Vinood B. Patel • Victor R. Preedy
Editors

Handbook of Nutrition, Diet, and Epigenetics

Volume 3

With 416 Figures and 143 Tables

Editors
Vinood B. Patel
School of Life Sciences
University of Westminster
London, UK

Victor R. Preedy
Diabetes and Nutritional Sciences Research Division
Faculty of Life Sciences and Medicine
King's College London
London, UK

ISBN 978-3-319-55529-4 ISBN 978-3-319-55530-0 (eBook)
ISBN 978-3-319-59052-3 (print and electronic bundle)
https://doi.org/10.1007/978-3-319-55530-0

Library of Congress Control Number: 2018953716

This Springer imprint is published by the registered company Springer Nature Switzerland AG
The registered company address is: Gewerbestrasse 11, 6330 Cham, Switzerland

Preface

The well-being of humankind is not only dependent upon individuals receiving adequate nutrition but also upon their genetic makeup. Genes may encode proteins responsible for structural components (e.g., membranes, subcellular organelles) and dynamics (e.g., enzymes, receptor–postreceptor cascades). Many of these components will require, from the outset, an adequate diet. For example, some antioxidant enzymes are critically dependent on the diet. This is illustrated by the role of dietary selenium which is necessary for glutathione peroxidase activities, while copper and zinc are necessary for superoxide dismutase activities. However, there is an increasing body of evidence to suggest that nutrition itself may alter the way in which genes are expressed via the process of epigenetics. Definitions of epigenetics vary and include modifications in the functional expression of DNA. This may involve changes in, or the influence of, DNA methylation, noncoding RNA, chromatin, histone acetylation or methylation, genomic imprinting, and other processes. There are many dietary components that impose epigenetic changes including folate, B vitamins, betaine, choline, and other extracts from plants, foods, and beverages. In fact, the knowledge base of how dietary components impact epigenetic processes has increased markedly over the past few years. As a prelude to understanding the role of epigenetics, it is also necessary to understand the basics of cellular and molecular biology, as well as the clinical basis of health and disease. However, marshaling all the information on the complex relationships between cellular and molecular biology, diet and nutrition, health and disease, and epigenetic processes is somewhat difficult due to the myriad of material. To address this, the editors have compiled the *Handbook of Nutrition, Diet, and Epigenetics*.

The book is divided into the following parts:

Part I. Introductory Material and Foundations
Part II. Organs, Disease, and Life Stages
Part III. Influence of Diet and Nutrition on Epigenetics
Part IV. Practical Techniques and Applications

Part I Introductory Material and Foundations covers biology of the cell, overviews, and comparative epigenetics. **Part II Organs, Disease, and Life Stages** covers weight control, metabolic syndrome and obesity, diabetes, insulin and glucose, the cardiovascular system, the nervous system, cancers and immune function, the intestinal tract, kidney, muscle and bone, life stages, pregnancy, development and programming, transgenerational effects, and aging. **Part III Influence of Diet and Nutrition on Epigenetics** covers energy, general treatments and nutritional modifications, lipids and proteins as macronutrients and their components, vitamins and minerals, combinations (mixtures of components), specific foods and nutraceuticals, and nutritional toxicology and adverse effects. **Part IV Practical Techniques and Applications** covers multilocus methylation assays, beadchips, bioinformatics databases, microRNAs, mass spectrometry, embryonic stem cells, and molecular pathways and resources.

It is difficult to list all the chapters as there are just over 120, and some cover numerous analytical or disease-based domains. The editors recognize the difficulties in assigning chapters to specific parts of the book as some chapters may well be suitable for two or more sections. Nevertheless, there is a wide breadth of material available. There are also unique features in this handbook, whereby each of the chapters includes the following sections:

- Dictionary of Terms
- Key Facts
- Summary Points

These features enable the transdisciplinary and transintellectual divides to be bridged.

Contributors are authors of international and national standing, leaders in the field, and trendsetters. Emerging fields of epigenetics in relation to diet and nutrition are also incorporated in the *Handbook of Nutrition, Diet, and Epigenetics*. This represents essential reading for nutritionists, dietitians, health care professionals, research scientists, molecular and cellular biochemists, physicians, general practitioners, public health practitioners, as well as those interested in health in general.

Vinood B. Patel
Victor R. Preedy

Contents

Volume 1

Volume 2

Volume 3

About the Editors

Vinood B. Patel
School of Life Sciences
University of Westminster
London, UK

Vinood B. Patel (B.Sc., Ph.D., FRSC) is a Reader in Clinical Biochemistry at the University of Westminster and honorary Fellow at King's College London. Dr. Patel graduated from 1992 the University of Portsmouth with a Degree in Pharmacology and completed his Ph.D. in protein metabolism from King's College London in 1997. His postdoctoral work was carried out at Wake Forest University Baptist Medical School studying structural-functional alterations to mitochondrial ribosomes, where he developed novel techniques to characterize their biophysical properties. Dr. Patel is a nationally and internationally recognized scientist, and in 2014 he was elected as a Fellow to the Royal Society of Chemistry. He presently directs studies on metabolic pathways involved in diabetes and liver disease, particularly related to mitochondrial energy regulation and cell death. He is currently conducting research to study the role of nutrients, antioxidants, phytochemicals, iron, alcohol, and fatty acids. His other areas of interest include identifying new biomarkers that can be used for the diagnosis and prognosis of liver disease and understanding mitochondrial oxidative stress in Alzheimer's disease and gastrointestinal dysfunction in autism. Dr. Patel has edited biomedical books in the areas of nutrition and health prevention and biomarkers and has published over 150 articles.

Victor R. Preedy (B.Sc., Ph.D., D.Sc., FRSB, FRSPH, FRCPath, FRSC) is a senior member of King's College London, where he is also Director of the Genomics Centre and a member of the Faculty of Life Sciences and Medicine.

Professor Preedy graduated in 1974 with an Honors Degree in Biology and Physiology with Pharmacology. He gained his University of London Ph.D. in 1981. In 1992, he received his Membership of the Royal College of Pathologists, and in 1993 he gained his second doctoral degree for his outstanding contribution to protein metabolism in health and disease. Professor Preedy was elected as Fellow to the Institute of Biology in 1995 and to the Royal College of Pathologists in 2000. Since then, he has been elected as Fellow to the Royal Society for the Promotion of Health (2004) and the Royal Institute of Public Health (2004). In 2009, Professor Preedy became Fellow of the Royal Society for Public Health and, in 2012, Fellow of the Royal Society of Chemistry. In his career, Professor Preedy has carried out research at the National Heart Hospital (part of Imperial College London), the School of Pharmacy (now part of University College London), and the MRC Centre at Northwick Park Hospital. He has collaborated with research groups in Finland, Japan, Australia, the USA, and Germany. He is a leading expert in the science of health and has a long-standing interest in food and nutrition for over 30 years, especially related to tissue pathology and cellular and molecular biology. He has lectured nationally and internationally. To his credit, Professor Preedy has published over 600 articles, which include peer-reviewed manuscripts based on original research, abstracts and symposium presentations, reviews, and numerous books and volumes.

Contributors

Rosita Aitoro Department of Translational Medical Science, University of Naples "Federico II", Naples, Italy

Aftab Alam Chromatin and Disease Biology Lab (# 08), National Centre for Cell Science, NCCS Complex, Savitribai Phule Pune University Campus, Pune, Maharashtra, India

Luigi Albano Department of Translational Medical Sciences, University of Naples "Federico II", Naples, Italy

David Albuquerque Department of Life Sciences, Research Centre for Anthropology and Health (CIAS), University of Coimbra, Coimbra, Portugal

Nezam Altorok Division of Rheumatology, Department of Internal Medicine, University of Toledo Medical Center, Toledo, OH, USA

Fawaz Alzaïd Sorbonne Universités, Université Pierre et Marie-Curie, Institut National de la Santé et de la Recherche Médicale (INSERM), UMR_S 1138 Cordeliers Research, Paris, France

Gerard Aragonès Nutrigenomics Research Group, Department of Biochemistry and Biotechnology, Universitat Rovira i Virgili (URV), Tarragona, Spain

Juan Arguelles Departamento de Biología Funcional, Área de Fisiología, Facultad de Medicina y Ciencias de la Salud, Universidad de Oviedo, España, Oviedo, Asturias, Spain

Takahiro Arima Department of Informative Genetics, Tohoku University Graduate School of Medicine, Aoba-ku, Sendai, Japan

Lluís Arola Nutrigenomics Research Group, Department of Biochemistry and Biotechnology, Universitat Rovira i Virgili (URV), Tarragona, Spain

Nutrition and Health Research Group, Technological Center for Nutrition and Health (EURECAT-CTNS), Tecnio, Campus of International Excellence Southern Catalonia (CEICS), Reus, Spain

Anna Arola-Arnal Nutrigenomics Research Group, Department of Biochemistry and Biotechnology, Universitat Rovira i Virgili (URV), Tarragona, Spain

Stella Aslibekyan Department of Epidemiology, University of Alabama at Birmingham, Birmingham, AL, USA

Paulo Pimentel Assumpção Núcleo de Pesquisas em Oncologia, Universidade Federal do Pará, Belém, Pará, Brazil

Susana Astiz Comparative Physiology Group. SGIT-INIA, Avda. Puerta de Hierro s/n, Madrid, Spain

Aneta Balcerczyk Faculty of Biology and Environmental Protection, Department of Molecular Biophysics, University of Lodz, Lodz, Poland

Donatella Barisani School of Medicine and Surgery, University of Milano Bicocca, Monza, Italy

Nicolas Barnich M2iSH, Université Clermont Auvergne, Inserm U1071, USC-INRA 2018, Clermont-Ferrand, France

Emma L. Beckett School of Medicine and Public Health, The University of Newcastle, Ourimbah, NSW, Australia

Megan Beetch Land and Food Systems, Food, Nutrition and Health, University of British Columbia, Vancouver, BC, Canada

Jomer Bernardo Texas A&M Irma Lerma Rangel College of Pharmacy, Kingsville, TX, USA

Cinta Bladé Nutrigenomics Research Group, Department of Biochemistry and Biotechnology, Universitat Rovira i Virgili (URV), Tarragona, Spain

Priscilla Boccia Department of Technological Innovation and Safety of Plants, Product and Anthropic Settlements (DIT), Italian Workers' Compensation Authority (INAIL), Rome, Italy

Loys Bodin GenPhySE, Université de Toulouse, INRA, INPT, ENVT, Castanet Tolosan, France

Tiziana Bonaldi Department of Experimental Oncology, European Institute of Oncology, Milan, Italy

Nikolay M. Borisov Centre for Convergence of Nano-, Bio-, Information and Cognitive Sciences and Technologies, National Research Centre "Kurchatov Institute", Moscow, Russia

Department of Personalized Medicine, First Oncology Research and Advisory Center, Moscow, Russia

John D. Bowman Texas A&M Irma Lerma Rangel College of Pharmacy, Kingsville, TX, USA

Sofia Braverman Department of Immunology, Weizmann Institute of Science, Rehovot, Israel

Francisca I. Bravo Nutrigenomics Research Group, Department of Biochemistry and Biotechnology, Universitat Rovira i Virgili (URV), Tarragona, Spain

Jean-Michel Brun GenPhySE, Université de Toulouse, INRA, INPT, ENVT, Castanet Tolosan, France

Gunnar Brunborg Department of Molecular Biology, Infection Control and Environmental Health, Norwegian Institute of Public Health, Oslo, Norway

Scott J. Bultman Department of Genetics, Lineberger Comprehensive Cancer Center, University of North Carolina at Chapel Hill, Chapel Hill, NC, USA

Rommel Rodríguez Burbano Laboratório de Biologia Molecular, Hospital Ophir Loyola, Belém, Pará, Brazil

Markus Burkard Department of Vegetative and Clinical Physiology, Institute of Physiology, University Hospital Tuebingen, Tuebingen, Germany

Anton A. Buzdin Centre for Convergence of Nano-, Bio-, Information and Cognitive Sciences and Technologies, National Research Centre "Kurchatov Institute", Moscow, Russia

OmicsWay Corporation, Walnut, CA, USA

Group for Genomic Regulation of Cell Signaling Systems, Shemyakin-Ovchinnikov Institute of Bioorganic Chemistry, Moscow, Russia

Lars Olov Bygren Department of Biosciences and Nutrition, Karolinska Institutet, Huddinge, Sweden

Demin Cai Key Laboratory of Animal Physiology and Biochemistry, Ministry of Agriculture, College of Veterinary Medicine, Nanjing Agricultural University, Nanjing, China

Danielle Queiroz Calcagno Núcleo de Pesquisas em Oncologia, Universidade Federal do Pará, Belém, Pará, Brazil

Luísa Camacho Division of Biochemical Toxicology, National Center for Toxicological Research, Food and Drug Administration, Jefferson, AR, USA

Roberto Berni Canani Department of Translational Medical Science, University of Naples "Federico II", Naples, Italy

European Laboratory for the Investigation of Food Induced Diseases (ELFID), University of Naples "Federico II", Naples, Italy

CEINGE Advanced Biotechnologies, University of Naples "Federico II", Naples, Italy

Andres Cardenas Department of Population Medicine, Harvard Medical School, Boston, MA, USA

Emily Chapman Department of Environmental Medicine and Public Health, Icahn School of Medicine at Mount Sinai, New York, NY, USA

Sybil Charriere Claude Bernard University, Lyon, France

Department of Endocrinology, Diabetology, Metabolic Diseases and Nutrition, Cardiovascular Hospital Louis Pradel, Bron Cedex, France

Samit Chattopadhyay Indian Institute of Chemical Biology, Kolkata, West Bengal, India

Chromatin and Disease Biology Lab (# 08), National Centre for Cell Science, NCCS Complex, Savitribai Phule Pune University Campus, Pune, Maharashtra, India

Qiao Yi Chen Department of Environmental Medicine, New York University School of Medicine, Tuxedo, NY, USA

Elizabeth Suchi Chen Division of Genetics, Department of Morphology and Genetics, Universidade Federal de São Paulo (UNIFESP), São Paulo, SP, Brazil

Guohua Chen Department of Pathology, Wayne State University School of Medicine, Detroit, MI, USA

Jia Chen Department of Environmental Medicine and Public Health, Icahn School of Medicine at Mount Sinai, New York, NY, USA

Department of Pediatrics, Icahn School of Medicine at Mount Sinai, New York, NY, USA

Department of Oncological Sciences, Icahn School of Medicine at Mount Sinai, New York, NY, USA

Department of Medicine, Hematology and Medical Oncology, Icahn School of Medicine at Mount Sinai, New York, NY, USA

Jin-Ran Chen Arkansas Children's Nutrition Center and the Department of Pediatrics, University of Arkansas for Medical Sciences, Little Rock, AR, USA

Alfred Sze-Lok Cheng School of Biomedical Sciences, The Chinese University of Hong Kong, Shatin, N.T., Hong Kong

Wen-Hsing Cheng Department of Food Science, Nutrition and Health Promotion, Mississippi State University, Mississippi State, MS, USA

Arpankumar Choksi Chromatin and Disease Biology Lab (# 08), National Centre for Cell Science, NCCS Complex, Savitribai Phule Pune University Campus, Pune, Maharashtra, India

Mahua Choudhury Department of Pharmaceutical Sciences, Irma Lerma Rangel College of Pharmacy, Texas A&M Health Science Center, College Station, TX, USA

Sayantani Chowdhury Division of Molecular Medicine, Bose Institute, Kolkata, India

Sabrina Chriett INSERM U1060, Oullins, France

Steven A. Claas College of Public Health, University of Kentucky, Lexington, KY, USA

Juan Francisco Codocedo CARE UC Biomedical Research Center, Faculty of Biological Sciences, Pontificia Universidad Católica de Chile, Santiago, Chile

Anne Collin INRA – URA, INRA, Nouzilly, France

Linda Cosenza Department of Translational Medical Science, University of Naples "Federico II", Naples, Italy

Max Costa Department of Environmental Medicine, New York University School of Medicine, Tuxedo, NY, USA

Vincent Coustham INRA – URA, INRA, Nouzilly, France

Anna Crescenti Nutrition and Health Research Group, Technological Center for Nutrition and Health (EURECAT-CTNS), Tecnio, Campus of International Excellence Southern Catalonia (CEICS), Reus, Spain

Juan Cui Department of Computer Science and Engineering, University of Nebraska-Lincoln, Lincoln, NE, USA

Alessandro Cuomo Department of Experimental Oncology, European Institute of Oncology, Milan, Italy

Lidia Daimiel Nutritional Genomics of the Cardiovascular Disease and Obesity, Foundation IMDEA Food CEI UAM + CSIC, Madrid, Spain

Department of Nutrition and Bromatology, Facultad de Farmacia, Universidad San Pablo-CEU, CEU universities, Boadilla del Monte, Madrid, Spain

Aline de Conti Division of Biochemical Toxicology, National Center for Toxicological Research, Food and Drug Administration, Jefferson, AR, USA

Flávia Ramos de Siqueira Department of Internal Medicine, Laboratory of Experimental Hypertension, School of Medicine, University of Sao Paulo, Sao Paulo, SP, Brazil

Department of Internal Medicine, University of São Paulo School of Medicine, Butantã, São Paulo, SP, Brazil

Department of Internal Medicine, Nephrology Division, School of Medicine, University of São Paulo, São Paulo, SP, Brazil

Stefanie Braga Maia de Sousa Núcleo de Pesquisas em Oncologia, Universidade Federal do Pará, Belém, Pará, Brazil

Christiana A. Demetriou Neurology Clinic D, The Cyprus Institute of Neurology and Genetics, Ayios Dhometios, Nicosia, Cyprus

The Cyprus School of Molecular Medicine, The Cyprus Institute of Neurology and Genetics, Ayios Dhometios, Nicosia, Cyprus

Jérémy Denizot M2iSH, Université Clermont Auvergne, Inserm U1071, USC-INRA 2018, Clermont-Ferrand, France

Maya A. Deyssenroth Department of Environmental Medicine and Public Health, Icahn School of Medicine at Mount Sinai, New York, NY, USA

Margherita Di Costanzo Department of Translational Medical Science, University of Naples "Federico II", Naples, Italy

Carmen di Scala Department of Translational Medical Science, University of Naples "Federico II", Naples, Italy

Svetlana Dinić Department of Molecular Biology, Institute for Biological Research Siniša Stanković, University of Belgrade, Belgrade, Serbia

Cristina Domenichelli Largo Alessandria del Carretto, Rome, Italy

Daoyin Dong Department of Obstetrics, Gynecology and Reproductive Sciences, University of Maryland School of Medicine, Baltimore, MD, USA

Jörg Dötsch Department of Pediatrics, University of Cologne, Cologne, Germany

Nur Duale Department of Molecular Biology, Infection Control and Environmental Health, Norwegian Institute of Public Health, Oslo, Norway

Thomas Eggermann Institute of Human Genetics, University Hospital, RWTH Technical University Aachen, Aachen, Germany

Eran Elinav Department of Immunology, Weizmann Institute of Science, Rehovot, Israel

Eduard Escrich Multidisciplinary Group for the Study of Breast Cancer, Department of Cell Biology, Physiology and Immunology, Physiology Unit, Faculty of Medicine, Universitat Autònoma de Barcelona, Barcelona, Spain

Dyfed Lloyd Evans School of Life Sciences, University of KwaZulu-Natal, Westville Campus, Durban, KwaZulu Natal, South Africa

Mingzhu Fang Department of Environmental and Occupational Health, School of Public Health Environmental and Occupational Health Sciences Institute, Rutgers, The State University of New Jersey, Piscataway, NJ, USA

Maria Rosaria Faraone-Mennella Department of Biology, University of Naples "Federico II", Naples, Italy

Yannick Faulconnier Herbivore Research Unit -Biomarkers Team, French Institut of Agricultural Research (INRA), St Genès-Champanelle, France

Christopher D. Faulk Department of Animal Sciences, University of Minnesota, St. Paul, MN, USA

Robert Feil Institute of Molecular Genetics (IGMM), Centre National de Recherche Scientifique (CNRS), UMR-5535, University of Montpellier, Montpellier, France

Michael Fenech CSIRO Health and Biosecurity, Personalised Nutrition and Healthy Ageing, Adelaide, SA, Australia

Carla Ferreri Institute of Organic Synthesis and Photoreactivity (ISOF), CNR, Bologna, Italy

Maxime François CSIRO Health and Biosecurity, Personalised Nutrition and Healthy Ageing, Adelaide, SA, Australia

Department of Molecular and Cellular Biology, School of Biological Sciences, University of Adelaide, Adelaide, SA, Australia

Laure Frésard Department of Pathology, Stanford University, Stanford, CA, USA

Luzia Naôko Shinohara Furukawa Department of Internal Medicine, Laboratory of Experimental Hypertension, School of Medicine, University of Sao Paulo, Sao Paulo, SP, Brazil

Department of Internal Medicine, University of São Paulo School of Medicine, Butantã, São Paulo, SP, Brazil

Andrea Fuso Department of Surgery "P. Valdoni", Sapienza University of Rome, Rome, Italy

Uma Gaur Institute of Animal Genetics and Breeding, Sichuan Agricultural University, Chengdu, Sichuan, China

Bruce German Department of Food Science and Technology, University of California Davis, Davis, CA, USA

Meenu Ghai School of Life Sciences, University of KwaZulu-Natal,Westville Campus, Durban, KwaZulu Natal, South Africa

Jyotirmoy Ghosh Department of Chemistry, Banwarilal Bhalotia College, Ushagram Asansol, West Bengal, India

Carolina Oliveira Gigek Division of Genetics, Department of Morphology and Genetics, Universidade Federal de São Paulo (UNIFESP), São Paulo, SP, Brazil

Division of Surgical Gastroenterology, Department of Surgery, Universidade Federal de São Paulo (UNIFESP), São Paulo, SP, Brazil

Elodie Gimier M2iSH, Université Clermont Auvergne, Inserm U1071, USC-INRA 2018, Clermont-Ferrand, France

Silvia Gioiosa Institute of Biomembranes and Bioenergetics, National Research Council, Bari, Italy

Toshinao Goda Graduate School of Nutritional and Environmental Sciences, University of Shizuoka, Shizuoka, Japan

Department of Nutrition, School of Food and Nutritional Sciences, The University of Shizuoka, Shizuoka, Japan

Xiaoming Gong Department of Pediatrics, Paul L. Foster School of Medicine, Texas Tech University Health Sciences Center El Paso, El Paso, TX, USA

Antonio Gonzalez-Bulnes Comparative Physiology Group. SGIT-INIA, Avda. Puerta de Hierro s/n, Madrid, Spain

Viviana Granata Department of Translational Medical Science, University of Naples "Federico II", Naples, Italy

Steven G. Gray Thoracic Oncology Research Group, Trinity Translational Medical Institute, St James's Hospital, Dublin, Ireland

Nevena Grdović Institute for Biological Research, University of Belgrade, Belgrade, Serbia

Michael Paul Greenwood School of Clinical Sciences, University of Bristol, Bristol, UK

Harvest F. Gu Department of Clinical Science, Intervention and Technologies, Karolinska Institutet, Karolinska University Hospital, Stockholm, Sweden

Center for Molecular Medicine, Karolinska Institutet, Karolinska University Hospital, Stockholm, Sweden

Sandrine Le Guillou Génétique Animale et Biologie Intégrative, French Institut of Agricultural Research (INRA), Jouy-en-Josas, France

Daniel B. Hardy Departments of Obstetrics and Gynecology and Physiology and Pharmacology, The Children's Health Research Institute and The Lawson Health Research Institute, The University of Western Ontario, London, ON, Canada

Natsuyo Hariya Department of Nutrition, Faculty of Health and Nutrition, Yamanashi Gakuin University, Kofu, Yamanashi, Japan

Hiromitsu Hattori Department of Informative Genetics, Tohoku University Graduate School of Medicine, Aoba-ku, Sendai, Japan

Trine B. Haugen Faculty of Health Sciences, Oslo and Akershus University College of Applied Sciences, Oslo, Norway

Christine Heberden Micalis Institute, INRA, AgroParisTech, Université Paris-Saclay, Jouy-en-Josas, France

Renato Heidor Laboratory of Diet, Nutrition and Cancer, Department of Food and Experimental Nutrition, Faculty of Pharmaceutical Sciences, University of São Paulo, São Paulo, SP, Brazil

Joel Claudio Heimann Department of Internal Medicine, Laboratory of Experimental Hypertension, School of Medicine, University of Sao Paulo, Sao Paulo, SP, Brazil

Hitoshi Hiura Department of Informative Genetics, Tohoku University Graduate School of Medicine, Aoba-ku, Sendai, Japan

Tara L. Hogenson Schulze Center for Novel Therapeutics, Mayo Clinic, Rochester, MN, USA

Kazue Honma Graduate School of Nutritional and Environmental Sciences, University of Shizuoka, Shizuoka, Japan

Department of Nutrition, School of Food and Nutritional Sciences, The University of Shizuoka, Shizuoka, Japan

Dave S. B. Hoon Department of Translational Molecular Medicine, John Wayne Cancer Institute at Providence Saint John's Health Center, Santa Monica, CA, USA

Willie Hsu Department of Chemistry, Texas A&M University, College Station, TX, USA

Yun Hu Key Laboratory of Animal Physiology and Biochemistry, Ministry of Agriculture, College of Veterinary Medicine, Nanjing Agricultural University, Nanjing, China

Kun Huang Tongji School of Pharmacy, Huazhong University of Science and Technology, Wuhan, China

Tyler C. Huff Dr. John T. Macdonald Foundation Department of Human Genetics, John P. Hussman Institute for Human Genomics, University of Miami Miller School of Medicine, Miami, FL, USA

Yuko Ibuki Graduate Division of Nutritional and Environmental Sciences, University of Shizuoka, Shizuoka, Japan

Nibaldo C. Inestrosa CARE UC Biomedical Research Center, Faculty of Biological Sciences, Pontificia Universidad Católica de Chile, Santiago, Chile

Centre for Healthy Brain Ageing, School of Psychiatry, Faculty of Medicine, University of New South Wales, Sydney, NSW, Australia

Centro de Excelencia en Biomedicina de Magallanes (CEBIMA), Universidad de Magallanes, Punta Arenas, Chile

Tomas Jakobsson Department of Laboratory Medicine, Karolinska Institutet, Huddinge, Sweden

Gopabandhu Jena Facility for Risk Assessment and Intervention Studies, Department of Pharmacology and Toxicology, National Institute of Pharmaceutical Education and Research (NIPER), Mohali, Punjab, India

Yuqian Jiang Department of Biochemistry and Molecular Medicine, University of California at Davis, Sacramento, CA, USA

Jordi Jiménez-Conde Department of Neurology, Hospital del Mar; Neurovascular Research Group, IMIM (Institut Hospital del Mar d'Investigacions Mèdiques), Barcelona, Spain

Ruta Jog Department of Pathology, Wayne State University School of Medicine, Detroit, MI, USA

Bjorn Johansson Centre of Molecular and Environmental Biology (CBMA), Department of Biology, University of Minho, Braga, Portugal

Joan Jory Guelph, ON, Canada

Sadhana Joshi Department of Nutritional Medicine, Interactive Research School for Health Affairs, Bharati Vidyapeeth Deemed University, Pune, India

Shailesh Joshi School of Life Sciences, University of KwaZulu-Natal,Westville Campus, Durban, KwaZulu Natal, South Africa

Bonnie R. Joubert Population Health Branch, National Institute of Environmental and Health Sciences, Durham, NC, USA

Jelena Arambašić Jovanović Department of Molecular Biology, Institute for Biological Research Siniša Stanković, University of Belgrade, Belgrade, Serbia

Tomasz P. Jurkowski Institute of Biochemistry, University of Stuttgart, Stuttgart, Germany

Gunnar Kaati Department of Biosciences and Nutrition, Karolinska Institutet, Huddinge, Sweden

Bashar Kahaleh Division of Rheumatology, Department of Internal Medicine, University of Toledo Medical Center, Toledo, OH, USA

Foteini Kakulas School of Medicine and Pharmacology, The University of Western Australia, Crawley, WA, Australia

Neerja Karnani Singapore Institute for Clinical Sciences (SICS), A*STAR, Brenner Centre for Molecular Medicine, Singapore, Singapore

Yukihiko Kato Department of Dermatology, Tokyo Medical University Hachioji Medical Center, Tokyo, Japan

Jaspreet Kaur Department of Biotechnology, University Institute of Engineering and Technology, Panjab University, Chandigarh, India

Pushpinder Kaur USC Keck School of Medicine, University of Southern California, Los Angeles, CA, USA

Sabbir Khan Facility for Risk Assessment and Intervention Studies, Department of Pharmacology and Toxicology, National Institute of Pharmaceutical Education and Research (NIPER), Mohali, Punjab, India

Yong Jin Kim Department of Obstetrics and Gynecology, Korea University Medical College, Korea University Guro Hospital, Seoul, South Korea

Yoon Young Kim Department of Obstetrics and Gynecology, Seoul National University College of Medicine, Seoul National University Hospital, Seoul, South Korea

Anna K. Kiss Department of Pharmacognosy and Molecular Basis of Phytotherapy, Faculty of Pharmacy, Medical University of Warsaw, Warsaw, Poland

Akane Kitamura Department of Informative Genetics, Tohoku University Graduate School of Medicine, Aoba-ku, Sendai, Japan

Jelena Knežević-Vukčević Microbiology, Center for Genotoxicology and Ecogenotoxicology, Faculty of Biology, University of Belgrade, Belgrade, Serbia

Norio Kobayashi Department of Informative Genetics, Tohoku University Graduate School of Medicine, Aoba-ku, Sendai, Japan

Robert A. Koza Center for Molecular Medicine, Maine Medical Center Research Institute, Scarborough, ME, USA

Seung-Yup Ku Department of Obstetrics and Gynecology, Seoul National University College of Medicine, Seoul National University Hospital, Seoul, South Korea

Takeo Kubota Faculty of Child Studies, Seitoku University, Matsudo, Chiba, Japan

Kyriacos Kyriacou Department of Electron Microscopy/Molecular Pathology, The Cyprus Institute of Neurology and Genetics, The Cyprus School of Molecular Medicine, The Cyprus School of Molecular, Nicosia, Cyprus

Zachary M. Laubach Department of Integrative Biology and Program in Ecology, Evolutionary Biology, and Behavior, Michigan State University, East Lansing, MI, USA

Minh Le Texas A&M Irma Lerma Rangel College of Pharmacy, Kingsville, TX, USA

Daniel Leclerc Departments of Human Genetics and Pediatrics, McGill University, The Research Institute of the McGill University Health Centre, Montreal, QC, Canada

Todd Leff Department of Pathology, Wayne State University School of Medicine, Detroit, MI, USA

Wayne Leifert CSIRO Health and Biosecurity, Personalised Nutrition and Healthy Ageing, Adelaide, SA, Australia

Department of Molecular and Cellular Biology, School of Biological Sciences, University of Adelaide, Adelaide, SA, Australia

Christian Leischner Department of Vegetative and Clinical Physiology, Institute of Physiology, University Hospital Tuebingen, Tuebingen, Germany

Alexey A. Leontovich Division of Biomedical Statistics and Informatics, Mayo Clinic, Rochester, MN, USA

Christine Leroux Herbivore Research Unit -Biomarkers Team, French Institut of Agricultural Research (INRA), St Genès-Champanelle, France

Department of Food Science and Technology, University of California Davis, Davis, CA, USA

Nancy Lévesque Departments of Human Genetics and Pediatrics, McGill University, The Research Institute of the McGill University Health Centre, Montreal, QC, Canada

Chuan Li Department of Immunology, School of Medicine, University of Connecticut Health Center, Farmington, CT, USA

Ting Lian Institute of Animal Genetics and Breeding, Sichuan Agricultural University, Chengdu, Sichuan, China

Ives Y. Lim Singapore Institute for Clinical Sciences (SICS), A*STAR, Brenner Centre for Molecular Medicine, Singapore, Singapore

Chengqi Lin Institute of Life Sciences, The Key Laboratory of Developmental Genes and Human Disease, Southeast University, Nanjing, Jiangsu, China

Xinyi Lin Singapore Institute for Clinical Sciences (SICS), A*STAR, Brenner Centre for Molecular Medicine, Singapore, Singapore

Birgitte Lindeman Department of Toxicology and Risk, Infection Control and Environmental Health, Norwegian Institute of Public Health, Oslo, Norway

Chengyu Liu Tongji School of Pharmacy, Huazhong University of Science and Technology, Wuhan, China

Haoyu Liu Department of Medical Cell Biology, Uppsala University, Uppsala, Sweden

Wenshe R. Liu Department of Chemistry, Texas A&M University, College Station, TX, USA

Ilana Livyatan Department of Genetics, The Hebrew University of Jerusalem, Jerusalem, Israel

Pang-Kuo Lo Greenebaum Cancer Center, Department of Biochemistry and Molecular Biology, University of Maryland School of Medicine, Baltimore, MD, USA

Gaspar Jesus Lopes-Filho Division of Surgical Gastroenterology, Department of Surgery, Universidade Federal de São Paulo (UNIFESP), São Paulo, SP, Brazil

Hsin-Yi Lu Department of Food Science, Nutrition and Health Promotion, Mississippi State University, Mississippi State, MS, USA

Mark Lucock School of Environmental and Life Sciences, The University of Newcastle, Ourimbah, NSW, Australia

Xiaoping Luo Department of Pediatrics, Tongji Hospital, Tongji Medical College, Huazhong University of Science and Technology, Wuhan, China

Zhuojuan Luo Institute of Life Sciences, The Key Laboratory of Developmental Genes and Human Disease, Southeast University, Nanjing, Jiangsu, China

Sonia Métayer-Coustard INRA – URA, INRA, Nouzilly, France

Qingyi Ma Center for Perinatal Biology, Division of Pharmacology, Department of Basic Sciences, Loma Linda University, School of Medicine, Loma Linda, CA, USA

Paolo Emidio Macchia Department of Clinical Medicine and Surgery, University of Naples "Federico II", Naples, Italy

Licínio Manco Department of Life Sciences, Research Centre for Anthropology and Health (CIAS), University of Coimbra, Coimbra, Portugal

Krishna Prahlad Maremanda Facility for Risk Assessment and Intervention Studies, Department of Pharmacology and Toxicology, National Institute of Pharmaceutical Education and Research (NIPER), Mohali, Punjab, India

J. Alfredo Martinez Department of Nutrition, Food Science and Physiology, and Centre for Nutrition Research, University of Navarra, Pamplona, Spain

Department of Nutrition and Dietetics, King's College London, London, UK

Madrid Institute of Advanced Studies (IMDEA Food), Madrid, Spain

David J. Martino Gastro and Food Allergy, Murdoch Children's Research Institute, The University of Melbourne Department of Paediatrics, The Royal Children's Hospital, Melbourne, VIC, Australia

Nina Nayara Ferreira Martins Núcleo de Pesquisas em Oncologia, Universidade Federal do Pará, Belem, Brazil

Diego M. Marzese Department of Translational Molecular Medicine, John Wayne Cancer Institute at Providence Saint John's Health Center, Santa Monica, CA, USA

Annalisa Masi Institute of Organic Synthesis and Photoreactivity (ISOF), CNR, Bologna, Italy

Petra Matoušková Department of Biochemical Sciences, Charles University, Faculty of Pharmacy, Hradec Králové, Czech Republic

Elise Maximin Micalis Institute, INRA, AgroParisTech, Université Paris-Saclay, Jouy-en-Josas, France

Andre Souza Mecawi Department of Physiological Sciences, Institute of Medical and Biological Sciences, Federal Rural University of Rio de Janeiro, Seropedica, RJ, Brazil

Claudia Meconi Research Organization CRF (Cooperativa Ricerca Finalizzata Sc), Tor Vergata University Science Park, Rome, Italy

Bodo C. Melnik Department of Dermatology, Environmental Medicine and Health Theory, University of Osnabrück, Osnabrück, Germany

Isabel Mendizabal School of Biological Sciences, Georgia Institute of Technology, Atlanta, GA, USA

Department of Genetics, Physical Anthropology and Animal Physiology, University of the Basque Country UPV/EHU, Leioa, Spain

Eran Meshorer Department of Genetics and the Edmond and Lily Safra Center for Brain Sciences (ELSC), The Hebrew University of Jerusalem, Jerusalem, Israel

Alessandra Mezzelani National Research Council, Institute of Biomedical Technologies (CNR-ITB), Segrate (MI), Italy

Mirjana Mihailović Department of Molecular Biology, Institute for Biological Research Siniša Stanković, University of Belgrade, Belgrade, Serbia

Fermin I. Milagro Department of Nutrition, Food Science and Physiology, and Centre for Nutrition Research, University of Navarra, Pamplona, Spain

Dragan Milenkovic School of Medicine, Division of Cardiovascular Medicine, University of California Davis, Davis, CA, USA

Department of Human Nutrition, French Institut of Agricultural Research (INRA), St Genès-Champanelle, France

Francis Minvielle UMR INRA/AgroParisTech – GABI, Jouy-en-Josas, France

Brett M. Mitchell Texas A&M, Department of Medical Physiology, College Station, TX, USA

Dragana Mitić-Ćulafić Microbiology, Center for Genotoxicology and Ecogenotoxicology, Faculty of Biology, University of Belgrade, Belgrade, Serbia

Naoko Miyauchi Department of Informative Genetics, Tohoku University Graduate School of Medicine, Aoba-ku, Sendai, Japan

Lenha Mobuchon Herbivore Research Unit -Biomarkers Team, French Institut of Agricultural Research (INRA), St Genès-Champanelle, France

Kazuki Mochizuki Department of Local Produce and Food Sciences, Faculty of Life and Environmental Sciences, University of Yamanashi, Kofu, Yamanashi, Japan

Myth Tsz-Shun Mok School of Biomedical Sciences, The Chinese University of Hong Kong, Shatin, N.T., Hong Kong

Simon G. Møller Department of Biological Sciences, St. John's University, New York, NY, USA

Norwegian Center for Movement Disorders, Stavanger University Hospital, Stavanger, Norway

Raquel Moral Multidisciplinary Group for the Study of Breast Cancer, Department of Cell Biology, Physiology and Immunology, Physiology Unit, Faculty of Medicine, Universitat Autònoma de Barcelona, Barcelona, Spain

Angélica Morales-Miranda Department of Reproductive Biology, Instituto Nacional de Ciencias Médicas y Nutrición "Salvador Zubirán", Mexico City, Mexico

Fabiano Cordeiro Moreira Núcleo de Pesquisas em Oncologia, Universidade Federal do Pará, Belém, Pará, Brazil

Fernando Salvador Moreno Laboratory of Diet, Nutrition and Cancer, Department of Food and Experimental Nutrition, Faculty of Pharmaceutical Sciences, University of São Paulo, São Paulo, SP, Brazil

Sumiko Morimoto Department of Reproductive Biology, Instituto Nacional de Ciencias Médicas y Nutrición "Salvador Zubirán", Mexico City, Mexico

Mireille Morisson GenPhySE, Université de Toulouse, INRA, INPT, ENVT, Castanet Tolosan, France

Elizangela Rodrigues da Silva Mota Núcleo de Pesquisas em Oncologia, Universidade Federal do Pará, Belém, Pará, Brazil

Philippe Moulin Claude Bernard University, Lyon, France

Department of Endocrinology, Diabetology, Metabolic Diseases and Nutrition, Cardiovascular Hospital Louis Pradel, Bron Cedex, France

Meltem Muftuoglu Department of Molecular Biology and Genetics, Acibadem University, Istanbul, Turkey

Begoña Muguerza Nutrigenomics Research Group, Department of Biochemistry and Biotechnology, Universitat Rovira i Virgili (URV), Tarragona, Spain

Eiji Munetsuna Department of Biochemistry, Fujita Health University School of Medicine, Toyoake, Aichi, Japan

Daniele Musiani Department of Experimental Oncology, European Institute of Oncology, Milan, Italy

Vivek Nagaraja Division of Rheumatology, Department of Internal Medicine, University of Toledo Medical Center, Toledo, OH, USA

Varun Sasidharan Nair Department of Pathology, Hallym University, College of Medicine, Chuncheon, Gangwon-Do, Republic of Korea

Kazuhiko Nakabayashi Division of Developmental Genomics, Department of Maternal-Fetal Biology, National Research Institute for Child Health and Development, Tokyo, Japan

Chau Nguyen Texas A&M Irma Lerma Rangel College of Pharmacy, Kingsville, TX, USA

Luciano Nicosia Department of Experimental Oncology, European Institute of Oncology, Milan, Italy

Biljana Nikolić Microbiology, Center for Genotoxicology and Ecogenotoxicology, Faculty of Biology, University of Belgrade, Belgrade, Serbia

Roberta Noberini Center for Genomic Science of IIT@SEMM, Istituto Italiano di Tecnologia, Milan, Italy

Clévio Nóbrega Department of Biomedical Sciences and Medicine, Center for Biomedical Research (CBMR), University of Algarve, Faro, Portugal

Center for Neuroscience and Cell Biology, University of Coimbra, Coimbra, Portugal

Rita Nocerino Department of Translational Medical Science, University of Naples "Federico II", Naples, Italy

Eva Nüsken Department of Pediatrics, Pediatric Nephrology, University of Cologne, Cologne, Germany

Kai-Dietrich Nüsken Department of Pediatrics, Pediatric Nephrology, University of Cologne, Cologne, Germany

Kwon Ik Oh Department of Pathology, Hallym University, College of Medicine, Chuncheon, Gangwon-Do, Republic of Korea

Koji Ohashi Department of Clinical Biochemistry, Fujita Health University School of Health Sciences, Toyoake, Aichi, Japan

Hiroaki Okae Department of Informative Genetics, Tohoku University Graduate School of Medicine, Aoba-ku, Sendai, Japan

Daniela Oliveira Centre for the Research and Technology of Agro-Environmental and Biological Sciences (CITAB), Department of Biology, University of Minho, Braga, Portugal

Rui Oliveira Centre for the Research and Technology of Agro-Environmental and Biological Sciences (CITAB), Department of Biology, University of Minho, Braga, Portugal

Centre of Biological Engineering (CEB), Department of Biology, University of Minho, Braga, Portugal

Kelly Cristina da Silva Oliveira Núcleo de Pesquisas em Oncologia, Universidade Federal do Pará, Belem, Pará, Brazil

Javier I. J. Orozco Department of Translational Molecular Medicine, John Wayne Cancer Institute at Providence Saint John's Health Center, Santa Monica, CA, USA

Ana Laura Ortega-Márquez Department of Reproductive Biology, Instituto Nacional de Ciencias Médicas y Nutrición "Salvador Zubirán", Mexico City, Mexico

Richa Pant Chromatin and Disease Biology Lab (# 08), National Centre for Cell Science, NCCS Complex, Savitribai Phule Pune University Campus, Pune, Maharashtra, India

Lorella Paparo Department of Translational Medical Science, University of Naples "Federico II", Naples, Italy

Sonal Patel Chromatin and Disease Biology Lab (# 08), National Centre for Cell Science, NCCS Complex, Savitribai Phule Pune University Campus, Pune, Maharashtra, India

Vinood B. Patel School of Life Sciences, University of Westminster, London, UK

Rakesh Pathak Institute of Molecular Genetics (IGMM), Centre National de Recherche Scientifique (CNRS), UMR-5535, University of Montpellier, Montpellier, France

Ketan S. Patil Department of Biological Sciences, St. John's University, New York, NY, USA

Wei Perng Department of Nutritional Sciences, Department of Epidemiology, University of Michigan School of Public Health, Ann Arbor, MI, USA

Meirav Pevsner-Fischer Department of Immunology, Weizmann Institute of Science, Rehovot, Israel

Luciano Pirola INSERM U1060, Oullins, France

Frédérique Pitel GenPhySE, Université de Toulouse, INRA, INPT, ENVT, Castanet Tolosan, France

Igor P. Pogribny Division of Biochemical Toxicology, National Center for Toxicological Research, Food and Drug Administration, Jefferson, AR, USA

Goran Poznanović Department of Molecular Biology, Institute for Biological Research Siniša Stanković, University of Belgrade, Belgrade, Serbia

Victor R. Preedy Diabetes and Nutritional Sciences Research Division, Faculty of Life Science and Medicine, King's College London, London, UK

Fabienne Le Provost Génétique Animale et Biologie Intégrative, French Institut of Agricultural Research (INRA), Jouy-en-Josas, France

Rajkumar Rajendram Department of Medicine, King Abdulaziz Medical City, Ministry of National Guard Health Affairs, Riyadh, Saudi Arabia

Diabetes and Nutritional Sciences Research Division, Faculty of Life Science and Medicine, King's College London, London, UK

Omar Ramos-Lopez Department of Nutrition, Food Science and Physiology, and Centre for Nutrition Research, University of Navarra, Pamplona, Spain

Department of Molecular Biology in Medicine, Civil Hospital of Guadalajara "Fray Antonio Alcalde", Guadalajara, Jalisco, Mexico

Richa Rathod Department of Nutritional Medicine, Interactive Research School for Health Affairs, Bharati Vidyapeeth Deemed University, Pune, India

Mirunalini Ravichandran Institute of Biochemistry, University of Stuttgart, Stuttgart, Germany

E. Albert Reece Department of Obstetrics, Gynecology and Reproductive Sciences, University of Maryland School of Medicine, Baltimore, MD, USA

Caroline Relton MRC Integrative Epidemiology Unit, School of Social and Community Medicine, University of Bristol, Bristol, UK

Institute of Genetic Medicine, Newcastle University, Newcastle upon Tyne, UK

Jose Ignacio Riezu-Boj Department of Nutrition, Food Science and Physiology, and Centre for Nutrition Research, University of Navarra, Pamplona, Spain

Alberto PG Romagnolo University of Arizona Cancer Center, College of Medicine, University of Arizona, Tucson, AZ, USA

Donato F. Romagnolo Department of Nutritional Sciences, University of Arizona, Tucson, AZ, USA

University of Arizona Cancer Center, College of Medicine, University of Arizona, Tucson, AZ, USA

Jaume Roquer Head of the Department of Neurology, Hospital del Mar; Neurovascular Research Group, IMIM (Institut Hospital del Mar d'Investigacions Mèdiques), Barcelona, Spain

Department of Medicine, Universitat Autònoma de Barcelona, Barcelona, Spain

Cheryl S. Rosenfeld Department of Biomedical Sciences, Bond Life Sciences Center Investigator, Thompson Center for Autism and Neurobehavioral Disorders, University of Missouri, Columbia, MO, USA

Rima Rozen Departments of Human Genetics and Pediatrics, McGill University, The Research Institute of the McGill University Health Centre, Montreal, QC, Canada

Lewis P. Rubin Departments of Pediatrics and Biomedical Sciences, Paul L. Foster School of Medicine, Texas Tech University Health Sciences Center El Paso, El Paso, TX, USA

Hitaji Sanford Texas A&M Irma Lerma Rangel College of Pharmacy, Kingsville, TX, USA

Michael P. Sarras Jr. Department of Cell Biology and Anatomy, Chicago Medical School, Rosalind Franklin University of Medicine and Science, North Chicago, IL, USA

E. R. Sauter Hartford HealthCare Cancer Institute, Hartford, CT, USA

Luís Fernando Schütz Department of Pharmaceutical Sciences, Irma Lerma Rangel College of Pharmacy, Texas A&M Health Science Center, College Station, TX, USA

Tanya L. Schwab Department of Biochemistry and Molecular Biology, Mayo Clinic, Rochester, MN, USA

Ornella I. Selmin Department of Nutritional Sciences, University of Arizona, Tucson, AZ, USA

University of Arizona Cancer Center, College of Medicine, University of Arizona, Tucson, AZ, USA

Parames C. Sil Division of Molecular Medicine, Bose Institute, Kolkata, India

Minati Singh Department of Pediatrics, The University of Iowa, Iowa City, IA, USA

Marilia Arruda Cardoso Smith Division of Genetics, Department of Morphology and Genetics, Universidade Federal de São Paulo (UNIFESP), São Paulo, SP, Brazil

Monica Soldi Department of Experimental Oncology, European Institute of Oncology, Milan, Italy

Berna Somuncu Department of Molecular Biology and Genetics, Acibadem University, Istanbul, Turkey

Carolina Soriano-Tárraga Neurovascular Research Group, IMIM (Institut Hospital del Mar d'Investigacions Mèdiques), Barcelona, Spain

Valeria Spadotto Department of Experimental Oncology, European Institute of Oncology, Milan, Italy

Barbara Stefanska Land and Food Systems, Food, Nutrition and Health, University of British Columbia, Vancouver, BC, Canada

Silvia Stringhini Institute of Social and Preventive Medicine (IUMSP), Lausanne University Hospital, Lausanne, Switzerland

Elena Sturchio Department of Technological Innovation and Safety of Plants, Product and Anthropic Settlements (DIT), Italian Workers' Compensation Authority (INAIL), Rome, Italy

Manuel Suárez Nutrigenomics Research Group, Department of Biochemistry and Biotechnology, Universitat Rovira i Virgili (URV), Tarragona, Spain

Souta Takahashi Department of Informative Genetics, Tohoku University Graduate School of Medicine, Aoba-ku, Sendai, Japan

Junji Takaya Department of Pediatrics, Kawachi General Hospital, Higashi-Osaka, Osaka, Japan

Department of Pediatrics, Kansai Medical University, Moriguchi, Osaka, Japan

Qihua Tan Epidemiology and Biostatistics, Department of Public Health, Unit of Human Genetics, Department of Clinical Research, Faculty of Health Sciences, University of Southern Denmark, Odense, Denmark

Paniz Tavakoli CSIRO Health and Biosecurity, Personalised Nutrition and Healthy Ageing, Adelaide, SA, Australia

Department of Molecular and Cellular Biology, School of Biological Sciences, University of Adelaide, Adelaide, SA, Australia

Tincy Thomas Department of Pharmaceutical Sciences, Irma Lerma Rangel College of Pharmacy, Texas A&M Health Science Center, College Station, TX, USA

Anja Tolić Institute for Biological Research, University of Belgrade, Belgrade, Serbia

Eckardt Treuter Department of Biosciences and Nutrition, Karolinska Institutet, Huddinge, Sweden

Paola Ungaro Institute for Experimental Endocrinology and Oncology, "G. Salvatore"(IEOS), National Research Council (CNR), Naples, Italy

Aleksandra Uskoković Department of Molecular Biology, Institute for Biological Research Siniša Stanković, University of Belgrade, Belgrade, Serbia

Karin van Veldhoven Department of Epidemiology and Biostatistics, School of Public Health, Imperial College London, London, UK

Ernesto Vargas-Mendez Laboratory of Diet, Nutrition and Cancer, Department of Food and Experimental Nutrition, Faculty of Pharmaceutical Sciences, University of São Paulo, São Paulo, SP, Brazil

Anthony T. Vella Department of Immunology, School of Medicine, University of Connecticut Health Center, Farmington, CT, USA

Nicolas Venteclef Sorbonne Universités, Université Pierre et Marie-Curie, Institut National de la Santé et de la Recherche Médicale (INSERM), UMR_S 1138 Cordeliers Research, Paris, France

Sascha Venturelli Department of Vegetative and Clinical Physiology, Institute of Physiology, University Hospital Tuebingen, Tuebingen, Germany

Martin Veysey School of Medicine and Public Health, The University of Newcastle, Gosford Hospital, Gosford, NSW, Australia

Mark H. Vickers Liggins Institute, University of Auckland, Auckland, New Zealand

Melita Vidaković Department of Molecular Biology, Institute for Biological Research Siniša Stanković, University of Belgrade, Belgrade, Serbia

Paolo Vineis Department of Epidemiology and Biostatistics, School of Public Health, Imperial College London, London, UK

Branka Vuković-Gačić Microbiology, Center for Genotoxicology and Ecogenotoxicology, Faculty of Biology, University of Belgrade, Belgrade, Serbia

Danyang Wan College of Life Sciences, Wuhan University, Wuhan, China

Zhipeng A. Wang Department of Chemistry, Texas A&M University, College Station, TX, USA

Gaofeng Wang Dr. John T. Macdonald Foundation Department of Human Genetics, John P. Hussman Institute for Human Genomics, University of Miami Miller School of Medicine, Miami, FL, USA

Department of Human Genetics, Dr. Nasser Ibrahim Al-Rashid Orbital Vision Research Center, Bascom Palmer Eye Institute, University of Miami Miller School of Medicine, Miami, FL, USA

Jian Wang Department of Pathology, Wayne State University School of Medicine, Detroit, MI, USA

Cardiovascular Research Institute, Wayne State University School of Medicine, Detroit, MI, USA

Inga Wessels Institute of Immunology, RWTH Aachen University Hospital, Aachen, Germany

Oliwia Witczak Faculty of Health Sciences, Oslo and Akershus University College of Applied Sciences, Oslo, Norway

Thomas E. Witzig Division of Hematology, Mayo Clinic, College of Medicine, Rochester, MN, USA

Benjamin Wolfson Greenebaum Cancer Center, Department of Biochemistry and Molecular Biology, University of Maryland School of Medicine, Baltimore, MD, USA

Vincent Wai-Sun Wong Department of Medicine and Therapeutics, The Chinese University of Hong Kong, Shatin, N.T., Hong Kong

Xiaosheng Wu Division of Hematology, Mayo Clinic, College of Medicine, Rochester, MN, USA

Xinhua Xiao Department of Endocrinology, Peking Union Medical College Hospital, Chinese Academy of Medical Sciences and Peking Union Medical College, Beijing, China

Xuemei Xie Department of Pediatrics, Tongji Hospital, Tongji Medical College, Huazhong University of Science and Technology, Wuhan, China

Department of Endocrinology, The First Affiliated Hospital of Guangxi Medical University, Nanning, China

Hiroya Yamada Department of Hygiene, Fujita health University School of Medicine, Toyoake, Aichi, Japan

Mingyao Yang Institute of Animal Genetics and Breeding, Sichuan Agricultural University, Chengdu, Sichuan, China

Peixin Yang Department of Obstetrics, Gynecology and Reproductive Sciences, University of Maryland School of Medicine, Baltimore, MD, USA

Department of Biochemistry and Molecular Biology, University of Maryland School of Medicine, Baltimore, MD, USA

Zoe Yates School of Biomedical Sciences and Pharmacy, The University of Newcastle, Ourimbah, NSW, Australia

Soojin V. Yi School of Biological Sciences, Georgia Institute of Technology, Atlanta, GA, USA

Wei Ying Department of Medicine, Division of Endocrinology and Metabolism, University of California, San Diego, La Jolla, CA, USA

Yangmian Yuan College of Life Sciences, Wuhan University, Wuhan, China

Miriam Zanellato Department of Technological Innovation and Safety of Plants, Product and Anthropic Settlements (DIT), Italian Workers' Compensation Authority (INAIL), Rome, Italy

Diana Zapata Department of Pharmaceutical Sciences, Irma Lerma Rangel College of Pharmacy, Texas A&M Health Science Center, College Station, TX, USA

Helmut Zarbl Department of Environmental and Occupational Health, School of Public Health Environmental and Occupational Health Sciences Institute, Rutgers, The State University of New Jersey, Piscataway, NJ, USA

Tatiana Zerjal UMR INRA/AgroParisTech – GABI, Jouy-en-Josas, France

Lubo Zhang Center for Perinatal Biology, Division of Pharmacology, Department of Basic Sciences, Loma Linda University, School of Medicine, Loma Linda, CA, USA

Xiaofei Zhang Fudan University Shanghai Cancer Center, Shanghai, China

Ruqian Zhao Key Laboratory of Animal Physiology and Biochemistry, Ministry of Agriculture, College of Veterinary Medicine, Nanjing Agricultural University, Nanjing, China

Jia Zheng Department of Endocrinology, Peking Union Medical College Hospital, Chinese Academy of Medical Sciences and Peking Union Medical College, Beijing, China

Ling Zheng College of Life Sciences, Wuhan University, Wuhan, China

Beiyan Zhou Department of Immunology, School of Medicine, University of Connecticut Health Center, Farmington, CT, USA

Qun Zhou Greenebaum Cancer Center, Department of Biochemistry and Molecular Biology, University of Maryland School of Medicine, Baltimore, MD, USA

Jianhong Zhu Department of Preventive Medicine, Department of Geriatrics and Neurology at the Second Affiliated Hospital, and Key Laboratory of Watershed Science and Health of Zhejiang Province, Wenzhou Medical University, Wenzhou, Zhejiang, China

Niv Zmora Department of Immunology, Weizmann Institute of Science, Rehovot, Israel

Part III

Influence of Diet and Nutrition on Epigenetics

Energy Metabolism and Epigenetics 76

Scott J. Bultman

Contents

Abstract

The mammalian genome is packaged as chromatin, which influences all enzyme-catalyzed processes that utilize DNA as a template including transcription, DNA replication, the DNA damage response and repair, and meiotic recombination.

S. J. Bultman (✉)
Department of Genetics, Lineberger Comprehensive Cancer Center, University of North Carolina at Chapel Hill, Chapel Hill, NC, USA
e-mail: Scott_Bultman@med.unc.edu

V. B. Patel, V. R. Preedy (eds.), *Handbook of Nutrition, Diet, and Epigenetics*,
https://doi.org/10.1007/978-3-319-55530-0_87

This is why epigenetic writers, readers, and erasers are so crucial for the regulation of genome function. Epigenetic mechanisms are tightly integrated with energy metabolism. For example, methyltransferase and acetyltransferase enzymes that modify DNA and histones require intermediary energy metabolites such as S-adenosylmethionine and acetyl-CoA as essential co-factors. To extend the analogy, epigenetic writers require epigenetic ink. The underlying logic of this relationship is that nutrient availability and cellular metabolism can regulate gene expression and genome function to maintain homeostasis. This review will focus on the functional interplay between epigenetic and metabolic mechanisms and discuss why this is relevant for health and disease prevention with an emphasis on cancer.

Keywords
DNA methylation · Histone methylation · Histone acetylation · Energy metabolites · SAM · Acetyl CoA · NAD^+ · TETs · Sirtuins · HATs

List of Abbreviations

CpG	Cytosine-guanine dinucleotide with phosphodiester (p) linkage
DNMT	DNA methyltransferase
SAM	S-Adenosylmethionine
1C	1 carbon
DMR	Differentially methylated region
ChIP	Chromatin immunoprecipitation
SNP	Single-nucleotide polymorphism
IGF	Insulin-like growth factor
ROS	Reactive oxygen species
TET	Ten-eleven translocation
α-KG	α Ketoglutarate
TCA	Tricarboxylic acid
JMJD	Jumonji C domain
FAD	Flavin adenine dinucleotide
NAD^+	Nicotinamide adenine dinucleotide
ATP	Adenosine triphosphate
PTM	Posttranslational modification
HAT	Histone acetyltransferase
HDAC	Histone deacetylase
SIRT	Sirtuin
AMPK	5′ adenosine monophosphate (AMP)-activated protein kinase
PARP	Poly-ADP-ribose polymerase
mTOR	Mechanistic target of rapamycin

Introduction

Diet and energy metabolism regulate gene expression to influence human health. This link is mediated by epigenetic mechanisms, which fit a paradigm whereby gene-environment interactions converge at the level of the epigenome to regulate

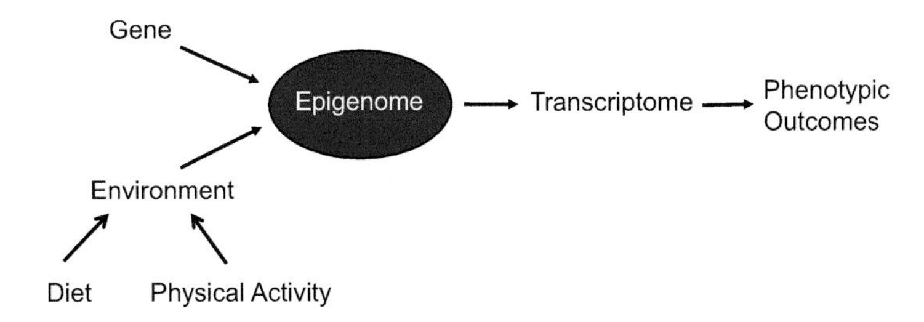

Fig. 1 **Energy balance affects epigenetics.** Many gene-environment interactions converge on the epigenome, which regulates the transcriptome to influence phenotypic outcomes. Diet and physical activity are environmental factors that modulate energy metabolites that can affect DNA methylation, histone posttranslational modifications (PTMs), and ATPase chromatin-remodeling complexes

transcriptome profiles and phenotypic outcomes (Fig. 1). Mechanistically, a number of energy metabolites function as co-factors for epigenetic enzymes (Table 1), which allows changes in nutrient availability to regulate expression and maintain homeostasis. Not surprisingly, chronic perturbations in energy balance and epigenetic mechanisms drive multiple disease states, and the obesity epidemic is increasing the prevalence of type-2 diabetes, cardiovascular disease, and cancer.

To prevent overlap with other chapters in this book, this chapter will not include a general review of epigenetic modifications such as DNA methylation, histone posttranslational modifications, and ATPase chromatin-remodeling complexes. Instead, it will focus on how diet and energy metabolites can modulate the function of epigenetic writers, readers, and erasers. Mouse models will be emphasized because their genetics and environment can be tightly controlled to elucidate molecular mechanisms. To highlight the translational relevance of these mechanisms to cancer and other disease states, human clinical cases will be discussed where there is supporting data.

1C Metabolism and SAM-Mediated Mechanism of DNA Methylation

In the mammalian genome, DNA methylation primarily occurs at cytosines in the context of CpG dinucleotides (Bird 2002). CpG methylation is a widespread and relatively stable epigenomic mark of considerable importance, in part, because promoter hypermethylation often confers transcriptional silence. The de novo DNA methyltransferases (DNMT3A, DNMT3B, and DNMT3C) and the only maintenance enzyme (DNMT1) require S-adenosylmethionine (SAM) as a co-factor that functions as the methyl-group donor (Table 1). For every methyltransferase reaction, the methyl group of SAM is catalytically transferred by a particular DNMT to the 5 position of cytosine to generate 5-methylcytosine. SAM is a product of 1-carbon (1C) metabolism that involves a number of micronutrients including zinc, methionine, and vitamin B family members such as choline and folate/folic acid (Fig. 2;

Table 1 Energy metabolites that are co-factors for epigenetic enzymes

Energy metabolite	Source (energetic process or pathway)	Affected epigenetic enzyme(s)	Effect of metabolite on enzymatic activity
SAM	1C metabolism	DNA methyltransferases (DNMTs)	Increase (methyl group donor)
SAM	1C metabolism	Histone methyltransferases	Increase (methyl group donor)
α-KG	TCA cycle	DNA "demethylases" (TETs)	Increase (R-2HG metabolite decreases)
α-KG	TCA cycle	Histone demethylases with JmjC domain	Increase (R-2HG metabolite decreases)
Succinate	TCA cycle	DNA "demethylases" (TETs)	Decrease
Succinate	TCA cycle	Histone demethylases with JmjC domain	Decrease
Fumarate	TCA cycle	DNA "demethylases" (TETs)	Decrease
Fumarate	TCA cycle	Histone demethylases with JmjC domain	Decrease
Acetyl-CoA	TCA cycle-citrate shuttle-ACLY; ACSS2; nuclear PDH	Histone acetyltransferases (HATs)	Increase (acetyl group donor)
NAD^+	TCA cycle/electron transport chain	Class III histone deacetylases (Sirtuins)	Increase
NAD^+	TCA cycle/electron transport chain	Poly-ADP-ribose polymerases (PARPs)	Increase (ADP-ribose group donor)
FADH2	TCA cycle/electron transport chain	LSD histone demethylases	Increase
ATP	Electron transport chain/oxidative phosphorylation	Chromatin-remodeling complexes with ATPase catalytic subunits	Increase
ATP	Electron transport chain/oxidative phosphorylation	AMP kinase (AMPK)	Decrease (AMP/ATP ratio)

Bultman 2015). Folate is a natural component of leafy vegetables and certain other foods, whereas folic acid is a synthetic form used as a dietary supplement. They are considered equivalent because each one is metabolized to 5-methyltetrahydrofolate (5-methylTHF) in the 1C pathway.

The epigenetic effects of 1C micronutrients can be readily observed in the viable-yellow agouti mouse model where genetically identical mice (when they are maintained on an inbred genetic background) can range in coat color from yellow, through various degrees of mottling where patches of yellow are mixed with agouti (brown), to completely agouti depending on the extent of DNA methylation near one of the transcriptional start sites at the agouti locus (Morgan et al. 1999). At the two extremes, yellow coat color is associated with hypomethylation and higher levels of

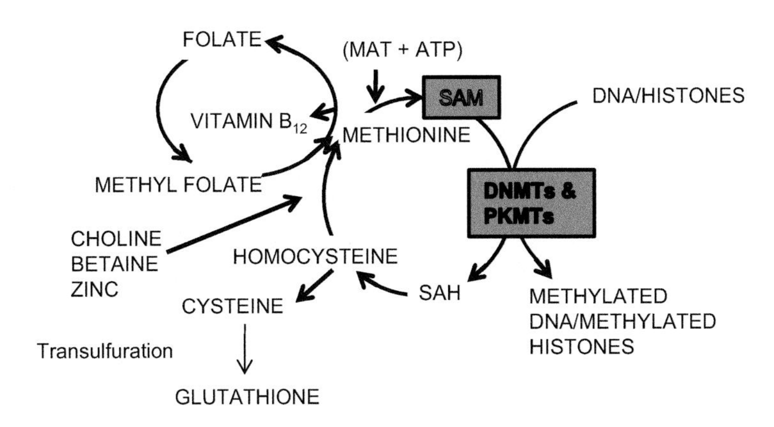

Fig. 2 1C metabolism and SAM influences the methylation of DNA and histones. A schematic of 1-carbon (1C) metabolism. A number of micronutrients including certain B vitamins are used to synthesize methionine, which is converted to S-adenosylmethionine (SAM) by the enzyme MAT in an ATP-dependent manner. SAM is an essential co-factor for DNA methyltransferases (DNMTs) and histone methyltransferases [also known as lysine methyltransferases (KMTs)] because it is the methyl-group donor. In methyltransferase reactions, SAM is converted to S-adenosylhomocysteine (SAH). High levels of SAH can inhibit methyltransferase reactions via feedback inhibition, and SAH is converted to homocysteine, which is recycled to make more methionine. Carbons from SAH can also be redirected to glutathione (via transulfuration) to counteract ROS

transcriptional activity, while agouti is associated with hypermethylation and lower levels of transcription. When viable-yellow females are pregnant and provided a diet with a ~ 3-fold increase in several 1C methyl donors including folic acid and choline, the coat-color distribution of their progeny shifts from yellow towards agouti (Fig. 3; Cooney et al. 2002; Waterland and Jirtle 2003). These studies clearly demonstrate that micronutrients such as vitamin B family members can be rate-limiting factors in SAM production and DNA methylation reactions in vivo. The viable-yellow phenotype is pleiotropic, and a shift toward agouti coat color is correlated with decreased incidence of obesity, type-2 diabetes, and cancer that is observed in all-yellow mice but not all-agouti mice. Dietary supplementation in viable-yellow mice also has multigenerational epigenetic effects because yellow-to-agouti skewing can be apparent in the next generation without any further supplementation although the effect size is smaller (Cropley et al. 2006). The basis for this multigenerational effect is that increased levels of SAM in utero result in increased methylation of the agouti gene in developing female germ cells as well as developing skin/hair follicles within the fetus (Cropley et al. 2006).

The viable-yellow mouse model serves as a good reporter to observe the effects of DNA methylation, but it is not unique. Folic acid or choline supplementation during pregnancy has been reported to alter the DNA methylation status and expression of other genes in both mice and humans including a differentially methylated region (DMR) of the imprinted gene *IGF2* (insulin-like growth factor 2) (Steegers-Theunissen et al. 2009; Waterland et al. 2006; Zeisel 2009). DNA methylation profiling and transcriptome profiling experiments are beginning to identify diverse

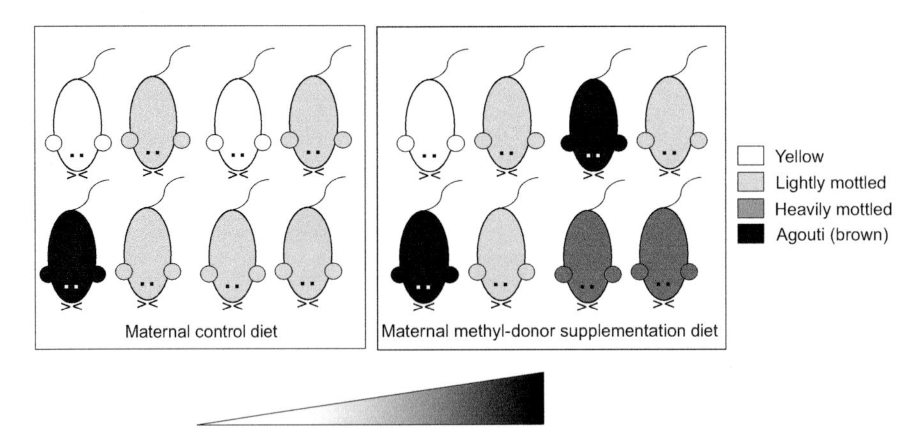

Fig. 3 **Maternal diet can influence the coat-color phenotype of agouti viable-yellow mice.** Shown is a schematic of two litters of viable-*yellow* mice with their coat colors indicated. On the *left* are offspring of a mother provided a control diet. The *yellow* and lightly *mottled* coat colors are well represented (see key on far *right*). On the right are offspring of a mother provided a diet with a three-fold increase in several methyl donors such as folic acid and choline while pregnant. The heavily mottled and agouti coat colors are well represented (see key on far *right*). At the very *bottom*, a gradient-colored *triangle* shows that the coat-color distribution of the offspring shifts based on maternal diet

genes (i.e., not associated with transposable elements as are viable-yellow agouti and axin-fused) that are regulated by these factors via DNA methylation in a tissue-specific manner (Amarasekera et al. 2014; Binder and Michels 2013; Fryer et al. 2011; Schaible et al. 2011). However, the affected genes can be either hyper-methylated or hypomethylated, which indicates the situation is not as simple as methyl donors increasing SAM and DNA methylation at all loci. One potential explanation for differential effects is that SAM might not be randomly distributed throughout the nucleus. Chromatin immunoprecipitation (ChIP) experiments have demonstrated that MAT (methionine adenosyltransferase), which catalyzes the final step of SAM production, is physically associated with chromatin and enriched at certain loci (Katoh et al. 2011), suggesting that SAM levels spike at specific sites that are methylated and silenced. And silencing of a transcriptional repressor could lead to activation (and hypomethylation) of downstream target genes.

1C Metabolism in Human Health: Genetics and Precision Medicine

1C metabolism is important for human health. Epidemiological experiments have demonstrated reproducibly that adequate folate/folic acid intake is important during pregnancy and can prevent neural tube defects such as spina bifida from occurring (Czeizel and Dudas 1992). The importance of these findings from a public health perspective led to flour and other grain-based food products being fortified with folic acid because the neural tube closes in the fetus during the first trimester before many women realize they are pregnant and have had an opportunity to modify their diets. However, this approach is potentially complicated by other epidemiologic findings, suggesting that too high of folate/folic acid intake during pregnancy is associated with several disease states in children after they are born including atopic dermatitis and asthma (Dunstan et al. 2012; Haberg et al. 2009; Kiefte-de Jong et al. 2012; Smith et al. 2008; Whitrow et al. 2009). If confirmed, this discrepancy would be compatible with folate/folic acid affecting many genes in a tissue-specific manner.

Genetics is an important variable when it comes to supplementation with different individuals having different dietary requirements (Bultman 2015). For example, women who are carriers of a *MTHFD1* (5,10-methylenetetrahydrofolate dehydrogenase) single-nucleotide polymorphism (SNP) have >15-fold increased risk of developing signs of choline deficiency when on a low-choline diet plus an increased risk of having children with neural tube defects (Zeisel 2009). This nonsynonymous SNP (resulting in G1958A) is predicted to affect the availability of 5-methylTHF for homocysteine remethylation and may therefore increase demand for choline as a methyl-group donor. Additionally, a C677T SNP in *MTHFR* (5, 10-methylenetetrahydrofolate reductase), which results in a single amino-acid substitution (A222V), also results in neural tube defects (Friso and Choi 2005). MTHFR is a key enzyme in 1C metabolism that catalyzes the production of 5-methylTHF and maintains the flux of methyl groups for the remethylation of homocysteine to methionine, and the A222V variant decreases this activity. Collectively, these genetic variants suggest precision medicine will be relevant to 1C metabolism. To add another layer of complexity on top of genetic heterogeneity, certain pharmaceuticals, including some drugs used to treat seizures, are associated with folate deficiency, suggesting that folate depletion is an adverse side effect. It seems likely that other environmental exposures will affect SAM levels and DNA methylation.

Despite the large body of literature regarding 1C metabolism, the precise mechanism of how it impacts human health is unclear. Although methyl-donor depletion/ supplementation and SNPs in 1C enzymes modulate global DNA methylation levels (Friso and Choi 2005; Zeisel 2009), the target genes that contribute to neural tube defects and other disease states have not been identified. More importantly, some studies have been underpowered in terms of the number of samples and loci analyzed, which has led to some findings not being replicated in subsequent studies (Crider et al. 2012). One challenge is that DNA methylation is highly cell-type dependent so cellular heterogeneity within samples can be a confounding factor. Another challenge is that there is a considerable amount of normal variation that occurs even in homogeneous samples such as cell lines and flow-sorted cells (which is evident from bisulfite sequencing analyses). Finally, the mechanism might not be entirely due to DNA methylation because SAM is also a co-factor for enzymes that methylate histones and nonhistone proteins as well as RNAs and certain lipids. It should also be noted that 1C metabolism influences nucleotide pools for DNA synthesis and also affects amino-acid metabolism. Determining the relative importance of these different processes that are dependent on 1C metabolism will probably not be trivial.

Developmental Windows of Epigenetic Susceptibility for Disease & Transgenerational Effects

The importance of maternal diet, which was discussed in the two previous sections, is supported by cohort studies that evaluated the health status of >2000 people at ~50 years of age who were conceived during the Dutch famine of 1944–1945 (Painter et al. 2005). In Nazi-occupied Netherlands, calorie consumption dropped to 400–800 per day, and children that were conceived during this period became adults with a higher incidence of schizophrenia and heart disease than their siblings. Although it is unclear whether this higher incidence was mediated through epigenetic alterations, it is known that dynamic changes in DNA methylation occur during the earliest stages of mammalian embryogenesis (Messerschmidt et al. 2014), and these studies identified a correlation between disease state and altered CpG methylation in certain genes including *IGF2* (Painter et al. 2005; Tobi et al. 2009). It is plausible that periconceptual folate deficiency contributed to this change (Lumey et al. 2011; Steegers-Theunissen et al. 2009; Sinclair et al. 2007). However, one caveat of the Dutch famine studies is that CpG methylation profiles of blood cells are being correlated with schizophrenia, and it is unclear whether these cells are functionally relevant or if they serve as a reliable proxy for CpG methylation profiles in the CNS. Prenatal malnutrition during the Great Leap Forward famine of 1959–1961 in China also has been associated with increased risk of adult schizophrenia (Xu et al. 2009), but this study did not analyze DNA methylation. A number of similar studies have reported increased incidence of cardiovascular problems, metabolic syndrome, and other health ailments due to the effects of various famines during different postconception stages of pregnancy (Lumey et al. 2011).

Many dietary factors and environmental agents will undoubtedly affect 1C metabolism and/or DNA methylation in general and during fetal development in particular. In fact, it is tempting to speculate that many teratogens perturb DNA methylation and other epigenomic marks. As a possible precedent, high levels of alcohol consumption can cause genome-wide changes in CpG methylation in mouse embryos that are associated with neural tube defects and other teratogenic effects (Liu et al. 2009). Therefore, fetal alcohol syndrome may have an epigenetic component (Haycock 2009). Although the mechanism is not clear, one possibility is that alcohol-induced oxidative stress leads to glutathione turnover, which is replenished by the transulfuration pathway that pulls carbons away from the 1C cycle at the level of cysteine and homocysteine (Fig. 2). This idea is supported by work where acute glutathione depletion affected methionine and led to CpG hypomethylation (Lertratanangkoon et al. 1997).

Recent mouse studies have also revealed links between paternal or maternal diet and DNA methylation in their offspring. One study showed that a paternal low-protein diet led to many modest (~10–20%) changes in DNA methylation status in the liver with altered expression of genes involved in lipid metabolism (Carone et al. 2010). A second study showed that paternal high-fat diets diminished the DNA methylation status of the *Il13ra2* gene in the pancreatic islets of female offspring (Ng et al. 2010). There is probably more evidence for maternal diets than paternal diets including one study linking poor maternal nutrition (low protein or high fat) during pregnancy and nursing with methylation of DNA encoding rRNA in their progeny that was correlated with poor growth (Holland et al. 2016). Yet another study demonstrated that maternal calorie restriction during pregnancy gave rise to males with aberrant DNA methylation of *Lxra* (a gene involved in lipogenesis) in their sperm that was transmitted to a somatic tissue (liver) the next generation (Martinez et al. 2014). This is an example of a transgenerational effect (as opposed to a multigenerational effect) because it persisted for two generations to animals that were never exposed to the stimulus (caloric restriction in this case) as a fetus and were not derived from a germ cell exposed to the stimulus. Another example of a transgenerational effect comes from a gene-trap line of mice carrying a hypomorphic mutation of *Mtrr* (methionine synthase reductase), which encodes an enzyme involved in 1C metabolism (Padmanabhan et al. 2013). This mutation results in global hypomethylated DNA profiles and confers intrauterine growth restriction, developmental delay, and congenital malformations including cardiac and neural tube defects. These defects are dependent on the *Mtrr* genotypes of the maternal grandparents, rather than the maternal or embryonic genotypes, and persist for five generations, possibly through transgenerational epigenetic inheritance.

The above examples demonstrate that what mom or grandma ate or where exposed to while pregnant can influence the epigenetics and phenotype of their children/grandchildren. These observations support the idea that there are windows of susceptibility for dietary factors or environmental exposures during development that contribute to subsequent health outcomes and transgenerational effects. Support for this idea comes from studies of endocrine disruptors such as bisphenol A (BPA) and vinclozolin. BPA is a chemical used to manufacture polycarbonate plastic that has been implicated in autism. When viable-yellow mice are exposed to BPA during fetal or neonatal development, their coat-color distribution is shifted toward yellow, opposite of methyl-donor supplementation, due to decreased CpG methylation (Dolinoy et al. 2007). Interestingly, maternal supplementation of methyl donors including folic acid or the phytoestrogen genistein (an isoflavine derived from soy beans that is abundant in soy-based foods) negates the effect of BPA in viable-yellow mice (Dolinoy et al. 2007). This result suggests that maternal diet might counteract certain adverse environmental exposures. Human case-control studies are currently being performed to investigate whether exposure to dietary factors or environmental agents during pregnancy or postnatal development can alter the epigenome to influence the incidence of autism spectrum disorders and other conditions that have been increasing in incidence in recent years and are therefore believed to have a strong environmental component that involve epigenetic perturbations.

Methyl-Donor Supplementation outside of Pregnancy

The influence of methyl donor supplementation on 1C metabolism is likely to be important outside the context of pregnancy, but this has been more difficult to establish. For example, there is widespread interest in reversing the changes in DNA methylation (and other epigenomic marks) that occur during carcinogenesis. Tumors often exhibit global hypomethylation, particularly in pericentric repeats which can lead to chromosomal instability, while also displaying focal hyper-methylation that can silence tumor-suppressor genes. Unfortunately, studies that have evaluated the effect of folic acid supplementation on DNA methylation status in tumors have been inconsistent (Crider et al. 2012). This could be due to limited sample sizes, different tissues/tumor types, and different DNA methylation assays performed at different loci. Genetics could also be a complicating factor as discussed above for *MTHFR* CC versus TT genotypes. Folic acid supplementation has been implicated in colorectal cancer prevention, but this is somewhat controversial and there is evidence that it may actually increase the progression of existing colonic adenomas or precancerous lesions in the prostate and possibly other sites (Kim 2007a, b; Smith et al. 2008; Ulrich and Potter 2007). Therefore, genetic background, dose (optimal versus too high), and the timing of supplementation may be crucial in distinguishing beneficial and detrimental effects. Both genetics and diet can influence the basal levels of a particular micronutrient, and this is an important factor based on a study of 672 individuals with histories of benign colorectal tumors where half of the participants received 1 mg of folic acid each day for $\geq$3 years and the other half received placebo (Wu et al. 2009). Although folic acid supplementation did not have a significant effect when all of the participants were analyzed, it did have a significant effect reducing the risk of recurrence in those individuals with the lowest levels of folate/folic acid intake prior to the onset of the trial (Wu et al. 2009). Additional work will be necessary to determine whether methyl-donor supplementation and 1C metabolism can modify CpG methylation profiles in adult tissues with cancer-preventive or other beneficial properties. Generally speaking, stratifying cases and controls will probably reveal novel relationships between diet (including methyl-donor supplementation), epigenomic profiles, and health outcomes in human studies.

A number of environmental agents including air pollution, benzene exposure, and particulate pollution have been reported to change DNA methylation in adults (Bollati et al. 2007; Crider et al. 2012; Madrigano et al. 2011). Although the mechanism(s) is not clear, it could involve reactive oxygen species (ROS), glutathione depletion, and the transulfuration pathway to affect 1C flux and DNA methylation as speculated above for alcohol exposure. Estrogen-replacement therapy has been associated with decreased plasma homocysteine levels and increased DNA methylation (Friso et al. 2007). An important area of future research will be to learn more about how the "exposome" impacts the epigenome to influence health outcomes.

DNA Demethylation and Oncometabolites

CpG methylation profiles are determined by DNMT- and SAM-mediated methylation as well as TET-mediated demethylation. There are three TET (ten-eleven translocation) proteins that convert 5-methylcytosine (5mC) to 5-hydroxylmethylcytosine (5hmC) and can further oxidize this base to 5-formylcytosine (5fC) and 5-carboxylcytosine (5caC) (Tan and Shi 2012). Although these 5-methylcytosine derivatives may have significant epigenetic properties of their own, they are clearly intermediates in the process of demethylation as they can be removed by glycosylases/base-excision repair enzymes and then be replaced by unmethylated cytosines (Bhutani et al. 2011; Nabel and Kohli 2011).

The TETs are dioxygenase enzymes that require oxygen, iron (Fe^{2+}), and α-ketoglutarate (α-KG) (Table 1). α–KG plays a crucial role in oxidative metabolism as an intermediate in the tricarboxylic acid (TCA) cycle within mitochondria (Fig. 4). In addition to being produced by isocitrate dehydrogenase (IDH) in the TCA cycle, α–KG can also be generated from glutamine, which is the most abundant amino acid in the body, where it enters the TCA via anapleurosis mediated by two enzymatic reactions (Fig. 4). Intracellular α–KG levels are estimated to be in the low millimolar range, well above the TET K_m values, suggesting that they are not normally rate limiting (Chowdhury et al. 2011; Thirstrup et al. 2011). However, neomorphic IDH mutations, which are common in certain tumors, deplete α–KG (also known as 2-oxyglutarate), and convert it to the R enantiomer of 2-hydroxyglutarate (R-2HG) (Kaelin and McKnight 2013; Yang et al. 2013). R-2HG is referred to as an oncometabolite because it is present at trace levels in normal tissues but accumulates to millimolar levels in IDH mutant tumors. Importantly, R-2HG inhibits TET activity and is both necessary and sufficient to promote certain aspects of tumorigenesis (Losman et al. 2013; Xu et al. 2011). Therefore, normal tissues have high levels of α–KG, which stimulates TET activity, whereas IDH mutant tumors have high levels of R-2HG that inhibit the same enzymes (Kaelin and McKnight 2013; Table 1, Fig. 4). As one would expect, this imbalance results in IDH mutant tumors having significantly increased global levels of DNA methylation (and histone methylation as mentioned in the next section) (Figueroa et al. 2010; Turcan et al. 2012; Xu et al. 2011).

There is additional evidence to support the above mechanism. In other tumors, loss-of-function mutations in two other genes encoding TCA cycle enzymes (Fig. 4), succinate dehydrogenase (SDH), and fumarate hydrase (FH) result in the accumulation of succinate and fumarate, respectively (Kaelin and McKnight 2013; Yang et al. 2013). These TCA cycle intermediates are also oncometabolites that inhibit TETs and cause global hypermethylation of CpGs (Killian et al. 2013; Letouze et al. 2013; Xiao et al. 2012; Fig. 4). Furthermore, evidence is emerging that this mechanism is also applicable to a much larger subset of tumors that do not have IDH, SDH, or FH mutations. Because TETs are oxygen dependent, they exhibit diminished activity in hypoxic tumor environments, which also leads to DNA.

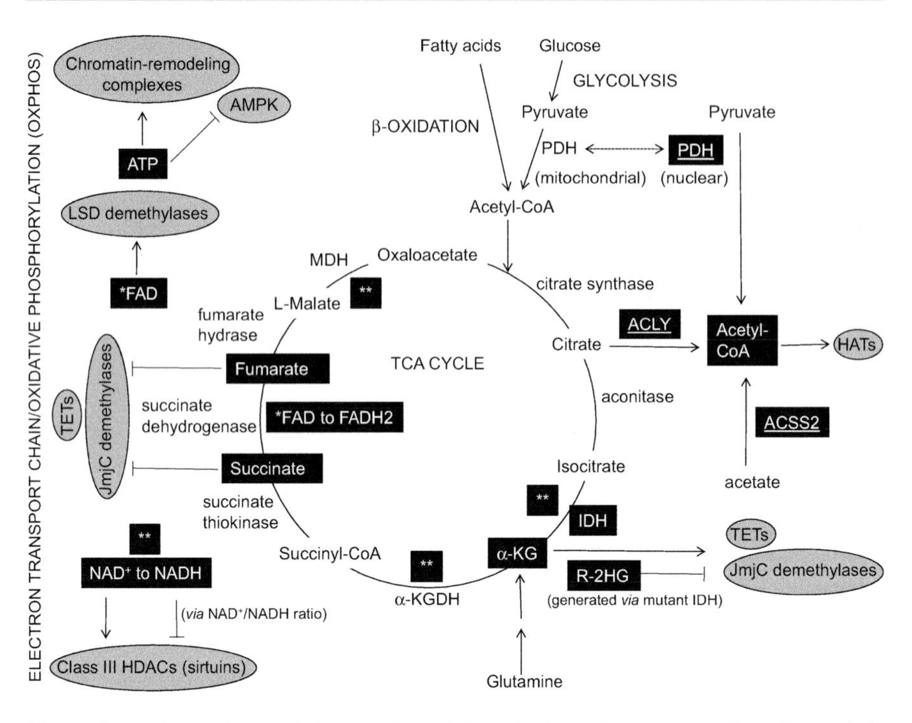

Fig. 4 The TCA cycle can influence the activity of epigenetic enzymes. The TCA cycle is shown along with inputs and outputs at different positions as indicated. Labeling of the TCA cycle and other metabolic processes/pathways are capitalized. Individual enzymes (IDH, ACL, ACS, PDH), energy metabolites (α-KG, succinate, fumarate, acetyl-CoA, ATP), and electron carriers modulated by redox state (NAD^+, NADH, FAD) that activate (*arrows*) or inhibit (*perpendicular lines*) epigenetic enzymes are highlighted by white text in *black boxes*. All of the metabolic enzymes that produce these energy metabolites and electron carries are mitochondrial (TCA cycle and OXPHOS) except for several nucleocytoplasmic enzymes that are *underlined* (ACLY, ACSS2, PDH). The epigenetic enzymes that are regulated by energy metabolites and electron carries are shown in *gray ovals*. See text for details

RNA Methylation and the Epitranscriptome

1C metabolism and SAM influence the methylation of macromolecules other than DNA. An emerging area of great interest is epitranscriptomics, which characterizes chemical modifications of RNAs with a current emphasis on adenosine methylation (Cao et al. 2016). N^6-methyladenosine (m^6A) of mRNAs is enriched near stop codons and in the 5′ and 3′ untranslated regions (UTRs) where it has been implicated in splicing, translation, and degradation (Cao et al. 2016). Methylation also occurs in noncoding (e.g., linc) RNAs, rRNAs, tRNAs, and snoRNAs. N^1-methyladenosine (m^1A) is less abundant but is induced by stress conditions and positively correlates with the level of gene product (Dominissini et al. 2016). Similar to DNA demethylation, RNA demethylation is catalyzed by α-KG-dependent dioxygenases with their activity linked to the TCA cycle. Furthermore, one RNA demethylating enzyme, FTO (fat mass and obesity-associated protein), is associated with human obesity. In the next few sections, there will be a shift from the methylation of nucleic acids to the methylation and acetylation of histones.

Energy Metabolites Involved in Histone Methylation and Demethylation

Genomic DNA is packaged as chromatin, and the nucleosome is the most fundamental unit. Each nucleosome consists of a 147-bp segment of DNA wrapped around a histone octamer (2 copies each of histones H2A, H2B, H3, and H4). Histones undergo a variety of posttranslational modifications (PTMs) including methylation and acetylation that alter nucleosome function. Methylation occurs at arginine (R) and lysine (K) residues and has differential effects on transcription. For example, histone 3, lysine 4 methylation (H3K4me) is associated with active transcription, whereas H3K9me and H3K27me are associated with transcriptional repression. Protein arginine methyltransferases (PRMTs) and protein lysine methyltransferases (PKMTs) catalyze histone methylation and also methylate other proteins. The 9 PRMTs are relatively promiscuous because they methylate multiple R residues on multiple histones, whereas the ~50 PKMTs exhibit target specificity for histone substrates (Greer and Shi 2012). For example, MLL family members methylate H3K4, G9a and SUV39H1 methylate H3K9, and EZH2 methylates H3K27. Although PKMTs do exhibit target specificity, there are many instances of overlap (e.g., G9a and SUV39H1 both target H3K9) with differences in their ability to mono/di/trimethylate (me1, me2, me3) or to function in pericentric heterochromatic regions and different euchromatic loci. All of the PKMTs except DOT1L (which methylates H3K79) contain a SET domain that catalyzes the methyltransferase reaction.

All of the PRMT and PKMT enzymes are dependent on SAM as a methyl donor co-factor similar to DNMTs and RNA methyltransferases such as METTL3/14 (Table 1, Fig. 2). Methyl-donor supplementation/deficiency and genetic variants of enzymes involved in 1C metabolism such as MTHFR will undoubtedly affect histone methylation levels, but relatively few studies have been published so far (Dobosy et al. 2008; Mehedint et al. 2010). The metabolism of the amino acid threonine and SAM are coupled to regulate H3K4me3 and pluripotency in mouse embryonic stem (ES) cells (Shyh-Chang et al. 2013). Although it is not clear why other histone methylation marks are apparently not regulated by this pathway, H3K4me3 is particularly abundant and turns over rapidly in this cell type. This finding suggests that other dietary factors and metabolic pathways will impinge on 1C metabolism. However, as mentioned above, it will be a challenge to determine the relative importance of methyl donor availability, 1C genetics, and SAM levels on DNA methylation versus histone methylation, especially since there is crosstalk between these epigenomic marks.

Methylthioadenosine phosphorylase (*MTAP*) encodes a ubiquitously expressed enzyme in the methionine salvage pathway that is recurrently deleted in human cancers because of its close proximity to the *CDKN2A* tumor-suppressor gene. As a result, methylthioadenosine (MTA) accumulates as an oncometabolite in tumors that inhibits PRMT5 methyltransferase activity (Kryukov et al. 2016; Mavrakis et al. 2016). The viability and growth of *MTAP*-deficient cancer cell lines are impaired by administration of either MTA or PRMT5, which suggests that this vulnerability might be exploited for therapy of *MTAP*/*CDKN2A*–deleted tumors (Kryukov et al. 2016; Mavrakis et al. 2016).

Similar to DNA and RNA, histone methylation is reversed by demethylation. Two families of lysine demethylase (KDM) enzymes have been characterized, but the identification of arginine demethylases has been more elusive. The larger family of KDMs contain a catalytic jumonji C domain (JMJD) and consist of ~30 members that demethylate-specific K residues. Multiple KDMs with JMJDs can act on the same residue with differences in their ability to cleave me1/me2/me3 as well as which loci they act upon in the genome. They also can demethylate nonhistone proteins. Similar to TETs, these KDM/JMJD enzymes are dioxygenases that require oxygen, iron (Fe^{2+}), and the TCA cycle intermediate α-KG (Table 1, Fig. 3). Consequently, neomorphic IDH mutations that generate the oncometabolite R-2HG in tumors result in increased histone methylation in addition to increased CpG methylation (Lu et al. 2012). Another feature of dioxygenases is that they produce formaldehyde as an enzymatic byproduct of demethylation. Endogenous formaldehyde is present in the body at surprisingly high levels (~100 μM) and potentially contributes to carcinogenic protein-DNA crosslinks and ROS, particularly in individuals with Fanconi anemia or possibly even carriers of *Fanc* mutations.

The second family of histone demethylase enzymes (although discovered first) consists of KDM1A and KDM1B (also known as lysine-specific demethylases LSD1 and LSD2). The former acts on HeK4me1/2 and H3K9me1/2, while the later has only been shown to act on H3K4me1/2. Both can also demethylate nonhistone proteins. These enzymes are structurally distinct from the JMJD family and are catalyzed by an amine oxidase domain. KDM1A and KDM1B require flavin adenine dinucleotide (FAD) as a co-factor (Table 1), which is an energy carrier derived from riboflavin (vitamin B2) and ADP that is dependent on the redox state of the cell. Their activities are stimulated by FAD when it is oxidized but not when it is reduced to $FADH_2$ such as occurs in the TCA cycle to carry high-energy electrons used in the electron transport chain of mitochondria that culminates in oxidative phosphorylation and ATP production (Fig. 4). KDM1A/LSD1 directly regulates PPAR function and mitochondrial respiration (Hino et al. 2012), which could potentially contribute to additional epigenomic alterations.

Acetyl-CoA and Histone Acetylation

Another intensively studied PTM is histone acetylation. Unlike methylation, it only occurs at K residues and is usually associated with transcriptional activity. Histone acetyltransferases (HATs) (also called lysine acetyl transferases or KATs because they acetylate nonhistone protein substrates) require acetyl-CoA as a co-factor because it is the acetyl group donor (Table 1). Acetyl-CoA is central to energy metabolism. It is generated in the mitochondria as a product of both glucose metabolism (glycolysis followed by pyruvate dehydrogenase) and lipid metabolism (β oxidation) and condenses with oxaloacetate to form citrate in the first step of the TCA cycle (Fig. 4). Although acetyl-CoA cannot exit the mitochondria by diffusing across membranes, a citrate antiporter system can shuttle citrate from the mitochondria to the cytosol, where ATP citrate lyase (ACLY) reconverts it to acetyl-CoA (Fig. 4). ATP and CoA are consumed by this ACLY-catalyzed reaction, and oxaloacetate is another reaction product. Cytosolic acetyl-CoA is the building block of de novo lipid biogenesis that involves reactions catalyzed by acetyl-CoA carboxylase and fatty acid synthase. Phospholipids are important for doubling the biomass of the plasma membrane every time a cell divides, and triglycerides are utilized for energy storage, especially in adipocytes and hepatocytes. ACLY is also present in the nucleus, and citrate can diffuse through nuclear pores into the nucleus (Wellen et al. 2009). Acetyl-CoA generated by ACLY in the cytosol can also enter the nucleus through nuclear pores. Thus, the citrate shuttle and ACLY represent a major pathway for the production of nuclear acetyl-CoA. Depending on nutritional or metabolic conditions, intracellular acetyl-CoA levels can fluctuate by ~10-fold and are within the K_m range of many HATs, suggesting acetyl-CoA availability can be rate-limiting for histone acetylation (Cai et al. 2011; Takamura and Nomura 1988). The importance of this pathway is supported by RNAi-mediated knockdown studies demonstrating that ACLY is crucial for maintaining global histone acetylation levels (Wellen et al. 2009). It is believed that the citrate shuttle-ACLY pathway is utilized more prominently when nutrients meet or surpass the energy needs of the cell, and ACLY is overexpressed in several tumor types and important for tumor growth (Bauer et al. 2005; Hatzivassiliou et al. 2005). Not surprisingly, this pathway is subject to feed-forward activation as glucose and insulin signaling upregulate ACLY activity via phosphorylation by AKT (Lee et al. 2014a).

The importance of nucleocytosolic acetyl-CoA is underscored by it being produced not only by ACLY but also by other mechanisms. One of two acetyl-CoA synthetase enzymes condenses acetate and coenzyme A in the nucleus and cytosol (Table 1, Fig. 4). As one might expect, RNAi-mediated knockdown this enzyme (ACSS2) in addition to ACLY results in particularly low levels of global histone acetylation (Wellen et al. 2009). ACSS2 is overexpressed in many tumors whose growth is dependent on acetate (Comerford et al. 2014). In addition, pyruvate dehydrogenase (PDH) can translocate to the nucleus instead of being exclusively localized to the mitochondria. This mitochondrial-to-nuclear translocation of PDH occurs when cells are stimulated with serum or epidermal growth factor (EGF), and RNAi knockdown studies demonstrate that PDH also contributes to nuclear pools of acetyl-CoA histone acetylation (Sutendra et al. 2014; Table 1, Fig. 4).

It is tempting to speculate that acetyl-CoA links energy status and gene expression to regulate cell-cycle progression. When cells are serum starved or lack enough energy metabolites to complete the next round of cell division, they do not pass a G_1 checkpoint called the restriction point and become quiescent because it would be deleterious to begin DNA replication without being able to complete it and the ensuing mitosis. This checkpoint protects against genomic instability and cancer. Re-addition of serum or nutrients allows cells to move beyond the restriction point and the G_1/S phase transition and also induces histone acetylation. RB phosphorylation is involved in this process, but acetyl-CoA is also a good candidate because it is influenced by energy balance, helps to build new cells via lipid biogenesis, and regulates gene expression as a HAT co-factor. In support of this idea, acetyl-CoA levels can fluctuate by six-fold in yeast depending on their metabolic cycle (oxidative versus reductive growth), and this correlates with H3K9ac levels at certain loci (Kaelin and McKnight 2013). Furthermore, when carbon sources are added back to nutrient-deprived yeast that are quiescent, acetyl-CoA levels increase and induce histone acetylation (via the GCN5 HAT) and transcription of growth-related genes (Cai et al. 2011). It is also likely that acetyl-CoA-driven HAT activity triggers the G_1/S phase transition by facilitating DNA replication because origins of replication are enriched in histone acetylation (Goren et al. 2008). Finally, it should be reiterated that HATs/KATs acetylate nonhistone proteins, and the acetylated proteome is enriched for metabolic enzymes and cell-cycle regulators including ACLY (Lin et al. 2013; Zhao et al. 2010).

NAD^+ and Histone Deacetylation

Histone deacetylases (HDACs) are divided into several classes, and the class III HDACs (also known as sirtuins) require nicotinamide adenine dinucleotide (NAD^+) as an essential co-factor (Table 1). NAD^+ is an energy carrier that accepts electrons from other molecules as it is reduced to NADH. Similar to FAD and $FADH_2$, this change in redox status links the TCA cycle to the electron transport chain and oxidative phosphorylation for ATP production in mitochondria (Fig. 4). A relatively low NAD^+/NADH ratio is therefore a measure of favorable energetics that is "sensed" by sirtuins (SIRTs) to downregulate their deacetylase activity (Guarente 2011). However, in response to prolonged fasting and diminished glucose intake, this ratio is elevated, which leads to increased sirtuin activity including SIRT1-mediated deacetylation of H3K9ac and H4K16ac at ribosomal DNA (rDNA) repeats that downregulates rDNA transcription, ribosome biogenesis, and consequently protein synthesis. SIRT1 and other sirtuins also regulate FoxO transcription factors and various aspects of autophagy (Hariharan et al. 2010; Lee et al. 2008; Ng and Tang 2013). These findings are compatible with sirtuins playing a role in cellular growth when nutrients are abundant and autophagy when nutrients are scarce.

Not only can diet elevate NAD^+ levels and the NAD^+/NADH ratio (Canto et al. 2012), but alcohol intake can have the opposite effect when it is metabolized by alcohol dehydrogenase (ADH) to acetaldehyde and then by acetaldehyde dehydrogenase (ALDH) to acetate because both reactions involve the reduction of NAD^+ to NADH (Shepard and Tuma 2009). This is expected to increase the NAD^+/NADH ratio and inhibit sirtuins, which is consistent with observations that that excessive alcohol exposure can increase histone acetylation (Haycock 2009). In this scenario, increased histone acetylation could also be driven by a second mechanism where the acetate end product is converted to acetyl-CoA by ACSS2 to increase HAT activity. Although alcohol-induced histone acetylation should occur primarily in the liver, it does have effects in the CNS (Pascual et al. 2012) and might contribute to the teratogenic effects of alcohol in addition to decreased DNA methylation (as well as epigenetic-independent mechanisms).

Similar to other HDACs, the sirtuins also deacetylate nonhistone proteins. In fact, most of the 7 sirtuins are localized outside of the nucleus with SIRT2 being cytosolic and SIRT3, SIRT4, and SIRT5 being mitochondrial (Houtkooper et al. 2012). This subcellular localization is consistent with the sirtuins regulating energy metabolism and supports proteomic data that metabolic enzymes are regulated by acetylation to a greater extent than other proteins. Only SIRT1, SIRT6, and SIRT7 are localized to the nucleus, and SIRT1 is also localized in the cytosol while SIRT7 is primarily found in the nucleolus (Houtkooper et al. 2012). SIRT1 plays an important role in the maintenance of circadian rhythms and part of this driven by the oscillation of NAD^+ levels (Nakahata et al. 2009; Ramsey et al. 2009; Sassone-Corsi 2012). Energy metabolism and autophagy play a central role in carcinogenesis, and several sirtuins including SIRT6 are tumor suppressors that coordinate glycolytic and TCA cycle activity (Sebastian et al. 2012). Part of this mechanism involves SIRT6 deacetylation of H3K9 at the promoters genes involved in glycolysis (Zhong et al. 2010).

Caloric Restriction, Lifespan, and Histone Acetylation

Reducing caloric intake to ~60% of ad libitum is known to diminish certain age-related diseases and increase lifespan up to 150% in organisms ranging from yeast to humans, and much of this effect has been attributed to SIRT1 activity. Although there has been some controversy regarding SIRT1 and lifespan extension in *C. elegans* and *Drosophila* (Burnett et al. 2011), it has not been challenged in mammals. For example, caloric restriction increases NAD^+ levels in certain tissues (muscle, liver, white adipose tissue) and extends the lifespan of wild-type mice, but it does not have this effect in SIRT1-deficient mice (Herranz and Serrano 2010). Although this indicates that the effect of caloric restriction is SIRT1-dependent, a caveat is that SIRT1-deficient mice are short lived regardless of diet with many dying in the neonatal period. Further support comes from the observation that transgenic mice overexpressing SIRT1 are protected from glucose intolerance and hyper-insulinemia in response to a high-fat diet although longevity is unaffected (Herranz and Serrano 2010). Similarly, resveratrol, which is a naturally occurring polyphenol present in the skin of red grapes (but present at low levels in red wine) and other fruits, activates SIRT1 and can improve insulin sensitivity and extend lifespan of mice provided a high-fat diet. However, resveratrol probably activates SIRT1 indirectly via AMPK, and the relative importance of SIRT1 compared to other targets is not known. Nevertheless, these findings support the idea that SIRT1 protects against metabolic stress and promotes "healthspan." It is not clear to what extent this is due to deacetylation of histones versus nonhistone targets. However, caloric restriction does lead to changes in histone acetylation and other epigenomic marks at various loci (Li et al. 2011). Other energy metabolites involved in epigenetics will probably be involved in mammalian aging as suggested by work in more experimentally tractable eukaryotic model systems. For example, α-KG extends the lifespan of *C. elegans* by inhibiting ATP synthase, and this is dependent on target of rapamycin (TOR) and also associated with increased autophagy (Chin et al. 2014). Furthermore, RNAi-mediated depletion of acetyl-CoA synthetase in yeast (*S. cerevisiae*) and fruit flies (*D. melanogaster*) alters nucleocytosolic acetyl-CoA levels to stimulate autophagy and prolong lifespan, and part of this is mediated by H4K16ac and upregulation of autophagy (atg) genes (Eisenberg et al. 2014).

Prolonged fasting during caloric restriction results in the liver producing ketone bodies such as β-hydroxybutyrate (β-OHB) as an energy source. This also occurs when glucose levels drop due to ketogenic diets (i.e., high fat/low carbohydrate diets such as the Atkins diet) or prolonged exercise. Serum levels of β-OHB are induced to the millimolar range compared to basal levels in the micromolar range. In addition to being important for energy metabolism, millimolar concentrations of β-OHB function as an inhibitor of class I HDACs (distinct from the sirtuins) and increase H3K9ac and H3K14ac (Shimazu et al. 2013). This is the case globally and at specific genes such as FOXO3A that are upregulated and may account for some of the homeostatic effects (Shimazu et al. 2013). It should be noted that β-OHB production is activated in response to fasting by mitochondrial SIRT3 deacetylating an enzyme (HMGCS2) that regulates ketone body production (Shimazu et al. 2010).

Additional Histone PTMs

Histones undergo additional PTMs that are derived from NAD^+ and other energy metabolites. One or more ADP-ribose moieties can be added to K, R, and several other amino-acid residues by 18 different poly-ADP-ribose polymerase (PARP) enzymes (Messner and Hottiger 2011). PARP1 has been investigated in the most detail and catalyzes the majority of ADP-ribose chains. Similar to sirtuins, the PARPs require NAD^+. NAD^+ is consumed because it donates the ADP-ribose group (Table 1). The poly-ADP-ribose chain length can be determined by NAD^+ concentrations (Huletsky et al. 1985), which is influenced by nutritional status and redox state (Fig. 4). Histone poly-ADP ribosylation plays a role in transcriptional regulation and DNA replication but is particularly important for DNA repair (Messner and Hottiger 2011). Because PARPs and sirtuins compete for the same NAD^+ pools, extensive DNA damage may result in PARP-mediated depletion of NAD^+ to an extent that attenuates sirtuin-mediated histone deacetylation. NAD^+ levels decline during normal aging, and this could be due to increased PARP1 DNA repair activity in response to an accumulation of abasic sites and single-strand DNA breaks. Dietary administration of nicotinamide riboside to replenish NAD^+ levels in aging mice ameliorates loss of muscle stem cells, which is one sign of aging, although this may involve SIRT-mediated effects on mitochondria rather than histone acetylation (Zhang et al. 2016). It is tempting speculate that PARP inhibitors, which are used as chemotherapeutic agents, might affect sirtuin activity. Either of these scenarios could involve nonhistone proteins as supported by the observation that DNA damage and mitochondrial SIRT4 tumor-suppressive function involve glutamine metabolism/anaplerosis and carbon flux into the TCA cycle (Jeong et al. 2013). Like other epigenomic marks, poly-ADP-ribosylation is reversible and PARPs are counteracted by ADP-ribosylhydrolases and PAR glycohydrolases.

Glucose induces histone GlcNacylation, and GlcNAc is derived from glycolysis via the hexosamine pathway (Fujiki et al. 2011). Glucose also induces histone monoubiquitination (H2B-K120ub), which is an active mark that is not targeted for proteosome-mediated degradation. New marks have recently been identified by mass spectrometry, and it will not be surprising if some are influenced by energy metabolism.

ATP Levels, Chromatin-Remodeling Complexes, and AMPK

Four families of chromatin-remodeling complexes (SWI/SNF, ISWI, NuRD/Mi2/CHD, and INO80) have been characterized that utilize catalytic subunits with ATPase activity to slide or evict nucleosomes at many sites in the genome. Much attention has focused on their occupancy at promoters and their role in nucleosome depletion (as detected by ATAC-seq) and transcriptional initiation. Because each one has a catalytic ATPase subunit, these complexes have activities that are dependent on adenosine triphosphate (ATP) (Table 1), which is produced by oxidative phosphorylation in the mitochondria and serves as the energy currency of the cell (Fig. 4). Theoretically, suboptimal energetic conditions could result in ATP dropping below normal steady-state levels such that it becomes rate limiting for chromatin remodeling. However, there is not any evidence to support this notion at the current time. For example, SWI/SNF complexes have robust activity in cell-free assays when 1–2 millimolar ATP is used, and this ATP concentration is toward the lower end of the normal intracellular concentration range (0.5–10 millimolar). However, ATP levels could conceivably drop below this range in the nucleus under extreme conditions (without causing cell death) or normal ATP levels might be insufficient to meet nuclear demand under certain circumstances that require extensive chromatin remodeling. There is evidence suggesting that mitochondrial-derived ATP might not always be sufficient for chromatin remodeling under nonbasal conditions. PARP hydrolysis yields ATP and represents a local source of nuclear ATP that is essential for hormone-induced chromatin remodeling (Wright et al. 2016). For mitochondrial- and nuclear-derived ATP, it is also possible that there are differences in the local concentration of ATP within different regions of the nucleus (e.g., nucleolus, nuclear periphery, pericentric heterochromatin), which could diminish chromatin remodeling at a subset of genomic loci.

In contrast to chromatin-remodeling complexes, the enzyme AMPK [5′ adenosine monophosphate (AMP)-activated protein kinase] is highly sensitive to ATP levels in the cell. It is activated by high AMP/ATP ratios, which are indicative of low energy status, and activates many substrates in multiple pathways including suppression of the pro-growth mTOR pathway and induction of autophagy to adapt to energetic stress and enable cell survival (Table 1, Fig. 4). Among its direct targets is H2B-S36. AMPK-mediated phosphorylation of this residue is important for transcriptional regulation and homeostasis (Bungard et al. 2010). Activation of AMPK alters the NAD^+/NADH ratio to modulate SIRT1 activity and the deacetylation of PPARs and FOXO transcription factors as another way to restore energy homeostasis (Canto et al. 2009). This suggests that AMPK may deacetylate histones indirectly via SIRT1 or SIRT6.

Not surprisingly, this pathway is perturbed in certain cancers. The upstream activator of AMPK (LKB1 [liver kinase B1] or STK11 [serine-threonine kinase 11]) is a tumor suppressor mutated in Peutz-Jeghers syndrome. In a mouse model of pancreatic cancer and primary pancreatic epithelial cells, mutation of LKB1 in KRAS G12D driven tumors prevents AMPK activation to increase glycolysis, divert glucose into serine biosynthesis, and increase SAM levels and DNMT-catalyzed DNA methylation (Kottakis et al. 2016).

Lack of Physical Activity, the Obesity Epidemic, Cancer, and Epigenetics

Physical activity and diet influence energy metabolites and epigenetics (Fig. 1). Although our knowledge of the former is not as strong as the latter, strenuous exercise can affect NAD^+ levels, AMPK activation, and certain epigenomic marks including DNA methylation (Kirchner et al. 2013). A more sedentary lifestyle combined with the Western diet (energy-rich, processed foods) is probably responsible for most or all of the increased adiposity of human populations, and the obesity epidemic will continue to have profound deleterious consequences on cardiovascular disease, diabetes, cancer, and other disease states. Many of the underlying disease etiologies will involve direct effects on energy balance that culminate in increased adiposity and metabolic syndrome, which are associated with atherosclerosis, insulin resistance, inflammation, and a tumor-permissive environment. However, chronic "overnutrition" is likely to have other effects including altered gene expression due to perturbed epigenomic marks that arise because of altered energy metabolites that function as co-factors for epigenetic enzymes (via mechanisms discussed in above sections). For example, the enzyme NNMT (nicotinamide N-methyltransferase) methylates nicotinamide (vitamin B3) using SAM as a methyl donor and is upregulated in white adipose tissue and liver from obese humans and rodents. NNMT can impact sirtuins because nicotinamide is a precursor of NAD^+, while it can simultaneously affect DNA methylation and histone methylation because SAM is converted to SAH. NNMT-depleted mice are resistant to diet-induced obesity and insulin resistance, and this phenotype is associated with altered SAM, NAD^+, and H3K4me levels in white adipose tissue (Kraus et al. 2014).

In human epidemiologic studies, a strong association between obesity and increased incidence of various cancers has been very reproducible (Willett 2008). This link is compatible with metformin, which includes AMPK as a target and has long used to treat type-2 diabetes, being repurposed for cancer treatment with considerable success (Quinn et al. 2013). This link may involve obesity leading to epigenomic perturbations that precede oncogenic transformation and contribute to carcinogenesis. For example, diet-induced obesity in mice lacking any oncogenic stimuli alters H3K27ac profiles at putative enhancers and transcriptome profiles in colonocytes, and these obesity-associated profiles resemble a colorectal cancer signature (Li et al. 2014). However, the relative contribution of epigenomics must still be determined for the obesity-cancer connection. Another challenge is that, unlike obesity, epidemiologic studies that have evaluated individual dietary factors such as fruits, vegetables, fiber, and red meat in cancer prevention have yielded inconsistent findings (Willett 2008).

Due to the Warburg effect (Vander Heiden et al. 2009), tumors primarily utilize aerobic glycolysis instead of oxidative metabolism. Although this is inefficient from an energetic perspective (2–4 ATPs versus 36 ATPs per glucose molecule, respectively), carbons and nitrogens from glycolytic intermediates and the pentose phosphate pathway are diverted into nucleotide and amino-acid salvage pathways (instead of the carbon "escaping" as CO_2), which enables the rapidly dividing cell to double its biomass in as little as 6–8 h. Similarly, the citrate shuttle-ACLY pathway is crucial for increased for lipid biogenesis (to double the plasma membrane content), but this also increases histone acetylation. This represents one way in which the transcriptome profile of the cancer cell diverges from the original cell-of-origin without requiring mutations in the promoter regions of the affected genes. To compensate for their inefficient ATP production, tumors upregulate the expression of their glucose transporters (GLUTs) and take up $>$10X glucose than normal cells, and this is the basis for tumor imaging in the clinic (via 2-FDG and PET scanning). Metabolic reprogramming is not exclusive to cancer. A mechanism similar or identical to the Warburg effect is responsible for the rapid proliferation and differentiation of activated T cells during an immune response (Peng et al. 2016). However, this process is deregulated in cancer, and it is reinforced by mutations in proto-oncogenes and tumor-suppressor genes that play an important role in the overexpression of GLUTs and other aspects of metabolic reprogramming (Soga 2013; Lee et al. 2014b).

Bioactive Food Components that Inhibit Epigenetic Enzymes

Certain naturally occurring bioactive food components can inhibit epigenetic enzymes by mechanisms other than 1C metabolism (Huang et al. 2011; Ong et al. 2011; Supic et al. 2013). Biochemical analysis of plant-derived extracts, particularly those used as traditional medicines in China and India, has led to a number of bioactive molecules. For example, several polyphenols can inhibit DNMTs: epigallocatechin-3-gallate (EGCG), a catechin abundant in green tea; genistein, an isoflavone from soybeans; curcumin from the roots of tumeric herb; resveratrol from grape skins; and caffeic acid and chlorogenic acid from coffee. Naturally occurring HDAC inhibitors include sulforaphane, an isothiocyanate abundant in cruciferous vegetables; diallyl disulfide as well as allyl mercaptan, organosulfur compounds abundant in garlic; and butyrate, a short-chain fatty acid produced by bacterial fermentation of fiber in the gut. The gut microbiome has prodigious metabolic capacity and will likely produce many metabolites other than butyrate and choline that affect host epigenetics (Donohoe et al. 2011, 2012; Spencer et al. 2011). Other dietary compounds inhibit HATs such as anacardic acid, salicylic acids with alkyl chains present in certain nuts; curcumin (listed above for DNMTs); and plumbagin, a hydroxynaphthoquinone from the roots of a leadwort plant. Other epigenetic enzymes can also be targeted. For example, omega-3 fatty acids, which are common in fish oils, can inhibit the EZH2 histone methyltransferase.

Depending on which specific HDACs or HATs (or other epigenetic enzyme) are affected by each dietary agent, different target genes will be affected in different tissues. This is relevant because many studies have been performed in cell lines and need to be extended to mouse models and human subjects. A benefit of in vivo studies is that they will more rigorously test the hypothesis that some of these factors are beneficial for human health, especially for cancer prevention. However, even in cases where these links are confirmed, it will be important to determine how much of the beneficial effect is mediated by altering epigenomic marks versus other possible mechanisms such as affecting nonhistone proteins or antioxidant activity. Many other bioactive food components and micronutrients may affect histone acetylation or other epigenomic marks that still await identification. Considering the size of the nutritional supplement industry and how little we know about many agents, it would be useful to learn more about their potential epigenetic effects by performing large-scale ChIP-seq, BS-seq, and ATAC-seq screens to evaluate histone PTMs, DNA methylation, and nucleosome occupancy/density, respectively.

Dictionary of Terms

- **Agouti viable yellow** – a mouse model of epigenetics where the effects of DNA methylation can be readily observed based on coat-color pigmentation.
- **Anaplerosis** – a process where TCA cycle intermediates are replenished by biochemical reactions such as the conversion of glutamine to glutamate and then α-KG (catalyzed by glutaminase and glutamate dehydrogenase, respectively).
- **Autophagy** – an intracellular process where organelles and macromolecules are broken down by autophagasomes vesicles fusing with lysosomes for energetics or quality-control purposes.
- **Cataplerosis** – a process where TCA cycle intermediates exit the mitochondria and are utilized for biosynthetic anabolic reactions and epigenetic reactions instead of ATP production. For example, the citrate transporter that allows ACLY to generate nucleocytoplasmic acetyl-CoA.
- **Histone modifications** – covalent posttranslational modifications of histones, such as acetylation or methylation, that are added by enzymatic "writers," interpreted by "reader" proteins, and removed by enzymatic "erasers."

- **Nucleosome** – the fundamental unit of chromatin that consists of 147-bp segments of DNA wrapped around histone octamers.
- **Redox** – a reduction-oxidation reaction where a molecule is reduced while another is oxidized and is a complimentary process with gain and loss of electrons, respectively. For example, NAD^+ and NADH where NADH serves as an electron carrier to link the TCA cycle to the electron transport chain in mitochondria.
- **Transgenerational epigenetic inheritance** – a form of epigenetic cellular memory where DNA methylation or another epigenetic mark at a particular gene is passed from one generation to subsequent generations by being stably maintained during gametogenesis.

Key Facts

- The majority of epigenetic enzymes utilize energy metabolites as essential co-factors.
- To extend the reader-writer-eraser analogy, epigenetic writers require epigenetic ink. For example, the energy metabolites SAM and acetyl-CoA are methyl-group and acetyl-group donors required for methylation and acetylation, respectively.
- In many cases, the energy metabolites that moonlight as co-factors for epigenetic enzymes are not rate limiting and link nutrient availability with regulation of gene expression to achieve homeostasis. This is particularly true for reactions that follow Michaelis-Menten kinetics and is compatible with the reversible nature of epigenetic reactions.
- The 1C cycle and TCA cycle are hubs for energy metabolism that also control carbon flux and redirect energy intermediates into biosynthetic and/or epigenetic reactions depending on the energetic state of the cell.
- Certain naturally occurring bioactive food components and metabolites of the gut microbiome can inhibit epigenetic enzymes.
- Perturbations in the interplay between metabolism and epigenetics drive tumorigenesis.

Summary Points

- The majority of epigenetic enzymes utilize energy metabolites as essential co-factors.
- To extend the reader-writer-eraser analogy, epigenetic writers require epigenetic ink. For example, the energy metabolites SAM and acetyl-CoA are methyl-group and acetyl-group donors required for methylation and acetylation, respectively.
- In many cases, the energy metabolites that moonlight as co-factors for epigenetic enzymes are not rate limiting and link nutrient availability with regulation of gene expression to achieve homeostasis. This is particularly true for reactions that

follow Michaelis-Menten kinetics and is compatible with the reversible nature of epigenetic reactions.
- The 1C cycle and TCA cycle are hubs for energy metabolism that also control carbon flux and redirect energy intermediates into biosynthetic and/or epigenetic reactions depending on the energetic state of the cell.
- Certain naturally occurring bioactive food components and metabolites of the gut microbiome can inhibit epigenetic enzymes.
- Perturbations in the interplay between metabolism and epigenetics drive tumorigenesis.

Permissions

Caister Academic Press has kindly granted permission for the author to re-use text and figures from a chapter previously published in "Epigenetics: Current Research and Emerging Trends" (edited by Dr. Brian Chadwick). The previously published chapter was cited multiple times and appears in the References as: Bultman (2015).

References

Amarasekera M, Martino D, Ashley S, Harb H, Kesper D, Strickland D, Saffery R, Prescott SL (2014) Genome-wide DNA methylation profiling identifies a folate-sensitive region of differential methylation upstream of ZFP57-imprinting regulator in humans. FASEB J 28:4068

Bauer DE, Hatzivassiliou G, Zhao F, Andreadis C, Thompson CB (2005) ATP citrate lyase is an important component of cell growth and transformation. Oncogene 24:6314–6322

Bhutani N, Burns DM, Blau HM (2011) DNA demethylation dynamics. Cell 146:866–872

Binder AM, Michels KB (2013) The causal effect of red blood cell folate on genome-wide methylation in cord blood: a Mendelian randomization approach. BMC Bioinf 14:353

Bird A (2002) DNA methylation patterns and epigenetic memory. Genes Dev 16:6–21

Bollati V, Baccarelli A, Hou L, Bonzini M, Fustinoni S, Cavallo D, Byun HM, Jiang J, Marinelli B, Pesatori AC et al (2007) Changes in DNA methylation patterns in subjects exposed to low-dose benzene. Cancer Res 67:876–880

Bultman SJ (2015) Metabolic inputs into epigenetics. In: Chadwick BP (ed) Epigenetics: current research and emerging trends. Caister Academic Press, Norfolk, pp 307–326

Bungard D, Fuerth BJ, Zeng PY, Faubert B, Maas NL, Viollet B, Carling D, Thompson CB, Jones RG, Berger SL (2010) Signaling kinase AMPK activates stress-promoted transcription via histone H2B phosphorylation. Science 329:1201–1205

Burnett C, Valentini S, Cabreiro F, Goss M, Somogyvari M, Piper MD, Hoddinott M, Sutphin GL, Leko V, McElwee JJ et al (2011) Absence of effects of Sir2 overexpression on lifespan in C. elegans and Drosophila. Nature 477:482–485

Cai L, Sutter BM, Li B, Tu BP (2011) Acetyl-CoA induces cell growth and proliferation by promoting the acetylation of histones at growth genes. Mol Cell 42:426–437

Canto C, Gerhart-Hines Z, Feige JN, Lagouge M, Noriega L, Milne JC, Elliott PJ, Puigserver P, Auwerx J (2009) AMPK regulates energy expenditure by modulating NAD+ metabolism and SIRT1 activity. Nature 458:1056–1060

Canto C, Houtkooper RH, Pirinen E, Youn DY, Oosterveer MH, Cen Y, Fernandez-Marcos PJ, Yamamoto H, Andreux PA, Cettour-Rose P et al (2012) The NAD(+) precursor nicotinamide riboside enhances oxidative metabolism and protects against high-fat diet-induced obesity. Cell Metab 15:838–847

Cao G, Li HB, Yin Z, Flavell RA (2016) Recent advances in dynamic m6A RNA modification. Open Biol 6:160003

Carone BR, Fauquier L, Habib N, Shea JM, Hart CE, Li R, Bock C, Li C, Gu H, Zamore PD et al (2010) Paternally induced transgenerational environmental reprogramming of metabolic gene expression in mammals. Cell 143:1084–1096

Chin RM, Fu X, Pai MY, Vergnes L, Hwang H, Deng G, Diep S, Lomenick B, Meli VS, Monsalve GC et al (2014) The metabolite alpha-ketoglutarate extends lifespan by inhibiting ATP synthase and TOR. Nature 509:397–401

Chowdhury R, Yeoh KK, Tian YM, Hillringhaus L, Bagg EA, Rose NR, Leung IK, Li XS, Woon EC, Yang M et al (2011) The oncometabolite 2-hydroxyglutarate inhibits histone lysine demethylases. EMBO Rep 12:463–469

Comerford SA, Huang Z, Du X, Wang Y, Cai L, Witkiewicz AK, Walters H, Tantawy MN, Fu A, Manning HC et al (2014) Acetate dependence of tumors. Cell 159:1591–1602

Cooney CA, Dave AA, Wolff GL (2002) Maternal methyl supplements in mice affect epigenetic variation and DNA methylation of offspring. J Nutr 132:2393S–2400S

Crider KS, Yang TP, Berry RJ, Bailey LB (2012) Folate and DNA methylation: a review of molecular mechanisms and the evidence for folate's role. Adv Nutr 3:21–38

Cropley JE, Suter CM, Beckman KB, Martin DI (2006) Germ-line epigenetic modification of the murine A vy allele by nutritional supplementation. Proc Natl Acad Sci U S A 103: 17308–17312

Czeizel AE, Dudas I (1992) Prevention of the first occurrence of neural-tube defects by periconceptional vitamin supplementation. N Engl J Med 327:1832–1835

Dobosy JR, Fu VX, Desotelle JA, Srinivasan R, Kenowski ML, Almassi N, Weindruch R, Svaren J, Jarrard DF (2008) A methyl-deficient diet modifies histone methylation and alters Igf2 and H19 repression in the prostate. Prostate 68:1187–1195

Dolinoy DC, Huang D, Jirtle RL (2007) Maternal nutrient supplementation counteracts bisphenol A-induced DNA hypomethylation in early development. Proc Natl Acad Sci U S A 104:13056–13061

Dominissini D, Nachtergaele S, Moshitch-Moshkovitz S, Peer E, Kol N, Ben-Haim MS, Dai Q, Di Segni A, Salmon-Divon M, Clark WC et al (2016) The dynamic N(1)-methyladenosine methylome in eukaryotic messenger RNA. Nature 530:441–446

Donohoe DR, Garge N, Zhang X, Sun W, O'Connell TM, Bunger MK, Bultman SJ (2011) The microbiome and butyrate regulate energy metabolism and autophagy in the mammalian colon. Cell Metab 13:517–526

Donohoe DR, Collins LB, Wali A, Bigler R, Sun W, Bultman SJ (2012) The Warburg effect dictates the mechanism of butyrate-mediated histone acetylation and cell proliferation. Mol Cell 48:612–626

Dunstan JA, West C, McCarthy S, Metcalfe J, Meldrum S, Oddy WH, Tulic MK, D'Vaz N, Prescott SL (2012) The relationship between maternal folate status in pregnancy, cord blood folate levels, and allergic outcomes in early childhood. Allergy 67:50–57

Eisenberg T, Schroeder S, Andryushkova A, Pendl T, Kuttner V, Bhukel A, Marino G, Pietrocola F, Harger A, Zimmermann A et al (2014) Nucleocytosolic depletion of the energy metabolite acetyl-coenzyme a stimulates autophagy and prolongs lifespan. Cell Metab 19:431–444

Figueroa ME, Abdel-Wahab O, Lu C, Ward PS, Patel J, Shih A, Li Y, Bhagwat N, Vasanthakumar A, Fernandez HF et al (2010) Leukemic IDH1 and IDH2 mutations result in a hypermethylation phenotype, disrupt TET2 function, and impair hematopoietic differentiation. Cancer Cell 18:553–567

Friso S, Choi SW (2005) Gene-nutrient interactions in one-carbon metabolism. Curr Drug Metab 6:37–46

Friso S, Lamon-Fava S, Jang H, Schaefer EJ, Corrocher R, Choi SW (2007) Oestrogen replacement therapy reduces total plasma homocysteine and enhances genomic DNA methylation in postmenopausal women. Br J Nutr 97:617–621

Fryer AA, Emes RD, Ismail KM, Haworth KE, Mein C, Carroll WD, Farrell WE (2011) Quantitative, high-resolution epigenetic profiling of CpG loci identifies associations with cord blood plasma homocysteine and birth weight in humans. Epigenetics 6:86–94

Fujiki R, Hashiba W, Sekine H, Yokoyama A, Chikanishi T, Ito S, Imai Y, Kim J, He HH, Igarashi K et al (2011) GlcNAcylation of histone H2B facilitates its monoubiquitination. Nature 480:557–560

Goren A, Tabib A, Hecht M, Cedar H (2008) DNA replication timing of the human beta-globin domain is controlled by histone modification at the origin. Genes Dev 22:1319–1324

Greer EL, Shi Y (2012) Histone methylation: a dynamic mark in health, disease and inheritance. Nat Rev Genet 13:343–357

Guarente L (2011) The logic linking protein acetylation and metabolism. Cell Metab 14:151–153

Haberg SE, London SJ, Stigum H, Nafstad P, Nystad W (2009) Folic acid supplements in pregnancy and early childhood respiratory health. Arch Dis Child 94:180–184

Hariharan N, Maejima Y, Nakae J, Paik J, Depinho RA, Sadoshima J (2010) Deacetylation of FoxO by Sirt1 plays an essential role in mediating starvation-induced autophagy in cardiac myocytes. Circ Res 107:1470–1482

Hatzivassiliou G, Zhao F, Bauer DE, Andreadis C, Shaw AN, Dhanak D, Hingorani SR, Tuveson DA, Thompson CB (2005) ATP citrate lyase inhibition can suppress tumor cell growth. Cancer Cell 8:311–321

Haycock PC (2009) Fetal alcohol spectrum disorders: the epigenetic perspective. Biol Reprod 81:607–617

Herranz D, Serrano M (2010) SIRT1: recent lessons from mouse models. Nat Rev Cancer 10:819–823

Hino S, Sakamoto A, Nagaoka K, Anan K, Wang Y, Mimasu S, Umehara T, Yokoyama S, Kosai K, Nakao M (2012) FAD-dependent lysine-specific demethylase-1 regulates cellular energy expenditure. Nat Commun 3:758

Holland ML, Lowe R, Caton PW, Gemma C, Carbajosa G, Danson AF, Carpenter AA, Loche E, Ozanne SE, Rakyan VK (2016) Early-life nutrition modulates the epigenetic state of specific rDNA genetic variants in mice. Science 353:495–498

Houtkooper RH, Pirinen E, Auwerx J (2012) Sirtuins as regulators of metabolism and healthspan. Nat Rev Mol Cell Biol 13:225–238

Huang J, Plass C, Gerhauser C (2011) Cancer chemoprevention by targeting the epigenome. Curr Drug Targets 12:1925–1956

Huletsky A, Niedergang C, Frechette A, Aubin R, Gaudreau A, Poirier GG (1985) Sequential ADP-ribosylation pattern of nucleosomal histones. ADP-ribosylation of nucleosomal histones. Eur J Biochem 146:277–285

Jeong SM, Xiao C, Finley LW, Lahusen T, Souza AL, Pierce K, Li YH, Wang X, Laurent G, German NJ et al (2013) SIRT4 has tumor-suppressive activity and regulates the cellular metabolic response to DNA damage by inhibiting mitochondrial glutamine metabolism. Cancer Cell 23:450–463

Kaelin WG Jr, McKnight SL (2013) Influence of metabolism on epigenetics and disease. Cell 153:56–69

Katoh Y, Ikura T, Hoshikawa Y, Tashiro S, Ito T, Ohta M, Kera Y, Noda T, Igarashi K (2011) Methionine adenosyltransferase II serves as a transcriptional corepressor of Maf oncoprotein. Mol Cell 41:554–566

Kiefte-de Jong JC, Timmermans S, Jaddoe VW, Hofman A, Tiemeier H, Steegers EA, de Jongste JC, Moll HA (2012) High circulating folate and vitamin B-12 concentrations in women during pregnancy are associated with increased prevalence of atopic dermatitis in their offspring. J Nutr 142:731–738

Killian JK, Kim SY, Miettinen M, Smith C, Merino M, Tsokos M, Quezado M, Smith WI Jr, Jahromi MS, Xekouki P et al (2013) Succinate dehydrogenase mutation underlies global epigenomic divergence in gastrointestinal stromal tumor. Cancer Discov 3:648–657

Kim YI (2007a) Folate and colorectal cancer: an evidence-based critical review. Mol Nutr Food Res 51:267–292

Kim YI (2007b) Folic acid fortification and supplementation – good for some but not so good for others. Nutr Rev 65:504–511

Kirchner H, Osler ME, Krook A, Zierath JR (2013) Epigenetic flexibility in metabolic regulation: disease cause and prevention? Trends Cell Biol 23:203–209

Kottakis F, Nicolay BN, Roumane A, Karnik R, Gu H, Nagle JM, Boukhali M, Hayward MC, Li YY, Chen T et al (2016) LKB1 loss links serine metabolism to DNA methylation and tumorigenesis. Nature 539:390–395

Kraus D, Yang Q, Kong D, Banks AS, Zhang L, Rodgers JT, Pirinen E, Pulinilkunnil TC, Gong F, Wang YC et al (2014) Nicotinamide N-methyltransferase knockdown protects against diet-induced obesity. Nature 508:258–262

Kryukov GV, Wilson FH, Ruth JR, Paulk J, Tsherniak A, Marlow SE, Vazquez F, Weir BA, Fitzgerald ME, Tanaka M et al (2016) MTAP deletion confers enhanced dependency on the PRMT5 arginine methyltransferase in cancer cells. Science 351:1214–1218

Lee IH, Cao L, Mostoslavsky R, Lombard DB, Liu J, Bruns NE, Tsokos M, Alt FW, Finkel T (2008) A role for the NAD-dependent deacetylase Sirt1 in the regulation of autophagy. Proc Natl Acad Sci U S A 105:3374–3379

Lee JV, Carrer A, Shah S, Snyder NW, Wei S, Venneti S, Worth AJ, Yuan ZF, Lim HW, Liu S et al (2014a) Akt-dependent metabolic reprogramming regulates tumor cell histone acetylation. Cell Metab 20:306–319

Lee Y, Dominy JE, Choi YJ, Jurczak M, Tolliday N, Camporez JP, Chim H, Lim JH, Ruan HB, Yang X et al (2014b) Cyclin D1-Cdk4 controls glucose metabolism independently of cell cycle progression. Nature 510:547–551

Lertratanangkoon K, Wu CJ, Savaraj N, Thomas ML (1997) Alterations of DNA methylation by glutathione depletion. Cancer Lett 120:149–156

Letouze E, Martinelli C, Loriot C, Burnichon N, Abermil N, Ottolenghi C, Janin M, Menara M, Nguyen AT, Benit P et al (2013) SDH mutations establish a hypermethylator phenotype in paraganglioma. Cancer Cell 23:739–752

Li Y, Daniel M, Tollefsbol TO (2011) Epigenetic regulation of caloric restriction in aging. BMC Med 9:98

Li R, Grimm SA, Chrysovergis K, Kosak J, Wang X, Du Y, Burkholder A, Janardhan K, Mav D, Shah R et al (2014) Obesity, rather than diet, drives epigenomic alterations in colonic epithelium resembling cancer progression. Cell Metab 19:702–711

Lin R, Tao R, Gao X, Li T, Zhou X, Guan KL, Xiong Y, Lei QY (2013) Acetylation stabilizes ATP-citrate lyase to promote lipid biosynthesis and tumor growth. Mol Cell 51:506–518

Liu Y, Balaraman Y, Wang G, Nephew KP, Zhou FC (2009) Alcohol exposure alters DNA methylation profiles in mouse embryos at early neurulation. Epigenetics 4:500–511

Losman JA, Looper RE, Koivunen P, Lee S, Schneider RK, McMahon C, Cowley GS, Root DE, Ebert BL, Kaelin WG Jr (2013) (R)-2-hydroxyglutarate is sufficient to promote leukemogenesis and its effects are reversible. Science 339:1621–1625

Lu C, Ward PS, Kapoor GS, Rohle D, Turcan S, Abdel-Wahab O, Edwards CR, Khanin R, Figueroa ME, Melnick A et al (2012) IDH mutation impairs histone demethylation and results in a block to cell differentiation. Nature 483:474–478

Lumey LH, Stein AD, Susser E (2011) Prenatal famine and adult health. Annu Rev Public Health 32:237–262

Madrigano J, Baccarelli A, Mittleman MA, Wright RO, Sparrow D, Vokonas PS, Tarantini L, Schwartz J (2011) Prolonged exposure to particulate pollution, genes associated with glutathione pathways, and DNA methylation in a cohort of older men. Environ Health Perspect 119:977–982

Martinez D, Pentinat T, Ribo S, Daviaud C, Bloks VW, Cebria J, Villalmanzo N, Kalko SG, Ramon-Krauel M, Diaz R et al (2014) In utero undernutrition in male mice programs liver lipid

metabolism in the second-generation offspring involving altered lxra DNA methylation. Cell Metab 19:941–951
Mavrakis KJ, McDonald ER 3rd, Schlabach MR, Billy E, Hoffman GR, deWeck A, Ruddy DA, Venkatesan K, Yu J, McAllister G et al (2016) Disordered methionine metabolism in MTAP/CDKN2A-deleted cancers leads to dependence on PRMT5. Science 351:1208–1213
Mehedint MG, Niculescu MD, Craciunescu CN, Zeisel SH (2010) Choline deficiency alters global histone methylation and epigenetic marking at the Re1 site of the calbindin 1 gene. FASEB J 24:184–195
Messerschmidt DM, Knowles BB, Solter D (2014) DNA methylation dynamics during epigenetic reprogramming in the germline and preimplantation embryos. Genes Dev 28:812–828
Messner S, Hottiger MO (2011) Histone ADP-ribosylation in DNA repair, replication and transcription. Trends Cell Biol 21:534–542
Morgan HD, Sutherland HG, Martin DI, Whitelaw E (1999) Epigenetic inheritance at the agouti locus in the mouse. Nat Genet 23:314–318
Nabel CS, Kohli RM (2011) Molecular biology. Demystifying DNA demethylation. Science 333:1229–1230
Nakahata Y, Sahar S, Astarita G, Kaluzova M, Sassone-Corsi P (2009) Circadian control of the NAD+ salvage pathway by CLOCK-SIRT1. Science 324:654–657
Ng F, Tang BL (2013) Sirtuins' modulation of autophagy. J Cell Physiol 228:2262–2270
Ng SF, Lin RC, Laybutt DR, Barres R, Owens JA, Morris MJ (2010) Chronic high-fat diet in fathers programs beta-cell dysfunction in female rat offspring. Nature 467:963–966
Ong TP, Moreno FS, Ross SA (2011) Targeting the epigenome with bioactive food components for cancer prevention. J Nutrigenet Nutrigenomics 4:275–292
Padmanabhan N, Jia D, Geary-Joo C, Wu X, Ferguson-Smith AC, Fung E, Bieda MC, Snyder FF, Gravel RA, Cross JC et al (2013) Mutation in folate metabolism causes epigenetic instability and transgenerational effects on development. Cell 155:81–93
Painter RC, Roseboom TJ, Bleker OP (2005) Prenatal exposure to the Dutch famine and disease in later life: an overview. Reprod Toxicol 20:345–352
Pascual M, Do Couto BR, Alfonso-Loeches S, Aguilar MA, Rodriguez-Arias M, Guerri C (2012) Changes in histone acetylation in the prefrontal cortex of ethanol-exposed adolescent rats are associated with ethanol-induced place conditioning. Neuropharmacology 62:2309–2319
Peng M, Yin N, Chhangawala S, Xu K, Leslie CS, Li MO (2016) Aerobic glycolysis promotes T helper 1 cell differentiation through an epigenetic mechanism. Science 354:481–484
Quinn BJ, Kitagawa H, Memmott RM, Gills JJ, Dennis PA (2013) Repositioning metformin for cancer prevention and treatment. Trends Endocrinol Metab 24:469–480
Ramsey KM, Yoshino J, Brace CS, Abrassart D, Kobayashi Y, Marcheva B, Hong HK, Chong JL, Buhr ED, Lee C et al (2009) Circadian clock feedback cycle through NAMPT-mediated NAD+ biosynthesis. Science 324:651–654
Sassone-Corsi P (2012) Minireview: NAD+, a circadian metabolite with an epigenetic twist. Endocrinology 153:1–5
Schaible TD, Harris RA, Dowd SE, Smith CW, Kellermayer R (2011) Maternal methyl-donor supplementation induces prolonged murine offspring colitis susceptibility in association with mucosal epigenetic and microbiomic changes. Hum Mol Genet 20:1687–1696
Sebastian C, Zwaans BM, Silberman DM, Gymrek M, Goren A, Zhong L, Ram O, Truelove J, Guimaraes AR, Toiber D et al (2012) The histone deacetylase SIRT6 is a tumor suppressor that controls cancer metabolism. Cell 151:1185–1199
Shepard BD, Tuma PL (2009) Alcohol-induced protein hyperacetylation: mechanisms and consequences. World J Gastroenterol 15:1219–1230
Shimazu T, Hirschey MD, Hua L, Dittenhafer-Reed KE, Schwer B, Lombard DB, Li Y, Bunkenborg J, Alt FW, Denu JM et al (2010) SIRT3 deacetylates mitochondrial 3-hydroxy-3-methylglutaryl CoA synthase 2 and regulates ketone body production. Cell Metab 12:654–661

Shimazu T, Hirschey MD, Newman J, He W, Shirakawa K, Le Moan N, Grueter CA, Lim H, Saunders LR, Stevens RD et al (2013) Suppression of oxidative stress by beta-hydroxybutyrate, an endogenous histone deacetylase inhibitor. Science 339:211–214

Shyh-Chang N, Locasale JW, Lyssiotis CA, Zheng Y, Teo RY, Ratanasirintrawoot S, Zhang J, Onder T, Unternaehrer JJ, Zhu H et al (2013) Influence of threonine metabolism on S-adenosylmethionine and histone methylation. Science 339:222–226

Sinclair KD, Allegrucci C, Singh R, Gardner DS, Sebastian S, Bispham J, Thurston A, Huntley JF, Rees WD, Maloney CA et al (2007) DNA methylation, insulin resistance, and blood pressure in offspring determined by maternal periconceptional B vitamin and methionine status. Proc Natl Acad Sci U S A 104:19351–19356

Smith AD, Kim YI, Refsum H (2008) Is folic acid good for everyone? Am J Clin Nutr 87:517–533

Soga T (2013) Cancer metabolism: key players in metabolic reprogramming. Cancer Sci 104:275–281

Spencer MD, Hamp TJ, Reid RW, Fischer LM, Zeisel SH, Fodor AA (2011) Association between composition of the human gastrointestinal microbiome and development of fatty liver with choline deficiency. Gastroenterology 140:976–986

Steegers-Theunissen RP, Obermann-Borst SA, Kremer D, Lindemans J, Siebel C, Steegers EA, Slagboom PE, Heijmans BT (2009) Periconceptional maternal folic acid use of 400 microg per day is related to increased methylation of the IGF2 gene in the very young child. PLoS One 4: e7845

Supic G, Jagodic M, Magic Z (2013) Epigenetics: a new link between nutrition and cancer. Nutr Cancer 65:781–792

Sutendra G, Kinnaird A, Dromparis P, Paulin R, Stenson TH, Haromy A, Hashimoto K, Zhang N, Flaim E, Michelakis ED (2014) A nuclear pyruvate dehydrogenase complex is important for the generation of acetyl-CoA and histone acetylation. Cell 158:84–97

Takamura Y, Nomura G (1988) Changes in the intracellular concentration of acetyl-CoA and malonyl-CoA in relation to the carbon and energy metabolism of Escherichia coli K12. J Gen Microbiol 134:2249–2253

Tan L, Shi YG (2012) Tet family proteins and 5-hydroxymethylcytosine in development and disease. Development 139:1895–1902

Thienpont B, Steinbacher J, Zhao H, D'Anna F, Kuchnio A, Ploumakis A, Ghesquiere B, Van Dyck L, Boeckx B, Schoonjans L et al (2016) Tumour hypoxia causes DNA hypermethylation by reducing TET activity. Nature 537:63–68

Thirstrup K, Christensen S, Moller HA, Ritzen A, Bergstrom AL, Sager TN, Jensen HS (2011) Endogenous 2-oxoglutarate levels impact potencies of competitive HIF prolyl hydroxylase inhibitors. Pharmacol Res 64:268–273

Tobi EW, Lumey LH, Talens RP, Kremer D, Putter H, Stein AD, Slagboom PE, Heijmans BT (2009) DNA methylation differences after exposure to prenatal famine are common and timing- and sex-specific. Hum Mol Genet 18:4046–4053

Turcan S, Rohle D, Goenka A, Walsh LA, Fang F, Yilmaz E, Campos C, Fabius AW, Lu C, Ward PS et al (2012) IDH1 mutation is sufficient to establish the glioma hypermethylator phenotype. Nature 483:479–483

Ulrich CM, Potter JD (2007) Folate and cancer – timing is everything. JAMA 297:2408–2409

Vander Heiden MG, Cantley LC, Thompson CB (2009) Understanding the Warburg effect: the metabolic requirements of cell proliferation. Science 324:1029–1033

Waterland RA, Jirtle RL (2003) Transposable elements: targets for early nutritional effects on epigenetic gene regulation. Mol Cell Biol 23:5293–5300

Waterland RA, Dolinoy DC, Lin JR, Smith CA, Shi X, Tahiliani KG (2006) Maternal methyl supplements increase offspring DNA methylation at Axin Fused. Genesis 44:401–406

Wellen KE, Hatzivassiliou G, Sachdeva UM, Bui TV, Cross JR, Thompson CB (2009) ATP-citrate lyase links cellular metabolism to histone acetylation. Science 324:1076–1080

Whitrow MJ, Moore VM, Rumbold AR, Davies MJ (2009) Effect of supplemental folic acid in pregnancy on childhood asthma: a prospective birth cohort study. Am J Epidemiol 170:1486–1493

Willett W (2008) Nutrition and cancer: the search continues. Nutr Cancer 60:557–559

Wright RH, Lioutas A, Le Dily F, Soronellas D, Pohl A, Bonet J, Nacht AS, Samino S, Font-Mateu J, Vicent GP et al (2016) ADP-ribose-derived nuclear ATP synthesis by NUDIX5 is required for chromatin remodeling. Science 352:1221–1225

Wu K, Platz EA, Willett WC, Fuchs CS, Selhub J, Rosner BA, Hunter DJ, Giovannucci E (2009) A randomized trial on folic acid supplementation and risk of recurrent colorectal adenoma. Am J Clin Nutr 90:1623–1631

Xiao M, Yang H, Xu W, Ma S, Lin H, Zhu H, Liu L, Liu Y, Yang C, Xu Y et al (2012) Inhibition of alpha-KG-dependent histone and DNA demethylases by fumarate and succinate that are accumulated in mutations of FH and SDH tumor suppressors. Genes Dev 26:1326–1338

Xu MQ, Sun WS, Liu BX, Feng GY, Yu L, Yang L, He G, Sham P, Susser E, St Clair D et al (2009) Prenatal malnutrition and adult schizophrenia: further evidence from the 1959–1961 Chinese famine. Schizophr Bull 35:568–576

Xu W, Yang H, Liu Y, Yang Y, Wang P, Kim SH, Ito S, Yang C, Wang P, Xiao MT et al (2011) Oncometabolite 2-hydroxyglutarate is a competitive inhibitor of alpha-ketoglutarate-dependent dioxygenases. Cancer Cell 19:17–30

Yang M, Soga T, Pollard PJ (2013) Oncometabolites: linking altered metabolism with cancer. J Clin Invest 123:3652–3658

Zeisel SH (2009) Epigenetic mechanisms for nutrition determinants of later health outcomes. Am J Clin Nutr 89:1488S–1493S

Zhang H, Ryu D, Wu Y, Gariani K, Wang X, Luan P, D'Amico D, Ropelle ER, Lutolf MP, Aebersold R et al (2016) NAD(+) repletion improves mitochondrial and stem cell function and enhances life span in mice. Science 352:1436–1443

Zhao S, Xu W, Jiang W, Yu W, Lin Y, Zhang T, Yao J, Zhou L, Zeng Y, Li H et al (2010) Regulation of cellular metabolism by protein lysine acetylation. Science 327:1000–1004

Zhong L, D'Urso A, Toiber D, Sebastian C, Henry RE, Vadysirisack DD, Guimaraes A, Marinelli B, Wikstrom JD, Nir T et al (2010) The histone deacetylase Sirt6 regulates glucose homeostasis via Hif1alpha. Cell 140:280–293

Milk Exosomes and MicroRNAs: Potential Epigenetic Regulators

77

Bodo C. Melnik and Foteini Kakulas

Contents

B. C. Melnik (✉)
Department of Dermatology, Environmental Medicine and Health Theory, University of Osnabrück, Osnabrück, Germany
e-mail: melnik@t-online.de

F. Kakulas
School of Medicine and Pharmacology, The University of Western Australia, Crawley, WA, Australia
e-mail: foteini.kakulas@bigpond.com

V. B. Patel, V. R. Preedy (eds.), *Handbook of Nutrition, Diet, and Epigenetics*,
https://doi.org/10.1007/978-3-319-55530-0_86

Abstract
The scientific perception of the biological functions of milk, mammal's secretory product of mammary gland epithelial cells during lactation, has dramatically changed in recent years from a simple food for the newborn mammal to a most sophisticated signaling system between the mother and her infant, in addition to nourishment, protection, and development it provides. From the wide range of extracellular vesicles found in milk, this review focuses primarily on milk exosomes and exosome-delivered microRNAs that emerge as an epigenetic regulatory software of milk decreasing genome methylation of the milk recipient. Thus, milk-derived exosomes are regarded as critical signalosomes for mother-to-child transmission of microRNAs that affect epigenetic regulatory circuits of the milk recipient. Evidence accumulates that epigenetic signaling of milk promotes the development of the infant's gastrointestinal tract, immune system, osteogenesis, myogenesis, adipogenesis, and neurogenesis. According to the *functional hypothesis*, milk exosomal microRNAs may reach the systemic circulation of the infant and the adult human milk consumer. Human and bovine milk exosomes and their associated microRNAs are found in the fat fraction of milk and in skim milk. microRNAs are also present in large numbers in the cellular fraction of milk, making it a microRNA-rich medium, likely the richest microRNA source of all body fluids in humans. Human milk and commercial cow's milk provides abundant amounts of microRNA-148a, microRNA-152, microRNA-29b, and microRNA-21, which all target DNA methyltransferases (DNMTs) that potentially affect whole genome DNA methylation patterns. DNA CpG demethylation upregulates the expression of many genes including the m^6A RNA demethylase *fat mass- and obesity-associated protein* (FTO). In this regard, milk exosomal microRNAs may function as potential epigenetic modifiers of DNA- and RNA methylation of the milk recipient, who under physiological conditions is the suckling infant, but also the human consumer of commercial cow's milk. Whereas milk exosome-driven epigenetic signaling appears to be indispensable for adequate postnatal growth and programming of the infant, this microRNA transmitter is almost absent in artificial formula, potentially leading to faulty or immature epigenetic programming of formula-fed infants. Furthermore, persistent consumption of commercial cow's milk that still contains bioactive microRNAs, some of which exhibit high complementarity to human milk microRNAs, has currently unknown consequences to human health and may bear a health risk for humans of developed milk-consuming civilizations.

Keywords
Breastfeeding · DNA methyltransferase · Epigenetic regulation · Epitranscriptome · Growth · Exosome · FoxP3 · FTO · Infant formula · Lactation · microRNA · Milk · NRF2 · NR4A3 · Western diseases

List of Abbreviations

AGO2	Argonaute 2
AKT	V-AKT murine thymoma viral oncogene homolog

CMA	Cow's milk allergy
D2R	Dopamine receptor type 2
DNMT	DNA methyltransferase
EV	Extracellular vesicle
FoxP3	Forkhead box P3
FTO	Fat mass- and obesity-associated gene
GLUT1	Glucose transporter 1
IEC	Intestinal epithelial cell
IgE	Immunoglobulin E
IGF-1	Insulin-like growth factor-1
IGF1R	Insulin-like growth factor-1 receptor
IL	Interleukin
ILV	Intralumial vesicle
IκBα	Nuclear factor κB inhibitor α
KO	Knockout
LCT	Lactase gene
m^6A	*N6*-methyladenosine
MEC	Mammary epithelial cell
MEF	Mouse embryonic fibroblast
MFG	Milk fat globule
MFGM	Milk fat globule membrane
microRNA	Micro-ribonucleic acid
mTORC1	Mechanistic target of rapamycin complex 1
MVB	Multi-vesicular body
NEC	Necrotizing enterocolitis
NFKBI	Nuclear factor of κ light chain gene enhancer in B cells inhibitor α
NF-κB	Nuclear factor κB
NR4A3	Nuclear receptor subfamily 4, group A, member 3
NRF2	Nuclear factor erythroid 2-related factor 2
p53	Transformation-related protein 53
PBMCs	Peripheral blood mononuclear cells
PGC1α	PPAR-γ coactivator 1-α
PI3K	Phosphatidylinositol 3-kinase
PPARγ	Peroxisome proliferator-activated receptor-γ
RASGRP1	Ras guanyl nucleotide-releasing protein-1
RISC	RNA silencing complex
RNA	Ribonucleic acid
ROCK1	Rho-associated coiled-coil containing protein kinase 1
RUNX1T1	Runt-related transcription factor 1, translocated to, 1
S6 K1	Ribosomal protein S6 kinase
SIDT1	Systemic RNA interference-defective-1 transmembrane family member 1
SNP	Single nucleotide polymorphisms
SREBP-1	Sterol regulatory element-binding protein 1
SRSF2	Serine/arginine-rich splicing factor-2

TCR	T cell receptor
TGFβ	Transforming growth factor-β
TNF	Tumor necrosis factor
TSDR	Treg-specific demethylated region
UTR	Untranslated region
VEC	Vascular endothelial cell

Introduction

It is now widely recognized that the environment in early life has important influence on human growth and development, including the "epigenetic programming" of far-reaching effects on the risk of developing common metabolic and noncommunicable diseases in later life (Godfrey et al. 2016). Milk is the very first postnatal nutritional environment of all mammals, who eventually wean off mother's milk to move into an adult diet. However, in humans who persistently consume commercial cow's milk or other mammal's milks after weaning off mother's own milk, now this new microRNA source continues to influence their epigenome throughout life. There is accumulating interest in the role of early nutrition for the epigenetic basis of developmental programming. Epigenetic processes play a central role in regulating tissue-specific gene expression and hence altering processes that can induce long-term changes in gene expression and metabolism which persist throughout life course (Godfrey et al. 2016).

Milk is a highly specialized, complex, and dynamic nutrient system developed by mammalian evolution to promote postnatal growth (Hassiotou et al. 2013). Furthermore, it has been speculated that milk may function as an epigenetic imprinting system (Melnik et al. 2013; Perge et al. 2017). There is compelling evidence suggesting that exosome-derived microRNAs of milk affect gene regulation of the milk recipient (Melnik et al. 2013, 2016; Baier et al. 2014; Alsaweed et al. 2015; Zempleni et al. 2017). This review illustrates the potential epigenetic impact of milk exosome-mediated microRNA transfer to the newborn infant and the adult human milk consumer.

It is generally accepted that human milk allows appropriate metabolic programming and protects against diseases of civilization in later life. Human milk is considered the optimal food for term infants in the first 6 months of life. Milk not only supplies multiple macromolecular compounds required for protein translation, growth, and development such as essential amino acids that activate the nutrient sensor mechanistic target of rapamycin complex 1 (mTORC1), but also transfers a signaling software to the growing infant ensuring mother-child communication for lifelong metabolic programming of the infant (Melnik et al. 2013, 2016c).

Milk Exosomes

Extracellular vesicles (EVs) have been classified into three main groups: (1) exosomes (30–100 nm in diameter) formed via the endocytic pathway, (2) microvesicles (100–2000 nm) formed by budding out of the plasma membrane in a calcium-dependent process, and (3) apoptotic bodies (>1000 nm) formed by blebbing of the plasma membrane during the process of apoptosis (Yáñez-Mó et al. 2015). Exosomes are assembled in intraluminal vesicles (ILVs) contained in multivesicular bodies (MVBs) that are released by fusing with the cell membrane (Yáñez-Mó et al. 2015). The significance of exosomes lies in their capacity to transfer information to the recipient cell thereby modulating gene functions (Yáñez-Mó et al. 2015). Thus, exosomes transfer functional RNAs from a donor to an acceptor cell, analogous to hormones that can signal in paracrine and autocrine modes.

Milk exosomes are pivotal components of mammalian milk and are regarded as most important signalosomes mediating cellular communication between the mother and the nursing infant. In 2007, exosomes from human colostrum and mature human milk have been isolated and characterized for the first time (Admyre et al. 2007). Later on, milk-derived exosomes have been detected in colostrum and mature milk of humans, cows, buffalos, goats, pigs, marsupial tammar wallabies, and rodents (Admyre et al. 2007; Kosaka et al. 2010; Hata et al. 2010; Reinhardt et al. 2012; Gu et al. 2012; Zhou et al. 2012; Izumi et al. 2012, 2015; Sun et al. 2013; Modepalli et al. 2014; Chen et al. 2014, 2016; Na et al. 2015; Baddela et al. 2016).

In the past decade, exosomes have been recognized as potent mediators of intercellular communication. Exosome-like vesicles have recently been observed within cytoplasmic crescents of human milk fat globules (MFGs) (Gallier et al. 2015). There are at least four major cellular sources of milk exosomes: (1) direct exosome release via mammary epithelial cells (MECs) during the different stages of lactation, (2) exosome sequestration from MFGs, (3) exosome release by immune cells present in milk, and (4) exosome release via other, nonimmune milk cells, such as milk epithelial cells and milk stem cells (Hassiotou et al. 2012, 2013; Reinhardt et al. 2012, 2013; Gallier et al. 2015; Sun et al. 2013; Alsaweed et al. 2015, 2016b). It is noteworthy to mention that the majority of human milk microRNAs primarily originate from MECs resulting in unique lactation-specific microRNA profiles (Alsaweed et al. 2016b, c).

Milk Exosomes Resist Intestinal Degradation

Exosomal package of microRNAs serves as an important biological feature protecting and stabilizing microRNA transfer between cells. Compared with exogenous synthetic microRNAs, exosomal immune-related microRNAs of human milk are resistant to harsh conditions (Zhou et al. 2012). Commercial bovine milk microRNAs are stable under degradative conditions, such as acidic environments and RNase treatment, boiling, and freezing, but were degraded by the addition of detergent and bacterial fermentation (Benmoussa et al. 2016; Izumi et al. 2012, 2015; Pieters et al. 2015; Howard et al. 2015). Pasteurization and homogenization of commercial cow's milk caused a 63% loss of microRNA-200c, whereas a 67% loss observed for microRNA-29b was statistically significant in skim milk (Howard et al. 2015). Buffalo milk microRNA-21 and microRNA-500 have also been shown to be stable even under household storage conditions (Baddela et al. 2016). Collectively, these data suggest that human and bovine milk exosomes are highly resistant against degradative conditions in the gastrointestinal tract, a very important requirement for their uptake by intestinal epithelial cells (IECs) and potential transfer into the systemic circulation.

Milk Exosomal microRNAs

Secreted microRNAs represent a newly recognized most important layer of gene regulation in eukaryotes, which plays a relevant role for intercellular communication (Yáñez-Mó et al. 2015). MicroRNAs are part of the epigenetic machinery and are predicted to regulate 30–60% of all genes in humans (Ambros 2004). MicroRNAs bind through partial sequence homology to the 3′-UTR of target mRNAs located in the RNA silencing complex (RISC) and cause either translational block or less frequently mRNA degradation (Ambros 2004). Human milk contains the highest concentration of total RNAs including microRNAs in comparison to other human body fluids like plasma (Weber et al. 2010). Exosomes can transfer microRNAs to recipient cells subsequently modifying target gene expression by specific microRNAs. According to our *functional hypothesis,* milk exosomes apparently travel the longest distance of all mammalian exosomes in order to transfer gene regulatory microRNAs from maternal MECs to the cells of the newborn infant (Melnik et al. 2016c; Fig. 1). They reach the infant's intestine and most likely the infant's blood circulation (Melnik et al. 2013, 2016c; Alsaweed et al. 2016c).

Milk Exosome Uptake

Cells are able to take up exosomes by a variety of endocytic pathways, including clathrin-dependent endocytosis and clathrin-independent pathways such as caveolin-mediated uptake, macropinocytosis, phagocytosis, and lipid raft-mediated internalization (Mulcahy et al. 2014).

Human as well as bovine milk exosomes including their microRNAs are taken up by human macrophages (Izumi et al. 2015; Pieters et al. 2015). The intestinal uptake of bovine milk exosomes is mediated by temperature-dependent endocytosis and depends on cell and exosome surface glycoproteins in IECs (Arntz et al. 2015; Wolf et al. 2015). PKH67-labeled bovine milk exosomes have been detected in IECs of the ileum and isolated splenocytes of mice that received bovine milk exosomes by daily oral gavage (Arntz et al. 2015). Recent evidence has been presented that porcine milk-derived exosomes are taken up by murine IECs and promoted IEC proliferation (Chen et al. 2016). In vivo, porcine exosomes significantly raised mice' villus height, crypt depth, and ratio of villus length to crypt depth of intestinal tissues (Chen et al. 2016). This study clearly demonstrated that milk-derived exosomal microRNAs exerted gene regulatory functions in the recipient IECs (Fig. 1).

Human vascular endothelial cells (VECs) have also been shown to take up bovine milk exosomes by endocytosis (Kusuma et al. 2016), which supports the notion that

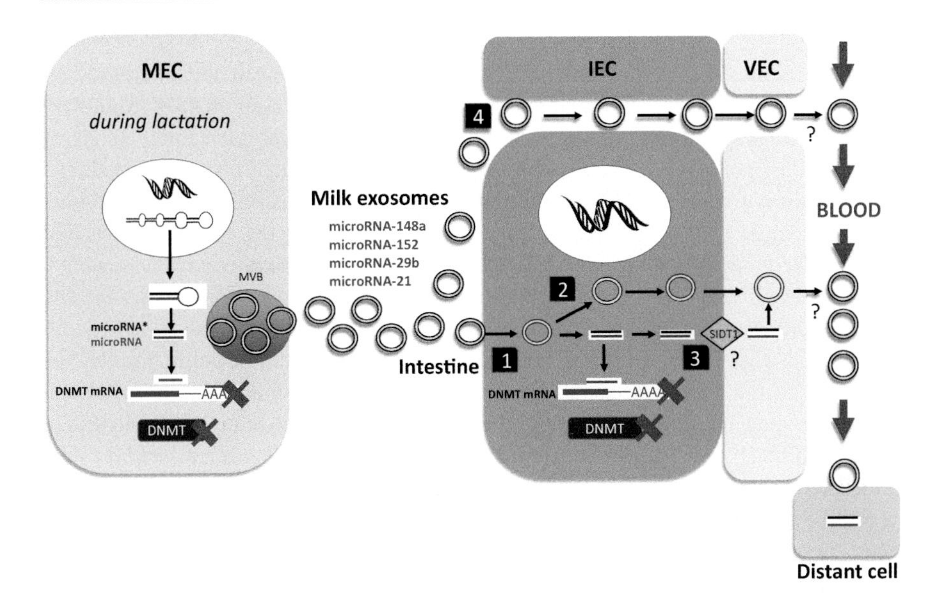

Fig. 1 Exosome transfer of lactation-promoting DNA methyltransferases-targeting microRNAs to the milk recipient. Exosome transfer of lactation-specific microRNAs (microRNA-148a, −152, −29b, and −21) that target DNA methyltransferases (DNMT) of the milk recipient. Mammary gland epithelial cells (MEC) secrete these microRNAs via exosomes. (1) Milk exosomes are taken up by intestinal epithelial cells (IEC) via endocytosis. The released microRNAs suppress intestinal DNMTs and thereby promote intestinal growth. (2) Milk exosomes may be transported to adjacent vascular endothelial cells (VEC) via transendocytosis. (3) Milk-derived microRNAs such as microRNA-21 may be transferred to VEC by the action of systemic RNA interference-defective-1 transmembrane family member 1 (SIDT1) protein, where they may be incorporated into VEC-derived exosomes. (4) During the postnatal period with associated high intestinal permeability, milk exosomes may move through IEC intercellular spaces and VEC fenestrae thereby easily reaching the systemic circulation. Milk exosomes may thus approach distant cells of the milk recipient, where they may reduce DNA methylation

milk-derived exosomes and their microRNA cargo may reach peripheral tissues of the milk recipient (Melnik et al. 2013, 2016c). In fact, it has been reported that bovine milk microRNAs (microRNA-29b and microRNA-200c) are absorbed in biologically meaningful amounts from nutritionally relevant doses of cow's milk and affect gene expression in peripheral blood mononuclear cells (PBMCs) of human volunteers (Baier et al. 2014).

Highly expressed lactation-specific microRNAs have been detected in the serum of the neonate wallaby (*Macropus eugenii*) compared to blood levels of adult wallabies, suggesting systemic uptake of milk-derived microRNAs during lactation (Modepalli et al. 2014). Immune-related porcine microRNAs, which are enriched in milk exosomes, have been detected in higher numbers in colostrum compared to mature porcine milk (Gu et al. 2012). Higher serum concentrations of these immune-related microRNAs of colostrum-only fed piglets have been measured compared to piglets fed mature milk, which further indicates a dose-dependent uptake of milk-derived microRNAs into the bloodstream of the piglet (Gu et al. 2012). An integrated genomics and computational analysis has characterized the likelihood of milk-derived microRNAs to get transferred into the systemic circulation (Shu et al. 2015). Remarkably, predicted target genes of 14 highly expressed microRNAs of bovine milk fractions are related with organismal development favoring a systemic gene-regulatory role of milk-derived microRNAs (Melnik et al. 2013, 2016c; Alsaweed et al. 2016c; Li et al. 2016). Table 1 summarizes recent studies that support the view that milk exosome-derived microRNAs may reach the systemic circulation and may exert gene-regulatory functions in the newborn infant and the adult consumer of commercial cow's milk.

Theoretically, there are three possible routes for lactation-specific microRNA transfer into the systemic circulation of the infant or the human milk consumer. Exosomes may travel along intercellular spaces of enterocytes and VEC fenestrae direct into the bloodstream of the milk recipient, a potential mechanism that may preferentially operate during the postnatal period associated with increased intestinal permeability. Alternatively, due to the high exosome gradient during lactation, milk exosomes may intracellularly saturate IECs, enter IEC's MVBs for exocytosis toward adjacent VECs. IECs as well as VECs are able to release exosomes (Kusuma et al. 2016). Finally, lactation-specific microRNAs may be shuttled between IECs

Table 1 Evidence for milk exosome microRNA transfer

The majority of milk exosomes is secreted by mammary epithelial cells	Alsaweed et al. 2016b
Milk exosomes resist intestinal degradation	Benmoussa et al. 2016; Hata et al. 2010; Izumi et al. 2012, 2015, Pieters et al. 2015; Zhou et al. 2012
Milk exosomes are taken up by intestinal epithelial cells	Arntz et al. 2015; Wolf et al. 2015; Chen et al. 2016
Milk exosomes are taken up by vascular endothelial cells	Kusuma et al. 2016
Increased serum levels of milk-derived microRNAs during the period of lactation	Gu et al. 2012; Modepalli et al. 2014
Dose-dependent increase of microRNA-29b and microRNA-200c in the serum of cow's milk consumers	Baier et al. 2014
Increase of microRNA-29b and microRNA-200c in peripheral blood mononuclear cells of human volunteers 6 h after commercial milk intake	Baier et al. 2014
Increased gene expression of RUNX2, a regulatory target of microRNA-29b, in PBMCs of healthy humans after cow's milk consumption	Baier et al. 2014
Detection of bovine milk exosomes in murine splenocytes	Arntz et al. 2015
Predicted role of milk microRNAs in organismal development and organ maturation	Alsaweed et al. 2015, 2016c; Li et al. 2016; Melnik et al. 2014; Shu et al. 2015

and VECs via the transmembrane protein systemic RNA interference-defective-1 transmembrane family member 1 (SIDT-1), which has been shown to transfer microRNA-21 between adjacent cells (Elhassan et al. 2012; Fig. 1).

Milk Exosomes Transfer Immune-Regulatory microRNAs

There is accumulating evidence from various mammalian species that milk exosomes transfer immune-regulatory microRNAs to the newborn infant (Admyre et al. 2007; Kosaka et al. 2010; Alsaweed et al. 2015). Admyre et al. (2007) showed for the first time that the addition of isolated human milk exosomes to PBMCs increased the number of FoxP3+ regulatory T cells (Tregs). Tregs present a unique T-cell lineage that plays a key role for the initiation and maintenance of immunological tolerance. Tregs express the forkhead box transcription factor FoxP3, which acts as a lineage-specifying factor that determines the unique properties of these immunosuppressive cells (Huehn and Beyer 2015). It has recently been speculated that milk-derived exosomes deliver tolerogenic signals to the offspring (Melnik et al. 2014; Parigi et al. 2015). FoxP3, the master transcription factor of Tregs, plays a key role in Treg function, which is strongly related to the induction of tolerance against self-antigens (prevention of autoimmunity) and environmental allergens (prevention of allergy) (Melnik et al. 2014, 2016). Milk exosomes via induction of FoxP3+ Tregs may thus play a crucial role in shaping intestinal and systemic immunity (Melnik et al. 2014, 2016a; Parigi et al. 2015).

Milk microRNAs: Epigenetic Inducers of Treg-Mediated Immune Tolerance

Epigenetic modifications in the CpG-rich *Treg-specific demethylated region* (TSDR) in the *FOXP3* locus are associated with stable FoxP3 expression (Huehn and Beyer 2015). Notably, stable FoxP3 expression was found only for cells displaying enhanced TSDR demethylation (Huehn and Beyer 2015). Recently, a linear correlation between FoxP3 expression and the degree of TSDR demethylation was demonstrated (Paparo et al. 2016). TSDR is an important methylation-sensitive element regulating FoxP3 expression. Epigenetic imprinting in the TSDR is thus critical for the establishment of a stable Treg lineage (Huehn and Beyer 2015). Notably, atopic individuals express lower numbers of demethylated FoxP3+ Tregs than nonatopic controls (Hinz et al. 2012).

Both DNA methyltransferases DNMT1 and DNMT3b are associated with the *FOXP3* locus in $CD4^{+}$ T cells (Josefowicz et al. 2009). DNMT1 deficiency resulted in highly efficient FoxP3 induction following T-cell receptor (TCR) stimulation (Josefowicz et al. 2009). Importantly, DNMT1 and DNMT3b are direct targets of microRNA-148a (Duursma et al. 2008; Pan et al. 2010). MicroRNA-148a is abundant in bovine colostrum, mature cow's milk, and human milk (Chen et al. 2010; Gu et al. 2012; Izumi et al. 2012; Zhou et al. 2012; Kirchner et al. 2016). MicroRNA-

148a is highly expressed in bovine milk fat and MFGs of human milk (Munch et al. 2013; Kirchner et al. 2016). MicroRNA-21, another abundant microRNA of cow's milk, indirectly inhibits DNMT1 expression by targeting Ras guanyl nucleotide-releasing protein-1 (RASGRP1) (Pan et al. 2010). MicroRNA-152 also targets DNMT1 and is involved in the development of lactation in mammary gland of the dairy cow (Wang et al. 2014). MicroRNA-29b, which increases dose-dependently in human serum after intake of pasteurized cow's milk (Baier et al. 2014), targets DNMT3a and DNMT3b (Fabbri et al. 2007). Notably, the seed sequences of microRNA-148a, microRNA-152, microRNA-29b, and microRNA-21 of *Homo sapiens* (hsa) and *Bos taurus* (bta) are identical (Table 2). It has been suggested that biologically active microRNAs of unprocessed cow's milk mediate the allergy preventive farm milk effect (Melnik et al. 2014, 2016; Kirchner et al. 2016). The transfer of DNMT-targeting microRNAs via unprocessed farm milk enhances FoxP3 expression, whereas boiling of raw cow's milk significantly reduced milk microRNA-148a levels (Kirchner et al. 2016) and abolished the allergy-preventive farm milk effect (Loss et al. 2011). Boiling of milk may disrupt the protective lipid bilayer of milk exosomes and MFGs, accelerating the degradation of critical DNMT-targeting milk-derived microRNAs. Furthermore, heat-induced alterations of

Table 2 High degree of complementarity between human and bovine microRNAs targeting DNA methyltransferases (mirbase.org)

<table>
<tr><th>Human microRNAs targeting DNMTs</th><th>Bovine microRNAs targeting DNMTs</th></tr>
<tr><td>hsa-miR-21 stem loop
5' gucgg agcuuauc gacug uguug cugu g a
||||| |||||||| ||||| ||||| |||| | u
3' caguc ucggguag cugac acaac ggua c c</td><td>bta-mir-21 stem loop
5' gucgg agcuuauc gacug uguug cugu g a
||||| |||||||| ||||| ||||| |||| | u
3' caguc ucggguag cugac acaac ggua c c</td></tr>
<tr><td>hsa-mir-148a stem loop
5' gaggcaaaguucug ag cacu gacu cug u
|||||||||||||| || |||| |||| ||| a
3' cucuguuucaagac uc guga cuga gau u</td><td>bta-mir-148a stem loop
5' gaggcaaaguucug ag cacu gacu cug u
|||||||||||||| || |||| |||| ||| a
3' cucuguuucaagac uc guga cuga gau u</td></tr>
<tr><td>hsa-mir-152 stem loop
5' gucc cc ggcccagguucugu au cacu gacu
|||| || |||||||||||||| || |||| |||
3' cagg gg ccggguucaagaca ua guga cuga</td><td>bta-mir-152 stem loop
5' gucc ucc ggcccagguucugu au cacu gacu
|||| ||| |||||||||||||| || |||| |||
3' cagg agg ccggguucaagaca ua guga cuga</td></tr>
<tr><td>hsa-mir-29b-1 stem loop
5' cuucaggaa gcugguuuca auggug uuagau u
||||||||| |||||||||| |||||| |||||| a
3' gggguucuu ugacuaaagu uaccac gaucug g</td><td>bta-mir-29b-1 stem loop
5' cuucaggaa gcugguuuca auggug uuagau u
||||||||| |||||||||| |||||| |||||| a
3' gggguucuu ugacuaaagu uaccac gaucug g</td></tr>
<tr><td>hsa-mir-29b-2
5' cuucuggaa gcugguuuca auggug cu agau a
||||||||| |||||||||| |||||| || ||||
3' gaggauuuu ugacuaaagu uaccac ga ucua u</td><td>bta-mir-29b-2
5' cuucuggaa gcugguuuca auggug cu agau a
||||||||| |||||||||| |||||| || ||||
3' gaggauuuu ugacuaaagu uaccac ga ucua u</td></tr>
</table>

exosomal membrane proteins may disturb intestinal exosome uptake. Thus, native milk-derived microRNAs via suppressing DNMTs may provide pivotal epigenetic signals stabilizing FoxP3 expression and Treg differentiation either in the intestine (inducible Tregs) or via the circulation in the neonatal thymus (natural Tregs) (Melnik et al. 2014). A recent epigenetic study confirmed general DNA hypermethylation in a group of children with cow's milk allergy (CMA) compared to the healthy control group (Petrus et al. 2016). *FOXP3* TSDR demethylation was significantly lower in children with active IgE-mediated CMA than in either children who outgrew CMA or in healthy children (Paparo et al. 2016). These observations indicate that milk and especially milk exosomal microRNAs shape the epigenetic environment for FoxP3-driven tolerance induction, a key mechanism for the prevention of allergy and autoimmunity (Melnik et al. 2014, 2016b).

The NR4A subfamily of orphan nuclear receptors belongs to the larger nuclear receptors superfamily of eukaryotic transcription factors. Recently, NR4A1, NR4A2, and NR4A3 are reported to regulate Treg cell development through transcriptional activation of *FOXP3* (Won and Hwang 2016). NR4A-deficient Tregs exhibited reduced FoxP3 expression (Sekiya et al. 2016). It is of critical importance that NR4A directly activates the *FOXP3* promoter (Sekiya et al. 2016). Intriguingly, NR4A3 was epigenetically silenced by NR4A3 promoter methylation, whereas NR4A3 promoter demethylation increased its expression (Yeh et al. 2016). It is thus conceivable that milk microRNA-mediated DNMT suppression enhances further FoxP3 expression via epigenetic upregulation of NR4A3, further promoting Treg differentiation (Fig. 1).

Milk Exosomal microRNAs: Activators of mTORC1-Dependent Translation

mTORC1 plays a key role for global and specific mRNA translation. Milk signaling promotes mTORC1-dependent translation (Melnik 2015a). Milk-derived exosomal microRNA-21 is one of milk's microRNAs that may enhance PI3K/AKT-mediated activation of mTORC1 via targeting the crucial PI3K inhibitor PTEN (Melnik 2015a).

There are further epigenetic pathways involving microRNA-mediated DNMT suppression that may enhance mTORC1 signaling. For instance, caveolin-1 (Cav-1) is under the epigenetic control of and interacts with insulin receptors (IRs) and IGF-I receptors (IGF-IR) and stimulates IR kinase activities, which are of critical importance for mTORC1 activation. Cav-1 interacts with low-density lipoprotein receptor-related protein 6 (LRP6) and the β-subunit of IR to generate an integrated signaling module that leads to the activation of IGF-IR/IR and results in stimulation of AKT-mTORC1 signaling (Palacios-Ortega et al. 2014). Exon 1 and first intron of the Cav-1 gene (*CAV1*) undergoes a demethylation process that is accompanied by a strong induction of Cav-1 expression (Palacios-Ortega et al. 2014). Notably, the expression of Cav-1 is enhanced by DNA demethylation during adipocyte differentiation (Palacios-Ortega et al. 2014). The exosomal transfer of milk's DNMT-targeting microRNAs to the exosome recipient cell via epigenetic enhancement of Cav-1 expression may further promote Cav-1-mediated exosome endocytosis.

Nuclear factor erythroid 2-related factor 2 (NRF2) is another important transcription factor, which is under epigenetic control. DNMT inhibition by 5-aza-2′-deoxycytidine increased NRF2 at both messenger RNA and protein levels via downregulating the expression of DNMTs (Cao et al. 2016). Remarkably, increasing cellular NRF2 results in direct transcriptional activation of the *MTOR* gene (Bendavit et al. 2016; Fig. 1). NRF2 also upregulates RagD, a small G-protein activator of the mTOR pathway (Sasaki et al. 2012). Intriguingly, NRF2 is a critical transcription factor activating microRNA-29-coding genes (Kurinna et al. 2014), which may attenuate DNMT3b expression causing further epigenetic upregulation of NRF2.

Milk Exosomal microRNAs: Potential Regulators of Inflammation

Human milk has important effects on neonatal intestinal health and reduces the risk of necrotizing enterocolitis (NEC) in preterm infants. Raw bovine milk improved gut responses to feeding relative to infant formula in preterm piglets. The maturational and protective effects on the immature intestine decreased in the order bovine colostrum > bovine milk > whole milk powder (Li et al. 2014). In an experimental model, in which NEC was mimicked by exposing Caco-2 cells to hypoxia/reoxygenation, human milk supernatant inhibited the expression of proinflammatory cytokines IL-1β, IL-6, and TNFα (Ruan et al. 2016), which are upregulated by activated NF-κB signaling. Nuclear factor κ light polypeptide gene enhancer in B-cells inhibitor-α (IκBα) is a member of a family of cellular proteins that function to inhibit the NF-κB transcription factor. IκBα inhibits NF-κB by masking the nuclear localization signals of NF-κB proteins and keeping them sequestered in an inactive state in the cytoplasm. The CpG methylation status in the promoter of IκBα exerts strong influences on NF-κB signaling (O'Gorman et al. 2010). Hypomethylation of the promoter of *NFKBI* increases IκBα expression, whereas methylation downregulates IκBα expression (O'Gorman et al. 2010). It is thus conceivable that milk microRNA-mediated downregulation of DNMT expression represents an anti-inflammatory epigenetic mechanism that enhances IκBα-mediated suppression of proinflammatory NF-κB signaling, which may protect against the development of NEC.

Milk Exosomal microRNAs: Potential Regulators of FTO Expression

Milk microRNAs may not only modify epigenetic regulation of DNA but may also have an impact on the epigenetic regulation of RNAs and thus the epitranscriptome (Melnik et al. 2015b). Milk plays a key regulatory role in the initiation of translation (Melnik 2015a) and provides abundant amounts of leucine, the predominant amino acid activator of mTORC1. FTO connects leucine availability to the activation of mTORC1 (Melnik 2015a, b). Importantly, FTO-deficient cells have reduced global translational rates, and FTO-dependent control of translation clearly depends on FTO enzymatic activity. Thus, there is already a close interaction between leucine, FTO, mTORC1, and translation. In concert with leucine-mTORC1-driven activation of translation, milk-derived exosomal microRNAs may further enhance mRNA transcription via epigenetic interactions.

To understand milk's potential impact on transcription, it is important to be familiar with the regulation of *N6*-methyladenosine (m^6A) in mRNA, which adds another layer of regulation at the posttranscriptional level in gene expression (Maity and Das 2016). Over 12,000 m^6A sites characterized by a typical consensus in the transcripts of more than 7000 human genes have been identified (Dominissini et al. 2012). FTO and alkylation repair homolog 5 (ALKBH5) are the two major m^6A demethylases that selectively remove the m^6A code from target mRNAs (Maity and Das 2016). Notably, loss-of-function mutation of FTO is associated with postnatal growth retardation in humans (Boissel et al. 2009). In accordance, FTO knockdown in mice decreased body weight, altered metabolism, and retarded growth (Fischer et al. 2009), whereas overexpression of FTO in mice led to a dose-dependent increase in body and fat mass, increased food intake resulting in obesity (Church et al. 2010). Overexpression of FTO led to a global decrease of m^6A in RNAs (Jia et al. 2011). The dynamic and reversible m^6A modification on RNA may also serve as an important epigenetic marker of profound biological significances.

The transfer of milk exosomal microRNAs targeting DNMT expression may serve as an important epigenetic mechanism that upregulates FTO expression of the milk recipient (Melnik 2015b). Hypomethylation of specific CpG sites of the FTO gene have been reported to enhance FTO expression (Liu et al. 2014). DNMT-targeting microRNAs of milk may synergize in reducing FTO CpG methylation thereby increasing FTO expression required for enhanced postnatal transcription (Fig. 2). In this regard, milk microRNAs may serve as epigenetic activators of FTO with crucial impact on the mRNA epitranscriptome. Milk specific microRNAs may thereby shape mRNA transcription, splicing, nuclear export, localization, translation, and especially mRNA stability (Maity and Das 2016).

FTO-Mediated Adipogenesis

Recent evidence links FTO overexpression to enhanced expression of the pro-adipogenic *short isoform* of the transcription factor RUNX1T1. FTO controls mRNA splicing by regulating the ability of the serine/arginine-rich splicing factor-2 (SRSF2) to bind to mRNA in an m^6A-dependent manner (Zhao et al. 2014). Moreover, the proadipogenic short isoform of RUNX1T1 stimulates mitotic clonal expansion of mouse embryonic fibroblasts (MEFs) and thus enhances adipocyte numbers (Merkestein et al. 2015). In mice fed with high-fat diet, an m^6A hypomethylation state was associated with increased FTO expression, whereas the supplementation of the methyl donor betaine prevented these changes (Chen et al. 2015).

FTO-Mediated Ghrelin Expression, Orexigenic Signaling, and Feeding Reward

Epigenetic FTO-mediated mechanisms of milk may regulate appetite and reward signals in order to guarantee adequate calorie intake during the postnatal growth phase. Increased expression of FTO in mice with two additional copies of FTO (FTO-4 mice) exhibited increased adiposity and hyperphagia (Merkestein et al. 2014). Intriguingly, FTO overexpression reduced ghrelin mRNA m^6A methylation, concomitantly increasing ghrelin mRNA and protein levels (Karra et al. 2013; Fig. 3). The ghrelin receptor is expressed in the brain to regulate feeding, including hypothalamic nuclei involved in energy balance regulation and reward-linked areas such as the ventral tegmental area (Perello and Dickson 2015). Ghrelin signaling at the level of the mesolimbic system is one of the key molecular substrates that provides a physiological signal connecting gut and reward pathways (Perello and Dickson 2015). Milk intake via epigenetic upregulation of FTO may thus trigger ghrelin

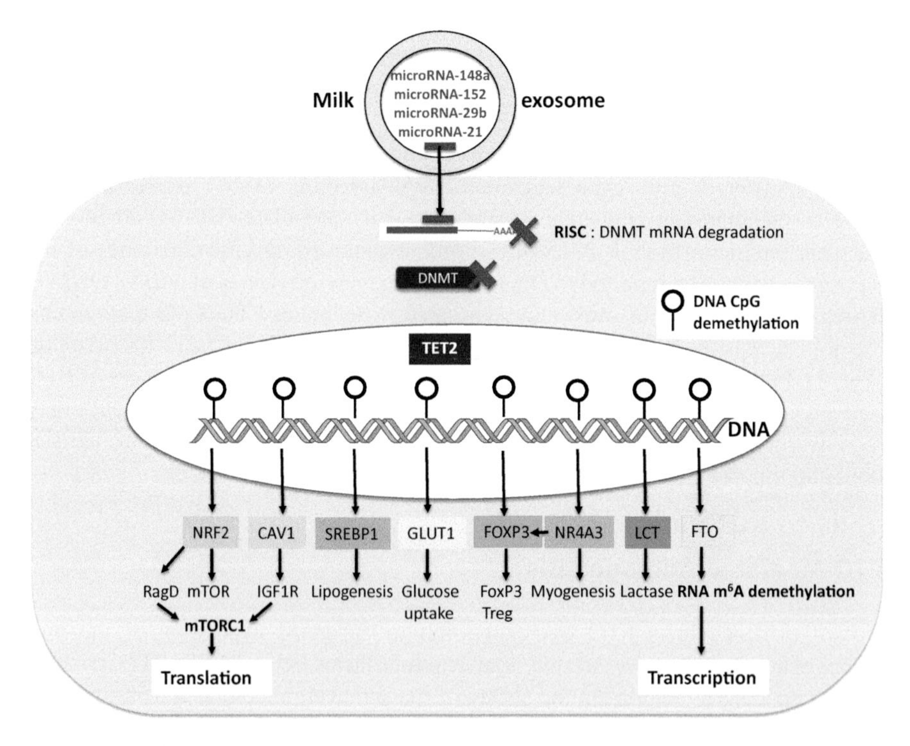

Fig. 2 Epigenetic regulation of gene expression via milk-derived DNA methyltransferases-targeting microRNAs. Milk exosome-derived DNMT-targeting microRNAs enhance DNA demethylation of critical CpG islets involved in the regulation of gene expression. The expression of pivotal transcription factors (NRF2, SREBP1, FOXP3, NR4A3) and key metabolic regulators (CAV1, GLUT1), the lactase gene (LCT), and the RNA m^6A demethylase FTO are upregulated via DNA promoter demethylation. Milk-derived DNMT-targeting microRNAs may thus play a fundamental role in epigenetic enhancement of transcription and translation

signaling enhancing appetite and suckling reward to secure postnatal food intake required for postnatal growth and development. Human milk via epigenetic FTO-mediated m^6A-demethylation of ghrelin mRNA may maintain specific orexigenic and reward signals, while ensuring appropriate appetite regulation, a developmental requirement adjusting physiological growth trajectories of the human infant.

FTO-Mediated Dopaminergic Signaling and Neuronal Development

Activation of dopaminergic signaling plays an important role in the midbrain and frontal cortex during postnatal development. Midbrain dopaminergic neurons in the substantia nigra pars compacta and ventral tegmental area regulate extrapyramidal movement and important cognitive functions, including motivation, reward associations, and habit learning (Luo and Huang 2016). The FTO gene regulates the activity of the dopaminergic midbrain circuitry (Hess et al. 2013). Inactivation of the FTO gene impairs dopamine receptor type 2 (D2R) and type 3 (D3R)-dependent control of neuronal activity and behavioral responses (Hess et al. 2013). Analysis of global m^6A modification of mRNAs in the midbrain and striatum of FTO-deficient mice revealed increased m^6A levels in a subset of mRNAs important for neuronal signaling, including many in the dopaminergic signaling pathway (Hess et al. 2013). Conversely, milk-mediated upregulation of FTO stimulated via DNMT-targeting exosomal microRNAs may increase FTO expression and m^6A demethylation of specific mRNAs promoting dopaminergic transmission (Fig. 3).

Milk Exosomal microRNAs: Enhancers of Intestinal Growth and Maturation

It has been shown that porcine milk exosomes transfer microRNAs into IECs and modify target gene expression promoting intestinal proliferation (Chen et al. 2016). Furthermore, DNMT inhibition promoted the differentiation of human induced pluripotent stem cells into enterocytes (Kodama et al. 2016).

Milk microRNAs: Epigenetic Inducers of Osteogenesis

Recent evidence indicates that the transcription factor NRF2 represents a key pathway in regulating bone metabolism. NRF2, which is upregulated by DNMT inhibition (Cao et al. 2016), is required for normal postnatal bone acquisition in mice (Kim et al. 2014). Mice lacking NRF2 exhibited a marked deficit in postnatal bone acquisition, which was most severe at 3 weeks of age when osteoblast numbers were 12-fold less than observed in control animals (Kim et al. 2014). Thus, milk exosomal microRNA-mediated suppression of DNMTs may promote NRF2-driven postnatal osteogenesis.

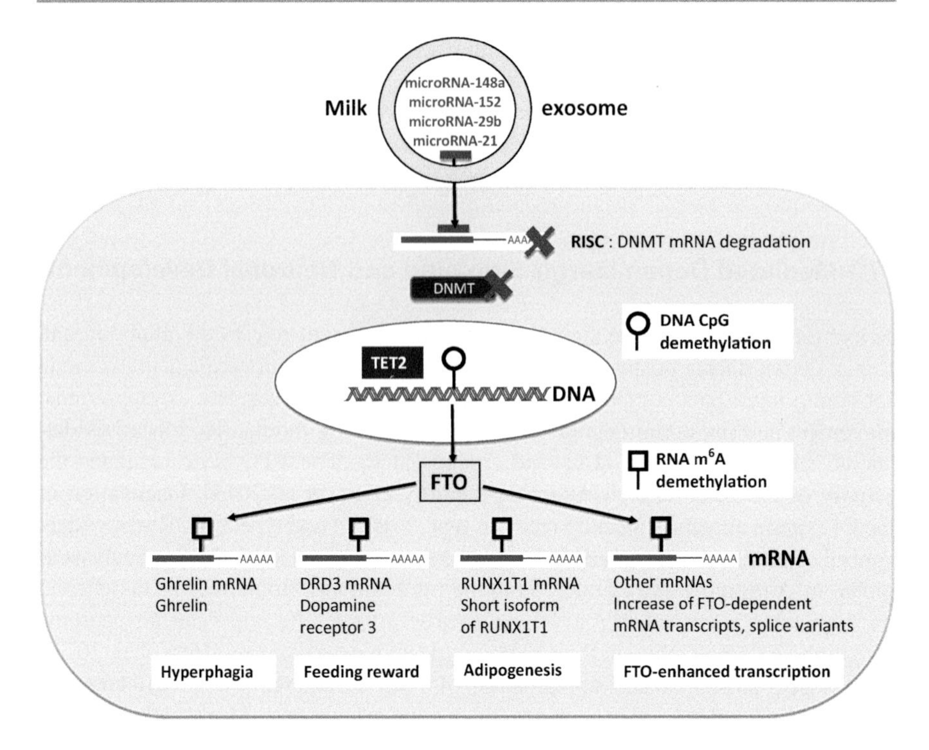

Fig. 3 Epigenetic upregulation of RNA transcription by exosomal milk-derived microRNAs enhancing the expression of the RNA m^6A demethylase FTO. Milk exosome-derived DNMT-targeting microRNAs apparently reduce FTO CpG DNA methylation. Demethylated DNA attracts the enzyme TET2 (ten-eleven-translocation), which catalyzes complete DNA demethylation. Demethylation of critical DNA CpG islets of the FTO gene increases FTO expression. FTO is an RNA m^6A demethylase, which erases m^6A marks on mRNAs, thereby increasing mRNA expression and changing RNA splice variant generation. The mRNAs of ghrelin and dopamine receptor 3 (DRD3) are targets of FTO-mediated upregulation. The resulting hyperphagia and feeding rewards support milk intake for infant growth requirements. The formation of the FTO-dependent short form of RUNX1T1 enhances adipogenesis

The transcription factor RUNX2 is a global regulator of osteogenesis. MicroRNA-29b promotes osteogenesis by directly targeting mRNA of HDAC4, which is an inhibitor of RUNX2 (Li et al. 2009). Notably, consumption of commercial cow's milk dose-dependently increased RUNX2 expression in PBMCs of healthy milk consumers (Baier et al. 2014).

Both, microRNA-148a and microRNA-21, abundant microRNA of milk, have been shown to promote osteogenic differentiation of human bone marrow-derived stem cells (Xu et al. 2014; Chen and Wang 2017). Bovine milk-derived EVs have recently been reported to promote the formation of small osteoclasts without reductions in bone resorption (Oliveira et al. 2017). Thus, a whole regulatory network of milk exosome-derived microRNAs in concert with bone-derived exosomal microRNAs may epigenetically enhance osteogenesis.

Milk microRNAs: Epigenetic Inducers of Myogenesis

The metabolic transcription factor NR4A3 is involved in myogenesis and postnatal development, and its expression critically depends on nutritional status (Pérez-Sieira et al. 2014). Milk-derived exosomal microRNAs may provide the required epigenetic signals for muscle cell differentiation and appropriate muscle protein acquisition. Notably, NR4A3 expression is epigenetically induced by NR4A3 promoter demethylation (Yeh et al. 2016). Genome demethylation with either 5-azacytidine treatment or overexpression of the antisense RNA against DNMT1 induced trans-differentiation of mouse fibroblasts into myoblasts (Taylor and Jones 1979; Szyf et al. 1992). Furthermore, it has been shown that demethylation of the distal enhancer in MyoD gene and the MyoG promoter is essential for the initiation of the myogenic differentiation program (Lucarelli et al. 2001; Palacios and Puri 2006). Micro-RNA-148a also promotes myogenic differentiation. Rho-associated coiled-coil containing protein kinase 1 (ROCK1), a known inhibitor of myogenesis, has been identified as a target of microRNA-148a (Zhang et al. 2012). Thus, multiple epigenetic mechanisms induced by milk exosomal microRNAs targeting DNMT expression may contribute to myogenesis.

Milk Exosomal microRNAs: Promotors of Skin Barrier Function

NRF2 expression plays a fundamental role in skin development and homeostasis. Epidermal barrier function of the skin depends on NRF2 signaling. Late cornified envelope 1 genes are transcriptional targets of NRF2 (Ishitsuka et al. 2016). After birth, milk microRNAs targeting DNMTs may support epigenetic upregulation of NRF2 augmenting the expression of cornified envelope proteins, which are crucial components for skin barrier function.

Milk microRNAs: Epigenetic Enhancers of Lactase Expression

It is not precisely known how the lactase gene (*LCT*) is downregulated after weaning. Recent evidence has been presented that epigenetically controlled regulatory elements account for the differences in lactase mRNA levels among individuals, intestinal cell types, and species (Labrie et al. 2016). In Europeans, a single nucleotide substitution ($-13910C > T$) with intron 13 of *MCM6*, the gene adjacent to *LCT*, is strongly associated with lactase persistence. This substitution resides within an enhancer of *LCT* increasing promoter activity. Seven epigenetically regulated regions have been identified that include the *LCT* enhancer located within intron 13 of *MCM6*, which house the functional DNA variants (Labrie et al. 2016). Low *LCT* transcription observed with the lactase nonpersistence haplotype was associated with a higher methylation status of intron 13, in contrast to a low methylation status of intron 13 in the lactase persistence haplotype resulting in high lactase expression (Labrie et al. 2016). We speculate that milk-derived microRNAs targeting DNMTs may maintain the low methylation status of intron 13, thus promoting lactase production during the period of breastfeeding. After weaning, the disappearance of milk microRNA-mediated DNMT suppression might downregulate intestinal lactase expression, which is no more required after the lactation period. Thus, milk epigenetics may assist to control intestinal lactase expression of the milk recipient.

Enhancing Commercial Milk Yield May Affect Consumer Health

Enhanced milk quality and quantity has become a major selection criterion for the genetic modification of livestock. Epigenetic regulations play a major role in bovine mammary gland development and lactation performance (Bian et al. 2015). Efforts of veterinary medicine to increase the milk yield of dairy cows may affect their milk exosome microRNA content. As the majority of milk microRNAs are derived from MECs, an increase of intracellular bovine MEC microRNA levels may enhance their concentration in exported MEC-derived exosomes taken up by consumers of commercial milk (Melnik 2015b). Lactation performance critically depends on microRNAs such as microRNA-148a, microRNA-152, microRNA-29b, and microRNA-21 that control the methylation status of lactation-specific genes (Wang et al. 2014; Bian et al. 2015; Chen et al. 2017). In the mammary gland of dairy cows producing high quality milk compared with cows producing low quality milk, the expression of microRNA-152 was significantly increased during lactation (Wang et al. 2014). DNMT1, which is a target of microRNA-152, was inversely correlated with the expression level of microRNA-152 in the mammary gland of dairy cows (Wang et al. 2014). Forced expression of microRNA-152 resulted in a marked reduction of DNMT1 at both the mRNA and protein levels (Wang et al. 2014). This in turn led to a decrease in global DNA methylation and increased the expression of two lactation-related genes such as AKT and PPARγ. Furthermore, microRNA-152 enhanced the viability and multiplication capacity of dairy cow MECs (Wang et al. 2014). Whereas microRNA-152 suppresses DNMT1, the microRNA-29 family regulates the DNA methylation level by inversely targeting both DNMT3A and DNMT3B in dairy cow MECs. Overexpression of microRNA-29a, microRNA-29b, and microRNA-29c increased the expression of casein-α s1 (CSN1S1), E74-like factor 5 (ELF5), PPARγ, SREBP1, and glucose transporter 1 (GLUT1) (Bian et al. 2015).

The link between microRNA-148a/152 and microRNA-29 s and DNMTs plays an important role in the epigenetic regulation of lactation performance but confers a yet unpredictable health risk for the human consumer of milk produced by high performance dairy cows (Melnik 2015b). Since the 1950s with the widespread distribution of refrigeration technology, exosomal microRNAs of cow's milk have been introduced in a large scale into the human food chain. In commercial pasteurized cow's milk the majority of DNMT-targeting microRNAs are still detectable in comparison to raw unprocessed milk (Howard et al. 2015; Baddela et al. 2016). In contrast, artificial infant formula contains only minor amounts of exosomal microRNAs most likely due to a loss of exosome integrity during formula processing (Chen et al. 2010; Sun et al. 2013; Alsaweed et al. 2015).

Increased transfer of exosomal microRNAs generated and exported by high performance, microRNA-148a/152 and microRNA-29b-overexpressing dairy cows, may thus have adverse effects on the human epigenome explaining the rising prevalence rates of noncommunicable diseases.

Conclusion

The vast majority of scientific data confirm the transfer of milk-derived exosomes and their exosomal microRNAs to IECs of the milk recipient, who under physiological conditions is the breastfed infant. Indirect evidence based on animal models, cell culture systems, computerized simulations, and translational research indicates that lactation-specific milk-derived exosomal microRNAs may reach the systemic circulation of the breastfed infant and human consumer of commercial cow's milk, respectively. Milk, the exclusive earliest nutritional environment of newborn mammals, may shape the infant's epigenome and epitranscriptome. Milk exosome-derived microRNAs may represent the critical epigenetic linkers between the mother and her infant. Abundant microRNAs of milk are microRNA-148a, microRNA-152, microRNA-29b, and microRNA-21, which all suppress the activity of DNA methyltransferases critically involved in the epigenetic regulation of gene expression. Important candidate genes for milk-microRNA-mediated epigenetic regulation are *FOXP3*, *NR4A*, *NRF2*, *FTO*, and *LCT*, respectively. Via induction of increased FoxP3 expression, milk-microRNAs may contribute to a tolerogenic intestinal environment during early nutrition and may be an important factor for the introduction and tolerance of solid food. Future studies should clarify whether milk-derived exosomes reach the thymus and modulate thymic FoxP3 expression critical for natural Treg differentiation. It is thus possible that milk-derived exosomal microRNAs function as a tolerogenic conditioner during early TCR-mediated antigen contacts for both inducible Tregs in the intestine and natural Tregs in the thymus. In this respect, the period of breastfeeding should last at least 1 year to ensure a perennial tolerogenic environment for all seasonal allergens.

Milk-microRNA controlled epigenetic activation of FTO expression may ensure sustained food intake via lactation-driven hyperphagia and feeding rewards during the lactation period. Milk via epigenetic enhancement of the expression of the m^6A demethylase FTO may increase critical mRNA transcription of the infant to promote adipogenesis, osteogenesis, myogenesis, and neurogenesis. Thus, adequate kinetics of signal transduction of milk-regulated FTO expression may be of critical importance for lifelong metabolic programming of the newborn infant. In all mammals except Neolithic humans, the beneficial transfer of milk's epigenetic machinery is restricted to the lactation/breastfeeding period. Since 1955, the widespread availability of household refrigerators associated with the storage of pasteurized cow's milk allowed a large-scale introduction of bioactive bovine milk exosomal microRNAs into the human food chain.

It is of considerable concern that persistent abuse of milk's exosomal microRNAs by consumption of commercial milk may disturb epigenetic homeostasis, which in the long run promotes adipogenic, diabetogenic, cancerogenic, and neurodegenerative

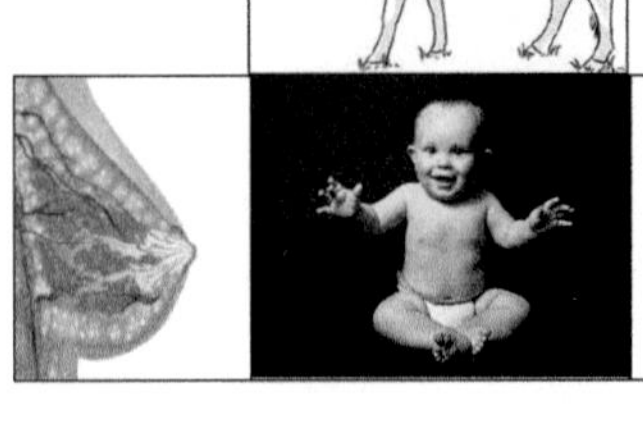
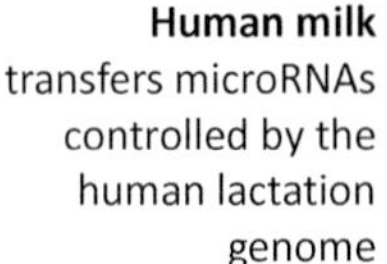

Fig. 4 Comparison of different milk microRNA exposures of human infants by natural breastfeeding, artificial formula feeding, and cow's milk consumption. In contrast to all other mammals, which only receive the epigenetic programming machinery of milk of their own species, the newborn human infant can be exposed to several modes of epigenetic regulation: natural and lactation-genome controlled breastfeeding, artificial formula feeding associated with severe deficiencies of the lactation-specific microRNA signaling, and in later infancy exposure to farm milk or commercial cow's milk. The correct milk-mediated epigenetic programming and imprinting during the postnatal period is a fundamental biological phenomenon tightly controlled by the species' lactation genome, a natural birth right of all mammals

pathways. As veterinary medicine's efforts to increase milk yield are associated with increments of DNMT-targeting bovine microRNAs, whose seed sequences are complementary to their human analogues (Table 2), subsequent disturbances of epigenetic homeostasis of the human milk consumer may be inevitable.

There are still many open questions that have to be proven experimentally as our perception of the epigenetic signaling of milk for infant development and non-communicable diseases is at the very beginning. For the growing infant, the tight lactation-genome-controlled milk environment appears to be of critical importance for lifelong epigenetic imprinting (Fig. 4). A deeper understanding of milk's epigenetic layers of communication will help to interfere therapeutically and to prevent adverse nutrigenomic effects on human health. Future research has to unravel whether it was a wise behavior modification of *Homo sapiens* to expose himself permanently to milk's epigenetic machinery.

Dictionary of Terms

- **FoxP3** – The master transcription factor of regulatory T cells (Tregs), which is activated via demethylation of the Treg-specific demethylation region (TSDR) of the *FOXP3* gene.
- **DNMTs** – Three DNA methyltransferases are important for the methylation of CpG islets of mammalian DNA: DNMT1, DNMT3a, and DNMT3b. De novo methylation patterns are established during embryonic development by DNMT3a and DNMT3b. They are maintained by a DNMT1-mediated copying mechanism

when cells divide. The heritability of DNA methylation patterns provides an epigenetic marking of the genome that is stable through multiple cell divisions and therefore constitutes a form of cellular memory. Milk may modify DNA methylation of the milk recipient via transfer of DNMTs-targeting microRNAs promoting postnatal epigenetic imprinting.

- **CpG islands** – CpG islands are regions of high CpG density (>50%), usually 200 bp–2 kb in length that lack CpG methylation, found at promoters of most human genes. Long-term silencing of the gene can be insured by methylation of the CpG region.
- **5-Azacytidine** – 5-Azacytidine is a nucleoside analogue that inhibits DNA methylation. It is incorporated into DNA in place of cytidine and forms a covalent adduct with DNMTs preventing further DNA methylation.
- **FTO** – Fat mass- and obesity-associated protein is a demethylase of m^6A methylation marks on mRNAs. The FTO-catalyzed m^6A erasing mechanism is of critical importance for enhancing mRNA transcription, mRNA stability, generation of splicing factors, and microRNA biogenesis.
- **Milk exosome** – Milk exosomes are secretory products of mammary epithelial cells. The majority of these extracellular vesicles found in human and bovine milk are 30–100 nm sized spherical particles covered by a resistant lipid bilayer membrane which contains tetraspanins and other proteins important for exosome adhesion and trafficking. Milk exosomes transport a complex cargo including high amounts of RNAs including microRNAs, which target DNMTs.
- **TSDR** – The Treg-cell specific demethylation region (also named CNS2: conserved noncoding sequence 2) is a CpG-rich element that lies in the first intron of the FOXP3 gene. The selective demethylation of this CpG motif is critical for the stabilization of FoxP3 expression.

Key Facts

- Milk contains abundant amounts of extracellular vesicles. Exosomes are a special class of extracellular vesicles in the size range of 30–100 nm, equipped with a unique proteome allowing exosome endocytosis.
- The majority of milk exosomes are secreted by mammary gland epithelial cells during lactation.
- Milk exosomes contain highest amounts various RNAs and microRNAs in comparison to other fluids of the human body.
- An abundant type of milk exosome-derived microRNAs are microRNA-148a, microRNA-152, microRNA-29b, and microRNA-21, which all target DNA methyltransferases (DNMTs).
- There is compelling evidence that milk exosomes are taken up by intestinal epithelial cells of the milk recipient and thereby transfer milk-derived microRNAs resulting in gene-regulatory changes in the recipient cells.

- Indirect translational evidence indicates that milk exosomes and their lactation-specific microRNAs may reach the systemic circulation of the milk recipient and exert distant gene-regulatory functions (*functional hypothesis*).
- The seed sequences of human and bovine DNMT-targeting microRNAs are identical, which means that bovine milk-derived microRNAs may modify gene regulation of the human consumer of commercial milk.
- Milk microRNA-mediated downregulation of DNMTs appears as a new epigenetic imprinting mechanism increasing global DNA demethylation thereby enhancing general transcription required for postnatal growth and adequate metabolic programming during the lactation period.
- Gene expression of several important transcription factors such as NRF2, NR4A3, SREBP1, PPARγ, FOXP3 as well as crucial metabolic regulators such as GLUT1, CAV1, LCT, and FTO are upregulated via DNMT suppression resulting in DNA CpG demethylation.
- There is good reason to assume that milk intake increases the expression and activity of the m^6A demethylase FTO thereby decreasing m^6A marks of specific RNAs leading to increased expression of ghrelin, dopamine, and adipogenic signaling.
- Artificial formula misses epigenetically active microRNAs of exosomal origin bearing the risk of inappropriate immunological and metabolic programming of the newborn infant.
- In mammalian physiology, milk exosome-mediated epigenetic imprinting is restricted to the lactation period and ceases with weaning.
- The persistence of milk-exosome-mediated epigenetic signaling into adulthood of Neolithic humans, who continuously consume commercial milk containing the epigenetic machinery of *Bos taurus*, may be an overlooked most critical risk factor promoting diseases of Western civilization.
- Efforts of dairy research to increase milk yield of dairy cows are associated with increased expression of DNMT-targeting lactation-specific microRNAs, which may further raise microRNA levels that reach the human food chain.
- With the widespread distribution of refrigeration technology and forced expression of lactation-promoting microRNAs secreted by high performance dairy cows, the consumer of commercial milk is exposed to an unnoticed change of their epigenetic nutritional environment with potential adverse effects on human health.

Summary Points

1. Milk-derived exosomes are transmitters of mother-to-child communication affecting epigenetic regulation of the infant.
2. Milk exosomes equipped with a special proteome allow exosome trafficking and endocytosis.
3. The majority of milk exosomes are secretory products of mammary gland epithelial cells during lactation.

4. Milk exosomes contain highest amounts RNAs and microRNAs.
5. Abundant milk exosome microRNA-148a, microRNA-29b, and microRNA-21 target DNA methyltransferases (DNMTs).
6. Milk exosomes are taken up by intestinal epithelial and vascular endothelial cells.
7. Milk exosome microRNAs reach the systemic circulation and modify gene regulation of the milk recipient.
8. Seed sequences of human and bovine DNMT-targeting microRNAs are identical.
9. Milk microRNA-mediated downregulation of DNMTs is a new epigenetic imprinting mechanism increasing global DNA demethylation enhancing transcription.
10. Fat mass- and obesity-associated gene (FTO) may be upregulated via DNMT-targeting microRNAs of milk.
11. The m^6A demethylase FTO may decrease m^6A marks of specific RNAs enhancing RNA transcription.
12. Artificial formula misses milk's epigenetically active microRNAs bearing the risk of inappropriate epigenetic programming of the newborn infant.
13. Milk exosome-mediated epigenetic imprinting is restricted to the lactation period.
14. Persistence of milk-mediated epigenetic signaling into adulthood may represent a critical risk factor promoting diseases of civilization.
15. Efforts to enhance milk yield of dairy cows increases the expression of DNMT-targeting microRNAs such as microRNA-148a accumulating in the human food chain.
16. MicroRNA preservation by refrigeration and overexpression by high performance dairy cows exposes the human milk consumer to epigenetic changes that may adversely affect human health.

References

Admyre C, Johansson SM, Qazi KR et al (2007) Exosomes with immune modulatory features are present in human breast milk. J Immunol 179:1969–1978

Alsaweed M, Hartmann PE, Geddes DT, Kakulas F (2015) MicroRNAs in breastmilk and the lactating breast: potential immunoprotectors and developmental regulators for the infant and the mother. Int J Environ Res Public Health 12:13981–14020

Alsaweed M, Lai CT, Hartmann PE et al (2016a) Human milk cells and lipids conserve numerous known and novel miRNAs, some of which are differentially expressed during lactation. PLoS One 11:e0152610

Alsaweed M, Lai CT, Hartmann PE et al (2016b) Human milk miRNAs primarily originate from the mammary gland resulting in unique miRNA profiles of fractionated milk. Sci Rep 6:20680

Alsaweed M, Lai CT, Hartmann PE et al (2016c) Human milk cells contain numerous miRNAs that may change with milk removal and regulate multiple physiological processes. Int J Mol Sci 17: E956

Ambros V (2004) The functions of animal microRNAs. Nature 431:350–355

Arntz OJ, Pieters BC, Oliveira MC et al (2015) Oral administration of bovine milk derived extracellular vesicles attenuates arthritis in two mouse models. Mol Nutr Food Res 59:1701–1712
Baddela VS, Nayan V, Rani P et al (2016) Physicochemical biomolecular insights into buffalo milk-derived nanovesicles. Appl Biochem Biotechnol 178:544–557
Baier SR, Nguyen C, Xie F et al (2014) MicroRNAs are absorbed in biologically meaningful amounts from nutritionally relevant doses of cow milk and affect gene expression in peripheral blood mononuclear cells, HEK-293 kidney cell cultures, and mouse livers. J Nutr 144:1495–1500
Bendavit G, Aboulkassim T, Hilmi K et al (2016) Nrf2 transcription factor can directly regulate mTOR; linking cytoprotective gene expression to a major metabolic regulator that generates redox activity. J Biol Chem 291:25476–25488
Benmoussa A, Lee CH, Laffont B et al (2016) Commercial dairy cow milk microRNAs resist digestion under simulated gastrointestinal tract conditions. J Nutr 146:2206–2215
Bian Y, Lei Y, Wang C et al (2015) Epigenetic regulation of miR-29s affects the lactation activity of dairy cow mammary epithelial cells. J Cell Physiol 230:2152–2163
Boissel S, Reish O, Proulx K et al (2009) Loss-of-function mutation in the dioxygenase-encoding FTO gene causes severe growth retardation and multiple malformations. Am J Hum Genet 85:106–111
Cao H, Wang L, Chen B et al (2016) DNA demethylation upregulated Nrf2 expression in Alzheimer's disease cellular model. Front Aging Neurosci 7:244
Chen D, Wang Z (2017) Adrenaline inhibits osteogenesis via repressing miR-21 expression. Cell Biol Int 41:8–15
Chen X, Gao C, Li H et al (2010) Identification and characterization of microRNAs in raw milk during different periods of lactation, commercial fluid, and powdered milk products. Cell Res 20:1128–1137
Chen T, Xi QY, Ye RS et al (2014) Exploration of microRNAs in porcine milk exosomes. BMC Genomics 15:100
Chen J, Zhou X, Wu W et al (2015) FTO-dependent function of N6-methyladenosine is involved in the hepatoprotective effects of betaine on adolescent mice. J Physiol Biochem 71:405–413
Chen T, Xie MY, Sun JJ et al (2016) Porcine milk-derived exosomes promote proliferation of intestinal epithelial cells. Sci Rep 6:33862
Chen Z, Luo J, Sun S et al (2017) miR-148a and miR-17-5p synergistically regulate milk TAG synthesis via PPARGC1A and PPARA in goat mammary epithelial cells. RNA Biol 14:326–338
Church C, Moir L, McMurray F et al (2010) Overexpression of Fto leads to increased food intake and results in obesity. Nat Genet 42:1086–1092
Dominissini D, Moshitch-Moshkovitz S, Schwartz S et al (2012) Topology of the human and mouse m6A RNA methylomes revealed by m6A-seq. Nature 485:201–206
Duursma AM, Kedde M, Schrier M et al (2008) miR-148 targets human DNMT3b protein coding region. RNA 14:872–877
Elhassan MO, Christie J, Duxbury MS (2012) Homo sapiens systemic RNA interference-defective-1 transmembrane family member 1 (SIDT1) protein mediates contact-dependent small RNA transfer and microRNA-21-driven chemoresistance. J Biol Chem 287:5267–5277
Fabbri M, Garzon R, Cimmino A et al (2007) MicroRNA-29 family reverts aberrant methylation in lung cancer by targeting DNA methyltransferases 3A and 3B. Proc Natl Acad Sci U S A 104:15805–15810
Fischer J, Koch L, Emmerling C et al (2009) Inactivation of the Fto gene protects from obesity. Nature 458:894–898
Gallier S, Vocking K, Post JA et al (2015) A novel infant milk formula concept: mimicking the human milk fat globule structure. Colloids Surf B Biointerfaces 136:329–339
Godfrey KM, Costello PM, Lillycrop KA (2016) Development, epigenetics and metabolic programming. Nestle Nutr Inst Workshop Ser 85:71–80

Gu Y, Li M, Wang T et al (2012) Lactation-related microRNA expression profiles of porcine breast milk exosomes. PLoS One 7:e43691

Hassiotou F, Beltran A, Chetwynd E et al (2012) Breastmilk is a novel source of stem cells with multilineage differentiation potential. Stem Cells 30:2164–2174

Hassiotou F, Geddes DT, Hartmann PE (2013) Cells in human milk: state of the science. J Hum Lact 29:171–182

Hata T, Murakami K, Nakatani H et al (2010) Isolation of bovine milk-derived microvesicles carrying mRNA and microRNAs. Biochem Biophys Res Commun 396:528–533

Hess ME, Hess S, Meyer KD et al (2013) The fat mass and obesity associated gene (Fto) regulates activity of the dopaminergic midbrain circuitry. Nat Neurosci 16:1042–1048

Hinz D, Bauer M, Röder S et al (2012) Cord blood Tregs with stable FOXP3 expression are influenced by prenatal environment and associated with atopic dermatitis at the age of one year. Allergy 67:380–389

Howard KM, Jati Kusuma R, Baier SR et al (2015) Loss of miRNAs during processing and storage of cow's (Bos taurus) milk. J Agric Food Chem 63:588–592

Huehn J, Beyer M (2015) Epigenetic and transcriptional control of Foxp3+ regulatory T cells. Semin Immunol 27:10–18

Ishitsuka Y, Huebner AJ, Rice RH et al (2016) Lce1 family members are Nrf2-target genes that are induced to compensate for the loss of loricrin. J Invest Dermatol 136:1656–1663

Izumi H, Kosaka N, Shimizu T et al (2012) Bovine milk contains microRNA and messenger RNA that are stable under degradative conditions. J Dairy Sci 95:4831–4841

Izumi H, Tsuda M, Sato Y et al (2015) Bovine milk exosomes contain microRNA and mRNA and are taken up by human macrophages. J Dairy Sci 98:2920–2933

Jia G, Fu Y, Zhao X et al (2011) N6-methyladenosine in nuclear RNA is a major substrate of the obesity-associated FTO. Nat Chem Biol 7:885–887

Josefowicz SZ, Wilson CB, Rudensky AY (2009) Cutting edge: TCR stimulation is sufficient for induction of Foxp3 expression in the absence of DNA methyltransferase 1. J Immunol 182:6648–6652

Karra E, O'Daly OG, Choudhury AI et al (2013) A link between FTO, ghrelin, and impaired brain food-cue responsivity. J Clin Invest 123:3539–3551

Kim JH, Singhal V, Biswal S et al (2014) Nrf2 is required for normal postnatal bone acquisition in mice. Bone Res 2:14033

Kirchner B, Pfaffl MW, Dumpler J et al (2016) microRNA in native and processed cow's milk and its implication for the farm milk effect on asthma. J Allergy Clin Immunol 137:1893–1895.e13

Kodama N, Iwao T, Kabeya T et al (2016) Inhibition of mitogen-activated protein kinase kinase, DNA methyltransferase, and transforming growth factor-β promotes differentiation of human induced pluripotent stem cells into enterocytes. Drug Metab Pharmacokinet 31:193–200

Kosaka N, Izumi H, Sekine K, Ochiya T (2010) microRNA as a new immune-regulatory agent on breast milk. Silence 1:7

Kurinna S, Schäfer M, Ostano P et al (2014) A novel Nrf2-miR-29-desmocollin-2 axis regulates desmosome function in keratinocytes. Nat Commun 5:5099

Kusuma RJ, Manca S, Friemel T et al (2016) Human vascular endothelial cells transport foreign exosomes from cow's milk by endocytosis. Am J Physiol Cell Physiol 310:C800–C807

Labrie V, Buske OJ, Oh E et al (2016) Lactase nonpersistence is directed by DNA-variation-dependent epigenetic aging. Nat Struct Mol Biol 23:566–573

Li Z, Hassan MQ, Jafferji M et al (2009) Biological functions of miR-29b contribute to positive regulation of osteoblast differentiation. J Biol Chem 284:15676–15684

Li Y, Jensen ML, Chatterton DE et al (2014) Raw bovine milk improves gut responses to feeding relative to infant formula in preterm piglets. Am J Physiol Gastrointest Liver Physiol 306: G81–G90

Li R, Dudemaine PL, Zhao X et al (2016) Comparative analysis of the miRNome of bovine milk fat, whey and cells. PLoS One 11:e0154129

Liu ZW, Zhang JT, Cai QY et al (2014) Birth weight is associated with placental fat mass- and obesity-associated gene expression and promoter methylation in a Chinese population. J Matern Fetal Neonatal Med 10:1–6
Loss G, Apprich S, Waser M et al (2011) The protective effect of farm milk consumption on childhood asthma and atopy: the GABRIELA study. J Allergy Clin Immunol 128:766–773
Lucarelli M, Fuso A, Strom R, Scarpa S (2001) The dynamics of myogenin site-specific demethylation is strongly correlated with its expression and with muscle differentiation. J Biol Chem 276:7500–7506
Luo SX, Huang EJ (2016) Dopaminergic neurons and brain reward pathways: from neurogenesis to circuit assembly. Am J Pathol 186:478–488
Maity A, Das B (2016) N6-methyladenosine modification in mRNA: machinery, function and implications for health and diseases. FEBS J 283:1607–1630
Melnik BC (2015a) Milk – a nutrient system of mammalian evolution promoting mTORC1-dependent translation. Int J Mol Sci 16:17048–17087
Melnik BC (2015b) Milk: an epigenetic amplifier of FTO-mediated transcription? Implications for Western diseases. J Transl Med 13:385
Melnik BC, John SM, Schmitz G (2013) Milk is not just food but most likely a genetic transfection system activating mTORC1 signaling for postnatal growth. Nutr J 12:103
Melnik BC, John SM, Schmitz G (2014) Milk: an exosomal microRNA transmitter promoting thymic regulatory T cell maturation preventing the development of atopy? J Transl Med 12:43
Melnik BC, John SM, Carrera-Bastos P, Schmitz G (2016a) Milk: a postnatal imprinting system stabilizing FoxP3 expression and regulatory T cell differentiation. Clin Transl Allergy 6:18
Melnik BC, John SM, Schmitz G (2016b) Milk: an epigenetic inducer of FoxP3 expression. J Allergy Clin Immunol 138:937–938
Melnik BC, Kakulas F, Geddes DT et al (2016c) Milk miRNAs: simple nutrients or systemic functional regulators? Nutr Metab (Lond) 13:42
Merkestein M, McTaggart JS, Lee S et al (2014) Changes in gene expression associated with FTO overexpression in mice. PLoS One 9:e97162
Merkestein M, Laber S, McMurray F et al (2015) FTO influences adipogenesis by regulating mitotic clonal expansion. Nat Commun 6:6792
Modepalli V, Kumar A, Hinds LA et al (2014) Differential temporal expression of milk miRNA during the lactation cycle of the marsupial tammar wallaby (Macropus eugenii). BMC Genomics 15:1012
Mulcahy LA, Pink RC, Carter DR (2014) Routes and mechanisms of extracellular vesicle uptake. J Extracell Vesicles 3:1
Munch EM, Harris RA, Mohammad M et al (2013) Transcriptome profiling of microRNA by next-gene deep sequencing reveals known and novel miRNA species in the lipid fraction of human breast milk. PLoS One 8:e50564
Na RS, GX E, Sun W et al (2015) Expressional analysis of immune-related miRNAs in breast milk. Genet Mol Res 14:11371–11376
O'Gorman A, Colleran A, Ryan A et al (2010) Regulation of NF-kappaB responses by epigenetic suppression of IkappaBalpha expression in HCT116 intestinal epithelial cells. Am J Physiol Gastrointest Liver Physiol 299:G96–G105
Oliveira MC, Di Ceglie I, Arntz OJ et al (2017) Milk-derived nanoparticle fraction promotes the formation of small osteoclasts but reduces bone resorption. J Cell Physiol 232:225–233
Palacios D, Puri PL (2006) The epigenetic network regulating muscle development and regeneration. J Cell Physiol 207:1–11
Palacios-Ortega S, Varela-Guruceaga M, Milagro FI et al (2014) Expression of caveolin 1 is enhanced by DNA demethylation during adipocyte differentiation. Status of insulin signaling. PLoS One 9:e95100
Pan W, Zhu S, Yuan M et al (2010) MicroRNA-21 and microRNA-148a contribute to DNA hypomethylation in lupus CD4+ T cells by directly and indirectly targeting DNA methyltransferase 1. J Immunol 184:6773–6781

Paparo L, Nocerino R, Cosenza L et al (2016) Epigenetic features of FoxP3 in children with cow's milk allergy. Clin Epigenetics 8:86
Parigi SM, Eldh M, Larssen P et al (2015) Breast milk and solid food shaping intestinal immunity. Front Immunol 6:415
Perello M, Dickson SL (2015) Ghrelin signalling on food reward: a salient link between the gut and the mesolimbic system. J Neuroendocrinol 27:424–434
Pérez-Sieira S, López M, Nogueiras R, Tovar S (2014) Regulation of NR4A by nutritional status, gender, postnatal development and hormonal deficiency. Sci Rep 4:4264
Perge P, Nagy Z, Decmann Á et al (2017) Potential relevance of microRNAs in inter-species epigenetic communication, and implications for disease pathogenesis. RNA Biol 14:391–401
Petrus NC, Henneman P, Venema A et al (2016) Cow's milk allergy in Dutch children: an epigenetic pilot survey. Clin Transl Allergy 6:16
Pieters BC, Arntz OJ, Bennink MB et al (2015) Commercial cow milk contains physically stable extracellular vesicles expressing immunoregulatory TGF-β. PLoS One 10:e0121123
Reinhardt TA, Lippolis JD, Nonnecke BJ, Sacco RE (2012) Bovine milk exosome proteome. J Proteomics 75:1486–1492
Reinhardt TA, Sacco RE, Nonnecke BJ, Lippolis JD (2013) Bovine milk proteome: quantitative changes in normal milk exosomes, milk fat globule membranes and whey proteomes resulting from Staphylococcus aureus mastitis. J Proteomics 82:141–154
Ruan WY, Bi MY, Feng WW et al (2016) Effect of human breast milk on the expression of proinflammatory cytokines in Caco-2 cells after hypoxia/re-oxygenation. Rev Invest Clin 68:105–111
Sasaki H, Shitara M, Yokota K et al (2012) RagD gene expression and NRF2 mutations in lung squamous cell carcinomas. Oncol Lett 4:1167–1170
Sekiya T, Nakatsukasa H, Lu Q, Yoshimura A (2016) Roles of transcription factors and epigenetic modifications in differentiation and maintenance of regulatory T cells. Microbes Infect 18:378–386
Shu J, Chiang K, Zempleni J, Cui J (2015) Computational characterization of exogenous micro-RNAs that can be transferred into human circulation. PLoS One 10:e0140587
Sun Q, Chen X, Yu J et al (2013) Immune modulatory function of abundant immune-related microRNAs in microvesicles from bovine colostrum. Protein Cell 4:197–210
Szyf M, Rouleau J, Theberge J, Bozovic V (1992) Induction of myogenic differentiation by an expression vector encoding the DNA methyltransferase cDNA sequence in the antisense orientation. J Biol Chem 267:12831–12836
Taylor SM, Jones PA (1979) Multiple new phenotypes induced in 10T1/2 and 3T3 cells treated with 5-azacytidine. Cell 17:771–779
Wang J, Bian Y, Wang Z et al (2014) MicroRNA-152 regulates DNA methyltransferase 1 and is involved in the development and lactation of mammary glands in dairy cows. PLoS One 9: e101358
Weber JA, Baxter DH, Zhang S et al (2010) The mircoRNA spectrum in 12 body fluids. Clin Chem 56:1733–1741
Wolf T, Baier SR, Zempleni J (2015) The intestinal transport of bovine milk exosomes is mediated by endocytosis in human colon carcinoma Caco-2 cells and rat small intestinal IEC-6 cells. J Nutr 145:2201–2206
Won HY, Hwang ES (2016) Transcriptional modulation of regulatory T cell development by novel regulators NR4As. Arch Pharm Res 39:1530–1536
Xu JF, Yang GH, Pan XH et al (2014) Altered microRNA expression profile in exosomes during osteogenic differentiation of human bone marrow-derived mesenchymal stem cells. PLoS One 9:e114627
Yáñez-Mó M, Siljander PR, Abdreu Z et al (2015) Biological properties of extracellular vesicles and their physiological functions. J Extracell Vesicles 4:27066
Yeh CM, Chang LY, Lin SH et al (2016) Epigenetic silencing of the NR4A3 tumor suppressor, by aberrant JAK/STAT signaling, predicts prognosis in gastric cancer. Sci Rep 6:31690

Zempleni J, Aguilar-Lozano A, Sadri M et al (2017) Biological activities of extracellular vesicles and their cargos from bovine and human milk in humans and implications for infants. J Nutr 147:3–10
Zhang J, Ying ZZ, Tang ZL et al (2012) MicroRNA-148a promotes myogenic differentiation by targeting the ROCK1 gene. J Biol Chem 287:21093–21101
Zhao X, Yang Y, Sun BF et al (2014) FTO and obesity: mechanisms of association. Curr Diab Rep 14:486
Zhou Q, Li M, Wang X et al (2012) Immune-related microRNAs are abundant in breast milk exosomes. Int J Biol Sci 8:118–123

Nutritional Regulation of Mammary miRNome: Implications for Human Studies

78

Christine Leroux, Dragan Milenkovic, Lenha Mobuchon, Sandrine Le Guillou, Yannick Faulconnier, Bruce German, and Fabienne Le Provost

Contents

C. Leroux (✉)
Herbivore Research Unit -Biomarkers Team, French Institut of Agricultural Research (INRA), St Genès-Champanelle, France

Department of Food Science and Technology, University of California Davis, Davis, CA, USA
e-mail: christine.leroux@inra.fr

D. Milenkovic
School of Medicine, Division of Cardiovascular Medicine, University of California Davis, Davis, CA, USA

Department of Human Nutrition, French Institut of Agricultural Research (INRA), St Genès-Champanelle, France
e-mail: dmilenkovic@ucdavis.edu; dragan.milenkovic@inra.fr

L. Mobuchon · Y. Faulconnier
Herbivore Research Unit -Biomarkers Team, French Institut of Agricultural Research (INRA), St Genès-Champanelle, France
e-mail: lmobuchon@hotmail.fr; yannick.faulconnier@inra.fr

S. Le Guillou · F. Le Provost
Génétique Animale et Biologie Intégrative, French Institut of Agricultural Research (INRA), Jouy-en-Josas, France
e-mail: sandrine.le-guillou@inra.fr; fabienne.le-provost@inra.fr

B. German
Department of Food Science and Technology, University of California Davis, Davis, CA, USA
e-mail: jbgerman@ucdavis.edu

V. B. Patel, V. R. Preedy (eds.), *Handbook of Nutrition, Diet, and Epigenetics*,
https://doi.org/10.1007/978-3-319-55530-0_88

Abstract

Mammary gland is the organ of milk component synthesis that provides the nutrients for growth and development of the mammalian neonate. In addition to macronutrients like proteins, carbohydrates, and lipids known for their roles in providing substrate and energy, a new class of components has been identified notably microRNA that have signaling roles regulating a large set of biological processes. MicroRNAs, short noncoding RNAs, have been reported to act on the mammary tissues, influencing mammary development and milk component biosynthesis, and evidence is now assembling that they also signal to the infant. The expression profile of these miRNAs can be under nutritional regulation. Their presence in milk and their relative persistency through industrial treatment open new way of investigations to use them as biomarkers of animal health, as well as to evaluate their effects on the health of those consuming them. Due to the role of miRNAs on human health and diseases, their transfer from milk or milk products to infants and adults is being actively researched, though their bioavailability is not known. Research is defining their distribution in the different fractions of milk (such as cells, exosomes, fat globule, or skim milk). Indeed, the unique packaging of miRNAs could be crucial for their action through the intestinal tract. The value of milk miRNAs to diverse aspects of human health is now an emerging field of science.

Keywords

MicroRNA · miRNome · Nutritional regulation · Expression · Bovine · Caprine · Human · Mammary gland · Milk · Healthy food

List of Abbreviations

3′UTR	Three prime untranslated region
ACACA	Acetyl-CoA carboxylase 1
ACSL1	Acyl-CoA synthetase long-chain family member 1
CSN1S1	Casein alpha S1
CSN2	Casein beta
DNA	Deoxyribonucleic acid
ESR1	Estrogen receptor 1
FASN	Fatty acid synthase
GLUT1	Glucose transporter 1
GMEC	Goat mammary epithelial cells
HER2	Human epidermal growth factor receptor 2
MEC	Mammary epithelial cells

MG	Mammary gland
miRNA	Microribonucleic acid
NGS	Next-generation sequencing
PPARG	Peroxisome proliferator-activated receptor gamma
RISC	RNA-induced silencing complex
SCD1	Stearoyl-CoA desaturase
SREBP1	Sterol regulatory element-binding protein 1

Introduction

Milk is a complex secretion of the mammary gland (**MG**) providing the nutrients for growth and development for mammalian neonates. MG development and functioning are one of the most important biological phenomena of evolution undergoing remarkable physiological changes during the reproductive cycles. MG is a dynamic organ changing from fetus development to complete functioning during lactation with repeating cycles of mammary epithelium establishment during each pregnancy, followed by a dry period (Anderson et al. 2007). During lactation, mammary epithelial cells (**MEC**) secrete prodigious quantities of proteins, lipids, minerals, lactose, and other milk components in a polarized fashion from their apical surface into the alveolar lumen that they surround. Milk is in large part responsible for the conspicuous success of mammalia in evolution by providing a complete system of nourishment to mammalian neonates. During the past 10,000 years, domestication of lactating mammals became agricultural practice and over the past century has become a global agricultural enterprise nourishing humans of all ages. Milk components have considerable effects on the nutritional, technological, and sensorial properties of milk and its derived products, cheese, butter, yogurt, etc. Scientific research on human milk has revealed a complex source of bioactive components beyond nutrition, providing protective and developmental benefits. For example, while fat in milk is recognized as a source of energy for the neonate of each species, detailed examinations are revealing new structures and functions of different lipid components (German and Dillard 2010). The nutritional quality of milk, which is a central concern for health, has been investigated to give objective data for human consumption recommendations. In particular, the milk lipid fraction has been studied due to its saturated fatty acid composition with the hypothesis that all of the effects are not positive to the health of all humans (Givens et al. 2010; German and Dillard 2010). More recently, milk oligosaccharides have been shown to have an influence on gut microbiota of consumers (Chichlowski et al. 2011). Based on this breadth of prior experience, other components less abundant are coming under attention due to their potential role on human health of which a new target is microRNA (**miRNA**).

miRNAs which are small noncoding RNA (~22 nucleotides) are crucial regulators of various biological processes such as proliferation, growth and cell death,

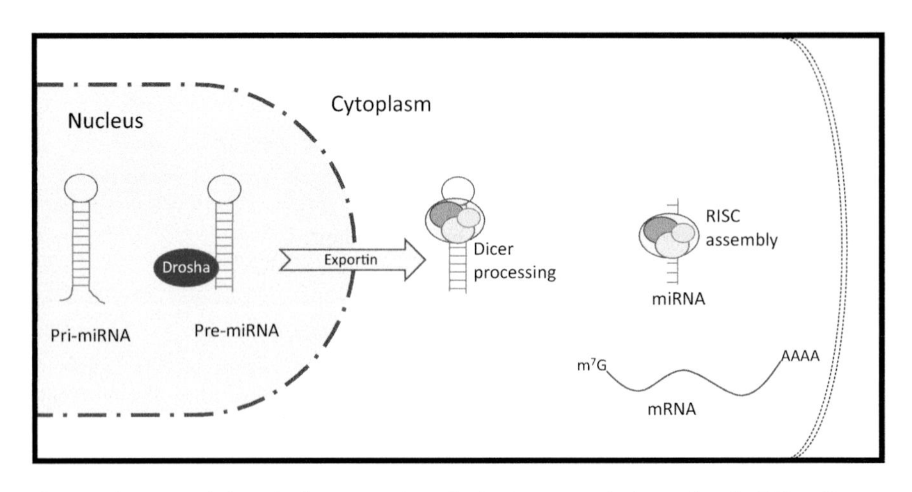

Fig. 1 After transcription, the first maturation of miRNAs occurs in the nucleus by RNase III-type Drosha with its cofactor DiGeorge syndrome critical region gene 8 (DGCR8). The pre-miRNAs are supported by a protein (Exportin 5) which carries them out the nucleus to the cytoplasm. In the cytoplasm, the pre-miRNAs are submitted to new cleavages. This maturation is accomplished by Dicer constituted of several domains including two RNase III regions. Dicer associated with protein (such as Argonautes) constitutes the effector complex RISC that enables miRNA to bind their mRNA targets (adapted from Kim et al. 2009).

immune response, and metabolism (Bartel 2004). Their activities are thought to regulate over 60% of genes, controlling gene expression at the posttranscription, pre-translational level (Bartel 2004). Their biogenesis is complex; this chapter only describes briefly the major steps of their biosynthesis in mammals (Fig. 1). miRNAs are transcribed from genes that can be found within the genome as isolated genes or linked in clusters (Griffiths-Jones et al. 2008). These clusters can include up to 40 miRNA genes, as in the *Bos taurus* chromosome BTA19 (Mobuchon et al. 2015a). Approximately half of the total known miRNA genes are located within intergenic regions with the balance within introns or exons of protein-coding transcription units (Kim et al. 2009). miRNAs are transcribed as long RNAs (Lee et al. 2002; Cai et al. 2004;) and then submitted to different steps of maturation (Kim et al. 2009) within the nucleus (involving the RNase III-type Drosha with its cofactor DCGR8 (Han et al. 2004)). After their transport from the nucleus to the cytoplasm by exportin 5, ribonuclease III, Dicer1, manages another maturation step to load the miRNAs in the RNA-induced silencing complex (**RISC**) leading to mature and functional miRNA (Rana 2007). miRNA can hybridize to their target mRNA by base pair complementarity and thus regulate gene expression through their actions on mRNA. Mismatch in the double strand leads to translation inhibition of the targeted mRNA, whereas complete hybridization leads to targeted mRNA degradation. The complexity of action of miRNA is heightened by the fact that one miRNA could target different mRNAs and one mRNA could be targeted by several miRNAs. The documented roles of miRNA in the regulation of gene expression and hence development and functioning of numerous organs are compelling reasons to investigate the role of miRNA in MG.

The objective of this chapter is to give an overview of the nutritional regulation of miRNAs with a focus, in particular, on MG and its potential influence on human health through milk consumption.

Characterization of miRNA Expression in Mammary Gland

Advanced technologies, microarray expression methods, and more recently high-throughput sequencing (next-generation sequencing: **NGS**) analyses led to a rapid expansion in our knowledge of miRNA expression in the last decade. In humans, miRNA have been studied notably for the crucial role they play in healthy individuals, and their dysregulations have also been identified as implicated in a wide range of diseases, including diabetes, cardiovascular disease, kidney disease, and cancer (Paul et al. 2017; Sayed and Abdellatif 2011) but also aging (Lin et al. 2016). miRNAs can also play an important role in prevention and therapeutics of non-contagious diseases (Nouraee and Mowla 2015). Within this context, the role of miRNA in MG biology has been investigated in relation to breast cancer. Studies have shown miRNAs to be involved through their actions on targeted mRNA (such as *ESR1* mRNA) in both tumorigenesis and metastasis (Sayed and Abdellatif 2011) or the epithelial to mesenchymal transition (Koutsaki et al. 2014). Recently, it has been suggested that miRNAs may act as endocrine and paracrine signaling messengers to facilitate information transfer between breast carcinoma and their neighboring cells (Jahagirdar et al. 2016). In other aspects, miRNAs have been used to complete the classical indicators predicting the performance of endocrine and biological therapies (such as *ESR1* and *HER2*) which were considered far from perfect (Gao et al. 2016). Thus, sets of miRNAs have been proposed as biomarkers for early breast cancer diagnosis, prognosis, and therapy prediction (Bertoli et al. 2015; Gao et al. 2016; Nassar et al. 2016). In addition, the loss of Drosha or Dicer proteins was observed to be associated with breast cancer development providing further evidence that alterations in miRNA regulation influence tumor behavior (Khoshnaw et al. 2012, 2013).

Currently, the function of miRNAs is only partially known in the healthy MG. Indeed, few studies, essentially performed in mouse mammary epithelial cell line or in transgenic mice, have reported the role of few miRNA in the development of the MG (Ibarra et al. 2007; Le Guillou et al. 2012). Further investigations on the regulation of their expression will be necessary to understand their involvement in MG development and lactation. In livestock, due to their economic importance, lactation function has been investigated mostly in ruminant species and consequently ruminant mammary miRNomes have been established (Table 1; Gu et al. 2007; Le Guillou et al. 2014; Li et al. 2016; Mobuchon et al. 2015a). Furthermore, the regulation of mammary miRNA expression was investigated by following the dynamic modifications occurring at each cycle of lactation (from dry, early, mid-, to

Table 1 Studies reporting miRNome in mammary gland or milk in cows and goat

Species	Status	Strategies of global analyses	References
Cow	Lactation vs non-lactation	RNAseq	Li et al. 2012b
	Milk exosomes, Infected vs non-infected	RNAseq	Sun et al. 2015
	Lactation, peak	RNAseq	Le Guillou et al. 2014
	Postpubertal Holstein vs Limousin	RNAseq	Li et al. 2015
	Lactation, peak vs late	Microarrays	Wicik et al. 2016
	Lactation, lipid supplementation	RNAseq	Li et al. 2015
	Lactation, lipid supplementation	RNAseq	Mobuchon et al. in press
	Milk, different fractions	RNAseq	Li et al. 2016
Goat	Lactation, early	RNAseq	Ji et al. 2012
	Lactation, peak vs late	RNAseq	Ji et al. 2012
	Lactation, dry vs peak	RNAseq	Li et al. 2012a
	Lactation, peak	RNAseq	Mobuchon et al. 2015a

late lactation periods). The expression of 100 miRNAs has been identified as differentially regulated between pregnancy and lactation, in ovine MG (Galio et al. 2013). In bovine, the expression of several miRNAs is also modified by the stage of lactation (Wang et al. 2012; Li Z et al. 2012b). For example, during early lactation, the upregulation of expression of *miR-221* and *miR-33b* has been observed, suggesting their involvement in the control of cell proliferation or angiogenesis and in lipogenesis in mammary tissue. This observation is consistent with the co-localization and the co-expression of *miR-33b* and SREBP-1 coding gene and their role in lipid metabolism in human (Najafi-Shoushtari et al. 2010). In the last 2 years, an increasing number of publications report the use of in vitro goat or bovine MEC in studies to decipher the role of miRNAs in the milk component biosynthesis. As the objective of this chapter is not to provide an exhaustive list of the corresponding publications, only a few examples will be described below. For example, in bovine MEC, the highly abundant *miR-29* has been showed to regulate DNA methylation levels by targeting the two DNA methyltransferases (DNMT3A and DNMT3B) crucial for epigenetic signature (Bian et al. 2015). The inhibition of *miR-29* led to global DNA hypermethylation including an increase of promoter methylation of genes (such as *CSN1S1*, *PPARG*, *SREBP1*, or *GLUT*1) known to be involved in milk component biosynthesis (Bian et al. 2015). In the same way, *miR-181a* or *miR-15a* has been reported to regulate milk biosynthesis genes such as *ACSL1*, involved in fat synthesis (Lian et al. 2016), and growth hormone receptor, involved in mammary cell viability (Li HM et al. 2012a). Recent in vitro studies using goat mammary epithelial cells (GMEC) have revealed potential roles of *miR-103* and *miR-27a* in the regulation of milk fat synthesis in these cells (Lin et al. 2013a, b). The highly abundant *miR-26(a/b)* and *miR-24* have also been reported to regulate the expression of genes related to fatty acid synthesis (like *PPARG, SREBF1, FASN, ACACA, or SCD1*) (Wang et al. 2015, 2016). Taken together, the increase of the number of publications on the miRNA effects on the regulation of gene expression involved in milk biosynthesis underlines the crucial role of miRNA in MG and in lactation.

Nutritional Regulation of miRNA Expression

The diverse roles of miRNA in numerous biological processes have made them candidate levers for the therapeutic regulation of these processes. The variation of miRNA gene expression in response to nutritional regulation has been investigated. The nutritional regulation of miRNA expression on human has begun to be studied recently and the data are still rare on this topic. It has been shown that an 8-week Mediterranean-based nutritional intervention was able to induce changes in the expression of miRNAs in circulating white blood cells from patients with metabolic syndrome (Marques-Rocha et al. 2016). Interestingly, the miRNAs whose expression has been affected by the diet have been associated with inflammatory gene regulation and human diseases linked to metabolic syndrome such as cancer, atherogenic and adipogenic processes and other inflammatory conditions. Specific foods can also modulate the expression of miRNA in the human following their consumption. For example, 4-week high-red meat consumption resulted in altered miRNA expression in rectal mucosa tissue of healthy volunteers, particularly levels of oncogenic mature miRNAs, including *miR-17–92* cluster miRNAs and *miR-21* and therefore increase or decrease risk for colorectal cancer (Humphreys et al. 2014). Interestingly, in the same volunteers, intake of butyrylated-resistant starch together with red meat restored the level of *miR-17– 92* miRNAs to baseline levels, suggesting that increased resistant starch consumption reduce risk associated with a red meat diet by, at least partially, counteracting the effect on miRNA expression. More recently, it has been observed that 4-week consumption of nondigestible carbohydrates (resistant starch and polydextrose) affected the expression of over 100 miRNAs in macroscopically normal human rectal epithelium (Malcomson et al. 2017). Some of the differentially expressed miRNA have been described to be involved in cellproliferation, migration, and invasion in colorectal and gastric cancer and reduce apoptosis and inflammation. Together with whole diets or food, it has also been reported that specific nutrients present in food can modulate the expression of miRNAs in human. For example, it has been shown that polyphenols, micronutrients present in fruits and vegetables, can modulate expression of miRNAs in vivo in animal models (Milenkovic et al. 2012, 2013). But only few studies have been done to evaluate their impact on miRNA expression in humans. Another example showed that 1-year daily intake of a resveratrol, polyphenol found in fruits like grape, by hypertensive male patients with type 2 diabetes mellitus affected the expression of miRNA in peripheral blood mononuclear cells of volunteers (Tomé-Carneiro et al. 2013). Among these miRNAs is a group involved in the regulation of the inflammatory response: *miR-21*, *miR-181b*, *miR-663*, *miR-30c2*, *miR-155*, and *miR-34a*. Several studies have also shown that supplementation with vitamin D can affect the expression of miRNA in different tissues in human suggesting that the anticancer effects of vitamin D could be exerted through alteration of miRNA expression (Zeljic et al. 2017). Taken together, these studies suggest that food and macro- or micronutrients can exert nutrigenomic effect on miRNA that present relevant mechanisms involved in development or prevention of different chronic diseases. Attention has been also paid to prevention of the age-related declines in muscle and the positive effect on lifespan (Orom et al. 2012). In these studies, the effects of caloric restriction on miRNA expression regulation have been investigated. The effects of 40% food restriction for 10 days induced, in the cartilaginous growth center of prepubertal rats, a decrease in expression of *miR-140-3p* targeting *SIRT1* (a NAD-dependent protein deacetylase of histones) mRNA was associated with an increase of *SIRT1* mRNA abundance. These results suggest that mRNA and miRNA expressions are responsive to nutritional cues and could both impact the two parts of epigenetic response: miRNA expression and histone deacetylases (Pando et al. 2012). Moreover, caloric restriction has been also shown to reverse the age-association miRNA alterations in skeletal muscle from monkeys by rescuing the levels of *miR-181b* and limiting the age-induced increase of *miR-451* and *miR-144* (Mercken et al. 2013). The effects of caloric restriction (a 70% control diet during 6 months) on miRNA expression profiles in mouse mammary tissue have been also demonstrated (Orom et al. 2012). The most significant changes were detected for *miR-29c*, *miR-203*, *miR-150*, and *miR-30*. *miR-29* and *miR-30* are known to be involved in senescence (Martinez et al. 2011) and *miR-203* in the regulation of several cellular processes (cholesterol homeostasis, signal transduction) and cancers including breast cancers (Orom et al. 2012).

In parallel with studies of human breast, studies on regulation of miRNA expression have been performed on livestock MG (Fig. 2; Table 1). With the aim to

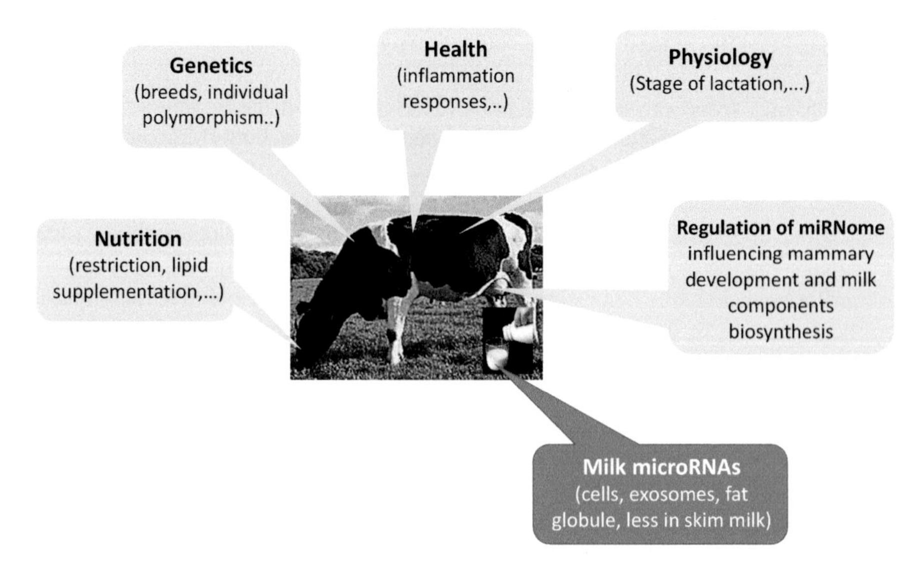

Fig. 2 Different factors regulating miRNA expression in mammary gland or milk

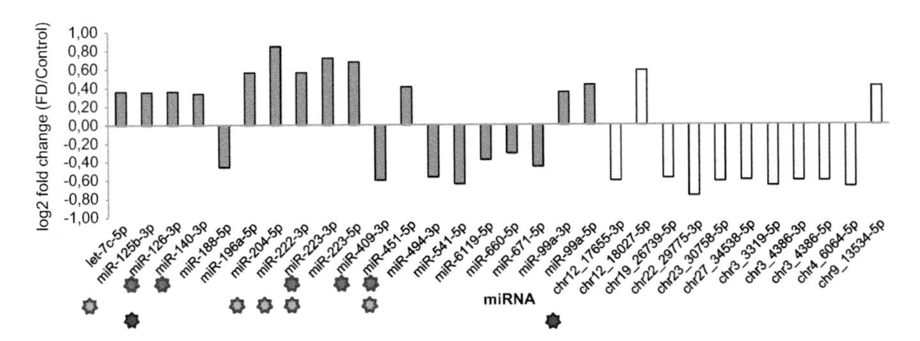

Fig. 3 Identification of 30 miRNAs differentially expressed in mammary gland of goats receiving either an ad libitum diet or food deprived (Mobuchon et al. 2015b). *Dark gray* bars indicate the known miRNA and *light gray* new predicted miRNAs. *Orange stars*, miRNAs already described in others restriction model in different tissues; *green stars*, miRNAs involved in lipid metabolism; *blue stars*, miRNAs involved in mammary epithelial cells proliferation.

decipher nutritional regulation of mammary miRNome, a first study has been achieved on an extreme model with 48 h of food deprivation applied to lactating goats. The expression of 30 miRNAs was highlighted as modified in MG by food deprivation (Fig. 3; Mobuchon et al. 2015b). Among them, nine were highly expressed in MG such as *miR-6119-5p* (considered up to now to be ruminant specific), *miR-126-3p*, and *miR-451-5p*. This last miRNA has been also identified as regulated by caloric restriction in monkey muscles (Mercken et al. 2013). Moreover, this miRNA may inhibit cell proliferation and target mRNA crucial for milk composition such as *CSN2* mRNA coding beta casein which is one of the six major milk proteins or for milk biosynthesis by targeting progesterone receptor (*PGR*) mRNA (Cui et al. 2011). The pathways most significantly targeted by the 30 nutri-regulated miRNAs are those regulating cellular proliferation and remodeling of MG and are most likely to be changed through the action of these miRNAs (Mobuchon et al. 2015b). Several miRNAs regulated by food deprivation have been identified as potentially to target mRNA involved in lipids, proteins, or lactose biosynthesis. Five known nutri-regulated miRNAs (*miR-222-3p*, *miR-188-5p*, *miR-541-5p*, *miR-494-3p*, and *chr-3319-5p*) may target phosphatase and tensin homologue (*PTEN*) mRNA which plays a crucial role in the secretion of beta casein, triglycerides, and lactose (Wang et al. 2014). In accordance with the way of regulation, 19 mRNA (detected in the same animals as differentially expressed by 48 h of food deprivation, Ollier et al. 2007) were identified as potentially targeted by at least one of the 30 nutri-regulated miRNAs. For example, the downregulation of the transcription factor *ESR1* (coding for an estrogen receptor) involved in the cell proliferation and growth in MG could be due to the upregulation of two known (*miR-125b-3p* and *miR-222-3p*) and one new miRNA (*Chr19_26739-5p*) targeting its 3′UTR. Finally, two miRNAs (*miR-204-5p* and *miR-223-3p*) were highlighted as affected by food deprivation and in turn target several mRNAs resulting in their being differentially expressed (Mobuchon et al. 2015b). Due to the large conservation of miRNAs through the evolution understanding their functions in animal will help to understand their importance to human biology.

miRNAs Are Milk Components

In addition to the nutri-regulation of miRNome in MG, the studies on livestock species have implications to human health through their transfer to consumers via milk. The realization that miRNAs are a signaling system from mother to infant raise the potential for these same molecules to act across species and influence milk's health effects for consumers. To answer to a public health question of contaminations of baby formula milk powder, the presence of miRNA in bovine milk has been studied. Chen et al. (2010) showed, for the first time, that miRNAs are an intrinsic component of cow milk and are stably present in milk. A comparison between MG and milk fat fraction showed a global similarity among all miRNomes although some miRNA have been detected in mammary tissue but lacking in bovine milk fat fraction (Lago-Novais et al. submitted). For example two (*miR-126* and *miR-204*) were not detected in milk fat fraction, whereas they were abundant in MG. In addition to the differences in the miRNAs distribution between milk and MG, there is difference depending to the stages of lactation. Thus, mature milk miRNomes and those identified in colostrum present ca. 90% of miRNAs in both stages, whereas 10% of miRNA are present only in one of these two stages. As expected, three highly expressed miRNAs (*miR-181a*, *miR-223*, and *miR-150*) known to be involved in immune response and immune system development were more abundant in colostrum than in mature milk.

For the same stage of lactation, miRNAs are differentially distributed in milk compartments. A study comparing the miRNomes of bovine milk fat, whey, and cells showed that more than 90% of miRNA were common in the three sources of miRNA (Li et al. 2016) but with some differences. Thus, a comparison between miRNome of exosomes obtained after centrifugation and of supernatant showed a differential repartition of miRNAs between these two fractions. miRNAs in milk were detected for a third both in exosomes and supernatant, a third only in exosomes, and the last third only in supernatant (Izumi et al. 2015).

The presence and the distribution of miRNAs are particularly interesting due to the miRNA resistance under degradative conditions, and therefore they are suspected to be unaffected by treatments that milk or milk-related commercial products can undergo before human consumption (Chen et al. 2010). However, the resistance is not complete. Indeed, the loss of miRNAs during processing and storage of cow's milk has been examined through a monitoring of two miRNAs (*miR-200c* and *miR-29b*) that are relatively abundant in bovine milk (Howard et al. 2015). This study showed a 63% loss of *miR-200c* by pasteurization and homogenization, whereas skim milk saw a 67% loss of *miR-29b* (Howard et al. 2015). Such results point out the differences of comportment of miRNA in front treatments underlining the importance of further investigation in this point.

As in ruminant milk, the presence of miRNA has been demonstrated in human breast milk with a diverse distribution depending to the lactation stages with a high expression level of immune-related miRNAs in the first 6 months of lactation (Kosaka et al. 2010). In addition, the miRNA expression profiles differ during lactation with more pronounced upregulation in month 4 than in month 2 or 6 (Alsaweed et al. 2016). The miRNA profiles differ depending on milk fraction

(Munch et al. 2013; Alsaweed et al. 2015b) despite great similarities between human milk cell and lipid fractions (Alsaweed et al. 2016). These authors suggested that miRNA expression reflect the remodeling of the MG in response to infant feeding patterns.

Role of miRNA in Healthy Food

The concept of milk as a healthy food has opened the way of studies on milk components, including macro- and micronutrients, as well as diverse biomolecules such as miRNA. Indeed, first evidences of the presence of *miR-168a*, a rice miRNA acquired orally through food intake, in sera of Chinese human subjects consuming a great quantity of rice have been reported as its role demonstrated in the regulation of the expression of target genes (Zhang et al. 2012). This result pointed out that the food-borne miRNAs have the potential to be both bioavailable and active affecting consumers' gene expression. Nevertheless, the studies on the effects of endogenous miRNAs or of consumed miRNAs are complex due to the sequence similarities of miRNA across animal species which render it difficult to distinguish those unique to diet (Zempleni et al. 2015). These similarities emphasize the potential role of dietary miRNAs from animals on consumers. Consequently, the presence of miRNAs in large quantities in milk (including commercial milk) has led the scientific community to focus their attention on the potential role of milk on human health through this mechanism (Fig. 4). In one study, a pathway-enrichment analysis was performed for the most abundant miRNAs in human milk cells and fat portion (Alsaweed et al.

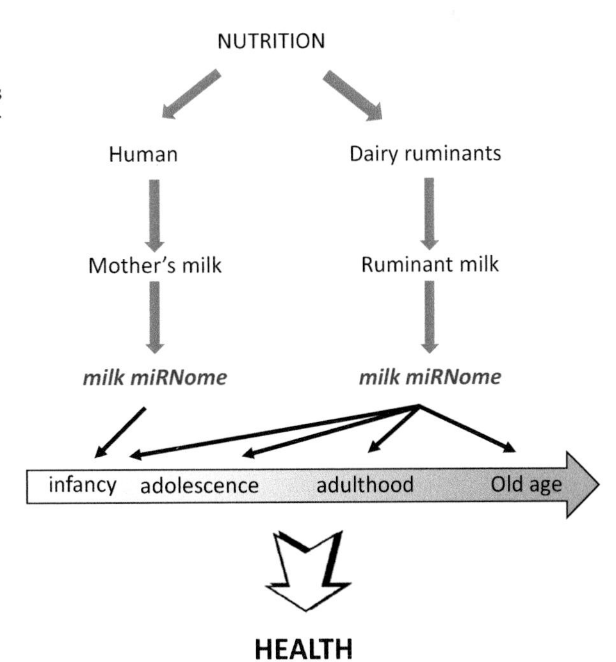

Fig. 4 The presence of miRNAs in human and ruminant milk open new ways of investigations, in particular in the field of human health

2015b). The analysis suggested that miRNAs in human milk could affect cell functions such as immunity, growth and development, cell proliferation, and apoptosis. It has also been reported that most of the microRNAs in human milk are known for their immunocompetence by being involved in mechanisms regulating the immune system, such as regulation of B and T cell differentiation and development, and innate/adaptive immune responses. They can also play key roles in autoimmune conditions, such as inflammatory bowel disease, and potentially regulate the development or prevention of these diseases (Alsaweed et al. 2015a). Nevertheless, very little is known about the biological relevance of miRNA present in milk on infant and adult health. The stability of miRNA including those in milk under various harsh conditions has led investigators to suggest that their packaging in milk cells or milk exosomes has the net effect to protect them (Zhou et al. 2012) thus allowing their survival in the gut intestinal tract of offspring (Zhang et al. 2012; Laubier et al. 2015). In particular, exosomes containing miRNAs protect them against degradation and facilitating uptake by endocytosis (Izumi et al. 2012; Zempleni et al. 2017). However, the bioavailability of miRNAs in milk for consumers remains controversial. Indeed, the high level of *miR-30b* in transgenic mouse milk was not associated with an increase of this miRNA in tissues of pups fed by transgenic females compared to pups fed wild-type females (Laubier et al. 2015). In addition, the lack of bioavailability of milk-derived miRNAs has been also suggested based on the evidence of a lack of detection of two miRNAs (*miR-375* and *miR-200c/141*) in these two miRNA knockout mice models receiving milk from wild-type mothers (Title et al. 2015). The distribution of miRNA in different fraction of milk could have a crucial role on the importance of effects after consumption (Baier et al. 2014). Thus, further investigations are needed to clarify this controversy in particular taking into account of the packaging of studied miRNAs. This structural dimension is crucial to evaluating the health consequences of milk-derived miRNAs linked with the nutritional regulation of milk miRNAs. Moreover, emergent studies have concluded that there is a link between the abundance of some miRNAs (involved in adipogenesis) in milk and maternal weight and infant gender (Xi et al. 2016).

Conclusions and Perspectives

Milk is a source of nutrients for neonates and adults, and its composition influences the health of consumers in both the short and long term. This composition varies depending on the genetics or environmental conditions including nutrition that modify mammary development and function. Among the main variables of biological signaling responsible for these modifications, there is a novel class of molecules with broad regulatory properties: miRNAs. Indeed, studies on food restriction or deprivation in ruminants demonstrated the nutritional regulation of miRNA expression profiles in MG leading to suggestions of their intimate involvement in milk components synthesis. Moreover, their transfer in milk opens a line of investigation to use miRNAs as indicators (biomarkers) for downstream consequences of health, physiological or metabolic status, in human and animals. The growing knowledge on the regulation of the miRNAs profiles in MG and in milk related to specific conditions (such as undernutrition, mastitis in ruminants, etc.) is compelling evidence to use them as biomarkers in humans and animals.

In addition to their direct actions on the regulation of mammary development, structure, and milk composition, the transfer of miRNAs into milk directly has prompted novel interest in their potential influence on consumers. These molecules are now considered to act specifically on neonates whatever the considered source of milk (mother's milk, cow's milk, or formula). The resistance of miRNAs during the treatment conditions applied to milk and milk products before their commercialization and consumption means that they have the potential to influence the health of consumers. However, the influence of miRNAs in milk and milk products must be clarified as their packaging into exosomes or milk fat globules provides a protective role for their miRNAs cargos. The interaction with intestinal tract must be investigated in different ways including their role on the nature of gut microbial as well as their endocytosis by epithelial and vascular endothelial cells providing their deliverability in blood. Thus, the role of miRNAs from milk in consumer health is actually emerging, and the nutri-regulation of mammary as milk miRNomes is an opportunity to modify milk nutritional quality.

Key Facts of Nutritional Regulation of Mammary miRNome: Implication for Human Studies

- Milk is a source of nutrients for neonates and adults, and its composition influences the health of consumers in short and long term.
- The transfer of miRNAs from mammary gland to milk, their regulation (e.g., by nutrition) as their relative resistance, open ways for investigations.
- miRNAs have been proposed as biomarkers of specific conditions (in particular diseases).
- Their role in nutritional quality of milk for consumers is a promising consideration.

Dictionary of Terms

- **MicroRNA (or miRNA)** – small RNA (~22 nucleotides), regulating various biological processes through base pairing targeted mRNA leading to their degradation or inhibiting their translation.
- **miRNomes** – global atlas of miRNAs presents in cells, tissues, or organs.
- **Nutrigenomics** – global analyses of the effects of nutrition on genes expression profile.
- **Exosomes** – small extracellular vesicles containing different constituents such as proteins and RNA.

Summary Points

This chapter presents recent data on the nutritional regulation of miRNAs with a focus on mammary gland (**MG**). miRNAs are involved in the regulation of mammary development and function as well as in the regulation of milk component biosynthesis. The presence of miRNAs in milk has opened new ways of investigations in two directions. Due to their relative resistance and their presence in milk and since miRNAs are clearly a fundamental mechanism of biological regulation, they have been related to development of diseases such as cancer in human or inflammation in ruminants, and they have been proposed as biomarkers of health or physiological status. Therefore, miRNA in milk including after treatments for commercialization can greatly influence the nutritional quality of milk. The structures by which miRNAs transfer from MG to milk are under investigation such as their packaging which could play a role in their functionality in milk consumers. Their role in the regulation of numerous biological processes suggests their potential as components of food for health.

References

Alsaweed M, Hartmann PE, Geddes DT, Kakulas F (2015a) MicroRNAs in Breastmilk and the lactating breast: potential immunoprotectors and developmental regulators for the infant and the mother. Int J Environ Res Public Health 12:13981–14020

Alsaweed M, Hepworth AR, Lefevre C, Hartmann PE et al (2015b) Human milk MicroRNA and Total RNA differ depending on milk fractionation. J Cell Biochem 116:2397–2407

Alsaweed M, Lai CT, Hartmann PE, Geddes DT, Kakulas F (2016) Human milk cells and lipids conserve numerous known and novel miRNAs, some of which are differentially expressed during lactation. PLoS One 11:e0152610

Anderson SM, Rudolph MC, Mcmanaman JL, Neville MC (2007) Key stages in mammary gland development – secretory activation in the mammary gland: it's not just about milk protein synthesis! Breast Cancer Res 9:204

Baier SR, Nguyen C, Xie F, Wood JR, Zempleni J (2014) MicroRNAs are absorbed in biologically meaningful amounts from nutritionally relevant doses of cow milk and affect gene expression in peripheral blood mononuclear cells, HEK-293 kidney cell cultures, and mouse livers. J Nutr 144:1495–1500

Bartel DP (2004) MicroRNAs: genomics, biogenesis, mechanism, and function. Cell 116:281–297

Bertoli G, Cava C, Castiglioni I (2015) MicroRNAs: new biomarkers for diagnosis, prognosis, therapy prediction and therapeutic tools for breast cancer. Theranostics 5:1122–1143

Bian Y, Lei Y, Wang C, Wang J et al (2015) Epigenetic regulation of miR-29s affects the lactation activity of dairy cow mammary epithelial cells. J Cell Physiol 230:2152–2163

Cai X, Hagedorn CH, Cullen BR (2004) Human microRNAs are processed from capped, polyadenylated transcripts that can also function as mRNAs. RNA 10:1957–1966

Chen X, Gao C, Li H, Huang L et al (2010) Identification and characterization of microRNAs in raw milk during different periods of lactation, commercial fluid, and powdered milk products. Cell Res 20:1128–1137

Chichlowski M, German JB, Lebrilla CB, Mills DA (2011) The influence of milk oligosaccharides on microbiota of infants: opportunities for formulas. Annu Rev Food Sci Technol 2:331–351

Cui W, Li Q, Feng L, Ding W (2011) MiR-126-3p regulates progesterone receptors and involves development and lactation of mouse mammary gland. Mol Cell Biochem 355:17–25

Galio L, Droineau S, Yeboah P, Boudiaf H et al (2013) MicroRNA in the ovine mammary gland during early pregnancy: spatial and temporal expression of miR-21, miR-205, and miR-200. Physiol Genomics 45:151–161
Gao Y, Cai Q, Huang Y, Li S et al (2016) MicroRNA-21 as a potential diagnostic biomarker for breast cancer patients: a pooled analysis of individual studies. Oncotarget 7:34498–34506
German JB, Dillard CJ (2010) Saturated fats: a perspective from lactation and milk composition. Lipids 45:915–923
Givens DI (2010) Milk and meat in our diet: good or bad for health? Animal 4:1941–1952
Griffiths-Jones S, Saini HK, van Dongen S, Enright AJ (2008) miRBase: tools for microRNA genomics. Nucleic Acids Res 36:D154–D158
Gu Z, Eleswarapu S, Jiang H (2007) Identification and characterization of microRNAs from the bovine adipose tissue and mammary gland. FEBS Lett 581:981–988
Han J, Lee Y, Yeom KH, Kim YK et al (2004) The Drosha-DGCR8 complex in primary microRNA processing. Genes Dev 18:3016–3027
Howard KM, Jati Kusuma R, Baier SR, Friemel T et al (2015) Loss of miRNAs during processing and storage of cow's (Bos taurus) milk. J Agric Food Chem 63:588–592
Humphreys KJ, Mckinnon RA, Michael MZ (2014) miR-18a inhibits CDC42 and plays a tumour suppressor role in colorectal cancer cells. PLoS One 9:e112288
Ibarra I, Erlich Y, Muthuswamy SK, Sachidanandam R, Hannon GJ (2007) A role for microRNAs in maintenance of mouse mammary epithelial progenitor cells. Genes Dev 21:3238–3243
Izumi H, Tsuda M, Sato Y, Kosaka N et al (2015) Bovine milk exosomes contain microRNA and mRNA and are taken up by human macrophages. J Dairy Sci 98:2920–2933
Jahagirdar D, Purohit S, Jain A, Sharma NK (2016) Export of microRNAs: a bridge between breast carcinoma and their neighboring cells. Front Oncol 6:147
Ji Z, Wang G, Xie Z, Zhang C, Wang J (2012) Identification and characterization of microRNA in the dairy goat (Capra hircus) mammary gland by Solexa deep-sequencing technology. Mol Biol Rep 39(10):9361–71
Khoshnaw SM, Rakha EA, Abdel-fatah TM, Nolan CC et al (2012) Loss of dicer expression is associated with breast cancer progression and recurrence. Breast Cancer Res Treat 135:403–413
Khoshnaw SM, Rakha EA, Abdel-fatah T, Nolan CC et al (2013) The microRNA maturation regulator Drosha is an independent predictor of outcome in breast cancer patients. Breast Cancer Res Treat 137:139–153
Kim VN, Han J, Siomi MC (2009) Biogenesis of small RNAs in animals. Nat Rev Mol Cell Biol 10:126–139
Kosaka N, Izumi H, Sekine K, Ochiya T (2010) microRNA as a new immune-regulatory agent in breast milk. Silence 1:7–7
Koutsaki M, Spandidos DA, Zaravinos A (2014) Epithelial-mesenchymal transition-associated miRNAs in ovarian carcinoma, with highlight on the miR-200 family: prognostic value and prospective role in ovarian cancer therapeutics. Cancer Lett 351:173–181
Lago-Novais D, Pawlowski K, Pires J, Mobuchon L, et al (2016) Milk fat globules as a source of mammary microRNA. In: 2016 ADSA/ASAS Joint Annual Meeting. p 401.
Laubier J, Castille J, Le Guillou S, Le Provost F (2015) No effect of an elevated miR-30b level in mouse milk on its level in pup tissues. RNA Biol 12:26–29
Le Guillou S, Sdassi N, Laubier J, Passet B et al (2012) Overexpression of miR-30b in the developing mouse mammary gland causes a lactation defect and delays involution. PLoS One 7:e45727
Le Guillou S, Marthey S, Laloe D, Laubier J et al (2014) Characterisation and comparison of lactating mouse and bovine mammary gland miRNomes. PLoS One 9:e91938
Lee Y, Jeon K, Lee JT, Kim S, Kim VN (2002) MicroRNA maturation: stepwise processing and subcellular localization. EMBO J 21:4663–4670
Li HM, Wang CM, Li QZ, Gao XJ (2012a) MiR-15a decreases bovine mammary epithelial cell viability and lactation and regulates growth hormone receptor expression. Molecules 17:12037–12048

Li Z, Liu H, Jin X, Lo L, Liu J (2012b) Expression profiles of microRNAs from lactating and non-lactating bovine mammary glands and identification of miRNA related to lactation. BMC Genomics 13:731

Li R, Dudemaine PL, Zhao X, Lei C, Ibeagha-awemu EM (2016) Comparative analysis of the miRNome of bovine milk fat, whey and cells. PLoS One 11:e0154129

Lian S, Guo JR, Nan XM, Ma L et al (2016) MicroRNA Bta-miR-181a regulates the biosynthesis of bovine milk fat by targeting ACSL1. J Dairy Sci 99:3916–3924

Lin X, Luo J, Zhang L, Wang W, Gou D (2013a) MiR-103 controls milk fat accumulation in goat (Capra Hircus) mammary gland during lactation. PLoS One 8:e79258

Lin XZ, Luo J, Zhang LP, Wang W et al (2013b) MiR-27a suppresses triglyceride accumulation and affects gene mRNA expression associated with fat metabolism in dairy goat mammary gland epithelial cells. Gene 521:15–23

Lin X, Zhan JK, Wang YJ, Tan P et al (2016) Function, role, and clinical application of MicroRNAs in vascular aging. Biomed Res Int 2016:6021394

Malcomson FC, Willis ND, McCallum I, Xie L et al (2017) Non-digestible carbohydrates supplementation increases miR-32 expression in the healthy human colorectal epithelium: a randomized controlled trial. Mol Carcinog. https://doi.org/10.1002/mc.22666

Marques-Rocha JL, Milagro FI, Mansego ML, Zulet MA et al (2016) Expression of inflammation-related miRNAs in white blood cells from subjects with metabolic syndrome after 8 wk of following a Mediterranean diet-based weight loss program. Nutrition 32:48–55

Martinez I, Cazalla D, Almstead LL, Steitz JA, Dimaio D (2011) miR-29 and miR-30 regulate B-Myb expression during cellular senescence. Proc Natl Acad Sci USA 108:522–527

Mercken EM, Majounie E, Ding J, Guo et al (2013) Age-associated miRNA alterations in skeletal muscle from rhesus monkeys reversed by caloric restriction. Aging (Albany NY) 5:692–703

Milenkovic D, Deval C, Gouranton E, Landrier JF et al (2012) Modulation of miRNA expression by dietary polyphenols in apoE deficient mice: a new mechanism of the action of polyphenols. PLoS One 7:e29837

Milenkovic D, Jude B, Morand C (2013) miRNA as molecular target of polyphenols underlying their biological effects. Free Radic Biol Med 64:40–51

Mobuchon L, Marthey S, Boussaha M, Le Guillou S et al (2015a) Annotation of the goat genome using next generation sequencing of microRNA expressed by the lactating mammary gland: comparison of three approaches. BMC Genomics 16:285

Mobuchon L, Marthey S, Le Guillou S, Laloe D et al (2015b) Food deprivation affects the miRNome in the lactating goat mammary gland. PLoS One 10:e0140111

Mobuchon L, Le Guillou S, Marthey S, Laubier J et al (2017) Sunflower oil supplementation affects the expression of miR-20a-5p and miR-142-5p in the lactating bovine mammary gland. PLoS One. in press

Munch EM, Harris RA, Mohammad M, Benham AL et al (2013) Transcriptome profiling of microRNA by next-gen deep sequencing reveals known and novel miRNA species in the lipid fraction of human breast milk. PLoS One 8:e50564

Najafi-Shoushtari SH, Kristo F, Li Y, Shioda T et al (2010) MicroRNA-33 and the SREBP host genes cooperate to control cholesterol homeostasis. Science 328:1566–1569

Nassar FJ, Nasr R, Talhouk R (2016) MicroRNAs as biomarkers for early breast cancer diagnosis, prognosis and therapy prediction. Pharmacol Ther 172:34–49

Nouraee N, Mowla SJ (2015) miRNA therapeutics in cardiovascular diseases: promises and problems. Front Genet 6:232

Ollier S, Robert-Granie C, Bernard L, Chilliard Y, Leroux C (2007) Mammary transcriptome analysis of food-deprived lactating goats highlights genes involved in milk secretion and programmed cell death. J Nutr 137:560–567

Orom UA, Lim MK, Savage JE, Jin L et al (2012) MicroRNA-203 regulates caveolin-1 in breast tissue during caloric restriction. Cell Cycle 11:1291–1295

Pando R, Even-Zohar N, Shtaif B, Edry L et al (2012) MicroRNAs in the growth plate are responsive to nutritional cues: association between miR-140 and SIRT1. J Nutr Biochem 23:1474–1481

Paul S, Lakatos P, Hartmann A, Schneider-Stock R, Vera J (2017) Identification of miRNA-mRNA modules in colorectal cancer using rough hypercuboid based supervised clustering. Sci Rep 7:42809

Rana TM (2007) Illuminating the silence: understanding the structure and function of small RNAs. Nat Rev Mol Cell Biol 8:23–36

Sayed D, Abdellatif M (2011) Micrornas in development and disease. Physiol Rev 91:827–887

Sun J, Aswath K, Schroeder SG, Lippolis JD et al (2015) MicroRNA expression profiles of bovine milk exosomes in response to Staphylococcus aureus infection. BMC Geno 16;16:806

Title AC, Denzler R, Stoffel M (2015) Uptake and function studies of maternal milk-derived MicroRNAs. J Biol Chem 290:23680–23691

Tome-Carneiro J, Larrosa M, Yanez-Gascon MJ, Davalos A et al (2013) One-year supplementation with a grape extract containing resveratrol modulates inflammatory-related microRNAs and cytokines expression in peripheral blood mononuclear cells of type 2 diabetes and hypertensive patients with coronary artery disease. Pharmacol Res 72:69–82

Wang M, Moisa S, Khan MJ, Wang J et al (2012) MicroRNA expression patterns in the bovine mammary gland are affected by stage of lactation. J Dairy Sci 95:6529–6535

Wang Z, Hou X, Qu B, Wang J et al (2014) Pten regulates development and lactation in the mammary glands of dairy cows. PLoS One 9:e102118

Wang H, Luo J, Chen Z, Cao WT et al (2015) MicroRNA-24 can control triacylglycerol synthesis in goat mammary epithelial cells by targeting the fatty acid synthase gene. J Dairy Sci 98:9001–9014

Wang H, Luo J, Zhang T, TIAN H et al (2016) MicroRNA-26a/b and their host genes synergistically regulate triacylglycerol synthesis by targeting the INSIG1 gene. RNA Biol 13:500–510

Wicik Z, Gajewska M, Majewska A, Walkiewicz D et al (2016) Characterization of microRNA profile in mammary tissue of dairy and beef breed heifers. J Anim Breed Genet 133(1):31–42

Xi Y, Jiang X, Li R, Chen M et al (2016) The levels of human milk microRNAs and their association with maternal weight characteristics. Eur J Clin Nutr 70:445–449

Zeljic K, Supic G, Magic Z (2017) New insights into vitamin D anticancer properties: focus on miRNA modulation. Mol Genet Genomics 292:511–524

Zempleni J, Baier SR, Howard KM, Cui J (2015) Gene regulation by dietary microRNAs. Can J Physiol Pharmacol 93:1097–1102

Zempleni J, Aguilar-Lozano A, Sadri M, Sukreet S et al (2017) Biological activities of extracellular vesicles and their cargos from bovine and human milk in humans and implications for infants. J Nutr 147:3–10

Zhang L, Hou D, Chen X, Li D et al (2012) Exogenous plant MIR168a specifically targets mammalian LDLRAP1: evidence of cross-kingdom regulation by microRNA. Cell Res 22:107–126

Zhou Q, Li M, Wang X, Li Q et al (2012) Immune-related microRNAs are abundant in breast milk exosomes. Int J Biol Sci 8:118–123

Diet-Induced Epigenetic Modifications and Implications for Intestinal Diseases

79

Elodie Gimier, Nicolas Barnich, and Jérémy Denizot

Contents

Abstract

Epigenetic modifications, such as post-translational modifications of histones, DNA methylation, and microRNA expression, are involved in gene transcription

E. Gimier · N. Barnich · J. Denizot (✉)
M2iSH, Université Clermont Auvergne, Inserm U1071, USC-INRA 2018, Clermont-Ferrand, France
e-mail: elodie.gimier@uca.fr; nicolas.barnich@uca.fr; jeremy.denizot@uca.fr

V. B. Patel, V. R. Preedy (eds.), *Handbook of Nutrition, Diet, and Epigenetics*,
https://doi.org/10.1007/978-3-319-55530-0_117

changes in the cells in response to environmental signals. It is now clear that particular phenotypes are the consequences of environmental effects on epigenetic marks. In this chapter, we describe the interactions existing between environment and epigenetic marks. Among the environmental factors, we'll specially focus on diet, through different examples such as the effect of diet on bee cast formation, on microbiota composition and short-chain fatty acid concentration in the gut, and the consequences on epigenetic marks. Finally, we describe the link that exists between diet and epigenetic modifications in the context of inflammatory bowel disease (IBD). So far, epigenetic marks have been poorly investigated in the context of IBD, but it has recently become an expanding field of research since new data raise crucial role for epigenetic modifications in the etiology of IBD.

Keywords
DNA methylation · DNMT3A · Microbiota · Inflammatory bowel disease · Folate · Short-chain fatty acids · Butyrate · Dysbiosis · Crohn's disease · Nutrition · GPR43

List of Abbreviations

BPA	Bisphenol A
CD	Crohn's Disease
DNMT	DNA-Methyltransferase
DSS	Dextran Sodium Sulfate
EWAS	Epigenome Wide Association Study
GPR43	G-protein Coupled Receptors 43
HC	Healthy Controls
HDAC	Histone Deacetylase
HFD	High-Fat Diet
HPTM	Histone Post-Translational Modification
IAP	Intracisternal A Particle
IBD	Inflammatory Bowel Disease
IEC	Intestinal Epithelial Cells
KAT2B	Lysine Acetyltransferase 2B
SCFA	Short-Chain Fatty Acid
SNP	Single-Nucleotide Polymorphism
UC	Ulcerative Colitis

Introduction

Epigenetic regulation of gene expression is essential to determine cell differentiation, cell morphology, and reproductive behavior phenotype. Epigenetic processes are at the interface between environmental signals and transcriptional responses, allowing rapid adaptation of the organism to environmental changes. Nutrition is one of the environmental factors that influence the most epigenetic marks through modification of epigenetics-related enzyme activity. Dietary factors are essential for normal biological processes and metabolic regulation but its imbalance can profoundly modify the transcriptional pattern of cells, contributing to specific disease development. Here, we review the effects of diet on epigenetic marks with an emphasis on the interaction between diet, microbiota, and epigenetic marks. We also discuss the role of epigenetic modifications in the onset of inflammatory bowel diseases (IBDs), such as Crohn's disease (CD) and ulcerative colitis (UC).

Effects of Diet on Epigenetic Marks

Environmental factors and especially the diet influence health throughout adulthood, but organisms are particularly sensitive during the early life. A well-known example of early nutrition influence on epigenetic marks is the honey bee queen/worker developmental process.

Example of Worker and Queen Bees

Queens and workers differ in morphology, reproductive capability, behavior, and life span despite being genetically identical. All the larvae bees are fed royal jelly during early life; however, worker bees are weaned earlier than the future queens. The former will switch their diet to pollen and nectar, whereas the latter will be fed royal jelly until they are adult (Mao et al. 2015) (Fig. 1). It is now well admitted that the

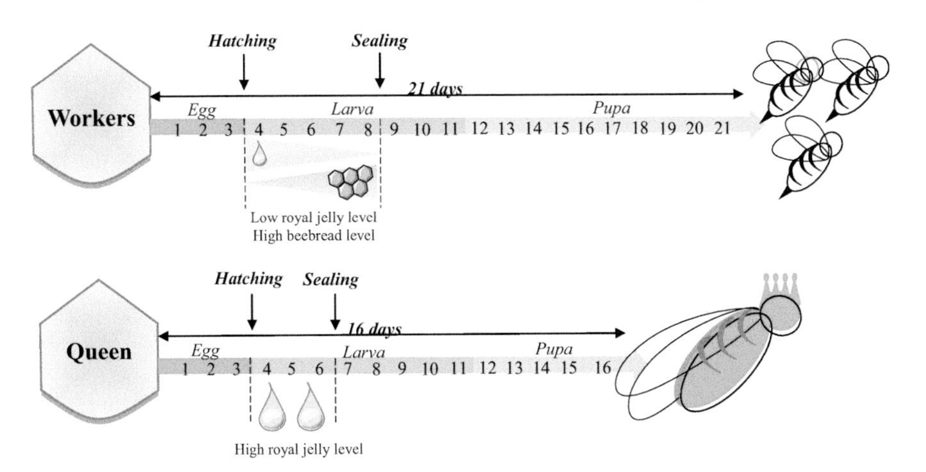

Fig. 1 Epigenetics in honey bee queen/worker developmental process. Whereas honey bee queens and workers are genetically identical, they differ in morphology, reproductive capabilities, behavior, and life span. These differences between adult queen and worker bees result from different diets in the early stages of the development. After egg hatching, all receive royal jelly: high level of royal jelly for future queens and beebread for future worker bees. This specific early nutritional exposure leads to epigenetics particularities between these two castes that explain the two adult phenotypes

nutritional composition of royal jelly affects the DNA methylation process in the developing bees. A bee fed royal jelly for a long period presents a unique brain methylation profile specific to the queen's caste, which is very different from the worker bee's caste. More than 550 genes show differential methylation levels between workers and queens in brain cells. A strong correlation between methylation pattern and splicing sites was observed, suggesting that alternative splicing is one mechanism by which DNA methylation could be linked to gene regulation in the honey bee. DNA methylation status is essential for caste differentiation as modulation of pathways involved in DNA methylation processes leads to a shift in bee's phenotype. A fundamental study revealed that the silencing of DNA methyltransferase-3 (DNMT3) (essential enzyme for global DNA methylation reprogramming) mimics the royal jelly diet directing the development of bees into queens (Kucharski et al. 2008). Another study revealed the inhibitory effect of royal jelly on activity of histone deacetylases (HDACs) in the bees, suggesting that histone post-translational modifications (HPTMs) could be involved in the fate of bees. In addition to the effect of royal jelly on honey bee caste formation, a new mechanism has recently been uncovered. A plant microRNA (miRNA) more enriched in worker bees' food (beebread) than in royal jelly is able to postpone larval development inducing sterile worker bees. The identified plant miRNA (miR-162a) delays development and decreases body and ovary size in honey bees, preventing larval differentiation into queens and inducing development into worker bees (Fig. 2) (Zhu et al. 2017).

Taken together, these findings show that multiple epigenetic mechanisms (HPTMs, DNA methylation, and miRNA expression) concur to the caste formation in honey bees, through the diet during early development. An early exposure to specific diet largely influences the development, the phenotype, and the adult physiology in bees; thus, it is likely that this phenomenon could also be found in mammalian organisms, with specific effects of the diet on gene expression and phenotype.

Folate and DNA Methylation

The diet composition modulates the concentration of molecules involved in DNA methylation. These molecules are part of the one-carbon metabolism involved in the synthesis of methyl group used by DNMT to maintain DNA methylation during replication (mostly DNMT1) or to methylate DNA de novo (DNMT3). Folate (vitamin B9) is the most extensively studied molecule for its role of methyl donor during the DNA methylation process (Fig. 3). Studies on animal models have revealed that folate depletion in the diet decreases global DNA methylation, and folate supplementation increases DNA methylation. However, such observations are difficult to reproduce in humans because folate is not the only methyl group donor molecule involved in DNA methylation (Keyes et al. 2007). Other molecules considered as methyl-donor molecules are choline, vitamin B12, methionine, and betaine.

Intestinal epithelial cells (IECs) are in direct contact with intra-luminal methyl-donor molecules, suggesting that DNA methylation could be modified in these cells in response to changes in folate concentrations in the lumen. Many studies have

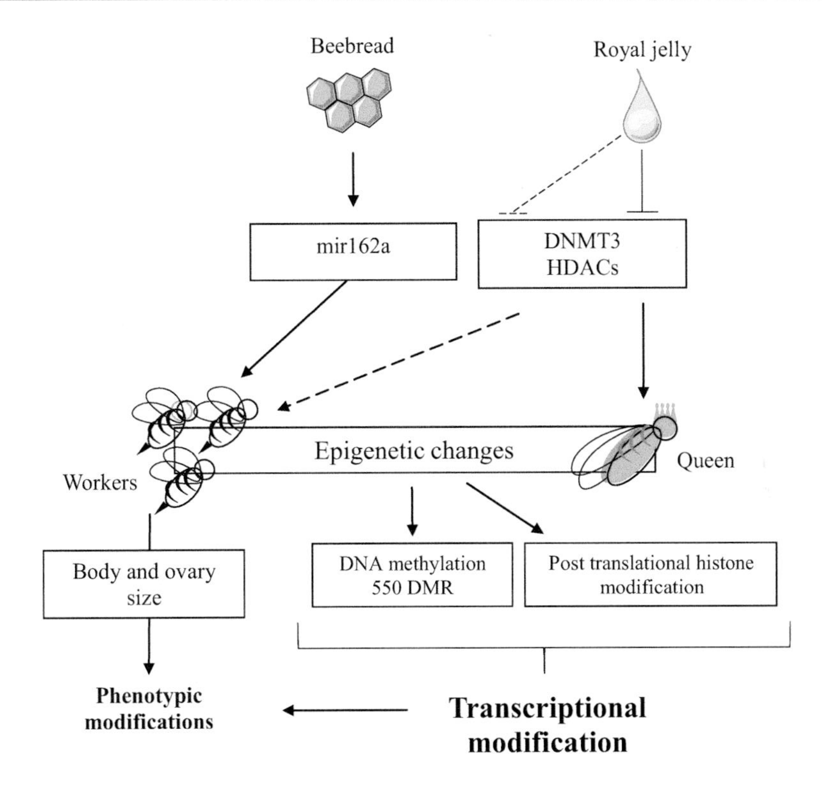

Fig. 2 Molecular effects of diet on honey bee development. DNA methylation and histone post-translational modification are essential for caste differentiation. Studies highlighted several properties of royal jelly and beebread as epigenetic modulators. Royal jelly has an inhibitory effect on DNA methyltransferase-3 (DNMT3) and histone desacetylase (HDAC) leading to histones acetylation and DNA modification. In addition to royal jelly effect, beebread that is enriched in miR-162a causes body and ovary size decrease

highlighted the effect of folate depletion or supplementation on intestinal mucosa in mouse models in different contexts such as intestinal inflammation and colorectal cancer in the offspring of treated-mothers. However, the effect of methyl donor supplementation is minor on pediatric or adult mice, suggesting the diet has a more pronounced effect on DNA methylation of the offspring during fetal life than in early life (Schaible et al. 2011).

Diet Effect on the Offspring's Epigenetic Marks

Effect of Maternal Diet on Epigenetics

Maternal diet is an important factor that determines offspring health during their entire life, but critical different periods of time can be defined. In general, the most sensitive periods to environmental changes are the early embryonic period and the

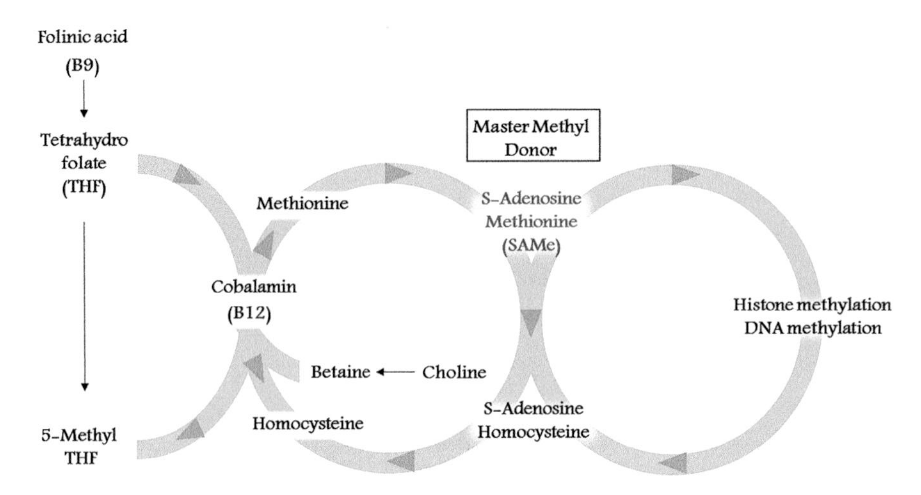

Fig. 3 **Methyl donor cycle involved in DNA and histone methylation**. S-Adenosine methionine (SAMe) is the master methyl-donor molecule used for DNA and histone methylation reactions. This molecule is produced by the catabolic pathways of the "methyl donor molecules" methionine, betaine, choline, folinic acid, and vitamin B12

fetus period. The weaning time is also an important window in contrast to the post-weaning period, which is less sensitive to diet changes. This moderate effect of the post-weaning diet is supported by the fact that a normal post-weaning diet does not reverse an in utero exposure to nutritional adverse effects. Interestingly, the effects of maternal diet persists through next generations even with a normal post-weaning diet. This suggests that in utero exposure to specific nutrition (during very early development) alters the epigenome in the somatic and germ line cells of the offspring allowing the persistence of some phenotype abnormalities in the next generations (Lee 2015). Environmental influences, such as diet, during pregnancy lead to epigenetic modification, metabolic and phenotype alterations in the offspring. Several animal models have been developed to study the impact of maternal diet on progeny.

One of the most studied mouse models used to investigate the impact of diet on DNA methylation is the viable yellow agouti (A^{vy}). This model has coat color variation directly correlated with the degree of CpG DNA methylation of intracisternal A particle (IAP) found in the A^{vy} allele. The A^{vy} allele is the result of the insertion of the IAP upstream of the transcription start site of the wild-type *Agouti* gene (Fig. 4a). This insertion results in the constitutive ectopic expression of the *Agouti* gene in all cells, leading to yellow fur coloration, and also to behavior modification as higher appetite, and physiological alterations such as obesity and high blood glucose concentrations (Dolinoy 2008). This mouse model was used to identify the effect of environmental factors on DNA methylation by studying the effect of different diets on coat color. Using this strategy, it has been found that xenobiotics such as bisphenol A (BPA), commonly used in food containers composed of plastics and resins, modulates the levels of DNA methylation. Maternal dietary exposure with BPA prior mating, gestation, and lactation results in a shift of

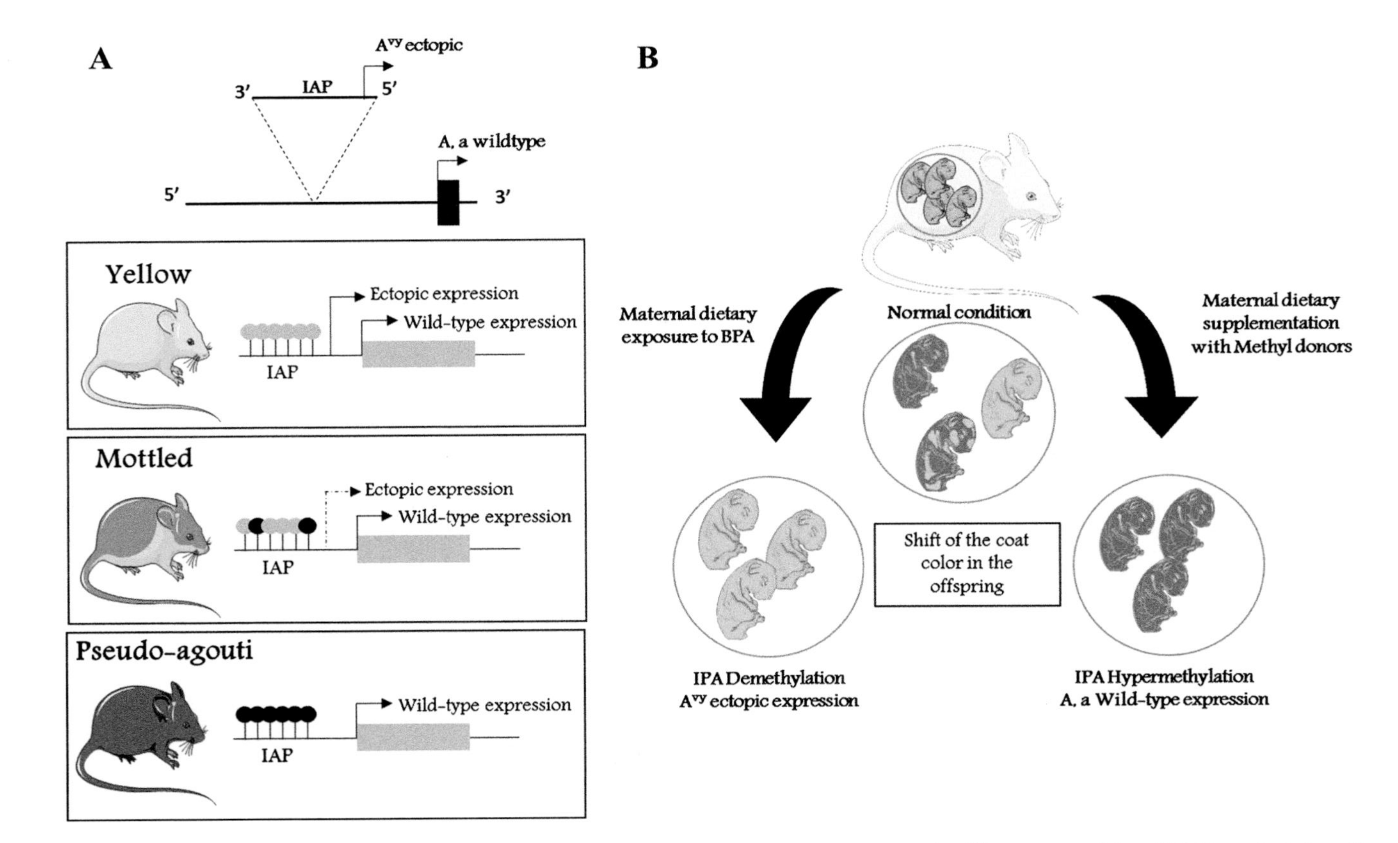

Fig. 4 **The viable yellow agouti mouse model (A^{vy}) and maternal diet**. (A) **The viable yellow agouti (A^{vy}) mouse model and DNA methylation**. The A^{vy} allele results of the insertion of the intracisternal A particle (IAP) upstream of the transcription start site of the wild-type agouti gene allele. This mouse model

the coat color in the offspring toward yellow, associated with a hypomethylation of CpG within the IAP of A^{vy} allele. Hypomethylation and yellow fur color subsequent to BPA exposure can be abolished by the maternal dietary supplemented in methyl donor (Dolinoy et al. 2007) (Fig. 4b).

Another study used this model to highlight the consequences of supplementation of maternal diet in methyl-donor molecules (folic acid, vitamin B12, betaine, choline, and methionine supplementation). When Agouti pregnant females were fed a methyl donor–supplemented diet, a high percentage of the offspring born with a brown coat color compared to the progeny of females fed a control diet. This shift in color coat of offspring is associated with hypermethylation of the IAP of *Agouti* gene (Cooney et al. 2002). These studies, and others, clearly showed that maternal diet composition can modulate DNA methylation levels on a specific gene in the offspring, leading to a specific phenotype.

The consequences of in utero undernutrition and overnutrition on epigenetic marks have also been extensively studied. High-fat diet during pregnancy leads to alterations of nutritional behavior and physiology of the offspring. The offspring of high-fat diet (HFD)-fed mice presents an increase in food intake, persistent insulin resistance throughout life associated with epigenetic alteration such as a decrease in H3 and H4 acetylation, and an increase in H3K9me3, H3K20me3, and in DNA methylation, leading to transcriptomic profile modification of genes involved in lipid metabolism. Remarkably, the opposite diet, a caloric-restricted diet, also leads to the same adverse effect on the offspring health and also promoted metabolic syndrome development. Whether the mothers were fed a low or high caloric diet, the offspring from these two groups presented high blood insulin and glucose levels, triglyceride, and low-density lipoprotein-cholesterol persistent to adulthood. These data show that early developmental exposure to undernutrition and overnutrition promotes metabolic disorders with a high susceptibility to chronic diseases associated with obesity (Reynolds et al. 2015).

Effect of Paternal Diet on Epigenetics

The maternal contribution in the persistence of characteristics acquired during environmental exposures, including diet, is widely studied. However, recent studies

Fig. 4 (continued) presents coat color variation correlated with the degree of DNA methylation of the IAP. A hypomethylation on CpG of the IAP allows ectopic expression of the agouti protein leading to a yellow fur phenotype. A hypermethylation of IAP prevents the ectopic expression of the A^{vy} allele, and mice present a pseudo-agouti phenotype. An intermediate level of DNA methylation leads to a molted phenotype. **(B) Maternal diet effect on DNA methylation level in the offspring**. In normal conditions, these mice give birth to offspring with different phenotypes associated with a variable level of methylation in the IAP. Maternal dietary exposure to xenobiotics like bisphenol A (BPA) leads to a shift in the coat color of offspring toward yellow because of the IPA demethylation by the BPA exposure. However, when the IAP is hypermethylated by maternal dietary supplementation of methyl donor, the offspring presents a shift in their phenotype toward the pseudo-agouti phenotype

highlighted that a paternal contribution to the phenotype and epigenome transmission to the next generation could also exist. Accumulating evidence indicates that paternal environment may affect the health of offspring. Paternal nutrition is a well-documented factor that alters sperm epigenetic marks which can be inherited through the paternal germ line in the next generation. Paternal HFD alters the DNA methylation profile in sperm and noncoding RNAs pool, allowing the delivery of epigenetic information through the paternal germ line to the offspring (Chen et al. 2016). Paternal HFD modifies expression of specific noncoding RNAs in sperm and confers paternally acquired disorder to the F1 generation, like glucose intolerance and insulin resistance. The noncoding RNA composition of sperm influences gene expression from the embryo to adulthood by transcriptional cascade deregulation at the early development stage, which can explain the transmission of a phenotype from adult to offspring (Chen et al. 2016). Paternal low-protein diet also alters the abundance of small noncoding RNAs in sperm and modifies gene expression in zygote, leading to a lower growth and hepatic metabolism perturbation in offspring (Sharma et al. 2016).

Another example of paternal transmission of specific phenotypes has been described in a recent paper, which reported that an aberrant DNA methylation pattern induced by dextran sodium sulfate (DSS) treatment (chemically induced colitis) can be transmitted to the offspring through paternal lineage. Alteration of DNA methylation can be detected in the sperm of the father and in the colonic cells of the offspring. These transmitted changes render the offspring more susceptible to DSS-induced colitis when compared to offspring of an untreated male. These observations give weight to the hypothesis that IBDs can have, in addition to genetic polymorphisms, an inheritable character through epigenetic mechanisms. (Tschurtschenthaler et al. 2016).

To conclude, parental history and early life diet modifications can have effects on epigenetic landscape of the offspring. However, even if many mechanisms involved in these changes are well described, it is not always clear whether the observed effect of the diet on epigenetic marks is direct or is a consequence of other changes in the organism. In the past 10 years, the new sequencing technologies allowed the analysis of the intestinal microbiota, opening a new field in the study of chromatin biology in the context of host–microbe interaction in the gut.

Diet–Microbiota Crosstalk in the Intestine: Effect on Epigenetic Marks

Intestinal Microbiota in Health

The human gut is lined by IECs organized in crypts and villus in the small intestine, whereas only crypts are found in the colonic epithelium. The luminal side of the gut is protected by a mucus layer (composed by a dense accumulation of mucin proteins), which acts as a barrier avoiding direct contact between the intestinal epithelium and the microorganisms composing the intestinal microbiota. Breach in the protective mucus layer allows commensal bacteria to be in contact with the intestinal epithelium, leading to inflammatory response. The microbiota is mostly composed of bacteria; however, viruses, yeast, and fungi are also found in the intestinal microbiota. The number of microorganisms colonizing the gastrointestinal tract is estimated to 3.8×10^{13}, which is equivalent to the number of cells in a human body (Sender et al. 2016). Sequencing approaches have identified 1000 different bacterial species composing the gut microbiota, and metagenomic data estimate that the microbiota possesses 150 times more genes than the human genome (Qin et al. 2010). This impressive number opens the research field to the analysis of the effect of bacterial gene expression on the host with a special emphasis on the effect of bacterial active products on epigenetic marks in IECs, in immune cells, and in other organs of the body. It is now clear that the interaction between microbiota and IECs regulates different important biological processes such as host immune response, nutrient uptake, angiogenesis, and neuronal development. However, the molecular mechanisms involved in these processes are not yet clear (Tremaroli and Bäckhed 2012). Epigenetic modifications, such as HPTM and DNA methylation, could be the missing link between the microbiota metabolic activity and host gene regulation explaining how microbiota can modulate epigenetic profiles in host cells. Epigenetic regulation represents a powerful way for the microbiota to communicate with the host genome and to regulate gene expression. This communication is made possible by metabolites produced by bacterial species having an effect on the activity of human enzymes, which evolved to sense microbial products involved in the regulation of the epigenome.

Effect of Diet on Microbiota-Derived Metabolic Product Synthesis

The commensal bacteria not only contribute to the regulation of host immune response and homeostasis, but also participate in the digestion of fibers and energy metabolism. Gut bacteria have enzymes that host cells lack for breaking down carbohydrates, turning them into different useful metabolites (Sun et al. 2017). The fermentation of carbohydrates by commensal bacteria such as Clostridia and Bifidobacteria leads to the synthesis of short-chain fatty acids (SCFAs). The main SCFAs found in the gut are acetate, propionate, and butyrate. The most important butyrate producer is *Faecalibacterium prausnitzii*, which belongs to the cluster of Firmicutes, comprised of *Clostridium leptum*, *Eubacterium rectale*, and *Clostridium coccoides* (Louis et al. 2010). SCFAs can be absorbed by IECs and transported to other organs through blood circulation. The level of SCFAs is very low in axenic mice confirming the necessary role of bacteria in the synthesis of these molecules (Maslowski et al. 2009). The concentration of SCFAs in the lumen is regulated by the concentration of SCFA-producing bacteria. The presence or absence of these bacteria is directly linked to the composition of the diet. For example, mice fed an occidental diet (high-fat, high-sugar low-fiber diet) present a decrease in butyrate, acetate, and propionate concentration in stools, associated with an increase in inflammatory markers and increased susceptibility to DSS-induced colitis and

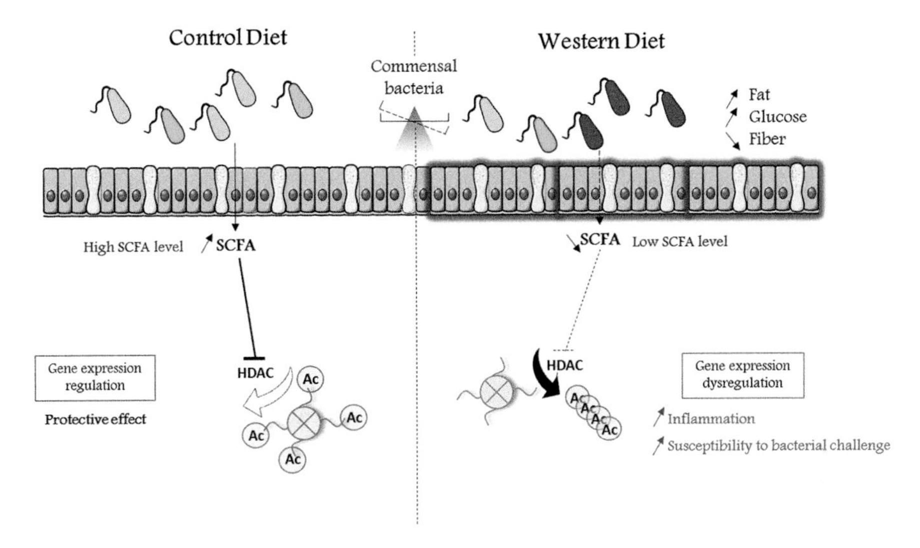

Fig. 5 Effect of diet on SCFA-producing bacteria and epigenetic modulations and intestinal homeostasis. The interaction between microbiota and intestinal epithelial cells regulates different important biological processes essential for intestinal homeostasis. Microbiota effect is mediated through bacteria active products such as short-chain fatty acids (SCFAs). SCFAs are HDAC inhibitors and their concentration modulates epigenetic marks in host's cells. SCFAs (acetate, butyrate, and propionate) concentration in the lumen is regulated by the concentration of SCFA-producing bacteria. The presence or absence of these bacteria is directly linked to the composition of the diet. Western diet leads to a decrease in SCFA concentration subsequent to microbiota alteration. This decrease of SCFA is associated with an increase in inflammatory markers and to increase susceptibility to bacterial challenge

bacterial challenge. This was mostly due to a shift in microbiota composition during the treatment (decrease in Firmicutes and increase in Proteobacteria) (Agus et al. 2016). This observation indicates that environmental factors, such as the diet, can drastically modify the microbiota and the metabolic compound concentrations synthetized by the microbiota (Fig. 5).

Effect of SCFA on Epigenetic Marks

Low concentrations of SCFAs have been associated with many diseases such as Parkinson's disease and IBD, suggesting that SCFAs could be important in the prevention of diseases (Sun et al. 2017). Butyrate and propionate are well-known HDAC inhibitors, the former being more potent than the latter as butyrate accumulates much more in colonocytes (Waldecker et al. 2008). Based on these observations, SCFAs have emerged as a central player mediating the crosstalk between microbiota and the host by acting on epigenetic marks in host's cells (Macfarlane and Macfarlane 2003). Butyrate treatment of human cells results in hyperacetylation of histones, thereby increasing the accessibility of transcription factors to chromatin and leading to active transcription of target genes. However, it is likely that butyrate has other (indirect) intracellular targets, including hyperacetylation of nonhistone proteins and alteration of DNA methylation, leading to transcriptomic changes in the cell (Hinnebusch et al. 2002). Butyrate exerts a very broad range of effects on many biological pathways via its inhibitory activity on HDAC such as inhibition of cell proliferation, induction of apoptosis, control of inflammatory response, and development and function of hematological system (Li et al. 2016). These observations shed light on the importance of SCFAs in gene regulation in the intestine and in the control of intestinal homeostasis.

The exact mechanism behind the beneficial effects of SCFAs on the intestine is not clear yet. It is well admitted that SCFAs act as HDAC inhibitors, which leads to enhanced histone acetylation and modified transcriptome. Another way of action of SCFAs could be the activation of G protein-coupled receptors (GPCRs) such as GPR43. A very interesting study revealed that acetate protects mice against colitis (in an acute model of colitis), in a GPR43-dependent manner. This clearly shows that the activation of GPCRs plays a key role in the protective effects of SCFAs in the intestine (Maslowski et al. 2009). However, the authors of this study did not analyze the HDAC activity and epigenetic changes induced in response to acetate treatment in the context of colitis, which could contribute to the protection against colitis.

In addition to the effects of microbiota-derived SCFAs on the intestine, SCFAs can reach the blood circulation and influence epigenetic marks in distant organs. A recent study revealed that gut microbiota composition affects host-tissue epigenetic status in a SCFA-dependent manner, especially in the liver and the adipose tissue. The authors also showed that Western diet suppresses microbiota-driven SCFA production. More importantly, SCFA supplementation in germ-free mice restores, partially, the histone PTM profile and gene expression profile, comparable to what is observed in a microbiota-colonized mouse. These observations demonstrate that SCFAs can induce aspects of a "colonized"-like state in host chromatin and gene expression in liver. These data suggest that "gut microbial community composition and metabolite production are important factors that connect gut microbiota and host chromatin states" (Krautkramer et al. 2016).

Epigenetics in Inflammatory Bowel Disease

IBDs, such as CD and UC, are characterized by uncontrolled inflammation of the intestinal mucosa (all segments of the digestive tract can be affected in CD, whereas only rectal and colonic mucosas are affected in UC). IBDs are relapsing conditions of the gastrointestinal tract characterized by chronic inflammation leading to abdominal pain, diarrhea, and alteration of the quality of life. The etiology of IBD is not clearly understood. However, it is widely accepted that IBD is a multifactorial disease. Genetic, infectious, and environmental factors have been involved in the development of the disease. A hypothesis is that CD arises in a genetically susceptible host (harboring single-nucleotide polymorphisms in specific genes) living in a specific environment (Western countries).

Genome-wide association's studies (GWAS) have identified more than 160 loci associated with an increased risk of developing CD. Among them, a risk variant of DNMT3A, a major actor of DNA methylation process, has been identified (Franke et al. 2010). Importantly, higher levels of expression of DNMT1 and DNMT3B were reported in the inflamed mucosa of UC patients (in comparison to non-inflamed controls) (Saito et al. 2011). These findings support the role of epigenetic mechanisms in the onset and progression in intestinal inflammation and open a new field of study in gastroenterology. Moreover, the environment-mediated intestinal dysbiosis (frequently observed in IBD) could influence epigenetic marks in patients, leading to increased susceptibility to intestinal inflammation.

Environmental Factors and Intestinal Dysbiosis in IBD

Among the environmental factors involved in the etiology of IBD, lifestyle has been proposed as an aggravating factor for the development of CD, because of the high incidence of the disease in developed countries, compared to developing countries (Rogler et al. 2016). More interestingly, it has been observed that the migrant population from Asia (having a very low prevalence of CD) to the United States (having a very high incidence of CD) reaches a similar incidence in CD after one generation. This observation gives weight to the influence of the environment and food habits in the development of the disease (Pinsk et al. 2007). Western lifestyle and human habits have been implicated in the rise of IBD incidence in developed countries (Bernstein and Shanahan 2008). This theory has been reinforced since the incidence of IBD has dramatically increased in the last half century, a period associated with huge changes in human's habits such as usage of antibiotics, sanitary conditions, diet rich in fat and sugar, consumption of processed food with their ubiquitous food additives. Rodent and human studies have shown that all these lifestyle habits are associated with dramatic modification in the composition of gut microbiota, which could explain, in part, the dysbiosis observed in IBD patients (Chassaing et al. 2016).

Dysbiotic microbiota is frequently observed in IBD patients, reinforcing the role of microbiota in IBD pathogenesis. However, it is not clear whether the dysbiosis triggers inflammation in the gut or if it is a consequence of chronic inflammation. CD patients present a decrease in the diversity of bacterial species associated with an increase in Bacteroides and Proteobacteria and a significant decrease in Firmicutes abundance (Sokol et al. 2008). At the genus level, *F. prausnitzii* was observed as significantly reduced in the microbiota of CD patients. *F. prausnitzii* is an anaerobic butyrate-producing bacterium. It has anti-inflammatory activity in a mouse model of colitis, a protective effect on barrier function (intestinal permeability), and modulates T-cell response by induction of anti-inflammatory cytokine IL-10 in human dendritic cells (Rossi et al. 2016). Importantly, a reduction of *F. prausnitzii* in the gut section of CD patients at the time of surgery (resection of an inflamed gut section) is associated with a higher risk of postoperative recurrence of ileal CD (Sokol et al. 2008). However, the molecular mechanisms behind this protective role are not yet clear. A 15-kDa protein having anti-inflammatory properties has been recently identified in *F. prausnitzii* culture supernatant. This protein inhibits the NF-κB signaling pathway and protects animals from DSS-induced colitis (Quévrain et al. 2016). In complement of the activity of this recently identified protein, the amount of butyrate produced by *F. prausnitzii* could also play a global role in the protection of the intestine against inflammation *via* its HDAC inhibitor activity.

Epigenetics Marks in IBD

HPTM in IBD

Histone post-translational modification remains poorly studied in the context of IBD. However, it was observed that the inflamed mucosa and Peyer's patches of CD patients show increase in histone H4 acetylation on the residues K8 and K12 when compared to the non-inflamed mucosa. Hyperacetylation seems to be a hallmark of inflammatory tissues as the same observations were obtained in a rat model of chemically induced colitis (Tsaprouni et al. 2011). Other studies revealed that HDAC inhibitors have a protective effect in different mouse models of colitis by inducing apoptosis and decreasing pro-inflammatory cytokine expression (Glauben et al. 2006). A recent study showed that a HDAC-6 selective inhibitor (LTB2) protects mice from DSS-induced colitis, suggesting that HDAC inhibitors could potentially be used for therapy of IBD, but this warrants further investigations. Data on the association of increased histone acetylation with disease are controversial with conflicting reports, and further investigations are needed. A recent study highlighted the downregulation of lysine acetyltransferase 2B (KAT2B) in inflamed mucosa from CD and UC patients compared to uninflamed mucosa. The decrease in KAT2B was associated with a loss in H4K5 acetylation in the inflamed mucosa and with decreased expression of the key anti-inflammatory cytokine IL-10 in the mucosa (Bai et al. 2016). Another study linked other key proteins in the deacetylation process of histone and nonhistone proteins. Sirt2 is a member of the NAD^+-dependent deacetylases, which binds and deacetylates the p65 subunit of NF-κB. Sirt2 deficiency induces hyperacetylation of NF-κB, leading to an increase in the level of pro-inflammatory cytokines. In a mouse model of colitis, Sirt2 deficiency is associated with a more severe DSS-induced colitis. These data confirm a protective role for SIRT2 against the development of inflammation. However, the role of Sirt2 has not been assessed in IBD patients and deserves further attention in the future.

DNA Methylation in IBD

A GWAS meta-analysis revealed an association between IBD and a genetic polymorphism in the *DNMT3A* gene. Moreover, higher expression levels of *DNMT1* and *DNMT3B* were reported in actively inflamed UC mucosa compared to patients in quiescent phase of the disease. These data suggest the potential important role of DNA methylation in the pathogenesis of IBD. Similar to genetic studies and GWAS, the identification of differentially methylated genes potentially involved in the pathogenesis of IBD is critical to improve our knowledge on molecular mechanisms leading to the disease.

A study highlighted, for the first time in 2011, differentially methylated regions in IBD patients compared to healthy controls, providing a new insight into the etiology of the disease. This work highlighted seven CpG sites differentially methylated in intestinal tissues of IBD patients. The authors have also identified changes in DNA methylation associated with the two major IBD subtypes, CD and UC. This study reports that the DNA methylation pattern could be disease-specific (Lin et al. 2011). Another study from the same group highlighted the altered methylation marks in B cells of CD patients when compared to siblings without IBD. Using this approach, the authors identified 11 IBD-associated CpG sites, 14 CD-specific CpG sites, and 24 UC-specific CpG sites with methylation changes in B cells. Many of the genes showing differential methylation have important immune and inflammatory response functions including several loci within the interleukin (IL)-12/IL-23 pathway (Lin et al. 2012). Another important study revealed 1117 sites differentially methylated (hypo or hypermethylated) in ileal mucosa of CD patients when compared to healthy controls. The genes harboring these modifications are involved in different cellular functions, such as immune activation, defense response to bacteria, and dendritic cell activity. Interestingly, the authors identified a significant enrichment of methylation changes within 50 kb of CD GWAS loci including IL-27, IL-19, TNF, MST1, and NOD2. It is important to note that methylation status was predictive of disease status, suggesting a central role of methylation of DNA in the onset and progression of ileal CD. Interestingly, *CARD9* gene, which is an IBD-associated locus identified through GWAS approaches, has been found to be differentially methylated in peripheral blood mononuclear cells (PBMC) (Nimmo et al. 2012) and in inflamed mucosa from UC patients compared with normal mucosa from HCs (Cooke et al. 2012).

A more recent study aimed at correlating the observed altered methylation patterns with specific bacterial populations found in the microbiota. The authors highlight a correlation between an increased abundance of Bacteroides and Roseburia genders with the DNA methylation of *KHDC3L* gene level in UC patients (Harris et al. 2016).

The factors involved in these mechanisms are not identified yet, but many studies focus on the effect of the diet and more specifically on methyl-donor molecules on DNA methylation in the intestine and on the composition in bacterial species of the microbiota. Studies on mouse models have shown that an enriched diet in methyl-donor molecules (folate, betaine, choline...) of a pregnant mouse leads to an increased sensitivity to chemically induced colitis in the offspring, suggesting DNA methylation and microbiota composition could be key factors controlling inflammation in the gut (Schaible et al. 2011). Our group also revealed that depletion of methyl-donor molecules in the diet leads to demethylation of the promoter of the gene *CEACAM6*, which encodes a receptor for CD-associated adherent-invasive *E. coli* (AIEC). This diet-mediated demethylation results in an increase in the expression of CEACAM6 in the colonic mucosa of genetically modified mice (CEABAC10 transgenic mice) and enhances gut colonization by AIEC and inflammatory response in the intestinal mucosa (Denizot et al. 2015). These data on mouse models clearly state the central role played by DNA methylation in the control of inflammation in the gut, in a diet-dependent manner, but further investigations are needed to better understand the underlying mechanisms to identify epigenetics-related therapeutic targets for the treatment of CD.

Conclusion

To conclude, the composition of the diet can have profound effects on epigenetic marks either through a direct effect, as observed in honey bee development during royal jelly or beebread feeding, or an indirect effect through modification of bacterial species colonizing the intestine. Diet modification can lead to disequilibrium in bacterial species colonizing the host (dysbiosis), as frequently observed in IBD patients, decreasing SCFA concentration in the gut and SCFA beneficial-associated effects. One of the consequences of SCFA decrease in the gut is the hyperacetylation of histones as SCFAs have direct HDAC inhibition activity (Fig. 6). One strategy to restore histone modification profile in IBD could be to supplement the diet in molecules having a HDAC inhibitor activity (butyrate, acetate...) or favor the growth of SCFA-producing bacteria through an increase in the ingestion of fibers (prebiotics). The latter strategy could potentially be used for prevention of IBD in Western countries where the incidence of IBD is very high.

Key Facts

Key Facts on Honey Bees

- The diet of bees (royal jelly or beebread) has profound effect on DNA methylation marks, on histone desacetylase activities, and on miRNA expression during the development of bees, leading to the growth of a worker or a queen bee.
- In honey bees, DNA methylation status, histone desacetylase inhibitors contained in royal jelly, and miRNA found in beebread regulate the development of bees to become a worker or a queen.

Key Facts on Microbiota and Inflammatory Bowel Disease

- The human microbiota is composed of more than 1000 different bacterial species. Metagenomics data estimate that the microbiota encodes for 150 times more genes than the human genome. These bacteria synthetize metabolically active products that can act on epigenetics-related proteins in the host. SCFAs, such as butyrate, are synthetized by the microbiota and can inhibit histone desacetylase activity resulting in changes in genes expression. The composition of the microbiota depends mostly on the composition of the diet.

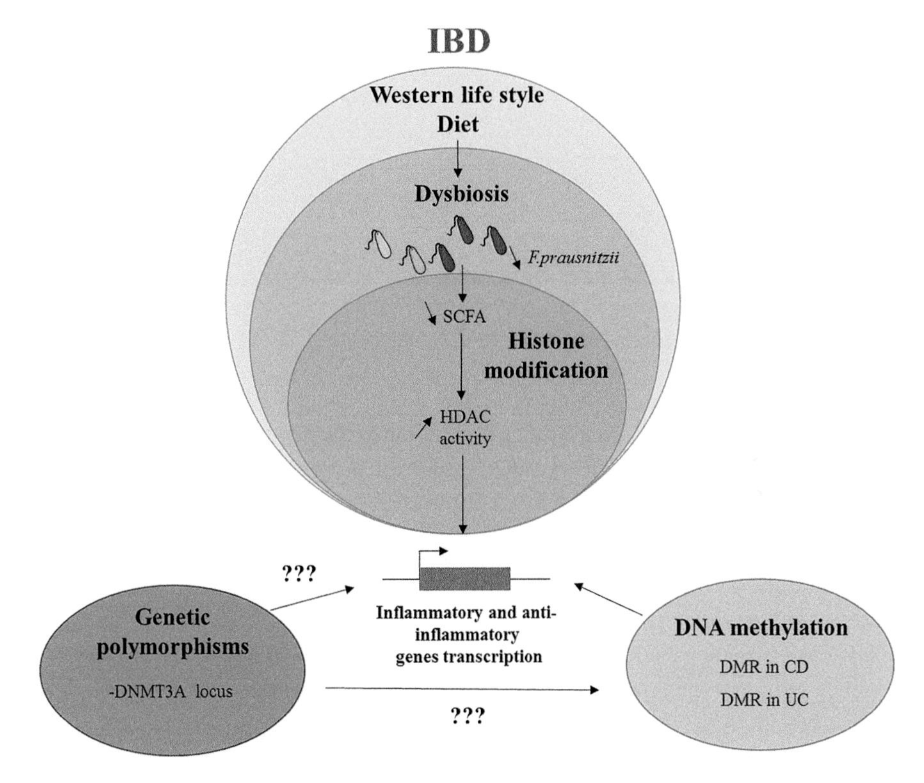

Fig. 6 **Epigenetic in IBD context**. Inflammatory bowel diseases (IBDs), such as Crohn's disease and ulcerative colitis, are characterized by uncontrolled inflammation of intestinal mucosa. These pathologies are multifactorial diseases and present a complex etiology. Western lifestyle is wildly known to increase the risk of developing IBD. The harmful effect of Western diet is, in part, mediated by microbiota alteration. Dysbiosis observed in IBD patients is characterized by a higher concentration of pro-inflammatory bacteria and a decrease of protective bacteria such as SCFA-producing bacteria (*F.prausnitzzi*). Decrease in SCFAs concentration can lead to modulation of histone acetylation level and transcription level of specific genes. Genome-wide association's studies have identified several loci associated with IBD such as DNMT3a that can explain altered DNA methylation profiles observed in IBD patients. DMR, differentially methylated regions

- Crohn's disease is an IBD for which the etiology remains unknown. It is a multifactorial disease with genetic, infectious, and environmental factors involved in the development of the disease. The western lifestyle, such as high-fat/high-sugar diet, is a key factor in the development of the disease. Recent studies focus on the role of epigenetic mechanisms in the onset of Crohn's disease.

Key Facts on Epigenetics in Inflammatory Bowel Disease

- Histones acetylation has been poorly studied in the context of Crohn's disease. However, recent data suggest abnormal regulation of KAT2B in Crohn's disease and others reveal that HDAC inhibitors protect mice against inflammation.

- DNA methylation status in IBD is becoming an expending field of research as many loci show differential methylation levels in IBD when compared to healthy controls.

Dictionary of Terms

- **Folate** – Folate is a vitamin B that is naturally present in some foods or added to others. Folate is involved in the synthesis of S-adenosyl methionine, an important methyl-donor molecule involved in the methylation of DNA and epigenetic regulation of genes expression.
- **Short-chain fatty acid** – Short-chain fatty acids, such as acetate, propionate, and butyrate, are the end products of fermentation of dietary fibers, mostly found in the colon. Butyrate is a well-characterized histone desacetylase inhibitor and has anti-inflammatory effects on intestinal mucosa.
- **Dysbiosis** – Dysbiosis refers to disequilibrium in the composition of microbial species in the gut. Dysbiosis is frequently observed in pathological conditions, such as IBD.
- **Crohn's disease** – Crohn's disease is an IBD, for which the exact etiology remains unknown. It is a multifactorial disease involving genetics, environmental, microbial, and epigenetic factors. To date, no curative treatment exists for this disease.

Summary Points

- Silencing of DNA methyltransferase-3 (DNMT3) (an essential enzyme for global DNA methylation reprogramming) mimics the royal jelly diet directing the development of bees into queens.
- Paternal diet can affect DNA methylation in the sperm and alters the offspring DNA methylation pattern and phenotype. Aberrant DNA methylation pattern induced by dextran sodium sulfate (DSS) treatment is associated with alteration of DNA methylation in the colonic cells of the offspring. These paternally transmitted changes render the offspring more susceptible to DSS-induced colitis.
- Epigenetic regulation represents a powerful way for the microbiota to communicate with the host genome and to regulate genes expression through molecules synthetized by microbiota.
- Short-chain fatty acids (SCFAs) have anti-inflammatory activities on intestinal mucosa, not only depending on their HDAC inhibitor activities, but also depending on SCFA receptors GPR43, which is necessary to mediate the anti-inflammatory effect of acetate and to protect against colitis.
- Many single-nucleotide polymorphisms have been associated with increased risk of developing the disease. *DNMT3A* gene harbors a risk variant for Crohn's disease (CD), suggesting the important role of DNA methylation in IBD

development. However, no functional study was performed to determine the effect of the variant on DNA methylation or susceptibility to inflammation.

- SCFAs are found in lower concentrations in CD, mainly due to the decreased abundance in butyrate-producing bacteria *Faecalibacterium prausnitzii.* The absence of *F. prausnitzii* is correlated with a higher risk of postoperative recurrence of ileal CD.
- Histone desacetylase inhibitors have anti-inflammatory effects in mouse models of colitis, suggesting the central role of histones acetylation in the inflammatory process.
- Lysine acetyl transferase 2B is downregulated in CD patients' inflamed mucosa and is associated with a loss in H4K5 acetylation mark.
- DNA methylation is modified in CD patient's mucosa and blood samples, compared to healthy controls. EWAS revealed differentially methylated loci located within 50 kb of one of the CD-associated loci.

References

Agus A, Denizot J, Thévenot J, Martinez-Medina M, Massier S, Sauvanet P, Bernalier-Donadille A, Denis S, Hofman P, Bonnet R et al (2016) Western diet induces a shift in microbiota composition enhancing susceptibility to adherent-invasive *E. coli* infection and intestinal inflammation. Sci Rep 6:19032

Bai AHC, Wu WKK, Xu L, Wong SH, Go MY, Chan AWH, Harbord M, Zhang S, Chen M, Wu JCY et al (2016) Dysregulated lysine acetyltransferase 2B promotes inflammatory bowel disease pathogenesis through transcriptional repression of interleukin-10. J Crohns Colitis 10:726–734

Bernstein CN, Shanahan F (2008) Disorders of a modern lifestyle: reconciling the epidemiology of inflammatory bowel diseases. Gut 57:1185–1191

Chassaing B, Koren O, Goodrich JK, Poole AC, Srinivasan S, Ley RE, Gewirtz AT (2016) Corrigendum: dietary emulsifiers impact the mouse gut microbiota promoting colitis and metabolic syndrome. Nature 536:238

Chen Q, Yan M, Cao Z, Li X, Zhang Y, Shi J, Feng G, Peng H, Zhang X, Zhang Y et al (2016) Sperm tsRNAs contribute to intergenerational inheritance of an acquired metabolic disorder. Science 351:397–400

Cooke J, Zhang H, Greger L, Silva A-L, Massey D, Dawson C, Metz A, Ibrahim A, Parkes M (2012) Mucosal genome-wide methylation changes in inflammatory bowel disease. Inflamm Bowel Dis 18:2128–2137

Cooney CA, Dave AA, Wolff GL (2002) Maternal methyl supplements in mice affect epigenetic variation and DNA methylation of offspring. J Nutr 132:2393S–2400S

Denizot J, Desrichard A, Agus A, Uhrhammer N, Dreux N, Vouret-Craviari V, Hofman P, Darfeuille-Michaud A, Barnich N (2015) Diet-induced hypoxia responsive element demethylation increases CEACAM6 expression, favouring Crohn's disease-associated *Escherichia coli* colonisation. Gut 64:428–437

Dolinoy DC (2008) The agouti mouse model: an epigenetic biosensor for nutritional and environmental alterations on the fetal epigenome. Nutr Rev 66(Suppl 1):S7–S11

Dolinoy DC, Huang D, Jirtle RL (2007) Maternal nutrient supplementation counteracts bisphenol A-induced DNA hypomethylation in early development. Proc Natl Acad Sci U S A 104:13056–13061

Franke A, McGovern DPB, Barrett JC, Wang K, Radford-Smith GL, Ahmad T, Lees CW, Balschun T, Lee J, Roberts R et al (2010) Genome-wide meta-analysis increases to 71 the number of confirmed Crohn's disease susceptibility loci. Nat Genet 42:1118–1125

Glauben R, Batra A, Fedke I, Zeitz M, Lehr HA, Leoni F, Mascagni P, Fantuzzi G, Dinarello CA, Siegmund B (2006) Histone hyperacetylation is associated with amelioration of experimental colitis in mice. J Immunol 176:5015–5022

Harris RA, Shah R, Hollister EB, Tronstad RR, Hovdenak N, Szigeti R, Versalovic J, Kellermayer R (2016) Colonic mucosal epigenome and microbiome development in children and adolescents. J Immunol Res 2016:9170162

Hinnebusch BF, Meng S, Wu JT, Archer SY, Hodin RA (2002) The effects of short-chain fatty acids on human colon cancer cell phenotype are associated with histone hyperacetylation. J Nutr 132:1012–1017

Keyes MK, Jang H, Mason JB, Liu Z, Crott JW, Smith DE, Friso S, Choi S-W (2007) Older age and dietary folate are determinants of genomic and p16-specific DNA methylation in mouse colon. J Nutr 137:1713–1717

Krautkramer KA, Kreznar JH, Romano KA, Vivas EI, Barrett-Wilt GA, Rabaglia ME, Keller MP, Attie AD, Rey FE, Denu JM (2016) Diet-microbiota interactions mediate global epigenetic programming in multiple host tissues. Mol Cell 64:982–992

Kucharski R, Maleszka J, Foret S, Maleszka R (2008) Nutritional control of reproductive status in honeybees via DNA methylation. Science 319:1827–1830

Lee H-S (2015) Impact of maternal diet on the epigenome during in utero life and the developmental programming of diseases in childhood and adulthood. Nutrients 7:9492–9507

Li C-J, Li RW, Baldwin RL, Blomberg LA, Wu S, Li W (2016) Transcriptomic sequencing reveals a set of unique genes activated by butyrate-induced histone modification. Gene Regul Syst Biol 10:1–8

Lin Z, Hegarty JP, Cappel JA, Yu W, Chen X, Faber P, Wang Y, Kelly AA, Poritz LS, Peterson BZ et al (2011) Identification of disease-associated DNA methylation in intestinal tissues from patients with inflammatory bowel disease. Clin Genet 80:59–67

Lin Z, Hegarty JP, Yu W, Cappel JA, Chen X, Faber PW, Wang Y, Poritz LS, Fan J-B, Koltun WA (2012) Identification of disease-associated DNA methylation in B cells from Crohn's disease and ulcerative colitis patients. Dig Dis Sci 57:3145–3153

Louis P, Young P, Holtrop G, Flint HJ (2010) Diversity of human colonic butyrate-producing bacteria revealed by analysis of the butyryl-CoA: acetate CoA-transferase gene. Environ Microbiol 12:304–314

Macfarlane S, Macfarlane GT (2003) Regulation of short-chain fatty acid production. Proc Nutr Soc 62:67–72

Mao W, Schuler MA, Berenbaum MR (2015) A dietary phytochemical alters caste-associated gene expression in honey bees. Sci Adv 1:e1500795

Maslowski KM, Vieira AT, Ng A, Kranich J, Sierro F, Yu D, Schilter HC, Rolph MS, Mackay F, Artis D et al (2009) Regulation of inflammatory responses by gut microbiota and chemoattractant receptor GPR43. Nature 461:1282–1286

Nimmo ER, Prendergast JG, Aldhous MC, Kennedy NA, Henderson P, Drummond HE, Ramsahoye BH, Wilson DC, Semple CA, Satsangi J (2012) Genome-wide methylation profiling in Crohn's disease identifies altered epigenetic regulation of key host defense mechanisms including the Th17 pathway. Inflamm Bowel Dis 18:889–899

Pinsk V, Lemberg DA, Grewal K, Barker CC, Schreiber RA, Jacobson K (2007) Inflammatory bowel disease in the South Asian pediatric population of British Columbia. Am J Gastroenterol 102:1077–1083

Qin J, Li R, Raes J, Arumugam M, Burgdorf KS, Manichanh C, Nielsen T, Pons N, Levenez F, Yamada T et al (2010) A human gut microbial gene catalogue established by metagenomic sequencing. Nature 464:59–65

Quévrain E, Maubert MA, Michon C, Chain F, Marquant R, Tailhades J, Miquel S, Carlier L, Bermúdez-Humarán LG, Pigneur B et al (2016) Identification of an anti-inflammatory protein from *Faecalibacterium prausnitzii*, a commensal bacterium deficient in Crohn's disease. Gut 65:415–425

Reynolds CM, Gray C, Li M, Segovia SA, Vickers MH (2015) Early life nutrition and energy balance disorders in offspring in later life. Nutrients 7:8090–8111

Rogler G, Zeitz J, Biedermann L (2016) The search for causative environmental factors in inflammatory bowel disease. Dig Dis Basel Switz 34(Suppl 1):48–55

Rossi O, van Berkel LA, Chain F, Tanweer Khan M, Taverne N, Sokol H, Duncan SH, Flint HJ, Harmsen HJM, Langella P et al (2016) Faecalibacterium prausnitzii A2-165 has a high capacity to induce IL-10 in human and murine dendritic cells and modulates T cell responses. Sci Rep 6:18507

Saito S, Kato J, Hiraoka S, Horii J, Suzuki H, Higashi R, Kaji E, Kondo Y, Yamamoto K (2011) DNA methylation of colon mucosa in ulcerative colitis patients: correlation with inflammatory status. Inflamm Bowel Dis 17:1955–1965

Schaible TD, Harris RA, Dowd SE, Smith CW, Kellermayer R (2011) Maternal methyl-donor supplementation induces prolonged murine offspring colitis susceptibility in association with mucosal epigenetic and microbiomic changes. Hum Mol Genet 20:1687–1696

Sender R, Fuchs S, Milo R (2016) Revised estimates for the number of human and bacteria cells in the body. PLoS Biol 14:e1002533

Sharma U, Conine CC, Shea JM, Boskovic A, Derr AG, Bing XY, Belleannee C, Kucukural A, Serra RW, Sun F et al (2016) Biogenesis and function of tRNA fragments during sperm maturation and fertilization in mammals. Science 351:391–396

Sokol H, Pigneur B, Watterlot L, Lakhdari O, Bermúdez-Humarán LG, Gratadoux J-J, Blugeon S, Bridonneau C, Furet J-P, Corthier G et al (2008) Faecalibacterium prausnitzii is an anti-inflammatory commensal bacterium identified by gut microbiota analysis of Crohn disease patients. Proc Natl Acad Sci U S A 105:16731–16736

Sun M, Wu W, Liu Z, Cong Y (2017) Microbiota metabolite short chain fatty acids, GPCR, and inflammatory bowel diseases. J Gastroenterol 52:1–8

Tremaroli V, Bäckhed F (2012) Functional interactions between the gut microbiota and host metabolism. Nature 489:242–249

Tsaprouni LG, Ito K, Powell JJ, Adcock IM, Punchard N (2011) Differential patterns of histone acetylation in inflammatory bowel diseases. J Inflamm Lond Engl 8:1

Tschurtschenthaler M, Kachroo P, Heinsen F-A, Adolph TE, Rühlemann MC, Klughammer J, Offner FA, Ammerpohl O, Krueger F, Smallwood S et al (2016) Paternal chronic colitis causes epigenetic inheritance of susceptibility to colitis. Sci Rep 6:31640

Waldecker M, Kautenburger T, Daumann H, Busch C, Schrenk D (2008) Inhibition of histone-deacetylase activity by short-chain fatty acids and some polyphenol metabolites formed in the colon. J Nutr Biochem 19:587–593

Zhu K, Liu M, Fu Z, Zhou Z, Kong Y, Liang H, Lin Z, Luo J, Zheng H, Wan P et al (2017) Plant microRNAs in larval food regulate honeybee caste development. PLoS Genet 13:e1006946

Nutritional and Epigenetics Implications in Esophageal Cancer

80

Danielle Queiroz Calcagno, Kelly Cristina da Silva Oliveira, and Nina Nayara Ferreira Martins

Contents

Abstract

The effects of diet on the epigenome and its modifications on gene expression can provide a window to understanding carcinogenesis. In this chapter, an update of the most recent studies of nutritional epigenetics and how dietary nutrients affect these target mechanisms in esophageal cancer were performed. One-carbon metabolism nutrients and how a variety of nutrients such as folate, methionine,

D. Q. Calcagno (✉) · K. C. d. S. Oliveira
Núcleo de Pesquisas em Oncologia, Universidade Federal do Pará, Belém, Pará, Brazil
e-mail: danicalcagno@gmail.com; daniellequeirozcalcagno@gmail.com; kellynut01@gmail.com

N. N. F. Martins
Núcleo de Pesquisas em Oncologia, Universidade Federal do Pará, Belem, Brazil
e-mail: nnmartinsnutricao@gmail.com

V. B. Patel, V. R. Preedy (eds.), *Handbook of Nutrition, Diet, and Epigenetics*,
https://doi.org/10.1007/978-3-319-55530-0_44

cobalamin, riboflavin, and vitamin B6 has the potential to perturb DNA and histone methylation patterns in esophageal cancer are reviewed. Additionally, the roles of nutrients such as genistein, epigallocatechin-3-gallate, resveratrol and curcumin as DNA methyltransferase inhibitors and modulators of histone modifications are also discussed in esophageal cancer.

Keywords
Esophageal squamous cell carcinoma · Esophageal adenocarcinoma · One-carbon metabolism · Folate · Methionine · Cobalamin · Riboflavin · Vitamin B6 · Genistein · Epigallocatechin-3-gallate · Resveratrol · Curcumin

List of Abbreviations

5mTHF	5,10-Methylenetetrahydrofolate
5-MTHF	5-Methylenetetrahydrofolate
BE	Barrett's esophagus
COX-2	Cytochrome c oxidase subunit II
DHF	Dihydrofolate
DNMTs	DNA methyltransferases
EAC	Esophageal adenocarcinoma
ESCC	Esophageal squamous cell carcinoma
ERK1/ERK2	Extracellular signal-regulated kinase 1
GERD	Gastroesophageal reflux disease
GSTP1	Glutathione S-transferase Pi 1
HATs	Histone acetyltransferases
HDACs	Histone deacetylases
HDMs	Histone demethylases
HMTs	Histone methyltransferases
hMLH1	Human mutL homolog 1
JUN	Jun proto-oncogene
MAT	Methionine adenosyltransferase
MTR	Methionine synthase
MTR	Methionine synthase
MTHFR	Methylenetetrahydrofolate reductase
MTRR	Methionine synthase reductase
MGMT	O^6-Methylguanine DNA methyltransferase
NOTCH-1	Notch homolog 1 gene
PTMs	Posttranslational modifications
RARß	Retinoid acid receptor-p
SAM	S-Adenosylmethionine
SAH	S-Adenosylhomocysteine
SAHH	S-Adenosylhomocysteine hydrolase
SHMT	Serine hydroxymethyltransferase
SNPs	Single nucleotide polymorphisms
THF	Tetrahydrofolate

Background

In the last decades, the incidence of esophageal cancer has risen, coinciding with a shift in histologic type and primary tumor location. Two histologic types account for most esophageal cancer: esophageal squamous cell carcinoma (ESCC) and esophageal adenocarcinoma (EAC). ESCC can develop mainly in the upper third of the esophagus, whereas EAC occurs in the lower esophagus (Reim and Friess 2016). ESCC is the predominant histologic type worldwide and is associated with lower socioeconomic status and alcohol and tobacco consumption. In contrast, EAC is more prevalent in the United States and Western Europe and is associated with obesity, gastroesophageal reflux disease (GERD) and Barrett's esophagus (BE) (Olefson and Moss 2015; Falk 2009).

The esophagus serves as a food conduit to the gastrointestinal tract. Due to early occlusion of the gastrointestinal passage, patients with esophageal cancer suffer involuntary weight loss and malnutrition. The generally curative intervention is esophagectomy (Cohen and Leichman 2015), and for selected cases of ESCC, chemoradiotherapy is the treatment choice. Most patients have advanced disease at diagnosis; consequently, the overall survival is under 30% at 5 years (Rice et al. 2010). Therefore, all these observations highlight the importance of prevention, as the treatment options are restricted and poor outcome is the rule.

Recent advances in personalized nutrition emphasize the role of nutrients in the modulation of epigenetic mechanisms as a possibly beneficial intervention for the prevention and treatment of different types of cancer (Ferguson et al. 2016).

Epigenetic modifications, such as DNA methylation and histone modification, draw on reversible modifications of DNA molecule and chromatin structure without altering the nucleotide sequence. These processes can occur in a gene-specific or in a global manner, depending on the presence of enzymes and dietary nutrients (Anderson et al. 2012).

DNA methylation refers to the addition of a methyl group at the 5′ position of the cytosine in a CpG dinucleotide by DNA methyltransferases (DNMTs) and required S-adenosylmethionine (SAM) as the methyl group donor, which is obtained from diet through the one-carbon metabolism. In cancer, DNA hypermethylation occurs in gene promoter regions and represses the expression of genes involved in DNA repair, apoptosis, and cell cycle control. In contrast, hypomethylation can result in genome instability and proto-oncogene upregulation and/or activation (Baylin and Jones 2016; Calcagno et al. 2015).

In addition, histone modification occurs in a cell's chromatin conformations; hence, the tighter the association between histone and other proteins, the greater will be the influence on gene expression or repression. The N-terminal regions of histone proteins are posttranslationally modified by various enzymes mainly acetylases and methylases. Histone acetylation, associated with an increase in gene transcription, is catalyzed by both histone acetyltransferases (HATs) and histone deacetylases (HDACs). On the other hand, histone methylation is carried by histone methyltransferases (HMTs) and histone demethylases (HDMs) and results in transcription repression or activation, depending on the target sites (Baylin and Jones 2016).

The effects of diet on the epigenome and its modifications on expression gene can provide a window to understanding carcinogenesis (Martinez et al. 2014).

Hence, this chapter focuses on dietary nutrients that affect epigenetic mechanisms in esophageal cancer, including DNA methylation and histone modifications. An overview of dietary compounds that modulate methylation and histone modification in esophageal cancer is divided into two categories: (i) one-carbon metabolism nutrients and (ii) dietary agents that act as DNMT inhibitors and modulators of histone modifications.

One-Carbon Metabolism Nutrients

One-carbon metabolism consists of folate and methionine cycle and utilizes a variety of nutrients capable of perturbing DNA and histone methylation patterns (Fig. 1). Diets deficient in folate, methionine, cobalamin, or vitamin B6 induce global hypomethylation and site-specific hypermethylation and have been linked to increased cancer development (Obeid 2013).

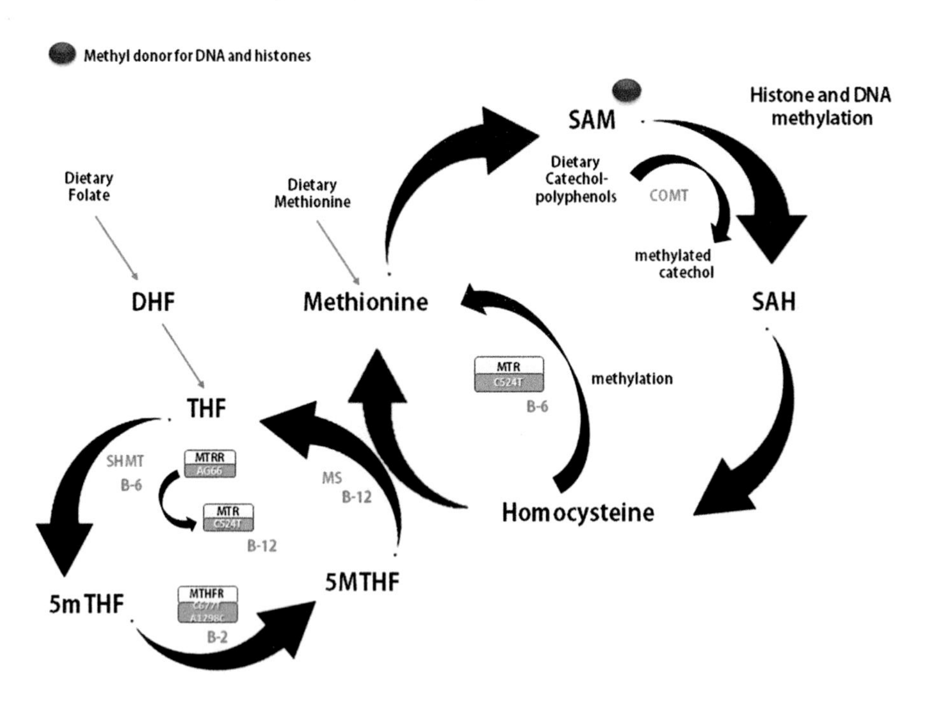

Fig. 1 Representation of one-carbon metabolism and its cofactors, enzymes and genetic polymorphisms. *MTRR* methionine synthase reductase (regenerates a functional MTR via reductive methylation), *MTR* methionine synthase (catalyzes the remethylation of homocysteine to methionine), *MTHFR* methylenetetrahydrofolate reductase (catalyzes the conversion of 5mTHF to 5-MTHF), *DHF* dihydrofolate, *THF* tetrahydrofolate, *5mTHF* 5,10-methylenetetrahydrofolate, *5-MTHF* 5-methyltetrahydrofolate, *SAM* S-adenosylmethionine, *SAH* S-adenosylhomocysteine, *COMT* catechol-*O*-methyltransferase

Folate is the natural form of vitamin B9 (bovine liver and green leafy vegetables) or in its synthetic form folic acid, used in human supplementation and fortified foods (Mudryj et al. 2016).

Before entering the folate cycle, folic acid must be reduced to dihydrofolate (DHF) and tetrahydrofolate (THF), which is then converted to 5,10-methylenetetrahydrofolate (5mTHF) by serine hydroxymethyltransferase (SHMT). Folate-derived 5mTHF is crucial for the formation of 5-methylcytosine. The enzyme methylenetetrahydrofolate reductase (MTHFR) irreversibly catalyzes the conversion of 5mTHF to 5-methyltetrahydrofolate (5-MTHF), which is the primary form of circulating folate in the peripheral blood (Ducker and Rabinowitz 2017).

In addition to DNA methylation, derivatives of folate maintain DNA stability by regulating DNA biosynthesis and DNA repair (Duthie 2011). Thus, inadequate folate intake has been implicated in the development of many types of cancer, mainly cancer associated with folate-dependent DNA replication and rapid tissue renewal, such as esophageal cancer (Sharp et al. 2013).

Several studies have suggested that low folate consumption is a significant risk factor for ESCC (Zhao et al. 2011; Ibiebele et al. 2011; Xiao et al. 2014). Furthermore, a significant interaction has been observed between folate and alcohol intake in esophageal cancer. Alcohol ingestion is a primary cause of folate deficiency, as alcohol may decrease internal folate levels through intestinal malabsorption, decreased hepatic storage, and increased renal excretion. Ethanol in alcoholic beverages is oxidized by alcohol dehydrogenase to form acetaldehyde (Seitz and Stickel 2010). Acetaldehyde is the most toxic ethanol metabolite in alcohol-associated carcinogenesis, including carcinogenesis in the esophagus (Liu et al. 2015a). Ethanol intake stimulates tumor development by inhibiting enzymes of the one-carbon metabolism, and folate deficiency facilitates the adverse effects of alcohol in methionine metabolism (Peng et al. 2016). Thus, ethanol may contribute to ESCC through aberrant gene methylation.

In an Australian case-control study, individuals with increased alcohol and decreased folate intake demonstrated a greater risk of ESCC compared with individuals with decreased alcohol and increased folate intake. Interestingly, this study also hypothesized that higher folate intake or synthetic supplementation can reduce EAC risk in patients with Barrett's esophagus (Ibiebele et al. 2011).

Methionine is an essential amino acid acquired principally from animal protein sources (Schweinberger and Wyse 2016). Methionine cycle begins with the addition of adenosyl group to methionine by methionine adenosyltransferase (MAT) to form SAM, which enters one-carbon metabolism as a methyl donor for DNA methylation and histone modifications. After transfer of the methyl group from SAM to a compatible acceptor, S-adenosylhomocysteine (SAH), a potent inhibitor of DNMTs and histone methyltransferases (HMTs), is produced (Mentch and Locasale 2015). Finally, hydrolysis of SAH yields adenosine and homocysteine, which can be remethylated by 5-MTHF to methionine (Gruber 2016). Methionine is also a precursor of cysteine and glutathione (Brown-Borg and Buffenstein 2016). Deficiency in SAM limits the ability to maintain DNA methylation and results from a low supply of necessary precursors or cofactors in the one-carbon metabolism pathway (Locasale 2013).

As inadequate folate intake is a possible risk factor for esophageal cancer, studies have demonstrated that a higher consumption of methionine in food might increase the risk of EAC (Ibiebele et al. 2011). Furthermore, Schweinberger and Wyse (2016) suggested that a misbalance in the methionine cycle results in high rate formation of its principal product homocysteine, increases the inflammatory status and aggravates preexisting lesions contributing to EAC carcinogenesis.

Su et al. (2016) reported that either the high or low intake of methionine was non-beneficial, causing alterations in the SAM/SAH concentration ratio and leading to epigenetic effects. Li et al. (2014) also reported that S-adenosylhomocysteine hydrolase (SAHH), an enzyme that catalyzes the hydrolysis of SAH to homocysteine, is downregulated in ESCC cells compared with normal esophageal epithelial cells. In addition, these authors observed that SAHH overexpression in ESCC cells promoted cell apoptosis and inhibited cell migration and adhesion, but it did not affect cell proliferation and cell cycle. All these findings suggest that SAHH may be involved in esophageal carcinogenesis.

Although it has been well established that nutrients in the one-carbon metabolism are involved in cancer development, limited studies have investigated the intakes of its cofactors and esophageal cancer. Table 1 summarizes the one-carbon metabolism nutrients and esophageal cancer risk.

Sharp et al. (2013) hypothesized that dietary methyl group factors such as folate and B6 and B12 vitamins are implicated in the etiology of EAC and its precursors – reflux esophagitis leading to metaplasia BE – as well as progression through the inflammation-metaplasia-adenocarcinoma sequence. Intriguingly, no association was observed between riboflavin and esophageal cancer although this vitamin has an important role as cofactor of one-carbon metabolism.

Found in dietary sources including meat, fish, cereals, eggs, and vegetables, vitamin B6 is a coenzyme of serine hydroxymethyltransferase required for the conversion of homocysteine to cystathionine (Sharp et al. 2013). An inverse relation between vitamin B6 intake and EAC was described in two different studies (Ibiebele et al. 2011; Mayne et al. 2001). However, Fanidi et al. (2015) and Xiao et al. (2014) found no association between vitamin B6 intake and the risk of developing esophageal cancer. In addition, Balbuena and Casson (2010) explore the effect of dietary vitamin B6 intake on p53 in the molecular

Table 1 One-carbon metabolism nutrients and esophageal cancer risk

	Intake	Esophageal cancer risk	References
Folate	↓↓↓	ESCC	Zhao et al. (2011), Ibiebele et al. (2011), and Xiao et al. (2014)
Methionine	↑↑↑	EAC	Su et al. (2016)
B6	↓↓↓	EAC	Ibiebele et al. (2011) and Mayne et al. (2001)
B12	↓↓↓	EAC	Chang et al. (2015)

↓ low intake, ↑ high intake, *ESCC* esophageal squamous cell carcinoma, *EAC* esophageal adenocarcinoma

pathogenesis of EAC and showed that vitamin B6 does not appear to influence p53, suggesting the existence of alternative molecular mechanisms underlying carcinogenesis of EAC.

Another cofactor, vitamin B12 (cobalamin), is essential for methionine synthase (MTR) activity that transforms homocysteine to methionine. Thus, B12 regulates the levels of SAM and SAH, which are inhibitors of enzymes catalyzing DNA methylation (Stefanska et al. 2012).

Major sources of vitamin B12 are liver, meat, milk, eggs, fish, oysters, and clams. Although vitamin B12 is known for its absence in plant source foods, edible species of mushrooms contain noticeable amounts (Nakos et al. 2017).

In a recent study, Chang et al. (2015) found a positive association between plasma vitamin B12 levels and esophageal cancer in a population-based case-control study through standardized epidemiologic questionnaires concomitantly with blood sample collections during the interviews.

Single Nucleotide Polymorphisms of Genes in the One-Carbon Metabolism

Many studies have described single nucleotide polymorphisms (SNPs) in genes involved in the one-carbon metabolism associated with cancer susceptibility, including methylenetetrahydrofolate reductase (*MTHFR*), methionine synthase (*MTR*), and methionine synthase reductase (*MTRR*) (Chang et al. 2014) (Fig. 1). Table 2 summarizes the SNPs of genes in the one-carbon metabolism and esophageal cancer risk.

MTHFR encodes an enzyme involved in DNA synthesis and responsible for the production of SAM. In esophageal cancer, the most commonly studied SNP of *MTHFR* gene, C677T (rs1801133), results in an alanine to valine substitution, leading to reduced MTHFR enzyme activity. The homozygote and heterozygote forms of C667T mutation yield 30% and 65% of wild-type activity, respectively. Moreover, the reduction of MTHFR activity decreased 5-MTHF and increased

Table 2 Single nucleotide polymorphisms of genes in the one-carbon metabolism and esophageal cancer risk

Gene	SNP	Esophageal cancer risk	References
MTHFR	*C677T* (rs1801133)	ESCC	Keld et al. (2014)
	A1298C (rs1801131)		Chang et al. (2014)
			Zhao et al. (2011)
			Li et al. (2011)
			Tang et al. (2014b)
MTR	*A2756G* (rs1805087)	No association	Chang et al. (2014)
MTRR	*A66G* (rs1801394)	No association	Chang et al. (2014)
	C524T (rs1532268)		

MTHFR methylenetetrahydrofolate reductase, *MTR* methionine synthase, *MTRR* methionine synthase reductase, *ESCC* esophageal squamous cell carcinoma

5mTHF levels in blood (Zhao et al. 2015). *MTHFR* C677T polymorphism increased the risk of developing ESCC in different populations (Mayne et al. 2001; Keld et al. 2014; Chang et al. 2014; Zhao et al. 2011).

The second more common *MTHFR* polymorphism characterized in esophageal cancer is a missense mutation at *A1298C* (rs1801131) that causes a glutamate to alanine substitution. Several case-control studies, mainly from China, have reported that both MTHFR 677TT and 1298CC variant genotypes were associated with ESCC (Tang et al. 2014b; Chang et al. 2014; Fang et al. 2011).

MTR and MTRR are two other important enzymes involved in one-carbon metabolism. MTR catalyzes the methylation of homocysteine to methionine. *MTR* A2756G (rs1805087), a substitution of aspartic acid to glycine, has been largely studied with regard to cancer risk. However, no apparent associations have been observed between rs1805087 and esophageal cancer (Ding et al. 2013; Chang et al. 2014). On the other hand, MTRR regenerates a functional MTR via reductive methylation. Two common polymorphisms, *MTRR* A66G (rs1801394) and C524T (rs1532268), have been found to regenerate MTR less efficiently (Amigou et al. 2012). Consistent with previous findings, Chang et al. (2014) reported a stronger association among individuals with lower plasma folate levels and alcohol consumption appeared to have modified the odds ratios relating *MTRR* (rs1801394) conversion of isoleucine to methionine in esophageal cancer, suggesting potential interactions between alcohol intake and polymorphism of genes of the one-carbon metabolism.

Agents That Act as DNMT Inhibitors and Modulators of Histone Modifications

The roles of nutrients such as genistein, epigallocatechin-3-gallate, resveratrol, and curcumin as DNA methyltransferase inhibitors and/or modulators of histone modifications are observed in Fig. 2. Additionally, Table 3 summarizes the dietary agents that act as DNMT inhibitors and modulators of histone modifications in esophageal cancer.

Genistein

Genistein is a well-known isoflavonoid, which belongs to the flavonoid subgroup of polyphenols, reported as the main anticancer element found in soybean and soy food derivatives (Maru et al. 2016). Genistein is also classified as a phytoestrogen, having a significant value in preventing hormone-dependent cancers, especially because of its property as an estrogen receptor (ER) antagonist (Lee et al. 2016).

Similarly, genistein's chemopreventive effects have also been reported in non-hormone-dependent cancers, including many cellular functions such as cell proliferation, apoptosis, cell cycle progression, migration, invasion, and metastasis (Lee et al. 2016; Pavese et al. 2014). Tang et al. (2014a) described that habitual

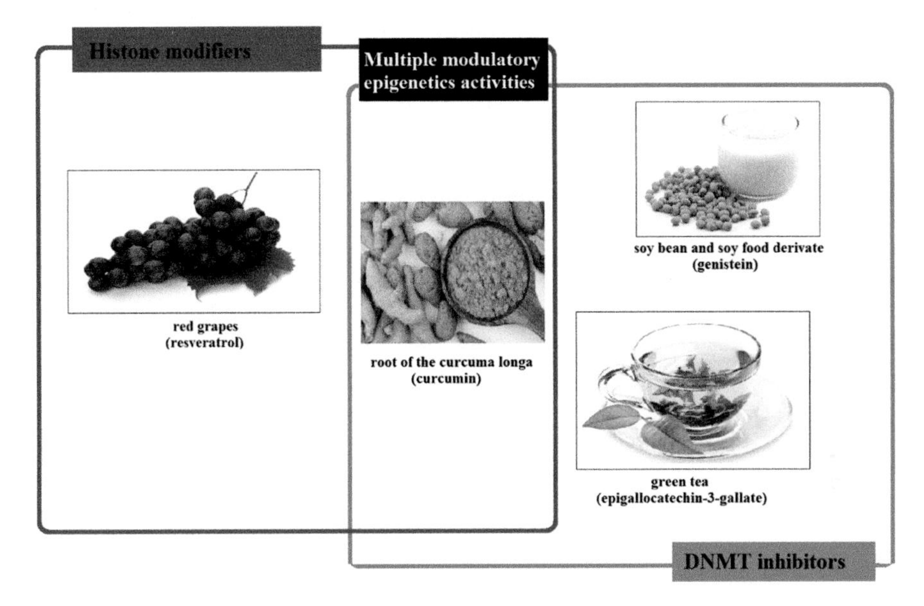

Fig. 2 Nutrients directly associated with esophageal cancer and their respective epigenetic effects. Nutrients directly associated with esophageal cancer and their principal dietary sources, classified by their respective epigenetic effects represented by histone modifiers (resveratrol) in *red frame*, DNMT inhibitors (genistein, epigallocatechin-3-gallate) in *blue frame*, multiple epigenetic modulatory activities (curcumin) at the intersection of both frames

consumption of genistein appears to be associated with reduced risk of esophageal cancer in northwestern China.

In recent years, a large number of the studies have found genistein capable of modulating epigenetic mechanisms such as DNA methylation and histone acetylation in many types of cancer (Kavoosi et al. 2016; Lynch 2016; Mukherjee et al. 2015; Hwang and Choi 2015).

In esophageal cancer, a preclinical study examined the effects of genistein in an ESCC cell line (KYSE 510) and described that genistein was capable of inhibiting DNMT, progressively reversing the hypermethylation status of *RARβ*, $p16^{INK4a}$, and *MGMT* genes. In addition, the reactivation of the expression of these genes may also be a consequence of the HDAC inhibitory effect of genistein (Fang et al. 2005a,). Moreover, the treatment of KYSE 510 cells with a combination of genistein (5 μmol/L) and trichostatin A (TSA), an organic selective inhibitor of the class I and II mammalian HDACs, synergistically increased the mRNA levels of $p16^{INK4a}$, *RARβ*, and *MGMT*. The level of unmethylation-specific *RARβ* DNA band was increased by genistein but not by TSA (Fang et al. 2007). These results suggest that hypermethylation of *RARβ*, *p16INK4a*, and *MGMT* genes may be prevented or reversed by genistein for the prevention of ESCC, and the epigenetic mechanism of genistein needs to be further elucidated in esophageal cancer.

Table 3 Dietary agents that act as DNMT inhibitors and modulators of histone modifications in esophageal cancer

	Effect	Mechanism	Model	References
Genistein	↓ Cell proliferation	Inhibition of DNMT activity	ESCC cell line (KYSE-510)	Fang et al. (2005a)
	↓ Colonies' formation	Reversion of DNA hypermethylation reactivation of *RARβ*, *p16*^{INK4a}, and *MGMT* genes		
	↑ Capacity to repair O^{6}-methylguanine	Reversion of hypermethylation and reactivation of the *MGMT* gene	ESCC cell line (KYSE-510)	Fang et al. (2005b)
	↓ Carcinogenesis			
	↓ Carcinogenesis	Increase of mRNA levels of *p16*^{INK4a}, *RARβ*, and *MGMT*	ESCC cell line (KYSE-510)	Fang et al. 2007
	↑ Capacity to repair O6-methylguanine			
	↓ Colonies' formation	Inhibition of HDAC activity		
	↓ Cell proliferation			
Epigallocatechin-3-gallate	Cell cycle arrest at the G1 phase	Increasing the expression of cleaved caspase-3 and decreasing VEGF protein levels	ESCC cell lines (Eca-109 and TE-1)	Liu et al. (2015b)
	↑ Apoptosis			
	↓ Cellular concentrations of 5-methylcytosine	Inhibits DNMT activity associated with promoter demethylation and reactivation of *p16*^{INK4a}, *RAR*β, *MGMT*, human mutL homolog 1 (*hMLH1*) and *GSTP1*	ESCC cell line (KYSE-150) and *in vivo*	Fang et al. (2007)

(*continued*)

Table 3 (continued)

	Effect	Mechanism	Model	References
Resveratrol	Cell cycle arrest at the sub-G1 phase	Elevation of LC3-II in autophagosomes, upregulation of multiple key autophagosome-regulatory proteins, such as Beclin-1 and ATG5 and the formation of AVO	ESCC cell lines (EC109, EC9706)	Tang et al. (2013)
	↑ Apoptosis			
	↑ Cytotoxicity			
	↓ Colonies' formation			
	↑ Autophagy			
		AMPK/mTOR pathway independent		
	↑ Apoptosis ↓ Adhesion	Upregulation of GRP78, ATF6, p-PERK, p-eIF2α, and CHOP protein expression downregulation Bcl-2 expression	ESCC cell lines (EC109 and TE-1) and *in vivo*	Feng et al. (2016)
	↓ Migration			
	↓ Colonies' formation			
		Upregulation PUMA expression		
Curcumin	↓ Colonies' formation	Downregulated of Notch-1 signaling	EAC cell lines (TE-7, TE-10 and ESO-1)	Subramaniam et al. (2012)
	↑ Apoptosis			
	↓ Viability	ERK1/ERK2, JUN, and COX-2 signaling	ESCC (TE-8) and EAC cell lines (SKGT-4) and *in vivo*	Ye et al. (2012)
	↓ Invasion			
	↓ Tumor growth			

↓ inhibition, ↑ stimulation, *DNMT* DNA methyltransferase, *ESCC* esophageal squamous cell carcinoma, *HDACs* histone deacetylases, *hMLH1* human mutL homolog 1, *NOTCH-1* Notch homolog 1 gene, *MGMT* O^6-methylguanine DNA methyltransferase, *RARβ* retinoid acid receptor-p, *GSTP1* glutathione S-transferase Pi 1, *ERK1/ERK2* extracellular signal-regulated kinase 1, *JUN* Jun proto-oncogene, *COX-2* cytochrome c oxidase subunit II, *LC3-II* LC3-phosphatidylethanolamine conjugate, *ATG5* autophagy related 5, *AVO* acidic vesicular organelles, *AMPK/mTOR* adenosine monophosphate-activated protein kinase/mechanistic target of rapamycin, *GRP78* glucose-regulated protein 78 kDa, *ATF6* activating transcription factor 6, *p-PERK* phosphorylated-PKR-like ER kinase, *p-Eif2α* phosphorylated eukaryotic translational initiation factor, *CHOP* transcription factor C/EBP homologous protein, *Bcl-2* B-cell lymphoma-2, *PUMA* p53 upregulation modulator of apoptosis.

Epigallocatechin-3-Gallate

The epigallocatechin-3-gallate (EGCG) is the most abundant flavonoid in green tea (Gilbert and Liu 2010). Their regular consumption may decrease the risk of many cancer types, including esophageal cancer (Khan and Mukhtar 2010). However, few studies concerning the role and mechanism of action of EGCG in esophageal cancer are available.

Liu et al. (2015) showed that EGCG inhibited the proliferation of ESCC cell lines (Eca-109 and TE-1) in a time- and dose-dependent manner. Tumor cells were arrested in the G1 phase, and apoptosis was accompanied by ROS production and caspase-3 cleavage. Moreover, EGCG significantly inhibited the growth of Eca-109 tumors by increasing the expression of cleaved caspase-3 and decreasing VEGF protein levels in a mouse model. Taken together, their findings suggest that EGCG may have future clinical applications for novel approaches in ESCC treatment.

EGCG has been shown capable of directly inhibiting DNMT activity resulting in decreased cellular concentrations of 5-methylcytosine (Meeran et al. 2011). In human cell lines, EGCG (20 μmol/L) inhibits DNMT activity in different cancer types, including esophageal (KYSE-150), colon (HT-29), prostate (PC-3), and breast (MCF-7 and MDA-MB-231) cancer cells (Ong et al. 2011).

Moreover, an indirect DNMT inhibition by flavan-3-ols is mediated by an increase in SAH concentration. This increase in SAH may be caused by catechol-*O*-methyltransferase (COMT)-mediated methylation of these flavonoids (Saavedra et al. 2009; Busch et al. 2015).

Resveratrol

Resveratrol (Res) is a natural polyphenolic compound present in numerous plant species, including the skin of red grapes, peanuts, berries, pistachio nuts, dried roots of the Japanese knotweed (*Polygonum cuspidatum*), and others (Patel et al. 2011; Sales and Resurreccion 2014).

According to Tang et al. (2013), Res induced ESCC cell cycle arrest at the sub-G1 phase leading to subsequent apoptosis. However, mechanistically, resveratrol-induced autophagy in the ESCC cells is AMPK/mTOR pathway independent.

Feng et al. (2016) showed that pterostilbene (PTE), a natural dimethylated resveratrol analogue, exerts an effective anticancer activity against human esophageal cancer cells (EC109 and TE-1) both in vitro and in vivo with possible activation of endoplasmic reticulum stress signaling pathways.

Interestingly, Yang et al. (2013) showed that cancer cells treated with Res underwent cell cycle arrest at the G1 phase, leading to cellular senescence, and the depletion of sirtuin 1 (SIRT1) reversed the Res effects. These data suggest that Res inhibits the proliferation of cancer cells in a SIRT1-dependent manner. SIRT1, a class III HDAC, has been reported to be a key target of Res in several tumor models (Frazzi et al. 2013; Ulrich et al. 2006). Taken together, these findings indicate that

Res exerts SIRT1-dependent inhibitory effects on cancer cells and suggests a potential therapeutic value for Res in cancer.

In esophageal cancer, the expression levels of SIRT1 protein were significantly correlated with TNM stage and lymph node status of ESCC patients. Upregulation of SIRT1 expression has been associated with poor clinical prognosis. However, the effect of Res in SIRT1 inhibition in esophageal cancer cells is still unknown.

Curcumin

Curcumin (diferuloylmethane) is a natural product extracts from the rhizomes (root) of *Curcuma longa* L., commonly known as turmeric, used to overcome inflammatory conditions (Lestari and Indrayanto 2014).

Among the pharmacological properties of curcumin, reduction or inhibition of cellular proliferation and enhancement of apoptosis in various cancer cell lines and xenograft tumors, including esophageal cancer, have been confirmed (Mahran et al. 2017). Subramaniam et al. (2012) found that efficacious inhibition of esophageal cancer cell line (TE-7, TE-10, and ESO-1) growth by curcumin targets Notch-1. These data suggest that Notch signaling inhibition is a novel mechanism of action for curcumin during therapeutic intervention in esophageal cancers.

In addition, Ye et al. (2012) established that curcumin individually (40 μmol/L) and combined with EGCG and lovastatin treatment significantly reduced the viability and invasion capacity of esophageal cancer cells (TE-8 and SKGT-4). At the molecular level, inhibition of the expression of phosphorylated ERK1/ERK2, JUN, and COX-2 and induced caspase-3 expression were demonstrated in vitro. Moreover, 49.4% and 77.6% of esophageal cancer tissue samples expressed phosphorylated ERK1/ERK2 and COX-2 protein, respectively. These results suggest that curcumin is a strong candidate for therapeutic applications for esophageal cancer.

The molecular effects of curcumin may be a consequence of epigenetic modulation, since curcumin exerts multiple epigenetic modulatory activities as inhibitors of DNMT, HDAC, and HAT. Although the actions of HATs and HDACs are antagonistic, both have been associated with cancer, and curcumin may allow suppression of tumor growth (Khan et al. 2016). However, the direct epigenetic effect of curcumin in esophageal carcinogenesis still remains unclear.

Recent studies have been conducted to overcome the problems of curcumin bioavailability. Preclinical and clinical studies have reported success in approaches combining curcumin with other treatments (Imran et al. 2016). Milano et al. (2013) utilized a highly absorbed form of curcumin, named Theracurmin, to investigate its effect against esophageal cancer. As a result, they found that nanocurcumin significantly decreased the proliferation of EAC cells (OE33 and OE19), but did not affect a normal esophageal cell line (HET-1A). Therefore, the use of nanocurcumin is possibly more effective than its free form in the treatment of esophageal cancer. Its epigenetic effects need to be further examined.

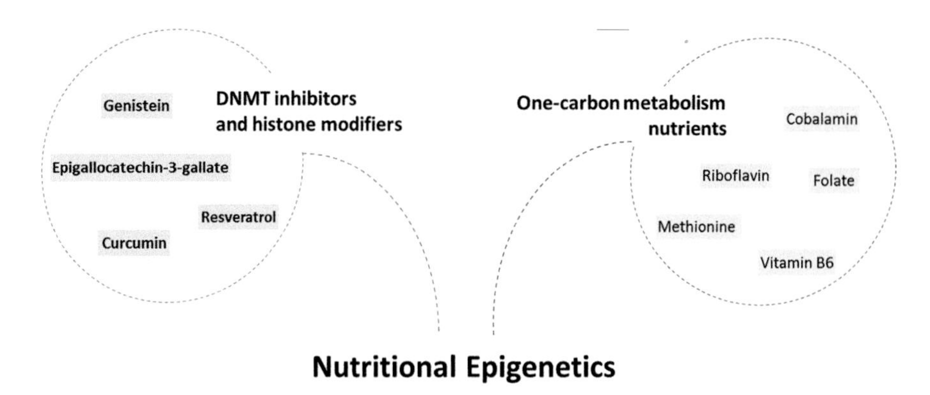

Fig. 3 Epigenetic events and the modulatory effects of nutrients. Nutritional epigenetics showing the epigenetic event regulated by dietary components inserting bioactive compounds in the light of new approaches on how they can alter epigenetic patterns and restore gene function

Conclusions

Dietary components contributing to epigenetic alterations and nutritional epigenetic interventions have been demonstrated to be useful tools in cancer prevention and treatment. Therefore, identifying specific nutrients that might influence epigenetic events is important, especially their implications in esophageal carcinogenesis that could be essential for developing personalized medicine approaches (Fig. 3).

There are still numerous questions to be answered about the exact role of epigenetics in esophageal cancer and the modulatory effects of nutrients. Nutritional epigenetic studies may help answer many questions such as how dietary components regulate epigenetic events, alter epigenetic patterns, and restore gene function and which gene-specific epigenetic inhibitors can be developed.

Finally, considering nutritional epigenetics in the era of increased consumption of nutrients and food supplementation highlights the importance of the differential effects of medicinal and toxic doses of nutrients, as both high and low intake can lead to the development of esophageal cancer as well as other types of cancer.

Dictionary of Terms

- **TNM stage** – Classification system which uses an alphanumeric method for indicating size (T) of original tumor, invasion (N) of proximal or distal lymph node, and affection of other areas of the body (M) causing metastasis.
- **Personalized nutrition** – Recommendations for individuals, personalized or individualized nutrition, based on epigenetic profile.
- **Esophagectomy** – Removal of the entire or part of the esophagus and often nearby areas as lymph node and/or part of the stomach by surgery.

- **Malnutrition** – An unhealthy condition caused by unbalanced consumption of nutrients or food.
- **Chemopreventive effects** – The use of drugs, nutrients, or other agents that could reduce the risk of or delay the development recurrence of cancer.
- **Bioactive food components** – Biocompounds present in foods that could modulate depending on the amount of diary intake, metabolic processes resulting in the improvement of health state.

Key Facts

Precision nutrition

- Treatment of the individual giving to their phenotype and genetic characteristics, pointed at both treatment and prevention of disease.
- Nutrigenetics studies the effect of individual genetic variation on differences in response to dietary components, nutrient requirements, and predisposition to disease.
- The genetic information, combined with anthropometric, biochemical, and dietary assessments, can improve recommendations to a personalized/individualized diet.
- Novel tools are required to support personalized nutrition to recognize new biomarkers in order for this method to be useful to individuals.
- Personalized nutrition is helping the development of functional foods specific at improvement of health status.

Summary Points

- Dietary components contributing to epigenetic alterations and nutritional epigenetic interventions have been shown to be useful tools in cancer prevention and treatment.
- One-carbon metabolism consists of the folate and methionine cycle and utilizes a variety of nutrients capable of perturbing DNA and histone methylation patterns.
- Inadequate folate intake has been implicated mainly in the development of cancer associated with rapid tissue renewal, such as esophageal cancer.
- Nutrients in the one-carbon metabolism may be involved in cancer development, based upon limited studies of the intakes of its cofactors and esophageal cancer.
- Studies have described single nucleotide polymorphisms of genes involved in the one-carbon metabolism associated with cancer susceptibility, including *MTHFR*, *MTR*, and *MTRR*.
- The roles of nutrients as DNMT inhibitors and modulators of histone modifications have also been reported in esophageal cancer, such as genistein, epigallocatechin-3-gallate, resveratrol, and curcumin.

- Hypermethylation of *RARβ*, *p16INK4a*, and *MGMT* may be avoided or reversed by genistein for the treatment of ESCC.
- Epigallocatechin-3-gallate has been shown to directly inhibit DNMT activity leading to decreased cellular concentrations of 5-methylcytosine in cancer cell lines, including esophageal cancer.
- The Sirt1-dependent inhibitory effects of Res on cancer cells suggest a therapeutic role for Res in cancer.
- Curcumin exerts multiple epigenetic modulator activities as DNMT, HDAC, and HAT inhibitors.
- Curcumin proves to be a strong candidate for therapeutic applications for esophageal cancer.
- Nutritional epigenetics in the era of increased consumption of nutrients and food supplementation highlights the importance of the differential effects of medicinal and toxic doses of nutrients.

References

Amigou A, Jérémie R, Laurent O et al (2012) Folic acid supplementation, MTHFR and MTRR polymorphisms, and the risk of childhood leukemia: the ESCALE study (SFCE). Cancer Causes Control 23(8):1265–1277

Anderson OS, Sant KE, Dolinoy DC (2012) Nutrition and epigenetics: an interplay of dietary methyl donors, one-carbon metabolism and DNA methylation. J Nutr Biochem 23(8):853–859

Balbuena L, Casson AG (2010) Dietary folate and vitamin B6 are not associated with p53 mutations in esophageal adenocarcinoma. Mol Carcinog 49(3):211–214

Baylin SB, Jones PA (2016) Epigenetic determinants of cancer. Cold Spring Harb Perspect Biol 8(9):a019505

Brown-Borg HM, Buffenstein R (2016) Cutting back on the essentials: can manipulating intake of specific amino acids modulate health and lifespan? Age Res Rev. pii: S1568–1637 (16):30204–30205

Busch C, Burkard M, Leischner C et al (2015) Epigenetic activities of flavonoids in the prevention and treatment of cancer. Clin Epigenetics 7(1):64

Calcagno DQ, de Arruda Cardoso Smith M, Burbano RR (2015) Cancer type-specific epigenetic changes: gastric cancer. Methods Mol Biol 1238:79–101

Chang S, Chang PY, Butler B et al (2014) Single nucleotide polymorphisms of one-carbon metabolism and cancers of the esophagus, stomach, and liver in a Chinese population. PLoS One 9(10):e109235

Chang SC, Goldstein BY, Mu L et al (2015) Plasma folate, vitamin B12, and homocysteine and cancers of the esophagus, stomach, and liver in a Chinese population. Nutr Cancer 67(2):212–223

Cohen DJ, Leichman L (2015) Controversies in the treatment of local and locally advanced gastric and esophageal cancers. J Clin Oncol 33(16):1754–1759

Ding W, Dong-lei Z, Xun J et al (2013) Methionine synthase A2756G polymorphism and risk of colorectal adenoma and cancer: evidence based on 27 studies. PLoS One 8(4):e60508

Ducker GS, Rabinowitz JD (2017) One-carbon metabolism in health and disease. Cell Metab 25(1):27–42

Duthie SJ (2011) Folate and cancer: how DNA damage, repair and methylation impact on colon carcinogenesis. J Inherit Metab Dis 34(1):101–109

Falk GW (2009) Risk factors for esophageal cancer development. S Oncol Clin N Am 18(3):469–485

Fang MZ, Chen D, Sun Y et al (2005a) Reversal of hypermethylation and reactivation of p16 INK4a, RAR b and MGMT genes by genistein and other isoflavones from soy. Clin Cancer Res 11(19):7033–7041

Fang MZ, Jin Z, Wang Y et al (2005b) Promoter hypermethylation and inactivation of O(6)-methylguanine-DNA methyltransferase in esophageal squamous cell carcinomas and its reactivation in cell lines. In J Oncol 26(3):615–622

Fang M, Chen D, Yang CS (2007) Dietary polyphenols may affect DNA methylation. J Nutr 137(Suppl 1):223S–228S

Fang Y, Xiao F, An Z et al (2011) Systematic review on the relationship between genetic polymorphisms of methylenetetrahydrofolate reductase and esophageal squamous cell carcinoma. Asian Pac J Cancer Prev 12:1861–1866

Fanidi A, Relton C, Ueland PM et al (2015) A prospective study of one-carbon metabolism biomarkers and cancer of the head and neck and esophagus. Int J Cancer 136(4):915–927

Feng Y, Yang Y, Fan C et al (2016) Pterostilbene inhibits the growth of human esophageal cancer cells by regulating endoplasmic reticulum stress. Cell Phys Biochem 38(3):1226–1244

Ferguson LR, De Caterina R, Gorman U et al (2016) Guide and position of the International Society of nutrigenetics/nutrigenomics on personalised nutrition: part 1 – fields of precision nutrition. J Nutrigenet Nutrigenomics 9(1):12–27

Frazzi R, Valli R, Tamagnini I et al (2013) Resveratrol-mediated apoptosis of Hodgkin lymphoma cells involves SIRT1 inhibition and FOXO3a hyperacetylation. In J Cancer 132(5):1013–1021

Gilbert ER, Liu D (2010) Flavonoids influence epigenetic-modifying enzyme activity: structure – function relationships and the therapeutic potential for cancer. Curr Med Chem 17(17):1756–1768

Gruber BM (2016) B-group vitamins: chemoprevention? Adv Clin Exp Med 25(3):561–568

Hwang KA, Choi KC (2015) Anticarcinogenic effects of dietary phytoestrogens and their chemopreventive mechanisms. Nutr Cancer 67(5):796–803

Ibiebele TI, Highes MC, Pandeya N et al (2011) High intake of folate from food sources is associated with reduced risk of esophageal cancer in an Australian population. J Nutr 141:274–283

Imran M, Saeed F, Nadeem M, et al (2016) Cucurmin; anticancer and antitumor perspectives – a comprehensive review. Crit Rev Food Sci Nutr 22:0

Kavoosi F, Dastjerdi MN, Valiani A et al (2016) Genistein potentiates the effect of 17-beta estradiol on human hepatocellular carcinoma cell line. Adv Biom Res 5(1):133

Keld R, Thian M, Hau C et al (2014) Polymorphisms of MTHFR and susceptibility to oesophageal adenocarcinoma in a Caucasian United Kingdom population. World J Gastroenterol 20(34):12212–12216

Khan N, Mukhtar H (2010) Cancer and metastasis: prevention and treatment by green tea. Cancer Metastasis Rev 29(3):435–445

Khan S, Karmokar A, Howells L et al (2016) Targeting cancer stem-like cells using dietary-derived agents – where are we now? Mol Nutr Food Res 60(6):1295–1309

Lee GA, Hwang KA, Choi KC (2016) Roles of dietary phytoestrogens on the regulation of epithelial-mesenchymal transition in diverse cancer metastasis. Toxicology 8(6):162

Lestari ML, Indrayanto G (2014) Curcumin. Profiles Drug Subst Excip Relat Methodol 39:113–204

Li Y, Wicha MS, Schwartz SJ et al (2011) Implications of cancer stem cell theory for cancer chemoprevention by natural dietary compounds. J Nutr Biochem 22(9):799–806

Li Q, Mao L, Wang R (2014) Overexpression of S-adenosylhomocysteine hydrolase (SAHH) in esophageal squamous cell carcinoma (ESCC) cell lines: effects on apoptosis, migration and adhesion of cells. Mol Biol Rep 41:2409–2417

Liu Y, Chen H, Sun Z et al (2015a) Molecular mechanisms of ethanol-associated oro-esophageal squamous cell carcinoma. Cancer Lett 361(2):164–173

Liu L, Hou L, Gu S et al (2015b) Molecular mechanism of epigallocatechin-3-gallate in human esophageal squamous cell carcinoma in vitro and in vivo. Oncol Rep 33(1):297–303
Locasale JW (2013) Serine, glycine and one-carbon units: cancer metabolism in full circle. Nat Rev Cancer 3(8):572–583
Lynch KL (2016) Is obesity associated with Barrett's esophagus and esophageal adenocarcinoma? Gastroenterol Clin N Am 45(4):615–624
Mahran RI, Hagras MM, Sun D et al (2017) Bringing curcumin to the clinic in cancer prevention: a review of strategies to enhance bioavailability and efficacy. AAPS J 19(1):54–81
Martinez JA, Navas-Carretero S, Saris WH et al (2014) Personalized weight loss strategies – the role of macronutrient distribution. Nat Rev Endocrinol 10(12):749–760
Maru GB, Hudlikar RR, Kumar G et al (2016) Understanding the molecular mechanisms of cancer prevention by dietary phytochemicals: from experimental models to clinical trials. World J Biol Chem 7(1):88–100
Mayne ST, Risch HA, Dubrow R et al (2001) Nutrient intake and risk of subtypes of esophageal and gastric cancer. Cancer Epidemiol Biomarkers Prev 10(10):1055–1062
Meeran SM, Patel SN, Tak-Hang C et al (2011) A novel prodrug of epigallocatechin-3-gallate: differential epigenetic *hTERT* repression in human breast cancer cells. Cancer Prev Res 4(8):1243–1254
Mentch SJ, Locasale JW (2015) One-carbon metabolism and epigenetics: understanding the specificity. Ann N Y Acad Sci 1363:91–98
Milano F, Mari L, Van de Luijtgaarden W et al (2013) Nano-curcumin inhibits proliferation of esophageal adenocarcinoma cells and enhances the T cell mediated immune response. Front Oncol 3:137
Mudryj AN, de Groh M, Aukema HM et al (2016) Folate intakes from diet and supplements may place certain Canadians at risk for folic acid toxicity. Br J Nutr 16(7):1236–1245
Mukherjee N, Kumar AP, Ghosh R (2015) DNA methylation and flavonoids in genitourinary cancers. Curr Pharmacol Rep 1(2):112–120
Nakos M, Pepelanova I, Beutel S et al (2017) Isolation and analysis of vitamin B12 from plant samples. Food Chem 216:301–308
Obeid R (2013) The metabolic burden of methyl donor deficiency with focus on the betaine homocysteine methyltransferase pathway. Nutrition 5(9):3481–3495
Olefson S, Moss SF (2015) Obesity and related risk factors in gastric cardia adenocarcinoma. Gastric Cancer 18(1):23–32
Ong TP, Moreno FS, Ross SA (2011) Targeting the epigenome with bioactive food components for cancer prevention. J Nutrigenet Nutrigenomics 4(5):275–292
Patel KR, Scott E, Brown VA et al (2011) Clinical trials of resveratrol. Ann N Y Acad Sci 1215(1):161–169
Pavese JM, Krishna SN, Bergan RC (2014) Genistein inhibits human prostate cancer cell detachment, invasion, and metastasis. Am J Clin Nutr 100(Suppl 1):431S–436S
Peng Q, Chen H, Huo JR (2016) Alcohol consumption and corresponding factors: a novel perspective on the risk factors of esophageal cancer (review). Oncol Lett 11(5):3231–3239
Reim D, Friess H (2016) Feeding challenges in patients with esophageal and gastroesophageal cancers. Gastrointest Tumors 2(4):166–177
Rice TW, Blackstone EH, Rusch VW (2010) 7th edition of the AJCC cancer staging manual: esophagus and esophagogastric junction. Ann Surg Oncol 17(7):1721–1724
Saavedra OM, Isakovic L, Llewellyn DB et al (2009) SAR around (l)-S-adenosyl-L-homocysteine, an inhibitor of human DNA methyltransferase (DNMT) enzymes. Bioorg Med Chem Lett 19(10):2747–2751
Sales JM, Resurreccion AVA (2014) Resveratrol in peanuts. Crit Rev Food Sci Nutr 54(6):734–770
Schweinberger BM, Wyse AT (2016) Mechanistic basis of hypermethioninemia. Amino Acids 48(11):2479–2489
Seitz HK, Stickel F (2010) Acetaldehyde as an underestimated risk factor for cancer development: role of genetics in ethanol metabolism. Genes Nutr 5(2):121–128

Sharp L, Carsin AE, Cantwell MM et al (2013) Intakes of dietary folate and other B vitamins are associated with risks of esophageal adenocarcinoma, Barrett's esophagus, and reflux esophagitis. J Nutr 143(12):1966–1973

Stefanska B, Karlic H, Varga F (2012) Epigenetic mechanisms in anti-cancer actions of bioactive food components – the implications in cancer prevention. Br J Pharmacol 167(2):279–297

Su X, Wellen KE, Rabinowitz JD (2016) Metabolic control of methylation and acetylation. Curr Opin Chem Biol 30:52–60

Subramaniam D, Ponnurangam S, Ramamoorthy P et al (2012) Curcumin induces cell death in esophageal cancer cells through modulating Notch signaling. PLoS One 7(2):1–11

Tang Q, Li G, Wei X et al (2013) Resveratrol-induced apoptosis is enhanced by inhibition of autophagy in esophageal squamous cell carcinoma. Cancer Lett 336(2):325–337

Tang L, Lee AH, Xu F et al (2014a) Soya and isoflavone intakes associated with reduced risk of oesophageal cancer in north-west China. Public Health Nutr 18(1):130–134

Tang M, Wang SQ, Liu BJ et al (2014b) The methylenetetrahydrofolate reductase (MTHFR) C677T polymorphism and tumor risk: evidence from 134 case – control studies. Mol Biol Rep 2014:4659–4673

Ulrich S, Rau O, Loitsch SM et al (2006) Peroxisome proliferator-activated receptor gamma as a molecular target of resveratrol-induced modulation of polyamine metabolism. Cancer Res 66(14):7348–7354

Xiao Q, Freedman ND, Ren J et al (2014) Intakes of folate, methionine, vitamin B6 and vitamin B12 with risk of esophageal and gastric cancer in a large cohort study. Br J Cancer 110(5):1328–1333

Yang Q, Wang B, Zang W et al (2013) Resveratrol inhibits the growth of gastric cancer by inducing G1 phase arrest and senescence in a sirt1-dependent manner. PLoS One 8(11):e70627. D Heymann, ed

Ye F, Zhang GH, Guan BX et al (2012) Suppression of esophageal cancer cell growth using curcumin, (−)-epigallocatechin-3-gallate and lovastatin. World J Gastroenterol 18(2):126

Zhao P, Fengsong L, Li Z et al (2011) Folate intake, methylenetetrahydrofolate reductase polymorphisms and risk of esophageal cancer. Asian Pac J Cancer Prev 12(8):019–023

Zhao T, Gu D, Xu Z et al (2015) Polymorphism in one-carbon metabolism pathway affects survival of gastric cancer patients: large and comprehensive study. Oncotarget 6(11):9564–9576

The Methyl-CpG-Binding Domain (MBD) Protein Family: An Overview and Dietary Influences

81

Carolina Oliveira Gigek, Elizabeth Suchi Chen, Gaspar Jesus Lopes-Filho, and Marilia Arruda Cardoso Smith

Contents

Abstract

The machinery involved in controlling gene expression via DNA methylation includes DNA methyltransferases and seven methyl-CpG-binding domain (MBD) proteins. By binding to methyl-CpG marks, the MBD proteins can recruit the correct epigenetic machinery to read and to interpret the epigenetic landscape. The MBD family comprises seven proteins: MeCP2 and the MBD1–6, each one

C. O. Gigek (✉)
Division of Genetics, Department of Morphology and Genetics, Universidade Federal de São Paulo (UNIFESP), São Paulo, SP, Brazil

Division of Surgical Gastroenterology, Department of Surgery, Universidade Federal de São Paulo (UNIFESP), São Paulo, SP, Brazil
e-mail: carolina.gigek@unifesp.br

E. S. Chen · M. A. C. Smith
Division of Genetics, Department of Morphology and Genetics, Universidade Federal de São Paulo (UNIFESP), São Paulo, SP, Brazil
e-mail: eschen@unifesp.br; macsmith@unifesp.br

G. J. Lopes-Filho
Division of Surgical Gastroenterology, Department of Surgery, Universidade Federal de São Paulo (UNIFESP), São Paulo, SP, Brazil
e-mail: gasparlopes@uol.com.br

V. B. Patel, V. R. Preedy (eds.), *Handbook of Nutrition, Diet, and Epigenetics*,
https://doi.org/10.1007/978-3-319-55530-0_79

with a main function and specific characteristics. The family grouping characteristic is the conserved methyl-binding domain sequence that can recognize a single CpG. Dietary intake contributes to the epigenetic equilibrium by preventing or promoting DNA methylation and influencing epigenetic machinery proteins; expression of transgenerational effects of epigenetic marks and regulators is also addressed.

Keywords
DNA methylation · MeCP2 · MBD1 · MBD2 · MBD3 · MBD4 · MBD5 · MBD6 · Methionine · Choline · Folate · High fat · Transgenerational epigenetics · Offspring

List of Abbreviations

BAZ2A	Bromodomain adjacent to zinc finger domain 2A
BAZ2B	Bromodomain adjacent to zinc finger domain 2B
BR	Black raspberries
BRCA1	Breast cancer 1 gene
CpG	5′-Citosine–phosphate–Guanine–3′
DNA	Deoxyribonucleic acid
DNMT	DNA methyltransferases
Glut4	Glucose transporter gene
HAT	Histone acetyltransferases
HDAC	Histone deacetylases
HDM	Histone demethylases
HF	High fat
HMT	Histone methyltransferases
ICR	Imprinting control region
MBD	Methyl-CpG-binding domain
MeCP2	Methyl-CpG-binding protein 2
miRNA	MicroRNAs
mRNA	Messenger ribonucleic acid
NuRD	Nucleosome remodeling and histone deacetylase
RefSeq	Reference sequence database
RING	Really interesting new gene
SET	Suppressor of variegation 3–9, enhancer of zeste, and trithorax group
SETDB1	SET domain bifurcated 1
SETDB2	SET domain bifurcated 1
SRA	SET and RING finger-associated domain

Introduction

Epigenetics was first introduced as a field to study mechanisms and processes between genotype and phenotype during embryogenesis (Waddington 1942). Waddington's concept about cell fate describes that "ridges and valleys" (landscape) will define the cell fate; therefore, the genetic information in the cell is read and processed by epigenetics processes, shaping cell differentiation and specialization. Nowadays, the landscape pattern/epigenetic marks are the mechanisms controlling gene expression: DNA methylation, histone modifications, and microRNAs (da Silva Oliveira et al. 2016). In order to maintain cell homeostasis, the correct epigenetic marks must be read and interpreted by the epigenetic machinery.

The machinery involved in controlling gene expression via DNA methylation includes four types of DNA methyltransferases (DNTM), each one with a specific role and time of action and seven methyl-CpG-binding domain (MBD) proteins. Histone modifications include methylation and acetylation, among others, catalyzed by histone methyltransferases (HMT) and demethylases (HDM) and by histone acetyltransferases (HAT) and deacetylases (HDAC), respectively (Hardy and Tollefsbol 2011). Moreover, the microRNAs (miRNA) control gene expression after the mRNA is transcribed, preventing the translation through binding to the target protein-coding mRNAs (da Silva Oliveira et al. 2016).

The epigenetic marks and maintenance are essential for homeostasis and regulate several biological processes, including cell differentiation and genomic imprinting. Therefore, alterations in a cell environment can lead to epigenetic alterations affecting transcriptional regulation. Many dietary compounds and nutrients are active molecules of the epigenetic machinery and/or chemical reaction donors, such as methionine to DNA methylation. Ingestion of nutrients may affect epigenetic processes, including cancer and other disease pathways, via inhibition of tumor suppressor genes, apoptosis and cell survival, and inflammatory signaling (Hardy and Tollefsbol 2011).

The MBD Family

The members of the MBD family are the proteins responsible for reading the DNA methylation marks. By binding to methyl-CpG marks, they can recruit the correct epigenetic machinery to read and to interpret the epigenetic landscape. In order to maintain homeostasis, precise levels of DNA methylation along with accurate expression of the MBD proteins are expected in healthy conditions, comparable to a symphony (Fig. 1a). All the cells in our body using the same DNA code throughout life are comparable to musicians playing different instruments reading the musical score in a concert. The epigenetic landscape and the musical arrangements are exclusive to each cell type and musical instrument. Thus, just like in a symphony score, the epigenetic code is complex and precisely arranged; therefore, if the wrong epigenetic mark/musical note is read and interpreted, the balance is lost (Fig. 1b). The same can occur if the reader (MBD protein/musician) is not able to accurately interpret the code (Fig. 1c) (Smith 2015).

The MBD family comprises seven proteins: MeCP2 and the MBD1–6. The main grouping family characteristic is the possession of a conserved methyl-binding domain sequence, consisting of about 70 amino acid residues that can recognize a single CpG (Fig. 2), as reviewed in Gigek et al. (2015) and Marchler-Bauer et al. (2015).

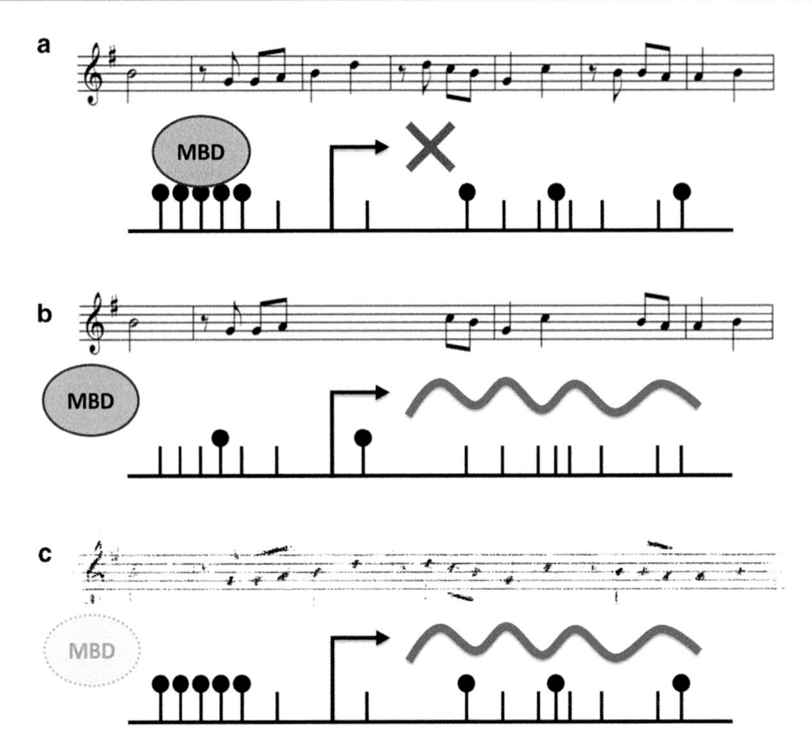

Fig. 1 Symphony and epigenetics. The musical score read during a concert is comparable to the correct functioning of both epigenetic marks and epigenetic machinery; in this illustration, (**a**) the musical score is read correctly, as the MBD proteins read the DNA methylation marks, silencing the gene expression. (**b**) The missing musical notes are comparable to missing/incorrect DNA methylation marks, leading to the abnormal gene expression. (**c**) The musical score cannot be read properly and therefore cannot be interpreted; similarly, the lost expression of the MBD proteins cannot act on the correct DNA methylation marks to silence gene expression. *mCpG* methylated CpG, | *uCpG* unmethylated CpG (Adapted from the proposed analogy by Smith (2015))

Biochemically, the domain is structured as wedge shaped, composed by four β-sheets, an α-helix and hairpin loop between two β-strands, thus forming a monomer that binds to its recognition site (Wade and Wolffe 2001). Despite all MBDs being structurally able to recognize methylation, only MeCP2 and MBD1, MBD2, and MBD4 have the ability to bind to methylated DNA and suppress gene expression; conversely, MBD3, MBD5, and MBD6 do not bind to methylated DNA because of particularities that will be discussed later in this topic.

It is worthy to mention that not all methyl-binding proteins have the MBD domain. Two structural families, the SET and RING finger-associated domain (SRA), are reported to bind to hemimethylated DNA (Arita et al. 2008), and the zinc finger family with a binding site to both methylated and non-methylated DNA (Daniel et al. 2002; Gigek et al. 2015). In addition, four other proteins can also read DNA methylation and carry an MBD-like domain (SETDB1, SETDB2, BAZ2A, and BAZ2B); however, they are not the focus of this review.

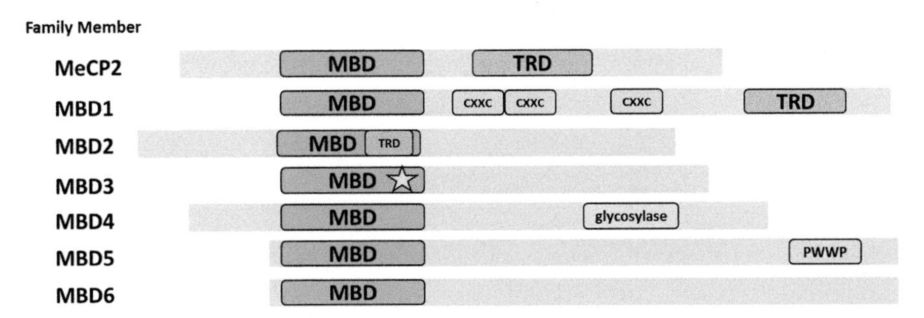

Fig. 2 A scheme of the structure of the seven components of the methyl-CpG-binding domain (*MBD*) protein family and other protein interaction domains. The *yellow star* represents the two distinct amino acid residues in the MBD of MBD3. *MBD* methyl-CpG-binding domain, *TRD* transcriptional repression domain, *CXXC* Cys-x-x-Cys domain, *PWWP* proline-tryptophan-tryptophan-proline domain (Adapted with permission from "Methyl-Cpg-Binding Protein (MBD) Family: Epigenomic Read-Outs Functions And Roles In Tumorigenesis And Psychiatric Diseases" by Gigek et al. 2015, J Cell Biochem., for more details please check the publication)

MeCP2. The first and the best characterized family member, MeCP2, can bind to a single methylated CpG. Upon binding, the recruitment of corepressor protein Sin3 and HDAC leads to chromatin condensation and gene silencing (Jones et al. 1998; Gigek et al. 2015). Conversely, MeCP2 was found in promoter regions of active genes in human neuroblastoma cell lines (Yasui et al. 2007). Moreover, MeCP2 has also been described as a splice regulator, interacting with a ribonucleoprotein in the brain (Young et al. 2005).

MeCP2 is highly expressed in all tissues, except for the oral mucosa, colon, small intestine, and rectum that show lower expression (Uhlen et al. 2015). Twenty-one transcripts are formed after splice variants, yet nine are protein coding, but only two are the main isoforms of the protein (Uhlen et al. 2015; Herrero et al. 2016). *MeCP2* mutations lead to autistic and neurological disorders, especially the Rett syndrome, and to metabolic disorders (Guy et al. 2011; Pitcher et al. 2013; Gigek et al. 2015) (Table 1).

MBD1. Among the 27 described transcripts of MBD1, 20 are protein coding, and 13 isoforms are the result of alternative splicing listed in RefSeq database (Herrero et al. 2016). MBD1 protein binds to methylated and unmethylated CpG and is involved in heterochromatin formation (Hameed et al. 2014), while MBD2 has strong preferences for highly methylated regions. MBD1 is implied in controlling cell division and differentiation. Therefore, deregulated MBD1 is described in several types of cancer, including colorectal, leukemia, breast, bladder, gastric, esophageal, and others (Gigek et al. 2015; Herrero et al. 2016). Additionally, knockout of MBD1 in vivo was associated with autism spectrum disorders (Allan et al. 2008) (Table 1).

MBD2. Another well-studied protein, MBD2, presents two functional transcripts, out of five, and has the ability to bind to promoter regions, to transcriptional start sites, and also to gene body (Table 1) (Chatagnon et al. 2011; Menafra et al. 2014). Despite the fact MBD2 is mostly recruited to highly methylated sites, levels of gene

Table 1 An overview of the MBD family members

Protein	Transcripts	Protein-coding isoforms (RefSeq)	Main binding	Main function	Main interactions
MeCP2	21	2	Binds to a single mCpG	Important to neurodevelopment	Promoter regions, corepressor Sin3 and HDAC
MBD1	27	13	Methylated and unmethylated CpG	Cell division and differentiation	HDAC
MBD2	5	2	Binds to highly methylated CpG	Stem cell pluripotency, gene expression repression	Transcriptional start sites, gene body, NuRD complex
MBD3	13	2	Binds to 5-hydroxi-mCpG	Normal development, cell reprograming	NuRD complex, DNA damage repair, chromatin assembly
MBD4	10	5	Binds to 5mCpG/TpG mismatch	Identification and excision of deaminated CpG	DNA methyltransferase 1
MBD5	33	9	Binds to chromocenters and methylated pericentric heterochromatin	Development, neurogenesis, and neuronal gene regulation	Epigenetic machinery
MBD6	14	9	Binds to chromatin	Cell differentiation and proliferation	MBD2, MBD4, MBD6, repressor complex

expression along with the presence of active histone marks were found in MBD2-binding regions (Menafra et al. 2014). MBD2 is a very abundant epigenetic protein in all tissues, shown to play a role in silencing tumor suppressor genes and participate in body formation in mice (Hendrich et al. 2001; Gigek et al. 2015; Uhlen et al. 2015).

MBD3. MBD2 and MBD3 can form a heterodimer in the NuRD complex context and interact with CpG domain, regulating gene transcription, DNA damage repair, maintenance of genome stability, and chromatin assembly (Allen et al. 2013). Regardless of being the smallest MBD protein, MBD3 has the ability to recognize hydroxymethylcytosines instead of methylated CpGs (Yildirim et al. 2011). MBD3 has been proved to have an essential role in embryonic development (Hendrich et al. 2001), is expressed in all tissues (Uhlen et al. 2015), and is reported to regulate gene expression via NuRD complex (Table 1) (Gigek et al. 2015). Eight MBD3 transcripts are protein coding, although only two are in the RefSeq database and another three transcripts are not translated (Herrero et al. 2016).

MBD4. A specific domain for deamination reactions of methylcytosines is involved in DNA repair and confers to MBD4 a special role among the MBDs.

The five protein-coding transcripts among the ten (Herrero et al. 2016) have the ability to bind to methylated DNA through the N-terminal MBD and repair DNA mismatch (C → T) in the C-terminal domain (Table 1) (Gigek et al. 2015; Uhlen et al. 2015). MBD4 transcriptional repression activity is linked to cell death signaling pathway, and it is reported to interact with DNMT to control gene expression (Laget et al. 2014). For that reason, its absence was shown to promote tumor formation in vivo (Millar et al. 2002).

MBD5. The remaining two family members, MBD5 and MBD6, were recently described by Laget et al. (2010) and cannot bind to methylated CpG (Laget et al. 2010). MBD5 has 33 transcripts, but only 9 are translated into protein (Table 1) (Herrero et al. 2016). Furthermore, MBD5 is expected to contribute to epigenetic machinery in neurons, since its deficiency results in severe intellectual disability (Talkowski et al. 2011; Gigek et al. 2015).

MBD6. The nine protein-coding MBD6 among 14 transcripts are predicted intracellular protein (Uhlen et al. 2015; Herrero et al. 2016), although not shown to bind to methylated DNA in vitro (Table 1) (Laget 2010). The number of studies with MBD6 is still modest, but mutations in the MBD6 gene have been reported in gastrointestinal tumors (Choi et al. 2015) and in autistic patients (Cukier et al. 2012), suggesting an important role in the brain, like other MBD family members. MBD6 is demonstrated to interact with MBD2–4 and MBD6 itself and to cell cycle-related genes, associated to its cell proliferation and differentiation function reported in vitro (Jung et al. 2013). Moreover, MBD5 and MBD6 are both necessary to interact with other protein in a repressor complex although exact function remains unknown. In addition, MBD6 is recruited to sites of DNA damage (Table 1) (Baymaz et al. 2014).

Thus, disruption of the balance between the correct DNA methylation marks and the status of the epigenetic machinery can lead to neurodevelopmental disease as illustrated here and several types of cancer such as breast, prostate, gastric, and colon among others (Gigek et al. 2015). A hypothesis that dietary intake will contribute to equilibrium by preventing or promoting DNA methylation is encouraged by the description of chemoprevention by common nutrients in several types of cancer (Arasaradnam et al. 2008) (Gigek et al., Part 2 of this handbook, chapter ▶ "Epigenetic Alterations in Stomach Cancer: Implications for Diet and Nutrition").

Nutrients and Diet Influences

The biochemical reactions for DNA methylation process encompass methyl-containing nutrients, the lipotropes, acting as methyl donors and cofactors, besides the DNMT enzymes. Methionine, choline, folate, and vitamin B12 are lipotropes present in the one-carbon metabolism, a process that provides methyl groups for epigenetic marks, as shown in Table 2 (Friso et al. 2016; Cho et al. 2012).

Maternal and paternal diet has been proven to influence development and differentiation of the offspring. High doses of lipotropes in maternal diet reduced mammary carcinogenesis in the offspring. Although not associated with alteration in the global DNA methylation pattern, the high doses of methyl diet decreased gene

Table 2 Dietary compound, food source, and outcomes of nutrients

Dietary compound/ style	Class	Food source	Outcomes
Choline (lipotrope)	Vitamin	Cauliflower, broccoli, wheat germ	Methylation maintenance
			Cancer prevention
			Deficiency: high MBD2 levels
Methionine (lipotrope)	Amino acid	Egg, sesame seed, cheese, Brazilian nuts	Cancer prevention
			Deficiency: high MBD2 levels
Folate (lipotrope)	Vitamin	Vegetables (dark green leafy vegetables), fruits, nuts, beans	Cancer prevention
			Deficiency: high MBD2 levels
Anthocyanins and ellagitannins		Black raspberry	Antioxidant
			Anti-inflammatory
Casein	Amino acid carbohydrate	Milk, cheese	Maintain correct epigenetic patterns
Genistein	Isoflavone		DNA methylation, decrease activity of DNMT and MBD2
Polyphenols resveratrol	Phenols	Fruits, vegetable, green tea	Decrease MBD1, MBD4, MeCP2, DNMT1, and HDAC1 levels
			Altered DNA methylation
High fat	Triglycerides	Oil, butter fat	Aberrant DNA methylation and miRNA profile in offspring
Alcohol	Ethanol	Alcoholic beverages	Decreases folate absorption and transportation
			Epigenetic dysregulation
Caloric restriction	–	Dietary regimen with low ingest	MeCP2 recruitment
High carbohydrate	Carbohydrate	Sugars, oligosaccharides and polysaccharides	Increased DNA methylation and recruitment of MeCP2

expression of *Hdac1* and *Mecp2*, previously implied in tumor initiation, and are suggested to prevent breast cancer (Cho et al. 2012). Besides, a high fat (HF) diet (high oil and butter fat) during pregnancy in animal model led to aberrant expression of MBD1 and MBD3 in the offspring with mammary tumor (Govindarajah et al. 2016). Breast cancer results from genetic and epigenetic alterations and displays aberrant DNA methylation, driving to oncogenesis in this tissue.

Additionally, low-protein (low-casein) diet during pregnancy period was shown to induce epigenetic alterations in the offspring rat liver. Gong et al. (2010) reported higher MBD2 and *Dnmt1* and *Dnmt3a* expression and no changes in global DNA methylation pattern, although hypermethylation of an important imprinting control region (ICR) was observed. Moreover, in the group of maternal low-protein diet supplemented with folic acid, ICR hypermethylation was attenuated compared to non-supplemented, and mRNA levels of MBD2 and Dnmts were similar to those observed in the control group (Gong et al. 2010). Another study observed miRNA alterations in the offspring of maternal HF diet during pregnancy and lactation. One of the most abundantly expressed miRNA in the liver, the miR-709, was reduced in the offspring. Also, an in silico analysis predicted MeCP2 and MBD6 as targets of miR-709 (Zhang et al. 2009), possibly resulting in higher protein levels.

Furthermore, alcohol intake can provoke epigenetic alterations by affecting methionine and other lipotrope pathway. Long-term intake of alcohol is linked to decreased absorption of folate and impaired transportation of folate to the fetus (Resendiz et al. 2013). Also, choline deficiency is implied to increase DNA methylation and inactivated epigenetic machinery genes, including MBD3 and *DNMT*. Newborns affected by fetal alcohol syndrome have neurological deficits correspondent to epigenetic dysregulation, besides growth retardation. The nutritional supplementation with lipotropes has been reported to lighten some effects of maternal alcohol intake to the offspring (Resendiz et al. 2013).

Nutrient deficiency in general also has an impact on DNA methylation and MBD protein action. Timely caloric restriction during pregnancy also reduced placental glucose transporter gene expression by enhancing its DNA methylation and recruiting Mecp2 (Ganguly et al. 2014).

Paternal diet habits also have an impact on offspring health. Paternal alcohol intake changed DNA methylation and *DNMT* expression levels in sperm, in animal model (Resendiz et al. 2013). Offspring of paternal low-protein diet mice displayed low histone methylation levels and specific promoter DNA hypermethylation in the liver (Carone et al. 2010). Besides, HF in mice and rats led to aberrant DNA methylation and miRNA profile in male offspring (Fullston et al. 2013) and lower gene-specific DNA methylation along with aberrant expression of more than 600 genes in the pancreas of female offspring (Ng et al. 2010), respectively. A more detailed review of how paternal diet affects sperm quality and offspring health is provided elsewhere (Schagdarsurengin and Steger 2016).

A number of evidences demonstrated that early-life and adulthood diet also affects epigenetic processes. Newborn rats that were submitted to high carbohydrate diet developed insulin resistance and obesity in adulthood. A molecular analysis revealed increased DNA methylation in the promoter of insulin-responsive glucose transporter gene (G*lut4*) and enhanced recruitment of MeCP2 and Dnmts, reducing *Glut4* expression in skeletal muscle (Raychaudhuri et al. 2014).

Methyl-deficient diet is known to induce hepatocellular carcinoma. In adult rat liver, the effects of a low methionine and folate and choline absent diet promoted higher levels of both MBD2 RNA and protein and low Mecp2 protein levels in the liver in vivo (Esfandiari et al. 2003). The reduction of Mecp2 level is implied to initiate hepatic cancer. Likewise, in early stages of hepatocellular carcinoma, animal model submitted to the same methyl-deficient diet had increased levels of MBD1–3 expression and MBD1, MBD2, and MBD4 protein levels (Ghoshal et al. 2006).

On the other hand, some diet compounds may help symptoms and attenuate disease status, toward healthy. Anthocyanins and ellagitannins are protective antioxidant and anti-inflammatory compounds present in black raspberries (BR) (Table 2). Beneficial effects to precancerous colon conditions were demonstrated in mice fed 5% BR diet. Genetic model of ulcerative colitis is knockout for interleukin-10 gene; mice fed with BR had less ulceration compared to the knockout in control diet (Wang et al. 2013b). Moreover, authors' conclusions point to inhibition of aberrant epigenetic events, since there was a recovery of promoter demethylation of tumor suppressor genes, consequently increasing mRNA expression. The epigenetic machinery proteins MBD2, DNMTs, and HDAC were decreased in bone marrow of BR-fed mice with chemo-induced ulcerative colitis (Wang et al. 2013b).

Genistein, a dietary isoflavone, has also been shown to affect DNA methylation patterns at specific genes in cancer animal models and in cancer cell lines (Table 2) (Zhang and Chen 2011; Li and Tollefsbol 2010). A transgenerational effect of a genistein-supplemented mother was observed to alter DNA methylation pattern by shifting the coat color in agouti offspring – a well document gene responsible for coat color distribution controlled by methylation in mice (Dolinoy et al. 2006). Besides, in renal cancer cells, genistein administration induced promoter demethylation of *BTG3*, a frequently hypermethylated gene. Moreover, the genistein treatment administrated with a demethylation agent (5-aza-C) led to decrease activity of DNMT and MBD2, and increased HAT activity, being a promising candidate for epigenetic therapy (Majid et al. 2009).

Chemopreventive polyphenols from fruits and vegetables may occur through epigenetic mechanisms. A particular polyphenol, from green tea, has an effect attributed to the antioxidant and anti-inflammatory properties (Table 2). An in vitro assay of prostate cancer cells exposed to the major constitute of green tea decreased mRNA and protein levels of MBD1, MBD4, and MeCP2 and altered DNA methylation (Pandey et al. 2010). The resveratrol mostly abundant in grapes, but found in several plants, is also a polyphenol known to affect epigenetics. Silencing of *BRCA1* expression due to promoter hypermethylation is a common finding in sporadic breast cancer. In vitro studies demonstrated, resveratrol administration reduced BRCA1 methylation, and, therefore, DNMT and MBD2 recruitment (Papoutsis et al. 2012). Curcumin from turmeric plant also has antioxidant and anti-inflammatory effects. Interestingly, neurological protective effects of curcumin have been reported to stimulate proliferation of neural stem cells in hippocampus and to have cardiovascular protection role (Aggarwal and Harikumar 2009). Importantly, breast cell cancer treated with resveratrol, green tea polyphenol, curcumin, and other polyphenols had decreased protein levels of DNMT1, HDAC1, and MeCP2, but not MBD2 (Mirza et al. 2013). Therefore, dietary compound may be a cancer preventive alternative to prevent epigenetic modifications.

MeCP2 mutation in animal model is characterized by neurological disorder and by alterations in endothelial vessels. These alterations, as well as the pathological stereotyped movements that characterize the pathological animal, were partially reversed by curcumin administration, attenuating circulatory problems (Panighini et al. 2013).

Dietary nutrients have been proved to act as epigenetic modifiers, stimulating favorable epigenetic landscape, resulting in a positive outcome in healthy and even improving disease conditions. Since epigenetic effects can be transgenerational, parent's better-quality dietary habits and lifestyle should be recommended by healthcare providers to a healthy offspring.

Dictionary of Terms

- **Epigenetic landscape** – A term coined to describe cell fate. The analogy is drawn to compare epigenetic modifications shaping the cell phenotype to ridges and valleys shaping a landscape.
- **Homeostasis** – Balance of a living organism in healthy conditions.
- **RefSeq database** – Open-access database with annotated sequences of DNA, RNA, and proteins.
- **Imprinting control region** – A chromosomal region with regulatory elements involved in controlling gene expression.
- **Fetal alcohol syndrome** – Babies of women who drink alcohol during pregnancy may present a range of mental and/or physical defects, including microcephaly, neurological impairment, characteristic facial features, and growth deficiency.
- **Chemopreventive** – Nutrients or other agents that can reduce cancer risk.

Key Facts of MBD Family and Nutrition

- The epigenetic landscape coined by Waddington is a metaphor for the embryonic development.
- There are seven MBD family members.
- MeCP2 is the first family member described in 1992 and is the only one that does not have MBD in the name.
- Absence or excess of some nutritional compounds affects MBD protein recruitment, therefore disturbing the epigenetic homeostasis.
- Transgenerational effects of both maternal and paternal diet exert an influence on the epigenetic regulation in the offspring.

Summary Points

- The MBD family is responsible for reading the DNA methylation marks.
- Each protein has a distinctive characteristic and can be recruited to a specific function.
- All MBDs are structurally able to recognize methylation; however, MBD3, MBD5, and MBD6 do not bind to methylated DNA.
- Lipotropes are chemopreventive, and deficiency in these nutrients can influence levels of MBD family members.
- Parental dietary habits can lead to epigenetic alterations in the offspring. These inherited epigenetic marks have an influence in cancer risk.

References

Aggarwal BB, Harikumar KB (2009) Potential therapeutic effects of curcumin, the anti-inflammatory agent, against neurodegenerative, cardiovascular, pulmonary, metabolic, autoimmune and neoplastic diseases. Int J Biochem Cell Biol 41:40–59. https://doi.org/10.1016/j.biocel.2008.06.010

Allan AM, Liang X, Luo Y, Pak C, Li X, Szulwach KE et al (2008) The loss of methyl-CpG binding protein 1 leads to autism-like behavioral deficits. Hum Mol Genet 17:2047–2057. https://doi.org/10.1093/hmg/ddn102

Allen HF, Wade PA, Kutateladze TG (2013) The NuRD architecture. Cell Mol Life Sci 70:3513–3524. https://doi.org/10.1007/s00018-012-1256-2

Arasaradnam RP, Commane DM, Bradburn D, Mathers JC (2008) A review of dietary factors and its influence on DNA methylation in colorectal carcinogenesis. Epigenetics 3:193–198

Arita K, Ariyoshi M, Tochio H, Nakamura Y, Shirakawa M (2008) Recognition of hemi-methylated DNA by the SRA protein UHRF1 by a base-flipping mechanism. Nature 455:818–821. https://doi.org/10.1038/nature07249

Baymaz HI, Fournier A, Laget S, Ji Z, Jansen PW, Smits AH et al (2014) MBD5 and MBD6 interact with the human PR-DUB complex through their methyl-CpG-binding domain. Proteomics 14:2179–2189. https://doi.org/10.1002/pmic.201400013

Carone BR, Fauquier L, Habib N, Shea JM, Hart CE, Li R et al (2010) Paternally induced transgenerational environmental reprogramming of metabolic gene expression in mammals. Cell 143:1084–1096. https://doi.org/10.1016/j.cell.2010.12.008

Chatagnon A, Perriaud L, Nazaret N, Croze S, Benhattar J, Lachuer J et al (2011) Preferential binding of the methyl-CpG binding domain protein 2 at methylated transcriptional start site regions. Epigenetics 6:1295–1307. https://doi.org/10.4161/epi.6.11.17875

Cho K, Mabasa L, Bae S, Walters MW, Park CS (2012) Maternal high-methyl diet suppresses mammary carcinogenesis in female rat offspring. Carcinogenesis 33:1106–1112. https://doi.org/10.1093/carcin/bgs125

Choi YJ, Yoo NJ, Lee SH (2015) Mutation and expression of a methyl-binding protein 6 (MBD6) in gastric and colorectal cancers. Pathol Oncol Res 21:857–858. https://doi.org/10.1007/s12253-015-9904-0

Cukier HN, Lee JM, Ma D, Young JI, Mayo V, Butler BL et al (2012) The expanding role of MBD genes in autism: identification of a MECP2 duplication and novel alterations in MBD5, MBD6, and SETDB1. Autism Res 5:385–397. https://doi.org/10.1002/aur.1251

da Silva Oliveira KC, Thomaz Araujo TM, Albuquerque CI, Barata GA, Gigek CO, Leal MF et al (2016) Role of miRNAs and their potential to be useful as diagnostic and prognostic biomarkers in gastric cancer. World J Gastroenterol 22:7951–7962. https://doi.org/10.3748/wjg.v22.i35.7951

Daniel JM, Spring CM, Crawford HC, Reynolds AB, Baig A (2002) The p120(ctn)-binding partner Kaiso is a bi-modal DNA-binding protein that recognizes both a sequence-specific consensus and methylated CpG dinucleotides. Nucleic Acids Res 30:2911–2919

Dolinoy DC, Weidman JR, Waterland RA, Jirtle RL (2006) Maternal genistein alters coat color and protects Avy mouse offspring from obesity by modifying the fetal epigenome. Environ Health Perspect 114:567–572

Esfandiari F, Green R, Cotterman RF, Pogribny IP, James SJ, Miller JW (2003) Methyl deficiency causes reduction of the methyl-CpG-binding protein, MeCP2, in rat liver. Carcinogenesis 24:1935–1940. https://doi.org/10.1093/carcin/bgg163

Friso S, Udali S, De Santis D, Choi SW (2016) One-carbon metabolism and epigenetics. Mol Asp Med. https://doi.org/10.1016/j.mam.2016.11.007

Fullston T, Ohlsson Teague EM, Palmer NO, DeBlasio MJ, Mitchell M, Corbett M et al (2013) Paternal obesity initiates metabolic disturbances in two generations of mice with incomplete penetrance to the F2 generation and alters the transcriptional profile of testis and sperm microRNA content. FASEB J 27:4226–4243. https://doi.org/10.1096/fj.12-224048

Ganguly A, Chen Y, Shin BC, Devaskar SU (2014) Prenatal caloric restriction enhances DNA methylation and MeCP2 recruitment with reduced murine placental glucose transporter isoform 3 expression. J Nutr Biochem 25:259–266. https://doi.org/10.1016/j.jnutbio.2013.10.015

Ghoshal K, Li X, Datta J, Bai S, Pogribny I, Pogribny M et al (2006) A folate- and methyl-deficient diet alters the expression of DNA methyltransferases and methyl CpG binding proteins involved in epigenetic gene silencing in livers of F344 rats. J Nutr 136:1522–1527

Gigek CO, Chen ES, Ota VK, Maussion G, Peng H, Vaillancourt K et al (2015a) A molecular model for neurodevelopmental disorders. Transl Psychiatry 5:e565. https://doi.org/10.1038/tp.2015.56

Gigek CO, Chen ES, Smith MA (2015b) Methyl-Cpg-binding protein (MBD) family: epigenomic read-outs functions and roles in tumorigenesis and psychiatric diseases. J Cell Biochem. 2016;117(1):29–38. https://doi.org/10.1002/jcb.25281

Gong L, Pan YX, Chen H (2010) Gestational low protein diet in the rat mediates Igf2 gene expression in male offspring via altered hepatic DNA methylation. Epigenetics 5:619–626

Govindarajah V, Leung YK, Ying J, Gear R, Bornschein RL, Medvedovic M et al (2016) In utero exposure of rats to high-fat diets perturbs gene expression profiles and cancer susceptibility of prepubertal mammary glands. J Nutr Biochem 29:73–82. https://doi.org/10.1016/j.jnutbio.2015.11.003

Guy J, Cheval H, Selfridge J, Bird A (2011) The role of MeCP2 in the brain. Annu Rev Cell Dev Biol 27:631–652. https://doi.org/10.1146/annurev-cellbio-092910-154121

Hameed UF, Lim J, Zhang Q, Wasik MA, Yang D, Swaminathan K (2014) Transcriptional repressor domain of MBD1 is intrinsically disordered and interacts with its binding partners in a selective manner. Sci Rep 4:4896. https://doi.org/10.1038/srep04896

Hardy TM, Tollefsbol TO (2011) Epigenetic diet: impact on the epigenome and cancer. Epigenomics 3:503–518. https://doi.org/10.2217/epi.11.71

Hendrich B, Guy J, Ramsahoye B, Wilson VA, Bird A (2001) Closely related proteins MBD2 and MBD3 play distinctive but interacting roles in mouse development. Genes Dev 15:710–723. https://doi.org/10.1101/gad.194101

Herrero J, Muffato M, Beal K, Fitzgerald S, Gordon L, Pignatelli M et al (2016) Ensembl comparative genomics resources. Database (Oxford) 2016. https://doi.org/10.1093/database/bav096

Jones PL, Veenstra GJ, Wade PA, Vermaak D, Kass SU, Landsberger N et al (1998) Methylated DNA and MeCP2 recruit histone deacetylase to repress transcription. Nat Genet 19:187–191. https://doi.org/10.1038/561

Jung JS, Jee MK, Cho HT, Choi JI, Im YB, Kwon OH et al (2013) MBD6 is a direct target of Oct4 and controls the stemness and differentiation of adipose tissue-derived stem cells. Cell Mol Life Sci 70:711–728. https://doi.org/10.1007/s00018-012-1157-4

Laget S, Joulie M, Le Masson F, Sasai N, Christians E, Pradhan S et al (2010) The human proteins MBD5 and MBD6 associate with heterochromatin but they do not bind methylated DNA. PLoS One 5:e11982. https://doi.org/10.1371/journal.pone.0011982

Laget S, Miotto B, Chin HG, Esteve PO, Roberts RJ, Pradhan S et al (2014) MBD4 cooperates with DNMT1 to mediate methyl-DNA repression and protects mammalian cells from oxidative stress. Epigenetics 9:546–556. https://doi.org/10.4161/epi.27695

Li Y, Tollefsbol TO (2010) Impact on DNA methylation in cancer prevention and therapy by bioactive dietary components. Curr Med Chem 17:2141–2151

Majid S, Dar AA, Ahmad AE, Hirata H, Kawakami K, Shahryari V et al (2009) BTG3 tumor suppressor gene promoter demethylation, histone modification and cell cycle arrest by genistein in renal cancer. Carcinogenesis 30:662–670. https://doi.org/10.1093/carcin/bgp042

Marchler-Bauer A, Derbyshire MK, Gonzales NR, Lu S, Chitsaz F, Geer LY et al (2015) CDD: NCBI's conserved domain database. Nucleic Acids Res 43:D222–D226. https://doi.org/10.1093/nar/gku1221

Menafra R, Brinkman AB, Matarese F, Franci G, Bartels SJ, Nguyen L et al (2014) Genome-wide binding of MBD2 reveals strong preference for highly methylated loci. PLoS One 9:e99603. https://doi.org/10.1371/journal.pone.0099603

Millar CB, Guy J, Sansom OJ, Selfridge J, MacDougall E, Hendrich B et al (2002) Enhanced CpG mutability and tumorigenesis in MBD4-deficient mice. Science 297:403–405. https://doi.org/10.1126/science.1073354

Mirza S, Sharma G, Parshad R, Gupta SD, Pandya P, Ralhan R (2013) Expression of DNA methyltransferases in breast cancer patients and to analyze the effect of natural compounds on DNA methyltransferases and associated proteins. J Breast Cancer 16:23–31. https://doi.org/10.4048/jbc.2013.16.1.23

Ng SF, Lin RC, Laybutt DR, Barres R, Owens JA, Morris MJ (2010) Chronic high-fat diet in fathers programs beta-cell dysfunction in female rat offspring. Nature 467:963–966. https://doi.org/10.1038/nature09491

Pandey M, Shukla S, Gupta S (2010) Promoter demethylation and chromatin remodeling by green tea polyphenols leads to re-expression of GSTP1 in human prostate cancer cells. Int J Cancer 126:2520–2533. https://doi.org/10.1002/ijc.24988

Panighini A, Duranti E, Santini F, Maffei M, Pizzorusso T, Funel N et al (2013) Vascular dysfunction in a mouse model of Rett syndrome and effects of curcumin treatment. PLoS One 8:e64863. https://doi.org/10.1371/journal.pone.0064863

Papoutsis AJ, Borg JL, Selmin OI, Romagnolo DF (2012) BRCA-1 promoter hypermethylation and silencing induced by the aromatic hydrocarbon receptor-ligand TCDD are prevented by resveratrol in MCF-7 cells. J Nutr Biochem 23:1324–1332. https://doi.org/10.1016/j.jnutbio.2011.08.001

Pitcher MR, Ward CS, Arvide EM, Chapleau CA, Pozzo-Miller L, Hoeflich A et al (2013) Insulinotropic treatments exacerbate metabolic syndrome in mice lacking MeCP2 function. Hum Mol Genet 22:2626–2633. https://doi.org/10.1093/hmg/ddt111

Raychaudhuri N, Thamotharan S, Srinivasan M, Mahmood S, Patel MS, Devaskar SU (2014) Postnatal exposure to a high-carbohydrate diet interferes epigenetically with thyroid hormone receptor induction of the adult male rat skeletal muscle glucose transporter isoform 4 expression. J Nutr Biochem 25:1066–1076. https://doi.org/10.1016/j.jnutbio.2014.05.011

Resendiz M, Chen Y, Ozturk NC, Zhou FC (2013) Epigenetic medicine and fetal alcohol spectrum disorders. Epigenomics 5:73–86. https://doi.org/10.2217/epi.12.80

Schagdarsurengin U, Steger K (2016) Epigenetics in male reproduction: effect of paternal diet on sperm quality and offspring health. Nat Rev Urol 13:584–595. https://doi.org/10.1038/nrurol.2016.157

Smith K (2015) Epigenome: the symphony in your cells. Retrieved Jan 2017. https://doi.org/10.1038/nature.2015.16955

Talkowski ME, Mullegama SV, Rosenfeld JA, van Bon BW, Shen Y, Repnikova EA et al (2011) Assessment of 2q23.1 microdeletion syndrome implicates MBD5 as a single causal locus of intellectual disability, epilepsy, and autism spectrum disorder. Am J Hum Genet 89:551–563. https://doi.org/10.1016/j.ajhg.2011.09.011

Uhlen M, Fagerberg L, Hallstrom BM, Lindskog C, Oksvold P, Mardinoglu A et al (2015) Proteomics. Tissue-based map of the human proteome. Science 347:1260419. https://doi.org/10.1126/science.1260419

Waddington CH (1942) The epigenotype. Endeavour 1:18–20

Wade PA, Wolffe AP (2001) ReCoGnizing methylated DNA. Nat Struct Biol 8:575–577. https://doi.org/10.1038/89593

Wang LS, Kuo CT, Huang TH, Yearsley M, Oshima K, Stoner GD et al (2013a) Black raspberries protectively regulate methylation of Wnt pathway genes in precancerous colon tissue. Cancer Prev Res (Phila) 6:1317–1327. https://doi.org/10.1158/1940-6207.CAPR-13-0077

Wang LS, Kuo CT, Stoner K, Yearsley M, Oshima K, Yu J et al (2013b) Dietary black raspberries modulate DNA methylation in dextran sodium sulfate (DSS)-induced ulcerative colitis. Carcinogenesis 34:2842–2850. https://doi.org/10.1093/carcin/bgt310

Yasui DH, Peddada S, Bieda MC, Vallero RO, Hogart A, Nagarajan RP et al (2007) Integrated epigenomic analyses of neuronal MeCP2 reveal a role for long-range interaction with active genes. Proc Natl Acad Sci U S A 104:19416–19421. https://doi.org/10.1073/pnas.0707442104

Yildirim O, Li R, Hung JH, Chen PB, Dong X, Ee LS et al (2011) Mbd3/NURD complex regulates expression of 5-hydroxymethylcytosine marked genes in embryonic stem cells. Cell 147:1498–1510. https://doi.org/10.1016/j.cell.2011.11.054

Young JI, Hong EP, Castle JC, Crespo-Barreto J, Bowman AB, Rose MF et al (2005) Regulation of RNA splicing by the methylation-dependent transcriptional repressor methyl-CpG binding protein 2. Proc Natl Acad Sci U S A 102:17551–17558. https://doi.org/10.1073/pnas.0507856102

Zhang Y, Chen H (2011) Genistein, an epigenome modifier during cancer prevention. Epigenetics 6:888–891

Zhang J, Zhang F, Didelot X, Bruce KD, Cagampang FR, Vatish M et al (2009) Maternal high fat diet during pregnancy and lactation alters hepatic expression of insulin like growth factor-2 and key microRNAs in the adult offspring. BMC Genomics 10:478. https://doi.org/10.1186/1471-2164-10-478

Epigenetic Effects of N-3 Polyunsaturated Fatty Acids

82

Christine Heberden and Elise Maximin

Contents

Abstract

N-3 polyunsaturated fatty acids (PUFAs) are essential components of the cell membrane. They display a wide range of effects, acting on brain functions and cardiovascular physiology. N-3 long-chain PUFAs have to be ingested from the diet and are nowadays quasi restricted to fish or seafood and are thus consumed in unsatisfactory low proportions compared to n-6 PUFAs. Experimental animal models have shown that n-3 PUFAs could act through epigenetic effects. For instance, maternal n-3 uptakes correlate negatively with offspring body weight in rodents. Such an association has been debated in humans. Yet, there is a negative correlation with n-3 maternal intakes and child's sensitivity to allergies or asthma. In experimental animals or in vitro, n-3 supplementation or incubation can

C. Heberden (✉) · E. Maximin
Micalis Institute, INRA, AgroParisTech, Université Paris-Saclay, Jouy-en-Josas, France
e-mail: christine.heberden@inra.fr; christine.heberden@gmail.com; elise.maximin@inra.fr

V. B. Patel, V. R. Preedy (eds.), *Handbook of Nutrition, Diet, and Epigenetics*,
https://doi.org/10.1007/978-3-319-55530-0_45

modify DNA methylation state, either globally or locally on some gene promoters. The promoter methylation of fatty acid desaturase-2 (FADS2), a desaturase involved in PUFA synthesis, is increased by DHA. The transcriptome of micro-RNAs (miR) is also modified by n-3 PUFAs. The compounds derived from n-3 PUFAs, such as ResolvinD1, are also able to alter miR synthesis to resolve inflammation. Therefore, these still fragmentary observations open the road to deepen the knowledge of the varied molecular mechanisms of the n-3 PUFAs.

Keywords

N-3 polyunsaturated fatty acids · Linoleic acid · Alpha-linolenic acid · Arachidonic acid · EPA · DHA · Resolvin D1 · FADS2 · Brain · Neurodevelopment · Offspring · Maternal · miR · DNA methylation

List of Abbreviations

AA	Arachidonic acid
ALA	Alpha-linolenic acid
BDNF	Brain-derived neurotrophic factor
BMI	Body mass index
DHA	Docosahexaenoic acid
ELOVL5	Fatty acid elongase 5
EPA	Eicosapentaenoic acid
FADS1	Fatty acid desaturase-1
FADS2	Fatty acid desaturase-2
LA	Linoleic acid
n-3 PUFAs	N-3 polyunsaturated fatty acids
RvD1	Resolvin D1

Introduction

The essential n-3 polyunsaturated fatty acids (n-3 PUFAs) are cell membrane constituents and are derived from diet. They have been the focus of numerous studies for their diverse beneficial roles in brain and cardiovascular physiology, or in the prevention of colon cancer, for instance. This diversity of actions is paralleled by a myriad of effects on several cell compartments: as membrane components they can modify the presence of proteins in the lipid rafts or the membrane properties such as membrane fluidity (Langelier et al. 2010). They can bind to membrane and nuclear receptors, and regulate gene expression. They are also metabolized in active compounds such as autacoids and others, as we will describe below (Piomelli et al. 2007). Besides this already large complexity, another aspect of their abilities is now emerging with the possibility that they could also play a role in epigenetics.

Before gathering the experimental evidences, we will briefly address the origin and metabolism of n-3 PUFAs and explain the importance of these dietary constituents.

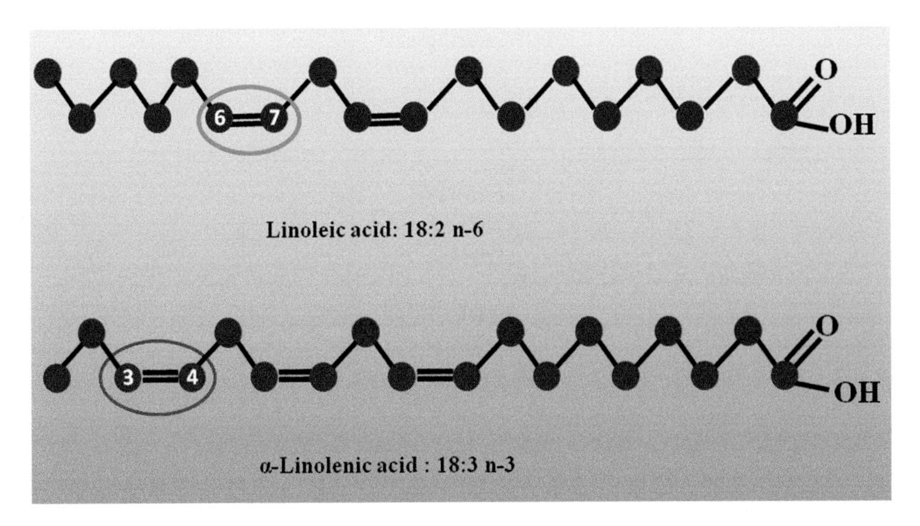

Fig. 1 Chemical structures of the precursors of the n-6 (or ω6) and the n-3 (or ω3) families. The double bond after which the family is named has been circled

Polyunsaturated Families

The most abundant PUFAs in the cell membrane phospholipids belong to two families, named n-6 and n-3. This denomination comes from the place of the first double bond in the aliphatic chain, i.e., carbon 6 and carbon 3 for the n-6 and n-3 family, respectively (Fig. 1).

Although these families are distinct molecularly and functionally, they share common characteristics. The precursors of both families cannot be synthesized by mammals and have to be ingested from the diet; thus they are essential. Both precursors are metabolized by the same set of enzymes and as a consequence compete with each other. Linoleic acid (LA, C18:2 n-6) is the precursor of the n-6 PUFAs and α-linolenic acid (ALA, C18:3 n-3) the precursor of the n-3 (Fig. 2).

They are metabolized through a succession of elongases and desaturases (Fig. 2) in the liver with two key enzymes, the fatty acid desaturase-2 (FADS2) and fatty acid desaturase-1 (FADS1), also named Δ6 and Δ5 desaturase, respectively. These enzymes are considered to determine the rate of conversion (Baker et al. 2016).

LA and ALA are converted into long-chain PUFAs, which are the active compounds of the families: Arachidonic acid (AA, C20:4 n-6), eicosapentaenoic acid (EPA, C20:5 n-3), and docosahexaenoic acid (DHA, C22:6 n-3) are the most active among long-chain PUFAs. The rate of conversion into long-chain PUFAs is low, and consequently, n-3 long-chain PUFAs have to be ingested from the diet to be present and stored in significant amounts in the cell phospholipids. A balanced nutrition demands a certain proportion of n-3 in the diet, and nutritional studies show that western diet is unfortunately characterized by a lack of n-3 PUFAs. The origin of n-3 PUFAs in modern diet comes almost exclusively from seafood and fish, and

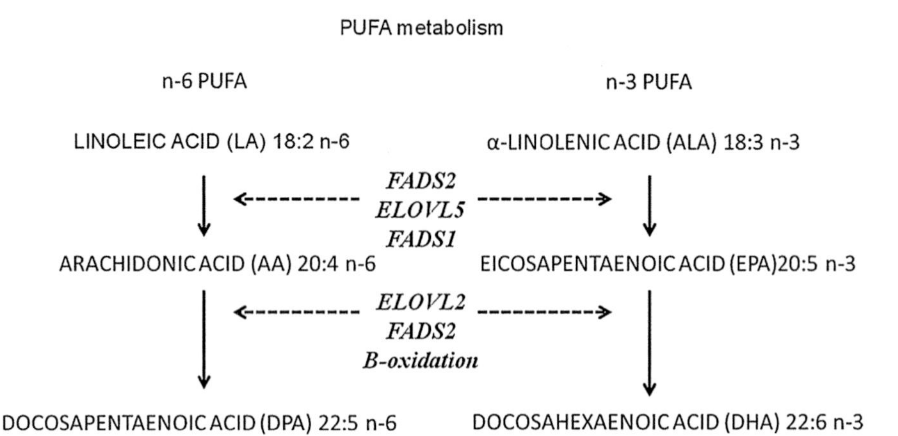

Fig. 2 Metabolism of the polyunsaturated fatty acids. The two families share the same set of enzymes, which explains the competition between them

therefore, the resulting ratio n-6/n-3 is close to 15, while 5 would be an ideal one (Baker et al. 2016).

Brain and retina are the organs in which the highest proportions of EPA and DHA can be found. A low abundance of these PUFAs in the cell membrane promotes several disorders in brain functions, and conversely, n-3 PUFAs abundance is favorable to brain development and physiology. Experimental animal models have amply demonstrated that impaired structure establishment and delayed cell migration in the brain are associated with a deficiency in DHA (Coti Bertrand et al. 2006). In humans, a recent study has shown that the maternal n-6/n-3 ratio had direct consequences on the child neurodevelopment (Bernard et al. 2013) and that higher DHA levels associated with low LA were favorable for preterm babies, particularly for the retina and visual acuity (Tam et al. 2016, Liu et al. 2010).

Heritable Effects of Maternal Exposure to N-3

The idea that n-3 could display epigenetic effects came from animal experiments mainly.

The fetus relies on maternal PUFA supply since its bioconversion of long-chain PUFAs is very limited (Haggarty 2010). The access through the placenta of different levels of n-3 PUFAs can directly affect molecule and protein synthesis, and processes involved in neurodevelopment, which could explain the delay in the fetal neurodevelopment of deficient pups already mentioned. Yet there exist other long-term consequences in the offspring which persist during adulthood and cannot be explained by the direct perinatal effects of these compounds.

Referring to experimental animal models, multiple examples can be cited. Hirabara et al. (2013) showed that feeding rats with fish oil for 2 generations resulted in a lower basal glycemia and an increase in insulin sensitivity in the G2 generation.

The maternal n-3 PUFA intake is correlated negatively with the offspring body weight in rodents and a higher n-3 maternal intake correlated positively with an alleviation of the metabolic syndrome (Kasbi-Chadli et al. 2014) although this association has not been confirmed in humans (Vidakovic et al. 2016). In piglets, maternal DHA favored offspring social behavior after weaning and improved sickness behavior after an inflammatory challenge (Clouard et al. 2015).

Similarly, we showed that in vitro cultures of neural stem cells (NSC) from newborn rats were deeply and durably modified by a maternal diet rich in DHA and EPA, functionally and molecularly (Goustard-Langelier et al. 2013): pups NSC from n-3-supplemented mothers proliferate more rapidly than cells from deficient animals (Fig. 3), but their differentiation was not enhanced by addition of DHA or AA in the culture medium, while the differentiation of deficient cells was increased, as assessed by the measurement of neurite lengths (Figs 4 and 5).

These observations point to the existence of persistent effects of n-3 PUFAs and, therefore, suggest an epigenetic mechanism of action. In humans, the evidence is less unequivocal, since observational or interventional studies have reached inconsistent results. While there seems to finally be no connection with maternal n-3 PUFA ingestion levels and child body mass index (BMI), fat deposition, or obesity in humans (Vidakovic et al. 2016), there seems to be a beneficial effect of maternal n-3 PUFAs on the outcome of allergic diseases and asthma in childhood (Best et al. 2016). A study also showed that ingestion of n-3 PUFAs during pregnancy correlated with lower systolic blood pressure, and a reduction of aortic stiffness in children, unrelated to the children's own n-3 PUFAs consumption (Bryant et al. 2015).

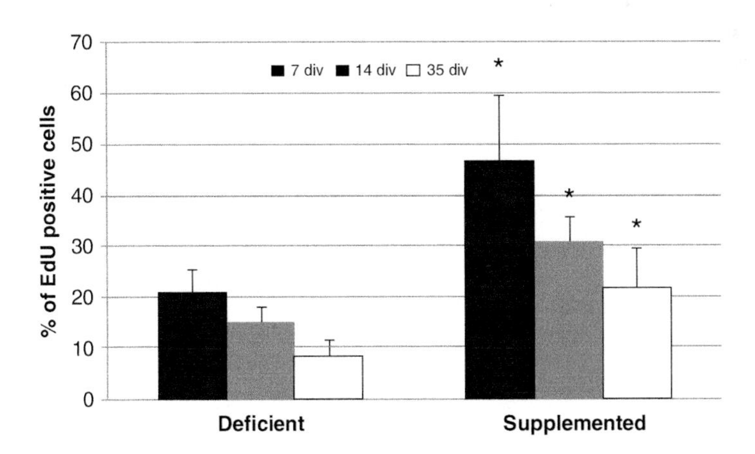

Fig. 3 Proliferation of neural stem cells derived either from n-3-deficient or n-3-supplemented rat pups. The dividing cells were labeled with 10 μM EdU for 6 h, allowed to adhere on glass coverslips coated with polyornithine and laminin, and fixed with paraformaldehyde. EdU labeling was realized per manufacturer's instructions, and the cells were counterstained with DAPI. Cells positive for EdU were counted, and the numbers were expressed as percentage of total cell counts. The cells were counted using Image J; at least 500 cells were counted for each coverslip. DIV, days in vitro (*Reprinted with permission from Elsevier from Goustard-Langelier* et al. *J. Nutr. Biochem. 24 (2013) 380–387*)

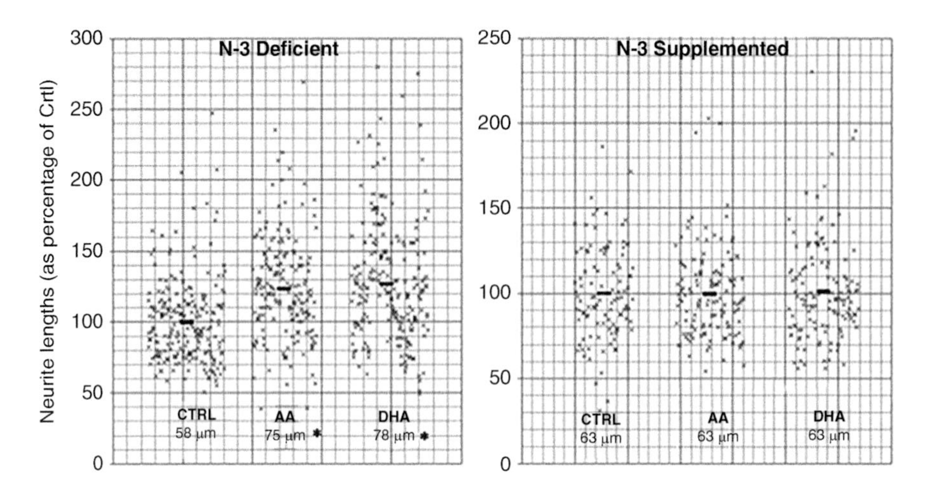

Fig. 4 Expansions of neurites in differentiated neurons from deficient or supplemented rat pups, supplemented or not with 5 μM AA or DHA. The cells from deficient or supplemented pups were allowed to differentiate on P/L glass coverslips for 8 days in the absence of growth factors. The cells were fixed with 4% paraformaldehyde and immunolabeled with β-III tubulin antibody, and the nuclei were counterstained with DAPI. The neurite lengths were measured using Image J. The figure compiles the results from cells that were examined from 20 to 35 days in vitro. The numbers at the bottom of the figure express the average length in μm recorded for cells in each treatment condition. The *asterisk* denotes a significant difference at $P < 0.05$ (*Reprinted with permission from Elsevier from Goustard-Langelier* et al. *J. Nutr. Biochem. 24 (2013) 380–387*)

The discrepancy of outcomes from experimental animal models and human observational studies may come from the fact that human diet is more complex than experimental diets for animal models and presents a variability in terms of quality, quantities, and ingestion pattern that can blur the effects of PUFAs. Also, as we will develop below, genetic variants in humans can influence the end result of PUFAs' actions.

In spite of the inconsistent and indefinite results obtained in humans, that have to be confirmed, a few studies have turned to the analysis of epigenetic effects induced by n-3 PUFAs.

Since this topic is just emerging, we will address the two domains that have been mainly developed so far, i.e., DNA methylation and micro-RNA synthesis.

N-3 PUFAs and DNA Methylation

There are several studies showing that n-3 PUFAs can modify global DNA methylation, either in vivo or in vitro, or the methylation state of the promoter of particular genes.

An association study in 2014 showed that in Alaska native Yup'ik people, 27 differentially methylated CpG sites correlated with n-3 PUFA intakes (Aslibekyan et al. 2014). The study was realized on 185 individuals, from both

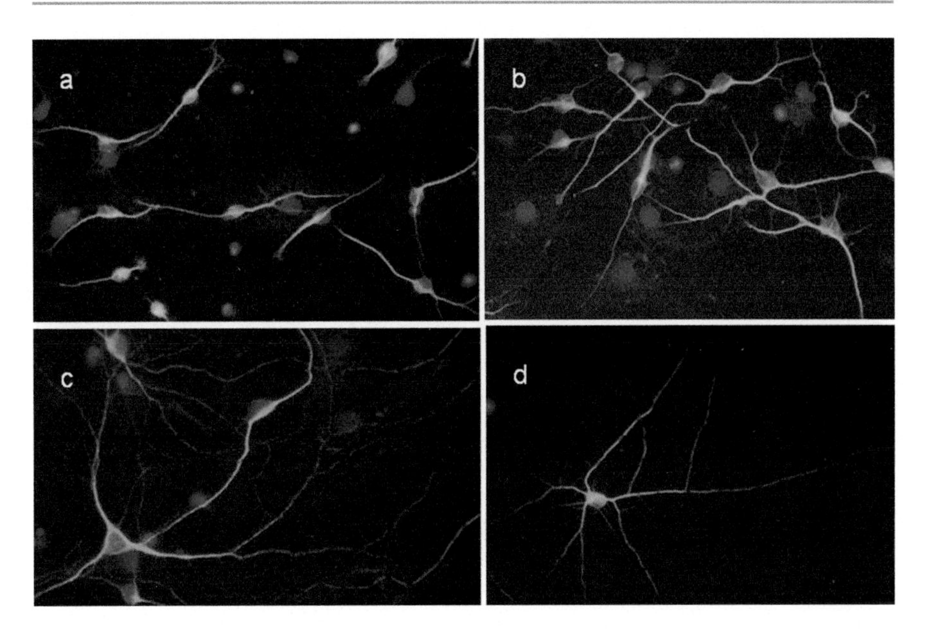

Fig. 5 Cells from deficient or supplemented rat pups differentiated in the presence of 5 μM DHA. (**a**) Neurons from n-3-deficient NSC cultures differentiated in the absence of 5 μM DHA. (**b**) Neurons from n-3-deficient NSC cultures differentiated in the presence of 5 μM DHA. (**c**) Neurons from n-3-supplemented NSC cultures differentiated in the absence of 5 μM DHA. (**d**) Neurons from n-3-supplemented NSC cultures differentiated in the presence of 5 μM DHA (*Unpublished (C. Heberden)*)

sex, and a total of approximately 470,000 CpG sites were examined. N-3 PUFAs modified the DNA methylation state in cord blood monocytes when the mothers had been supplemented with 400 mg/day DHA during the pregnancy from 8 to 12 weeks of gestation until parturition. The methylation levels of the promoters of T-relevant genes were modified by n-3 PUFA supplementation, and the Th1/Th2 balance in infants was modified as well (Lee et al. 2013).

In the same vein, in the rat fed a diet rich in fat and supplemented or not with n-3 PUFAs, n-3 PUFAs could modify the global methylation state and prevent the decrease of methylation induced by the high fat diet in Pparg2 promoter (Amaral et al. 2014).

Hirabara et al. (2013) showed that in mice supplemented with fish oil for 2 generations, the global DNA methylation was decreased.

The modifications of the methylation are more often studied on specific genes. For instance, in patients with renal diseases from either gender, 4 g per day of n-3 PUFAs for 8 weeks modified FADS2 and ELOVL5 methylation promoters in peripheral blood monocytes. The modifications were significant but different according to the sex. These modifications correlated with the modulation of the synthesis of PUFAs (Hoile et al. 2014).

Experimental animal models have also shown that supplementation with n-3 PUFAs can modify the methylation level of specific genes: when dams were supplemented with ALA during gestation and lactation, the maternal liver FADS2 gene promoter and intron I were hypermethylated, and a negative association was noted between hypermethylation and FADS2 expression. Similarly, the FADS2 promoter was hypermethylated in the offspring livers, to an extent comparable to the mothers' (Niculescu et al. 2013).

He et al. (2014) showed that in the rat daily supplementation of ALA during gestation and lactation modified the expressions of several genes in brain offspring and decreased the methylation of the promoter of methyl-CpG binding protein 2. In Sprague Dawley rats, n-3 PUFA supplementation reduced the methylation of the *BDNF* exon IV promoter induced by western diet. This methylation was associated with anxiety like in behavior in open arms equipment and correlated with gene expression and protein content (Tyagi et al. 2015).

In vitro as well, the impact of n-3 PUFAs on DNA methylation has been described: Ceccarelli et al. (2011) demonstrated that an incubation of leukemia U937 cells with EPA induces a demethylation of a single CpG site, and this demethylation correlates with a decrease in cell proliferation. More recently, it has been shown that EPA could, in the same cells, induce the demethylation of intron 1 of *HRas* (Ceccarelli et al. 2014).

If experimental models and in vitro studies confirm the existence of the effect of n-3 PUFAs on DNA methylation, the conclusions are less clear-cut in human studies. A recent study has pointed out the intricacy of the question by investigating the effects of n-3 PUFAs on the methylation of interleukin-6 promoter (IL6) (Ma et al. 2016). They showed that the effect of n-3 PUFAs was dependent on genetic variants: in heterozygotes for an allele of IL6, higher n-3 PUFAs correlated with lower methylation at the IL6 promoter, but did not in the homozygotes. Therefore, the effects of n-3 PUFAs could be modified by single-nucleotide polymorphisms.

The complexity of the human situation, either because of its genetics or its nutrition, results in somewhat contradictory results, but should not mask or exclude the possibility that n-3 PUFAs can act on DNA methylation state.

N-3 PUFAs and miR

The miR can regulate the transcriptome and consequently cell physiology. In view of the wide range of effects of fatty acids, the hypothesis that these components could affect miR expression has been tested on several models covering different fields. The next paragraph will sum up a few of the recent studies, and Table 1 recapitulates the identities of the miR involved and the putative functions that have been ascribed to them.

In vivo, Shah et al. (2011) demonstrated that a diet supplemented with fish oil had chemoprotective properties in the rat colon and that the diet had a specific miR signature. In obese Sprague Dawley rats, a diet enriched in marine alga oil, rich in DHA, was shown to reduce the amounts of miR-33 and miR-122, both involved in the enhancement of lipid metabolism genes related with obesity (Baselga-Escudero et al. 2013). Recently, it was shown that brown thermogenesis, beneficial against

Table 1 N-3 PUFAs and miR expressions in the literature

Ref	Organism	Treatment	miR	Function
Shah et al. 2011	Rat colon	Diet, fish oil	↑18a, 27b, 93, 200c, 497 ↓21	Chemoprevention
Baselga-Escudero et al. 2013	Obese rat liver	Marine alga oil	↓33, 122	Lipid metabolism
Goustard-Langelier et al. 2013	Newborn rat NSC cultures (ex vivo)	Diet during pregnancy, fish oil	↑210	Cell proliferation, BDNF?
Kim et al. 2016	Mice	Diet, fish oil	↑30c	thermogenesis
Farago et al. 2011	Glioma	DHA	↑20b, 17, 26a, 29c	apoptosis
Mandal et al. 2012	Breast cancer	DHA	↓21	Induction of PTEN and decrease of CSF1
Gil-Zamorano et al. 2014	Caco2	DHA	↑192, 30c	Lipid metabolism
Antal et al. 2014	Glioma	DHA + ionization	↑146, 181	Apoptosis
Daimiel-Ruiz et al. 2015	Caco2	DHA	↑107	Circadian rhythms
Yao et al. 2015	Human cholangiocarcinoma cells	DHA	↓26 a 26 b	↑15 PGDH

obesity and promoted by EPA, was induced through increased binding of EPA to the free fatty acid receptor 4 (FFAR4) leading to miR-30b and miR378 upregulation. The in vitro association was confirmed in vivo in fish-oil-fed mice (Kim et al. 2016).

In neural stem cell cultures derived from rat newborns born to deficient or supplemented n-3 PUFA mothers, we observed a significant increase in miR-210 expression. This miR is involved in cell proliferation, a parameter that was indeed increased in the supplemented-derived cultures (Goustard-Langelier et al. 2013).

In vitro also, Farago et al. showed that the cytotoxic action of AA and DHA on glioma cells was the result of specific changes of miR, the putative targets of which were apoptosis related. DHA has been often studied for its anticarcinogenic properties: in breast cancer cells, DHA treatment downregulates miR-21, which in turn upregulates the phosphatase and tensin homolog (PTEN) and which results in the decrease of colony-stimulating factor 1 (Mandal et al. 2012). In Caco2 cells, DHA upregulates miR associated with lipid metabolism and circadian rhythms (Gil-Zamorano et al. 2014; Daimiel-Ruiz et al. 2015). In human cholangiocarcinoma cells, DHA downregulates miR-26a and miR-26b, which consequently promotes the expression of 15-hydroxyprostaglandin dehydrogenase (*15 PGDH*), the enzyme which catalyzes the synthesis of prostaglandin E2 (Yao et al. 2015).

The PUFAs can act either on their own or in association with another treatment: for instance, PUFA associated with ionizing radiations induced the overexpression of miR-146 in human glioma cells in vitro (Antal et al. 2014). Also, PUFAs associated with fermentable fibers can synergize to play a protective role in colon cancer and modify the expression of several miR (Shah et al. 2011). Table 1 recapitulates the different miR modifications.

Unlike the work by Kim et al. cited above and related to brown thermogenesis, it has been not always determined whether the PUFAs act themselves by binding to their specific receptors or through their multiple derived compounds. Yet in the field of inflammation, it has been now demonstrated that n-3 PUFA derivatives are acting as mediators and are able to modify the panel of miR. Acute inflammation is indeed terminated by effective resolution which then allows tissue repair. If the resolution process is inefficient, a chronic inflammation takes place and leads to cellular damages. This process is coordinated by anti-inflammatory and pro-resolving mediators. Resolvins, protectins, or lipoxins are specialized mediators which are derived from n-3 PUFAs and are actively involved in the disruption of persistent inflammation. The history of the identification of resolvins has been retraced by Serhan et al. (2004), and for deeper knowledge, the reader is invited to refer to this excellent review. Resolvin D1 (7*S*,8,17*R*-Trihydroxy-docosa-4*Z*,9*E*,11*E*,13*Z*,15*E*,19*Z*-hexaenoic acid, RvD1) was identified in resolution exudates of mice treated with aspirin and DHA. Recchiuti et al. (2011) showed that an injection of 300 ng/Kg in mice could upregulate a set of 6 specific miR, the targets of which were involved in pro-inflammatory pathways. The results were reproduced in vitro on human macrophage cultures. RvD1 binds to two G-protein-coupled receptors, and the same laboratory (Krishnamoorthy et al. 2012) showed that through the binding to one of the GPCR, RvD1 was able to upregulate miR-208a, an miR upregulating the anti-inflammatory cytokine IL10 in human macrophages. In the rat, an intravitreal injection of RvD1 induced a downregulation of specific micro-RNAs related to the synthesis of Sirtuin1 (Rossi et al. 2015).

In the obesity field, a recent article has shown that in mice developing non-alcoholic steatohepatitis (NASH), RvD1 coupled to calorie restriction could improve multiple parameters deteriorated in NASH and induce a specific miR signature.

Thus through direct action or their derivatives, n-3 PUFAs can modify miR transcriptome and, therefore, modulate numerous target mRNAs and proteins.

Conclusion

In many ways, the epigenetic field is still in its infancy, and more work is needed to define its overall influence in physiology. Although the classical actions already described and recognized cannot be disregarded, epigenetic seems to be part of their abilities, and the effects of PUFAs are multiple. It is obvious that PUFAs represent more than mere membrane constituents. The molecular mechanisms of the PUFAs need to be better described and understood, since they are brought by the diet but also are consumed as supplements which are available over the counter. Therefore, the pioneering studies cited in this chapter deserve to be thoroughly dealt with and open the road to many more studies related to these essential food constituents.

Dictionary of Terms

- **Autacoids** – Substances released locally that can act as either messengers or hormones.
- **Body mass index** – Value obtained by dividing the weight of an individual by the square of his height. For an adult, a BMI lower than 25 kg/m^2 is normal, while higher than is considered overweight, and higher than 30 is considered obese.
- **CpG sites** – In the DNA sequence, a cytosine (C) immediately followed by a guanine (G). These sites are susceptible to methylation.
- **Desaturase** – Enzyme that creates a double bond by removing two carbon atoms from the aliphatic chain of a fatty acid.
- **Elongase** – Enzyme that adds two carbon atoms at the end of the aliphatic chain of a fatty acid.
- **Lipid rafts** – Microdomains of the plasma membrane, characterized by their lipid compositions. Compared to the rest of the plasma membrane, the lipid rafts are abundant in cholesterol and sphingolipids. Due to the lipid composition, these domains tend to be less fluid than the rest of the membrane. They serve as signaling platforms.
- **Polyunsaturated fatty acids** – Fatty acids with more than one double bond in their chemical structure. Fatty acids are divided in two categories: saturated, without double bond (e.g., palmitic acid, C16:0) or unsaturated. The unsaturated fatty acids can be monounsaturated (one double bond, e.g., oleic acid C18:1 n-9) or polyunsaturated (more than one double bond, e.g., linoleic acid, C18:2 n-6).

Key Facts

Key facts of linoleic and α-linolenic fatty acids

- Abundant in vegetables, not synthetized by mammals.
- Metabolized by the same set of enzymes.
- Esterified in membrane phospholipids.
- Linoleic acid is abundant in peanuts and sunflower.
- α-linolenic abundant in rapeseed and soybean.

Key facts of arachidonic acid

- Most abundant derivative of linoleic acid metabolism
- Ubiquitous

- Released by phospholipase A2
- Metabolized in eicosanoids: prostaglandins, leukotrienes
- Mainly involved in pro-inflammatory processes

Key facts of docosahexaenoic acid

- Most abundant derivative of α-linolenic acid metabolism
- Not ubiquitous, abundant in CNS
- Metabolized in resolvin, protectin, etc.
- Involved in anti-inflammatory processes
- Mainly present in seafood and pumpkin seeds

Key facts of neural stem cells

- Neural stem cells are present in defined areas of the brain: the subventricular zone, the hippocampus, and the hypothalamus.
- They are dividing during the whole life although at a slower pace with age.
- They are localized in proximity of ventricles or capillaries, which make them sensitive to the signals of the environment.
- They have different functions, related to the niche to which they belong: In the subventricular zone, they give rise to olfactory neurons; in the hippocampus, the new neurons impact memory, learning abilities, and mood; and in the hypothalamus, food intake and energy homeostasis.
- Nutrition can modulate adult neurogenesis by modifying neural stem cell properties.

Summary Points

1. Long-chain PUFAs belong to two different families, the n-6 and n-3 families.
2. N-3 long-chain PUFAs are inefficiently metabolized from ALA.
3. In the western diet, n-6 PUFA ingestion overrides n-3 PUFAs.
4. The main dietary source of long-chain n-3 PUFAs is fish and seafood.
5. The effects of dietary n-3 PUFAs are abundant and rely on varied molecular mechanisms.
6. Experimental models demonstrate that some effects are heritable in offspring.
7. In humans, heritable effects have been shown as well, such as aortic stiffness in children, related with n-3 maternal levels.
8. N-3 PUFAs modify the DNA methylation of the FADS2 gene promoter, a key enzyme for PUFA synthesis.
9. EPA and DHA can regulate miR synthesis in different organs.
10. Resolvin D1, a DHA derivative, acts through miR modifications to terminate inflammation.

References

Amaral CL, Crisma AR, Masi LN et al (2014) DNA methylation changes induced by a high-fat diet and fish oil supplementation in the skeletal muscle of mice. J Nutrigenet Nutrigenomics 7:314–326

Antal O, Hackler L Jr, Shen J et al. (2014) Combination of unsaturated fatty acids and ionizing radiation on human glioma cells: cellular, biochemical and gene expression analysis. Lipids Health Dis 13:article142

Aslibekyan S, Wiener HW, Havel PJ et al (2014) DNA methylation associated with n-3 fatty acids uptake in Yup'ik people. J Nutr 144:425–430

Baker EJ, Miles EA, Burdge GC et al (2016) Metabolism and functional effects of plant-derived omega-3 fatty acids in humans. Prog Lipid Res 60:30–56

Baselga-Escudero L, Arola-Arnal A, Pascual-Serrano A et al (2013) Chronic administration of proanthocyanidins or docosahexaenoic acid reverses the increase of miR-33a and miR-122 in dyslipidemic obese rats. PLoS One 8:e69817

Bernard JY, De Agostini M, Forhan A et al (2013) The dietary n-6: n-3 fatty acid ratio during pregnancy is inversely associated with child neurodevelopment in the EDEN mother-child cohort. J Nutr 143:1484–1488

Best KP, Gold M, Kennedy D et al (2016) Omega-3 long-chain PUFA intake during pregnancy and allergic disease outcomes in the offspring: a systematic review and meta-analysis of observational studies and randomized controlled trials. Am J Clin Nutr 103:128–143

Bryant J, Hanson M, Peebles C et al (2015) Higher oily fish consumption in late pregnancy is associated with reduced aortic stiffness in the child at age 9 years. Circ Res 116:1202–1205

Ceccarelli V, Racanicchi S, Martelli MP et al (2011) Eicosapentaenoic acid demethylates a single CpG that mediates expression of tumor suppressor CCAAT/enhancer-binding protein delta in U937 leukemia cells. J Biol Chem 286:27092–27102

Ceccarelli V, Nocentini G, Billi M et al (2014) Eicosapentaenoic acid activates RAS/ERK/C/EBPβ pathway through H-Ras intron 1 CpG island demethylation in U937 leukemia cells. PLoS One 9:e85025

Clouard C, Souza AS, Gerrits WJJ et al (2015) Maternal fish oil supplementation affects the social behavior, brain fatty acid profile, and sickness response of piglets. J Nutr 145: 2176–2184

Coti Bertrand P, O'Kusky JR, Innis SM (2006) Maternal dietary (n-3) fatty acid deficiency alters neurogenesis in the embryonic rat brain. J Nutr 136:1570–1575

Daimiel-Ruiz L, Klett-Mingo M, Konstantinidou V et al (2015) Dietary lipids modulate the expression of miR-107, an miRNA that regulates the circadian system. Mol Nutr Food Res 59:552–565

Farago N, Feher LZ, Kitajka K et al (2011) MicroRNA profile of polyunsaturated fatty acid treated glioma cells reveal apoptosis-specific expression changes. Lipids Health Dis 10. Article Number: 173

Gil-Zamorano J, Martin R, Daimiel L et al (2014) Docosahexaenoic acid modulates the enterocyte Caco-2 cell expression of microTNAs involved in lipid metabolism. J Nutr 144:575–585

Goustard-Langelier B, Koch M, Lavialle M et al (2013) Rat neural stem cell proliferation and differentiation are durably altered by the in utero polyunsaturated fatty acid supply. J Nutr Biochem 24:380–387

Haggarty P (2010) Fatty acid supply to the human fetus. Annu Rev Nutr 30:237–255

He F, Lupu DS, Niculescu MD (2014) Perinatal α-linolenic acid availability alters the expression of genes related to memory and to epigenetic machinery, and the Mecp2 DNA methylation in the whole brain of mouse offspring. Int J Dev Neurosci 36:38–44

Hirabara SM, Folador A, Fiamoncini J et al (2013) Fish oil supplementation for two generations increases insulin sensitivity in rats. J Nutr Biochem 24:1136–1145

Hoile SP, Clarke-Harris R, Huang RC et al (2014) Supplementation with N-3 long-chain polyunsaturated fatty acids or olive oil in men and women with renal disease induces differential changes in the DNA methylation of FADS2 and ELOVL5 in peripheral blood mononuclear cells. PLoS One 17:e109896

Kasbi-Chadli F, Boquien CY, Simard G et al (2014) Maternal supplementation with n-3 long chain polyunsaturated fatty acids during perinatal period alleviates the metabolic syndrome disturbances in adult hamster pups fed a high-fat diet after weaning. J Nutr Biochem 25:726–733

Kim J, Okla M, Erickson A et al (2016) Eicosapentaenoic acid potentiates Brown thermogenesis through FFAR4-dependent up-regulation of miR-30b and miR-378. J Biol Chem 291:20551–20562

Krishnamoorthy S, Recchiuti A, Chiang N et al (2012) Resolvin D1 receptor Stereoselectivity and regulation of inflammation and Proresolving MicroRNAs. Am J Pathol 180:2018–2027

Langelier B, Linard A, Bordat C et al (2010) Long chain-polyunsaturated fatty acids modulate membrane phospholipid composition and protein localization in lipid rafts of neural stem cell cultures. J Cell Biochem 110:1356–1366

Lee HS, Barraza-Villarreal A, Hernandez-Vargas H et al (2013) Modulation of DNA methylation states and infant immune system by dietary supplementation with ω-3 PUFA during pregnancy in an intervention study. Am J Clin Nutr 98:480–487

Liu A, Chang J, Lin Y, Shen Z, Bernstein PS (2010) Long-chain and very long-chain polyunsaturated fatty acids in ocular aging and age-related macular degeneration. J Lipid Res 51:3217–3229

Ma Y, Smith CE, Lai CQ et al (2016) The effects of omega-3 polyunsaturated fatty acids and genetic variants on methylation levels of the interleukin-6 gene promoter. Mol Nutr Food Res 60:410–419

Mandal CC, Ghosh-Choudhury T, Dey N et al (2012) miR-21 is targeted by omega-3 polyunsaturated fatty acid to regulate breast tumor CSF-1 expression. Carcinogenesis 33:1897–1908

Niculescu MD, Lupu DS, Craciunescu CN (2013) Perinatal manipulation of α-linolenic acid intake induces epigenetic changes in maternal and offspring livers. FASEB J 27:350–358

Piomelli D, Astarita G, Rapaka R (2007) A neuroscientist's guide to lipidomics. Nat Rev Neurosci 8:743–754

Recchiuti A, Krishnamoorthy S, Fredman G et al (2011) MicroRNAs in resolution of acute inflammation: identification of novel resolvin D1-miRNA circuits. FASEB J 25:544–560

Rossi S, Di Filippo C, Gesualdo C et al (2015) Interplay between intravitreal RvD1 and local endogenous Sirtuin-1 in the protection from endotoxin-induced uveitis in rats. Mediat Inflamm 2015:126408

Serhan CN, Gotlinger K, Hong S et al (2004) Resolvins, docosatrienes, and neuroprotectins, novel omega-3-derived mediators, and their aspirin-triggered endogenous epimers: an overview of their protective roles in catabasis. Prostaglandins Other Lipid Mediat 73:155–172

Shah MS, Schwartz SL, Zhao C et al (2011) Integrated microRNA and mRNA expression profiling in a rat colon carcinogenesis model: effect of a chemo-preventive diet. Physiol Genomics 43:640–654

Tam EWY, Chau V, Barkovich AJ et al (2016) Early postnatal docosahexaenoic acid levels and improved preterm brain development. Pediatr Res 79:723–730

Tyagi E, Zhuang Y, Agrawal R et al (2015) Interactive actions of Bdnf methylation and cell metabolism for building neural resilience under the influence of diet. Neurobiol Dis 73:307–318

Vidakovic AJ, Gishti O, Voortman T et al (2016) Maternal plasma PUFA concentrations during pregnancy and childhood adiposity: the generation R study. Am J Clin Nutr 103:1017–1025

Yao L, Han C, Song K et al (2015) Omega-3 polyunsaturated fatty acids upregulate 15-PGDH expression in cholangiocarcinoma cells by inhibiting miR-26a/b expression. Cancer Res 75:1388–1398

Threonine Catabolism: An Unexpected Epigenetic Regulator of Mouse Embryonic Stem Cells

83

Ruta Jog, Guohua Chen, Todd Leff, and Jian Wang

Contents

R. Jog · G. Chen · T. Leff
Department of Pathology, Wayne State University School of Medicine, Detroit, MI, USA
e-mail: rjog@med.wayne.edu; gche@med.wayne.edu; tleff@wayne.edu

J. Wang (✉)
Department of Pathology, Wayne State University School of Medicine, Detroit, MI, USA

Cardiovascular Research Institute, Wayne State University School of Medicine, Detroit, MI, USA
e-mail: jianwang@med.wayne.edu

V. B. Patel, V. R. Preedy (eds.), *Handbook of Nutrition, Diet, and Epigenetics*,
https://doi.org/10.1007/978-3-319-55530-0_103

Abstract

Mouse embryonic stem cells (mESCs) are prototypical *in vitro* models of pluripotent stem cells. They are characterized by a capacity for infinite self-renewal while retaining the ability to differentiate into each of the cell types of the embryo. The maintenance of their pluripotent state relies on a complex regulatory network involving cytokine signaling and transcriptional controls at genetic and epigenetic levels. More recently, it has become evident that mESC pluripotency requires a specific nutritional environment. We now understand that mESC pluripotency is critically dependent on threonine catabolism for provision of one- and two-carbon donors for pluripotency-related chromatin modifications. In this chapter, we provide a comprehensive overview of the cellular processes required for the maintenance of mESC pluripotency, including signaling pathways, transcriptional networks, and epigenetic regulation. In addition, we discuss the latest developments concerning the unique dependence of mESC on threonine and the role of the amino acid in establishing the epigenetic status required for mESC self-renewal.

Keywords

Embryonic stem cell · Pluripotency · Transcription regulation · Epigenetic regulation · Histone methylation · Bivalent domain · Threonine metabolism · Glycine metabolism · One carbon metabolism · Threonine dehydrogenase

List of Abbreviations

CDK	Cyclin dependent kinase
DNMT	DNA methyltransferases
EC	Embryonal carcinoma
Erk	Extracellular signal-related kinase
ESC	Embryonic stem cells
GCAT	2-amino-3-oxobutyrate coenzyme A ligase
GCS	Glycine cleavage system
GLDC	Glycine decarboxylase
GSK3	Glycogen synthase kinase 3
HAT	Histone acetyltransferase
hCys	Homocysteine
HMT	Histone methyltransferase
ICM	Inner cell mass
iPSC	Induced pluripotent stem cell
JAK	Janus-associated kinases
LIF	Leukemia inhibitory factor
LIFR	Leukemia inhibitory factor receptor
mESC	Mouse embryonic stem cells

Met	Methionine
pRB	Phosphorylated retinoblastoma protein
RB	Retinoblastoma protein
SAH	S-adenosyl homocysteine
SAM	S-adenosylmethionine
TCA	Tricarboxylic acid
TDH	Threonine dehydrogenase
TrxG	Trithorax group

Introduction

On the third day after fertilization, a developing mouse embryo progresses to the blastocyst stage, a hollow spherical embryonic structure. The outer layer of the blastocyst is composed of trophoectoderm cells that provide nutritional support to the growing embryo and eventually give rise to a large part of the placenta. The inner cell mass (ICM) of the blastocyst is composed of pluripotent embryonic stem cells (ESCs) that ultimately develop into all cell types that comprise the three embryonic germ layers – ectoderm, mesoderm, and endoderm. ESCs of the ICM exist transiently in the blastocyst stage and progressively disappear over the course of embryonic development. In 1980s, the first ESC line was isolated from the 3.5-day mouse embryos by Evans and Kaufman (1981). This landmark study demonstrated that pluripotency could be captured and preserved in a culture dish. Since then, ESCs have been widely used in many research areas and have been invaluable tools for understanding cell differentiation and for modeling genetic diseases. In addition, they became an essential tool in the generation of genetically modified mouse lines. The isolation of human ESCs in 1998 (Shamblott et al. 1998; Thomson et al. 1998) has generated tremendous interest in the potential utility of these pluripotent stem cells (PSCs) for regenerative medicine applications.

Under proper culture conditions, ESCs can undergo unlimited self-renewal through symmetric cell division, while maintaining their pluripotency, i.e., the ability to differentiate into all cell types of a developing embryo upon stimulation by the differentiation signals. Interestingly, ESCs exhibit an unconventional mode of cell division and a unique dependence on genetic and epigenetic control of the expression of genes involved in the maintenance of pluripotency. This dependence imposes stringent demands on nutritional support and cytokine signaling, which collectively contribute to their preservation in the naïve, or “ground state,” of pluripotency.

In this chapter, we provide a historical outline of the milestones of ESC research, highlighting the advances in our understanding of the cell cycle, cytokine signaling, and transcriptional control of ESCs. We will then discuss the findings that demonstrate how the epigenetic and metabolic state shape the self-renewal and differentiation of ESCs, with a focus on threonine catabolism in mouse ESCs (mESCs). Figure 1 presents a timeline of the key discoveries in ESC biology. Readers are encouraged to refer to the following reviews for broader summaries on amino acid

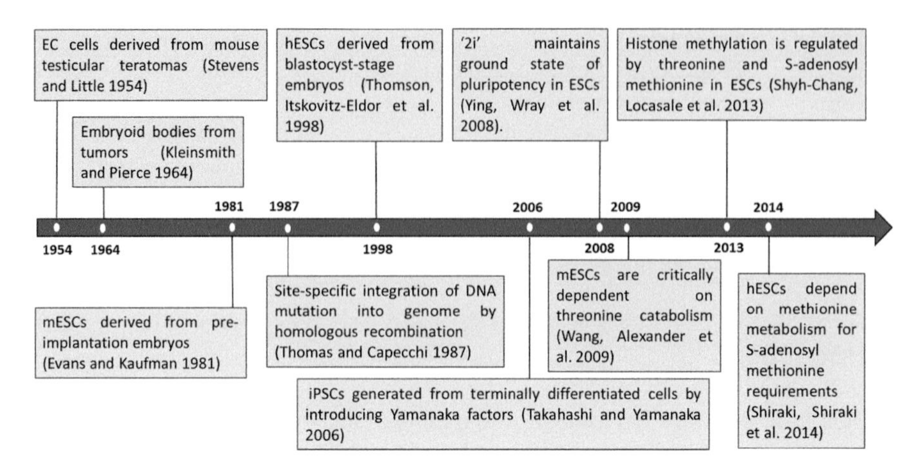

Fig. 1 A timeline illustrates the key events in the study of embryonic stem cell and the recent development in understanding regulation of epigenetic state by amino acid metabolism in embryonic stem cells

metabolism, epigenetics, and stem cell maintenance (Kaelin and Mcknight 2013; Kilberg et al. 2016; Martello and Smith 2014; Mentch and Locasale 2016).

Brief Historical Background of Embryonic Stem Cell Research

Embryonal Carcinoma Cells *In Vitro* and *In Vivo*

A seminal discovery made in 1954 by Stevens and Little set the stage for the emergence of the field of stem cell research. They observed that unlike most tumors, spontaneously occurring teratomas in mouse testes were composed of a mixture of cells that resembled a wide variety of cell and tissue types from both embryonic and adult tissues. Interestingly, when these teratomas were transplanted into new host animals, the resulting tumor contained undifferentiated cells (Stevens and Little 1954). These undifferentiated cells were called embryonal carcinoma (EC) cells.

Kleinsmith and Pierce developed a technique to successfully clone EC cells by enzymatically dissociating tumors into single cells from fresh tumors, which were then transplanted into new host animals. The resulting teratomas were composed of a heterogeneous mass of multipotential EC cells that were organized into an embryoid body (Kleinsmith and Pierce 1964). This method of growing teratoma cells served as a useful model for studying mammalian cell fate determination that in many ways was more experimentally amenable than normal embryos (Martin and Evans 1975).

Derivation of Embryonic Stem Cells

In 1981, mouse embryonic stem cells (mESCs) were successfully derived from pre-implantation embryos (Evans and Kaufman 1981). ESCs are in vitro equivalents of the inner cell mass of blastocyst-stage embryos. In 1998, Thomson et al. successfully derived human ESCs from donated blastocyst-stage embryos from in vitro fertilization procedures. These early-stage embryos contained pluripotent cells that could give rise to all three germ layers. The ESCs derived from these embryos had a normal karyotype and contained high levels of telomerase activity which conferred an immortality to these cell lines that other normal diploid human somatic cells do not possess. These ESCs expressed cell surface markers specific for undifferentiated pluripotent cells that distinguish them from other lineage-specific stem cells (Thomson et al. 1998).

The derivation of ESCs opened new avenues of investigation that not only provided insights into basic developmental processes (Nichols and Smith 2012) but also served as the central component of methodologies developed for manipulating gene expression profiles through genetic targeting (Thomas and Capecchi 1987). These methods allowed the development of new disease models and provided insights into the potential use of ESCs for regenerative medicine. Despite the enormous scientific and medical value of human ESC research, ethical concerns posed impact on research into the application of ESC-based approaches to the development of therapies to treat a variety of human diseases.

Genetic Manipulation of Embryonic Stem Cells

The discovery of pluripotent stem cells, and the development of methods for their manipulation, formed the basis for what became one of the most valuable genetic research tools of the last several decades. Thomas and Capecchi (1987) made use of the endogenous enzymatic machinery for homologous recombination to generate site-specific integration of a DNA molecule into the genome of ESCs. This provided the means to produce experimental animal models by introduction of germline deletions or mutations in specific gene(s). This technique was subsequently refined to allow for the mutation or deletion of a gene to be limited to a specific cell-type or tissue, as opposed to the "global" alteration of a gene in all tissues. This was accomplished by using a Cre/loxP based approach in which the expression of the recombinase is limited to specific cell or tissue types. Further refinements allowed for temporal regulation of the recombinase, providing the means to induce a gene knockout, or disease-related mutation into a specific tissue and at a specific developmental stage (Baron et al. 1999).

Reprogramming of Somatic Cells to Induced Pluripotent Stem Cells

In 2006, the pioneering work of Takahashi and Yamanaka demonstrated that PSCs could be generated from terminally differentiated fibroblasts simply by the introduction of four transcription factors: Oct4, Sox2, Klf4, and c-Myc (later designated as Yamanaka factors). These cells were termed "induced pluripotent stem cells" or iPSCs. The iPSCs exhibited the full spectrum of molecular and developmental properties associated with ESCs (Takahashi and Yamanaka 2006). This was a defining moment in stem cell biology as iPSCs could now be generated easily from terminally differentiated cells, broadening the accessibility and utility of this approach in the study and treatment of human disease. For example, iPSCs made from individuals affected by genetic diseases have been used to generate unique animal and cell models that are useful in probing the molecular mechanisms of disease (Romito and Cobellis 2016). Recent development of methods utilizing benign, noninsertional gene transfer vectors (Okita et al. 2010), together with epigenetic (Liang and Zhang 2013) and nutritional (Romito and Cobellis 2016) modulation, have improved the safety and efficacy of iPSC reprogramming.

Fundamentals of Embryonic Stem Cell Self-Renewal

The property of pluripotency is transferred from one generation of ESCs to the next through self-renewal. Unlocking the molecular basis for ESC self-renewal has been a major research focus in the field. Cell division of ESC is symmetrical, producing two progeny cells that are structurally and functionally equivalent to the parent cell. ESCs are small in size, with the nucleus making up a major portion of the cell volume (Thomson et al. 1998). Replication of genetic contents imposes a major task on ESC self-renewal. Recent developments have defined a naïve, pluripotent ground state, and the transcriptional network critical for maintenance of this pluripotent state.

ESC Self-Renewal Couples with Unconventional Rapid Cell Proliferation

The nature of ESC self-renewal is highlighted by rapid cell proliferation and a unique cell cycle structure (Fig. 2). Under standard culture conditions, the doubling time of mESCs can be as short as 8 h (Savatier et al. 1994). The eukaryotic cell cycle is divided into four chronological phases: G1 (Gap 1), an interphase in which cells synthesize mRNA and proteins in preparation for the next phase; S (synthesis), in which DNA is replicated; G2 (Gap 2), a second interphase in which cells synthesize mRNA and proteins in preparation for mitosis; and M (mitosis), in which cellular contents of a parent cell are segregated into two daughter cells. The main differences of the cell cycle organization between renewing and differentiated ESCs are the relative lengths of the G1 and S phases. Normally, differentiated somatic cells have relatively longer G1 (around 18 h) and a shorter S phase, whereas renewing ESCs spend a much shorter amount of time in the G1 phase (less than 2 h) and have an extended S-phase (Hindley and Philpott 2013). In spite of these differences, the total time of genome duplication is not significantly different between the two cell types (Ahuja et al. 2016; Savatier et al. 1994). The relatively long S phase satisfies the demands of ESCs DNA replication. Moreover, the robustness and plasticity of ESCs is related to their short G1 phase (Festuccia et al. 2016); only ESCs in G1 phase

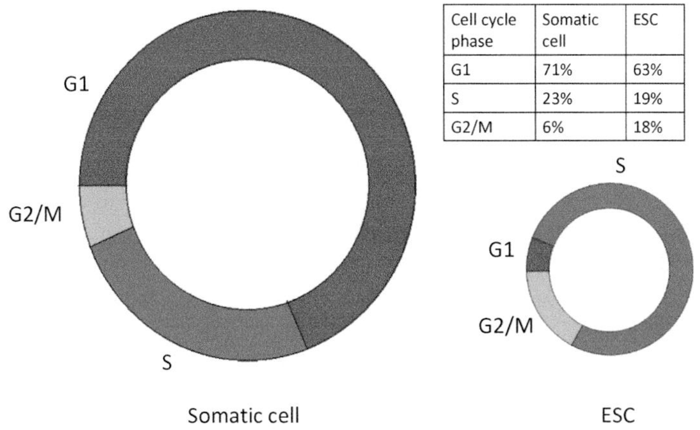

Cell cycle phase	Somatic cell	ESC
G1	71%	63%
S	23%	19%
G2/M	6%	18%

G1
G2/M
S
Somatic cell

S
G1
G2/M
ESC

Fig. 2 A schematic representation of cell cycle structures in renewing and differentiated embryonic stem cells. Circular diagrams illustrate the relative length of time for somatic (*Left*) or embryonic stem (*Right*) cells to spend in each phase during cell cycle progression, with the percentage of total cells in each phase shown in the inset table (Savatier et al. 1994). *G1* Gap1, *S* DNA synthesis, *G2* Gap2, *M* Mitosis

display developmental plasticity in response to differentiation signals (Coronado et al. 2013).

Studies on the mechanisms for maintaining rapid proliferation of ESCs revealed important roles for key modulators of cell cycle progression, such as retinoblastoma protein (Rb) and cyclin and cyclin-dependent kinases (CDKs). Savatier et al. analyzed the difference between rapidly renewing mESCs and regularly proliferating mouse embryonic fibroblasts by examining phosphorylated Rb protein expression levels in these cell types. They demonstrated that, while total Rb levels were similar in differentiated versus renewing cells, renewing ESCs sustained higher levels of the hyperphosphorylated form of Rb (Savatier et al. 1994). This high level of hyper-phosphorylated Rb "jumpstarts" the ESC cell cycle by promoting the transition from G1 to S phase. Due to negligible expression levels of CDK inhibitors, ESCs exhibit high CDK activities throughout cell cycle (Koledova et al. 2010). Particularly, high levels of CDK2, as a result of the continuous expression of both cyclin E and A, contribute to the rapid G1-S transition in ESCs (Stead et al. 2002). Loss-of-function studies demonstrated that inhibition of CDK2 extends G1 phase and sensitizes ESCs to differentiation signals (Neganova et al. 2009). While this rapid cell cycle promotes pluripotency, it imposes specific demand for cellular metabolism, which will be discussed in detail below.

"Naïve" ESCs Are Self-Autonomous

The search for exogenous cytokines critical for ESC self-renewal revealed that the myeloid leukemia inhibitory factor (LIF), a soluble factor secreted by the inactive feeder layer of fibroblasts, is required for preventing ESC differentiation in culture (Smith et al. 1988; Williams et al. 1988). LIF, a secreted glycoprotein in the form of a single polypeptide chain, belongs to interleukin-6 cytokine family (Metcalf 1991). LIF receptor (LIFR) forms a complex with the coreceptor glycoprotein 130 (Gp130) on the surface of ESCs (Hilton et al. 1988). Binding of LIF to LIFR/Gp130 heterodimer activates the Janus kinase (JAK) pathway, leading to sustained ESC pluripotency mediated by Stat3-dependent transcriptional activation (Niwa et al. 1998). However, the downstream of LIF signaling bifurcates into two pathways with opposing activities. The extracellular signal-related kinase (Erk) promotes ESC differentiation (Burdon et al. 1999), while Stat3 activation is required for ES cell self-renewal (Niwa et al. 1998).

Later, the nature of ESC self-renewal was further defined by the delineation of the "ground state" of naïve pluripotency. Martello and Smith (2014) explain ground state as "robust self-renewal of a biologically homogeneous population of cells, each with unbiased potential to form all somatic cell lineages and germ cells." Practically, the "ground state" of pluripotency is stabilized in culture using "2i" media, a chemically defined serum-free formula supplemented with two key small molecule kinase inhibitors – PD0325901 and CHIR99021, inhibitors for Erk and glycogen synthase kinase 3 (GSK3) signaling, respectively (Ying et al. 2008). Under conventional serum/LIF culture condition, ESCs exist in a "metastable" state, cycling in and out of naïve pluripotency, because serum-derived and autocrine FGF4/ErK (Kunath et al. 2007) differentiation signals continuously counteract ESC self-renewal. In contrast, ESCs maintained in "2i" culture uniformly exist in the naïve pluripotent state, showing consistent spherical colony morphology and, importantly, enhanced clonogenicity and chimerization capacity (Ying et al. 2008). At the molecular level, "2i" culture locks ESCs in a distinct transcriptional and epigenetic state that consists of the uniform expression of key pluripotent transcription factors, as well as specific DNA and histone methylation patterns (Ying et al. 2008, 2013). The discovery of a factor that acts to maintain naïve pluripotency by blocking differentiation signals reveals the self-autonomous nature of ESC self-renewal and confirms the importance of intrinsic cellular processes, such as a pluripotency-related transcriptional hierarchy, for the maintenance of ESC plasticity.

Master Transcriptional Regulators of Pluripotency: The Oct4, Sox2, and Nanog Triad

Cell fate decisions depend on precise control of gene expression, largely achieved through the regulation of gene expression by specific transcription factors. Accumulated evidence has shown that Oct4, Sox2, and Nanog are the core transcriptional factors of pluripotency, summarized in the review by Boyer et al. (2005). They are highly expressed in renewing ESCs and control an indispensable regulatory circuitry within the transcriptional hierarchy in ESCs. For their molecular features refer to Table 1.

Table 1 Molecular features of the master pluripotent transcription factors Oct4, Sox2, and Nanog

Transcription factor	Size (# amino acids)	Expression	DNA binding sites	Functional features
Oct4	360	ICM ESC	HMG-domain: 5′.. CTTTGTT.0.3′ POU-domain: 5′.. ATGCATCT.0.3′	Essential for pluripotency Expression rapidly decreases during embryonic development (Scholer et al. 1990)
Sox2	320	ICM Adult stem or progenitor cells (Arnold et al. 2011) ESC	HMG-domain: 5′.. CATTGTG.0.3′ POU-domain: 5′.. ATGCATAT.0.3′	Essential for pluripotency Deletion is embryonic lethal (Avilion et al. 2003)
Nanog	300	ICM Epiblast ESC	HMG-domain: 5′.. CATTGTA.0.3′ POU-domain: 5′.. ATGCAAAA.0.3′	Deletion is embryonic lethal (Mitsui et al. 2003) Regulatory role in naïve and primed PSCs Fluctuating levels provide ESCs with variable resistance to differentiation (Chambers et al. 2007)

The core transcription factors (Oct4, Sox2, and Nanog) form an interconnected network in which they autoregulate their own transcription also bind to elements in the promoters of the other pluripotency genes and regulate their expression in a highly coordinated manner. The core triad activates a set of genes that maintain pluripotent state (protein-coding genes and miRNA genes) while repressing genes that promote differentiation or lineage specification of the ESCs (Young 2011; Cole and Young 2008). Oct4 and Sox2 can function as a dimer, with the POU-domain of Oct4 directly interacting with the HMG-domain of Sox2 (Ambrosetti et al. 1997; Masui et al. 2007). Nanog is a target of Oct4/Sox2 and acts in turn to transactivate Oct4/Sox2, forming a feedforward regulatory loop that consolidates and stabilizes the pluripotent state (Loh et al. 2006). By means of genome-scale location analysis and high-throughput expression profiling (Boyer et al. 2005; Loh et al. 2006), it was determined that almost half of the target gene promoters which are occupied by Oct4 also have Sox2 binding sites. Interestingly, more than 90% these Oct4-Sox2 complex bound promoters were also occupied by Nanog at proximity forming an interconnected autoregulatory loop. This core set of several hundred genes that are regulated by Oct4/Sox2/Nanog triad is enriched for epigenetic regulators that presumably establish gene regulatory networks that promote pluripotency and repress lineage specification (Boyer et al. 2005; Cole and Young 2008; Sharov et al. 2008). Epigenetic regulation through DNA methylation at the CpG islands and histone post-translational modifications, in conjunction with other epigenetic modifiers, contributes to the precise control of pluripotent gene expression program.

DNA and Histone Methylation: Epigenetic Regulation of ESC Self-Renewal

In conjunction with the action of transcription factors, epigenetic regulation of gene expression, achieved by chemical modifications of genomic DNA and histones, plays a critical role in defining cellular phenotype (Goldberg et al. 2007). Among a variety of epigenetic mechanisms, such chromatin modifications mainly involve methylation of cytosine residues in the CpG islands on genomic DNA and methylation or acetylation of the lysine residues on histone tails. Both classes of modifications can dynamically alter gene expression patterns in response to developmental cues. In principle, the DNA methylation (a marker of heterochromatin) represses, and the histone acetylation (a marker of euchromatin) activates gene expression. However, the regulatory effects by histone methylation are complex and variable depending on which lysine residue is modified, and the number of methyl group attached to it. For example, trimethylation of histone H3 at lysine 9 (H3K9me3), lysine 27 (H3K27me3), or lysine 36 (H3K36me3) is associated with transcriptional repression, while the same modification at lysine 4 (H3K4me3) activates gene expression. For molecular details of epigenetic regulation, readers are referred to these excellent reviews (Jaenisch and Bird 2003; Rice and Allis 2001). ESCs have an enriched euchromatin that is required for self-renewal. Below, we describe the feature and functional importance of DNA and histone methylation in ESCs.

DNA Methylation in Early Development and in Embryonic Stem Cells

Genome-wide profiling of DNA methylation patterns have demonstrated the dynamic range of DNA methylation in early embryonic development by shedding light upon its role in regulating cell fate decision of pluripotent stem cells. Immediately after fertilization, the paternal genome of the zygote undergoes a rapid global demethylation through an active mechanism mediated in part by the DNA demethylase Tet3 (Shen et al. 2014). In contrast, the maternal genome undergoes demethylation passively, mainly through DNA replication in cell divisions during the cleavage stage of development (Santos et al. 2002). Together, these result in global hypomethylation in the ICM of the epiblast-stage embryo where the pluripotent stem cells exist. Coincident with embryo implantation, when cell differentiation enters the center stage, DNA methylation patterns start to be reset via the upregulation of the DNA methyltransferases (DNMTs) that catalyze the addition of methyl groups to DNA (Santos et al. 2002). The temporal order of these events indicates that DNA methylation is crucial for cell lineage specification but dispensable for establishment of pluripotency. This concept has been validated using ESC models. ESCs at the naïve pluripotent state in "2i" media exhibit uniform global DNA hypomethylation (Leitch et al. 2013). Importantly, complete erasure of DNA methylation by deletion of all DNMT paralogs does not alter the ESC morphology and self-renewal but does block ESC differentiation (Jackson et al. 2004). Thus, DNA methylation is a key epigenetic mechanism for facilitating the exit of ESCs from the pluripotent state.

Histone Methylation and Stem Cell Histone Bivalent Domains

Congruent with the global DNA hypomethylation, ESCs are enriched with histone modifications associated with active chromatin. These include acetylation of histones H3 and H4, and trimethylation of lysine 4 on histone 3 (H3K4me3) (Kimura et al. 2004; Lee et al. 2004). Exit from the naïve state and commencement of cellular differentiation is associated with a decrease in the global levels of these modifications, rendering the epigenetic landscape relatively suppressive. Although acetylation of histones is of clear importance for the regulation of the pluripotent state, we will focus primarily on the functional importance of H3K4me3 in maintaining pluripotency. The major enzymatic mediators of the H3K4me3 modification belong to the Trithorax group (TrxG) family of histone methyltransferase complexes (Shilatifard 2012). Studies have demonstrated that knockdown of different subunits of TrxG protein complexes strongly affect the expression of pluripotent genes and result in defective self-renewal and pluripotency (Clouaire et al. 2012; Wysocka et al. 2005; Jiang et al. 2011). Sustaining high levels of H3K4me3 is clearly required for the establishment and maintenance of pluripotency.

Interestingly, ESC epigenome contains many specialized regions called "bivalent domains" (Azuara et al. 2006; Bernstein et al. 2006). These domains are enriched in genes that establish cell identity. They were termed "bivalent" because they contain counterbalancing histone modifications: H3K4me3 (activating) and H3K27me3 (repressing). The presence of these two modifications function to maintain these developmental genes in a crucial state: silenced but also poised for induction from differentiating cues, if required. Upon receipt of differentiation cues, the epigenetic state is altered and the bivalent domains are resolved at the developmental genes (Fig. 3). The specific epigenetic pattern provides cells with their functional identity. This cellular identity is retained after cell division in the resultant daughter cells through multiple generations, with DNA and histone modification playing an important part in the regulation at epigenetic level.

Linking of Cellular Metabolism to Epigenetic Regulation

Metabolism refers to the collection of biochemical reactions through which cells convert nutrients, such as carbohydrates, lipids, and amino acids, into small metabolic intermediates that fuel the energy production and macromolecule biosynthesis essential for cellular homeostasis. Certain small metabolic intermediates, such as acetyl-CoA, S-adenosylmethionine (SAM), NAD^{+}, and ATP, serve dual functions as both substrates for biosynthetic reactions and as regulators of gene expression. Fluctuations in their amounts relay information about the metabolic fitness of the cell to multiple regulatory processes at virtually all levels, acting as critical feedback links between metabolic pathways and regulatory controls of the cell (Kaelin and Mcknight 2013; Janke et al. 2015).

In this regard, the profound effect and functional importance of the intrinsic links between metabolism and cellular regulation is seen most clearly in the influence that

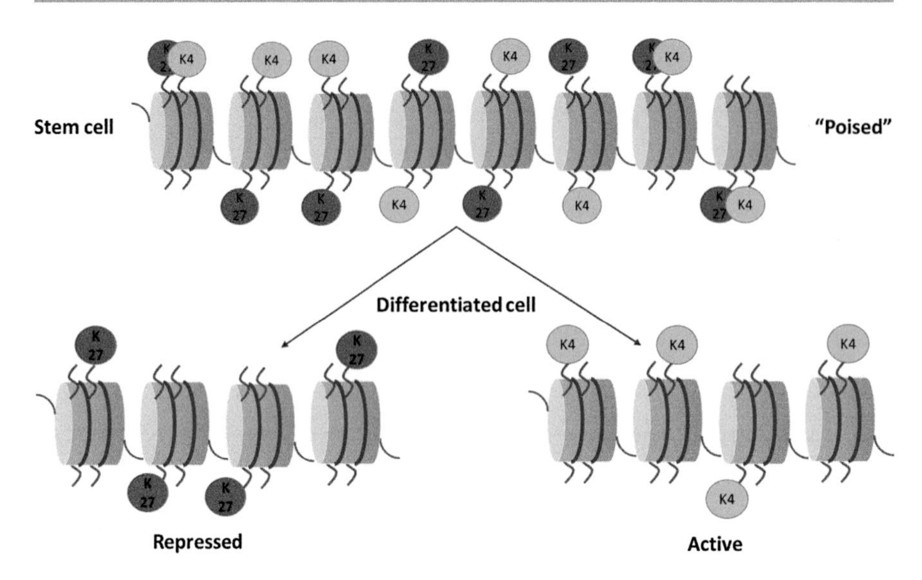

Fig. 3 **Bivalent domain of histone methylation in embryonic stem cell**. In renewing embryonic cells (*Top*), the gene promoters of lineage-specification genes are enriched with the bivalent domains that are characterized by the co-existence of repressive H3K27me3 (*Red*) and activating H3K4me3 (*Green*) histone modification makers. These bivalent epigenetic markers render these lineage-specification genes remain repressed but poised for activation in response to development cues. In differentiated cells, the bivalent domains are resolved to either a repressed, which enriches H3K27Me3 (*Bottom left*), or an active state, which enriches H3K4me3 (*Bottom right*). The cell lineage is thus established

cellular metabolic status has on epigenetic states. DNMTs and histone methyltransferases (HMTs) consume SAM when adding methyl marks on DNA and histones, while histone acetyltransferases (HATs) consume acetyl-CoA to modify histones. Thus, the cellular levels of acetyl-CoA and SAM can affect epigenetic status and thereby influence cell fate decisions (Su et al. 2016; Bertolo and Mcbreairty 2013; Cai et al. 2011; Moussaieff et al. 2015).

Essential Role of Threonine Catabolism in Mouse ESCs

Threonine Catabolism Generates Key Anabolic Metabolites and Epigenetic Modifiers

Threonine is an essential amino acid. Apart from its role in protein synthesis, threonine catabolism generates glycine and acetyl-CoA, key metabolic intermediates participating in the biosynthesis of nucleotides and the production of epigenetic modifiers, respectively (Fig. 4). In eukaryotic cells, degradation of threonine takes place in mitochondria via a two-step biochemical pathway. In the first step, threonine is oxidized by the rate-limiting threonine dehydrogenase (TDH) to 2-amino-3-

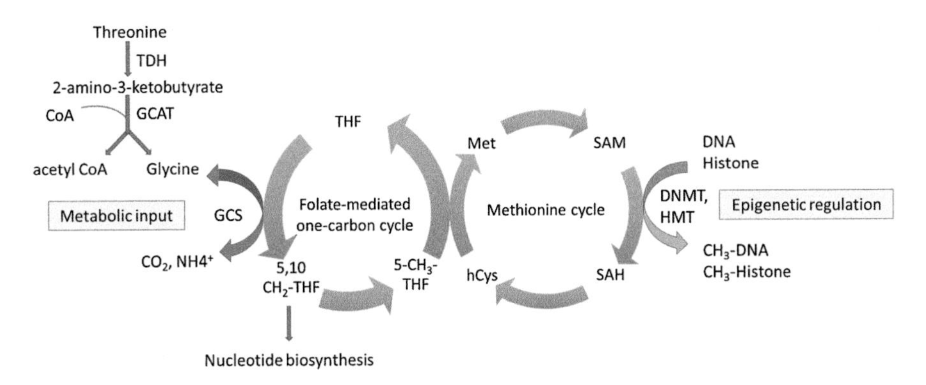

Fig. 4 Contribution of TDH-mediated threonine catabolism to the nucleotide biosynthesis and epigenetic regulation of embryonic stem cell. Threonine catabolism generates acetyl-CoA and glycine that in turn fuels nucleotide biosynthesis and DNA/histone methylation via interconnected folate-mediated one-carbon cycle and methionine cycle. *TDH* threonine dehydrogenase, *GCAT* 2-amino-3-ketobutyrate CoA ligase, *GCS* glycine cleavage system, *DNMT* DNA methyltransferase, *HMT* histone methyltransferase, *THF* tetrahydrofolate, *Met* methionine, *hCys* homocysteine, *SAM* S-adenosyl-methionine, *SAH* S-adenosyl-homocysteine

oxobutyrate, which is then split by 2-amino-3-oxobutyrate coenzyme A ligase (GCAT) in the reaction with coenzyme A to form glycine and acetyl-CoA (Dale 1978). The importance of threonine catabolism is highlighted by the metabolic routes of these threonine catabolites. Via the glycine cleavage system (GCS), glycine is catabolized to methylene-tetrahydrofolate, the activated methyl donor that provides one-carbon groups for nucleotide synthesis and ultimately for histone/DNA methylation (Fig. 4). Of equal importance is the acetyl-CoA produced by TDH which can be used either as metabolic fuel for the tricarboxylic acid (TCA) cycle, or it can contribute to far-reaching effects on gene expression by participation in acetylation of histones (Cai et al. 2011). It was demonstrated that fast-growing cancer cells require high level of glycine consumption (Jain et al. 2012), and that upregulation of glycine decarboxylase (GLDC), the rate-limiting enzyme of GCS, drives proliferation of cancer cells (Zhang et al. 2012). Interestingly, in bacteria, the breakdown of threonine into glycine and acetyl-CoA forms a high-flux metabolic backbone that contributes to anabolism during periods of rapid proliferation (Almaas et al. 2004). Thus, threonine catabolism constitutes important metabolic routes for supporting both cell proliferation and epigenetic regulation.

Dependence of Mouse ESCs on Threonine

In search for the metabolic basis for ESC self-renewal, McKnight and coworkers performed mass-spectrometry-based metabolomics studies to compare the intracellular metabolite abundance between renewing and differentiated mouse ESCs (Wang et al. 2009, 2011). It was observed that renewing ESCs were characterized by high levels of nucleotide biosynthetic intermediates compared to differentiated cells, indicating an increased demand of nucleotide biosynthesis for supporting the extensive genome replication of ESCs. Measurement of folate species demonstrated that renewing ESCs express low methylated but high uncharged levels of folate, indicating that the metabolic flux of one-carbon groups is a constraint of the self-renewal state. Furthermore, low threonine but high acetyl-CoA levels were found in renewing ESCs, ranking threonine metabolism at the top list of the candidates for modulating self-renewal. Indeed, threonine deficiency completely inhibited the DNA synthesis of ESC. Likewise, exposing ESCs to media deficient of one of each of the 20 amino acids demonstrated the selective dependence of ESC colony formation on threonine. Transcription profiling revealed that ESCs express extremely high level of the threonine-catabolic enzyme TDH. A key role for TDH in pluripotency is strongly supported by the observation that multiple pluripotent transcriptional factors bind to the TDH promoter (Kim et al. 2008). Importantly, chemical (Alexander et al. 2011) and genetic (Shyh-Chang et al. 2013) inhibition of TDH blocked DNA synthesis and stopped renewal of ESCs. Conversely, ectopically forced expression of TDH in somatic cells significantly enhanced the success of iPSC reprogramming (Han et al. 2013). Together these experiments clearly demonstrate the crucial role of threonine catabolism in maintenance of ESC self-renewal and further establish the importance of its effects on promoting flux through the one-carbon cycle.

Threonine Catabolism Selectively Maintains the Level of H3K4me3 Modification

In a follow-up study, Cantley, Daley, and coworkers deepened the understanding of the mechanism by which threonine catabolism fuels self-renewal of ESCs by linking threonine catabolic activity to the modulation of methionine cycle and the control of histone methylation (Shyh-Chang et al. 2013). It was observed that by stimulating the one carbon cycle, TDH-mediated threonine catabolism also promoted high-flux synthesis of methionine, which in turn elevated production of SAM, a universal donor for methylation reactions. This elevated the cellular methylation potential as judged by an increased SAM/SAH ratio. Assessment of the influence on the epigenetic status by threonine deprivation or TDH silencing demonstrated that inhibiting threonine catabolism does not influence global DNA methylation, but strikingly, selectively erases H3K4me3 and H3K4me2 marks, which (as described above) are the crucial modifications controlling the stem cell bivalent domain of epigenome. Maintaining H3K4me3 is critical for ESC self-renewal (Ang et al. 2011). TDH-SAM axis thus links threonine catabolism to the epigenetic regulation important for establishing pluripotency. Recently, the high sensitivity of H3K4me3 to the fluctuation of SAM/SAH ratio has been observed in many other cells (Liu and Pile 2017; Mentch et al. 2015), clearly exemplifying the functional importance of the connection between cellular metabolism and epigenetic regulation.

However, it is important to point out that human ESCs do not express a functional TDH. This may partly explain why these cells grow much more slowly and are more susceptible to differentiation than mouse ESCs. In spite of these difference in threonine metabolism between mouse and human cells, methionine supplementation is required to maintain H3K4me3 level and self-renewal of human ESCs (Kilberg et al. 2016; Shiraki et al. 2014), indicating that the overall pathways of one-carbon metabolism are essential components of the regulatory pathways that control ESC pluripotency.

Conclusion Remarks

The power of endless self-renewal and pluripotency makes embryonic stem cells the excellent models in biomedical research and underlies their potential utility in the field of regenerative medicine. The highly proliferative nature and unique epigenetic state associated with pluripotency imposes specific nutritional demands for successful ESC self-renewal. In this regard, discovery of the dependence of mESCs on threonine catabolism represents a breakthrough in our understanding of how specific nutritional parameters are crucial for successful self-renewal and safeguarding pluripotency. Threonine catabolism by TDH simultaneously supplies acetyl-CoA and methylene-THF, the crucial one- and two-carbon intermediates of cellular metabolism that fuel macromolecule biosynthesis, energy production, and epigenetic modification. The unexpected discovery that threonine-mediated one-carbon metabolism guarantees self-renewal by supporting both nucleotide synthesis and histone methylation has clearly advanced our understanding of the pluripotent state. We expect further exciting advancements in our understanding of nutritional influences on pluripotency when the impact of the acetyl-CoA branch of threonine catabolism on ESC self-renewal is further elucidated.

Dictionary of Terms

- **Pluripotency** – The ability of the cells to differentiate into all three germ layers – endoderm, mesoderm, and ectoderm – eventually forming all organs in the animal.
- **Self-renewal** – The maintenance of the undifferentiated stem cell pool through cell divisions.
- **iPSC** – Adult/terminally differentiated cells reprogrammed by introducing certain transcription factors to express embryonic stem cell specific markers and properties.
- **Cell cycle** – The accurate duplication of the genomic material and its precise segregation into two identical daughter cells through a series of events.
- **Naïve pluripotency** – The ability of the cells of the inner cell mass of the preimplantation embryo to give rise to all somatic cell lineages in vivo.
- **Clonogenicity** – Colony-forming ability of cells.
- **Chimerization** – The process in which cells from two or more species are combined to give rise to a chimeric embryo.

- **Bivalent domain** – Chromatin segments which contain epigenetic regulators both, activating and repressing modifications, in the same histone protein bound chromatin region.
- **One carbon cycle** – Nutritional input in the form of glucose, amino acids, and vitamins are integrated by the folate and methionine cycle reactions to generate biosynthetic output, epigenetic modifiers and maintain redox balance.
- **Methionine cycle** – Coupled to the folate cycle in one-carbon metabolism. The methionine cycle begins with methyl THF being donated from the folate cycle and its conversion to methionine. The main output of the methionine cycle is the production of S-adenosyl methionine which is a methyl group donor.

Key Facts of Embryonic Stem Cells

- Embryonic stem cells are the *in vitro* equivalents to the pluripotent stem cells that transiently express during early embryonic development.
- Embryonic stem cells can infinitely self-renew and differentiate into all cell types of an adult.
- Self-renewing embryonic stem cells present a highly proliferative feature. This requires specific nutritional support.
- The epigenome of embryonic stem cells is usually characterized by DNA hypomethylation, histone hyperacetylation, and enriched trimethylation of lysine 4 on histone 3 (H3K4me3) that is favorable for pluripotent gene expression.
- The promoter regions of lineage-specification genes of embryonic stem cells usually contain both active H3K4me3 and repressive H3K27Me3 epigenetic histone modifications (bivalent domains).
- Bivalent domains render the lineage-specification genes remain repressed but poised for activation in response to developmental cues.
- Embryonic stem cells provide unlimited source for stem cell research and hold great promise for implication in regenerative medicine.

Summary Points

- This chapter focuses on the metabolic and epigenetic features of mouse embryonic stem cells.
- The self-renewal of embryonic stem cells imposes unique metabolic demands for supporting the unusually high rate of proliferation.
- The epigenome of embryonic stem cells is enriched of active histone modifications including trimethylation of lysine 4 on histone H3 (H3K4me3) that are required for sustained pluripotent gene expression.
- TDH-mediated threonine catabolism is highly active in embryonic stem cells, simultaneously producing acetyl-CoA and glycine.
- Glycine produced by threonine catabolism feeds one-carbon metabolism that maintains nucleotide synthesis and H3K4me3 modification, playing essential

roles in supporting cell proliferation and epigenetic regulation of stem cells renewal.

- Currently, it is not clear how the acetyl-CoA produced from threonine catabolism contributes to the self-renewal of stem cells.

References

Ahuja AK, Jodkowska K, Teloni F, Bizard AH, Zellweger R, Herrador R, Ortega S, Hickson ID, Altmeyer M, Mendez J, Lopes M (2016) A short G1 phase imposes constitutive replication stress and fork remodelling in mouse embryonic stem cells. Nat Commun 7:10660

Alexander PB, Wang J, Mcknight SL (2011) Targeted killing of a mammalian cell based upon its specialized metabolic state. Proc Natl Acad Sci 108:15828–15833

Almaas E, Kovacs B, Vicsek T, Oltvai Z, Barabási A-L (2004) Global organization of metabolic fluxes in the bacterium Escherichia coli. Nature 427:839–843

Ambrosetti DC, Basilico C, Dailey L (1997) Synergistic activation of the fibroblast growth factor 4 enhancer by Sox2 and Oct-3 depends on protein-protein interactions facilitated by a specific spatial arrangement of factor binding sites. Mol Cell Biol 17:6321–6329

Ang Y-S, Tsai S-Y, Lee D-F, Monk J, Su J, Ratnakumar K, Ding J, Ge Y, Darr H, Chang B (2011) Wdr5 mediates self-renewal and reprogramming via the embryonic stem cell core transcriptional network. Cell 145:183–197

Arnold K, Sarkar A, Yram MA, Polo JM, Bronson R, Sengupta S, Seandel M, Geijsen N, Hochedlinger K (2011) Sox2(+) adult stem and progenitor cells are important for tissue regeneration and survival of mice. Cell Stem Cell 9:317–329

Avilion AA, Nicolis SK, Pevny LH, Perez L, Vivian N, Lovell-Badge R (2003) Multipotent cell lineages in early mouse development depend on SOX2 function. Genes Dev 17:126–140

Azuara V, Perry P, Sauer S, Spivakov M, Jørgensen HF, John RM, Gouti M, Casanova M, Warnes G, Merkenschlager M (2006) Chromatin signatures of pluripotent cell lines. Nat Cell Biol 8:532–538

Baron U, Schnappinger D, Helbl V, Gossen M, Hillen W, Bujard H (1999) Generation of conditional mutants in higher eukaryotes by switching between the expression of two genes. Proc Natl Acad Sci U S A 96:1013–1018

Bernstein BE, Mikkelsen TS, Xie X, Kamal M, Huebert DJ, Cuff J, Fry B, Meissner A, Wernig M, Plath K (2006) A bivalent chromatin structure marks key developmental genes in embryonic stem cells. Cell 125:315–326

Bertolo RF, Mcbreairty LE (2013) The nutritional burden of methylation reactions. Curr Opin Clin Nutr Metab Care 16:102–108

Boyer LA, Lee TI, Cole MF, Johnstone SE, Levine SS, Zucker JP, Guenther MG, Kumar RM, Murray HL, Jenner RG, Gifford DK, Melton DA, Jaenisch R, Young RA (2005) Core transcriptional regulatory circuitry in human embryonic stem cells. Cell 122:947–956

Burdon T, Stracey C, Chambers I, Nichols J, Smith A (1999) Suppression of SHP-2 and ERK signalling promotes self-renewal of mouse embryonic stem cells. Dev Biol 210:30–43

Cai L, Sutter BM, Li B, Tu BP (2011) Acetyl-CoA induces cell growth and proliferation by promoting the acetylation of histones at growth genes. Mol Cell 42:426–437

Chambers I, Silva J, Colby D, Nichols J, Nijmeijer B, Robertson M, Vrana J, Jones K, Grotewold L, Smith A (2007) Nanog safeguards pluripotency and mediates germline development. Nature 450:1230–1234

Clouaire T, Webb S, Skene P, Illingworth R, Kerr A, Andrews R, Lee J-H, Skalnik D, Bird A (2012) Cfp1 integrates both CpG content and gene activity for accurate H3K4me3 deposition in embryonic stem cells. Genes Dev 26:1714–1728

Cole MF, Young RA (2008) Mapping key features of transcriptional regulatory circuitry in embryonic stem cells. Cold Spring Harb Symp Quant Biol 73:183–193

Coronado D, Godet M, Bourillot PY, Tapponnier Y, Bernat A, Petit M, Afanassieff M, Markossian S, Malashicheva A, Iacone R, Anastassiadis K, Savatier P (2013) A short G1 phase is an intrinsic determinant of naive embryonic stem cell pluripotency. Stem Cell Res 10:118–131
Dale RA (1978) Catabolism of threonine in mammals by coupling of L-threonine 3-dehydrogenase with 2-amino-3-oxobutyrate-CoA ligase. Biochim Biophys Acta Gen Subj 544:496–503
Evans MJ, Kaufman MH (1981) Establishment in culture of pluripotential cells from mouse embryos. Nature 292:154–156
Festuccia N, Gonzalez I, Navarro P (2016) The epigenetic paradox of pluripotent ES cells. J Mol Biol 429:1476
Goldberg AD, Allis CD, Bernstein E (2007) Epigenetics: a landscape takes shape. Cell 128:635–638
Han C, Gu H, Wang J, Lu W, Mei Y, Wu M (2013) Regulation of L-threonine dehydrogenase in somatic cell reprogramming. Stem Cells 31:953–965
Hilton DJ, Nicola NA, Metcalf D (1988) Specific binding of murine leukemia inhibitory factor to normal and leukemic monocytic cells. Proc Natl Acad Sci U S A 85:5971–5975
Hindley C, Philpott A (2013) The cell cycle and pluripotency. Biochem J 451:135–143
Jackson M, Krassowska A, Gilbert N, Chevassut T, Forrester L, Ansell J, Ramsahoye B (2004) Severe global DNA hypomethylation blocks differentiation and induces histone hyperacetylation in embryonic stem cells. Mol Cell Biol 24:8862–8871
Jaenisch R, Bird A (2003) Epigenetic regulation of gene expression: how the genome integrates intrinsic and environmental signals. Nat Genet 33:245–254
Jain M, Nilsson R, Sharma S, Madhusudhan N, Kitami T, Souza AL, Kafri R, Kirschner MW, Clish CB, Mootha VK (2012) Metabolite profiling identifies a key role for glycine in rapid cancer cell proliferation. Science 336:1040–1044
Janke R, Dodson AE, Rine J (2015) Metabolism and epigenetics. Annu Rev Cell Dev Biol 31:473–496
Jiang H, Shukla A, Wang X, Chen W-Y, Bernstein BE, Roeder RG (2011) Role for Dpy-30 in ES cell-fate specification by regulation of H3K4 methylation within bivalent domains. Cell 144:513–525
Kaelin WG Jr, Mcknight SL (2013) Influence of metabolism on epigenetics and disease. Cell 153:56–69
Kilberg MS, Terada N, Shan J (2016) Influence of amino acid metabolism on embryonic stem cell function and differentiation. Adv Nutr 7:780s–789s
Kim J, Chu J, Shen X, Wang J, Orkin SH (2008) An extended transcriptional network for pluripotency of embryonic stem cells. Cell 132:1049–1061
Kimura H, Tada M, Nakatsuji N, Tada T (2004) Histone code modifications on pluripotential nuclei of reprogrammed somatic cells. Mol Cell Biol 24:5710–5720
Kleinsmith LJ, Pierce GB Jr (1964) Multipotentiality of single embryonal carcinoma cells. Cancer Res 24:1544–1551
Koledova Z, Krämer A, Kafkova LR, Divoky V (2010) Cell-cycle regulation in embryonic stem cells: centrosomal decisions on self-renewal. Stem Cells Dev 19:1663–1678
Kunath T, Saba-El-Leil MK, Almousailleakh M, Wray J, Meloche S, Smith A (2007) FGF stimulation of the Erk1/2 signalling cascade triggers transition of pluripotent embryonic stem cells from self-renewal to lineage commitment. Development 134:2895–2902
Lee JH, Hart SR, Skalnik DG (2004) Histone deacetylase activity is required for embryonic stem cell differentiation. Genesis 38:32–38
Leitch HG, Mcewen KR, Turp A, Encheva V, Carroll T, Grabole N, Mansfield W, Nashun B, Knezovich JG, Smith A (2013) Naive pluripotency is associated with global DNA hypomethylation. Nat Struct Mol Biol 20:311–316
Liang G, Zhang Y (2013) Genetic and epigenetic variations in iPSCs: potential causes and implications for application. Cell Stem Cell 13:149–159
Liu M, Pile LA (2017) The transcriptional corepressor SIN3 directly regulates genes involved in methionine catabolism and affects histone methylation, linking epigenetics and metabolism. J Biol Chem 292:1970–1976

Loh Y-H, Wu Q, Chew J-L, Vega VB, Zhang W, Chen X, Bourque G, George J, Leong B, Liu J (2006) The Oct4 and Nanog transcription network regulates pluripotency in mouse embryonic stem cells. Nat Genet 38:431–440

Martello G, Smith A (2014) The nature of embryonic stem cells. Annu Rev Cell Dev Biol 30:647–675

Martin GR, Evans MJ (1975) Differentiation of clonal lines of teratocarcinoma cells: formation of embryoid bodies in vitro. Proc Natl Acad Sci U S A 72:1441–1445

Masui S, Nakatake Y, Toyooka Y, Shimosato D, Yagi R, Takahashi K, Okochi H, Okuda A, Matoba R, Sharov AA, Ko MS, Niwa H (2007) Pluripotency governed by Sox2 via regulation of Oct3/4 expression in mouse embryonic stem cells. Nat Cell Biol 9:625–635

Mentch SJ, Locasale JW (2016) One-carbon metabolism and epigenetics: understanding the specificity. Ann N Y Acad Sci 1363:91–98

Mentch SJ, Mehrmohamadi M, Huang L, Liu X, Gupta D, Mattocks D, Padilla PG, Ables G, Bamman MM, Thalacker-Mercer AE (2015) Histone methylation dynamics and gene regulation occur through the sensing of one-carbon metabolism. Cell Metab 22:861–873

Metcalf D (1991) The leukemia inhibitory factor (LIF). Int J Cell Cloning 9:95–108

Mitsui K, Tokuzawa Y, Itoh H, Segawa K, Murakami M, Takahashi K, Maruyama M, Maeda M, Yamanaka S (2003) The homeoprotein Nanog is required for maintenance of pluripotency in mouse epiblast and ES cells. Cell 113:631–642

Moussaieff A, Rouleau M, Kitsberg D, Cohen M, Levy G, Barasch D, Nemirovski A, Shen-Orr S, Laevsky I, Amit M, Bomze D, Elena-Herrmann B, Scherf T, Nissim-Rafinia M, Kempa S, Itskovitz-Eldor J, Meshorer E, Aberdam D, Nahmias Y (2015) Glycolysis-mediated changes in acetyl-CoA and histone acetylation control the early differentiation of embryonic stem cells. Cell Metab 21:392–402

Neganova I, Zhang X, Atkinson S, Lako M (2009) Expression and functional analysis of G1 to S regulatory components reveals an important role for CDK2 in cell cycle regulation in human embryonic stem cells. Oncogene 28:20–30

Nichols J, Smith A (2012) Pluripotency in the embryo and in culture. Cold Spring Harb Perspect Biol 4:a008128

Niwa H, Burdon T, Chambers I, Smith A (1998) Self-renewal of pluripotent embryonic stem cells is mediated via activation of STAT3. Genes Dev 12:2048–2060

Okita K, Hong H, Takahashi K, Yamanaka S (2010) Generation of mouse-induced pluripotent stem cells with plasmid vectors. Nat Protoc 5:418–428

Rice JC, Allis CD (2001) Histone methylation versus histone acetylation: new insights into epigenetic regulation. Curr Opin Cell Biol 13:263–273

Romito A, Cobellis G (2016) Pluripotent stem cells: current understanding and future directions. Stem Cells Int 2016:9451492

Santos F, Hendrich B, Reik W, Dean W (2002) Dynamic reprogramming of DNA methylation in the early mouse embryo. Dev Biol 241:172–182

Savatier P, Huang S, Szekely L, Wiman KG, Samarut J (1994) Contrasting patterns of retinoblastoma protein expression in mouse embryonic stem cells and embryonic fibroblasts. Oncogene 9:809–818

Scholer HR, Dressler GR, Balling R, Rohdewohld H, Gruss P (1990) Oct-4: a germline-specific transcription factor mapping to the mouse t-complex. EMBO J 9:2185–2195

Shamblott MJ, Axelman J, Wang S, Bugg EM, Littlefield JW, Donovan PJ, Blumenthal PD, Huggins GR, Gearhart JD (1998) Derivation of pluripotent stem cells from cultured human primordial germ cells. Proc Natl Acad Sci U S A 95:13726–13731

Sharov AA, Masui S, Sharova LV, Piao Y, Aiba K, Matoba R, Xin L, Niwa H, Ko MS (2008) Identification of Pou5f1, Sox2, and Nanog downstream target genes with statistical confidence by applying a novel algorithm to time course microarray and genome-wide chromatin immunoprecipitation data. BMC Genomics 9:269

Shen L, Inoue A, He J, Liu Y, Lu F, Zhang Y (2014) Tet3 and DNA replication mediate demethylation of both the maternal and paternal genomes in mouse zygotes. Cell Stem Cell 15:459–470

Shilatifard A (2012) The COMPASS family of histone H3K4 methylases: mechanisms of regulation in development and disease pathogenesis. Annu Rev Biochem 81:65–95

Shiraki N, Shiraki Y, Tsuyama T, Obata F, Miura M, Nagae G, Aburatani H, Kume K, Endo F, Kume S (2014) Methionine metabolism regulates maintenance and differentiation of human pluripotent stem cells. Cell Metab 19:780–794

Shyh-Chang N, Locasale JW, Lyssiotis CA, Zheng Y, Teo RY, Ratanasirintrawoot S, Zhang J, Onder T, Unternaehrer JJ, Zhu H, Asara JM, Daley GQ, Cantley LC (2013) Influence of threonine metabolism on S-adenosylmethionine and histone methylation. Science 339:222–226

Smith AG, Heath JK, Donaldson DD, Wong GG, Moreau J, Stahl M, Rogers D (1988) Inhibition of pluripotential embryonic stem cell differentiation by purified polypeptides. Nature 336:688–690

Stead E, White J, Faast R, Conn S, Goldstone S, Rathjen J, Dhingra U, Rathjen P, Walker D, Dalton S (2002) Pluripotent cell division cycles are driven by ectopic Cdk2, cyclin A/E and E2F activities. Oncogene 21:8320

Stevens LC, Little CC (1954) Spontaneous testicular Teratomas in an inbred strain of mice. Proc Natl Acad Sci U S A 40:1080–1087

Su X, Wellen KE, Rabinowitz JD (2016) Metabolic control of methylation and acetylation. Curr Opin Chem Biol 30:52–60

Takahashi K, Yamanaka S (2006) Induction of pluripotent stem cells from mouse embryonic and adult fibroblast cultures by defined factors. Cell 126:663–676

Thomas KR, Capecchi MR (1987) Site-directed mutagenesis by gene targeting in mouse embryo-derived stem cells. Cell 51:503–512

Thomson JA, Itskovitz-Eldor J, Shapiro SS, Waknitz MA, Swiergiel JJ, Marshall VS, Jones JM (1998) Embryonic stem cell lines derived from human blastocysts. Science 282:1145–1147

Wang J, Alexander P, Wu L, Hammer R, Cleaver O, Mcknight SL (2009) Dependence of mouse embryonic stem cells on threonine catabolism. Science 325:435–439

Wang J, Alexander P, Mcknight S (2011) Metabolic specialization of mouse embryonic stem cells. Cold Spring Harb Symp Quant Biol 76:183–193. Cold Spring Harbor Laboratory Press

Williams RL, Hilton DJ, Pease S, Willson TA, Stewart CL, Gearing DP, Wagner EF, Metcalf D, Nicola NA, Gough NM (1988) Myeloid leukaemia inhibitory factor maintains the developmental potential of embryonic stem cells. Nature 336:684–687

Wysocka J, Swigut T, Milne TA, Dou Y, Zhang X, Burlingame AL, Roeder RG, Brivanlou AH, Allis CD (2005) WDR5 associates with histone H3 methylated at K4 and is essential for H3 K4 methylation and vertebrate development. Cell 121:859–872

Yamaji M, Ueda J, Hayashi K, Ohta H, Yabuta Y, Kurimoto K, Nakato R, Yamada Y, Shirahige K, Saitou M (2013) PRDM14 ensures naive pluripotency through dual regulation of signaling and epigenetic pathways in mouse embryonic stem cells. Cell Stem Cell 12:368–382

Ying QL, Wray J, Nichols J, Batlle-Morera L, Doble B, Woodgett J, Cohen P, Smith A (2008) The ground state of embryonic stem cell self-renewal. Nature 453:519–523

Young RA (2011) Control of the embryonic stem cell state. Cell 144:940–954

Zhang WC, Shyh-Chang N, Yang H, Rai A, Umashankar S, Ma S, Soh BS, Sun LL, Tai BC, Nga ME, Bhakoo KK, Jayapal SR, Nichane M, Yu Q, Ahmed DA, Tan C, Sing WP, Tam J, Thiruganananam A, Noghabi MS, Pang YH, Ang HS, Mitchell W, Robson P, Kaldis P, Soo RA, Swarup S, Lim EH, Lim B (2012) Glycine decarboxylase activity drives non-small cell lung cancer tumor-initiating cells and tumorigenesis. Cell 148:259–272

The Role and Epigenetic Modification of the Retinoic Acid Receptor

84

Yukihiko Kato

Contents

Abstract

Dietary vitamin A is converted to retinoids, including retinal, retinol, and retinoic acid, in the metabolic pathway. Two types of retinoic acid exist in the cell nucleus along with two types of receptor, the retinoic acid receptor (RAR) and retinoic X receptor (RXR) each of which has three subtypes, α, β, and γ. Retinoic acid receptors are involved in a wide variety of functions including mediating cell differentiation, tissue growth, blood vessel formation, the emotional and cognitive functions, and tumor suppression.

RARβ, a tumor suppressor gene, is epigenetically suppressed in most cancers. Epigenetic modification of RAR in many cancers includes DNA methylation and histone hypoacetylation. Histone deacetylase inhibitor and RA restore RAR expression and have shown a strong antitumor effect. The epigenetic modification of RAR could have clinical applications such as in diagnosing malignancies.

Y. Kato (✉)
Department of Dermatology, Tokyo Medical University Hachioji Medical Center, Tokyo, Japan
e-mail: y-kato@tokyo-med.ac.jp; y2011ka@gmail.com

V. B. Patel, V. R. Preedy (eds.), *Handbook of Nutrition, Diet, and Epigenetics*,
https://doi.org/10.1007/978-3-319-55530-0_114

In the future, as many other types of epigenetic modifications become better understood, we can expect their diagnostic and therapeutic applications to be greatly expanded.

Keywords
Vitamin A · All-trans retinoic acid · Retinoic acid receptor · 9-cis-retinoic acid · Retinoic X receptor · Retinoid · Tumor suppressor gene · RARβ2 · Histone deacetylase · Hypoacetylation

List of Abbreviations

ATRA	All-trans retinoic acid
CRA	9-cis-retinoic acid
HDAC	Histone deacetylase
RA	Retinoic acid
RAR	Retinoic acid receptor
RARE	Receptor responsive element
RXR	Retinoic X receptor

Introduction

Vitamin A and Retinoid

The importance of vitamin A in vision was first noticed in the 1950s. Since then, vitamin A has attracted considerable attention as a strong inducer of cell differentiation. Nearly two decades of studies have revealed that the gene expression of this lipophilic vitamin was regulated epigenetically via the chromatin structure.

Dietary vitamin A is converted to all-trans retinoic (ATRA) acid and 9-cis-retinoic acid (CRA) via retinal in the metabolic pathway. Two types of retinoic acid receptor exist in the cell nucleus along with two types of receptor, RAR and retinoic X receptor (RXR), each of which has three subtypes, α, β, and γ. The heterodimer of RAR and RXR binds to the retinoic acid receptor responsive element (RARE) and regulates the transcriptional activity of the target genes (Fig. 1).

A "retinoid," a derivative of a monocyclic compound, functions in a manner similar to retinoic acid via specific receptors. This current definition is remarkable in that it anticipated the discovery of retinoic acid receptor (RAR). The best-known retinoids are retinal, retinol, and retinoic acid (Fig. 2).

Retinoids perform various functions throughout the body and are involved in the maintenance of eye and skin health, the formation of sperm, normal functioning of the immune system, the activation of tumor suppressor genes, and cell differentiation. Retinoic acid is involved in the switching between proliferation and differentiation in various neural stem cells in vitro. Retinoids are also used for the treatment of a number of dermatological diseases including psoriasis, lymphoma of the skin, acne, photoaging, and skin wrinkles.

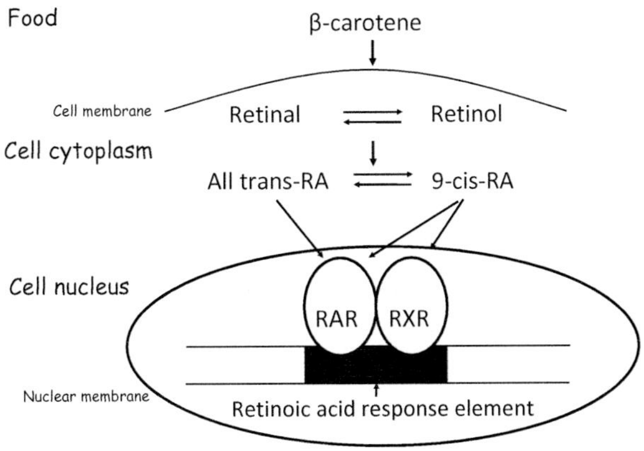

Fig. 1 Intracellular translocation of vitamin A. Dietary vitamin A, which results from the transformation of provitamin A (β-carotene), becomes all-trans retinoic acid (*ATRA*) and 9-cis-retinoic acid (*CRA*) via retinal in the metabolic pathway. Two types of retinoic acid exist in the cell nucleus together with two types of receptor, namely, the retinoic acid receptor (*RAR*) and retinoic X receptor (*RXR*). The heterodimer of RAR and RXR binds to the retinoic acid receptor responsive element (*RARE*) and regulates the transcriptional activity of the target genes

Fig. 2 Chemical structure of vitamin A

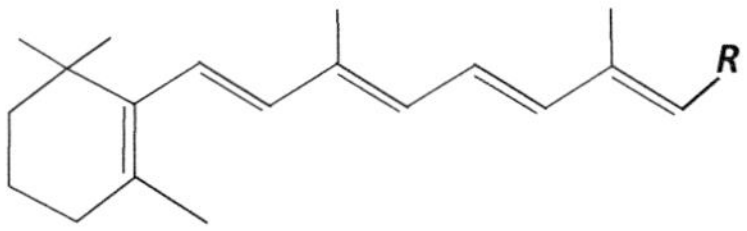

R	Retinoids
OH	Retinol
CHO	Retinal
COOH	Retinoic acid

The Retinoic Acid Receptor

Retinoic acid receptors mediate a wide variety of functions including tissue growth, blood vessel formation, the regulation of the emotional and cognitive functions, and tumor suppression.

Angiogenesis of Blood Vessels

The RAR agonists, all-trans retinoic acid (ATRA) and Am580, upregulate angiogenesis by producing nitric oxide via the phosphoinositide 3-kinase (PI3K)/Akt pathway (Uruno et al. 2005). According to one study, this angiogenic effect was mediated in vitro by the production of vascular endothelial growth factor (VEGF) via the expression of VEGF receptor 2 on endothelial cells (Saito et al. 2007). Similarly tazarotene, which is a RAR agonist but not a RXR agonist, was found to promote angiogenesis and wound healing (Al Haj Zen et al. 2016).

Emotional and Cognitive Functions

A study by Etchamendy et al. (2001) found that RAR normalized age-related memory deficits in mice. Similarly, Mingaud et al. (2008) also found that RAR improved short-term/working memory organization and long-term declarative memory encoding in mice. Vitamin A deficiency causes amyloid β deposition in the adult rat brain by disrupting the retinoid signaling pathway (Corcoran et al. 2004). RA rescued memory deficits in an Alzheimer's disease transgenic mouse model by downregulating the amyloid precursor protein and tau protein in the brain (Ding et al. 2008).

Tumor Suppressive Function

RA is involved in cell differentiation as described above. However, RARβ, a tumor suppressor gene, is epigenetically suppressed in most cancers. This next chapter will discuss a number of studies that have been done on the epigenetics of RARβ activity and review some of the possible clinical implications.

Epigenetics of RAR

Most studies of RAR involving epigenetics have been performed in the field of oncology. Sirchia et al. (2000) first reported that the loss of RARβ in breast cancer was due to DNA methylation of the promoter site and suggested the possibility of a repressive mechanism other than DNA methylation after a demethylating agent failed to restore the expression of RARβ. DNA methylation has been found to suppress RAR expression (Moison et al. 2013) in gastric cancer (Hayashi et al. 2001), esophageal squamous cell carcinoma (Liu et al. 2005), cervical cancer (Zhang et al. 2007), breast cancer (Fang et al. 2015; Sun et al. 2011), and bladder cancer (Berrada et al. 2012). Furthermore, the tobacco smoke carcinogen, 4-(methylnitrosamino)-1-(3-pyridyl)-1-butanone, induced RARβ hypermethylation in esophageal squamous epithelial cells (Wang et al. 2012).

Some reports have demonstrated that the epigenetic modification of RAR included DNA methylation and hypoacetylation of histone in prostate cancer (Nakayama et al. 2001; Qian et al. 2005), renal cell carcinoma (Wang et al. 2005), and melanoma (Kato et al. 2007). In these studies, histone deacetylase (HDAC) inhibitor and RA restored the expression of RAR and the antitumor effect of RA. Kato et al. showed that histone acetylation was superior to DNA methylation in restoring RAR expression in cutaneous T cell lymphoma and demonstrated the double antitumor effect of HDAC inhibitor and RA (Fig. 3) (Kato et al. 2016). These findings might explain why the demethylating agent failed to restore the expression of RAR in the study by Sirchia et al. (2000).

A number of studies have demonstrated that the deacetylation, rather than the methylation, of the RAR promoter repressed RAR expression via the

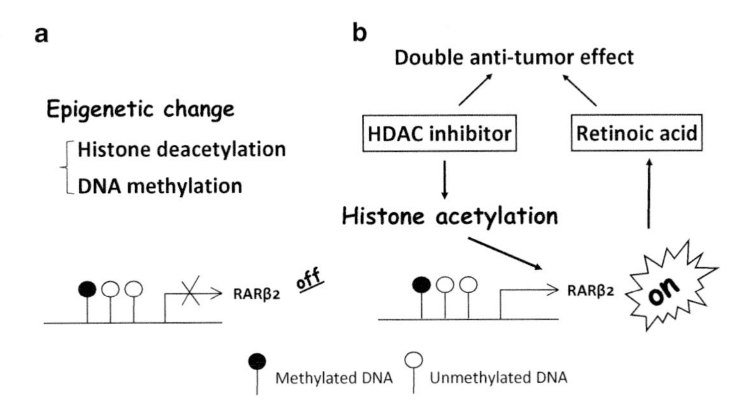

Fig. 3 Reexpression of RARβ2 induced by histone deacetylase (*HDAC*) inhibitor. (**a**) Retinoic acid receptor (*RAR*) β2 was repressed epigenetically. (**b**) RARβ2 expression was restored by HDAC inhibitor and had a double antitumor effect via HDAC inhibitor and retinoic acid

hypomethylation of the promoter itself. Histone deacetylase inhibitor alone, which does not affect the methylation status, restored RAR in thyroid cancer cells (Cras et al. 2007) and cervical cancer cells (Zhang et al. 2007).

Some studies have treated RARβ2 as a tumor suppressor gene. Widschwendter was the first to observe that RARβ2 was apparently methylated in breast cancer and unmethylated in nonneoplastic breast tissue or normal tissue. Widschwendter thus suggested that the treatment of cancer patients with demethylating agents followed by retinoic acid might constitute a new chemopreventive therapeutic method (Widschwendter et al. 2001). However, Kato et al. pointed that RARβ2 was a tumor suppressor gene regulated by histone acetylation rather than by DNA methylation (Kato et al. 2016, p. 58).

Other studies have pointed that RAR is regulated epigenetically in concert with other genes. He et al. (2009) showed that inducing the overexpression of Myc, an oncogene, inhibited RARβ expression by methylation in a benign prostate epithelial cell line (He et al. 2009). One of the target genes of promyelocytic leukemia zinc finger-retinoic acid receptor alpha (PLZF-RARalpha), an oncogenic transcriptional repressor, is an RAR element in leukemogenesis (Choi et al. 2014). RARβ and p16 were downregulated in epithelial ovarian carcinomas and low-malignant potential tumors by methylation while no methylation was observed in normal ovarian tissue (Bhagat et al. 2014).

Epigenetic modification of RAR could have clinical applications. Since the promoter hypermethylation of RARβ2 was closely associated with prostate cancer rather than prostate hypertrophy, the methylation status of RARβ2 might be useful for clinical diagnosis (Dumache et al. 2012). Pirouzpanah et al. reported that hypermethylated RARβ2 was associated with younger age in breast cancer patients, and that hypermethylation of estrogen receptor (ER) alpha was associated with smoking, duration of estradiol exposure, ER-negative tumor, and body mass index (Pirouzpanah et al. 2010). Cassinat interestingly found that in an acute promyelocytic leukemia (APL) patient, granulocyte colony-stimulating factor (G-CSF) restored RAR alpha expression via the activation of the extracellular signal-regulated kinase (ERK)/mitogen-activated protein kinase (MAPK) pathway and subsequent histone phosphorylation, histone acetyltransferase activity, and histone acetylation (Cassinat et al. 2011).

The relationship between viral infections and RARβ2 is complicated. Human papilloma virus type 16 (HPV16) increased RARβ mRNA and protein in vitro and in vivo in cervical cancer cells (Gutierrez et al. 2015). On the other hand, in a model of HPV-18-positive HeLa cells, ectopic expression of RARβ2 decreased HPV-18 transcription, induced p53, p21, and p27, and stopped cell proliferation (De-Castro Arce et al. 2007). The hepatitis C virus core protein inhibited cell growth by suppressing RARβ2 via DNA methylation in hepatocellular carcinoma (Lee et al. 2013). Latent membrane protein 1, an Epstein-Barr virus (EBV) oncogene product, induced hypermethylation of RARβ2 and EBV-mediated tumorigenesis in nasopharyngeal carcinoma (Seo et al. 2008). Hepatitis B virus X protein-induced hypermethylation of RARβ2 and suppressed RA-induced cell growth (Jung et al. 2010). HCV, EBV, and HBV, but not necessarily HPVs, produced oncogenic effects via epigenetic modification of RARβ2. Further studies on the relationship between viral infection and RARβ as well as RAR are needed to shed more light on this interesting topic.

Epigenetic modification of RAR also has implications for mental function and development. Neonatal maternal separation in rats weakened adult hippocampal neural differentiation by decreasing RAR alpha via methylation of its promoter (Boku et al. 2015). Vitamin A deficiency induced learning and memory impairment in rats by decreasing RAR alpha via suppression of H3 and H4 histone acetylation (Hou et al. 2015). Mature retinal ganglion cells in rats, considered to be irreplaceable, could exert a neuritogenic (regenerative) effect on optic axons by inducing RARβ via nitrosylation of histone deacetylase (HDAC) 2 and the acetylation of histone H3 (Koriyama et al. 2013). RAR was also found to preserve the stemness of adult tendon stem cells (TSCs) during expansion in vitro by histone methylation (Webb et al. 2016).

The epigenetic modifications of RAR are intimately associated with the pathogenesis of various diseases. Up to now the study of RAR epigenetic modification has focused mainly on DNA methylation and histone acetylation and has been led chiefly by oncologists. In the future, many other types of epigenetic modifications will undoubtedly come to light and improve our understanding of RAR, thereby also creating promising diagnostic and therapeutic applications.

Dictionary of Terms

- **Vitamin A** – The name of a group of fat-soluble retinoids. Vitamin A supports cell growth and differentiation, playing a critical role in the normal formation and maintenance of the heart, lungs, kidneys, and other organs.

- **The retinoic acid receptor (RAR)** – A type of nuclear receptor which can also act as a transcription factor that is activated by both all-trans retinoic acid and 9-cis retinoic acid.
- **Angiogenesis** – The physiological process through which new blood vessels form from preexisting vessels. Angiogenesis is a normal and vital process in growth and development, as well as in wound healing but the angiogenesis inhibitors are used in the treatment of cancer.
- **Oncogene** – a gene with the ability to cause a normal cell to become cancerous.
- **Tumor suppressor gene** – a gene that inhibits uncontrolled cell proliferation in normal cells. When this gene becomes inactivated, the cell is at increased risk of malignant proliferation.

Key Facts

1. Vitamin A is necessary for the body's growth and health.
2. Vitamin A is especially important for the maintenance of vision, skin health, normal cognition, the formation of sperm, and the prevention of malignancies.
3. Vitamin A is converted to retinoids, which perform various important functions.
4. Retinoic acid, a type of retinoid, binds to the retinoic acid receptor (RAR).
5. RARβ2, a type of RAR, works as a tumor suppressor gene to prevent cancer progression.
6. RARβ2 is lost in most cancers by epigenetic modification.
7. If the epigenetic modification is reversed, RARβ2 is reexpressed and suppresses tumorigenesis and tumor progression.

Summary Points

1. Dietary vitamin A is converted to retinoids including retinal, retinol, and retinoic acid, in the metabolic pathway.
2. Two types of retinoic acid exist in the cell nucleus along with two types of receptor, the retinoic acid receptor (RAR) and retinoic X receptor (RXR). Each of these receptors has three subtypes, α, β, and γ.
3. Retinoic acid receptors are involved in a wide variety of functions including mediating tissue growth, blood vessel formation, the emotional and cognitive functions, and tumor suppression.
4. Retinoic acid is involved in cell differentiation. However, RARβ, a tumor suppressor gene, is epigenetically suppressed in most cancers.
5. Epigenetic modification of RAR in many cancers includes DNA methylation and histone hypoacetylation.
6. Histone deacetylase inhibitor and RA restore RAR expression and have shown a strong antitumor effect.

7. Some studies have suggested that RAR is regulated epigenetically in concert with other genes.
8. The relationship between viral infections and RARβ2 is complicated.
9. Epigenetic modification of RAR has implications for mental function and development.
10. The epigenetic modification of RAR could have clinical applications such as in diagnosing malignancies.
11. In the future, as many other types of epigenetic modifications become better understood, we can expect their diagnostic and therapeutic applications to be greatly expanded.

References

Al Haj Zen A et al (2016) The retinoid agonist Tazarotene promotes angiogenesis and wound healing. Mol Ther 24(10):1745–1759

Berrada N et al (2012) Epigenetic alterations of adenomatous polyposis coli (APC), retinoic acid receptor beta (RARbeta) and survivin genes in tumor tissues and voided urine of bladder cancer patients. Cell Mol Biol (Noisy-le-Grand) (Suppl 58):OL1744–OL1751

Bhagat R et al (2014) Epigenetic alteration of p16 and retinoic acid receptor beta genes in the development of epithelial ovarian carcinoma. Tumour Biol 35(9):9069–9078

Boku S et al (2015) Neonatal maternal separation alters the capacity of adult neural precursor cells to differentiate into neurons via methylation of retinoic acid receptor gene promoter. Biol Psychiatry 77(4):335–344

Cassinat B et al (2011) New role for granulocyte colony-stimulating factor-induced extracellular signal-regulated kinase 1/2 in histone modification and retinoic acid receptor alpha recruitment to gene promoters: relevance to acute promyelocytic leukemia cell differentiation. Mol Cell Biol 31(7):1409–1418

Choi WI et al (2014) Promyelocytic leukemia zinc finger-retinoic acid receptor alpha (PLZF-RARalpha), an oncogenic transcriptional repressor of cyclin-dependent kinase inhibitor 1A (p21WAF/CDKN1A) and tumor protein p53 (TP53) genes. J Biol Chem 289(27):18641–18656

Corcoran JP, So PL, Maden M (2004) Disruption of the retinoid signalling pathway causes a deposition of amyloid beta in the adult rat brain. Eur J Neurosci 20(4):896–902

Cras A et al (2007) Epigenetic patterns of the retinoic acid receptor beta2 promoter in retinoic acid-resistant thyroid cancer cells. Oncogene 26(27):4018–4024

De-Castro Arce J, Gockel-Krzikalla E, Rosl F (2007) Retinoic acid receptor beta silences human papillomavirus-18 oncogene expression by induction of de novo methylation and hetero-chromatinization of the viral control region. J Biol Chem 282(39):28520–28529

Ding Y et al (2008) Retinoic acid attenuates beta-amyloid deposition and rescues memory deficits in an Alzheimer's disease transgenic mouse model. J Neurosci 28(45):11622–11634

Dumache R et al (2012) Retinoic acid receptor beta2 (RARbeta2): nonivasive biomarker for distinguishing malignant versus benign prostate lesions from bodily fluids. Chirurgia (Bucur) 107(6):780–784

Etchamendy N et al (2001) Alleviation of a selective age-related relational memory deficit in mice by pharmacologically induced normalization of brain retinoid signaling. J Neurosci 21(16):6423–6429

Fang C et al (2015) Promoter methylation of the retinoic acid receptor Beta2 (RARbeta2) is associated with increased risk of breast cancer: a PRISMA compliant meta-analysis. PLoS One 10(10):e0140329

Gutierrez J et al (2015) Human papillomavirus type 16 E7 oncoprotein upregulates the retinoic acid receptor-beta expression in cervical cancer cell lines and K14E7 transgenic mice. Mol Cell Biochem 408(1–2):261–272

Hayashi K et al (2001) Inactivation of retinoic acid receptor beta by promoter CpG hypermethylation in gastric cancer. Differentiation 68(1):13–21

He M et al (2009) Epigenetic regulation of Myc on retinoic acid receptor beta and PDLIM4 in RWPE1 cells. Prostate 69(15):1643–1650

Hou N et al (2015) Vitamin A deficiency impairs spatial learning and memory: the mechanism of abnormal CBP-dependent histone acetylation regulated by retinoic acid receptor alpha. Mol Neurobiol 51(2):633–647

Jung JK, Park SH, Jang KL (2010) Hepatitis B virus X protein overcomes the growth-inhibitory potential of retinoic acid by downregulating retinoic acid receptor-beta2 expression via DNA methylation. J Gen Virol 91(2):493–500

Kato Y et al (2007) Antitumor effect of the histone deacetylase inhibitor LAQ824 in combination with 13-cis-retinoic acid in human malignant melanoma. Mol Cancer Ther 6(1):70–81

Kato Y et al (2016) Combination of retinoid and histone deacetylase inhibitor produced an anti-tumor effect in cutaneous T-cell lymphoma by restoring tumor suppressor gene, retinoic acid receptorβ2, via histone acetylation. J Dermatol Sci 81(1):17–25

Koriyama Y et al (2013) Requirement of retinoic acid receptor β for genipin derivative-induced optic nerve regeneration in adult rat retina. PLoS One 8(8):e71252

Lee H et al (2013) Hepatitis C virus Core protein overcomes all-trans retinoic acid-induced cell growth arrest by inhibiting retinoic acid receptor-β2 expression via DNA methylation. Cancer Lett 335(2):372–379

Liu Z et al (2005) 5-Aza-2′-deoxycytidine induces retinoic acid receptor-beta(2) demethylation and growth inhibition in esophageal squamous carcinoma cells. Cancer Lett 230(2): 271–283

Mingaud F et al (2008) Retinoid hyposignaling contributes to aging-related decline in hippocampal function in short-term/working memory organization and long-term declarative memory encoding in mice. J Neurosci 28(1):279–291

Moison C et al (2013) DNA methylation associated with polycomb repression in retinoic acid receptor beta silencing. FASEB J 27(4):1468–1478

Nakayama T et al (2001) The role of epigenetic modifications in retinoic acid receptor beta2 gene expression in human prostate cancers. Lab Investig 81(7):1049–1057

Pirouzpanah S et al (2010) The effect of modifiable potentials on hypermethylation status of retinoic acid receptor-beta2 and estrogen receptor-alpha genes in primary breast cancer. Cancer Causes Control 21(12):2101–2111

Qian DZ et al (2005) In vivo imaging of retinoic acid receptor beta2 transcriptional activation by the histone deacetylase inhibitor MS-275 in retinoid-resistant prostate cancer cells. Prostate 64(1):20–28

Saito A et al (2007) All-trans retinoic acid induces in vitro angiogenesis via retinoic acid receptor: possible involvement of paracrine effects of endogenous vascular endothelial growth factor signaling. Endocrinology 148(3):1412–1423

Seo SY, Kim EO, Jang KL (2008) Epstein-Barr virus latent membrane protein 1 suppresses the growth-inhibitory effect of retinoic acid by inhibiting retinoic acid receptor-beta2 expression via DNA methylation. Cancer Lett 270(1):66–76

Sirchia SM et al (2000) Evidence of epigenetic changes affecting the chromatin state of the retinoic acid receptor beta2 promoter in breast cancer cells. Oncogene 19(12):1556–1563

Sun J et al (2011) Epigenetic regulation of retinoic acid receptor β2 gene in the initiation of breast cancer. Med Oncol 28(4):1311–1318

Uruno A et al (2005) Upregulation of nitric oxide production in vascular endothelial cells by all-trans retinoic acid through the phosphoinositide 3-kinase/Akt pathway. Circulation 112(5):727–736

Wang XF et al (2005) Epigenetic modulation of retinoic acid receptor beta2 by the histone deacetylase inhibitor MS-275 in human renal cell carcinoma. Clin Cancer Res 11(9):3535–3542

Wang J et al (2012) 4-(Methylnitrosamino)-1-(3-pyridyl)-1-butanone induces retinoic acid receptor β hypermethylation through DNA methyltransferase 1 accumulation in esophageal squamous epithelial cells. Asian Pac J Cancer Prev 13(5):2207–2212

Webb S et al (2016) Retinoic acid receptor signaling preserves tendon stem cell characteristics and prevents spontaneous differentiation in vitrox. Stem Cell Res Ther 7:45

Widschwendter M et al (2001) Epigenetic downregulation of the retinoic acid receptor-beta2 gene in breast cancer. J Mammary Gland Biol Neoplasia 6(2):193–201

Zhang Z et al (2007) Retinoic acid receptor beta2 is epigenetically silenced either by DNA methylation or repressive histone modifications at the promoter in cervical cancer cells. Cancer Lett 247(2):318–327

Histone Deacetylase Inhibitor Tributyrin and Vitamin A in Cancer

85

Renato Heidor, Ernesto Vargas-Mendez, and Fernando Salvador Moreno

Contents

Abstract

Bioactive food compounds like vitamin A and the butyrate's prodrug tributyrin have preventive activities against different types of cancer, and their use in association could represent a promising strategy for cancer treatment and chemoprevention. Both compounds can induce cell differentiation and apoptosis of neoplastic and preneoplastic cells by means of modulation of gene transcription, yet they act through different but interconnected mechanisms. Vitamin A acts through nuclear receptors that are tightly regulated by histone modifications such as acetylation and DNA methylation. Tributyrin modulates transcription of genes by HDACs inhibition and histone hyperacetylation. This chapter describes how epigenetics mediates the antineoplastic and chemopreventive activity of vitamin A, tributyrin, and their derivatives and how their combination can be used to help overcome current limitations in cancer treatment and prevention. We also show

R. Heidor · E. Vargas-Mendez · F. S. Moreno (✉)
Laboratory of Diet, Nutrition and Cancer, Department of Food and Experimental Nutrition, Faculty of Pharmaceutical Sciences, University of São Paulo, São Paulo, SP, Brazil
e-mail: rheidor@usp.br; evargas@usp.br; rmoreno@usp.br

V. B. Patel, V. R. Preedy (eds.), *Handbook of Nutrition, Diet, and Epigenetics*,
https://doi.org/10.1007/978-3-319-55530-0_72

how the mechanisms of action of vitamin A and tributyrin have aided in the development of synthetic and bioengineered compounds like synthetic retinoids and structured lipids, respectively.

Keywords
Cancer chemoprevention · Hepatocarcinogenesis · Bioactive food compounds · Butyrate · Tributyrin · Vitamin A · Retinoids · ATRA · DNA methylation · Histone acetylation · HDAC inhibitors

List of Abbreviations

5-aza	5-azacytidine
9cRA	9-*cis* retinoic acid
ALA	α linolenic acid
APL	Acute promyelocytic leukemia
ATRA	All-*trans* retinoic acid
b.w.	Body weight
BA	Butyric acid
BFC	Bioactive food compound
Crm-1	Chromosomal region maintenance 1
CRP	Chromatin remodeling complex
DNMT	DNA methyltransferase
GST-P	Placental glutathione-S-transferase
H3K18ac	Histone 3 acetylated in lysine residue 18
H3K27me3	Histone 3 trimethylated in lysine residue 27
H3K4ac	Histone 3 acetylated in lysine residue 4
H3K9me3	Histone 3 trimethylated in lysine residue 9
H4K12ac	Histone 4 acetylated in lysine residue 12
H4K16ac	Histone 4 acetylated in lysine residue 16
HAT	Histone acetyltransferase
HDAC	Histone deacetylase
LINE-1	Long interspersed nucleotide element 1
Mdm2	Human homologue of the mouse double minute 2
NES	Nuclear exportation signal
PPAR	Peroxisomal proliferator activated receptor
RARE	Retinoic acid response elements
RARα	Retinoic acid receptor alfa
RARβ2	Retinoic acid receptor beta isoform 2
RH	Resistant hepatocyte model
ROL	Retinol
RXR	Retinoid X receptors
RXRE	RXR response elements
TB	Tributyrin
UBE1L	Ubiquitin-activating enzyme E1-like
VDR	Vitamin D receptor

Introduction

The concept of diet and nutritional status as potential risk factors for non-transmissible chronic diseases such as cancer is not new (Michels 2005). Several epidemiological studies conducted during the 1970s investigated the relationship between diet and cancer incidence concluding that nutritional factors play, indeed, an important role in neoplasm prevention. In this regard, in vitro and in vivo studies showed effects of bioactive food compounds (BFCs) in the prevention of different types of cancer, mainly due to their pleiotropic properties and to their putative capability to act selectively on neoplastic cells (Fig. 1) (Alcantara and Speckmann 1976; Gori 1978; Ahmad et al. 2012; Kotecha et al. 2016).

The prevention of carcinogenesis involves mechanisms related to inhibition of proliferation as well as induction of apoptosis and differentiation. The molecular pathways associated with those processes could be controlled genetically or they can even show epigenetic regulation. This depends on functionality and cooperation of multiple factors such as the activity of chromatin remodeling complexes (CRPs) or metabolism of methyl groups as well as post-transcriptional modulation of the miRNAs' genetic expression (Huang et al. 2011). Nevertheless, the inhibition in

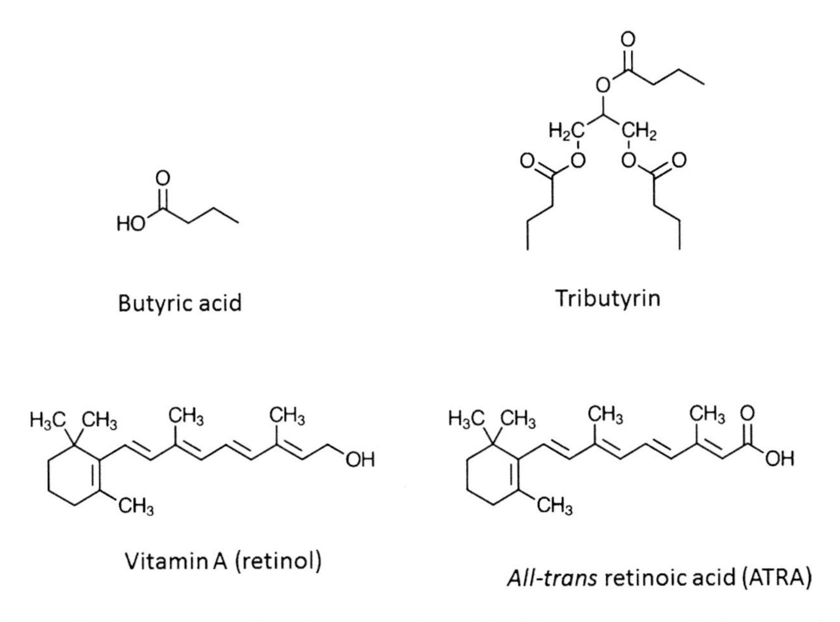

Fig. 1 Chemical structures. Butyric acid is a short-chain fatty acid composed of a four-carbon backbone and a carboxylic group attached to carbon 1 (C1) or carbon alfa (α). Three molecules of butyric acid bind to one molecule of glycerol through ester bonds to form a tributyrin molecule. Vitamin A and its main biological effector all-trans retinoic acid (*ATRA*) share a common structure that comprises a cyclic group known as β-ionone ring in one end and a polar group in the other end of a nine-carbon backbone. In vitamin A, the polar group is a primary alcohol and in the case of retinoic acid it comprises a carboxylic acid

the expression of a gene or the inhibition of a molecular pathway could not be sufficient to affect the overall survival of neoplastic cell populations. An alternative approach targeting different genes or molecular pathways simultaneously seems to be promising against cancer (Steward and Brown 2013). In this regard, there is an interest in the development of new strategies combining drugs or BFCs with different mechanisms of action and capable of targeting several carcinogenic pathways simultaneously (Banudevi et al. 2015). Therefore, translational studies analyzing the mechanisms of action of BFCs for the inclusion of this combinatorial approach into clinical practice are necessary.

This chapter discusses the role of vitamin A and its analogs as well as butyrate and its prodrug tributyrin in the context of cancer prevention. Epigenetic aspects related to these BFCs are discussed, and future perspectives of the use of these compounds for the prevention and control of cancer are highlighted.

Butyrate and Tributyrin

Butyric acid (BA), a short chain fatty acid (Fig. 1) produced during anaerobic fermentation of dietary fibers by colonic microbiota, has been studied since 1970. It has been proposed that the primary butyrate-producing bacteria in humans are *Faecalibacterium prausnitzii* and *Eubacterium rectale/Roseburia* spp. (Louis and Flint 2009), both residents of the proximal part of the large intestine. Due to the flux of intestinal mucus and the peristaltic movements, the butyrate concentration in the proximal and distal portions of the colon tends to vary. Likewise, in the colonic crypts, a gradient-like pattern of butyrate concentration ranging from 0.5 mM in the basal cells to 5 mM in the apical cells close to the lumen has been described (Donohoe et al. 2012).

It has been described that BA, or its ionized form butyrate, shows, besides morphological changes on cells, an inhibitory activity on DNA synthesis and cellular proliferation (Leder and Leder 1975; Prasad and Sinha 1976; Rephaeli et al. 1994). Both upregulation and downregulation of gene expression has also been observed in cells treated with butyrate (Berni Canani et al. 2012). The antineoplastic activity of this fatty acid was verified in vitro and in vivo in experimental models of prostate, breast, stomach, and liver cancer. This anticarcinogenic activity could be associated mainly with induction of apoptosis or with cellular differentiation (Heidor et al. 2012). In this regard, it was observed that treatment with butyrate increased the susceptibility of hepatoma cells to apoptosis through stimulation of the TNF signaling pathway (Ogawa et al. 2004) as well as induction of differentiation of several cell lines from different cancers including colon, pancreas, cervix, prostate, liver, and leukemias (Heidor et al. 2012).

Butyrate acts as a weak ligand of histone deacetylase (HDAC) enzymes, inhibiting HDACs members of the classes I and II as well as HDAC4, HDAC5, HDAC7, and HDAC9. This activity has been related to its structure which facilitates access to the active site of the enzyme and allows the formation of a coordinate bond with the enzymatic cofactor Zn^{2+} (Davie 2003). Therefore, butyrate promotes histone acetylation leading to an increased expression of genes associated with apoptosis and differentiation (Heidor et al. 2012). It was observed in vitro that butyrate can both inhibit the expression of HDAC4 and increase the levels of acetylated histone H3 (H3K9ac) in the hepatocellular carcinoma cell lines SMMC-7721 and HepG2. Besides, butyrate inhibited cellular migration/invasion of the cells increasing the expression of E-cadherin, considered an epithelial marker, and decreasing the expression of vimentin, a mesenchymal marker (Wang et al. 2013).

Histone acetylation by butyrate occurs preferentially in neoplastic cells where glucose metabolism is the main source of energy as a consequence of the Warburg effect (Van der Heiden et al. 2009). Thus, butyrate is metabolized inefficiently and then accumulated inside the cell. Such accumulation increases butyrate diffusion to the nucleus and favors its HDACs' ligand activity. There is an alternative mechanism that could be also associated with butyrate capacity of promoting histone acetylation. In this case, butyrate is first metabolized to acetyl-CoA, an essential factor for histone acetyl transferases (HAT) activity. Then, through the increased activity of HATs, epigenetic modulation of gene expression takes place (Donohoe and Bultman 2012). The critical factor determining which mechanism of acetylation prevails is the metabolic status of the cell. In this regard, it has been observed that butyrate shows specificity toward epigenetic activity in neoplastic cells but not in normal cells where butyrate is used as an energetic substrate (Donohoe et al. 2012).

Other epigenetic modifications such as histone methylation can be promoted by butyrate. For instance, histone H3 trimethylation in lysine 27 (H3K27me3) has been observed in SW480 colon cancer cells treated with butyrate at 1.25 mM. In addition, the increased levels of H3K27me3 led to a reduction in the expression of NFκB both at transcriptional and translational level. These findings were confirmed in vivo using KK.Cg-Ay/a, a mice strain predisposed to metabolic syndrome, fed with insoluble modified fiber as butyrate source (Liu et al. 2016). Changes in DNA methylation patterns also occurred following treatment with butyrate. In this regard, colon cancer cells HT-29 treated with 4 mM butyrate showed a reduction in DNA methylation of the promoter region of RARβ2 gene. In the same study, it was observed that there is a difference between the demethylation effect induced by butyrate and Aza-2′deoxycytine. The former modifies the methylation pattern of specific loci and the latter induces a broader genomic hypomethylation associated with DNMTs activity (Spurling et al. 2008). In this sense, the ultimate mechanisms by which butyrate modulate DNA methylation are still unknown.

Antineoplastic mechanisms induced by butyrate such as apoptosis and cellular differentiation can also be regulated by miRNAs. During carcinogenesis, altered expression of several miRNAs occurs by mechanisms still unclear. Nonetheless, it has been suggested that DNA methylation and histone modifications could contribute to the modulation of those expression patterns of miRNAs observed in neoplasms (Calin and Croce 2006). Epigenetic regulation of miRNA expression has been reported in several neoplastic cell lines as well as in tissues from colon, prostate, pancreas, kidney, bladder, and lung cancer (Wang 2014). The treatment of the colon cancer cell lines HT29 and HCT116 with 5 mM butyrate reduced the expression of the miRNAs cluster miR-17-92 and increased the expression of their target genes such as PTEN, BCL2L11, and CDKN1A (Humphreys et al. 2013). Likewise, treatment of LT97 cells derived from colon cancer with 2–10 mM butyrate modulated the expression of miRNAs involved in cell cycle and proliferation like miR-135a, miR-135b, miR-24, miR-106b, and miR-let-7a (Schlörmann et al. 2015).

Given its promising antineoplastic activity, sodium butyrate was included in a clinical trial conducted in acute leukemia patients. This trial resulted in the partial remission of the illness without evidence of toxicity under a 15 g/kg/day intravenous regime for up to 17 days (Rephaeli et al. 1994). A sustained plasma concentration of butyrate of at least 0.5 mM has been associated with a greater therapeutic activity of the compound (Miller et al. 1987). The biological half-life of butyrate is 5 min when administered in rodents. In humans, butyrate's half-life shows a biphasic behavior with an accelerated excretion in the first 30s and followed by a peak at 14 min, when butyrate binds to plasma proteins (Daniel et al. 1989). Despite its low toxicity, the fast clearance of butyrate is considered a major limitation for its use in clinics.

Butyrate's half-life can be increased significantly if administered in the form of tributyrin (TB), a triacylglycerol constituted by three molecules of BA esterified to a molecule of glycerol (Fig. 1). The complete hydrolysis of one mole of TB by lipases yields two moles of butyrate and one mole of 2-monobutyrylglycerol. Butyrate's half-live after oral administration in rodents is 40 min (Su et al. 2006) and even doses up to 8.2 g/kg b.w. in rodents did not show evidence of toxic effects (Egorin et al. 1999). The fact that TB can be administered orally has advantages. Besides its favorable pharmacokinetics, TB's organoleptic characteristics are better than butyrate's. The treatment of mice C57BL/6 with TB under hyperlipidemic (59.1% fat) diet resulted in a hypolipidemic effect with partial reversion of hepatic steatosis and stimulated production of adiponectin by adipose tissue (Vinolo et al. 2012).

This triacylglycerol showed chemopreventive activity in rats subjected to a hepatocarcinogenesis (Fig. 2) model (Kuroiwa-Trzmielina et al. 2009; de Conti et al. 2012, 2013; Guariento et al. 2014; Heidor et al. 2016). Besides apoptosis induction, TB increased p21 expression and H3 (Kuroiwa-Trzmielina et al. 2009; de Conti et al. 2012) as well as H3K18ac and H4K16ac levels (de Conti et al. 2013). Trimethylation of histone H3 in lysine 9 (H3K9me3) was observed in Wistar rats treated with TB and subjected to a model of hepatocarcinogenesis, a key event in the repression of gene transcription (Heidor et al. 2016).

It has been observed that TB treatment of rats subjected to a hepatocarcinogenesis model promotes the nuclear accumulation of p53 and a corresponding reduction of its cytoplasmic levels. This event is critical in the control of apoptosis/proliferation equilibrium mediated by p53 (Kuroiwa-Trzmielina et al. 2009; Ortega et al. 2016). Further in vitro experimentation with JM-1 and PLC/PRF/5 hepatocarcinoma cell lines confirmed the nuclear localization of p53 after treatment with butyrate (Ortega et al. 2016). p53 protein contains three nuclear localization signals and one nuclear exportation signal (NES) in its C-terminal extremity, as well as one additional NES inside the transactivation domain. Proteins such as p53 containing at least one NES are recognized by the exporter receptor Crm-1 (chromosomal-region-maintenance 1), a member of the cariofenin-β family of receptor proteins responsible for the translocation of molecules from inside the nucleus to the cytoplasm. In neuroblastoma cells

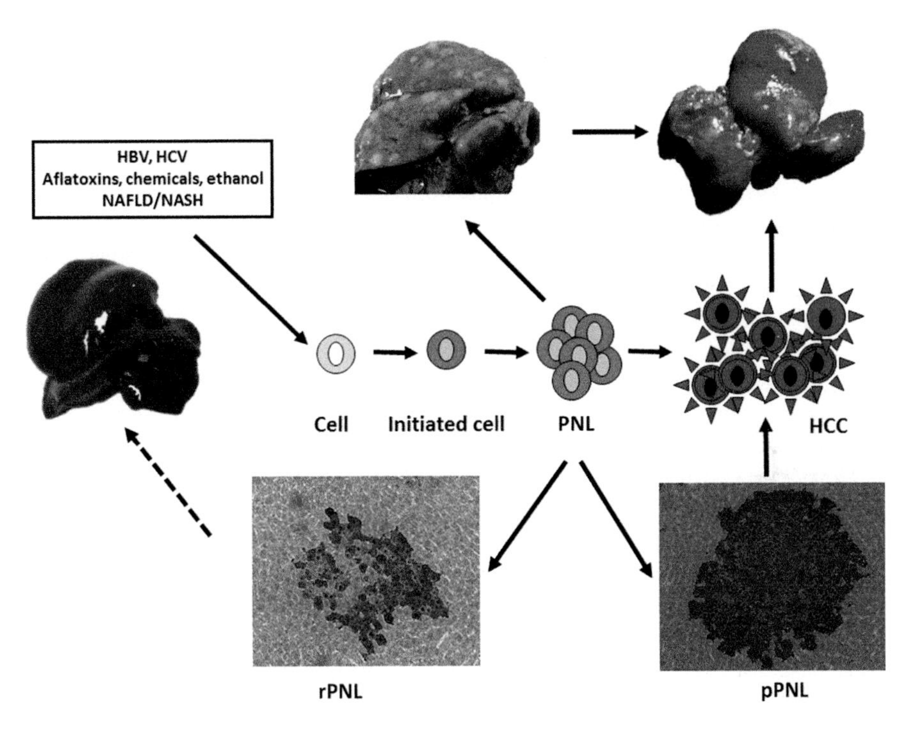

Fig. 2 Hepatocarcinogenesis. Etiologic factors of HCC are associated with genetic alterations in hepatocyte's DNA. If those alterations are not repaired, the cell becomes initiated. After several cycles of cellular proliferation, preneoplastic lesions (PNLs) are originated from foci of initiated cells. PNLs can be classified as remodeling (rPNLs) or persistent (pPNLs) and can even be macroscopically visible when those lesions form nodules. Remodeling lesions tend to reverse their phenotype and become apparently normal or they can persist as a lesion and then evolve toward HCC

the cytoplasmic localization of p53 could be attributed to alterations in its retention inside the nucleus because of Crm-1 exportation activity (Stommel et al. 1999). Likewise, preneoplastic lesions, hepatic nodules, and hepatocellular carcinomas derived from F344 rats submitted to the model of hepatocarcinogenesis known as "Resistant Hepatocyte" (RH) showed increased levels of Crm-1 compared to normal livers (Pascale et al. 2005). In this regard, Wistar rats subjected to the same model (RH) but treated with TB showed a nuclear accumulation of Crm-1 in hepatic foci stained for placental glutathione-S-transferase (GST-P) as observed in JM-1 and PLC/PRF/5 cells treated with butyrate (Ortega et al. 2016).

Mdm2 (human homologue of the mouse double minute 2) is a key regulator of p53 signaling pathway acting as a transcriptional inhibitor and as a degradation stimulator. The balance between mdm2 and p53 is determined by an autoregulatory negative feedback loop in which p53 increases mdm2 expression through its binding to response elements located in mdm2 gene promoter, a process that can be controlled by HDACs inhibitors like butyrate (Xie et al. 2016). In this regard, rats treated with TB and subjected to the RH model showed reduced levels of mdm2 expression compared to its corresponding isocaloric controls (de Conti et al. 2013).

Autophagy is a catabolic process in which cellular harmful constituents are degraded. This cellular mechanism is essential for differentiation, survival, and homeostasis of the cell and its regulation has been related to p53 activity. Cytoplasmic localization of p53 associates with suppression of autophagy, while autophagy itself has been associated with tumor progression (Cordani et al. 2016). In the case of liver carcinogenesis, augmented levels of autophagy and concomitant expression of the autophagy markers LC3B and BECN1 were identified in rats submitted to the RH-model (de Conti et al. 2013).

Despite TB's antineoplastic and chromatin-modulating activity, few studies have evaluated its antitumoral potential in humans. In one of these studies, carried out in patients with different types of cancer, it was observed that oral administration of 200 mg/kg of body weight TB resulted in serum concentrations of 0.45 mM butyrate. Given that those blood levels were detectable for only 4 h after administration, serial doses of 200 mg/kg each 8 h were necessary (Conley et al. 1998). A following study further demonstrated that TB at that same dose was sufficient to stabilize the disease without any additional toxicity. Nevertheless, to achieve the desired concentration of 200 mg/kg, 28 capsules containing 1 g each of TB would be necessary per day (Edelman et al. 2003).

To increase the therapeutic efficiency of TB, structural modifications of its molecule or a vehicle-mediated delivery should be implemented. In this regard, a lipid emulsion containing phosphatidylcholine, cholesteryl oleate, and TB was tested orally in rats. Blood levels of butyrate increased significantly compared to the same concentration of TB without any vehicle (Su et al. 2006). Another emulsion containing Tween 80®, phosphatidylcholine, and TB induced apoptosis in B16-F10 melanoma cells and prevented the generation of lung tumors in mice inoculated with those cells (Kang et al. 2011).

With the current advances in biotechnology, it is possible to combine the properties of different fatty acids in a single molecule of triacylglycerol by means of lipid modification techniques. Hence, the development of functional foods or pharmaceuticals for the treatment of diseases will be achievable (Farfán et al. 2013). One of those lipid modification techniques consists in the production of structured lipids from preexisting triacylglycerols. Such modification alters the composition and/or distribution of fatty acids along the glycerol backbone in a controlled manner (Fig. 3). A good candidate to produce structured lipids is alfa-linolenic acid (ALA; C18:3n-3), a precursor of metabolites with anti-inflammatory properties and preventive potential of cardiovascular diseases, arthritis, and cancer (Rajaram 2014; Kim et al. 2014). Among the most important sources of ALA, linseed oil (*Linum usitatissimum*) has high levels of unsaturated fatty acids localized, mainly, in the sn-2 position of glycerol molecules (Kim et al. 2014). In this regard, the treatment of rats subjected to the RH model with structured lipids obtained by enzymatic interesterification between TB and linseed oil showed a reduction in the number of GST-P positive preneoplastic lesions, similarly to rats treated with TB. However, given the lack of a strong proapoptotic response of the structured lipid compared to

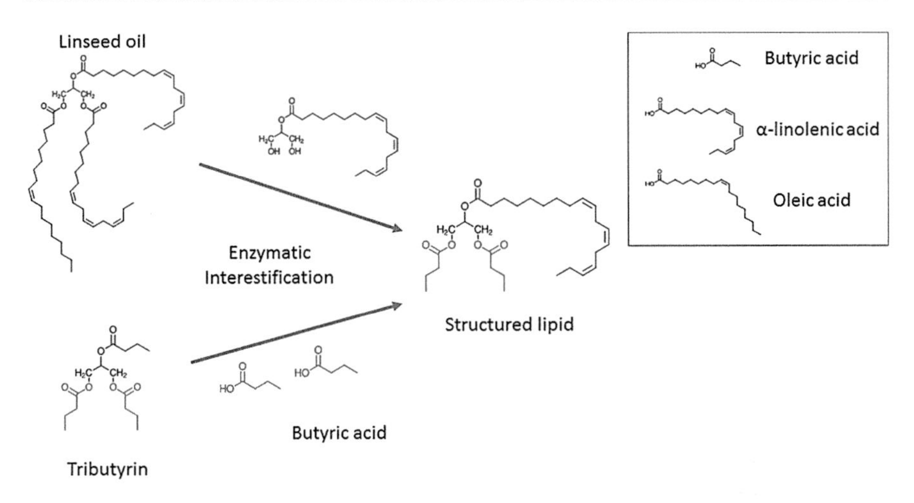

Fig. 3 Synthesis of structured lipids. Structured lipids are the product of enzymatic interesterification of triacylglycerols. In this case, tributyrin is interesterified with linseed oil to produce new triacylglycerols containing both butyric acid and unsaturated fatty acids. Given that alfa-linolenic acid and oleic acid are the most abundant fatty acids in linseed oil, the interesterification process of tributyrin yields triacylglycerols with butyric and alfa-linolenic acid in their molecules as exemplified in this figure

TB, it was concluded that both compounds have different mechanisms of action in hepatocarcinogenesis prevention. Moreover, structured lipids increased the levels of histone H3K9me3 and histone H3 trimethylated in lysine 27 (H3K27me3) compared to controls. Considering that this event is associated with transcription repression, the structured lipids-induced downregulation of the hepatic oncogenes *Ccnd-1*, *Myc*, and *Stat3* were confirmed. Accordingly, increased levels of H3K9me3 and H3K27me3 were found in the promoter regions of *Ccnd-1* and *Myc* genes (Heidor et al. 2016).

Tributyrin as a prodrug of the butyric acid found in foods shows several mechanisms of action including chromatin-modifying activity. Given that high doses of this compound are required to achieve its therapeutic activity, incorporation of TB in emulsions or elaboration of butyrate molecular carriers from TB could be considered promissory alternatives for its use in the prevention of chronic diseases. It should be also highlighted the potential of combinatorial treatments of TB along with other BFCs or drugs able to act through different mechanisms in such a way that the overall chemopreventive or therapeutic activity against cancer is increased. An example of this approach is the association of TB (1 g/Kg b.w.) with folic acid (0.8 g/Kg) that resulted in chemopreventive activity in rats subjected to the RH model. In this study, the association or combinatorial treatment was more effective than the isolated compounds in the reduction of preneoplastic lesions (Guariento et al. 2014). Recent unpublished findings of our group also demonstrated that association of TB with sorafenib shows anticarcinogenic activity in rats with syngeneic transplants of JM-1 and GP7TB cells. The former developed carcinomas with hepatoblastic pattern and the latter developed carcinomas following a sarcomatous morphology. Clinical and preclinical trials have also evaluated the association of HDACs inhibitors with retinoids since this is a promissory strategy for the prevention and therapy of cancer (Tang and Gudas 2011).

Vitamin A and Its Analogs

Vitamin A, also known as retinol (ROL), was the first liposoluble vitamin discovered by experiments conducted in rats. Treatment of these animals with regular chow plus butter, egg yolk extracts, alfalfa leaves, liver, or kidney led to an improved growth (McCollum and Davis 1913). The relevance of this vitamin was then confirmed by several clinical and preclinical studies, in which the essential role of ROL in physiological processes like embryonic development, vision, immunity, reproduction (Al Tanoury et al. 2013), and stem cell differentiation was demonstrated (Gudas 2013). Recently, the involvement of ROL in chronic diseases such as diabetes (Trasino et al. 2016) and metabolic syndrome (Wei et al. 2016) has been investigated. Besides, the modulatory effect of vitamin A in different types of neoplasia has been increasingly studied (Tang and Gudas 2011).

Some pleiotropic activities of ROL can be associated to its metabolites. That is the case of 11-cis retinaldehyde, which is a key factor in retina's phototransduction as well as some retinoic acid isomers like all-trans retinoic acid (ATRA) which is responsible for the modulation of genes associated with cell proliferation, differentiation, and apoptosis (Al Tanoury et al. 2013; Li et al. 2014). ROL, its metabolites, and the synthetic analogs derived from their structures are grouped in a family of compounds known as retinoids (Fig. 1) (Uray et al. 2016). These retinoids can exert regulatory activity of gene expression acting as ligands of nuclear receptors colocalized in transcription complexes. There are three types of retinoic acid receptors (RARα, RARβ, and RARγ) and several isoforms of each type. These RARs act along with retinoid X receptors (RXR) forming heterodimers RAR/RXR (Chambon 1996). The heterodimers recognize specific sequences in several gene promoters denominated retinoic acid response elements (RAREs). Binding of ATRA to one of these RARs initiate the transcription of the corresponding gene (Mangelsdorf and Evans 1995). Transcription induction mediated by ATRA occurs at physiologic concentrations (1×10^{-9} M) (Lo-Coco and Ammatuna 2006) and is associated with increased levels of coactivator proteins like HATs (p300 and p300/cAMP response element–binding protein–associated factor), pCIP, and RNA polymerase II in the RAREs. It has also been observed that repressor proteins such as NCoR and HDACs are associated with RAREs, but they dissociate in response to retinoic acid (Urvalek et al. 2014). Moreover, histone modifications are produced in response to retinoic acid signaling, for instance, increased levels of H3K27me3, H3K9ac, and H3K4ac. These epigenetic modifications have been associated with stem cell differentiation (Gudas 2013).

It has been observed a reduced expression of RARs in different types of neoplasia. Thus, it can be hypothesized that genes encoding these proteins have tumor-suppressor properties (Li et al. 2014). In the case of acute promyelocytic leukemia (APL), the RARα gene has suffered a rearrangement due to a chromosomal translocation. The rearranged gene expresses a fusion protein known as PML-RARα. This fusion protein binds to the transcription inhibitor complex recruited by HDACs and promotes the hypermethylation of genes involved in cell differentiation and apoptosis due to its affinity for the DNA methyltransferases DNMT 1 and DNMT 2 (Lo-Coco and Ammatuna 2006). Another RAR whose expression has been reported as altered during carcinogenesis is RARβ2. Considered the main receptor subtype induced by retinoic acid, RARβ2 has been observed downregulated or completely silenced in several types of cancer (Li et al. 2014). Silencing of RARβ2 could be associated with epigenetic events as has been suggested in the case of prostate and lung cancer where high frequency of hypermethylation in RARβ2 promoter have been observed compared to surrounding tissue (di Masi et al. 2015). Still, there are other epigenetic mechanisms such as histone deacetylation that could be also responsible for the repression of RARβ2 expression (Raif et al. 2009).

Regarding RXRs, 9-cis retinoic acid (9cRA) has showed in vitro binding capability toward RXR's binding domain (Gilardi and Desvergne 2014). Following this idea, synthetic compounds sharing structural features with retinoids have been produced. Those compounds known as rexinoids can inhibit cellular proliferation and induce differentiation and apoptosis of neoplastic cells (Uray et al. 2016). RXRs can also form dimers with other nuclear receptors like peroxisomal proliferator activated receptors (PPARs), vitamin D receptors (VDRs), or retinoid-related orphan nuclear receptors. In this case, the signal transduction is independent of retinoids. The competition among receptors to form stable dimers with the RXRs constitutes an important level of regulation of gene expression (Uray et al. 2016). The formation of homodimers RXR/RXR is also possible. In this case, the homodimer associates with DNA sequences named RXR response elements (RXRE) and activates gene transcription following ligand binding to the receptor. However, when RXR forms heterodimers like RAR/RXR, RXR's agonists cannot trigger their signal transduction possibly due to its association with ATRA (Uray et al. 2016).

Retinol and its analogs show chemopreventive potential against cancer as in vitro and in vivo studies have already demonstrated (Hayden and Satre 2002; Alizadeh et al. 2014). In this regard, evidence suggests that ROL interferes with carcinogenesis through the modulation of cellular processes like inhibition of proliferation, differentiation induction, and triggering of apoptosis. Moreover, retinoids' anticarcinogenic effects seem to be related not only with neoplastic cell dynamics but also with angiogenesis and metastasis (Li et al. 2014). Another potential target for retinoids is the catalytic activity of the senescence-related enzyme telomerase. Increased levels of this enzyme have been observed in vitro after treatment of breast cancer cell lines BT549 and MDA-MB435 as well as prostate cancer cell lines DU145 and PC3 with ATRA at 80uM during 6 h (Xiao et al. 2005). In vivo experiments with rats subjected to the RH model and treated with ATRA, 9cRA, or ROL resulted in chemopreventive activity in the initial phases of hepatocarcinogenesis. In this experiment, retinoids reduced cellular proliferation but only ROL reduced the level of damaged DNA (Fonseca et al. 2005).

Since nuclear receptors are the key mediators of vitamin A biological activity, clinical trials have been conducted to evaluate the relevance of RARs and RXRs in the retinoid-dependent signaling pathway observed during carcinogenesis. In the context of APL, retinoic acid induces expression of the protein UBE1L (ubiquitin-activating enzyme E1-like), leading to degradation of the fusion protein PML-RARα and apoptosis (Kitareewan et al. 2002). Degradation of this fusion protein has been observed at pharmacological concentrations of retinoic acid (Lo-Coco and Ammatuna 2006). In some cases of APL, patients resistant to ATRA treatment show other rare types of fusion proteins, all of them involving RARα gene like STAT5B-RARα and PLZF-RARα (Alizadeh et al. 2014). The treatment of cutaneous T-cell lymphoma with the rexinoid bexarotene induced apoptosis through a mechanism involving selective cleavage of caspase-3. Following this finding, preclinical studies with bexarotene showed promissory results for the treatment of breast, prostate, and non-small-cell lung cancer (Qi et al. 2016).

Modulation of epigenetic-dependent gene expression can be also associated with retinoids in the prevention of chronic diseases (Ong et al. 2011). In this way, rats treated with retinyl palmitate, ATRA, or 13-cis retinoic acid at 30 μmol/kg showed increased activity of the hepatic enzyme glycine-*N*-methyltransferase involved in the maintenance of adenosylmethionine/s-adenosylhomocystein equilibrium and thus suggesting an increase in transmethylation rates (Rowling et al. 2002). Alterations in global DNA methylation, measured as long interspersed nucleotide element (LINE-1) methylations in peripheral white blood cells, have been linked with the risk of cardiovascular diseases and cancer. Owing to LINE-1, methylation is susceptible to environmental factors such as diet, it has been demonstrated that this epigenetic alteration is inversely proportional to plasma levels of vitamin A, and it can be related with modulation of DNMTS activity or expression (Perng et al. 2012). In this regard, T47D cells from breast cancer showed a reduction in DNMT1, DNMT3a, and DNMT3b expression when treatment with 10 μM ATRA was applied during 12 days. This event was associated with the possible loss of carcinogenic potential of those cells after treatment with ATRA (Hansen et al. 2007). In another report, colon cancer cells from the lineages SW48, DLD1, RKO, SW837, HCT-116, SW480, Colo-320, and HT29 silencing of CRBP1 and RARbeta2 by promoter hypermethylation was confirmed (Esteller et al. 2002). CRBP1, responsible for the intracellular transport of ROL and RARβ2, has also been reported to be silenced in human samples of colon but, in these cases, additional information about diet led to the conclusion that patients that consumed high levels of vitamin A showed lower frequencies of promoter methylation in both CRBP1 and RARβ2 genes (Esteller et al. 2002). During the promotion phase of hepatocarcinogenesis, rats treated with retinyl palmitate (1 mg/100 g b. w.) exhibited lower incidence of hepatic neoplasia as well as an inhibition of cellular proliferation. However, the treatment with this retinoid did not produce alterations in the methylation patterns of either HMG-CoA reductase or c-myc (Moreno et al. 2002).

Regarding cellular differentiation, it has been demonstrated that human embryonic cell lines CHA3-hES, CHA4-hES, and SNUhES3 treated with 50 μM of retinoic acid displayed greater levels of global DNA methylation compared to nontreated cells. Moreover, around 2000 genes exhibited differences in expression when cells were treated with retinoic acid and half of them revealed hypoexpression status. Combinatorial analyses of methylation and gene expression unveiled that *THY1*, a gene involved in the undifferentiated status of the cell, was hypermethylated (Cheong et al. 2010). Besides promoter hypermethylation, retinoic acid induced downregulation of genes related to DNA methylation such as *Dnmt3b*, *Dnmt3l*, *Tet1*, and *Tet2* in embryonic cells from J1 mice and thus suggested a global DNA hypomethylation as confirmed by 5-methylcytosine immunofluorescence. Lower expression of *Hdac2* and *Hdac8*, genes involved in histone acetylation, was observed in cells treated with retinoic acid. Yet, treatment with retinoic acid failed to show increasing levels of H3K9ac. Therefore, it was suggested that the main epigenetic modification modulated by retinoic acid in the differentiation of those cells was DNA methylation, even though dimethylation of histone H3 in lysine 4 (H3K4me2) was also observed. Treatment with retinoic acid also modulated the expression of microRNAs in cells of J1 mice. miRNAs associated with cellular differentiation such as miR-135, miR-302, miR-449a, miR-200b, miR-200c, miR193b, miR-130, and miR-141 were downregulated while miR-10a, miR-181, and miR-480 were upregulated (Zhang et al. 2015). miR-10a was also studied in human samples of breast cancer where its expression was otherwise downregulated (Khan et al. 2015). Likewise, cells HGC-27 and MGC-803 derived from gastric adenocarcinomas showed the same pattern of miR-10a downregulation. When these cells were treated with the DNA demethylating agent 5-azacitidine (5-aza) it was confirmed the epigenetic repression of miR-10a due to its increased expression following treatment with 5-aza (Jia et al. 2014).

Toxicity is the main limitation of vitamin A and retinoids for their use in prevention and treatment of cancer. In this regard, some strategies have been suggested to reduce retinoid toxicity without reducing its therapeutic efficacy. For instance, intermittent administration of high doses of 13cis RA could be more effective than reduced doses administered continually as has been demonstrated in neuroblastoma patients (Veal et al. 2007). Structural modifications of retinoid molecules could be also a good strategy to reduce retinoid toxicity. Synthetic analogs of ATRA such as all-trans retinoic acid α-tocoferil-ester is 150 times less toxic than ATRA and it does not show teratogenic activity at therapeutic concentrations (Makishima and Honma 1997).

Combinatorial treatment with associations of retinoids with other compounds represents another plausible strategy to both reduce toxicity and increase the treatment efficiency. Several studies suggested the association of ATRA with other compounds that show epigenetic modulatory activity like DNMTs and HDACs inhibitors (Schenk et al. 2014). Thus, the therapeutic potential of combining retinoids with butyrate or its prodrug tributyrin is here emphasized.

Tributyrin Association with Vitamin A

Evidence suggest that methyl-CpG binding proteins such as MeCP2 and MBD2 interact with HDACs and that they recruit corepressor proteins of transcription to the promoter site, thus establishing an association between two key epigenetic modulation mechanisms of gene transcription: DNA methylation and histone acetylation. Combination of HDACs and DNMTs inhibitors could modulate chromatin architecture and reestablish the expression of genes silenced by repressive complexes. Therefore, this strategy would increase the therapeutic efficacy against cancer (Savickiene et al. 2012).

Vitamin A is being used for both prevention and treatment of breast cancer. However, the main limiting factor for this kind of approach is the deregulation of ROL metabolism that takes place in those breast cancer cells. For instance, impeded synthesis of retinyl esters interferes with vitamin A intracellular stoking which could be a consequence of a downregulation in the expression of RARβ and CRBP-1. In this regard, MCF-7 cells treated with 10 μM vitamin A in association with 1 mM sodium butyrate showed an increased expression of RARβ but no alteration in CRBP-1. Even though the global DNA methylation pattern and the specific level of methylation of RARβ and CRBP-1 gene promoters did not change after treatment, levels of H3K9ac did raise (Andrade et al. 2012). In vivo, association of vitamin A with TB was evaluated in the context of hepatocarcinogenesis chemoprevention. In this study, rats submitted to the RH model and treated with vitamin A plus TB during the promotion phase of carcinogenesis showed higher levels of H3K9ac and apoptotic bodies compared to controls. As observed in vitro, rats submitted to the RH-model did not show changes in CRBP-1 promoter methylation pattern, suggesting that epigenetic silencing of CRBP-1 is not associated with alterations of ROL metabolism in early phases of hepatocarcinogenesis (de Conti et al. 2012).

Vitamin A analogs have also been tested in association with TB as shown in Table 1. For instance, TB and ATRA induced cellular differentiation in human promyelocytic leukemia cells HL60. This result was observed when HL60 cells were treated with 12–38 nM ATRA, mimicking the ATRA blood levels found in healthy patients, along with TB in a concentration 10 times lower than those used to induced differentiation alone (Chen and Breitman 1994). In another study where the authors used NB4 cells as human promyelocytic leukemia model, treatment with

Table 1 Effects of the combinatorial treatment between tributyrin and retinoids

Combinatorial treatment effects	Model	Reference
Cellular differentiation	Human promyelocytic leukemia cells HL60 Human promyelocytic leukemia cells NB4	Chen and Breitman 1994 Taimi et al. 1998; Chen et al. 1999
Fetal hemoglobin expression	Human erythroleukemia cell K562	Witt et al. 2000
Inhibition of proliferation and upregulation of sodium-iodide symporter	Human follicular thyroid carcinoma cell FTC-133	Zhang et al. 2011
Increased expression of RARβ and H3K9ac	Human breast cancer cell MCF7	Andrade et al. 2012
Apoptosis and increased expression of H3K9ac	Wistar rats subjected to a hepatocarcinogenesis model	de Conti et al. 2012

retinoic acid plus TB also leads to cellular differentiation (Taimi et al. 1998; Chen et al. 1999). Likewise, human erythroleukemia cells K562 expressed fetal hemoglobin after treatment with retinoic acid along with TB following once again a cellular differentiation pathway. In FTC-133 cells, a model for human thyroid follicular carcinoma, association of ATRA with TB inhibited cellular proliferation and increased the expression of NIS (sodium-iodide symporter). The latter resulted in higher radioiodine uptake compared to cells treated with each compound alone. Nanostrutured carrier systems produced from ATRA and TB together also showed efficacy inhibiting cellular proliferation in neoplastic cells (Su et al. 2008; Silva et al. 2015). However, the epigenetic mechanisms involved with these antineoplastic activities were not elucidated.

In the context of arterial hypertension, C57/BL6 mice bearing an interrupted copy of the gene *Npr1* (GC-A/natriuretic peptide receptor-A) showed an increase in *Npr1* expression after treatment with ATRA plus sodium butyrate. This change in *Npr1* expression occurred due to changes in chromatin architecture mediated by HDACs and HATs and the consequent increase in H3K9ac and H4K12ac levels and the decrease in H3K9me2, H3K9me3, and H3K27me3 levels. In this regard, association of ATRA with sodium butyrate stimulated the recruitment of proteins associated with transcription initiation such as HATs, RARα, activator complex containing E26 transformation-specific 1 to Npr1 promoter (Kumar et al. 2014). The efficacy of the association of retinoids with HDACs depends on the inhibition of HDAC1 and the activation of RARs (Spurling et al. 2008). It has been demonstrated that HDAC1 and HDAC2 bind several RAREs. However, HDAC1 is associated with RARβ2-RARE (Urvalek et al. 2014). Strategies combining retinoic acid with HDACs inhibitors and with DNA methyltransferases are also being tested. Association of retinoic acid with valproic acid with 5-Aza induced the expression of RARβ and inhibited proliferation in breast cancer and promyelocytic leukemia cell lines (Mongan and Gudas 2005).

Although drugs with modulating properties of chromatin structure and synthetic retinoids have demonstrated more efficacy than tributyrin and retinol in cancer treatment, bioactive natural products are of particular interest for the chemoprevention of early neoplastic lesions. As long as chemopreventive agents feature a broad action in epigenetic processes, their effects on global DNA methylation, activation of tumor suppressor genes silenced by DNA methylation, modulation of histone modification patterns, and the expression of miRNAs evidence their relevance in several molecular pathways activated during carcinogenesis (Huang et al. 2011). Figure 4 shows the epigenetic mechanisms of carcinogenesis in which TB and ROL have a modulatory action. More studies are required to identify whether those chemopreventive activities described in this chapter for the combinatorial treatment of TB and ROL are mediated by epigenetic regulation of gene expression or by other mechanisms. Epigenetic regulation of gene expression by means of the association of TB and ROL seems to show specificity for some neoplastic cells in vitro and for preneoplastic lesions in vivo (de Conti et al. 2012). Considering that inhibition of HDACs by BFCs is a transitory process (Huang et al. 2011), whether the occasional consumption of natural sources of butyrate like dietary fibers and dairy products

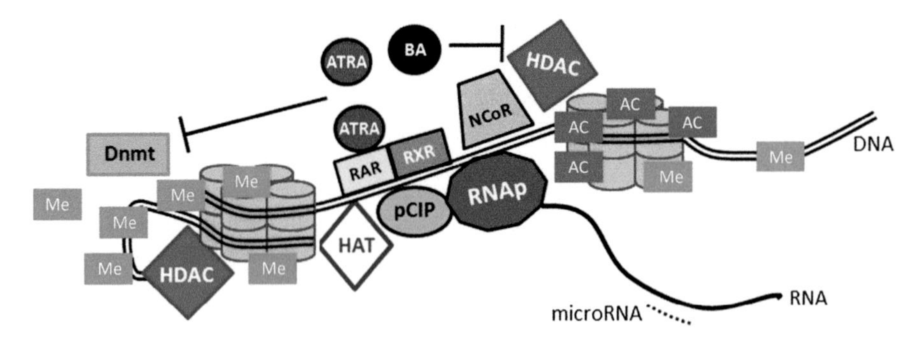

Fig. 4 Epigenetic mechanisms of TB and retinoids in carcinogenesis. Nuclear signaling of ATRA (all-*trans* retinoic acid) depends on the binding to RARs (retinoic acid receptor). These receptors form heterodimers RAR/RXR that recognize specific DNA sequences in the promoter region of several genes. ATRA-dependent transcription is associated with increasing levels of coactivator proteins like HATs (p300 and p300/cAMP response element-binding protein-binding-associated factor), pCIP, and RNA polymerase II. There are also transcription corepressor proteins like NCoR and HDACs that associates to RAR/RXR dimers in the absence of the ligand ATRA but dissociates following binding of ATRA to the corresponding RAR. A reduction in DNMT expression is observed after treatment with retinoic acid. Then, global DNA hypomethylation and miRNA modulation occur due to DNMT downregulation. TB inhibits HDACs leading to an increase in histone acetylation and modulation of miRNAs expression

results in actual histone hyperacetylation is still unclear. In the case of vitamin A, the intracellular storage allows for a continuous supply of ROL for its biological functions, for instance, regulation of gene expression. Moreover, chemopreventive (Fig. 5) approaches could be more effective if combinatorial treatments with BFCs were applied. It is important to highlight that the association of TB with vitamin A in the chemoprevention of animal models may resemble the complex interactions between BFCs in some human diets. Specifically, in the case of diets with high content of fruits and vegetables, where the relationship with a reduced risk for cancer have been well established (World Cancer Research Fund/American Institute for Cancer Research 2007).

Dictionary of Terms

- **Neoplasia** – An abnormal mass of tissue with excessive growth that may present invasive behavior such as cancer.
- **Bioactive food compounds** – Dietary constituents that in addition to macro and micronutrients promotes health benefits.
- **Carcinogenesis** – The process that culminates with the development of cancer.
- **Prodrug** – A compound that requires biotransformation to show pharmacological effects.
- **Hepatocarcinogenesis** – The process that culminates with the development of liver cancer.

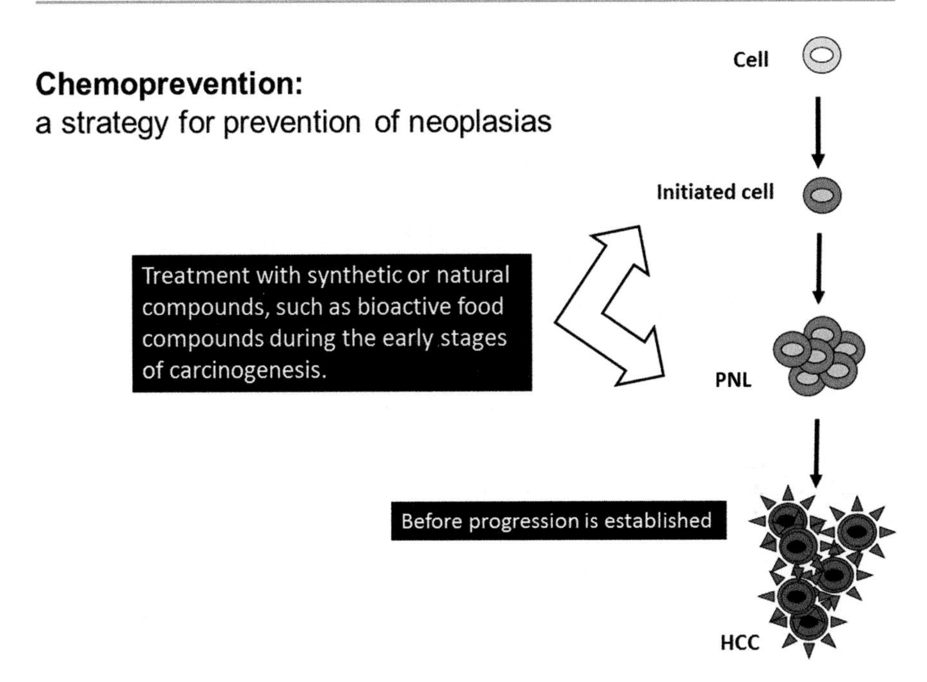

Fig. 5 Chemoprevention. Chemoprevention can be considered an important strategy for the cancer prevention when the intervention occurs in early phases of carcinogenesis like initiation and promotion. In these phases, chemopreventive agents can act on initiated cells and preneoplastic lesions (*PNLs*), inhibiting cellular proliferation and inducing apoptosis or cellular differentiation. These mechanisms can delay the onset of cancer progression when cells acquire aggressive behavior and invasive potential

- **5-Aza-2′deoxycytidine** – A compound that inhibits DNA methyltransferase resulting in DNA hypomethylation.
- **Placentary gluthatione-S-transferase** – A placentary enzyme whose expression is upregulated in pre- and neoplastic lesions and that is used as a biomarker of chemical carcinogenesis.
- **Sorafenib** – A multikinase inhibitor drug used for the treatment of liver cancer.
- **Valproic acid** – An anticonvulsant drug that inhibits HDACs.

Key Facts of Chemoprevention of Cancer with Tributyrin and Vitamin A

- Cancer chemoprevention is a strategy for the prevention of neoplasia occurrence based on the administration of natural or synthetic compounds during the early phases of carcinogenesis.
- Evidence suggests that chemopreventive agents may act specifically on neoplastic cells.

- Butyrate and HDAC have been evaluated as potential chemopreventive agents given their inhibitory properties of cellular proliferation, induction of cellular differentiation, and apoptosis.
- Tributyrin shows a better pharmacokinetic behavior than butyrate, allowing its oral administration.
- Vitamin A also presents cancer chemopreventive activities acting by epigenetic mechanisms that mostly involve changes in DNA methylation patterns.
- Retinoids toxicity may be a limiting factor for their use in cancer chemoprevention.
- Approaches using chemopreventive agents with different mechanisms of action such as tributyrin and retinoids may consist in a potential strategy for the chemoprevention of carcinogenesis.

Summary Points

- Nutritional factors play an important role in cancer prevention.
- Butyric acid derived from dietary fibers fermentation shows in vitro and in vivo anticarcinogenic activities.
- Butyric acid acts as a HDAC inhibitor promoting histones acetylation, and it may also methylate these proteins and regulate the expression of miRNAs.
- Compared to butyric acid, tributyrin is a triacylglycerol that shows an increased half-life, acting also in acetylation and methylation of histones.
- Given that high doses of tributyrin are required to achieve its therapeutic activity, elaboration of butyrate molecular carriers from this triacylglycerol could be considered promissory alternatives for its use in the prevention of chronic diseases.
- The association of HDAC inhibitors with retinoids may consists in an important chemopreventive strategy against cancer.
- Retinoids can exert regulatory activity of gene expression acting as ligands of nuclear receptors colocalized in transcription complexes.
- Vitamin A derived compounds may modify gene expression by epigenetics mechanisms, including DNA methylation and the expression of microRNAs.
- The association of a HDAC inhibitor with a DNA methylation modulator, such as tributyrin and retinoids, respectively, may have a potential cancer chemopreventive action.
- Because epigenetic events are transitory, the importance of an adequate and diversified dietary pattern should be stressed.

References

Ahmad A, Sakr WA, Rahman KM (2012) Novel targets for detection of cancer and their modulation by chemopreventive natural compounds. Front Biosci (Ellite Ed) 4:410–425

Al Tanoury Z, Piskunov A, Rochette-Egly C (2013) Vitamin a and retinoid signaling: genomic and nongenomic effects. J Lipid Res 54:1761–1775

Alcantara EN, Speckmann EW (1976) Diet, nutrition, and cancer. Am J Clin Nutr 29:1035–1047
Alizadeh F, Bolhassani A, Khavari A et al (2014) Retinoids and their biological effects against cancer. Int Immunopharmacol 18:43–49
Andrade FO, Nagamine MK, Conti AD et al (2012) Efficacy of the dietary histone deacetylase inhibitor butyrate alone or in combination with vitamin a against proliferation of MCF-7 human breast cancer cells. Braz J Med Biol Res 45:841–850
Banudevi S, Swaminathan S, Maheswari KU (2015) Pleiotropic role of dietary phytochemicals in cancer: emerging perspectives for combinational therapy. Nutr Cancer 67:1021–1048
Berni Canani R, Di Costanzo M, Leone L (2012) The epigenetic effects of butyrate: potential therapeutic implications for clinical practice. Clin Epigenetics 4:4
Calin GA, Croce CM (2006) MicroRNA signatures in human cancers. Nat Rev Cancer 6:857–866
Chambon P (1996) A decade of molecular biology of retinoic acid receptors. FASEB J 10:940–954
Chen ZX, Breitman TR (1994) Tributyrin: a prodrug of butyric acid for potential clinical application in differentiation therapy. Cancer Res 54:3494–3499
Chen Z, Wang W, Pan J et al (1999) Combination of all-trans retinoic acid with butyric acid and its prodrugs markedly enhancing differentiation of human acute promyelocytic leukemia NB4 cells. Chin Med J 112:352–355
Cheong HS, Lee HC, Park BL et al (2010) Epigenetic modification of retinoic acid-treated human embryonic stem cells. BMB Rep 43:830–835
Conley BA, Egorin MJ, Tait N et al (1998) Phase I study of the orally administered butyrate prodrug, tributyrin, in patients with solid tumors. Clin Cancer Res 4:629–634
Cordani M, Butera G, Pacchiana R et al (2016) Molecular interplay between mutant p53 proteins and autophagy in cancer cells. Biochim Biophys Acta 1867:19–28
Daniel P, Brazier M, Cerutti I et al (1989) Pharmacokinetic study of butyric acid administered in vivo as sodium and arginine butyrate salts. Clin Chim Acta 181:255–263
Davie JR (2003) Inhibition of histone deacetylase activity by butyrate. J Nutr 133:2485S–2493S
de Conti A, Kuroiwa-Trzmielina J, Horst MA et al (2012) Chemopreventive effects of the dietary histone deacetylase inhibitor tributyrin alone or in combination with vitamin a during the promotion phase of rat hepatocarcinogenesis. J Nutr Biochem 23:860–866
de Conti A, Tryndyak V, Koturbash I et al (2013) The chemopreventive activity of the butyric acid prodrug tributyrin in experimental rat hepatocarcinogenesis is associated with p53 acetylation and activation of the p53 apoptotic signaling pathway. Carcinogenesis 34:1900–1906
di Masi A, Leboffe L, De Marinis E et al (2015) Retinoic acid receptors: from molecular mechanisms to cancer therapy. Mol Asp Med 41:1–115
Donohoe DR, Bultman SJ (2012) Metaboloepigenetics: interrelationships between energy metabolism and epigenetic control of gene expression. J Cell Physiol 227:3169–3177
Donohoe DR, Collins LB, Wali A et al (2012) The Warburg effect dictates the mechanism of butyrate-mediated histone acetylation and cell proliferation. Mol Cell 48:612–626
Edelman MJ, Bauer K, Khanwani S et al (2003) Clinical and pharmacologic study of tributyrin: an oral butyrate prodrug. Cancer Chemother Pharmacol 51:439–444
Egorin MJ, Yuan ZM, Sentz DL et al (1999) Plasma pharmacokinetics of butyrate after intravenous administration of sodium butyrate or oral administration of tributyrin or sodium butyrate to mice and rats. Cancer Chemother Pharmacol 43:445–453
Esteller M, Guo M, Moreno V et al (2002) Hypermethylation-associated inactivation of the cellular retinol-binding-protein 1 gene in human cancer. Cancer Res 62:5902–5905
Farfán M, Villalón MJ, Ortíz ME et al (2013) The effect of interesterification on the bioavailability of fatty acids in structured lipids. Food Chem 139:571–577
Fonseca EMAV, Chagas CE, Mazzantini RP et al (2005) All-trans and 9-cis retinoic acids, retinol and beta-carotene chemopreventive activities during the initial phases of hepatocarcinogenesis involve distinct actions on glutathione S-transferase positive preneoplastic lesions remodeling and DNA damage. Carcinogenesis 26:1940–1946
Gilardi F, Desvergne B (2014) RXRs: collegial partners. Subcell Biochem 70:75–102
Gori GB (1978) Role of diet and nutrition in cancer cause, prevention and treatment. Bull Cancer 65:115–126

Guariento AH, Furtado KS, de Conti A et al (2014) Transcriptomic responses provide a new mechanistic basis for the chemopreventive effects of folic acid and tributyrin in rat liver carcinogenesis. Int J Cancer 135:7–18

Gudas LJ (2013) Retinoids induce stem cell differentiation via epigenetic changes. Semin Cell Dev Biol 24:701–705

Hansen NJ, Wylie RC, Phipps SM et al (2007) The low-toxicity 9-cis UAB30 novel retinoid down-regulates the DNA methyltransferases and has anti-telomerase activity in human breast cancer cells. Int J Oncol 30:641–650

Hayden LJ, Satre MA (2002) Alterations in cellular retinol metabolism contribute to differential retinoid responsiveness in normal human mammary epithelial cells versus breast cancer cells. Breast Cancer Res Treat 72:95–105

Heidor R, Ortega JF, de Conti A et al (2012) Anticarcinogenic actions of tributyrin, a butyric acid prodrug. Curr Drug Targets 13:1720–1729

Heidor R, de Conti A, Ortega JF et al (2016) The chemopreventive activity of butyrate-containing structured lipids in experimental rat hepatocarcinogenesis. Mol Nutr Food Res 60:420–429

Huang J, Plass C, Gerhauser C (2011) Cancer chemoprevention by targeting the epigenome. Curr Drug Targets 12:1925–1956

Humphreys KJ, Cobiac L, Le Leu RK et al (2013) Histone deacetylase inhibition in colorectal cancer cells reveals competing roles for members of the oncogenic miR-17-92 cluster. Mol Carcinog 52:459–474

Jia H, Zhang Z, Zou D et al (2014) MicroRNA-10a is down-regulated by DNA methylation and functions as a tumor suppressor in gastric cancer cells. PLoS One 9:e88057

Kang SN, Lee E, Lee MK et al (2011) Preparation and evaluation of tributyrin emulsion as a potent anti-cancer agent against melanoma. Drug Deliv 18:143–149

Khan S, Wall D, Curran C et al (2015) MicroRNA-10a is reduced in breast cancer and regulated in part through retinoic acid. BMC Cancer 15(1):345

Kim KB, Nam YA, Kim HS et al (2014) α Linolenic acid: nutraceutical, – pharmacological and toxicological evaluation. Food Chem Toxicol 70:163–178

Kitareewan S, Pitha-Rowe I, Sekula D et al (2002) UBE1L is a retinoid target that triggers PML/RARalpha degradation and apoptosis in acute promyelocytic leukemia. Proc Natl Acad Sci USA 99:3806–3811

Kotecha R, Takami A, Espinoza JL (2016) Dietary phytochemicals and cancer chemoprevention: a review of the clinical evidence. Oncotarget 7:52517–52529

Kumar P, Periyasamy R, Das S et al (2014) All-trans retinoic acid and sodium butyrate enhance natriuretic peptide receptor a gene transcription: role of histone modification. Mol Pharmacol 85:946–957

Kuroiwa-Trzmielina J, de Conti A, Scolastici C et al (2009) Chemoprevention of rat hepatocarcinogenesis with histone deacetylase inhibitors: efficacy of tributyrin, a butyric acid prodrug. Int J Cancer 124:2520–2527

Leder A, Leder P (1975) Butyric acid, a potent inducer of erythroid differentiation in cultured erythroleukemic cells. Cell 5:319–322

Li M, Sun Y, Guan X et al (2014) Advanced progress on the relationship between RA and its receptors and malignant tumors. Crit Rev Oncol Hematol 91:271–282

Liu Y, Upadhyaya B, Fardin-Kia AR et al (2016) Dietary resistant starch type 4-derived butyrate attenuates nuclear factor-kappa-B1 through modulation of histone H3 trimethylation at lysine 27. Food Funct 7:3772–3781

Lo-Coco F, Ammatuna E (2006) The biology of acute promyelocytic leukemia and its impact on diagnosis and treatment. Hematology Am Soc Hematol Educ Program 1:156–161

Louis P, Flint HJ (2009) Diversity, metabolism and microbial ecology of butyrate-producing bacteria from the human large intestine. FEMS Microbiol Lett 294:1–8

Makishima M, Honma Y (1997) Tretinoin tocoferil as a possible differentiation-inducing agent against myelomonocytic leukemia. Leuk Lymphoma 26:43–48

Mangelsdorf DJ, Evans RM (1995) The RXR heterodimers and orphan receptors. Cell 83:841–850
McCollum EV, Davis M (1913) The necessity of certain lipids during growth. J Biol Chem 15:167–175
Michels KB (2005) The role of nutrition in cancer development and prevention. Int J Cancer 114:163–165
Miller AA, Kurschel E, Osieka R et al (1987) Clinical pharmacology of sodium butyrate in patients with acute leukemia. Eur J Cancer Clin Oncol 23:1283–1287
Mongan NP, Gudas LJ (2005) Valproic acid, in combination with all-trans retinoic acid and 5-aza-2′-deoxycytidine, restores expression of silenced RARbeta2 in breast cancer cells. Mol Cancer Ther 4:477–486
Moreno FS, S-Wu T, Naves MM et al (2002) Inhibitory effects of beta-carotene and vitamin a during the progression phase of hepatocarcinogenesis involve inhibition of cell proliferation but not alterations in DNA methylation. Nutr Cancer 44:80–88
Ogawa K, Yasumura S, Atarashi Y et al (2004) Sodium butyrate enhances Fas-mediated apoptosis of human hepatoma cells. J Hepatology 40:278–284
Ong TP, Moreno FS, Ross SA (2011) Targeting the epigenome with bioactive food components for cancer prevention. J Nutrigenet Nutrigenomics 4:275–292
Ortega JF, de Conti A, Tryndyak V et al (2016) Suppressing activity of tributyrin on hepatocarcinogenesis is associated with inhibiting the p53-CRM1 interaction and changing the cellular compartmentalization of p53 protein. Oncotarget 7:24339–24347
Pascale RM, Simile MM, Calvisi DF et al (2005) Role of HSP90, CDC37, and CRM1 as modulators of P16(INK4A) activity in rat liver carcinogenesis and human liver cancer. Hepatology 42:1310–1319
Perng W, Rozek LS, Mora-Plazas M et al (2012) Micronutrient status and global DNA methylation in school-age children. Epigenetics 7:1133–1141
Prasad KN, Sinha PK (1976) Effect of sodium butyrate on mammalian cells in culture: a review. In Vitro 12:125–132
Qi L, Guo Y, Zhang P et al (2016) Preventive and therapeutic effects of the retinoid X receptor agonist bexarotene on tumors. Curr Drug Metab 17:118–128
Raif A, Marshall GM, Bell JL et al (2009) The estrogen-responsive B box protein (EBBP) restores retinoid sensitivity in retinoid-resistant cancer cells via effects on histone acetylation. Cancer Lett 277:82–90
Rajaram S (2014) Health benefits of plant-derived α-linolenic acid. Am J Clin Nutr 100(Suppl 1):443S–448S
Rephaeli A, Nordenberg J, Aviram A et al (1994) Butyrate-induced differentiation in leukemic myeloid cells in vitro and in vivo studies. Int J Oncol 4:1387–1391
Rowling MJ, McMullen MH, Schalinske KL (2002) Vitamin a and its derivatives induce hepatic glycine N-methyltransferase and hypomethylation of DNA in rats. J Nutr 132:365–369
Savickiene J, Treigyte G, Borutinskaite VV et al (2012) Antileukemic activity of combined epigenetic agents, DNMT inhibitors zebularine and RG108 with HDAC inhibitors, against promyelocytic leukemia HL-60 cells. Cell Mol Biol Lett 17:501–525
Schenk T, Stengel S, Zelent A (2014) Unlocking the potential of retinoic acid in anticancer therapy. Br J Cancer 111:2039–2045
Schlörmann W, Naumann S, Renner C et al (2015) Influence of miRNA-106b and miRNA-135a on butyrate-regulated expression of p21 and cyclin D2 in human colon adenoma cells. Genes Nutr 10:50
Silva EL, Carneiro G, Caetano PA et al (2015) Nanostructured lipid carriers loaded with tributyrin as an alternative to improve anticancer activity of all-trans retinoic acid. Expert Rev Anticancer Ther 15:247–256
Spurling CC, Suhl JA, Boucher N et al (2008) The short chain fatty acid butyrate induces promoter demethylation and reactivation of RARbeta2 in colon cancer cells. Nutr Cancer 60:692–702
Steward WP, Brown K (2013) Cancer chemoprevention: a rapidly evolving field. Br J Cancer 109:1–7

Stommel JM, Marchenko ND, Jimenez GS et al (1999) A leucine-rich nuclear export signal in the p53 tetramerization domain: regulation of subcellular localization and p53 activity by NES masking. EMBO J 18:1660–1672

Su J, He L, Zhang N et al (2006) Evaluation of tributyrin lipid emulsion with affinity to low-density lipoprotein: pharmacokinetics in adult male Wistar rats and cellular activity on Caco-2 and HepG2 cell lines. J Pharmacol Exp Ther 316:62–70

Su J, Zhang N, Ho PC (2008) Evaluation of the pharmacokinetics of all-trans-retinoic acid (ATRA) in Wistar rats after intravenous administration of ATRA loaded into tributyrin submicron emulsion and its cellular activity on Caco-2 and HepG2 cell lines. J Pharm Sci 97:2844–2853

Taimi M, Chen ZX, Breitman TR (1998) Potentiation of retinoic acid-induced differentiation of human acute promyelocytic leukemia NB4 cells by butyric acid, tributyrin, and hexamethylene bisacetamide. Oncol Res 10:75–84

Tang XH, Gudas LJ (2011) Retinoids, retinoic acid receptors, and cancer. Annu Rev Pathol 6:345–364

Trasino SE, Tang XH, Jessurun J et al (2016) Retinoic acid receptor β2 agonists restore glycaemic control in diabetes and reduce steatosis. Diabetes Obes Metab 18:142–151

Uray IP, Dmitrovsky E, Brown PH (2016) Retinoids and rexinoids in cancer prevention: from laboratory to clinic. Semin Oncol 43:49–64

Urvalek A, Laursen KB, Gudas LJ (2014) The roles of retinoic acid and retinoic acid receptors in inducing epigenetic changes. Subcell Biochem 70:129–149

Van der Heiden MG, Cantley LC, Thompson CB (2009) Understanding the Warburg effect: the metabolic requirements of cell proliferation. Science 324:1029–1033

Veal G, Rowbotham S, Boddy A (2007) Pharmacokinetics and pharmacogenetics of 13-cis-retinoic acid in the treatment of neuroblastoma. Therapie 62:91–93

Vinolo MA, Rodrigues HG, Festuccia WT et al (2012) Tributyrin attenuates obesity-associated inflammation and insulin resistance in high-fat-fed mice. Am J Physiol Endocrinol Metab 303: E272–E282

Wang H (2014) Predicting cancer-related MiRNAs using expression profiles in tumor tissue. Curr Pharm Biotechnol 15:438–444

Wang HG, Huang XD, Shen P et al (2013) Anticancer effects of sodium butyrate on hepatocellular carcinoma cells in vitro. Int J Mol Med 31:967–974

Wei X, Peng R, Cao J et al (2016) Serum vitamin a status is associated with obesity and the metabolic syndrome among school-age children in Chongqing. China Asia Pac J Clin Nutr 25:563–570

Witt O, Schmejkal S, Pekrun A (2000) Tributyrin plus all-trans-retinoic acid efficiently induces fetal hemoglobin expression in human erythroleukemia cells. Am J Hematol 64:319–321

World Cancer Research Fund/American Institute for Cancer Research (2007) Food, nutrition, physical activity, and the prevention of cancer: a global perspective. AICR, Washington, DC

Xiao X, Sidorov IA, Gee J et al (2005) Retinoic acid-induced downmodulation of telomerase activity in human cancer cells. Exp Mol Pathol 79:108–117

Xie C, Wu B, Chen B et al (2016) Histone deacetylase inhibitor sodium butyrate suppresses proliferation and promotes apoptosis in osteosarcoma cells by regulation of the MDM2-p53 signaling. Onco Targets Ther 9:4005–4013

Zhang M, Guo R, Xu H et al (2011) Retinoic acid and tributyrin induce in-vitro radioiodine uptake and inhibition of cell proliferation in a poorly differentiated follicular thyroid carcinoma. Nucl Med Commun 32:605–610

Zhang J, Gao Y, Yu M et al (2015) Retinoic acid induces embryonic stem cell differentiation by altering both encoding RNA and microRNA expression. PLoS One 10:e0132566

Association Between MicroRNA Expression and Vitamin C in Ovarian Cells

86

Yong Jin Kim, Yoon Young Kim, and Seung-Yup Ku

Contents

Abstract

Vitamin C (L-ascorbic acid) is an essential, water-soluble micronutrient that exists predominantly as the ascorbate anion under physiological pH conditions and has been implicated in several processes of reproduction of reproductive organ cells. Several studies described the regulatory roles of vitamin C in various cellular developmental processes, via microRNA mechanism, using in vivo and in vitro animal models. To date, some specific microRNAs have been regarded as candidates that have regulatory roles in vitamin C metabolism. Many questions should

Y. J. Kim
Department of Obstetrics and Gynecology, Korea University Medical College, Korea University Guro Hospital, Seoul, South Korea
e-mail: zinigo@gmail.com

Y. Y. Kim · S.-Y. Ku (✉)
Department of Obstetrics and Gynecology, Seoul National University College of Medicine, Seoul National University Hospital, Seoul, South Korea
e-mail: yoonykim96@gmail.com; jyhsyk@snu.ac.kr; jyhsyk@gmail.com

V. B. Patel, V. R. Preedy (eds.), *Handbook of Nutrition, Diet, and Epigenetics*,
https://doi.org/10.1007/978-3-319-55530-0_80

be further investigated, such as the following: whether and how vitamin C directly regulates epigenetic modifiers, whether vitamin C regulates gene and miRNA promoters through specific signaling pathways, and whether vitamin C-induced DNA demethylation occurs.

Keywords
microRNA · Vitamin C · Follicle development

List of Abbreviations

μM	Micro molarity
G-cell	Granulosa cell
GLUT	Glucose transporters
Gulo	L-gulono-γ-lactone oxidase
HDL3	High-density lipoprotein 3
IGF-1	Insulin-like growth factor 1
IOM	Institute of Medicine
IVM	In vitro maturation
mg	Milligram
MII	Meiosis II
miRNAs	MicroRNAs
ODS	Osteogenic disorder Shionogi
RDA	Recommended dietary allowances
SVCT	Sodium-dependent vitamin C transporters

MicroRNAs are noncoding 21–23 nucleotide RNA molecules that regulate the expressions of other genes by inhibiting translation or cleaving complementary target mRNAs (Bartel 2009). Although the exact function has not yet been determined, microRNAs have been revealed to play important roles in intracellular signaling (Lai et al. 2005), apoptosis (Xu et al. 2003), metabolism (Poy et al. 2004), and organogenesis (Sokol and Ambros 2005) in addition to clinical implication (Kim et al. 2016). Because of their importance on cell biology, microRNA may have significant impacts on the metabolism of trace elements such as vitamins.

Vitamin C (L-ascorbic acid) is an essential, water-soluble micronutrient that exists predominantly as the ascorbate anion under physiological pH conditions. It is well established that ascorbate is an antioxidant and free radical scavenger and, further, an essential cofactor in numerous enzymatic reactions (Chan 1993). This essential, water-soluble micronutrient exists predominantly as the ascorbate anion (L-ascorbic acid) under physiological pH conditions. Vitamin C is involved in a wide variety of biological processes as an antioxidant and free radical scavenger, including hormonal action and cell behavior (Lee et al. 2008b; Patak et al. 2004).

Recently, some studies showed relationship between vitamin C and microRNA profiles in various cell. In this chapter, the evidence and potential molecular mechanism of vitamin C regulating microRNA expression are reviewed.

Metabolism of Vitamin C

Vitamin C is a chain breaking agent that stops the propagation of the peroxidative process, helping recycle oxidized vitamin E and glutathione (Chan 1993). This essential, water-soluble micronutrient exists predominantly as the ascorbate anion (L-ascorbic acid) under physiological pH conditions. Vitamin C is involved in a wide variety of biological processes as an antioxidant and free radical scavenger, including hormonal action and cell behavior (Lee et al. 2008a; Patak et al. 2004). Also, vitamin C has an essential role in collagen crosslinking. Severe vitamin C deficiency can cause scurvy due to incomplete collagen crosslinking (Gould and Woessner 1957). Most mammals, such as rodents, synthesize ascorbate de novo in the liver from glucose through a biosynthetic pathway. In contrast, humans – as well as primates, guinea pigs, and fruit bats – no longer can synthesize ascorbate due to a mutant and nonfunctional enzyme, L-gulonolactone oxidase (Gulo), which catalyzes the last step of ascorbate biosynthesis (Linster and Van Schaftingen 2007). For these mammalian species, ascorbate is a vitamin that needs to be supplied through dietary sources and supplements. From either dietary sources or synthesis in the liver, ascorbate enters cells primarily through sodium-dependent vitamin C transporters (SVCTs). SVCT1 is the transporter of high capacity and low affinity, primarily responsible for ascorbate absorption and reabsorption in intestinal and renal epithelial cells. SVCT2 is a more ubiquitous transporter that has high affinity and low capacity, and it distributes ascorbate to most tissues (Wilson 2005). The average concentration of ascorbate in the plasma of healthy humans or mice is ~50 μM. Currently, the recommended dietary allowance (RDA) by the Institute of Medicine (IOM) is 90 mg for adult males and 75 mg for adult females, although the tolerable upper intake level for adults is 2000 mg per day.

L-Gulono-γ-Lactone Oxidase Knockout Mouse

Whereas ascorbic acid is endogenously produced in most animals, higher primates of the suborder Haplorhini (which includes prosimians, Old World monkeys, apes, and New World monkeys) are among the few groups that lack this ability (Pollock and Mullin 1987). This is due to the loss of Gulo [NCBI Entrez Gene ID 2989], which encodes GULO, responsible for the last step of ascorbate synthesis (Nishikimi et al. 1988). The transport of vitamin C requires the expression of glucose transporters (GLUT) for facilitated diffusion of dehydroascorbic acid and sodium-dependent vitamin C transporters (SVCT) for active transport of ascorbic acid (Yu and Schellhorn 2013). The guinea pig, the first animal known to develop scurvy, also lacks GULO due to an evolutionary loss-of-function mutation event (Burns et al. 1956). The osteogenic disorder Shionogi (ODS) rat was then isolated with GULO deficiency (Kawai et al. 1992) before mice became widely accepted as the vertebrate model organism.

Humans and guinea pigs express GLUT1 on erythrocyte membranes, specifically associated with the uptake of dehydroascorbic acid, whereas ascorbic acid-synthesizing animals such as mice express the GLUT4 transporter (Montel-Hagen et al. 2008). Active transport of ascorbic acid against a concentration gradient occurs in intestinal, renal, and liver epithelial cells expressing SVCT1, the bulk ascorbic acid transporter. Specialized and usually metabolically active cells, such as brain, retinal, and placental cells, express SVCT2 for ascorbic acid uptake (Li and Schellhorn 2007).

L-gulono-γ-lactone oxidase knockout mouse (Gulo-/-) was constructed by deleting exons 3 and 4 from Gulo in a C57BL/6 background (Maeda et al. 2000). This renders Gulo fully nonfunctional, and vitamin C supplementation is required to maintain viability in these mice. Interestingly, the plasma ascorbate concentration of Gulo-/- mice fed with a vitamin C-deficient diet does not reach zero upon dietary restriction and is instead maintained at 15% of wild-type concentrations, suggesting an uncharacterized pathway to generate a small amount of ascorbate. Consistently, vitamin C concentrations in unsupplemented Gulo-/- mice after 5 weeks of deficiency are still 25–30% of those of the wild type in major organ stores such as the adrenal gland, lungs, and brain (Kim et al. 2012). Similar to observations made with guinea pigs, vitamin C concentrations increased before endpoint in several tissues, including the intestines, heart, adrenal gland, and spleen. As the genomics and genetics of the mouse have been extensively studied, the Gulo-/- mouse has become the model of choice in studying the role of vitamin C in complex diseases. Vitamin C production has been successfully restored in Gulo-/- mice using a gene therapeutic adenovirus (Li et al. 2008), making it possible to robustly manipulate physiological ascorbate concentrations in an inbred mouse. Unlike guinea pigs treated with modified GULO enzyme (Hadley and Sato 1989), it is not necessary to provide exogenous L-gulonolactone to virus-treated Gulo2/2 mice (Li et al. 2008), because L-gulonolactone is produced at sufficient levels in Gulo-/- mice for GULO activity.

Vitamin C and Reproduction

Vitamin C has been implicated in several processes of reproduction by follicular, luteal development and maintenance of healthy pregnancy for female. Vitamin C is essential for collagen biosynthesis, and it is especially important for the growth of the ovarian follicle, ovulation and the luteal phase (Murray et al. 2001).

Ovarian tissue contains high concentrations of vitamin C (Luck et al. 1995). Several in vitro ovarian follicle maturation studies showed that vitamin C improves the quality of follicles and their survival (Jimenez et al. 2016, Kim et al. 2010; Thomas et al. 2001). Although it has not fully revealed, this regulatory role of vitamin C may be due to the influence on MMP production.

Vitamin C deficiency is rarely reported in individuals with a healthy standard diet. However, during pregnancy, vitamin C requirements are increased as vitamin C is actively transported across the placenta (Streeter and Rosso 1981). As a result, maternal plasma vitamin C levels fall during pregnancy and the recommended dietary intake for vitamin C is increased to 60 mg per day during pregnancy. During lactation the recommended dietary intake for vitamin C is increased to 85 mg per day due to loss through breast milk.

A key feature in the development of complications in pregnancy like preeclampsia, intrauterine growth restriction, and premature rupture of membranes (PROM) is oxidative stress (Beach et al. 1999; Myatt and Cui 2004). Oxidative stress has also been implicated in many of the disorders common to preterm infants including chronic lung disease, intraventricular hemorrhage, periventricular leukomalacia, retinopathy of prematurity, necrotizing enterocolitis, and bronchopulmonary dysplasia (Saugstad 1988, 2001). Having an increased dietary intake of vitamin C in early pregnancy has been associated with small increases in birthweight and placental weight (Mathews et al. 1999).

For male, ordinary antioxidants in semen consist of vitamin C, vitamin E, superoxide dismutase, glutathione, and thioredoxin. Increasing vitamin C concentration in seminal plasma may prohibit DNA damage of sperm of human (Colagar and Marzony 2009; Greco et al. 2005a) and improve outcome of ICSI/IVF for in-/subfertile male (Greco et al. 2005a). However, clinical evidence on the efficacy of vitamin C supplementation is yet unclear.

MicroRNAs and Vitamin C in Ovarian Follicle Cells

Vitamin C or ascorbic acid has been found to play an important role in reproductive processes. L-gulono-γ-lactone oxidase (Gulo-/-) knockout mouse is a vitamin C-deficient mouse line with a targeted deletion of the gene that codes for the enzyme Gulo, which catalyzes the final step of vitamin C biosynthesis (Maeda et al. 2000). This mouse is an optimal experimental model for the study on effects of oxidative stress because a deficient status of vitamin C makes the capacity opposing oxidative stress weak (Lee et al. 2008a).

In a previous study using in vitro ovarian follicle maturation model (Fig. 1) to investigate the expression profiles of microRNAs after adding hCG and to assess the effects of vitamin C-deficient status, vitamin C-deficient mice showed altered microRNA profiles in matured oocytes and granulosa cells (Kim et al. 2015).

The follicles of Gulo-/- mice expressed high level of mmu-let-7b, miR-16, miR-30a, miR-126, miR-143, miR-322, and miR-721 and less let-7c than those of wild-type mice before in vitro culture (Fig. 2a). After supplementation of hCG, the oocytes of Gulo-/- mice expressed less mmu-let-7b and miR-142 than those of wild-type mice (Fig. 2b). In the G cells, mmu-let-7b, let-7c, miR-16, miR-27a, miR-30a, miR-126, miR-143, and miR-721 of Gulo-/- mice were downregulated compared to those of wild-type mice after supplementation of hCG (Fig. 2c). The other miR expression in oocytes or G cells of Gulo-/- mice was similar to that of wild-type mice.

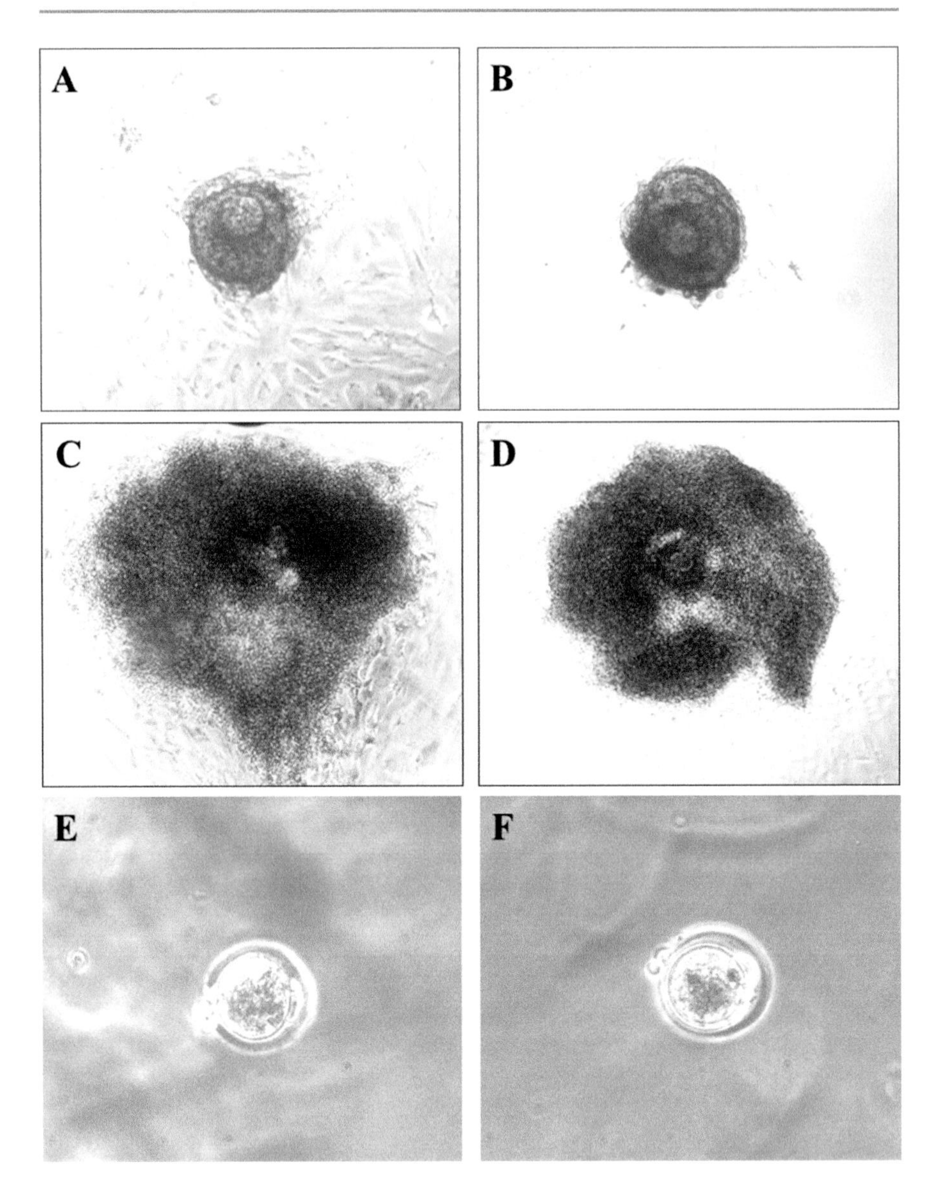

Fig. 1 In vitro ovarian follicle maturation model. Pre-antral ovarian follicle of wild-type (**a**) and Gulo-/- (**b**) mice before IVM, matured ovarian follicle of wild-type (**c**) and Gulo-/- (**d**) mice by IVM, and MII oocyte of wild-type (**e**) and Gulo-/- (**f**) mice. Key: *Gulo* L-gulono-γ-lactone oxidase, *MII* Meiosis II

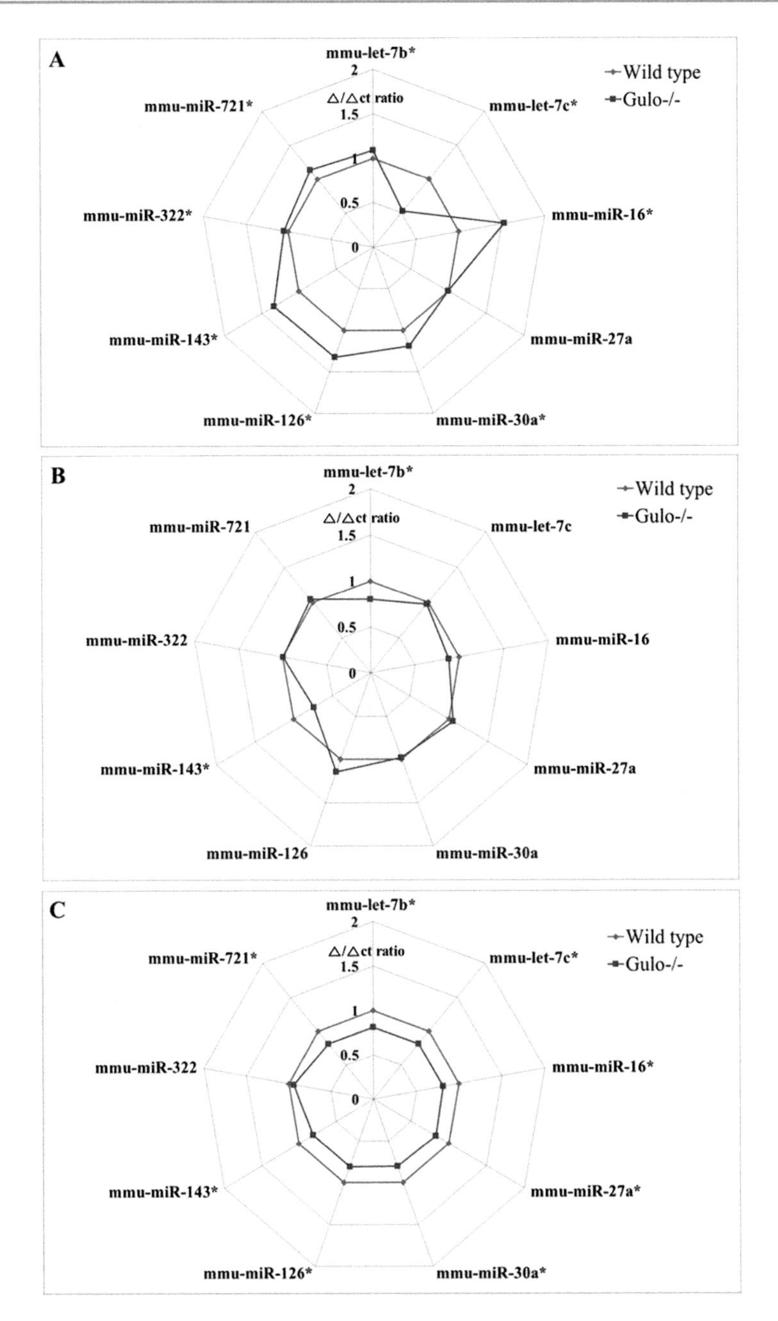

Fig. 2 Difference in microRNAs profile between wild-type and Gulo-/- mice. Relative expression of microRNAs in ovarian follicle before in vitro culture (**a**), oocyte (**b**) and G-cell (**c**) of Gulo-/- mice compared to wild type in in vitro follicle culture model ($^{*}P < 0.05$ in Gulo-/- mice vs. wild type). Key: *G-cell* Granulosa cell

Table 1 Studies on the relationship between vitamin C and microRNAs

Author	Year	Results	References
Kim et al.	2010	Compared to wild type, vitamin C-deficient mice showed altered microRNA profiles in matured oocytes and granulosa cells	Kim et al. (2010)
Shima et al.	2011	The muscle-specific miRNAs were barely detectable in the cells cultured without vitamin C and IGF-I at 30 °C, though they expressed with vitamin C and IGF-I at levels similar to the cells cultured at 38 °C	Shima et al. (2011)
Venturelli et al.	2014	Vitamin C alters the expression of 151 of microRNA in human melanoma cells according to dose	Venturelli et al. (2014)
Gao et al.	2015	Vitamin C promoted mouse embryonic stem cell-specific miRNAs	Gao et al. (2015)
Kim et al.	2015	The expression level of microRNA 155 in HDL3 was reduced by 49% and 75% in nonsmokers and smokers with consumption of a high-dose daily vitamin C (1,250 mg) for 8 weeks, respectively	Kim et al. (2015)

Key: *miRNAs* MicroRNAs, *IGF-1* Insulin-like growth factor 1, *HDL3* High-density lipoprotein 3

Regulation of MicroRNA in Other Cells

To date, various cells were studied to evaluate the function of vitamin C on microRNA profiles (Tables 1 and 2). In an experiment, supplementation with vitamin C and IGF-I promoted the expression of microRNAs that are involved in the regulation of myogenic differentiation or muscle regeneration (Shima et al. 2011). In study of mouse embryonic stem cells, vitamin C promoted ESC-specific microRNAs (Gao et al. 2015). Then, differentiation and development genes were repressed by ESC-enriched miRNAs, which maintained the stem cell state. This study suggested that epigenetic modifiers have a central role during vitamin C-induced reprogramming and pluripotency maintenance. In the study of human melanoma cells, vitamin C alters the expression of 151 of microRNA according to dose (Venturelli et al. 2014). A previous report showed the daily consumption of a high dose of vitamin C for 8 weeks resulted in enhanced anti-senescence and anti-atherosclerotic effects via an improvement of lipoprotein parameters and microRNA expression through antioxidation and anti-glycation, especially in smokers (Kim et al. 2015). Regulation of important gene expression can be regulated by microRNA transfection (Kim et al. 2013).

Conclusions

The microRNA expression profiles in oocytes and granulosa cells are altered by hCG supplementation during IVM of mouse follicles. These microRNA expression profiles differ between oocytes and granulosa cells. These alterations show different features between vitamin C-deficient and wild-type mice. Taken together, the impacts of hormone and oxidative stress may have an influence on the expression

Table 2 Candidate miRNA related to vitamin C with published known target genes (The European Bioinformatics Institute database; http://www.ebi.ac.uk)

Species	Gene name	Transcript	Description	Length	Total sites	No. cons species	No. miRNAs
Os taurus	LOC505492	ENSBTAT00000010823	Solute carrier family 23 member 1 (sodium-dependent vitamin C transporter 1) (hSVCT1) (Na(+)/L-ascorbic acid transporter 1) (yolk sac permease-like molecule 3) [Source:Uniprot/SWISSPROT;Acc:Q9UHI7] BY ORTHOLOGY TO: SINFRUT00000150933	600	7	2	hsa-miR-154, hsa-miR-188-5p, hsa-miR-339-3p, hsa-miR-515-3p, hsa-miR-519e, hsa-miR-599, hsa-miR-672, mmu-miR-350, mmu-miR-434-3p, mmu-miR-465a-3p
Canis familiaris	478025	ENSCAFT00000009175	Solute carrier family 23 member 1 (sodium-dependent vitamin C transporter 1) (hSVCT1) (Na(+)/L-ascorbic acid transporter 1) (yolk sac permease-like molecule 3) [Source:Uniprot/SWISSPROT;Acc:Q9UHI7] BY	500	10	1	bta-miR-103, bta-miR-107, bta-miR-200b. bta-miR-200c, bta-miR-545, hsa-miR-129-3p, hsa-miR-216a, hsa-miR-384, hsa-miR-488, hsa-miR-623, hsa-miR-758, mmu-miR-677,

(continued)

Table 2 (continued)

Species	Gene name	Transcript	Description	Length	Total sites	No. cons species	No. miRNAs
			ORTHOLOGY TO: SINFRUT00000150933				mmu-miR-743b-3p
Danio rerio	SLC23A2	ENSDART00000021715	Solute carrier family 23 member 2 (sodium-dependent vitamin C transporter 2) (hSVCT2) (Na(+)/L-ascorbic acid transporter 2) (yolk sac permease-like molecule 2) (nucleobase transporter-like 1 protein) [Source: Uniprot/SWISSPROT; Acc:Q9UGH3]	700	10	2	dre-miR-10a, dre-miR-10b, dre-miR-10d, dre-miR-18b, dre-miR-19b, dre-miR-19d, dre-miR-23a, dre-miR-23b, dre-miR-27d, dre-miR-153a, dre-miR-153b, dre-miR-153c, dre-miR-155, dre-miR-181a, dre-miR-204, dre-miR-216a, dre-miR-216b, dre-miR-221, dre-miR-451, dre-miR-458, dre-miR-722, dre-miR-726

Gallus gallus	SLC23A2	ENSGALT00000000261	Solute carrier family 23 member 2 (sodium-dependent vitamin C transporter 2) (hSVCT2) (Na(+)/L-ascorbic acid transporter 2) (yolk sac permease-like molecule 2) (nucleobase transporter-like 1 protein) [Source: Uniprot/SWISSPROT; Acc:Q9UGH3]	406	18	4	bta-miR-425-3p, gga-miR-23b, gga-miR-29a, gga-miR-29b, gga-miR-29c, gga-miR-142-5p, gga-miR-155, gga-miR-183, gga-miR-184, gga-miR-466, gga-miR-489, hsa-miR-28-3p, hsa-miR-197, hsa-miR-220b, hsa-miR-338-3p, hsa-miR-379, hsa-miR-421, hsa-miR-519c-3p, hsa-miR-543, hsa-miR-548a-5p, hsa-miR-548b-3p, hsa-miR-548b-5p, hsa-miR-548c-3p, hsa-miR-607, hsa-miR-610, hsa-miR-621, hsa-miR-652, hsa-miR-654-3p, hsa-miR-760, hsa-miR-871, hsa-miR-873, mmu-miR-467a, mmu-miR-467b,

(*continued*)

Table 2 (continued)

Species	Gene name	Transcript	Description	Length	Total sites	No. cons species	No. miRNAs
							mmu-miR-467d, mmu-miR-685, mmu-miR-704, mmu-miR-883a-3p, mmu-miR-883b-3p
Gallus gallus	SLC23A1	ENSGALT00000003962	Solute carrier family 23 member 1 (sodium-dependent vitamin C transporter 1) (hSVCT1) (Na(+)/L-ascorbic acid transporter 1) (yolk sac permease-like molecule 3) [Source:Uniprot/SWISSPROT;Acc:Q9UHI7]	79	4	3	bta-miR-93, bta-miR-532, gga-miR-17-5p, gga-miR-20a, gga-miR-106, gga-miR-142-5p, gga-miR-20b, hsa-miR-371-3p, hsa-miR-512-3p, hsa-miR-515-3p, hsa-miR-518a-3p, hsa-miR-519d, hsa-miR-520a-3p, hsa-miR-520d-3p, hsa-miR-644, hsa-miR-675, mmu-miR-290-3p, mmu-miR-291a-3p, mmu-miR-292-3p, mmu-miR-294, mmu-miR-295, mmu-miR-467a, mmu-miR-676

Gallus gallus	SLC23A2	ENSGALT00000041086	Solute carrier family 23 member 2 (sodium-dependent vitamin C transporter 2) (hSVCT2) (Na(+)/L-ascorbic acid transporter 2) (yolk sac permease-like molecule 2) (nucleobase transporter-like 1 protein) [Source: Uniprot/SWISSPROT; Acc:Q9UGH3]	600	11	1	bta-miR-369-3p, gga-miR-19a, gga-miR-19b, gga-miR-122, gga-miR-155, gga-miR-206, hsa-miR-338-3p, hsa-miR-379, hsa-miR-543, hsa-miR-654-3p, hsa-miR-648, hsa-miR-873, hsa-miR-935, mmu-miR-685, mmu-miR-695, mmu-miR-883b-3p
Gasterosteus aculeatus	SLC23A2	ENSGACT00000007527	Solute carrier family 23 member 2 (sodium-dependent vitamin C transporter 2) (hSVCT2) (Na(+)/L-ascorbic acid transporter 2) (yolk sac permease-like molecule 2) (nucleobase transporter-like 1 protein) [Source: Uniprot/SWISSPROT; Acc:Q9UGH3]	500	6	1	dre-miR-461, dre-miR-726, fru-miR-27b, fru-miR-27c, fru-miR-27e, dre-miR-150, fru-miR-193, fru-miR-200b, fru-miR-221

(*continued*)

Table 2 (continued)

Species	Gene name	Transcript	Description	Length	Total sites	No. cons species	No. miRNAs
Gasterosteus aculeatus	SLC23A2	ENSGACT00000007535	Solute carrier family 23 member 2 (sodium-dependent vitamin C transporter 2) (hSVCT2) (Na(+)/L-ascorbic acid transporter 2) (yolk sac permease-like molecule 2) (nucleobase transporter-like 1 protein) [Source: Uniprot/SWISSPROT; Acc:Q9UGH3]	500	2	1	dre-miR-150, fru-miR-27b, fru-miR-27c, fru-miR-205
Gasterosteus aculeatus	SLC23A1	ENSGACT00000027710	Solute carrier family 23 member 1 (sodium-dependent vitamin C transporter 1) (hSVCT1) (Na(+)/L-ascorbic acid transporter 1) (yolk sac permease-like molecule 3) [Source:Uniprot/ SWISSPROT;Acc: Q9UHI7]	500	5	1	dre-miR-462, fru-miR-30b, fru-miR-30c, fru-miR-30d, fru-miR-100, fru-miR-122, fru-miR-460

Homo sapiens	SLC23A1	ENST00000353963	Solute carrier family 23 member 1 (sodium-dependent vitamin C transporter 1) (hSVCT1) (Na(+)/L-ascorbic acid transporter 1) (yolk sac permease-like molecule 3) [Source:Uniprot/ SWISSPROT;Acc: Q9UHI7]	446	19	4	gga-miR-460, hsa-let-7d, hsa-miR-10a, hsa-miR-10b, hsa-miR-20a, hsa-miR-33b, hsa-miR-135b, hsa-miR-150, hsa-miR-19b-2, hsa-miR-27a, hsa-miR-29b-1, hsa-miR-29b-2, hsa-miR-33a, hsa-miR-103, hsa-miR-107, hsa-miR-133a, hsa-miR-133b, hsa-miR-376a, hsa-miR-376b, hsa-miR-384, hsa-miR-431, hsa-miR-491-3p, hsa-miR-499-5p, hsa-miR-509-3p, hsa-miR-516a-3p, hsa-miR-520a-5p, hsa-miR-525-5p, hsa-miR-573, hsa-miR-582-3p, hsa-miR-586, hsa-miR-601, hsa-miR-627, hsa-miR-628-5p,

(*continued*)

Table 2 (continued)

Species	Gene name	Transcript	Description	Length	Total sites	No. cons species	No. miRNAs
							hsa-miR-629, hsa-miR-656, hsa-miR-744, hsa-miR-921, hsa-miR-934, mmu-miR-465a-5p, mmu-miR-465c-5p, mmu-miR-471, mmu-miR-547, mmu-miR-710, mmu-miR-743b-5p
Homo sapiens	SLC23A2	ENST00000379321	Solute carrier family 23 member 2 (sodium-dependent vitamin C transporter 2) (hSVCT2) (Na(+)/L-ascorbic acid transporter 2) (yolk sac permease-like molecule 2) (nucleobase transporter-like 1 protein) [Source: Uniprot/SWISSPROT; Acc:Q9UGH3]	290	14	1	hsa-miR-17, hsa-miR-27a, hsa-miR-29b-2, hsa-miR-30a, hsa-miR-30b, hsa-miR-30c, hsa-miR-30d, hsa-miR-30e, hsa-miR-34a, hsa-miR-34b, hsa-miR-34c-5p, hsa-miR-92a-1, hsa-miR-92a-2, hsa-miR-93, hsa-miR-122,

							hsa-miR-130b, hsa-miR-132, hsa-miR-140-5p, hsa-miR-142-5p, hsa-miR-212, hsa-miR-223, hsa-miR-299-5p, hsa-miR-361-5p, hsa-miR-370, hsa-miR-449a, hsa-miR-449b, hsa-miR-483-5p, hsa-miR-499-5p, hsa-miR-548a-5p, hsa-miR-548b-5p, hsa-miR-548c-5p, hsa-miR-548d-5p, hsa-miR-574-5p, hsa-miR-559, hsa-miR-597, hsa-miR-598, hsa-miR-624, hsa-miR-639, hsa-miR-662, hsa-miR-802, hsa-miR-924, mmu-miR-341, mmu-miR-669b, mmu-miR-683, mmu-miR-699, mmu-miR-707, mmu-miR-714, mmu-miR-717,

(*continued*)

Table 2 (continued)

Species	Gene name	Transcript	Description	Length	Total sites	No. cons species	No. miRNAs
							mmu-miR-721, mmu-miR-879, rno-miR-333
Homo sapiens	SLC23A2	ENST00000379333	Solute carrier family 23 member 2 (sodium-dependent vitamin C transporter 2) (hSVCT2) (Na(+)/L-ascorbic acid transporter 2) (yolk sac permease-like molecule 2) (nucleobase transporter-like 1 protein) [Source: Uniprot/SWISSPROT; Acc:Q9UGH3]	4,616	1	2	hsa-miR-487a
Macaca mulatta	SLC23A1	ENSMMUT00000001429	Solute carrier family 23 member 1 (sodium-dependent vitamin C transporter 1) (hSVCT1) (Na(+)/L-ascorbic acid transporter 1) (yolk sac permease-like molecule 3) [Source:Uniprot/ SWISSPROT;Acc: Q9UHI7]	800	10	2	hsa-miR-10a, hsa-miR-10b, hsa-miR-324-5p, hsa-miR-367, hsa-miR-495, hsa-miR-499-5p, hsa-miR-553, hsa-miR-566, hsa-miR-623, hsa-miR-628-5p, hsa-miR-672, mml-miR-34a, mml-miR-188, mmu-miR-673-5p

Macaca mulatta	SLC23A2	ENSMMUT00000028117	Solute carrier family 23 member 2 (sodium-dependent vitamin C transporter 2) (hSVCT2) (Na(+)/L-ascorbic acid transporter 2) (yolk sac permease-like molecule 2) (nucleobase transporter-like 1 protein) [Source: Uniprot/SWISSPROT; Acc:Q9UGH3]	800	8	1	hsa-miR-151-3p, hsa-miR-203, hsa-miR-544, hsa-miR-556-3p, hsa-miR-769-5p, hsa-miR-921, mmu-miR-466f-5p, mmu-miR-466 g, mmu-miR-468, mmu-miR-696
Macaca mulatta	SLC23A2	ENSMMUT00000044526	Solute carrier family 23 member 2 (sodium-dependent vitamin C transporter 2) (hSVCT2) (Na(+)/L-ascorbic acid transporter 2) (yolk sac permease-like molecule 2) (nucleobase transporter-like 1 protein) [Source: Uniprot/SWISSPROT; Acc:Q9UGH3]	4,040	2	1	hsa-miR-193b, hsa-miR-365
Monodelphis domestica	SLC23A2	ENSMODT00000014836	Solute carrier family 23 member 2 (sodium-dependent vitamin C transporter 2) (hSVCT2) (Na(+)/L-ascorbic acid transporter 2) (yolk sac permease-like molecule 2)	500	15	4	bta-miR-99b, bta-miR-148a, bta-miR-148b, bta-miR-532, hsa-miR-28-3p, hsa-miR-155, hsa-miR-197, hsa-miR-220b,

(*continued*)

Table 2 (continued)

Species	Gene name	Transcript	Description	Length	Total sites	No. cons species	No. miRNAs
			(nucleobase transporter-like 1 protein) [Source: Uniprot/SWISSPROT; Acc:Q9UGH3]				hsa-miR-512-5p, hsa-miR-519c-3p, hsa-miR-543, hsa-miR-548a-3p, hsa-miR-548b-3p, hsa-miR-555, hsa-miR-607, hsa-miR-675, hsa-miR-708, hsa-miR-760, hsa-miR-871, mdo-miR-17-3p, mdo-miR-100, mdo-miR-101, mdo-miR-152, mdo-miR-181a, mdo-miR-181b, mdo-miR-200c, mdo-miR-338, mmu-miR-467b, mmu-miR-467d, mmu-miR-719, mmu-miR-680, mmu-miR-682

Monodelphis domestica	SLC23A1	ENSMODT00000015171	Solute carrier family 23 member 1 (sodium-dependent vitamin C transporter 1) (hSVCT1) (Na(+)/L-ascorbic acid transporter 1) (yolk sac permease-like molecule 3) [Source:Uniprot/SWISSPROT;Acc:Q9UHI7]	500	14	1	hsa-miR-154, hsa-miR-324-5p, hsa-miR-409-5p, hsa-miR-488, hsa-miR-516a-5p, hsa-miR-517b, hsa-miR-525-3p, hsa-miR-588, hsa-miR-647, hsa-miR-668, mdo-miR-221, mdo-miR-449, mmu-miR-434-3p, mmu-miR-676, mmu-miR-677, mmu-miR-686, mmu-miR-701, mmu-miR-763
Ornithorhynchus anatinus	SLC23A2	ENSOANT00000021807	Solute carrier family 23 member 2 (sodium-dependent vitamin C transporter 2) (hSVCT2) (Na(+)/L-ascorbic acid transporter 2) (yolk sac permease-like molecule 2) (nucleobase transporter-like 1 protein) [Source:Uniprot/SWISSPROT;Acc:Q9UGH3]	500	8	1	bta-miR-132, bta-miR-200b, bta-miR-345, bta-miR-487a, hsa-miR-212, hsa-miR-371-3p, hsa-miR-371-5p, hsa-miR-377, hsa-miR-498, hsa-miR-513-3p, hsa-miR-518c, hsa-miR-518d-3p, hsa-miR-518e, hsa-miR-518f, hsa-miR-598,

(*continued*)

Table 2 (continued)

Species	Gene name	Transcript	Description	Length	Total sites	No. cons species	No. miRNAs
							hsa-miR-615-3p, hsa-miR-646, hsa-miR-767-3p, hsa-miR-888, mmu-miR-292-3p, mmu-miR-292-5p, mmu-miR-471
Oryzias latipes	SLC23A2	ENSORLT00000003611	Solute carrier family 23 member 2 (sodium-dependent vitamin C transporter 2) (hSVCT2) (Na(+)/L-ascorbic acid transporter 2) (yolk sac permease-like molecule 2) (nucleobase transporter-like 1 protein) [Source: Uniprot/SWISSPROT; Acc:Q9UGH3]	500	1	1	fru-miR-96, fru-miR-101a, fru-miR-183
Oryzias latipes	SLC23A2	ENSORLT00000003616	Solute carrier family 23 member 2 (sodium-dependent vitamin C transporter 2) (hSVCT2) (Na(+)/L-ascorbic acid transporter 2) (yolk sac permease-like molecule 2) (nucleobase transporter-	500	2	1	dre-miR-459, dre-miR-462, fru-miR-194, fru-miR-429

			like 1 protein) [Source: Uniprot/SWISSPROT; Acc:Q9UGH3]				
Oryzias latipes	SLC23A1	ENSORLT00000015917	Solute carrier family 23 member 1 (sodium-dependent vitamin C transporter 1) (hSVCT1) (Na(+)/L-ascorbic acid transporter 1) (yolk sac permease-like molecule 3) [Source:Uniprot/SWISSPROT;Acc: Q9UHI7]	500	6	1	dre-miR-34, dre-miR-34b, fru-miR-10d, fru-miR-17, fru-miR-20, fru-miR-30c, fru-miR-100, fru-miR-130, fru-miR-216b, xtr-miR-425-5p
Oryzias latipes	SLC23A1	ENSORLT00000015919	Solute carrier family 23 member 1 (sodium-dependent vitamin C transporter 1) (hSVCT1) (Na(+)/L-ascorbic acid transporter 1) (yolk sac permease-like molecule 3) [Source:Uniprot/SWISSPROT;Acc: Q9UHI7]	500	5	1	dre-miR-93, fru-miR-137, fru-miR-200b, fru-miR-218a, fru-miR-218b, fru-miR-365, fru-miR-429
Pan troglodytes	SLC23A2	ENSPTRT00000024573	Solute carrier family 23 member 2 (sodium-dependent vitamin C transporter 2) (hSVCT2) (Na(+)/L-ascorbic acid transporter 2) (yolk sac permease-	4,647	1	2	hsa-miR-487a

(continued)

Table 2 (continued)

Species	Gene name	Transcript	Description	Length	Total sites	No. cons species	No. miRNAs
			like molecule 2) (nucleobase transporter-like 1 protein) [Source: Uniprot/SWISSPROT; Acc:Q9UGH3]				
Pan troglodytes	SLC23A1	ENSPTRT00000043127	Solute carrier family 23 member 1 (sodium-dependent vitamin C transporter 1) (hSVCT1) (Na(+)/L-ascorbic acid transporter 1) (yolk sac permease-like molecule 3) [Source:Uniprot/SWISSPROT;Acc: Q9UHI7]	446	17	4	gga-miR-460, hsa-miR-376a, hsa-miR-376b, hsa-miR-384, hsa-miR-431, hsa-miR-491-3p, hsa-miR-499-5p, hsa-miR-509-3p, hsa-miR-516a-3p, hsa-miR-520a-5p, hsa-miR-525-5p, hsa-miR-573, hsa-miR-582-3p, hsa-miR-586, hsa-miR-601, hsa-miR-627, hsa-miR-628-5p, hsa-miR-629, hsa-miR-656, hsa-miR-744, hsa-miR-921, hsa-miR-934,

							mmu-miR-434-5p, mmu-miR-465a-5p, mmu-miR-465c-5p, mmu-miR-471, mmu-miR-547, mmu-miR-710, mmu-miR-743b-5p, ptr-miR-33, ptr-miR-103, ptr-miR-107
Rattus norvegicus	Slc23a1	ENSRNOT00000027048	Solute carrier family 23 member 1 (sodium-dependent vitamin C transporter 1) (Na(+)/L-ascorbic acid transporter 1) [Source:Uniprot/SWISSPROT;Acc:Q9WTW7]	651	14	4	hsa-miR-515-3p, hsa-miR-516a-3p, hsa-miR-516a-5p, hsa-miR-517a, hsa-miR-517b, hsa-miR-517c, hsa-miR-518b, hsa-miR-518d-3p, hsa-miR-518e, hsa-miR-519c-3p, hsa-miR-519e, hsa-miR-520f, hsa-miR-526b, hsa-miR-581, hsa-miR-610, hsa-miR-635, hsa-miR-636, hsa-miR-656, mmu-miR-295, mmu-miR-465a-3p, mmu-miR-

(*continued*)

Table 2 (continued)

Species	Gene name	Transcript	Description	Length	Total sites	No. cons species	No. miRNAs
							465a-5p, mmu-miR-465b-5p, mmu-miR-465c-5p, mmu-miR-721, rno-miR-19b, rno-miR-27a, rno-miR-130b, rno-miR-145, rno-miR-200b, rno-miR-218, rno-miR-329, rno-miR-376a, rno-miR-376b-3p, rno-miR-429, rno-miR-543
Rattus norvegicus	Slc23a2	ENSRNOT00000028885	Solute carrier family 23 member 2 (sodium-dependent vitamin C transporter 2) (Na(+)/L-ascorbic acid transporter 2) [Source:Uniprot/SWISSPROT;Acc:Q9WTW8]	948	8	4	hsa-miR-885-5p, hsa-miR-935, mmu-miR-197, mmu-miR-685, mmu-miR-704, mmu-miR-719, rno-miR-34c, rno-miR-100, rno-miR-183, rno-miR-365,

							rno-miR-652, rno-miR-873
Rattus norvegicus	Slc23a2	ENSRNOT00000041570	Solute carrier family 23 member 2 (sodium-dependent vitamin C transporter 2) (Na(+)/L-ascorbic acid transporter 2) [Source:Uniprot/SWISSPROT;Acc:Q9WTW8]	700	8	1	hsa-miR-620, hsa-miR-649, mmu-miR-679, mmu-miR-704, mmu-miR-714, mmu-miR-719, rno-let-7b, rno-let-7d, rno-let-7e, rno-miR-34c
Rattus norvegicus	Slc23a2	ENSRNOT00000046491	Solute carrier family 23 member 2 (sodium-dependent vitamin C transporter 2) (Na(+)/L-ascorbic acid transporter 2) [Source:Uniprot/SWISSPROT;Acc:Q9WTW8]	700	11	1	hsa-miR-371-5p, hsa-miR-561, hsa-miR-607, mmu-miR-362-3p, mmu-miR-653, mmu-miR-675-3p, mmu-miR-675-5p, mmu-miR-683, rno-miR-141, rno-miR-200a, rno-miR-200b, rno-miR-200c, rno-miR-208, rno-miR-324-3p, rno-miR-330, rno-miR-421, rno-miR-451
Takifugu rubripes	SLC23A2	SINFRUT00000128512	Solute carrier family 23 member 2 (sodium-dependent vitamin C transporter 2)	400	8	2	dre-miR-150, dre-miR-363, fru-let-7a, fru-let-7b, fru-let-7d,

(*continued*)

Table 2 (continued)

Species	Gene name	Transcript	Description	Length	Total sites	No. cons species	No. miRNAs
			(hSVCT2) (Na(+)/L-ascorbic acid transporter 2) (yolk sac permease-like molecule 2) (nucleobase transporter-like 1 protein) [Source: Uniprot/SWISSPROT; Acc:Q9UGH3]				fru-let-7e, fru-let-7 g, fru-let-7 h, fru-let-7j, fru-miR-9, fru-miR-10b, fru-miR-10c, fru-miR-25, fru-miR-29a, fru-miR-29b, fru-miR-92, fru-miR-129, fru-miR-183, fru-miR-217, fru-miR-223, fru-miR-489, xtr-miR-33a, xtr-miR-367
Takifugu rubripes	SLC23A1	SINFRUT00000150933	Solute carrier family 23 member 1 (sodium-dependent vitamin C transporter 1) (hSVCT1) (Na(+)/L-ascorbic acid transporter 1) (yolk sac permease-like molecule 3) [Source:Uniprot/SWISSPROT;Acc: Q9UHI7]	400	4	2	dre-miR-93, fru-miR-15b, fru-miR-192, xtr-miR-215

Tetraodon nigroviridis	GSTENG00014656001	GSTENT00014656001	Solute carrier family 23 member 1 (sodium-dependent vitamin C transporter 1) (hSVCT1) (Na(+)/L-ascorbic acid transporter 1) (yolk sac permease-like molecule 3) [Source:Uniprot/SWISSPROT;Acc:Q9UHI7]Solute carrier family 23 member 1 (sodium-dependent vitamin C transporter 1) (hSVCT1) (Na(+)/L-ascorbic acid transporter 1) (yolk sac permease-like molecule 3) [Source:Uniprot/SWISSPROT;Acc:Q9UHI7] BY ORTHOLOGY TO: SINFRUT00000150933	400	5	2	tni-miR-29a, tni-miR-29b, tni-miR-103, tni-miR-122, tni-miR-192, tni-miR-200a, xtr-miR-215, xtr-miR-367, xtr-miR-449
Xenopus tropicalis	SLC23A1	ENSXETT00000018417	Solute carrier family 23 member 1 (sodium-dependent vitamin C transporter 1) (hSVCT1) (Na(+)/L-ascorbic acid transporter 1) (yolk sac permease-like molecule 3) [Source:Uniprot/	700	2	1	dre-miR-735, xtr-miR-182

(*continued*)

Table 2 (continued)

Species	Gene name	Transcript	Description	Length	Total sites	No. cons species	No. miRNAs
			SWISSPROT;Acc: Q9UHI7]				
Xenopus tropicalis	SLC23A2	ENSXETT00000036732	Solute carrier family 23 member 2 (sodium-dependent vitamin C transporter 2) (hSVCT2) (Na(+)/L-ascorbic acid transporter 2) (yolk sac permease-like molecule 2) (nucleobase transporter-like 1 protein) [Source: Uniprot/SWISSPROT; Acc:Q9UGH3]	700	5	1	dre-miR-220, xtr-let-7 g, xtr-miR-25, xtr-miR-17-3p, xtr-miR-92a, xtr-miR-92b, xtr-miR-142-5p, xtr-miR-199a, xtr-miR-199b
Xenopus tropicalis	SLC23A2	ENSXETT00000036734	Solute carrier family 23 member 2 (sodium-dependent vitamin C transporter 2) (hSVCT2) (Na(+)/L-ascorbic acid transporter 2) (yolk sac permease-like molecule 2) (nucleobase transporter-like 1 protein) [Source: Uniprot/SWISSPROT; Acc:Q9UGH3]	700	2	1	dre-miR-220, dre-miR-726, xtr-miR-99, xtr-miR-100

Key: *Cons* Conserved, *hsa* Home sapiens, *Mmu* Mus musculus, *Bta* Bos taurus, *Dre* Danio rerio, *Gga* Gallus gallus, *Fru* Fugu rubripes

of microRNAs in germ cells during maturation in vitro. The relationship between developmental competence of germ cells and changes in the microRNA expression needs to be evaluated in further studies.

In spite of emerging researches on regulating function of microRNA in vitamin C metabolism, many questions should be further investigated, such as the following: whether vitamin C directly regulates epigenetic modifiers and how, whether vitamin C regulates gene and miRNA promoters via signaling, and whether vitamin C-induced DNA demethylation mainly occurs. More in-depth research should be conducted to guide people in effectively using vitamin C.

Dictionary of Terms

- **In vitro ovarian follicle maturation** – A technology by which ovarian follicles were cultured in vitro to obtain mature oocyte.
- **L-gulono-γ-lactone oxidase knockout mouse** – Mice in a C57BL/6 background, which is required to maintain viability of vitamin C supplementation, by deleting exons 3 and 4 from Gulo.

Key Facts of MicroRNA Expression and Vitamin C

- To date, some studies showed relationship between vitamin C and microRNA profiles in many cells.
- L-gulono-γ-lactone oxidase knockout mouse is a useful animal model for vitamin C study.
- Vitamin C alters the profile of microRNAs in various cells.
- Deficiency of vitamin C affects the cell function through altering the profiles of microRNAs.
- Vitamin C or ascorbic acid has been found to play an important role in reproductive processes.

Summary Points

- L-gulono-γ-lactone oxidase knockout mouse is a useful animal model for vitamin C study and used for the analyses of vitamin C in development.
- MicroRNA expression is affected by vitamin C in many cell types.
- Vitamin C has a significant role in reproductive biology.
- Deficiency of vitamin C affects the function of various cells and proved the correlation between the microRNA and vitamin C.
- Further studies are necessary for the investigation on functional action of microRNA altered by vitamin C.

References

Bartel DP (2009) MicroRNAs: target recognition and regulatory functions. Cell 136:215–233
Beach MJ, Addiss DG, Roberts JM, Lammie PJ (1999) Treatment of trichuris infection with albendazole. Lancet 353:237–238
Burns JJ, Moltz A, Peyser P (1956) Missing step in guinea pigs required for the biosynthesis of L-ascorbic acid. Science 124:1148–1149
Chan AC (1993) Partners in defense, vitamin E and vitamin C. Can J Physiol Pharmacol 71:725–731
Colagar AH, Marzony ET (2009) Ascorbic acid in human seminal plasma: determination and its relationship to sperm quality. J Clin Biochem Nutr 45:144–149
Gao Y, Han Z, Li Q, Wu Y, Shi X, Ai Z, Du J, Li W, Guo Z, Zhang Y (2015) Vitamin C induces a pluripotent state in mouse embryonic stem cells by modulating microRNA expression. FEBS J 282:685–699
Gould BS, Woessner JF (1957) Biosynthesis of collagen; the influence of ascorbic acid on the proline, hydroxyproline, glycine, and collagen content of regenerating guinea pig skin. J Biol Chem 226:289–300
Greco E, Romano S, Iacobelli M, Ferrero S, Baroni E, Minasi MG, Ubaldi F, Rienzi L, Tesarik J (2005a) ICSI in cases of sperm DNA damage: beneficial effect of oral antioxidant treatment. Hum Reprod 20:2590–2594
Greco E, Iacobelli M, Rienzi L, Ubaldi F, Ferrero S, Tesarik J (2005b) Reduction of the incidence of sperm DNA fragmentation by oral antioxidant treatment. J Androl 26:349–353
Hadley KB, Sato PH (1989) Catalytic activity of administered gulonolactone oxidase polyethylene glycol conjugates. Enzyme 42:225–234
Jimenez CR, Araujo VR, Penitente-Filho JM, de Azevedo JL, Silveira RG, Torres CA (2016) The base medium affects ultrastructure and survival of bovine preantral follicles cultured in vitro. Theriogenology 85:1019–1029
Kawai T, Nishikimi M, Ozawa T, Yagi K (1992) A missense mutation of L-gulono-gamma-lactone oxidase causes the inability of scurvy-prone osteogenic disorder rats to synthesize L-ascorbic acid. J Biol Chem 267:21973–21976
Kim YJ, Ku SY, Rosenwaks Z, Liu HC, Chi SW, Kang JS, Lee WJ, Jung KC, Kim SH, Choi YM et al (2010) MicroRNA expression profiles are altered by gonadotropins and vitamin C status during in vitro follicular growth. Reprod Sci 17:1081–1089
Kim H, Bae S, Yu Y, Kim Y, Kim HR, Hwang YI, Kang JS, Lee WJ (2012) The analysis of vitamin C concentration in organs of gulo(-/-) mice upon vitamin C withdrawal. Immune Netw 12:18–26
Kim YJ, Ku S-Y, Kim YY, Liu HC, Chi SW, Kim SH, Choi YM, Kim JG, Moon SY (2013) MicroRNAs transfected into granulosa cells may regulate oocyte meiotic competence during in vitro maturation of mouse follicles. Hum Reprod 28:3050–3061
Kim SM, Lim SM, Yoo JA, Woo MJ, Cho KH (2015) Consumption of high-dose vitamin C (1250 mg per day) enhances functional and structural properties of serum lipoprotein to improve anti-oxidant, anti-atherosclerotic, and anti-aging effects via regulation of anti-inflammatory microRNA. Food Funct 6:3604–3612
Kim YJ, Ku SY, Kim YY, Suh CS, Kim SH, Choi YM (2016) MicroRNA profile of granulosa cells after ovarian stimulation differs according to maturity of retrieved oocytes. Geburtshilfe Frauenheilkd 76:704–708
Lai EC, Tam B, Rubin GM (2005) Pervasive regulation of *Drosophila* notch target genes by GY-box-, Brd-box-, and K-box-class microRNAs. Genes Dev 19:1067–1080
Lee CW, Wang XD, Chien KL, Ge Z, Rickman BH, Rogers AB, Varro A, Whary MT, Wang TC, Fox JG (2008a) Vitamin C supplementation does not protect L-gulono-gamma-lactone oxidase-deficient mice from Helicobacter pylori-induced gastritis and gastric premalignancy. Int J Cancer 122:1068–1076
Lee SK, Kang JS, Jung da J, Hur DY, Kim JE, Hahm E, Bae S, Kim HW, Kim D, Cho BJ et al (2008b) Vitamin C suppresses proliferation of the human melanoma cell SK-MEL-2 through the

inhibition of cyclooxygenase-2 (COX-2) expression and the modulation of insulin-like growth factor II (IGF-II) production. J Cell Physiol 216:180–188
Li Y, Schellhorn HE (2007) New developments and novel therapeutic perspectives for vitamin C. J Nutr 137:2171–2184
Li Y, Shi CX, Mossman KL, Rosenfeld J, Boo YC, Schellhorn HE (2008) Restoration of vitamin C synthesis in transgenic Gulo-/- mice by helper-dependent adenovirus-based expression of gulonolactone oxidase. Hum Gene Ther 19:1349–1358
Linster CL, Van Schaftingen E (2007) Vitamin C. Biosynthesis, recycling and degradation in mammals. FEBS J 274:1–22
Luck MR, Jeyaseelan I, Scholes RA (1995) Ascorbic acid and fertility. Biol Reprod 52:262–266
Maeda N, Hagihara H, Nakata Y, Hiller S, Wilder J, Reddick R (2000) Aortic wall damage in mice unable to synthesize ascorbic acid. Proc Natl Acad Sci U S A 97:841–846
Mathews F, Yudkin P, Neil A (1999) Influence of maternal nutrition on outcome of pregnancy: prospective cohort study. BMJ (Clinical research ed) 319:339–343
Montel-Hagen A, Kinet S, Manel N, Mongellaz C, Prohaska R, Battini JL, Delaunay J, Sitbon M, Taylor N (2008) Erythrocyte Glut1 triggers dehydroascorbic acid uptake in mammals unable to synthesize vitamin C. Cell 132:1039–1048
Murray AA, Molinek MD, Baker SJ, Kojima FN, Smith MF, Hillier SG, Spears N (2001) Role of ascorbic acid in promoting follicle integrity and survival in intact mouse ovarian follicles in vitro. Reproduction 121:89–96
Myatt L, Cui X (2004) Oxidative stress in the placenta. Histochem Cell Biol 122:369–382
Nishikimi M, Koshizaka T, Ozawa T, Yagi K (1988) Occurrence in humans and guinea pigs of the gene related to their missing enzyme L-gulono-gamma-lactone oxidase. Arch Biochem Biophys 267:842–846
Patak P, Willenberg HS, Bornstein SR (2004) Vitamin C is an important cofactor for both adrenal cortex and adrenal medulla. Endocr Res 30:871–875
Pollock JI, Mullin RJ (1987) Vitamin C biosynthesis in prosimians: evidence for the anthropoid affinity of Tarsius. Am J Phys Anthropol 73:65–70
Poy MN, Eliasson L, Krutzfeldt J, Kuwajima S, Ma X, Macdonald PE, Pfeffer S, Tuschl T, Rajewsky N, Rorsman P et al (2004) A pancreatic islet-specific microRNA regulates insulin secretion. Nature 432:226–230
Saugstad OD (1988) Hypoxanthine as an indicator of hypoxia: its role in health and disease through free radical production. Pediatr Res 23:143–150
Saugstad OD (2001) Update on oxygen radical disease in neonatology. Curr Opin Obstet Gynecol 13:147–153
Shima A, Pham J, Blanco E, Barton ER, Sweeney HL, Matsuda R (2011) IGF-I and vitamin C promote myogenic differentiation of mouse and human skeletal muscle cells at low temperatures. Exp Cell Res 317:356–366
Sokol NS, Ambros V (2005) Mesodermally expressed *Drosophila* microRNA-1 is regulated by twist and is required in muscles during larval growth. Genes Dev 19:2343–2354
Streeter ML, Rosso P (1981) Transport mechanisms for ascorbic acid in the human placenta. Am J Clin Nutr 34:1706–1711
Thomas FH, Leask R, Srsen V, Riley SC, Spears N, Telfer EE (2001) Effect of ascorbic acid on health and morphology of bovine preantral follicles during long-term culture. Reproduction 122:487–495
Venturelli S, Sinnberg TW, Berger A, Noor S, Levesque MP, Bocker A, Niessner H, Lauer UM, Bitzer M, Garbe C et al (2014) Epigenetic impacts of ascorbate on human metastatic melanoma cells. Front Oncol 4:227
Wilson JX (2005) Regulation of vitamin C transport. Annu Rev Nutr 25:105–125
Xu P, Vernooy SY, Guo M, Hay BA (2003) The *Drosophila* microRNA Mir-14 suppresses cell death and is required for normal fat metabolism. Curr Biol 13:790–795
Yu R, Schellhorn HE (2013) Recent applications of engineered animal antioxidant deficiency models in human nutrition and chronic disease. J Nutr 143:1–11

Rewriting the Script: The Story of Vitamin C and the Epigenome 87

Tyler C. Huff and Gaofeng Wang

Contents

Abstract

Vitamin C is a vital micronutrient in the maintenance of numerous cellular functions and the development of mammalian systems. Vitamin C predominantly exists physiologically as the ascorbate anion, an antioxidant classically linked to the prevention of scurvy. Current research has shown that ascorbate plays an

T. C. Huff
Dr. John T. Macdonald Foundation Department of Human Genetics, John P. Hussman Institute for Human Genomics, University of Miami Miller School of Medicine, Miami, FL, USA
e-mail: tch33@miami.edu

G. Wang (✉)
Dr. John T. Macdonald Foundation Department of Human Genetics, John P. Hussman Institute for Human Genomics, University of Miami Miller School of Medicine, Miami, FL, USA

Department of Human Genetics, Dr. Nasser Ibrahim Al-Rashid Orbital Vision Research Center, Bascom Palmer Eye Institute, University of Miami Miller School of Medicine, Miami, FL, USA
e-mail: gwang@med.miami.edu

V. B. Patel, V. R. Preedy (eds.), *Handbook of Nutrition, Diet, and Epigenetics*,
https://doi.org/10.1007/978-3-319-55530-0_46

additional role critical in DNA demethylation by acting as a cofactor for the ten-eleven translocation (TET) family of methylcytosine dioxygenase enzymes. TET enzymes hydroxylate 5-methylcytosine (5mC) to 5-hydroxymethylcytosine (5hmC), an epigenetic marker whose further processing results in cleavage of the methylated cytosine and subsequent repair via the base excision repair pathway, resulting in completion of active DNA demethylation. Recent work has also speculated ascorbate's role in mediating histone demethylation dynamics via Jumonji C domain (JmjC) demethylase enzymes belonging to the same enzyme family as TET dioxygenases. Although these roles in demethylation are of principal importance, the need for ascorbate initially evolved in early photosynthetic eukaryotes who required a reducing agent to protect themselves from photodamage generated by the chloroplast, a role that ultimately affected the evolutionary paths of insects and herbivorous animals. Altogether, the wide-reaching functions of ascorbate play a critical role in the maintenance of mammalian demethylation dynamics and organismal development.

Keywords
Vitamin C · Ascorbate · DNA demethylation · TET · 5hmC · JmjC · GULO · Histone demethylation · Eusociality · Iron

List of Abbreviations

2O	2-Oxoglutarate
5caC	5-Carboxylcytosine
5fC	5-Formylcytosine
5hmC	5-Hydroxymethylcytosine
5mC	5-Methylcytosine
DNMTs	DNA methyltransferases
DSBH	Double-stranded beta helix
GLDH	L-galactonolactone dehydrogenase
GULO	Gulonolactone (L-) oxidase
Jmj	Jumanji
TDG	Thymine-DNA glycosylase
TET	Ten-eleven translocation

Introduction

Vitamin C (L-ascorbic acid) is a water-soluble dietary micronutrient that plays a variety of roles in the maintenance of numerous physiological processes. Under normal physiological conditions, vitamin C predominantly exists as the ascorbate anion, a well-established antioxidant which facilitates a number of enzymatic reactions and is classically known for its role in collagen crosslinking implicated in the prevention of scurvy (Van Robertson and Schwartz 1953). Although this is the classic clinical relevance of ascorbate, this is only one of many physiological roles that ascorbate plays. Recent discoveries have shown that ascorbate is a vital mediator of the epigenome, serving as a key promotional element to the process of DNA demethylation. As a result, the role of ascorbate in health and disease from the epigenetic perspective has become a hot topic of recent interest and is extensively reviewed elsewhere (Camarena and Wang 2016). This chapter will discuss the molecular mechanisms of ascorbate-mediated DNA and histone demethylation, the biosynthesis of ascorbate, and its evolutionary relevance.

Vitamin C and DNA Demethylation

The epigenome is the cellular interface of a complex relationship between the seemingly static genome and the ever-changing environment. Orchestrating the modulation of gene expression involves a common repertoire of epigenetic events that include chromatin remodeling, noncoding RNA regulation, and modification of both histones and nucleotides. Of these hallmark epigenetic events, methylation at the C^5 position of cytosine [5-methylcytosine (5mC)] is the best-characterized major covalent modification of DNA. As the most widespread epigenetic modification in vertebrates, cytosine methylation has an important impact on the transcriptional regulation of cell-specific proteins and the transcription factors that govern cell identity (Schübeler 2015). DNA methyltransferases (DNMTs) are responsible for creating this epigenetic landmark by transferring a methyl group from the donor S-adenosylmethionine to the nucleotide's C^5 position, thus forming 5mC. After the formation of 5mC, especially within the context of the CpG dinucleotide pattern, methyl-CpG-binding proteins can recognize and bind to these patterns to mediate transcriptional regulation of the associated DNA. Although 5mC is considered a relatively stable epigenetic hallmark of DNA, it can be lost via dilution during DNA replication due to lack of maintenance by DNMT1, thus resulting in passive demethylation (Bhutani et al. 2011). Until recently, the question of whether an active demethylation process existed and how this putative process was implemented remained largely unclear.

TET Dioxygenases Facilitate Active DNA Demethylation

This question of active DNA demethylation was addressed after the discovery of the ten-eleven translocation (TET) gene family. This conserved family of genes consists of three members (TET1, TET2, and TET3) and was named after the discovery that a chromosomal translocation of 10q22 and 11q23 resulting in the fusion of the TET1 and mixed-lineage leukemia (MLL) gene occur in acute myeloid leukemia (Lorsbach et al. 2003). Soon after their initial discovery, somatic mutations in TET2 had been identified in around 15% of patients suffering from myeloid leukemia, but the role that TETs play in the pathogenesis of the disease was far from clear as was the basic biological function of this newly discovered gene family (Delhommeau et al. 2009).

The two seemingly unrelated questions regarding the existence of active DNA demethylation and the enigmatic biological function of TETs became inextricably linked after the breakthrough discovery that TET1 enzymatic activity promoted the hydroxylation of 5-methylcytosine to 5-hydroxymethylcytosine (5hmC), thus demonstrating the existence of active DNA demethylation (Tahiliani et al. 2009). Cultured cells expressing wild-type TET1 showed a drastic decrease of 5mC signal, a trend not observed in mutant TET1-overexpressing cells. Subsequent experiments showed that TET enzymes further oxidize 5hmC to 5-formylcytosine (5fC) and 5-carboxylcytosine (5caC), two transient DNA tags which can be excised by the DNA repair enzyme thymine-DNA glycosylase (TDG) (Ito et al. 2011) (Fig. 1). TDG base excision results in an abasic position that is eventually repaired by the DNA base excision repair pathway, a process which ultimately yields an unmodified cytosine and thus completes active DNA demethylation (He et al. 2011).

TET enzymes belong to the Fe(II)- and 2-oxoglutarate (2OG, also known as α-ketoglutarate)-dependent dioxygenase superfamily, a diverse class of enzymes containing several well-conserved subfamilies found widely across the evolutionary gamut, including prokaryotes, metazoans, and plants (Salminen et al. 2015) (Table 1). Although their function can vary across taxa, their undeviating feature is their catalytic mechanism: each member requires both 2OG and molecular oxygen as co-substrates and utilizes Fe(II) as a cofactor. TET enzymes have been demonstrated to operate in the same manner, as mutations affecting putative iron-binding sites eliminate enzymatic activity as can the binding of competitive inhibitors such as 2-hydroxyglutarate (Xu et al. 2011b). The TET enzymes share very similar structural features; TET1 and TET3 both contain an N-terminal CXXC zinc finger domain that binds clustered unmethylated CpG dinucleotides with high affinity. TET2 appears to have also shared the CXXC domain before losing it to an evolutionary gene inversion event that caused the motif to separate from the TET2 coding sequence and become a separate gene, now called CXXC4 or IDAX (Ko et al. 2013).

TETs regulate DNA demethylation via two mechanisms: First, binding of TET to CpG-rich genomic regions inherently prevents unwarranted DNA methyltransferase activity. However, the principal method by which TETs mediate DNA demethylation is by active conversion of 5mC to 5hmC through hydroxylation (Xu et al. 2011a). In the oxidation reaction, TET is initially activated by the binding of 2OG and ferrous iron Fe(II) to their respective binding sites within the catalytic domain (Fig. 2). Upon binding, TET coordinates molecular oxygen (O_2) to bind Fe(II) which stimulates the decarboxylation of 2OG and causes subsequent release of CO_2 and succinic acid. This reaction stimulates the formation of a ferryl iron intermediate, Fe(IV), which then hydroxylates 5mC to form 5hmC before converting to Fe(III) (Rose et al. 2011; Li et al. 2015). In the reaction's entirety, the importance of iron's role becomes ever so clear: iron must first be in ferrous form in order to promote the decarboxylation of 2OG and the subsequent hydroxylation of the substrate. However, following the reaction's completion, iron is predominantly left in its ferric state Fe(III) and in this form is unable to initialize or mediate further hydroxylation of 5mC substrates. Without the conversion of iron species back to their previous ferrous form, TET-mediated DNA demethylation could not be sustained.

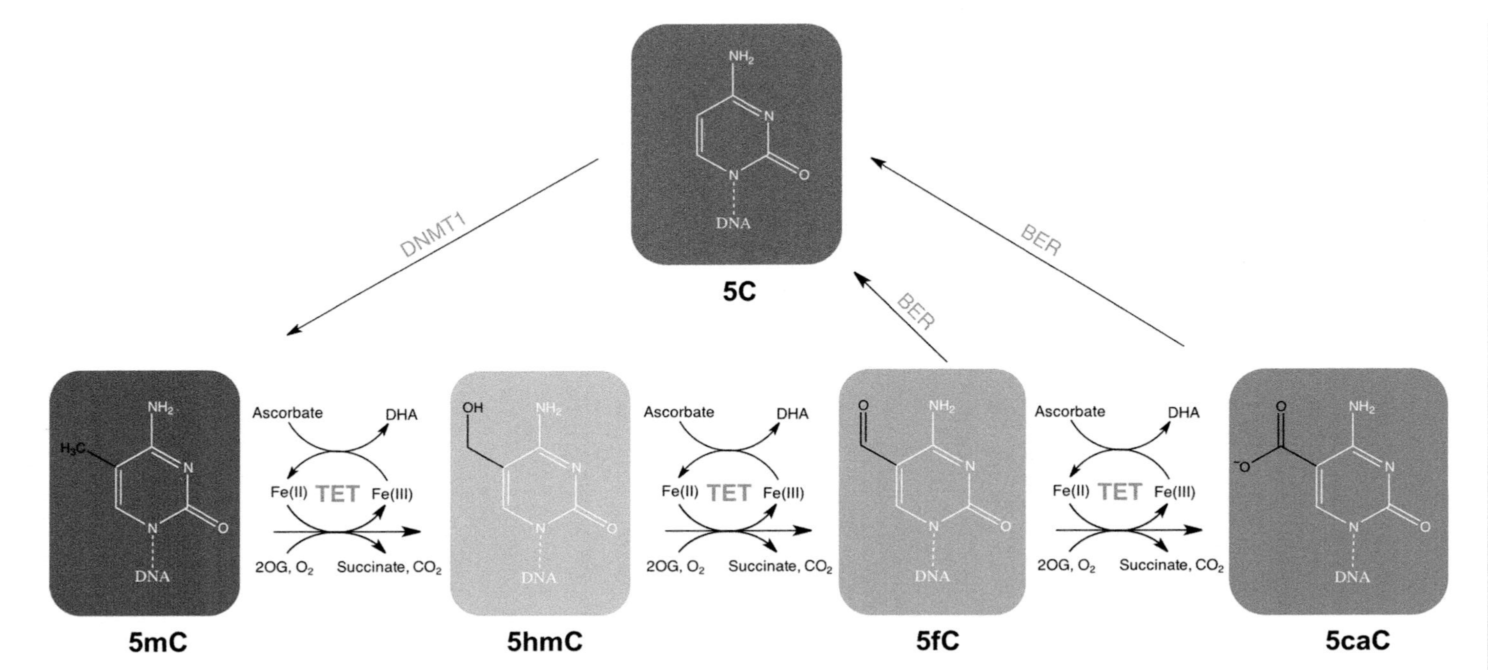

Fig. 1 TET-mediated DNA demethylation. DNMT1 can methylate unmodified cytosine at the 5′ position (5C) to form 5mC. Ascorbate promotes the TET-mediated oxidative cascade of 5mC to form 5hmC, 5fC, and 5caC. Ascorbate is required for continuous oxidation of 5mC intermediates by TET enzymes. Ascorbate replenishes catalytically active Fe(II)by reducing Fe(III) produced as a byproduct of TET oxidation. This allows for sustainment of the 5mC oxidative cascade and TET-mediated DNA demethylation. Both 5fC and 5caC are transient DNA tags that can be excised by thymine-DNA glycosylase in the base excision DNA repair (BER) pathway to form an abasic position that is eventually replaced by unmodified cytosine, thus completing active DNA demethylation. Key: 5mC = 5-methylcytosine. 5hmC = 5-hydroxymethylcytosine. 5fC = 5-formylcytosine. 5caC = 5-carbocylcytosine. 2OG = 2-oxoglutarate

Table. 1 Fe(II)- and 2OG-dependent enzyme family members listed by function. The Fe(II)- and 20G-dependent enzyme family is a large and functionally diverse class of enzymes. Family members were selected by McDonough et al. (2010) and Martinez and Hausinger (2015)

Hydroxylase		Demethylase		Other/unknown function	
Group	Members	Group	Members	Group	Members
BBOX1/ TMLHE	TMLHE	**ALKBH**	ALKBH1	**C17orf101**	C17orf101
	BBOX1 (GBBH)		ALKBH2		
			ALKBH3	**CarC**	CarC (saturation and epimerization)
C-P3H	LEPRE1		ALKBH4		
	LEPREL1		ALKBH5	**CAS**	CAS1 (oxidative ring cyclization)
	LEPREL2		ALKBH6		CAS1 (oxidative ring cyclization)
			ALKBH7		
C-P4H	P4HA1		ALKBH8	**DAOCS**	DAOCS (oxidative ring expansion)
	P4HA2				
	P4HA3	**FTO**	FTO	**FtmOx1**	FtmOx1 (endoperoxidase)
EGFH	ASPH	**JARID2**	JARID2	**H6H**	H6H (epoxidase)
	ASPHD1				
	ASPHD2	**KDM2 (FBXL)**	FBXL11	**JmjC domain**	JMJD4
					JMJD5
JmjC domain	JMJD6	**KDM3 (JMJD1)**	HR		JMJD7
	HSPBAP1		JMJD1A		JMJD8
	HIF1AN (FIH)		JMJD1B		MINA
	C2orf60 (TYW5)		JMJD1C		C14orf169 (NO66)
P4HTM	P4HTM	**KDM4 (JMJD2)**	JMJD2A	**OGFOD1**	OGFOD1
			JMJD2B		
PHD	PHD1 (EGLN2)		JMJD2C	**OGFOD2**	OGFOD2
	PHD2 (EGLN1)		JMJD2D		
	PHD3 (EGLN3		JMJD2E	**SyrB2**	SyrB2 (halogenase)
PHYH	PHYH (PAHX)	**KDM5 (JARID)**	JARID1A		
			JARID1B		

(continued)

Table. 1 (continued)

Hydroxylase		Demethylase		Other/unknown function	
PLOD	PLOD1		JARID1C		
	PLOD2		JARID1D		
	PLOD3				
		KDM6 (UTX/Y)	UTX		
TauD	TauD		UTY		
			JMJD3		
TET	TET1				
	TET2	**KDM7 (PHF)**	PHF2		
	TET3		PHF8		

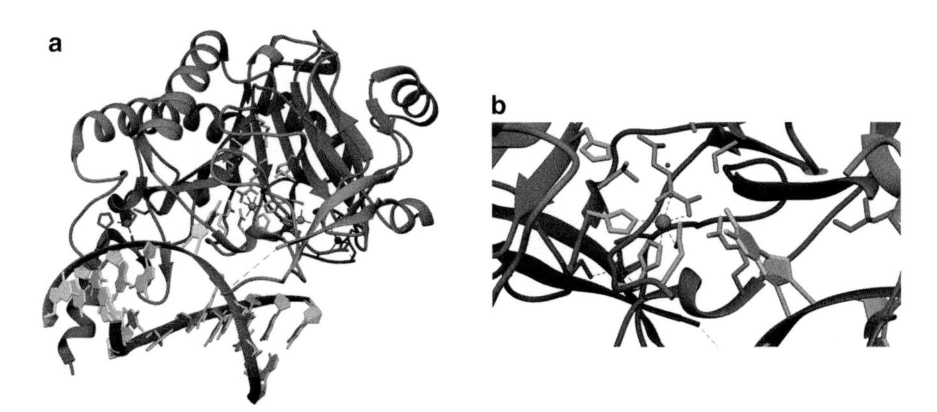

Fig. 2 Crystal structure of the TET2-DNA complex. **(a)** Here a truncated, yet catalytically active, TET2 crystal structure (*red*) is shown bound to a 12 bp segment of DNA that includes a methylated cytosine base (*yellow*). Methylated cytosine is flipped out of guanine base pairing and is inserted into the double-stranded beta helix (DSBH) catalytic core of TET2. **(b)** Inside the DBSH catalytic core lie residues critical to TET2 catalytic activity and substrate coordination (residues shown in *blue*). H1382, D1384, and H1881 chelate Fe^{2+} (*purple*) into the catalytic core and subsequently coordinates O_2 (shown here as H_2O, *red*) and 2OG (*brown*) which are required for TET2 catalytic activity. This structure was resolved by Hu et al. (2013) and can be accessed through the Protein Data Bank with the PDB code 4NM6

Ascorbate is a Likely Cofactor Needed for Continued TET DNA Demethylation

It is now well understood that ascorbate is responsible for the conversion of iron back to its reactive ferrous form and is a critical component for the continuation of TET-mediated DNA demethylation (Young et al. 2015). Recent work has demonstrated that ascorbate enhances 5hmC generation in cultured cells, most likely acting as an additional cofactor for TET-mediated 5mC hydroxylation (Minor et al. 2013; Dickson et al. 2013). This finding was later validated in many other differing cell types and animal models generated by other groups (Blaschke et al. 2013; Yin et al. 2013). Evidence for ascorbate's role as an additional cofactor for TET arises from a series of elegant experiments: siRNA knockdown of TETs largely inhibited the effect of ascorbate on 5hmC generation, demonstrating what had previously been known about TET involvement in active DNA demethylation (Tahiliani et al. 2009; Minor et al. 2013). Treatment with other native reducers such as glutathione did not alter 5hmC level, thus suggesting that the role of ascorbate in TET-mediated demethylation is specific and not due to its capacity as a general reducing agent. Inhibitors of ascorbate transporters such as phloretin and sulfinpyrazone diminished the effect of ascorbate treatment on 5hmC generation, suggesting that entry and accumulation of ascorbate within the cell is important for promoting TET catalytic activity and subsequent DNA demethylation. Although it has been previously shown that ascorbate enhances the uptake of iron by cells and initially stirred speculation that ascorbate's effect on 5hmC may be indirectly caused by increased iron accumulation, experiments where iron was removed from the cell culture medium demonstrated that iron withdrawal did not affect ascorbate-induced generation of 5hmC, thus suggesting that ascorbate has a direct effect on 5hmC generation (Lane et al. 2013; Dickson et al. 2013). Altogether, these results suggest that ascorbate actively participates in TET-mediated conversion of 5mC to 5hmC, most likely as an additional cofactor.

In retrospect, the role of ascorbate in DNA demethylation isn't all too unexpected. Previous work had shown that ascorbate induces sweeping DNA demethylation of nearly 2000 genes in human embryonic stem cells (Chung et al. 2010). Additionally, ascorbate was demonstrated to promote the generation of induced pluripotent stem cells (iPSCs) from terminally differentiated cells, a transition that is often followed by widespread DNA demethylation (Stadtfeld et al. 2012; Esteban et al. 2010). With hindsight it becomes clear that the effect of ascorbate on the cell was widespread and pronounced despite the lack of knowledge regarding the underlying mechanism. In recent years the breadth of knowledge regarding ascorbate's role in Fe(II)- and 2OG-dependent dioxygenase reactions continues to be expanded, with work now demonstrating that ascorbate is required for a number of enzymes in this family to maintain their catalytically active forms (McDonough et al. 2010). However, although the role that TET enzymes play in DNA demethylation is one of noteworthy importance, it is only one variety of a vast repertoire of functions that Fe(II)- and 2OG-dependent dioxygenases can biologically serve. With such a large family that exhibits a fine capacity for oxidation, the reach of their biological functions is widely distributed from functions in DNA/RNA repair to fatty acid metabolism, oxygen sensing, and even the biosynthesis of antibiotics (Martinez and Hausinger 2015). With such wide-reaching hands spanning the gamut of biological roles and a proclivity for ascorbate, it soon begs the question as to what is the true effective scope of vitamin C: Is vitamin C limited to assisting only TET-mediated DNA demethylation, or does the breadth of vitamin C's influence extend to other epigenetic domains as does its partner dioxygenase family?

Vitamin C and Histone Demethylation

The answer to this question lies with the discovery of the founding member of an important Fe(II)- and 2OG-dependent dioxygenase subfamily. While using gene trap approaches to discover novel regulators of mouse embryonic development, researchers identified jmj (JARID2 in humans) as an important class of transcriptional modulators (Johansson et al. 2014). The class name is derived from the Japanese cruciform-shaped character "Jumanji" (十), a prevalent shape found in the neural groove development of jmj mutant mice. Following analysis of the jmj primary gene structure and domain architecture was the identification of an ARID/Bright domain, a domain prevalent in numerous DNA-binding proteins and whose identification in jmj strongly suggested a chromatin-associating function. Second, and most pertinent, was the subsequent discovery of the novel Jmj N and C domains (named JmjN and JmjC, respectively), of which JmjC was found to be a vastly conserved protein fold whose origin has been traced to the cupin metalloenzyme family and found in every domain of life from the simplest prokaryotes to the highest order animal species (Takeuchi et al. 2006; Clissold and Ponting 2001). Most interestingly, despite low sequence similarity, JmjC enzymes also share the familiar Fe(II)- and 2OG-dependent catalytic core as seen in TETs and other Fe(II)- and 2OG-dependent enzymes, thus prompting its addition to the so-named superfamily (Johansson et al. 2014). Members of JmjC subfamily share a common structural topology composed of a double-stranded beta helix (DSBH) fold that is comprised of eight antiparallel beta sheets to form a barrel-like structure, a pattern typical of cupin metalloenzymes (Acari and Fisher 2015). This barrel is the structural motif that contains the binding sites for Fe(II) and 2OG and is responsible for coordinating these two cofactors to carry out the functional mechanisms of this enzyme family. Soon after the shared structural components of these enzymes were beginning to be resolved, great interest in determining their function was to subsequently follow.

The seminal work on histone methyl-antagonizing proteins PADI and LSD1 led to the groundbreaking discovery of JmjC domain-containing histone demethylases (JHDMs), comprising the largest enzyme family responsible for cellular histone demethylation dynamics (Klose et al. 2006). This family is divided into seven subfamilies based on structural motifs and the methylated histone substrate they specifically target (Accari and Fisher 2015) (Table 2). The family is further separated into two distinct functional classes: one with the capacity to remove methyl groups from histone lysine residues (KDM2-KDM7) and one with currently unknown function that only structurally is composed of the JmjC domain (JmjC domain-only subgroup). JHDMs are fully functional and versatile enzymes that are capable of removing all three histone lysine methylation states, unlike LSD1 which can only antagonize the mono- and dimethylated states (Klose et al. 2006). The roles of JHDMs are critical in organismal development and have been linked to various mammalian disease states. KDM5 (also known as JARID2) has been shown to associate with Polycomb group (PcG) proteins in embryonic stem cells, among many others, and is critical for embryonic stem cell differentiation (Shen et al. 2009). KDM3A (also known as JHDM2) is highly expressed and is critical in

Table 2 JmjC domain-containing histone demethylases and their substrates. Members of JmjC histone demethylase families are shown along with the histone lysine residue modifications they antagonize

Subfamily	Members (synonym)	Substrate
JmjC only	JMJD6 (PTDSR)	H3R2me2, H4R3me2
	NO66	H3K3me2/H3K3me3, H3K36me2/H3K36me3, H3K4me1/H3K4me2/H3K4me3
KDM2	KDM2A (JHDM1A, FBXL11, Ndy2)	H3K36me1/me2, H3K4me3
	KDM2B (JHDM1B, FBXL10, Ndy1)	H3K36me1/H3K36me2
KDM3	KDM3A (JHDM2A,TSGA, JMJD1A)	H3K9me1/H3K9me2
	KDM3B (JHDM2B, 5qCNA, JMJD1B)	H3K9me1/H3K9me2
	KDM3C (JHDM2C, TRIP8, JMJD1C)	H3K9me1/H3K9me2
	HR	ND
KDM4	KDM4A (JHDM3A, JMJD2A)	H3K9me2/H3K9me3, H3K36me2/H3K36me3
	KDM4B (JMJD2B)	H3K9me2/H3K9me3, H3K36me2/H3K36me3
	KDM4C (JMJD2C, GASC1)	H3K9me2/H3K9me3, H3K36me2/H3K36me3
	KDM4D (JMJD2D)	H3K9me1/H3K9me2/H3K9me3, H3K36me2/H3K36me3
KDM5	KDM5A (JARID1A, RBP2)	H3K4me2/H3K4me3
	KDM5B (JARID1B, PLU-1)	H3K4me2/H3K4me3
	KDM5C (JARID1C, SMCX)	H3K4me2/H3K4me3
	KDM5D (JARID1D, SMCY)	H3K4me2/H3K4me3
	JARID2 (Jumonji)	ND
KDM6	KDM6A (UTX)	H3K27me2/H3K27me3
	KDM6B (JMJD3)	H3K27me2/H3K27me3
	UTY	ND
KDM7	KDM7A (KIAA1718)	H3K9me1/H3K9me2, H3K27me1/H3K27me2, H4K20me1
	KDM7B (PHF8)	H3K9me2, H4K20me1
	KDM7C (PHF2)	H3K9me1/H3K9me2, H3K27me1/H3K27me2, H4K20me3

Key: *H* histone. *K* Lysine. *Me* methyl group. *ND* not determined

mouse spermatogenesis, thus granting it its alternative moniker of testis-specific gene A (TSGA). Male knockout mice develop with small testes, acute reduction in sperm count, and develop to be infertile (Okada et al. 2007). Additionally, these mice develop an obesity phenotype into adulthood (Tateishi et al. 2009). These are only but a few examples that attest to the wide-reaching and critical influence of this enzyme family in mammalian development. The full breadth of knowledge regarding the role of JHDMs in disease is profound and is well reviewed elsewhere (Johansson et al. 2014).

Demethylation by JHDM enzymes initiates via N-methyl group hydroxylation of the substrate by using 2-OG, Fe(II), and molecular oxygen to form a hydroxylated methyllysine intermediate, succinate, and CO_2 (Marmorstein and Treivel 2009). This intermediate form is unstable and subsequently undergoes spontaneous fragmentation to produce demethylated lysine and formaldehyde. This precise series of reactions is critical for the function of the enzyme, as this catalytic sequence is responsible for the JHDM family's ability to remove all three methyllysine states. As one may notice, the molecular steps of this reaction proceed in a very familiar fashion akin to that of other classic Fe(II)- and 2OG-dependent dioxygenase enzymes. It follows that since JHDM enzymes belong to this dioxygenase superfamily, just as TET enzymes, they may also require ascorbate as a cofactor to engage in full catalytic activity. This inquiry spurs great interest, as the answer implicates whether ascorbate's function extends not only to DNA demethylation but also into the regulation of histone tail demethylation dynamics, a topic of profound interest as it relates to the epigenetics of disease. As it currently stands, it appears that ascorbate may indeed be vital to the functionality of JHDM enzymes. Tsukada et al. (2006) demonstrated in vitro that ascorbate is required for the optimal activity of JHDM1 (KDM2), likely due to its reductive capacity to regenerate Fe(II) from Fe(III), just as it does in TET-mediated DNA demethylation. Mutation of H212, the histidine residue highly conserved in JmjC domains and responsible for binding iron in the classic Fe(II)-dependent dioxygenase FIH, completely abolished enzymatic activity (Elkins et al. 2002; Tsukada et al. 2006). Although the relationship between ascorbate and JHDM enzymes is still largely unexplored, it is a topic of great significance to the field and to those seeking to bolster the connection between the epigenome and disease.

The Evolution of Ascorbate Biosynthesis and Associated DNA Demethylating Systems

The overwhelming work of the past decade has made it plainly clear that the active exchange between ascorbate, TET, and demethylation dynamics is vital to the development and survival of mammalian species. TET-mediated DNA and histone demethylation are essential cogs within the epigenetic machinery that make it possible for organismal diversity and the development of complex life. Though evolutionary forces have forged their vital association, the current roles of TET and ascorbate emerged from distinct biological circumstances that fortuitously developed in closely associating systems.

Evolution of Ascorbate Biosynthesis

In countless lineages connecting many species across all living things, ascorbate is produced via closely related, endogenous biosynthetic pathways (Wheeler et al. 2015) (Fig. 3). However, despite a relatively conserved mechanism of synthesis and the great need for ascorbate, many animal lineages including haplorhine primates, guinea pigs, fruit bats, and other species have lost this ability and instead have

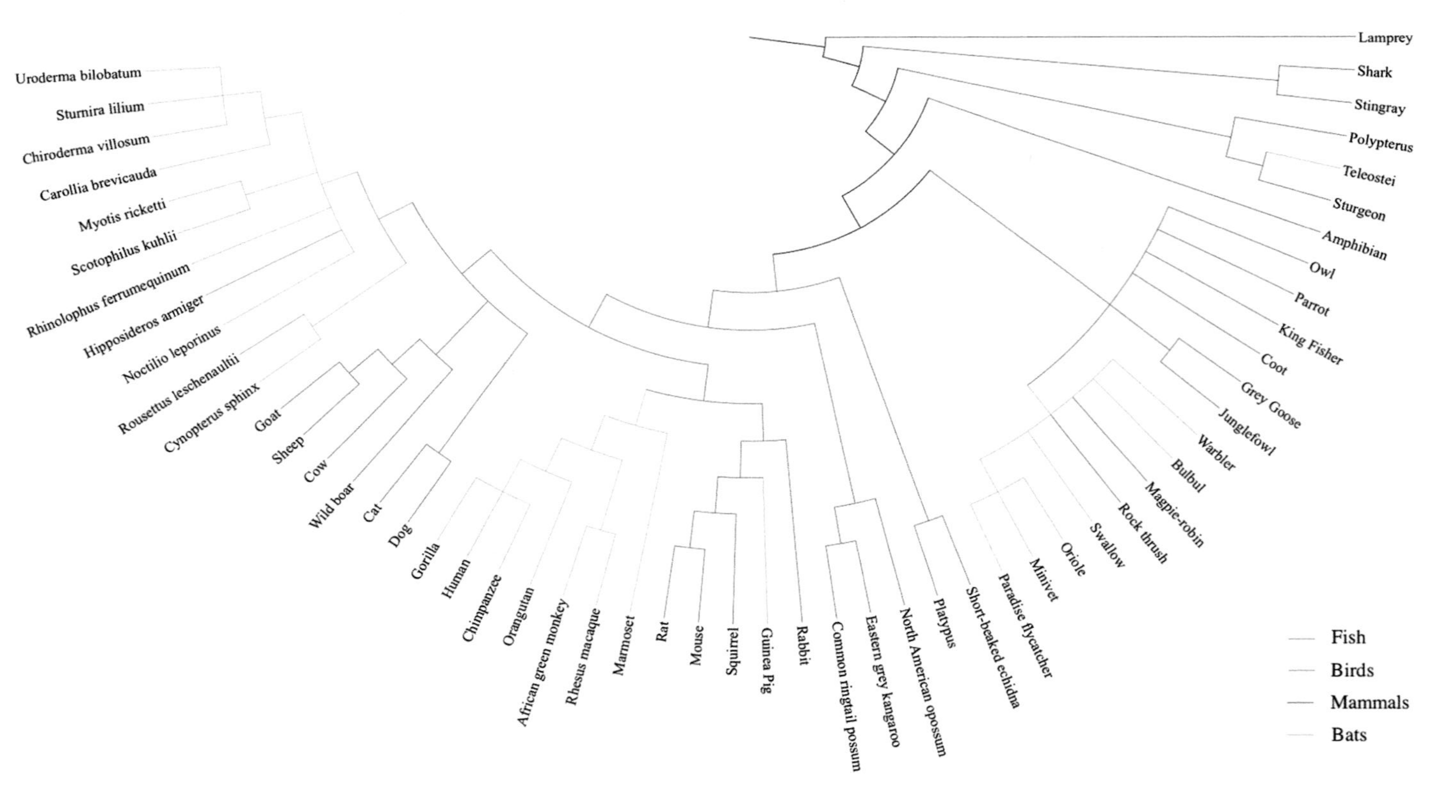

Fig. 3 Vitamin C production across the animal kingdom. A functional biosynthetic pathway for vitamin C is well conserved among species within all major animal classes. *Colored lines* indicate animals capable of endogenous vitamin C biosynthesis, while *gray lines* indicate auxotrophic species who must acquire the nutrient through dietary means. Auxotrophic taxa include teleost fish, many species of Passeriforme birds, anthropoid primates, and most species of bat. Species were selected from Drouin et al. (2011). The phylogenetic tree was created using PhyloT and visualized in iTOL

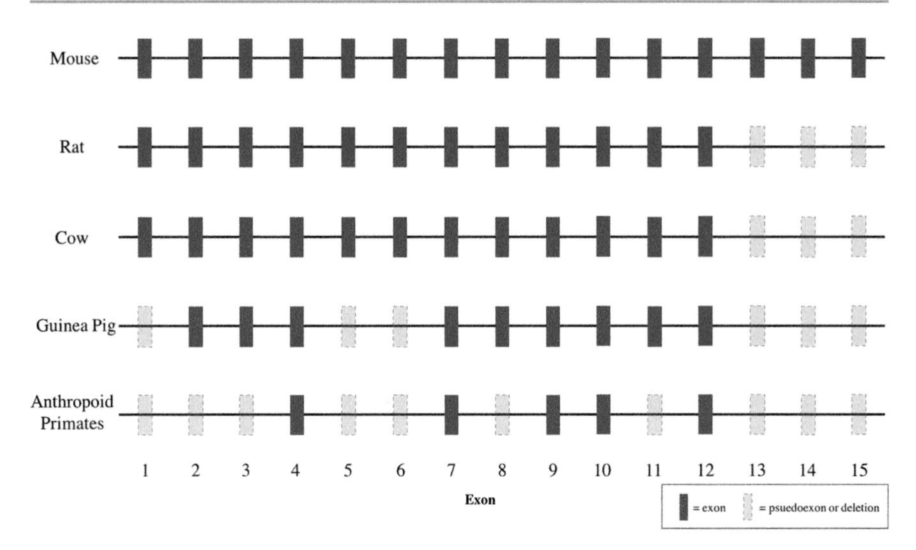

Fig. 4 Schematic representation of the GULO gene in related mammal species. Exons are indicated by *rectangles* (*blue*) while introns are indicated by *lines* (*black*). Many mammal species capable of vitamin C production share the same exonic regions. Mice, rats, and cows endogenously produce vitamin C, while guinea pigs and anthropoid primates, including humans, are ascorbate auxotrophs who have lost many exonic regions shared among other mammalian species (*opaque rectangles, dashed lines*)

become ascorbate auxotrophs that must consume vitamin C through dietary means. In all animal ascorbate auxotrophs that have been studied, the gene encoding gulonolactone (L-) oxidase (GULO) has been either lost or, in the case of humans and bats, has become a pseudogene (Drouin et al. 2011) (Fig. 4). GULO is the terminal enzyme in the animal ascorbate biosynthesis pathway that is responsible for converting L-gulonolactone into L-ascorbate by using molecular oxygen as its electron acceptor (Wheeler et al. 2015). However, the use of molecular oxygen for this purpose results in the production of H_2O_2 reactive oxygen species (ROS), a harmful cellular inhabitant that may have shaped the selective pressures for animal ascorbate auxotrophs to lose this enzyme (Mandl et al. 2009).

In contrast, plants and nearly all photosynthetic eukaryotes have lost GULO but instead use L-galactonolactone dehydrogenase (GLDH) to synthesize ascorbate from aldonolactone (Wheeler et al. 2015). GULO and GLDH are very closely related enzymes, both sharing striking sequence similarity and both belonging to the flavoprotein vanillyl-alcohol oxidase (VAO) family (Leferink et al. 2008). However, despite much similarity and few differences regarding substrate specificity and the localization of these enzymes, a critical distinction lies with the byproduct of their associated ascorbate biosynthesis pathways: While GULO generates ROS by utilizing molecular oxygen as an electron acceptor, GLDH instead uses cytochrome C to convert aldonolactone to ascorbate, a transition that does not produce H_2O_2 or any such ROS waste product (Smirnoff 2001; Wheeler et al. 2015). This distinction, besides inherently seeming beneficial to organisms who can produce vitamin C without the consequence of generating harmful ROS, becomes much more impressive and comprehensible within the context of photosynthetic systems.

For the first Archaeplastida species that acquired an endosymbiotic cyanobacteria and would bring the gift of photosynthesis to plants and algae, the danger of photodamage was imminent. The photosynthetic cyanobacteria that was the ancestor to the chloroplast secreted H_2O_2, thus putting the cell at risk of damage for becoming autotrophic. It is due to these circumstances that a strong antioxidant system that wouldn't subject the cell to further ROS risk was needed. Thus, for this purpose it is believed that the GLDH-driven biosynthesis of ascorbate was recruited (Wheeler et al. 2015). The acquisition of the plastid increased the antioxidant demands of the cell and therefore ascorbate was used to fulfill this role. Ascorbate's role in protecting photosynthetic cells from photodamage is now well understood. Through the ascorbate-glutathione cycle, ascorbate peroxidase (APX), localized to the cytosol and chloroplast, removes H_2O_2 produced by photosystem I, a catalytic mechanism that may account for 10% of photosynthetic electron transport (Smirnoff 2011). This process, along with ascorbate's role in preventing lipid peroxidation of the chloroplast thylakoid membranes, serves as a vital measure to protect photosynthetic cells from accumulating ROS damage.

As it happens, the evolutionary forces that bestowed certain eukaryotes with photosynthesis also consequently shaped the ascorbate biosynthetic pathways of other herbivorous taxa. Recent data has shown that in all documented animal ascorbate auxotrophs the primary source of dietary ascorbate comes from GLDH derived from land plants (Wheeler et al. 2015). This holds true even for insectivorous animals and insects who also cannot synthesize ascorbate and must acquire it through dietary means, most notably from land plants. Thus, the loss and subsequent replacement of GULO with GLDH in photosynthetic eukaryotes may ultimately have driven the loss of GULO in many herbivorous animals and insects (Wheeler et al. 2015). Due to the self-harming nature of ROS byproducts derived from GULO ascorbate synthesis and the abundance of ascorbate being produced in plant prey species, the continued use of GULO to synthesize ascorbate seems superfluous, if not deleterious, from the evolutionary perspective. Thus, the evolutionary history of auxotrophic animals and humankind has been intimately shaped by the hands of ascorbate and its narrative.

Evolution of TET Genes

In contrast to ascorbate, TET enzymes developed far more narrowly within the tree of life. This arises from the fact that DNA demethylation operates differently between plants and animals. As described previously, the animal DNA demethylation pathway begins with methylated cytosine that is converted to 5hmC, oxidized several more times into 5fC and 5caC intermediates, and then excised by TDG DNA glycosylase before the resulting abasic position is fixed by the DNA base excision repair pathway. In plants, however, this process remained much simpler than in its mammalian counterpart. As studied thoroughly in Arabidopsis, plant DNA glycosylases such as ROS1 are able to directly excise 5mC, leaving an abasic position that is also subsequently fixed and filled with unmodified cytosine just as in animals (Zhang and Zhu 2012). This pathway skips all oxidative intermediate steps, most notably the oxidation of 5mC to 5hmC whose existence in plants, as expected, has yet to be demonstrated (Jang et al. 2014). This demethylation pathway leaves no room for TET involvement; thus, the bulk of work studying the evolutionary history and function of these enzymes has been conducted in mammalian systems.

All three TET genes are widely distributed amongst mammals, a characteristic that suggests the TET gene family putatively resulted from two successive gene duplication events before diversifying in this class (Iyer et al. 2009; Pastor et al. 2013). As mentioned previously, TET1 and TET3 share the common DNA-binding motif CXXC domain while TET2 underwent an evolutionary chromosomal inversion event that separated its CXXC domain into a separate functional gene now called IDAX (Ko et al. 2013). Recent work has found that evolutionary forces shaped TET enzyme family members differentially. The overall evolutionary trend for these enzymes was purifying selection (negative selection), though TET1 and TET2 were found to have undergone positive selection more frequently than TET3 (Akahori et al. 2015). The overall trend of negative selection emphasizes the physiological importance of these enzymes throughout mammalian history. Eukaryotic TET enzymes were found to originate from ancestral genes encoded by bacteriophages who produce a diversity of DNA modifications to evade host cell restriction systems (Aravind et al. 2012). Therefore, TET enzymes may have been recruited by bacteriophages as a tactic to avoid host cell surveillance through DNA modification, a strategy that was later adopted and repurposed by eukaryotes for transcriptional regulation.

Additionally, TET enzymes may also have a surprising role in maintaining hierarchical phenotypes within eusocial insect species. A rising trend within the field of epigenetics is the use of eusocial species as epigenetic models, including honey bees, ants, and social wasps (Yan et al. 2015). This trend arises from the rationale that these organisms form hives composed of thousands of members all containing the same genetic content, yet these genetically identical members can further differentiate into social castes (e.g., queen, worker) that exhibit gross phenotypic and behavioral differences. Thus, these species serve as excellent epigenetic models for studying the functional consequences of histone and DNA methylation dynamics. It has now been well demonstrated that both histone and DNA methylation are integral to regulating social castes and behavior within eusocial species (Yan et al. 2015). This finding instantly raises questions regarding how methylation status within these castes is maintained, particularly by active demethylation systems such as that of TET enzymes. It has been demonstrated that honey bees have a single TET orthologue, AmTET, that is capable of converting 5mC to 5hmC in vitro (Wojciechowski et al. 2014). Although TET's involvement in regulating social caste status and behavior has not been directly demonstrated, further research is needed to determine the prevalence, function, and significance of TET orthologues in eusocial species.

Conclusions

Presently, at a time of great innovation and prospect for the field of epigenetics, the role that ascorbate plays in DNA and histone demethylation is one of fascinating interest. As evolutionary history has made clear, ascorbate is an integral component to the development of biological systems and has taken on diverse critical roles in both the photoprotection of plants and regulating demethylation dynamics in the animal kingdom. Continuing to explore the relationship between ascorbate and its diverse biological relevance in these systems can lead to greater understanding of DNA methylation dynamics and the potential maladies that arise from its disruption.

Dictionary of Terms

- **Gulonolactone (L) oxidase** – An enzyme encoded by the GULO gene and is the terminal enzyme in the animal ascorbate biosynthesis pathway that uses molecular oxygen to convert L-gulonolactone into L-ascorbate.
- **DNA methylation** – The addition of methyl groups to DNA cytosine and adenine bases. Cytosine methylation is an important epigenetic modification implicated in many important developmental processes including gene transcriptional regulation and embryonic stem cell differentiation.
- **5-hydroxymethylcytosine (5hmC)** – A stable epigenetic modification resulting from the hydroxylation of 5-methylcytosine (5mC) that affects transcriptional regulation. The hydroxylation reaction is performed by the TET methylcytosine dioxygenases and is the first step in active DNA demethylation.
- **Ten-eleven translocation (TET) methylcytosine dioxygenases** – An enzyme family consisting of three members responsible for DNA demethylation in mammals. TET dioxygenases sequentially convert 5-methylcytosine to 5-hydroxymethylcytosine, 5-formylcytosine, and 5-carboxylcytosine.
- **Eusociality** – An organizational schema within animals where species members are divided into distinct behavioral castes that perform different reproductive and labor functions. This occurs within crustaceans, insects, and some mammals and has been best studied in the insect order Hymenoptera (bees, wasps, and ants).

Key Facts of Ascorbate

- Vitamin C (ascorbate) is a reducing agent clinically linked to the prevention of scurvy.
- Fruits and vegetables were known by sailors to prevent scurvy for hundreds of years before vitamin C was eventually discovered.

- Albert Szent-Györgyiwas the first to synthesize vitamin C in the early 1930s and link its role to the prevention of scurvy, a seminal finding that awarded him the Nobel Prize in 1937.
- Ascorbate plays an important role in catecholamine neurotransmitter synthesis.
- Ascorbate biosynthesis occurs via different, yet closely related, biochemical pathways in both plants and animals.

Summary Points

- Vitamin C (ascorbate) is a dietary nutrient classically implicated in the prevention of scurvy.
- 5-Methylcytosine (5mC) is the best-characterized DNA modification of the mammalian genome.
- Although the existence of DNA methylation was known for many years, the question of whether methylation states could be actively altered or removed was debated.
- The ten-eleven translocation (TET) family of enzymes was demonstrated to be responsible for active DNA demethylation in mammals.
- TET enzymes hydroxylate 5-methylcytosine (5mC) to 5-hydroxymethylcytosine (5hmC). TET enzymes further convert 5hmC to 5-formylcytosine (5fC) and 5-carboxylcytosine (5caC), two transient DNA tags excised by DNA repair enzyme thymine-DNA glycosylase (TDG). This results in an abasic position that is replaced by an unmodified cytosine via the base excision repair pathway, thus completing DNA demethylation.
- TET enzymes belong to the Fe(II)- and 2-oxoglutarate(2OG)-dependent dioxygenase superfamily that utilize 2OG and molecular oxygen as co-substrates and require ferrous iron as a cofactor.
- Ascorbate is an additional cofactor for TET enzymes and is responsible for converting Fe(III) to Fe(II) to sustain TET activity.
- JmjC demethylases, responsible for removing methylated histone modifications, may also use ascorbate as an additional cofactor.
- Humans, fruit bats, guinea pigs, and other animals have lost a functional GULO gene, a gene that encodes the terminal enzyme in the animal ascorbate biosynthesis pathway, thus making these species unable to endogenously produce ascorbate.
- Plants synthesize ascorbate via an alternative pathway that requires the gene GLDH. GLDH encodes the terminal enzyme in the plant ascorbate biosynthesis pathway and was substituted for GULO after the acquisition of the chloroplast, a switch that both served to protect photosynthetic plants and ultimately shaped the evolutionary forces guiding ascorbate auxotrophy.

Acknowledgements The work on the epigenomic regulation by vitamin C in the Wang lab is supported by grants (R01NS089525, R21CA191668) from the National Institutes of Health.

References

Accari SL, Fisher PR (2015) Emerging roles of JmjC domain-containing proteins. Int Rev Cell Mol Biol 319:165–220

Akahori H, Guindon S, Yoshizaki S, Muto Y (2015) Molecular evolution of the TET gene family in mammals. Int J Mol Sci 16:28472–28485. https://doi.org/10.3390/ijms161226110

Aravind L, Anantharaman V, Zhang D, de Souza R, Iyer L (2012) Gene flow and biological conflict systems in the origin and evolution of eukaryotes. Front Cell Infect Microbiol 2. https://doi.org/10.3389/fcimb.2012.00089

Bhutani N, Burns DM, Blau HM (2011) DNA demethylation dynamics. Cell 146:866–872. https://doi.org/10.1016/j.cell.2011.08.042

Blaschke K, Ebata KT, Karimi MM, Zepeda-Martínez JA, Goyal P, Mahapatra S, Tam A, Laird DJ, Hirst M, Rao A, Lorincz MC, Ramalho-Santos M (2013) Vitamin C induces Tet-dependent DNA demethylation and a blastocyst-like state in ES cells. Nature 500:222–226. https://doi.org/10.1038/nature12362

Camarena V, Wang G (2016) The epigenetic role of vitamin C in health and disease. Cell Mol Life Sci 73:1645–1658. https://doi.org/10.1007/s00018-016-2145-x

Chung TL, Brena RM, Kolle G, Grimmond SM, Berman BP, Laird PW, Pera MF, Wolvetang EJ (2010) Vitamin C promotes widespread yet specific DNA demethylation of the epigenome in human embryonic stem cells. Stem Cells 28:1848–1855. https://doi.org/10.1002/stem.493

Clissold PM, Ponting CP (2001) JmjC: cupin metalloenzyme-like domains in jumonji, hairless and phospholipase A2β. Trends Biochem Sci 26:7–9. https://doi.org/10.1016/s0968-0004(00)01700-x

Delhommeau F, Dupont S, Della Valle V, James C, Trannoy S, Massé A, Kosmider O, Le Couedic JP, Robert F, Alberdi A, Lécluse Y, Plo I, Dreyfus FJ, Marzac C, Casadevall N, Lacombe C, Romana SP, Dessen P, Soulier J, Viguié F, Fontenay M, Vainchenker W, Bernard OA (2009) Mutation in TET2 in myeloid cancers. New England Journal of Medicine 360:2289–301. https://doi.org/10.1056/NEJMoa0810069

Dickson KM, Gustafson CB, Young JI, Züchner S, Wang G (2013) Ascorbate-induced generation of 5-hydroxymethylcytosine is unaffected by varying levels of iron and 2-oxoglutarate. Biochem Biophys Res Commun 439:522–527. https://doi.org/10.1016/j.bbrc.2013.09.010

Drouin G, Godin JR, Page B (2011) The genetics of vitamin C loss in vertebrates. Curr Genomics 12:371–378. https://doi.org/10.2174/138920211796429736

Elkins JM, Hewitson KS, Mcneill LA, Seibel JF, Schlemminger I, Pugh CW, Ratcliffe PJ, Schofield CJ (2002) Structure of factor-inhibiting hypoxia-inducible factor (HIF) reveals mechanism of oxidative modification of HIF-1. J Biol Chem 278:1802–1806. https://doi.org/10.1074/jbc.c200644200

Esteban MA, Wang T, Qin B, Yang J, Qin D, Cai J, Li W, Weng Z, Chen J, Ni S, Chen K, Li Y, Liu X, Xu J, Zhang S, Li F, He W, Labuda K, Song Y, Peterbauer A, Wolbank S, Redl H, Zhong M, Cai D, Zeng L, Pei D (2010) Vitamin C enhances the generation of mouse and human induced pluripotent stem cells. Cell Stem Cell 6:71–79. https://doi.org/10.1016/j.stem.2009.12.001

He YF, Li BZ, Li Z, Liu P, Wang Y, Tang Q, Ding J, Jia Y, Chen Z, Li L, Sun Y, Li X, Dai Q, Song C-X, Zhang K, He C, Xu G-L (2011) Tet-mediated formation of 5-Carboxylcytosine and its excision by TDG in mammalian DNA. Science 333:1303–1307. https://doi.org/10.1126/science.1210944

Hu L, Li Z, Cheng J, Rao Q, Gong W, Liu M, Shi Y, Zhu J, Wang P, Xu Y (2013) Crystal structure of TET2-DNA complex: insight into TET-mediated 5mC oxidation. Cell 155:1545–1555. https://doi.org/10.1016/j.cell.2013.11.020

Ito S, Shen L, Dai Q, SC W, Collins LB, Swenberg JA, He C, Zhang Y (2011) Tet proteins can convert 5-methylcytosine to 5-formylcytosine and 5-carboxylcytosine. Science 333:1300–1303. https://doi.org/10.1126/science.1210597

Iyer LM, Tahiliani M, Rao A, Aravind L (2009) Prediction of novel families of enzymes involved in oxidative and other complex modifications of bases in nucleic acids. Cell Cycle 8:1698–1710. https://doi.org/10.4161/cc.8.11.8580

Jang H, Shin H, Eichman BF, Huh JH (2014) Excision of 5-hydroxymethylcytosine by DEMETER family DNA glycosylases. Biochem Biophys Res Commun 446:1067–1072. https://doi.org/10.1016/j.bbrc.2014.03.060

Johansson C, Tumber A, Che K, Cain P, Nowak R, Gileadi C, Oppermann U (2014) The roles of Jumonji-type oxygenases in human disease. Epigenomics 6:89–120. https://doi.org/10.2217/epi.13.79

Klose RJ, Kallin EM, Zhang Y (2006) JmjC-domain-containing proteins and histone demethylation. Nature Reviews Genetics 7:715–727. https://doi.org/10.1038/nrg1945

Ko M, An J, Bandukwala HS, Chavez L, Äijö T, Pastor WA, Segal MF, Li H, Koh KP, Lähdesmäki H, Hogan PG, Aravind L, Rao A (2013) Modulation of TET2 expression and 5-methylcytosine oxidation by the CXXC domain protein IDAX. Nature 497:122–126. https://doi.org/10.1038/nature12052

Lane DJ, Chikhani S, Richardson V, Richardson DR (2013) Transferrin iron uptake is stimulated by ascorbate via an intracellular reductive mechanism. Biochim Biophys Acta (BBA) Mol Cell Res 1833:1527–1541. https://doi.org/10.1016/j.bbamcr.2013.02.010

Leferink NG, Heuts DP, Fraaije MW, Berkel WJV (2008) The growing VAO flavoprotein family. Arch Biochem Biophys 474:292–301. https://doi.org/10.1016/j.abb.2008.01.027

Li D, Guo B, Wu H, Tan L, Lu Q (2015) TET family of dioxygenases: crucial roles and underlying mechanisms. Cytogenet Genome Res 146:171–180. https://doi.org/10.1159/000438853

Lorsbach RB, Moore J, Mathew S, Raimondi SC, Mukatira ST, Downing JR (2003) TET1, a member of a novel protein family, is fused to MLL in acute myeloid leukemia containing the t (10;11)(q22;q23). Leukemia 17:637–641. https://doi.org/10.1038/sj.leu.2402834

Mandl J, Szarka A, Bánhegyi G (2009) Vitamin C: update on physiology and pharmacology. Br J Pharmacol 157:1097–1110. https://doi.org/10.1111/j.1476-5381.2009.00282.x

Marmorstein R, Trievel RC (2009) Histone modifying enzymes: structures, mechanisms, and specificities. Biochim Biophys Acta (BBA) Gene Regul Mech 1789:58–68. https://doi.org/10.1016/j.bbagrm.2008.07.009

Martinez S, Hausinger R (2015) Catalytic mechanisms of Fe(II)- and 2-oxoglutarate-dependent oxygenases. J Biol Chem 290:20702–20711. https://doi.org/10.1074/jbc.r115.648691

McDonough M, Loenarz C, Chowdhury R, Clifton I, Schofield C (2010) Structural studies on human 2-oxoglutarate dependent oxygenases. Curr Opin Struct Biol 20:659–672. https://doi.org/10.1016/j.sbi.2010.08.006

Minor EA, Court BL, Young JI, Wang G (2013) Ascorbate induces ten-eleven translocation (Tet) methylcytosine dioxygenase-mediated generation of 5-hydroxymethylcytosine. J Biol Chem 288:13669–13674. https://doi.org/10.1074/jbc.c113.464800

Okada Y, Scott G, Ray MK, Mishina Y, Zhang Y (2007) Histone demethylase JHDM2A is critical for Tnp1 and Prm1 transcription and spermatogenesis. Nature 450:119–123. https://doi.org/10.1038/nature06236

Pastor WA, Aravind L, Rao A (2013) TETonic shift: biological roles of TET proteins in DNA demethylation and transcription. Nat Rev Mol Cell Biol 14:341–356. https://doi.org/10.1038/nrm3589

Rose NR, McDonough MA, King ON, Kawamura A, Schofield CJ (2011) Inhibition of 2-oxoglutarate dependent oxygenases. Chem Soc Rev 40:4364

Salminen A, Kauppinen A, Kaarniranta K (2015) 2-Oxoglutarate-dependent dioxygenases are sensors of energy metabolism, oxygen availability, and iron homeostasis: potential role in the regulation of aging process. Cell Mol Life Sci 72:3897–3914. https://doi.org/10.1007/s00018-015-1978-z

Schübeler D (2015) Function and information content of DNA methylation. Nature 517:321–326. https://doi.org/10.1038/nature14192

Shen X, Kim W, Fujiwara Y, Simon MD, Liu Y, Mysliwiec MR, Yuan G-C, Lee Y, Orkin SH (2009) Jumonji modulates polycomb activity and self-renewal versus differentiation of stem cells. Cell 139:1303–1314. https://doi.org/10.1016/j.cell.2009.12.003

Smirnoff N (2001) L-ascorbic acid biosynthesis. Vitam Horm 61:241–266. https://doi.org/10.1016/s0083-6729(01)61008-2

Smirnoff N (2011) Vitamin C. Adv Bot Res Biosynthesis Vitam Plants Part B 107–177. https://doi.org/10.1016/b978-0-12-385853-5.00003-9

Stadtfeld M, Apostolou E, Ferrari F, Choi J, Walsh RM, Chen T, Ooi SSK, Kim SY, Bestor TH, Shioda T, Park PJ, Hochedlinger K (2012) Ascorbic acid prevents loss of Dlk1-Dio3 imprinting and facilitates generation of all–iPS cell mice from terminally differentiated B cells. Nat Genet 44:398–405. https://doi.org/10.1038/ng.1110

Tahiliani M, Koh KP, Shen Y, Pastor WA, Bandukwala H, Brudno Y, Agarwal S, Iyer LM, Liu DR, Aravind L, Rao A (2009) Conversion of 5-Methylcytosine to 5-Hydroxymethylcytosine in mammalian DNA by MLL partner TET1. Science 324:930–935. https://doi.org/10.1126/science.1170116

Takeuchi T, Watanabe Y, Takano-Shimizu T, Kondo S (2006) Roles of jumonji and jumonji family genes in chromatin regulation and development. Dev Dyn 235:2449–2459. https://doi.org/10.1002/dvdy.20851

Tateishi K, Okada Y, Kallin EM, Zhang Y (2009) Role of Jhdm2a in regulating metabolic gene expression and obesity resistance. Nature 458:757–761. https://doi.org/10.1038/nature07777

Tsukada Y, Fang J, Erdjument-Bromage H, Warren ME, Borchers CH, Tempst P, Zhang Y (2006) Histone demethylation by a family of JmjC domain-containing proteins. Nature 439:811–6

Van Robertson WB, Schwartz B (1953) Ascorbic acid and the formation of collagen. J Biol Chem 201:689–696

Wheeler G, Ishikawa T, Pornsaksit V, Smirnoff N (2015) Evolution of alternative biosynthetic pathways for vitamin C following plastid acquisition in photosynthetic eukaryotes. elife 4. https://doi.org/10.7554/elife.06369

Wojciechowski M, Rafalski D, Kucharski R, Misztal K, Maleszka J, Bochtler M, Maleszka R (2014) Insights into DNA hydroxymethylation in the honeybee from in-depth analyses of TET dioxygenase. Open Biology 4:140110–140110. https://doi.org/10.1098/rsob.140110

Xu W, Yang H, Liu Y, Yang Y, Wang P, Kim SH, Ito S, Yang C, Wang P, Xiao MT, Liu LX, Jiang WQ, Liu J, Zhang JY, Wang B, Frye S, Zhang Y, YH X, Lei QY, Guan KL, Zhao SM, Xiong Y (2011a) Oncometabolite 2-hydroxyglutarate is a competitive inhibitor of α-ketoglutarate-dependent dioxygenases. Cancer Cell 19:17–30. https://doi.org/10.1016/j.ccr.2010.12.014

Xu Y, Wu F, Tan L, Kong L, Xiong L, Deng J, Barbera AJ, Zheng L, Zhang H, Huang S, Min J, Nicholson T, Chen T, Xu G, Shi Y, Zhang K, Shi YG (2011b) Genome-wide regulation of 5hmC, 5mC, and gene expression by Tet1 hydroxylase in mouse embryonic stem cells. Mol Cell 42:451–464. https://doi.org/10.1016/j.molcel.2011.04.005

Yan H, Bonasio R, Simola DF, Liebig J, Berger SL, Reinberg D (2015) DNA methylation in social insects: how epigenetics can control behavior and longevity. Annu Rev Entomol 60:435–452. https://doi.org/10.1146/annurev-ento-010814-020803

Yin R, Mao SQ, Zhao B, Chong Z, Yang Y, Zhao C, Zhang D, Huang H, Gao J, Li Z, Jiao Y, Li C, Liu S, Wu D, Gu W, Yang YG, GL X, Wang H (2013) Ascorbic acid enhances tet-mediated 5-methylcytosine oxidation and promotes DNA demethylation in mammals. J Am Chem Soc 135:10396–10403. https://doi.org/10.1021/ja4028346

Young JI, Züchner S, Wang G (2015) Regulation of the epigenome by vitamin C. Annu Rev Nutr 35:545–564. https://doi.org/10.1146/annurev-nutr-071714-034228

Zhang H, Zhu JK (2012) Active DNA demethylation in plants and animals. Cold Spring Harb Symp Quant Biol 77:161–173. https://doi.org/10.1101/sqb.2012.77.014936

Vitamin C and DNA Demethylation in Regulatory T Cells

88

Varun Sasidharan Nair and Kwon Ik Oh

Contents

Abstract

For the immune homeostasis, regulation of effector T cells is indispensable. This is performed by a distinct subclass of $CD4^+$ T cell called "regulatory T cells" (Tregs). The Tregs express the canonical transcription factor called Forkhead box P3 (Foxp3) throughout their life span for their proper development and suppressive function, and the expression of Foxp3 is regarded as a reliable marker of Tregs. Tregs can be generated in the thymus, peripheral tissues, and even in vitro. Thus, Treg populations are divided into three groups. The first one is the Tregs generated in the thymus (thymic Treg, tTreg) and occupies the major fraction of the total Treg population in vivo. The second one is the minor fraction generated in periphery from naïve $CD4^+$ T cells, when they meet cognate antigen under

V. S. Nair · K. I. Oh (✉)
Department of Pathology, Hallym University, College of Medicine, Chuncheon, Gangwon-Do, Republic of Korea
e-mail: varuns1982@gmail.com; varanbio@gmail.com; kwonik@hallym.ac.kr; kwonikoh@gmail.com

V. B. Patel, V. R. Preedy (eds.), *Handbook of Nutrition, Diet, and Epigenetics*,
https://doi.org/10.1007/978-3-319-55530-0_30

tolerogenic conditions (peripheral Treg, pTreg). Tregs can also be generated in vitro upon TCR activation in the presence of TGF-β (induced Treg, iTreg). Although all three Treg populations have suppressive activity in common, each population shows distinct genetic and epigenetic features. For instance, in Foxp3 gene there is a unique evolutionarily conserved intronic region with several CpG motifs, which is called CNS2 (conserved non-coding sequence 2). The CpG motifs in CNS2 are fully methylated in almost all Foxp3^{-} T cells including CD4 single positive thymocytes (tTreg precursors), naïve CD4^{+} T cells (pTreg and iTreg precursors) and CD8^{+} T cells, and some Foxp3^{+} T cells such as iTregs. In contrast, they are fully demethylated in Tregs generated in vivo (tTregs and pTregs). This dichotomic pattern seen in CNS2 (de-)methylation has attracted researchers' attention. In this chapter, we are going to discuss the underlying mechanisms of CNS2 demethylation in various types of Tregs and how vitamin C contributes to this process.

Keywords
CNS2 demethylation · CpG motifs · Foxp3 · Histone demethylation · Iron- and 2-oxoglutarate-dependent dioxygenases · Jmjd2 · Regulatory T cell (Treg) · Stability · Tet · Vitamin C

List of Abbreviations

5hmC	5-hydroxymethylcytosine
5mC	5-methylcytosine
CNS2	Conserved non-coding sequence 2
Dnmt	DNA methyl transferase
Foxp3	Forkhead box P3
IL	Interleukin
iTreg	Induced Treg
pTreg	Peripheral Treg
SVCT	Sodium-dependent vitamin C transporter
Tet	Ten-eleven-translocation
TGF-β	Transforming growth factor-β
Treg	Regulatory T
tTreg	Thymic Treg

DNA Methylation

The epigenetic marks of a genome are transferred to the daughter cells through replication without changing the DNA sequences (Reik et al. 2001, 1089), and one of these epigenetic marks is DNA methylation which occurs in the 5th position of the pyrimidine ring of cytosine residue in a cytosine (C)–guanine (G) sequence context, which is called CpG ("p" stands for phosphate group) (Fig. 1a). In our genome there are CpG-rich regions which are called CpG islands. They are usually seen near to gene regulatory elements such as promoters and enhancers and known to be

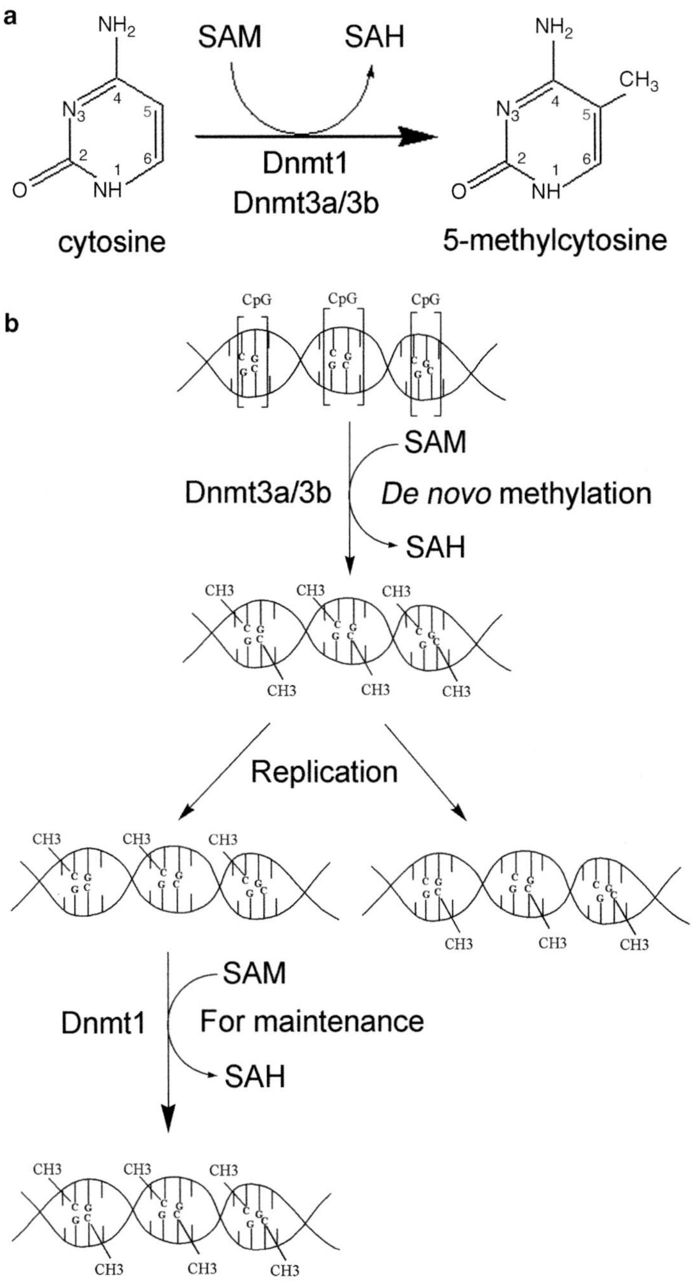

Fig. 1 DNA methylation. (**a**) DNA methylation happens in a CpG context. *S*-adenosyl-L-methionine (*SAM*) donates methyl group to the fifth position of cytosine and converts cytosine to 5-methylcytosine by means of DNMT. (**b**) DNMT3a and DNMT3b help de novo methylation, and DNMT1 helps to prevent the loss of methylation of cytosine in the nascent DNA and maintains the preexisting methylation pattern. *SAM* *S*-adenosyl-L-methionine, *SAH* *S*-adenosyl-L-homocysteine

involved in transcriptional repression. The methylation of CpG motifs is carried out by three different enzymes which belong to DNA methyltransferase (Dnmt) family named Dnmt1, Dnmt3a, and Dnmt3b by using S-adenosylmethionine as a methyl group donor. The de novo methylation of DNA is done by Dnmt3a and Dnmt3b, and after replication the methyl group of nascent DNA strand in the daughter cells is maintained by Dnmt1 (Cheng 1995, 293; Jones and Liang 2009, 805) (Fig. 1b).

DNA Demethylation

For gene transcription DNA needs to be accessible to various molecules such as transcription factors and converted from the heterochromatin (condensed) to euchromatin (relaxed) structure. In the heterochromatin structure, the DNA is methylated, but in the euchromatin form, the DNA becomes demethylated (or unmethylated) and transcriptionally permissive (Allis and Jenuwein 2016, 487). For the DNA demethylation, there are two different pathways, active and passive (Fig. 2a). Dnmt1 has an essential role in the maintenance of methyl group throughout replication, but sometimes due to the rapid proliferation, Dnmt1 could be diluted and fail to secure its methyl hallmark to corresponding CpGs. Thereby DNA would be demethylated. This process is called passive demethylation (Messerschmidt et al. 2014, 812). For understanding the active demethylation, we need to understand a class of enzymes which belong to an iron- and 2-oxoglutarate-dependent dioxygenases family, called ten-eleven translocation (Tet), comprising three members (Tet1, Tet2, and Tet3). Tet enzymes can recognize 5-methylcytosine (5mC), oxidize 5mC into 5-hydroxymethylcytosine (5hmC), and iteratively modify the oxidized cytosine products, leading to the generation of 5-formylcytosine (5fC) and 5-carboxylcytosine (5caC). The oxidized cytosine is removed by thymine DNA glycosylase and eventually replaced with unmodified cytosine by means of base excision repair mechanisms (Pastor et al. 2013, 341) (Fig. 2b). During this process, iron and 2-oxoglutarate act as a cofactor and a substrate.

Importance of CNS2 Demethylation in Tregs

Forkhead box P3$^+$ (Foxp3$^+$) Tregs are a dedicated CD4$^+$ T cell population that holds excessive immune responses in check and prevents autoimmune diseases (Sakaguchi and Powrie 2007, 317). Majority of Tregs found in peripheral secondary lymphoid organs like lymph nodes arise from the thymus through the high-affinity interaction between T cell receptors (TCRs) and self-antigens and retain the Foxp3 expression and immune regulatory phenotype stably. Thereby, Tregs can be activated by cognate antigens and differentiated into effector cells while maintaining the expression of immune suppressive genes including Foxp3, implying that Treg is a distinct T cell lineage. Apart from thymic generation, Tregs can also be generated de novo in

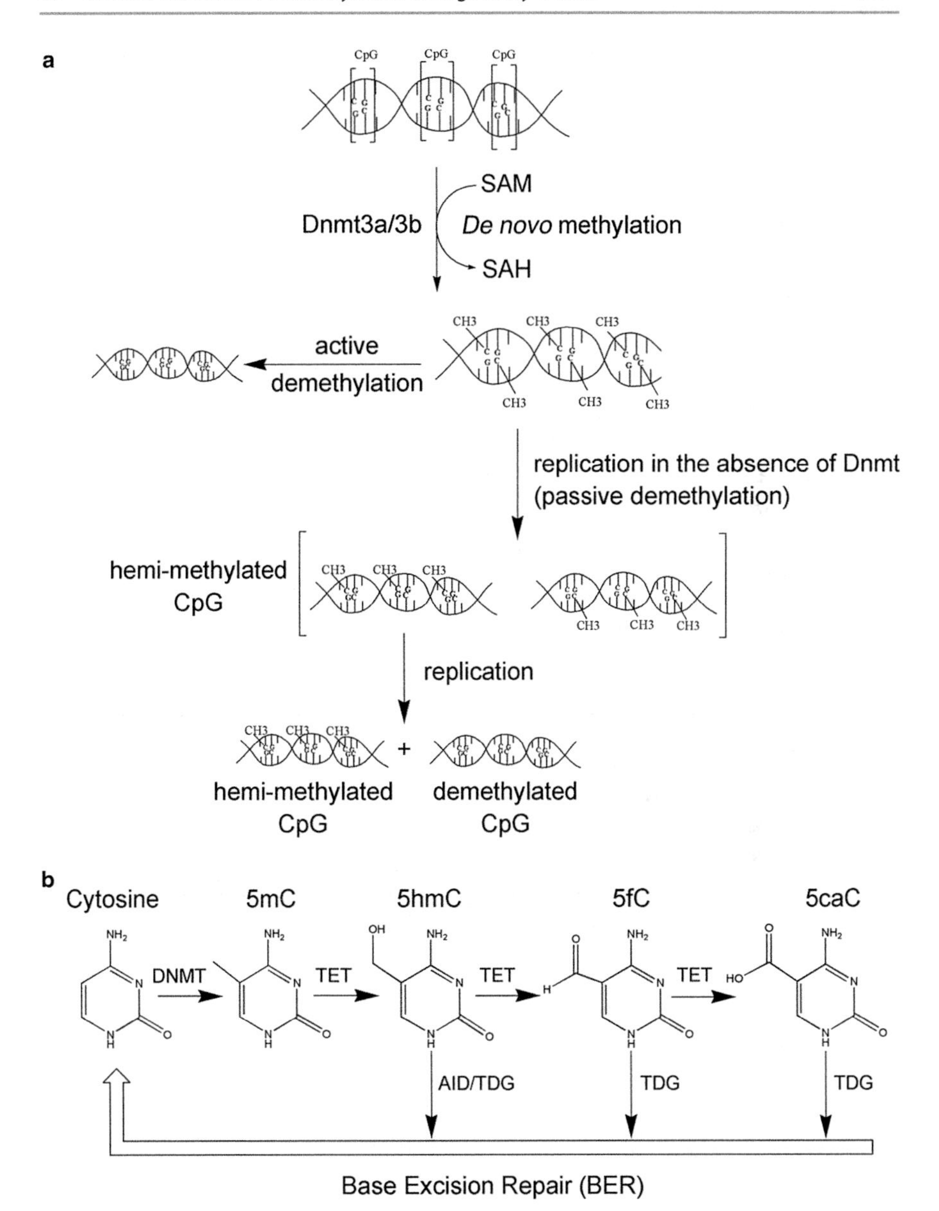

Fig. 2 Active and passive demethylation. (**a**) Methylated DNA can be demethylated in two ways, active and passive demethylations. The active DNA demethylation is done by Tet enzymes, and the passive demethylation can occur by DNA replication in the absence of Dnmt. The hemi-methylated DNA produced in the absence of Dnmt1 is finally demethylated upon consecutive replications. (**b**) 5-Methylcytosine (5mC) can be converted to 5-hydroxymethylcytosine (5hmC), 5-formylcytosine (5fC), and 5-carboxylcytosine (5caC) by Tet enzymes. Finally by means of base excision repair (*BER*), the oxidized cytosine is removed and replaced with unmodified cytosine

the peripheral organs like colons upon antigen stimulation and even in vitro in the presence of transforming growth factor-β (TGF-β) from naïve $CD4^+$ T cells. What attracts our attention in Tregs generated in vitro by TGF-β is their unstable phenotype. When these Tregs are stimulated by cognate antigens, most of them lose Foxp3 expression unless TGF-β is supplied continuously. This finding indicates that antigen plus TGF-β stimulation can upregulate the expression of Foxp3, but as-yet-unknown factors are required to differentiate naïve $CD4^+$ T cells into stable Foxp3-expressing Treg lineage. Collectively, Treg populations can be divided into three groups, namely, thymus-derived Tregs (tTreg), peripheral Tregs (pTregs), and Tregs induced in vitro (induced Tregs, iTregs). Although all three Treg populations have suppressive activity in common, each type of Treg shows unique genetic and epigenetic profiles (Feuerer et al. 2010, 5919; Ohkura et al. 2012, 785), suggesting that each develops under different environments.

Foxp3 is the essential transcription factor for the immune regulatory functions of Tregs (d'Hennezel et al. 2009, 1710). Therefore, the stable expression of Foxp3 is critical and Foxp3 instability was also reported to be pathologically relevant (Hori 2011, 295; Komatsu et al. 2014, 62; Miyao et al. 2012, 262; Rubtsov et al. 2010, 1667; Zhou et al. 2009, 1000). Stable expression of Foxp3 is known to be accompanied by epigenetic modulation of the CpG-rich enhancer (which is called conserved non-coding sequence 2, CNS2) within the first intron of the Foxp3 locus (Floess et al. 2007, e38; Huehn and Beyer 2015, 10; Huehn et al. 2009, 9; Kim and Leonard 2007, 1543). Demethylation of CpG motifs enables critical transcription factors involved in the Foxp3 expression like Runt-related transcription factor 1 (Runx1) and Foxp3 itself to bind to the CNS2 region and keep the transcription of Foxp3 active in the progeny of dividing Tregs (Feng et al. 2014, 749; Li et al. 2014, 734; Zheng et al. 2010, 808). Accordingly, CNS2 is demethylated in tTregs and pTregs expressing Foxp3 stably but fully methylated in iTregs expressing Foxp3 transiently (Miyao et al. 2012, 262; Polansky et al. 2008, 1654) (Fig. 3).

Vitamin C Accelerates CNS2 Demethylation in iTregs in a Tet-Dependent Manner

Demethylation of CpG motifs in the CNS2 enhancer (referred to as CNS2 demethylation hereafter) occurs gradually during the thymic development stages, through an active DNA demethylation process (Toker et al. 2013, 3180). Recently, it was also reported that Tet1 and Tet2 double-deficient Tregs lost 5hmC on the Foxp3 regulatory elements including CNS2 enhancer, leading to a reduced Treg population and systemic autoimmunity (Yang et al. 2015b, 251). These studies clearly showed the role of Tet proteins in CNS2 demethylation in tTregs. However, the question of how iTregs expressing Tet proteins exhibit CNS2 methylation still remained unsolved.

Vitamin C is known to be not only a general antioxidant but also a cofactor for a large enzyme family known as the iron- and 2-oxoglutarate-dependent dioxygenases (Young et al. 2015, 545). A typical example is the collagen prolyl-4-hydroxylase in collagen maturation and scurvy (Gorres and Raines 2010, 106). In the absence of

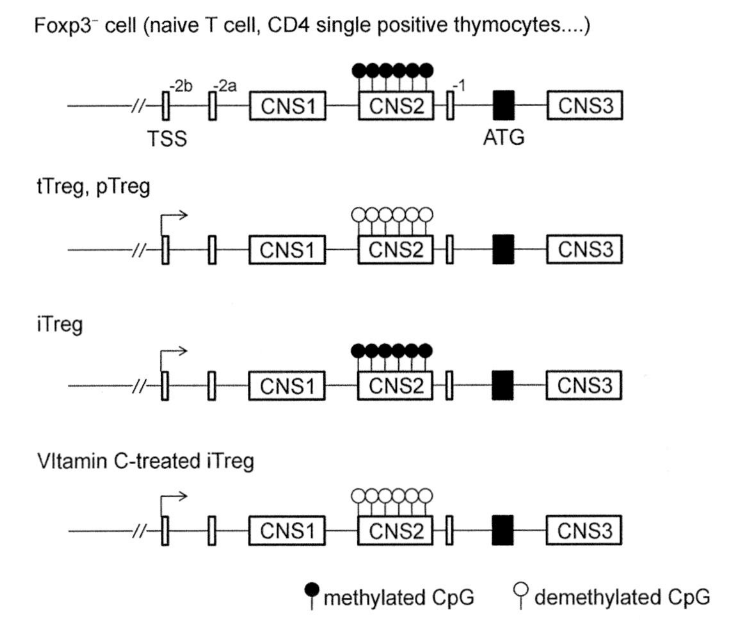

Fig. 3 Methylation pattern of CpG motifs in Foxp3 CNS2 region. A schematic picture of the first part of the mouse Foxp3 locus. *Small white boxes* (−2b, −2a, and −1) indicate untranslated exons. Three CNS elements are also shown. First CpG motif is located 4275 bases downstream of the TSS (transcription start site, +1)

vitamin C, the initial hydroxylation catalyzed by the collagen prolyl-4-hydroxylase proceeds albeit less efficiently; however, the catalytically inactive oxidized iron species accumulates soon and lowers the activity of collagen prolyl-4-hydroxylase, leading to an incomplete hydroxylation of residues in collagen (Gorres and Raines 2010, 106; Kuiper and Vissers 2014, 359). These studies reveal that vitamin C plays a role in reducing inactive iron and maintaining continued enzyme cycling. Since Tet enzymes also belong to an iron- and 2-oxoglutarate-dependent dioxygenases family (Ito et al. 2010, 1129; Tahiliani et al. 2009, 930), it is tempting to speculate that vitamin C might be involved in active CNS2 demethylation in Tregs by enhancing the activity of Tet enzymes. Indeed, vitamin C was reported to promote DNA demethylation of Foxp3 CNS2 region in iTregs (Sasidharan Nair et al. 2016, 2119; Yue et al. 2016, 377) (Fig. 3). Interestingly, vitamin C treatment facilitated the conversion of 5mC into 5hmC, and subsequently the methyl group of the CpG motifs was erased rapidly from the Foxp3 CNS2 enhancer. In other words, although the hydroxylation reaction of 5mC into 5hmC could be initiated in both iTregs, further oxidation reaction was not completed (therefore, oxidized cytosine like 5hmC remained) in conventional iTregs as mentioned in the case of collagen hydroxylation and scurvy (Fig. 4).

Vitamin C induces CNS2 demethylation in a Tet-dependent manner. When Tet2-deficient naïve $CD4^+$ cells were cultured under iTreg conditions (anti-CD3/anti-CD28

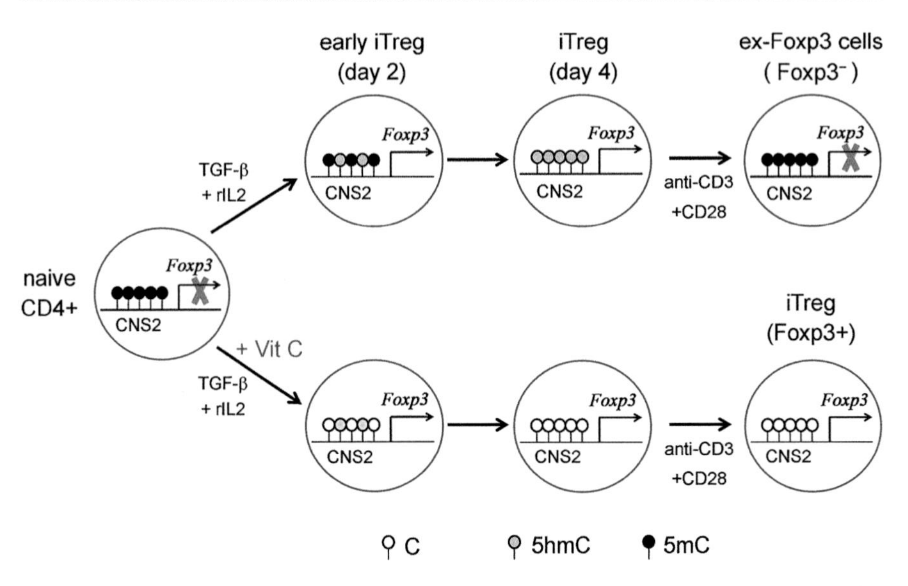

Fig. 4 CNS2 demethylation was accelerated by vitamin C in iTregs. The majority of naïve CD4$^+$ T cells cultured under iTreg conditions (in the presence or absence of vitamin C) start to express Foxp3 in 2 days, and the methylation status was checked on day 0 (naïve CD4$^+$ T cells), day 2 (early iTreg), and day 4 (late iTreg). While 5hmC was gradually enriched in the iTregs which were not treated with vitamin C, it was enriched and then disappeared abruptly in the vitamin C-treated iTregs. These findings indicate that vitamin C helps to complete the active demethylation of CNS2 more quickly, which otherwise would proceed slowly and inefficiently

antibodies plus polarizing cytokines like interleukin-2 (IL-2) and TGF-β) in the presence of low doses (5–10 μg/ml, the physiological level of vitamin C in serum) of vitamin C, CpG motifs in CNS2 remained methylated (WT iTregs were demethylated). However, a higher dose (100 μg/ml) of vitamin C induced CNS2 demethylation even in Tet2-deficient iTregs, implying that other Tet proteins like Tet1 and Tet3 might be substituted for Tet2 in vitamin C-rich environments. Indeed, CpG motifs in CNS2 of Tet2/Tet3 double-deficient iTregs did not become demethylated even after the high dose of vitamin C treatment (Fig. 5). Furthermore, iTregs overexpressing Tet2 showed fully demethylated CNS2 without vitamin C treatment and maintained Foxp3 expression even without TGF-β. Altogether, these findings suggest that vitamin C promotes CNS2 demethylation in a Tet2-/Tet3-dependent fashion, but could be dispensable in cells with a high level of Tet2 like tTregs (Toker et al. 2013, 3180).

Foxp3 Stability in iTregs Treated with Vitamin C

Although it is well known that CNS2 demethylation is required for maintaining the expression of Foxp3 in tTregs, is the expression of Foxp3 maintained stably in iTregs treated with vitamin C and having the demethylated CNS2 enhancer? Surprisingly,

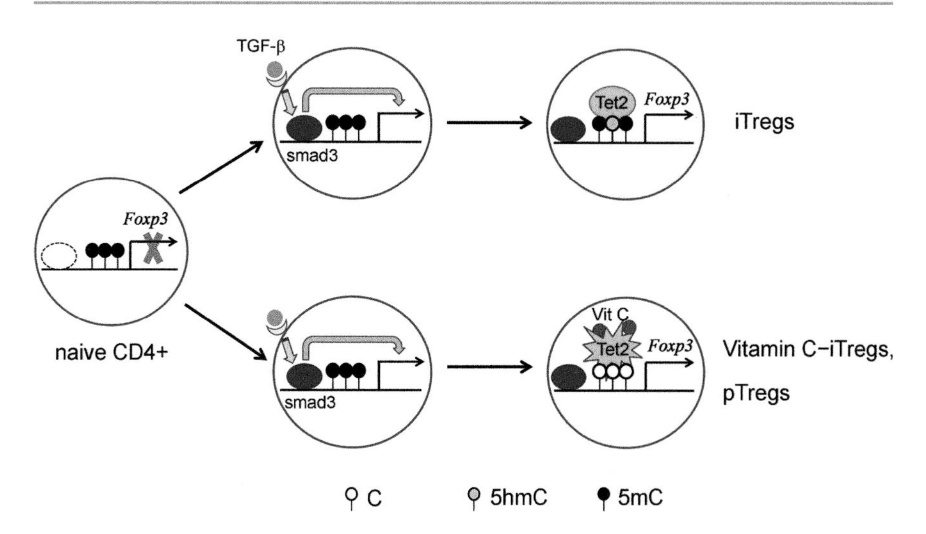

Fig. 5 Schematics of Foxp3 expression and CNS2 demethylation in iTregs and pTregs. Vitamin C promotes Tet-mediated active DNA demethylation of *Foxp3* CNS2 region in iTregs and pTregs

iTregs generated in the presence of vitamin C expressed Foxp3 stably and did not lose it even after TCR stimulation in vitro and in vivo. When vitamin C-treated iTregs were stimulated in vitro with anti-CD3/anti-CD28 antibodies (which mimic TCR and costimulatory signals) plus IL-2 (without vitamin C and TGF-β) or in vivo with the cognate antigen plus complete Freund's adjuvant, Foxp3 expression was maintained stably even in proliferating cells (Fig. 4). Furthermore, the suppressive activity of vitamin C-treated iTregs was superior to that of iTregs.

Recently it was reported that the numbers of genes whose expression was significantly altered by the treatment of vitamin C were only about 200 genes in embryonic stem cells (Blaschke et al. 2013, 222), implying that DNA demethylation in the regulatory elements per se induced by vitamin C might not be sufficient to change the gene expression, probably because of the availability of transcription factors and other required epigenetic processes. Consistent with the above findings, a small number of (approximately 50) genes were up- or downregulated by more than twofold in vitamin C-treated iTregs, compared with iTregs, and only 5 genes (CDKN1A, C3, TBX21, FCER1G, LAG3) were related to "positive regulation of immune system process," suggesting that vitamin C does not alter the gene expression profiles and related immune functions in iTregs (Sasidharan Nair et al. 2016, 2119).

The Roles of Vitamin C and Vitamin C Transporters in CNS2 Demethylation of pTregs and tTregs

Vitamin C has been known to enter cells through a sodium-dependent active vitamin C transporter. Two isoforms of the transporters were identified a couple of decades ago and called sodium-dependent vitamin C transporter 1 and 2 (SVCT1/2,

SLC23A1/2). Several chemical inhibitors for SVCT have also been described, and sulfinpyrazone is one of the well-established inhibitors of SVCT (May 2011, 1793). Interestingly, consistent with the previous report that the effect of vitamin C on 5hmC was prevented in fibroblast cells by the treatment of sulfinpyrazone (Dickson et al. 2013, 522), sulfinpyrazone reduced the effect of vitamin C on CNS2 demethylation during the generation of iTregs, suggesting that the uptake of vitamin C through SVCT is critical in CNS2 demethylation of iTregs. Given the developmental similarity between iTregs and pTreg, it is not quite surprising that vitamin C and Tet play roles in the CNS2 demethylation of pTregs as well. Indeed, while pTregs are generated in vivo, the treatment of sulfinpyrazone dramatically prevents CNS2 demethylation in pTregs. More interestingly, the effects of sulfinpyrazone were much greater in Tet2-deficient cells. Sulfinpyrazone restricted not only CNS2 demethylation but also the pTreg development in Tet2-deficient T cells. Altogether, these findings show that the cooperation of Tet2 and vitamin C is indispensable for CNS2 demethylation and Foxp3 expression in both iTregs and pTregs (Fig. 5, Sasidharan Nair et al. 2016, 2119).

Exceptionally, tTregs are resistant to sulfinpyrazone treatment. Why is vitamin C (or the entry of vitamin C) dispensable for the CNS2 demethylation in tTregs? Since CNS2 of Tet2-deficient tTregs treated with sulfinpyrazone remains methylated, a lack of sensitivity of tTregs to vitamin C seems to be caused by the high expression levels of Tet genes (Toker et al. 2013, 3180), which is also supported by the finding that overexpression of Tet2 rendered the CNS2 of iTregs fully demethylated without vitamin C. These results raise a question whether Tet2-deficient mice treated with sulfinpyrazone developed the inflammatory diseases like CNS2-deficient or Tet1/Tet2 double-deficient mice. Unexpectedly, autoimmune diseases containing tissue inflammations and immune cell activations were not developed spontaneously, indicating that Tet1 and Tet2 might contribute to the developments and functions of Tregs in a vitamin C-independent manner or vitamin C can be involved in T cell activation (as discussed below).

Roles of Vitamin C in Th17 Differentiation

T helper type 17 (Th17) cell is an important cellular mediator in diverse immune responses (Korn et al. 2009, 485) and can be generated from naïve $CD4^+$ T cells with stimulations via CD3/CD28 and polarizing cytokines like IL-6 and TGF-β (Mangan et al. 2006, 231). CD3/CD28 stimulations open up and make the target gene chromatin structures permissive to IL-6-associated transcription factor, signal transducer, and activator of transcription 3 (STAT3). Then, STAT3 acts together with various modifiers to promote the expression of retinoic acid receptor-related orphan receptor gamma-t (RORγt) (Durant et al. 2010, 605; Ivanov et al. 2006, 1121), leading to the lineage specification of Th17 cells. Th17 cells produce a group of proinflammatory cytokines including IL-17, which mediate diverse autoimmune and inflammatory diseases (Patel and Kuchroo 2015, 1040). Here, we are going to discuss the role of vitamin C in Th17 differentiation and IL-17 expression briefly.

When naïve $CD4^+$ T cells were cultured under Th17 conditions in the presence of vitamin C, the generation of IL-17-expressing cells was increased significantly. Since CpG motifs in IL17 promoter are less methylated in Th17 cells (Yang et al. 2015a, 1537) and 5hmC and Tet2 play essential roles in Th17 cells (Ichiyama et al. 2015, 613), it is tempting to hypothesize that vitamin C might contribute to IL-17 expression as a cofactor of Tet enzymes. However, the expression levels of IL-17 and the methylation status of CpG motifs in IL17 promoter were not altered significantly by Tet proteins or vitamin C. Instead, vitamin C treatment reduced the trimethylation of histone H3 lysine 9 (H3K9me3) in the IL17 promoter and enhancer in a Jumonji domain-containing proteins 2 (Jmjd2) in a dependent fashion (Song et al. 2016). These results suggest that vitamin C can affect the expression of IL-17 through the modulation of histone demethylase activity.

In conclusion vitamin C looks an intriguing nutrient which can regulate the epigenetic pattern and be used clinically. For example, iTregs treated with vitamin C could be used for various inflammatory diseases including transplantation rejection as a substitute for tTregs which are being tested in clinical trials. These findings also indicate that nutritional factors can bring about changes in immune homeostasis through epigenetic mechanisms.

Dictionary of Terms

- **Foxp3** – Foxp3 is a transcription factor which specifies the phenotypes and regulates the functions of Tregs by promoting and repressing the transcription of key genes in Tregs. Although it can be expressed transiently in activated T cells, Foxp3 is widely used as a specific marker of Tregs.
- **H3K9me3** – H3K9me3 (trimethylation of histone 3 lysine 9) is a histone modification related with heterochromatin formation and transcriptional silencing. This modification can be removed by histone lysine demethylase enzymes.
- **Iron- and 2-oxoglutarate-dependent dioxygenases** – They indicate a big enzyme family whose activities depend on key nutrients including oxygen, iron, 2-oxoglutarate (an intermediate in TCA cycle), and, in some cases, a reducing agent like vitamin C. They can be divided by their target molecules such as DNA demethylases (Tet family), histone demethylases (Jmjd or KDM family), and prolyl hydroxylase (PHD proteins).
- **Naïve T cell** – Naïve T cell is a mature T cell that has finished the developmental process within the thymus, comes into periphery, and, unlike memory T cells, has not encountered its cognate antigen yet. Under appropriate condition, naïve T cells can be differentiated into various types of effector T cells like Th17 cells.
- **T helper type 17 (Th17) cell** – Th17 cell is a subset of T helper cells defined by the production of proinflammatory cytokine, IL17. Th17 cells play an essential role in clearing pathogens at mucosal surfaces, but they are also implicated in autoimmune and inflammatory disorders. RORγt functions as a master transcription factor of Th17 cell differentiation and IL17 expression.

- **TCR** – TCR (T cell receptor) is a molecule found on the surface membrane of T cells and is responsible for recognizing antigens presented as peptides bound to major histocompatibility complex (MHC) molecules. When the TCR engages with its cognate antigen peptide and MHC complex, very complicated signal transduction pathways are activated, leading to T cell proliferation and differentiation.
- **Transcription factor** – Transcription factor is a sequence-specific DNA binding protein whose activity is to help initiate or inhibit gene transcription. One of the features of transcription factor is that it contains at least one DNA binding domain which binds to corresponding sequence of DNA in promoter or enhancer. The DNA sequence that the transcription factor binds is usually located close to the gene that the transcription factor regulates.

Key Facts

Key facts of self-/immune tolerance

1. One of the features in the immune system is the diversity of lymphocyte receptors like T cell receptors (TCR) and B cell receptors (BCR), which helps protect our body from rapidly evolving pathogens.
2. The diversity of lymphocyte receptors is generated during the developmental periods through the random recombination process.
3. The above strategy inevitably produces lymphocytes having receptors which recognize and attack self-tissues, which are called self-reactive lymphocytes.
4. The self-reactive lymphocytes need to be restrained properly.
5. "Self-/immune tolerance" means the state that the immune systems do not respond to self-tissues, which is usually accomplished by actively repressing (or deleting) self-reactive lymphocytes.
6. Excessive and potentially harmful immune responses can be induced by normal lymphocytes as well as self-reactive lymphocytes unless well controlled. The phenomenon that the immune system itself regulates excessive immune responses by using various methods is called "immune regulation."
7. Foxp3^{+} Treg is an essential cellular component in self-tolerance and immune regulation.

Summary Points

- Demethylation of CpG motifs in Foxp3 CNS2 enhancer is indispensable for the stable expression of Foxp3 in Tregs.
- Vitamin C induces CNS2 demethylation in iTregs and pTregs in a Tet2-/Tet3-dependent manner.
- Vitamin C is essential for CNS2 demethylation in tTregs when Tet is downregulated.

- Vitamin C was required for the CNS2 demethylation mediated by Tet proteins, which was essential for Foxp3 expression.
- Vitamin C also induces the upregulation of IL17 through Jmjd2 histone demethylase enzymes.
- Nutritional factors like vitamin C contribute to immune homeostasis through epigenetic mechanisms.

References

Allis CD, Jenuwein T (2016) The molecular hallmarks of epigenetic control. Nat Rev Genet 17:487–500

Blaschke K, Ebata KT, Karimi MM, Zepeda-Martinez JA, Goyal P, Mahapatra S, Tam A, Laird DJ, Hirst M, Rao A et al (2013) Vitamin C induces Tet-dependent DNA demethylation and a blastocyst-like state in ES cells. Nature 500:222–226

Cheng X (1995) Structure and function of DNA methyltransferases. Annu Rev Biophys Biomol Struct 24:293–318

d'Hennezel E, Ben-Shoshan M, Ochs HD, Torgerson TR, Russell LJ, Lejtenyi C, Noya FJ, Jabado N, Mazer B, Piccirillo CA (2009) FOXP3 forkhead domain mutation and regulatory T cells in the IPEX syndrome. N Engl J Med 361:1710–1713

Dickson KM, Gustafson CB, Young JI, Zuchner S, Wang G (2013) Ascorbate-induced generation of 5-hydroxymethylcytosine is unaffected by varying levels of iron and 2-oxoglutarate. Biochem Biophys Res Commun 439:522–527

Durant L, Watford WT, Ramos HL, Laurence A, Vahedi G, Wei L, Takahashi H, Sun HW, Kanno Y, Powrie F et al (2010) Diverse targets of the transcription factor STAT3 contribute to T cell pathogenicity and homeostasis. Immunity 32:605–615

Feng Y, Arvey A, Chinen T, van der Veeken J, Gasteiger G, Rudensky AY (2014) Control of the inheritance of regulatory T cell identity by a cis element in the Foxp3 locus. Cell 158:749–763

Feuerer M, Hill JA, Kretschmer K, von Boehmer H, Mathis D, Benoist C (2010) Genomic definition of multiple ex vivo regulatory T cell subphenotypes. Proc Natl Acad Sci USA 107:5919–5924

Floess S, Freyer J, Siewert C, Baron U, Olek S, Polansky J, Schlawe K, Chang HD, Bopp T, Schmitt E et al (2007) Epigenetic control of the foxp3 locus in regulatory T cells. PLoS Biol 5:e38

Gorres KL, Raines RT (2010) Prolyl 4-hydroxylase. Crit Rev Biochem Mol Biol 45:106–124

Hori S (2011) Regulatory T cell plasticity: beyond the controversies. Trends Immunol 32:295–300

Huehn J, Beyer M (2015) Epigenetic and transcriptional control of Foxp3+ regulatory T cells. Semin Immunol 27:10–18

Huehn J, Polansky JK, Hamann A (2009) Epigenetic control of FOXP3 expression: the key to a stable regulatory T-cell lineage? Nat Rev Immunol 9:83–89

Ichiyama K, Chen T, Wang X, Yan X, Kim BS, Tanaka S, Ndiaye-Lobry D, Deng Y, Zou Y, Zheng P et al (2015) The methylcytosine dioxygenase Tet2 promotes DNA demethylation and activation of cytokine gene expression in T cells. Immunity 42:613–626

Ito S, D'Alessio AC, Taranova OV, Hong K, Sowers LC, Zhang Y (2010) Role of Tet proteins in 5mC to 5hmC conversion, ES-cell self-renewal and inner cell mass specification. Nature 466:1129–1133

Ivanov II, McKenzie BS, Zhou L, Tadokoro CE, Lepelley A, Lafaille JJ, Cua DJ, Littman DR (2006) The orphan nuclear receptor ROR gammat directs the differentiation program of proinflammatory IL-17+ T helper cells. Cell 126:1121–1133

Jones PA, Liang G (2009) Rethinking how DNA methylation patterns are maintained. Nat Rev Genet 10:805–811

Kim HP, Leonard WJ (2007) CREB/ATF-dependent T cell receptor-induced FoxP3 gene expression: a role for DNA methylation. J Exp Med 204:1543–1551

Komatsu N, Okamoto K, Sawa S, Nakashima T, Oh-hora M, Kodama T, Tanaka S, Bluestone JA, Takayanagi H (2014) Pathogenic conversion of Foxp3+ T cells into TH17 cells in autoimmune arthritis. Nat Med 20:62–68

Korn T, Bettelli E, Oukka M, Kuchroo VK (2009) IL-17 and Th17 cells. Annu Rev Immunol 27:485–517

Kuiper C, Vissers MC (2014) Ascorbate as a co-factor for fe- and 2-oxoglutarate dependent dioxygenases: physiological activity in tumor growth and progression. Front Oncol 4:359

Li X, Liang Y, LeBlanc M, Benner C, Zheng Y (2014) Function of a Foxp3 cis-element in protecting regulatory T cell identity. Cell 158:734–748

Mangan PR, Harrington LE, O'Quinn DB, Helms WS, Bullard DC, Elson CO, Hatton RD, Wahl SM, Schoeb TR, Weaver CT (2006) Transforming growth factor-beta induces development of the T(H)17 lineage. Nature 441:231–234

May JM (2011) The SLC23 family of ascorbate transporters: ensuring that you get and keep your daily dose of vitamin C. Br J Pharmacol 164:1793–1801

Messerschmidt DM, Knowles BB, Solter D (2014) DNA methylation dynamics during epigenetic reprogramming in the germline and preimplantation embryos. Genes Dev 28:812–828

Miyao T, Floess S, Setoguchi R, Luche H, Fehling HJ, Waldmann H, Huehn J, Hori S (2012) Plasticity of Foxp3(+) T cells reflects promiscuous Foxp3 expression in conventional T cells but not reprogramming of regulatory T cells. Immunity 36:262–275

Ohkura N, Hamaguchi M, Morikawa H, Sugimura K, Tanaka A, Ito Y, Osaki M, Tanaka Y, Yamashita R, Nakano N et al (2012) T cell receptor stimulation-induced epigenetic changes and Foxp3 expression are independent and complementary events required for Treg cell development. Immunity 37:785–799

Pastor WA, Aravind L, Rao A (2013) TETonic shift: biological roles of TET proteins in DNA demethylation and transcription. Nat Rev Mol Cell Biol 14:341–356

Patel DD, Kuchroo VK (2015) Th17 cell pathway in human immunity: lessons from genetics and therapeutic interventions. Immunity 43:1040–1051

Polansky JK, Kretschmer K, Freyer J, Floess S, Garbe A, Baron U, Olek S, Hamann A, von Boehmer H, Huehn J (2008) DNA methylation controls Foxp3 gene expression. Eur J Immunol 38:1654–1663

Reik W, Dean W, Walter J (2001) Epigenetic reprogramming in mammalian development. Science 293:1089–1093

Rubtsov YP, Niec RE, Josefowicz S, Li L, Darce J, Mathis D, Benoist C, Rudensky AY (2010) Stability of the regulatory T cell lineage in vivo. Science 329:1667–1671

Sakaguchi S, Powrie F (2007) Emerging challenges in regulatory T cell function and biology. Science 317:627–629

Sasidharan Nair V, Song MH, Oh KI (2016) Vitamin C facilitates demethylation of the Foxp3 enhancer in a Tet-dependent manner. J Immunol 196:2119–2131

Song MH, Nair VS, Oh KI (2016) Vitamin C enhances the expression of IL17 in a Jmjd2-dependent manner. BMB Rep 50:49–54

Tahiliani M, Koh KP, Shen Y, Pastor WA, Bandukwala H, Brudno Y, Agarwal S, Iyer LM, Liu DR, Aravind L et al (2009) Conversion of 5-methylcytosine to 5-hydroxymethylcytosine in mammalian DNA by MLL partner TET1. Science 324:930–935

Toker A, Engelbert D, Garg G, Polansky JK, Floess S, Miyao T, Baron U, Duber S, Geffers R, Giehr P et al (2013) Active demethylation of the Foxp3 locus leads to the generation of stable regulatory T cells within the thymus. J Immunol 190:3180–3188

Yang BH, Floess S, Hagemann S, Deyneko IV, Groebe L, Pezoldt J, Sparwasser T, Lochner M, Huehn J (2015a) Development of a unique epigenetic signature during in vivo Th17 differentiation. Nucleic Acids Res 43:1537–1548

Yang R, Qu C, Zhou Y, Konkel JE, Shi S, Liu Y, Chen C, Liu S, Liu D, Chen Y et al (2015b) Hydrogen sulfide promotes Tet1- and Tet2-mediated Foxp3 demethylation to drive regulatory T cell differentiation and maintain immune homeostasis. Immunity 43:251–263

Young JI, Zuchner S, Wang G (2015) Regulation of the epigenome by vitamin C. Annu Rev Nutr 35:545–564

Yue X, Trifari S, Aijo T, Tsagaratou A, Pastor WA, Zepeda-Martinez JA, Lio CW, Li X, Huang Y, Vijayanand P et al (2016) Control of Foxp3 stability through modulation of TET activity. J Exp Med 213:377–397

Zheng Y, Josefowicz S, Chaudhry A, Peng XP, Forbush K, Rudensky AY (2010) Role of conserved non-coding DNA elements in the Foxp3 gene in regulatory T-cell fate. Nature 463:808–812

Zhou X, Bailey-Bucktrout SL, Jeker LT, Penaranda C, Martinez-Llordella M, Ashby M, Nakayama M, Rosenthal W, Bluestone JA (2009) Instability of the transcription factor Foxp3 leads to the generation of pathogenic memory T cells in vivo. Nat Immunol 10:1000–1007

Cobalamin, Microbiota and Epigenetics 89

Joan Jory

Contents

Abstract

Functional cobalamin (B12) status and assessment are inextricably intertwined with the human microbiome. Small bowel bacterial overgrowth can both cause and result from gastritis and alter dietary cobalamin absorption. Some bacterial species may produce human-inaccessible cobalamin corrinoids and may create competition for human-accessible cobalamin. Increased human-inaccessible corrinoids from bacterial production may raise the total corrinoid level assessed by the serum total cobalamin, limiting diagnostic utility and masking a deficiency of human-accessible cobalamin. Anaerobic bacteria may reverse the propionic to succinic acid pathway, converting methylmalonic acid back to propionic acid to release CO2; this could raise propionic acid and lower methylmalonic acid levels, limiting its diagnostic utility. Cobalamin deficiency limits enzymatic conversion

J. Jory (✉)
Guelph, ON, Canada
e-mail: joanjory.2012@gmail.ca; jjory@uoguelph.ca

V. B. Patel, V. R. Preedy (eds.), *Handbook of Nutrition, Diet, and Epigenetics*,
https://doi.org/10.1007/978-3-319-55530-0_47

of homocysteine to methionine and increases homocysteine levels. Increased homocysteine can be reduced by diversion into the transsulfuration pathway, limiting the diagnostic power of this metabolite. Finally, in the delicate balance between folate and cobalamin which regulates DNA synthesis, excess synthetic folate from public health policies can combine with bacterial folate production to mask the macrocytic anemia of cobalamin deficiency.

Small bowel bacterial overgrowth can increase propionic acid production and reduce cobalamin bioavailability. Both propionic acid administration and cobalamin deficiency can alter brain fatty acid levels and brain function and cause autistic symptomology. Essential fatty acid ratios can modify gut bacterial species which can, in turn, modify fatty acid composition and inflammation. Omega-3 supplementation can reverse many of the symptoms of propionic acid neurotoxicity. Cobalamin supplementation can raise omega-3 fatty acid levels in the brain and can improve autism symptomology. Therefore, there are strong epigenetic interrelationships among cobalamin and its enzymatic activity, propionic acid, essential fatty acids, folate, and the human bacterial microbiome.

Keywords
Autism spectrum disorder · B12 · Brain-derived neurotrophic factor · Cobalamin · Dysbiosis · Methylation · Methylmalonic acid · Propionic acid · Micronucleated lymphocytes · Polyunsaturated fatty acid · Short-chain fatty acids · Small bowel bacterial overgrowth

List of Abbreviations

AA	Arachidonic acid
ADHD	Attention deficit hyperactivity disorder
ASD	Autism spectrum disorder; B12 = cobalamin
BDNF	Brain-derived neurotrophic factor
DHA	Docosahexaenoic acid
EPA	Eicosapentaenoic acid
IF	Intrinsic factor
MCM	Methylmalonyl-CoA mutase
MDA	Malondialdehyde
MetH	Methionine synthase
MMA	Methylmalonic acid
MNL	Micronucleated lymphocytes MTHF = methyltetrahydrofolate
PPA	Propionic acid
PUFA	Polyunsaturated fatty acid
SAH	S-adenosylhomocysteine
SAM-e	S-adenosylmethionine
SCFA	Short-chain fatty acids
SIBO	Small bowel bacterial overgrowth
THF	Tetrahydrofolate

Introduction

Vitamin B12 (cobalamin) is essential and cannot be synthesized by humans. Cobalamin refers to a corrin ring-containing cobalt complex belonging to a larger group of cobalamin corrinoids (Krautler 2012). The richest dietary cobalamin sources include red meat, liver, and organ meat (Gille and Schmid 2015; Williams 2007). In industrialized countries, red and organ meat intake has declined amid increasing vegetarian and vegan trends (Clonan et al. 2016; Daniel et al. 2011). Generally then, there has been an overall decline in cobalamin intakes among industrialized populations, which can have negative multigenerational impacts. Low cobalamin status increases neural tube risk (Molloy et al. 2009). Cobalamin deficiency is associated with a higher rate of miscarriages (Bennett 2001). Breastfed babies of vegan or vegetarian mothers can exhibit profound neurological delays and evidence of under-myelination on MRI (Lovblad et al. 1997; Kocaoglu et al. 2014; Guez et al. 2012).

Cobalamin is a component of key enzymes that influence both the homocysteine-methionine pathway and the propionic-succinic acid pathway, potentially altering DNA and cellular energy production. Equally, propionic acid-producing small bowel intestinal bacteria can modify cobalamin absorption and host cobalamin availability as well as influence fatty acid-mediated inflammatory responses and brain fatty acids. Cobalamin, therefore, demonstrates a unique interplay with the gut microbiome, with methylation cofactors, and with short-chain (propionic acid) and polyunsaturated fatty acids (PUFA) in an epigenetic manner.

Cobalamin Absorption and Metabolism: Interrelationships with Gut Microbiota and PPA

Once cobalamin is consumed in the diet, it must undergo a complex process of absorption (Fig. 1), including hydrochloric acid-mediated isolation from protein in the stomach, bonding with parietal cell-secreted haptocorrin-binding proteins and intrinsic factor in the duodenum, receptor-mediated absorption of the intrinsic factor (IF)-cobalamin complex in the terminal ileum, and transformation via the binding protein transcobalamin into the biologically active holotranscobalamin which is delivered through the blood to all the cells in the body (Schjonsby 1989).

Only 1% of cobalamin intake can be passively absorbed; 99% must be IF facilitated, and IF is produced by the gastric parietal cells (Muyshondt and Schwartz 1964). Thus gastrointestinal (GI) health can have a profound effect on the bioavailability and absorption of dietary cobalamin intakes. At the first juncture, insufficiency of gastric acid can alter the ability to cleave cobalamin from dietary protein. Gastric hypochlorhydria can reduce parietal cell secretions (McKoll et al. 1998) and contribute to atrophic gastritis (Sipponen et al. 1996) and small bowel bacterial overgrowth (SIBO) (Belitsos et al. 1992). Paradoxically, hypochlorhydria can lead to

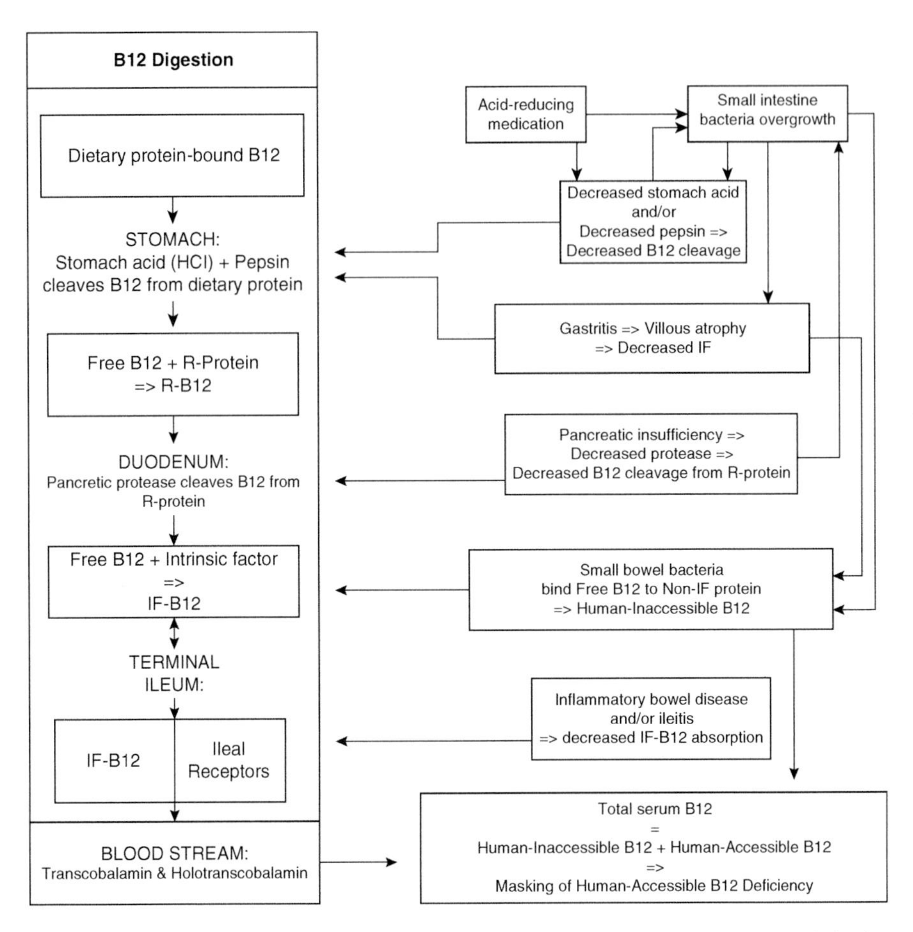

Fig. 1 B12 digestion and metabolism with associated pathway modifying factors. *B12* cobalamin, *HCl* hydrochloric acid, *IF* intrinsic factor, *IF-B12* intrinsic factor-bound B12, *R-protein* haptocorrin, *R-B12* R-protein-bound B12

heartburn and high acid symptoms (Bredenoord et al. 2006). Treatment for heartburn symptoms includes acid-reducing medications such as proton pump inhibitors and H2 antagonists. Use of acid-reducing medication is associated with cobalamin deficiency and increased methylmalonic acid (MMA)(Ruscin et al. 2002; Saltzman et al. 1994). Proton pump inhibitor (PPI) use is also associated with SIBO (Fujimori 2015; Lewis et al. 1996); in one study, 53% of patients on omeprazole demonstrated SIBO (Thorens et al. 1996). Acid-reducing medications are increasingly prescribed to pregnant women, children, and infants (Diav-Citrin et al. 2005; Smith et al. 2013; Van der Pol et al. 2011) and may have a multigenerational impact on the human microbiome.

At the next juncture in cobalamin digestion, alterations in GI epithelial health can influence the ability to produce and release IF, a rate-limiting step in absorption. The incidence of atrophic gastritis, celiac disease, non-gluten enteropathies, and inflammatory bowel disease is rising in industrialized countries. Gastritis is associated with villous atrophy, impaired IF production, and compromised cobalamin absorption (Lebwohl et al. 2015; Meyniel et al. 1981; Wood et al. 1964). SIBO is associated with villous atrophy (Lappinga et al. 2010) and may be reversible with antibiotic treatments (Haboubi et al. 1991). Patients with atrophic gastritis and bacterial overgrowth absorb significantly less protein-bound cobalamin, which is reversed by antibiotics (Suter et al. 1991).

At the third juncture of cobalamin absorption, exocrine pancreatic insufficiency may preferentially transfer the cobalamin to alternative proteins than IF, rendering the cobalamin functionally inaccessible to the human host (Marcoullis et al. 1980). Pancreatic insufficiency can also contribute to the development of SIBO (Therrien et al. 2016).

Thus, there is considerable overlap between conditions contributing to cobalamin malabsorption and to SIBO. Further, once established, large colonies of small bowel bacteria may have unique needs for cobalamin. Although some bacterial species can produce their own cobalamin (Morita et al. 2008; LeBlanc et al. 2013), excessive bacterial growth may set up an environment of competition for cobalamin between bacterial residents and the human host (Brandt et al. 1977; Degnan et al. 2014; Schjonsby, 1989). Some small bowel bacteria can selectively couple dietary cobalamin to alternative binding proteins, forming non-IF complexes which are inaccessible to the human host and deprive the host of this essential cobalamin source. At the same time, small bowel bacteria may themselves produce cobalamin corrinoids whose molecular structure does not allow them to be used by humans (Schjonsby 1989). A decrease in IF-bound cobalamin availability to the human host amid extensive bacterial production of human-inaccessible corrinoid forms may not only contribute to cobalamin deficiency development but also mask the diagnosis of cobalamin deficiency if only serum cobalamin is tested. Serum cobalamin measures only total corrinoids and is not able to distinguish between human-specific and bacteria-specific cobalamin corrinoids (Degnan et al. 2014). Overgrowth of small bowel bacteria may therefore contribute to a deceivingly normal or even elevated total cobalamin corrinoid status while obscuring human-specific cobalamin deficiency.

Small bowel bacterial overgrowth is associated with increased bacterial production of short-chain fatty acids (SCFA) such as butyrate, propionate, and acetate. Although these SCFAs may be largely anti-inflammatory at normal physiological levels (Rios-Covian et al. 2016), excess production of propionic acid (PPA) may have negative implications in SIBO, particularly in association with the neurodevelopmental disorder of autism spectrum disorder (ASD) (Frye et al. 2015). Elevated concentrations of several propionate-producing bacterial species have been isolated in stool samples of children with ASD, as have elevated levels of PPA (Finegold et al. 2010; Song et al. 2004; Wang et al. 2012). Though genetically mediated propionic acidemia does not usually elicit symptoms of ASD, cerebral infusions of PPA in the rodent model do (Al-Owain et al. 2013; McFabe et al. 2007). A multifactorial association between elevated PPA levels and the complex symptomology of autism has been evocatively formulated by McFabe et al. Pediatric

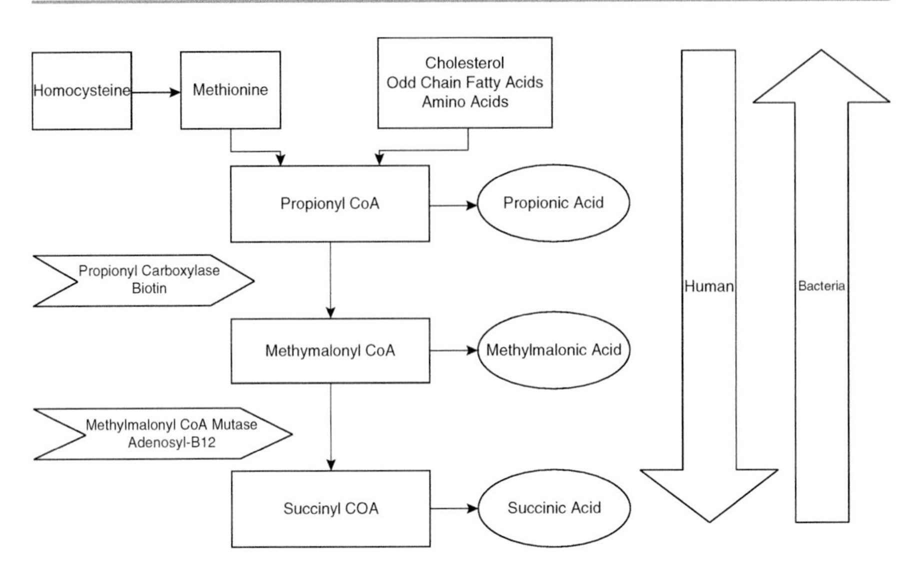

Fig. 2 Bidirectional propionic-succinic acid pathway. Bacterial reconversion of methylmalonyl-CoA to propionyl-CoA may mask potential methylmalonic acidemia of B12 deficiency. *B12* cobalamin, *CoA* coenzyme A

antibiotic treatment and high intakes of carbohydrates and PPA additive-rich foods have been identified as possible contributing causes to both small bowel bacteria overgrowth and elevated PPA concentrations in ASD (McFabe 2012).

In humans, propionic acid is a first-step product of cholesterol, odd-chain fatty acid, and select amino acid metabolism along the succinic acid pathway (Fig. 2), leading into the Krebs energy cycle. It is of note that methionine, one of the amino acid precursors for PPA production, is an end product of the cobalamin-dependent homocysteine methylation cycle via the cobalamin-dependent enzyme methionine synthase (MetH). In the second step, the propionyl-CoA is converted to methylmalonyl-CoA via the biotin-dependent enzyme propionyl carboxylase, followed by conversion to succinyl-CoA via the cobalamin-dependent enzyme methylmalonyl-CoA mutase (MCM) (Takahashi-Iniguez et al. 2012).

Thus, human-accessible cobalamin corrinoids are critically implicated in both the production and degradation of propionic acid. Methionine synthase and methylmalonyl-CoA mutase are the only two cobalamin-dependent enzymes in human metabolism. However, within the bacterial microbiome of the human gut, there are many cobalamin corrinoid-dependent enzymes, among which MCM and MetH are the most prominent (Degnan et al. 2014). In some bacteria, the conversion of PPA to succinic acid via cobalamin-specific MMA can also be a particularly useful bidirectional pathway (Fig. 2); anaerobic bacteroides can capitalize on the production of CO2 from the decarboxylation of MMA by running the pathway in reverse, thereby also increasing the pool of PPA (Fischbach and Sonnenburg 2011).

Cobalamin and PPA in Autism: Interrelationships with MMA and the Microbiome

Much recent attention has been given to a potential association between small bowel dysbiosis and elevated propionic acid in the pathophysiology of autism. However, the production of PPA as part of a cobalamin-dependent continuum involving MMA must also be taken into account, especially given the interrelatedness of human cobalamin species with the pathway substrates, products, and essential enzyme activities. MMA is emerging as an important metabolite biomarker of functional cobalamin deficiency since the bioactive corrinoid adenosylcobalamin is a rate-limiting ingredient in the MMA conversion to succinic acid. Thus, in many conditions, MMA levels may rise before serum cobalamin levels drop (Klee 2000). It is also of note that in cobalamin-deficient patients, antibiotic treatment can lower MMA levels without altering homocysteine levels, indicating that PPA produced by anaerobic gut bacteria may be a precursor to MMA in cobalamin-deficient patients (Lindenbaum et al. 1990). Equally, however, early work on organic acidemias raised the question of whether the presence of excess PPA could mask a concurrent elevation in MMA (Duran et al. 1973). Bacteria can produce PPA from non-SCFA precursors moving forward along the PPA to succinic acid pathway. However, anaerobic bacteria are also able to produce PPA from MMA moving backward along the PPA to succinic acid pathway, in order to increase CO_2 production (Fischbach and Sonnenburg 2011). This latter cobalamin-independent conversion of MMA to PPA, such as in established small intestinal bacterial overgrowth (SIBO), could conceivably reduce the hallmark elevations of MMA which are diagnostically important to the confirmation of cobalamin deficiency and, at the same time, contribute to the elevations in PPA theoretically linked to the pathophysiology of autism.

Symptoms of cobalamin deficiency resemble those of autism (Agrawal and Nathani 2009), and levels of methylcobalamin, adenosylcobalamin, and methionine are lower in the autistic brain (Zhang et al. 2016). Symptoms associated with PPA administration to rats also resemble autism (McFabe 2012). However, documentation of PPA and MMA levels in human autism research is limited. One study of 58 children with autism found unexpectedly lower stool PPA than among controls (Adams et al. 2011a), while stool PPA was significantly elevated among a group of 23 children with autism compared to controls (Wang et al. 2012). In a retracted Lancet study (Wakefield et al. 1998), urinary MMA levels were elevated among 8 children with autism and GI symptomology. By contrast, no significant elevations in plasma MMA were found among a group of 55 children with ASD; however, there were also no indications of significant GI symptomology or SIBO among this group (Adams et al. 2011b). In a case report of a young child with ASD and evidence of small bowel bacterial overgrowth, there was a high-normal MMA which was nonresponsive to oral cobalamin intervention (Fitzgerald et al. 2012). However, when subcutaneous cobalamin injections were initiated, bypassing the SIBO issue, there were dramatic improvements in behavior and development. Thus, children with autism and concurrent significant GI symptoms may have higher MMA levels than children with autism and minimal GI symptoms, potentially confirming that SIBO could contribute to elevated MMA levels. Further, in SIBO, bacteria may preferentially sequester supplemental cobalamin travelling through the GI tract and limit the effectiveness of oral cobalamin treatment. The reportedly dramatic impact of subcutaneous cobalamin treatment on behavior and development in an ASD child previously nonresponsive to oral supplementation would appear to confirm the preferential sequestering of oral cobalamin by small bowel bacteria and the need to bypass the gut for effective redress of functional cobalamin deficiency during SIBO.

Cobalamin and Epigenetics: Interrelationships with Folate and Gut Microbiota

Alterations in human-accessible cobalamin availability in conditions of small bowel bacterial overgrowth will have effects on other biochemical pathways, with potential epigenetic implications. Notably, the methylation of homocysteine to methionine is both cobalamin and folate dependent and directly impacts DNA synthesis. If there is insufficient cobalamin, or if the cobalamin is oxidized during oxidative stress, methyltetrahydrofolate (MTHF) will not be converted to tetrahydrofolate (THF) for recycling, and methionine synthase production (MetH) will drop with reduced conversion of homocysteine to methionine. This reduction in methionine synthesis will then impact the methylation of DNA via the S-adenosylmethionine (SAM-e) to S-adenosylhomocysteine (SAH) pathway (Chiang et al. 1996; Waterland). Changes in DNA methylation are principal epigenetic mediators. Rodent experiments supplementing methylation cofactors before and during pregnancy have demonstrated increased DNA methylation in both early and mid-gestation (Waterland 2006; Waterland et al. 2006). The increased DNA methylation in response to gestational supplementation does not appear to be multigenerational in nature (Waterland et al. 2007).

Targeted cobalamin and folate supplementation in humans has also been shown to alter DNA, as determined by the frequency of micronucleated lymphocytes (MNL). In both adolescent and adult males, the frequency of MNL demonstrated a significant negative relationship with serum cobalamin levels but not folate status, despite the absence of clinical cobalamin deficiency. Although the adolescents were each supplemented with 3.5 and 10 times the RDI for both folate and cobalamin, the magnitude of serum cobalamin increase was much smaller than that of the RBC folate increase. Further, the increases in folate status were not associated with the improvements in MNL frequency seen with improved cobalamin status. Overall, the greatest reduction in MNL frequency was achieved at post-supplementation cobalamin levels >300 pmol/L (Fenech et al. 1997, 1998). The differences in cobalamin and folate response to supplementation may reflect differences in barriers to absorption between dietary folate and cobalamin and underline the challenges in addressing the epigenetic consequences of cobalamin deficiency through gut-mediated fortification or supplementation.

Children with Down syndrome exhibit altered cobalamin- and folate-mediated homocysteine-to-methionine conversion, and a higher frequency of MNL, than do controls (Youness et al. 2016). Levels of micronucleated lymphocytes are also elevated among young mothers of children with Down syndrome, which may indicate a multigenerational effect (Coppede et al. 2016). Children with autism who also have altered homocysteine-methionine function may or may not have similar increased rates of micronucleated cells: among a small number of sibling pairs (6), indicators of DNA damage (including micronuclei frequency) in response to hydrogen peroxide challenge were higher among the children with autism, but failed to reach statistical significance. However, the children with autism did demonstrate higher rates of lymphoblast necrosis (Main et al. 2013).

In theory, this altered homocysteine-to-methionine conversion associated with cobalamin deficiency should result in increased homocysteine levels and provide a potentially useful diagnostic index of cobalamin status. However, homocysteine is not always elevated in cobalamin deficiency (Fig. 3); in some cases, extra homocysteine is diverted into the transsulfuration pathway to produce cysteine and glutathione (Stipanuk and Ueki 2011). In other cases, betaine (trimethylglycine) may upregulate the homocysteine-methionine pathway via betaine homocysteine methyltransferase (BHMT) and bypass limitations imposed by reduced cobalamin-dependent methionine synthase activity (Kim and Kim 2005).

There is a unique dance between cobalamin and folate availability (which impacts methylation reactions throughout the human body) and their related epigenetic effects. While some human gut bacteria can produce cobalamins which are inaccessible to the human host, other flora can synthesize folates which appear to be accessible to the host (Leblanc et al. 2013; Rossi et al. 2011). Rat studies have

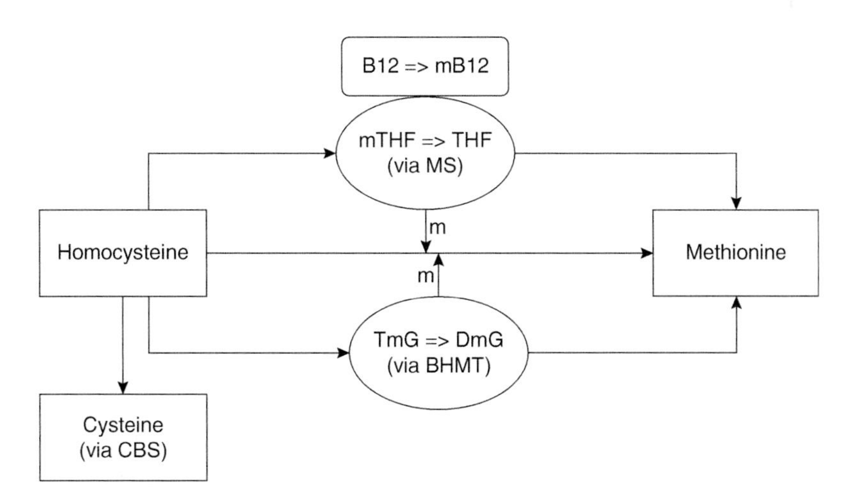

Fig. 3 Homocysteine metabolism and B12-independent pathways. *B12* cobalamin, *BHMT* betaine homocysteine methyl transferase, *CBS* cystathionine beta synthase, *DMG* dimethylglycine, *m* methyl group (CH_3), *MS* methionine synthase, *MTHF* methyltetrahydrofolate, *THF* tetrahydrofolate, *TMG* trimethylglycine

identified significantly higher fecal folate levels among Bifidobacterium models which correlate with serum folate levels and hemoglobin and mean cell volume status (Sugahara et al. 2015). The balance between bacteria variants in the gut may thus influence the balance between cobalamin and folate in human.

On a larger scale, relative ratios between dietary intakes of cobalamin and folate may also influence methylation pathways and DNA methylation. Since the late 1990s, numerous countries have elected to add synthetic folic acid to flour to reduce neural tube defects and to encourage the use of synthetic folic acid-containing supplements, particularly among women who are pregnant or of child-bearing age. Levels of total RBC folate and unmetabolized folic acid are elevated in the populations of several folate-fortified nations. Unmetabolized folic acid has also been detected in the cord blood of infants whose mothers did not receive maternal folate supplements (Obeid et al. 2010). By contrast, serum cobalamin levels are not elevated (McFarlane et al. 2011). Among Canadian women, cobalamin levels were lowest among adolescents and women of child-bearing age; lower cobalamin status was associated with higher homocysteine levels (McFarlane et al. 2011).

The biological significance of supraphysiological total folate and high levels of unmetabolized folate, particularly in the context of cobalamin- and folate-related polymorphisms, remains to be clarified. Supraphysiological folate may mask the macrocytic anemia of cobalamin deficiency, further complicating a complicated diagnosis (Wyckoff and Ganji 2007). There may also be subpopulations who are more sensitive than others to the accumulation of synthetic folate, as evidenced by differences between Canadian women with and without detectable unmetabolized folic acid levels following supplementation (Tam et al. 2012). However, an elevated ratio of folate to cobalamin has been associated with adverse neurological effects among the elderly (Morris et al. 2007) and with altered hippocampal microstructure and memory performance (Kobe et al. 2016) in adults. An increased folate-to-cobalamin ratio is also associated with embryonic delay and growth retardation in mice (Pickell et al. 2011) and with reduced neonatal growth anthropometrics in humans (Gadgil et al. 2014). Intake of synthetic folic acid supplements during pregnancy has been positively correlated with risk of autism in analyses of the Rochester (MN) Epidemiology Project and the Centers for Disease Control and Prevention pediatric dataset, respectively (Beard et al. 2011; DeSoto et al. 2012). High folate in the presence of genetic alterations to folate/homocysteine pathways may also increase the risk of Down syndrome (Coppede 2009).

Cobalamin and Brain Function: Interrelationships with PUFA, Microbiota, and PPA

The ratio of cobalamin to folate may also interact with the composition of polyunsaturated fatty acids, which are important methyl group acceptors, in the brain. Among older persons with mild cognitive impairment, cobalamin and folic acid supplementation appears to slow the rate of brain atrophy among patients with high baseline omega-3 fatty acids. However, there was no benefit from folate and cobalamin supplements among patients with lower omega-3 levels (Jerneren et al. 2015). The ratio of folic acid to cobalamin during pregnancy can have direct impacts on fetal brain docosahexaenoic acid (DHA) and arachidonic acid (AA) accretion. During pregnancy, maternal DHA levels are primary regulators of brain DHA in offspring, while fetal DHA status influences neurogenesis and neuron survival during pregnancy (Dhobale and Joshi 2012). However, in rat studies, excess folate supplementation of cobalamin-deficient mothers decreased DHA concentrations in both mothers and offspring. Omega-3 supplementation of cobalamin-deficient rats receiving high folate improved DHA levels in both mother and offspring, but AA levels were reduced. Rates of malondialdehyde (MDA), indicative of lipid peroxidation, were also elevated in B12-deficient rats and offspring receiving high folate (Roy et al. 2012). Further study identified alterations in the mRNA of brain-derived neurotrophic factors (BDNF) in offspring of rats fed with high-folate, cobalamin-deficient diets; in rats receiving prenatal supplements of DHA and eicosapentaenoic acid (EPA), abnormal mRNA of BDNF was normalized (Sable et al. 2014).

DHA levels are lower in children with autism (Brigandi et al. 2015) and attention deficit hyperactivity disorder (ADHD) (Parletta et al. 2016). DHA supplementation of children with autism failed to demonstrate significant positive effects in core behavioral symptoms (Voigt et al. 2014). However, interactions between DHA and methylation nutrients such as folate and cobalamin were not accounted for. It is possible that correction of potential folate-to-cobalamin imbalances may be required before a therapeutic response to DHA supplementation can be demonstrated in autism. Similar limitations to study protocols and therapeutic outcomes are evident in omega-3 supplementation research with ADHD (Gillies et al. 2012).

Gut microbiota can alter the impact of omega-3 and omega-6 fatty acid ratios in mice in vivo, and vice versa (Kaliannan et al. 2015; Wall et al. 2009). Mice fed an omega-6-enriched diet demonstrate elevated levels of gastrointestinal inflammatory biomarkers; this effect is not seen in genetically altered mice that intrinsically produce omega-3 fatty acids. Treatment with antibiotics eliminates the effect. Microbiome analyses indicate omega-3-producing mice have higher counts of anti-inflammatory bacterial species including bifidobacterium and lactobacillus strains; mice with high omega-6 and inflammatory biomarker levels demonstrate bacterial overgrowth with pro-inflammatory proteobacteria. Cohousing of the mouse types leads to fecal transfer of bacteria from the omega-3 mice, which increased intestinal zinc-dependent alkaline phosphatase levels and altered bacterial growth in the omega-6 mice toward anti-inflammatory species (Kaliannan et al. 2015).

In humans, fish oil supplementation in combination with breastfeeding cessation altered the gut bacterial species of infants, compared to sunflower oil supplements (Andersen et al. 2011). Among older adults, a high omega-6 diet altered gut flora and caused gut dysbiosis; this effect was reversed by fish oil supplementation (Ghosh et al. 2013). A further mouse study found that in utero and early life omega-3 supplementation beneficially altered not only gut microbiota but also depressive, social, and cognitive behaviors during later life (Robertson et al. 2016); however, some of the effects of DHA supplementation, and the subsequently altered gastric microbiome, appear to be gender specific and stronger among males (Davis et al.

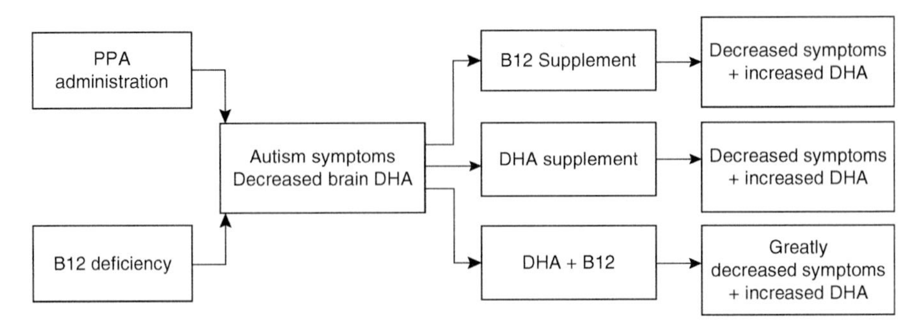

Fig. 4 Overlapping epigenetic impacts of PPA and B12 on brain DHA and symptoms of autism. *B12* cobalamin, *DHA* docosahexaenoic acid, *PPA* propionic acid

2017). Rates of autism spectrum disorder and ADHD are also higher among males (CDC).

Administration of gut dysbiosis-related PPA (Fig. 4) to rats decreased absolute levels of omega-3 fatty acids, AA, and fatty acid ratios, as well as induced autistic behaviors (El-Ansary and Al-Ayadhi 2014). However, omega-3 fatty acid supplementation conferred protection against the neurotoxic effects of PPA administration, suggesting the potential to modify the symptoms of the PPA-induced autism and gut dysbiosis model (El-Ansary et al. 2011). Like PPA administration, cobalamin deficiency also decreased DHA levels in the rat model and was associated with symptoms of autism (Kulkarni et al. 2011). However, maternal cobalamin supplementation alone increased DHA levels, BDNF, and cognition in rat offspring, thereby demonstrating power to modify brain fatty acids and neurodevelopment even in the absence of preexisting cobalamin deficiency. This effect was amplified by concomitant omega-3 supplementation (Rathod et al. 2014) and by multigenerational supplementation (Rathod et al. 2016), suggesting an important potential epigenetic impact of cobalamin supplementation on gut dysbiosis-mediated alterations in brain development and function.

Conclusions

Gut microbiota (Fig. 5) can influence fatty acid composition which, in turn, can alter gut microbiota. Elevated PPA is associated with gut dysbiosis, and gut dysbiosis is associated with altered cobalamin bioavailability. Both PPA administration and cobalamin deficiency can decrease essential fatty acid levels and cause symptoms of autism. Cobalamin supplementation can raise DHA levels and improve autistic symptoms, while omega-3 supplementation can improve the symptoms of PPA administration. If the neurotoxic effects of PPA associated with gut dysbiosis can be reversed by omega-3 supplementation, and if cobalamin supplementation can raise levels of brain omega-3 and neurotrophic growth factor, it is possible that the pathologies of cobalamin deficiency and PPA neurotoxicity are intricately

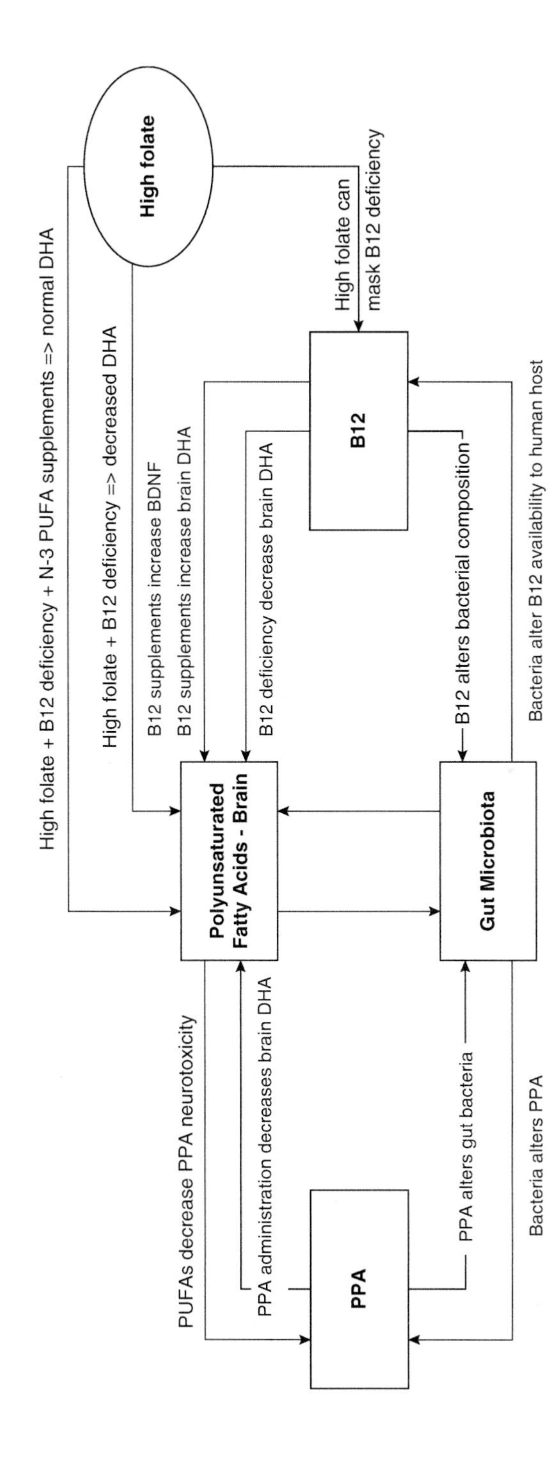

Fig. 5 Unifying hypothesis of cobalamin and microbiota in epigenetics. *B12* cobalamin, *BDNF* brain-derived neurotrophic factors, *DHA* docosahexaenoic acid, *N-3 PUFA* omega-3 polyunsaturated fatty acids, *PPA* propionic acid

interrelated with each other, with polyunsaturated fatty acids and with the gut microbiome in a strongly epigenetic manner. Within this theoretical framework, stepwise alterations of cobalamin, folate, and essential fatty acid levels in a PPA-gut dysbiosis model may be able to modify gut microbiota, PPA production, core methylation pathways, as well as brain fatty acid distributions, myelination patterns, and DNA integrity.

Dictionary of Terms

- **Autism** – a pediatric neurological disorder affecting communication, behavior, and cognition.
- **Cobalamin** – a cobalt-based vitamin and essential enzymatic cofactor for pathways affecting DNA and RNA.
- **Gut dysbiosis** – bacterial imbalance causing inflammation and nutrient malabsorption.
- **Polyunsaturated fatty acids** – double-bond fatty acids important for brain function including DHA, EPA, and AA.
- **Propionic acid** – a short-chain fatty acid produced by bacteria, fermentation, and metabolism of some amino and fatty acids.

Key Facts About Vitamin Cobalamin

- Cobalamin supports normal production of white blood cells and platelets, affecting immune responses and blood clotting.
- Cobalamin assists in red blood cell and hemoglobin production to carry oxygen to tissues and organs.
- Cobalamin assists in myelin formation, the insulating sheath that protects nerves throughout the brain and body.
- Cobalamin influences production of neurotransmitters such as dopamine and serotonin.
- Cobalamin influences hormonal regulation of bone growth and strength.

Summary Points

- Some bacteria compete for cobalamin, divert human-accessible cobalamin into human-inaccessible forms, and alter the gut environment, influencing cobalamin absorption, bioavailability, metabolism, and assessment.
- Cobalamin deficiency can raise methylmalonic acid (MMA) and alter the propionic → methylmalonic → succinic acid pathway.
- Small bowel bacterial overgrowth can increase propionic acid (PPA) from precursors; some bacteria reverse MMA to PPA, increasing total PPA.

- Both cobalamin deficiency and PPA can cause autistic symptoms and alter brain function.
- Cobalamin-folate ratios modify PPA and MMA production, alter methylation pathways and DNA, and impact pregnancy outcomes and brain function.
- Cobalamin-folate ratios alter brain polyunsaturated fatty acid (PUFA) levels and response to PUFA supplementation.
- Gut bacteria alter PUFA levels; PUFA supplementation can alter gut bacterial species.
- Both cobalamin deficiency and PPA administration can alter brain PUFA levels, and altered brain PUFA levels are associated with autism.
- PUFA supplementation can reverse the neurotoxicity of PPA.
- Cobalamin supplementation can raise PUFA levels.
- Cobalamin and PUFA supplementation, with controlled cobalamin-folate ratios, may epigenetically modify the PPA model of autism.

References

Adams JB, Johansen LJ, Powell LD et al (2011a) Gastrointestinal flora and gastrointestinal status in children with autism – comparisons to typical children and correlation with autism severity. BMC Gastroenterol 11:22

Adams JB, Audhya T, McDonough-Means S et al (2011b) Nutritional and metabolic status of children with autism vs. neurotypical children, and the association with autism severity. Nutr Metab (London) 8(34):1–32

Agrawal S, Nathani S (2009) Neuro-regression in vitamin B12 deficiency. BMJ Cas Rep 2009: bcr06.2008.0235

Al-Owain M, Kaya N, Al-Shamrani H et al (2013) Autism spectrum disorder in a child with propionic acidemia. JIMD Rep 7:63–66

Andersen AD, Molbak L, Michaelsen KF et al (2011) Molecular fingerprints of the human fecal microbiota from 9 to 18 months old and the effect of fish oil supplementation. J Pediatr Gastroenterol Nutr 53(3):303–309

Beard CM, Panser LA, Katusic SKI (2011) Folic acid supplementation a risk for autism? Med Hypoth 77(1):15–17

Belitsos PC, Greenson JK, Yardley JH et al (1992) Association of gastric hypoacidity with opportunistic enteric infections in patients with AIDS. J Infect Dis 166(2):277–284

Bennett M (2001) Vitamin B12 deficiency, infertility and fetal loss. J Reprod Med 46(3):209–212

Brandt LJ, Berstein LH, Wagle A (1977) Production of vitamin B12 analogues in patients with small-bowel bacterial overgrowth. Ann Intern Med 87(5):546–551

Bredenoord AJ, Baron A, Smout AJPM (2006) Symptomatic gastro-oesophageal reflux in a patient with achlorhydria. Gut 55(7):1054–1055

Brigandi SA, Shao H, Qian SY et al (2015) Autistic children exhibit decreased levels of essential fatty acids in red blood cells. Int J Mol Sci 16(5):10061–10076

Chiang PK, Gordon RK, Tal J et al (1996) S-Adenosylmethionine and methylation. FASEB J 10:471–480

Clonan A, Roberts KE, Holdsworth M (2016) Socioeconomic and demographic drivers of red and processed meat consumption: implications for health and environmental sustainability. Proc Nutr Soc 75(3):367–373

Coppede F (2009) The complex relationship between folate/homocysteine metabolism and risk of down Syndrome. Mutat Res 682(1):54–70

Coppede F, Denaro M, Tannorella P, Migliore L (2016) Increased MTHFR promotor methylation in young mothers of down Syndrome individuals. Mutat Res 787:1–6

Daniel CR, Cross AJ, Koebnick C et al (2011) Trends in meat consumption in the United States. Public Health Nutr 14(4):575–583

Davis DJ, Hecht PM, Jasarevic E et al (2017) Sex-specific effects of docosahexaenoic acid (DHA) on the microbiome and behaviour of socially-isolated mice. Brain Behav Immun 59:38–48

Degnan PH, Taga ME, Goodman AL (2014) Vitamin B_{12} as a modulator of gut microbial ecology. Cell Metab 20(5):769–778

DeSoto MC, Hitlan RT (2012) Synthetic folic acid supplementation during pregnancy may increase risk of developing autism. J Ped Biochemist 2(4):251–261

Dhobale M, Joshi S (2012) Altered maternal micronutrients (folic acid, vitamin B12) and omega 3 fatty acids through oxidative stress may reduce neurotrophic factors in preterm pregnancy. J Matern Fetal Neonatal Med 25(4):317–325

Diav-Citrin O, Arnon J, Schectman S (2005) The safety of proton pump inhibitors in pregnancy: a multicenter prospective controlled study. Aliment Pharmacol Ther 21(3):269–275

Duran M, Ketting D, Wadman SK et al (1973) Propionic acid, an artifact which can leave methylmalonic acidemia undiscovered. Clinica Chemica Acta 49:177–179

El-Ansary A, Al-Ayadhi L (2014) Relative abundance of short chain and polyunsaturated fatty acids in propionic acid-induced autistic features in rat pups as potential markers in autism. Lipids Health Dis 13:140

El-Ansary AK, Al-Daihan SK, El-Gezeery AR (2011) On the protective effect of omega-3 against propionic acid-induced neurotoxicity in rat pups. Lipids Health Dis 10:142

Fenech MF, Dreosti IE, Rinaldi JR (1997) Folate, vitamin B12, homocysteine status and chromosome damage rate in lymphocytes of older men. Carcinogenesis 18(7):1329–1336

Fenech M, Aitken C, Rinaldi J (1998) Folate, vitamin B12, homocysteine status and DNA damage in young Australian adults. Carcinogenesis 19(7):1163–1171

Finegold SM, Dowd SE, Gontcharova V et al (2010) Pyrosequencing study of fecal microflora of autistic and control children. Anaerobe 16(4):444–453

Fischbach MA, Sonnenburg JL (2011) Eating for two: how metabolism establishes interspecies interactions in the gut. Cell Host Microbe 10(4):336–347

Fitzgerald K, Hyman M, Swift K (2012) Autism Spectrum disorders. Glob Adv Health Med 1(4):62–74

Frye RE, Rose S, Slattery J et al (2015) Gastrointestinal dysfunction in autism spectrum disorder: the role of the mitochondria and the enteric microbiome. Microb Ecol Health Dis 26:2748

Fujimori S (2015) What are the effects of proton pump inhibitors on the small intestine? World J Gastroenterol 21(22):6817–6819

Gadgil M, Joshi K, Pandit A et al (2014) Imbalance of folic acid and vitamin B12 is associated with birth outcome: an Indian pregnant women study. Eur J Clin Nutr 68(6):726–729

Ghosh S, Molcan E, DeCoffe D et al (2013) Diets rich in n-6 PUFA induce intestinal microbial dysbiosis in aged mice. Brit J Nutr 110(3):515–523

Gille D, Schmid A (2015) Vitamin B12 in meat and dairy products. Nutr Rev 73(2):106–115

Gillies D, Sinn JK, Lad SS et al (2012) Polyunsaturated fatty acids (PUFA) for attention deficit hyperactivity disorder (ADHD) in children and adolescents. Cochrane Database Syst Rev 7: CD007986

Guez S, Chiarelli G, Menni F et al (2012) Severe vitamin B12 deficiency in an exclusively breastfed 5-month-old Italian infant born to a mother receiving multivitamin supplementation during pregnancy. BMC Pediatr 12:85

Haboubi NY, Lee GS, Montgomery RD (1991) Duodenal mucosal morphometry of elderly patients with small intestinal bacterial overgrowth: response to antibiotic treatment. Age Ageing 20(1):29–32

Jerneren F, Elshorbagy AK, Oulhaj A (2015) Brain atrophy in cognitively impaired elderly: the importance of long-chain w-3 fatty acids and B vitamin status in a randomized controlled trial. Am J Clin Nutr 102(1):215–221

Kaliannan K, Wang B, Li X-Y et al (2015) A host-microbiome interaction mediates the opposing effects of omega-6 and omega-3 fatty acids on endotoxemia. Sci Rep 5:11276

Kim SK, Kim YC (2005) Effects of betaine supplementation on hepatic metabolism of sulphur-containing amino acids in mice. J Hepatol 42(6):907–913
Klee GG (2000) Cobalamin and folate evaluation: measurement of methylmalonic acid and homocysteine vs vitamin B12 and folate. Clin Chem 46(8):1277–1283
Kobe T, Witte AV, Schnelle A et al (2016) Vitamin B-12 concentration, memory performance, and hippocampal structure in patients with mild cognitive impairment. Am J Clin Nutr 103(4):1045–1054
Kocaoglu C, Akin F, Caksen H et al (2014) Cerebral atrophy in a vitamin B12-deficient infant of a vegetarian mother. J Health Popul Nutr 32(2):367–371
Krautler B (2012) Biochemistry of B12-cofactors in human metabolism. Subcell Biochem 56:323–346
Kulkarni A, Dangat K, Kale A et al (2011) Effects of altered maternal folic acid, vitamin B12 and docosahexaenoic acid on placental global DNA methylation patterns in Wistar rats. PLoS One 6(3):e17706
Lappinga PJ, Abraham SC, Murray JA et al (2010) Small intestinal bacterial growth: histopathological features and clinical correlates in an underrecognized entity. Arch Pathol Lab Med 134(2):264–270
LeBlanc JG, Milani C, de Giori GS et al (2013) Bacteria as vitamin suppliers to their host: a gut microbiota perspective. Curr Opin Biotechnol 24(2):160–168
Lebwohl B, Green PH, Genta RM (2015) The coeliac stomach: gastritis in patients with coeliac disease. Alimen Pharmaco Ther 42(2):180–187
Lewis SJ, Franco S, Young G et al (1996) Altered bowel function and duodenal bacterial overgrowth in patients treated with omeprazole. Aliment Pharmacol Ther 10(4):557–561
Lindenbaum J, Savage DG, Stabler SP et al (1990) Diagnosis of cobalamin deficiency: II. Relative sensitivities of serum cobalamin, methylmalonic acid, and total homocysteine concentrations. Am J Hematol 34:99–107
Lovblad K, Ramelli G, Remonda L et al (1997) Retardation of myelination due to dietary B12 deficiency: cranial MRI findings. Pediatr Radiol 27(2):155–158
MacFabe DF (2012) Short-chain fatty acid fermentation products of the gut microbiome: implications in autism spectrum disorders. Microb Ecol Health Dis 23:1–24
MacFabe DF, Cain DP, Rodriguez-Capote K et al (2007) Neurobiological effects of intraventricular propionic acid in rats: possible role of short chain fatty acids on the pathogenesis and characteristics of autism spectrum disorders. Behav Brain Res 176(1):149–169
MacFarlane AJ, Greene-Finestone LS, Shi Y (2011) Vitamin B-12 and homocysteine status in a folate-replete population: results from the Canadian health measures survey. Am J Clin Nutr 94(4):1079–1087
Main PA, Thomas P, Esterman A et al (2013) Necrosis is increased in lymphoblastoid cell lines from children with autism compared with their non-autistic siblings under conditions of oxidative and nitrosative stress. Mutagenesis 28(4):475–484
Marcoullis G, Parmentier Y, Nicolas JP et al (1980) Cobalamin malabsorption due to non-degradation of R proteins in the human intestine. Inhibited cobalamin absorption in exocrine pancreatic dysfunction. J Clin Invest 66(3):430–440
McKoll KE, el-Omar E, Gillen D (1998) Interactions between H. Pylori infection, gastric secretion and anti-secretory therapy. Mr Med Bull 54(1):121–138
Meyniel D, Petit J, Bodin F et al (1981) Vitamin B12 deficiency in chronic atrophic gastritis. 3 cases (author's transl). Nouvelle Presse Med 10(27):2281–2284
Molloy AM, Kirke PN, Troendle JF et al (2009) Maternal vitamin B12 status and risk of neural tube defects in a population with high neural tube defect prevalence and no folic acid fortification. Pediatrics 123(3):917–923
Morita H, Toh H, Fukuda S et al (2008) Comparative genome analysis of lactobacillus reuteri and lactobacillus fermentum reveal a genomic island for reuterin and cobalamin production. DNA Res 15(3):151–161
Morris MS, Jacques PF, Rosenberg IJ et al (2007) Folate and vitamin B-12 status in relation to anemia, macrocytosis, and cognitive impairment in older Americans in the age of folic acid fortification. Am J Clin Nutr 85(1):193–2000

Muyshondt E, Schwartz SI (1964) Vitamin B12 absorption following vagectomy and gastric surgery. Ann Surg 160(5):788–792
Obeid R, Kasoha M, Kirsch SH et al (2010) Concentrations of unmetabolized folic acid and primary folate forms in pregnant women at delivery and in umbilical cord blood. Am J Clin Nutr 92(6):1416–1422
Parletta N, Niyonsenga T, Duff J (2016) Omega-3 and omega-6 polyunsaturated fatty acid levels and correlations with symptoms in children with attention deficit hyperactivity disorder, autistic spectrum disorder and typically developing controls. PLoS One 11(5):e0156432
Pickell L, Brown K, Li D et al (2011) High intakes of folic acid disrupts embryonic development in mice. Birth Defects Res A Clin Mol Teratol 91(1):8–19
Rathod R, Khaire A, Kemse N et al (2014) Maternal omega-3 fatty acid supplementation on vitamin B12 rich diet improves brain omega-3 fatty acids, neurotrophins and cognition in the Wistar rat offspring. Brain Dev 36(10):853–863
Rathod RS, Khaire AA, Aa K et al (2016) Effect of vitamin B12 and omega-3 fatty acid supplementation on brain neurotrophins and cognition in rats: a multigeneration study. Biochimie 128–129:201–208
Ríos-Covián D, Ruas-Madiedo P, Margolles A et al (2016) Intestinal short chain fatty acids and their link with diet and human health. Front Microbiol 7:185
Robertson RC, Seira Oriach C, Murphy K et al (2016) Omega-3 polyunsaturated fatty acids critically regulate behaviour and gut microbiota development in adolescence and adulthood. Brain Behav Immun 59:21–37
Rossi M, Amaretti A, Raimondi S (2011) Folate production by probiotic bacteria. Forum Nutr 3(1):118–134
Roy S, Kale A, Dangat K et al (2012) Maternal micronutrients (folic acid and vitamin B12) and omega 3 fatty acids: implications for neurodevelopmental risk in the rat offspring. Brain and Development 34(1):64–71
Ruscin JM, Page RL 2nd, Valuck RJ (2002) Vitamin B(12) deficiency associated with histamine(2)-receptor antagonists and a proton-pump inhibitor. Ann Pharmacother 36(5):812–816
Sable P, Kale A, Joshi A et al (2014) Maternal micronutrient imbalance alters gene expression of BDNF, NGF, TrkB and CREB in the offspring at an adult age. Int J Dev Neurosci 34:24–32
Saltzman JR, Kemp JA, Golner BB et al (1994) Effect of hypochlorhydria due to omeprazole treatment or atrophic gastritis on protein-bound vitamin B12 absorption. J Am Coll Nutr 13(6):584–591
Schjonsby H (1989) Vitamin B12 absorption and malabsorption. Gut 30:1686–1691
Sipponen P, Kekki M, Seppala K et al (1996) The relationships between chronic gastritis and gastric acid secretion. Aliment Pharmacol Ther 10(1):103–118
Smith CH, Israel DM, Schreiber R et al (2013) Proton pump inhibitors for irritable infants. Can Fam Physician 59(2):153–156
Song Y, Liu C, Finegold SM (2004) Real-time PCR quantitation of clostridia in feces of autistic children. Appl Environ Microbiol 70(11):6459–6465
Stipanuk MH, Ueki I (2011) Dealing with methionine/homocysteine sulfur: cysteine metabolism to taurine and inorganic sulfur. J Inherit Metab Dis 34(1):17–32
Sugahara H, Odamaki T, Hashikura N et al (2015) Differences in folate production by bifidobacteria of different origins. Biosci Microbiota Food Health 34(4):87–93
Suter PM, Golner BB, Goldin BR et al (1991) Reversal of protein-bound vitamin B12 malabsorption with antibiotics in atrophic gastritis. Gastroenterology 101(4):1039–1045
Takahashi-Iñiguez T, García-Hernandez E, Arreguín-Espinosa R et al (2012) Role of vitamin B_{12} on methylmalonyl-CoA mutase activity. J Zhejiang Univ Sci B 13(6):423–437
Tam C, O'Connor D, Koren G (2012) Circulating unmetabolized folic acid: relationship to folate status and effect of supplementation. Obstet Gynecol Int 2012:485179
Therrien A, Bouchard S, Sidani S et al (2016) Prevalence of small intestinal bacterial overgrowth among chronic pancreatitis patients: a case-control study. Can J Gastroenterol Hepatol 2016:7424831–7424837

Thorens J, Froehlich F, Schwizer W et al (1996) Bacterial overgrowth during treatment with omeprazole compared with cimetidine: a prospective, randomized double blind study. Gut 39(1):54–59
Van der Pol RJ, Smits MJ, van Wijk MP et al (2011) Efficacy of proton-pump inhibitors in children with gastroesophageal reflux disease: a systematic review. Pediatr 127(5):925–935
Voigt RG, Mellon MW, Katusic SK et al (2014) Dietary docosahexaenoic acid supplementation in children with autism. J Pediatr Gastroenterol Nutr 58(6):715–722
Wakefield AJ, Murch SH, Anthony A et al (1998) Ileal-lymphoid-nodular hyperplasia, non-specific colitis, and pervasive developmental disorder in children (Retracted). Lancet 351(9103):637–641
Wall R, Ross RP, Shanahan F et al (2009) Metabolic activity of the enteric microbiota influences the fatty acid composition of murine and porcine liver and adipose tissues. Am J Clin Nutr 89(5):1393–1401
Wang L, Christophersen C, Sorich M et al (2012) Elevated fecal short chain fatty acid and ammonia concentrations in children with autism spectrum disorder. Dig Dis Sci 57(8):2096–2102
Waterland RA (2006) Assessing the effects of high methionine intake on DNA methylation. J Nutr 136(6):1706S–1710S
Waterland RA, Dolinoy DC, Lin JR et al (2006) Maternal methyl supplements increase offspring DNA methylation at Axin fused. Genesis 44(9):401–406
Waterland RA, Travisano M, Tahiliani KG (2007) Diet-induced hypermethylation at agouti viable yellow is not inherited transgenerationally through the female. FASEB J 21(12):3380–3385
Williams PG (2007) Nutritional composition of red meat. Nutr Dietet 64(4):S113–S119
Wood IJ, Ralston M, Ungar B et al (1964) Vitamin B12 deficiency in chronic gastritis. Gut 5:27–37
Wyckoff KF, Ganji V (2007) Proportion of individuals with low serum vitamin B-12 concentrations without macrocytosis is higher in the post folic acid fortification period than in the pre folic acid fortification period. Am J Clin Nutr 86(4):1187–1192
Youness ER, Aly HF, El-Bassyouni HT et al (2016) Micronucleus assay and pro-oxidant status of patients with known chromosomal aneuploidy. Der Pharma Chemica 8(13):158–164
Zhang Y, Hodgson NW, Trivedi MS et al (2016) Decreased brain levels of vitamin B12 in aging, autism and schizophrenia. PLoS One 11(1):e0146797

Maternal Folate and DNA Methylation in Offspring

90

Emma L. Beckett, Mark Lucock, Martin Veysey, and
Bonnie R. Joubert

Contents

E. L. Beckett (✉)
School of Medicine and Public Health, The University of Newcastle, Ourimbah, NSW, Australia
e-mail: emma.beckett@newcastle.edu.au

M. Lucock
School of Environmental and Life Sciences, The University of Newcastle, Ourimbah, NSW, Australia
e-mail: mark.lucock@newcastle.edu.au

M. Veysey
School of Medicine and Public Health, The University of Newcastle, Gosford Hospital, Gosford, NSW, Australia
e-mail: martin.veysey@newcastle.edu.au

B. R. Joubert
Population Health Branch, National Institute of Environmental and Health Sciences, Durham, NC, USA
e-mail: bonnie.joubert@nih.gov

V. B. Patel, V. R. Preedy (eds.), *Handbook of Nutrition, Diet, and Epigenetics*,
https://doi.org/10.1007/978-3-319-55530-0_3

Abstract

Folate plays a critical role in DNA methylation as it is a key source of methyl donors via the one-carbon metabolism cycle. Folate supplementation is recommended during the periconceptional period for the prevention of neural tube defects in offspring. However, maternal folate levels during pregnancy may also influence the risk of many other conditions in offspring, but the underlying mechanisms involved are unclear. As such, it is important to investigate the possible association between maternal folate status and disease risk that act via modulation of the methylome. Improving methods and technologies available for profiling DNA methylation has allowed for rapidly expanding investigations in this field; however, limitations in study design remain. On the available evidence, global DNA methylation does not appear to be associated with maternal folate status in cord blood samples, but this response may be tissue specific as correlations have been found in fetal brains and adult murine intestines. Several studies have shown differential locus-specific methylation in response to maternal folate status. However, results may vary depending on the assay methods employed, including different assessments of the methylome, different measures of folate status, and cohort composition. Although maternal folate status is linked to disease risk, additional research is required to link this modulation of the methylome to altered health and disease outcomes.

Keywords

Folate/folic acid · Maternal diet · Methylation · Methylome · Methyl · One-carbon metabolism · Offspring · Newborn · Epigenome

List of Abbreviations

5CaC	5-Carboxylcytosine
1CM	One-carbon metabolism
5fmC	5-Formylcytosine
5hmC	5-Hydroxymethylcytosine
5mC	5-Methylcystosine
AID	Activation-induced deaminase
DHF	Dihydrofolate
DHFR	Dihydrofolate reductase
DMR	Differentially methylated region
DNMT	DNA methyltransferase
DOHaD	Developmental origins of health and disease
IAP	Intracisternal A particle

MS-PCR	Methylation-specific polymerase chain reaction
MS-qPCR	Methylation-specific quantitative polymerase chain reaction
MTHFR	Methylenetetrahydrofolate reductase
MTR	Methionine synthase
MTRR	Methionine synthase reductase
NTDs	Neural tube defects
SAH	S-Adenosylhomocysteine
SAM	S-Adenosylmethionine
SHMT	Serine hydroxymethyltransferase
Tet	Ten-eleven translocation
THF	Tetrahydrofolate

Introduction

Folate, used as a generic term for food folate and folic acid, is a water-soluble vitamin. Folate occurs naturally in some foods, is added to folate fortified foods, or can be consumed as a dietary supplement. The fully oxidized monoglutamate form is used in dietary supplements and fortified foods, while food folates exist in a variety of reduced polyglutamate forms (Bailey and Gregory 2006).

Folate plays an important role in one-carbon metabolism (1CM), acting as a coenzyme or co-substrate in DNA synthesis and repair reactions, serine-glycine interconversions, histidine catabolism, and the de novo generation of methionine for provision of methyl donors for non-genomic and genomic methylation reactions, including DNA methylation (Bailey and Gregory 2006). Folate is also instrumental in the conversion of homocysteine to methionine. As such folate levels are associated with levels of homocysteine in the systemic circulation, an independent risk factor in multiple conditions including vascular disease (Selhub et al. 2000).

Folate plays a vital role in fetal development, and the importance of adequate periconceptional folate for the prevention of neural tube defects is well known. However, the mechanism involved, including the role of the epigenome, in this protective role is not well understood. The periconceptional period is a vital window of plasticity, during which fetal growth and development may be significantly influenced by maternal exposures. Epigenetics, in particular DNA methylation, may be modulated by nutrients, and nutrition may play an important role in programming of the epigenome during development (Burdge and Lillycrop 2010). Therefore, in the context of the developmental origins of health and disease (DOHaD) hypothesis, periconceptional maternal folate status could have consequences for health and disease of offspring in later life via epigenetics (Burdge and Lillycrop 2010).

Recent advances in techniques available for studying the methylome have expanded the potential to investigate the influence of maternal nutrition, including periconceptional folate status, on methylation status of offspring. As such, the potential for maternal folate status to influence the methylome in the offspring and the potential consequences for disease risk are becoming apparent.

The Importance of Adequate Maternal Periconceptional Folate

Folic acid supplementation ($\geq$0.4 mg per day) is recommended prior to, and in the early stages of, pregnancy. Over 50 countries have implemented programs to fortify foods with folic acid to increase folate levels in women of childbearing age (Miller and Ulrich 2013), with the primary aim of reducing the incidence of NTDs in offspring. It is clear that rates of NTDs have decreased following fortification (Miller and Ulrich 2013), and there is increasing interest in the potential for higher supplementation to reduce the risk of additional birth defects including oral clefts and cardiac defects (Bortolus et al. 2014).

It is poorly understood how folate prevents NTDs, and potentially other birth defects and later-life health outcomes (Nakouzi and Nadeau 2014), but there is growing evidence that epigenetic changes are involved. As folate is a critical micronutrient in DNA methylation, periconceptional maternal folate levels may modulate methylation patterns established in utero that are vital for fetal development, thus impacting later health outcomes in the offspring. Although current recommendations for maternal folate intake in the periconceptional period are based on the prevention of NTDs (Nakouzi and Nadeau 2014), there is a broad scope for maternal folate status to influence methylation at a number of genes and in a range of tissues, and the potential for this to influence offspring health is an active area of investigation.

Folate as a Methyl-Group Donor

Folate is critical for DNA methylation, feeding the folate-dependent 1CM cycle. Synthetic folate vitamers are first reduced to dihydrofolate (DHF) and then converted to tetrahydrofolate (THF) by dihydrofolate reductase (DHFR). THF is further converted by serine hydroxymethyltransferase (SHMT) and methylenetetrahydrofolate reductase (MTHFR) into 5-methyl THF, where natural food folates also enter the cycle. 5-Methyl THF feeds into the 1CM cycles by donating its methyl group to homocysteine, converting it into methionine, catalyzed by methionine synthase (MS), in a vitamin B_{12}-dependent manner. Methionine is then converted to SAM (S-adenosylmethionine), which donates methyl groups to biological targets (Fig. 1). After methyl-group transfer, SAM converts to S-adenosylhomocysteine (SAH), an inhibitor of methyltransferases. Choline and betaine can also serve as methyl donors, and additional B vitamins act as cofactors for these transformations (Choi and Friso 2010).

Therefore, consumption of folate and other methyl-group donors partially dictates the availability of methyl groups for methylation reactions. As such, folate supplementation is thought to be associated with increased levels of methylation, and

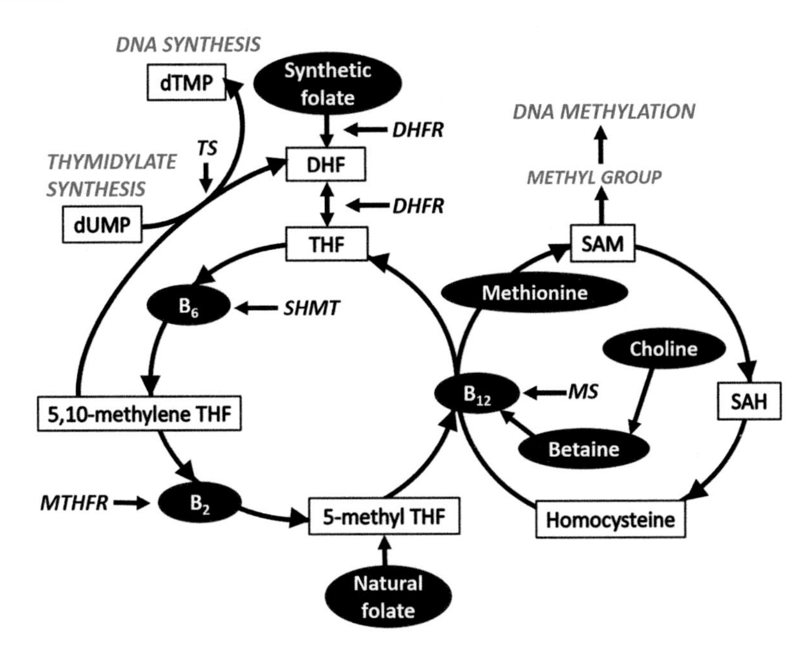

Fig. 1 Involvement of dietary micronutrients in one-carbon metabolism (*1CM*) and the generation of methyl groups for DNA methylation. Synthetic folate is converted to dihydrofolate (*DHF*) and then tetrahydrofolate (*THF*) by dihydrofolate reductase (*DHFR*). Vitamin B_6 is a cofactor in the conversion of THF to 5,10-methylene THF. Vitamin B_2 (riboflavin) serves as a cofactor for methylenetetrahydrofolate reductase (*MTHFR*), which converts 5,10-methylene THF to 5-methyl THF. Vitamin B_{12} is a cofactor for methionine synthase (MTR), which is involved in remethylation of homocysteine to form methionine. Methionine is then converted to SAM (S-adenosyl-methionine), which donates methyl groups to biological targets. After transferring the methyl group, SAM is converted to S-adenosylhomocysteine (SAH). Betaine and choline may also act as methyl donors. *dTMP* deoxythymidine monophosphate, *dUMP* deoxyuridine monophosphate

deficiency has been thought to be associated with decreased methylation levels (Friso and Choi 2005).

Methods for Studying the Methylome: Considerations for Offspring and Folate Studies

In mammals, DNA methylation occurs when a methyl group is added to a cytosine residue in DNA. This primarily occurs where a cytosine residue is followed by a guanine residue, referred to as a “CpG” dinucleotide. CpG dinucleotides are distributed throughout the genome, but are often found in clusters, known as CpG islands. When CpG islands occur in the gene promoter or other regulatory regions, methylation status can act to regulate gene expression. Approximately half of all human genes have CpG islands in their promoter regions. In a large proportion of genes studied so far, differential methylation has been found to occur in the 5′ untranslated region and is commonly inversely correlated with gene expression (Mathers 2008).

There are multiple strategies for assessing DNA methylation, and approaches may be locus specific (genome wide or candidate gene) or global. Approaches include bisulfite conversion, differential enzymatic digestion, and affinity enrichment, which can be paired with array, locus-specific amplification, plate based, or sequencing analysis for the detection of the methylation patterns. Each of the resulting combinations has their own unique set of features and drawbacks, including the ability to discriminate between 5-methylcytosine (5mC) and other cytosine derivatives (5-carboxylcytosine (5CaC), 5-formylcytosine (5fmC), and 5-hydroxymethylcytosine (5hmC)) and input DNA required, cost, issues with statistical analysis, and allele specificity and sensitivity (Shen and Waterland 2007). Global methylation is simple to measure and can indicate a change in the methylome; however, it offers no information on the genomic positioning of these changes. Locus-specific analyses offer this information, but can present challenges in selecting relevant sites for inclusion (Shen and Waterland 2007).

Analysis of the methylome in neonatal and fetal development presents additional limitations, including access to relevant tissues. Human studies are often restricted to cord blood, peripheral blood cells isolated from whole blood after birth, or aborted fetal tissue. As methylation patterns are often tissue specific, this may present a limitation. However, cord blood does contain mesenchymal stem cells and vascular tissue and so may be considered useful when looking for associations with later life anthropometry (Godfrey et al. 2011). Animal models overcome the issues involved in accessing tissue, but present additional limitations due to the different features of epigenetic marking and patterns of reprogramming in model systems, relative to humans (Smith et al. 2012).

Additionally, studying the influence of maternal exposures on the offspring methylome is further complicated by the difficulty involved in distinguishing what is transgenerational inheritance and what is a consequence of the in utero environment directing the placement of epigenetic marks in the offspring (Heard and Martienssen 2014). Studies on the effect of maternal diet on offspring often focus on the periconceptional period; however, it is important to note that maternal exposures that are temporally removed from conception may influence the epigenome via influence on the oocytes or dietary modulation of other phenotypes. Location is also an important consideration when conducting studies of folate supplementation, as provision of a supplement in a country with a mandatory or voluntary folic acid fortification policy may produce less noticeable modulation from baseline effects, compared to supplementation in a deficient population.

DNA Methylation and the Life Cycle

DNA methylation is dynamic during the life cycle, with distinct phases of reprogramming and de novo methylation. DNA methylation is important in the regulation of key developmental processes including genomic imprinting, transposon silencing, genomic stability, gene regulation, and the direction and maintenance of cell lineage (Messerschmidt et al. 2014; Tang et al. 2016). However, these marks pose barriers to sexual reproduction, and so mass erasure occurs at key points to return cells to a totipotent state. The first reprogramming occurs in primordial germ cells, followed by the establishment of epigenetic signatures involved in meiotic maturation and fertilization. A second round of reprogramming occurs post-fertilization (Messerschmidt et al. 2014; Tang et al. 2016).

Demethylation during reprogramming can occur via passive or active mechanisms. Passive demethylation is mediated by the suppression of the DNA methylation maintenance machinery during replication. Conversely, active demethylation occurs in a replication-independent manner via enzymatically driven actions, including the action of ten-eleven translocation (tet) enzymes, which oxidize 5mC to 5hmC, 5fC and 5caC, and potentially other enzymes such as activation-induced deaminase (AID) (Messerschmidt et al. 2014; Amouroux et al. 2016).

Interestingly, in human and murine zygotes, it appears that the paternal genome undergoes rapid active demethylation within hours of fertilization (Mayer et al. 2000; Oswald et al. 2000; Guo et al. 2014b), followed by subsequent replication-dependent passive dilution (Iqbal et al. 2011; Wossidlo et al. 2011). Conversely, the majority of the demethylation of the maternal genome appears to occur more slowly during subsequent mitotic divisions, dominated by passive mechanisms (Rougier et al. 1998; Guo et al. 2014b). Epigenetic marks are then re-established during development to direct cell lineage (Guo et al. 2014a).

However, it is apparent that some methylation marks are able to escape this reprogramming or are reestablished following reprogramming. A poignant example is genetic imprinting, which silences one parental allele (Heard and Martienssen 2014; Tang et al. 2016). Additionally, it is becoming clear that some epigenetic marks can be inherited transgenerationally, although the mechanisms are not yet fully elucidated (Heard and Martienssen 2014). However, it is becoming clear that maternal nutrition, along with other lifestyle and environmental exposures, can influence patterns of epigenetic inheritance and genetic imprinting.

The Relationships Between Maternal Folate Status and DNA Methylation in Newborns

Recently, the influence of maternal folate status on DNA methylation in offspring has been studied in multiple models. Human studies (Tables 1 and 2) are largely restricted to cord blood or peripheral blood samples. Animal models (Table 3) allow the analyses in additional tissues, but lack the dietary and genetic background diversity of humans. Additionally, the restriction or supplementation of multiple methyl donors, including folate, may be used in animal models, whereas human models are restricted to blood levels, intake estimates, and supplementation status. Available results are mixed, but multiple associations between folate and DNA methylation in offspring have been reported.

Table 1 Human observational studies of global DNA methylation profiles in relation to maternal blood folate levels. No associations have been found between blood levels of maternal folate and global DNA methylation in cord blood. Maternal folate status may be linked to global methylation in other tissues

Nutrients assessed	Maternal observation	Offspring tissue	Method of epigenome assessment	Influence on offspring epigenome	Study size	References
Folate only	Highest versus lowest quartile serum folate	Cord blood	LINE-1; Sequenom EpiTyper	No association	23	Amarasekera et al. (2014)
Folate only	Serum folate (continuous)	Cord blood	LINE-1; pyrosequencing	No association	24	Fryer et al. (2009)
Folate only	Erythrocyte folate (continuous)	Cord blood and whole blood (3–5 postnatal)	Enzyme digestion	No associations	37	Schlinzig et al. (2009)
Folate only	Serum folate	Aborted fetuses (with and without NTDs)	HPLC	Maternal serum folate lower in NTD fetuses than controls. Maternal serum folate correlated with density of methylation in the fetal brain. Hypomethylation of brain DNA in NTD fetuses and hypermethylation of the skin and heart in NTD fetuses, relative to controls	20	Chang et al. (2011)

[*] *FDR* false discovery rate, *NTDs* neural tube defects, *No association* indicates no statistically significant association was reported by the authors

The most dramatic demonstration of the epigenetic impact of folate, along with other methyl-group donors, involves using the viable yellow agouti (A^{vy}) mouse, in which coat color variation is correlated to epigenetic marks established early in development. The A^{vy} mutation is caused by retrotransposition of an intracisternal A particle (IAP) upstream of the *agouti* gene, which regulates the production of yellow pigment in hair follicles (Duhl et al. 1994) A cryptic promoter in the IAP promotes agouti expression, causing a yellow-colored coat. Yellow mice are also larger, hyperinsulinemic, obese, more susceptible to cancers, and experience shorter average lifespans, compared to non-yellow offspring. The agouti gene is an example of a metastable epiallele (Cooney et al. 2002), where epigenotype, including DNA methylation, is established stochastically in the early embryo and maintained

Table 2 Human observational studies of loci-specific DNA methylation profiles in relation to maternal blood folate levels. Results are mixed regarding the relationships between loci-specific methylation and maternal blood folate levels

Nutrients assessed	Maternal observation	Offspring tissue	Method of epigenome assessment	Influence on offspring epigenome	Study size	References
Folate only	Plasma folate (continuous)	Cord blood	Genome-wide (BeadChip array)	443 CpGs (at 320 genes) significantly associated with maternal plasma folate during pregnancy at a FDR 5%	1988	Joubert et al. (2016)
Folate only	Highest versus lowest quartile serum folate	Cord blood	Genome-wide (bisulfite conversion and array)	High-folate group had lower methylation at regions including ZFP57 DMR	23	Amarasekera et al. (2014)
Folate and vitamin D	Plasma folate (continuous)	Cord blood	Genome wide (BeadChip array)	No association	200	Mozhui et al. (2015)
Folate only	Erythrocyte folate (quartiles)	Cord blood	Candidate gene: DMRs of IGF2/H19, DLK1/MEG3, PEG3, MEST, PEG10, SGCE, NNAT, PLAGL1 (pyrosequencing)	DNA methylation at the PLAGL1, SGCE, DLK1/MEG3, and IGF2/H19 DMRs was associated with maternal folate levels	496	Hoyo et al. (2014)
Folate and homocysteine	Plasma folate (continuous).	Cord blood	11 regions of the seven genes: NR3C1, DRD4, 5-HTT, IGF2DMR, H19, KCNQ1OT1, and MTHFR (bisulfite conversion and MassARRAY	No association	463	van Mil et al. (2014)
Folate only	Serum and cord blood folate	Cord blood	P2 and P3 region of IGF2 promotor (MS-qPCR)	Methylation patterns of both promoters in cord blood not associated with serum folate levels in either cord or maternal blood	99	Ba et al. (2011)

(*continued*)

Table 2 (continued)

Nutrients assessed	Maternal observation	Offspring tissue	Method of epigenome assessment	Influence on offspring epigenome	Study size	References
Folate and homocysteine	Serum folate (continuous)	Cord blood	Genome-wide (bisulfite conversion and array)	Cluster analysis revealed a difference in DNA methylation between clusters; however, serum folate levels did not differ	12	Fryer et al. (2011)

**FDR* false discovery rate, *DMR* differentially methylated region, *No association* indicates no statistically significant association was reported by the authors

throughout life. Therefore, isogenic agouti mice display a range of phenotypes, from yellow (hypomethylated at A^{vy}) to brown coats (hypermethylated at A^{vy}). Supplementation with methyl-group donors, including folic acid, in agouti mothers significantly influences A^{vy} methylation and phenotype of offspring, despite having no impact on methylation status at the same gene in the mother (Cooney et al. 2002).

Whether this maternal diet-induced epigenetic change can be inherited to the next generation is still a matter of contention. Some studies have demonstrated that methyl-donor supplementation of the pregnant dam shifts A^{vy} phenotypes not only in exposed fetuses but also in the next generation of offspring (Cooney 2006; Cropley et al. 2006). However, a recent study reported that diet-induced hypermethylation at A^{vy} is not inherited in the female germ line (Waterland et al. 2007).

Another murine metastable epiallele, axin fused *(Axin^{Fu})*, has also been assessed in the context of maternal methyl-donor diet (Waterland et al. 2006). DNA methylation at the $Axin^{Fu}$ IAP is negatively linked with tail-kinking. Supplementation with methyl donors, including folate, during gestation led to increased $Axin^{Fu}$ methylation and silenced the expression, decreasing the incidence of tail kinking in offspring (Waterland et al. 2006). This demonstrates that maternal diets may epigenetically influence multiple loci and influence several of phenotypes in offspring.

Global Methylation

Studies assessing the relationships between maternal blood folate levels or intake and global DNA methylation in cord blood collected from offspring at birth have commonly reported no association (Fryer et al. 2009; Schlinzig et al. 2009; Boeke et al. 2012; Amarasekera et al. 2014), despite multiple methods being utilized. However, it is likely that these relationships are tissue specific. Due to the difficulties associated with accessing fetal tissue, cord blood is more commonly studied.

The one study published, to date, that does assess aborted fetal tissue found that global DNA methylation in brain tissue was positively correlated to maternal serum folate (Chang et al. 2011). The same study compared DNA methylation levels and

Table 3 Animal studies examining changes in offspring epigenome following alteration of maternal folate status. Animal studies of supplementation suggest a link between dietary methyl-donor levels and DNA methylation in offspring tissues

Dietary manipulation	Model animal	Tissue assessed	Method of epigenome assessment	Influence on offspring epigenome	References
Folic acid, B12, and methionine-deficient diet	Sheep	Liver	Restriction landmark genome scanning (1,400 CpG islands)	4% of the CpG islands assessed had altered methylation (majority hypomethylated) relative to controls	Sinclair et al. (2007)
Folic acid, B12, methionine, betaine (medium and high supplementation)	Mouse	Liver and kidney	Candidate gene: long terminal repeat of agouti gene (various)	Increased methylation of long terminal repeat and expression of agouti gene (and healthier phenotype) in high methyl-donor-supplemented diet	Wolff et al. (1998), Cooney et al. (2002), and Waterland et al. (2007)
Folic acid (supplementation) in low-protein diet	Rat	Liver	Candidate gene: GR, PPAR (MS-PCR)	Supplementation protected against the hypomethylation of GR and PPAR observed in protein restriction alone	Lillycrop et al. (2005)
Folic acid (supplementation) in low-protein diet	Rat	Liver	Candidate gene: IGF2, H19; MS-PCR	Supplementation protected against the hypermethylation of IGF2 and H19 observed in protein restriction alone	Gong et al. (2010)
Folic acid (low vs. high)	Mice	Brain	Sequencing (bisulfite converted)	High maternal folate led to differential methylation of 16% of CpG sites	Barua et al. (2014a)
Folic acid (low vs. normal)	Mice	Intestines	Candidate gene: p53 (MS-qPCR)	Reduced p53 methylation in adult mice born from mothers on low-folate diet	McKay et al. (2011a)

(*continued*)

Table 3 (continued)

Dietary manipulation	Model animal	Tissue assessed	Method of epigenome assessment	Influence on offspring epigenome	References
Folic acid (low vs. normal)	Mice	Intestines	Global (enzyme digestion)	Low maternal folate intake during pregnancy resulted in global hypomethylation of the adult offspring	McKay et al. (2011b)
Folic acid (low vs. normal)	Mice	Intestines	Candidate gene: Slc394a, Esr1, and Igf2 DMR (pyrosequencing)	Low folate reduced methylation at Slc394a, but did not influence Esr1 or Igf2 DMR	McKay et al. (2011c)

MS-PCR methyl-specific PCR (polymerase chain reaction), *MS-qPCR* methyl-specific quantitative PCR, *DMR* differentially methylated region

maternal serum folate between fetuses with NTDs and those without and found that those with NTDs had lower global methylation in the brain, which was also associated with lower maternal serum folate (Chang et al. 2011). Interestingly, in this study maternal serum folate was not associated with DNA methylation in skin, liver, or kidney tissue (Chang et al. 2011).

However, global hypomethylation in the intestines of offspring has also been demonstrated in a mouse model of low maternal folate (McKay et al. 2011a). DNA methylation was not influenced if maternal folate was low only during lactation, rather than during gestation. This suggests that early life folate depletion affects epigenetic markings, that this effect is not modulated by postweaning folate supply, and that altered epigenetic marks persist into adulthood. Altered methylation profiles in the gastrointestinal tract may influence the expression of genes involved in nutrient absorption, and this may represent a pathway for additional secondary modulation of the epigenome in other tissues.

Locus-Specific Methylation

Imprinted Regions

Perinatal maternal folate status has also been associated with differential methylation in specific imprinted genes in cord blood; however, results are mixed. In a small study of genome-wide methylation, maternal folate was linked to differential methylation upstream of the gene *ZFP57*, which plays a central role in the regulation and maintenance of imprinting (Amarasekera et al. 2014). However, a candidate gene analysis found no relationship between maternal folate intake in the last trimester and DNA methylation at *ZAC1*, another gene involved in the regulation of imprinting (Azzi et al. 2014). However, in this study, the coenzyme vitamin B_2 was related to *ZAC1* methylation density. Additional and larger studies are required to assess the impacts of altered DNA methylation in regulatory regions on other imprinted genes.

The best studied imprinted DMRs are *IGF2* and *H19*, the two are closely linked genes expressed together during fetal development (Leighton et al. 1995). A study of nearly 500 mother-baby pairs found that methylation at the DMRs of *IGF2* and *H19* varies by maternal erythrocyte folate (Hoyo et al. 2014), when the highest and lowest quartiles of folate levels are compared. In this study, a variance in methylation density was also reported at multiple additional imprinted DMRs, including *DLK1/MEG3* and *PLAG1* (Hoyo et al. 2014).

The above findings have been supported for *H19*, but not *IFG2*, in a similarly sized study, when maternal folate was assessed based on the use of a folate supplement prior to or during pregnancy (Hoyo et al. 2011). In peripheral blood taken from offspring at 17 months of age, methylation at the *IGF2* DMR was higher in those whose mothers had taken a supplement (Steegers-Theunissen et al. 2009). Additionally, a murine study using liver tissue found that folate supplementation was able to protect against the hypermethylation that occurred at *Igf2* and *H19*, and the hypomethylation that occurred at *GR* and *PPAR*, in a model of protein restriction (Lillycrop et al. 2005; Gong et al. 2010). However, in a mouse model, low maternal folate did not influence *Igf2* methylation in the intestines of offspring (McKay et al. 2011c). Interestingly, a randomized control trial in Gambia using a multivitamin including folate found that *IGF2R* methylation in cord blood was linked to supplementation in female offspring only (Cooper et al. 2012).

However, others have reported no association between *IGF2* and *H19* DMR methylation in cord blood in a large study of reported supplementation during pregnancy and serum folate levels (van Mil et al. 2014) or a smaller study assessing both maternal serum folate (Ba et al. 2011) and cord blood folate (Ba et al. 2011). Additionally, a mouse model showed no change in *Igf2* methylation in the small intestines of offspring from folate mothers who were folate deficient (McKay et al. 2011b). Lack of consistency between these studies may be explained by differences between erythrocyte folate, serum folate, cord blood folate, intake estimates, timing of sample collection, baseline folate levels or genetic background of participant, and methods of methylation assessment. The studies finding no associations used methylation-specific PCR (Ba et al. 2011) and mass array methods (van Mil et al. 2014), while those reporting differences used pyrosequencing methods (Hoyo et al. 2011, 2014).

Genome-Wide Methylation and Candidate Gene Studies

Two small genome-wide studies found no association between maternal plasma or serum folate and methylation status at any loci in cord blood (Fryer et al. 2011; Mozhui et al. 2015). However, the null results may be due to the small sample sizes (200 or less participants) being unable to withstand correction for multiple testing.

Corrections for multiple testing are required in genome-wide analyses to prevent the identification of false-positive associations. A more recent and larger genome-wide analysis revealed that methylation status of 443 CpGs in cord blood DNA was significantly associated with maternal plasma folate at a false discovery rate of 5%. Forty-eight of sites remained significant after Bonferroni correction, a strict method of multiple testing adjustment (Joubert et al. 2016). Most of the genes identified as differentially methylated in this study are not previously known to be related to folate biology and include *APC2*, *GRM8*, *SLC16A12*, *OPCML*, *PRPH*, *LHX1*, *KLK4*, and *PRSS21*. Some relate to birth defects other than neural tube defects, neurological functions, or varied aspects of embryonic development (Joubert et al. 2016). Additional investigations are required to assess the functional consequences of this differential methylation on gene expression and disease and health outcomes.

Animal studies are able to assess the influence of maternal folate on DNA methylation in a broader range of tissues. In a sheep model of a maternal methyl-donor-deficient diet (folate, B12, and methionine deficient) on genome-wide methylation in the livers of offspring, 4% of the CpG sites assessed were differentially methylated. A majority of these were hypomethylated compared to the control diet, as would be expected in the methyl-donor-deficient environment (Sinclair et al. 2007). Similarly, 16% of the CpG sites assessed in the brains of mice were differentially methylated when their mothers were fed a high-folate diet (Barua et al. 2014a).

Mice born from mothers on a folate-deficient diet in utero had reduced methylation of *p53* (McKay et al. 2011b) and *slc39a4* (McKay et al. 2011c) and in the small intestine. Hypomethylation of the *p53* gene has been associated with colon tumorigenesis in rat models (Kim et al. 1996) and with increased mutation rates within p53 itself (Tornaletti and Pfeifer 1995). These pathways may be mechanistic links between folate status and risk for colorectal cancer.

Summary of Findings

Based on the data available to date, it appears that maternal folate has no influence on global methylation in cord blood, but may influence global methylation in other organs, such as the brain and intestines of the offspring. Although results are mixed, it appears that maternal folate can modulate locus-specific methylation in the cord blood and potentially the peripheral blood cells of human offspring. However, additional studies are required to assess how the changes in cord blood are relevant to other organ systems in the offspring. Animal models have already demonstrated some gene specific effects in tissues, but additional studies are required to apply these findings to humans.

Further studies on the relationship between maternal folate and DNA methylation in the offspring are needed to expand our understanding of tissue specificity and interactions with other nutrients and to assess the role of polymorphism in the genes involved in the 1CM cycles and epigenetic regulation in modulating these effects. Future efforts should be globally directed to clarify the impact of baseline folate prior to supplementation and any potentially differential effect between intakes from fortification, supplementation, and food folates. Furthermore, longitudinal studies are required to link the observed changes to DNA methylation with health outcomes.

Maternal Folate, DNA Methylation, and Health Outcomes

In addition to the prevention of NTDs, higher maternal folate levels may have additional health benefits for offspring. Other potential beneficial effects of high maternal folate in humans include reduced risk of low birthweight, preterm delivery, language delay, leukemia, childhood brain tumors, and autism and increased cognitive performance, although the evidence is inconsistent (Barua et al. 2014b; O'Neill et al. 2014; Veena et al. 2016). However, higher folate intake during pregnancy has also been associated with an increased risk of some diseases, including childhood retinoblastoma and early respiratory illness (Barua et al. 2014b), Many of these conditions are also linked with aberrant DNA methylation (Smith 2011; Schiepers et al. 2012; Turan et al. 2012; Nardone et al. 2014; Schoofs et al. 2014; Vidal et al. 2014); however, the extent to which folate-dependent epigenetics underlies these diseases in humans is unknown.

Although maternal folate status is linked to both disease risks and DNA methylation, it is difficult to determine if folate-induced modulation of DNA methylation is a causative factor in altered disease risk. Few studies are yet to be conducted which examine both maternal folate, offspring methylation, and disease risk in the same cohort, and this will require future longitudinal investigations. Additionally, while folate is a critical player in the generation of methyl donors, it is also involved in multiple other cellular processes, including DNA replication and repair, and is inversely associated with homocysteine, an independent risk factor for cardiovascular and neurodegenerative diseases. Therefore, these mechanisms may add or interact with DNA methylation aberration, or other genetic or lifestyle risk factors, to modulate disease risk.

One human study that unites disease outcomes with both maternal folate and DNA methylation profiling used aborted fetal tissue and compared fetuses with and without NTDs. This study observed both a lower maternal serum folate and lower global DNA methylation in the brain tissue of those with NTDs, compared to those without (Chang et al. 2011).

Additionally, animal models provide evidence of modulation of DNA methylation and disease markers in offspring, as a response to maternal folate. The agouti mouse model provides a clear proof of principle whereby methyl-donor diet influences phenotype via a DNA methylation-dependent pathway (Wolff et al. 1998; Cooney et al. 2002; Waterland et al. 2007). Waterland et al. have proposed that either the reduced methyl-donor availability lead to altered DNA methylation at metastable epialleles by affecting ICM or the activity of Dnmt1 or that the repression of critical genes may occur during de novo DNA methylation in early fetal development

(Waterland and Michels 2007). Additionally, a methyl-deficient diet during early development in female mature sheep results in alterations of promoter DNA methylation and leads offspring to obesity, altered immune responses, insulin resistance, and elevated blood pressure (Sinclair et al. 2007). However, it is unclear whether these outcomes are DNA methylation signature dependent or related to additional cellular functions of folate.

Maternal folate status may be linked to global DNA methylation in the gastrointestinal tracts of offspring (McKay et al. 2011a). Methyl-donor deficiency has also been linked the impaired release of ghrelin into the blood (Bossenmeyer-Pourie et al. 2010). This may be a mechanism through which maternal folate and DNA methylation link to the modulation of birthweight and growth in offspring.

In utero deprivation of methyl donors, including folate, has been linked to metabolic syndrome, liver disease, and heart disease in rodents (Garcia et al. 2011; Pooya et al. 2012). These changes relate to decreased expression of PPAR-α (Garcia et al. 2011; Sun et al. 2015). Maternal folate supplementation protects against the aberrant methylation of *Ppar* in a rodent model of protein restriction. Hypomethylation of the *PPAR-α*, and other loci in the liver, affects nuclear receptor activity and results in the dysregulation of mitochondrial fatty acid oxidation (Garcia et al. 2011; Pooya et al. 2012) and thus may provide a epigenetic link between maternal folate status and these conditions.

Conclusions

The link between maternal folate status and disease is well established for NTDs and is becoming established for numerous other conditions in offspring. Due to the critical role of folate in the provision of methyl-groups for DNA methylation via the1CM cycle, it is logical that the role of DNA methylation in modulating these relationships is now a burgeoning area of investigation. Maternal folate status does not appear to be related to global DNA methylation in cord blood; however, it may be related in other tissues. Further studies are required to assess additional tissues.

Multiple studies have now found an association between maternal folate status and altered DNA methylation at multiple loci in cord blood and other tissues; however, more research is needed to understand the consequences of modulated methylation at these loci in terms of gene expression and disease outcomes. Although many diseases linked to aberrant DNA methylation are also linked to suboptimal maternal folate, additional research is required to fully elucidate the mechanisms involved. This endeavor is hampered by the multifactorial etiology of many of the diseases of interest and the fact that folate is critical in multiple cellular processes in addition to the provision of methyl donors for DNA methylation. Studies in humans are hampered by differences in underlying genetics, variations in folate fortification between countries, and the difficulty involved in accessing fetal or offspring tissues, other than blood and cord blood. However, given the established recommendation of folate supplementation during the periconceptional period to prevent NTDs, elucidating the impacts of maternal folate status on DNA methylation and risk of other diseases remains important to ensure dosing is optimal and that there are limited off target adverse effects.

Dictionary of Terms

- **CpG** – Cytosine and guanine separated by only one phosphate; phosphate links any two nucleosides together in DNA.
- **Developmental origins of health and disease (DOHaD) hypothesis** – A hypothesis that proposes that environment in utero predicts the risk of postnatal complex diseases.
- **Erythrocyte** – Red blood cell.
- **Methylome** – The sum of the methylation modifications on the nucleic acids of an organism's genome.
- **Metastable epiallele** – Where epigenotype established stochastically in the early embryo and maintained throughout life.

Key Facts of Maternal Folate

- Supplementation with folate is recommended during pregnancy worldwide to prevent neural tube defects.
- Many countries have mandatory or voluntary folate fortification in the food chain to ensure adequate levels in women of childbearing age.
- Supplementation and fortification have been successful in reducing the incidence of neural tube defects in offspring.
- Folate is critical in DNA methylation as the key source of methyl donors via the one-carbon metabolism cycle.
- Maternal folate status is related to the risk of multiple other conditions in offspring and is involved related to differential modulation of methylation at numerous of CpG sites.

Summary Points

- Folate is critical in DNA methylation as the key source of methyl donors via the one-carbon metabolism cycle.
- Maternal folate supplementation and fortification has successfully reduced the incidence of neural tube defects; it is not clear if this is DNA methylation dependent or independent.
- Maternal folate appears to be related to locus-specific methylation, but not global methylation in offspring.
- Many diseases linked to aberrant DNA methylation are also linked to suboptimal maternal folate; additional research is required to fully elucidate the mechanisms involved.

- Multifactorial etiology of many of the diseases of interest, and the fact that folate is critical in multiple cellular processes in addition to the provision of methyl donors for DNA methylation, makes studying these relationships complex.

References

Amarasekera M et al (2014) Genome-wide DNA methylation profiling identifies a folate-sensitive region of differential methylation upstream of ZFP57-imprinting regulator in humans. FASEB J 28(9):4068–4076

Amouroux R et al (2016) De novo DNA methylation drives 5hmC accumulation in mouse zygotes. Nat Cell Biol 18(2):225–233

Azzi S et al (2014) Degree of methylation of ZAC1 (PLAGL1) is associated with prenatal and post-natal growth in healthy infants of the EDEN mother child cohort. Epigenetics 9(3):338–345

Ba Y et al (2011) Relationship of folate, vitamin B12 and methylation of insulin-like growth factor-II in maternal and cord blood. Eur J Clin Nutr 65(4):480–485

Bailey L, Gregory J (2006) Folate: present knowledge in nutrition, vol I. International Life Sciences Institute, Washington, DC, pp 278–301

Barua S et al (2014a) Single-base resolution of mouse offspring brain methylome reveals epigenome modifications caused by gestational folic acid. Epigenetics Chromatin 7(1):3

Barua S et al (2014b) Folic acid supplementation in pregnancy and implications in health and disease. J Biomed Sci 21:77

Boeke CE et al (2012) Gestational intake of methyl donors and global LINE-1 DNA methylation in maternal and cord blood: prospective results from a folate-replete population. Epigenetics 7(3):253–260

Bortolus R et al (2014) Prevention of congenital malformations and other adverse pregnancy outcomes with 4.0 mg of folic acid: community-based randomized clinical trial in Italy and the Netherlands. BMC Pregnancy Childbirth 14:166

Bossenmeyer-Pourie C et al (2010) Methyl donor deficiency affects fetal programming of gastric ghrelin cell organization and function in the rat. Am J Pathol 176(1):270–277

Burdge GC, Lillycrop KA (2010) Nutrition, epigenetics, and developmental plasticity: implications for understanding human disease. Annu Rev Nutr 30:315–339

Chang H et al (2011) Tissue-specific distribution of aberrant DNA methylation associated with maternal low-folate status in human neural tube defects. J Nutr Biochem 22(12):1172–1177

Choi S, Friso S (2010) Epigenetics: a new bridge between nutrition and health. Adv Nutr 1(1):8–16

Cooney CA (2006) Germ cells carry the epigenetic benefits of grandmother's diet. Proc Natl Acad Sci U S A 103(46):17071–17072

Cooney CA et al (2002) Maternal methyl supplements in mice affect epigenetic variation and DNA methylation of offspring. J Nutr 132(8 Suppl):2393s–2400s

Cooper WN et al (2012) DNA methylation profiling at imprinted loci after periconceptional micronutrient supplementation in humans: results of a pilot randomized controlled trial. FASEB J 26(5):1782–1790

Cropley JE et al (2006) Germ-line epigenetic modification of the murine A vy allele by nutritional supplementation. Proc Natl Acad Sci U S A 103(46):17308–17312

Duhl DM et al (1994) Neomorphic agouti mutations in obese yellow mice. Nat Genet 8(1):59–65

Friso S, Choi SW (2005) Gene-nutrient interactions in one-carbon metabolism. Curr Drug Metab 6(1):37–46

Fryer AA et al (2009) LINE-1 DNA methylation is inversely correlated with cord plasma homocysteine in man: a preliminary study. Epigenetics 4(6):394–398

Fryer AA et al (2011) Quantitative, high-resolution epigenetic profiling of CpG loci identifies associations with cord blood plasma homocysteine and birth weight in humans. Epigenetics 6(1):86–94

Garcia MM et al (2011) Methyl donor deficiency induces cardiomyopathy through altered methylation/acetylation of PGC-1alpha by PRMT1 and SIRT1. J Pathol 225(3):324–335
Godfrey KM et al (2011) Epigenetic gene promoter methylation at birth is associated with child's later adiposity. Diabetes 60(5):1528–1534
Gong L et al (2010) Gestational low protein diet in the rat mediates Igf2 gene expression in male offspring via altered hepatic DNA methylation. Epigenetics 5(7):619–626
Guo F et al (2014a) Active and passive demethylation of male and female pronuclear DNA in the mammalian zygote. Cell Stem Cell 15(4):447–458
Guo H et al (2014b) The DNA methylation landscape of human early embryos. Nature 511(7511):606–610
Heard E, Martienssen RA (2014) Transgenerational epigenetic inheritance: myths and mechanisms. Cell 157(1):95–109
Hoyo C et al (2011) Methylation variation at IGF2 differentially methylated regions and maternal folic acid use before and during pregnancy. Epigenetics 6(7):928–936
Hoyo C et al (2014) Erythrocyte folate concentrations, CpG methylation at genomically imprinted domains, and birth weight in a multiethnic newborn cohort. Epigenetics 9(8):1120–1130
Iqbal K et al (2011) Reprogramming of the paternal genome upon fertilization involves genome-wide oxidation of 5-methylcytosine. Proc Natl Acad Sci U S A 108(9):3642–3647
Joubert BR et al (2016) Maternal plasma folate impacts differential DNA methylation in an epigenome-wide meta-analysis of newborns. Nat Commun 7:10577
Kim YI et al (1996) Exon-specific DNA hypomethylation of the p53 gene of rat colon induced by dimethylhydrazine. Modulation by dietary folate. Am J Pathol 149(4):1129–1137
Leighton PA et al (1995) An enhancer deletion affects both H19 and Igf2 expression. Genes Dev 9(17):2079–2089
Lillycrop KA et al (2005) Dietary protein restriction of pregnant rats induces and folic acid supplementation prevents epigenetic modification of hepatic gene expression in the offspring. J Nutr 135(6):1382–1386
Mathers J (2008) Session 2: personalised nutrition. Epigenomics: a basis for understanding individual differences? Proc Nutr Soc 67(4):390–394
Mayer W et al (2000) Demethylation of the zygotic paternal genome. Nature 403(6769):501–502
McKay JA et al (2011a) Folate depletion during pregnancy and lactation reduces genomic DNA methylation in murine adult offspring. Genes Nutr 6(2):189–196
McKay JA et al (2011b) Effect of maternal and post-weaning folate supply on gene-specific DNA methylation in the small intestine of weaning and adult apc and wild type mice. Front Genet 2:23
McKay JA et al (2011c) Maternal folate supply and sex influence gene-specific DNA methylation in the fetal gut. Mol Nutr Food Res 55(11):1717–1723
Messerschmidt DM et al (2014) DNA methylation dynamics during epigenetic reprogramming in the germline and preimplantation embryos. Genes Dev 28(8):812–828
Miller JW, Ulrich CM (2013) Folic acid and cancer – where are we today? Lancet 381 (9871):974–976
Mozhui K et al (2015) Ancestry dependent DNA methylation and influence of maternal nutrition. PLoS One 10(3):e0118466
Nakouzi GA, Nadeau JH (2014) Does dietary folic acid supplementation in mouse NTD models affect neural tube development or gamete preference at fertilization? BMC Genet 15:91
Nardone S et al (2014) DNA methylation analysis of the autistic brain reveals multiple dysregulated biological pathways. Transl Psychiatry 4:e433
O'Neill RJ et al (2014) Maternal methyl supplemented diets and effects on offspring health. Front Genet 5:289
Oswald J et al (2000) Active demethylation of the paternal genome in the mouse zygote. Curr Biol 10(8):475–478
Pooya S et al (2012) Methyl donor deficiency impairs fatty acid oxidation through PGC-1alpha hypomethylation and decreased ER-alpha, ERR-alpha, and HNF-4alpha in the rat liver. J Hepatol 57(2):344–351

Rougier N et al (1998) Chromosome methylation patterns during mammalian preimplantation development. Genes Dev 12(14):2108–2113
Schiepers OJ et al (2012) DNA methylation and cognitive functioning in healthy older adults. Br J Nutr 107(5):744–748
Schlinzig T et al (2009) Epigenetic modulation at birth – altered DNA-methylation in white blood cells after caesarean section. Acta Paediatr 98(7):1096–1099
Schoofs T et al (2014) Origins of aberrant DNA methylation in acute myeloid leukemia. Leukemia 28(1):1–14
Selhub J et al (2000) Relationship between plasma homocysteine and vitamin status in the Framingham study population. Impact of folic acid fortification. Public Health Rev 28(1–4): 117–145
Shen L, Waterland RA (2007) Methods of DNA methylation analysis. Curr Opin Clin Nutr Metab Care 10(5):576–581
Sinclair KD et al (2007) DNA methylation, insulin resistance, and blood pressure in offspring determined by maternal periconceptional B vitamin and methionine status. Proc Natl Acad Sci U S A 104(49):19351–19356
Smith SD (2011) Approach to epigenetic analysis in language disorders. J Neurodev Disord 3(4):356–364
Smith ZD et al (2012) A unique regulatory phase of DNA methylation in the early mammalian embryo. Nature 484(7394):339–344
Steegers-Theunissen RP et al (2009) Periconceptional maternal folic acid use of 400 microg per day is related to increased methylation of the IGF2 gene in the very young child. PLoS One 4(11): e7845
Sun C et al (2015) Potential epigenetic mechanism in non-alcoholic fatty liver disease. Int J Mol Sci 16(3):5161–5179
Tang WW et al (2016) Specification and epigenetic programming of the human germ line. Nat Rev Genet 17:585–600
Tornaletti S, Pfeifer GP (1995) Complete and tissue-independent methylation of CpG sites in the p53 gene: implications for mutations in human cancers. Oncogene 10(8):1493–1499
Turan N et al (2012) DNA methylation differences at growth related genes correlate with birth weight: a molecular signature linked to developmental origins of adult disease? BMC Med Genet 5:10
van Mil NH et al (2014) Determinants of maternal pregnancy one-carbon metabolism and newborn human DNA methylation profiles. Reproduction 148(6):581–592
Veena SR et al (2016) Association between maternal nutritional status in pregnancy and offspring cognitive function during childhood and adolescence; a systematic review. BMC Pregnancy Childbirth 16:220
Vidal AC et al (2014) Maternal stress, preterm birth, and DNA methylation at imprint regulatory sequences in humans. Genet Epigenet 6:37–44
Waterland RA, Michels KB (2007) Epigenetic epidemiology of the developmental origins hypothesis. Annu Rev Nutr 27:363–388
Waterland RA et al (2006) Maternal methyl supplements increase offspring DNA methylation at Axin fused. Genesis 44(9):401–406
Waterland RA et al (2007) Diet-induced hypermethylation at agouti viable yellow is not inherited transgenerationally through the female. FASEB J 21(12):3380–3385
Wolff GL et al (1998) Maternal epigenetics and methyl supplements affect agouti gene expression in Avy/a mice. FASEB J 12(11):949–957
Wossidlo M et al (2011) 5-hydroxymethylcytosine in the mammalian zygote is linked with epigenetic reprogramming. Nat Commun 2:241

Modulation of microRNA by Vitamin D in Cancer Studies

91

Emma L. Beckett, Martin Veysey, Zoe Yates, and Mark Lucock

Contents

E. L. Beckett (✉)
School of Medicine and Public Health, The University of Newcastle, Ourimbah, NSW, Australia
e-mail: emma.beckett@newcastle.edu.au

M. Veysey
School of Medicine and Public Health, The University of Newcastle, Gosford Hospital, Gosford, NSW, Australia
e-mail: martin.veysey@health.nsw.gov.au

Z. Yates
School of Biomedical Sciences and Pharmacy, The University of Newcastle, Ourimbah, NSW, Australia
e-mail: zoe.yates@newcastle.edu.au

M. Lucock
School of Environmental and Life Sciences, The University of Newcastle, Ourimbah, NSW, Australia
e-mail: mark.lucock@newcastle.edu.au

V. B. Patel, V. R. Preedy (eds.), *Handbook of Nutrition, Diet, and Epigenetics*,
https://doi.org/10.1007/978-3-319-55530-0_4

Abstract

Vitamin D, a steroid hormone, is well known for its influence in regulating gene expression via the action of the vitamin D receptor, in addition to its classical roles in maintaining calcium homeostasis and bone health. Recently, vitamin D status has been linked to a number of additional nonskeletal diseases, including cancers. Aberrant miRNA profiles have been demonstrated in malignant tissues and in the serum and plasma of cancer patients, leading to investigations into the potential that vitamin D-dependent modulation of miRNA profiles is involved in determining the risk and progression of malignancy. A number of studies, mostly in cell culture models, have demonstrated the modulation of a number of miRNA in a number of cancers; however, results vary depending on the cell line, stimulation concentration, and time of treatment. Additional studies are needed to assess similar relationships in other diseases where risk is linked to vitamin D status. While few studies have been conducted in humans, differences in serum profiles relative to vitamin D levels have been demonstrated. miRNA may provide a link between vitamin D status and disease risk, and this may offer a potential therapeutic avenue. Evidence exists to show that vitamin D can modulate miRNA levels by altering expression of the enzymes involved in miRNA biogenesis and direct and indirect induction of miRNA transcription. However, additional studies are needed to fully elucidate the genetic pathways resulting in modulation of miRNA and to understand the complex interactions between miRNA and vitamin D-related targets.

Keywords

Vitamin D · miRNA · Calcitriol · Cancer · Vitamin D receptor · VDR · VDRE

List of Abbreviations

$1,25(OH)_2D$	1,25-dihydroxycholecalciferol/calcitriol
25(OH)D	25-hydroxycholecaliferol/calcidiol
Ago	Argonaute
AML	Acute myeloid leukemia
CRC	Colorectal cancer
LPS	Lipopolysaccharide
mRNA	Messenger RNA
miRNA	microRNA
pre-miRNA	Precursor microRNA
pri-miRNA	Primary microRNA
RISC	RNA-induced silencing complex
RXR	Retinoic acid receptor
VDR	Vitamin D receptor
VDRE	Vitamin D response element

Introduction

Vitamin D is a fat-soluble steroid vitamin and prohormone. It can endogenously be synthesized in the epidermis following UV-B exposure or consumed from dietary or supplementary sources as cholecalciferol (vitamin D_3) or ergocalciferol (vitamin D_2) (Prosser and Jones 2004).

The metabolism of vitamin D is a multistep enzymatic process. Ergocalciferol and cholecalciferol are converted by calciol-25-hydroxylase into 25-hydroxyergocalciferol and 25-hydroxycholecaliferol, respectively (25(OH)D), which is the circulating storage form of the vitamin. Calcidiol-1α-hydroxylase then converts 25(OH)D into calcitriol (1,25-dihydroxycholecalciferol; 1,25 $(OH)_2D$) which can then act on the vitamin D receptor (VDR). A 24-hydroxylase enzyme is responsible for the inactivation of both 25(OH)D and 1,25$(OH)_2D$ via hydroxylation (Prosser and Jones 2004) (Fig. 1).

Vitamin D exerts its pleotropic effects via genomic and non-genomic pathways. Non-genomic, rapid, actions occur in a wide range of cells via the activation of cell membrane receptors (Carmeliet et al. 2015). The genomic actions of 1,25$(OH)_2D$ occur via its action on the ubiquitously expressed VDR nuclear receptor which acts

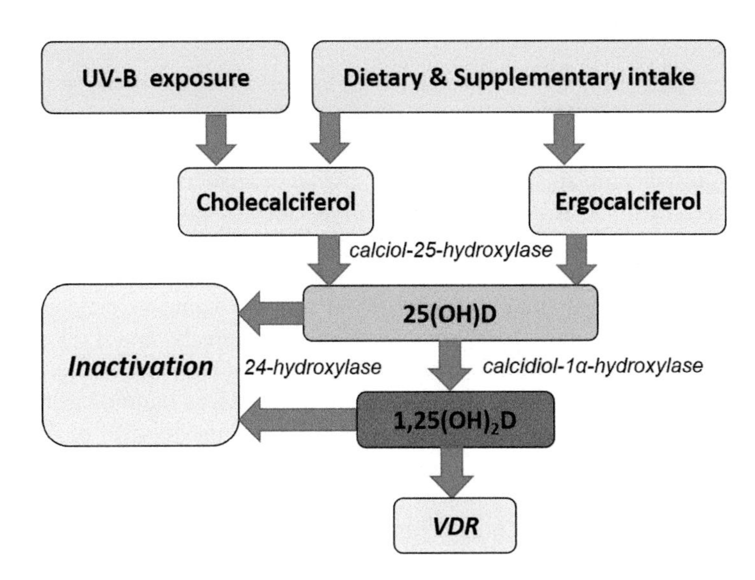

Fig. 1 Simplified flowchart of vitamin D metabolism. Vitamin D is obtained as cholecalciferol or ergocalciferol from foods and supplements or synthesized in the skin following sun exposure. Calciol-25-hydroxylase converts cholecalciferol and ergocalciferol to 25-hydroxycholecaliferol and 25-hydroxyergocalciferol, respectively (25(OH)D). Calcidiol-1α-hydroxylase then converts 25 (OH)D into 1,25-dihydroxycholecalciferol (1,25$(OH)_2D$), the biologically active form. The 24-hydroxylase enzyme inactivates 25(OH)D and 1,25$(OH)_2D$. The vitamin D receptor (VDR) is activated by 1,25$(OH)_2D$

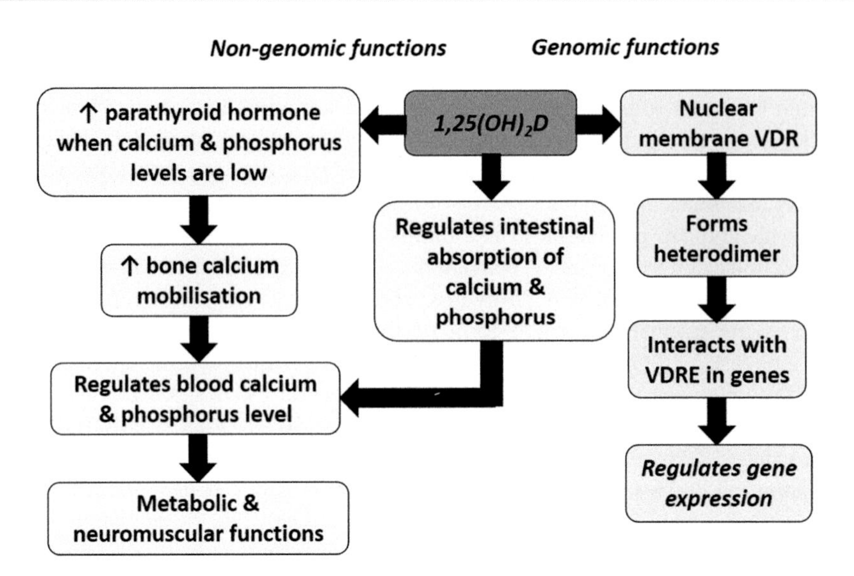

Fig. 2 Simple overview of the genomic and non-genomic functions of vitamin D. Storage and metabolism pathways are involved in maintaining bone mineralization and calcium homeostasis. Action via the vitamin D receptor (*VDR*) has additional consequences for gene transcription

as a nuclear transcription factor to activate the expression of genes with a vitamin D response element (VDRE) (Wang et al. 2012).

Vitamin D metabolism is inextricably linked to calcium homeostasis and bone health (Carmeliet et al. 2015) (Fig. 2). Evidence also is now mounting that suboptimal vitamin D levels are involved in a range of extra-skeletal conditions including cardiovascular disease (Wang et al. 2008), autoimmune and inflammatory disease (Adorini and Penna 2008), and cancers (Lappe et al. 2007). Vitamin D supplementation and vitamin D optimization are now being investigated in the treatment and prevention of these conditions (Lamprecht and Lipkin 2003). Interestingly, aberrant microRNA (miRNA) signatures have been noted in these same conditions. This, as well as the presence of VDRE at multiple miRNA loci (Guan et al. 2013), has led to a recent increase in the investigation of the role of vitamin D status in modulating miRNA expression and the impact this has on disease risk.

miRNA is a class of short (~18–22 nucleotides long) noncoding RNA. They are involved in maintenance of complex posttranscriptional networks essential to the regulation of gene expression. miRNA acts via complementarity with messenger RNA (mRNA) sequences, primarily to silence gene expression through either the degradation or inhibition of translation of target mRNA (Bartel 2004) (Fig. 3). It is predicted that miRNA regulates a large portion of coding genes, and through these actions, miRNA is thought to be involved in the fine-tuning of the transcriptional regulation of gene expression (Garzon et al. 2006). However, miRNA may also play distinct roles in the regulation of the proliferation, differentiation, and function of

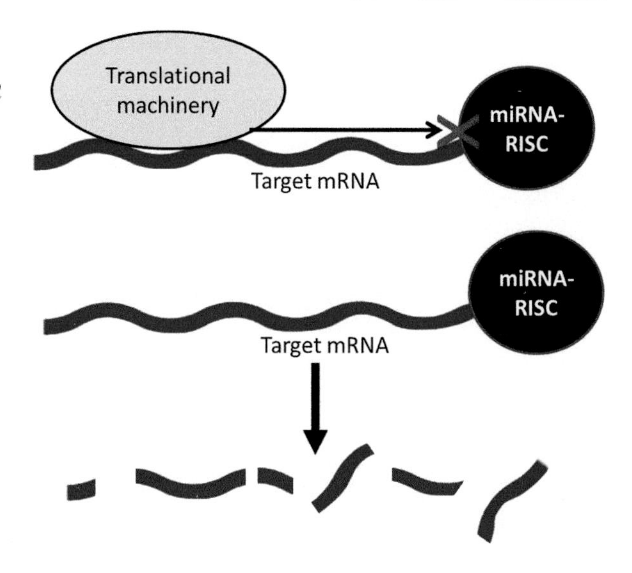

Fig. 3 MicroRNA mechanisms of action. Binding of the miRNA-RISC (RNA-induced silencing complex) can act on target mRNA (represented by the *blue line*) by (**a**) blocking gene translational machinery or (**b**) inducing mRNA instability or degradation

specific cell types. Therefore, miRNA regulatory networks may be particularly important for the regulation of the effects of signaling molecules that exert pleiotropic effects on numerous tissues throughout the body, such as the active form of vitamin D, $1,25(OH)_2D$.

miRNA is transcribed from genomic DNA, via the same mechanisms as protein-coding RNA (Lee et al. 2002). As such, the expression of miRNA can be regulated by other genetic or epigenetic mechanisms, including by nuclear receptor-response element interactions. A number of miRNA loci have VDRE in or upstream from transcription start sites (Fetahu et al. 2014). Following transcription, several processing steps are required to produce functional miRNA, allowing a number of potential points of modulation.

Immature primary precursors (pri-miRNA) are processed into precursor miRNA (pre-miRNA) by the enzyme drosha and the cofactor DGCR8 (Yi et al. 2003). After export to the cytoplasm, the enzyme dicer cleaves the pre-miRNA loop leaving a miRNA duplex ~18–22 nucleotides in length. In some cases, the protein argonaute (Ago) 2 cleaves the mature miRNA from the precursor (Cheloufi et al. 2010). Ago proteins then select a strand to be incorporated into the RNA-induced silencing complex (RISC); the other is discarded or incorporated into another RISC (Kim and Kim 2012). This process may result in two unique mature miRNA, being generated from the same precursor. Conversely, identical mature miRNA may be generated from unique pri-miRNA (Biasiolo et al. 2011).

Each mature miRNA can regulate multiple genes, and each gene may be targeted by a number of miRNA. The complexity of higher eukaryotic organisms requires the coordinated expression of multiple genes for housekeeping and specific functions in specialized tissues. Noncoding regulatory sequences, such as miRNA, are important in these processes.

Vitamin D: A Genetically Active Nutrient

Thousands of VDRE exist in the human genome (Ramagopalan et al. 2010; Heikkinen et al. 2011). However, 1,25(OH)$_2$D-sensitive genes differ by tissue and cell type. This may be due to ligand-induced and DNA binding-induced alterations in VDR structure and activity between cell types (Zhang et al. 2011). The activated VDR, incorporating co-modulator proteins, regulates gene transcription via well-documented mechanisms involving interaction with the VDRE at single gene loci or as part of networks of genes (Fig. 4).

VDR-VDRE interactions can regulate expression of protein-coding genes directly or indirectly, and posttranscriptional regulatory mechanisms for vitamin D have also been proposed. Regulation of mRNA levels via miRNA signaling is now becoming recognized as a potential additional mechanism for the action of vitamin D, and consequences of these interactions are now being investigated. Similar to protein-coding genes, miRNA expression may be regulated by VDR-VDRE interactions at miRNA loci (Fetahu et al. 2014). VDR-induced regulation of miRNA via VDRE has been demonstrated for several miRNA whose pri-miRNA have multiple VDR/RXR binding sites, suggesting that these miRNA could potentially be directly regulated by vitamin D metabolites (Guan et al. 2013). Additionally, miRNA profiles may be modulated indirectly by altered expression of other genes, including those involved in miRNA biogenesis, via modulation of

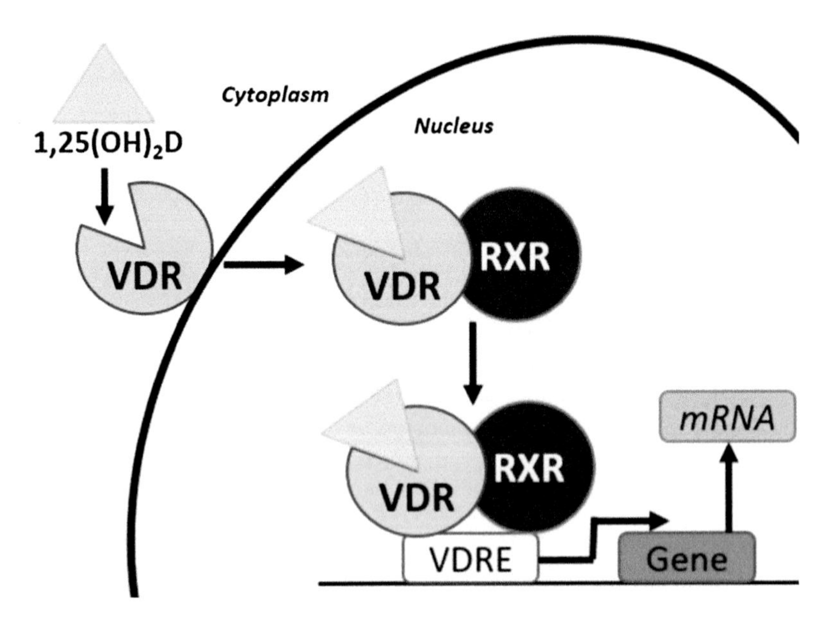

Fig. 4 Schematic of vitamin D regulation of gene transcription. The active vitamin D metabolite, calcitriol (1,25-dihydroxycholecalciferol, 1,25(OH)$_2$D, represented by the *triangle*), binds with the vitamin D receptor (*VDR*). Following ligand binding the VDR forms a heterodimer, often with the retinoid X receptor (*RXR*). Interaction of this heterodimer with the vitamin D response element (*VDRE*) triggers gene transcription

intermediate gene expression or through the modulation of other epigenetic marks (Fetahu et al. 2014).

Negative feedback loops can exist between miRNA and VDR signaling. One example is miR-125b, where overexpression can reduce VDR/RXR expression. As miR-125b levels are often reduced in malignant tissues, it has been proposed that such a decrease in miR-125b may result in the upregulation of VDR and increasing antitumor effects driven by vitamin D in cancer cell models (Mohri et al. 2009). The interacting and overlapping nature of these mechanisms allows for regulation of miRNA levels and thus their genetic targets via complex feedback loops.

Vitamin D Links to miRNA in Cancer Studies

The dysregulation of miRNA signatures in malignant tissues is well established. The interactions between vitamin D metabolites and miRNA expression have been well studied in the context of a number of cancers (Table 1). The majority of studies utilize cell culture models; however, there is significant epidemiological evidence for the association between vitamin D and risk of a number of cancers.

Prostate Cancer

The potential anticancer and chemopreventive properties of vitamin D in prostate tissue are supported by cell culture and animal models (Wang et al. 2009, 2011); however, epidemiological evidence is mixed (Giovannucci 2005). Low serum 1,25 $(OH)_2D$ has been linked to an increased risk of mortality (Fang et al. 2011) and recurrent disease (Trump et al. 2009) in those with prostate cancer but does not appear to be associated with increased risk for primary disease (Giovannucci 2005), suggesting that vitamin D may be an important disease progression or protection against more aggressive forms of the disease. Prostate cancer cells also have reduced levels of CYP24A activity, which may alter 1,25$(OH)_2D$ levels in prostate tissue independently of systemic levels (Hsu et al. 2001).

Unique miRNA signatures and increased dicer levels have been demonstrated in prostate cancer cells (Chiosea et al. 2006), and these profiles may be modified by vitamin D treatment. 1,25$(OH)_2D$ treatment of multiple prostate cancer cell lines led to the upregulation of miR-100 and miR-125b, both known tumor-suppressor miRNA. Furthermore, miR-100 and miR-125b expression was lower in primary prostate cancer cells than in cells derived from benign tissue, and prostatic concentrations of 1,25$(OH)_2D$ positively correlated with levels of both miRNA prostate cancer and benign prostate tissue (Giangreco et al. 2013).

PLK1 and *E2F3*, target genes of miR-100 and miR-125b, were downregulated by 1,25$(OH)_2D$ in a VDR-dependent manner (Giangreco et al. 2013). miR-125b also inhibits VDR expression (Mohri et al. 2009), and in turn VDR can downregulate miR-125b (Zhuo de et al. 2010; Iosue et al. 2013). Other targets of miR-125b include NCOR2/SMRT (Yang et al. 2012), which is itself a VDR target gene, suggesting the

Table 1 Summary of associations between vitamin D and microRNA in cancer. Results are mixed and likely to be dependent on type of malignancy, tissue, cell line, and time and dose of vitamin D treatment

miRNA	Related cancers/cell line	Observation[a]	Refs/notes
Prostate cancer			
miR-100 miR-125b	RWPE-1, RWPE-2, PrEC, and PrE cells	↑ By 1,25(OH)$_2$D treatment	Target genes include *PLK1*, *E2F3*, NCOR2/ SMRT, and *VDR*; Giangreco et al. (2013)
	Malignant versus benign prostate tissue	↓ In malignant prostate tissue, relative to control	
	Local prostatic 1,25 (OH)$_2$D levels in prostate cancer patients given oral vitamin D3 for 3–8 weeks	↑ By 1,25(OH)$_2$D treatment	Bartoszewski et al. (2011)
miR-1911* miR-106b	RWPE-1 cells	↑ By 1,25(OH)$_2$D treatment	4 additional miRNA upregulated and 5 downregulated, however, by <0.5-fold; Singh et al. (2015)
miR-494 miR-1973 miR-200b	HPr1 cells	↑ By 1,25(OH)$_2$D treatment	8 additional miRNA upregulated and 10 downregulated, however, by <0.5-fold; Singh et al. (2015)
miR-198 miR-1915*	HPr1AR cells	↑ By 1,25(OH)$_2$D treatment	5 additional miRNA upregulated and 6 downregulated, however, by <0.5-fold; Singh et al. (2015)
miR-138-1* miR-98 miR-1274b	LNCaP cells	↑ By 1,25(OH)$_2$D treatment	9 additional miRNA upregulated and 5 downregulated, however, by <0.5-fold; Singh et al. (2015)
miR-143 miR-616	LNCaP-C4–2 cells	↑ By 1,25(OH)$_2$D treatment	3 additional miRNA upregulated and 4 downregulated, however, by <0.5-fold; Singh et al. (2015)
miR-125b-1*	PC3 cells	↑ By 1,25(OH)$_2$D treatment	19 additional miRNA upregulated and 21 downregulated, however, by <0.5-fold; Singh et al. (2015)
miR-520 g miR-933		↓ By 1,25(OH)$_2$D treatment	
miR-106b	RWPE-1 cells	↑ By 1,25(OH)$_2$D treatment	Contributes to p21-mediated cell-cycle arrest; Sakaki et al. (2005) and Thorne et al. (2011)
	Local prostatic 1,25 (OH)$_2$D levels in prostate cancer patients given oral vitamin D3 for 3–8 weeks	↑ By 1,25(OH)$_2$D treatment	Bartoszewski et al. (2011)

(*continued*)

Table 1 (continued)

miRNA	Related cancers/cell line	Observation[a]	Refs/notes
miR-98	LNCaP cells	↑ By 1,25(OH)$_2$D treatment	Wang et al. (2009)
miR-663	LNCaP cells Malignant prostate tissue	↑ By 1,25(OH)$_2$D treatment correlates with Gleeson score	Wang et al. (2011), Liu et al. (2012), and Jiao et al. (2014)
miR-371-5p	LNCaP cells Malignant prostate tissue	↑ By 1,25(OH)$_2$D treatment also correlates with cancer progression	Wang et al. (2011), Liu et al. (2012), and Jiao et al. (2014)
miR-22 miR-29ab miR-134 miR-371-5p miR-663 miR-1207-5p	LNCaP cells	↑ By 1,25(OH)$_2$D treatment	Wang et al. (2011)
miR-17/92 Cluster miR-20a	LNCaP cells	↓ By 1,25(OH)$_2$D treatment	miR-17/92 cluster known oncogenic function; Wang et al. (2011), He et al. (2005), and Volinia et al. (2006)
miR-141 miR-331-3p miR-103 let-7a/b	Local prostatic 1,25 (OH)$_2$D levels in prostate cancer patients given oral vitamin D for 3–8 weeks	↑ By 1,25(OH)$_2$D treatment	Bartoszewski et al. (2011)
Breast cancer			
miR-26b miR-182 miR-200b miR-200c Let-7 (family)	MCF12F cells	↑ In conditions of serum starvation prevented by 1,25(OH)$_2$D treatment	Peng et al. (2010)
miR-125b	MCF7 cells	miR-125b downregulates VDR expression. These cells are resistant to the antiproliferative effects of 1,25(OH)$_2$D treatment seen in other cell lines	Mohri et al. (2009)
Colorectal cancer			
miR-22 miR-146a miR-222	SW80-ADH and HCT116 cells	↑ By 1,25(OH)$_2$D treatment	miR-222 and miR-22 target *DNMT3* and *HDAC4*, respectively; Sonkoly et al. (2012), Zhang et al. (2010), and Lee et al. (2013)

(*continued*)

Table 1 (continued)

miRNA	Related cancers/cell line	Observation[a]	Refs/notes
miR-203	SW80-ADH and HCT116 cells	↓ By $1,25(OH)_2D$ treatment	miR-203 suppresses the proto-oncogene *JUN*; Sonkoly et al. (2012)
miR-627	HT29 cells	↑ By $1,25(OH)_2D$ treatment	Correlated with a downregulation of expression of *JMJD1A*, a histone demethylase gene; Padi et al. (2013)
Leukemia			
miR-32	HL60-G and U937 cells	↑ By $1,25(OH)_2D$ treatment	Suppresses pro-apoptotic factor Bcl-2-like protein 11. Decreased expression of miRNA-32 can sensitize AML cells to the cytotoxic agents; Gocek et al. (2011)
Let-7e-5p, miR-22-3p, miR-146a-5p	HL60 cells	↑ By $1,25(OH)_2D$ treatment	Additional miRNA modulated but not confirmed by qPCR; Fontemaggi et al. (2015)
miR-96-5p, miR-17-5p	HL60 cells	↓ By $1,25(OH)_2D$ treatment	Additional miRNA modulated but not confirmed by qPCR; Fontemaggi et al. (2015)
miR-181a/b	HL60 and U937	↓ By $1,25(OH)_2D$ treatment	Associated with arrest of cell-cycle progression in G_1 phase; Wang et al. (2009)
miR-17-5p/20a/106a cluster	HL60 PGK and siAgo2 cells	↓ By $1,25(OH)_2D$ treatment	These miRNA target AML1, VDR, and C/EBPβ; Iosue et al. (2013)
miR-125b, miR-155			
miR-22	HeLa and SiHa cells	↑ By $1,25(OH)_2D$ treatment	Dicer expression also upregulated; Gonzalez-Duarte et al. (2015)
miR-296-3p, miR-498			
miR-26a	HL60 and NB4 cells	↑ By $1,25(OH)_2D$ treatment	Salvatori et al. (2011)
Lung cancer			
miR-27b	Lung fibroblasts	↓ By $1,25(OH)_2D$ treatment	Li et al. (2015)
let-7a-2	A549 cells	↑ By $1,25(OH)_2D$ treatment	Guan et al. (2013)

(*continued*)

Table 1 (continued)

miRNA	Related cancers/cell line	Observation[a]	Refs/notes
Bladder cancer			
miR-17	253 J cells	Modulated by 1,25 $(OH)_2D$ treatment	Direction of modulation not reported; Ma et al. (2015)
let-7a			
miR-1201			
miR-10a	253 J-BV cells	Modulated by 1,25 $(OH)_2D$ treatment	Direction of modulation not reported; Ma et al. (2015)
miR-22			
miR-29a			
miR-30d			
miR-96			
miR-125b-1			
miR-126			
miR-130a			
miR-147			
miR-147b			
miR-193b, miR-335			
miR-421			
miR-454			
miR-542-5p			
miR-1237			
Gastric cancer			
miR-145	SGC-7901 and AGS cells	↑ By 1,25$(OH)_2D$ treatment	Suppressed *E2F3*, *CDK6*, *CDK2*, and *CCNA2* expression, genes involved in proliferation; Chang et al. (2015)

[a]↑ indicates upregulation, ↓ indicates downregulation

existence of multiple levels of co-regulation and an interdependent relationship between the VDR, the mRNA and miRNA transcriptomes, and perhaps the wider epigenome in regulation carcinogenesis.

In another study using seven prostate cancer cell lines, stimulation with 1,25 $(OH)_2D$ led to the modulation of 111 miRNA; however, there was limited correlation in expression between each cell line, suggesting that control of miRNA expression is flexible and multifaceted (Singh et al. 2015). In another study, miR-106b alone was upregulated by 1,25$(OH)_2D$ in prostate cells and contributed to *p21*-mediated cell-cycle arrest (Sakaki et al. 2005). miR-106b expression is required for the 1,25 $(OH)_2D$-induced feed-forward loop regulating *p21* expression in nonmalignant RWPE-1 cells (Sakaki et al. 2005; Thorne et al. 2011).

In LNCaP cells, 1,25$(OH)_2D$ treatment induced expression of miR-98, a tumor-suppressor miRNA, and suppressed proliferation. This occurred both via direct induction through the enhanced VDR-VDRE interaction in the miR-98 promoter

(Ting et al. 2013). In another study of LNCaP cells, multiple miRNAs were differentially regulated by both vitamin D and testosterone, with similar regulatory effects following treatment with either vitamin D or testosterone and an additive effect in combination. Among these, miR-22, miR-29ab, miR-134, miR-371-5p, miR-663, and miR-1207-5p are upregulated, while miR-17 and miR-20a, members of the miR-17/92 cluster, are downregulated. Multiple genes implicated in cell-cycle progression, lipid synthesis and accumulation, and calcium homeostasis are among targets of these miRNA (Wang et al. 2011). Elevated miR-371-5p and miR-663 expression have been correlated with cancer progression, and miR-663 expression positively associates with Gleason scores (Liu et al. 2012; Jiao et al. 2014). Conversely, the miR-17/92 cluster is known to be oncogenic, and its expression is higher in more advanced prostate cancers (He et al. 2005; Volinia et al. 2006).

Some of these cell culture findings have been paralleled in a study of prostate cancer patients given oral vitamin D3 prior to prostatectomy. 1,25$(OH)_2$D levels in prostate tissue positively correlated with the expression of multiple miRNA (miR-100, miR-125b, miR-106b, miR-141, miR-331-3p, miR-103, let-7a, and let-7b) in both normal and malignant prostate tissue (Bartoszewski et al. 2011). These results show that vitamin D may globally augment miRNA in both benign and malignant prostate tissue.

Breast Cancer

Low serum 1,25$(OH)_2$D is associated with decreased breast cancer risk, recurrence, and mortality, although results are mixed and may be dependent on breast cancer type (Carlberg and Campbell 2013; Jacobs et al. 2016). Unique miRNA signatures and differential dicer expression occur in cancerous, relative to benign breast tissue (Friedman et al. 2009; Wickramasinghe et al. 2009).

Vitamin D may play a role in protection from cellular stress, which may be relevant to carcinogenesis. In the breast epithelial cell line (MCF12F), 1,25 $(OH)_2$D treatment protected cells against death in models of starvation, oxidative stress, hypoxia, and apoptosis induction. Serum starvation led to significant increases in the expression of multiple miRNA including miR-26b, miR-182, miR-200b, miR-200c, and the let-7 family. However, this was reversed following 1,25$(OH)_2$D treatment, indicating that miRNA expression may be a mechanism that protects against cellular stress via a vitamin D-related process (Peng et al. 2010).

CYP24A has also been reported to be overexpressed in various cancers, including breast cancer. miR-125b, which is often underexpressed in cancer tissues (Townsend et al. 2005), targets CYP24A, decreasing expression (Komagata et al. 2009). miR-125b has also been implicated in the resistance to 1,25$(OH)_2$D described in MCF7 breast cancer cells by suppressing endogenous levels of VDR protein (Mohri et al. 2009).

Colorectal Cancer

There appears to be an inverse correlation between serum vitamin D levels and risk for colorectal cancer (CRC) (Gandini et al. 2011). Mice fed a vitamin D-deficient diet have larger tumors and decreased VDR and CYP27B1 levels, relative to mice fed a normal diet (Tangpricha et al. 2005). Mice fed a low calcium and vitamin D, and high fat diets, develop colonic tumors, while mice receiving calcium and vitamin D supplementation had reduced tumor progression (Newmark et al. 2009).

Vitamin D-dependent suppression of proliferation has been demonstrated in malignant CRC cell lines. In SW80-ADH and HCT116 cells, the antiproliferative effects of 1,25$(OH)_2$D treatment correlated with the induction of miR-22, miR-146a, and miR-222 and a reduction in miR-203 expression. miR-203 is a known suppressor of the proto-oncogene *JUN* (Sonkoly et al. 2012). miR-222 and miR-22 are known epi-miRNA, targeting *DNMT3* and *HDAC4*, respectively (Zhang et al. 2010; Lee et al. 2013). Moreover, miR-22 expression is lower in human colon cancer tissue compared to surrounding normal tissue, correlating with changes in VDR expression levels (Alvarez-Díaz et al. 2012). This suggests miR-22 may have an anticancer effect that is modulated by vitamin D and the VDR.

However, in another CRC cell line (HT29), miR-627 was the only miRNA significantly upregulated by 1,25$(OH)_2$D treatment (Padi et al. 2013). Overexpression of miR-627 reduced proliferation of CRC cells in culture and growth of xenograft tumors in mice. Conversely, blocking miR-627 activity inhibited the effects of 1,25$(OH)_2$D. The biological plausibility of these interactions is further evidenced by the decrease of miR-627 in human colon adenocarcinoma samples, relative to healthy controls (Padi et al. 2013).

More complex feedback loops may exist between miRNA- and vitamin D-regulated pathways. In HT29 and HCT116 cells, miR-346 has been shown to target VDR. Blocking of miR-346 expression prevented the inhibition of VDR by TNF-α, confirming a role for miR-346 in modulating the relationship between VDR and inflammation. This may, therefore, have implications for carcinogenesis (Chen et al. 2014).

Leukemia

1,25$(OH)_2$D modulates cellular differentiation and proliferation in leukemia cell lines (Miyaura et al. 1981; Munker et al. 1986), and vitamin D status may be predictive of leukemia risk (Mohr et al. 2011). However, vitamin D therapy trials have had limited success (Jorde et al. 2012). Several studies have demonstrated aberrant miRNA profiles in leukemia (Schotte et al. 2012).

1,25$(OH)_2$D treatment of acute myeloid leukemia (AML) cells led to the induction of features of normal monocytes, paralleled by increased miR-32 expression. miR-32 suppresses the pro-apoptotic factor Bcl-2-like protein 11 (also known as Bim). Suppression of drosha, dicer, or miR-32 activity led to Bim suppression

(Gocek et al. 2011). Treatment of pro-myeloblastic (HL60) and promonocytic leukemia (U937) cells with low concentrations of 1,25(OH)$_2$D also led to decreased expression of miR-181a and miR-181b in a dose- and time-dependent manner and the arrest of the cell-cycle progression in the G$_1$ phase. Transfection of pre-miR-181a attenuated these effects, suggesting that aberrant expression may contribute to the malignant phenotype and that this can be moderated by vitamin D treatment (Wang et al. 2009). miR-181a inhibition by 1,25(OH)$_2$D results in increased p27Kip1 mRNA and protein, which, in turn, leads to G$_1$/S blockade (Wang et al. 2009).

Treatment of HL60 cells with 1,25(OH)$_2$D also induces differentiation from myoblast to monocytes (Fontemaggi et al. 2015). Vitamin D-induced differentiation is associated with modulated expression of 31 miRNA related to the ribosomal machinery (9 miRNA upregulated and 22 miRNA downregulated). let-7e-5p, miR-22-3p, and miR-146a-5p were validated by qPCR as upregulated by vitamin D treatment, and miR-96-5p and miR-17-5p were downregulated (Fontemaggi et al. 2015).

1,25(OH)$_2$D treatment in AML cells has also downregulated the miR-17-5p/20a/106a cluster, miR-125b, and miR-155, which target AML1, VDR, and C/EBPβ (Iosue et al. 2013), and upregulated mir-26a (Salvatori et al. 2011), which targets the transcriptional repressor E2F7 (Salvatori et al. 2012). E2F7 repression contributes to the increased expression of p21Cip1/Waf1 observed during 1,25(OH)$_2$D-induced monocytic differentiation of AML cells. Moreover, silencing of E2F7 results in inhibition of c-Myc activity and downregulation of its transcriptional target, the oncogenic miR-17-92 cluster (Salvatori et al. 2012). This suggests that miR-26a may be involved in monocytic differentiation and leukemogenesis.

In VDR expressing HeLa and SiHa cells, but not in VDR-deficient C33-A cells, treatment with 1,25(OH)$_2$D upregulated the expression of dicer, with an associated modification of tumor-suppressor miRNA such as miR-22, miR-296-3p, and miR-498 (Gonzalez-Duarte et al. 2015).

Lung Cancer

1,25(OH)$_2$D has been implicated in regulating cellular differentiation in lung tissue. 1,25(OH)$_2$D treatment of fibroblasts downregulates miR-27b expression and inhibits differentiation. Overexpression of miR-27b reduced VDR protein levels via direct targeting of the VDR 3′ untranslated region, while inhibition had the opposite effect (Li et al. 2015).

VDR-VDRE interaction has been shown to upregulate let-7a-2 expression in human lung cancer (A549) cells (Guan et al. 2013). In non-squamous cell carcinoma, lung cells, synthetic inhibitors of miR-92a and miR-1226*, caused toxicity, associated with loss of p53 expression and sequence-specific downregulation of the miR-17-92 cluster. Downregulation of the miR-17-92 cluster was shown to be selectivity toxic due to derepression of vitamin D signaling via suppression of

CYP24A1. Interestingly, high *CYP24A1* expression was also significantly correlated with poor patient outcome in multiple lung cancer cohorts (Borkowski et al. 2015).

Bladder Cancer

The bladder cancer cell lines, 253J and 253J-BV, both express functional VDR. 1,25 $(OH)_2D$ differentially regulated miRNA expression profiles in these cells in a dynamic manner; however, profiles were unique between each cell line. Pathway analysis of the miRNA target genes revealed distinct patterns of contribution to the molecular functions and biological processes in the two cell lines (Ma et al. 2015).

Gastric Cancer

In gastric cancer cells, miR-145 expression was induced by $1,25(OH)_2D$ treatment in a dose-dependent manner. Inhibition of miR-145 also reverses the antiproliferative effect of $1,25(OH)_2D$. Furthermore, miR-145 expression was lower in tumor tissue, relative to normal tissue. miR-145 expression also correlated with increased E2F3 transcription factor expression. miR-145 overexpression reduced cell viability and induced cell arrest in S-phase by targeting E2F3 and CDK6. Conversely, miR-145 inhibition resulted in the suppression of $1,25(OH)_2D$-mediated downregulation of *E2F3*, *CDK6*, *CDK2*, and *CCNA2* expression. Therefore, it is apparent that miR-145 mediates the antiproliferative and gene regulatory effects of vitamin D in gastric cancer cells, and as such, it might hold promise for prognosis and therapeutic strategies for gastric cancer treatment (Chang et al. 2015).

Potential Mechanisms

While a wealth of data now exists to show that vitamin D metabolites are able to modulate miRNA levels, the mechanisms of modulation are not fully clear. Direct VDR-VDRE interactions have been demonstrated for some miRNA (Griffin et al. 2001), but not all. However, the heterogeneity in VDRE consensus sequences and for VDREs to be very distal, up to 100 kb away from the transcription start site (Giangreco and Nonn 2013), may pose difficulties to the elucidation of which relationships are directly VDR-VDRE interaction dependent.

Vitamin D stimulation may also modulate the activity or expression of factors involved in miRNA biogenesis. In myeloid cell lines and primary blasts, it has been demonstrated that Ago2 has a key role in human monocytic cell fate determination and in lipopolysaccharide-induced inflammatory response of $1,25(OH)_2D$)-treated cells. The silencing of Ago2 impairs the vitamin D-dependent downregulation of the miR-17-5p/20a/106a cluster, miR-125b and miR-155; the accumulation of their translational targets AML1, VDR, and C/EBPb; and the monocytic cell differentiation (Iosue et al. 2013). Additional mechanisms may exist including the alteration of chromatin states following VDR stimulation, which may alter transcriptional activity (Disanto et al. 2012).

Modulation of inflammation may provide a link between vitamin D status and disease processes. VDR signaling attenuates toll-like receptor-mediated inflammation by enhancing the negative feedback inhibition. VDR inactivation leads to hyperinflammatory responses in mice and macrophage cultures when challenged with lipopolysaccharide (LPS), due to miR-155 overproduction that suppresses SOCS1, a key regulator that enhances the negative feedback loop. SOCS inhibits pro-inflammatory pathways that release cytokines and can inhibit the LPS-induced inflammatory response by directly blocking TLR4 signaling by targeting the IL-1R-associated kinases (IRAK) 1 and 4 (Kinjyo et al. 2002; Chen et al. 2013). Deletion of miR-155 in mice attenuates the vitamin D suppression of lipopolysaccharide-induced inflammation via this pathway (Chen et al. 2013; Li et al. 2014). miR-146 has also been shown to modulate inflammatory responses mediated by TLR4 (Ye and Steinle 2016). These results suggest that vitamin D can orchestrate miRNA diversity involved in TLR signaling, thereby regulating inflammatory responses and activation of immune responses.

1,25(OH)2D also promotes the immaturity of dendritic cells (DCs) by inducing downregulation of co-stimulatory molecules and interleukin (IL)-12 with enhanced expression of IL-10 production resulting in decreased T-cell activation and responsiveness (Griffin et al. 2001). miR-378 levels are increased in antigen-presenting cells following vitamin D treatment (Mohri et al. 2009; Pedersen et al. 2009; Lee et al. 2014). In patients with chronic hepatitis B, plasma levels of hepatitis B viral DNA were inversely correlated with plasma 25(OH)D levels and positively correlated with miR-378 levels. Serum miR-378 is proposed as a biomarker for hepatocellular carcinoma (Mohamadkhani et al. 2015). This may play a role in the relationship between inflammatory responses and cancer.

$1,25(OH)_2D$ stimulation enhances the susceptibility of malignant cells to the cytotoxicity of natural killer cells. This occurs partly via the downregulation of miR-302c and miR-520c, which leads to the upregulation of the NKG2D ligands. NKG2D is a receptor that recognizes several ligands, including polymorphic major histocompatibility complex class I chain-related proteins A and B (MICA/B) and unique long 16-binding proteins. These ligands are present on cancer cells and are recognized by NKG2D in a cell-structure-sensing manner, triggering natural killer cell cytotoxicity (Min et al. 2013). Vitamin D stimulation also upregulates let-7 expression in thymus cells and may play a role in differentiation and effector cell function (Pobezinsky et al. 2015).

miRNA is differentially expressed in serum and plasma of cancer patients and may serve as potential biomarkers for cancer diagnosis and prognosis (Beckett et al. 2014). While limited studies have been conducted on the role of vitamin D in regulating these biomarkers, the two small studies which have been conducted demonstrate that vitamin D supplementation and plasma 25(OH)D levels correlate with modulation of miRNA profiles (Enquobahrie et al. 2011; Jorde et al. 2012). However, additional studies are needed due to the small cohorts used in the available studies.

Conclusions

While elucidating the role of nutrition and diet in the modulation of miRNA profiles remains a relatively new field of investigation, there is mounting evidence of numerous relationships between vitamin D stimulation and miRNA modulation.

While there is significant epidemiological evidence linking both miRNA and vitamin D to the risk for, and progression of, a number of cancers, the majority of the studies conducted to date investigating the two simultaneously rely on cell culture systems, with varying time frames and doses of stimulation, and as such results can be highly variable. Additional observational and interventional studies in appropriately sized human cohorts are needed to elucidate if these observations translate directly into the more complex environment of exposures and interactions that occur in the human body and at normal physiological doses.

Furthermore, the implications of identified associations between vitamin D and miRNA signatures cannot be fully appreciated without further research to identify the mRNA targets of more miRNA, as the majority remain poorly defined. Additional functional studies are required in future research to understand if there are physiological consequences for modulation of miRNA by vitamin D and if these relationships can be harnessed for therapeutic or diagnostic purposes. The increasing affordability and practicality of miRNA assays are likely to contribute to an increased volume of investigation in coming years.

While associations have been identified between vitamin D, miRNA, and a number of cancers, additional investigations are needed for other conditions such as cardiovascular disease and diabetes that have also been linked to vitamin D status to determine the role of miRNA in these diseases.

Dictionary of Terms

- **miRNA** – microRNA – a class of short noncoding miRNA involved in the regulation of gene expression.
- **Transcription** – the production of gene mRNA from DNA.
- **Translation** – the production of protein from gene mRNA.
- **Transcription factor** – a protein that binds to specific DNA sequences, thereby controlling the rate of transcription of genetic information from DNA to messenger RNA.
- **Nuclear membrane** – the phospholipid bilayer which surrounds the genetic material and nucleolus in eukaryotic cells.

Key Facts of Vitamin D

- Vitamin D can be obtained from food, supplements, or synthesis in the skin.
- Nuclear membrane expression of the vitamin D receptor allows 1,25(OH)$_2$D, the active vitamin D metabolite, to regulate gene expression via interaction of the receptor with vitamin D response elements in genes.

- Suboptimal vitamin D status has been linked to increased risk, mortality, and recurrence in a number of diseases, which included many cancers.
- This may occur via modulation of miRNA signatures.
- Vitamin D treatment modulates miRNA expression in a number of malignant and normal cell lines.

Summary Points

- Vitamin D is well known for its influence of the regulation of gene expression.
- As well as bone health, vitamin D has been linked to risk for a number of additional nonskeletal diseases, including cancer.
- Aberrant miRNA signatures are well established in cancer tissues and in plasma of cancer patients.
- Cell culture models have linked vitamin D to miRNA profiles in studies of bone development, inflammation regulation, and a number of cancers; however, results vary depending on the cells studied and dose and duration of treatment.
- miRNA may provide a link between vitamin D status and disease risk, and this may offer a potential therapeutic avenue.
- Expression of miRNA can be altered by vitamin D receptor-vitamin D response element interactions directly or indirectly via other vitamin D-dependent mechanisms.

References

Adorini L, Penna G (2008) Control of autoimmune diseases by the vitamin D endocrine system. Nat Clin Pract Rheumatol 4:404–412

Alvarez-Díaz S et al (2012) MicroRNA-22 is induced by vitamin D and contributes to its antiproliferative, antimigratory and gene regulatory effects in colon cancer cells. Hum Mol Genet 21(10):2157–2165

Bartel DP (2004) MicroRNAs: genomics, biogenesis, mechanism, and function. Cell 116 (2):281–297

Bartoszewski R et al (2011) The unfolded protein response (UPR)-activated transcription factor X-box-binding protein 1 (XBP1) induces microRNA-346 expression that targets the human antigen peptide transporter 1 (TAP1) mRNA and governs immune regulatory genes. J Biol Chem 286(48):41862–41870

Beckett EL et al (2014) The role of vitamins and minerals in modulating the expression of microRNA. Nutr Res Rev 27(1):94–106

Biasiolo M et al (2011) Impact of host genes and strand selection on miRNA and miRNA* expression. PLoS One 6(8):e23854

Borkowski R et al (2015) Genetic mutation of p53 and suppression of the miR-17 approximately 92 cluster are synthetic lethal in non-small cell lung cancer due to upregulation of vitamin D signaling. Cancer Res 75(4):666–675

Carlberg C, Campbell MJ (2013) Vitamin D receptor signaling mechanisms: integrated actions of a well-defined transcription factor. Steroids 78(2):127–136

Carmeliet G et al (2015) Vitamin D signaling in calcium and bone homeostasis: a delicate balance. Best Pract Res Clin Endocrinol Metab 29(4):621–631

Chang S et al (2015) miR-145 mediates the antiproliferative and gene regulatory effects of vitamin D3 by directly targeting E2F3 in gastric cancer cells. Oncotarget 6(10):7675–7685

Cheloufi S et al (2010) A dicer-independent miRNA biogenesis pathway that requires ago catalysis. Nature 465:584–589

Chen Y et al (2013) 1,25-Dihydroxyvitamin D promotes negative feedback regulation of TLR signaling via targeting microRNA-155-SOCS1 in macrophages. J Immunol 190(7):3687–3695

Chen Y et al (2014) MicroRNA-346 mediates tumor necrosis factor alpha-induced downregulation of gut epithelial vitamin D receptor in inflammatory bowel diseases. Inflamm Bowel Dis 20(11):1910–1918

Chiosea S et al (2006) Up-regulation of dicer, a component of the MicroRNA machinery, in prostate adenocarcinoma. Am J Pathol 169(5):1812–1820

de Zhuo X et al (2010) Vitamin D3 up-regulated protein 1(VDUP1) is regulated by FOXO3A and miR-17-5p at the transcriptional and post-transcriptional levels, respectively, in senescent fibroblasts. J Biol Chem 285(41):31491–31501

Disanto G et al (2012) Vitamin D receptor binding, chromatin states and association with multiple sclerosis. Hum Mol Genet 21(16):3575–3586

Enquobahrie D et al (2011) Global maternal early pregnancy peripheral blood mRNA and miRNA expression profiles according to plasma 25-hydroxyvitamin D concentrations. J Matern Fetal Neonatal Med 24(8):1002–1012

Fang F et al (2011) Prediagnostic plasma vitamin D metabolites and mortality among patients with prostate cancer. PLoS One 6(4):e18625

Fetahu IS et al (2014) Vitamin D and the epigenome. Front Physiol 5:164

Fontemaggi G et al (2015) Identification of post-transcriptional regulatory networks during myeloblast-to-monocyte differentiation transition. RNA Biol 12(7):690–700

Friedman RC et al (2009) Most mammalian mRNAs are conserved targets of microRNAs. Genome Res 19(1):92–105

Gandini S et al (2011) Meta-analysis of observational studies of serum 25-hydroxyvitamin D levels and colorectal, breast and prostate cancer and colorectal adenoma. Int J Cancer 128(6):1414–1424

Garzon R et al (2006) MicroRNA expression and function in cancer. Trends Mol Med 12(12):580–587

Giangreco AA, Nonn L (2013) The sum of many small changes: microRNAs are specifically and potentially globally altered by vitamin D3 metabolites. J Steroid Biochem Mol Biol 136: 86–93

Giangreco A et al (2013) Tumor suppressor microRNAs, miR-100 and -125b, are regulated by 1,25-dihydroxyvitamin D in primary prostate cells and in patient tissue. Cancer Prev Res (Phila) 6(5):483–494

Giovannucci E (2005) The epidemiology of vitamin D and cancer incidence and mortality: a review (United States). Cancer Causes Control 16(2):83–95

Gocek E et al (2011) MicroRNA-32 upregulation by 1,25-dihydroxyvitamin D3 in human myeloid leukemia cells leads to Bim targeting and inhibition of AraC-induced apoptosis. Cancer Res 71(19):6230–6239

Gonzalez-Duarte RJ et al (2015) Calcitriol increases Dicer expression and modifies the microRNAs signature in SiHa cervical cancer cells. Biochem Cell Biol 93(4):376–384

Griffin MD et al (2001) Dendritic cell modulation by 1alpha,25 dihydroxyvitamin D3 and its analogs: a vitamin D receptor-dependent pathway that promotes a persistent state of immaturity in vitro and in vivo. Proc Natl Acad Sci U S A 98(12):6800–6805

Guan H et al (2013) 1,25-Dihydroxyvitamin D3 up-regulates expression of hsa-let-7a-2 through the interaction of VDR/VDRE in human lung cancer A549 cells. Gene 522(2):142–146

He L et al (2005) A microRNA polycistron as a potential human oncogene. Nature 435 (7043):828–833

Heikkinen S et al (2011) Nuclear hormone 1alpha,25-dihydroxyvitamin D3 elicits a genome-wide shift in the locations of VDR chromatin occupancy. Nucleic Acids Res 39(21):9181–9193

Hsu JY et al (2001) Reduced 1alpha-hydroxylase activity in human prostate cancer cells correlates with decreased susceptibility to 25-hydroxyvitamin D3-induced growth inhibition. Cancer Res 61(7):2852–2856

Iosue I et al (2013) Argonaute 2 sustains the gene expression program driving human monocytic differentiation of acute myeloid leukemia cells. Cell Death Dis 4:e926

Jacobs ET et al (2016) Vitamin D and colorectal, breast, and prostate cancers: a review of the epidemiological evidence. J Cancer 7(3):232–240

Jiao L et al (2014) miR-663 induces castration-resistant prostate cancer transformation and predicts clinical recurrence. J Cell Physiol 229(7):834–844

Jorde R et al (2012) Plasma profile of microRNA after supplementation with high doses of vitamin D3 for 12 months. BMC Res Notes 5(1):245

Kim Y, Kim V (2012) MicroRNA factory: RISC assembly from precursor MicroRNAs. Mol Cell 46(4):384–386

Kinjyo I et al (2002) SOCS1/JAB is a negative regulator of LPS-induced macrophage activation. Immunity 17(5):583–591

Komagata S et al (2009) Human CYP24 catalyzing the inactivation of calcitriol is post-transcriptionally regulated by miR-125b. Mol Pharmacol 76(4):702–709

Lamprecht S, Lipkin M (2003) Chemoprevention of colon cancer by calcium, vitamin D and folate: molecular mechanisms. Nat Rev Cancer 3:601–614

Lappe J et al (2007) Vitamin D and calcium supplementation reduces cancer risk: results of a randomized trial1,2. Am J Clin Nutr 85(6):1586–1591

Lee Y et al (2002) MicroRNA maturation: stepwise processing and subcellular localization. EMBO J 21(17):4663–4670

Lee J et al (2013) Hypermethylation and post-transcriptional regulation of DNA methyltransferases in the ovarian carcinomas of the laying hen. PLoS One 8(4):e61658

Lee HJ et al (2014) Low 25(OH) vitamin D3 levels are associated with adverse outcome in newly diagnosed, intensively treated adult acute myeloid leukemia. Cancer 120(4):521–529

Li YC et al (2014) MicroRNA-mediated mechanism of vitamin D regulation of innate immune response. J Steroid Biochem Mol Biol 144(Pt A):81–86

Li F et al (2015) 1alpha,25-Dihydroxyvitamin D3 prevents the differentiation of human lung fibroblasts via microRNA-27b targeting the vitamin D receptor. Int J Mol Med 36 (4):967–974

Liu PT et al (2012) MicroRNA-21 targets the vitamin D-dependent antimicrobial pathway in leprosy. Nat Med 18(2):267–273

Ma Y et al (2015) 1alpha,25(OH)2D3 differentially regulates miRNA expression in human bladder cancer cells. J Steroid Biochem Mol Biol 148:166–171

Min D et al (2013) Downregulation of miR-302c and miR-520c by 1,25(OH)2D3 treatment enhances the susceptibility of tumour cells to natural killer cell-mediated cytotoxicity. Br J Cancer 109(3):723–730

Miyaura C et al (1981) 1 alpha,25-Dihydroxyvitamin D3 induces differentiation of human myeloid leukemia cells. Biochem Biophys Res Commun 102(3):937–943

Mohamadkhani A et al (2015) Negative association of plasma levels of vitamin D and miR-378 with viral load in patients with chronic hepatitis B infection. Hepat Mon 15(6):e28315

Mohr SB et al (2011) Ultraviolet B and incidence rates of leukemia worldwide. Am J Prev Med 41(1):68–74

Mohri T et al (2009) MicroRNA regulates human vitamin D receptor. Int J Cancer 125(6):1328–1333

Munker R et al (1986) Vitamin D compounds. Effect on clonal proliferation and differentiation of human myeloid cells. J Clin Invest 78(2):424–430

Newmark HL et al (2009) Western-style diet-induced colonic tumors and their modulation by calcium and vitamin D in C57Bl/6 mice: a preclinical model for human sporadic colon cancer. Carcinogenesis 30(1):88–92
Padi S et al (2013) MicroRNA-627 mediates the epigenetic mechanisms of vitamin D to suppress proliferation of human colorectal cancer cells and growth of xenograft tumors in mice. Gastroenterology 145(2):437–446
Pedersen AW et al (2009) Phenotypic and functional markers for 1alpha,25-dihydroxyvitamin D(3)-modified regulatory dendritic cells. Clin Exp Immunol 157(1):48–59
Peng X et al (2010) Protection against cellular stress by 25-hydroxyvitamin D3 in breast epithelial cells. J Cell Biochem 110(6):1324–1333
Pobezinsky LA et al (2015) Let-7 microRNAs target the lineage-specific transcription factor PLZF to regulate terminal NKT cell differentiation and effector function. Nat Immunol 16(5):517–524
Prosser DE, Jones G (2004) Enzymes involved in the activation and inactivation of vitamin D. Trends Biochem Sci 29(12):664–673
Ramagopalan SV et al (2010) A ChIP-seq defined genome-wide map of vitamin D receptor binding: associations with disease and evolution. Genome Res 20(10):1352–1360
Sakaki T et al (2005) Metabolism of vitamin D3 by cytochromes P450. Front Biosci 10:119–134
Salvatori B et al (2011) Critical role of c-Myc in acute myeloid leukemia involving direct regulation of miR-26a and histone methyltransferase EZH2. Genes Cancer 2(5):585–592
Salvatori B et al (2012) The microRNA-26a target E2F7 sustains cell proliferation and inhibits monocytic differentiation of acute myeloid leukemia cells. Cell Death Dis 3:e413
Schotte D et al (2012) MicroRNAs in acute leukemia: from biological players to clinical contributors. Leukemia 26(1):1–12
Singh PK et al (2015) VDR regulation of microRNA differs across prostate cell models suggesting extremely flexible control of transcription. Epigenetics 10(1):40–49
Sonkoly E et al (2012) MicroRNA-203 functions as a tumor suppressor in basal cell carcinoma. Oncogene 1:e3
Tangpricha V et al (2005) Vitamin D deficiency enhances the growth of MC-26 colon cancer xenografts in Balb/c mice. J Nutr 135(10):2350–2354
Thorne JL et al (2011) Epigenetic control of a VDR-governed feed-forward loop that regulates p21 (waf1/cip1) expression and function in non-malignant prostate cells. Nucleic Acids Res 39(6):2045–2056
Ting H et al (2013) Identification of microRNA-98 as a therapeutic target inhibiting prostate cancer growth and a biomarker induced by vitamin D. J Biol Chem 288(1):1–9
Townsend K et al (2005) Autocrine metabolism of vitamin D in normal and malignant breast tissue. Clin Cancer Res 11(9):3579–3586
Trump DL et al (2009) Vitamin D deficiency and insufficiency among patients with prostate cancer. BJU Int 104(7):909–914
Volinia S et al (2006) A microRNA expression signature of human solid tumors defines cancer gene targets. Proc Natl Acad Sci U S A 103(7):2257–2261
Wang T et al (2008) Vitamin D deficiency and risk of cardiovascular disease. Circulation 117:503–511
Wang X et al (2009) MicroRNAs181 regulate the expression of p27Kip1 in human myeloid leukemia cells induced to differentiate by 1,25-dihydroxyvitamin D3. Cell Cycle 8(5): 736–741
Wang WL et al (2011) Effects of 1alpha,25 dihydroxyvitamin D3 and testosterone on miRNA and mRNA expression in LNCaP cells. Mol Cancer 10:58
Wang Y et al (2012) Where is the vitamin D receptor? Arch Biochem Biophys 523(1):123–133
Wickramasinghe NS et al (2009) Estradiol downregulates miR-21 expression and increases miR-21 target gene expression in MCF-7 breast cancer cells. Nucleic Acids Res 37 (8):2584–2595

Yang X et al (2012) miR-125b regulation of androgen receptor signaling via modulation of the receptor complex co-repressor NCOR2. Biores Open Access 1(2):55–62

Ye EA, Steinle JJ (2016) miR-146a attenuates inflammatory pathways mediated by TLR4/NF-kappaB and TNFalpha to protect primary human retinal microvascular endothelial cells grown in high glucose. Mediat Inflamm 2016:3958453

Yi R et al (2003) Exportin-5 mediates the nuclear export of pre-microRNAs and short hairpin RNAs. Genes Dev 17(24):3011–3016

Zhang J et al (2010) microRNA-22, downregulated in hepatocellular carcinoma and correlated with prognosis, suppresses cell proliferation and tumourigenicity. Br J Cancer 103(8):1215–1220

Zhang J et al (2011) DNA binding alters coactivator interaction surfaces of the intact VDR-RXR complex. Nat Struct Mol Biol 18(5):556–563

Epigenetics and Minerals: An Overview

92

Inga Wessels

Contents

Abstract

Epigenetics plays a decisive role in gene regulation and is vulnerable to environmental challenges, including supply with nutritional factors, such as minerals. The observation that the function of epigenetically active enzymes requires cofactors such as minerals supports this hypothesis. Data are available that reveal

I. Wessels (✉)
Institute of Immunology, RWTH Aachen University Hospital, Aachen, Germany
e-mail: iwessels@ukaachen.de

V. B. Patel, V. R. Preedy (eds.), *Handbook of Nutrition, Diet, and Epigenetics*,
https://doi.org/10.1007/978-3-319-55530-0_48

direct and indirect effects of essential minerals on the methylation status of the DNA, on epigenetic modifications of histones, and on the regulation of RNA interference. As is true for most epigenetically active factors, the mineral balance mostly effects the epigenome generation during embryonic development, but changes can be induced throughout life as part of lifelong epigenome editing. It has indeed been suggested that changes induced by minerals cumulate during aging and can be passed on to the next generation. Together, this suggests the use of mineral supplementation to prevent dysplasia originating from errors in establishing the epigenome or correct epigenetic disturbances. Despite immense advances in recent years, literature on the impact of minerals on the epigenome is still scarce compared to our general knowledge in nutritional epigenetics. This chapter provides an overview over epigenetic activities of calcium, chromium, manganese, magnesium, iron, selenium, and zinc and briefly mentions data for molybdenum and mineral mixes. An association between disturbed mineral balance, epigenetics, and certain types of diseases will be addressed as well.

Keywords
Calcium · Chromium · Manganese · Magnesium · Iron · Selenium · Zinc · Mineral balance · Nutritional epigenetics · Interactome

List of Abbreviations

BDNF	Brain-derived neurotrophic factor
HDAC	Histone deacetylase
hTMA	Hair tissue mineral analysis
ICP-AES	Inductively coupled plasma-atomic emission spectroscopy

Introduction

As broadly discussed within this book, gene expression is not only regulated by the DNA sequence but largely depends on the epigenome. Dysplasia originating from all developmental stages and a high number of diseases have so far been connected to epigenetic abnormalities, triggered by environmental factors including nutritional aspects (Kanherkar et al. 2014).

The association of nutrition-dependent epigenetics with the decision over healthy development or disease is generally accepted. One of the best illustrations for the epigenetic effects of early-life nutrition is the development of the bee. If larvae are fed with a royalactin-containing diet, they will develop a queen phenotype; if not, they will become workers. In this simple example, decision on the phenotype is primarily made by the DNA methylation status (Lyko et al. 2010). In case of humans, mechanisms are more complex than in the bee, not only because of a more complex (epi)genome but also regarding the composition of human nutrition and other aspects varying between individuals such as lifestyle and genetic makeup (Fig. 1; Lyko et al. 2010; Zhang et al. 2011). However, studies indicate that a lot of epigenetic mechanisms are conserved over a range of species. Earliest findings in

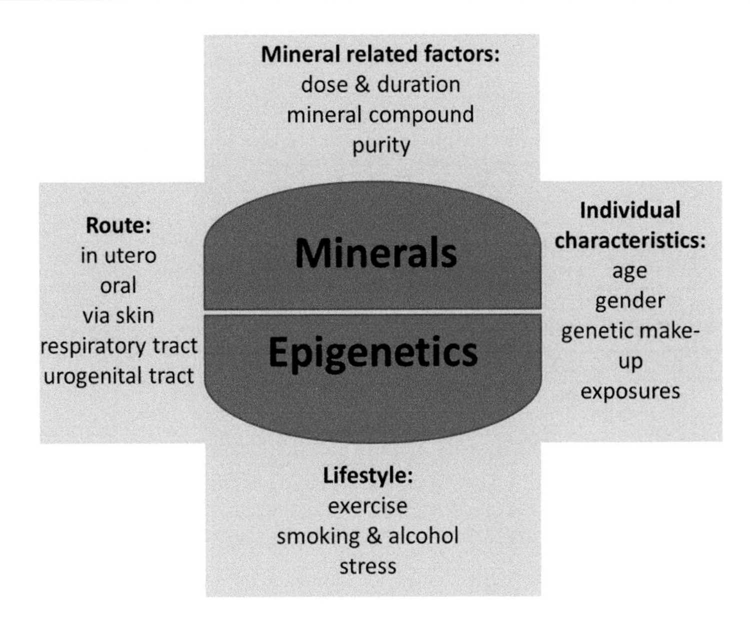

Fig. 1 Illustration of the interactome. The figure illustrates the interaction of mineral-related factors, mineral dose, characteristics of the individual exposed to the mineral, as well as their lifestyle on the effect a certain mineral can have on the epigenome. All factors together decide over health and disease

this research area mostly indicated that undernutrition affected epigenetics, especially during development (Brown and Susser 2008). Recently, investigations on single element level shed light in that the balance of individual minerals affects the setup of a functioning epigenome during development and also influences the epigenetic setup later in life.

For humans, most important and essential minerals include calcium, magnesium, iron, manganese, copper, iodine, selenium, and chromium. The need for the latter is still discussed controversially. As shown in Table 1, the amounts of minerals essential for the human body are rather low, ranging from a few μg to mg per day (Wessels 2015). The amount depends on developmental stage but also daily activity (Alegria-Torres et al. 2011). Epigenetic influences are suggested for most of the abovementioned elements, as will be described in the upcoming paragraphs.

As with most nutrition-based epigenetic effects, the effects of altered mineral supply or metabolism start to become apparent in utero (infantile window), especially if changes affect stem cells. Thus, embryonic to late adult stages can be influenced and promote diseases. Specific to minerals, their epigenetic effects display majorly in neurological disturbances (Keen et al. 2003; Wright and Baccarelli 2007; Wang et al. 2013; Tarale et al. 2016).

This chapter will describe some basic concepts of our understanding of how minerals alter the epigenome, including specification of the "interactome" (Ho et al. 2012; Watts 2015). Important research techniques, challenges, and pitfalls in this area of research will be briefly discussed, and major data for each mineral will be

Table 1 RDAs for epigenetically active trace minerals

Element	Age [years]	Ca mg	Mg mg	Fe mg	Zn mg	Mn mg	Cu μg	I μg	Se μg	Cr[b] μg
Infants	0–0.5[a]	200	40	6	2	0.003	200	40	15	0.2
	0.5–1	260	75	11	3	0.6	220	50	20	5.5
Children	1–3	700	80	7	3	1.2	340	90	20	11
	4–6	1000	130	10	5	1.5	440	90	25	13
	7–10	1100	170	8	7	1.7	550	120	30	15
Male	11–14	1300	270	11	8	1.9	700	150	40	25
	15–18	1300	400	11	11	2.2	890	150	55	35
	19–24	1100	400	8	11	2.3	900	150	70	35
	25–50	1000	420	8	11	2.3	900	150	70	35
	50+	1000	420	8	11	2.3	900	150	55	30
Female	11–14	1300	280	15	8	1.6	700	150	40	21
	15–18	1300	360	15	9	1.6	890	150	55	24
	19–24	1100	310	15	8	1.8	900	150	55	25
	25–50	1000	320	18	8	1.8	900	150	55	25
	50+	1200	320	10	8	1.8	900	150	55	20
Pregnant		1300	360	27	12	2.0	1000	220	60	29
Nursing	14–18	1300	360	10	13	2.6	1300	290	75	44
Nursing	19–50	1100	320	9	12	2.6	1300	290	70	45

Sources: These reports may be accessed via www.nap.edu; Copyright 2000/2001 by the National Academies

The table is adapted from the Dietary Recommended Intake (DRI) reports, see www.nap.edu

[a]Recommended adequate intake (RAI)

[b]RDA for chromium might have to be adjusted as recent literature suggests lower values

described and associated to disease. Open questions regarding an association of mineral metabolism to epigenetics will be formulated and should encourage deeper research in this area.

Health and Disease Depend on the Interactome

The effects of minerals on the epigenome in a complex organism depend on a variety of aspects as illustrated in Fig. 1. First and most obviously, the dose of the mineral in combination with its purity or contamination with other minerals or trace elements decides about its effect on the epigenome. A lot of minerals are available as different compounds as in case of zinc as zinc oxide, zinc glutamate, zinc aspartate, zinc chloride, and many more (Ollig et al. 2016). For most minerals, “the dose makes the poison” is true, which should be extended by the aspect of duration; some elements can induce either DNA hyper- or hypomethylation depending on the length of the exposure (Ho et al. 2012). Minerals can enter the body via different routes. This does not only alter the amount absorbed by the body but also if a systemic effect or a contained local effect can be anticipated. A zinc lotion, for example, applied once to

Fig. 2 Effect of mineral deficiencies during pregnancy and early childhood. The figure illustrates the consequences of deficiencies in calcium (*Ca*), copper (*Cu*) iron (*Fe*), magnesium (*Mg*), manganese (*Mn*), selenium (*Se*), and zinc (*Zn*) during pregnancy and early childhood. Included are all reported phenotypes that are based on metal-deficiency-induced epigenetic changes (references see text)

the skin, will most probably not induce lifelong epigenetic changes, while long-term suboptimal supply with oral zinc might induce chronic inflammation via epigenetic mechanisms (Kahmann et al. 2008; Wessels 2014, 2015). Age or stage of life when exposure occurs and gender are also decisive in this regard. Table 1 underlines this point, showing that the need for a certain mineral strongly differs depending on gender, age, and activity.

Epigenetics serves as key mechanisms during embryonic and fetal development as they control which genes are activated or silenced, thereby controlling the differentiation of cells and tissues. Alterations of the embryonal epigenome development due to mineral excess or deficiency can become apparent directly after birth, but also later in life, which is in line with Barker hypothesis of the early origin of adult diseases (Fig. 2; Godfrey and Barker 2001; Keen et al. 2003). Also, subsequent stages of life always have critical windows of differentiation, more sensitive to mineral imbalance than others, defining mature phenotypes of cells and tissues (Kanherkar et al. 2014). Finally, due to lifelong editing, aberrant gene expression and diseases can be induced via epigenetic mechanisms, caused by accumulation of minerals or chronic deficiencies. The same elevated dose (compared to RDA) of a mineral that changes severely the epigenome in the embryo might not have any consequence in an adult. For an excellent review, see Ho et al. (2012). As generally true for nutrition-mediated epigenetics, lifestyle including the amount and kind of exercises, smoking, and consuming alcohol can alter the magnitude of an epigenetic effect of a mineral (Fig. 1; Alegria-Torres et al. 2011; Kanherkar et al. 2014). Finally, inherited information or the genetic makeup, carried on DNA, might influence mineral-dependent epigenetic effects, as has been suggested for nutrients in general. This point has however not been investigated much, yet.

In summary, factors illustrated in Fig. 1 form a multidimensional "interactome" that decides about health and disease outcomes of an individual, as a function of mineral balance (Alegria-Torres et al. 2011; Ho et al. 2012).

Reasons to Study the Association of Minerals with the Epigenome

First, it was assumed that the influence of minerals in diseases was mostly due to DNA damage (MacGregor 1990), but soon evidence for epigenetic changes became available. Mineral-associated epigenetic changes are broad and have been described for all areas of epigenetics (Fig. 3; Ho et al. 2012). DNA methylation, for instance, requires more than a dozen of nutrients which include zinc, magnesium, selenium, and copper (Watts 2015). Most mineral-associated effects on the epigenome are mild but can be drastic and persistent, depending on the context (Fig. 1).

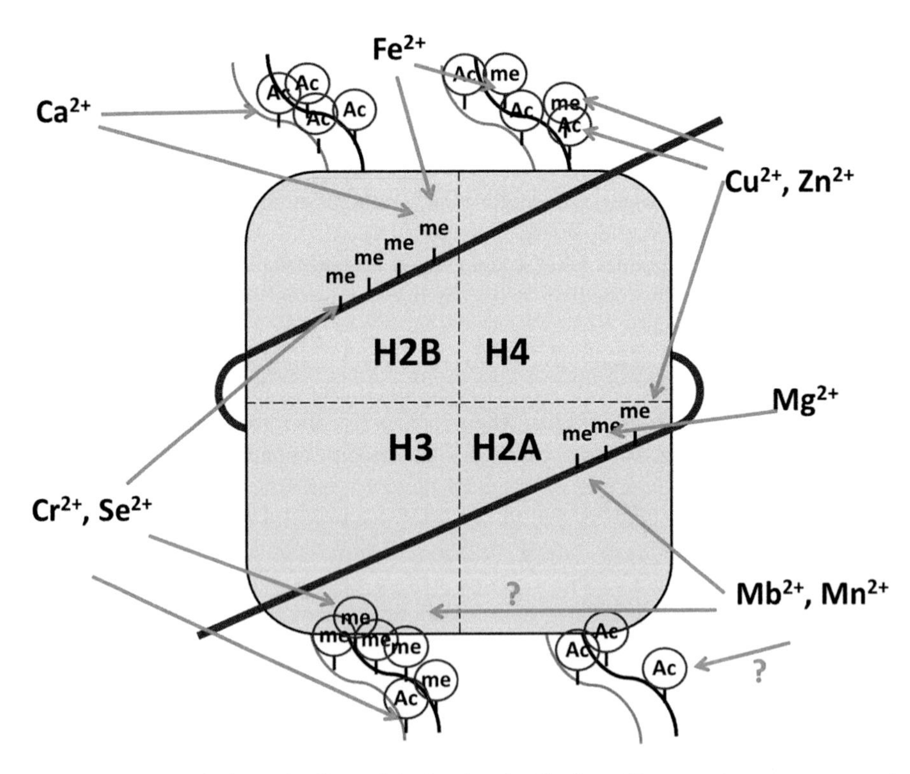

Fig. 3 Summary of epigenetic effects of single minerals. The figure illustrates the consequences of deficiencies in calcium (Ca^{2+}), Copper (Cu^{2+}), iron (Fe^{2+}), molybdenum (Mb^{2+}), magnesium (Mg^{2+}), manganese (Mn^{2+}), selenium (Se^{2+}), and zinc (Zn^{2+}) during pregnancy and during aging. DNA methylation as well as histone methylation and acetylation are affected by metal deficiency. *DNAme* DNA methylation, (Ac) *HAc* histone acetylation, (me) *HM* histone methylation

Chronic consequences of an unbalanced mineral status include changes in the susceptibility and the response to infection, due to the importance of epigenetic mechanisms in the regulation of the immune response (Wang et al. 2012; Wessels 2014, 2015).

Recent evidence from nutritional studies on diabetes, neuronal diseases, cardiovascular diseases, cancer, or adverse reproductive and developmental abnormalities provides implications that the epigenetic changes that form the basis of those diseases are associated to an altered mineral balance (Christensen and Marsit 2011; Zhitkovich 2011). The amount of studies connecting mineral inbalance to epigenetic changes and disease is constantly increasing. More data could discover new options to treat and prevent certain diseases in a mostly natural way, accepted by the majority of cultures and religions, as an alternative to current medications.

Tools for Investigating the Association of Minerals with Epigenetic Changes

One of the key techniques in mineral research is hair tissue mineral analysis (hTMA) (Watts 2015). Calcium, magnesium, potassium, sodium, zinc, copper, and phosphate belong to the regular panel of analytes as well as toxic elements such as lead, cadmium, mercury, arsenic, aluminum, and nickel. This method is firmly established, uses noninvasive sample generation, and is comparably cost-effective especially compared to genetic testing. It provides not only a record of current trace element levels but also of past exposures, making it a great tool for research in this area. This is especially interesting to approach our scarce knowledge of the consequences of mineral deficiencies in adults and during aging (Fig. 4). However, not for all elements and research questions, hTMA is well suited. In case of zinc, for example, a reliable biomarker is still lacking (Ollig et al. 2016). Another very important tool to accurately measure up to 60 elements in one sample in less than a minute is inductively coupled plasma-atomic emission spectroscopy (ICP-AES). Samples can range from particulate matter, filter extracts, and water samples to urine and blood. Researchers have to choose their techniques carefully according to the research question to answer and the element(s) to investigate. Techniques for studying genome and epigenome are broadly described throughout the book and will not be explored here again. Obviously, delineation of the effects of individual minerals is one of the most difficult tasks.

Calcium

The most abundant mineral in the human body is calcium, which is mostly found in the bone (98–99%), but also in intracellular or extracellular fluids and inside the cell membrane (1–2%). Calcium is naturally available from a wide variety of foods, added to others, available as supplement and contained in certain medications. Main functions of calcium include its importance during blood coagulation as well as

Fig. 4 Effect of mineral deficiencies in adults and their accumulation in the elderly. The figure illustrates the consequences of deficiencies in calcium (*Ca*), chromium (*Cr*), copper (*Cu*) iron (*Fe*), magnesium (*Mg*), manganese (*Mn*), selenium (*Se*), and zinc (*Zn*) in adults and during aging. Included are all reported phenotypes that are based on metal-deficiency-induced epigenetic changes (references see text)

growth and mineralization of skeletal muscle. Overload has been connected to neuronal depression, while deficiency can result in neuromuscular defects especially in preterms and children. Later in life, calcium dyshomeostasis elicits far less symptoms (Uusi-Rasi et al. 2013).

Targeting calcium signaling has been suggested to reverse epigenetic silencing of tumor suppressor genes in colonic cancer cells. Drugs, known to reactivate silenced tumor genes, lead to this alteration in calcium signaling and resulted in exclusion of methyl-CpG-binding protein 2 from the nucleus, hereby causing methylation and silenced CpG island promoters (Raynal et al. 2016).

Increased intracellular calcium causes translocation of HDAC from the nucleus to the cytoplasm of cardiomyocytes. This resulted in histone hyperacetylation and alternative splicing. Underlying mechanisms have to be explored more deeply (Sharma et al. 2014). A great review by Graff and Tsai broadly discusses the role of calcium in regulating histone modification, which will therefore not be specified here in more detail (Graff and Tsai 2013). Calcium deficiency was connected to hypomethylation (Takaya et al. 2013). For more information, also see other chapters in this book.

Chromium

As multiple industries use chromium and as it arises from automobile catalytic converters, the exposure of humans is widespread. Absorbed chromium is stored in the liver, spleen, soft tissue, and bone. In the 1960s chromium was proposed to be an essential element for mammals with a role in maintaining proper carbohydrate and lipid metabolism. Inadequacies of chromium are associated with malfunctions, malformations, and acute and chronic diseases including diabetes-like symptoms (Chen et al. 2016). Chromium supplements for diabetics, for correcting blood lipid levels, for promoting weight loss, and during pregnancy are however discussed controversial as well as the classification of this mineral as essential (Chen et al. 2016). The necessary amounts should be met by including chromium-rich foods such as whole grains, meat, broccoli, and herbs in the diet, as supplements might result in malformation and fetal toxicity, respiratory conditions, and cancer that have been connected to epigenetic changes (Martinez-Zamudio and Ha 2011).

Chronic exposure to chromium in workers of chromate industry leads to global alteration in DNA methylation responsible for silencing of tumor suppressor and DNA-repair genes, as was detected for example in lung cancer specimens (Ali et al. 2011).

Similar but also slightly deviating data were obtained in studies using cell culture models (A549 lung cells and B lymphoblastoid cells). Chromium induced global hypomethylation of DNA, which resulted in cell cycle arrest, but effects on certain promoters were different than reported for workers (Lou et al. 2013). Because of those discrepancies between in vivo and in vitro data, caution has to be taken when extrapolating data.

Histone modifications are affected by the chromium levels. Phosphorylation of H2AX was induced by chromium, and biotinylation as well as acetylation of histones has been suggested (Xia et al. 2011). Underlying mechanisms include the chromium-dependent crosslinking of DNMT1 and HDAC1. Also, an effect of chromium on the expression of histone-modifying enzymes has been proposed, and first data for increased H3K9-specific methyltransferase G9a resulting in higher di- and trimethylated H3K9 are available (Sun et al. 2009). Increased H3K4me3 has been observed globally and is promoter specific. For a more detailed discussion of the epigenetic effects of chromium, please see review by Martinez-Zamudio and Ha (2011). Finally, there are first indications that chromium is associated with miRNA expression (miR-146a, miR-143) in humans (He et al. 2013). Analyses that are more detailed are necessary here.

Copper

About 100 mg of copper can be found in healthy individuals. Roughly 75% is stored in the skeleton and muscle. Especially the liver, brain, heart, and kidney are high in copper content. Food with high copper levels includes shellfish, nuts, legumes, organ meats, and chocolate.

Copper plays a critical role in metabolism as a cofactor for several cupric enzymes. It is especially important for adequate use of iron by the body and for a functioning redox system. The latter makes it potentially toxic. Copper deficiencies as well as overload have been associated with diseases and malfunctioning especially during embryonic development. Consequences include severe connective tissue abnormalities, skeletal defects, lung abnormalities, and even embryotoxicity (Keen et al. 2003; Kambe et al. 2008).

Aging has not been associated with significant changes in the requirement for copper, suggesting that copper metabolism is a robust system throughout life. Due to the difficulty in measurement of copper status, RDA could not be established for copper, and instead a safe and adequate range of copper intake is recommended (Table 1).

Decreased global DNA methylation seen in diabetic mice was normalized by copper supplementation (Ergaz et al. 2014). Copper chelation by penicillamine could restore DNA methylation levels. Copper induced deregulation of miRNA expression in the zebra fish olfactory system (Wang et al. 2013), which represents the only data for this research area.

A study in adrenal gland cells suggests that Dnmt1 and Sirt-1 are expressed differently depending on copper levels, added to the cells. Underlying mechanisms were not discussed (Sun et al. 2014). Kang et al. investigated the effect of elevated copper levels during Alzheimer's disease on histone acetylation in human hepatoma cells. They found a significant decrease in histone acetylation, especially for H3 and H4 (Kang et al. 2004). Chelation of copper by bathocuproine disulfonate mostly reversed copper-induced alteration. In human promyeloid HL-60 cells as well as in human Hep3B hepatoma cells, copper inhibited histone acetylation and increased apoptosis in HL-60 cells; an increase in generation of H_2O_2 and O_2^- in a concentration and time-dependent manner was noted. Interestingly, this could be attenuated by treatment with HDAC inhibitors, linking epigenetics to the cytotoxic function of copper (Chen et al. 2008). As similar effects for H4 acetylation in yeast have been described as well, this might be a general function of copper (Kang et al. 2004).

Another proposed mechanism for the interference of copper with histone acetylation is a direct binding of copper to histidine residues in histone tails. As those interactions have only been observed in cell-free systems (Karavelas et al. 2005), more research is necessary at this point. The suggestion that there might be an endoplasmic-reticulum-localized protein with copper-dependent nuclease activity (reticulon) that can inhibit HDACs needs to be evaluated as well (Nepravishta et al. 2010).

Iodine

An epigenetic activity of iodine could not be found.

Iron

Iron is another major essential mineral for humans, with about 4–5 g total body content for a 70-kg adult. Iron-rich foods include red meat, fish, beans, and whole wheat. It plays important roles in cell replication and differentiation, oxygen metabolism, and transport. Iron deficiency affects two billion people worldwide, especially pregnant women and their offspring. Anemia, fatigue, poor work performance and endurance, neurological disturbances, and decreased immunity are common symptoms (Brunner and Wuillemin 2010).

Literature suggests a role of iron in epigenetics, which remains to be investigated in more detail. The notion that key enzymes of DNA duplication, repair, and epigenetics have iron sulfur clusters as prosthetic groups could explain their dysregulation during iron deficiency (Camaschella 2013). Another example for an iron-dependent enzyme is JmjC-domain-containing histone demethylase-2, which regulates histone H3 lysine 9 (H3K9) mono- and dimethylation, known to be critical marks for long-term gene silencing. If the central iron ion is replaced by nickel, enzyme activity is inhibited, resulting in higher global H3K9me1 and H3K9me2 (Chervona et al. 2012). In case of a loss of the central iron ion during iron deficiency, the same consequences are likely, explaining epigenetic changes during iron shortage.

Iron deficiency at early stages of life results in neurological dysfunctions that include decreased speed of processing, learning, and memory and can cause schizophrenia. One possible explanation may be changes in brain-derived neurotrophic factor (BDNF) signaling, forming the basis for learning and memory. Recent data show that BDNF can be epigenetically modified (Jiang et al. 2008). Pronounced iron deficiency in utero (as part of overall nutritional deficiency) was connected to the development of schizophrenia, when samples from the Dutch Hunger Winter and the China Famine were analyzed (Brown and Susser 2008). However, as in other studies, direct evidence for an association of early-life iron deficiencies with epigenetic changes is missing.

Magnesium

Total magnesium content in the adult human body is about 22–26 g, majorly located in the bone and muscle. Food, rich in this mineral, includes legumes, grain, cereal, nuts, and some kinds of mineral water. Magnesium is essential for neurotransmission, energy generation, the stability of bones, and a variety of cellular functions. Embryonic development is severely affected during magnesium deficiency, which presents later in life as metabolic disturbances and diseases including diabetes, suggesting epigenetic involvement (Komiya et al. 2014).

Also, magnesium is important for the function of MycE, which is involved in the mycinamicin biosynthetic pathway, so that during deficiency enzymatic functioning is disrupted, which results in changes in methylation patterns (Akey et al. 2011). Much more research is necessary in this area.

Manganese

From oxidation states ranging between −III and +VII, the only form absorbed by humans is Mn^{2+}, which is oxidized to Mn^{3+} over time in plasma. Approximately 2.5–5 mg of Manganese can be found in the human body, with 25–40% present in the bone. In addition to exposure via contaminated water and inhalation from the atmosphere, dietary manganese is the main route of exposure under normal circumstances for most people. Manganese-rich foods are whole grain, nuts, rice, and tees (Aschner and Aschner 2005). As essential mineral, manganese is needed for proper fetal development, brain functioning, and other important aspects of metabolism. Manganese deficiency is associated with impaired growth, skeletal defects, reduced reproduction, abnormal glucose, and lipid and carbohydrate metabolism. Manganese excess can induce adverse reproductive, respiratory, and neurological effects including diseases such as Parkinson's disease especially in the elderly (Wood 2009; Tarale et al. 2016) and neurotoxicity in infants.

A connection of manganese to epigenetics has been found recently. Manganese causes overexpression of a-synuclein, which interacts with histones and thereby regulates apoptosis. As a-synuclein can sequester Dnmts to the cytoplasm, hypomethylation of DNA was also induced (Tarale et al. 2016). Maternal manganese exposure caused sustained promoter hypermethylation and transcript downregulation in offspring through postnatal day 77 (Wang et al. 2013). Other important changes due to manganese imbalance are the downregulation of miR-7 and miR-153 and interaction with histone protein p300. However, the epigenetic effects of manganese are supposedly mediated via ROS generation (Tarale et al. 2016) and need to be evaluated.

Molybdenum

Molybdenum plays are role in sulfate metabolism, important for the generation of precursors for DNA and histone methylations. However, as a direct effect has not been described yet, see paragraph on selenium for more information on substrate generation, which might also be related to alterations in molybdenum homeostasis (Wang et al. 2012).

Selenium

The presence of selenium in the human body depends strongly on the region where a person lives. Same is true for its food content as broadly discussed in Navarro-Alarcon and Cabrera-Vigue (2008). Through selenoproteins, selenium functions in thyroid hormone metabolism, reproduction, synthesis of DNA, and formation of the epigenome (Sunda 2012). Concerning the latter, DNA methylation and also histone methylation and acetylation are altered, mostly via effects on the methionine-homocytein cycle. Here, methionine serves as methyl donor and can act inhibitory for HDACs (Uthus and Ross 2009; Metes-Kosik et al. 2012). An effect of selenium deficiency during pregnancy on embryonic development and generation of the epigenome can therefore be assumed. However, data on this link are scarce, yet. In the liver of adult rats as well as in cell culture experiments using intestinal Caco-2 cells, selenium deficiency causes hypomethylation while hypermethylation has been observed in selenium-deficient mice (Metes-Kosik et al. 2012), and an inverse correlation of serum selenium with DNA methylation in leukocytes has been described in humans (Pilsner et al. 2011). This underlines the complexity of effects and the need for further research. In various cell lines, selenium supplementation reduced Dnmt activity causing decreased global DNA methylation and H3K9 methylation. Histone acetylation was altered after selenium supplementation in murine macrophages (Narayan et al. 2015). An increase in histone acetylation has been found as well as changes in global DNA methylation (Metes-Kosik et al. 2012). For a more detailed view, see the other chapters in this book.

Taken together, there are a lot of open points regarding the epigenetic effect of altered selenium levels. However, the importance of selenium for a balanced redox and methylation metabolism should encourage the use of selenium as a therapeutic option to prevent and treat certain diseases (Wessels 2014).

Zinc

The total body content of zinc is between 2 and 4 g in the human body and 12–16 μM in plasma. Zinc-rich foods include oysters, meat, and whole meal flour. Absorption and distribution of zinc within body tissues is accomplished via 24 different zinc transporters (Kloubert and Rink 2015). Apoptosis, signal transduction, transcription, differentiation, and replication are all functions involving changes in zinc homeostasis and known to be regulated strongly via epigenetic processes. The fact that several epigenetically active enzymes such as Dnmts, methyl-binding proteins, and histone-modifying enzymes (acetylases, deacetylases, methylases) are zinc dependent suggests that alterations and pathologies due to disturbed zinc homeostasis might be established through epigenetic mechanisms (Black 2001; Wessels 2014). The notion of global DNA hypomethylation and increasing numbers of hypomethylated genes during zinc deficiency underlines this hypothesis (Wallwork and Duerre 1985; Wessels et al. 2013). Actually, there is evidence that, during zinc deficiency, IL-1β, IL-6, and TNFα expressions are increased in vitro mainly due to changes in the chromatin structure of adjacent genes (Wessels et al. 2013; Kessels et al. 2016). However, deeper insights into the underlying cellular mechanisms are still missing. HDAC1 and HDAC2 are also influenced, and chromatin condensation is impaired so that the cell remains in the prometaphase (Wallwork and Duerre 1985). As part of the methionine synthase reaction, zinc causes changes in the generation of substrates for DNA and histone methylation, as well (Wong et al. 2015).

Malformation and poor health in adulthood belong to the detrimental consequences of zinc deficiency during pregnancy. Zinc deficiency increases DNA and histone methylation thereby disturbing regular follicular development and resulting in disorganized chromatin and modified gene expression. Consequently, oocyte fertilization and preimplantation development are impaired, which was discovered in vitro. However, extrapolation to the in vivo situation has to be carefully investigated as fertilization in vivo was not affected as dramatically as during in vitro studies (Tian and Diaz 2013). On the other hand, zinc deficiency during pregnancy leads to disease at later stages of life. Hypertension and impaired learning are consequences of pre- and postnatal zinc deficiency in pregnant rats (Kim et al. 2010). Immunodeficiency as a consequence of lack of zinc during pregnancy even persisted for three succeeding generations, suggesting the importance of zinc in establishing an epigenetic memory (Kong et al. 2012).

In addition to the establishment of the epigenome in the embryo, zinc is important for epigenome maintenance throughout life as well. Long-term zinc deficiencies can cause accumulation of disturbances in the epigenome in the elderly, especially regarding DNA methylation and histone acetylation that cause dysfunctional activation of the immune response. Chronic zinc deficiency, for example, leads to a basal increase of inflammatory markers. In combination with altered expression of zinc transporters, which are epigenetically regulated, this creates a vicious circle, adding to the disturbed zinc homeostasis during aging (Maywald and Rink 2015). This connection has been nicely explored recently by Wong et al., who found age-related hypermethylation of Zip6, ZnT1, and ZnT5, amplifying the age-related decline in zinc status (Wong et al. 2015).

Later-in-life diseases connected to an effect of zinc on the epigenome are especially neuronal diseases such as schizophrenia, autism, depression, various kinds of impaired learning, and anxieties (Fig. 4; Wang et al. 2012). This suggests that an epigenetic decision to favor regular structural and functional brain development over development of mood disorders might be zinc dependent. Although additional clinical studies are advisable to prove those observations for in vivo situations in humans, data point to a high importance of a well-balanced zinc homeostasis throughout life.

Mineral Mixes, Open Questions, and Challenges

More data on epigenetic effects of minerals are available. However, often mixes are investigated such as Diesel exhaust. Weeklong workplace exposure to particulate matter containing arsenic, iron, nickel, lead, cadmium, chromium, and manganese led to increases in miRNAs; the combination of nickel, cadmium, and chromium VI has been associated to epigenetic changes. Global hypomethylation and specific hypermethylation of certain promoters have been observed after exposure to diesel exhaust particles, containing among others arsenic, chromium, manganese mercury, selenium, and phosphorus. Affected genes included *interferon gamma* and *foxp3* (Brunst et al. 2013), the letter being essential for development of regulatory T cells which are responsible for immune tolerance (Rosenkranz et al. 2016). Lots of research will be necessary to clearly define the roles of individual minerals in epigenetic processes.

In general, the importance of minerals for a functioning epigenome just started to emerge. Nowadays, a lot of metabolites can be analyzed with great accuracy simultaneously using, for example, AES-MS. Modern techniques should be combined with established measurements such as hTMA to improve accuracy of results. While the amount of data on mineral effects during embryogenesis has largely increased during recent years, studies on the accumulation of minerals throughout life have been touched much less. However, the degree of interactions multiplies over time and could be infinite. Also, interaction can be attenuating, antagonistic, combinatory, synergistic, and more. Careful planning and literature research are necessary. Estimating an "exposome" as suggested by Wild (2005) including frequency and duration of exposures, dose and route, and other factors similar to what is illustrated in Fig. 1 sounds like an interesting approach. Epigenetic changes might be good biomarkers, to assess the status of an individual concerning health and susceptibility to certain diseases on a genetic level. For this, persistent epigenetic memory and temporal changes must be explored more deeply to probably define changes that are specific for a certain kind of exposure. A panel of epigenetic signatures that depend on certain mineral deficiencies or their combination would be helpful. Understanding mechanisms of interaction between different minerals will add parts to the puzzle, help to better understand changes, and can finally path the way for more personalized approaches containing health and fighting disease. Obtaining and analyzing best-suited samples is another obstacle to climb but could provide important data for simple treatment strategies.

Dictionary of Terms

- **The Interactome** – This term is used to summarize the various influences that decide if a change in mineral balance has an effect on the epigenome or not.
- **hTMA** – Hair tissue mineral analyses is a non-invasive technique, where a representative hair sample is analyzed. The amount of chemical elements deposited in the cells and between them during growth (3 – 4 months) can be measured. Total body load of a certain element cannot be determined.
- **ICP-AES** – Inductively coupled plasma-atomic emission spectroscopy is a nowadays common technique, which can be used to analyze all kinds of dissolved samples. The sample is excited to a higher energy level, typically with a plasma source, causing its dissociation into single ions. When the sample "returns" to its starting point of energy level, it emits photons. The wavelength of the latter is characteristic for the element that is present. It is great to analyze up to 60 elements in only one minute, but it is one of the most expensive methods as well.

Key Facts

Keyfacts on Exposome

- Term chosen by Christopher Paul Wild (2005) to describe "the other 50%," as it gets more and more gets clear, that the predisposition to disease not solely lies in the genetic background of an individual.
- Adds up all exposures from conception onwards that complement and alter the genome.
- Does not take into account genetic make-up itself including mutations or polymorphisms, which distinguishes it from the interactome.

Keyfacts on the Barker Hypothesis of the Early Origin of Adult Diseases

- Originally, genetic predisposition and lifestyle were held responsible for disease in grown-ups.
- According to the hypothesis of Dr. David Barker (1986), nutritional supply to the growing fetus can form the basis for certain diseases later in life, as has been widely accepted nowadays.

Keyfacts on Critical Window

- A certain timeframe that is sensitive for alterations (in this case of the epigenome).
- Can vary in length including, for example, certain week during pregnancy, the whole period of puberty, and phases of compromised immunity during infection.

Summary Points

- Mineral imbalance can induce epigenetic changes in utero, during early-life reprogramming, windows of susceptibility, continuous alterations during life, and even the transmission to the next generation.
- Minerals with well-described epigenetic effects include calcium, chromium, manganese, magnesium, iron, selenium, and zinc.
- In addition to embryonic malformation, epigenetic effects of mineral imbalance majorly lead to neuronal and heart diseases.
- "The dose makes the poison" is true for epigenetic effects of minerals.
- Mineral imbalance can alter DNA methylation and histone modifications and affects RNA interference.
- A well-balanced diet during pregnancy and early life is necessary to assure healthy epigenome establishment and prevention of diseases in the adult.
- Modern techniques such as AES-MS should be combined with established measurements such as hTMA.

References

Akey DL, Li S, Konwerski JR et al (2011) A new structural form in the SAM/metal-dependent omethyltransferase family: MycE from the mycinamicin biosynthetic pathway. J Mol Biol 413:438–450
Alegria-Torres JA, Baccarelli A, Bollati V (2011) Epigenetics and lifestyle. Epigenomics 3:267–277
Ali AH, Kondo K, Namura T et al (2011) Aberrant DNA methylation of some tumor suppressor genes in lung cancers from workers with chromate exposure. Mol Carcinog 50:89–99
Aschner JL, Aschner M (2005) Nutritional aspects of manganese homeostasis. Mol Asp Med 26:353–362
Black RE (2001) Micronutrients in pregnancy. Br J Nutr 85(Suppl 2):S193–S197
Brown AS, Susser ES (2008) Prenatal nutritional deficiency and risk of adult schizophrenia. Schizophr Bull 34:1054–1063
Brunner C, Wuillemin WA (2010) Iron deficiency and iron deficiency anemia – symptoms and therapy. Ther Umsch 67
Brunst KJ, Leung YK, Ryan PH et al (2013) Forkhead box protein 3 (FOXP3) hypermethylation is associated with diesel exhaust exposure and risk for childhood asthma. J Allergy Clin Immunol 131:592–594
Camaschella C (2013) Iron and hepcidin: a story of recycling and balance. Hematol Am Soc Hematol Educ Program 2013:1–8
Chen J, Du C, Kang J et al (2008) Cu2+ is required for pyrrolidine dithiocarbamate to inhibit histone acetylation and induce human leukemia cell apoptosis. Chem Biol Interact 171:26–36
Chen WY, Mao FC, Liu CH et al (2016) Chromium supplementation improved post-stroke brain infarction and hyperglycemia. Metab Brain Dis 31:289–297
Chervona Y, Arita A, Costa M (2012) Carcinogenic metals and the epigenome: understanding the effect of nickel, arsenic, and chromium. Metallomics 4:619–627
Christensen BC, Marsit CJ (2011) Epigenomics in environmental health. Front Genet 2:84
Ergaz Z, Guillemin C, Neeman-Azulay M et al (2014) Placental oxidative stress and decreased global DNA methylation are corrected by copper in the Cohen diabetic rat. Toxicol Appl Pharmacol 276:220–230
Godfrey KM, Barker DJ (2001) Fetal programming and adult health. Public Health Nutr 4:611–624
Graff J, Tsai LH (2013) Histone acetylation: molecular mnemonics on the chromatin. Nat Rev Neurosci 14:97–111
He J, Qian X, Carpenter R et al (2013) Repression of miR-143 mediates Cr (VI)-induced tumor angiogenesis via IGF-IR/IRS1/ERK/IL-8 pathway. Toxicol Sci 134:26–38
Ho SM, Johnson A, Tarapore P et al (2012) Environmental epigenetics and its implication on disease risk and health outcomes. ILAR J 53:289–305
Jiang X, Tian F, Du Y et al (2008) BHLHB2 controls Bdnf promoter 4 activity and neuronal excitability. J Neurosci 28:1118–1130
Kahmann L, Uciechowski P, Warmuth S et al (2008) Zinc supplementation in the elderly reduces spontaneous inflammatory cytokine release and restores T cell functions. Rejuvenation Res 11:227–237
Kambe T, Weaver BP, Andrews GK (2008) The genetics of essential metal homeostasis during development. Genesis 46:214–228
Kang J, Lin C, Chen J et al (2004) Copper induces histone hypoacetylation through directly inhibiting histone acetyltransferase activity. Chem Biol Interact 148:115–123
Kanherkar RR, Bhatia-Dey N, Csoka AB (2014) Epigenetics across the human lifespan. Front Cell Dev Biol 2:49
Karavelas T, Mylonas M, Malandrinos G et al (2005) Coordination properties of cu(II) and Ni (II) ions towards the C-terminal peptide fragment -EL. J Inorg Biochem 99:606–615
Keen CL, Hanna LA, Lanoue L et al (2003) Developmental consequences of trace mineral deficiencies in rodents: acute and long-term effects. J Nutr 133:1477S–1480S

Kessels JE, Wessels I, Haase H et al (2016) Influence of DNA-methylation on zinc homeostasis in myeloid cells: regulation of zinc transporters and zinc binding proteins. J Trace Elem Med Biol 37:125–133

Kim AM, Vogt S, O'Halloran TV et al (2010) Zinc availability regulates exit from meiosis in maturing mammalian oocytes. Nat Chem Biol 6:674–681

Kloubert V, Rink L (2015) Zinc as a micronutrient and its preventive role of oxidative damage in cells. Food Funct 6:3195–3204

Komiya Y, Su LT, Chen HC et al (2014) Magnesium and embryonic development. Magnes Res 27:1–8

Kong BY, Bernhardt ML, Kim AM et al (2012) Zinc maintains prophase I arrest in mouse oocytes through regulation of the MOS-MAPK pathway. Biol Reprod 87:11,1–11,12

Lou J, Wang Y, Yao C et al (2013) Role of DNA methylation in cell cycle arrest induced by Cr (VI) in two cell lines. PLoS One 8:e71031

Lyko F, Foret S, Kucharski R et al (2010) The honey bee epigenomes: differential methylation of brain DNA in queens and workers. PLoS Biol 8:e1000506

MacGregor JT (1990) Dietary factors affecting spontaneous chromosomal damage in man. Prog Clin Biol Res 347:139–153

Martinez-Zamudio R, Ha HC (2011) Environmental epigenetics in metal exposure. Epigenetics 6:820–827

Maywald M, Rink L (2015) Zinc homeostasis and immunosenescence. J Trace Elem Med Biol 29:24–30

Metes-Kosik N, Luptak I, Dibello PM et al (2012) Both selenium deficiency and modest selenium supplementation lead to myocardial fibrosis in mice via effects on redox-methylation balance. Mol Nutr Food Res 56:1812–1824

Narayan V, Ravindra KC, Liao C et al (2015) Epigenetic regulation of inflammatory gene expression in macrophages by selenium. J Nutr Biochem 26:138–145

Navarro-Alarcon M, Cabrera-Vique C (2008) Selenium in food and the human body: a review. Sci Total Environ 400:115–141

Nepravishta R, Bellomaria A, Polizio F et al (2010) Reticulon RTN1-C(CT) peptide: a potential nuclease and inhibitor of histone deacetylase enzymes. Biochemistry 49:252–258

Ollig J, Kloubert V, Wessels I et al (2016) Parameters influencing zinc in experimental systems in vivo and in vitro. Metals 6

Pilsner JR, Hall MN, Liu X et al (2011) Associations of plasma selenium with arsenic and genomic methylation of leukocyte DNA in Bangladesh. Environ Health Perspect 119:113–118

Raynal NJ, Lee JT, Wang Y et al (2016) Targeting calcium signaling induces epigenetic reactivation of tumor suppressor genes in cancer. Cancer Res 76:1494–1505

Rosenkranz E, Metz CH, Maywald M et al (2016) Zinc supplementation induces regulatory T cells by inhibition of Sirt-1 deacetylase in mixed lymphocyte cultures. Mol Nutr Food Res 60:661–671

Sharma A, Nguyen H, Geng C et al (2014) Calcium-mediated histone modifications regulate alternative splicing in cardiomyocytes. Proc Natl Acad Sci U S A 111:E4920–E4928

Sun Y, Liu C, Liu Y et al (2014) Changes in the expression of epigenetic factors during copper-induced apoptosis in PC12 cells. J Environ Sci Health A Tox Hazard Subst Environ Eng 49:1023–1028

Sun H, Zhou X, Chen H et al (2009) Modulation of histone methylation and MLH1 gene silencing by hexavalent chromium. Toxicol Appl Pharmacol 237:258–266

Sunda WG (2012) Feedback interactions between trace metal nutrients and phytoplankton in the ocean. Front Microbiol 3:204

Takaya J, Iharada A, Okihana H et al (2013) A calcium-deficient diet in pregnant, nursing rats induces hypomethylation of specific cytosines in the 11beta-hydroxysteroid dehydrogenase-1 promoter in pup liver. Nutr Res 33:961–970

Tarale P, Chakrabarti T, Sivanesan S et al (2016) Potential role of epigenetic mechanism in manganese induced neurotoxicity. Biomed Res Int 2016:2548792

Tian X, Diaz FJ (2013) Acute dietary zinc deficiency before conception compromises oocyte epigenetic programming and disrupts embryonic development. Dev Biol 376:51–61
Uthus EO, Ross S (2009) Dietary selenium (Se) and copper (Cu) interact to affect homocysteine metabolism in rats. Biol Trace Elem Res 129:213–220
Uusi-Rasi K, Karkkainen MU, Lamberg-Allardt CJ (2013) Calcium intake in health maintenance – a systematic review. Food Nutr Res 57
Wallwork JC, Duerre JA (1985) Effect of zinc deficiency on methionine metabolism, methylation reactions and protein synthesis in isolated perfused rat liver. J Nutr 115:252–262
Wang L, Shiraki A, Itahashi M et al (2013) Aberration in epigenetic gene regulation in hippocampal neurogenesis by developmental exposure to manganese chloride in mice. Toxicol Sci 136:154–165
Wang J, Wu Z, Li D et al (2012) Nutrition, epigenetics, and metabolic syndrome. Antioxid Redox Signal 17:282–301
Watts DL (2015) Nutrition, epigenetics and hair tissue mineral analysis (HTMA). Trace Elements Newsl 26:1–3
Wessels I (2014) Epigenetics and metal deficiencies. Curr Nutr Rep 3
Wessels I. (2015) Metal homeostastis during development, maturation, and aging, SFR 16:ISBN 978-0-262-02919-3
Wessels I, Haase H, Engelhardt G et al (2013) Zinc deficiency induces production of the proinflammatory cytokines IL-1beta and TNFalpha in promyeloid cells via epigenetic and redox-dependent mechanisms. J Nutr Biochem 24:289–297
Wild CP (2005) Complementing the genome with an "exposome": the outstanding challenge of environmental exposure measurement in molecular epidemiology. Cancer Epidemiol Biomark Prev 14:1847–1850
Wong CP, Rinaldi NA, Ho E (2015) Zinc deficiency enhanced inflammatory response by increasing immune cell activation and inducing IL6 promoter demethylation. Mol Nutr Food Res 59:991–999
Wood RJ (2009) Manganese and birth outcome. Nutr Rev 67:416–420
Wright RO, Baccarelli A (2007) Metals and neurotoxicology. J Nutr 137:2809–2813
Xia B, Yang LQ, Huang HY et al (2011) Chromium (VI) causes down regulation of biotinidase in human bronchial epithelial cells by modifications of histone acetylation. Toxicol Lett 205:140–145
Zhang FF, Cardarelli R, Carroll J et al (2011) Physical activity and global genomic DNA methylation in a cancer-free population. Epigenetics 6:293–299
Zhitkovich A (2011) Chromium in drinking water: sources, metabolism, and cancer risks. Chem Res Toxicol 24:1617–1629

Calcium-Deficient Diets in Pregnancy and Nursing: Epigenetic Change in Three Generations of Offspring

93

Junji Takaya

Contents

Abstract

Prenatal malnutrition can affect the phenotype of offspring by changing epigenetic regulation. Calcium (Ca) plays an important role in the pathogenesis of insulin resistance syndrome. We previously reported that feeding a Ca-restricted diet to pregnant rats results in hypomethylation and decreased expression from the *11β-hydroxysteroid dehydrogenase-1* promoter in the liver of offspring at day 21. These findings show that a maternal Ca deficiency during pregnancy can affect the regulation of non-imprinted genes by altering the epigenetic regulation of gene expression, thereby inducing different metabolic phenotypes. The epigenome is an important target of environmental modification. In addition,

J. Takaya (✉)
Department of Pediatrics, Kawachi General Hospital, Higashi-Osaka, Osaka, Japan

Department of Pediatrics, Kansai Medical University, Moriguchi, Osaka, Japan
e-mail: takaya@takii.kmu.ac.jp; takaya@kawati.or.jp

V. B. Patel, V. R. Preedy (eds.), *Handbook of Nutrition, Diet, and Epigenetics*,
https://doi.org/10.1007/978-3-319-55530-0_61

we determined the effects of a Ca deficiency during pregnancy and/or lactation on insulin resistance and secretion in at least three generations. Female Wistar rats consumed either a Ca-deficient or control diet ad libitum from three weeks preconception to 21 days postparturition and were mated with control males. Randomly selected first (F1)- and second-generation (F2) females were mated with males of each generation on postnatal day 70. F1 and F2 dams were fed with a control diet ad libitum during pregnancy and lactation. On 180 days, homeostasis model assessment of beta cell function (HOMA-β%) gradually decreased in F1 through F3 and that in F2 and F3 males and females was significantly lower than control. These findings indicated that maternal Ca restriction during pregnancy and/or lactation influences insulin secretion in three generations of offspring.

Keywords
Calcium · Glucocorticoid receptor · HOMA-IR · HOMA-β% · Insulin resistance · Metabolic syndrome · Pregnancy · Pyrosequencing · Rat · 11β-Hydroxysteroid dehydrogenase-1

List of Abbreviations

Ca	Calcium
F1	First generation
F2	Second generation
F3	Third generation
GR	Glucocorticoid receptor
HOMA-IR	Homeostasis model assessment of insulin resistance
HOMA-β%	Homeostasis model assessment of beta cell function
11β-HSD1	11β-hydroxysteroid dehydrogenase-1
11β-HSD2	11β-hydroxysteroid dehydrogenase-2
Hsd11b1	11β-hydroxysteroid dehydrogenase-1 gene
Hsd11b2	11β-hydroxysteroid dehydrogenase-2 gene
Nr3c1	glucocorticoid receptor gene
PEPCK	phosphoenolpyruvate carboxykinase
Pck1	phosphoenolpyruvate carboxykinase gene
PPARα	peroxisome proliferator-activated receptor α
Ppara	rat peroxisome proliferator-activated receptor α gene

Introduction

Epidemiologic studies have reported a link between calcium (Ca) intake and insulin resistance in obesity and metabolic syndrome (Pikilidou et al. 2009; Schrager 2005). The results of several observational prospective studies have shown a relationship between low/insufficient oral Ca intake and the incidence of type 2 diabetes mellitus (Liu et al. 2005; Pittas et al. 2007) or metabolic syndrome (Pereira et al. 2002). Conjugated linoleic acid and Ca supplementation modify the methylation pattern of fatty-acid-related genes under a high-fat diet in adult mice (Chaplin et al. 2017).

A 2-week Ca-deficient diet in rats was associated with upregulated hepatic expression of 11β-hydroxysteroid dehydrogenase-1 (11β-HSD1) mRNA, which occurred before the animals developed obesity or overt features of metabolic syndrome. Overactivity of 11β-HSD1 is associated with increased intracellular active glucocorticoids (Cooper and Stewart 2009; Tomlinson et al. 2004; Walker 2006). Our finding of insulin resistance and upregulation of 11β-HSD1 mRNA suggests that hepatic 11β-HSD1 regulates insulin resistance in Ca-deficient animals (Takaya et al. 2011). Genetic studies in rodents also suggest that increased 11β-HSD1 expression or activity increases the risk of several components of metabolic syndrome [Masuzaki et al. 2003; Morton and Seckl 2008].

Maternal undernutrition and the consequent low birth weight predispose offspring to various diseases, including adult-onset insulin resistance syndrome (Valdez et al. 1994; Warner and Ozanne 2010). Maternal protein restriction during rat pregnancy produces offspring with hypertension, hyperglycemia, altered hepatic enzyme profiles, or some combination of these conditions (Desai et al. 1997). Lillycrop et al. reported that feeding a protein-restricted diet to pregnant rats resulted in hypomethylation and increased expression from the peroxisome proliferator-activated receptor α (PPARα, *Ppara*) and glucocorticoid receptor (GR, *Nr3c1*) promoters in the liver of the offspring (Lillycrop et al. 2005, Lillycrop et al. 2008). This finding shows that maternal nutrition during pregnancy can affect regulation of non-imprinted genes via altered epigenetic regulation of gene expression, thereby inducing different metabolic phenotypes. We investigated the methylation of individual cytosine-guanine (CpG) dinucleotides in glucocorticoid-related genes in liver tissue of neonatal offspring from Ca-deficient rat dams. In addition, we determined the effects of a Ca deficiency during pregnancy and/or lactation on insulin resistance and secretion in at least three generations.

A Calcium-Deficient Diet in Pregnant, Nursing Rats Affects the Methylation of Specific Cytosines in the *11β-Hydroxysteroid Dehydrogenase-1* Promoter Within Pup Liver

Female Wistar rats consumed either a Ca-deficient (D: 0.008% Ca) or control (C: 0.90% Ca) diet ad libitum from 3 weeks preconception to 21 days postparturition (Table 1). Pups were allowed to nurse from their original mothers and then killed on day 21. Methylation of CpG dinucleotides in the phosphoenolpyruvate carboxykinase (PEPCK, *Pck1*) (Hanson and Patel 1994), PPARα, (*Ppara*), GR (*Nr3c1*), 11β-HSD1 (*Hsd11b1*), and 11β-HSD2 (*Hsd11b2*) promoters was measured in liver tissue using pyrosequencing (Table 2)(Takaya et al. 2013). The methylation levels of all genes did not differ between groups, except for that of *Hsd11b1*, which was lower in the Ca-deficient group (Fig. 1). Serum corticosterone levels of male pups from the Ca-deficient dams were higher than that from the control (Table 3). Expression of

Table 1 Diet ingredients

Ingredients	Control (g/100 g diet)	Calcium deficient (g/100 g diet)
Milk casein	24.50	24.50
Corn starch	45.50	45.50
Granulated sugar	10.00	10.00
Corn oil	6.00	6.00
Cellulose powder	5.00	5.00
α-Starch	1.00	1.00
Vitamin mix	1.00	1.00
Mineral mix[a]	7.00	7.00
(Calcium)	(0.90)	(0.008)
Total	100.00	100.00

[a]Mineral mix is free of calcium

Table 2 Percentage of methylation of mean each gene

	Control		Ca Deficit	
	Male	Female	Male	Female
Hsd11b1	17.8 ± 0.7	17.0 ± 0.4	15.5 ± 0.5**	15.8 ± 0.3*
Hsd11b2	11.0 ± 3.7	10.0 ± 4.9	10.5 ± 0.5	11.9 ± 1.9
Pck1	18.6 ± 4.5	18.5 ± 1.7	13.6 ± 1.6	15.3 ± 1.9
Nr3c1	4.22 ± 1.09	6.85 ± 0.58	3.97 ± 1.05	4.34 ± 0.93
Ppara	2.06 ± 2.06	1.16 ± 1.16	3.52 ± 2.15	2.73 ± 1.67

*$p < 0.05$; **$p < 0.01$ compared to control

Pck1 and *Nr3c1* except *Hsd11b1*, *Hsd11b2*, and *Ppara* was lower in the Ca-deficient group than the control group (Table 4) (Takaya et al. 2013).

Methylation level of hepatic *Hsd11b1* was altered in the offspring as a consequence of the maternal dietary manipulation, but the epigenetic changes were not reflected in corresponding alterations in transcription. The nuclear receptor corepressor complex is affected by environmental factors such as nutrients and hormones that can result in altered DNA methylation, acetylation, histone modification, other epigenetic changes, or some combination thereof; such epigenetic changes can and do alter the activity of DNA. These factors can alter feedback loops involving nuclear receptors that normally balance repression (Kaelin Jr and McKnight 2013).

Downregulation of *Hsd11b1* suggests that a compensatory mechanism may diminish cortisol production in the liver. Reduced hepatic glucocorticoid exposure also represents a compensatory mechanism that limits the metabolic complications of insulin resistance. In our study, no significant difference was found among the groups in serum 11β-HSD1 levels; however, this may have been due to tissue-specific differences between serum and liver. Whether glucocorticoids modulate *Hsd11b1* expression is unknown; hepatic *Hsd11b1* expression is very different from that in other tissues (Lindsay et al. 2003; Paulmyer-Lacrox et al. 2002; Rask et al. 2001). Obese rodents exhibit tissue-specific dysregulation of 11β-HSD1,

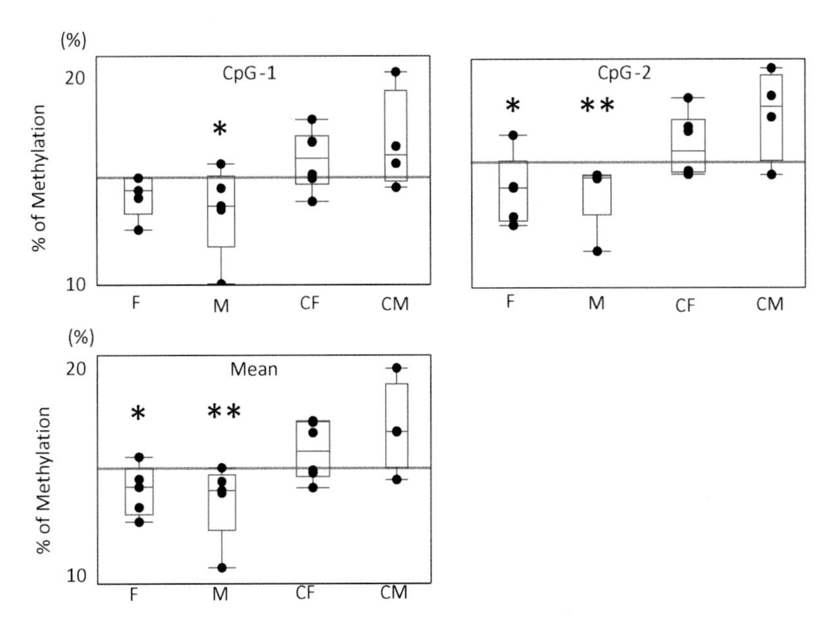

Fig. 1 Bisulfite pyrosequencing analysis of *11β-hydroxysteroid dehydrogenase-1* (*Hsd11b1*) in liver tissue from offspring; overall methylation levels and position-specific methylation levels were sensitive to the effect of calcium deficiency. The experimental samples were prepared and analyzed as described in Materials and Methods. The methylation levels for *Hsd11b1* (position 1–2) were calculated by averaging the methylation levels of each dietary group (male n = 5 and female n = 5 of each group). The results were analyzed by ANOVA followed by a post hoc Tukey test. Total and core methylation levels were significantly lower in calcium-deficient animals than in the controls (*$P < 0.05$, **$P < 0.01$). Box plots show percentage of methylation of individual CpG dinucleotides (central horizontal line, median; box 25th and 75th percentiles; whiskers, minimum and maximum). F, female pups from the Ca-deficient group; M, male pups from the Ca-deficient group; CF, female pups from the control group; CM, male pups from the control group

Table 3 Profile and serum biochemical data in pups

	Control		Ca deficit	
	Male	Female	Male	Female
Body weight (g)	36 ± 1	30 ± 4	29 ± 3	26 ± 5
Heart rate (bpm)	510 ± 13	530 ± 15	500 ± 7	520 ± 23
Systolic BP (mmHg)	76 ± 14	81 ± 9	63 ± 4	83 ± 5
Diastolic BP (mmHg)	28 ± 10	35 ± 9	18 ± 3	42 ± 14
Ionized Ca (nmol/L)	1.44 ± 0.03	1.43 ± 0.04	1.47 ± 0.05	1.45 ± 0.03
Ionized Mg (nmol/L)	0.70 ± 0.07	0.74 ± 0.04	0.69 ± 0.04	0.68 ± 0.05
Corticosterone (ng/mL)	380 ± 67	530 ± 63	530 ± 29*	530 ± 29
Insulin (ng/mL)	4.90 ± 0.95	5.59 ± 1.48	3.35 ± 0.31	2.35 ± 0.45*

BP, blood pressure; Ca, calcium; Mg, magnesium
*$P < 0.05$

Table 4 mRNA expression of each gene

	Control		Ca Deficit	
	Male	Female	Male	Female
Hsd11b1	1.15 ± 0.23	1.17 ± 0.15	0.64 ± 0.14	0.65 ± 0.07
Hsd11b2	1.35 ± 0.16	1.30 ± 0.09	1.18 ± 0.17	1.20 ± 0.13
Pck1	1.07 ± 0.24	1.04 ± 0.18	0.45 ± 0.18*	0.37 ± 0.05*
Nr3c1	0.90 ± 0.20	0.94 ± 0.14	0.41 ± 0.08*	0.42 ± 0.02*
Ppara	0.88 ± 0.15	0.82 ± 0.08	0.61 ± 0.08	0.56 ± 0.01

*$p < 0.05$ compared to each sex control

usually with upregulation in adipose tissue and downregulation in the liver (Hemanowski-Vosatka et al. 2005; Liu et al. 2003). In both obese Zucker rats and obese humans, 11β-HSD1 activity is high in adipose tissue, but low in the liver (Livingstone et al. 2001; Paulmyer-Lacrox et al. 2002; Rask et al. 2001). In adipose tissue and smooth muscle cells, glucocorticoid induces *Hsd11b1* mRNA expression, but controversial results have been obtained in the liver (Livingstone et al. 2001; Rask et al. 2001). Hamo et al. reported that liver-specific 11β-HSD1 knockout mice given low-dose 11-dehydrocorticosterone do not show any of the adverse metabolic effects seen in wild-type mice (Hamo et al. 2013). This result implies that liver-derived intra-tissue glucocorticoids, rather than circulating glucocorticoids, contribute to the development of metabolic syndrome and suggest that local action within hepatic tissue mediates these effects. In contrast, Morgan et al. reported that adipose-specific 11β-HSD1 knockout mice given higher-dose glucocorticoids are protected from hepatic steatosis and circulating fatty acid excess, whereas liver-specific 11β-HSD1 knockout mice develop full metabolic syndrome phenotypes (Morgan et al. 2014). This result demonstrates that 11β-HSD1, particularly in adipose tissue, is key to the development of the adverse metabolic profile associated with circulating glucocorticoid excess.

A Ca-deficient diet during pregnancy and nursing induced hypomethylation of specific CpG dinucleotides in the *Hsd11b1* promoter in liver tissue of neonatal offspring. These changes in *Hsd11b1* expression probably contribute to marked increases in glucocorticoid hormone action in liver tissue (Draper and Stewart 2005) and potentiate induction of insulin resistance during adult life (Anagnostis et al. 2009).

A Calcium-Deficient Diet in Rat Dams During Gestation and Nursing Affects Hepatic *11β-Hydroxysteroid Dehydrogenase-1* Expression in the Offspring

Next, we determined the effects of nursing in addition to Ca deficiency during pregnancy on hepatic *Hsd11b1* expression in offspring. Female Wistar rats ate either a Ca-deficient or control diet (Table 1) from 3 weeks preconception to 21 days post-parturition. On postnatal day 1, pups were cross fostered to the same or opposite

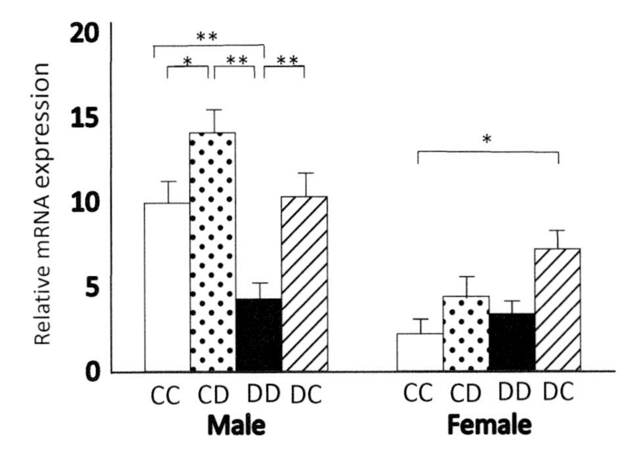

Fig. 2 Hepatic mRNA expression levels of *Hsd11b1* after delivery, foster mothers fed with the same or different diets: CC, DD, CD, and DC (first letter, diet of original mother; second letter, diet of nursing mother; C, control diet; D, calcium deficient diet). Data are represented as the means ± standard error. Statistically significant differences among the groups were determined with two-way analysis of variance (ANOVA), with diet and sex as the main factors. If interactions were found, the data were split and analyzed with one-way ANOVA. $^{*}P < 0.05$, $^{**}P < 0.01$

dams and divided into the following four groups: CC, DD, CD, and DC (first letter, original mother's diet; second letter, nursing mother's diet; C, control diet; D, Ca-deficient diet). All offspring were fed a control diet beginning at weaning (day 21) and were killed on day 200 (Takaya et al. 2014).

In males, mean levels of insulin, glucose, and homeostasis model assessment of insulin resistance (HOMA-IR) (Matthews et al. 1985) were higher in the DD and DC groups than in the CC group (Fig. 2). We found no difference in HOMA-IR between the CC and CD groups in either males or females. Expression of *Hsd11b1* was lower in male DD rats than in CC rats. *Hsd11b1* expression in male offspring nursed by cross fostered dams was higher than that in those nursed by dams fed with the same diet: CC vs. CD and DD vs. DC. In females, *Hsd11b1* expression in DC rats was higher than that in CC rats.

Male offspring developed insulin resistance after nursing from Ca-deficient dams. Sex differences in metabolism are well established, and females have been shown to be more insulin sensitive and secrete more insulin compared with males (Geer and Shen 2009; Macotela et al. 2009). Previous reports have shown that the epigenetic effect induced by intrauterine growth restriction is gender specific (Fu et al. 2009; Gong et al. 2010; Ng et al. 2010). Hall et al. reported that chromosome-wide and gene-specific sex differences in DNA methylation associated with altered expression, microRNA, and insulin secretion in human pancreatic islets (Hall et al. 2014). Testosterone may affect epigenetic regulation; however, further investigation is necessary. In our study, serum ionized Ca showed pronounced sexual dimorphism, with higher levels in females ($P < 0.05$). 11β-HSD1 also shows pronounced sexual dimorphism in the liver, with lower levels in females (Lax et al. 1978).

Our data show that metabolic programming effects are modified by nutrition in the immediate postnatal period. Hepatic *Hsd11b1* expression in cross fostered male offspring (DC and CD) was higher than that in offspring (DD and CC) nursed by dams fed with the same diet. Mismatched nutrition after birth, i.e., cross fostered nursing, disrupts the adaptation that has been programmed in the fetus. Although hepatic expression of *Hsd11b1* in offspring from Ca-deficient dams may have been originally upregulated by epigenetic mechanisms, *Hsd11b1* was likely down-regulated by other mechanisms during the early postnatal period. If the nutritional environment established before birth is mismatched after birth, the balance established by this compensation collapses. Metabolic programming effects translate into an adverse intrauterine environment during pregnancy. Altered mismatched nutritional experience during the suckling period can impact adult health in the offspring (Patel and Srinivasan 2011; Srinivasan et al. 2013).

In summary, maternal Ca restriction during pregnancy alters postnatal growth, *Hsd11b1* expression, and insulin resistance in a sex-specific manner (Takaya et al. 2014). Further work is required to support for the hypothesis that early postnatal lactation plays a sexually divergent role in programming the phenotype later in life.

Calcium-Deficient Diets in Pregnancy and Nursing: Epigenetic Change in Three Generations of Offspring

Dearden and Balthasar reported that increased female vulnerability to develop metabolic disturbance in response to maternal diet propagates a vicious cycle of obesity and type 2 diabetes in subsequent generations (Dearden and Balthasar 2014). Next, we postulated that metabolic dysregulation induced by a Ca deficiency during pregnancy in F0 rat models can be transmitted to more than one generation of offspring. We investigated whether alterations in insulin resistance and secretion induced in F1 offspring by feeding dams (F0) a Ca-deficient diet during pregnancy and lactation are passed on with or without further change to two subsequent generations.

Twelve-week-old virgin Wistar rats (F0) were fed with either a control (C) diet containing 0.90% Ca or a Ca-deficient (D) diet containing 0.008% Ca ad libitum for three weeks (Table 1). Three weeks later, the rats were mated with males (F0 generation) who were fed with the control diet. F0 dams continued to consume their respective diets throughout gestation and lactation. F1 and F2 females were mated with males of each generation on postnatal day 70. F1 and F2 dams were fed with the control diet during pregnancy and lactation, and offspring were weaned onto the control diet.

Homeostasis model assessment of beta cell function (HOMA-β %) (Hermans et al. 1999; Matthews et al. 1985; Wallace et al. 2004), insulin secretion parameter, was decreased in the F1 through F3 offspring of Ca-deficient F0 dams, and male F1 offspring from these dams developed insulin resistance. However, insulin resistance recovered in F2 male offspring, and HOMA-IR was significantly lower, and serum ionized Ca was higher in F3 offspring. HOMA-IR was affected only in the F1 generation of males, and it was corrected after F2. Although insulin secretion basically worsened, insulin resistance did not develop in F2 and F3 offspring because of sex-specific modifications (Takaya et al. 2016). Our findings indicated the insulin resistance had apparently been corrected, whereas, in fact, the ability to secrete insulin decreased with each passing generation. A mechanism to correct a tendency toward insulin resistance might function through the generations. The surrounding environment and adaptation by compensation might be involved. The intergenerational transmission of developmentally programmed insulin resistance is determined in part by the relative insulin sensitivity of the mother during pregnancy/lactation (Benyshek et al. 2008). Benyshek et al. reported that insulin sensitivity is normalized in the F3 offspring of developmentally programmed insulin-resistant (F2) rats fed with an energy-restricted diet. The F0 maternal Ca deficiency did not consistently develop from the point of insulin resistance in offspring over three consecutive generations. A gestating F0 female can generate F1 embryos and an F2 germ line that has been directly exposed to an environmental factor, such that F3 becomes the first generation that is not directly exposed to the factor (Heard and Martienssen 2014; Skinner 2008).

In conclusion, maternal Ca restriction during pregnancy alters postnatal insulin secretion for three generations. This condition also sex-specifically affects postnatal insulin resistance.

Key Facts

Key Facts of Glucocorticoid and Metabolic Syndrome

- Dysregulation of glucocorticoid action has been proposed to be one of the central features of metabolic syndrome.
- In the major metabolic organs, tissue sensitivity and exposure to glucocorticoids are determined by levels of intracellular peroxisome proliferator-activated receptor α (PPARα), glucocorticoid receptor (GR), and the activity of the microsomal enzyme 11β-hydroxysteroid dehydrogenase type 1 (11β-HSD1).

Key Facts of 11β-Hydroxysteroid Dehydrogenase-1

- 11β-HSD1 is highly expressed in liver and adipose tissue where glucocorticoids reduce insulin sensitivity and action.
- Activity of 11β-HSD1 in liver and adipose tissue might contribute to the development of several features of insulin resistance or metabolic syndrome.
- 11β-HSD1 regulates key hepatic gluconeogenic enzymes, including phosphoenolpyruvate carboxykinase (PEPCK), through the amplification of GR-mediated tissue glucocorticoid action.

Dictionary of Terms

- **HOMA-IR** – Currently, insulin resistance can be estimated with the use of several biological measurements that evaluate different aspects of this complex situation. The homeostasis model assessment of insulin resistance (HOMA-IR) has proved to be a robust tool for the surrogate assessment of insulin resistance. HOMA-IR = [glucose] × [insulin]/22.5 was used as a measure of insulin resistance.
- **HOMA-β%** – A surrogate estimate of pancreatic β-cell function based on measurements of fasting plasma glucose and insulin concentrations has continued to be used as an assessment of insulin secretory function. HOMA-β% = 20 × [insulin]/([glucose] - 3.5)%. HOMA-IR and HOMA-β% are the most popular methods of evaluating insulin resistance and insulin secretion even in rodent study as well as in epidermal studies.
- **11β-HSD1** – 11β-HSD1 converts inactive glucocorticoids (cortisone in humans, 11-dehydrocorticosterone in rodents) to their active forms (cortisol and corticosterone, respectively).
- **11β-HSD2** – 11β-hydroxysteroid dehydrogenase-2 (11β-HSD2) converts excess cortisol to inactive cortisone.
- **Phosphoenolpyruvate Carboxykinase** – Phosphoenolpyruvate carboxykinase, one of the key hepatic gluconeogenic enzymes, simultaneously decarboxylates and phosphorylates oxaloacetate into phosphoenolpyruvate in one of the earliest rate-limiting steps of gluconeogenesis.

Summary Points

- A calcium-deficient diet for a dam during gestation and early nursing may alter glucocorticoid metabolism and lead to higher intracellular glucocorticoid concentration in hepatic cells of her offspring, and this abnormal glucocorticoid metabolism may induce the metabolic complications associated with calcium deficiency.
- Maternal calcium restriction during pregnancy alters postnatal growth, *11β-hydroxysteroid dehydrogenase-1* expression, and insulin resistance in a sex-specific manner.
- Maternal calcium restriction during pregnancy and/or lactation influences insulin secretion in three generations of offspring.
- Prenatal nutrition affected the glucocorticoid metabolism in the offspring in part by affecting the epigenome of offspring.

References

Anagnostis P, Athyros VG, Tziomalos K, Karagiannis A, Mikhailidis D (2009) The pathogenetic role of cortisol in the metabolic syndrome: a hypothesis. J Clin Endocrinol Metab 94:2692–2701
Benyshek DC, Johnston CS, Martin JF (2008) Insulin sensitivity is normalized in the third generation (F3) offspring of developmentally programmed insulin resistant (F2) rats fed an energy-restricted diet. Nutr Metab (Lond) 5:26

Chaplin A, Palou A, Serra F (2017) Methylation analysis in fatty-acid related genes reveals their plasticity associated with conjugated acid and calcium supplementation in adult mice. Eur J Nutr 56:879–891

Cooper MS, Stewart PM (2009) 11β-hydroxysteroid dehydrogenase type 1 and its role in the hypothalamus-pituitary-adrenal axis, metabolic syndrome, and inflammation. J Clin Endocrinol Metab 94:4645–4654

Dearden L, Balthasar N (2014) Sexual dimorphism in offspring glucose-sensitive hypothalamic Gene expression and physiological responses to maternal high-fat diet feeding. Endocrinology 155:2144–2154

Desai M, Byrne CD, Zhang J, Petry CJ, Lucas A, Hales CN (1997) Programming of hepatic insulin-sensitive enzymes in offspring of rat dams fed a protein-restricted diet. Am J Phys 272: G1083–G1090

Draper N, Stewart PM (2005) 11β-hydroxysteroid dehydrogenase and the pre-receptor regulation of corticosteroid hormone action. J Endocrinol 186:251–271

Fu Q, Yu X, Callaway CW, Lane RH, McKnight RA (2009) Epigenetics: intrauterine growth retardation (IUGR) modifies the histone code along the rat hepatic IGF-1 gene. FASEB J 23:2438–2449

Geer EB, Shen W (2009) Gender differences in insulin resistance, body composition, and energy balance. Gend Med 6(Suppl 1):60–75

Gong L, Pan YX, Chen H (2010) Gestational low protein diet in the rat mediates Igf2 gene expression in male offspring via altered hepatic DNA methylation. Epigenetics 5:619–626

Hall E, Volkov P, Tasnim Dayeh T, Esguerra JLS, Salö S et al (2014) Sex differences in the genome-wide DNA methylation pattern and impact on gene expression, microRNA levels and insulin secretion in human pancreatic islets. Genome Biol 15:522

Hamo E, Cottrell EC, Keevil BG, DeSchoolmeester J, Bohlooly- YM et al (2013) 11-Dehydrocorticosterone causes metabolic syndrome, which is prevented when 11β-HSD1 is knocked out in livers of male mice. Endocrinology 154:3599–3609

Hanson RW, Patel YM (1994) Phosphoenolpyruvate carboxykinase (GTP) gene. Adv Enzymol Relat Areas Mol Biol 69:203–281

Heard E, Martienssen RA (2014) Transgenerational epigenetic inheritance: myths and mechanisms. Cell 157:95–109

Hemanowski-Vosatka A, Balkovec JM, Cheng K, Chen HY, Hernandez M et al (2005) 11beta-HSD1 inhibition ameliorates metabolic syndrome and prevents progression of atherosclerosis in mice. J Exp Med 202:517–527

Hermans MP, Levy JC, Morris RJ, Turner RC (1999) Comparison of tests of beta-cell function across a range of glucose tolerance from normal to diabetes. Diabetes 48:1779–1786

Kaelin Jr WG, McKnight SL (2013) Influence of metabolism on epigenetics and disease. Cell 153:56–69

Lax ER, Ghraf R, Schriefers H (1978) The hormonal regulation of hepatic microsomal 11beta-hydroxtsteroid dehydrogenase activity in the rat. Acta Endocrinol 89:352–357

Lillycrop KA, Phillips ES, Torrens C, Hanson MA, Jackson AA, Burdge GC (2008) Feeding pregnant rats a protein-restricted diet persistently alters the methylation of specific cytosines in the hepatic PPAR alpha promoter of the offspring. Br J Nutr 100:278–282

Lillycrop KA, Phillips ES, Jackson AA, Hanson MA, Burdge GC (2005) Dietary protein restriction of pregnant rats induces and folic acid supplementation prevents epigenetic modification of hepatic gene expression in the offspring. J Nutr 135:1382–1386

Lindsay RS, Wake DJ, Nair S, Bunt J, Livingstone DE et al (2003) Subcutaneous adipose 11 beta-hydroxysteroid dehydrogenase type 1 activity and messenger ribonucleic acid levels are associated with adiposity and insulinemia in pima Indians and Caucasians. J Clin Endocrinol Metab 88:2738–2744

Liu S, Song Y, Ford ES, Manson JE, Buring JE, Ridker PM (2005) Dietary calcium, vitamin D, and the prevalence of metabolic syndrome in middle-aged and older U.S. women. Diabetes Care 28:2926–2932

Liu Y, Nakagawa Y, Wang Y, Li R, Li X et al (2003) Leptin activation of corticosterone production in hepatocytes may contribute to the reversal of obesity and hyperglycemia in leptin-deficient ob/ob mice. Diabetes 52:1409–1416

Livingstone DE, Jones GC, Smith K, Jamieson PM, Andrew R et al (2001) Understanding the role of glucocorticoids in obesity: tissue-specific alterations of corticosterone metabolism in obese Zucker rats. Endocrinology 141:560–563

Macotela Y, Boucher J, Tran TT, Kahn CR (2009) Sex and depot differences in adipocyte insulin sensitivity and glucose metabolism. Diabetes 58:803–812

Masuzaki H, Yamamoto H, Kenyon CJ, Elmquist JK, Morton NM, Paterson JM, Shinyama H, Sharp MG, Fleming S, Mullins JJ, Seckl JR, Flier JS (2003) Transgenic amplification of glucocorticoid action in adipose tissue causes high blood pressure in mice. J Clin Invest 112:83–90

Matthews DR, Hosker JP, Rudenski AS, Naylor BA, Treacher DF, Turner RC (1985) Homeostasis model assessment: insulin resistance and beta-cell function from fasting plasma glucose and insulin concentrations in man. Diabetologia 28:412–419

Morgan SA, McCabe EL, Gathercole LL, Hassan-Smith ZK, Larner DP et al (2014) 11β-HSD1 is the major regulator of the tissue-specific effects of circulating glucocorticoid excess. Proc Natl Acad Sci USA 111:E2482–E2491

Morton NM, Seckl JR (2008) 11beta-hydroxysteroid dehydrogenase type 1 and obesity. Front Horm Res 36:146–164

Ng SF, Lin RC, Laybutt DR, Barres R, Owens JA et al (2010) Chronic high-fat diet in fathers programs ß-cell dysfunction in female rat offspring. Nature 467:963–966

Patel MS, Srinivasan M (2011) Metabolic programming in the immediate postnatal life. Ann Nutr Metab 58(Suppl 2):18–28

Paulmyer-Lacrox O, Boullu S, Oliver C, Alessi MC, Grino M (2002) Expression of the mRNA coding for 11beta-hydroxysteroid dehydrogenase type 1 in adipose tissue from obese patients: an in situ hybridization study. J Clin Endocrinol Metab 87:2701–2705

Pereira MA, Jacobs DRJ, Van Horn L, Slattery ML, Kartashov AI, Ludwig DS (2002) Dairy consumption, obesity, and the insulin resistance syndrome in young adults: the CARDIA study. JAMA 287:2081–2089

Pikilidou MI, Lasaridis AN, Sarafidis PA et al (2009) Insulin sensitivity increase after calcium supplementation and change in intraplatelet calcium and sodium-hydrogen exchange in hypertensive patients with type 2 diabetes. Diabetes Med 26:211–219

Pittas AG, Lau J, FB H, Dawson-Hughes B (2007) The role of vitamin D and calcium in type 2 diabetes. A systematic review and meta-analysis. J Clin Endocrinol Metab 92:2017–2029

Rask E, Olsson T, Sodenberg S, Andrew R, Livingstone DE et al (2001) Tissue-specific dysregulation of cortisol metabolism in human obesity. J Clin Endocrinol Metab 86:1418–1421

Schrager S (2005) Dietary calcium intake and obesity. J Am Board Fam Pract 18:205–210

Skinner MK (2008) What is an epigenetic transgenerational phenotype? F3 or F2. Reprod Toxicol 25:2–6

Srinivasan M, Mahmood S, Patel MS (2013) Metabolic programming effects initiated in the suckling period predisposing for adult-onset obesity cannot be reversed by calorie restriction. Am J Physiol Endocrinol Metab 304:E486–E494

Takaya J, Yamanouchi S, Tanabe Y, Kaneko K (2016) A calcium-deficient diet in rat dams during gestation decreases HOMA-β% in 3 generations of offspring. J Nutrigenet Nutrigenomics 9:276–286

Takaya J, Yamanouchi S, Kaneko K (2014) A calcium-deficient diet in rat dams during gestation and nursing affects hepatic 11β-hydroxysteroid dehydrogenase-1 expression in the offspring. PLoS One 9:e84125

Takaya J, Iharada A, Okihana H, Kaneko K (2013) A calcium-deficient diet in pregnant, nursing rats induces hypomethylation of specific cytosines in the 11β-hydroxysteroid dehydrogenase-1 promoter in pup liver. Nutr Res 33:961–970

Takaya J, Iharada A, Okihana H, Kaneko K (2011) Upregulation of hepatic 11β-hydroxysteroid dehydrogenase-1 expression in calcium-deficient rats. Ann Nutr Metab 59:73–78

Tomlinson JW, Walker EA, Bujalska IJ, Draper N, Lavery GG, Cooper MS, Hewison M, Stewart PM (2004) 11β-hydroxysteroid dehydrogenase type 1: a tissue-specific regulator of glucocorticoid response. Endocr Rev 25:831–866

Valdez R, Athens MA, Thompson GH, Bradshaw BS, Stem MP (1994) Birthweight and adult health outcomes in a biethnic population in the USA. Diabetologia 37:624–631

Walker BR (2006) Cortisol—cause and cure for metabolic syndrome? Diabet Med 23:1281–1288

Wallace TM, Levy JC, Matthews DR (2004) Use and abuse of HOMA modeling. Diabetes Care 27:1487–1495

Warner MJ, Ozanne SE (2010) Mechanisms involved in the developmental programming of adulthood disease. Biochem J 427:333–347

Selenoproteins and Epigenetic Regulation in Mammals

94

Hsin-Yi Lu, Berna Somuncu, Jianhong Zhu, Meltem Muftuoglu, and Wen-Hsing Cheng

Contents

H.-Y. Lu · W.-H. Cheng (✉)
Department of Food Science, Nutrition and Health Promotion, Mississippi State University, Mississippi State, MS, USA
e-mail: HL619@msstate.edu; wc523@msstate.edu

B. Somuncu · M. Muftuoglu
Department of Molecular Biology and Genetics, Acibadem University, Istanbul, Turkey
e-mail: berna.somuncu@live.acibadem.edu.tr; meltem.muftuoglu@acibadem.edu.tr

J. Zhu
Department of Preventive Medicine, Department of Geriatrics and Neurology at the Second Affiliated Hospital, and Key Laboratory of Watershed Science and Health of Zhejiang Province, Wenzhou Medical University, Wenzhou, Zhejiang, China
e-mail: jhzhu@wmu.edu.cn

V. B. Patel, V. R. Preedy (eds.), *Handbook of Nutrition, Diet, and Epigenetics*,
https://doi.org/10.1007/978-3-319-55530-0_31

Abstract

Selenium is an essential mineral. There is a total of 25 mammalian selenoproteins that confer the majority of physiological and pathophysiological functions of selenium. All functionally characterized selenoproteins are oxidoreductases. In humans, extremely low levels of selenium in the body result in classic selenium deficiency diseases, and patients with mutations in genes involved in selenoprotein expression show selenoprotein deficiency and multisystem defects. Recent progress suggests important roles of certain selenoproteins in epigenetic regulation of promoter methylation, histone modifications, noncoding RNA expressions, and genome stability. Conversely, such epigenetic events can also influence selenoprotein expression. Understanding how selenoproteins function in epigenetic regulations will continue to offer positive impact on selenium regulation toward optimal health.

Keywords

Selenium · Selenocysteine · Selenoproteins · Mineral · Nutrition · Oxidative stress · Genome maintenance · CpG methylation · Histone · Noncoding RNA

List of Abbreviations

5-mC	5-methylcytosine
DIO	Iodothyronine deiodinase
DNMT	DNA methyltransferase
GPX	Glutathione peroxidase
lincRNA	Long intergenic noncoding RNA
miRNA	MicroRNA
ncRNA	Noncoding RNA
piRNA	Piwi-interacting RNA
SBP2	SECIS-binding protein 2
SECIS	Selenocysteine insertion sequence
SELENO	Selenoprotein
SEPHS2	Selenophosphate synthetase-2
SEPSECS	Selenocysteine synthase
siRNA	Small interfering RNA
TXNRD	Thioredoxin reductase

Selenium and Selenoproteins

Selenium is a nutritionally essential metalloid and exerts its physiological and pathophysiological functions mainly through selenoproteins. The best known selenium deficiency syndromes in humans are Keshan cardiomyopathy, Kashin-Beck bone degeneration, and cretinism. Through recoding the stop codon UGA, selenoproteins incorporate selenium in the form of selenocysteine at translational level. Based on their structures and functions, the 25 mammalian selenoproteins are classified as glutathione peroxidases (GPX1-4 and 6), thioredoxin reductases (TXNRD1-3), iodothyronine deiodinases (DIO1-3), the Rdx family of selenoproteins (SELENOW, SELENOT, SELENOH, and SELENOV), thioredoxin-like fold endoplasmic reticulum proteins (SELENOF and SELENOM), or others (methionine-*R*-sulfoxide reductase-1, selenophosphate synthetase-2 (SEPHS2), SELENOI, SELENOK, SELENOS, SELENOO, SELENON, and SELENOP). GPX6 is a selenoprotein in humans, but not rodents. All functionally characterized selenoproteins are known as redox enzymes. Some selenoproteins are directly involved in selenium metabolism. For example, SEPHS2 generates monoselenophosphate from selenide and ATP, and SELENOP transports selenium from liver to other tissues. Physiological and molecular roles of selenoproteins are comprehensively reviewed in Labunskyy et al. (2014). Interestingly, although selenoproteins are generally considered to be beneficial, paradoxical roles of selenoproteins such as GPX1-3, TXNRD1, and SELENOP in cancer and diabetes have also been reported (Lei et al. 2016).

In addition to UGA codon, translational incorporation of selenocysteine requires the *cis*-acting selenocysteine insertion sequence (SECIS) at the 3′-untranslated region of all selenoprotein mRNAs, as well as *trans*-acting factors including SECIS-binding protein 2 (SBP2, encoded by *SECISBP2*) and selenocysteinyl-tRNA (encoded by *TRSP* with additional modifications). The synthesis of mature selenocysteinyl-tRNA requires the sequential enzymatic reactions of seryl-tRNA synthetase, phosphoseryl-tRNA$^{[Ser]Sec}$ kinase, and selenocysteine synthase (encoded by *SEPSECS*) on the *TRSP* gene product. Noticeably, *TRSP* encodes a precursor of both selenocysteinyl and cysteinyl tRNAs. Selenocysteine synthase catalyzes the synthesis of selenocysteinyl or cysteinyl tRNA depending on whether monoselenophosphate ($H_2SePO_3^-$) or thiophosphate ($H_2SPO_3^-$) is incorporated into the amino acid backbone of the serine moiety on tRNA$^{[Ser]Sec}$. Although a small fraction of TXNRD1 is known to contain cysteine at the selenocysteine position, dietary selenium deficiency stimulates such misincorporation as the formation of both $H_2SePO_3^-$ and $H_2SPO_3^-$ are catalyzed by SEPHS2 (for details, see Labunskyy et al. (2014)). The essentiality of these 25 selenoproteins is exemplified by the fact that *Trsp*$^{-/-}$ mice are embryonically lethal (Bosl et al. 1997). Recent advances have identified a couple of human mutations in genes necessary for selenoprotein synthesis, including those on *TRSP* (Schoenmakers et al. 2016), *SECISBP2* (Schoenmakers et al. 2010), and *SEPSECS* (Agamy et al. 2010; Anttonen et al. 2015). Phenotypes and selenium markers of these mutations in humans are summarized in Table 1. Consistent with their essentiality in selenoprotein synthesis, these mutations result in significant decrease in plasma selenium concentrations and tissue selenoprotein expression (Table 1).

Tissue selenium concentrations differ, ranging from 1.5 nmol/g in brain or muscle to 10–18 nmol/g in kidney, liver, and testes (Burk et al. 2006). GPX1 accounts for 58% of total liver selenium in mice (Cheng et al. 1997), and the two selenium transporters, GPX3 and SELENOP, altogether represent 54% of total plasma selenium in healthy humans (Combs et al. 2011). In addition to selenoproteins, selenium can exist specifically and posttranslationally in selenium-binding proteins

Table 1 Human diseases known to be attributed to mutations in genes in association with selenoprotein synthesis. Mutations in *TRSP*, *SECISBP2*, and *SEPSECS* distinctly result in an array of phenotypes and selenoprotein deficiency. Of note, they are not selenoprotein genes, but they encode proteins necessary for selenoprotein synthesis. Plasma selenium concentration at 1.65 μmol/L (130 μg/L) is considered optimal in adult humans

Genes	Mutations	Phenotypes	Plasma selenium (μmol/L)	Selenoprotein deficiency	References
TRSP	Point mutation	Abdominal pain, fatigue, muscle weakness, and low plasma levels of selenium	0.28	SELENOP, GPX3 in plasma. GPX1 in fibroblasts. *SELENOW* mRNA in fibroblasts	Schoenmakers et al. (2016)
SECISBP2	Heterozygous for paternally inherited frameshift/ premature stop mutation; maternally inherited missense mutation	Azoospermia, muscular dystrophy, photosensitivity, abnormal cytokine secretion, and telomere shortening and replication stress in T lymphocytes	0.11–0.14	SELENOP and GPX3 in plasma; mGPX4, TXNRD3, and SELENOV in seminal plasma; SELENON, MSRB1, and GPX1 protein, and *SELENOH*, *SELENOW*, and *SELENOT* mRNAs in fibroblasts	Schoenmakers et al. (2010)
SEPSECS	Missense mutation	Autosomal recessive symptoms of progressive cerebellocerebral atrophy	Not reported	Not reported	Agamy et al. (2010)
	Missense mutation; nonsense mutation	Cortical laminar necrosis, loss of myelin and neurons, and astrogliosis	Not reported	GPX1, GPX4 TXNRD1 in brain	Anttonen et al. (2015)

presumably through their structural configurations (Bansal et al. 1989), as well as nonspecifically by replacement of sulfur with selenium in methionine or cysteine residues of all proteins (Fig. 1). Speciation and doses of selenium affect its incorporation into proteins. Numerous studies have compared selenate, selenite, and selenomethionine bioavailability and their impact on selenoprotein expression. Although both organic and inorganic forms of dietary selenium are highly bioavailable, inorganic forms of selenium are directly absorbed without the need of prior digestion and ready to be used for synthesis of selenocysteine (Sunde and Raines 2011). The efficacy of selenate for selenoprotein expression or supporting GPX

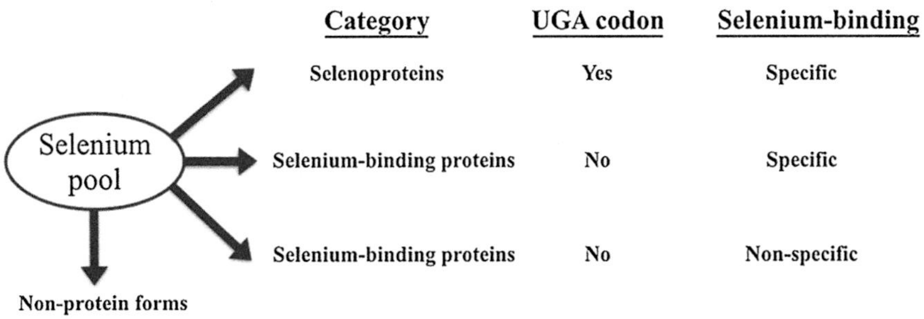

Selenium pool

Category	UGA codon	Selenium-binding
Selenoproteins	Yes	Specific
Selenium-binding proteins	No	Specific
Selenium-binding proteins	No	Non-specific

Non-protein forms

Fig. 1 The three forms of selenium-containing proteins. Selenium exists in proteins in the form of selenoproteins, selenium-binding proteins, and cysteine and methionine residues of any proteins. They are categorized by the presence of in-frame UGA codon and specificity in binding selenium. A small portion of body selenium is present as low molecular weight, nonprotein forms

activity is similar to, if not greater than, that of selenite (Bosse et al. 2010; Sunde and Raines 2011). While ^{75}Se from selenate can be readily transferred to the body during ileal absorption in rats, a substantial amount of ^{75}Se from selenite or selenomethionine is retained within the ileal tissues (Vendeland et al. 1992). Above nutritional level, selenium in the form of selenomethionine, but not selenate, can be nonspecifically incorporated into albumin (Burk et al. 2001). Besides proteins, a small fraction of selenium exists as low molecular weight metabolites (Combs 2015).

Dietrich Behne pioneered in development of the "hierarchy concept" through studies of ^{75}Se distribution in tissues and proteins of selenium-deficient rats (Behne et al. 1988). This notion implies that essential selenoproteins and tissues that prioritize selenium are maintained at the expense of others that store or does not necessarily need selenium. When the body is low in selenium status, GPX1, SELENOH, and SELENOW are greatly depleted and thus ranked at the bottom of selenoprotein hierarchy, whereas brain, thyroid, and reproductive organs (testes and ovaries) retain body selenium and are positioned at the top of tissue hierarchy (Behne et al. 1988; Sunde and Raines 2011). In particular, selenium concentrations are decreased by 80% in liver but increased by 3.3-fold and 80%, respectively, in brain and testes of mice on a selenium-deficient diet for 18 weeks (Burk and Hill 2009). Those ranked high in tissue hierarchy further complete for selenium, as castration increases selenium concentration in the brain of *Scly*$^{-/-}$*Sepp1*$^{-/-}$ mice carrying compromised selenium transport and unitization (Pitts et al. 2015). Over the last two decades, it is becoming clear that selenoprotein hierarchy is attributed to nonsense-mediated decay, CRL2 ubiquitin ligase-mediated degradation, isoforms of selenocysteinyl-tRNA, and SECIS-binding mediators such as the stimulatory nucleolin and the inhibitory eukaryotic initiation factor 4a3 (Budiman et al. 2009; Lin et al. 2015; Miniard et al. 2010; Moustafa et al. 2001; Seyedali and Berry 2014). Tissue hierarchy of selenium retention is mainly determined by the tissue-specific expression of the two low-density lipoprotein receptor family members, apolipoprotein E receptor-2 (apoER2) and megalin (Olson et al. 2007, 2008). The expression of a*poER2* is excessively high in testes and brain that are ranked high in tissue hierarchy and extremely low in kidney and liver that are ranked at the bottom. SELENOP delivers selenium from liver to tissues that express apoER2 or megalin to facilitate their endocytosis, followed by cleavage of selenocysteine residues on SELENOP by selenocysteine lyase to release selenide for tissue usage. SELENOP transfers selenium from liver to brain and testes through the nine selenocysteine residues in its C-terminal domain and to other tissues presumably through other domains (Hill et al. 2007; Schomburg et al. 2003). Such selenoprotein regulation of selenium homeostasis is consistent with the observation of selenoprotein hierarchy in testis, as requirements for *Gpx1* mRNA levels are high but *Selenop* mRNA levels low in this organ prioritized with body selenium (Sunde and Raines 2011).

Epigenetic Pathways

In addition to briefly discuss the well-studied epigenetic events on chromatin modifications (Du et al. 2015), here noncoding regulatory RNA and genome maintenance are also explored.

Histone Modifications

Chromatin packages nucleosomes and allows them to fit in the nucleus. Nucleosomes are composed of 146 base pairs of DNA wrapping around the core octamer of histone proteins (H2A, H2B, H3, and H4). Recent advances suggest that regulatory RNAs are also functionally present in chromatins (Zhang et al. 2016b). Nucleosomes are linked and stabilized through histone H1 binding on its linker DNA (Luger et al. 2012). The conserved *N*-terminal tails of core histone proteins are subject to various forms of covalent modifications that are crucial for the regulation of gene expression, replication, and DNA damage response. The histone code marked by phosphorylation, acetylation, and methylation on histones determines whether it is the compact heterochromatin in favor of transcription repression or the loose euchromatin promoting transcriptional activation. For example, acetylation of lysine residues on histone H3 and H4 by histone acetyltransferases neutralizes them and discourages binding with the negatively charged DNA, resulting in relaxation of the chromatin structure and transcriptional activation. In contrast, histone deacetylases promote chromatin condensation. Similarly, the compactness of chromatin is known to be regulated by histone methyltransferases on the arginine and lysine residues of histone H3 and H4. Furthermore, histone H3 phosphorylation during mitosis and meiosis may induce chromatin condensation.

Nucleic Acid Methylation

DNA methylation is a well-characterized form of epigenetic modification that leads to transcriptional silencing. DNA methyltransferases (DNMTs) catalyze the formation of 5-methylcytosine (5-mC) on genomic DNA. While more than 98% of 5-mC occurs in CpG dinucleotides in somatic cells, it is commonly observed in non-CpG context in embryonic stem cells. CpG islands are CpG dinucleotide rich at the proximity of transcriptional starting site and predominantly unmethylated. CpG island methylation is known to silence genes involved in genomic imprinting, X-chromosome inactivation and differentiation through direct inhibition of transcription factor binding, and recruitment of chromatin modifiers such as polycomb complexes, nucleosome positioning, noncoding RNA, and ATP-dependent chromatin remodeling proteins (Du et al. 2015; Fingerman et al. 2013). Interestingly, DNA methylation and histone modifications cooperate in the regulation of gene expression. For example, promoter methylation triggers histone H3 methylation, and DNA methyltransferases interact with histone deacetylases, histone methyltransferases, and methyl cytosine-binding proteins in a complex network (Du et al. 2015; Torres and Fujimori 2015). Recently, a reversible, posttranscriptional modification of N^1-methyladenosine on mRNAs for the promotion of translation has been identified in thousands of different eukaryotic mRNAs (Dominissini et al. 2016).

Noncoding RNAs

Noncoding RNAs (ncRNA) are functional RNA molecules that regulate translation, chromatin organization, histone modification, DNA methylation, RNA editing, and gene silencing. Except transfer and ribosomal RNAs, the long (>200 nt) and short (<30 nt) ncRNAs are the two main groups of ncRNAs (Collins et al. 2011; Wang and Chang 2011). Long ncRNAs are known to play important roles in chromatin remodeling, transcriptional regulation, X-chromosome inactivation, and posttranscriptional regulation. Large intergenic ncRNAs (lincRNAs) belong to a subgroup of long ncRNAs that can complex with chromatin-modifying proteins and thus regulate gene expression (Chen 2016). Short ncRNAs include microRNA (miRNA), small interfering RNA (siRNA) and piwi-interacting RNA (piRNA). Sequence-specific binding of miRNAs and siRNAs targets mRNAs for degradation. In addition, siRNAs can induce heterochromatin formation via an RNA-induced transcriptional silencing complex, and miRNAs and siRNAs can regulate DNA methylation and histone modifications. Conversely, miRNAs themselves can be regulated by these modifications (Carthew and Sontheimer 2009; Collins et al. 2011; MacFarlane and Murphy 2010). piRNAs are PIWI protein-associated short ncRNAs and known to be involved in chromatin regulation and transposon silencing in germline and somatic cells. The PIWI–piRNA pathway impacts epigenetics through histone modification and DNA methylation (Ku and Lin 2014).

Genome Maintenance

Genome maintenance safeguards cells carrying damaged DNA through cell cycle arrest and DNA repair. If the damage persists or is unrepairable, the cells can undergo senescence or various death pathways (Campisi and di Fagagna 2007). In particular, telomere dysfunction and attrition eventually expose the ends of linear chromosomes, resulting in persistent DNA damage response and cellular senescence (di Fagagna et al. 2003). Because the conserved telomeric TTAGGG repeats in vertebrates are rich in the oxidation-susceptible guanine, selenoproteins in principle may play important roles in the protection against oxidative telomere damage as all characterized selenoproteins are known to be oxidoreductases.

Selenoproteins and Chromatin Modifications

While epigenetic regulation by selenium metabolites and selenoproteins has been extensively reviewed (Cheng et al. 2014; Speckmann and Grune 2015), this chapter focuses on the latter. Evidence suggests that chromatin modifications and selenoproteins reciprocally interact in the pathogenesis of cancer and diabetes. CpG island methylation on the AP-2 regulatory region of *GPX1* promoter limits accessibility of the transcription factor TFAP2C, resulting in suppressed GPX1 expression at mRNA and protein levels in human primary breast cancer (Kulak et al. 2013). Similarly, *GPX1* promoter hypermethylation is associated with advanced gastric cancer in human biopsies (Min et al. 2012). Hypermethylation in *GPX3* promoter and the subsequent gene expression silencing are associated with increased risks to cancers of the prostate, ovary, stomach, endometrium, and Barrett's esophagus (Cheng et al. 2014). Conversely, selenoproteins may regulate chromatin modifications. Overexpression of *Gpx1* in mice induces hyperacetylation of histone H3 and H4 in *Pdx1* promoter and elevates *Pdx1* mRNA levels in pancreatic islets (Wang et al. 2008). Although specific selenoprotein(s) that confers such epigenetic regulation is unknown, results from a mouse study show marked reductions in histone H4 acetylation by selenoproteins in bone marrow-derived macrophages isolated from selenoprotein-deficient mice (Narayan et al. 2015).

Selenoproteins and Noncoding RNAs

As regulators of mRNA stability and translation, miRNAs have been proposed as biomarkers for a variety of diseases and physiological conditions, as well as mediators of intercellular communications when carried by extracellular vesicles in blood. Selenoprotein expression can be directly or indirectly regulated by specific miRNAs. In the DIO family, it has been reported that (1) miR-224 and miR-383 directly target 3′-untranslated region of *DIO1* and decrease its mRNA expression in clear renal cell carcinoma (Boguslawska et al. 2011); (2) the thyroid hormone-inactivating DIO3 stimulates the expression of miR-214 to negatively regulate its expression in mouse cardiomyocytes following myocardial infarction (Janssen et al. 2016); and (3) the genome-imprinted, epigenetic silencing *Dlk1-Dio3* region contains 54 miRNAs and this structure is associated with suppressed global DNA methylation in mouse lupus splenic cells (Dai et al. 2016). In the GPX and

TXNRD families and other selenoproteins, recent lines of evidence show that (1) oxidative stress leads to elevated miR-181a expression that may target *Gpx1* mRNA for degradation and then apoptosis in H9c2 rat cardiomyocytes (Wang et al. 2014); (2) miR-185 targets *GPX2* and *SEPHS2* mRNAs for degradation in Caco-2 cells (Maciel-Dominguez et al. 2013); (3) miR-17 targets and induces degradation of *GPX2* and *TXNRD2* mRNAs in the mitochondria of prostate cancer and blood mononuclear cells (Tian et al. 2016; Xu et al. 2010); (4) miR-17-3p downregulates the expression of *TXNRD2* in human retinal pigment epithelium cells (Curti et al. 2014); (5) argonaute-2 prevents senescence in adipose tissue-derived stem cells derived from humans through suppression of miR-10b and miR-23b that target *GPX3* and *SELENON* mRNAs (Kim et al. 2011); and (6) miR-7 targets 5′-untranslated region of *Selenop* variant 1b transcripts in N2A mouse neuroblastoma cells (Dewing et al. 2012). Selenium status may also alter the expression of miRNAs. It has been reported that selenium deficiency alters the expression of 12 miRNAs in Caco-2 cells (Maciel-Dominguez et al. 2013) and 8 miRNAs in short telomere mice (Wu et al. 2017), and, conversely, supplementation of selenite in human hepatocellular carcinoma cells reduces the expression of miR-544a that targets *SELENOK* mRNA at the 3′-untranslated region (Potenza et al. 2016). Thus, selenium status not only changes selenoprotein expression but also regulates miRNA profiles.

Selenoproteins and Genome Maintenance

Recent studies have indicated that selenium and selenoprotein deficiency play important roles in genome maintenance. Primary embryonic fibroblasts isolated from *Gpx1*$^{-/-}$ mice (de Haan et al. 2004) and human MRC-5 lung normal fibroblasts with *SELENOH* knockdown (Wu et al. 2014) show senescence-like features in a manner depending on reactive oxygen species. The protection of SELENOH against senescence is associated with inhibition of ATM pathway activation and repair of DNA breaks (Wu et al. 2014), but how GPX1 attenuates senescence and maintains genome integrity are less understood. Although *GPX1* knockdown increases the formation of UV-induced micronuclei, a marker of DNA damage, in LNCaP human prostate cancer cells (Baliga et al. 2008), the same group also reports induction of γH2AX formation, another marker of DNA damage, by GPX1 overexpression in MCF-7 human breast cancer cells (Jerome-Morais et al. 2013). Since additional selenium supplementation is necessary to support GPX1 overexpression, selenium metabolites in the presence of GPX1 may elicit DNA damage or chromatin remodeling that can induce ATM activation and thus γH2AX formation even in the absence of DNA breaks (Kim et al. 2009). Furthermore, heterozygous mutations of *SECISBP2*, a gene encoding SBP2 essential for the expression of all selenoproteins, in humans or knockdown of this gene in MSTO mesothelioma and SY5Y neuroblastoma cell lines lead to shortened telomeres, and such effect appears to be independent of telomerase (Schoenmakers et al. 2010; Squires et al. 2009). Similarly, long-term dietary selenium deprivation shortens telomeres by 22% in primary colonocytes isolated from generation 3 *Terc*$^{-/-}$ short telomere mice at 24 months of age (Wu et al. 2017).

Summary and Conclusions

That all functionally characterized selenoproteins are oxidoreductases supports a role of selenium at nutritional levels of intake in the protection against oxidative DNA damage. In addition, nuclear and mitochondrial selenoproteins may participate in genome maintenance through mechanism other than redox regulation (Zhang et al. 2016a). Considering that mitochondrial genome lacks histone proteins, contains very short intergenic regions, and is circular, epigenetic regulation by selenium and selenoproteins is expected to differ in nuclei and mitochondria. As selenium status is known to regulate histone modifications at various sites (Zhang et al. 2016a), it is also of future interests to identify selenoproteins that confer such epigenetic effect. Although recent progress has been made to enhance our understanding in selenoprotein regulation on nucleosome modifications and genome maintenance, unanswered questions include how nuclear and mitochondrial selenoproteins regulate (1) CpG site methylation and histone modifications; (2) epigenetic pathway cross-talks such as interactions between miRNA and senescence (Bilsland et al. 2013) or ncRNAs and DNA repair (Zhang et al. 2016b); (3) genomic tandem repeats such as telomeres and centromeres. Furthermore, the discovery of N^1-methyladenosine on mRNA (Dominissini et al. 2016) offers an interesting direction of selenoprotein studies on this new form of methylation.

Dictionary of Terms

- **Selenoproteins** – Selenoproteins carry selenocysteine residues being added co-translationally by decoding the UGA codon.
- **Essential mineral** – Minerals that are nutrients and essential for life. There are deficiency syndromes if dietary intake of essential minerals is low.
- **Selenoprotein hierarchy** – In response to selenium deficiency, certain selenoproteins are depleted faster and more significantly than others, and they are ranked low in the hierarchy.
- **Noncoding RNA** – A noncoding RNA is transcribed from the genome but is not translated into a protein. They are functional and examples include transfer RNA, ribosomal RNA, and microRNA.
- **Replicative senescence** – A form of senescence that is attributed to progressive telomere shortening by continuous cell proliferation, to a stage when telomeres at chromosome ends are too short to form the protective loops. Continuous exposure of chromosome ends results in persistent DNA damage response and senescence.
- **Genome maintenance** – In response to DNA damage, sensor, signaling, and repair proteins are recruited to the damaged sites that signal cell cycle to arrest.

If the damage is too severe to be repaired, the cells may subsequently undergo senescence, apoptosis, or other forms of cell death.

Key Facts

- Mutations in genes necessary for selenoprotein synthesis show multisystem defects in humans.
- The 25 mammalian selenoproteins are all experimentally verified or predicted as oxidoreductases, and many of them are nuclear or mitochondrial proteins.
- Reciprocal interactions between chromatin modifications and certain selenoprotein exist and impact the pathogenesis of certain chronic diseases including cancer and diabetes.
- Many miRNAs have been identified to target selenoprotein mRNAs for degradation under various physiological and pathophysiological conditions.
- A couple of selenoproteins have been found to impact genome maintenance including DNA damage response and telomere integrity.

Summary Points

- Selenium is an essential mineral.
- The physiological functions of selenium are mainly mediated by selenoproteins.
- Selenoproteins use the otherwise stop codon UGA to code selenocysteine, which is enabled by a couple of *cis*-acting and *trans*-acting factors.
- All functionally characterized selenoproteins are oxidoreductases.
- In humans, extremely low body selenium status results in classic deficiency diseases.
- Mutations in genes involved in selenoprotein expression show selenoprotein deficiency and multisystem defects.
- Recent progress suggests important roles of certain selenoproteins in epigenetic regulation of promoter methylation, histone modifications, noncoding RNA expression, and genome stability.
- Epigenetic regulation can impact selenoprotein expression.
- Long-term dietary selenium deficiency shortens telomeres in primary cells isolated from aged short telomere mice.
- Understanding how selenoproteins function in epigenetic regulations will be continued to offer positive impact on selenium regulation toward optimal health.

Acknowledgments This chapter was partially supported by the USDA National Institute of Food and Agriculture (Multistate NE1439, accession no. 1008124, project no. MIS-384050), the Scientific and Technological Research Council of Turkey (TUBITAK, grant no. 114Z875), Zhejiang Provincial Natural Science Foundation Distinguished Young Scholar Program (LR13H020002), and Wenzhou Science and Technology Bureau (Y20150005).

References

Agamy O, Zeev BB, Lev D, Marcus B, Fine D, Su D, Narkis G, Ofir R, Hoffmann C, Leshinsky-Silver E (2010) Mutations disrupting selenocysteine formation cause progressive cerebello-cerebral atrophy. Am J Hum Genet 87:538–544

Anttonen A-K, Hilander T, Linnankivi T, Isohanni P, French RL, Liu Y, Simonović M, Söll D, Somer M, Muth-Pawlak D (2015) Selenoprotein biosynthesis defect causes progressive encephalopathy with elevated lactate. Neurology 85:306–315

Baliga MS, Diwadkar-Navsariwala V, Koh T, Fayad R, Fantuzzi G, Diamond AM (2008) Selenoprotein deficiency enhances radiation-induced micronuclei formation. Mol Nutr Food Res 52:1300–1304

Bansal MP, Oborn CJ, Danielson KG, Medina D (1989) Evidence for two selenium-binding proteins distinct from glutathione peroxidase in mouse liver. Carcinogenesis 10:541–546

Behne D, Hilmert H, Scheid S, Gessner H, Elger W (1988) Evidence for specific selenium target tissues and new biologically important selenoproteins. Biochim Biophys Acta 966:12–21

Bilsland AE, Revie J, Keith W (2013) MicroRNA and senescence: the senectome, integration and distributed control. Crit Rev Oncog 18:373–390

Boguslawska J, Wojcicka A, Piekielko-Witkowska A, Master A, Nauman A (2011) MiR-224 targets the 3′ UTR of type 1 5′-iodothyronine deiodinase possibly contributing to tissue hypothyroidism in renal cancer. PLoS One. https://doi.org/10.1371/journal.pone.0024541

Bosl MR, Takaku K, Oshima M, Nishimura S, Taketo MM (1997) Early embryonic lethality caused by targeted disruption of the mouse selenocysteine tRNA gene (Trsp). Proc Natl Acad Sci U S A 94:5531–5534

Bosse AC, Pallauf J, Hommel B, Sturm M, Fischer S, Wolf NM, Mueller AS (2010) Impact of selenite and selenate on differentially expressed genes in rat liver examined by microarray analysis. Biosci Rep 30:293–306

Budiman ME, Bubenik JL, Miniard AC, Middleton LM, Gerber CA, Cash A, Driscoll DM (2009) Eukaryotic initiation factor 4a3 is a selenium-regulated RNA-binding protein that selectively inhibits selenocysteine incorporation. Mol Cell 35:479–489

Burk RF, Christensen JM, Maguire MJ, Austin LM, Whetsell WO, May JM, Hill KE, Ebner FF (2006) A combined deficiency of vitamins E and C causes severe central nervous system damage in guinea pigs. J Nutr 136:1576–1581

Burk RF, Hill KE (2009) Selenoprotein P – expression, functions, and roles in mammals. Biochim Biophys Acta 1790:1441–1447

Burk RF, Hill KE, Motley AK (2001) Plasma selenium in specific and non-specific forms. Biofactors 14:107–114

Campisi J, di Fagagna FDA (2007) Cellular senescence: when bad things happen to good cells. Nat Rev Mol Cell Biol 8:729–740

Carthew RW, Sontheimer EJ (2009) Origins and mechanisms of miRNAs and siRNAs. Cell 136:642–655

Chen L-L (2016) Linking long noncoding RNA localization and function. Trends Biochem Sci 41:761–772

Cheng W-H, Ho Y-S, Ross DA, Valentine BA, Combs GF, Lei XG (1997) Cellular glutathione peroxidase knockout mice express normal levels of selenium-dependent plasma and phospholipid hydroperoxide glutathione peroxidases in various tissues. J Nutr 127:1445–1450

Cheng W-H, Muftuoglu M, Wu RTY (2014) Selenium and epigenetic effects on histone marks and DNA methylation. In: Ho E, Domann F (eds) Nutrition and epigenetics. CRC Press, New York, pp 273–297

Collins LJ, Schönfeld B, Chen XS (2011) The epigenetics of non-coding RNA. In: Tollefsbol (ed) Handbook of epigenetics: the new molecular and medical genetics. Academic, Cambridge, pp 49–61

Combs F Jr (2015) Biomarkers of selenium status. Forum Nutr 7:2209–2236

Combs GF, Watts JC, Jackson MI, Johnson LK, Zeng H, Scheett AJ, Uthus EO, Schomburg L, Hoeg A, Hoefig CS (2011) Determinants of selenium status in healthy adults. Nutr J. https://doi.org/10.1186/1475-2891-10-75

Curti V, Capelli E, Boschi F, Nabavi SF, Bongiorno AI, Habtemariam S, Nabavi SM, Daglia M (2014) Modulation of human miR-17–3p expression by methyl 3-O-methyl gallate as explanation of its in vivo protective activities. Mol Nutr Food Res 58:1776–1784

Dai R, Lu R, Ahmed SA (2016) The upregulation of genomic imprinted DLK1-Dio3 miRNAs in murine lupus is associated with global DNA hypomethylation. PLoS One. https://doi.org/10.1371/journal.pone.0153509

de Haan JB, Bladier C, Lotfi-Miri M, Taylor J, Hutchinson P, Crack PJ, Hertzog P, Kola I (2004) Fibroblasts derived from Gpx1 knockout mice display senescent-like features and are susceptible to H_2O_2-mediated cell death. Free Radic Biol Med 36:53–64

Dewing AST, Rueli RH, Robles MJ, Nguyen-Wu ED, Zeyda T, Berry MJ, Bellinger FP (2012) Expression and regulation of mouse selenoprotein P transcript variants differing in non-coding RNA. RNA Biol 9:1361–1369

di Fagagna FDA, Reaper PM, Clay-Farrace L, Fiegler H, Carr P, von Zglinicki T, Saretzki G, Carter NP, Jackson SP (2003) A DNA damage checkpoint response in telomere-initiated senescence. Nature 426:194–198

Dominissini D, Nachtergaele S, Moshitch-Moshkovitz S, Peer E, Kol N, Ben-Haim MS, Dai Q, Di Segni A, Salmon-Divon M, Clark WC (2016) The dynamic N1-methyladenosine methylome in eukaryotic messenger RNA. Nature 530:441–446

Du J, Johnson LM, Jacobsen SE, Patel DJ (2015) DNA methylation pathways and their crosstalk with histone methylation. Nat Rev Mol Cell Biol 16:519–532

Fingerman IM, Zhang X, Ratzat W, Husain N, Cohen RF, Schuler GD (2013) NCBI epigenomics: what's new for 2013. Nucleic Acids Res 41:D221–D225

Hill KE, Zhou J, Austin LM, Motley AK, Ham A-JL, Olson GE, Atkins JF, Gesteland RF, Burk RF (2007) The selenium-rich C-terminal domain of mouse selenoprotein P is necessary for the supply of selenium to brain and testis but not for the maintenance of whole body selenium. J Biol Chem 282:10972–10980

Janssen R, Zuidwijk MJ, Muller A, van Mil A, Dirkx E, Oudejans CBM, Paulus WJ, Simonides WS (2016) MicroRNA 214 is a potential regulator of thyroid hormone levels in the mouse heart following myocardial infarction, by targeting the thyroid-hormone-inactivating enzyme deiodinase type III. Front Endocrinol (Lausanne). https://doi.org/10.3389/fendo.2016.00022

Jerome-Morais A, Bera S, Rachidi W, Gann PH, Diamond AM (2013) The effects of selenium and the GPx-1 selenoprotein on the phosphorylation of H2AX. Biochim Biophys Acta 183D:3399–3406

Kim Y-C, Gerlitz G, Furusawa T, Catez F, Nussenzweig A, Oh K-S, Kraemer KH, Shiloh Y, Bustin M (2009) Activation of ATM depends on chromatin interactions occurring before induction of DNA damage. Nat Cell Biol 11:92–96

Kim BS, Jung JS, Jang JH, Kang KS, Kang SK (2011) Nuclear Argonaute 2 regulates adipose tissue-derived stem cell survival through direct control of miR10b and selenoprotein N1 expression. Aging Cell 10:277–291

Ku H-Y, Lin H (2014) PIWI proteins and their interactors in piRNA biogenesis, germline development and gene expression. Natl Sci Rev 1:205–218

Kulak MV, Cyr AR, Woodfield GW, Bogachek M, Spanheimer PM, Li T, Price DH, Domann FE, Weigel RJ (2013) Transcriptional regulation of the GPX1 gene by TFAP2C and aberrant CpG methylation in human breast cancer. Oncogene 32:4043–4051

Labunskyy VM, Hatfield DL, Gladyshev VN (2014) Selenoproteins: molecular pathways and physiological roles. Physiol Rev 94:739–777

Lei XG, Zhu J-H, Cheng W-H, Bao Y, Ho Y-S, Reddi AR, Holmgren A, Arnér ESJ (2016) Paradoxical roles of antioxidant enzymes: basic mechanisms and health implications. Physiol Rev 96:307–364

Lin H-C, Ho S-C, Chen Y-Y, Khoo K-H, Hsu P-H, Yen H-CS (2015) CRL2 aids elimination of truncated selenoproteins produced by failed UGA/sec decoding. Science 349:91–95

Luger K, Dechassa ML, Tremethick DJ (2012) New insights into nucleosome and chromatin structure: an ordered state or a disordered affair? Nat Rev Mol Cell Biol 13:436–447

MacFarlane L-A, R Murphy P. (2010) MicroRNA: biogenesis, function and role in cancer. Curr Genomics 11:537–561

Maciel-Dominguez A, Swan D, Ford D, Hesketh J (2013) Selenium alters miRNA profile in an intestinal cell line: evidence that miR-185 regulates expression of GPX2 and SEPSH2. Mol Nutr Food Res 57:2195–2205

Min SY, Kim HS, Jung EJ, Jung EJ, Do Jee C, Kim WH (2012) Prognostic significance of glutathione peroxidase 1 (GPX1) down-regulation and correlation with aberrant promoter methylation in human gastric cancer. Anticancer Res 32:3169–3175

Miniard AC, Middleton LM, Budiman ME, Gerber CA, Driscoll DM (2010) Nucleolin binds to a subset of selenoprotein mRNAs and regulates their expression. Nucleic Acids Res 38:4807–4820

Moustafa ME, Carlson BA, El-Saadani MA, Kryukov GV, Sun Q-A, Harney JW, Hill KE, Combs GF, Feigenbaum L, Mansur DB (2001) Selective inhibition of selenocysteine tRNA maturation and selenoprotein synthesis in transgenic mice expressing isopentenyladenosine-deficient selenocysteine tRNA. Mol Cell Biol 21:3840–3852

Narayan V, Ravindra KC, Liao C, Kaushal N, Carlson BA, Prabhu KS (2015) Epigenetic regulation of inflammatory gene expression in macrophages by selenium. J Nutr Biochem 26:138–145

Olson GE, Winfrey VP, Hill KE, Burk RF (2008) Megalin mediates selenoprotein P uptake by kidney proximal tubule epithelial cells. J Biol Chem 283:6854–6860

Olson GE, Winfrey VP, NagDas SK, Hill KE, Burk RF (2007) Apolipoprotein E receptor-2 (ApoER2) mediates selenium uptake from selenoprotein P by the mouse testis. J Biol Chem 282:12290–12297

Pitts MW, Kremer PM, Hashimoto AC, Torres DJ, Byrns CN, Williams CS, Berry MJ (2015) Competition between the brain and testes under selenium-compromised conditions: insight into sex differences in selenium metabolism and risk of neurodevelopmental disease. J Neurosci 35:15326–15338

Potenza N, Castiello F, Panella M, Colonna G, Ciliberto G, Russo A, Costantini S (2016) Human MiR-544a modulates SELK expression in hepatocarcinoma cell lines. PLoS One. https://doi.org/10.1371/journal.pone.0156908

Schoenmakers E, Agostini M, Mitchell C, Schoenmakers N, Papp L, Rajanayagam O, Padidela R, Ceron-Gutierrez L, Doffinger R, Prevosto C (2010) Mutations in the selenocysteine insertion sequence–binding protein 2 gene lead to a multisystem selenoprotein deficiency disorder in humans. J Clin Invest 120:4220–4235

Schoenmakers E, Carlson B, Agostini M, Moran C, Rajanayagam O, Bochukova E, Tobe R, Peat R, Gevers E, Muntoni F (2016) Mutation in human selenocysteine transfer RNA selectively disrupts selenoprotein synthesis. J Clin Invest 126:992–996

Schomburg L, Schweizer U, Holtmann B, Flohé L, Sendtner M, Köhrle J (2003) Gene disruption discloses role of selenoprotein P in selenium delivery to target tissues. Biochem J 370:397–402

Seyedali A, Berry MJ (2014) Nonsense-mediated decay factors are involved in the regulation of selenoprotein mRNA levels during selenium deficiency. RNA 20:1248–1256

Speckmann B, Grune T (2015) Epigenetic effects of selenium and their implications for health. Epigenetics 10:179–190

Squires JE, Davy P, Berry MJ, Allsopp R (2009) Attenuated expression of SECIS binding protein 2 causes loss of telomeric reserve without affecting telomerase. Exp Gerontol 44:619–623

Sunde RA, Raines AM (2011) Selenium regulation of the selenoprotein and nonselenoprotein transcriptomes in rodents. Adv Nutr 2:138–150

Tian B, Maidana DE, Dib B, Miller JB, Bouzika P, Miller JW, Vavvas DG, Lin H (2016) miR-17-3p exacerbates oxidative damage in human retinal pigment epithelial cells. PLoS One. https://doi.org/10.1371/journal.pone.0160887

Torres IO, Fujimori DG (2015) Functional coupling between writers, erasers and readers of histone and DNA methylation. Curr Opin Struct Biol 35:68–75
Vendeland SC, Butler JA, Whanger PD (1992) Intestinal absorption of selenite, selenate, and selenomethionine in the rat. J Nutr Biochem 3:359–365
Wang KC, Chang HY (2011) Molecular mechanisms of long noncoding RNAs. Mol Cell 43:904–914
Wang L, Huang H, Fan Y, Kong B, Hu H, Hu K, Guo J, Mei Y, Liu W-L (2014) Effects of downregulation of microRNA-181a on H2O2-induced H9c2 cell apoptosis via the mitochondrial apoptotic pathway. Oxidative Med Cell Longev. https://doi.org/10.1155/2014/960362
Wang XD, Vatamaniuk MZ, Wang SK, Roneker CA, Simmons RA, Lei XG (2008) Molecular mechanisms for hyperinsulinaemia induced by overproduction of selenium-dependent glutathione peroxidase-1 in mice. Diabetologia 51:1515–1524
Wu RT, Cao L, Chen BPC, Cheng W-H (2014) Selenoprotein H suppresses cellular senescence through genome maintenance and redox regulation. J Biol Chem 289:34378–34388
Wu RT, Cao L, Mattson E, Witwer KW, Cao J, Zeng H, He X, Combs GF, Cheng WH (2017) Opposing impacts on healthspan and longevity by limiting dietary selenium in telomere dysfunctional mice. Aging Cell 16:125–135
Xu Y, Fang F, Zhang J, Josson S, Clair WHS, Clair DKS (2010) miR-17* suppresses tumorigenicity of prostate cancer by inhibiting mitochondrial antioxidant enzymes. PLoS One. https://doi.org/10.1371/journal.pone.0014356
Zhang Y, He Q, Hu Z, Feng Y, Fan L, Tang Z, Yuan J, Shan W, Li C, Hu X (2016b) Long noncoding RNA LINP1 regulates repair of DNA double-strand breaks in triple-negative breast cancer. Nat Struct Mol Biol 23:522–530
Zhang X, Zhang L, Zhu JH, Cheng WH (2016a) Nuclear selenoproteins and genome maintenance. IUBMB Life 68:5–12

DNA Methylation in Anti-cancer Effects of Dietary Catechols and Stilbenoids: An Overview of Underlying Mechanisms

95

Megan Beetch and Barbara Stefanska

Contents

M. Beetch · B. Stefanska (✉)
Land and Food Systems, Food, Nutrition and Health, University of British Columbia, Vancouver, BC, Canada
e-mail: meganbeetch@gmail.com; barbara.stefanska@ubc.ca

V. B. Patel, V. R. Preedy (eds.), *Handbook of Nutrition, Diet, and Epigenetics*,
https://doi.org/10.1007/978-3-319-55530-0_104

Abstract

Carcinogenesis involves an accumulation of genetic mutations and epigenetic alterations. DNA methylation, a dynamic epigenetic modification, may underlie genomic instability, silencing of genes with tumor suppressor functions, and activation of genes associated with cancer progression. Therefore, reversing DNA methylation patterns established during carcinogenesis constitutes a promising anti-cancer strategy. Interestingly, studies have indicated that certain dietary polyphenols, such as those from catechol and stilbenoid classes present in grapes, blueberries, and green tea, exert anti-cancer effects through epigenetic regulation of gene expression. A basis of evidence demonstrating the importance of DNA methylation in cancer formation and progression as well as the impact of catechol and stilbenoid compounds on these events are presented in this review. *In vitro* and *in vivo* evidence for the chemopreventive and therapeutic potential of polyphenols through their influence on DNA methylation is discussed. Current mechanistic insights on the changes in DNA methylation machinery upon exposure to polyphenols are further emphasized. Such studies are ongoing and crucially needed for transition into application of polyphenols as agents in cancer prevention and/or treatment in the clinical setting.

Keywords

DNA methylation · Epigenetics · Bioactive · Polyphenols · Catechols · Stilbenoids · Epigallocatechin gallate · Resveratrol · Pterostilbene · Chemoprevention · Cancer · Carcinogenesis

List of Abbreviations

AP-1	Activator protein 1
APC	Adenotamous polyposis coli
ATM	Ataxia telangiectasia mutated
ATR	Ataxia telangiectasia and rad3-related protein
BRCA1	Breast cancer 1
CDK	Cyclin-dependent kinase
CHK1/2	Checkpoint kinase 1/2
COMT	Catechol-O-methyltransferase
DNA	Deoxyribonucleic acid
DNMT	DNA methyltransferase
EGCG	Epigallocatechin gallate
EMT	Epithelial-to-mesenchymal transtion
ER	Estrogen receptor
EXOSC4	Exosome component 4
GSTP1	Glutathione S-transferase pi 1
HCC	Hepatocellular carcinoma
HDAC	Histone deacetylase
IGF2	Insulin-like growth factor 2

MAML2	Mastermind-like transcriptional coactivator 2
MAPK	Mitogen-activated protein kinase
MBD2	Methyl-CpG-binding domain 2
MMP	Matrix metalloproteinase
MTHFR	Methylenetetrahydrofolate reductase
NENF	Neuron-derived neurotrophic factor
NF-κB	Nuclear factor kappa B
OCT1	Octamer-binding transcription factor 1
PCNA	Proliferating cell nuclear antigen
PI3K/Akt	Phosphatidylinositol-4,5-bisphosphate 3-kinase/protein kinase B
PTEN	Phosphatase and tensin homolog
RASAL2	Ras-GTPase-activating protein 2
RASSF-1α	Ras association domain family member 1
RNA	Ribonucleic acid
RXRα	Retinoid X receptor A
SAH	S-adenosyl-L-homocysteine
SAM	S-adenosyl-L-methionine
SENP6	SUMO1/sentrin-specific peptidase 6
STAT3	Signal transducer and activator 3
TNF	Tumor necrosis factor
TRAMP	Transgenic adenocarcinoma of the mouse prostate
TYMS	Thymidylate synthase
VEGF	Vascular endothelial growth factor
WBSCR22	Williams-Beuren syndrome chromosome region 22

Introduction

The multistep process of carcinogenesis involves stages of initiation, promotion, and progression (Renan 1993). Throughout these stages, an accumulation of aberrant expression of oncogenes and tumor suppressor genes drives cancer development by massively dysregulating numerous signaling pathways and cellular functions including cell proliferation, apoptosis, angiogenesis, invasion, and metastasis to name a few (Hanahan and Weinberg 2011). Aberrations in gene expression can occur as a result of both genetic mutations and epigenetic alterations (Hanahan and Weinberg 2011; Baylin 2001). Overexpression and activation of oncogenes, like cyclin-dependent kinases (*CDK*s) controlling the cell cycle (Williams and Stoeber 2012) and *Ras*-related signaling involved in promoting cell growth and survival (Tsuchida et al. 2016) appears in cancer. Down-regulation of tumor suppressors, like *p53* whose function is essential for genomic stability, DNA repair, and cell cycle regulation (Williams and Stoeber 2012), and adenotamous polyposis coli (*APC*) which controls cell adhesion (Chiurillo 2015), also contributes to cancer. In addition to genetic mutations in these hallmark genes and others, abnormal gene expression as a result of epigenetic processes plays a direct role in carcinogenesis (Lao and Grady 2011).

The role of DNA methylation in cancer

Epigenetics refers to the control of gene expression without changes to the DNA sequence by coordinated components such as DNA methylation, covalent histone modifications, and noncoding RNA mechanisms (Jones and Takai 2001; Strahl and Allis 2000; Razin 1998; Liu and Gao 2016). Epigenetic components work together to activate or repress regions of the genome. DNA hypomethylation and histone acetylation within gene regulatory regions have been associated with open, active chromatin, whereas DNA hypermethylation and histone deacetylation have been associated with a closed chromatin state (Strahl and Allis 2000; Jones 2012). An intricate cross talk facilitates the recruitment of enzymes that catalyze these epigenetic processes. While all components of the epigenetic machinery are important to influence gene expression, this review will focus on DNA methylation. DNA methylation is thought to provide stable, long-term regulation by sustaining gene expression over time (Cedar and Bergman 2009).

DNA methylation is a covalent modification catalyzed by DNA methyltransferases (DNMTs) in mammalian cells. DNMTs transfer a methyl group donated by S-adenosyl-L-methionine (SAM) to the fifth position of the cytosine pyrimidine ring (Gruenbaum et al. 1981). Mammals have three active DNMTs that are classified into two categories: maintenance (DNMT1) and *de novo* DNA methyltransferases (DNMT3A and DNMT3B). DNMT1 targets hemimethylated DNA to maintain methylation patterns from mother to daughter strand during replication; thus propagating methylation patterns to the next generation (Berkyurek et al. 2014). DNMT3A and DNMT3B catalyze methylation of new, previously unmethylated regions of the DNA (Chen and Li 2004). Another member of the family, DNMT3L, has no methyltransferase activity but is essential in regulating DNMT3A and DNMT3B recruitment to DNA (Hata et al. 2002; Van Emburgh and Robertson 2011). Both maintenance and *de novo* DNMTs have been presented as vital players in carcinogenesis as their presence at the DNA dictates transcriptional activity of genes (reviewed in (Sharma et al. 2010)).

Early in carcinogenesis, tumor suppressor genes are hypermethylated most commonly at their promoters and silenced (reviewed in (Wilson et al. 2007; Esteller et al. 2001; Szyf et al. 2004)). Simultaneously, oncogenes and pro-metastatic genes are hypomethylated within regulatory regions and activated (Stefanska et al. 2011, 2013, 2014). In addition, DNA methylation levels decrease globally which occurs mainly in repetitive sequences and transposons and results in chromosomal rearrangements and genome instability (Hatada et al. 2006; Fig. 1). For instance, tumor suppressor gene *APC* was shown to be hypermethylated at its promoter and silenced in many types of cancer including non-small cell lung cancer (Liu et al. 2016), hepatocellular carcinoma (HCC) (Jain et al. 2011), and colorectal cancer (Galamb et al. 2016). Another example is *BRCA1* gene whose promoter hypermethylation arises during breast and ovarian carcinogenesis contributing to gene silencing (Romagnolo et al. 2015; Shariati-Kohbanani et al. 2016). Phosphatase and tensin homolog (*PTEN*) methylation and silencing are associated with more advanced tumor stage and metastatic capability in endometrial carcinomas (Salvesen et al. 2001). In contrast,

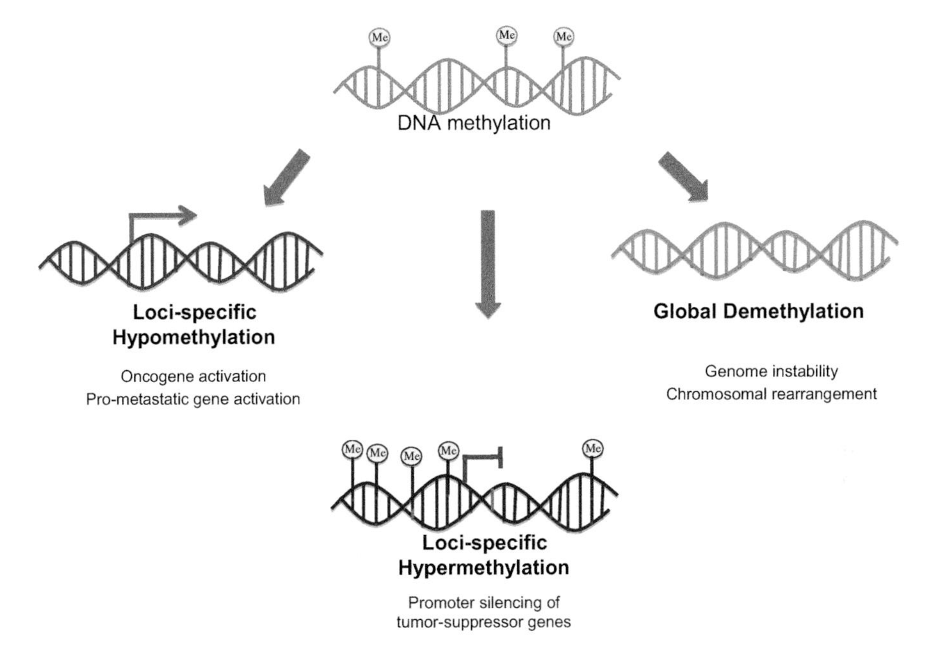

Fig. 1 Aberrations in DNA methylation result in cancer. Loci-specific hypomethylation of gene regulatory regions, like promoters and enhancers, can lead to activation of oncogenes and pro-metastatic genes. Loci-specific hypermethylation of certain gene promoters can lead to transcriptional silencing of tumor suppressor genes. A global demethylation event occurring during carcinogenesis promotes genome instability and chromosomal rearrangements

growth factor signaling is dysregulated by hypomethylation of insulin-like growth factor (*IGF2*) to induce downstream expression of pro-proliferative genes and acceleration of tumor formation in a liver cancer model (Martinez-Quetglas et al. 2016). Oncogenic players in the Wnt signaling pathway undergo hypomethylation and activation as colorectal cancer progresses (Galamb et al. 2016). Recent studies have provided more evidence on hypomethylation as a mechanism of activation of oncogenes and pro-metastatic genes (Stefanska et al. 2011, 2013, 2014). In our genome-wide investigation, we found that hypomethylation at gene promoters is as common as promoter hypermethylation and is associated with gene activation (Stefanska et al. 2011). The hypomethylated promoters appeared in clusters across the genome suggesting that a high-level organization underlies the epigenomic changes in cancer. The hypomethylated genes are mainly involved in cell growth, cell adhesion and communication, signal transduction, mobility, and invasion; functions that are essential for cancer progression and metastasis (Stefanska et al. 2011). We further delivered a mechanistic proof that aberrant DNA promoter hypomethylation in cancer leads to activation of genes that are required for continued cancer cell proliferation and invasiveness (Stefanska et al. 2014). We showed, for the first time, that silencing of genes hypomethylated in tumors, such as *RASAL2*, *NENF*, *EXOSC4*, *SENP6*, and *WBSCR22*, inhibits cell growth and invasive properties *in vitro*. Depletion of *RASAL2* and *NENF* also reduced human subcutaneous tumor xenograft growth in mice (Stefanska et al. 2014). Loci-specific DNA hypomethylation may be associated with overexpression of *MBD2* in cancer as we further mechanistically established (Stefanska et al. 2013). We discovered that MBD2 protein, that was previously shown to be involved in demethylation of specific human breast and prostate cancer genes, is upregulated in liver cancer patients. Whereas *MBD2* depletion in normal liver cells had little or no effect, we found that its depletion in human liver cancer cells resulted in suppression of cell growth, anchorage-independent growth and invasiveness, as well as an increase in promoter methylation and silencing of several of the genes that are hypomethylated in tumors and drive tumor growth (Stefanska et al. 2013).

As a whole, aberrant DNA methylation patterns play a major role in carcinogenesis through epigenetic regulation of transcription and reprogramming gene expression profiles. Although considered as a stable mark, DNA methylation is dynamic, reversible, and undergoes changes upon various environmental exposures, including diet. We will focus on the latter aspect in the current review.

Anti-Cancer Effects of Polyphenols

An accumulating body of evidence demonstrates a role for bioactive compounds present in foods in combating cancer as agents in both prevention and treatment. Epidemiological studies have correlated bioactive compounds from polyphenol class with prevention of degenerative diseases and establishment of anti-cancer activities (Arts and Hollman 2005; Williamson and Manach 2005). Dietary polyphenols are found in a wide variety of plant-based foods and beverages (Zamora-Ros et al. 2014). Polyphenols are chemically classified into two main subgroups: flavonoids and nonflavonoids. This review will highlight catechols from flavonoid compounds and stilbenoids from nonflavonoid compounds. Specifically, we will discuss (−)-epigallocatechin gallate (EGCG), (−)-epicatechin gallate (ECG), (−)-epigallocatechin (EGC), and (−)-epicatechin (EC) as major catechols found in green tea, with EGCG being the most abundant and bioactive (Yang et al. 2011). Resveratrol and pterostilbene, abundantly present in grapes and blueberries (Burns et al. 2002), respectively, will be discussed as major stilbenoid compounds (Table 1).

Epidemiological Evidence of Anti-Cancer Effects of Polyphenols

A number of epidemiological studies exploring the influence of green tea consumption on cancer outcomes have been reported (Yuan et al. 2011; Yuan 2013). Some of them showed a protective role and low risk of cancer in people regularly consuming green tea (Hanahan and Weinberg 2011; Yuan et al. 2011; Yuan 2013; Myung et al. 2009; Sun et al. 2006; Zhong et al. 2001; Tuveson and Hanahan 2011;

Table 1 **Structure, source, and characterization of epigenetic targets and mechanisms of selected catechol and stilbenoid compounds**. Catechols and stilbenoids of differing chemical structures are found within various dietary sources and influence many targets and pathways

Dietary Compound	Structure	Sources	Epigenetic target	Mechanism	Genes/pathways (epigenetically) affected
Catechols					
Epicatechin (EC)		Tea, apricots, apples, chocolate	DNMT inhibitor	COMT-mediated SAM/SAH ratio	
Epigallocatechin (EGC)		Tea, apricots, apples, chocolate	DNMT inhibitor	COMT-mediated SAM/SAH ratio	
Epicatechin gallate (ECG)		Tea, grapes (red wine), legumes	DNMT inhibitor	COMT-mediated SAM/SAH ratio	
Epigallocatechin gallate (EGCG)		Tea, chocolate, apricots, apples, grapes (red wine), legumes	DNMT inhibitor	COMT, influence SAM/SAH ratio, binding to catalytic site of DNMT, upregulation of p21	p16, p21, RARβ, MGMT ER, GSTP1, RXRα

(continued)

Table 1 (continued)

Dietary Compound	Structure	Sources	Epigenetic target	Mechanism	Genes/pathways (epigenetically) affected
Stilbenoids					
Resveratrol	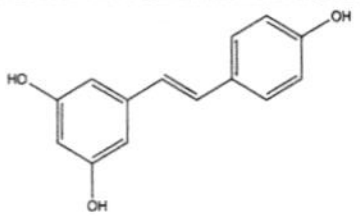	Grapes (red wine), mulberries, cocoa, peanuts	Inhibits DNMT expression and activity	Affect p21, AP-1, and PTEN expression to regulate DNMT, changes in transcription factor binding	PTEN p21, AP-1 NOTCH signaling STAT3 DNMT, HDAC
Pterostilbene	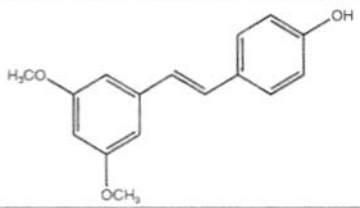	Blueberries, peanuts, almonds	Inhibits DNMT expression	Upregulation of p21, changes in transcription factor binding	p21 DNMT, HDAC

Herman and Baylin 2001; Yan et al. 2001). Yuan (2013; Yuan et al. 2011) reviews association studies evaluating green tea consumption and the risk of oral, esophageal, gastric, colorectal, lung, prostate, and breast cancers. Of note, green tea intake has been observed to have favorable effects on gastric and colorectal cancers (Myung et al. 2009; Sun et al. 2006). Inverse association has been observed between green tea intake and lung cancer risk in nonsmokers (Zhong et al. 2001). Prospective evaluation in a cohort of 69,710 Chinese women, 6 years of follow up demonstrates lower risk of colorectal cancer in women drinking green tea regularly compared with nondrinkers (Tuveson and Hanahan 2011). The higher the amount of green tea consumed and the duration of lifetime exposure, the lower the risk. Similar correlation has been established for breast cancer (Hanahan and Weinberg 2011). Weekly/daily green tea intake was inversely associated with breast cancer risk. This association was significantly strong in women with low folate intake and/or high-activity methylenetetrahydrofolate reductase (MTHFR) and thymidylate synthase (TYMS) genotypes. It would suggest that folate pathway inhibition might be a mechanism of green tea protective effects against breast cancer (Hanahan and Weinberg 2011). Case-control studies in China further confirm that green tea consumption is linked to reduced risk of breast and prostate cancers (Herman and Baylin 2001; Yan et al. 2001). In prostate cancer study, drinking green tea and simultaneous consumption of fruits and vegetables rich in lycopene were found to synergistically reduce risk of cancer (Yan et al. 2001). These studies suggest that regular consumption of green tea may reduce risk of cancer. Catechols present in tea may be at least partially responsible for observed effects. Interestingly, the presence of catechol- and phenolic acid-rich Annurca apple in the diet of a population in Southern Italy has been linked to a lower incidence of sporadic colorectal cancer compared to populations consuming foods characteristic of the Western diet (Fini et al. 2011). Although beneficial effects have been attributed to anti-oxidant properties of tea polyphenols, the precise mechanism of chemoprevention remain to be elucidated.

Populations consuming a Mediterranean diet consisting largely of fruits, vegetables, legumes and nuts, whole grains, fish, olive oil, and moderate amounts of wine have a decreased incidence of cancer compared to populations who do not consume whole, plant-based diet like the Western diet (Giacosa 2004). The Mediterranean diet differs from the Western diet because of its enrichment of bioactive compounds derived from plant sources (Grosso et al. 2013). One well-studied component of the Mediterranean diet is resveratrol, reported to have health benefits in cardiovascular, neurological, and cancer pathologies (reviewed in (Aires and Delmas 2015)).

Overall, epidemiological evidence presents a role for polyphenols in reduction of incidence of cancer. A consistent conclusion from population-based studies indicates that comprehensive and mechanistic studies are warranted to substantiate the protective effects of polyphenols in cancer and other diseases. Presence of mechanistic evidence would increase application of polyphenols in preventive strategies and presumably as support of anti-cancer therapy.

In Vitro and *In Vivo* Studies of Anti-Cancer Effects of Polyphenols

Associations between polyphenol intake and decreased cancer incidence stimulated interest into the beneficial cellular changes that take place after exposure to polyphenols. A number of *in vitro* and *in vivo* studies have identified suppressive effects on various types of cancer cells and models and protective roles mediating specific cellular pathways after introduction of stilbenoids and catechols. Since the discovery that polyphenols including catechols and stilbenes, especially resveratrol, have preventative effects in multiple stages of carcinogenesis in the late 1990s (Imai et al. 1997; Jang et al. 1997), the body of preclinical evidence has been growing rapidly. Oral administration, injection, and topical application have been used as methods to introduce polyphenols into animal systems. Generally, these methods have induced prevention or delay of carcinogenesis, but not all *in vitro* evidence has been confirmed in *in vivo* models, suggesting that factors such as dosage, diet, and cancer type can influence efficacy (Ziegler et al. 2004; Niles et al. 2006). Nevertheless, a large amount of data from animal models show results consistent with effects observed *in vitro*.

Catechols: Preclinical Evidence of Anti-Cancer Action

Studies have attempted to elucidate the pathways in cancer affected by catechol-containing compounds and the results are many. EGCG attenuates cell growth and invasive capacity of melanoma cells through binding to TNF receptor-associated factor 6 (TRAF6) whose role in signal transduction and activation of oncogenic pathways, like the NF-κB pathway, is critical (Zhang et al. 2016). Treatment with EGCG induces apoptosis via suppression of Wnt/β-catenin signaling in head and neck cancer cells (Shin et al. 2016) as well gastric cancer cells (Yang et al. 2016). EGCG stimulates apoptosis through a different mechanism in hepatocellular carcinoma cells; by inhibiting phosphofructokinase activity (Li et al. 2016). Xiang et al. (2016) review *in vitro* evidence of the suppressive effects of tea catechins in breast cancer cells extensively. Breast cancer cells treated with tea catechins may be protected through the following mechanisms: (1) alleviating reactive oxygen species, thus reducing oxidative stress and DNA damage, (2) competing with oncogenic pathways to inhibit their activation, (3) inhibiting vascular endothelial growth factor (VEGF) to promote anti-angiogenic activity, (4) regulating cyclins and cyclin-dependent kinases to stimulate apoptosis, (5) decreasing invasive potential by modulating proteolytic enzymes and epithelial-to-mesenchymal transition (EMT), and (6) altering the rate of DNA methylation reaction (Xiang et al. 2016). Catechols influence a wide range of cellular processes en route to inhibiting carcinogenesis.

Treatment with green tea polyphenols has shown promise in reducing cancer in several rodent models. In a chemically induced mouse model of HCC, administration of green tea prevented incidence of hepatic tumors (Umemura et al. 2003). Additionally, increased apoptosis and suppressed growth have been observed in HCC xenograft mice treated with EGCG. Decreased expression of *Bcl-xL*, *Akt*, and

VEGF receptor-2 is likely responsible for these changes (Nishikawa et al. 2006; Shirakami et al. 2009). EGCG treatment of the transgenic adenocarcinoma of the mouse prostate (TRAMP) model results in alterations in NF-κB, mitogen-activated protein kinases (MAPK), and IGF signaling pathways (reviewed in (Khan and Mukhtar 2013)).

Stilbenoids: Preclinical Evidence of Anti-Cancer Action

Accumulating evidence also demonstrates that stilbenoids, like resveratrol and pterostilbene, exert strong anti-cancer effects *in vitro* through modulation of various cellular functions. Resveratrol, a phytoestrogen, acts as an estrogen receptor (ER) antagonist to mitigate estrogen-related signaling pathways, such as phosphatidylinositol-4,5-bisphosphate 3-kinase/protein kinase B (PI3K/Akt), and enhance activation of p53- and p21-dependent pathways, in breast cancer cells. Resveratrol interferes with signal transducer and activator of transcription 3 (STAT3) as well as VEGF to induce antiproliferative and antiangiogenic effects in breast cancer cells. Resveratrol inhibits EMT by downregulating focal adhesion proteins and matrix metalloproteinases (MMPs) which contributes to reducing the ability of cancer cells to metastasize. (reviewed in (Sinha et al. 2016)). Similarly, resveratrol has been shown to inhibit STAT3 in ovarian and pancreatic cancer cells and the NF-κB pathway in pancreatic cancer cells, leading to decreased cell viability (Zhong et al. 2015; Duan et al. 2016).

Pterostilbene, an analogue of resveratrol, has been shown to be a more efficient and potent inhibitor of cancer cell proliferation (McCormack et al. 2011; Lubecka et al. 2016). One study identified that pterostilbene targets a p53-driven DNA damage response complex that consists of ataxia telangiectasia and rad3-related protein (ATR), ataxia telangiectasia mutated (ATM), and checkpoint kinase 1/2 (CHK1/2) in lung cancer cells (Lee et al. 2016).

As seen in *in vitro* studies, apoptosis and cell cycle interference has been observed in *in vivo* cancer models treated with resveratrol via increasing pro-apoptotic proteins, decreasing cyclins and CDKs, and inhibiting pro-proliferative signaling pathways like NF-κB and PI3K-Akt (reviewed in (Medina-Aguilar et al. 2016)). Other alterations in cellular processes like metastasis and invasion have been demonstrated in animal studies and linked to inhibition of MMPs (Lee et al. 2012a). Importantly, resveratrol has been successful in delaying carcinogenesis after introduction of carcinogens (Huderson et al. 2013) as well as enhancing the effectiveness of chemotherapeutic agents when used in combination *in vitro* and in animals models (El-Mowafy et al. 2010; Yong et al. 2016; Stefanska et al. 2010).

Summary

Altogether, preclinical evidence has established numerous direct and indirect molecular targets of catechols and stilbenoids. These compounds modulate many important cellular processes including proliferation, apoptosis, cell cycle arrest, angiogenesis, and invasion of the cancer cell. A question remains what are the mechanisms behind the observed effects and what are the molecular targets.

Modifying DNA Methylation Patterns by Catechols and Stilbenoids Contribute to Anti-Cancer Effects

Polyphenols exhibit anti-cancer effects by modulating many molecular pathways. The widespread cellular impact in response to polyphenol exposure may be, at least partially, attributed to the influence of polyphenols on DNA methylation events and DNA methylation machinery. Catechol and stilbenoid treatment in a number of *in vitro* and *in vivo* cancer models has strengthened the argument that upon exposure to dietary polyphenols, changes in DNA methylation occur, DNA methylation machinery is altered, and a link between these changes and anti-cancer actions exists. A compilation of studies highlighting the effects of catechol and stilbenoid compounds on DNA methylation status and anti-cancer outcomes follows.

Catechols: Impact on DNA Methylation in Cancer Models

The effect of green tea polyphenols on DNA methylation has been investigated in a wide variety of cancer types. Henning et al. (2013) review evidence of modifying DNA methylation in response to EGCG treatment in esophageal, skin, breast, lung, oral, and prostate cancer cells. There are consistent findings among studies including decreased global DNA methylation, reduced DNMT activity, and activation of methylation-silenced genes with antioncogenic and tumor suppressive functions (Henning et al. 2013). Of note, EGCG treatment for 3 days can reactivate estrogen receptor (ER) expression in ER-negative breast cancer cells and sensitizes them to hormonal therapies (Li et al. 2010). EGCG treatment of skin cancer cells for 6 days decreases global DNA methylation and reduces expression of DNMTs leading to increased expression of tumor suppressors *p16* and *p21* (Gallagher et al. 2001; Li et al. 2001). Treatment of prostate cancer cells with green tea extract was associated with hypomethylation at the promoter of glutathione S-transferase pi 1 (GSTP1) and restoration of expression of this gene whose silencing is implicated in cancer development (Pandey et al. 2010).

Fewer *in vivo* studies have been performed evaluating catechol compound-mediated changes of DNA methylation in cancer models. Rodent models of intestinal and skin cancer treated with EGCG have reported similar anti-cancer effects as *in vitro* studies. Administration of green tea solution in drinking water for 8–16 weeks to Apc(Min/−) mice resulted in decreased promoter methylation of the retinoid X receptor a (*RXRα*) and *RXRα* activation, which was concomitant with reduction of number of intestinal tumors (Volate et al. 2009). Topical application of EGCG-containing cream to hairless mice exposed to UVB reversed global decrease in DNA methylation, one of the hallmarks of cancer (Mittal et al. 2003).

Stilbenoids: Impact on DNA Methylation in Cancer Models

Resveratrol appears to regulate DNA methylation through interactions with epigenetic machinery like DNMTs, histone deacetylases (HDACs), and methyl binding domain proteins (Lee et al. 2012a). The methylation landscape of breast cancer cells changes in response to resveratrol treatment (Lubecka et al. 2016; Medina-Aguilar et al. 2016), causing an altered profile of gene expression. For example, resveratrol treatment induces a significant reduction in methylation of the tumor suppressor *PTEN* promoter region in noninvasive and invasive breast cancer cells concomitant with increased expression of *PTEN* (Yong et al. 2016), thus reactivating an important gene lost in cancer. Nine-day treatment with stilbenoids, like resveratrol and pterostilbene, leads to epigenetic inhibition of NOTCH oncogenic signaling pathway; the compounds reverse demethylation-mediated activation of *MAML2* oncogene that is a coactivator of NOTCH target genes (Lubecka et al. 2016). Combinatorial resveratrol and pterostilbene treatment of triple-negative breast cancer cells results in downregulation of *DNMT* and *HDAC* expression and a decrease in DNA damage response enzymes, leading to synergistic inhibition of cancer cell growth (Kala et al. 2015).

Twenty–one week administration of resveratrol to rodent model of estrogen-dependent mammary carcinoma resulted in differing levels of DNMT3B between normal and mammary tumor tissue. The decrease in DNMT3B seen in mammary tissue demonstrates that resveratrol differentially influences normal versus mammary tissue gene expression (Qin et al. 2014) highlighting its potential as an agent in cancer therapy.

Collectively, DNA methylation appears to play an important role in anti-cancer effects of polyphenol exposure. Alterations in DNA methylation, and possibly specific components of DNA methylation machinery, may be the basis for molecular changes observed after polyphenol treatment contributing to anti-cancer effects.

Underlying Mechanisms through which Catechols and Stilbenoids Elicit Effects on DNA Methylation

Research defining mechanisms through which polyphenols may exert effects on DNA methylation is ongoing. As a whole, it is thought that nutrition, specifically bioactive compounds such as polyphenols, can influence the DNA methylation process at multiple layers, which is believed to hamper carcinogenesis. First, polyphenols affect the main source of methyl groups, SAM, utilized for methylation reactions in the human body (Fig. 2). Second, polyphenols alter the activity/expression of DNA methyltransferases. Third, they change the affinity of methylating enzymes to DNA. Lastly, polyphenols change interactions of proteins such as transcription factors with DNA (Lubecka et al. 2016; Anderson et al. 2012; Stefanska et al. 2012). Modulation of the following mechanisms involving DNA methylation machinery occurs upon exposure to catechol and stilbenoid compounds and contributes to anti-cancer action.

Shifting the SAM/SAH Ratio

Inhibition of DNA methylation can occur through decrease in SAM pool available for methylation reactions and subsequent increase in S-adenosyl-L-

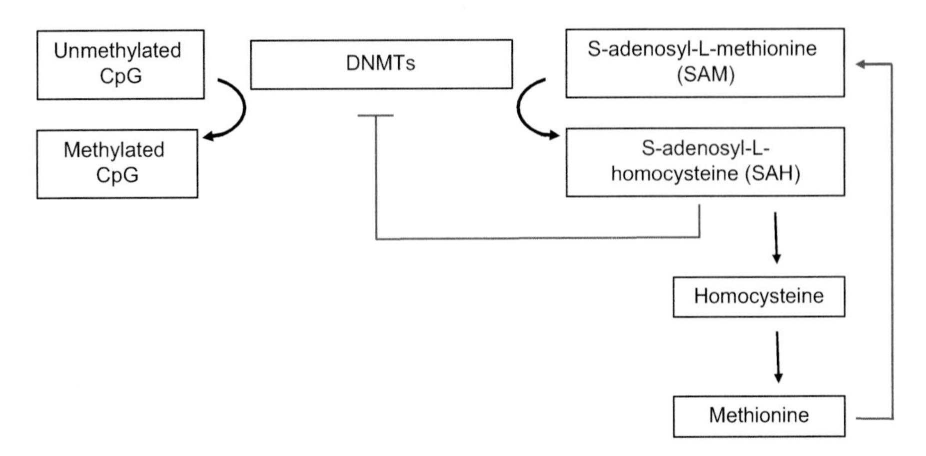

Fig. 2 One carbon metabolism and the importance of S-adenosyl-L-methionine (SAM) in methylation reactions. One carbon metabolism provides the methyl donor, S-adenosyl-L-methionine (*SAM*), for methylation reactions. A series of biochemical reactions leads to formation and reformation of SAM, resulting in multiple intermediate molecules. S-adenosyl-L-homocysteine (*SAH*) formation and accumulation after SAM utilization by DNA methyltransferases (*DNMTs*) directly inhibits DNMTs

homocysteine (SAH) levels. Catechol groups present in tea catechins are favorable substrates for methylation by catechol-O-methyltransferase (COMT) in the same way that DNA is an available substrate for methylation by DNMTs (Lee and Zhu 2006). COMT methylation of catechol groups on catechol-containing compounds uses SAM, the same methyl donor as for the DNA methylation reaction. This distraction of SAM from DNMT to COMT results in less methyl donor available for DNMTs to methylate DNA (Lee et al. 2005; Fig. 3a). Consequently, changes in DNA methylation machinery occur leading to decrease in methylation at certain loci (activation of tumor suppressor genes) and increase in methylation at other loci (repression of overactive oncogenes) (Lee et al. 2005; Renan et al. 1993).

High rate of COMT-mediated methylation of catechol compounds leads to formation of SAH (Lee et al. 2005) and increase plasma SAH concentrations (Olthof et al. 2001) (Fig. 3a). SAH, formed after SAM donates a methyl group during methylation reaction, is a potent noncompetitive inhibitor of DNMTs (James et al. 2003). Thus, catechol compounds impose indirect inhibition of DNMTs. Inhibition of DNMTs by SAH results in reactivation of hypermethylated tumor suppressor genes which was demonstrated in many cancers (Lee et al. 2005).

p21 and Competition with PCNA for the Same Binding Site on DNMT1 – Decrease in DNMT1 Activity

DNMTs activity may be affected by levels of *p21* expression, especially during DNA damage when *p21* is upregulated (Baylin 2001). p21 can disrupt DNMT1-PCNA

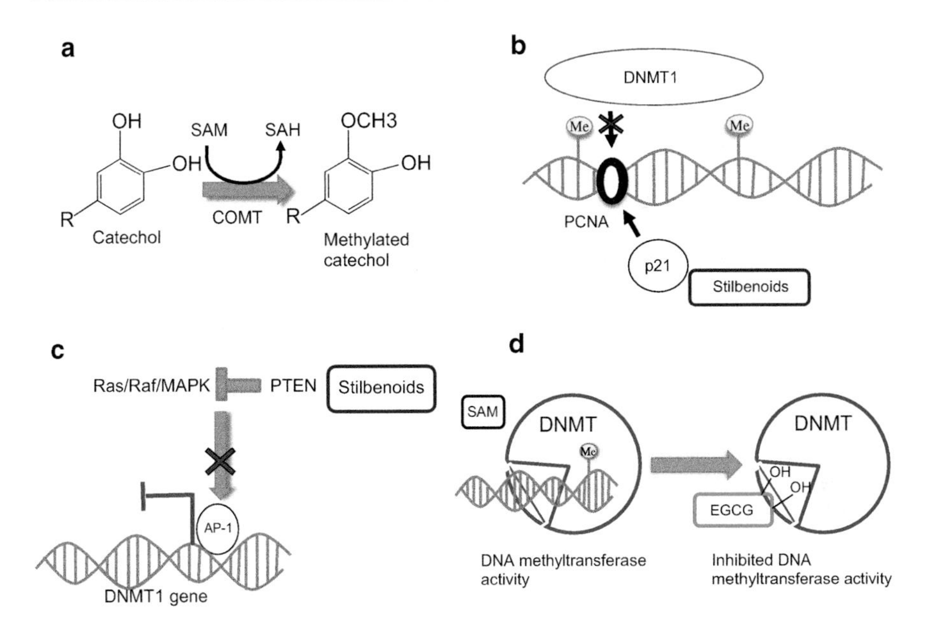

Fig. 3 Schematic representation of mechanisms used by catechol and stilbenoid compounds to induce changes to DNA methylation machinery. (**a**) Catechol-O-methyltransferase (*COMT*) uses catechol compounds as substrates for methylation reaction. COMT-mediated utilization of S-adenosyl-L-methionine (*SAM*) to methylate catechol compounds decreases a methyl donor pool available for methylating DNA. Further, S-adenosyl-L-homocysteine (*SAH*) accumulation directly inhibits DNA methyltransferases (*DNMTs*). (**b**) DNMT1 targets proliferating cell nuclear antigen (*PCNA*) on proliferating DNA to maintain DNA methylation during replication. *p21* competes with DNMT1 for the same binding site on PCNA. Upon exposure to catechol and stilbenoids compounds, *p21* is upregulated and its competitive binding to PCNA impairs DNMT1 activity. (**c**) Transcription factor AP-1 is required for transcription of the *DNMT1* gene. AP-1 is activated by the Ras/Raf/MAPK signaling pathway. Upon treatment with stilbenoids, Ras/Raf/MAPK pathway inhibitor PTEN is upregulated and disallows binding of AP-1 at the *DNMT1* gene, decreasing *DNMT1* expression. (**d**) Epigallocatechin gallate (*EGCG*) directly binds to the catalytic site of DNMTs, forming hydrogen bonds and inhibiting methyltransferase activity

(proliferating cell nuclear antigen) interaction by competing with DNMT1 for the same binding site on PCNA (Fig. 3b). DNMT1 is localized to replicating DNA by interaction with PCNA, a marker of proliferation; therefore, PCNA is crucial for DNMT1 activity (Chuang et al. 1997; Iida et al. 2002). As a result of competition for binding, the enzymatic activity of methyltransferases decreases which may at least partially contribute to decrease in methylation of regulatory regions of tumor suppressor genes causing subsequent gene activation (Yong et al. 2016; Stefanska et al. 2012). Resveratrol, as a phytoestrogen, was shown to increase expression of estrogen-responsive gene *p21* by acting as an agonist of the ER (Yong et al. 2016; Mandal and Davie 2010). Pterostilbene-mediated regulation of cell cycle arrest was also associated with *p21* upregulation (Esteller et al. 2001). Exposure to catechol compounds such as EGCG led to upregulation of *p21* in cancer (Dickinson et al. 2014) with evidence that the rise in expression is linked to hypomethylation of *p21* promoter (Li et al. 2001).

Increase in PTEN Reduces the Activity of AP-1 Transcription Factor – Decrease in DNMT1 Expression

DNMT1 is transcriptionally regulated by transcription factor AP-1 (Tsai et al. 2006; Bigey et al. 2000). A study in breast cancer cells identified activation of estrogen receptor (ER) in driving this process by inducing Ras/Raf/MAPK/AP-1 signaling and subsequently activating the *DNMT1* promoter (Bigey et al. 2000; Pethe and Shekhar 1999). Tumor suppressor PTEN negatively regulates AP-1 by inhibiting the aforementioned pathway (Chung et al. 2006). Interruption of AP-1 binding to *DNMT1* occurs upon resveratrol treatment (Fig. 3c). Specifically, resveratrol exposure leads to promoter hypomethylation and elevated expression of *PTEN* whose increased presence blocks AP-1 activity at the *DNMT1* gene promoter. As a result, *DNMT1* expression is downregulated (Yong et al. 2016; Manna et al. 2000).

Direct Binding to DNMT Catalytic Site

A mechanism of direct inhibition of DNA methylation has been described for several bioactive compounds including curcumin and catechol-containing EGCG (Renan et al. 1993; Dhar et al. 2016; Li et al. 2013; Singh et al. 2014; Ardaens and Renan 1993). Among catechol compounds, the activity of EGCG is 10–20 times higher in inhibiting human DNMT1 compared with catechin and epicatechin. EGCG shows strong inhibition of DNA methylation in the presence or absence of COMT, indicating that its inhibitory activity is not strictly COMT-mediated. Studies revealed that EGCG interacts directly with DNMT1 by forming hydrogen bonds at multiple sites within the catalytic pocket to inhibit the activity of the enzyme (Renan et al. 1993; Fig. 3d). A recent investigation in HeLa cells indicates that EGCG also decreases expression and activity of DNMT3B. Molecular modeling and docking site data support DNMT3B as another direct target of EGCG (Fig. 3d). EGCG-mediated inhibition of DNMT1 or DNMT3B leads to reversal of hypermethylation at promoters of tumor suppressors and reactivation of these important anti-cancer genes (Lee et al. 2005; Khan et al. 2015).

Changes in Dynamics of Interactions between DNA and Transcription Factors or DNMTs

In our recent study, stilbenoid treatment of breast cancer cells led to epigenetic changes, which conferred altered transcription factor binding and altered DNMT3B binding at specific DNA loci (Fig. 4). Hypermethylation of oncogene *MAML2* after resveratrol treatment was linked to increased occupancy of DNMT3B as well as reduced binding of transcription factor OCT1 at *MAML2* enhancer region. This reveals that enrichment of cancer promoting transcription factors like OCT1 might be modified by presence of methylation at its binding element on the DNA. Hindrance of OCT1 binding at enhancer region of *MAML2* led to downregulation of *MAML2* expression causing inhibition of the NOTCH signaling pathway (Lubecka

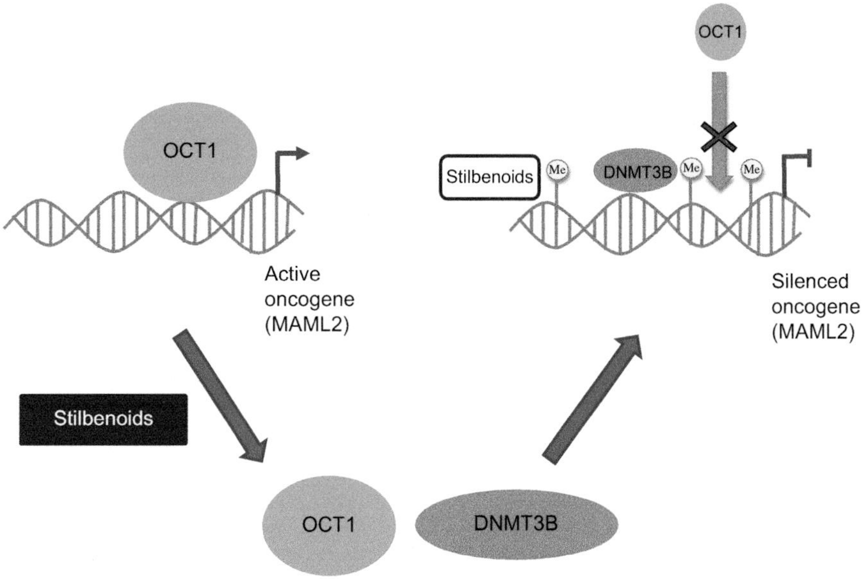

Fig. 4 **Stilbenoids epigenetically inhibit oncogenic signaling by altering dynamics of the interactions of DNA methyltransferases (DNMTs) and transcription factors with DNA**. Normally, oncogenes are active in cancer. Upon treatment with stilbenoids, *de novo* methyltransferase DNMT3B is being recruited and loci-specific increase in DNA methylation mediated by DNMT3B occurs within regulatory regions of oncogenes. Presence of hypermethylation sterically hinders transcription factor binding to recognized DNA binding elements, resulting in silencing of cancer-driving genes

et al. 2016; Fig. 5). These results suggest an important role for stilbenoids in loci-specific alterations of DNA methylation via changes in DNA methylation machinery and consequent changes in transcriptional activation by regulatory factors.

Another study identified oncogenic transcription factor STAT3 as a target of resveratrol to induce demethylation of promoters of tumor suppressor genes (Lee et al. 2012b). The presence of STAT3 acetylated at a lysine residue was observed in melanoma tumors. The presence of this acetylation mark was shown to mediate and facilitate a STAT3-DNMT1 interaction. Upon treatment with resveratrol that is considered to activate HDACs, STAT3 is deacetylated which impairs its interaction with DNMT1. It consequently reduces DNMT1 activity at promoters of tumor suppressor genes leading to their demethylation and activation as shown for *ERα* promoter (Lee et al. 2012b).

Conclusions and Future Prospects

The association of polyphenols from stilbenoid and catechol classes with reduced incidence of cancer has been reported in many epidemiological studies. *In vitro* and *in vivo* investigations have further delivered evidence on the effects in different cancer types and different animal models resulting often in inconsistent data. It is

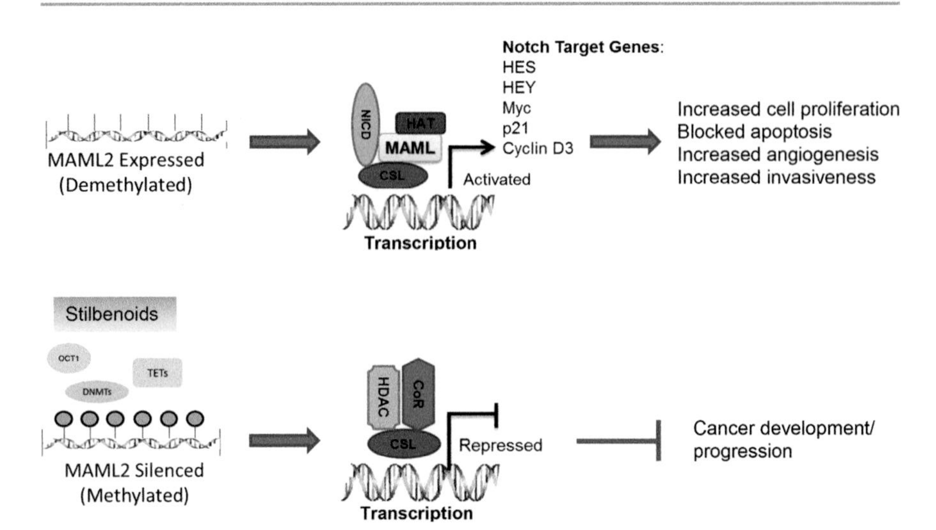

Fig. 5 Mechanistic proposal of how stilbenoids epigenetically inhibit the oncogenic Notch signaling pathway. Oncogenes, like *MAML2*, are expressed in cancer and induce further activation of target genes and tumorigenic properties. Stilbenoids promote an altered DNA methylation machinery which methylates and represses the *MAML2* gene and subsequently downregulates NOTCH signaling, leading to inhibition of cancer development and progression

apparent that dietary polyphenols exert effects in cancer cells and animal models but further mechanistic insights are needed before regular implementation in the clinical setting can be seriously considered. Clinical trials have proven challenging possibly due to bioavailability, dosage, and wide range of cancer types (Carter et al. 2014). The first piece of data from a randomized clinical trial investigating the effects of resveratrol in patients with colorectal cancer at the time of cancer resection surgery showed that the compound leads to a decrease in the activity of WNT oncogenic signaling in normal colon tissue but no effects were detected in colon cancer tissue (Nguyen et al. 2009). Another study on colon cancer noted minimal reduction of cell proliferation (Sodir et al. 2011). These pieces of evidence would suggest that resveratrol is more effective in reducing the risk of cancer rather than in treating/ attenuating established cancer. Indeed, several trials in healthy patients found that resveratrol decreases plasma levels of proteins associated with cancer development and metastasis (Pere et al. 2011; Allen et al. 2011). More recently, activation of tumor suppressor *RASSF-1α* was detected in women with increased risk of breast cancer exposed to resveratrol (Zhu et al. 2012).

Further trials in colon cancer patients showed that resveratrol increases apoptosis rate in hepatic metastases (Howells et al. 2011). In this study, resveratrol was used in a new formulation, SRT501, where the compound is micronized into very small uniform particles to maximize uptake into the body. Two years later, SRT501 was applied in patients with relapsed and refractory multiple myeloma (Mialon et al. 1993). Although promising, the study was prematurely terminated due to nephropathy and the death of one of the patients. It is probable that resveratrol did not directly cause kidney problems. Dehydration due to nausea and vomiting in patients on SRT501 could be the major cause of kidney toxicity. In addition, multiple myeloma patients have already higher risk of developing renal failure. In the clinical trials described above, the number of cancer patients enrolled was low (n = 8–24), and resveratrol was used at different doses and in different formulations. Thus, we need to be careful about conclusions and the results need to be confirmed in future studies on larger cohorts. It is also important that the cohorts are appropriately selected for a given study. For instance, resveratrol was claimed to have no beneficial metabolic effects in postmenopausal women (Collisson et al. 2011). However, the cohort consisted of nonobese healthy women with normal glucose tolerance. Hence, the study showed that resveratrol maintains metabolic health in postmenopausal women. It was not design appropriately to address a question whether resveratrol improves metabolic functions in postmenopausal women. Another aspect that requires attention in clinical trials is administration mode of a compound. Administration occurs after cancer formation and is limited by issues of toxicity, metabolic process, and bioavailability. To improve clinical success we need better understanding of mechanism of observed chemoprevention. Establishment of solid, comprehensive mechanisms is vitally needed in order to advance the field. Epigenetics, especially DNA methylation, if confirmed in humans, would constitute a powerful target of bioactive compounds due to reversibility of the alterations.

Dictionary of Terms

- **Angiogenesis** – Formation of new blood vessels.
- **Apoptosis** – Programmed cell death.
- **Carcinogens** – Cancer-causing agent.
- **Epithelial-to-mesenchymal transition** – Process cancer cells undergo to evade primary tumor site and travel to distant sites during metastasis.
- **Metastasis** – Tumor growths develop at sites distant from primary tumor site.
- **Phytoestrogen** – Plant-derived compound that mimics the hormone estrogen.
- **Polyphenols** – Bioactive compounds characterized by the presence of multiple phenol structures.
- **Xenograft** – Transplant of cells, tissue, or organ from donor to recipient of different species.

Key Facts of Carcinogenesis

- Broadly, adverse changes at molecular and cellular levels instigate carcinogenesis, leading to rampant cell proliferation, primary tumor formation, and potential to invade secondary sites.
- Stages of cancer initiation, promotion, and progression involve accumulation of genetic and epigenetic alterations within oncogenes and tumor suppressor genes.

- Specific genetic mutations, chromosomal translocations, and gene amplification occur during carcinogenesis.
- Aberrant epigenetic events involving DNA methylation such as loci-specific hypomethylation of oncogenes and pro-metastatic genes, loci-specific hypermethylation of tumor suppressor genes, and global demethylation resulting in genome instability appear in cancer.
- Carcinogenesis is characterized by dysregulation or misprogramming of many critical signaling pathways leading to altered cellular functions such as cell proliferation and growth, apoptosis, angiogenesis, invasion and metastasis.
- Chemopreventive and therapeutic strategies targeting altered pathways are traditionally used in treatment of cancer.
- Treatment challenges arise due to the complexity, heterogeneity, and diversity of the disease.
- Numerous side effects and resistance to current therapeutic options often arise when treating cancer, demonstrating a need for new, alternative strategies and an innovative outlook for combating cancer.

Key Facts of Polyphenols

- Polyphenols are bioactive compounds that can be obtained from a wide variety of dietary sources.
- Polyphenols can contribute to color, flavor, and odor of foods and beverages.
- Multiple phenol structural units characterize polyphenols and their structural composition classifies them into two distinct groups: flavonoids and nonflavanoids.
- Polyphenols are variably bioavailable and absorbed, and many are extensively metabolized.
- Consumption of polyphenols has been associated with protective effects against a wide variety of diseases including cardiovascular disease, diabetes, neurodegenerative diseases, and cancer.

Summary Points

- Genetic mutations and epigenetic alterations contribute to carcinogenesis.
- In cancer, DNA methylation events may underlie genomic instability, silencing of genes with tumor suppressor functions, and activation of genes associated with cancer progression.
- Targeting DNA methylation patterns established during carcinogenesis constitutes a promising anti-cancer strategy.
- Interestingly, epidemiological studies have found that consumption of certain dietary compounds known as polyphenols has profound beneficial health effects.
- Polyphenols have been shown to alter DNA methylation patterns in normal and disease states.

- Indeed, polyphenols, in particular those from catechol and stilbenoid classes, can exert anti-cancer effects through epigenetic regulation of gene expression.
- Mechanistic evidence of changes to the DNA methylation machinery by catechols and stilbenoids has surfaced over the past decade.
- Preclinical studies have presented chemopreventive and therapeutic potential of polyphenols through their influence on DNA methylation.
- Insights into mechanistic changes to the DNA methylation machinery upon exposure to polyphenols are limited but are under active investigation.
- Several mechanisms have been reported involving alterations to expression and activity of DNA methyltransferases (DNMTs), shifting of SAM/SAH ratio, and changes in interactions between DNA and transcription factors or DNMTs after polyphenol exposure.
- Further comprehensive mechanistic studies are ongoing and crucially needed for transition into application of polyphenols as agents in cancer prevention and/or treatment in the clinical setting.

References

Aires V, Delmas D (2015) Common pathways in health benefit properties of RSV in cardiovascular diseases, cancers and degenerative pathologies. Curr Pharm Biotechnol 16(3):219–244

Allen E, Walters IB, Hanahan D (2011) Brivanib, a dual FGF/VEGF inhibitor, is active both first and second line against mouse pancreatic neuroendocrine tumors developing adaptive/evasive resistance to VEGF inhibition. Clin Cancer Res 17(16):5299–5310

Anderson OS, Sant KE, Dolinoy DC (2012) Nutrition and epigenetics: an interplay of dietary methyl donors, one-carbon metabolism and DNA methylation. J Nutr Biochem 23(8):853–859

Ardaens Y, Renan CA (1993) Modern imaging of ovarian cysts. Contracep Fertil Sex 21(4): 321–324

Arts IC, Hollman PC (2005) Polyphenols and disease risk in epidemiologic studies. Am J Clin Nutr 81(1 Suppl):317S–325S

Baylin S (2001) DNA methylation and epigenetic mechanisms of carcinogenesis. Dev Biol 106:85–87. discussion 143–160

Berkyurek AC, Suetake I, Arita K et al (2014) The DNA methyltransferase Dnmt1 directly interacts with the SET and RING finger-associated (SRA) domain of the multifunctional protein Uhrf1 to facilitate accession of the catalytic center to hemi-methylated DNA. J Biol Chem 289(1): 379–386

Bigey P, Ramchandani S, Theberge J, Araujo FD, Szyf M (2000) Transcriptional regulation of the human DNA Methyltransferase (dnmt1) gene. Gene 242(1–2):407–418

Burns J, Yokota T, Ashihara H, Lean ME, Crozier A (2002) Plant foods and herbal sources of resveratrol. J Agric Food Chem 50(11):3337–3340

Carter LG, D'Orazio JA, Pearson KJ (2014) Resveratrol and cancer: focus on in vivo evidence. Endocr Relat Cancer 21(3):R209–R225

Cedar H, Bergman Y (2009) Linking DNA methylation and histone modification: patterns and paradigms. Nat Rev Genet 10(5):295–304

Chen T, Li E (2004) Structure and function of eukaryotic DNA methyltransferases. Curr Top Dev Biol 60:55–89

Chiurillo MA (2015) Role of the Wnt/beta-catenin pathway in gastric cancer: an in-depth literature review. World J Exp Med 5(2):84–102

Chuang LS, Ian HI, Koh TW, Ng HH, Xu G, Li BF (1997) Human DNA-(cytosine-5) methyltransferase-PCNA complex as a target for p21WAF1. Science 277(5334):1996–2000

Chung JH, Ostrowski MC, Romigh T, Minaguchi T, Waite KA, Eng C (2006) The ERK1/2 pathway modulates nuclear PTEN-mediated cell cycle arrest by cyclin D1 transcriptional regulation. Hum Mol Genet 15(17):2553–2559

Collisson EA, Sadanandam A, Olson P et al (2011) Subtypes of pancreatic ductal adenocarcinoma and their differin5g responses to therapy. Nat Med 17(4):500–503

Dhar S, Kumar A, Zhang L et al (2016) Dietary pterostilbene is a novel MTA1-targeted chemo-preventive and therapeutic agent in prostate cancer. Oncotarget 7(14):18469–18484

Dickinson D, Yu H, Ohno S et al (2014) Epigallocatechin-3-gallate prevents autoimmune-associated downregulation of p21 in salivary gland cells through a p53-independent pathway. Inflammation & allergy drug targets 13:15–24

Duan J, Yue W, JianYu E et al (2016) In vitro comparative studies of resveratrol and tri-acetylresveratrol on cell proliferation, apoptosis, and STAT3 and NFkappaB signaling in pancreatic cancer cells. Sci Rep 6:31672

El-Mowafy AM, El-Mesery ME, Salem HA, Al-Gayyar MM, Darweish MM (2010) Prominent chemopreventive and chemoenhancing effects for resveratrol: unraveling molecular targets and the role of C-reactive protein. Chemotherapy 56(1):60–65

Esteller M, Fraga MF, Guo M et al (2001) DNA methylation patterns in hereditary human cancers mimic sporadic tumorigenesis. Hum Mol Genet 10(26):3001–3007

Fini L, Piazzi G, Daoud Y et al (2011) Chemoprevention of intestinal polyps in ApcMin/+ mice fed with western or balanced diets by drinking annurca apple polyphenol extract. Cancer Prev Res 4(6):907–915

Galamb O, Kalmar A, Peterfia B et al (2016) Aberrant DNA methylation of WNT pathway genes in the development and progression of CIMP-negative colorectal cancer. Epigenetics 11(8): 588–602

Gallagher JC, Baylink DJ, Freeman R, McClung M (2001) Prevention of bone loss with tibolone in postmenopausal women: results of two randomized, double-blind, placebo-controlled, dose-finding studies. J Clin Endocrinol Metab 86(10):4717–4726

Giacosa A (2004) The Mediterranean diet and its protective role against cancer. Eur J Cancer Prev 13(3):155–157

Grosso G, Buscemi S, Galvano F et al (2013) Mediterranean diet and cancer: epidemiological evidence and mechanism of selected aspects. BMC Surg 13(Suppl 2):S14

Gruenbaum Y, Stein R, Cedar H, Razin A (1981) Methylation of CpG sequences in eukaryotic DNA. FEBS Lett 124(1):67–71

Hanahan D, Weinberg RA (2011) Hallmarks of cancer: the next generation. Cell 144(5):646–674

Hata K, Okano M, Lei H, Li E (2002) Dnmt3L cooperates with the Dnmt3 family of de novo DNA methyltransferases to establish maternal imprints in mice. Development 129(8):1983–1993

Hatada I, Fukasawa M, Kimura M et al (2006) Genome-wide profiling of promoter methylation in human. Oncogene 25(21):3059–3064

Henning SM, Wang P, Carpenter CL, Heber D (2013) Epigenetic effects of green tea polyphenols in cancer. Epigenomics 5(6):729–741

Herman JG, Baylin SB (2001) Methylation-specific PCR. Curr Protoc Hum Genet. Chapter 10: Unit 10 16

Howells LM, Berry DP, Elliott PJ et al (2011) Phase I randomized, double-blind pilot study of micronized resveratrol (SRT501) in patients with hepatic metastases – safety, pharmacokinetics, and pharmacodynamics. Cancer Prev Res 4(9):1419–1425

Huderson AC, Myers JN, Niaz MS, Washington MK, Ramesh A (2013) Chemoprevention of benzo (a)pyrene-induced colon polyps in ApcMin mice by resveratrol. J Nutr Biochem 24(4):713–724

Iida T, Suetake I, Tajima S et al (2002) PCNA clamp facilitates action of DNA cytosine methyltransferase 1 on hemimethylated DNA. Genes Cells 7(10):997–1007

Imai K, Suga K, Nakachi K (1997) Cancer-preventive effects of drinking green tea among a Japanese population. Prev Med 26(6):769–775

Jain S, Chang TT, Hamilton JP et al (2011) Methylation of the CpG sites only on the sense strand of the APC gene is specific for hepatocellular carcinoma. PLoS One 6(11):e26799

James SJ, Pogribny IP, Pogribna M, Miller BJ, Jernigan S, Melnyk S (2003) Mechanisms of DNA damage, DNA hypomethylation, and tumor progression in the folate/methyl-deficient rat model of hepatocarcinogenesis. J Nutr 133(11 Suppl 1):3740S–3747S

Jang M, Cai L, Udeani GO et al (1997) Cancer chemopreventive activity of resveratrol, a natural product derived from grapes. Science 275(5297):218–220

Jones PA (2012) Functions of DNA methylation: islands, start sites, gene bodies and beyond. Nat Rev Genet 13(7):484–492

Jones PA, Takai D (2001) The role of DNA methylation in mammalian epigenetics. Science 293(5532):1068–1070

Kala R, Shah HN, Martin SL, Tollefsbol TO (2015) Epigenetic-based combinatorial resveratrol and pterostilbene alters DNA damage response by affecting SIRT1 and DNMT enzyme expression, including SIRT1-dependent gamma-H2AX and telomerase regulation in triple-negative breast cancer. BMC Cancer 15:672

Khan N, Mukhtar H (2013) Modulation of signaling pathways in prostate cancer by green tea polyphenols. Biochem Pharmacol 85(5):667–672

Khan MA, Hussain A, Sundaram MK et al (2015) (−)-Epigallocatechin-3-gallate reverses the expression of various tumor-suppressor genes by inhibiting DNA methyltransferases and histone deacetylases in human cervical cancer cells. Oncol Rep 33(4):1976–1984

Lao VV, Grady WM (2011) Epigenetics and colorectal cancer. Nat Rev Gastroenterol Hepatol 8(12):686–700

Lee WJ, Zhu BT (2006) Inhibition of DNA methylation by caffeic acid and chlorogenic acid, two common catechol-containing coffee polyphenols. Carcinogenesis 27(2):269–277

Lee WJ, Shim JY, Zhu BT (2005) Mechanisms for the inhibition of DNA methyltransferases by tea catechins and bioflavonoids. Mol Pharmacol 68(4):1018–1030

Lee HS, Ha AW, Kim WK (2012a) Effect of resveratrol on the metastasis of 4T1 mouse breast cancer cells in vitro and in vivo. Nutr Res Pract 6(4):294–300

Lee H, Zhang P, Herrmann A et al (2012b) Acetylated STAT3 is crucial for methylation of tumor-suppressor gene promoters and inhibition by resveratrol results in demethylation. Proc Natl Acad Sci U S A 109(20):7765–7769

Lee H, Kim Y, Jeong JH, Ryu JH, Kim WY (2016) ATM/CHK/p53 pathway dependent chemo-preventive and therapeutic activity on lung cancer by Pterostilbene. PLoS One 11(9):e0162335

Li X, Mohan S, Gu W, Wergedal J, Baylink DJ (2001) Quantitative assessment of forearm muscle size, forelimb grip strength, forearm bone mineral density, and forearm bone size in determining humerus breaking strength in 10 inbred strains of mice. Calcif Tissue Int 68(6):365–369

Li Y, Yuan YY, Meeran SM, Tollefsbol TO (2010) Synergistic epigenetic reactivation of estrogen receptor-alpha (ERalpha) by combined green tea polyphenol and histone deacetylase inhibitor in ERalpha-negative breast cancer cells. Mol Cancer 9:274

Li K, Dias SJ, Rimando AM et al (2013) Pterostilbene acts through metastasis-associated protein 1 to inhibit tumor growth, progression and metastasis in prostate cancer. PLoS One 8(3):e57542

Li S, Wu L, Feng J et al (2016) In vitro and in vivo study of epigallocatechin-3-gallate-induced apoptosis in aerobic glycolytic hepatocellular carcinoma cells involving inhibition of phospho-fructokinase activity. Sci Rep 6:28479

Liu HT, Gao P (2016) The roles of microRNAs related with progression and metastasis in human cancers. Tumour Biol 37:15383-15397

Liu B, Song J, Luan J et al (2016) Promoter methylation status of tumor suppressor genes and inhibition of expression of DNA methyltransferase 1 in non-small cell lung cancer. Exp Biol Med (Maywood) 241(14):1531–1539

Lubecka K, Kurzava L, Flower K et al (2016) Stilbenoids remodel the DNA methylation patterns in breast cancer cells and inhibit oncogenic NOTCH signaling through epigenetic regulation of MAML2 transcriptional activity. Carcinogenesis 37(7):656–668

Mandal S, Davie JR (2010) Estrogen regulated expression of the p21 Waf1/Cip1 gene in estrogen receptor positive human breast cancer cells. J Cell Physiol 224(1):28–32

Manna SK, Mukhopadhyay A, Aggarwal BB (2000) Resveratrol suppresses TNF-induced activation of nuclear transcription factors NF-kappa B, activator protein-1, and apoptosis: potential role of reactive oxygen intermediates and lipid peroxidation. J Immunol 164(12):6509–6519

Martinez-Quetglas I, Pinyol R, Dauch D et al (2016) IGF2 is Upregulated by epigenetic mechanisms in hepatocellular carcinomas and is an actionable oncogene product in experimental models. Gastroenterology 151:1192

McCormack D, Schneider J, McDonald D, McFadden D (2011) The antiproliferative effects of pterostilbene on breast cancer in vitro are via inhibition of constitutive and leptin-induced Janus kinase/signal transducer and activator of transcription activation. Am J Surg 202(5):541–544

Medina-Aguilar R, Perez-Plasencia C, Marchat LA et al (2016) Methylation landscape of human breast cancer cells in response to dietary compound resveratrol. PLoS One 11(6):e0157866

Mialon MM, Camous S, Renand G, Martal J, Menissier F (1993) Peripheral concentrations of a 60-kDa pregnancy serum protein during gestation and after calving and in relationship to embryonic mortality in cattle. Reprod Nutr Dev 33(3):269–282

Mittal A, Piyathilake C, Hara Y, Katiyar SK (2003) Exceptionally high protection of photocarcinogenesis by topical application of (−)-epigallocatechin-3-gallate in hydrophilic cream in SKH-1 hairless mouse model: relationship to inhibition of UVB-induced global DNA hypomethylation. Neoplasia 5(6):555–565

Myung SK, Bae WK, Oh SM et al (2009) Green tea consumption and risk of stomach cancer: a meta-analysis of epidemiologic studies. Int J Cancer 124(3):670–677

Nguyen AV, Martinez M, Stamos MJ et al (2009) Results of a phase I pilot clinical trial examining the effect of plant-derived resveratrol and grape powder on Wnt pathway target gene expression in colonic mucosa and colon cancer. Cancer Manag Res 1:25–37

Niles RM, Cook CP, Meadows GG, Fu YM, McLaughlin JL, Rankin GO (2006) Resveratrol is rapidly metabolized in athymic (nu/nu) mice and does not inhibit human melanoma xenograft tumor growth. J Nutr 136(10):2542–2546

Nishikawa T, Nakajima T, Moriguchi M et al (2006) A green tea polyphenol, epigalocatechin-3-gallate, induces apoptosis of human hepatocellular carcinoma, possibly through inhibition of Bcl-2 family proteins. J Hepatol 44(6):1074–1082

Olthof MR, Hollman PC, Zock PL, Katan MB (2001) Consumption of high doses of chlorogenic acid, present in coffee, or of black tea increases plasma total homocysteine concentrations in humans. Am J Clin Nutr 73:532–538

Pandey M, Shukla S, Gupta S (2010) Promoter demethylation and chromatin remodeling by green tea polyphenols leads to re-expression of GSTP1 in human prostate cancer cells. Int J Cancer 126(11):2520–2533

Pere H, Montier Y, Bayry J et al (2011) A CCR4 antagonist combined with vaccines induces antigen-specific CD8+ T cells and tumor immunity against self antigens. Blood 118(18): 4853–4862

Pethe V, Shekhar PV (1999) Estrogen inducibility of c-Ha-ras transcription in breast cancer cells. Identification of functional estrogen-responsive transcriptional regulatory elements in exon 1/intron 1 of the c-Ha-ras gene. J Biol Chem 274(43):30969–30978

Qin W, Zhang K, Clarke K, Weiland T, Sauter ER (2014) Methylation and miRNA effects of resveratrol on mammary tumors vs. normal tissue. Nutr Cancer 66(2):270–277

Razin A (1998) CpG methylation, chromatin structure and gene silencing-a three-way connection. EMBO J 17(17):4905–4908

Renan MJ (1993) How many mutations are required for tumorigenesis? Implications from human cancer data. Mol Carcinog 7(3):139–146

Renan R, Freire VN, Auto MM, Farias GA (1993) Transmission coefficient of electrons through a single graded barrier. Phy Rev B Condens Matter 48(11):8446–8449

Romagnolo DF, Papoutsis AJ, Laukaitis C, Selmin OI (2015) Constitutive expression of AhR and BRCA-1 promoter CpG hypermethylation as biomarkers of ERalpha-negative breast tumorigenesis. BMC Cancer 15:1026

Salvesen HB, MacDonald N, Ryan A et al (2001) PTEN methylation is associated with advanced stage and microsatellite instability in endometrial carcinoma. Int J Cancer 91(1):22–26

Shariati-Kohbanani M, Zare-Bidaki M, Taghavi MM et al (2016) DNA methylation and microRNA patterns are in association with the expression of BRCA1 in ovarian cancer. Cell Mol Biol (Noisy-le-Grand, France) 62(1):16–23

Sharma S, Kelly TK, Jones PA (2010) Epigenetics in cancer. Carcinogenesis 31(1):27–36

Shin YS, Kang SU, Park JK et al (2016) Anti-cancer effect of (−)-epigallocatechin-3-gallate (EGCG) in head and neck cancer through repression of transactivation and enhanced degradation of beta-catenin. Phytomedicine 23(12):1344–1355

Shirakami Y, Shimizu M, Adachi S et al (2009) (−)-Epigallocatechin gallate suppresses the growth of human hepatocellular carcinoma cells by inhibiting activation of the vascular endothelial growth factor-vascular endothelial growth factor receptor axis. Cancer Sci 100(10):1957–1962

Singh B, Shoulson R, Chatterjee A et al (2014) Resveratrol inhibits estrogen-induced breast carcinogenesis through induction of NRF2-mediated protective pathways. Carcinogenesis 35(8):1872–1880

Sinha D, Sarkar N, Biswas J, Bishayee A (2016) Resveratrol for breast cancer prevention and therapy: preclinical evidence and molecular mechanisms. Semin Cancer Biol 40-41:209–232

Sodir NM, Swigart LB, Karnezis AN, Hanahan D, Evan GI, Soucek L (2011) Endogenous Myc maintains the tumor microenvironment. Genes Dev 25(9):907–916

Srivastava AK, MacFarlane G, Srivastava VP, Mohan S, Baylink DJ (2001a) A new monoclonal antibody ELISA for detection and characterization of C-telopeptide fragments of type I collagen in urine. Calcif Tissue Int 69(6):327–336

Srivastava AK, Bhattacharyya S, Li X, Mohan S, Baylink DJ (2001b) Circadian and longitudinal variation of serum C-telopeptide, osteocalcin, and skeletal alkaline phosphatase in C3H/HeJ mice. Bone 29(4):361–367

Stefanska B, Rudnicka K, Bednarek A, Fabianowska-Majewska K (2010) Hypomethylation and induction of retinoic acid receptor beta 2 by concurrent action of adenosine analogues and natural compounds in breast cancer cells. Eur J Pharmacol 638(1–3):47–53

Stefanska B, Huang J, Bhattacharyya B et al (2011) Definition of the landscape of promoter DNA hypomethylation in liver cancer. Cancer Res 71(17):5891–5903

Stefanska B, Karlic H, Varga F, Fabianowska-Majewska K, Haslberger A (2012) Epigenetic mechanisms in anti-cancer actions of bioactive food components – the implications in cancer prevention. Br J Pharmacol 167(2):279–297

Stefanska B, Suderman M, Machnes Z, Bhattacharyya B, Hallett M, Szyf M (2013) Transcription onset of genes critical in liver carcinogenesis is epigenetically regulated by methylated DNA-binding protein MBD2. Carcinogenesis 34(12):2738–2749

Stefanska B, Cheishvili D, Suderman M et al (2014) Genome-wide study of hypomethylated and induced genes in patients with liver cancer unravels novel anticancer targets. Clin Cancer Res 20(12):3118–3132

Strahl BD, Allis CD (2000) The language of covalent histone modifications. Nature 403(6765): 41–45

Sun CL, Yuan JM, Koh WP, Yu MC (2006) Green tea, black tea and colorectal cancer risk: a meta-analysis of epidemiologic studies. Carcinogenesis 27(7):1301–1309

Szyf M, Pakneshan P, Rabbani SA (2004) DNA methylation and breast cancer. Biochem Pharmacol 68(6):1187–1197

Tsai CL, Li HP, Lu YJ et al (2006) Activation of DNA methyltransferase 1 by EBV LMP1 involves c-Jun NH(2)-terminal kinase signaling. Cancer Res 66(24):11668–11676

Tsuchida N, Murugan AK, Grieco M (2016) Kirsten Ras* oncogene: significance of its discovery in human cancer research. Oncotarget 7:46717

Tuveson D, Hanahan D (2011) Translational medicine: cancer lessons from mice to humans. Nature 471(7338):316–317

Umemura T, Kai S, Hasegawa R et al (2003) Prevention of dual promoting effects of pentachlorophenol, an environmental pollutant, on diethylnitrosamine-induced hepato- and cholangiocarcinogenesis in mice by green tea infusion. Carcinogenesis 24(6):1105–1109

Van Emburgh BO, Robertson KD (2011) Modulation of Dnmt3b function in vitro by interactions with Dnmt3L, Dnmt3a and Dnmt3b splice variants. Nucleic Acids Res 39(12):4984–5002

Volate SR, Muga SJ, Issa AY, Nitcheva D, Smith T, Wargovich MJ (2009) Epigenetic modulation of the retinoid X receptor alpha by green tea in the azoxymethane-Apc Min/+ mouse model of intestinal cancer. Mol Carcinog 48(10):920–933

Williams GH, Stoeber K (2012) The cell cycle and cancer. J Pathol 226(2):352–364

Williamson G, Manach C (2005) Bioavailability and bioefficacy of polyphenols in humans. II. Review of 93 intervention studies. Am J Clin Nutr 81(1 Suppl):243S–255S

Wilson AS, Power BE, Molloy PL (2007) DNA hypomethylation and human diseases. Biochim Biophys Acta 1775(1):138–162

Xiang LP, Wang A, Ye JH et al (2016) Suppressive effects of tea catechins on breast cancer. Forum Nutr 8(8):458

Yan T, Wergedal J, Zhou Y, Mohan S, Baylink DJ, Strong DD (2001) Inhibition of human osteoblast marker gene expression by retinoids is mediated in part by insulin-like growth factor binding protein-6. Growth Horm IGF Res 11(6):368–377

Yang CS, Wang H, Li GX, Yang Z, Guan F, Jin H (2011) Cancer prevention by tea: evidence from laboratory studies. Pharmacol Res 64(2):113–122

Yang C, Du W, Yang D (2016) Inhibition of green tea polyphenol EGCG((−)-epigallocatechin-3-gallate) on the proliferation of gastric cancer cells by suppressing canonical wnt/beta-catenin signalling pathway. Int J Food Sci Nutr 67(7):818–827

Yong WS, Hsu FM, Chen PY (2016) Profiling genome-wide DNA methylation. Epigenetics Chromatin 9:26

Yuan JM (2013) Cancer prevention by green tea: evidence from epidemiologic studies. Am J Clin Nutr 98(6 Suppl):1676S–1681S

Yuan JM, Sun C, Butler LM (2011) Tea and cancer prevention: epidemiological studies. Pharmacol Res 64(2):123–135

Zamora-Ros R, Touillaud M, Rothwell JA, Romieu I, Scalbert A (2014) Measuring exposure to the polyphenol metabolome in observational epidemiologic studies: current tools and applications and their limits. Am J Clin Nutr 100(1):11–26

Zhang J, Lei Z, Huang Z et al (2016) Epigallocatechin-3-gallate(EGCG) suppresses melanoma cell growth and metastasis by targeting TRAF6 activity. Oncotarget 7:79557

Zhong L, Goldberg MS, Gao YT, Hanley JA, Parent ME, Jin F (2001) A population-based case-control study of lung cancer and green tea consumption among women living in shanghai. China Epidemiol 12(6):695–700

Zhong LX, Li H, Wu ML et al (2015) Inhibition of STAT3 signaling as critical molecular event in resveratrol-suppressed ovarian cancer cells. J Ovarian Res 8:25

Zhu W, Qin W, Zhang K et al (2012) Trans-resveratrol alters mammary promoter hypermethylation in women at increased risk for breast cancer. Nutr Cancer 64(3):393–400

Ziegler CC, Rainwater L, Whelan J, McEntee MF (2004) Dietary resveratrol does not affect intestinal tumorigenesis in Apc(Min/+) mice. J Nutr 134(1):5–10

Epigenetic Drivers of Resveratrol-Induced Suppression of Mammary Carcinogenesis: Addressing miRNAs, Protein, mRNA, and DNA Methylation

96

E. R. Sauter

Contents

Abstract

There are both *cis*- and *trans*-isomers of resveratrol, with the *trans*-isomer the more active of the two. *Trans*-resveratrol is present in a variety of foods, and preclinical data suggest that the agent is effective in both preventing the formation of mammary tumors and in shrinking tumors that have developed. The agent has mild to no human toxicity over a wide dose range. On the other hand, efficacy in humans has not been proven, and first-pass effects through conjugation in the liver have led people to question its clinical usefulness,

E. R. Sauter (✉)
Hartford HealthCare Cancer Institute, Hartford, CT, USA
e-mail: Edward.sauter@hhchealth.org

V. B. Patel, V. R. Preedy (eds.), *Handbook of Nutrition, Diet, and Epigenetics*,
https://doi.org/10.1007/978-3-319-55530-0_2

although evidence of deconjugation intracellularly has also been demonstrated. The epigenetic mechanisms of action of the agent are many, including effects on miRNAs, protein, mRNA, and DNA methylation. Findings to date will be reviewed, with a look to where investigations are headed to better assess clinical efficacy.

Keywords
Trans-resveratrol · *Cis*-resveratrol · miRNA · DNA methylation · mRNA expression · Protein expression · Glucuronide metabolite · Sulfate metabolite

List of Abbreviations

DNA	Deoxyribonucleic acid
miRNA	microRNA
mRNA	Messenger RNA
RNA	Ribonucleic acid

Introduction

There are both *cis*- and *trans*-isomers of resveratrol. *Trans*-resveratrol (resveratrol) is a polyphenol present in foods that has been touted as an anticancer agent (Subbaramaiah et al. 1998). The *cis*-isomer is generally present at lower levels in foods and thought to have less potent anticancer properties (2). *Trans*-resveratrol is found in high concentration in the skin of red grapes and in red wine, with lesser amounts in mulberries and peanuts. High levels are also found in the rhizome of *P. cuspidatum*, and most *trans*-resveratrol dietary supplements contain *P. cuspidatum* (Zhu et al. 2012).

One of the highly desirable attributes of resveratrol is that while it is effective in preventing cancer formation (Jang et al. 1997), it has little or no toxicity in normal tissues, even at doses which far exceed what is found in common foods (Zhu et al. 2012). It inhibits human breast proliferation in a dose- and time-dependent manner (Mgbonyebi et al. 1998) and inhibited mammary tumor incidence (by 45%) and multiplicity (by 55%) when administered as part of the diet (Jang et al. 1997). There is a dose-dependent effect, with higher doses of resveratrol being more effective in chemoprevention than lower doses (Qin et al. 2014).

The data suggest that resveratrol may have beneficial effects if used as a chemopreventive agent for breast cancer (Bhat et al. 2001). However, clinical efficacy is not yet proven, and poor bioavailability has raised concerns about its usefulness in clinical medicine (Ronghe et al. 2016). In short, the preclinical data regarding resveratrol are promising, but human data are limited and inconsistent. Herein we review preclinical and clinical findings suggesting that resveratrol works in large measure through epigenetic modification and discuss ongoing strategies to optimize efficacy in vitro, in animals and in humans.

Resveratrol as an Antiestrogen

Resveratrol is a phytoestrogen. It exerts antiproliferative effects on breast cancer cells via an estrogen receptor (ER)-dependent mechanism, even at low concentrations, but is also capable of maintaining the survival of normal breast cells via ER-independent or other mechanisms (Chen and Chien 2014). In the absence of estrogen, resveratrol has been found to exert mixed estrogen agonist/antagonist activities in vitro (Bhat et al. 2001). When estrogen is present, it functions as an antiestrogen. In animals, it inhibits mammary tumor formation (Qin et al. 2014), carcinogen-induced preneoplastic lesions, and mammary tumors. There is a dose response, with higher doses of resveratrol being more effective in preventing mammary tumor formation (Qin et al. 2014). Resveratrol treatment in the presence of estrogen upregulates the expression of nuclear factor erythroid 2-related factor 2 (NRF2) in mammary tissues (Singh et al. 2014). NRF2-regulated genes involved in protection against oxidative DNA damage are increased as well. Resveratrol treatment also induces apoptosis and inhibits estradiol (E2)-mediated increase in DNA damage in mammary tissues. Data suggest that resveratrol inhibits E2-induced breast carcinogenesis via induction of NRF2-mediated protective pathways (Singh et al. 2014).

ERα is frequently upregulated in breast cancer and can be modulated by methylation of the promoter (Hayashi et al. 2003). Resveratrol treatment increases promoter methylation of ESRl (Berner et al. 2010), the gene that encodes ERα, one of the two main ERs (the other being ERβ). Resveratrol has also been found to block aromatase, the rate-limiting enzyme in the conversion of androgens to estrogens (Subbaramaiah et al. 2013). A series of new resveratrol analogues have been designed and synthesized, with many of the compounds showing increased inhibitory activity in vitro compared to the parent resveratrol compound (Sun et al. 2010).

Effects on Gene Methylation

The most studied epigenetic alteration in human neoplasms is the hypermethylation of CpG islands in gene promoter regions, which has a profound role in gene regulation and carcinogenesis (Baylin and Ohm 2006). DNA methyltransferases (DNMTs 1, 3a, and 3b) influence DNA methylation patterns. DNMT1 maintains methylation patterns in newly formed DNA, whereas 3a and 3b introduce methylation at unmethylated CpG sites (Klose and Bird 2006). DNMTs silence gene expression both by repressing gene transcription and through the epigenetic modification of cytosine (Klose and Bird 2006).

There are extensive preclinical and limited human studies evaluating the effect of resveratrol on gene methylation. In a preclinical study, resveratrol was found to increase the methylation of genes with oncogenic function, including MAML2, a coactivator of NOTCH targets, leading to gene silencing (Lubecka et al. 2016). After resveratrol treatment, DNMT3b bound to the MAML2 enhancer, thereby increasing MAML2 methylation. We reported that resveratrol decreased the expression of DNMT1 and 3b in breast cancer cells in a dose-dependent fashion (Qin et al. 2005). In a rodent model,

DNMT3b decreased in mammary tumor but increased in normal mammary tissue (Qin et al. 2012). Hi-dose resveratrol and 5-Aza-2-deoxycytidine (Aza) (Qin et al. 2012) induced a differential effect on DNMT3b expression in tumor compared to normal tissue which was not observed for DNMT1 (Table 1). DNMT3b, which is overexpressed in 30% of breast cancers compared to 5% for DNMT1 and 3% for DNMT3a (Veeck and Esteller 2010), is believed to be the predominant methyltransferase in breast tumorigenesis (Veeck and Esteller 2010). While resveratrol did not significantly influence DNMT1 expression in normal tissue or mammary tumors, it did alter DNMT3b in a dose-dependent fashion (Qin et al. 2014), similar to Aza, a demethylating agent used in clinical medicine. In women at increased breast cancer risk, we observed that the tumor suppressor gene *RASSF-1a* decreased (Fig. 1) in

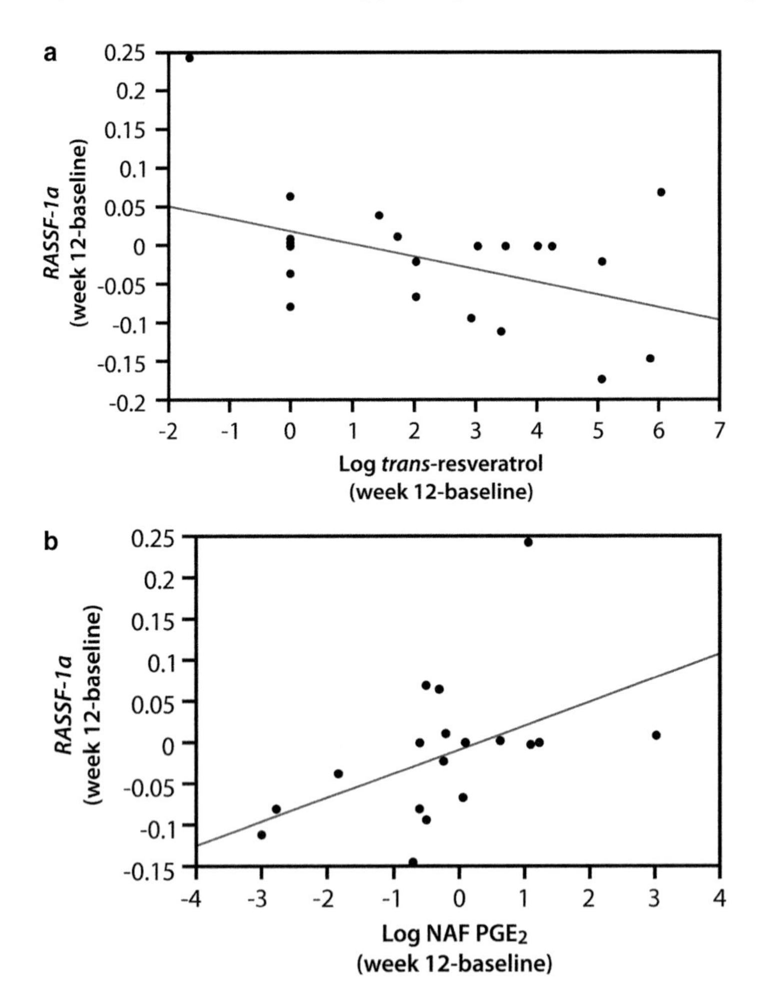

Fig. 1 *RASSF-1a* methylation change vs. change in (**a**) trans-resveratrol and (**b**) NAF PGE_2 concentration. *RASSF-1a* methylation significantly decreased as the concentration of total trans-resveratrol increased ($P = 0.047$, $r^2 = -0.14$) and as PGE_2 decreased ($P = 0.045$, $r^2 = 0.18$)

conjunction with increasing levels of circulating resveratrol ($P = 0.047$, $r^2 = -0.14$) and as PGE_2 decreased ($P = 0.045$, $r^2 = 0.18$) (Zhu et al. 2012).

Alterations in miRNAs

It is known that micro(mi)RNAs, 18–25 nucleotide stretches of single-stranded RNA which downregulate gene expression either by translational inhibition or messenger(m) RNA decay, are targets for chemopreventive agents, including resveratrol (Izzotti et al. 2012). Target genes of resveratrol-regulated miRNAs thus far reported, primarily from tissue culture studies, are related to apoptosis, cell cycle regulation, cell proliferation, and differentiation (Izzotti et al. 2012). Relatively little is known about the role of resveratrol in influencing gene methylation, including methylation related to miRNAs.

Influence on Cancer

miRNA machinery is well conserved among species (Izzotti et al. 2012), increasing the potential human relevance of animal studies. It has been reported (Dhar et al. 2011) that resveratrol significantly downregulated oncogenic miRNAs and upregulated tumor suppressor miRNAs in cancer cells. As previously noted, the compound has little or no toxicity in normal tissues, even at doses which far exceed what is found in common foods (Zhu et al. 2012).

Resveratrol impairs the expression of oncogenic miRNAs and upregulates tumor suppressor miRNAs (Lancon et al. 2012). Many miRNAs have been reported to be downregulated in cancer, including miR1, miR-10b, miR-129, miR-145, miR-199a, and miR-205 (Lopez-Serra and Esteller 2012; Suzuki et al. 2012). Hi-dose resveratrol increased miR-21, miR-129, miR-204, and miR-489 >twofold in hormone-sensitive mammary tumors and decreased the same miRNAs in normal tissue 10–50% compared to untreated control (Fig. 2). The effects of resveratrol on miRNAs appear to be dose related. Lo- but not hi-dose resveratrol increased miR-10a and miR-10b expression in tumor tissue, whereas hi- but not lo-dose resveratrol increased miR-21, miR-129, miR-204, and miR-489 (Fig. 2).

An important way in which miRNAs are inactivated in human cancer is through hypermethylation (Lopez-Serra and Esteller 2012). miRNAs have also been reported to regulate DNMT3b expression (Malumbres 2012). Of the six miRNAs influenced by resveratrol treatment in a rodent model of hormone-sensitive breast cancer, there was a significant ($P < 0.05$) inverse correlation with DNMT3b for three (129, $r^2 = 0.29$; 204, $r^2 = 0.20$; 489, $r^2 = 0.18$) in normal and one (489, $r^2 = 0.26$) in tumor tissue, suggesting their epigenetic regulation (Fig. 3) (Qin et al. 2014). The two that experienced the greatest increase with treatment in tumor tissue were miR-129 and miR-489. miR-129 is known to be aberrantly methylated and down-regulated in endometrial, colon, esophageal, and stomach cancer (Suzuki et al. 2012). The inverse correlation of miRNA-129 with DNMT3b in normal but not tumor tissue may indicate that its downregulation occurs early in mammary tumor

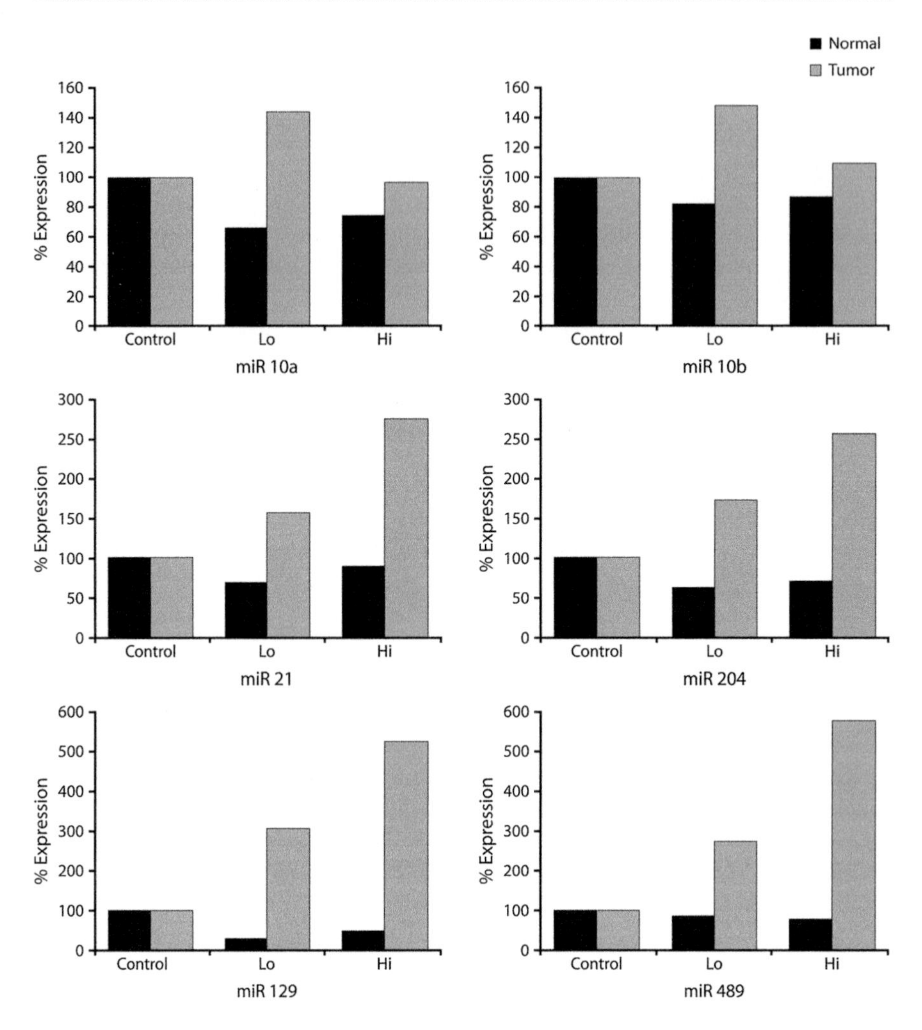

Fig. 2 Effect of low (*lo*)- and high (*hi*)-dose resveratrol on miRNA expression compared to control. Treatment with lo resveratrol significantly influenced miR-10a (**a**) and miR-10b (**b**) expression. There was a dose-dependent effect of resveratrol treatment on miR-21, miR-204, miR-129, and miR-489 (**c**–**f**)

development, before histopathologic evidence of malignancy. We are not aware of a report demonstrating that miR-489 is inversely correlated with DNA methylation or methyltransferase levels. The correlation in both normal and tumor tissues suggests that methylation may be important both in tumor initiation and progression.

Cardioprotection

Some of the medications used to treat breast cancer, including adriamycin and trastuzumab, are cardiotoxic (Brown et al. 2015). Patients also often receive

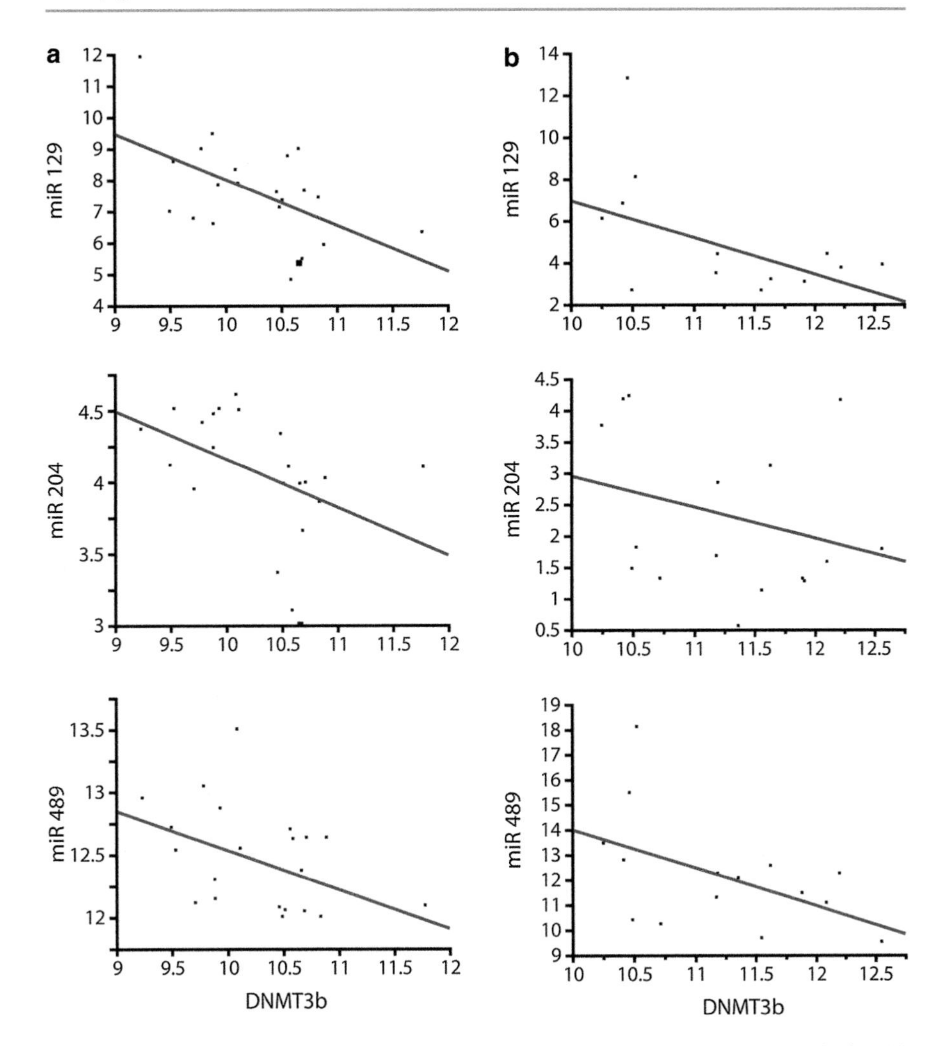

Fig. 3 Correlation of miRNAs with DNMT3b. ΔCt values for miR129, miR-204, and miR-489 were inversely correlated with ΔCt DNMT3b values in normal (**a**) and in tumor (**b**) tissue

radiation to the breast and/or chest wall to treat breast cancer, resulting in a portion of the heart being radiated. Resveratrol has been reported to exhibit cardioprotective effects (Fan et al. 2016), including reducing the severity of cardiac hypertrophy. One possible mechanism for this is resveratrol's ability to downregulate miR-155 in cardiomyocytes. Breast cancer susceptibility gene (BRCA) 1 inactivation can increase expression of miR-155, contributing to cardiac hypertrophy, and resveratrol, by downregulating miR-155, activates BRCA1. miR-155 is also known to be involved in breast cancer progression (Mattiske et al. 2012). Thus, the effect of resveratrol on miR-155 appears to decrease breast cancer risk/progression and also decrease cardiac hypertrophy.

Resveratrol Analogs to Increase Bioavailability and Efficacy

Human studies evaluating the efficacy of resveratrol have been mixed. Investigators who have measured resveratrol levels in the blood or urine have pointed to the high concentration of resveratrol metabolites compared to levels of free *trans*-resveratrol (Gambini et al. 2015) as a possible reason for inconsistent efficacy in humans, since the free form is generally considered the most active form. Nonetheless, evidence suggests that at least some resveratrol derivatives also have an antiproliferative effect (Colin et al. 2008).

Trans-resveratrol is efficiently absorbed after oral administration and rapidly metabolized by glucuronidation and sulfation in the liver (Zhu et al. 2012). In the blood, *trans*-resveratrol binds both to albumin and low-density lipoproteins (LDLs) which assist in its delivery to epithelial cells for uptake (Zhu et al. 2012). Once taken up, resveratrol metabolites can be cleaved or hydrolyzed back to its free (most active) form (Zhu et al. 2012).

To address concerns over poor efficacy and bioavailability, various groups have synthesized resveratrol analogs and tested them for efficacy. While many investigators report that some of the synthesized resveratrol analogs demonstrated efficacy superior to free *trans*-resveratrol in vitro and/or in animals (Ronghe et al. 2016), none have proven sufficiently convincing to be in common clinical use.

Resveratrol in Therapy

Combination Stilbene Therapy

5-Aza-2′-deoxycytidine is a demethylating agent approved to treat various blood disorders (Cheishvili et al. 2015). Its side effects, pancytopenia, nausea, and vomiting, often limit its use. The plant stilbenes resveratrol and pterostilbene have epigenetic activity and are well tolerated. Pterostilbene is found in high concentration in blueberries. It is not found in the skin of red grapes in appreciable amounts, unlike resveratrol. Pterostilbene is reported to have increased bioavailability in comparison to other stilbene compounds such as resveratrol (Mccormack and Mcfadden 2013), which may enhance its dietary and clinical benefit.

A recent preclinical study evaluating the potential additive or synergistic effects of the two compounds (Kala and Tollefsbol 2016) demonstrated that the combination decreased DNMT enzyme activity and 5-methylcytosine levels in MDA-MB-157 breast cancer cells, as well as reactivation of ERα expression. After treatment, the previously ERα-negative MDA-MB-157 breast cancer cells increased their proliferation in response to estradiol, with decreased proliferation after treatment with 4-hydroxytamoxifen.

Resveratrol and pterostilbene administered together at close to physiologically relevant doses resulted in synergistic growth inhibition of HCC1806 and MDA-MB-157 triple negative breast cancer cells (Kala et al. 2015). SIRT1, a type III histone deacetylase (HDAC), and DNA methyltransferase enzymes, were downregulated in response to treatment.

Resveratrol and Chemotherapy

Resveratrol has demonstrated synergistic/additive tumor regression when administered with the chemotherapeutic agents adriamycin (Rai et al. 2016) and cisplatin (Osman et al. 2015).

Summary

There is strong preclinical evidence that resveratrol suppresses tumor formation and shrinks already formed tumors. Resveratrol works primarily through epigenetic mechanisms to induce apoptosis, decrease cell proliferation, and silence gene transcription. There is in vivo evidence that these effects are selective for tumors, sparing normal cells, and that there is a dose response, supporting their specificity related to resveratrol treatment rather than a random effect. It has been suggested that resveratrol does not work in humans because of first-pass effects in the liver, where most free resveratrol (believed to be the most active form) is modified yielding less active glucuronide and sulfate metabolites. However, short-term (8 days) administration of resveratrol led to high concentrations of both free resveratrol and its glucuronide and sulfate metabolites in normal and malignant colon tissues (Patel et al. 2010). In the blood, *trans*-resveratrol binds both to albumin and low-density lipoproteins (LDLs) (Jannin et al. 2004) which assists in its delivery to the epithelial cell surface for cell membrane uptake (Jannin et al. 2004). Intracellularly, the glucuronide can be enzymatically cleaved (Wang et al. 2004) or sulfated *trans*-resveratrol hydrolyzed (Santner et al. 1984) to its free form. This intracellular "reactivation" provides rationale for the efficacy reported in animal and limited human studies that resveratrol is a molecule worth further study to prevent and/or treat breast cancer, as well as its side effects such as cardiotoxicity.

Key Facts

- Resveratrol has strong preclinical evidence for efficacy to prevent and treat mammary cancer.
- Resveratrol has only limited data supporting its efficacy to treat breast cancer in humans.
- Resveratrol is rapidly metabolized by the liver, such that the free form of resveratrol is difficult to detect in the circulation.
- There is evidence that resveratrol metabolites are cleaved or hydrolyzed intracellularly, increasing the percentage of intracellular free resveratrol.

- Additional human studies are needed to determine the potential usefulness of the agent in the prevention and/or treatment of human breast cancer, as well to protect humans from the side effects of breast cancer treatment.

Dictionary of Terms

- **Methylation** – alteration in the DNA of a gene which decreases the amount of RNA and protein that is produced.
- **Metabolism** – modification of the native (free) molecule by the addition of a chemical side group, such as a glucuronide or sulfate, which often alters the function of the native molecule.
- **Synergy** – when agents are combined, yielding more than an additive effect
- **Polyphenol** – a compound, often found in nature, containing large multiples of the chemical structure phenol. The structures are found widely in plants and components in foods, dyes, paints, varnishes, rubbing compounds, and many other uses.

Summary Points

- Resveratrol is found in two isometric forms, *cis* and *trans*, with the latter predominating.
- *Trans*-resveratrol (resveratrol) is a polyphenol present in foods that has shown promise in tissue culture and animal studies as an anticancer agent.
- Resveratrol has many non-mutational, or epigenetic, mechanisms of action demonstrating its preclinical evidence of efficacy.
- There is evidence that resveratrol may have cardioprotective effects during the treatment of breast cancer.
- Human studies to confirm the preclinical findings are limited.
- The metabolism of free resveratrol in the liver had led some investigators to question the potential of resveratrol to prevent or treat breast cancer.
- Resveratrol analogs have been synthesized to increase the bioavailability of the agent. None had been proven to have clinical usefulness.
- Combination therapy of resveratrol with another natural compound, or with a synthesized compound, has been investigated and shows some promise.
- More human studies are needed to determine the role of resveratrol in the prevention or treatment of human breast cancer or to mitigate the toxicity of breast cancer treatment.

References

Baylin SB, Ohm JE (2006) Epigenetic gene silencing in cancer – a mechanism for early oncogenic pathway addiction? Nat Rev Cancer 6:107–116

Berner C, Aumuller E, Gnauck A, Nestelberger M, Just A, Haslberger AG (2010) Epigenetic control of estrogen receptor expression and tumor suppressor genes is modulated by bioactive food compounds. Ann Nutr Metab 57:183–189

Bhat KP, Lantvit D, Christov K, Mehta RG, Moon RC, Pezzuto JM (2001) Estrogenic and antiestrogenic properties of resveratrol in mammary tumor models. Cancer Res 61:7456–7463

Brown SA, Sandhu N, Herrmann J (2015) Systems biology approaches to adverse drug effects: the example of cardio-oncology. Nat Rev Clin Oncol 12:718–731

Cheishvili D, Boureau L, Szyf M (2015) DNA demethylation and invasive cancer: implications for therapeutics. Br J Pharmacol 172:2705–2715

Chen FP, Chien MH (2014) Phytoestrogens induce differential effects on both normal and malignant human breast cells in vitro. Climacteric 17:682–691

Colin D, Lancon A, Delmas D, Lizard G, Abrossinow J, Kahn E, Jannin B, Latruffe N (2008) Antiproliferative activities of resveratrol and related compounds in human hepatocyte derived HepG2 cells are associated with biochemical cell disturbance revealed by fluorescence analyses. Biochimie 90:1674–1684

Dhar S, Hicks C, Levenson AS (2011) Resveratrol and prostate cancer: promising role for microRNAs. Mol Nutr Food Res 55:1219–1229

Fan Y, Liu L, Fang K, Huang T, Wan L, Liu Y, Zhang S, Yan D, Li G, Gao Y, Lv Y, Chen Y, Tu Y (2016) Resveratrol ameliorates cardiac hypertrophy by down-regulation of miR-155 through activation of breast cancer type 1 susceptibility protein. J Am Heart Assoc 5:e002648

Gambini J, Ingles M, Olaso G, Lopez-Grueso R, Bonet-Costa V, Gimeno-Mallench L, Mas-Bargues C, Abdelaziz KM, Gomez-Cabrera MC, Vina J, Borras C (2015) Properties of resveratrol: in vitro and in vivo studies about metabolism, bioavailability, and biological effects in animal models and humans. Oxidative Med Cell Longev 2015:837042

Hayashi SI, Eguchi H, Tanimoto K, Yoshida T, Omoto Y, Inoue A, Yoshida N, Yamaguchi Y (2003) The expression and function of estrogen receptor alpha and beta in human breast cancer and its clinical application. Endocr Relat Cancer 10:193–202

Izzotti A, Cartiglia C, Steele VE, Deflora S (2012) MicroRNAs as targets for dietary and pharmacological inhibitors of mutagenesis and carcinogenesis. Mutat Res Rev Mutat Res 751:287–303

Jang M, Cai L, Udeani GO, Slowing KV, Thomas CF, Beecher CW, Fong HH, Farnsworth NR, Kinghorn AD, Mehta RG, Moon RC, Pezzuto JM (1997) Cancer chemopreventive activity of resveratrol, a natural product derived from grapes. Science 275:218–220

Jannin B, Menzel M, Berlot JP, Delmas D, Lancon A, Latruffe N (2004) Transport of resveratrol, a cancer chemopreventive agent, to cellular targets: plasmatic protein binding and cell uptake. Biochem Pharmacol 68:1113–1118

Kala R, Tollefsbol TO (2016) A novel combinatorial epigenetic therapy using resveratrol and pterostilbene for restoring estrogen receptor-alpha (ERalpha) expression in ERalpha-negative breast cancer cells. PLoS One 11:e0155057

Kala R, Shah HN, Martin SL, Tollefsbol TO (2015) Epigenetic-based combinatorial resveratrol and pterostilbene alters DNA damage response by affecting SIRT1 and DNMT enzyme expression, including SIRT1-dependent gamma-H2AX and telomerase regulation in triple-negative breast cancer. BMC Cancer 15:672

Klose RJ, Bird AP (2006) Genomic DNA methylation: the mark and its mediators. Trends Biochem Sci 31:89–97

Lancon A, Kaminski J, Tili E, Michaille JJ, Latruffe N (2012) Control of microRNA expression as a new way for resveratrol to deliver its beneficial effects. J Agric Food Chem 60:8783–8789

Lopez-Serra P, Esteller M (2012) DNA methylation-associated silencing of tumor-suppressor microRNAs in cancer. Oncogene 31:1609–1622

Lubecka K, Kurzava L, Flower K, Buvala H, Zhang H, Teegarden D, Camarillo I, Suderman M, Kuang S, Andrisani O, Flanagan JM, Stefanska B (2016) Stilbenoids remodel the DNA methylation patterns in breast cancer cells and inhibit oncogenic NOTCH signaling through epigenetic regulation of MAML2 transcriptional activity. Carcinogenesis 37:656–668

Malumbres M (2012) miRNAs and cancer: An epigenetics view. Mol Aspects Med 34:863–874
Mattiske S, Suetani RJ, Neilsen PM, Callen DF (2012) The oncogenic role of miR-155 in breast cancer. Cancer Epidemiol Biomark Prev 21:1236–1243
Mccormack D, Mcfadden D (2013) A review of pterostilbene antioxidant activity and disease modification. Oxidative Med Cell Longev 2013:575482
Mgbonyebi OP, Russo J, Russo IH (1998) Antiproliferative effect of synthetic resveratrol on human breast epithelial cells. Int J Oncol 12:865–869
Osman AM, Telity SA, Damanhouri ZA, Al-Harthy SE, Al-Kreathy HM, Ramadan WS, Elshal MF, Khan LM, Kamel F (2015) Chemosensitizing and nephroprotective effect of resveratrol in cisplatin -treated animals. Cancer Cell Int 15:6
Patel KR, Brown VA, Jones DJ, Britton RG, Hemingway D, Miller AS, West KP, Booth TD, Perloff M, Crowell JA, Brenner DE, Steward WP, Gescher AJ, Brown K (2010) Clinical pharmacology of resveratrol and its metabolites in colorectal cancer patients. Cancer Res 70:7392–7399
Qin W, Zhu W, Sauter ER (2005) Resveratrol induced DNA methylation in ER+ breast cancer. Proc Am Assoc Cancer Res 96:2750A
Qin W, Zhu W, Zhang K, Clarke K, Sauter ER (2012) Resveratrol decreases tumor formation in a human relevant animal model of breast cancer. In: 11th annual conference, Frontiers of Cancer Prevention Research, A36
Qin W, Zhang K, Clarke K, Weiland T, Sauter ER (2014) Methylation and miRNA effects of resveratrol on mammary tumors vs. normal tissue. Nutr Cancer 66:270–277
Rai G, Mishra S, Suman S, Shukla Y (2016) Resveratrol improves the anticancer effects of doxorubicin in vitro and in vivo models: a mechanistic insight. Phytomedicine 23:233–242
Ronghe A, Chatterjee A, Singh B, Dandawate P, Abdalla F, Bhat NK, Padhye S, Bhat HK (2016) 4-(E)-{(p-tolylimino)-methylbenzene-1,2-diol}, 1 a novel resveratrol analog, differentially regulates estrogen receptors alpha and beta in breast cancer cells. Toxicol Appl Pharmacol 301:1–13
Santner SJ, Feil PD, Santen RJ (1984) In situ estrogen production via the estrone sulfatase pathway in breast tumors: relative importance versus the aromatase pathway. J Clin Endocrinol Metab 59:29–33
Singh B, Shoulson R, Chatterjee A, Ronghe A, Bhat NK, Dim DC, Bhat HK (2014) Resveratrol inhibits estrogen-induced breast carcinogenesis through induction of NRF2-mediated protective pathways. Carcinogenesis 35:1872–1880
Subbaramaiah K, Chung WJ, Michaluart P, Telang N, Tanabe T, Inoue H, Jang M, Pezzuto JM, Dannenberg AJ (1998) Resveratrol inhibits cyclooxygenase-2 transcription and activity in phorbol ester-treated human mammary epithelial cells. J Biol Chem 273:21875–21882
Subbaramaiah K, Sue E, Bhardwaj P, Du B, Hudis CA, Giri D, Kopelovich L, Zhou XK, Dannenberg AJ (2013) Dietary polyphenols suppress elevated levels of proinflammatory mediators and aromatase in the mammary gland of obese mice. Cancer Prev Res (Phila) 6:886–897
Sun B, Hoshino J, Jermihov K, Marler L, Pezzuto JM, Mesecar AD, Cushman M (2010) Design, synthesis, and biological evaluation of resveratrol analogues as aromatase and quinone reductase 2 inhibitors for chemoprevention of cancer. Bioorg Med Chem 18:5352–5366
Suzuki H, Maruyama R, Yamamoto E, Kai M (2012) DNA methylation and microRNA dysregulation in cancer. Mol Oncol 6:567–578
Veeck J, Esteller M (2010) Breast cancer epigenetics: from DNA methylation to microRNAs. J Mammary Gland Biol Neoplasia 15:5–17
Wang LX, Heredia A, Song H, Zhang Z, Yu B, Davis C, Redfield R (2004) Resveratrol glucuronides as the metabolites of resveratrol in humans: characterization, synthesis, and anti-HIV activity. J Pharm Sci 93:2448–2457
Zhu W, Qin W, Zhang K, Rottinghaus GE, Chen YC, Kliethermes B, Sauter ER (2012) Trans-resveratrol alters mammary promoter hypermethylation in women at increased risk for breast cancer. Nutr Cancer 64:393–400

PARylation, DNA (De)methylation, and Diabetes

97

Melita Vidaković, Anja Tolić, Nevena Grdović, Mirunalini Ravichandran, and Tomasz P. Jurkowski

Contents

Abstract

Diabetes and diabetic complications, autoimmunity and inflammatory diseases, have recently become the focus of epigenetic therapy, since with epigenetic drugs it is possible to reverse aberrant gene expression profiles associated with the disease states. For diabetes, the therapy challenges depend on identifying the most appropriate molecular target and its influence on a relevant gene product. This chapter summarizes the current view on the interplay between ten-eleven

M. Vidaković (✉)
Department of Molecular Biology, Institute for Biological Research Siniša Stanković, University of Belgrade, Belgrade, Serbia
e-mail: melita@ibiss.bg.ac.rs

A. Tolić · N. Grdović
Institute for Biological Research, University of Belgrade, Belgrade, Serbia
e-mail: anja.tolic@ibiss.bg.ac.rs; nevena@ibiss.bg.ac.rs

M. Ravichandran · T. P. Jurkowski
Institute of Biochemistry, University of Stuttgart, Stuttgart, Germany
e-mail: mirunalini.ravichandran@ibc.uni-stuttgart.de; tomasz.jurkowski@ibc.uni-stuttgart.de

V. B. Patel, V. R. Preedy (eds.), *Handbook of Nutrition, Diet, and Epigenetics*,
https://doi.org/10.1007/978-3-319-55530-0_55

translocation (TETs) and the poly(ADP-ribose) polymerase (PARPs) family of enzymes in regulating DNA methylation and how this interplay could be targeted to attenuate diabetes. This molecular interchange jigsaw puzzle is emerging as an important focus of research, and we can expect to see further advances in the elucidation of its role in diabetes as well as other pathologies. Moreover, the possibility for designating specific PARP-1 inhibitors as potential "EPI-drugs" for diabetes prevention/attenuation is also discussed. Understanding the epigenetic machinery and the differential roles of its components is essential for the development of targeted epigenetic therapies for diseases.

Keywords
Diabetes · DNA methylation · DNA demethylation · DNMT enzymes · Epigenetic drug targets · Chromatin architecture · PARylation · PARP-1 inhibitors · TET enzymes

List of Abbreviations

3AB	3-aminobenzamide
5caC	5-carboxylcytosine
5fC	5-formylcytosine
5hmC	5-hydroxymethylcytosine
5hmU	5-hydroxymethyluridine
5mC	5-methylcytosine
BER	base excision repair
C	cytosine
CpG	cytosine-phosphate-guanine
CRISPR/Cas9	clustered regularly interspaced short palindromic repeats/associated protein-9 nuclease
DNMTs	DNA methyltransferases
NAD^+	nicotinamide adenine dinucleotide
PARPs	poly(ADP-ribose) polymerase family of enzymes
PARs	poly(ADP-ribose) polymers
PARylation	poly(ADP-ribosyl)ation
PARG	poly(ADP-ribose) glycohydrolases
RO/NS	reactive oxygen/nitrogen species
T1D	type 1 diabetes
T2D	type 2 diabetes
TDG	thymine-DNA glycosylase
TETs	ten-eleven translocation family of enzymes
α-KG	α-ketoglutarate

Introduction

Epigenetic marks represent an additional stratum of information that can, in combination with genetic information, produce a plethora of phenotypic outcomes. As a result of its reversible nature, deregulation of epigenetic processes is at the root of

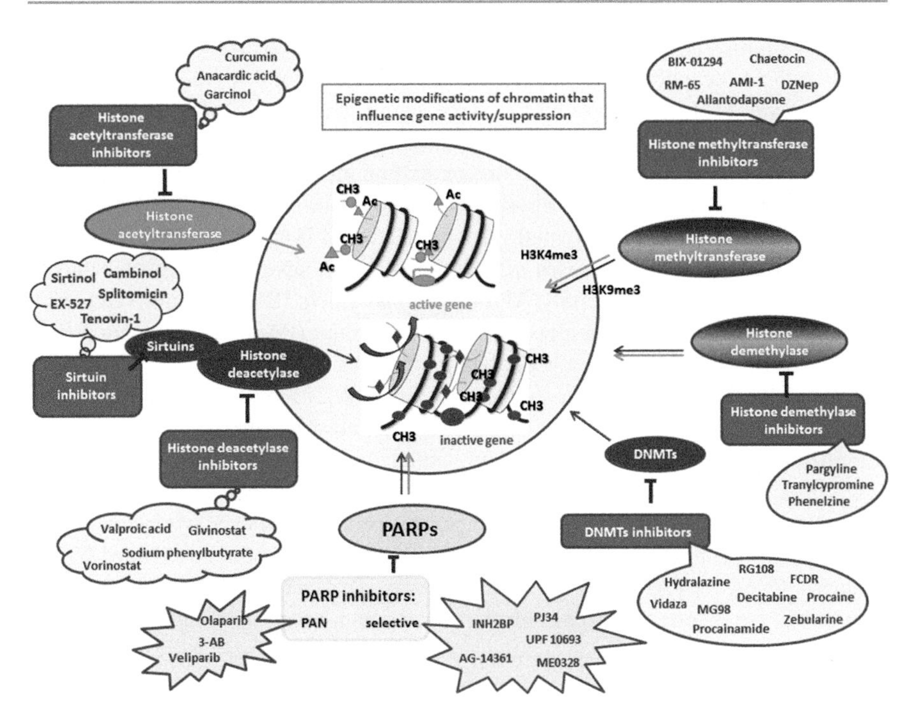

Fig. 1 Epigenetic drugs as an upgrade for the treatment of diabetes and obesity. Although there are no completely effective preventive or therapeutic strategies due to the multifactorial etiology of diabetes, the plastic nature of the epigenome represents a potential target in the design of future therapy and novel epigenetic drugs

several complex disorders, including diabetes. While these diseases cannot be effectively prevented by therapeutic strategies due to their complex etiologies and phenotypes, the epigenetic malleability of the genome renders it open to therapeutic drug targeting (Fig. 1). Thus research into new players that interact with and regulate epigenetic mechanisms presents a potential avenue for novel therapeutic approaches. As this field continues to develop, the epigenome could also become a biomarker for tracking therapeutics efficacy and toxicity of potential therapies.

Diabetes is a complex metabolic disorder characterized by hyperglycemia as a consequence of either impaired insulin secretion or insulin action, which presents in two major forms: type 1 diabetes (T1D) resulting from autoimmune destruction of insulin-producing pancreatic b-cells and type 2 diabetes (T2D) which develops as a result of insensitivity of target cells to insulin that can progressively compromise b-cell function and survival. Several studies have provided comprehensive information on the DNA methylomes of human pancreatic islets, underscoring epigenetic regulation as a mechanism driving the onset and development of diabetes (Agardh et al. 2015; Dayeh et al. 2014). Consequently DNA methylation has become an acceptable diagnostic tool for both diabetes types as changes in the methylation pattern are observed even prior to disease development.

While DNA methylation is the most investigated epigenetic modification, aside from the knowledge of passive DNA demethylation which has been linked to reduced activities or absence of DNA methyltransferases (DNMTs), the mechanism of active DNA demethylation has remained elusive for quite some time. Recently, the ten-eleven translocation (TET) enzyme family has been identified as the key initiator of active DNA demethylation through the iterative oxidation pathway of 5-methylcytosine (5mC) that leads to generation of three consecutive oxidized cytosine forms. The poly(ADP-ribose) polymerase family of enzymes (PARPs) is a focal point that links DNA methylation (via DNMTs) and demethylation (via TETs) by controlling both classes of enzymes. PARPs have the ability to catalytically add long and branched poly(ADP-ribose) polymers (PARs) using NAD^+ as a substrate, to target proteins in a poly(ADP-ribosyl)ation (PARylation) reaction (Bai 2015). At present, there is only one study that has incorporated the TET-PARP interplay in diabetes-related research (Dhliwayo et al. 2014).

To untangle this triangle, we first need to learn how to synchronize the enzyme activities of PARPs and TETs to ensure fine-tuned regulation of DNA (de)methylation. More stringent control of this equilibrium could attenuate the onset and progression of several complex diseases, including diabetes. This could be achieved by elucidating all the elements that maintain homeostasis of PARylation, a process determined by the balanced actions of PAR synthesizers and erasers. The fact that PARP inhibitors are included in therapeutic approaches for the prevention or reversal of diabetic complications (Burkart et al. 1999) lends additional support to further research efforts.

Establishment and Removal of DNA Methylation

DNA methylation is the most extensively studied epigenetic mark. As a major regulator of transcriptional activity, it participates in the control of the expression of a variety of genes. It plays an important role in maintaining genome integrity (Dodge et al. 2005), in genomic imprinting, in X chromosome inactivation, and in the silencing of retrotransposons, repetitive elements, and tissue-specific genes (Jones and Takai 2001). As once set, DNA methylation state can be faithfully maintained during each cycle of DNA replication and cell division.

In mammals, DNA methylation is the covalent addition of a methyl group to the carbon at the fifth position of cytosine (C) catalyzed by DNMTs. It is usually introduced in the context of cytosine-phosphate-guanine (CpG) dinucleotides located in regions that can differ in size and position within the gene: CpG islands, shores (regions 0–2 kb from CpG islands), shelves (regions 2–4 kb from CpG islands), and open sea (regions are isolated CpG sites with no specific position). De novo methyltransferases DNMT3a and DNMT3b are responsible for establishing de novo DNA methylation patterns (Yokochi and Robertson 2002), whereas the maintenance methyltransferase DNMT1 is responsible for renewing previously established methylation patterns after DNA replication (Hermann et al. 2004). Yet, this classical division for de novo and maintenance DNMTs seems to be more complex than initially assumed (Jeltsch and Jurkowska 2014). In addition to these enzymes, the availability of cofactors, such as S-adenosyl-L-methionine, is essential for the regulation of DNA and protein methylation (Stead et al. 2006).

A breakthrough in DNA (de)methylation research was made with the identification of the ten-eleven translocation (TET) family of DNA dioxygenases capable of generating 5-hydroxymethylcytosine (5hmC) through catalytic oxidation of 5mC (Ito et al. 2010). The TETs (TET1/2/3 enzymes) belong to a larger family of Fe [2]/α-ketoglutarate-dependent dioxygenases. TETs use molecular oxygen for oxidative decarboxylation of α-ketoglutarate (α-KG), thereby generating a high-valent Fe (IV)-oxo intermediate capable of converting 5mC to 5hmC. TETs can also generate 5-formylcytosine (5fC) and 5-carboxylcytosine (5caC) in successive oxidation reactions (Ito et al. 2011) (Fig. 2).

DNA methylation and 5mC formation within CpG islands could represent also the mutation hot spot, since the deamination of 5mC eventually leads to thymine formation, which is the primary cause of elevated mutagenesis. Since 5mC also represents a potential threat for genomic integrity, a cell needs to rely on a strictly controlled DNA demethylation mechanism that should be viewed as a part of the DNA repair process. In dividing cells, passive DNA demethylation occurs during replication and is caused by loss of DNMT1 activity, which leads to the dilution of 5mC throughout the genome (Kagiwada et al. 2013). Furthermore, TET- and replication-dependent pathways promote 5mC conversion to 5hmC (possibly further to 5fC and 5caC) which in turn dilutes methylation marks in successive DNA replications as DNMT1 does not recognize these modifications as location markers

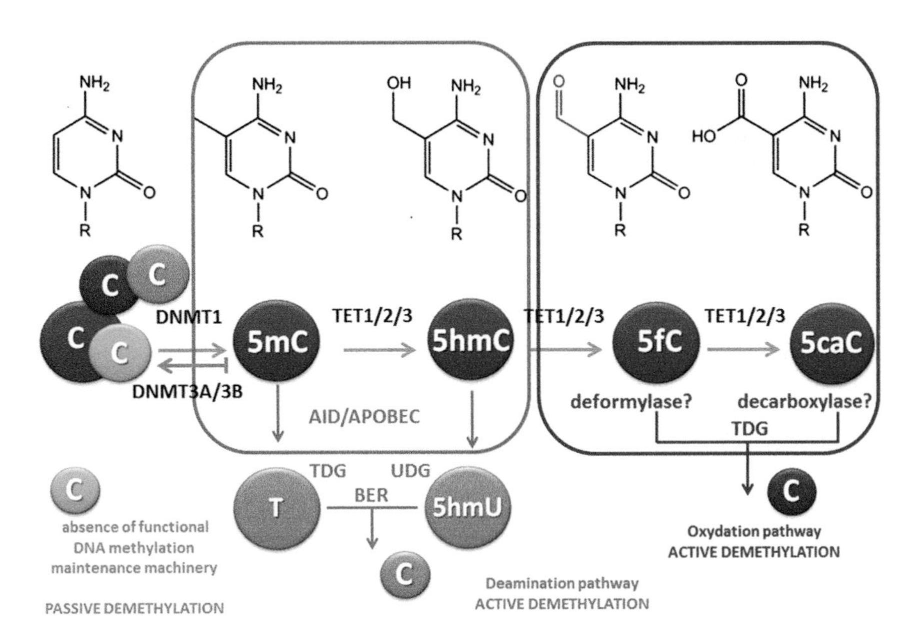

Fig. 2 Role of TETs in DNA demethylation pathways. DNA methylation is initiated by DNMTs. The 5mC can be further oxidized by TET1/2/3 proteins to generate 5hmC, 5fC, and 5caC. All modifications can be lost through passive demethylation (*green route*). Active demethylation can occur by the action of TDG that can excise 5fC and 5caC, which can be further repaired to cytosine (*red route*) or through deamination of 5mC or 5hmC in thymine or 5hmU (*orange route*)

for reintroduction of methylation (Kohli and Zhang 2013). An alternative, replication-independent, active DNA demethylation mechanism initiated by the oxidation of 5mC and further to 5fC and 5caC by TETs and finalized by base excision repair (BER), resulting in the replacement of a modified base with an unmodified C, was recently proposed (Fig. 2). One possibility is the deamination of 5hmC to 5-hydroxymethyluridine (5hmU) through the activity of AID/APOBEC enzymes, followed by BER (Guo et al. 2011). This model is controversial as there appear to be problems in replicating the results of the initial study. The other model that relies on BER has the best experimental confirmation and is generally accepted. An alternative mechanism is through direct 5caC decarboxylation; however, the enzyme catalyzing this reaction has not yet been identified. Finally, another BER-independent pathway has been proposed and incorporates the fact that DNMT3a/3b can function as DNA 5hmC-dehydroxymethylases in an oxidizing environment (Chen et al. 2012) yet not confirmed by independent study.

DNA demethylation, both active and passive, is indispensable for maintaining genome integrity and for its proper functioning. TET-mediated oxidation is an important epigenetic modification which leads to the creation of new epigenetic marks, such as 5hmC, 5fC, and 5caC. Several important questions need to be answered before we can obtain a full understanding of this vital process, especially in terms of cell-fate transition. Aside from the great importance of TET enzyme activities and DNA demethylation in cellular reprograming, the role of TETs in DNA methylation fidelity, modulation of chromatin architecture, and in transcriptional priming (Hill et al. 2014) should also be underlined. The role of TETs in faithful DNA methylation is realized by their ability to neutralize aberrant de novo DNA methylation and thereby preserve the proper unmethylated state of certain genomic regions and in transcriptional regulation through the formation of 5hmC that acts as an epigenetic mark allowing for gene activation or repression in combination with certain histone marks (Hill et al. 2014).

DNA (De)methylation and Diabetes

DNA methylation has been implicated in a number of biological processes and has been shown to be affected by environmental influences and aging (Christensen et al. 2009). However, much less is known about DNA methylation in diabetes.

Altered DNA methylation patterns observed in hyperglycemia, oxidative stress, and inflammation have been proven to be responsible for impaired gene regulation in diabetic individuals (Rakyan et al. 2011; Toperoff et al. 2012). There are only a limited number of studies regarding epigenetic changes in target tissues from patients with diabetes. It has been confirmed that the promoters of several genes involved in glucose metabolism exhibit differential DNA methylation. These are genes encoding for glucose transporter 4, the major glucose transporter in adipose and muscle tissues (Yokomori et al. 1999), and uncoupling protein 2 (Carretero et al. 1998), which is a candidate gene for the development of T2D. DNA methylation of another candidate gene, *PPARGC1A*, was increased in pancreatic islets from patients with T2D when compared to healthy control subjects (Ling et al. 2008). Accordingly, *PPARGC1A* expression was lower in diabetic islets and correlates inversely with the degree of DNA methylation. A recent study also discovered that p.Ile1762Val substitution in TET2 is associated with liver *PPARGC1A* methylation and transcription (Pirola et al. 2015). A significant association between p.Ile1762Val TET2 and T2D came as no surprise since it was previously shown that liver expression of *PPARGC1A* modulates insulin resistance (Sookoian et al. 2010). In light of the above findings, it was suggested that TET2 is involved in the modulation of the *PPARGC1A* (de)methylation balance, possible in response to changes in the cellular environment (Pirola et al. 2015). Further, analysis of skin fibroblasts derived from T1D-affected monozygotic twins cultured in media containing high levels of glucose demonstrated differential gene expression in twin pairs in a number of genes that are involved in epigenetic regulation (Caramori et al. 2012).

It has been suggested that errors in DNA methylation can lead to decreased gene responsiveness to external stimuli, thus contributing to the development of diabetes (Gallou-Kabani and Junien 2005). This assumption was confirmed by the first comprehensive attempt at DNA methylation profiling of pancreatic islets in T2D and nondiabetic individuals (Volkmar et al. 2012). The authors discovered that 276 CpG loci associated with promoters of 254 genes are differentially methylated in diabetic islets. These methylation patterns were not detected in blood cells from T2D individuals nor were they found in nondiabetic islets exposed to high glucose concentration (Volkmar et al. 2012). In a more recent study, researchers performed a genome-wide DNA methylation analysis of 479,927 CpG sites in human pancreatic islets from T2D donors (Dayeh et al. 2014). This work revealed the presence of differential DNA methylation patterns for 1649 CpG sites belonging to 853 genes in T2D patients. The authors published a detailed methylome map of the human pancreatic islets with 102 genes that showed differential DNA methylation and gene expression patterns in islets of T2D patients, demonstrating that changes in DNA methylation are responsible, at least in part, for altered insulin secretion and pathogenesis of T2D (Dayeh et al. 2014). In both of these large-scale DNA methylation profiling studies, the majority of the identified differentially methylated genes had decreased methylation and increased expression in T2D islets, while changes in global levels of DNA methylation have not been observed (Dayeh et al. 2014; Volkmar et al. 2012). Concerning T1D, a very comprehensive study in monozygotic twins has identified DNA methylation variable positions that arise very early prior to the development of overt T1D (Rakyan et al. 2011). The most recent epigenome-wide association study across 406,365 CpGs in 52 monozygotic twin pairs revealed enrichment of differentially variable CpG positions in T1D twins when compared with their healthy co-twins, at gene regulatory elements which, after integration with cell type-specific gene regulatory circuits, highlight the pathways involved in immune cell metabolism and the cell cycle, including mTOR signaling (Paul et al. 2016).

Since a common feature of both types of diabetes is a reduction in b-cell mass, therapeutic interventions that promote the growth and survival of functional b-cell are a valid approach for diabetes treatment. It is now clear that adult pancreatic b-cells replicate to maintain normoglycemia during body growth and after oxidative stress damage and inflammation (Bouwens and Rooman 2005). DNA methylation is proven to be important for maintenance of b-cell identity. Extreme loss of b-cells due to severe metabolic stress can induce transdifferentiation of pancreatic a- to b-cells, suggesting that b-cells deficient in DNMT1 are converted to pancreatic a-cells. Moreover, Aristaless-related homeobox gene was identified as methylated and repressed in b-cells but hypomethylated and expressed in a-cells and DNMT1-deficient b-cells (Dhawan et al. 2011). Another study has shown that adult human skin fibroblasts can be converted to insulin-secreting cells after they have been exposed to the DNA methyltransferase inhibitor 5-azacytidine (Pennarossa et al. 2013). Further research in this field also revealed that pancreatic islet dysfunction and development of diabetes in rats are associated with DNA methylation-dependent epigenetic silencing of the Pdx1 gene, which is a key transcriptional regulator of b-cell differentiation and insulin gene expression (Park et al. 2008).

Cell reprogramming requires viral transfection of appropriate transcription factors and use of many growth factors which limits its therapeutic potential. Therefore, alternative approaches need to be developed to overcome this limitation. The concept of epigenetic editing could be a vital alternative cell reprogramming approach. Recently, an epigenetic editing platform based on a programmable DNA-binding domain composed of CRISPR/Cas9 fused to engineered DNA methyltransferases was developed in order to change the DNA methylation status of the selected loci in cells (Stepper et al. 2016). These novel epigenetic editing tools could be employed for targeted DNA methylation of key transcription factors responsible for maintaining a- and b-cell identity (Fig. 3). Finally, any methodological success in increasing the number of insulin-producing b-cells is a promising therapeutic avenue for curing diabetes.

The crucial information obtained from these studies underlies that the profiles of differentially methylated genes determine b-cell survival, function, and adaptation to stressors and point to a tight link between DNA methylation, diabetes, and its complications. Diabetes is a complex disease with polygenic susceptibility, thereby depending on several environmental influences that through interplay between genome-epigenome factors influence its onset and further development (Fig. 4). Environmental factors that influence diabetes onset and progression have been well characterized in contrast to the details of their interaction with genetic factors and epigenetic modifications. While we are continuously acquiring knowledge about the interactions between these factors, the relationship between specific epigenetic modifications and genomic features remains poorly explained in the context of the diabetes phenotype. Thus, there is an urgent need for integrative approaches aimed at obtaining fundamental insight into how genetic and epigenetic factors underlie diabetic etiopathogenesis.

Poly(ADP-ribosyl)ation in Diabetes

PARylation has been implicated in numerous processes both in normal and pathological cell states and is involved in the regulation of epigenetic mechanisms and,

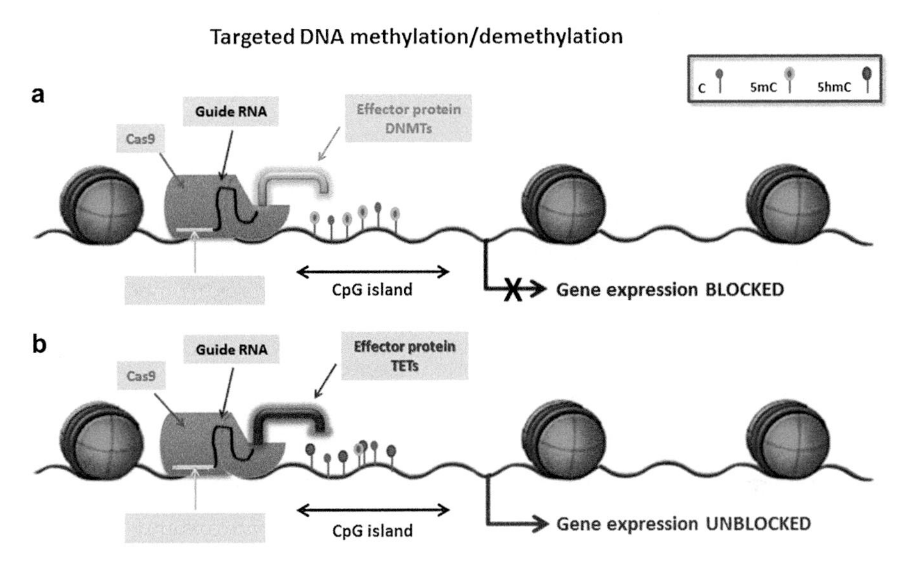

Fig. 3 The concept of epigenetic editing. The targeting device (CRISPRs/Cas9), a sequence-specific DNA-binding domain that can be redesigned to recognize desired sequences by the guided RNA, is fused to an effector domain (DNMTs or TETs) that can modify the epigenetic state of the targeted locus, leading to a long-lasting biological effect (gene repression (**a**) or activation (**b**)). *Blue lollipops*–cytosine; *green lollipops*–5mC; *red lollipops*–5hmC

accordingly, also in diabetes development and progression. The mammalian PARP superfamily of enzymes consists of 17 members all sharing a conserved ADP-ribosyl transferase catalytic domain, with PARP-1 and PARP-2 being the most prominent family members. PARPs have been implicated in many molecular and cellular processes, such as replication, DNA repair, genomic stability, chromatin remodeling, and transcriptional regulation (Schuhwerk et al. 2016). PARP-mediated PARylation is a reversible posttranslational modification that can be directed to both the enzymes carrying out the reaction (automodification) and to other target proteins (hetero-modification). In the cell, PARP activity in both basal and stimulated states is primarily exerted by PARP-1 and to lesser extent by PARP-2 (Bai 2015).

Historically, PARP-1 and PARP-2 were implicated in DNA repair as their activity is markedly stimulated by their binding to DNA strand breaks. Under basal conditions, PARP-1 activity is generally low, but excessive DNA damage or other pathophysiological conditions can induce PARP-1 hyperactivation, thus initiating an energy-consuming cycle leading to NAD^+ and ATP depletion and necrosis (Ha and Snyder 1999). On the other hand, oxidative stress resulting from an overload of the endogenous enzymatic and nonenzymatic antioxidant defenses with reactive oxygen/nitrogen species (RO/NS) plays an important role in the etiology of both types of diabetes. The prooxidant action in the hyperglycemic environment causes DNA damage and subsequent PARP-1/2 activation which in turn leads to a specific type of pancreatic b-cell death due to extreme energy consumption (Grdović et al. 2014). Since PAR synthesis is generally evoked by different stress signals,

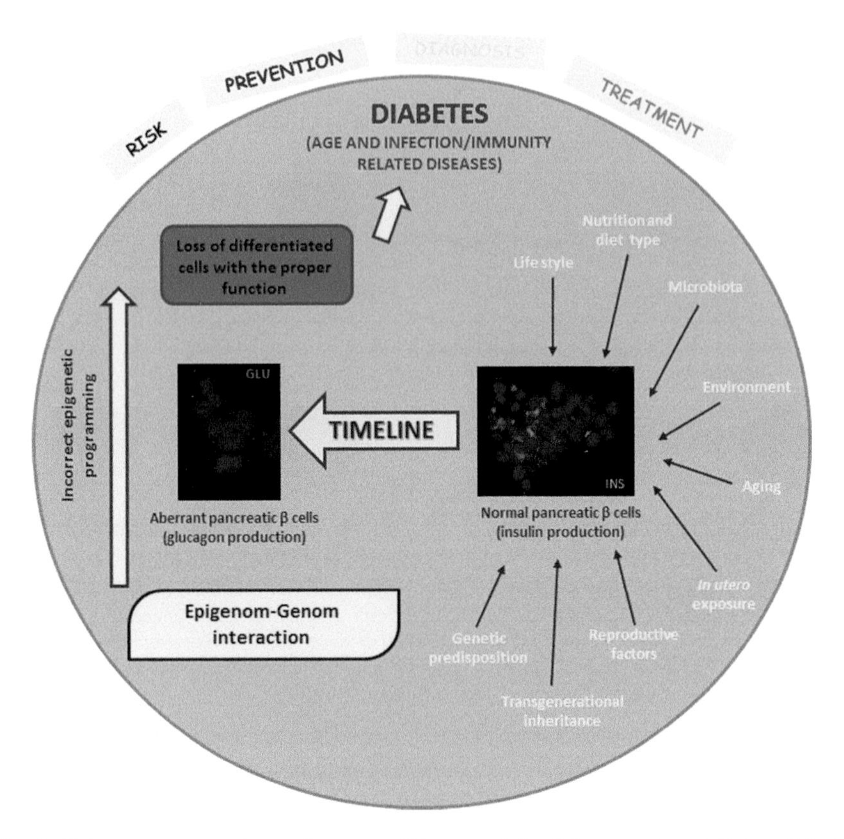

Fig. 4 Epigenome-genome interaction as a valid approach for diabetes treatment. Future diabetes research should be directed toward achieving a solid research link between genetics, epigenetics, and nongenetic factors in order to improve the quality of life in both health and disease state

genotoxic, hormonal, or metabolic, the regulation of their formation by PARPs and their degradation by poly(ADP-ribose) glycohydrolases (PARG) are an essential aspect of the adaptive response of the cell. Transient PAR formation/degradation coordinates chromatin remodeling and transcriptional regulation, thus allowing the cell to follow the proper survival pathway. Therefore, the precise regulation of the equilibrium between PAR synthesis and degradation is of utmost importance for cell fate in diabetes (Fig. 5). Additionally, PARP-1 can also induce the production of pro-inflammatory mediators via activation of NF-κB transcription, which is especially important during the autoimmune destruction of b-cells in T1D (Ba and Garg 2011). Other complications observed in diabetic patients appear to be also connected to these pathological consequences of PARP-1 hyperactivation. In models of T1D, it was shown that mice lacking PARP-1 gene expression are resistant to pancreatic b-cell destruction and diabetes development (Burkart et al. 1999). Taking into account the aforementioned, PARP-1 inhibition has been identified as a promising approach for the treatment of diabetes complications (Pacher and Szabo 2007). Several PARP-1 inhibitors have been used in clinical studies in susceptible individuals as agents capable of preventing the development of insulin-dependent diabetes (Pandya et al. 2010).

Up to now, several inhibitors of PARP-1 have been identified (Szabo et al. 2006). The first introduced was the competitive inhibitor of PARP-1, nicotinamide, a natural NAD^+-metabolizing enzyme competitor (Khan et al. 2007). A structurally similar compound, 3-aminobenzamide (3AB), is the foremost PARP inhibitor which is still in use. Both nicotinamide and 3AB have been shown to protect against diabetes development (Masiello et al. 1990). The administration of nicotinamide, a free-radical scavenger and a weak and nonselective PARP-1/2 inhibitor, prevents the development of diabetes in a spontaneous mouse model of autoimmune diabetes (Virag and Szabo 2002). Clinical trials assessing diabetes treatment by PARP-1/2 inhibition were performed with nicotinamide and, unfortunately, revealed no significant therapeutic effects. The problem with these inhibitors is their low potency (mM range) and low specificity, along with additional effects, such as antioxidant activity (Szabo et al. 1998). Other more potent PARP-1 inhibitors have been discovered, and some, such as PJ34 and INH2BP, show promising effects in diabetes models and also exert anti-inflammatory effects (Szabo et al. 2006). PARP-1 inhibitor, PD128763, is effective at protecting islet cells from NO, ROS, and streptozotocin at concentrations 100 times lower than required for nicotinamide (Wurzer et al. 2000).

In general, the involvement of PARP-1/2 proteins in the maintenance of genome integrity and major cellular functions anticipates that PARP-1 inhibition could be used in "next-generation" diabetes prevention therapy (Fig. 5). In the latest review that summarizes epigenetic drugs used for the treatment of diabetes and obesity (Arguelles et al. 2016), PARP inhibitors are not listed, and this should be corrected in any further work, since inhibitors of PARP activity due to all of the abovementioned deserve to be defined as a potential group of epigenetic drugs. Researchers should carefully contemplate on the effect they aim to induce by inhibiting PARPs. In some medical conditions, a global inhibition of PARPs could be more beneficial since it would stop any PAR synthesis and energy expenditure. Search for a selective inhibitor for a specific PARP enzyme will be more effective if the disease to be cured depends on a particular PARP molecule. Considering the importance of the equilibrium between PAR synthesis and degradation for cell fate in diabetes, aside from developing new, selective and potent PARP-1/2 inhibitors, researchers should also consider producing molecules capable of regulating PARG activity.

A Mingled Yarn: PARylation, DNA (De)methylation and Diabetes

During the last decade, it has become evident that PARPs and PARylation also have a role in epigenetic mechanisms, more precisely in the regulation of DNA

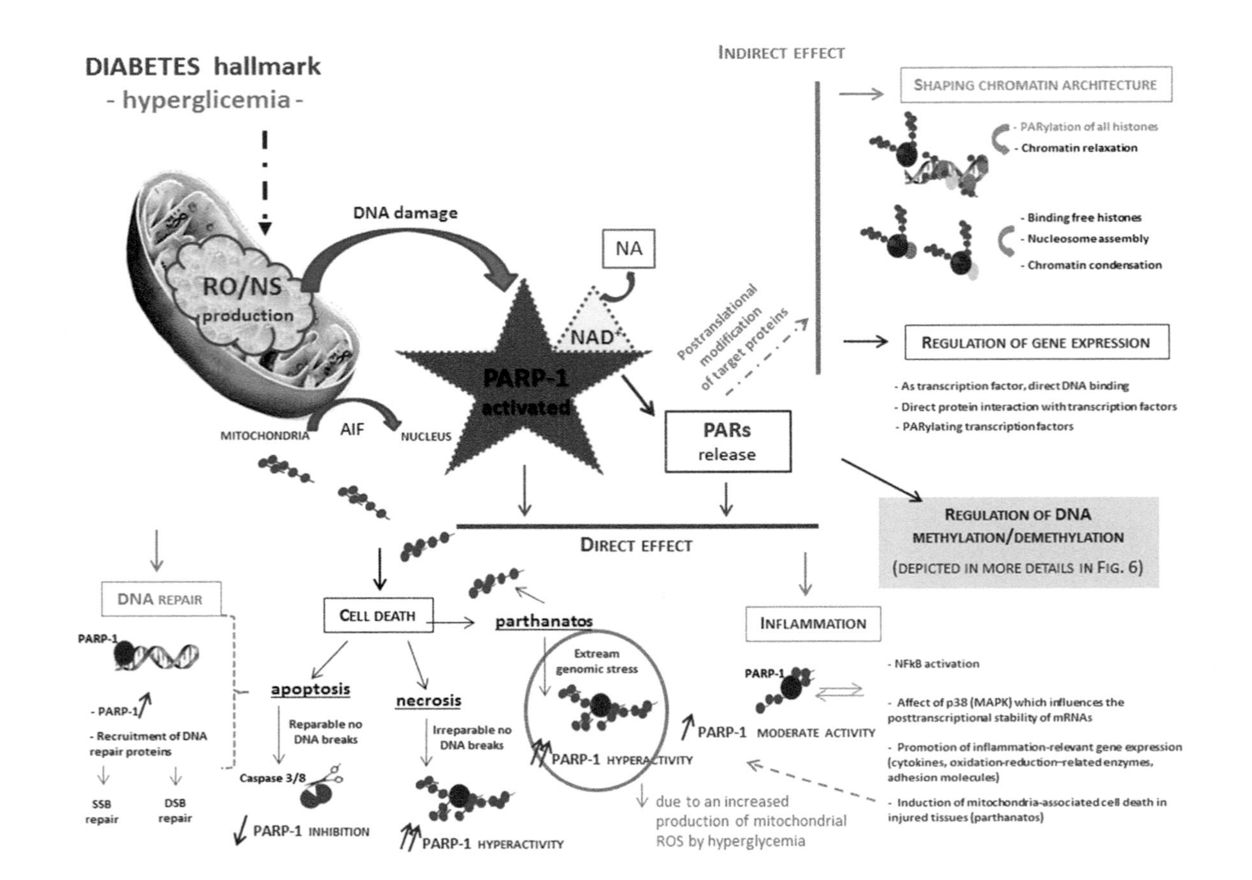

Fig. 5 (continued)

methylation (Fig. 6). The first information linking DNA methylation and PARP-1 activity was reported in 1997, where it was pointed out that the PARylated isoform of histone H1 and/or long and branched protein-free PARs could protect genomic DNA against methylation of the CpGs (Zardo et al. 1997). In the confirmation of this result in 2008, it was indicated that automodified, PARylated PARP-1 marks and protects sequences in the genome that should remain unmethylated, thus directly pointing to a role of PARP in the epigenetic regulation of gene expression (Zampieri et al. 2009). Caiafa and coworkers revealed that cells with hyperactivated PARP-1 are characterized by widespread DNA hypomethylation (Guastafierro et al. 2008). In vitro experiments showed that DNMT1 activity is inhibited by both free and PARs bound to PARP-1, while co-immunoprecipitation results confirmed that DNMT1 interacts with PARylated PARP-1 in vivo (Reale et al. 2005). Moreover, a new molecular participant in demethylation process, CTCF, as an insulator that binds preferentially to unmethylated target DNA sequences and whose insulator function is impaired by the inhibition of its N-terminal domain by PARylation, was identified (Yu et al. 2004). PARP-1, CTCF, and DNMT1 were identified as the main players in the cross talk between PARylation and DNA methylation. There is strong evidence that CTCF can activate PARP-1, leading to increased levels of nuclear PARs. Also, overexpression of CTCF can indirectly, through the agency of PARP-1, decrease DNMT1 activity by 70%, leading to widespread hypomethylation (Guastafierro et al. 2008). Recently it was shown that PAR depletion leads to the removal of CTCF from its target sites and to its perinuclear accumulation (Guastafierro et al. 2013). The consequence of both PAR depletion and CTCF silencing is increased DNA condensation and DNA hypermethylation, which emphasizes the importance of the cross talk between CTCF and PARylation in maintenance of chromatin structure and organization.

In 2014, Caiafa's group described the involvement of PARylation in the control of DNA and histone methylation in the TET1 gene (Ciccarone et al. 2014) and next year the importance of TET1 and PARP-1 interplay (Ciccarone et al. 2015). It was suggested that TET1 has the ability to stimulate PARP-1 activity independently of DNA damage, which can promote covalent PARylation of TET1, bringing about either an increase or inhibition of TET1 activity as a result of noncovalent interaction with PAR polymers (Ciccarone et al. 2015). It was shown that PARP-1 is also involved in active demethylation in mouse primordial germ cells by upregulation of TET1 transcription (Ciccarone et al. 2012). As TET enzymes are undoubtedly initiators of DNA demethylation, the search for factors that can influence them in a direct or indirect manner continues. PARPs are proteins that are increasingly

Fig. 5 PARP-1 in diabetes: an omnipresent molecule with great importance. PARP-1 is a key enzyme in DNA repair. PARP-1 is activated by hyperglycemia-induced RO/NS overproduction that causes DNA damage; PARP-1 overspends the energy pool of the cell, leading to cell death (apoptosis, necrosis, parthanatos); PARP-1 assumes the roles of transcriptional regulator, an important contributor to chromatin 3D reshaping and of a regulator of DNA methylation/demethylation. The role of PARP-1 in the regulation of inflammatory process is highlighted in diabetes

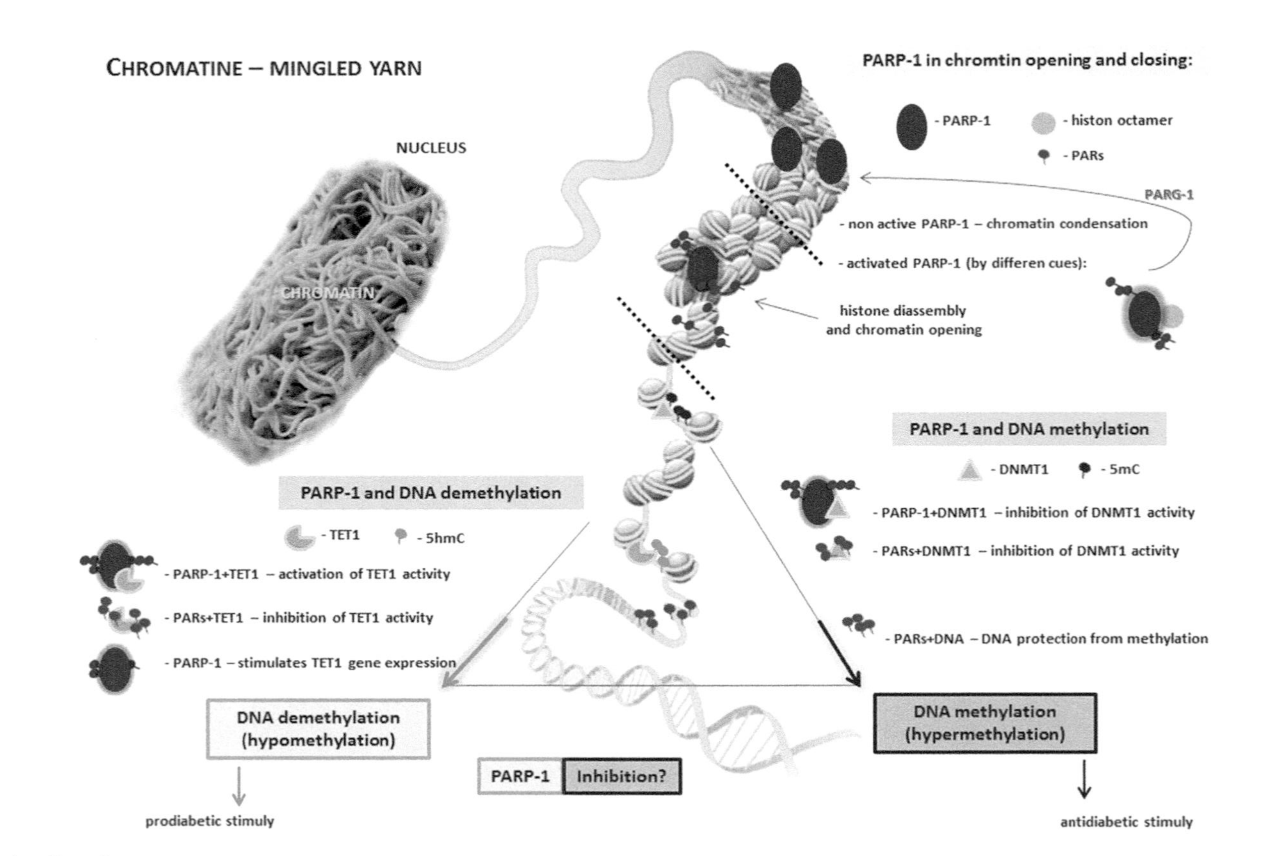

Fig. 6 (continued)

implicated in DNA demethylation (Fig. 6). As important components of the BER pathway, the inhibition of PARPs was used to confirm the involvement of BER in TET-initiated demethylation (Guo et al. 2011). Evidence suggesting that PARP-1 can also have a BER-independent role in DNA demethylation has emerged (Ciccarone et al. 2012). The same study showed that PARylation inhibition considerably reduces TET1 gene expression and moderately raises the TET3 gene expression in mouse primordial germ cells. Furthermore it was confirmed that all three TETs interact with PARP-1 (Muller et al. 2014). Additionally, it has been demonstrated that PARP-1 and TET2 are recruited to Nanog and Esrrb loci during somatic cell reprogramming and are important for downregulating 5mC modification and establishing 5hmC (Doege et al. 2012). Recent findings that TET1/2 can bind PAR polymers add yet another layer of complexity to the intricate interplay of TETs and PARP-1 in DNA demethylation (Fujiki et al. 2013). Specifically, the authors discovered that the Cys-rich domain exhibits strong affinity for binding PAR in both TET1/2, while the N-terminal region and CXXC domain of TET1 show somewhat weaker binding. In this study it was postulated that PARylation of sequence-dependent transcription factors serves to produce landmarks and docking site for TETs, directing targeted DNA demethylation (Fujiki et al. 2013). Additionally, PARylation of TET1 may lead to the formation of complexes containing PARylated and PAR-interacting proteins that participate in the regulation of DNA demethylation and transcription (Ciccarone et al. 2015). Surprisingly, no research has focused explicitly on the role of PARP-2 in the regulation of DNA (de)methylation, even though it contributes to overall PARylation levels in a cell.

Finally, describing molecular machinery that is responsible for the hyperglycemia-induced DNA demethylation, diabetes is included in this already complicated interplay of PARPs and TETs in regulating DNA demethylation process (Dhliwayo et al. 2014). It was proposed that PARP-1 can stimulate TET1-initiated demethylation in diabetes that eventually leads to persistent complications. Though it was previously documented that hyperglycemia induces DNA demethylation (Williams et al. 2008), this research went further by demonstrating that for most specific CpG islands, demethylation induced by hyperglycemia persists in metabolic memory which underlies diabetic complications, even when glycemic control is reestablished (Dhliwayo et al. 2014). This study demonstrated that PARP-1 inhibition can prevent hyperglycemia-induced demethylation and restore the regenerative capacity of the fin in the zebrafish model, giving hope for new treatment possibilities (Fig. 6).

Fig. 6 A mingled yarn: PARylation, DNA (de)methylation, and diabetes. Alterations in DNA methylation pattern are involved in different aspects of diabetes. Three enzyme families, DNMTs, TETs, and PARPs, have elevated this already complicated game to a higher level of complexity. PARP-1 is a connecting factor since it modifies chromatin structure, interacts with DNMT1 and TET1, and is capable of PARylating both enzymes and thereby influencing their activities. PARP inhibition could be the key for controlling widespread genomic hypomethylation that is observed in diabetes

Conclusion

Since diabetes is such a complex multifactorial disease, it is hard to pinpoint the precise pathways of its progression and development of diabetic complications. Nevertheless, it is important to gain as much knowledge of the possible mechanisms that underlie this widespread disease. Epigenetic marks and DNA methylation, in particular, represent a promising direction for future diabetes research since, as evinced in this chapter, a number of studies have implicated alterations in DNA methylation patterns in different aspects of diabetes. Therefore it is both important to gain knowledge from basic research examining the processes involved in the establishment, removal, and regulation of DNA methylation marks and to apply these insights in research focused on diabetes therapy.

DNA methylation is an important epigenetic mark, and thus regulation of its establishment and removal is crucial for a basic understanding of epigenetic regulatory mechanisms. The discovery of TET enzymes has opened a new path in the research of mechanism of active DNA demethylation. On the other hand, PARPs are implicated in a wide variety of cellular processes, including DNA methylation. PARPs are emerging as promising regulators of TETs' activities and possibly even in their localization in the genome. They thus represent molecules that link DNMTs and TETs and which, through fine-tuned activity, can influence DNA methylation/demethylation process. Therefore, studying TETs/PARPs and DNMTs/PARPs interactions and the influence of PARylation on TETs or DNMTs activity can provide new important insights into DNA metabolism. Research of basic molecular mechanisms is the first important step on the long road to the clinical application of epigenetics in diseases therapy.

Key Facts of DNA Methylation in Diabetes

- A common feature of diabetes is the reduction in b-cell mass; thus therapeutic interventions promoting functional b-cell growth and survival are a valid approach for diabetes treatment.
- DNA methylation has been implicated in diabetes onset and development of diabetic complications and is proven to be important for maintenance of pancreatic b-cell identity.
- To cure diabetes by triggering the transdifferentiation of pancreatic a- to b-cells, synthetic epigenetic editing tools could be employed for targeted DNA methylation of key transcription factors that maintain pancreatic cell phenotypes.
- PARP-1 activity and PARylation have been implicated in the regulation of DNA (de)methylation by influencing TET1 and DNMT1 activities and also in diabetes development and progression.
- PARP-1 inhibitors could be used as "next-generation" diabetes prevention drugs.

Dictionary of Terms

- **Diabetes** – a complex metabolic disorder marked by hyperglycemia due to either impaired insulin secretion or insulin action.
- **DNA methyltransferases (DNMTs)** – epigenetic enzymes that modify DNA by adding a methyl group on the fifth position of cytosine.
- **Poly(ADP-ribosyl) polymerases (PARPs)** – mammalian superfamily of enzymes capable of cleaving NAD^+-forming poly(ADP-ribose) polymers covalently attached to themselves or other target proteins.
- **Ten-eleven translocation (TET) family of DNA dioxygenases** – enzymes responsible for iterative oxidation of 5-methylcytosine.
- **Epigenetic drugs** – molecules with a potential to change the activity of druggable proteins involved in any aspect of epigenetic processes.

Summary Points

- Alterations in DNA methylation pattern are implicated in different aspects of diabetes.
- Diabetes and diabetic complications have recently become the focus of epigenetic therapy.
- PARP-1 serves as a pivot of the DNA methylation equilibrium in the genome.
- PARP-1 has emerged as a bridging molecule regulating the activities of both TETs and DNMTs.
- Efforts aimed at defining the interplay between TETs and PARPs in regulating DNA methylation are key for the development of novel diabetes attenuation strategies.
- Epigenetic editing concept could be an alternative cell reprogramming approach for diabetes cure.
- The epigenetic editing tools could be employed for targeted DNA methylation of key transcription factors responsible for maintaining pancreatic cell identity.
- "EPI-drugs" allow reversal of aberrant gene expression associated with different disease states.
- Specific PARP-1 inhibitors are viable candidates for potentially new "EPI-drugs" for diabetes prevention/attenuation.
- Understanding the epigenetic machinery and differential roles of its components is essential for developing more targeted epigenetic therapy for noncancerous diseases.

Acknowledgments This work was supported by the Alexander von Humboldt foundation, program for funding a Research Group Linkage (2014) and Ministry of Education, Science and Technological Development of the Republic of Serbia, Grant No. 173020. This article is based upon work from COST Action (CM1406), supported by COST (European Cooperation in Science and Technology), participants MV and TPJ.

References

Agardh E, Lundstig A, Perfilyev A et al (2015) Genome-wide analysis of DNA methylation in subjects with type 1 diabetes identifies epigenetic modifications associated with proliferative diabetic retinopathy. BMC Med 13:182

Arguelles AO, Meruvu S, Bowman JD et al (2016) Are epigenetic drugs for diabetes and obesity at our door step? Drug Discov Today 21:499–509

Ba X, Garg NJ (2011) Signaling mechanism of poly(ADP-ribose) polymerase-1 (PARP-1) in inflammatory diseases. Am J Pathol 178:946–955

Bai P (2015) Biology of poly(ADP-Ribose) polymerases: the factotums of cell maintenance. Mol Cell 58:947–958

Bouwens L, Rooman I (2005) Regulation of pancreatic beta-cell mass. Physiol Rev 85:1255–1270

Burkart V, Wang ZQ, Radons J et al (1999) Mice lacking the poly(ADP-ribose) polymerase gene are resistant to pancreatic beta-cell destruction and diabetes development induced by streptozocin. Nat Med 5:314–319

Caramori ML, Kim Y, Moore JH et al (2012) Gene expression differences in skin fibroblasts in identical twins discordant for type 1 diabetes. Diabetes 61:739–744

Carretero MV, Torres L, Latasa U et al (1998) Transformed but not normal hepatocytes express UCP2. FEBS Lett 439:55–58

Chen CC, Wang KY, Shen CK (2012) The mammalian de novo DNA methyltransferases DNMT3A and DNMT3B are also DNA 5-hydroxymethylcytosine dehydroxymethylases. J Biol Chem 287:33116–33121

Christensen BC, Houseman EA, Marsit CJ et al (2009) Aging and environmental exposures alter tissue-specific DNA methylation dependent upon CpG island context. PLoS Genet 5:e1000602

Ciccarone F, Klinger FG, Catizone A et al (2012) Poly(ADP-ribosyl)ation acts in the DNA demethylation of mouse primordial germ cells also with DNA damage-independent roles. PLoS One 7:e46927

Ciccarone F, Valentini E, Bacalini MG et al (2014) Poly(ADP-ribosyl)ation is involved in the epigenetic control of TET1 gene transcription. Oncotarget 5:10356–10367

Ciccarone F, Valentini E, Zampieri M et al (2015) 5mC-hydroxylase activity is influenced by the PARylation of TET1 enzyme. Oncotarget 6:24333–24347

Dayeh T, Volkov P, Salo S et al (2014) Genome-wide DNA methylation analysis of human pancreatic islets from type 2 diabetic and non-diabetic donors identifies candidate genes that influence insulin secretion. PLoS Genet 10:e1004160

Dhawan S, Georgia S, Tschen SI et al (2011) Pancreatic beta cell identity is maintained by DNA methylation-mediated repression of Arx. Dev Cell 20:419–429

Dhliwayo N, Sarras MP Jr, Luczkowski E et al (2014) Parp inhibition prevents ten-eleven translocase enzyme activation and hyperglycemia-induced DNA demethylation. Diabetes 63:3069–3076

Dodge JE, Okano M, Dick F et al (2005) Inactivation of Dnmt3b in mouse embryonic fibroblasts results in DNA hypomethylation, chromosomal instability, and spontaneous immortalization. J Biol Chem 280:17986–17991

Doege CA, Inoue K, Yamashita T et al (2012) Early-stage epigenetic modification during somatic cell reprogramming by Parp1 and Tet2. Nature 488:652–655

Fujiki K, Shinoda A, Kano F et al (2013) PPARgamma-induced PARylation promotes local DNA demethylation by production of 5-hydroxymethylcytosine. Nat Commun 4:2262

Gallou-Kabani C, Junien C (2005) Nutritional epigenomics of metabolic syndrome: new perspective against the epidemic. Diabetes 54:1899–1906

Grdović N, Dinic S, Mihailovic M et al (2014) CXC chemokine ligand 12 protects pancreatic beta-cells from necrosis through Akt kinase-mediated modulation of poly(ADP-ribose) polymerase-1 activity. PLoS One 9:e101172

Guastafierro T, Cecchinelli B, Zampieri M et al (2008) CCCTC-binding factor activates PARP-1 affecting DNA methylation machinery. J Biol Chem 283:21873–21880

Guastafierro T, Catizone A, Calabrese R et al (2013) ADP-ribose polymer depletion leads to nuclear Ctcf re-localization and chromatin rearrangement(1). Biochem J 449:623–630
Guo JU, Su Y, Zhong C et al (2011) Hydroxylation of 5-methylcytosine by TET1 promotes active DNA demethylation in the adult brain. Cell 145:423–434
Ha HC, Snyder SH (1999) Poly(ADP-ribose) polymerase is a mediator of necrotic cell death by ATP depletion. Proc Natl Acad Sci U S A 96:13978–13982
Hermann A, Goyal R, Jeltsch A (2004) The Dnmt1 DNA-(cytosine-C5)-methyltransferase methylates DNA processively with high preference for hemimethylated target sites. J Biol Chem 279:48350–48359
Hill PW, Amouroux R, Hajkova P (2014) DNA demethylation, Tet proteins and 5-hydroxymethylcytosine in epigenetic reprogramming: an emerging complex story. Genomics 104:324–333
Ito S, D'Alessio AC, Taranova OV et al (2010) Role of Tet proteins in 5mC to 5hmC conversion, ES-cell self-renewal and inner cell mass specification. Nature 466:1129–1133
Ito S, Shen L, Dai Q et al (2011) Tet proteins can convert 5-methylcytosine to 5-formylcytosine and 5-carboxylcytosine. Science 333:1300–1303
Jeltsch A, Jurkowska RZ (2014) New concepts in DNA methylation. Trends Biochem Sci 39:310–318
Jones PA, Takai D (2001) The role of DNA methylation in mammalian epigenetics. Science 293:1068–1070
Kagiwada S, Kurimoto K, Hirota T et al (2013) Replication-coupled passive DNA demethylation for the erasure of genome imprints in mice. EMBO J 32:340–353
Khan JA, Forouhar F, Tao X et al (2007) Nicotinamide adenine dinucleotide metabolism as an attractive target for drug discovery. Expert Opin Ther Targets 11:695–705
Kohli RM, Zhang Y (2013) TET enzymes, TDG and the dynamics of DNA demethylation. Nature 502:472–479
Ling C, Del Guerra S, Lupi R et al (2008) Epigenetic regulation of PPARGC1A in human type 2 diabetic islets and effect on insulin secretion. Diabetologia 51:615–622
Masiello P, Novelli M, Fierabracci V et al (1990) Protection by 3-aminobenzamide and nicotinamide against streptozotocin-induced beta-cell toxicity in vivo and in vitro. Res Commun Chem Pathol Pharmacol 69:17–32
Muller U, Bauer C, Siegl M et al (2014) TET-mediated oxidation of methylcytosine causes TDG or NEIL glycosylase dependent gene reactivation. Nucleic Acids Res 42:8592–8604
Pacher P, Szabo C (2007) Role of poly(ADP-ribose) polymerase 1 (PARP-1) in cardiovascular diseases: the therapeutic potential of PARP inhibitors. Cardiovasc Drug Rev 25:235–260
Pandya KG, Patel MR, Lau-Cam CA (2010) Comparative study of the binding characteristics to and inhibitory potencies towards PARP and in vivo antidiabetogenic potencies of taurine, 3-aminobenzamide and nicotinamide. J Biomed Sci 17(Suppl 1):S16
Park JH, Stoffers DA, Nicholls RD et al (2008) Development of type 2 diabetes following intrauterine growth retardation in rats is associated with progressive epigenetic silencing of Pdx1. J Clin Invest 118:2316–2324
Paul DS, Teschendorff AE, Dang MA et al (2016) Increased DNA methylation variability in type 1 diabetes across three immune effector cell types. Nat Commun 7:13555
Pennarossa G, Maffei S, Campagnol M et al (2013) Brief demethylation step allows the conversion of adult human skin fibroblasts into insulin-secreting cells. Proc Natl Acad Sci U S A 110:8948–8953
Pirola CJ, Scian R, Gianotti TF et al (2015) Epigenetic modifications in the biology of nonalcoholic fatty liver disease: the role of DNA hydroxymethylation and TET proteins. Medicine (Baltimore) 94:e1480
Rakyan VK, Beyan H, Down TA et al (2011) Identification of type 1 diabetes-associated DNA methylation variable positions that precede disease diagnosis. PLoS Genet 7:e1002300
Reale A, Matteis GD, Galleazzi G et al (2005) Modulation of DNMT1 activity by ADP-ribose polymers. Oncogene 24:13–19

Schuhwerk H, Atteya R, Siniuk K et al (2016) PARPing for balance in the homeostasis of poly (ADP-ribosyl)ation. Semin Cell Dev Biol. https://doi.org/10.1016/j.semcdb.2016.09.011
Sookoian S, Rosselli MS, Gemma C et al (2010) Epigenetic regulation of insulin resistance in nonalcoholic fatty liver disease: impact of liver methylation of the peroxisome proliferator-activated receptor gamma coactivator 1alpha promoter. Hepatology 52:1992–2000
Stead LM, Brosnan JT, Brosnan ME et al (2006) Is it time to reevaluate methyl balance in humans? Am J Clin Nutr 83:5–10
Stepper P, Kungulovski G, Jurkowska RZ et al (2016) Efficient targeted DNA methylation with chimeric dCas9-Dnmt3a-Dnmt3L methyltransferase. Nucleic Acids Res 45(4):1703–713
Szabo C, Virag L, Cuzzocrea S et al (1998) Protection against peroxynitrite-induced fibroblast injury and arthritis development by inhibition of poly(ADP-ribose) synthase. Proc Natl Acad Sci U S A 95:3867–3872
Szabo C, Biser A, Benko R et al (2006) Poly(ADP-ribose) polymerase inhibitors ameliorate nephropathy of type 2 diabetic Leprdb/db mice. Diabetes 55:3004–3012
Toperoff G, Aran D, Kark JD et al (2012) Genome-wide survey reveals predisposing diabetes type 2-related DNA methylation variations in human peripheral blood. Hum Mol Genet 21:371–383
Virag L, Szabo C (2002) The therapeutic potential of poly(ADP-ribose) polymerase inhibitors. Pharmacol Rev 54:375–429
Volkmar M, Dedeurwaerder S, Cunha DA et al (2012) DNA methylation profiling identifies epigenetic dysregulation in pancreatic islets from type 2 diabetic patients. EMBO J 31:1405–1426
Williams KT, Garrow TA, Schalinske KL (2008) Type I diabetes leads to tissue-specific DNA hypomethylation in male rats. J Nutr 138:2064–69
Wurzer G, Herceg Z, Wesierska-Gadek J (2000) Increased resistance to anticancer therapy of mouse cells lacking the poly(ADP-ribose) polymerase attributable to up-regulation of the multidrug resistance gene product P-glycoprotein. Cancer Res 60:4238–4244
Yokochi T, Robertson KD (2002) Preferential methylation of unmethylated DNA by Mammalian de novo DNA methyltransferase Dnmt3a. J Biol Chem 277:11735–11745
Yokomori N, Tawata M, Onaya T (1999) DNA demethylation during the differentiation of 3T3-L1 cells affects the expression of the mouse GLUT4 gene. Diabetes 48:685–690
Yu W, Ginjala V, Pant V et al (2004) Poly(ADP-ribosyl)ation regulates CTCF-dependent chromatin insulation. Nat Genet 36:1105–1110
Zampieri M, Passananti C, Calabrese R et al (2009) Parp1 localizes within the Dnmt1 promoter and protects its unmethylated state by its enzymatic activity. PLoS One 4:e4717
Zardo G, D'Erme M, Reale A et al (1997) Does poly(ADP-ribosyl)ation regulate the DNA methylation pattern? Biochemistry 36:7937–7943

Extra Virgin Olive Oil and Corn Oil and Epigenetic Patterns in Breast Cancer

98

Raquel Moral and Eduard Escrich

Contents

Abstract

Breast cancer is the leading neoplasia in women worldwide. Nutrition and especially dietary lipids can influence mammary carcinogenesis through multiple mechanisms. This works aims to get insight into the effects of two common oils,

R. Moral (✉) · E. Escrich
Multidisciplinary Group for the Study of Breast Cancer, Department of Cell Biology, Physiology and Immunology, Physiology Unit, Faculty of Medicine, Universitat Autònoma de Barcelona, Barcelona, Spain
e-mail: Raquel.Moral@uab.cat; raquel.moral1908@gmail.com; Eduard.Escrich@uab.cat; gr.mecm@uab.cat

V. B. Patel, V. R. Preedy (eds.), *Handbook of Nutrition, Diet, and Epigenetics*, https://doi.org/10.1007/978-3-319-55530-0_15

extra virgin olive oil (EVOO) and corn oil, on mammary carcinogenesis and the molecular mechanisms of such effects. The administration of a diet high in corn oil (HCO) from weaning had a clear stimulating effect on 7,12-dimethylbenz(a) anthracene-induced mammary carcinogenesis, increasing the morphological and clinical degree of tumor malignancy, while a high-EVOO diet has a weak tumor-enhancing effect. The HCO diet modified gene expression profiles in mammary gland and tumors, downregulating genes with a role in apoptosis and immune system. On the contrary, the high-EVOO diet mainly modulated genes with a role in metabolism. These effects may be a consequence of an influence on the epigenetic machinery. Thus, the high-EVOO diet increased global DNA methylation in the mammary gland, mainly around puberty, and also in experimental mammary tumors. In relation to gene-specific methylation, the HCO diet, but not the high EVOO one, increased the total DNA methyltransferase activity in mammary glands and tumors, concomitantly with the increase in Rassf1a and Timp3 promoter methylation. Both high-fat diets may influence the modification of histones (the levels of H3K4me2, H3K27me3, H4K16ac, and H4K20me3), especially in the mammary gland. Although there is little data reported at other epigenetic levels, the differential effects of the diets are likely to be also due to different modification of microRNA patterns. Considering the unspecific tumor-promoting effect of all high-fat diets, the results suggest some beneficial effect of EVOO that counteracts the deleterious influence of excessive fat intake. The EVOO minor components may have a key role in such beneficial effects modulating, at least in part, the epigenetic machinery.

Keywords

Breast cancer · Mediterranean diet · High-fat diets · Extra virgin olive oil · N-6 PUFA · DMBA · Experimental mammary tumors · Mammary gland · Global DNA methylation · DNMT activity · Rassf1a · Timp3 · Histone H3 · Histone H4

List of Abbreviations

DMBA 7,12-Dimethylbenz(a)anthracene
DNMT DNA methyltransferase
EVOO Extra virgin olive oil
H3K4me2 Dimethylation at lysine 4 of histone H3
H3K27me3 Trimethylation at lysine 27 of histone H3
H4K16ac Acetylation at lysine 16 of histone H4
H4K20me3 Trimethylation at lysine 20 of histone H4
HCO High corn oil
HDAC Histone deacetylase
HEVOO High extra virgin olive oil
LF Low fat
MUFA Monounsaturated fatty acid
PUFA Polyunsaturated fatty acid

Introduction

Breast cancer is the most frequent malignant neoplasia in women worldwide with increasing incidence rates in all countries (Ferlay et al. 2015). This neoplasia is a heterogeneous and multifactorial disease, with several factors acting simultaneously and/or sequentially in all the steps of the carcinogenesis process, i.e., genetic and epigenetic, endocrine, and environmental factors. Geographical variation of incidence rates suggests an important contribution of lifestyle, especially diet and nutrition, in its etiology (WCRF/AICR 2007).

The complex process by which a normal mammary cell becomes neoplastic involves profound changes in the function of a myriad of genes. Such changes are elicited by genetic and epigenetic alterations, which are the basis of the acquisition of the different capacities for cell transformation into malignant cancer (Hanahan and Weinberg 2011). In mammary carcinogenesis, there has been described disruption of epigenetic patterns at all levels, including aberrant DNA methylation, histone modifications, and microRNA profiles (Fraga et al. 2005; Veeck and Esteller 2010). The fact that epigenetic events are heritable and reversible provides a mechanistic link for the environmental influence on cell biology in health and disease. It has long been reported that dietary factors such as lipids may modify epigenetic events regulating metabolism genes (Burdge and Lillycrop 2014), but little is known about their implication in the control of genes with a role in breast cancer. This work is focused on the effects that diets rich in two commonly used oils (olive and corn oils) have on experimental mammary carcinogenesis and if such effects are accompanied by changes in the epigenetic profiles of mammary glands and tumors.

Breast Cancer and Dietary Lipids

Epidemiological and experimental studies have demonstrated the influence of nutritional factors, especially dietary lipids, on the development of some neoplasias including breast cancer (Escrich et al. 2006; WCRF/AICR 2007). Although some analyses in humans have generated conflicting results, prospective cohort studies associating food patterns with breast cancer risk have shown an effect of saturated and total fat intake (Schulz et al. 2008; Sieri et al. 2008). Furthermore, experimental assays have provided evidence of the modulatory influence of lipids on the susceptibility of the mammary gland to malignant transformation, and this influence depends not only on the total amount but also on the type of dietary fat. Hence, diets rich in n-6 polyunsaturated fatty acids (PUFA) have a strong stimulating effect on mammary carcinogenesis. Saturated fats and *trans*-fatty acids are also stimulators. On the contrary, mainly n-3 PUFA but also gamma-linolenic acid and conjugated linoleic acid have shown antiproliferative properties on cancer cells. The influence of monounsaturated fatty acids (MUFA) still remains unclear, and studies have reported from weak promoting to protective effects on experimental mammary carcinogenesis (Escrich et al. 2006, 2011). Olive oil is rich in the n-9 MUFA oleic acid, and there is a large body of evidence of its health benefits (Quiles et al. 2006).

Table 1 Minor components in extra virgin olive oil. Groups and classes of the representative minor components found in EVOO (detailed in Quiles et al. 2006)

Non-glyceryde esters and waxes
Aliphatic alcohols
Volatile compounds: aldehydes, ketones, alcohols, acids, esthers, etc.
Triterpene alcohols: erythrodiol, uvaol
Sterols: β-sitosterol, campesterol, stigmasterol, avenasterol
Hydrocarbons
Squalene
Carotenoids: β-carotene, lycopene
Volatile hydrocarbons: phenanthrene, pyrene, fluoranthene
Pigments
Chlorophylls
Pheophytins
Lipophilic phenolics
Tocopherols
Tocotrienols
Hydrophilic phenolics
Phenolic acids: gallic, vanillic, cinnamic, caffeic, coumanic acids
Phenolic alcohols: hydroxytyrosol, tyrosol, and their glucosides
Secoiridoids: oleuropein and ligstroside derivatives (such as oleocanthal)
Lignans: pinoresinol
Flavonoids: apigenin, luteolin

Such benefits have been related to its high MUFA content but also to its many minor but highly bioactive compounds (Table 1). Olive oil is the main source of fat in the Mediterranean diet, and this dietary pattern has been traditionally linked to a protective effect on some chronic diseases such as cancer, obesity, inflammatory, and cardiovascular diseases (Sofi et al. 2010; Couto et al. 2011). Actually, prospective studies have associated the Mediterranean dietary pattern with a reduction of the breast cancer risk (Couto et al. 2011).

One of the most widely used experimental models of mammary carcinogenesis is the one induced in female rats using dimethylbenz(a)anthracene (DMBA). The mammary tumors generated with this carcinogen are predominantly adenocarcinomas resembling in pathogenesis, morphology, hormone dependence, and molecular features to human breast tumors (Escrich 1987; Russo and Russo 1996; Costa et al. 2002). Using this model, it has been observed different effects of diets high in extra virgin olive oil (EVOO) or in corn oil (rich in n-6 PUFA) on clinical behavior and histopathological features of mammary tumors. The results obtained in 18 experimental series point to the conclusion that dietary fat would not modify key classification features of human breast cancer (e.g., hormone receptors, amplification of HER2). On the other hand, dietary lipids act through multiple, complex, and lipid-specific mechanisms (Fig. 1). Such mechanisms include advanced growth and sexual maturation (Moral et al. 2011), alteration of hepatic carcinogen detoxification (Manzanares et al. 2015), and molecular changes in mammary gland and tumor, such as in membrane composition, in the signaling pathways driving to modifications in the proliferation/apoptosis balance (Solanas et al. 2010), in differentiation (Escrich et al. 2004; Moral et al. 2008), in oxidative stress, in DNA damage (Solanas et al. 2010), and in gene expression profiles (Escrich et al. 2004; Moral et al. 2016).

Effects of Extra Virgin Olive Oil and Corn Oil on DMBA-Induced Carcinogenesis and Gene Expression Profile

Dietary intervention with a high extra virgin olive oil diet (HEVOO, with 3% corn oil +17% extra virgin olive oil -w/w-) and a high-corn oil diet (HCO, with 20% corn oil -w/w-), both containing 39.5% calories in the form of fat, affected morphological, clinical, and molecular development of DMBA-induced tumors (Table 2). Sprague-Dawley rats were fed with the experimental diets from weaning and induced with 5 mg of DMBA on 53 days of age. EVOO contained 73.7% of the MUFA oleic acid, while the corn oil contained 51.3% of the n-6 PUFA linoleic acid (Moral et al. 2011). Hence, adenocarcinomas from animals fed the HCO diet displayed anatomopathological characteristics of high malignancy (high nuclear and pattern grade, high mitotic activity, stromal reaction, necrotic areas, and overall histopathologic grade). Moreover, clinical parameters of the carcinogenesis showed an acceleration of the disease, i.e., the group fed with such n-6 PUFA diet showed the shortest latency time (earliest onset of tumor appearance), and the highest percentage of tumor-bearing animals and total mammary adenocarcinomas (Table 2, Fig. 2). In contrast, the high-EVOO diet had a weak stimulating effect. In previous experiments, animals fed the HEVOO diet showed clinical manifestations of the disease similar to those of control group or intermediate between control and HCO diet groups (Escrich et al. 2006; Moral et al. 2011, 2016; Solanas et al. 2010). Thus, while diets rich in n-6 PUFA had a strong and clear stimulating effect on experimental mammary carcinogenesis, the weaker and more variable effect (depending on the studied parameter) of diets rich in EVOO is probably related to the different varieties of this oil. In any case, considering that all high-fat diets have an unspecific promoting influence on carcinogenesis due to the high content of lipids (Escrich et al. 2006), the EVOO must have some health benefits that may partly counteract the effect of the intake of high amounts of fat.

One of the mechanisms by which dietary lipids may influence the susceptibility or resistance of the mammary gland to carcinogenesis is altering gene expression. In fact, the administration of HEVOO and HCO differentially modified gene expression profiles of the gland at different life stages (36, 51, 100, and 246 days of age). These ages are of interest when studying the malignant transformation and promotion of DMBA-induced breast tumors. At 36 days, just after the puberty onset, the mammary gland is actively proliferating and developing. Such active period extends

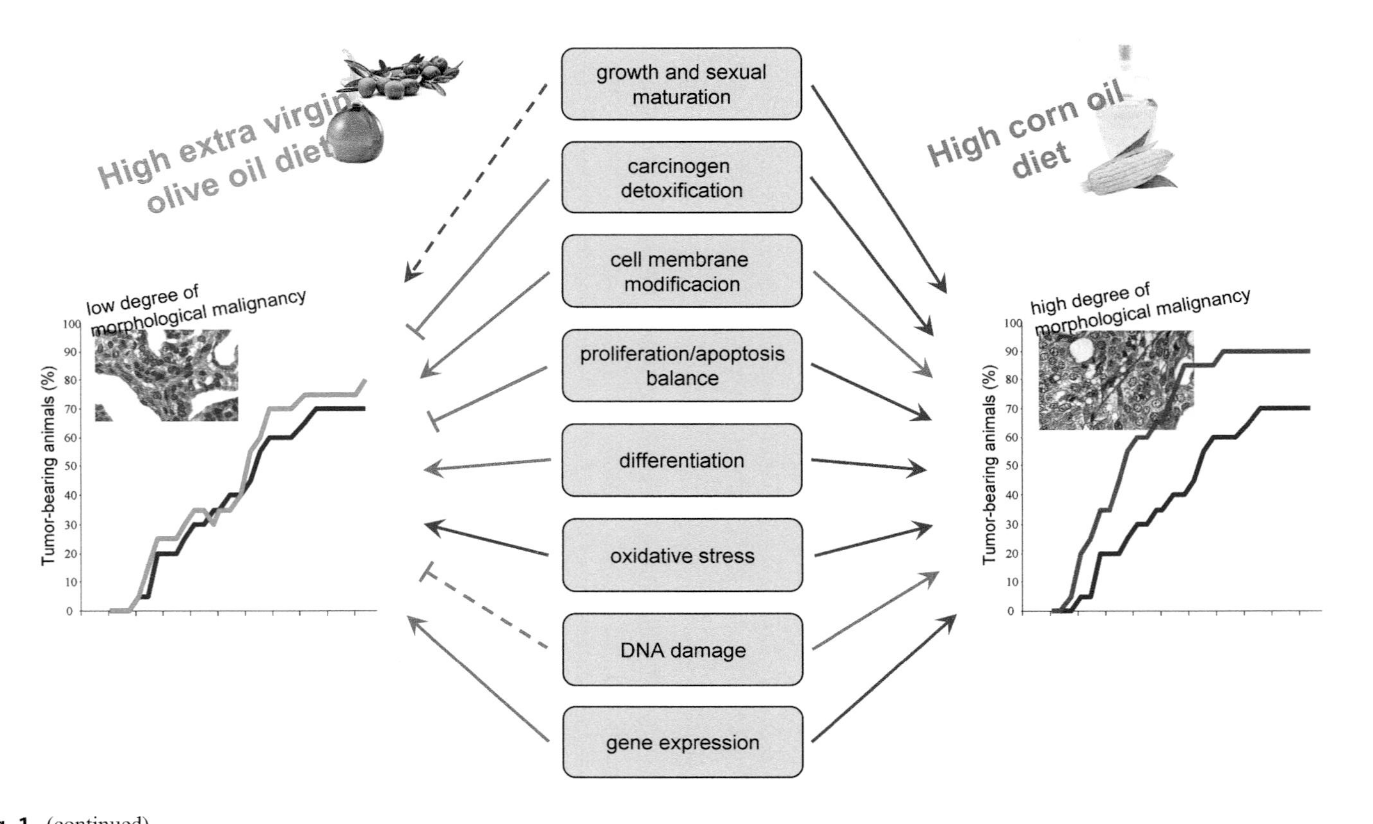

Fig. 1 (continued)

Table 2 Effects of experimental diets on morphological and clinical parameters of DMBA-induced mammary carcinogenesis. Histological degree of tumor malignancy was categorized using the modified Scarff-Bloom-Richardson (SBR) method (Costa et al. 2002). The group of animals fed with the high-corn oil diet (HCO) had the lowest percentage of low-degree tumors (SBR3/4/5) and the highest percentage of high-degree tumors (SBR9/10/11). Carcinogenesis parameters also indicated a more aggressive behavior in the group fed the HCO diet, since tumors appeared earlier (lowest latency time), and there were more affected animals and total number of tumors. *: $p < 0.05$ compared to low-fat (LF) diet group; Mann-Whitney U test

	LF	HEVOO	HCO
Histological degree (SBR score)			
Low (SBR3/4/5, % of tumors)	57.4	36.5	23
Medium (SBR6/7/8, % of tumors)	29.8	42.3	44
High (SBR9/10/11, % of tumors)	12.8	21.1	33
Latency time (days)	97	89	71.5
Tumor-bearing animals (%, n)	80 (16/20)	75 (15/20)	100 (20/20)*
Tumor yield (number of tumors)	46	58	100*

to 51 days, time point of particular relevance since it is within the window of maximum susceptibility of the gland to transformation (Russo and Russo 1996), and 2 days before the chemical induction with DMBA. Around 100 days of age (i.e., 47 days after induction), carcinogenesis starts manifesting clinically in all groups. In this experimental model, and with the dose used (5 mg of DMBA), the assays are classically extended to 200–250 days.

Analysis of the transcriptome profile of the mammary gland indicated a significant effect of both diets (stronger in the case of HCO) especially short after dietary intervention (36 days of age, 12–13 days on the experimental diets), while differences were smaller as intervention extends to the adaptation to chronic consumption. At different ages, both high-fat diets co-regulated genes related to metabolism. Regarding specific sequences, not commonly modified, the HCO diet downregulated genes with roles in immune system function and in apoptosis, while the HEVOO diet changed the expression of metabolism genes. In tumor tissues, there were also

Fig. 1 Mechanisms of the differential effects of a high-EVOO diet and a high-corn oil diet on DMBA-induced mammary carcinogenesis. Animals fed with a high-EVOO diet (HEVOO, *green line*) showed a similar degree of tumor morphological malignancy and percentage of affected animals than the controls fed with a low-fat diet (*blue line*), while animals fed with a high-corn oil diet (HCO, *red line*) presented more aggressive morphological and clinical manifestation of the disease. Mechanisms of such differential effects included advance in growth and sexual maturation, especially by the HCO diet, alteration in liver metabolism resulting in increased carcinogen detoxification by the HEVOO diet while decreased by the HCO diet, modifications in cell membrane composition by effect of both diets, changes in signaling pathways driving to increased apoptosis by the HEVOO diet while increased proliferation by the HCO diet, decreased expression of differentiation genes by the HCO diet, increased oxidative stress by both diets, changes in monoubiquitinated-PCNA suggesting lower DNA damage by the HEVOO diet, and modifications by both diets, especially the HCO, in gene expression profiles

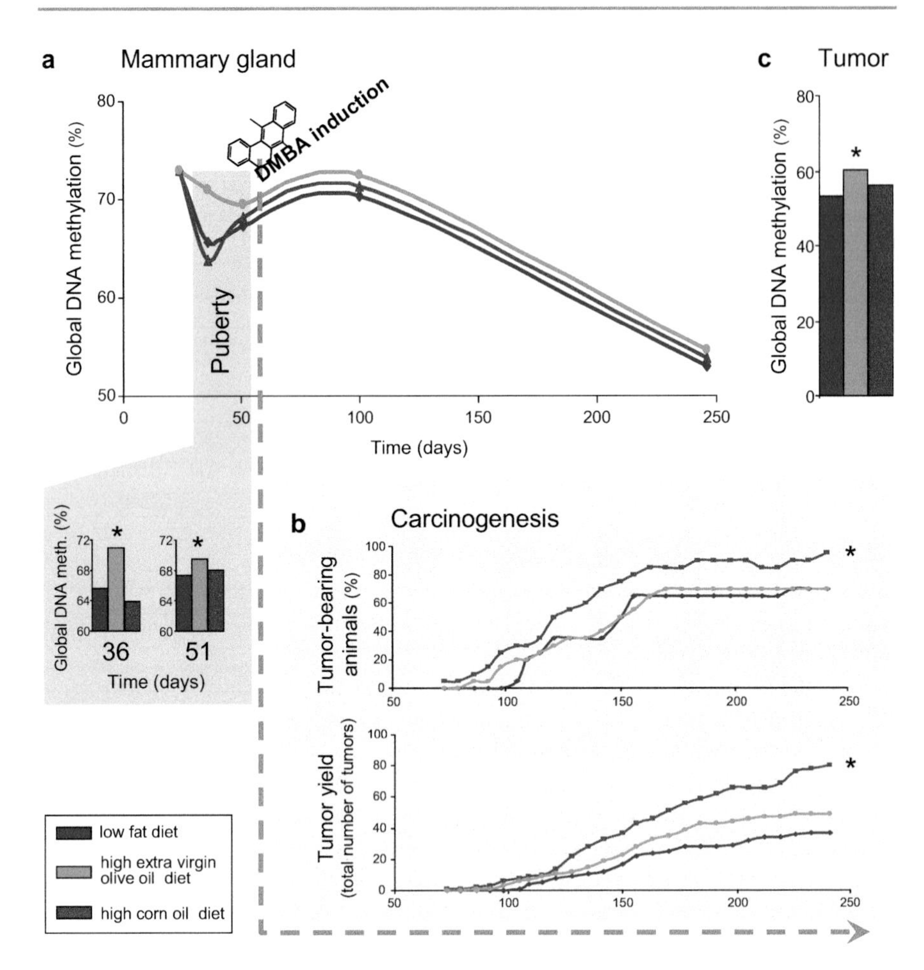

Fig. 2 Effects of high-EVOO and high-corn oil diets on global DNA methylation and carcinogenesis. (**a**) Global DNA methylation (%, median) in mammary gland at 24, 36, 51, 100, and 246 days of age. Histograms detail the values at ages around puberty (36 and 51 days). *: $p < 0.1$ compared to HCO (36 days) or compared to LF (51 days); Mann-Whitney U test. (**b**) Evolution of carcinogenesis parameters (tumor-bearing animals and tumor yield, medians) from day 74 to the end of the assay. *: $p < 0.05$ compared to LF and HEVOO groups; Friedman test. (**c**) Global DNA methylation in the experimental mammary tumors at 246 days. *: $p < 0.05$ compared to LF group; Mann-Whitney U test

evidence of the differential effect of these high-fat diets. Although the number of detected modified genes was low (probably due to the high variability among tumors), analysis of biological significance and functional clustering revealed similar enriched categories of genes compared to the mammary gland. Specifically the HCO diet downregulated genes associated with immune response and cell death, while the diet rich in EVOO upregulated genes with a role in proliferation and cell death (Moral et al. 2016).

Effects of Extra Virgin Olive Oil and Corn Oil on Global DNA Methylation in Mammary Gland and Experimental Tumors

The mammary gland undergoes profound remodeling through puberty, reproductive life, and aging (Russo and Russo 1996). Although there is evidence that gene expression profiles modify during physiologic mammary development and in malignant transformation, scarce data have been published in relation to the changes in the epigenetic profiles of this tissue through lifelong development.

The study of the global DNA methylation in rat mammary gland showed variations at different life stages, independently of the dietary administration (Fig. 2a). Global DNA methylation decreased around puberty (from postweaning age of 23 days to pubertal age of 36 days) and also with aging (from young adult of 100 days to 246 days). Puberty is a key developmental period of the mammary gland, and the decrease in methylation at this stage may be related to the high proliferation rate and the remodeling of the tissue. Interestingly, the end of this period (around 50 days of age) is the most susceptible age to induce mammary carcinogenesis in this model (Russo and Russo 1996). Moreover, the age-related decrease in mammary global DNA methylation in adulthood is in accordance with the gradual loss of methylation with aging reported in human tissues (Sierra et al. 2015), although no data have been published in this model.

In addition to the variations throughout lifetime, dietary intervention with high-fat diets modified global DNA methylation in mammary gland and tumors. Animals fed with the high-EVOO diet showed higher levels of methylated DNA, especially at puberty (close to significance compared to animals fed HCO at 36 days and compared to animals fed the low-fat diet at 51 days; Fig. 2a). As already mentioned, puberty is a critical window of susceptibility for mammary malignant transformation. Considering that global DNA hypomethylation is associated to chromosome instability, cell transformation, and tumor progression (Eden et al. 2003), the higher levels of global DNA methylation at puberty by the effect of the HEVOO diet may decrease the vulnerability of the mammary gland to chemically induced carcinogenesis. In accordance with this, clinical and histological features of tumors from the HEVOO group showed lower degree of malignancy than those from the HCO group (Fig. 2b, Table 2). The high-EVOO diet, despite being high in fat, elicited similar tumor behavior than the control low-fat diet, while the HCO clearly increased the percentage of tumor-bearing animals and tumor yield. Moreover, DNA methylation levels in tumors were also significantly increased in HEVOO group than in the control group (Fig. 2c). Since hypomethylation is considered a hallmark of cancer cells, the higher global DNA methylation in tumors from this group is also in accordance with a low degree of tumor aggressiveness.

This influence of EVOO decreasing global DNA hypomethylation in tumors has been observed with two different dietary interventions: administering the HEVOO diet from weaning or feeding the rats with the low-fat diet until carcinogen induction (day 53 of age) and thereafter with the HEVOO diet (Rodríguez-Miguel et al. 2015). Thus, the data obtained in mammary glands and tumors and with different timing of exposure indicated an influence of EVOO even after the carcinogenic insult occurred, suggesting a beneficial effect on malignant transformation but also on tumor progression.

Paradoxically, global methylation was not decreased in DMBA-induced tumor tissues in comparison with mammary glands. Although there is no published data using this model and the interpretation is unclear, these results may be related to other observations suggesting altered global methylation in experimentally induced rat mammary glands (Starlard-Davenport et al. 2010; Kutanzi and Kovalchuk 2013).

Effects of Extra Virgin Olive Oil and Corn Oil on Gene-Specific Methylation in Mammary Gland and Experimental Tumors

There is a wealth of evidence on the important role of gene-specific hypermethylation on breast cancer. Environmental factors affecting tumorigenesis processes are likely to disrupt methylation patterns of tumor suppressor genes. DNA methylation is catalyzed by the highly conserved family of enzymes DNA methyltransferases (DNMT). Thus, determination of total DNMT activity in mammary glands and DMBA-induced tumors showed a significant increase in tumors in comparison with the glands, resembling the increase in activity reported in human breast cancer (Veeck and Esteller 2010). Interestingly, the high-corn oil diet significantly increased the total DNMT activity in mammary gland and tumor, when compared with the low-fat diet, but also when compared with the high-EVOO diet (Fig. 3a). This increase in DNMT activity was not a consequence of the up-modulation of mRNA levels of the distinct isoforms with catalytic activity (DNMT1, DNMT3a, DNMT3b) (Rodríguez-Miguel et al. 2015), suggesting a stimulation of the activity rather than a modulation of the protein expression.

Epigenetic silencing of Ras-association domain family 1 isoform A (Rassf1a) is one of the most common molecular changes in human cancers and is considered a frequent and early event in breast cancer (Hesson et al. 2007). Rassf1a has an important role in cell cycle and apoptosis. Tissue inhibitor of metalloproteinase-3 (Timp3) avoids degradation of the extracellular matrix and is also a gene frequently hypermethylated in human breast cancer (Radpour et al. 2009). However, there is no data on the role of the silencing of these genes in DMBA-induced mammary carcinogenesis. Analysis of the methylation levels of their promoters in rat mammary glands and experimental tumors showed an increase in tumor tissues in relation to the gland for both genes, what suggests the importance of silencing these genes in experimental transformation. Interestingly, methylation levels were influenced by dietary intervention. The administration from weaning of the HCO diet significantly increased Rassf1a and Timp3 promoter methylation both in mammary gland and tumor, while the HEVOO diet only increased the promoter methylation of Rassf1a in tumors (Fig. 3b). Differences in the effects of both isocaloric high-fat diets suggest again a key role of the specific composition in fatty acids and minor compounds of corn oil and EVOO. In this sense, hydroxytyrosol and oleuropein, the most representative polyphenols in EVOO, are able to regulate by epigenetic mechanisms the expression of the tumor suppressor gene CB1 in Caco-2 cells and in rat colon

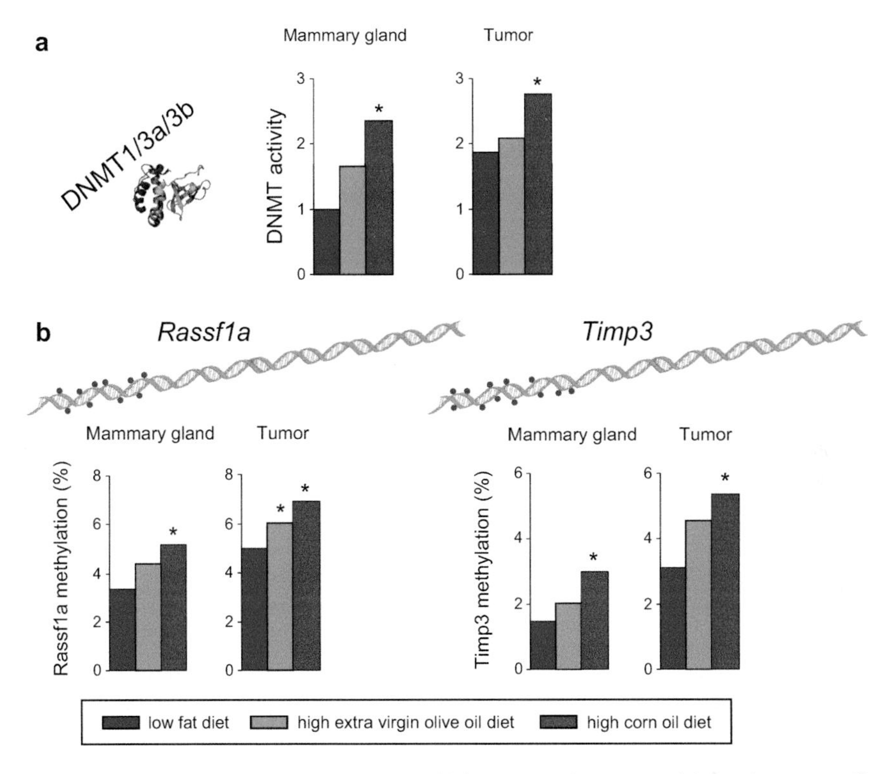

Fig. 3 Effects of high-EVOO and high-corn oil diets on total DNMT activity and gene-specific methylation. (**a**) Determination of total DNA methyltransferase (*DNMT*) activity (medians) in the mammary glands and tumors from animals fed the experimental diets. Values depicted are relative to the median level in the mammary gland from the control group. *: $p < 0.05$ compared to LF and HEVOO groups; Mann-Whitney U test. (**b**) Analysis by pyrosequencing of the methylation levels (%, median) of Rassf1a and Timp3 promoter methylation in mammary glands and tumors. *: $p < 0.05$ compared to LF group; Mann-Whitney U test

(Di Francesco et al. 2015). No data have been reported about the influence that oleic or linoleic acids may exert on gene-specific methylation in breast cancer cells.

Although the methylation levels of Rassf1a and Timp3 were differentially influenced by the high-EVOO or high-corn oil diet, further gene expression analysis by real-time PCR showed little effect of such diets in mammary glands and no clear correlation between promoter methylation and gene expression (Rodríguez-Miguel et al. 2015). Since epigenetic silencing of Rassf1a and Timp3 is considered early events in the carcinogenesis multistep process, methylation levels of these specific genes long after transformation may not reflect the degree of clinical and morphological tumor malignancy. In any case, these results suggest that the high-corn oil diet had a role in carcinogenesis increasing DNMT activity, which was reflected in the higher methylation of Rassf1a and Timp3, but should also be affecting other genes. The inhibition of different genes would contribute in some step to the acquisition of the hallmarks of cancer, such as resisting cell death or evading growth suppressors, although once the genes are silenced they may become less important in cancer progression.

Effects of Extra Virgin Olive Oil and Corn Oil on Histone Modifications in Mammary Gland and Experimental Tumors

Postraductional modifications in histones, specially affecting lysine (K) residues, have a key role in transformation. Aberrant histone modifications such as global decreases of acetylation at K16 and trimethylation at K20 of H4 are considered hallmarks of human cancer (Fraga et al. 2005). Moreover, global decrease in the methylation of H3 is frequently detected in human breast cancer, such as dimethylation at K4 and trimethylation at K27 (Greer and Shi 2012).

There is some evidence that different components of dietary oils may modulate the histone-modifying machinery, thus changing the global levels of histone modifications, but limited data have been reported in breast tumor cells. As an example, in MCF7, T47D, and MDA-MB-231 cells, treatment with n-3 PUFA, but not with n-6 PUFA, downregulated EZH2. This protein is the catalytic subunit of the polycomb repressive complex 2, responsible for the trimethylation of H3K27, and is frequently dysregulated in cancer cells (Dimri et al. 2010). Moreover, several minor components of extra virgin olive oil have demonstrated inhibition of histone deacetylase (HDAC) activity in breast cancer cells (Tseng et al. 2017).

In the rat DMBA-induced mammary cancer model, determination of posttranslational modifications of histone H3 (H3K4me2, H3K27me3) and histone H4 (H4K20me3, H4K16ac) in nuclear extracts showed lower levels of all four modifications in experimental tumors compared with mammary glands at 246 days of age (Fig. 4), indicating the aberrant histone modification also occurring in experimental tumorigenesis. In relation to the influence of the high-fat diets, in mammary glands a significant reduction in H3K27me3 was found by the effect of HCO diet. Although not statistically significant, alterations in H4K20me3 (increased) and H4K16ac (decreased) by the high-EVOO diet in mammary gland were observed. One of the enzymes with a role in deacetylation of H4K16 is Sirt6 (Han et al. 2015). Interestingly, oleoylethanolamide, an *N*-acylethanolamine derived from oleic acid, increases Sirt6 activity (Rahnasto-Rilla et al. 2016). In the experimental mammary tumors, no clear differences have been found by the effect of the high-EVOO or HCO diets when administered from weaning (Fig. 4). However, when the EVOO diet was administered from induction onward, a decrease in the global levels of H4K20me3 was observed (Rodríguez-Miguel et al. 2015). Thus, the results suggested a disruption in the histone modifications pattern by influence of high-fat diets on experimental carcinogenesis. These modifications, in the context of different DNA methylation levels, could contribute to the different gene expression profiles observed in the mammary gland and tumors of the animals subjected to dietary intervention (Moral et al. 2016).

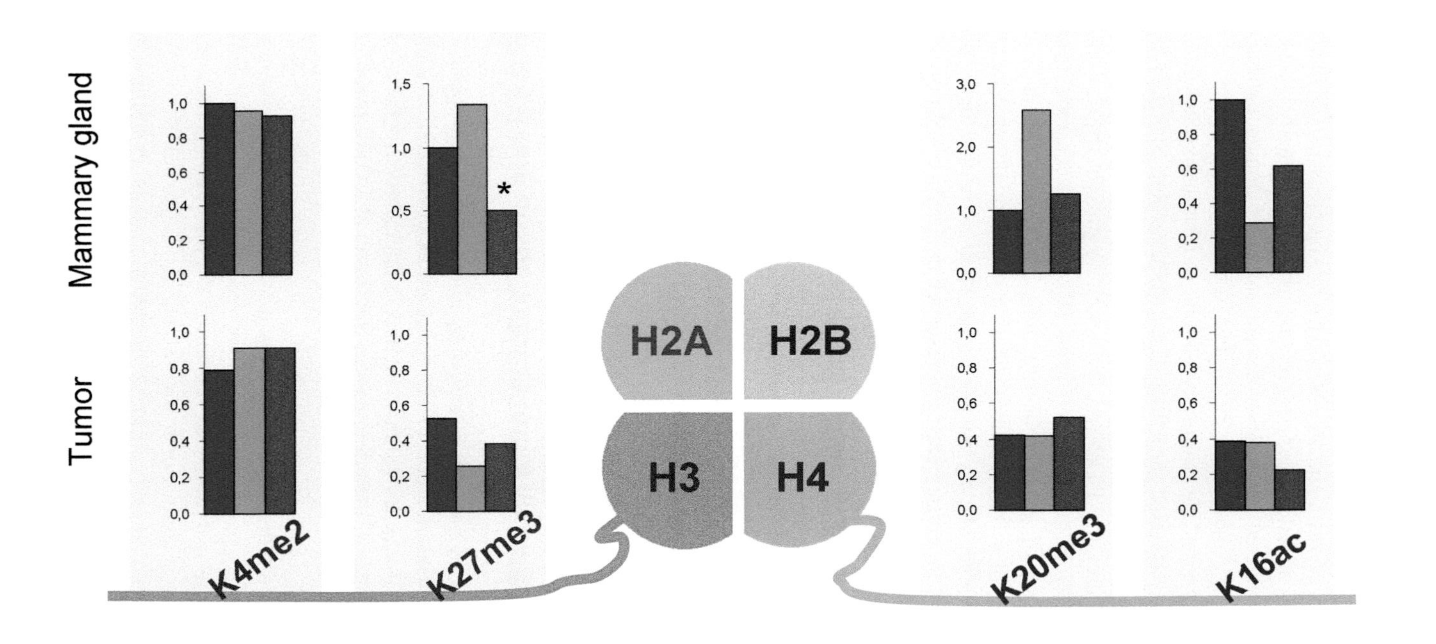

Fig. 4 Effects of high-EVOO and high-corn oil diets on H3 and H4 modifications. Analysis by Western blot of the levels of dimethylation at lysine 4 and trimethylation at lysine 27 of histone H3 (H3K4me2, H3K27me3) and trimethylation at lysine 20 and acetylation at lysine 16 of histone H4 (H4K20me3, H4K16ac). For each sample the levels of each modification have been relativized to the total level of the histone (H3 or H4). For each histone modification, values depicted are relative to the median level in the mammary gland from the control group. *: $p < 0.05$ compared to LF group; Mann-Whitney U test

Influence of Olive Oil and Other Dietary Lipids on MicroRNA Expression Patterns in Cancer

It is likely that the effects of olive oil and other dietary lipids on breast cancer risk are also mediated by changes in microRNA expression patterns. Actually, there is evidence that EVOO modulated microRNA transcriptome in human peripheral blood mononuclear cells (D'Amore et al. 2016) and in mouse brain (Luceri et al. 2017), but there is no data regarding mammary tumors. In relation to other dietary lipids, in vitro studies have assessed the effect of docosahexaenoic acid (DHA), a long-chain n-3 PUFA. In the breast cancer cell lines MDA-MB-231 and MCF-7, DHA blocked miR-21, increasing PTEN protein levels to prevent expression of CSF-1 (Mandal et al. 2012). In the same cell lines, DHA altered exosome microRNAs, especially the levels of let-7a, miR-23b, miR-27a/b, miR-21, let-7, and miR-320b, which are known to have antiproliferative or anti-angiogenic activity (Hannafon et al. 2015). Other dietary interventions, such as caloric restriction, altered microRNA patterns in mouse breast tissues, such as the levels of miR-29c, miR-203, miR-150, and miR-30 (Ørom et al. 2012).

There are also scarce data addressing the effects of olive oil and other dietary lipids in other types of cancers. As an example, the only study related to olive oil was performed with olive tree leaf extract (richer than olive oil in the phenolic-type oleuropein). This extract induced the expression of miR-153, miR-145, and miR-137 in stemlike cells of glioblastoma multiforme tumors (Tezcan et al. 2014). In addition, treatment of glioblastoma cells with different types of PUFAs (GLA, AA, and DHA) also resulted in altered expression of microRNAs (Faragó et al. 2011). On the other hand, diets rich in n-3 and in n-6 PUFA differentially modified the microRNA profile in experimental colon tumors (Davidson et al. 2009).

Epigenetic Effects of Olive Oil Minor Compounds on Breast Cancer

As already mentioned, there is a wealth of evidence on the favorable properties of olive oil minor compounds, in addition to its high content in MUFA, on health. Many bioactive components have antioxidant and anti-inflammatory properties both associated with chronic diseases, including cancer (Quiles et al. 2006). In general, compounds rich in antioxidants may reduce oxidative stress levels, thus altering histone modifications and DNA methylation. In this sense, cells exposed to reactive oxygen species presented altered activity of histone demethylases and acetyltransferases as well as decreased DNA demethylase activity of the family TET (Niu et al. 2015). More specifically, in relation to the minor compounds more abundant in EVOO (terpenes, sterols, or hydrocarbons like squalene), there are no studies on epigenetic effects on breast cancer. On the other hand, other components, especially polyphenols, have demonstrated antitumor properties concomitantly with modifications of epigenetic patterns in breast cancer cells (Table 3). Caffeic acid, a component of EVOO phenolic acids, inhibited in vitro DNA methylation catalyzed by DNMT1 and decreased RARβ2 promoter methylation in MCF-7 and

Table 3 Epigenetic effects of EVOO minor components on breast cancer cells. Polyphenols found in the hydrophilic phenolic class of EVOO minor compounds that have demonstrated an epigenetic effect on breast cancer cells

Phenolic class	Compound	In vitro model	Epigenetic action	References
Phenolic acids	Caffeic acid	MCF-7 and MDA-MB-231	Inhibition of DNMT1 activity	Lee and Zhu (2006)
Secoiridoids	Secoiridoids extract	JIMT-1	Hyperacetylation of H3 at lysine 18 (H3K18)	Oliveras-Ferraros et al. (2011)
Flavonoids	Apigenin	MDA-MB-231	Inhibition of HDAC activity, increased H3 acetylation	Tseng et al. (2017)
	Luteolin	MCF-7	Decreased H4 acetylation at PLK-1 promoter	Markaverich et al. (2011)

MDA-MB-231 breast cancer cells (Lee and Zhu 2006). In JIMT-1 breast cancer cells, an EVOO phenolic extract rich in secoiridoids regulated histone deacetylase (HDAC) activity, increasing the levels of acetylated H3 (Oliveras-Ferraros et al. 2011). Flavonoids such as apigenin and luteolin have also demonstrated in vitro effects on the epigenetics machinery. Apigenin induced histone H3 acetylation through inhibition of HDAC in MDA-MB-231 cells (Tseng et al. 2017), while luteolin blocked the acetylation of histone H4 associated with the promoter of the cell cycle gene PLK1 in MCF7 cells (Markaverich et al. 2011).

Other polyphenols, abundant in other vegetable sources although represented in little quantities in some olive oil varieties, had strong effects on epigenetic mechanisms in breast cancer cells. That is the case of quercetin inhibiting p300 histone acetyltransferase activity in MDA-MB-231 and MCF7 cells (Xiao et al. 2011) or ellagic acid inhibiting the DNA methyltransferase activity in MCF7 cells (Paluszczak et al. 2010).

Dictionary of Terms

- **Mediterranean diet** – This diet includes a variety of food patterns from the Mediterranean region and is characterized by the consumption of abundant and varied plant foods (fruits, vegetables, legumes, and nuts), dairy products, fish, and olive oil as the principal source of fat.
- **Extra virgin olive oil** – The juice obtained from the first pressing of olives, through mechanical processes and without chemical substances.
- **Extra virgin olive oil minor compounds** – Several compounds representing 1–2% of oil weight, including over 230 components some of them highly bioactive. They are predominantly present in extra virgin oil but not in refined oil.
- **7,12-Dimethylbenz(a)anthracene** – Polycyclic aromatic hydrocarbon that in a single dose by oral gavage is able to evoke mammary tumors in rats highly similar to human breast cancers

- **Carcinogenesis** – Multistep process by which cancer is originated and includes initiation, promotion, and progression. In experimental models, clinical evolution of cancer can be monitored by different carcinogenesis parameters (latency time, percentage of tumor-bearing animals, tumor yield, or tumor volume).
- **Saturated fatty acid (SFA)** – Fatty acid without double bonds. Foods rich in SFA are animal fat (dairy products and fatty meat) and some vegetable oils (coconut or palm kernel oil).
- **Monounsaturated fatty acid (MUFA)** – Fatty acid with one double bond. They are mainly n-9 (omega-9 or ω-9; the double bond is the omega-9 position, i.e., the ninth bond from the end of the fatty acid). Foods rich in MUFA include vegetable oils (olive or canola oils) and nuts.
- **Polyunsaturated fatty acid (PUFA)** – Fatty acid with more than one double bond, mainly classified in n-6 (last double bond in omega-6 position) and n-3 (last double bond in omega-3 position). Fats rich in n-6 PUFA are mainly vegetable oils (sunflower or corn oils), while the richest in n-3 PUFA are oily fish and some vegetable oils (flaxseed oil).
- ***Trans*-fatty acids** – For unsaturated fatty acids (MUFA and PUFA), the acid is a *cis*-isomer if hydrogen atoms are on the same side of the double bond and a *trans*-isomer if hydrogen atoms are on opposite sides of the double bond. *Trans*-fatty acids are uncommon in natural sources but produced industrially from vegetable fats.

Key Facts of Breast Cancer

- The mammary gland, unlike most organs, remains highly undifferentiated at birth. With puberty the gland develops and undergoes profound remodeling, but differentiation is only completed at the end of the first full-term pregnancy.
- The windows of susceptibility are the periods in which mammary gland is especially vulnerable to environmental factors that influence breast cancer risk, such as the pubertal period.
- Breast cancer is the most frequent malignancy among women worldwide, with more than 1.6 million new cases diagnosed in 2012 (25% of all newly cases in women).
- In 2012 this neoplasia was the first cause of cancer death in women in less developed regions, while the second leading cause of cancer death, after lung cancer, in more developed regions.
- Only a small percentage of breast cancers are linked to inherited high-susceptibility genes such as BRCA. Most cases are caused by the accumulation of spontaneous somatic mutations of DNA and epigenetic alterations.
- There are clear geographical differences in incidence and prevalence rates of breast cancer as well as changes in the incidence among immigrants, suggesting an important influence of environmental and lifestyle factors.

Summary Points

- The administration of a high-corn oil diet from weaning has a clear stimulating effect on 7,12-dimethylbenz(a)anthracene-induced mammary carcinogenesis, increasing the morphological and clinical degree of tumor malignancy, while a high extra virgin olive oil (EVOO) diet has a weak tumor-enhancing effect.
- The high-corn oil diet modifies gene expression profiles in mammary glands at different ages and in tumors, downregulating genes with a role in apoptosis and immune function. The high-EVOO diet mainly modulates genes with a role in metabolism.
- The high-EVOO diet increases global DNA methylation in the mammary gland, mainly around puberty, which is compatible with a lower vulnerability of the gland to malignant transformation.
- The high-EVOO diet increases global DNA methylation in experimental tumors, which is concordant with the low degree of tumor anatomopathological malignancy.
- The high-corn oil diet, but not the high EVOO one, increases total DNA methyltransferase activity in mammary glands at 246 days of age and tumors, concomitantly with an increase in Rassf1a and Timp3 promoter methylation.
- Both high-fat diets may influence the modification of histones (the levels of H3K4me2, H3K27me3, H4K16ac, and H4K20me3), especially in the mammary gland.
- Independently of dietary intervention, in 7,12-dimethylbenz(a)anthracene-induced mammary carcinogenesis, disruption of epigenetic patterns occurs similarly to human breast cancer (higher DNA methyltransferase activity and gene-specific methylation, decreased levels of key histone modifications).
- Although there are scarce data reported on other mechanisms, the effects of these diets are likely to be also due to different modification on microRNA patterns.
- Considering the unspecific tumor-promoting effect of all high-fat diets, there is some beneficial effect of EVOO that counteracts the deleterious influence of excessive fat intake.
- The EVOO minor components may have a key role in the beneficial effects of this oil modulating, at least in part, the epigenetic machinery.

Acknowledgments Research in the authors' laboratory is funded by grants from "Plan Nacional de I+D+I" (AGL2006-07691; AGL2011-24778); "Fundación Patrimonio Comunal Olivarero (FPCO)" (FPCO2008-165.396; FPCO2013-CF611.084); "Agencia para el Aceite de Oliva del Ministerio de Medio Ambiente y de Medio Rural y Marino" (AAO2008-165.471); "Organización Interprofesional del Aceite de Oliva Español (OIAOE)" (OIP2009-CD165.646); "Departaments d'Agricultura, Alimentació i Acció Rural, i de Salut de la Generalitat de Catalunya" (GC2010-165.000); and FPCO and OIAOE (FPCO-OIP2016-CF614.087). The sponsors had no role in study designs, data collection and analyses, interpretation of results, preparation of the manuscript, and the decision to submit the manuscript for publication or the writing of the manuscript. The authors are grateful to I. Costa, R. Escrich, C. Rodríguez-Miguel, M.C. Ruiz de Villa, M. Solanas, and E. Vela for their collaboration in these studies.

References

Burdge GC, Lillycrop KA (2014) Fatty acids and epigenetics. Curr Opin Clin Nutr Metab Care 17:156–161

Costa I, Solanas M, Escrich E (2002) Histopathologic characterization of mammary neoplastic lesions induced with 7,12-dimethylbenz(alpha)anthracene in the rat: a comparative analysis with human breast tumors. Arch Pathol Lab Med 126:915–927

Couto E, Boffetta P, Lagiou P et al (2011) Mediterranean dietary pattern and cancer risk in the EPIC cohort. Br J Cancer 104:1493–1499

D'Amore S, Vacca M, Cariello M et al (2016) Genes and miRNA expression signatures in peripheral blood mononuclear cells in healthy subjects and patients with metabolic syndrome after acute intake of extra virgin olive oil. Biochim Biophys Acta 1861:1671–1680

Davidson LA, Wang N, Shah MS et al (2009) n-3 polyunsaturated fatty acids modulate carcinogen-directed non-coding microRNA signatures in rat colon. Carcinogenesis 30:2077–2084

Di Francesco A, Falconi A, Di Germanio C et al (2015) Extravirgin olive oil up-regulates CB1 tumor suppressor gene in human colon cancer cells and in rat colon via epigenetic mechanisms. J Nutr Biochem 26:250–258

Dimri M, Bommi PV, Sahasrabuddhe AA et al (2010) Dietary omega-3 polyunsaturated fatty acids suppress expression of EZH2 in breast cancer cells. Carcinogenesis 31:489–495

Eden A, Gaudet F, Waghmare A et al (2003) Chromosomal instability and tumors promoted by DNA hypomethylation. Science 300:455

Escrich E (1987) Validity of the DMBA-induced mammary cancer model for the study of human breast cancer. Int J Biol Markers 2:197–206

Escrich E, Moral R, García G et al (2004) Identification of novel differentially expressed genes by the effect of a high-fat n-6 diet in experimental breast cancer. Mol Carcinog 40:73–78

Escrich E, Solanas M, Moral R (2006) Olive oil, and other dietary lipids, in cancer: experimental approaches. In: Quiles JL, Ramírez-Tortosa MC, Yaqoob P (eds) Olive oil and health. CABI Publishing, Oxford, pp 317–374

Escrich E, Solanas M, Moral R et al (2011) Modulatory effects and molecular mechanisms of olive oil and other dietary lipids in breast cancer. Curr Pharm Des 17:813–830

Faragó N, Fehér LZ, Kitajka K et al (2011) MicroRNA profile of polyunsaturated fatty acid treated glioma cells reveal apoptosis-specific expression changes. Lipids Health Dis 10:173

Ferlay J, Soerjomataram I, Dikshit R et al (2015) Cancer incidence and mortality worldwide: sources, methods and major patterns in GLOBOCAN 2012. Int J Cancer 136:E359–E386

Fraga MF, Ballestar E, Villar-Garea A et al (2005) Loss of acetylation at Lys16 and trimethylation at Lys20 of histone H4 is a common hallmark of human cancer. Nat Genet 37:391–400

Greer EL, Shi Y (2012) Histone methylation: a dynamic mark in health, disease and inheritance. Nat Rev Genet 13:343–357

Han L, Ge J, Zhang L et al (2015) Sirt6 depletion causes spindle defects and chromosome misalignment during meiosis of mouse oocyte. Sci Rep 5:15366

Hanahan D, Weinberg RA (2011) Hallmarks of cancer: the next generation. Cell 144:646–674

Hannafon BN, Carpenter KJ, Berry WL et al (2015) Exosome-mediated microRNA signaling from breast cancer cells is altered by the anti-angiogenesis agent docosahexaenoic acid (DHA). Mol Cancer 14:133

Hesson LB, Cooper WN, Latif F (2007) The role of RASSF1A methylation in cancer. Dis Markers 23:73–87

Kutanzi K, Kovalchuk O (2013) Exposure to estrogen and ionizing radiation causes epigenetic dysregulation, activation of mitogen-activated protein kinase pathways, and genome instability in the mammary gland of ACI rats. Cancer Biol Ther 14:564–573

Lee WJ, Zhu BT (2006) Inhibition of DNA methylation by caffeic acid and chlorogenic acid, two common catechol-containing coffee polyphenols. Carcinogenesis 27:269–277

Luceri C, Bigagli E, Pitozzi V et al (2017) A nutrigenomics approach for the study of anti-aging interventions: olive oil phenols and the modulation of gene and microRNA expression profiles in mouse brain. Eur J Nutr 56:865–877

Mandal CC, Ghosh-Choudhury T, Dey N et al (2012) miR-21 is targeted by omega-3 polyunsaturated fatty acid to regulate breast tumor CSF-1 expression. Carcinogenesis 33:1897–1908

Manzanares MA, Solanas M, Moral R et al (2015) Dietary extra-virgin olive oil and corn oil differentially modulate the mRNA expression of xenobiotic-metabolizing enzymes in the liver and in the mammary gland in a rat chemically induced breast cancer model. Eur J Cancer Prev 24:215–222

Markaverich BM, Shoulars K, Rodriguez MA (2011) Luteolin regulation of estrogen signaling and cell cycle pathway genes in MCF-7 human breast cancer cells. Int J Biomed Sci 7:101–111

Moral R, Solanas M, Garcia G et al (2008) High corn oil and high extra virgin olive oil diets have different effects on the expression of differentiation-related genes in experimental mammary tumors. Oncol Rep 20:429–435

Moral R, Escrich R, Solanas M et al (2011) Diets high in corn oil or extra-virgin olive oil provided from weaning advance sexual maturation and differentially modify susceptibility to mammary carcinogenesis in female rats. Nutr Cancer 63:410–420

Moral R, Escrich R, Solanas M et al (2016) Diets high in corn oil or extra-virgin olive oil differentially modify the gene expression profile of the mammary gland and influence experimental breast cancer susceptibility. Eur J Nutr 55:1397–1409

Niu Y, DesMarais TL, Tong Z et al (2015) Oxidative stress alters global histone modification and DNA methylation. Free Radic Biol Med 82:22–28

Oliveras-Ferraros C, Fernández-Arroyo S, Vazquez-Martin A et al (2011) Crude phenolic extracts from extra virgin olive oil circumvent de novo breast cancer resistance to HER1/HER2-targeting drugs by inducing GADD45-sensed cellular stress, G2/M arrest and hyperacetylation of histone H3. Int J Oncol 38:1533–1547

Ørom UA, Lim MK, Savage JE et al (2012) MicroRNA-203 regulates caveolin-1 in breast tissue during caloric restriction. Cell Cycle 11:1291–1295

Paluszczak J, Krajka-Kuźniak V, Baer-Dubowska W (2010) The effect of dietary polyphenols on the epigenetic regulation of gene expression in MCF7 breast cancer cells. Toxicol Lett 192:119–125

Quiles JL, Ramírez-Tortosa MC, Yaqoob P (eds) (2006) Olive oil and health. CABI Publishing, Oxford

Radpour R, Kohler C, Haghighi MM et al (2009) Methylation profiles of 22 candidate genes in breast cancer using high-throughput MALDI-TOF mass array. Oncogene 28:2969–2978

Rahnasto-Rilla M, Kokkola T, Jarho E et al (2016) *N*-acylethanolamines bind to SIRT6. Chembiochem 17:77–81

Rodríguez-Miguel C, Moral R, Escrich R et al (2015) The role of dietary extra virgin olive oil and corn oil on the alteration of epigenetic patterns in the rat DMBA-induced breast cancer model. PLoS One 10:e0138980

Russo IH, Russo J (1996) Mammary gland neoplasia in long-term rodent studies. Environ Health Perspect 104:938–967

Schulz M, Hoffmann K, Weikert C et al (2008) Identification of a dietary pattern characterized by high-fat food choices associated with increased risk of breast cancer: the European Prospective Investigation into Cancer and Nutrition (EPIC)-Potsdam study. Br J Nutr 100:942–946

Sieri S, Krogh V, Ferrari P et al (2008) Dietary fat and breast cancer risk in the European Prospective Investigation into Cancer and Nutrition. Am J Clin Nutr 88:1304–1312

Sierra MI, Fernández AF, Fraga MF (2015) Epigenetics of aging. Curr Genomics 16:435–440

Sofi F, Abbate R, Gensini GF et al (2010) Accruing evidence on benefits of adherence to the Mediterranean diet on health: an updated systematic review and meta-analysis. Am J Clin Nutr 92:1189–1196

Solanas M, Grau L, Moral R et al (2010) Dietary olive oil and corn oil differentially affect experimental breast cancer through distinct modulation of the p21Ras signaling and the proliferation-apoptosis balance. Carcinogenesis 31:871–879
Starlard-Davenport A, Tryndyak VP, James SR et al (2010) Mechanisms of epigenetic silencing of the Rassf1a gene during estrogen-induced breast carcinogenesis in ACI rats. Carcinogenesis 31:376–381
Tezcan G, Tunca B, Bekar A et al (2014) Olea europaea leaf extract improves the treatment response of GBM stem cells by modulating miRNA expression. Am J Cancer Res 4:572–590
Tseng TH, Chien MH, Lin WL et al (2017) Inhibition of MDA-MB-231 breast cancer cell proliferation and tumor growth by apigenin through induction of G2/M arrest and histone H3 acetylation-mediated p21WAF1/CIP1 expression. Environ Toxicol 32:434–444
Veeck J, Esteller M (2010) Breast cancer epigenetics: from DNA methylation to microRNAs. J Mammary Gland Biol Neoplasia 15:5–17
WCRF/AICR – World Cancer Research Fund/American Institute for Cancer Research (2007) Food, nutrition, physical activity, and the prevention of cancer: a global perspective. American Institute for Cancer Research, Washington, DC
Xiao X, Shi D, Liu L et al (2011) Quercetin suppresses cyclooxygenase-2 expression and angiogenesis through inactivation of P300 signaling. PLoS One 6:e22934

Natural Polyphenol Kaempferol and Its Epigenetic Impact on Histone Deacetylases: Focus on Human Liver Cells

99

Sascha Venturelli, Christian Leischner, and Markus Burkard

Contents

Abstract

The flavonol kaempferol, which is found in many vegetables and fruits, is suggested to exhibit various and promising beneficial health effects *in vitro* and *in vivo*. Although there is strong evidence for health-promoting effects and good tolerability of kaempferol as common ingredient of daily nutrition, only little is known about the underlying pharmacodynamics and especially kaempferol-mediated effects on liver gene expression, enzyme levels, and phase I metabolism. Noteworthy, recent studies revealed that kaempferol is

S. Venturelli · C. Leischner · M. Burkard (✉)
Department of Vegetative and Clinical Physiology, Institute of Physiology, University Hospital Tuebingen, Tuebingen, Germany
e-mail: sascha.venturelli@med.uni-tuebingen.de; sascha.venturelli@gmx.de; christian.leischner@uni-tuebingen.de; chlei@gmx.de; markus.burkard@uni-tuebingen.de; markus-burkard@gmx.de

V. B. Patel, V. R. Preedy (eds.), *Handbook of Nutrition, Diet, and Epigenetics*,
https://doi.org/10.1007/978-3-319-55530-0_62

an interesting inhibitor of histone deacetylases with high affinity toward all members of HDAC families I, II, and IV that were tested. Therefore, the epigenetic activity of kaempferol could, at least in part, be responsible for the promising health effects and remain to be intensively studied *in vivo*. Investigation of hepatotoxic effects and interactions with CYP450 enzymes is one of the major prerequisites for a possible clinical use of kaempferol. Therefore, preclinical evaluation of high doses of kaempferol was performed with primary human hepatocytes, which are widely used as valuable tools to predict toxic drug effects on the human liver. Additionally, an *in vivo* chicken embryotoxicity assay to check for embryotoxic effects yielded good tolerability of kaempferol. According to the promising preliminary results, it would be important to evaluate long-term effects of low physiological doses of kaempferol compared to interventions with high pharmacological doses in future experiments.

Keywords
Flavonoid · Flavonol · Kaempferol · Quercetin · Hepatocellular carcinoma · Primary human hepatocytes · Pan-HDAC inhibitor · Cytochrome p450 enzymes · Liver · Phase I metabolism

List of Abbreviations

B[a]P	Benzo[a]pyrene
CDK1	Cyclin-dependent kinase 1
CYP	Cytochrome P450
DMBA	7,12-Dimethylbenz[a]anthracene
DNMT	DNA methyl transferase
GSTP1–1	Glutathione S-transferase Pi 1 peptide 1
HAT	Histone acetyl transferase
HCC	Hepatocellular carcinoma
HDAC	Histone deacetylase
PHH	Primary human hepatocyte
P-PST	Phenol-sulfating form of phenol sulfotransferase
QR	Quinone reductase
SAHA	Suberoylanilide hydroxamic acid
SIRT	Sirtuin
SULT1A1	Sulfotransferase family 1A Member 1
SULT1E1	Sulfotransferase family 1E Member 1
TCDD	2,3,7,8-Tetrachlorodibenzo-p-dioxin
TSA	Trichostatin A
TS-PST	Thermostable phenol sulfotransferase
UDP	Uridine diphosphate
UGT	UDP-glucuronyl transferase

Introduction

To date, malignant diseases are still ranked as the second most disease-related death causes worldwide. Despite recent progress in some areas of oncology, several tumor entities are still characterized by poor prognosis and especially solid tumors like the hepatocellular carcinoma (HCC) that distinguish themselves by a lack of treatment options. Long absence of any HCC-related symptoms procures a delay in diagnosing the disease. Median survival time is hence only 3–6 months. Besides alcohol abuse, nonalcoholic fatty liver disease, and diabetes mellitus, viral infections with hepatitis B or C virus are the major causes for malignant transformation of human hepatocytes (El-Serag 2012). Surgical resection is the therapeutic option of choice but only applicable for the minority of the very diffuse-growing HCCs. On the other hand, sorafenib, which is the most effective pharmacological treatment option to date, can prolong median survival only for a few months, thereby often exhibiting severe toxicity and unwanted side effects. Therefore, there is an urgent need to find and develop new treatment options and anticancer compounds. Polyphenols derived from nutrition seem to be an increasingly attractive alternative to treat different tumor entities with a favorable toxicity profile, at once. Especially, kaempferol emerged as a promising plant polyphenol with strong antitumor effects on human HCC cell lines and a distinct epigenetic activity. Its clinical use in the treatment or prevention of HCC and eventually other chronic liver diseases requires careful examination of kaempferol effects on human liver cells and their metabolic activity.

Kaempferol and Corresponding Glycosides

Kaempferol (3,5,7,4′-tetrahydroxyflavone), a naturally occurring secondary plant metabolite, belongs to the flavonol (3-hydroxyflavone) subclass of flavonoids from a chemical point of view. Further flavonoid subclasses are flavan-3-ols, flavones, flavanones, isoflavones, and the positively charged anthocyanidins (Fig. 1). To date, the flavonoid family comprises more than 5000 individual compounds, and their number is still increasing. All of them share the flavan backbone as their basic chemical structure, which consists of two aromatic rings (ring A and B) linked to an oxygenated heterocyclic ring (C ring), which defines the different flavonoid subclasses by its structural variation.

The flavonol kaempferol possesses a double bond at the 2–3 position and a hydroxyl group at C3. Other common flavonol derivatives with di- and tri-hydroxylated B rings are, e.g., quercetin, myricetin, fisetin, and galangin (Fig. 2). Many of these polyphenols are part of the daily diet showing distinct bioactive properties, which are suggested to provide health benefits like reduced risk of developing chronic diseases such as cardiovascular issues (Liu 2004). In nature, flavonoids often appear as O-glycosylated or esterified forms preferably at the C3 position. The corresponding aglycones arise during cooking or are prevalent in food, which was otherwise processed. More than 80 naturally occurring different sugars account for

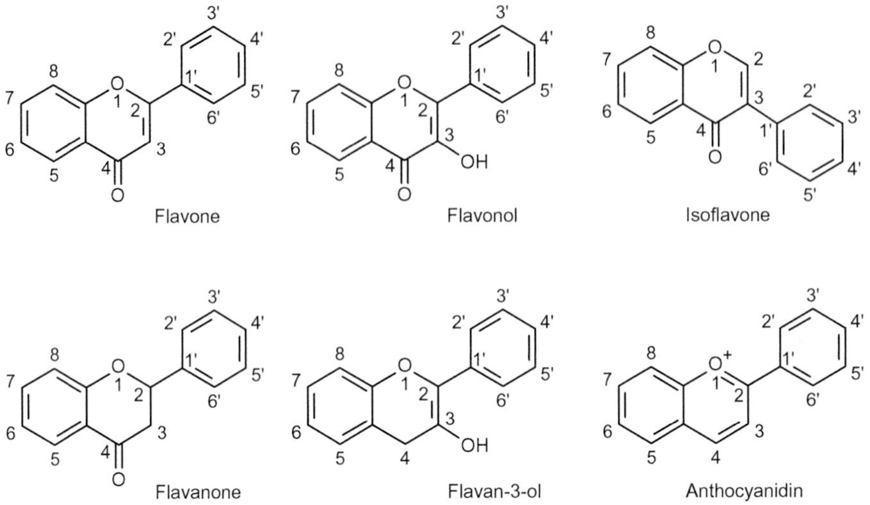

Flavone Flavonol Isoflavone

Flavanone Flavan-3-ol Anthocyanidin

Fig. 1 Chemical structures of the six major flavonoid subclasses. The flavonoid family is usually divided into flavones, flavonols (e.g., kaempferol), isoflavones, flavanones, flavan-3-ols, and anthocyanidins

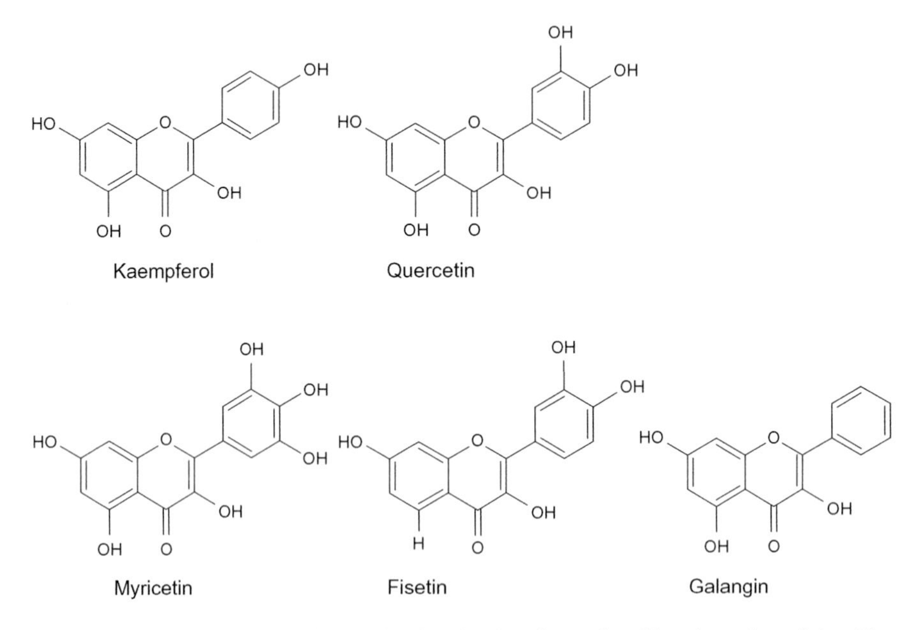

Fig. 2 Chemical structures of kaempferol and other flavonols with epigenetic activity. The kaempferol molecule is closely related to other naturally occurring flavonols like quercetin, myricetin, and fisetin. Interestingly, it can also be metabolized, e.g., to quercetin or isorhamnetin by hepatic phase I metabolism. The close structural relationship might explain that the displayed flavonols all exhibit epigenetic activity to certain degree

an exceptionally diversified repertoire of glycosides that results in an almost unmanageable multitude of different flavonoid derivatives when combined to the more than 5000 known flavonoid aglycones (Hollman and Arts 2000). Less frequently the C7 position and very seldom the 4′-, 3′- and 5-positions also serve as bonding sites (Herrmann 1988). The most common sugar found in flavonoid glycosides is glucose resulting in flavonoid glucosides. Other sugars that often contribute to flavonoid glycoside formation are, e.g., D-galactose, L-rhamnose, L-arabinose, D-xylose, D-apiose, or D-glucuronic acid. D-sugars generate β-glycosides, whereas sugars of the L-series result in the formation of the α-configuration (Herrmann 1988).

Occurrence of Kaempferol in Plants and Nutrition

Kaempferol, measured as free aglycone after hydrolysis (Table 1), is described as a bioactive plant polyphenol at varying concentrations notably in leek (~30–31 mg kg^{-1} fresh weight), endive (~46 mg kg^{-1}), broccoli (~60–72 mg kg^{-1}), and most abundantly in kale (~211–470 mg kg^{-1}). To a lesser extent, kaempferol is also found in French beans (<12 mg kg^{-1}), brussels sprouts (~7.4–9 mg kg^{-1}), black tea (~14–16 mg l^{-1}), strawberries (~5–12 mg kg^{-1}), and other fruits like tomatoes, apples, red grapes, as well as hop (Hertog et al. 1992, 1993b; Justesen et al. 1998). Peculiar seasonal dependence with great variations in kaempferol content by factor 3 to 5 was described, e.g., for endive and leek (Hertog et al. 1992). In addition, kaempferol can be found in honey (Jaganathan and Mandal 2009).

As mentioned above kaempferol predominantly occurs as glycosides in the biological context (as exemplified in Fig. 3). In broccoli, e.g., the main glycoside is kaempferol-3-O-sophoroside (166 mg kg^{-1} fresh weight) together with minor amounts of kaempferol-3-O-glucoside (14 mg kg^{-1}) and kaempferol-diglucoside

Table 1 Kaempferol occurrence and content of selected foods. Kaempferol is widely distributed throughout the flora. Some plants containing particular high levels of kaempferol and the according glycosides are displayed below. Contents of kaempferol were determined as aglycones after hydrolysis

Source	Content (aglycone after hydrolysis)	Reference
Leek	30–31 mg kg^{-1} fresh weight	(Hollman and Arts 2000; Hertog et al. 1992,1993b; Justesen et al. 1998)
Endive	46 mg kg^{-1}	
Broccoli	60–72 mg kg^{-1}	
Kale	211–470 mg kg^{-1}	
French beans	<12 mg kg^{-1}	
Brussels sprouts	7,4-9 mg kg^{-1}	
Black tea	14–16 mg l^{-1}	
Strawberries	5–12 mg kg^{-1}	

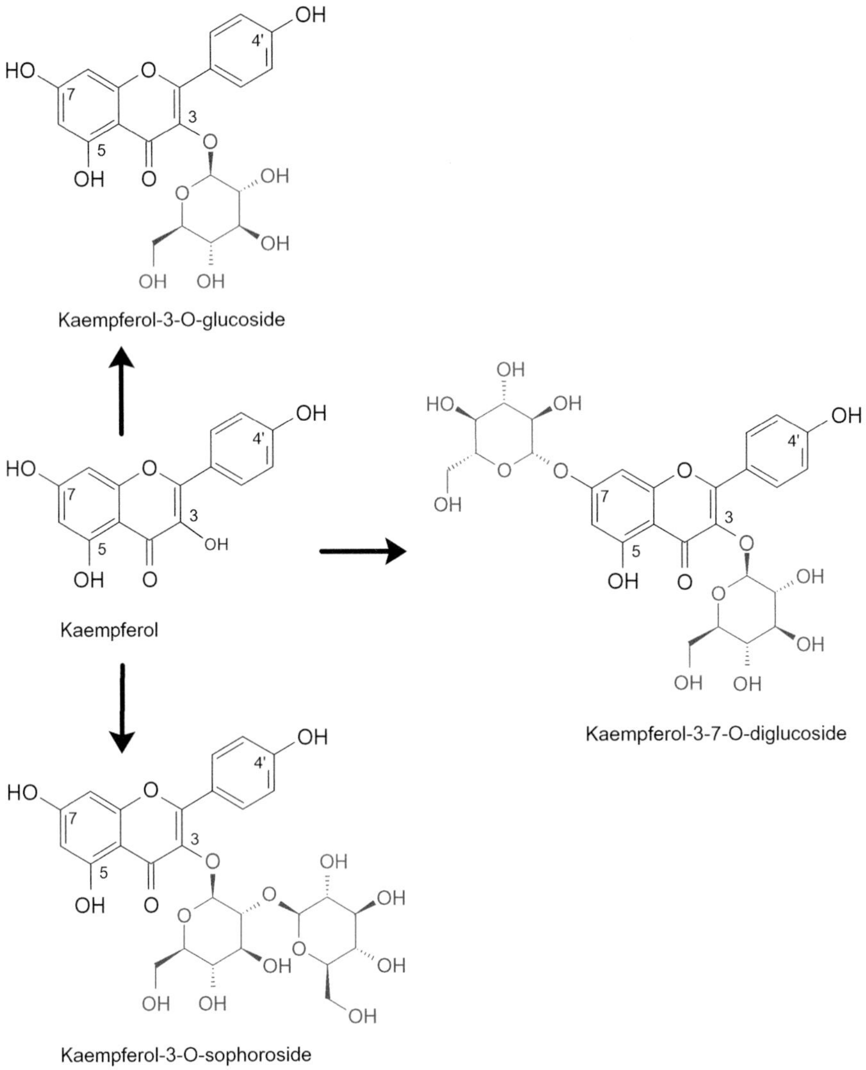

Kaempferol-3-O-glucoside

Kaempferol

Kaempferol-3-7-O-diglucoside

Kaempferol-3-O-sophoroside

Fig. 3 Kaempferol and according glycosides. In plants kaempferol is predominantly found as various glycosides with one or more sugar moieties, which influence the bioavailability of the polyphenol

(15 mg kg^{-1}) (Price et al. 1998a). Different tea infusions contain kaempferol-3-O-glucoside, kaempferol-3-O-rhamnosylglucoside, and kaempferol-3-O-glucosylrhamnosyl-glucoside (Price et al. 1998b).

Mean dietary intake of flavones and flavonols like quercetin, myricetin, and kaempferol was determined to be about 23 mg day^{-1} with a portion of kaempferol of 3.9 mg day^{-1} among a representative group of individuals in the Netherlands (Hertog et al. 1993a). Other cohort studies investigated flavonol intake in different countries and revealed the lowest values for Finland (4 mg day^{-1} (Knekt et al. 1996)) and the highest values for Japan (64 mg day^{-1} (Hertog et al. 1995)). This high flavonol intake, which is mainly caused by high green tea consumption containing kaempferol (Kim and Choi 2013) and high amounts of green tea catechins, could partially account for a phenomenon sometimes called "Asian paradox." It is characterized by a decreased prevalence of different cardiovascular and malignant diseases faced by an elevated cigarette consumption in comparison to many Western countries (Nugala et al. 2012).

After ingestion, unabsorbed or biliary secreted flavonol glycosides can be hydrolyzed by gut microbiota, and the aglycones can be further degraded through cleavage of the C ring. The resulting phenolic acids can be easily absorbed (Griffiths and Smith 1972). Some flavonol glycosides can also be taken up without prior hydrolysis, e.g., actively by glucose transporters suggesting the sugar moiety playing a role in the degree of absorption and bioavailability rate (Nemeth et al. 2003; Sesink et al. 2003; Wolffram et al. 2002; Cermak et al. 2003).

Following intravenous injection of kaempferol pharmacokinetic studies in a rat model revealed a declining plasma concentration to 50% 3–4 h after administration with no free kaempferol detectable after 24 h (Barve et al. 2009). When given orally, kaempferol concentration peaked at 1–1.5 h with no free kaempferol detectable after 6 h. It has been shown that kaempferol is extensively metabolized in the liver and finally renally excreted either as the parent compound or as corresponding glucuronidated or sulfated metabolites, respectively. Moreover, to a lesser extent, kaempferol is processed into quercetin or isorhamnetin through phase I detoxifying biotransformation reactions by O-methylation of the catechol group. Subsequently, quercetin and rhamnetin undergo phase II metabolism resulting in the corresponding glucuronidated counterparts (Barve et al. 2009). Hereby a portion of about 2.5% of the ingested kaempferol is directly renally excreted (Calderon-Montano et al. 2011). The peak plasma concentration after oral intake was found to be about of 150 nM in human after predefined endive ingestion (DuPont et al. 2004). Low bioavailability of kaempferol remains one of the major issues in regard to the interesting pharmacologic effects described below. Overall flavonoid (aglycones and conjugated metabolites) plasma levels of nutritionally relevant doses rarely exceed 1 μM (Hollman 2004).

Pharmacodynamics of Kaempferol

Kaempferol is suggested to exhibit cancer chemopreventive activity by inhibiting the phase I oxidative metabolism CYPP450 enzyme CYP1A1 (Table 2) and additionally by enhancing UDP-glucuronosyltransferase activity of phase II metabolism, which is responsible for the conjugation of xenobiotics to facilitate their renal excretion (Sun et al. 1998). Furthermore, kaempferol was found to downregulate cyclin-dependent kinase 1 (CDK1), which is also known as cell division cycle protein 2 homolog. Cyclin A and B are also downregulated by kaempferol, which together result in cell cycle arrest at the G2/M phase that is accompanied by enhanced

Table 2 Flavonols have strong influence on phase I and phase II metabolism. Kaempferol and other selected flavonols have strong impact on metabolic enzymes

	Kaempferol	Quercetin	Myricetin	Galangin	Fisetin	Reference
Phase I						
CYP1A1	↓ (activity),↓ (TCDD or B[a]P-induced transcription)	↑ (mRNA expression and activity)		↓ (activity),↑ (mRNA expression),↓ (TCDD or DMBA-induced transcription)		(Chang et al. 2006; Kang et al. 1999; Ciolino and Yeh 1999)
CYP1A2	↓ (activity)	↓ (activity)		↓ (activity)		(Chang et al. 2006; Tsyrlov et al. 1994; Zhai et al. 1998)
CYP1B2	↓ (activity)	↓ (activity)				(Chang et al. 2006)
CYP3A4		↓ (activity)	↓ (activity)			(Obach 2000; Ho et al. 2001)
CYP19				↓ (activity)		(Kao et al. 1998)
Phase II						
UGT1A1		↓ (activity)		↑ (mRNA expression)		(Williams et al. 2002; Walle and Walle 2002)
UGT	↑ (activity)	↑ (activity, in vivo)		↑ (activity)		(Sun et al. 1998; van der Logt et al. 2003)
SULT1A1 (P-PST or TS-PST)	↓ (activity)	↓ (activity)	↓ (activity)	↓ (activity)	↓ (activity)	(Eaton et al. 1996)
SULT1E1		↓ (activity)				(Ohkimoto et al. 2004)
GSTP1–1		↓ (activity)				(van Zanden et al. 2003)
QR	↑ (activity)	↑ (activity)		↑ (activity)		(Uda et al. 1997)

B[a]P benzo[a]pyrene, *CYP* cytochrome P450, *CYP1A1* cytochrome P450 family 1A member 1, *CYP1A2* cytochrome P450 family 1A member 2, *CYP1B2* cytochrome P450 family 1B member 2, *CYP3A4* cytochrome P450 family 3A member 4, *CYP19* cytochrome P450 family 19, *DMBA* 7,12-dimethylbenz[a]anthracene, *GSTP1–1* glutathione S-transferase Pi 1 peptide 1, *P-PST* phenol-sulfating form of phenol sulfotransferase, *QR* quinone reductase, *SULT1A1* sulfotransferase family 1A member 1, *SULT1E1* sulfotransferase family 1E member 1, *TCDD* 2,3,7,8-tetrachlorodibenzo-p-dioxin, *TS-PST* thermostable phenol sulfotransferase, *UDP* uridine diphosphate, *UGT* UDP-glucuronyl transferase

induction of apoptosis in human breast cancer cells (Choi and Ahn 2008). Other *in vitro* studies revealed antiproliferative and apoptosis-inducing properties of kaempferol for several other tumor entities such as non-small-cell lung cancer (Leung et al. 2007), leukemia (Marfe et al. 2009), prostate cancer (De Leo et al. 2006), oral cavity cancer (Kim et al. 2005), or colon cancer (Mutoh et al. 2000). The latter effects are partially caused by increased p53 expression with concomitant serine phosphorylation at position 15 followed by apoptosis (Choi and Ahn 2008), inactivation of serine/threonine protein kinase AKT with simultaneous activation of proapoptotic BAX protein and NAD-dependent deacetylase sirtuin (SIRT)3 (Marfe et al. 2009), or caspase-3-dependent apoptosis (Kim et al. 2005). In summary, kaempferol has been shown to exhibit pleiotropic effects on various tumor cell lines (Kang et al. 2010; Leung et al. 2007; Marfe et al. 2009; Huang et al. 2010).

Epigenetic Activity of Kaempferol

Epigenetic alterations give rise to changed gene expression patterns and among them especially oncogenes, which lead to uncontrolled cell cycle or suppressed apoptotic regulation, when dysregulated. Aberrations and changes in epigenetic DNA maintenance therefore often lead to the development of cancer by silencing important control genes, e.g., through overexpression of histone deacetylases (HDACs). These HDACs deacetylate histone proteins, and therefore the transcriptionally active euchromatin is condensed to the more tightly packed heterochromatin. Different classes of histone deacetylases (HDAC classes I, II, and IV; class III (sirtuins)) are considered to provide controlled activation or inactivation of genes by chromatin remodeling together with multiple other DNA-modulating enzymes including the histone acetyl transferases (HATs).

Epigenetic activity (inhibition of DNA methyl transferases (DNMT) and inhibition of HDACs) was described for compounds derived from plants and intriguingly for some distinct members of the flavonoid family. Examples are catechin and epigallocatechin-3-gallate (flavan-3-ols), apigenin and luteolin (flavones), hesperetin and naringenin (flavanones), genistein, and daidzein (isoflavones), as well as quercetin, myricetin, and fisetin (flavonols) (Busch et al. 2015, Gilbert and Liu 2010).

In this context, the flavonol kaempferol was intensively investigated and identified as an inhibitor of HDACs (classes I, II, and IV) leading to growth inhibition and cytotoxicity in human hepatocellular cancer cell lines (HepG3 and Hep3B) in a dose-dependent manner (Berger et al. 2013). First hints of an epigenetic mode of action by kaempferol were obtained by characterization of kaempferol with an *in silico* docking analysis that suggested a high suitability of the kaempferol molecule to fit in the binding pocket of several HDACs, where kaempferol could be able to interact with the required zinc ion of the catalytic site. Noteworthy, kaempferol-yielded GoldScores in docking analysis were higher in comparison to the GoldScores of trichostatin A (TSA) or suberoylanilide hydroxamic acid (SAHA) regarding HDAC isoenzymes 4, 7, and 8 (Berger et al. 2013). *In vitro* assays with nuclear human hepatoma cell line extract verified substantial HDAC inhibitory activity starting already at low micromolar concentration ranges. Further profiling with recombinant human HDACs of class I (HDAC1, 2, 3, and 8), class II (HDAC4, 5, 6, 7, 9, and 10), and class IV (HDAC11) proved an individual but overall inhibitory profile. The profiling was performed with 50 μM kaempferol to ensure a sufficient inhibitor concentration. 5 μM kaempferol was already found to significantly inhibit HDAC activity (Berger et al. 2013).

Kaempferol can therefore be regarded as a pan-HDACi with only low selectivity for certain HDAC isoenzymes. Thus, kaempferol caused hyperacetylation of histone H3 in human HCC cell lines (HepG2 and Hep3B) and in colon cancer cell line HCT-116, which was determined by western blot analysis (Berger et al. 2013). The kaempferol treatment that was performed with concentrations between 20 and 100 μM revealed significantly reduced cell proliferation for both HepG2 and Hep3B, unaffected by their respective state of p53. This is interesting due to the observation that some anticancer effects of kaempferol are attributed to a functional p53. Colon cancer as exemplified by HCT-116 cells is another possible indication for flavonols that show epigenetic activity like kaempferol because issues regarding bioavailability and biotransformation can be partially circumvented due to close contact between the tumor and the feces and/or nutrients, respectively. Interestingly, activation of NAD-dependent SIRT3 as a representative of HDAC class III was also described for kaempferol (Marfe et al. 2009). Treatment with 50 μM kaempferol induced oxidative stress in the K562 (human chronic myelogenous leukemia) and U937 (human lung lymphoblast) cell lines and increased the expression of BAX and SIRT3 in a time-dependent manner accompanied by decreased BCL-2 levels. These effects were ultimately followed by cancer cell decline associated with caspase-3 activation and cytochrome 3 release (Marfe et al. 2009).

Preclinical Evaluation of Kaempferol and Role in Hepatotoxicity

Kaempferol seems to display cytotoxic effects in healthy primary human hepatocytes starting at ~50 μM in *in vitro* assays (Berger et al. 2013). Higher concentrations of ~200 μM are necessary to show toxic effects in an *in vivo* chicken embryotoxicity assay. These results demonstrate a selective efficacy against malignant cells compared to nonmalignant healthy cells or tissue, respectively.

The pleasant therapeutic window in regard to human liver cells renders kaempferol to a promising compound or lead structure for cancer therapy and/or prevention. Protective activity of kaempferol on isoniazid- and rifampicin-induced hepatotoxicity in mice was also reported strengthening the hypothesis of good tolerability of kaempferol in regard to liver cells (Shih et al. 2013). Similar protective effects for kaempferol glycosides were reported in a mouse model of CCl_4-induced oxidative liver damage (Wang et al. 2015).

Despite the promising results in some preclinical models, suitable animal experiments with very high doses of kaempferol (pharmacological doses instead of physiological doses) are urgently needed.

Albeit a possible therapeutic use of kaempferol increased consumption of kaempferol (and other flavonoids) containing food in the daily diet seems to be desirable. Even though physiological kaempferol doses are low, there is rising evidence for the crucial role of nutrition in shaping the human epigenome (Busch et al. 2015). Therefore, kaempferol may be considered as nutrition-derived epigenome modifier with cumulative influence on human epigenome during life span at physiological doses, whereas it has also high therapeutic potential if used in pharmacological doses.

Role of Other Flavonols in Nutrition, Epigenetics, and Hepatotoxicity

Despite the distinct epigenetic activity of kaempferol, there is increasing evidence that other flavonols may also influence the human epigenome in various ways (Table 3). Quercetin, the most abundant flavonol, which is found, e.g., in tea, apples, berries, and wine, was described to inhibit DNMT1 activity *in vitro* with an IC_{50} value of only 1.6 μM, whereas fisetin that is contained in strawberries, apples, onions, wine, and tea had an IC_{50} value of 3.5 μM, while myricetin from grapes, berries, red wine, and tea showed the strongest inhibitory activity ($IC_{50} = 1.2$ μM) *in vitro* (Busch et al. 2015; Ong et al. 2011; Kim et al. 2005). The myricetin molecule that showed the strongest DNMT1 inhibition is characterized by a pyrogallol moiety similar to the gallic acid moiety of epigallocatechin-3-gallate (Gilbert and Liu 2010; Kim et al. 2005).

Furthermore, quercetin was found to activate HAT and/or inhibition of HDAC activity at high doses (100 μM for 6 h) (Lee et al. 2011), and fisetin and quercetin were described to activate sirtuins (Busch et al. 2015; Dashwood 2007).

Besides kaempferol, many other flavonols are also considered to exert different effects on phase I and phase II metabolism (Table 2). Quercetin seems to be rather protective than toxic toward liver cells as exemplified by the attenuated cadmium-induced oxidative damage to rat hepatocytes (Vicente-Sanchez et al. 2008). Quercetin supplementation was described for a beneficial effect in the context of chronic ethanol-induced hepatotoxicity in rats accompanied by an increase of glutathione levels (Vidhya and Indira 2009). Similar protective effects on alcoholic liver damage in mice were also reported for fisetin (Koneru et al. 2016). In a study with diabetic rats, myricetin showed not only antidiabetic activity but also no serious hepatotoxicity (Semwal et al. 2016; Ong and Khoo 2000). Therefore, supplementation with flavonols seems to be rather protective than deleterious on liver function at least in rodents.

Kaempferol and Its Derivatives in the Context of Other Epigenetic Modifiers

Even though kaempferol and some other flavonols show definite epigenetic effects, their activity is significantly less pronounced when compared, e.g., to the synthetic HDAC inhibitor suberoylanilide hydroxamic acid (SAHA), which is FDA approved for the treatment of certain hematological malignancies but not for the HCC and other solid tumors. These finding is substantiated by comprehensive profiling

Table 3 Epigenetic activity of kaempferol and the according flavonols. Kaempferol and some structurally closely related flavonoids are described for various very interesting epigenetic activities with remarkable effects on DNA methylation and/or histone deacetylation

Flavonoid	Treatment	Effects on DNA methylation	Effects on Histone acetylation	Reference
Kaempferol	20–100 μM Hep3B and HepG2 (HCC)	n.d.	Hyperacetylation of histone H3. Inhibition of HDAC1–11 in a HDAC profiling assay. Inhibition of HDAC1 in a HDAC inhibitor screening assay	(Berger et al. 2013)
Kaempferol	50 μM, up to 72 h. K562 and U937 cancer cells	n.d.	Increase of SIRT3 expression. Induction of cancer cell death	(Marfe et al. 2009)
Quercetin	$IC_{50} = 1.6$ μM	Inhibition of isolated SssI DNMT and DNMT1	n.d.	(Kim et al. 2005)
Quercetin	100 μM on human leukemia HL-60 cells	n.d.	HAT activation and HDAC inhibition	(Lee et al. 2011)
Myricetin	$IC_{50} = 1.2$ μM	Inhibition of isolated SssI DNMT and DNMT1	n.d.	(Kim et al. 2005)
Fisetin	$IC_{50} = 3.5$ μM	Inhibition of isolated SssI DNMT and DNMT1	n.d.	(Kim et al. 2005)

and Western blot analysis. Nevertheless, the therapeutic window of kaempferol and its derivatives seem to be extraordinary broad, and therefore kaempferol can be considered as a promising drug candidate for the treatment of malignant liver diseases and probably even their prerequisites.

Dictionary of Terms

- **Chicken embryotoxicity assay** – It is a fast and comparably sensitive *in vivo* assay for the preliminary determination of possible embryotoxic effects of test compounds using fertilized eggs of leghorn chickens.
- **Flavonoids** – Secondary plant polyphenols that are usually colored enabling their function as plant pigments.
- **GoldScore** – Scoring determinant that facilitates the *in silico* docking analysis and the prediction of ligand-binding positions in catalytic sites of enzymes including parameters like H-bonding energy, van der Waals energy, and metal interactions.

- **Hepatocellular carcinoma (HCC)** – The HCC is the most frequent malignant disease of liver cells in human and often caused by viral infections with the human hepatitis B or C virus as well as chronical alcohol abuse.
- **Hepatitis B virus** – The human hepatitis B virus can cause acute or chronic types of hepatitis in which chronic infection is often associated with liver cirrhosis and/or hepatocellular carcinoma.
- **Hep3B** – Human hepatocellular carcinoma cell line isolated from an 8-year-old male that is characterized by a defect variant of p53 and produces infectious hepatitis B virus particles.
- **HepG2** – Human hepatocellular carcinoma cell line isolated from a 15-year-old male Caucasian that is wildtype for p53 and negative for hepatitis B virus.
- **p53** – The tumor protein p53 or tumor suppressor p53 was found to be mutated in about 50% of human cancers and is seemingly one of the most important tumor suppressor genes, which is, e.g., involved in DNA repair, cell cycle regulation, and apoptosis induction.
- **Sorafenib** – Pharmacological standard therapy for the treatment of the non-resectable hepatocellular carcinoma.
- **Suberoylanilide hydroxamic acid (SAHA)** – SAHA, also known as vorinostat, which is a strong and reversible pan-HDAC inhibitor that was approved by the FDA for the treatment of distinct hematological malignant diseases.
- **Trichostatin A (TSA)** – TSA was originally described as an antifungal compound with antibiotic properties, which was subsequently found to be a very strong inhibitor for the majority of human HDACs except sirtuins with an inappropriate toxicity profile for the approval by the FDA.

Key Facts of Kaempferol

- The flavonol Kaempferol belongs to the class of flavonoids.
- Flavonoids are secondary plant metabolites that are ubiquitously found in fruits and vegetables fulfilling various tasks in plants such as coloration to attract pollinator animals, UV filtration, and possibly modulation of the cell cycle.
- Kaempferol mainly occurs as one of many different glycosides in plants.
- Kaempferol can be metabolized into quercetin and isorhamnetin by phase I metabolism.
- Kaempferol gained increasing attention for its very promising health-promoting effects and particularly its preventive potential in regard to cardiovascular and malignant diseases.
- Low bioavailability and/or fast metabolism of kaempferol and flavonoids in general are still great issues on the field of flavonoid research.
- Several members of the flavonoid family were recently described for their epigenetic activity such as inhibition of DNA methyltransferases or histone deacetylases.
- The underlying mechanisms for the observed beneficial health effects of kaempferol are vastly unclear and therefore have to be elucidated.

Summary Points

- This chapter focusses on kaempferol, a flavonol, which is found in several healthy foods and especially in vegetables and its effect on healthy as well as malignant liver cells.
- Kaempferol and other nutrition-derived flavonoids are suggested to be well tolerated *in vivo* after oral administration at physiological doses considering the high fruit and vegetable consumption all over the world and even high pharmacological doses are proposed to be well tolerated after oral administration.
- Kaempferol is present, e.g., in leek, endive, broccoli, kale, and many other vegetables and fruits and therefore readily available at low costs.
- The flavonol kaempferol was recently described as a novel inhibitor of histone deacetylase HDAC1–11 enzymes (HDAC families I, II, and IV) which can be considered as a pan-HDAC inhibitor due to its broad spectrum of inhibitory activity toward histone deacetylases.
- Epigenetic activity of kaempferol and other flavonoids was mostly determined in *in vitro* experiments, and only little is known about epigenetic effects of flavonoids *in vivo* to date.
- Kaempferol was also found to show strong anticancer activity in certain malignant human cell lines (hepatocellular carcinoma and colon carcinoma).
- The toxicity in regard to healthy primary human hepatocytes seems to be very low starting at concentrations above 50 μM kaempferol.
- Therefore, kaempferol could be a promising novel compound in the prevention and/or therapy of the hepatocellular carcinoma or at least an auspicious lead structure to develop such drugs.

References

Barve A, Chen C, Hebbar V et al (2009) Metabolism, oral bioavailability and pharmacokinetics of chemopreventive kaempferol in rats. Biopharm Drug Dispos 30:356–365

Berger A, Venturelli S, Kallnischkies M et al (2013) Kaempferol, a new nutrition-derived pan-inhibitor of human histone deacetylases. J Nutr Biochem 24:977–985

Busch C, Burkard M, Leischner C et al (2015) Epigenetic activities of flavonoids in the prevention and treatment of cancer. Clin Epigenetics 7:64

Calderon-Montano JM, Burgos-Moron E, Perez-Guerrero C et al (2011) A review on the dietary flavonoid Kaempferol. Mini-Rev Med Chem 11:298–344

Cermak R, Landgraf S, Wolffram S (2003) The bioavailability of quercetin in pigs depends on the glycoside moiety and on dietary factors. J Nutr 133:2802–2807

Chang TKH, Chen J, Yeung EYH (2006) Effect of Ginkgo biloba extract on procarcinogen-bioactivating human CYP1 enzymes: identification of isorhamnetin, kaempferol, and quercetin as potent inhibitors of CYP1B1. Toxicol Appl Pharmacol 213:18–26

Choi EJ, Ahn WS (2008) Kaempferol induced the apoptosis via cell cycle arrest in human breast cancer MDA-MB-453 cells. Nutr Res Pract 2:322–325

Ciolino HP, Yeh GC (1999) The flavonoid galangin is an inhibitor of CYP1A1 activity and an agonist/antagonist of the aryl hydrocarbon receptor. Br J Cancer 79:1340–1346

Dashwood RH (2007) Frontiers in polyphenols and cancer prevention. J Nutr 137:267S–269S

De Leo M, Braca A, Sanogo R et al (2006) Antiproliferative activity of Pteleopsis suberosa leaf extract and its flavonoid components in human prostate carcinoma cells. Planta Med 72:604–610

Dupont MS, Day AJ, Bennett RN et al (2004) Absorption of kaempferol from endive, a source of kaempferol-3-glucuronide, in humans. Eur J Clin Nutr 58:947–954

Eaton EA, Walle UK, Lewis AJ et al (1996) Flavonoids, potent inhibitors of the human P-form phenolsulfotransferase. Potential role in drug metabolism and chemoprevention. Drug Metab Dispos 24:232–237

El-Serag HB (2012) Epidemiology of viral hepatitis and hepatocellular carcinoma. Gastroenterology 142:1264–1273. e1

Gilbert ER, Liu D (2010) Flavonoids influence epigenetic-modifying enzyme activity: structure - function relationships and the therapeutic potential for cancer. Curr Med Chem 17:1756–1768

Griffiths LA, Smith GE (1972) Metabolism of myricetin and related compounds in the rat. Metabolite formation in vivo and by the intestinal microflora in vitro. Biochem J 130:141–151

Herrmann K (1988) On the occurrence of flavonol and flavone glycosides in vegetables. Zeitschrift für Lebensmittel-Untersuchung und Forschung 186:1–5

Hertog MGL, Hollman PCH, Katan MB (1992) Content of potentially anticarcinogenic flavonoids of 28 vegetables and 9 fruits commonly consumed in the Netherlands. J Agric Food Chem 40:2379–2383

Hertog MG, Hollman PC, Katan MB et al (1993a) Intake of potentially anticarcinogenic flavonoids and their determinants in adults in The Netherlands. Nutr Cancer 20:21–29

Hertog MGL, Hollman PCH, Van De Putte B (1993b) Content of potentially anticarcinogenic flavonoids of tea infusions, wines, and fruit juices. J Agric Food Chem 41:1242–1246

Hertog MG, Kromhout D, Aravanis C et al (1995) Flavonoid intake and long-term risk of coronary heart disease and cancer in the seven countries study. Arch Intern Med 155:381–386

Ho PC, Saville DJ, Wanwimolruk S (2001) Inhibition of human CYP3A4 activity by grapefruit flavonoids, furanocoumarins and related compounds. J Pharm Pharm Sci 4:217–227

Hollman PCH (2004) Absorption, bioavailability, and metabolism of flavonoids. Pharm Biol 42:74–83

Hollman PCH, Arts ICW (2000) Flavonols, flavones and flavanols – nature, occurrence and dietary burden. J Sci Food Agric 80:1081–1093

Huang WW, Chiu YJ, Fan MJ et al (2010) Kaempferol induced apoptosis via endoplasmic reticulum stress and mitochondria-dependent pathway in human osteosarcoma U-2 OS cells. Mol Nutr Food Res 54:1585–1595

Jaganathan SK, Mandal M (2009) Antiproliferative effects of honey and of its polyphenols: a review. J Biomed Biotechnol 830616:19

Justesen U, Knuthsen P, Leth T (1998) Quantitative analysis of flavonols, flavones, and flavanones in fruits, vegetables and beverages by high-performance liquid chromatography with photodiode array and mass spectrometric detection. J Chromatogr A 799:101–110

Kang ZC, Tsai SJ, Lee H (1999) Quercetin inhibits benzo[a]pyrene-induced DNA adducts in human Hep G2 cells by altering cytochrome P-450 1A1 gene expression. Nutr Cancer 35:175–179

Kang JW, Kim JH, Song K et al (2010) Kaempferol and quercetin, components of Ginkgo biloba extract (EGb 761), induce caspase-3-dependent apoptosis in oral cavity cancer cells. Phytother Res 24(Suppl 1):S77–S82

Kao YC, Zhou C, Sherman M et al (1998) Molecular basis of the inhibition of human aromatase (estrogen synthetase) by flavone and isoflavone phytoestrogens: a site-directed mutagenesis study. Environ Health Perspect 106:85–92

Kim SH, Choi KC (2013) Anti-cancer effect and underlying mechanism(s) of Kaempferol, a phytoestrogen, on the regulation of apoptosis in diverse cancer cell models. Toxicol Res 29:229–234

Kim KS, Rhee KH, Yoon JH et al (2005) Ginkgo biloba Extract (EGb 761) induces apoptosis by the activation of caspase-3 in oral cavity cancer cells. Oral Oncol 41:383–389

Knekt P, Jarvinen R, Reunanen A et al (1996) Flavonoid intake and coronary mortality in Finland: a cohort study. BMJ 312:478–481

Koneru M, Sahu BD, Kumar JM et al (2016) Fisetin protects liver from binge alcohol-induced toxicity by mechanisms including inhibition of matrix metalloproteinases (MMPs) and oxidative stress. J Funct Foods 22:588–601

Lee WJ, Chen YR, Tseng TH (2011) Quercetin induces FasL-related apoptosis, in part, through promotion of histone H3 acetylation in human leukemia HL-60 cells. Oncol Rep 25:583–591

Leung HW, Lin CJ, Hour MJ et al (2007) Kaempferol induces apoptosis in human lung non-small carcinoma cells accompanied by an induction of antioxidant enzymes. Food Chem Toxicol 45:2005–2013

Liu RH (2004) Potential synergy of phytochemicals in cancer prevention: mechanism of action. J Nutr 134:3479S–3485S

Marfe G, Tafani M, Indelicato M et al (2009) Kaempferol induces apoptosis in two different cell lines via Akt inactivation, Bax and SIRT3 activation, and mitochondrial dysfunction. J Cell Biochem 106:643–650

Mutoh M, Takahashi M, Fukuda K et al (2000) Suppression of cyclooxygenase-2 promoter-dependent transcriptional activity in colon cancer cells by chemopreventive agents with a resorcin-type structure. Carcinogenesis 21:959–963

Nemeth K, Plumb GW, Berrin JG et al (2003) Deglycosylation by small intestinal epithelial cell beta-glucosidases is a critical step in the absorption and metabolism of dietary flavonoid glycosides in humans. Eur J Nutr 42:29–42

Nugala B, Namasi A, Emmadi P et al (2012) Role of green tea as an antioxidant in periodontal disease: the Asian paradox. J Indian Soc Periodontol 16:313–316

Obach RS (2000) Inhibition of human cytochrome P450 enzymes by constituents of St. John's wort, an herbal preparation used in the treatment of depression. J Pharmacol Exp Ther 294:88–95

Ohkimoto K, Liu MY, Suiko M et al (2004) Characterization of a zebrafish estrogen-sulfating cytosolic sulfotransferase: inhibitory effects and mechanism of action of phytoestrogens. Chem Biol Interact 147:1–7

Ong KC, Khoo HE (2000) Effects of myricetin on glycemia and glycogen metabolism in diabetic rats. Life Sci 67:1695–1705

Ong TP, Moreno FS, Ross SA (2011) Targeting the epigenome with bioactive food components for cancer prevention. J Nutrigenet Nutrigenomics 4:275–292

Price KR, Casuscelli F, Colquhoun IJ et al (1998a) Composition and content of flavonol glycosides in broccoli florets (brassica olearacea) and their fate during cooking. J Sci Food Agric 77:468–472

Price KR, Rhodes MJC, Barnes KA (1998b) Flavonol glycoside content and composition of tea infusions made from commercially available teas and tea products. J Agric Food Chem 46:2517–2522

Semwal DK, Semwal RB, Combrinck S et al (2016) Myricetin: a dietary molecule with diverse biological activities. Forum Nutr 8:90

Sesink AL, Arts IC, Faassen-Peters M et al (2003) Intestinal uptake of quercetin-3-glucoside in rats involves hydrolysis by lactase phlorizin hydrolase. J Nutr 133:773–776

Shih TY, Young TH, Lee HS et al (2013) Protective effects of Kaempferol on isoniazid- and rifampicin-induced hepatotoxicity. AAPS J 15:753–762

Sun XY, Plouzek CA, Henry JP et al (1998) Increased UDP-glucuronosyltransferase activity and decreased prostate specific antigen production by biochanin a in prostate cancer cells. Cancer Res 58:2379–2384

Tsyrlov IB, Mikhailenko VM, Gelboin HV (1994) Isozyme- and species-specific susceptibility of cDNA-expressed CYP1A P-450s to different flavonoids. Biochim Biophys Acta 13:325–335

Uda Y, Price KR, Williamson G et al (1997) Induction of the anticarcinogenic marker enzyme, quinone reductase, in murine hepatoma cells in vitro by flavonoids. Cancer Lett 120:213–216

Van Der Logt EM, Roelofs HM, Nagengast FM et al (2003) Induction of rat hepatic and intestinal UDP-glucuronosyltransferases by naturally occurring dietary anticarcinogens. Carcinogenesis 24:1651–1656

Van Zanden JJ, Ben Hamman O, Van Iersel ML et al (2003) Inhibition of human glutathione S-transferase P1-1 by the flavonoid quercetin. Chem Biol Interact 145:139–148

Vicente-Sanchez C, Egido J, Sanchez-Gonzalez PD et al (2008) Effect of the flavonoid quercetin on cadmium-induced hepatotoxicity. Food Chem Toxicol 46:2279–2287

Vidhya A, Indira M (2009) Protective effect of quercetin in the regression of ethanol-induced hepatotoxicity. Indian J Pharm Sci 71:527–532

Walle UK, Walle T (2002) Induction of human UDP-glucuronosyltransferase UGT1A1 by flavonoids-structural requirements. Drug Metab Dispos 30:564–569

Wang Y, Tang CY, Zhang H (2015) Hepatoprotective effects of kaempferol 3-O-rutinoside and kaempferol 3-O-glucoside from Carthamus tinctorius L. on CCl4-induced oxidative liver injury in mice. J Food Drug Anal 23:310–317

Williams JA, Ring BJ, Cantrell VE et al (2002) Differential modulation of UDP-glucuronosyltransferase 1A1 (UGT1A1)-catalyzed estradiol-3-glucuronidation by the addition of UGT1A1 substrates and other compounds to human liver microsomes. Drug Metab Dispos 30:1266–1273

Wolffram S, Block M, Ader P (2002) Quercetin-3-glucoside is transported by the glucose carrier SGLT1 across the brush border membrane of rat small intestine. J Nutr 132:630–635

Zhai S, Dai R, Friedman FK et al (1998) Comparative inhibition of human cytochromes P450 1A1 and 1A2 by flavonoids. Drug Metab Dispos 26:989–992

Dietary Methylselenocysteine and Epigenetic Regulation of Circadian Gene Expression

100

Helmut Zarbl and Mingzhu Fang

Contents

Abstract

Selenium (Se) is an essential trace element. Methylselenocysteine (MSC) is an organic form of selenium obtained primarily through dietary ingestion. Selenium, especially MSC, showed significant inhibitory effect on mammary tumorigenesis, especially at early stages, in carcinogen-induced rat mammary tumor models.

H. Zarbl · M. Fang (✉)
Department of Environmental and Occupational Health, School of Public Health Environmental and Occupational Health Sciences Institute, Rutgers, The State University of New Jersey, Piscataway, NJ, USA
e-mail: fang@eohsi.rutgers.edu

V. B. Patel, V. R. Preedy (eds.), *Handbook of Nutrition, Diet, and Epigenetics*,
https://doi.org/10.1007/978-3-319-55530-0_63

However the underlying mechanisms are not fully understood. Accumulating evidence indicates that disruption of circadian rhythm by shift work or jet lag increases the risk of breast, prostate, and colon cancers. About 10% of genes, including many genes involved in hormone signaling and DNA damage response and repair (DDRR), are under circadian control and as a result show significant oscillation in expression across the day. Recent mechanism studies demonstrated that MSC restored and enhanced circadian expression of major clock genes, especially Period 2 (Per2), and circadian-controlled genes to inhibit mammary tumorigenesis induced by carcinogens in rats. Moreover, MSC restores and enhances circadian gene expression by increasing NAD^+/NADH and SIRT1 activity and modulated acetylation of circadian regulatory protein, BMAL1, and histone 3 associated with Per2 gene promoter. This chapter will focus on how the dietary chemopreventive regimen of MSC epigenetically modulates the circadian rhythm at molecular level and how it contributes to its chemopreventive activity.

Keywords

Methylselenocysteine · Nitrosomethylurea · Chemoprevention · Breast cancer · Circadian rhythm · NAD^+ · SIRT1 activity · Acetylation · Period 2 · DNA damage response and repair

List of Abbreviations

AcBMAL1	Acetylated BMAL1
AcH3K9	Acetylated histone 3 lysine 9
CCG	Circadian-controlled gene
CG	Circadian (clock) gene
DDRR	DNA damage response and repair
F344	Fisher 344 rat strain
MSC	Methylselenocysteine
NMU	Nitrosomethylurea
SCN	Suprachiasmatic nucleus

Introduction

Selenium (Se) is a micronutrient essential to numerous biological processes in our body, including antioxidant defense systems, thyroid hormone metabolism, and immune function (Jackson and Combs 2008). Elemental Se is incorporated into selenium amino acids by yeast and plants, especially allium and cruciferous vegetables. As reviewed by El-Bayoumy et al. (2006), the inhibitory effect of Se on mammary carcinogenesis was first found by Ip and coworkers in animals fed with garlic grown in high-selenium soil. Whereas exposure of virgin female rats to a variety of carcinogens during puberty results in close to 100% incidence of mammary carcinomas in many susceptible strains, animals fed with selenium-enriched garlic showed dramatic reductions in tumor incidence. The chemopreventive effects were independent of the carcinogen used or the strain of susceptible rats. Subsequent studies demonstrated that organic forms of Se, especially methylselenocysteine (MSC) , had the highest chemopreventive potential and its inhibitory effect on mammary carcinogenesis was most effective when MSC-enriched diet (3 ppm Se) was continued for 30 days following exposure to the carcinogen, suggesting that MSC mediates its effects during the early stage of mammary carcinogenesis (El-Bayoumy et al. 2006). Subsequent studies by Clark in selenium-deficient populations demonstrated chemopreventive activity of dietary selenium supplementation on several cancers.

The molecular mechanisms of selenium action were extensively investigated, and multiple mechanisms have been proposed, suggesting that the biological effects of this compound are likely highly pleiotropic. As reviewed in the Fang et al. recent publication (Fang et al. 2015a), selenium compounds protect against peroxynitrite-induced DNA damage through glutathione-mediated redox cycling; MSC induces cell senescence through activation of the ATM-mediated DNA repair upon exposure to genotoxic and oxidative stresses. These effects are supported by clinical intervention studies showing that selenium supplementation reduced oxidative DNA damage and cancer incidence in the patients who carry *BRCA1* mutations. Moreover, selenium modulates multiple transcription factor activities involved in circadian regulation, cellular responses to stress, and tumor suppressor function. Recently, selenium also showed epigenetic chemopreventive activity by upregulating tumor suppressor gene expression by inhibiting the activity of histone deacetylases (HDAC) (Hazane-Puch et al. 2016; Xiang et al. 2008) and DNA methyltransferases (DNMT) (Fiala et al. 1998). Particularly, MSC showed to modulate epigenetic marks (DNMT1 expression and H3K9Me3/K4K16Ac) specifically in breast cancer cells (de Miranda et al. 2014). However, the failures of several clinical intervention trials using organic selenium underscored the need for more mechanistic insight for development of rational and effective Se intervention strategies (Ganther 1999).

Dietary MSC Resets and Enhances Circadian Rhythm in Mammary Glands Exposed to Carcinogen

To elucidate the mechanism of MSC-mediated chemoprevention of mammary carcinogenesis, Dr. Zarbl and his graduate student confirmed its chemopreventive efficacy in Fischer 344 rats exposed to mammary tumor specific carcinogen, nitrosomethylurea (NMU) (Zhang and Zarbl 2008). Dietary supplementation with Se (3 ppm Se in the form of MSC) reduced the incidence of mammary carcinogenesis by ~60% compared to a standardized control diet (0.1 ppm Se). Mechanism study with the mammary glands of carcinogen-treated rats maintained on a control or MSC-enriched diet for 30 days showed that during the course of Se-induced chemoprevention, there were significant effects on both the steady-state levels and temporal oscillations of genes involved in circadian rhythm. Moreover, exposure of pubescent female rats to a single carcinogenic dose ablated the rhythmic expression of key circadian genes (CGs) and circadian-controlled genes (CCGs) in the target mammary tissue, but not in the liver. More importantly, a chemopreventive 30-day dietary regimen of MSC reset and enhanced the circadian expression of these genes, especially Period 2 (Per2), in the mammary glands of carcinogen-treated rats (Fang et al. 2010) (Fig. 1). Further, MSC-induced enhancement of circadian gene expression was not observed in the tumors that arose in MSC-treated animals, suggesting that loss of circadian control is required for carcinogenesis and restoration of circadian gene expression might play a critical role in inhibition of tumorigenesis (Zhang and Zarbl 2008). This was the first study to demonstrate a plausible mechanistic link between chemoprevention and enhancement of circadian rhythm.

Overview of Circadian Rhythm and Its Epigenetic Regulation

The circadian clock regulates a wide variety of fundamental biological, physiological, and cellular processes, ranging from gene expression to sleep behavior in a precise and sustained rhythm with a periodicity of ~24 hrs. The circadian clock comprises a biochemical oscillator that is present in the both the suprachiasmatic nucleus (SCN) of the hypothalamus, which functions as the central pacemaker, and in cells comprising most peripheral tissues. As reviewed (Haus and Smolensky 2013; Fu and Kettner 2013), the periodicity of the circadian clock is regulated through the activity of interconnected transcriptional/translational feedback loops that regulate circadian gene expression. Heterodimers of the Bmal1 and either Clock or Npas2 proteins regulate transcription of core circadian genes by binding to E-box elements in their promoter regions. The core circadian genes (CGs) including Per, Cry, ROR, and Rev-ErbAα in turn regulate the express numerous circadian-controlled genes (CCGs), including hormone receptors, growth-associated genes, and DNA damage response and repair (DDRR) genes. As they accumulate in the cytoplasms, heterodimers of Per:Cry proteins undergo posttranslational phosphorylation by casein kinase and are transported to the nucleus, where they repress the activity of the Clock:Bmal1 transcription factor. In this way, core circadian genes feedback and limit their own transcription to set up the rhythmic expression of CG and CCG that regulate numerous signaling pathways.

To synchronize the periodicity among cells in peripheral organs *in vivo* and adjust the phase to changing environmental signals, the intrinsic molecular oscillator is entrained by both external and internal signals. Entrainment is accomplished by a variety of physiological mediators that include light, energy balance, and hormones (melatonin and glucocorticoids) and neural inputs (Balsalobre 2002). In this way, SCN coordinates the behavior, physiological and biological functions of organisms and cells with light-dark cycles, food availability, and a variety of environmental signals. At molecular level, the key regulator of circadian gene expression is Clock

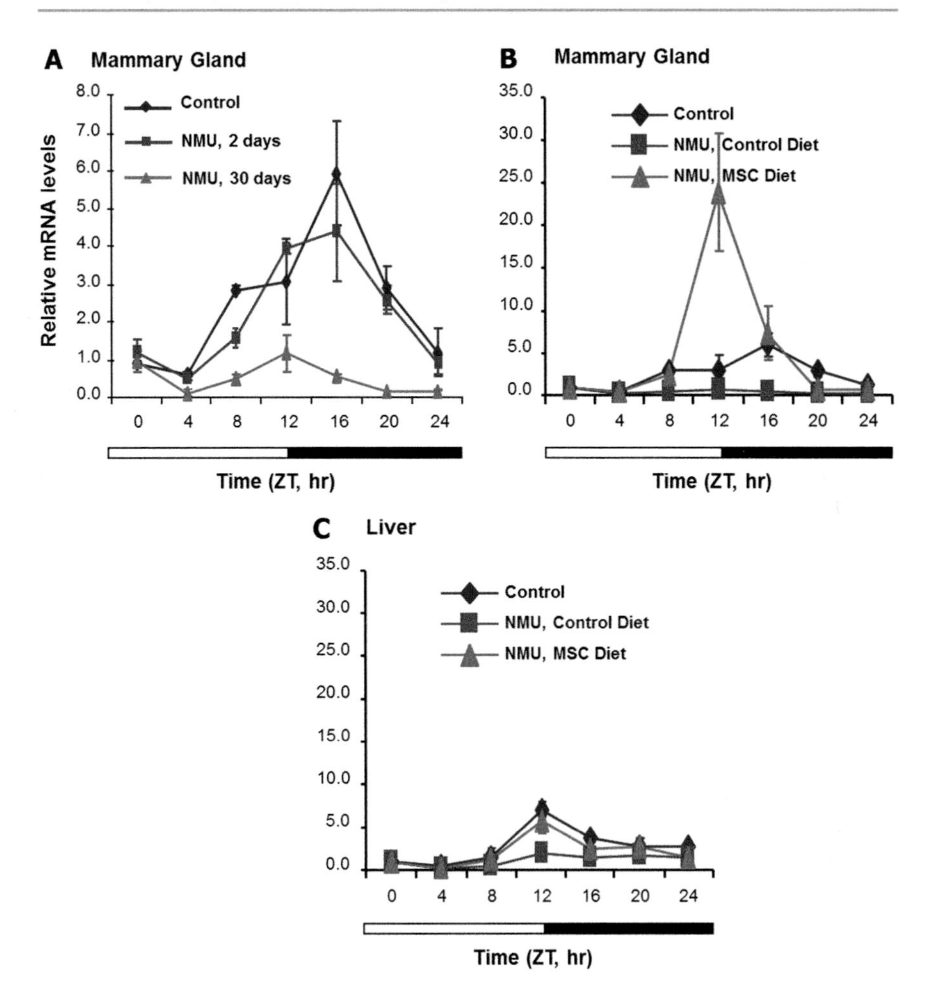

Fig. 1 Chemopreventive MSC reset and enhanced the circadian rhythm of Per2 mRNA expression in mammary gland of carcinogen (NMU)-treated rats. Three rats in each group were sacrificed every 4 h over a 24-h period, beginning at 7 AM. Per2 mRNA expression levels were determined using RT-qPCR with total RNA samples extracted from rat mammary glands, and the results were analyzed with a comparative Ct method. Results were normalized with endogenous control, β-actin, and then with the expression level of the first sample at ZT0. X-axis: Zeitgeber Time (ZT), empty bar indicates light-on from ZT0 (7 AM), and black bar indicates light-off from ZT12 (7 PM); Y-axis: relative mRNA level, shown in mean ± SE (n = 3). (**a**) NMU-treated rats were sacrificed at day 0 (control, *blue*), 2 (*red*), or 30 (*green*) during posttreatment. (**b**) NMU-treated rats were maintained on control diet (*pink*) or MSC-enriched diet (*green*) for 30 days postexposure to NMU; rats untreated and maintained on control diet were used as control group (*blue*). (**c**) The significant changes of Per2 mRNA expression observed in mammary gland (**b**) were not found in the liver. These results were previously published in *Cancer Prev Res.* (Fang et al. 2010) (**b** and **c**) and *Oncotarget* (Fang et al. 2015b) (**a**) (under license CC BY 2.0)

gene, which encodes a transcription factor with inherent protein acetyltransferase activity. Clock acetylates its binding partner Bmal1 and histone 3 on circadian gene promoters (Doi et al. 2006). SIRT1 protein deacetylase counterbalances Clock-directed acetylation (Nakahata et al. 2008). This epigenetic change modulates the enhancer binding activity of CLOCK:BMAL1 to regulate the circadian expression of CGs and CCGs. The regulation of SIRT1 activity by NAD^+ links cellular metabolism, and redox cycling links circadian transcription to control a wide variety of cellular functions, including DDRR and cell proliferation (Rutter et al. 2001; Nakahata et al. 2008). Intracellular NAD^+/NADH ratio and SIRT1 activity have thus emerged as key regulators of circadian gene expression. Therefore, intrinsic molecular oscillators in individual cells can be disrupted or reset through epigenetic reprogramming by external signals, including light, genotoxic stress, nutrients, hormones, and other environmental factors. In this way, circadian clocks integrate a wide variety of environmental and endogenous inputs to maintain normal cellular and physiological homeostasis under changing environmental conditions.

Epidemiological studies indicate that shift workers, including nurses and flight crews, have increased risks of breast, prostate, and colon cancers (Akerstedt et al. 2015; Parent et al. 2012; Knutsson et al. 2013). These findings are corroborated by rodent studies indicating that constant light, light at night, and jet lag increase the incidence of spontaneous and carcinogen-induced tumors and also accelerate tumor growth. Long-term shift work causes repeated changes of sleep and wake times over several months or years and results in delayed resetting of the circadian clock, leading to increased susceptibility to cancer. Chronic jet-lag protocols, which mimic long-term shift work, accelerate carcinogenesis by disrupting circadian expression of major CGs (e.g., Per2) and CCGs (e.g., tumor suppressor and DDRR genes) in rodents. Knocking out or mutating major CGs (e.g., Per2) increases cancer cell growth and accelerates spontaneous and carcinogen-induced tumor development in mice. By contrast, normal or ectopic expression of clock genes induces cell cycle arrest and sensitizes cancer cells to DNA damage-induced apoptosis. Genetic and epigenetic variation of CGs expression is significantly associated with breast cancer risk (Stevens and Zhu 2015). In addition, Per2 links circadian cycle to estrogen receptor signaling (Fang et al. 2010; Gery et al. 2007). These findings indicate that the normally functioning circadian clock is protective against carcinogenesis and that frequent disruption of circadian rhythm is an important tumor-promoting factor.

Given the fact that, in industrialized societies, up to 30% of the workforce (hospital staff, flight crews, janitorial staff, firefighters, police, soldiers, factory workers, etc.) is employed in occupations that require chronic rotation into work shifts that result in exposure to light at night, the health implications of circadian disruption may be significant. In addition, exposure to increasing amount of light due to light pollution, television and computer monitors, smartphones, and other electronics continue to extend the daily period of light exposure (Fang et al. 2015a). Therefore, the International Agency for Research on Cancer classified shift work as a probable human carcinogen (Type 2A) (IARC 2010). Shift work has also been associated with increased risk of metabolic syndrome including obesity and type II diabetes, cardiovascular diseases, and neuropsychological diseases (Fang et al.

2015a). Strategies that mitigate the effects of light at night on sleep patterns and on disruption of circadian rhythm, therefore, have the potential to significantly improve public health. Understanding the mechanisms by which MSC can restore circadian rhythm in those who serve the community by working at night therefore has significant public health implications.

Epigenetic Modulation of Circadian Gene Expression by Methylselenocysteine

In mammals, a key regulator of circadian rhythm is light. Blue light entering the eye activates specialized retinal melanopsin-expressing ganglion cells that innervate the SCN, the central circadian pacemaker. The incoming signals that are integrated are sent to the pineal gland, where they regulate the synthesis and release of the hormone, melatonin. Melatonin synchronizes the rhythms of peripheral cells by binding to and activating its receptor, MTNR1A. Recently, Fang et al. (2010) found that neither NMU nor MSC mediated their effects by altering melatonin secretion; however, both carcinogen and chemopreventive agent changed melatonin receptor expression in mammary gland, albeit in opposite directions. The melatonin receptor gene was also found to be a CCG, indicating that disruption of circadian rhythm in cells could also reduce their sensitivity to melatonin. In addition, neither NMU nor MSC significantly changed the expression of the core circadian regulators (Fang et al. 2010). These results indicate that NMU and MSC affected neither secretion of melatonin that regulate central clock nor expression of circadian regulatory genes, suggesting that the effects NMU and MSC on the expression CG and CCG might be mediated via epigenetic mechanisms (Fig. 2).

In mammary epithelial cells *in vitro*, NMU induced a dose-dependent decrease in intracellular redox cycling NAD^+/NADH and related SIRT1 activity; in contrast, MSC counteracted the effect of NMU, causing a dose-dependent increase of NAD^+/NADH and SIRT1 activity, while NMU decreased both of these epigenetic modulators (Fig. 3). To test the hypothesis that NMU and MSC mediate their epigenetic effects on circadian rhythm by modulating NAD^+-dependent SIRT1 activity, Dr. Fang developed an *in vitro* circadian reporter system in mammary epithelial cells transfected with human PER2 promoter-driven firefly luciferase expression vector. In this *in vitro* model, exposure to NMU induced a dose-dependent ablation of circadian gene expression. More importantly, MSC was able to restore circadian gene expression in NMU-treated cells. The results further indicated that SIRT1 inhibitors had the same effect on circadian cycling as NMU and their effects were also blocked by MSC (Fig. 4). It is important to note that NMU and MSC did not inhibit the activity of purified SIRT1 deacetylase in *ex vivo* assay, indicating that the NMU and MSC changed SIRT1 enzyme activity indirectly through modulating intracellular level of NAD^+, a substrate of SIRT1 activity. These results provided plausible mechanisms of circadian deregulation by NMU and its restoration by a chemopreventive regimen of MSC *in vivo*.

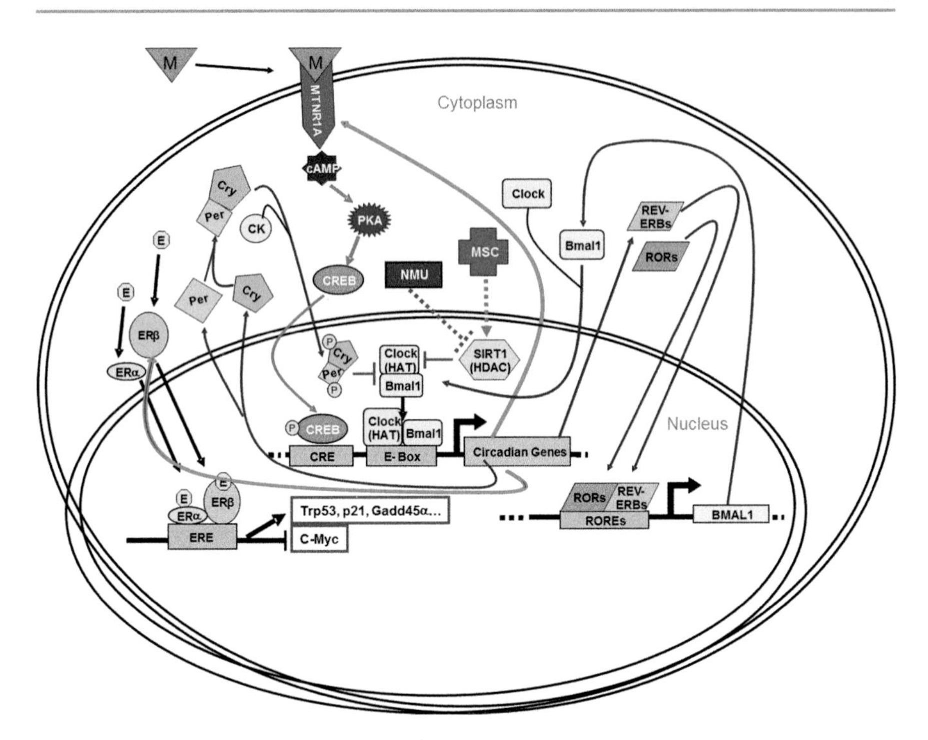

Fig. 2 Hypothetical model for the interaction of NMU and MSC with circadian control network during early stage of carcinogenesis. The proposed model combines the changes in gene expression observed in the Fang et al. (2010) study with known regulating network (*blue arrows*), comprising and modulating the activity of the molecular oscillator in peripheral mammalian cells. The rhythmic expression of the melatonin receptor (MTNR1A), itself a circadian-controlled gene that plays a crucial role in the entrainment and synchronization of peripheral circadian rhythm with central pacemaker, was abolished by NMU and reset by MSC without altering the expression of Bmal1 and Clock. These data are consistent with the possibility that NMU and MSC regulate rhythmic expressions of CGs and CCGs by modifying the activity of Clock:Bmal1. Altered Bmal1:Clock activity would also modulate the expression of ERβ and MTNR1A and other growth regulatory genes as well. Hypothetical positive or negative inputs from NMU and MSC are indicated by *dotted red* or *green lines*, respectively. *E* estrogen, *ERE* estrogen-response element, *M* melatonin, *CK* casein kinase, *PKA* protein kinase A, *CREB* cAMP-response element binding protein, *HAT* histone acetyltransferase, *HDAC* histone deacetylase. This diagram was previously published in *Cancer Prev Res* (Fang et al. 2010)

Mammary glands of rats exposed to vehicles, NMU, or NMU followed by a 30-day dietary regimen of MSC were therefore further examined for effects on circadian oscillation over a 24-h period in (1) the expression of Per2, (2) the ratio of NAD^+/NADH, (3) SIRT1 deacetylase activity, and (4) the acetylation status of circadian regulatory protein and histone tail on the Per2 promoter. The results indicated that the exposure to NMU decreased to circadian expression of Per2 as a function of time after exposure. The effect on circadian expression was first detectable 2 days after exposure and was ablated at 30 days (Fig. 1a). The decrease in circadian expression of Per2 coincided with a parallel decrease in intracellular

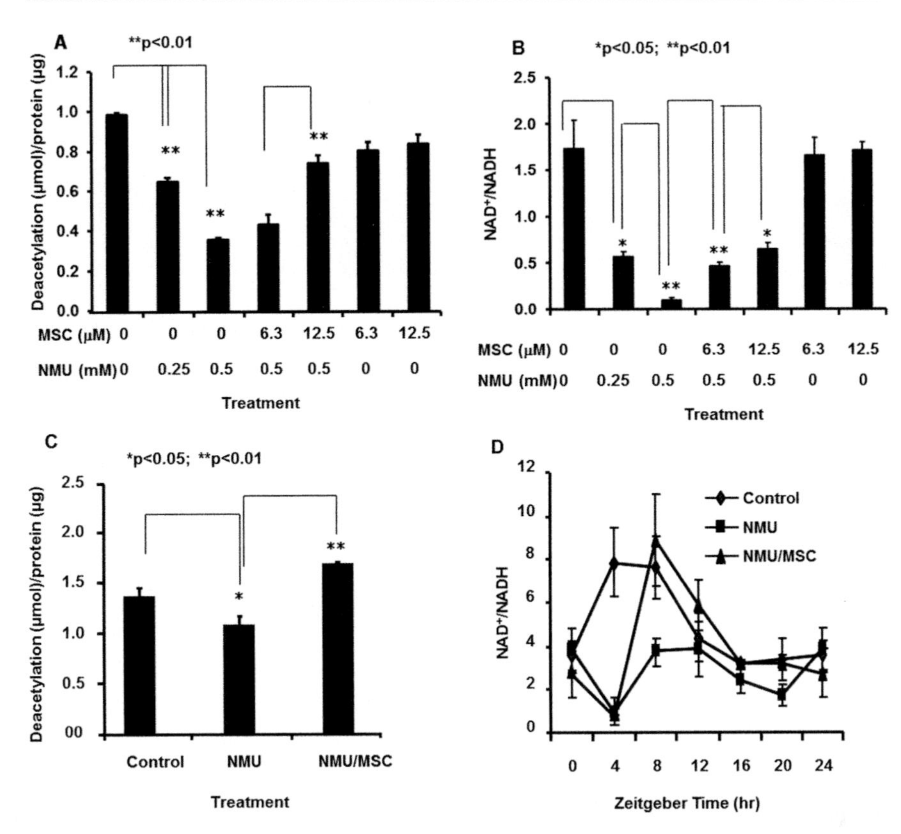

Fig. 3 MSC counteracts NMU to restore SIRT1 activity and NAD^+/NADH in mammary epithelial cells. (**a** and **b**) MCF10A cells were treated with NMU for 1 h, followed by treatment with MSC for 72 h at indicated concentrations. (**a**) Total protein samples extracted from cells were used in determination of SIRT1 activity. (**b**) Cell extracts were prepared and used in NAD^+/NADH quantification. (**c** and **d**) On day 30 postexposure to NMU, three rats per group (control, NMU, or NMU/MSC) were sacrificed every 4 h over a 24-h period, beginning right after lights-on at 7 AM (ZT0). (**c**) Total protein samples extracted from mammary tissues of rats sacrificed at ZT12 subjected to SIRT1 activity determination. (**d**) Tissue extracts prepared from mammary tissues of rats sacrificed at different time points over 24 h subjected to NAD^+/NADH quantification. X-axis, treatment group or zeitgeber time (ZT); Y-axis, SIRT1 activity (deacetylation) (**a** and **c**) or NAD^+/NADH (**b** and **d**), mean ± SE (n = 3). * and ** indicate statistical significance at $p \leq 0.05$ and $p \leq 0.01$, respectively. This result was previously published in *Oncotarget* by Fang et al. (2015b) and is licensed under CC BY 2.0

NAD^+/NADH ratios and a decrease of SIRT1 activity. Consistent with a causal role for these changes, there was also temporal correlation with increased acetylation of Bmal1 and decreased acetylation of histone 3 lysine 9 (H3K9), both of which have been associated with decreased circadian expression. Remarkably, the effects of NMU on NAD^+/NADH cycling, SIRT1 deacetylase activity, and acetylation of histone H3K9 and Bmal1 were reversed by the chemopreventive regimen MSC (Fig. 5) (Fang et al. 2015b).

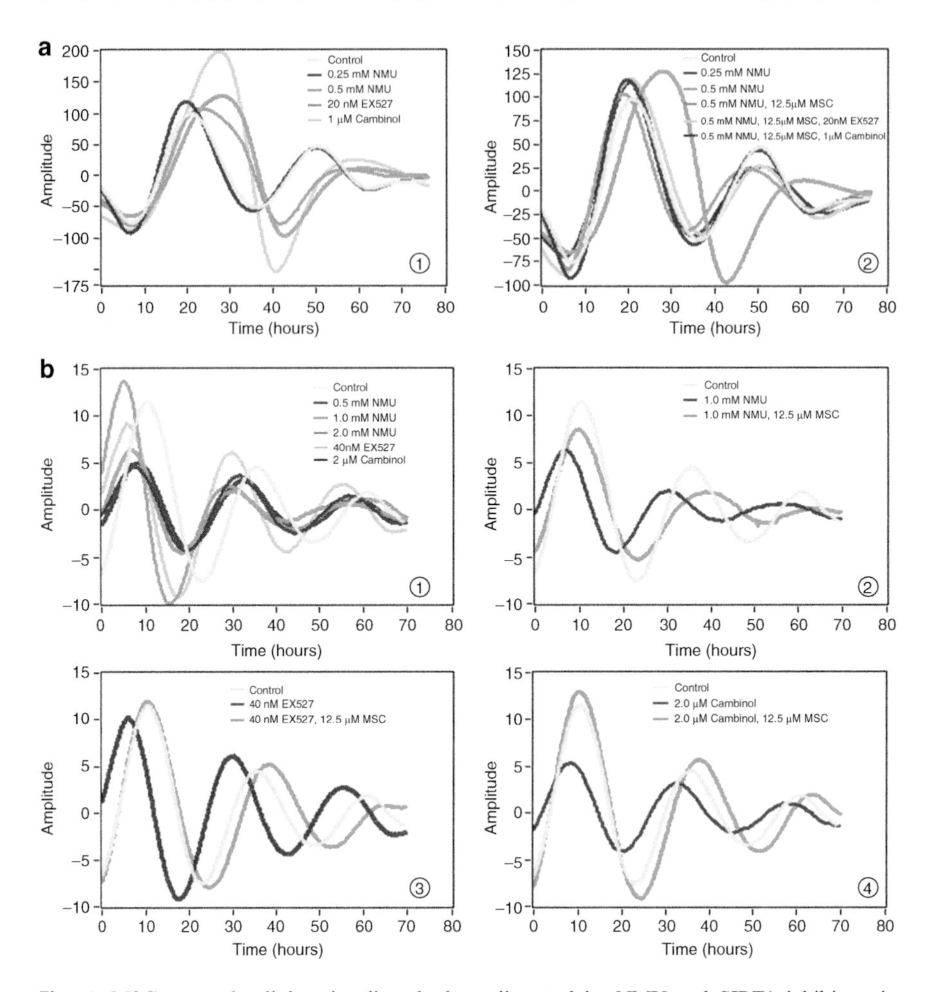

Fig. 4 MSC restored cellular circadian rhythms disrupted by NMU and SIRT1 inhibitors in mammary epithelial cells in vitro. Bioluminescence assays were performed on MCF10A/PER2-dLuc reporter cells after synchronization with 50% horse serum. X-axis, time (hours) (post-NMU treatment time); Y-axis, amplitude. A. Results from transiently transfected cells. ① Cells were treated with 0, 0.25, or 0.5 mM NMU, 20 nM EX527, or 1 μM cambinol for 1 h following synchronization. *Yellow* control, *red* 0.25 mM NMU, *green* 0.5 mM NMU, *blue* 20 nM EX527, *yellow green* 1 μM cambinol. ② Cells were treated with 12.5 μM MSC alone or in combination with 20 nM EX527 or 1 μM cambinol in recording medium following exposure to 0.5 mM NMU. *Yellow* control, *red* 0.25 mM NMU, *green* 0.5 mM NMU, *blue* 0.5 mM NMU + 12.5 μM MSC, *yellow green* 0.5 mM NMU + 12.5 μM MSC + 20 nM EX527, *purple* 0.5 mM NMU + 12.5 μM MSC + 1 μM cambinol. B. Results from stably transfected cells. ① Cells were treated with 0, 0.5, 1.0, or 2.0 mM NMU, 40 nM EX527, or 2 μM cambinol for 1 h after synchronization. *Yellow* control, *red* 0.5 mM NMU, *green* 1.0 mM NMU, *blue* 2.0 mM NMU, *yellow green* 40 nM EX527, *purple* 2 μM cambinol. ② Cells were treated with 12.5 μM MSC in recording medium following exposure to 1.0 mM NMU. *Yellow* control, *red* 1.0 mM NMU, *green* 1.0 mM NMU + 12.5 μM MSC. ③ Cells were treated with 12.5 μM MSC following exposure to 40 nM EX257. *Yellow* control, *red* 40 nM EX257, *green* 40 nM EX257 + 12.5 μM MSC. ④ Cells were treated with 12.5 μM MSC following exposure to 2 μM cambinol. *Yellow* control, *red* 2 μM cambinol, *green* 2 μM cambinol +25 μM MSC. This result was previously published in *Oncotarget* by Fang et al. (2015b) and licensed under CC BY 2.0

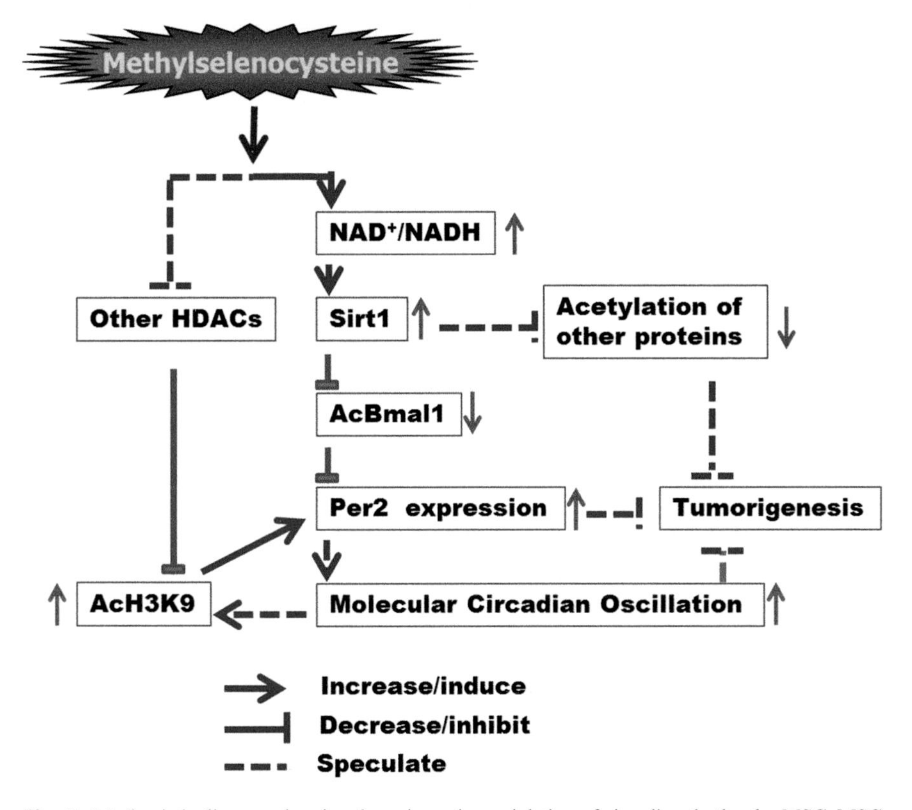

Fig. 5 Mechanistic diagram showing the epigenetic modulation of circadian rhythm by MSC. MSC increases NAD^+/NADH and Sirt1 activity, resulting in decrease of acetylated Bmal1 and enhancement of the circadian expression of Per2. MSC also increases acetylation of histone 3, which may further restore the circadian gene expression. The epigenetic modulation of circadian rhythm by MSC contributes to its chemopreventive activity at the early stage of tumorigenesis. This diagram was previously published in *Oncotarget* by Fang et al. (2015b) and is licensed under CC BY 2.0

Mechanistic Link Between Enhanced Circadian Rhythm and Chemopreventive Activity of MSC on Carcinogen-Treated Mammary Epithelial Cells

Restoration of Circadian Control to DNA Damage Response and Repair

As discussed above, disruption of circadian rhythm by lifestyle, occupational, experimental, and genetic factors increases risk of many chronic diseases, including breast cancer, in humans and other mammals. The most important question is how disruption of circadian rhythm contributes to mammary carcinogenesis. Recent studies indicated that the circadian clock regulates transcription, translation, and/or posttranslational modification of ~10% of genes and/or proteins, including most involved in DDRR signaling pathways and cell cycle progression (Rana and Mahmood 2010). Genotoxic agents can reset circadian rhythm through coordinated, interlocking regulation of CGs (e.g., Per2) and ATM/Chk2 in P53-dependent manner to protect against DNA damaging processes and carcinogenesis. Consistent with these observations, it was demonstrated that disruption of CG expression by a single dose of NMU ablated the rhythmic expression of major clock genes, especially Per2, and DDRR (e.g., p53, p21, and Gadd45α) in mammary gland of F344 rats (Fang et al. 2010). Significantly, dietary MSC restored and enhanced the circadian expression of clock (i.e., Per2) and DDRR genes during early state of carcinogenesis, leading to reduction of mammary tumor incidence by 63% in NMU-treated F344 rats (Fig. 2).

Restoration of Circadian Control to Melatonin and Estrogen Receptor Signalings

Melatonin receptor-mediated signaling and estrogen receptor signaling (especially ERα) are key regulators of mammary gland differentiation and proliferation. Animal studies indicated that disruption of circadian rhythm by exposure to constant light, jet-lagged light, and dark cycles or by ablation of melatonin by pinealectomy increases the incidence and enhances the growth of chemically induced mammary tumors in rat model (Shah et al. 1984). Melatonin was deemed the "chemical expression of darkness" and also known to be a potent antioxidant, suggesting that rhythmic melatonin signaling plays a critical role in protection from carcinogenesis by reducing oxidative stress under normal circadian control (Reiter 1991; Peyrot and Ducrocq 2008). Studies have shown that MSC does not affect systemic melatonin levels; rather it modulates cellular sensitivity to melatonin by enhancing the rhythmic expression of melatonin receptor (Fang et al. 2010).

The major clock gene, Per2, also binds to and accelerates degradation of ERα, suggesting that circadian control provides feedback mechanisms that allows diurnal responses to estrogen stimulation. Significantly, while exposure of rats to a carcinogenic dose of NMU disrupted circadian regulation of ERα, a chemopreventive regimen of MSC restored circadian expression that was temporally correlated with enhanced Per2 expression. Although ERβ is expressed in mammary cells at lower levels, it is exquisitely sensitive to circadian regulation. In contrast to the proliferative effects of ERα, activation of ERβ by estrogen binding is thought to arrest the cell cycle and induce mammary cell differentiation. Our results indicated that restoration of circadian rhythm contributes to the inhibitory effect of MSC on mammary carcinogenesis by enhancing DNA repair capacity and restoration of ERβ-mediated inhibition of cell proliferation and promotion of differentiation (Fang et al. 2010).

Role of MSC-Induced Enhancement of Redox Cycling on NAD^+/NADH and SIRT1 Activity in Chemoprevention

Cellular NAD^+ levels are maintained by cell metabolism but can be depleted by poly (ADP-ribose) polymerase (PARP)-mediated DNA repair after genotoxic stress (Druzhyna et al. 2000). Decreased NAD^+ reduces SIRT1 deacetylase activity and activates nicotinamide phosphoribosyltransferase (Nampt), the rate-limiting enzyme in NAD^+ biosynthesis (Ramsey et al. 2009).

SIRT1 is involved in several physiological functions, including control of gene expression, cell cycle regulation, apoptosis, DNA repair, metabolism, and circadian rhythm (Lavu et al. 2008; Houtkooper et al. 2012). Gene silencing by the SIRT1-mediated chromatin deacetylation is directly correlated with a longer life-span in yeast and worms (Yang and Sauve 2006). SIRT1 deacetylates various proteins that function as transcriptional factors, including the p53 tumor suppressor protein, a modulator of DDRR, and apoptosis (Smith 2002). SIRT1 removes acetyl group from lysine 537 of the Bmal1 transcription factor, regulating the transcriptional activity of the Clock:Bmal1 heterodimers associated with E-box elements in circadian gene promoters. SIRT1 also can regulate circadian gene expression by deacetylating H3 within promoter regions of CG (Nakahata et al. 2008), in turn, acetylation of histone is also controlled by circadian rhythm (Etchegaray et al. 2003). Importantly, a chemopreventive regimen of MSC-induced SIRT1 activity was correlated with restoration and enhancement of circadian histone acetylation (AcH3K9). Given the fact that the rhythmic AcH3K9 is also controlled by the circadian clock (Etchegaray et al. 2003), the enhanced AcH3K9 rhythm might thus be regulated by increased circadian rhythm that was modulated by decreased-AcBMAL1 resulting from increased NAD^+-dependent SIRT1 activity by MSC. Alternatively, the inhibitory effect of MSC on other histone deacetylases could also contribute to the increased level of AcH3K9 *in vivo* (Fang et al. 2015a). These results suggest that enhancement of NAD^+-dependent SIRT1 activity by MSC can restore circadian gene expression disrupted by exposure to carcinogen. The resulting increase in molecular circadian oscillation may resynchronize the AcH3K9 cycling in the promoters of CG and CCG to further amplify the rhythmic gene expression (Fig. 5).

In vitro mechanistic studies using SIRT1 inhibitors further support the hypothesis that inhibition of NAD^+-dependent SIRT1 activity is responsible for the disruptive effects of NMU on circadian control on DDRR in mammary glands of F344 rats (Fang et al. 2015b). Consistent with these findings, reduction of NAD^+ by decreased expression of Nampt showed to promote breast cancer metastasis (Santidrian et al. 2014); systemic enhancement of NAD^+/NADH through treatment with NAD^+ precursor (nicotinamide riboside) inhibited mammary tumor growth and metastasis and increased survival in rodent models (Santidrian et al. 2013). The beneficial effects of exercise, calorie restriction, and diet polyphenols (i.e., resveratrol) on the prevention of cancer and life-span extension also appear to be mediated by the NAD^+/NADH and NAD^+-dependent SIRT1 activity (Yang et al. 2007). Together, these findings suggest a mechanistic link between redox capacities, impaired circadian response to genotoxic stress, decreased DNA repair capacity, and increased susceptibility to carcinogenesis. Dietary factors (e.g., MSC) that increase NAD^+-dependent SIRT1 activity may therefore play a protective role during the early stages of carcinogenesis by restoration of circadian control on DDRR function.

Significance to Public Health

As mentioned previously, up to 30% of the workforce that regularly engages in shift work is at increased risk of developing cancers and other chronic diseases. Understanding of how circadian rhythm coordinates cellular responses to genotoxic stresses and of how it can be reversed or prevented by dietary compounds on circadian regulation thus has significant implications for public health. Our studies clearly demonstrated that epigenetic reprogramming of circadian gene expression with a 30-day dietary antioxidant, MSC, restored circadian expression of major clock genes (e.g., Per2) and recoupled Per2 to hormone receptor genes (e.g., ERβ) and DDRR genes during the early stage of carcinogenesis. Most importantly, it reduced mammary tumor incidence by ~63% at the later stage of rat mammary tumorigenesis. These findings are the first to link the restoration of circadian rhythm and chemoprevention and support the feasibility of reducing morbidity and mortality associated with chronic shift work by dietary MSC supplementation in shift workers.

To test this possibility, an intervention trial has been initiated in shift workers to determine if the effects of shift work on circadian gene expression in peripheral blood cells could be mitigated in shift workers by MSC. In the first phase of the study (phase I) in rotating shift workers, it has been found that Per2 mRNA expression in blood cells was significantly impacted by shift, suggesting that Per2 can be used to monitor the effects of MSC in shift workers' intervention trials (phase II) (Fang et al. 2015a). The second phase of this study is a 30-day blinded randomized placebo-controlled trial of MSC on indicators (e.g., Per2 and ERβ) of circadian rhythm in 100 night shift workers. The study was recently completed and results are currently being analyzed. It is expected to see a significant increase of Per2 and ERβ in shift workers in the MSC intervention arm relative to placebo. If this is the case, the results could form the basis for a prospective longitudinal intervention studies with dietary MSC in the restoration of peripheral circadian rhythm in shift workers.

Future Direction

Although the role of SIRT1 gene in carcinogenesis remains controversial, accumulating evidences indicated a significant association between the increases in NAD^+ level and SIRT1 activity by fasting, exercise, and other chemopreventive regimens and the prevention of aging-related diseases. The beneficial effects of various SIRT1-activating compounds on stress-related disease processes may be partially mediated through their effects on circadian rhythm. It will be interesting to test the effects of various toxicants and chemopreventive agents that modulate NAD^+ and SIRT1 activity on circadian gene expression. The circadian reporting vector that we recently established can be used in large panel of cell lines to determine if MSC-mediated restoration of circadian rhythm disrupted by environmental factors is general mechanism of its preventive role in different organs. The ability of dietary MSC to modulate circadian rhythm by enhancing SIRT1 activity may have therefore significant implication in other aging-related chronic diseases. In light of findings to date, further studies are warranted to assess whether the chemopreventive potential of MSC can be enhanced when combined with other agents that prevent NAD^+ depletion and activate SIRT1 activity during the early stage of tumorigenesis.

The most recent studies demonstrated that rat strains with differential genetic susceptibility to mammary carcinogenesis showed differential circadian regulation on DDRR. Genetic differences that affect NAD^+/NADH and SIRT1 activity contribute to this differential circadian control and differential susceptibility to mammary carcinogenesis (Fang et al. 2017). More in-depth mechanistic studies are needed to further study how the NAD^+-SIRT1 pathway integrates cellular responses from different internal conditions and external signals/stressors (Fig. 6). In addition, NAD^+/NADH and SIRT1 activity could also be developed as mechanistic biomarkers of disrupted circadian rhythm or as molecular targets of chemoprevention in shift workers. These biomarkers or molecular targets would be very useful in future intervention studies with dietary agents in women at elevated risk of breast cancer due to shift work.

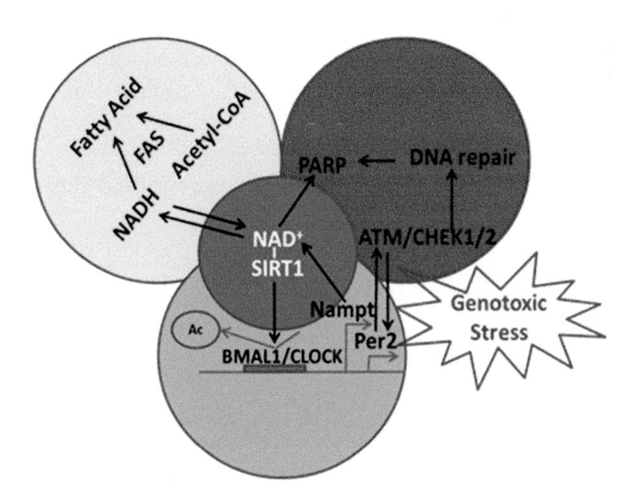

Fig. 6 Working hypothesis of how circadian rhythm integrates signals from endogenous factors and genotoxic stresses to couple DNA damage response and repair to circadian control. NAD^+-dependent SIRT1 deacetylase activity plays a central role in circadian regulation of DNA damage response and repair at differential metabolic conditions, contributing to differential susceptibility to breast cancer. *FAS* fatty acid synthesis, *PARP* poly (ADP-ribose) polymerase. This diagram was previously published in *Oncotarget* by Fang et al. (2017) and is under Creative Commons Attribution 3.0 License

Mini-dictionary of Terms

- **Chemoprevention** – refers to prevention of cancer with chemical compound.
- **Circadian rhythm** – about 24-h biobehavioral, physiological, and biochemical rhythm in living organisms. At molecular level, the expression of some genes also has approximately 24-h rhythm.
- **Epigenetics** – refers to changes in a chromosome that affect gene activity and expression without changing DNA sequence. Epigenetic changes may cause heritable phenotypic change that is not caused by a modification of the genome sequence.
- **Shift work** – is a working schedule designed to make use of, or provide service across, 24 h a day regardless day or night time. The day is typically divided into two or three shifts, set 12-h or 8-h time period during which different groups of workers perform their duties. The most common shift schedules are 12-h day or night shifts and 8-h rotating shift.
- **Jet lag** – refers to circadian dysrhythmia caused by frequent cross time-zone flight.
- **Circadian control** – refers to synchronization of different biological rhythm by central and peripheral circadian clock.
- **NAD^+/NADH** – reflects cellular redox cycling driven by cellular metabolism.

Key Facts of Methylselenocysteine (MSC)

- MSC is the most effective organic form of selenium on prevention of cancer.
- MSC increases cellular redox cycling and NAD^+ –protein deacetylating enzyme activity.
- MSC modulates acetylation of protein and histone to regulate circadian rhythm.
- MSC restores circadian expression of major clock genes and circadian-controlled genes.
- MSC enhances circadian control on DNA damage response and repair.
- These studies are the first linking the chemopreventive activity of MSC to circadian rhythm.

Summary Points

- This chapter focuses on epigenetic modulation of methylselenocysteine (MSC) on circadian rhythm.
- MSC is an organic form of selenium presenting in plants, especially allium and cruciferous vegetables.
- Chemopreventive regimens of MSC inhibit early stages of mammary carcinogenesis in carcinogen-induced rat models.
- MSC increases NAD^+/NADH and SIRT1 activity.

- SIRT1 modulates acetylation of BMAL1 and histone 3 on the Period 2 gene promoter.
- MSC restored and enhances circadian expression of major clock genes, especially Period 2.
- SIRT1 also restores and enhances circadian control on DNA damage response and repair, contributing to inhibitory effects of MSC on carcinogenesis.

References

Akerstedt T, Knutsson A, Narusyte J et al (2015) Night work and breast cancer in women: a Swedish cohort study. BMJ Open 5:e008127

Balsalobre A (2002) Clock genes in mammalian peripheral tissues. Cell Tissue Res 309:193–199

De Miranda JX, Andrade fde O, Conti A et al (2014) Effects of selenium compounds on proliferation and epigenetic marks of breast cancer cells. J Trace Elem Med Biol 28:486–491

Doi M, Hirayama J, Sassone-Corsi P (2006) Circadian regulator CLOCK is a histone acetyltransferase. Cell 125:497–508

Druzhyna N, Smulson ME, Ledoux SP et al (2000) Poly(ADP-ribose) polymerase facilitates the repair of N-methylpurines in mitochondrial DNA. Diabetes 49:1849–1855

El-Bayoumy K, Sinha R, Pinto J et al (2006) Cancer chemoprevention by garlic and garlic-containing sulfur and selenium compounds. J Nutr 136:864S–869S

Etchegaray JP, Lee C, Wade P et al (2003) Rhythmic histone acetylation underlies transcription in the mammalian circadian clock. Nature 421:177–182

Fang MZ, Zhang X, Zarbl H (2010) Methylselenocysteine resets the rhythmic expression of circadian and growth-regulatory genes disrupted by nitrosomethylurea in vivo. Cancer Prev Res (Phila) 3:640–652

Fang MZ, Ohman-Strickland P, Kelly-Mcneil K et al (2015a) Sleep interruption associated with house staff work schedules alters circadian gene expression. Sleep Med 16:1388–1394

Fang M, Guo WR, Park Y et al (2015b) Enhancement of NAD(+)-dependent SIRT1 deacetylase activity by methylselenocysteine resets the circadian clock in carcinogen-treated mammary epithelial cells. Oncotarget 6:42879–42891

Fang M, Ohman-Strickland PA, Kang H-G et al (2017) Uncoupling genotoxic stress responses from circadian control increases susceptibility to mammary carcinogenesis. Oncotarget. 16;8:32752–32768

Fiala ES, Staretz ME, Pandya GA et al (1998) Inhibition of DNA cytosine methyltransferase by chemopreventive selenium compounds, determined by an improved assay for DNA cytosine methyltransferase and DNA cytosine methylation. Carcinogenesis 19:597–604

Fu L, Kettner NM (2013) The circadian clock in cancer development and therapy. Prog Mol Biol Transl Sci 119:221–282

Ganther HE (1999) Selenium metabolism, selenoproteins and mechanisms of cancer prevention: complexities with thioredoxin reductase. Carcinogenesis 20:1657–1666

Gery S, Virk RK, Chumakov K et al (2007) The clock gene Per2 links the circadian system to the estrogen receptor. Oncogene 26:7916–7920

Haus EL, Smolensky MH (2013) Shift work and cancer risk: potential mechanistic roles of circadian disruption, light at night, and sleep deprivation. Sleep Med Rev 17:273–284

Hazane-Puch F, Arnaud J, Trocme C et al (2016) Sodium selenite decreased HDAC activity, cell proliferation and induced apoptosis in three human glioblastoma cells. Anti Cancer Agents Med Chem 16:490–500

Houtkooper RH, Pirinen E, Auwerx J (2012) Sirtuins as regulators of metabolism and healthspan. Nat Rev Mol Cell Biol 13:225–238

IARC (2010) Painting firefighting, and shiftwork. IARC Monogr Eval Carcinog Risks Hum 98:563–764

Jackson MI, Combs GF Jr (2008) Selenium and anticarcinogenesis: underlying mechanisms. Curr Opin Clin Nutr Metab Care 11:718–726
Knutsson A, Alfredsson L, Karlsson B et al (2013) Breast cancer among shift workers: results of the WOLF longitudinal cohort study. Scand J Work Environ Health 39:170–177
Lavu S, Boss O, Elliott PJ et al (2008) Sirtuins–novel therapeutic targets to treat age-associated diseases. Nat Rev Drug Discov 7:841–853
Nakahata Y, Kaluzova M, Grimaldi B et al (2008) The NAD+–dependent deacetylase SIRT1 modulates CLOCK-mediated chromatin remodeling and circadian control. Cell 134:329–340
Parent ME, El-Zein M, Rousseau MC et al (2012) Night work and the risk of cancer among men. Am J Epidemiol 176(9):751
Peyrot F, Ducrocq C (2008) Potential role of tryptophan derivatives in stress responses characterized by the generation of reactive oxygen and nitrogen species. J Pineal Res 45:235–246
Ramsey KM, Yoshino J, Brace CS et al (2009) Circadian clock feedback cycle through NAMPT-mediated NAD+ biosynthesis. Science 324:651–654
Rana S, Mahmood S (2010) Circadian rhythm and its role in malignancy. J Circadian Rhythms 8:3
Reiter RJ (1991) Melatonin: the chemical expression of darkness. Mol Cell Endocrinol 79: C153–C158
Rutter J, Reick M, LC W et al (2001) Regulation of clock and NPAS2 DNA binding by the redox state of NAD cofactors. Science 293:510–514
Santidrian AF, Matsuno-Yagi A, Ritland M et al (2013) Mitochondrial complex I activity and NAD +/NADH balance regulate breast cancer progression. J Clin Invest 123:1068–1081
Santidrian AF, Leboeuf SE, Wold ED et al (2014) Nicotinamide phosphoribosyltransferase can affect metastatic activity and cell adhesive functions by regulating integrins in breast cancer. DNA Repair (Amst) 23:79–87
Shah PN, Mhatre MC, Kothari LS (1984) Effect of melatonin on mammary carcinogenesis in intact and pinealectomized rats in varying photoperiods. Cancer Res 44:3403–3407
Smith J (2002) Human Sir2 and the 'silencing' of p53 activity. Trends Cell Biol 12:404–406
Stevens RG, Zhu Y (2015) Electric light, particularly at night, disrupts human circadian rhythmicity: is that a problem? Philos Trans R Soc Lond Ser B Biol Sci 370:20140120
Xiang N, Zhao R, Song G et al (2008) Selenite reactivates silenced genes by modifying DNA methylation and histones in prostate cancer cells. Carcinogenesis 29:2175–2181
Yang T, Sauve AA (2006) NAD metabolism and sirtuins: metabolic regulation of protein deacetylation in stress and toxicity. AAPS J 8:E632–E643
Zhang X, Zarbl H (2008) Chemopreventive doses of methylselenocysteine alter circadian rhythm in rat mammary tissue. Cancer Prev Res (Phila) 1:119–127

Proanthocyanidins and Epigenetics 101

Cinta Bladé, Anna Arola-Arnal, Anna Crescenti, Manuel Suárez, Francisca I. Bravo, Gerard Aragonès, Begoña Muguerza, and Lluís Arola

Contents

Abstract

Proanthocyanidins, also known as condensed tannins, are the most abundant flavonoids in the human diet. Proanthocyanidins provide beneficial health effects

C. Bladé (✉) · A. Arola-Arnal · M. Suárez · F. I. Bravo · G. Aragonès · B. Muguerza
Nutrigenomics Research Group, Department of Biochemistry and Biotechnology, Universitat Rovira i Virgili (URV), Tarragona, Spain
e-mail: mariacinta.blade@urv.cat; anna.arola@urv.cat; manuel.suarez@urv.cat; franciscaisabel.bravo@urv.cat; gerard.aragones@urv.cat; begona.muguerza@urv.cat

A. Crescenti
Nutrition and Health Research Group, Technological Center for Nutrition and Health (EURECAT-CTNS), Tecnio, Campus of International Excellence Southern Catalonia (CEICS), Reus, Spain
e-mail: anna.crescenti@ctns.cat

L. Arola
Nutrigenomics Research Group, Department of Biochemistry and Biotechnology, Universitat Rovira i Virgili (URV), Tarragona, Spain

Nutrition and Health Research Group, Technological Center for Nutrition and Health (EURECAT-CTNS), Tecnio, Campus of International Excellence Southern Catalonia (CEICS), Reus, Spain
e-mail: lluis.arola@urv.cat

V. B. Patel, V. R. Preedy (eds.), *Handbook of Nutrition, Diet, and Epigenetics*,
https://doi.org/10.1007/978-3-319-55530-0_16

by mitigating inflammation, oxidative stress, and risk factors associated with metabolic syndrome. Moreover, proanthocyanidins protect against cardiovascular disease and some cancers. Though several molecular mechanisms have been suggested to explain the beneficial effects of proanthocyanidins, epigenetic mechanisms have recently emerged as important mediators of the effects of proanthocyanidins. This chapter focuses on studies showing that proanthocyanidins can regulate cell functionality by modulation of miRNA expression, DNA methylation, and histone acetylation and methylation.

Keywords

Proanthocyanidins · Procyanidins · Flavonoids · miRNAs · DNA methylation · Histone acetylation

List of Abbreviations

CVD	Cardiovascular disease
DNMT	DNA methyltransferase
FXR	Farnesoid X receptor
HATs	Histone acetyltransferases
HDAC	Histone deacetylases
MAPK	Mitogen-activated protein kinases
MetS	Metabolic syndrome
miRNA	microRNA
NF-κB	Nuclear factor kappa B
PBMCs	Peripheral blood mononuclear cells

Introduction

Proanthocyanidins, also known as condensed tannins, are a group of polyphenols in the flavonoid family. Specifically, they belong to the flavanol subclass. This family of compounds is widespread among fruits and vegetables, and proanthocyanidins are considered among the most abundant phenolic compounds after lignins (Gu et al. 2003). Proanthocyanidins are highly distributed in plant foods; Table 1 shows the distribution of proanthocyanidins in different food sources.

Several studies have sought to estimate the intake of proanthocyanidin. For example, the American intake of proanthocyanidins, based on data from the USDA, has been estimated to be 90–300 mg/day (Vogiatzoglou et al. 2014; Wang et al. 2011b), while the daily intake in the Spanish population is approximately 570 mg (Arranz et al. 2010).

Structurally, proanthocyanidins are monomers or polymers composed of repetitions of flavanol units (Neilson et al. 2016). However, there is controversy over the inclusion of monomers in the definition, and several authors only include oligomers and polymers.

The most important proanthocyanidin monomers are catechin and epicatechin. Less abundant natural proanthocyanidins include (epi)afzelechin, (epi)gallocatechin,

Table 1 Estimated proanthocyanidin content in different food sources

	Proanthocyanidins (mg/100 g fresh weight foods or mg/L of beverages)							
	Dimers						Trimers	
Food	Monomer	B1	B2	B3	B4	C1	EEC	T2
Fruits								
Apple	8.4	5.7	10.0	0.2		7.0	1.4	
Apricots	6.4	0.1		0.05			0.01	
Blackberry	12.4	0.2	0.8	0.3			0.06	
Custard apple	6.3	1.3	7.0	0.3	2.5	3.5	1.0	
Grape (green)	2.2	0.6	0.06	0.2	0.4	0.07	0.13	
Nectarine, whole	4.7	9.9						
Persimmon	0.8	0.1		0.01			0.04	
Peaches	13.5	25.8	2.3	2.4	0.1	2.5	0.3	
Pears	3.5	0.01	0.7	0.005		0.2	0.01	
Plums	6.8	8.8	5.2	1.0	1.0	10.1	7.7	
Strawberries	6.8	0.6	0.03	1.1	0.1		0.5	
Cereals								
Barley	1.2			10.9				
Legumes								
Common beans, raw	5.3	1.2	0.1	0.8				
Green beans, raw	1.1	0.3	0.6	0.5				
Other								
Apple juice	9.7	0.9	7.9			30.0		
Grape (green) juice	2.5	0.1	0.3	0.9				
Cocoa, powder	271.1	112.0	71.6			23.8		
tea (black), infusion	49.5	3.7	2.5	0.5	1.8	0.8		
tea (green), infusion	65.7	0.6	0.8	0.4	1.8	1.1		
Wine (red)	11.5	4.1	5.0	9.5	7.3	2.6		0.1
Wine (white)	2.1	0.008	0.005	0.003	0.007			

Information obtained from the database Phenol-Rxplorer (Neveu et al. 2010; Rothwell et al. 2012; Rothwell et al. 2013)

(epi)fisetinidol, and (epi)robinetinidol (Rasmussen et al. 2005). Table 2 shows the nomenclature used for some proanthocyanidins depending on the monomer constituents. Figure 1 shows the basic structure of these monomers and the linkages between them.

The number of monomers included in each proanthocyanidin molecule (i.e., its degree of polymerization) can range from dimers to complex structures formed from multiple monomeric proanthocyanidin units (Dixon et al. 2005). In general, the most

Table 2 Nomenclature of the different types of proanthocyanidins based on their monomeric composition

Proanthocyanidins		Substitution pattern of monomers[a]					
(Monomers)$_n$	Monomers	R_1	R_2	R_3	R_4	R_5	R_6
Proapigeninidin	Apigeniflavan	H	OH	H	H	OH	H
Probutinidin	Butiniflavan	H	H	H	OH	OH	H
Procassinidin	Cassiaflavan	H	H	H	H	OH	H
Procyanidin	Catechin	OH	OH	H	OH	OH	H
Prodelphinidin	Gallocatechin	OH	OH	H	OH	OH	OH
Prodistenidin	Distenin	OH	OH	H	H	H	H
Profisetinidin	Fisetinidol	OH	H	H	OH	OH	H
Proguibourtinidin	Guibourtinidol	OH	H	H	H	OH	H
Proluteolinidin	Luteoliflavan	H	OH	H	OH	OH	H
Promelacacinidin	Mesquitol	OH	H	OH	OH	OH	H
Promopanidin	Mopanane	OCH_2-	H	H	OH	OH	H
Propelargonidin	Afzelechin	OH	OH	H	H	OH	H
Propeltogynidin	Peltogynane	OCH_2-	H	H	H	OH	OH
Prorobinetinidin	Robinetinidol	OH	H	H	OH	OH	OH
Proteracacinidin	Oritin	OH	H	H	H	OH	H
Protricetinidin	Tricetiflavan	H	OH	H	OH	OH	OH

[a]Side chain (R) are represented in Fig. 1a.

common degree of polymerization is from two to ten in fruits, grains, and nuts, whereas in other foods, this value can be as high as 190 (Neilson et al. 2016).

Proanthocyanidins can also be classified as A type and B type depending on the atoms involved in the linkage between the monomers. A-type proanthocyanidins are less abundant in nature than B type. A-type proanthocyanidins have been found in fruits (such as plums, cranberries, avocados, and peanuts) as well as in herbs such as cinnamon and curry (Gu et al. 2003). Some examples of A-type proanthocyanidins are procyanidin dimers A1 and A2. B-type proanthocyanidins can also be found in fruits such as grapes, banana, cocoa, and apricot, cereals such as sorghum and barley, and nuts such as almonds and pecans, among other sources (Gu et al. 2003). Within the B-type proanthocyanidins, several structures have been identified, such as dimers B1 to B8 and C1 (He et al. 2008).

Proanthocyanidins can undergo natural modifications such as methylation, acylation, or glycosylation (Rasmussen et al. 2005) thereby increasing their complexity. For example, proanthocyanidins in grape seeds can be esterified with gallic acid (He et al. 2008).

Biochemical and Physiological Effects of Proanthocyanidins

Epidemiological studies suggest that proanthocyanidin consumption is associated with a low incidence of cardiovascular disease CVD (McCullough et al. 2012) and some cancers (Wang et al. 2011a). Proanthocyanidins may protect from diseases

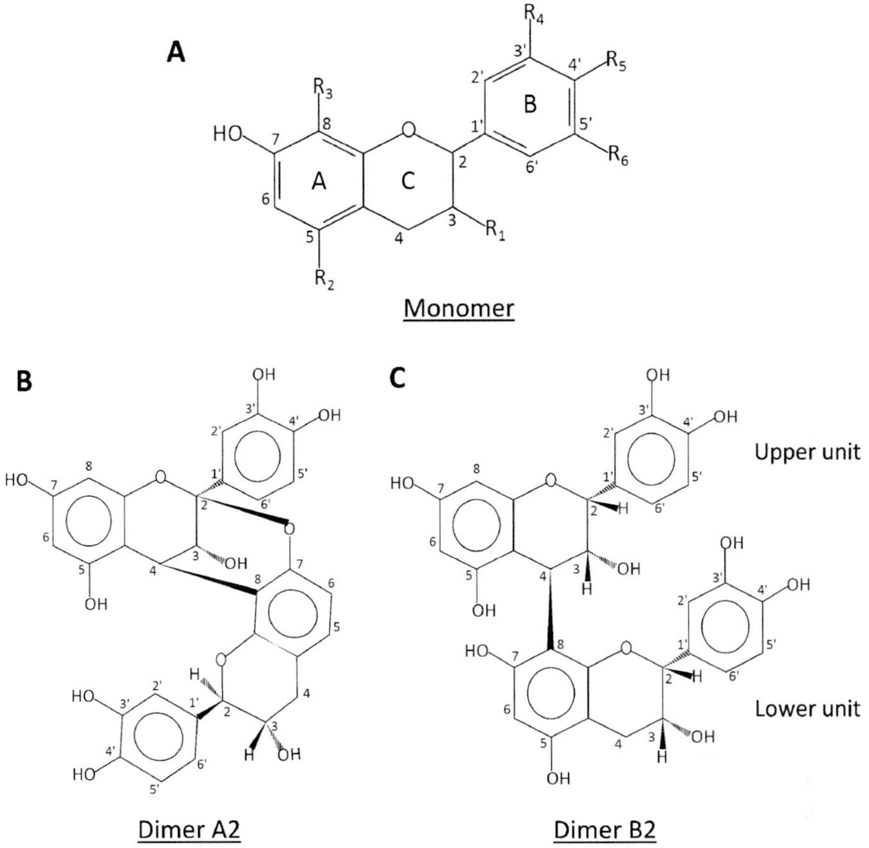

Fig. 1 Chemical structure of proanthocyanidins. Basic structure of proanthocyanidins monomers (a) and example of A-type (b) and B-type (c) structures

because they mitigate many "overarching processes" such as inflammation and metabolic and oxidative stresses (Bladé et al. 2016).

Related to oxidative stress, proanthocyanidins inhibit the oxidation of low-density lipoproteins (LDLs), exhibit antigenotoxic effects (Llópiz et al. 2004), modulate antioxidant enzyme activity (Puiggros et al. 2005), and have a clear antioxidant effect in the liver (Fernández-Iglesias et al. 2014).

The effects of proanthocyanidins on inflammation are well established. Proanthocyanidins modulate various mediators of inflammation (e.g., eicosanoids, cytokines, and nitric oxide (NO)) as well as the nuclear factor kappa B (NF-κB) and mitogen-activated protein kinase (MAPK) pathways (revised in Martinez-Micaelo et al. 2012). The regulation of NF-κB has also been involved in cell proliferation and oncogenic processes. Thus, several studies have suggested that proanthocyanidins may exert significant anticancer effects through the suppression of NF-κB.

In addition to the antioxidant and anti-inflammatory properties, proanthocyanidins also act as antihypertensive agents (Pons et al. 2015), improve lipid metabolism (Bladé et al. 2010), limit adipogenesis (Pinent et al. 2006), modulate glucose metabolism (Gonzalez-Abuin et al. 2015), and reestablish appropriate leptin signaling in both the hypothalamus and peripheral tissues (Ibars et al. 2016). Moreover, unabsorbed proanthocyanidins may control satiety and food intake through an incretin-like action, the stimulation of GLP-1/DPP4 activity (Gonzalez-Abuin et al. 2015), and by the regulation of gastrointestinal tract-brain signals (Salvadó et al. 2015).

Molecular Mechanisms by Which Proanthocyanidins Modulate Cell Functionality

For decades, proanthocyanidins have been considered "simple" antioxidant molecules. However, the low bioavailability of proanthocyanidins, even after the consumption of proanthocyanidin-rich foods, has led researches to question the correlation between their direct antioxidant activity and their beneficial health effects (Hollman et al. 2011). Thus, a direct antioxidant effect of proanthocyanidins as free radical scavengers is unlikely to be physiologically relevant in most tissues, except for those exposed to a high concentration of these compounds such as the gastrointestinal tract and liver.

Now, it is recognized that proanthocyanidins modulate cell functionality by modifying specific proteins and enzymes (Fraga et al. 2010). These specific interactions can result in biological effects depending on the function of the protein involved, including the modification of enzymatic activities, receptor-ligand binding, and the interaction between transcription factors and specific sites in DNA, among others. In vivo and in vitro studies have shown that proanthocyanidins regulate lipid homeostasis by the activation of several nuclear receptors, such as Farnesoid X receptor (FXR) and nuclear receptor small heterodimer partner (NR0B2/SHP) in hepatic cells (Del Bas et al. 2009, 2008). Several in vitro studies have demonstrated that proanthocyanidins can also interact with phospholipid moieties of cell membranes, thereby indirectly affecting cell function by modifying cell membrane structure, fluidity, density, and electrical properties.

Many studies support the capacity of proanthocyanidins to modulate cell functionality by an epigenetic mechanism. Thus, the next sections focus on the effects of proanthocyanidins on microRNAs (miRNAs), DNA methylation, and histone modifications. The majority of these studies have been performed in vitro using cell lines; a considerably smaller number of studies feature in vivo experiments on animals, and very few studies include human clinical trials. Furthermore, most of the studies have been performed with food polyphenol extracts with a variable concentration and composition of proanthocyanidins, including extracts with a high content of proanthocyanidins, such as grape seed or cocoa, and those with a low content of proanthocyanidins, such as green tea.

Modulation of miRNAs by Proanthocyanidins

Though many foods have been shown to modulate miRNAs, we describe herein only those studies employing proanthocyanidin-rich extracts and foods (Table 3). The modulation of miRNAs by proanthocyanidin-rich nuts, such as walnuts in mice with colorectal cancer (Tsoukas et al. 2015) and pistachios in human plasma (Hernández-Alonso et al. 2016), has been attributed to their fatty acid composition. Nevertheless, the role of proanthocyanidins in the modulation of miRNAs cannot be ruled out. Polyphenols extracted from red wine, which exhibit anti-inflammatory activity in human colon-derived CCD-18Co myofibroblasts cells, were demonstrated to increase the expression of miR-126 and decrease the expression of VCAM-1 (Angel-Morales et al. 2012). In this study, a luciferase assay employing miR-126-antagomir demonstrated that the repression of VCAM-1 was caused by the increase in miR-126 levels and that this regulatory response was the mechanism underling the anti-inflammatory capacity of red wine. More studies on the effect of wine polyphenols on miRNA expression have been conducted; in general, the effect is attributed to resveratrol. However, we believe that proanthocyanidins also contribute to the modulation of miRNAs by red wine.

Studies using proanthocyanidin-rich extracts are, therefore, a better model to study the modulation of miRNAs by proanthocyanidins. For example, a proanthocyanidin-rich cranberry extract was found to modulate miR-586, miR-520d-5p, miR-516a-3p, miR-410, and miR-202, which are deregulated in esophageal adenocarcinoma cells (EAC) (Kresty et al. 2011). These results suggest that the protective effect of cranberry proanthocyanidins against esophageal adenocarcinoma is at least partially due to the modulation of these miRNAs.

Most studies on the implication of miRNAs in the mechanism by which proanthocyanidins exert their beneficial effects on cancer and lipid and glucose metabolism have been conducted using extracts from grape seeds.

The carcinogenic miRNAs miR-135b, miR-196a, and miR-21, which are upregulated in colorectal cancer cells, have been shown to be downregulated by grape seed proanthocyanidins in a mouse model of human colorectal carcinoma (Derry et al. 2013). The miRNA miR-106b is overexpressed in melanoma cell lines and downregulated by grape seed proanthocyanidins in vitro and in vivo, suggesting that the inhibition of miR-106b by this extract may be the mechanism by which proanthocyanidins reduce melanoma proliferation (Prasad and Katiyar 2014). In human colorectal carcinoma, NF-κB, hypoxia-inducible factor (HIF)-1a, VEGF, and β-catenin pathways are upregulated and controlled by miR-19a, miR-20a, miR-205, and let-7a, respectively. In a mouse model of human colorectal carcinoma, treatment with grape seed proanthocyanidins modulated these miRNAs (Derry et al. 2013). The miRNAs let-7a, miR-205, and miR-103, which are modulated by grape seed proanthocyanidins, are known to act in the MAPK, NOTCH, and KRAS pathways, which have been shown to be involved in cancer development. Therefore, these miRNAs may be involved in the chemopreventive effect of proanthocyanidins against colorectal cancer. Grape seed proanthocyanidins downregulated miR-19a and miR-19b and upregulated the expression of their target genes encoding insulin-

Table 3 Studies evaluating the effect of proanthocyanidins on miRNAs

Type of study	Cell line/ animals/clinical study	PAs source	miRNAs	Target genes/ effect	Reference
In vitro	CCD-18Co myofibroblast cells	Red wine	↑ miR-126	VCAM-1/ inflammation	Angel-Morales et al. (2012)
In vitro	Esophageal adenocarcinoma cells	Cranberry	miR-586, miR-520d-5p, miR-516a-3p, miR-410 and miR-202	Anticarcinogenic	Kresty et al. (2011)
In vivo	Mice model of colorectal carcinoma	GSPs	↓ miR-135b, miR-196a and miR-21	Anticarcinogenic	Derry et al. (2013)
In vivo	Mice model of colorectal carcinoma	GSPs	miR-19a, miR-20a, miR-205 and let-7a	NF-KB, HIF-1a, VEGF, and β-catenin/ anticarcinogenic	Derry et al. (2013)
In vivo	Mice model of colorectal carcinoma	GSPs	↓ let-7a, miR-205 and miR-103	MAPK, NOTCH, and KRAS/ anticarcinogenic	Derry et al. (2013)
In vitro/ In vivo	Melanoma cell lines/mice model of melanoma	GSPs	↓ miR-106b	Anticarcinogenic	Prasad and Katiyar (2014)
In vitro/ In vivo	Lung cancer A549 cell line/ athymic nude mice	GSPs	↓ miR-19a and miR-19b	IGF-2R and PTEN/ anticarcinogenic	Mao et al. (2016)
In vitro/ In vivo	Hepatoma HepG2 cell line/ healthy and dyslipidemic rats	GSPs	↓ miR-33 and miR-122	ABCA1 and FAS/ hypolipidemic	Baselga-Escudero et al. (2012)
					Baselga-Escudero et al. (2013)
					Baselga-Escudero et al. (2014a)
					Baselga-Escudero et al. (2015)
In vivo	Dam and offspring rats	GSPs	↓ miR-33	ABCA1/ imprinting	Del Bas et al. (2015a)

(*continued*)

Table 3 (continued)

Type of study	Cell line/ animals/clinical study	PAs source	miRNAs	Target genes/ effect	Reference
In vivo	Rat	GSPs	↓ miR-1249, miR-483, miR-30c-1*	Pancreatic insulin secretion and antidiabetic	Castell-Auví et al. (2013)
			↑ miR-3544		
In vitro	Hepatoma HepG2 cell line	GSPs	↑ miR-1224-3p, miR-449b*, miR-197, miR-1249, miR-1234, miR-532-3p, miR-15b*, miR-522, miR-744*		Arola-Arnal and Bladé (2011)
			↓ miR-2110, miR-485-5p, miR-320c, miR-453, miR-1290, miR-30b*		
In vitro	Hepatoma HepG2 cell line	Cocoa	↑ miR-197, miR-1224-3p, miR-532-3p		Arola-Arnal and Bladé (2011)
			↓ miR-765, miR-187*, miR-30b*		
In vitro	3 T3-L1 adipocyte cell line	A-type dimers of ECG and EGCG	↑ miR-27a and miR-27b		Zhu et al. (2015)

PAs proanthocyanidins, *GSPs* grape seed polyphenols, ↑ upregulation, ↓ downregulated

like growth factor II receptor (IGF-2R) and phosphatase and tensin homolog (PTEN) in lung cancer model using A549 cells and athymic nude mice (Mao et al. 2016).

There is some controversy as to how proanthocyanidins modulate miRNAs in cancer. For instance, miR-19 is downregulated in lung cancer (Mao et al. 2016), whereas in colorectal cancer it is upregulated (Derry et al. 2013). Therefore, more studies may be necessary to elucidate the effect of proanthocyanidins on miRNAs to know whether this opposite effect is the consequence of the cancer type, the proanthocyanidin dose, or the exact type of proanthocyanidins present in the extracts.

Studies on the effect of proanthocyanidins on lipid metabolism have been focused on miR-33 and miR-122 in the liver, where these two miRNAs have key roles. The inhibition of miR-122, which is liver specific, has been associated with the deregulation of key genes that control lipid metabolism in the liver, including fatty acid synthase (FAS) (Chang et al. 2004). Moreover, miR-33 plays an important role in the regulation of genes involved in cholesterol homeostasis such as ABCA1, NPC1, and ABCG1, which control lipid and lipoprotein metabolism in the liver (Dávalos and Fernández-Hernando 2013). Studies in vitro and in vivo have demonstrated that grape seed proanthocyanidins downregulate miR-33 and miR-122 within1 h of administration, suggesting that the reduction of these miRNAs could be involved in the hypolipidemic effect of proanthocyanidins in absorptive and postabsorptive conditions in rats (i.e., 3–5 h after grape seed proanthocyanidin administration together with a lipid load) (Baselga-Escudero et al. 2012). A kinetic experiment in which a high dose of grape seed proanthocyanidins was administered once a day showed that proanthocyanidins are able to downregulate miR-33a and miR-122 for 24 h (Baselga-Escudero et al. 2014a). In addition to the effect of acute administration of grape seed proanthocyanidins on these miRNAs, chronic treatment also demonstrated a dose-response effect (Baselga-Escudero et al. 2014a). Specifically, a study using 5, 15, 25, and 50 mg/Kg of grape seed proanthocyanidins for 3 weeks in rats showed that only the higher doses (25 and 50 mg/Kg) significantly repressed liver miR-33a and miR-122. In concordance with the proanthocyanidin dose-response effect on miRNAs, a dose-response effect was also observed in the reduction of cholesterol and triglyceride plasma levels. Using a translational conversion for interspecies drug scaling (Reagan-Shaw et al. 2008), these doses correspond to 284 and 560 mg of proanthocyanidins/day in humans, levels that could easily be reached in proanthocyanidin-rich diets. Long-term consumption of grape seed proanthocyanidins has been shown to normalize miR-33a and miR-122 deregulation in the livers of obese and dyslipidemic rats (Baselga-Escudero et al. 2015). However, comparing these two metabolic situations, it appears that dyslipidemic obese rats are more sensitive to proanthocyanidins modulated miRNAs than lean animals. The changes of miR-33 and miR-122 expression by grape seed proanthocyanidins in all these studies correlate with the changes in ABCA1 and FAS, which are known target genes of these miRNAs. This result suggests that proanthocyanidins, through the repression of miR-33 and miR-122, increase liver cholesterol efflux to HDL formation and reduce fatty acid synthesis, improving lipid postprandial state.

Hence, grape seed proanthocyanidins have been shown to modulate miR-33 acutely and chronically, in healthy and dyslipidemic states, and may even have an imprinting effect. Emerging evidence has revealed that proanthocyanidins in the maternal diet can affect miRNA expression in the offspring. Specifically, when grape seed proanthocyanidins were administered during the perinatal period (gestation and lactation) to dams, liver miR-33 was reduced in pups (Del Bas et al. 2015a).

Interestingly, the levels of miR-33 and ABCA1 in peripheral blood mononuclear cells (PBMCs) reflected the modifications in the liver that were induced by diet and treatment (Baselga-Escudero et al. 2013). Thus, analyzing miR-33a levels in PBMCs can provide information about the state of miR-33a expression in the liver in a noninvasive manner, as well as information about the physiological state of humans.

To our knowledge, only one study has focused on the effect of proanthocyanidins on miRNAs involved in glucose homeostasis. Downregulation of miR-1249, miR-483, and miR-30c-1* and upregulation of miR-3544 were observed in the pancreas of rats fed chronically with grape seed proanthocyanidins (Castell-Auví et al. 2013). Because these modified miRNAs are related to insulin secretion, this miRNA modulation in the pancreas by proanthocyanidins may partially explain their antidiabetic effect. Additionally, miR-33 has also been shown to control the expression of AMP-activated kinase (AMPKα1) and sirtuin 6 (Sirt6) (Fernández-Hernando et al. 2013), which are involved in the regulation of both lipid and glucose metabolism.

Emerging evidences suggest that the composition of proanthocyanidins can influence the capacity of these extracts to modulate miRNAs. A miRNA array study employing the human hepatoma cell line HepG2 treated with a proanthocyanidin-rich cocoa or grape seed extract demonstrated that grape seed proanthocyanidins modulate 15 miRNAs, whereas cocoa proanthocyanidins showed altered only 6 miRNAs (Arola-Arnal and Bladé 2011). Specifically, grape seed proanthocyanidins upregulated miR-1224-3p, miR-449b*, miR-197, miR-1249, miR-1234, miR-532-3p, miR-15b*, miR-522, and miR-744* and downregulated miR-2110, miR-485-5p, miR-320c, miR-453, miR-1290, and miR-30b*, while cocoa proanthocyanidins upregulated miR-197, miR-1224-3p, and miR-532-3p and downregulated miR-765, miR-187*, and miR-30b*. This disparity could be attributed to the fact that these two extracts differ mainly in their content on proanthocyanidin gallic esters: grape seed extract is richer in gallated proanthocyanidins compared to cocoa extract (Margalef et al. 2016; Ortega et al. 2008). On the other hand, both proanthocyanidin extracts have the capacity to repress miR-30b* and upregulated miR-1224-3p, miR-197, and miR-532-3p, suggesting that proanthocyanidins can modulate these four miRNAs independent of the differences in their exact composition (Arola-Arnal and Bladé 2011).

However, the same study (Arola-Arnal and Bladé 2011) showed that the monomeric epigallocatechin gallate (EGCG) was less successful in modifying miRNA expression than proanthocyanidins extracts, suggesting that the degree of polymerization of the extracts is related to their capacity to modulate miRNA expression. Moreover, using two different proanthocyanidin extracts from grapes as well as their monomeric and oligomeric fractions, it was shown that miR-33 and miR-122 are distinctly modulated in liver cells (Baselga-Escudero et al. 2014b). Furthermore, a B2-containing extract repressed both miR-33 and miR-122, whereas pure B2 only modulated miR-122 in HepG2 cells (Baselga-Escudero et al. 2014b).

The isomeric form of the proanthocyanidin can affects its capacity to modulate miRNA expression. A-type dimers of ECG and EGCG were shown to upregulate miR-27a and miR-27b in the 3 T3-L1 adipocyte cell line, while A- and B-type EC dimers are not effective (Zhu et al. 2015). The proanthocyanidin chemical structure and the combination of proanthocyanidins in the extract clearly have an effect on miRNA modulation.

The exact mechanism by which proanthocyanidins modulate miRNA expression is becoming clearer. Some miRNAs are intronic genes, such as miR-33a which is in the intronic region SREBP2 (Horie et al. 2010). We have shown that the regulation of miR-33a by proanthocyanidins is independent of the expression of its host gene, SREBP2 (Baselga-Escudero et al. 2014b). This suggests that proanthocyanidins may modulate miRNAs during miRNA biogenesis and/or via the direct binding of proanthocyanidins to specific miRNAs.

There is evidence that polyphenols can bind to DNA and mRNAs (Kuzuhara et al. 2006), proteins (Xiao and Kai 2012), and lipids (Patra et al. 2008). We have determined by ^{1}H-NMR spectroscopy that EGCG, a proanthocyanidin monomer, binds directly to miR-33 and miR-122 (Baselga-Escudero et al. 2014b). This binding is very specific, as other polyphenols, such as resveratrol, interact differently with these miRNAs. Thus, the interaction of proanthocyanidins with specific miRNAs may influence miRNA functionality by altering the binding of these miRNAs to the seed sequence of their target genes.

Other authors have shown that proanthocyanidins induce Ago2 (Derry et al. 2013), which is involved in miRNA processing, and polyphenols can inhibit miRNA maturation by modifying the Dicer complex (Taliaferro et al. 2013). More studies are necessary to fully elucidate the mechanism by which proanthocyanidins modulate miRNAs.

Summing up, proanthocyanidins have the capacity to modulate miRNAs although the capacity of each extract depends of the type of proanthocyanidins present, which in turn is dependent on their botanical origin. Hence, it is necessary to take into account the food or extract that is used in each experiment.

Modulation of DNA Methylation by Proanthocyanidins

Table 4 summarizes in vitro, in vivo, and epidemiological studies evaluating the effect of proanthocyanidins on DNA methylation.

The effect of pure proanthocyanidins on DNA methylation was investigated by Shilpi et al. (2015) who used in silico analyses to demonstrate that procyanidin B2 was a potent inhibitor of the S-adenosyl-L-homocysteine (SAH)-binding pocket of DNA methyltransferases (DNMT). In vitro experiments confirmed this activity (Shilpi et al. 2015).

Grape seed proanthocyanidins have been shown to modulate DNA methylation in in vitro assays and in vivo studies. For example, grape seed proanthocyanidins modulate DNA methylation in A431 and ACC13 human skin cancer cells (Vaid et al. 2012) by decreasing the level of 5-methylcytosine (5-mC) and DNMT activity, as well as the mRNA and protein of DNMT1, DNMT3a, and DNMT3b, leading to the activation of tumor suppressor genes (Vaid et al. 2012). In a mouse model of UV-induced skin cancer, supplementation with grape seed proanthocyanidins reduces methylated DNA and 5-mC levels, represses DNMT activity, and restores the expression of tumor suppressor genes in both skin and tumor samples (Katiyar et al. 2012).

Proanthocyanidin-rich apple polyphenol extract has been studied for its capacity to induce epigenetic changes through DNA methylation. Apple polyphenols reduce DNMT1 and DNMT3b protein levels in RKO, SW48, and SW480 colon cancer cells, resulting in reduced DNA methylation of several tumor suppressor gene promoters, such as hMLH1, p14^{ARF}, and p16^{INK4a} that, in turn, restore the expression of these tumor suppressor genes (Fini et al. 2007). An apple polyphenol extract, rich in procyanidin B, was shown to increase the methylation of LINE-1 and age-related hypomethylated genes, such as Igf2-DMR1 and P2rx7 in mice with colorectal cancer (Fini et al. 2011). Apple polyphenols are also able to modulate the expression and DNA methylation of leptin and other adipose-specific genes, such as Aqp7 (Boqué et al. 2013). Interestingly, a significant correlation is observed between the degree of DNA methylation and expression for leptin and Aqp7.

Table 4 Studies evaluating the effect of proanthocyanidins on DNA methylation

Type of study	Cell line/animals/ clinical study	PA source	Epigenetic mechanism	Target effect	Reference
In vitro	MDA-MB-231 human cancer cell line	Pure Procyanidin B2	↓ total DNMT activity	↑ mRNA expression of *E-cadherin*, *Maspin*, and *BRCA1* genes	Shilpi et al. (2015)
In vitro	A431 and SCC13 human skin cancer cell lines	GSPs	↓ global DNA methylation, ↓ 5-mC, ↓ total DNMT activity, and ↓ mRNA and protein expression of DNMT1, DNMT3a, and DNMT3b genes	↑ mRNA and protein expression of tumor suppressor genes (RASSF1a, p16^{INK4a} and Cip1/p21)	Vaid et al. (2012)
In vivo	Mice model of UV-induced skin tumor growth	GSPs	↓ global DNA methylation levels, ↓5-mC, and ↓ total DNMT activity in the skin and in the skin tumor samples	↑ mRNA expression of tumor suppressor genes (*Cip1/p21* and *p16*^{INK4a})	Katiyar et al. (2012)
Humans	Randomized controlled clinical pilot study (13 subjects, 8 weeks)	GSPs (150 mg/ day)	↔ Gene-specific DNA methylation levels	No correlation with mRNA expression levels	Milenkovic et al. (2014)
In vitro	RKO, SW48, and SW480 human colon cancer cell lines	APE	↓ mRNA expression of *Dnmt1* and *Dnmt3b* genes and ↓ promoter DNA methylation of tumor suppressor genes (hMLH1, p14^{ARF}, p16^{INK4a})	↑ mRNA expression of tumor suppressor genes (*hMLH1*, *p14*^{ARF}, *p16*^{INK4a})	Fini et al. (2007)
In vivo	Apc$^{Min/+}$ mice model of colorectal cancer	APE	↑ LINE-1 DNA methylation levels and ↑ DNA methylation of age-related hypomethylated genes Igf2-DMR1 and P2rx7	Protection against hypomethylation	Fini et al. (2011)

(*continued*)

Table 4 (continued)

Type of study	Cell line/animals/ clinical study	PA source	Epigenetic mechanism	Target effect	Reference
In vivo	Rats fed with high-fat diet	APE	↑ promoter DNA methylation level of Aqp7 and ↓ promoter DNA methylation level of Lep genes in adipose tissue	↑ mRNA expression of *Aqp7* gene and ↓ mRNA expression of *Lep* gene	Boqué et al. (2013)
In vivo	LNCaP human prostate cancer cell lines	GTPs	↓ total DNMT activity, ↓ mRNA and protein expression of DNMT1 gene, ↓ promoter DNA methylation of GSTP1 gene	↑ mRNA and protein expression of GSTP1 gene	Pandey et al. (2010)
In vitro	MDA-MB-231 and MDA-MB-453 human breast cancer cell lines	GTPs	↓ promoter DNA methylation levels of tumor suppressor genes ($p21^{CIP/WAF1}$ and KLOTHO)	↑ mRNA and protein expression of tumor suppressor genes (*p21*$^{CIP/WAF1}$ and *KLOTHO*)	Sinha et al. (2015)
In vivo	Male severe immunodeficiency mice with androgen-dependent human LAPC4 prostate cancer cell subcutaneous xenografts	GTPs	↓ mRNA and protein expression of DNMT1 gene in tumor	↓ tumor volume and weight	Henning et al. (2012)
In vivo	$Apc^{Min/+}$ mice model of colorectal cancer	GTPs	↓ promoter DNA methylation level of RXRα gene in mouse intestinal tumors	↑ mRNA expression of *RXRα* gene in mouse intestinal tumors	Volate et al. (2009)
Humans	Retrospective study with 106 patients with primary gastric carcinoma	GTPs (≥ 7cups/day)	↓ DNA methylation level of CDX2 and BMP-2 genes	–	Yuasa et al. (2009)

(*continued*)

Table 4 (continued)

Type of study	Cell line/animals/ clinical study	PA source	Epigenetic mechanism	Target effect	Reference
Humans	Retrospective study with 106 patients with primary gastric carcinoma	GTPs (≥ 7cups/ day)	↓ DNA methylation level of CDX2 and BMP-2 genes	–	Yuasa et al. (2009)
Humans	Randomized controlled study with humans with CVD risk factors (2 weeks)	Cocoa (6 g/day)	↓ global DNA methylation level of peripheral leukocytes in humans with CVD risk factors and↓ mRNA expression of *Dnmt1*, *Dnmt3a*, *Dnmt3*b, *Mthfr*, and *Mtrr* genes in PBMCs	–	Crescenti et al. (2013)
Humans	Phase 1 pilot study with 20 colorectal cancer patients (4 weeks)	Black raspberries (20 g/day)	↓ protein expression of DNMT1 gene and ↓ promoter DNA methylation level of tumor suppressor genes (SFRP2, Pax6, p16, SFRP5, and WIF1 genes)	–	Wang et al. (2011a)

PAs proanthocyanidins, *GSPs* grape seed polyphenols, *GTPs* green tea polyphenols, *APE* apple polyphenol extract, *5-mC* 5-methylcytosine, *CVD* cardiovascular disease, *DNMT* DNA methyltransferase, *MTHFR* methylenetetrahydrofolate reductase, *MTRR* methionine synthase reductase, *PBMCs* peripheral blood mononuclear cells, ↓ reduction, inhibition, ↔ no effect, ↑ induction, stimulation, – not detected

Green tea polyphenols extract is by far the most studied plant extract in relation to its effects on DNA methylation. Green tea extracts are rich in monomeric proanthocyanidins; epigallocatechin gallate is the most abundant polyphenol. But, because the effects of food extracts may be due to the collective effects of all polyphenol and proanthocyanidin components, we have included the effects of green tea extracts on DNA methylation in this chapter. It has been shown that green tea polyphenol extract inhibits DNMT1 and induces the extensive demethylation of the GSTP1 gene, which correlates with the restorations of its expression, in human prostate cancer LNCaP cells (Pandey et al. 2010). Additionally, Sinha et al. (2015) demonstrated that the treatment of human breast cancer cells with a combination of green tea polyphenols and sulforaphane (a dietary HDAC inhibitor) synergistically leads to the reactivation of tumor-silenced genes, such as $p21^{CIP1/WAF1}$ and KLOTHO, through DNA demethylation and histone modifications in the promoters of these genes. In vivo studies have shown that brewed green tea inhibits DNMT1 expression in prostate xenograft LAPC4 tumor tissue (Henning et al. 2012) and significantly decreases the degree of methylation of the RXRα promoter in mouse intestinal tumors (Volate et al. 2009).

Human studies on the epigenetic effects of proanthocyanidins are scarce but demonstrate that proanthocyanidins from different origins are effective at modulating DNA methylation. A randomized controlled clinical study showed that cocoa consumption (6 g per day for 2 weeks) significantly reduces the global DNA methylation in peripheral leukocytes of people with cardiovascular disease risk factors (Crescenti et al. 2013). Interestingly, treating PBMCs from these subjects with a cocoa extract demonstrated that the cocoa significantly represses the expression of various enzymes involved in DNA methylation, including DNMTs, methylenetetrahydrofolate reductase (MTHFR), and methionine synthase reductase (MTRR) (Crescenti et al. 2013). A retrospective study with gastric cancer patients demonstrated that high consumption of green tea polyphenols (seven cups or more per day) is inversely related to the methylation status of CDX2 and BMP-2 genes, which are important for preventing gastric carcinogenesis and are frequently hypermethylated in gastric cancers (Yuasa et al. 2009). A human phase 1 pilot study with colorectal cancer patients showed that the consumption of black raspberries (60 g/day freeze-dried black raspberries for 9 weeks) reduces the degree of methylation of various gene promoters, including sFRP2, Pax6, and WIF1, which correlates with a lower expression of DNMT1 (Wang et al. 2011a). On the other hand, daily supplementation of proanthocyanidins from grape seeds (150 mg for 8 weeks) had no effect on gene methylation in an intervention study (Milenkovic et al. 2014). However, this lack of activity of proanthocyanidins on DNA methylation was attributed by the authors to the inclusion of subjects who smoked and who showed a strong interindividual variability on DNA methylation.

Histone Modifications Induced by Proanthocyanidins

Several in vitro and in vivo studies have shown that proanthocyanidins are able to modulate histone acetyltransferase (HAT) and histone deacetylase (HDAC) activities and, therefore, are able to induce histone modifications (Table 5). Specifically, procyanidin B3 inhibits HAT activity and p300-mediated androgen receptor (AR) acetylation, consequently suppressing p300-enhanced AR transcription in an acetylation-dependent manner in prostate cancer cells (Choi et al. 2011). Furthermore, procyanidin B2 activates the silent mating-type information regulation 2 homolog 1 (SIRT1), a class III HDAC, in rat mesangial cells (Bao et al. 2015).

Table 5 Studies evaluating the effect of proanthocyanidins on histone modifications

Type of study	Cell line/ animals/ clinical study	PAs source	Epigenetic mechanism	Target effect	Reference
In vitro	Prostate cancer cells	Pure Procyanidin B3	↓ total HAT activity and ↓ p300-mediated androgen receptor acetylation levels	↓ androgen receptor-dependent prostate cancer cell growth	Choi et al. (2011)
In vitro	Rat mesangial cells under high dose of glucosamine	pure Procyanidin B2	↑ SIRT1 protein expression	–	Bao et al. (2015)
In vitro	A431 and SCC13 human skin cancer cell lines	GSPs	↓ total HDAC activity and ↑ total HAT activity,↑ acetylated histones levels, and ↓methylated histone levels	↑ mRNA and protein expression of tumor suppressor genes (RASSF1a, $p16^{INK4a}$ and Cip1/p21)	Vaid et al. (2012)
In vivo	Male rats under high-carbohydrate/ high-fat diet	GSPs	↑ SIRT1 protein expression in renal tissue	Amelioration of podocyte injury in diabetic nephropathy	Bao et al. (2014)
In vivo	Healthy rats	GSPs	↑ mRNA *Sirt1* gene expression and ↑ SIRT1 activity in liver tissue	Protection against hepatic triglyceride accumulation	Aragonès et al. (2016)
In vitro	H_2O_2-induced cellular senescence model with human lung diploid fibroblasts	Persimmon	↑ nuclear SIRT1 protein expression	Protective effect against oxidative damage under the aging process	Lee et al. (2008)
In vivo	Senescence-accelerated mouse P8 model	Persimmon	↑ SIRT1 protein expression in the brain	Extension of the life span of animals	Yokozawa et al. (2009)
In vitro	LNCaP and PC-3 human prostate cancer cells	GTPs	↓ class I HDAC protein expression and activity, ↑ acetylation of histone H3	↑ Expression of proapoptotic genes (*p21/waf1* and *Bax* genes). Induce cell cycle arrest and apoptosis	Thakur et al. (2012)

(*continued*)

Table 5 (continued)

Type of study	Cell line/ animals/ clinical study	PAs source	Epigenetic mechanism	Target effect	Reference
In vitro	LNCaP human prostate cancer cell lines	GTPs	↓ mRNA and protein expression of HDAC 1–3; ↓ total HDAC activity, ↑ acetylation of histones H3 and H4	↑ mRNA and protein expression of GSTP1 gene	Pandey et al. (2010)
In vitro	A375, Hs294t and SK-Mel28 human melanoma cell lines	GTPs	↓ total HDAC activity, ↓ class I HDAC protein expression, ↑ HAT activity, and ↑ acetylation of histones H3	Inhibition of cell viability and induction of toxicity in melanoma cells	Prasad and Katiyar (2015)
In vitro	MCF-7 and MDA-MB-231 breast cancer cell lines	GTPs	↓ class I HDAC 1–3,8 protein expression, ↓ EZH2 protein expression, and ↓ H3K27 trimethylation, and ↑ histone H3K9/18 acetylation of TIMP-3 promoter gene	↑ mRNA and protein expression of TIMP-3 gene	Deb et al. (2015)

PAs proanthocyanidins, *GSPs* grape seed polyphenols, *GTPs* green tea polyphenols, *HDAC* histone deacetylase, *HAT* histone acetyltransferase, *SIRT1* sirtuin 1, *EZH2* histone methyltransferase enhancer of zeste homolog 2, ↓ reduction, inhibition, ↔ no effect, ↑ induction, stimulation; – not detected

Proanthocyanidin extracts also show histone modification activities in several in vitro studies. Grape seed proanthocyanidins decrease HDAC activity, increase histone acetylation, and decrease histone methylation in skin cancer cells, resulting in the reexpression of silenced tumor suppressor genes RASSF1A, $p16^{INK4a}$, and Cip1/p21 (Vaid et al. 2012). Additionally, grape seed proanthocyanidins reduce HDAC 1–3 expression (mRNA and protein) and increase histone H3 and H4 acetylation, inducing cell cycle arrest and apoptosis, in the human prostate cancer LNCap cell line (Pandey et al. 2010; Thakur et al. 2012). Grape seed proanthocyanidins reduce class I HDAC proteins and enhance HAT activity and histone acetylation in various human melanoma cell lines (Prasad and Katiyar 2015). In breast cancer cells, grape seed proanthocyanidins reduce EZH2 and HDAC I protein levels and induce TIMP-3 expression through the enrichment of H3K27 trimethylation and the increase of histone H3K9/18 acetylation at its gene promoter (Deb et al. 2015).

In vivo studies have demonstrated that grape seed proanthocyanidins increase SIRT1 expression and activity in renal (Lee et al. 2005) and hepatic (Aragonès et al. 2016) tissues. Proanthocyanidin extracts from other plants are also able to modulate SIRT1 activity. For instance, persimmon peel proanthocyanidins increase nuclear SIRT1 expression leading to a protective effect against oxidative damage under H_2O_2-induced cellular senescence in human lung diploid fibroblasts (Lee et al. 2008). The administration of persimmon proanthocyanidins to senescence-accelerated mouse P8 mice extended the life span of animals with elevated SIRT1 expression (Yokozawa et al. 2009).

Dictionary of Terms

- **Proanthocyanidin-rich extract** – mix of proanthocyanidins extracted from proanthocyanidin-rich foods or other sources, such as grape seeds.
- **Proanthocyanidin monomers** – smallest units of proanthocyanidins that are attached to one another forming proanthocyanidins. There are different monomers.
- **Degree of polymerization** – number of monomers included in each proanthocyanidin molecule. These can range from 2 to 190.
- **Gallated proanthocyanidins** – proanthocyanidins esterified with gallic acid.
- **Procyanidins** – proanthocyanidins composed exclusively by catechin and epicatechin monomers.

Key Facts of Proanthocyanidins

- Proanthocyanidins are polyphenols widespread in plant-derived foods; thus, they are very abundant in the human diet.
- Proanthocyanidins protect from diseases such as cardiovascular disease, some cancers, and from some risk factors of metabolic syndrome.
- Numerous studies, mainly in vitro, indicate that proanthocyanidins modulate miRNA expression, DNA methylation, and histone acetylation and methylation.
- Proanthocyanidins are food components, and the epigenetic effects of proanthocyanidins ingested in a healthy diet will not be as significant as they would be if they were administered in drug form.
- Long-term consumption of proanthocyanidins, which induces a moderate effect on specific epigenetic processes, can contribute to the health effects of these polyphenols, mainly in cancer and dyslipidemias.
- Not all proanthocyanidin-rich foods or extracts will have the same activity toward modulating epigenetic processes. The capacity of a proanthocyanidin-rich food or extract to modulate miRNA levels is dependent of their specific proanthocyanidin composition which, in turn, is dependent on their botanical origin.

Summary Points

- Many studies support the capacity of proanthocyanidins to modulate cell functionality by epigenetic mechanisms.
- Studies have been conducted using several proanthocyanidin-rich extracts, mainly grape seed extracts.
- In vitro studies have shown that proanthocyanidins modulate several miRNAs that are deregulated in colorectal and lung cancer.
- In vitro and in vivo studies have demonstrated that proanthocyanidins downregulate miR-33 and miR-122, suggesting that these miRNAs could be involved in the hypolipidemic effect of proanthocyanidins.
- The composition of proanthocyanidins (i.e., the degree of polymerization of proanthocyanidins, their isomeric form, as well as the presence of gallated proanthocyanidins) is related to the capacity of a specific proanthocyanidin-rich extract to modulate miRNA expression.
- Proanthocyanidins have been shown to modulate DNA methylation in in vitro assays and in animal and human studies through the modulation of the activity of different DNA methyltransferases.
- Several in vitro and in vivo studies have shown that proanthocyanidins are able to modulate the activity of histone acetyltransferases and histone deacetylases and, therefore, to induce histone modifications.

References

Angel-Morales G, Noratto G, Mertens-Talcott S (2012) Red wine polyphenolics reduce the expression of inflammation markers in human colon-derived CCD-18Co myofibroblast cells: potential role of microRNA-126. Food Funct 3:745–752

Aragonès G, Suárez M, Ardid-Ruiz A et al (2016) Dietary proanthocyanidins boost hepatic NAD+ metabolism and SIRT1 expression and activity in a dose-dependent manner in healthy rats. Sci Rep 6:24977

Arola-Arnal A, Bladé C (2011) Proanthocyanidins modulate microRNA expression in human HepG2 cells. PLoS One 6:e25982

Arranz S, Silván JM, Saura-Calixto F (2010) Nonextractable polyphenols, usually ignored, are the major part of dietary polyphenols: a study on the Spanish diet. Mol Nutr Food Res 54:1646–1658

Bao L, Cai X, Zhang Z et al (2015) Grape seed procyanidin B2 ameliorates mitochondrial dysfunction and inhibits apoptosis via the AMP-activated protein kinase–silent mating type information regulation 2 homologue 1–PPARγ co-activator-1α axis in rat mesangial cells under high-dose glucosamine. Br J Nutr 113:35–44

Baselga-Escudero L, Bladé C, Ribas-Latre A et al (2012) Grape seed proanthocyanidins repress the hepatic lipid regulators miR-33 and miR-122 in rats. Mol Nutr Food Res 56:1636–1646

Baselga-Escudero L, Arola-Arnal A, Pascual-Serrano A et al (2013) Chronic administration of proanthocyanidins or docosahexaenoic acid reverses the increase of miR-33a and miR-122 in dyslipidemic obese rats. PLoS One 8:e69817

Baselga-Escudero L, Blade C, Ribas-Latre A et al (2014a) Chronic supplementation of proanthocyanidins reduces postprandial lipemia and liver miR-33a and miR-122 levels in a dose-dependent manner in healthy rats. J Nutr Biochem 25:151–156

Baselga-Escudero L, Blade C, Ribas-Latre A et al (2014b) Resveratrol and EGCG bind directly and distinctively to miR-33a and miR-122 and modulate divergently their levels in hepatic cells. Nucleic Acids Res 42:882–892

Baselga-Escudero L, Pascual-Serrano A, Ribas-Latre A et al (2015) Long-term supplementation with a low dose of proanthocyanidins normalized liver miR-33a and miR-122 levels in high-fat diet-induced obese rats. Nutr Res 35:337–345

Bladé C, Arola L, Salvadó MJ (2010) Hypolipidemic effects of proanthocyanidins and their underlying biochemical and molecular mechanisms. Mol Nutr Food Res 54:37–59

Bladé C, Aragonès G, Arola-Arnal A et al (2016) Proanthocyanidins in health and disease. Biofactors 42:5

Boqué N, de la Iglesia R, de la Garza AL et al (2013) Prevention of diet-induced obesity by apple polyphenols in Wistar rats through regulation of adipocyte gene expression and DNA methylation patterns. Mol Nutr Food Res 57:1473–1478

Castell-Auví A, Cedó L, Movassat J et al (2013) Procyanidins modulate microRNA expression in pancreatic islets. J Agric Food Chem 61:355–363

Chang J, Nicolas E, Marks D et al (2004) miR-122, a mammalian liver-specific microRNA, is processed from hcr mRNA and may downregulate the high affinity cationic amino acid transporter CAT-1. RNA Biol 1:106–113

Choi K-C, Park S, Lim BJ et al (2011) Procyanidin B3, an inhibitor of histone acetyltransferase, enhances the action of antagonist for prostate cancer cells via inhibition of p300-dependent acetylation of androgen receptor. Biochem J 433:235–244

Crescenti A, Solà R, Valls RM et al (2013) Cocoa consumption alters the global DNA methylation of peripheral leukocytes in humans with cardiovascular disease risk factors: a randomized controlled trial. PLoS One 8:e65744

Dávalos A, Fernández-Hernando C (2013) From evolution to revolution: miRNAs as pharmacological targets for modulating cholesterol efflux and reverse cholesterol transport. Pharmacol Res 75:60–72

Deb G, Thakur VS, Limaye AM et al (2015) Epigenetic induction of tissue inhibitor of matrix metalloproteinase-3 by green tea polyphenols in breast cancer cells. Mol Carcinog 54:485–499

Del Bas JM, Ricketts ML, Baiges I et al (2008) Dietary procyanidins lower triglyceride levels signaling through the nuclear receptor small heterodimer partner. Mol Nutr Food Res 52:1172–1181

Del Bas JM, Ricketts M-L, Vaqué M et al (2009) Dietary procyanidins enhance transcriptional activity of bile acid-activated FXR in vitro and reduce triglyceridemia in vivo in a FXR-dependent manner. Mol Nutr Food Res 53:805–814

Del Bas JM, Crescenti A, Arola-Arnal A et al (2015a) Intake of grape procyanidins during gestation and lactation impairs reverse cholesterol transport and increases atherogenic risk indexes in adult offspring. J Nutr Biochem 26:1670–1677

Del Bas JM, Crescenti A, Arola-Arnal A et al (2015b) Grape seed procyanidin supplementation to rats fed a high-fat diet during pregnancy and lactation increases the body fat content and modulates the inflammatory response and the adipose tissue metabolism of the male offspring in youth. Int J Obes 39:7–15

Derry MM, Raina K, Balaiya V et al (2013) Grape seed extract efficacy against azoxymethane-induced colon tumorigenesis in a/j mice: interlinking miRNA with cytokine signaling and inflammation. Cancer Prev Res 6:625–633

Dixon RA, Xie D-Y, Sharma SB (2005) Proanthocyanidins – a final frontier in flavonoid research? New Phytol 165:9–28

Fernández-Hernando C, Ramírez CM, Goedeke L et al (2013) MicroRNAs in metabolic disease. Arterioscler Thromb Vasc Biol 33:178–185

Fernández-Iglesias A, Pajuelo D, Quesada H et al (2014) Grape seed proanthocyanidin extract improves the hepatic glutathione metabolism in obese Zucker rats. Mol Nutr Food Res 58:727–737

Fini L, Selgrad M, Fogliano V et al (2007) Annurca apple polyphenols have potent demethylating activity and can reactivate silenced tumor suppressor genes in colorectal cancer cells. J Nutr 137:2622–2628
Fini L, Piazzi G, Daoud Y et al (2011) Chemoprevention of intestinal polyps in ApcMin/+ mice fed with western or balanced diets by drinking annurca apple polyphenol extract. Cancer Prev Res (Phila) 4:907–915
Fraga CG, Galleano M, Verstraeten SV et al (2010) Basic biochemical mechanisms behind the health benefits of polyphenols. Mol Asp Med 31:435–445
Gonzalez-Abuin N, Pinent M, Casanova-Marti A et al (2015) Procyanidins and their healthy protective effects against type 2 diabetes. Curr Med Chem 22:39–50
Gu L, Kelm MA, Hammerstone JF et al (2003) Screening of foods containing proanthocyanidins and their structural characterization using LC-MS/MS and thiolytic. J Agric Food Chem 51:7513–7521
He F, Pan QH, Shi Y, Duan CQ (2008) Biosynthesis and genetic regulation of proanthocyanidins in plants. Molecules 13:2674–2703
Henning SM, Wang P, Said J et al (2012) Polyphenols in brewed green tea inhibit prostate tumor xenograft growth by localizing to the tumor and decreasing oxidative stress and angiogenesis. J Nutr Biochem 23:1537–1542
Hernández-Alonso P, Giardina S, Salas-Salvadó J et al (2016) Chronic pistachio intake modulates circulating microRNAs related to glucose metabolism and insulin resistance in prediabetic subjects. Eur J Nutr doi:10.1007/s00394-016-1262-5
Hollman PCH, Cassidy A, Comte B et al (2011) The biological relevance of direct antioxidant effects of polyphenols for cardiovascular health in humans is not established. J Nutr 141:989S–1009S
Horie T, Ono K, Horiguchi M et al (2010) MicroRNA-33 encoded by an intron of sterol regulatory element-binding protein 2 (Srebp2) regulates HDL in vivo. Proc Natl Acad Sci USA 107:17321–17326
Ibars M, Ardid-Ruiz A, Suárez M et al (2016) Proanthocyanidins potentiate hypothalamic leptin/ STAT3 signaling and Pomc gene expression in rats with diet-induced obesity signaling. Int J Obes 41:129
Katiyar SK, Singh T, Prasad R et al (2012) Epigenetic alterations in ultraviolet radiation-induced skin carcinogenesis: interaction of bioactive dietary components on epigenetic targets. Photochem Photobiol 88:1066–1074
Kresty LA, Clarke J, Ezell K et al (2011) MicroRNA alterations in Barrett's esophagus, esophageal adenocarcinoma, and esophageal adenocarcinoma cell lines following cranberry extract treatment: insights for chemoprevention. J Carcinog 10:34
Kuzuhara T, Sei Y, Yamaguchi K et al (2006) DNA and RNA as new binding targets of green tea catechins. J Biol Chem 281:17446–17456
Lee WJ, Shim J-Y, Zhu BT (2005) Mechanisms for the inhibition of DNA methyltransferases by tea catechins and bioflavonoids. Mol Pharmacol 68:1018–1030
Lee YA, Cho EJ, Yokozawa T (2008) Protective effect of persimmon (*Diospyros kaki*) peel proanthocyanidin against oxidative damage under H_2O_2-induced cellular senescence. Biol Pharm Bull 31:1265–1269
Lei Bao, Xiaxia Cai, Xiaoqian Dai, Ye Ding, Yanfei Jiang, Yujie Li, Zhaofeng Zhang, Yong Li (2014) Grape seed proanthocyanidin extracts ameliorate podocyte injury by activating peroxisome proliferator-activated receptor-Î³ coactivator 1Î± in low-dose streptozotocin-and high-carbohydrate/high-fat diet-induced diabetic rats. Food & Function 5(8):1872
Llópiz N, Puiggròs F, Céspedes E et al (2004) Antigenotoxic effect of grape seed procyanidin extract in Fao cells submitted to oxidative stress. J Agric Food Chem 52:1083–1087
Mao JT, Xue B, Smoake J et al (2016) MicroRNA-19a/b mediates grape seed procyanidin extract-induced anti-neoplastic effects against lung cancer. J Nutr Biochem 34:118–125
Margalef M, Pons Z, Iglesias-Carres L et al (2016) Gender-related similarities and differences in the body distribution of grape seed flavanols in rats. Mol Nutr Food Res 60:760–772

Martinez-Micaelo N, González-Abuín N, Ardèvol A et al (2012) Procyanidins and inflammation: molecular targets and health implications. Biofactors 38:257–265

McCullough ML, Peterson JJ, Patel R et al (2012) Flavonoid intake and cardiovascular disease mortality in a prospective cohort of US adults. Am J Clin Nutr 95:454–464

Milenkovic D, Vanden Berghe W, Boby C et al (2014) Dietary flavanols modulate the transcription of genes associated with cardiovascular pathology without changes in their DNA methylation state. PLoS One 9:e95527

Neilson AP, O'Keefe SF, Bolling BW (2016) High-molecular-weight proanthocyanidins in foods: overcoming analytical challenges in pursuit of novel dietary bioactive components. Annu Rev Food Sci 7:43–64

Neveu V, Perez-Jimenez J, Vos F, Crespy V, du Chaffaut L, Mennen L, Knox C, Eisner R, Cruz J, Wishart D, Scalbert A (2010) Phenol-Explorer: an online comprehensive database on polyphenol contents in foods. Database 2010 (0):bap024-bap024

Ortega N, Romero M-P, Macià A et al (2008) Obtention and characterization of phenolic extracts from different cocoa sources. J Agric Food Chem 56:9621–9627

Pandey M, Shukla S, Gupta S (2010) Promoter demethylation and chromatin remodeling by green tea polyphenols leads to re-expression of GSTP1 in human prostate cancer cells. Int J Cancer 126:2520–2533

Patra SK, Rizzi F, Silva A et al (2008) Molecular targets of (-)-epigallocatechin-3-gallate (EGCG): specificity and interaction with membrane lipid rafts. J Physiol Pharmacol 59:217–235

Pinent M, Bladé C, Salvadó MJ et al (2006) Procyanidin effects on adipocyte-related pathologies. Crit Rev Food Sci Nutr 46:543–550

Pons Z, Margalef M, Bravo FI et al (2015) Acute administration of single oral dose of grape seed polyphenols restores blood pressure in a rat model of metabolic syndrome: role of nitric oxide and prostacyclin. Eur J Nutr 55:749–758

Prasad R, Katiyar SK (2014) Down-regulation of miRNA-106b inhibits growth of melanoma cells by promoting G1-phase cell cycle arrest and reactivation of p21/WAF1/Cip1 protein. Oncotarget 5:10636–10649

Prasad R, Katiyar SK (2015) Polyphenols from green tea inhibit the growth of melanoma cells through inhibition of class I histone deacetylases and induction of DNA damage. Genes Cancer 6:49–61

Puiggros F, Llópiz N, Ardévol A et al (2005) Grape seed procyanidins prevent oxidative injury by modulating the expression of antioxidant enzyme systems. J Agric Food Chem 53:6080–6086

Rasmussen SE, Frederiksen H, Krogholm KS et al (2005) Dietary proanthocyanidins: occurrence, dietary intake, bioavailability, and protection against cardiovascular disease. Mol Nutr Food Res 49:159–174

Reagan-Shaw S, Nihal M, Ahmad N (2008) Dose translation from animal to human studies revisited. FASEB J 22:659–661

Rothwell JA, Urpi-Sarda M, Boto-Ordonez M, Knox C, Llorach R, Eisner R, Cruz J, Neveu V, Wishart D, Manach C, Andres-Lacueva C, Scalbert A (2012) Phenol-Explorer 2.0: a major update of the Phenol-Explorer database integrating data on polyphenol metabolism and pharmacokinetics in humans and experimental animals. Database 2012 (0):bas031–bas031

Rothwell JA, Perez-Jimenez J, Neveu V, Medina-Remon A, M'Hiri N, Garcia-Lobato P, Manach C, Knox C, Eisner R, Wishart DS, Scalbert A (2013) Phenol-Explorer 3.0: a major update of the Phenol-Explorer database to incorporate data on the effects of food processing on polyphenol content. Database 2013 (0):bat070–bat070

Salvadó MJ, Casanova E, Fernández-iglesias A et al (2015) Roles of proanthocyanidin rich extracts in obesity. Food Funct 6:1053–1071

Shilpi A, Parbin S, Sengupta D et al (2015) Mechanisms of DNA methyltransferase-inhibitor interactions: Procyanidin B2 shows new promise for therapeutic intervention of cancer. Chem Biol Interact 233:122–138

Sinha S, Shukla S, Khan S et al (2015) Epigenetic reactivation of p21CIP1/WAF1 and KLOTHO by a combination of bioactive dietary supplements is partially ERα-dependent in ERα-negative human breast cancer cells. Mol Cell Endocrinol 406:102–114

Taliaferro JM, Aspden JL, Bradley T et al (2013) Two new and distinct roles for Drosophila Argonaute-2 in the nucleus: alternative pre-mRNA splicing and transcriptional repression. Genes Dev 27:378–389

Thakur VS, Gupta K, Gupta S (2012) Green tea polyphenols causes cell cycle arrest and apoptosis in prostate cancer cells by suppressing class I histone deacetylases. Carcinogenesis 33:377–384

Tsoukas MA, Ko BJ, Witte TR et al (2015) Dietary walnut suppression of colorectal cancer in mice: mediation by miRNA patterns and fatty acid incorporation. J Nutr Biochem 26:776–783

Vaid M, Prasad R, Singh T et al (2012) Grape seed proanthocyanidins reactivate silenced tumor suppressor genes in human skin cancer cells by targeting epigenetic regulators. Toxicol Appl Pharmacol 263:122–130

Vogiatzoglou A, Mulligan AA, Luben RN et al (2014) Assessment of the dietary intake of total flavan-3-ols, monomeric flavan-3-ols, proanthocyanidins and the aflavins in the European Union. Br J Nutr 111:1463–1473

Volate SR, Muga SJ, Issa AY et al (2009) Epigenetic modulation of the retinoid X receptor alpha by green tea in the azoxymethane-Apc Min/+ mouse model of intestinal cancer. Mol Carcinog 48:920–933

Wang L-S, Arnold M, Huang Y-W et al (2011a) Modulation of genetic and epigenetic biomarkers of colorectal cancer in humans by black raspberries: a phase I pilot study. Clin Cancer Res 17:598–610

Wang Y, Chung S-J, Song WO et al (2011b) Estimation of daily proanthocyanidin intake and major food sources in the US diet. J Nutr 141:447–452

Xiao J, Kai G (2012) A review of dietary polyphenol-plasma protein interactions: characterization, influence on the bioactivity, and structure-affinity relationship. Crit Rev Food Sci Nutr 52:85–101

Yokozawa T, Lee YA, Zhao Q et al (2009) Persimmon oligomeric proanthocyanidins extend life span of senescence-accelerated mice. J Med Food 12:1199–1205

Yuasa Y, Nagasaki H, Akiyama Y et al (2009) DNA methylation status is inversely correlated with green tea intake and physical activity in gastric cancer patients. Int Cancer 124:2677–2682

Zhu W, Zou B, Nie R et al (2015) A-type ECG and EGCG dimers disturb the structure of 3T3-L1 cell membrane and strongly inhibit its differentiation by targeting peroxisome proliferator-activated receptor with miR-27 involved mechanism. J Nutr Biochem 26:1124–1135

Application of Nutraceuticals in Pregnancy Complications: Does Epigenetics Play a Role?

102

Luís Fernando Schütz, Jomer Bernardo, Minh Le, Tincy Thomas, Chau Nguyen, Diana Zapata, Hitaji Sanford, John D. Bowman, Brett M. Mitchell, and Mahua Choudhury

Contents

L. F. Schütz · T. Thomas · D. Zapata · M. Choudhury (✉)
Department of Pharmaceutical Sciences, Irma Lerma Rangel College of Pharmacy, Texas A&M Health Science Center, College Station, TX, USA
e-mail: schutz@pharmacy.tamhsc.edu; tthomas@pharmacy.tamhsc.edu; dzapata@pharmacy.tamhsc.edu; mchoudhury@pharmacy.tamhsc.edu

J. Bernardo · M. Le · C. Nguyen · H. Sanford · J. D. Bowman
Texas A&M Irma Lerma Rangel College of Pharmacy, Kingsville, TX, USA
e-mail: jbernardo@pharmacy.tamhsc.edu; mle@pharmacy.tamhsc.edu; cnguyen1@pharmacy.tamhsc.edu; hsanford@pharmacy.tamhsc.edu; bowman@pharmacy.tamhsc.edu

B. M. Mitchell
Texas A&M, Department of Medical Physiology, College Station, TX, USA
e-mail: bmitchell@medicine.tamhsc.edu

V. B. Patel, V. R. Preedy (eds.), *Handbook of Nutrition, Diet, and Epigenetics*,
https://doi.org/10.1007/978-3-319-55530-0_81

Abstract

Nutraceuticals provide the prevention or treatment of diseases through dietary supplementation. These become especially important during pregnancy to prevent disorders secondary to nutrient deficiency. In the light of research accomplished in the recent years, it is now established that maternal nutrition affects pregnancy outcomes and disorders through epigenetics, which are heritable gene expression modifications that occur without a change in the DNA sequence. The most studied epigenetic modifications are DNA methylation, histone modifications, and small noncoding RNAs (microRNAs). Recent research has started to unveil how nutraceuticals may prevent pregnancy complications such as preeclampsia, intrauterine growth restriction (IUGR), preterm delivery, and miscarriage through epigenetic mechanisms.

Keywords

Nutraceuticals · Pregnancy · Epigenetics · Methylation · Histones · MicroRNAs · Omega-3 · Vitamin D · Folic acid · Preeclampsia · Intrauterine growth restriction

List of Abbreviations

CYP27B1	Cytochrome P450 family 27 subfamily B member 1
CYP24A1	Cytochrome P450 family 24 subfamily B member 1
DHA	Docosahexaenoic acid
DNMTs	DNA methyltransferases
EPA	Eicosapentaenoic acid
HATs	Histone acetyltransferases
HDACs	Histone deacetylases
HMTs	Histone methyltransferases
IUGR	Intrauterine growth restriction
LDL	Low density lipoprotein
PPARγ	Peroxisome proliferator-activated receptor gamma
PTH	Parathyroid hormone
RXR	Retinoid X receptor
SAH	S-adenosylhomocysteine
SAM	S-adenosylmethionine
Setd8	SET domain containing (lysine methyltransferase) 8
TET	Ten-eleven translocation
UTRs	Untranslated regions
UVB	Ultraviolet B
VDRE	Vitamin D response element

Introduction

Nutrition has an essential role in the history and evolution of mankind. Factors such as the control of fire, the development of cuisine, the evolution of agricultural systems, and the industrialization have brought changes to alimentary habits with an enormous impact to human health (Armelagos 2014). Indeed, nutrition is believed to have both positive and negative effects on the level of human mortality (Gage and O'Connor 1994). An increased awareness of the public to the importance of nutrition for health allied to advances in food science and technology has led to the expansion of the market of innovative nutrition products (Hardy 2000). The search for healthy diets gave rise to the market of nutritionally or medicinally enhanced foods, the so-called nutraceuticals (Brower 1998).

The term *nutraceutical* was coined in 1979 by Stephen DeFelice, founder and chairman of the Foundation for Innovation in Medicine (Kalra 2003). According to DeFelice, nutraceutical means "a food, or part of a food, that provides medical or health benefits, including the prevention and/or treatment of a disease" (Kalra 2003). The list of nutraceutical compounds includes, among others, probiotics, vitamins, and antioxidants, which are generally sold in medicinal forms such as tablets, capsules, powders, solutions, or potions (Gul et al. 2016). Nutraceuticals, different than dietary supplements, not only complement the diet but should also help in the prevention and/or treatment of disorders or diseases (Kalra 2003). In addition to bringing benefits to overall health maintenance, nutraceuticals have also been used to prevent pregnancy complications (Agrawal et al. 2015; Omotayo et al. 2016).

Nutrients and nutraceuticals can influence gene expression, playing an important role in disease prevention by suppressing harmful genes (Tokunaga et al. 2013). Epigenetics, defined as heritable modifications in gene expression that occur without a change in the DNA sequence itself (Wolffe and Matzke 1999), is often the result of environmental and nutritional factors that alter phenotypic expression by modifying the DNA/histone (Handy et al. 2011; Tokunaga et al. 2013). It is now becoming more accepted that the nutrition of the mother during gestation has implications in the health of the offspring through epigenetic modifications (Handy et al. 2011). This chapter will focus on reviewing the role of nutraceuticals and epigenetics on preventing pregnancy complications.

Nutraceutical Supplementation During Pregnancy

Preeclampsia and intrauterine growth restriction (IUGR) are the leading causes of perinatal mortality and morbidity, affecting millions of women and their babies worldwide (Amhed et al. 2014). The pathophysiology of both preeclampsia and IUGR appears to be resultant from a failure in trophoblast invasion, which has critical consequences for the perfusion of the placenta with nutrients and oxygenated blood coming from the maternal blood (Huppertz et al. 2014). Importantly, these pregnancy complications not just affect the outcomes of pregnancy at delivery but also may affect the long-term cardiovascular health of the affected women and their offspring (Zhong et al. 2010).

The role of maternal nutrition in preventing diseases during pregnancy has been receiving an increasing level of attention throughout the recent decades (Fig. 1). Indeed, several studies have been conducted to determine the effectiveness of dietary

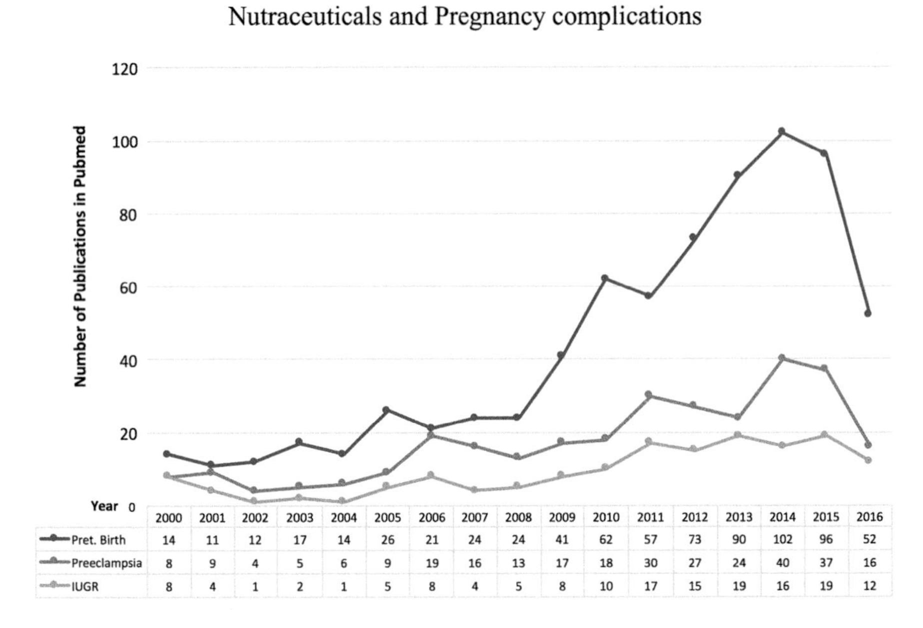

Year	2000	2001	2002	2003	2004	2005	2006	2007	2008	2009	2010	2011	2012	2013	2014	2015	2016
Pret. Birth	14	11	12	17	14	26	21	24	24	41	62	57	73	90	102	96	52
Preeclampsia	8	9	4	5	6	9	19	16	13	17	18	30	27	24	40	37	16
IUGR	8	4	1	2	1	5	8	4	5	8	10	17	15	19	16	19	12

Fig. 1 Number of PubMed publications with the terms "nutraceuticals" along with pregnancy complication such as "preeclampsia," "preterm birth," and "IUGR" (intrauterine growth restriction) over the last 16 years

supplementation with omega-3 polyunsaturated fatty acids, minerals, and vitamins, including folic acid and vitamin D, during gestation to reduce the risks of disorders such as preeclampsia and to prevent impaired fetal development (Agrawal et al. 2015; Al-Dughaishi et al. 2016; De-Regil et al. 2016; Omotayo et al. 2016). Some of these supplements, however, can also be detrimental to the offspring health when in excessive doses (Brown et al. 2014).

Omega-3 Polyunsaturated Fatty Acids

Omega-3 fatty acids are long-chain polyunsaturated fatty acids of great importance for human health (Gladyshev et al. 2013). Nevertheless, humans lack key enzymes necessary for the insertion of double bonds into the ω3 position of fatty acid chains and are unable to synthesize omega-3 fatty acids physiologically (Gladyshev et al. 2013). Therefore, humans can only obtain these nutrients from diet or via supplementation (Scorletti and Byrne 2013; Wiktorowska-Owczarek et al. 2015). Dietary sources of omega-3 fatty acids are mainly marine fish predators but also include green leafy vegetables, nuts, legumes, algae, and other seafood (Scorletti and Byrne 2013; Wiktorowska-Owczarek et al. 2015). In addition to food sources, dietary supplements containing omega-3 fatty acids are becoming popular worldwide (Wiktorowska-Owczarek et al. 2015).

Docosahexaenoic acid (DHA) and eicosapentaenoic acid (EPA) are physiologically important omega-3 fatty acids (Gladyshev et al. 2013; Wiktorowska-Owczarek et al. 2015). These fatty acids have some common biological actions (Table 1), including reduction of triglyceride plasma levels, decrease of total LDL cholesterol fraction, normalization of blood pressure, anti-inflammatory activity, and antithrombotic activity (Wiktorowska-Owczarek et al. 2015). Nevertheless, DHA and EPA have important distinctions in their biological effects. EPA provides healthy function of cardiovascular system (Gladyshev et al. 2013), whereas DHA, although also playing a role in cardiovascular health, is commonly found in the brain and in the retina. DHA plays a critical role in the functioning of the nervous system of adults, in nervous system development during fetal life and childhood, and has cytoprotective and anti-inflammatory activities in the eyes (Wiktorowska-Owczarek et al. 2015).

The use of omega-3 fatty acids as nutraceuticals during gestation has been intensively studied (Kar et al. 2016). The potential benefits of dietary omega-3 fatty acids during pregnancy rely on the fact that these nutrients could prevent pregnancy complications by reducing inflammation and oxidative stress. Indeed, increased placental inflammation and oxidative stress have been associated with placental disorders such as preeclampsia, IUGR, and gestational diabetes mellitus (Jones et al. 2014). Omega-3 fatty acids have been shown to reduce placental

Table 1 List of nutrients that are nutraceuticals available as supplements, their main food sources, and biological functions

Nutrients	Food sources	Biological functions
Omega-3 fatty acids	Marine fish Green leafy vegetables Nuts Legumes Algae	Plasma triglycerides reduction Total LDL cholesterol fraction reduction Blood pressure normalization Anti-inflammatory activity Antithrombotic activity Cardiovascular system function improvement[a] Nervous system development[b] Cytoprotective activity in the eyes[b] Anti-inflammatory activity in the eyes[b]
Vitamin D	Milk Eggs Oily fish Meat	Regulation of calcium and phosphorus homeostasis Bone development and maintenance Neuromuscular function Immune processes Cell growth and differentiation Fetal development
Folate/folic acid	Green leafy vegetables Legumes Yeast Liver Dairy products	Nucleotide synthesis Cell division and growth Growth and development of the placenta and fetus

[a]Mainly eicosapentaenoic acid (EPA)
[b]Mainly docosahexaenoic acid (DHA)

inflammation by either promoting the generation of anti-inflammatory eicosanoids or reducing pro-inflammatory eicosanoid generation (Scorletti and Byrne 2013; Jones et al. 2014; Poniedzialek-Czajkowska et al. 2014) and may limit placental oxidative stress by decreasing production or increasing scavenging of reactive oxygen species (Jones et al. 2014).

Another important aspect of maternal dietary omega-3 fatty acids is the importance of DHA to the neuronal development of the offspring. Because of its importance in nervous system development, DHA accumulates in the mammalian fetal brain during late gestation (Cetin et al. 2002). Interestingly, DHA concentrations are reduced in pregnancy complications such as preterm birth, preeclampsia, and IUGR, which may interfere with fetal development (Cetin et al. 2002; Dhobale and Joshi 2012). Since enzymes responsible for omega-3 fatty acid synthesis are undetectable in the placenta, the great majority of the omega-3 fatty acids must come from the diet, reinforcing the importance of the use of these nutrients as nutraceuticals during pregnancy (Jones et al. 2014). In fact, many studies aiming to verify the efficacy of omega-3 fatty acid supplementation in preventing preterm delivery in women have reported a reduction in the risk of preterm delivery (Mozurkewich and Klemens 2012; Poniedzialek-Czajkowska et al. 2014; Kar et al. 2016).

Vitamin D and Calcium

Vitamin D is known for its critical importance in the regulation of calcium and phosphorus homeostasis and is required for bone development and maintenance and neuromuscular function (DeLuca and Zierold 1998; Christakos et al. 2007). Calcium is critically important for several biological functions (Table 1), including regulation of muscle and nerve function, bone and teeth development and maintenance, and blood coagulation (Bouillon et al. 2003). The active form of vitamin D is a major regulator of intestinal calcium absorption and calcium reabsorption in the kidney, which regulates the calcium balance in the body (Bouillon et al. 2003; Sutton and MacDonald 2003). But the functions of this secosteroid hormone go beyond the regulation of mineral balance, playing roles in fetal development, immune processes, and cell growth and differentiation in a variety of tissues (Sutton and MacDonald 2003; Christakos et al. 2007; Pérez-López 2007).

Vitamin D itself is metabolically inactive and must undergo sequential hydroxylations to produce its active form (DeLuca and Zierold 1998). The production of the bioactive form of vitamin D ($1,25\text{-}(OH)_2D_3$) starts when vitamin D3 (cholecalciferol) is obtained from the diet or is produced in the skin upon exposure to ultraviolet B (UVB) radiation. Vitamin D3 is then transported to the liver and gets hydroxylated to produce 25-hydroxyvitamin D3 (25(OH)D3), the major circulating form of vitamin D in mammals. The second hydroxylation takes place in the kidney, catalyzed by 1α-hydroxylase and regulated by serum levels of calcium, phosphorus, and parathyroid hormone (PTH), resulting in the production of the bioactive vitamin D (Sutton and MacDonald 2003; Calvo et al. 2005). The active vitamin D goes to target tissues, where it binds to the vitamin D receptor (VDR), heterodimerizes with the retinoid X receptor (RXR), and binds to vitamin D response element (VDRE) in the promoter of target genes (Christakos et al. 2007). The best studied target organs for the hormonally active vitamin D are the intestine, the kidney, and the bone, but many other target tissues have also been recognized (Calvo et al. 2005).

Vitamin D3 can be obtained from food sources such as milk, eggs, oily fish, and meat (Calvo et al. 2005; Perez-Lopez 2007). Apart from diet and supplements, another way of maintaining adequate levels of vitamin D is from skin exposure to UVB light, which is influenced by geographical location, season, dark skin tone, use of sunscreen, and the air pollution (Zhang et al. 2015). Keeping adequate levels of vitamin D3 is critical to avoid health-related problems, including osteoporosis, diabetes, ischemic heart diseases, autoimmune diseases, cancer, multiple sclerosis, and rheumatoid arthritis (Calvo et al. 2005). Nevertheless, because of low consumption of vitamin D3 in the diet and inadequate exposure to the sun, several studies have reported a high prevalence of vitamin D deficiency (Calvo et al. 2005). This deficiency is also associated with obesity and sedentary activity (De-Regil et al. 2016).

During pregnancy, vitamin D and calcium are critical for fetal development (Lewis et al. 2010; De-Regil et al. 2016). Fetal skeletal development during the third trimester of gestation demands an increase in circulating calcium, which requires an increase in the levels of active form of vitamin D for enhanced calcium absorption (Fleet and Schoch 2010; Lewis et al. 2010). The fetus is totally dependent on the concentration of the active form of vitamin D that crosses the placental barrier, and, therefore, production of the 25(OH)D_3 during gestation becomes elevated as a result of increased renal hydroxylation and possibly due to placental production (Perez-Lopez 2007). This increased vitamin D requirement in pregnancy must be met through sun exposure, dietary intake, and supplements (Lewis et al. 2010; Zhang et al. 2015).

Vitamin D deficiency during pregnancy is common among women worldwide and may lead to pregnancy complications (Lewis et al. 2010; De-Regil et al. 2016). Maternal vitamin D deficiency during gestation has been associated with an increased risk of preeclampsia and gestational diabetes mellitus (Lewis et al. 2010; De-Regil et al. 2016). Interestingly, vitamin D has been reported to stimulate insulin production and to prevent insulin deficiency, which are central to the pathogenesis of gestational diabetes mellitus, and to decrease pro-inflammatory cytokines present in preeclamptic placentas (Lewis et al. 2010). Furthermore, women with preeclampsia appear to have an impaired calcium metabolism, and calcium supplementation reduces the risk of preeclampsia (Novakovic et al. 2009; Lewis et al. 2010). Thus, supplementation with vitamin D and calcium can be considered to prevent pregnancy complications.

Folic Acid

Folic acid, also known as vitamin B9, is the fully oxidized form of the folate vitamin that is synthetically produced for use in fortified foods and dietary supplements

(Saini et al. 2016). Humans cannot synthesize folate, which naturally occurs in an array of foods, including leafy green vegetables, legumes, yeast, liver, and dairy products (Lucock 2000; Dang et al. 2000; Stea et al. 2006; Winkels et al. 2007). The instability of folate does not allow it to be available at desired levels in processed and stored food, making the more stable folic acid a preferable component for dietary supplements (Winkels et al. 2007). Since folic acid is more stable than dietary folates, they have better bioavailability and are more rapidly absorbed across the intestine brush border than natural folates (Goh and Koren 2008).

Folic acid derivatives take part in specific biological activities (Table 1), including nucleotide synthesis, methionine regeneration from homocysteine, and oxidation and reduction of one-carbon units required for normal cell division and growth (Saini et al. 2016). Folates play critical roles in the regulation of several physiological functions. Disorders affected by folate status include macrocytic anemia, cardiovascular diseases, cancer, Alzheimer's disease, affective disorders, and Down's syndrome (Lucock 2000; Konings et al. 2001; Winkels et al. 2007).

A high folate demand exists during gestation, driven by the critical role of folate for DNA synthesis, cellular division and proliferation, and the growth and development of the placenta and fetus (Dhobale and Joshi 2012). In pregnant women, low levels of folic acids are associated with increased risks of neural tube defects, recurrent pregnancy loss, preeclampsia, preterm birth, and miscarriage (Lucock 2000; Dhobale and Joshi 2012; Balogun et al. 2016). The main known action performed by folic acid to prevent pregnancy complications is the reduction of homocysteine levels, which might reduce oxidative stress (Shahbazian et al. 2016).

Nutraceuticals and Epigenetics

Epigenetic modifications affect gene expression of current and future generations without changing the DNA sequence itself (Wolffe and Matzke 1999). To date, the better studied epigenetic mechanisms are DNA methylation, histone modifications, and microRNAs (Table 2). DNA methylation and histone modifications regulate gene transcription, whereas microRNAs act posttranscriptionally to regulate gene or protein expression (Tammen et al. 2013).

DNA Methylation

DNA methylation is characterized by the methylation of the 5′ position of a cytosine within the genome, mostly within CpG dinucleotides, and is catalyzed by the enzymes DNA methyltransferases (DNMTs), resulting in the formation of 5-methylcytosine (Tammen et al. 2013). On the other hand, the removal of methyl from 5-methylcytosine, called DNA demethylation, is intermediated by Ten-eleven translocation (TET) enzymes through the process of hydroxymethylation (Tammen et al. 2013). In general, hypermethylation of DNA prevents gene transcription,

Table 2 Epigenetic modifications and their characteristics

Epigenetic modification	Enzymes	Mechanisms	Effects
DNA methylation	DNMTs	Methylation of the 5′ position of cytosines	Prevention of gene transcription
DNA demethylation	TETs	Removal of methyl group from 5-methylcitosine	Increase in gene transcription
Histone acetylation	HATs	Acetylation of the ε-amino group of lysine residues in the tails of histones	Open chromatin for increase in gene transcription
Histone deacetylation	HDACs	Removal of acetyl groups from lysine residues in the tails of histones	Close chromatin for prevention of gene transcription
Histone methylation	HMTs	Methylation of lysine and arginine residues in the tails of histones	Chromatin remodeling for gene activation or repression
Histone demethylation	Histones demethylases	Removal of methyl groups from amino residues in the tails of histones	Chromatin remodeling for gene activation or repression
MicroRNAs	Dicer (for microRNA maturation)	Binding to the target mRNA 3′ untranslated region	Gene translation repression and/or mRNA cleavage

whereas hypomethylation is associated with an increase in gene transcription (Tammen et al. 2013).

Nutrition has an impact on gene expression via alterations in DNA methylation and histone modifications (Tammen et al. 2013). During pregnancy, maternal nutrition and the use of nutraceuticals may impact the epigenome of the offspring, bringing stable, long-term modifications to the offspring gene expression (Lillycrop et al. 2005; Tokunaga et al. 2013). With advances in our understanding of epigenetics, it is now becoming well accepted that the use of nutraceuticals has an impact on pregnancy complications through epigenetic mechanisms (Table 3).

Micronutrients such as folate and DHA affect fetal development and pregnancy outcomes through epigenetic mechanisms (Dhobale and Joshi 2012). These nutrients take part in the one-carbon metabolism (Fig. 2), a critical pathway for nucleotide biosynthesis and fetal development, where they reduce oxidative stress, inflammation, and the risk of preeclampsia and regulate global DNA methylation (Dhobale and Joshi 2012; Kemse et al. 2014).

Folate is required for the remethylation of homocysteine to methionine and provides the methyl group for methionine to be converted into S-adenosylmethionine (SAM), a major methyl donor for many reactions including the methylation of DNA, RNA, and proteins (Goh and Koren 2008; Dhobale and Joshi 2012). In addition, folate is used for the recycling of homocysteine back to methionine and for nucleotide biosynthesis, including purines, pyrimidines, and thymidines, and for the synthesis of DNA (Lucock 2000; Goh and Koren 2008). When folate concentrations are low, SAM levels are decreased, which has been

Table 3 Impacts of nutrients deficiency during pregnancy on epigenetic alterations and pregnancy complications

Nutrient deficiency during pregnancy	Reported epigenetic alterations	Reported pregnancy complications
Omega-3 fatty acid deficiency	Increased global DNA methylation, altered histone methylation, altered expression of circulating microRNAs	Preeclampsia, preterm delivery, and intrauterine growth restriction
Vitamin D deficiency	Altered DNA methylation, increased histone acetylation, altered expression of circulating microRNAs	Preeclampsia and gestational diabetes
Folate deficiency	Decreased global DNA and histone methylation, altered placental microRNA expression	Neural tube defects, recurrent pregnancy loss, preeclampsia, preterm birth, miscarriage

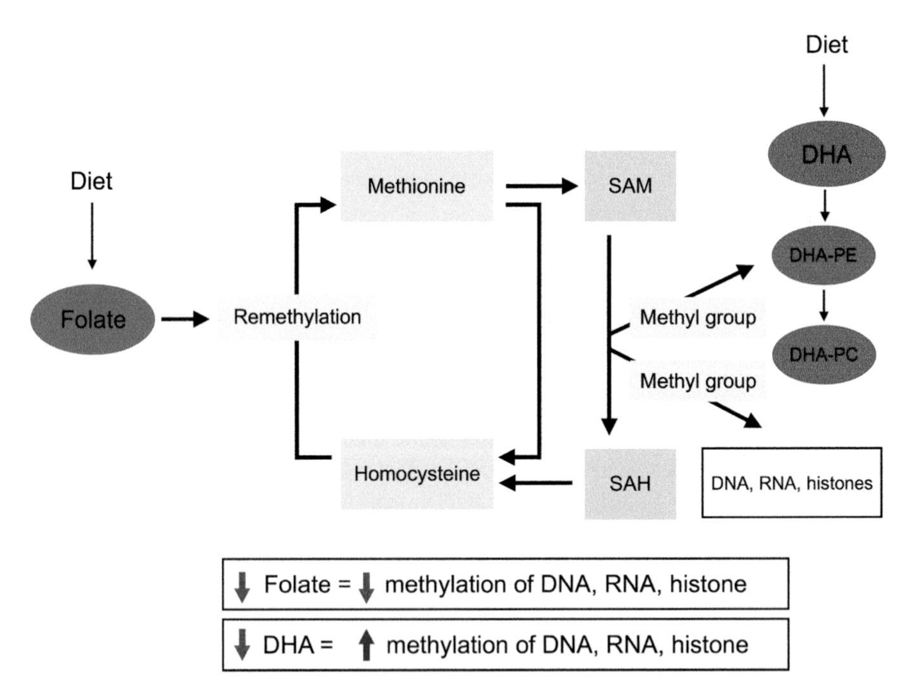

Fig. 2 Dietary status impact of folate and docosahexaenoic acid (*DHA*) on global methylation of DNA, RNA, and histones through the one-carbon metabolism. Folate is important to providing the methyl group to methionine, which will then be transferred to S-adenosylmethionine (*SAM*). SAM provides methyl groups for methylation of nucleic acids and histones and to DHA, incorporated into phosphatidyl ethanolamine (*PE*), which will be converted into phosphatidylcholine (*PC*). After donating the methyl group, SAM becomes S-adenosylhomocysteine (*SAH*)

associated with a decrease in genome-wide methylation in humans and animals and may lead to DNA strand damage and result in offspring genetic mutations (Choi and Mason 2002; Dhobale and Joshi 2012) and disorders such as autism (DeVilbiss et al. 2015). In addition, when SAM levels are decreased due to folic acid deficiency, there is an increase in the levels of homocysteine, a cytotoxic sulfhydryl-containing molecule. This increase in homocysteine enhances pro-inflammatory cytokine expression, induces oxidative stress through promotion of reactive oxygen species generation, and may lead to global DNA hypomethylation (Handy et al. 2011; Dhobale and Joshi 2012) and alterations in DNMTs (Kalani et al. 2014). Elevated maternal levels of homocysteine are associated with prematurity, low birth weight, preeclampsia, miscarriage, and IUGR (Aubard et al. 2000).

Similar to folic acid, dietary DHA, an important omega-3 fatty acid, can also alter global DNA methylation patterns through the one-carbon metabolism (Kulkarni et al. 2011; Kemse et al. 2014). But differences exist in how folic acid and DHA participate in this metabolism and how they play a role in epigenetics (Fig. 2). While folic acid is an important factor for the generation of methionine from homocysteine and can contribute to the generation of SAM, a ubiquitous methyl donor, DHA receives methyl groups from SAM through the actions of phosphatidyl ethanolamine-*N*-transferase (Kulkarni et al. 2011). This methyl transfer to DHA is important for the conversion of phosphatidyl ethanolamine into phosphatidylcholine, a phospholipid that is crucial for the delivery of polyunsaturated fatty acids such as DHA from the liver to the plasma for posterior distribution to peripheral tissues (Kulkarni et al. 2011). When maternal levels of DHA are low, however, there will be less requirement of methyl groups for the conversion of phosphatidyl ethanolamine into phosphatidylcholine, and the excessive influx of methyl groups will increase placental global DNA methylation, altering placental gene expression (Kulkarni et al. 2011). Therefore, while folic acid deficiency may lead to a decrease in global DNA methylation, low levels of dietary DHA may increase global DNA methylation patterns. It is possible that maternal dietary deficiencies of these micronutrients lead to maternal complications through epigenetic alterations.

The increased levels of the active form of vitamin D during pregnancy also provide an interesting example of how DNA methylation affects pregnancy outcomes and may be altered by diet (Fig. 3). Vitamin D is crucial for fetal development and for remodeling of the maternal vasculature during gestation (Novakovic et al. 2009). As a result, low levels of maternal vitamin D and calcium have been associated with pregnancy complications such as preeclampsia, which can be prevented by calcium supplementation (Novakovic et al. 2009; Lewis et al. 2010). Vitamin D exerts its biological effects through the binding and activation of VDR, which heterodimerizes with the RXR and regulates gene expression through the binding of target gene promoters (Christakos et al. 2007; Goyal et al. 2014). Interestingly, VDR and RXR have been reported to be downregulated in placental tissue from preeclamptic pregnancies through the process of DNA hypermethylation (Anderson et al. 2015). In addition, DNA methylation regulates the expression of two critical genes for regulation of vitamin D levels during

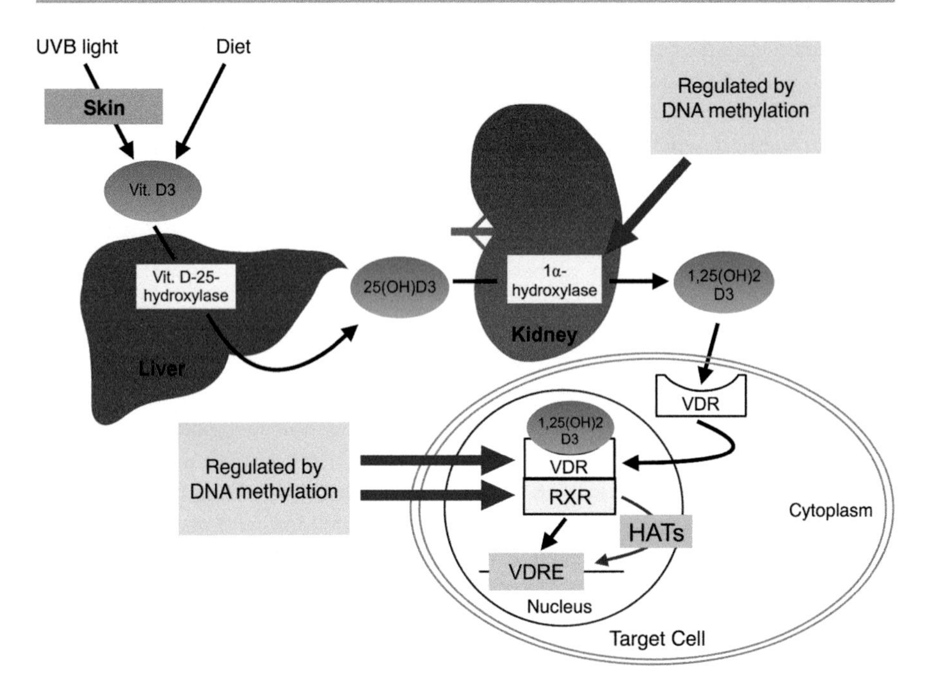

Fig. 3 Epigenetics impact on vitamin D synthesis and biological activity. In order to become active, dietary vitamin D undergoes successive hydroxylations by the enzymes vitamin D-25-hydroxylase and 1α-hydroxylase. When active, vitamin D binds to vitamin D receptor (*VDR*) and forms a heterodimer with retinoid X receptor (*RXR*), which is required for vitamin D to exert its effects through binding to vitamin D response element (*VDRE*) of target cells. These events are regulated by epigenetic mechanisms: DNA methylation regulates the expression of 1α-hydroxylase, VDR, and RXR, whereas the heterodimer VDR/RXR interacts with histone acetyltransferases (*HATs*) to stimulate histone acetylation and target gene transcription

pregnancy: 25(OH)-1α-hydroxylase (*CYP27B1*), which is required for the activation of vitamin D and must be up-regulated during pregnancy, and 24-hydroxylase (*CYP24A1*), which inactivates vitamin D and must be downregulated during pregnancy (Novakovic et al. 2009; Goyal et al. 2014). Indeed, the *CYP24A1* promoter has been reported to be hypomethylated, whereas *CYP27B1* is unmethylated in the normotensive human placenta (Novakovic et al. 2009) and hypermethylated in human preeclamptic placenta (Anderson et al. 2015). Therefore, we can state that metabolism of vitamin D can be altered through epigenetic mechanisms in women with pregnancy complications, but further studies will have to investigate the role of dietary vitamin D in these epigenetic alterations.

Histone Modifications

The eukaryotic genome is compacted within the cell in nucleosomes, which constitute the basic units of chromatin. Each nucleosome is composed of DNA wrapped nearly twice in an octameric protein core that contains two copies of the core histones H2A, H2B, H3, and H4 (Tsukiyama and Wu 1997; Peterson and Laniel 2004). Histone modifications such as acetylation, methylation, phosphorylation, ubiquitination, and sumoylation open or close the chromatin structure to transcription and, therefore, regulate gene expression (Tammen et al. 2013; Friso et al. 2016). These modifications take place in short amino acid sequences attached to the histones, which are called histone tails (Friso et al. 2016).

Histone acetylation of the ε-amino group of a lysine residue within the histone tail, catalyzed by histone acetyltransferases (HATs), opens the chromatin structure for transcription, while the removal of acetyl groups from histone tail lysine residues is catalyzed by histone deacetylases (HDACs) and is associated with repression of gene transcription (Tammen et al. 2013). Histone methylation can result in gene activation or repression, depending on the histone site, the degree of methylation, the amino residues affected, and the position of the amino residues in the histone tail (Fetahu et al. 2014). Histones can be methylated by the action of histone methyltransferases (HMTs), which transfer the methyl group from SAM, a ubiquitous methyl donor, mainly to specific histone lysine and arginine residues, generating S-adenosylhomocysteine (SAH) and regulating gene transcription through chromatin remodeling and recruitment of the transcriptional machinery (Handy et al. 2011; Mentch and Locasale 2016). Histone methylation is maintained by a balance between the action of HMTs and histone demethylases (Friso et al. 2016; Mentch and Locasale 2016). Lysine residues can be mono-, di-, or trimethylated, whereas arginine residues can be mono- or dimethylated (Friso et al. 2016).

Folate, an important micronutrient for preventing pregnancy complications, has been reported to affect histone modifications (Friso et al. 2016), and its dietary status provides a clear example of the association between nutrition and epigenetics (Fig. 2). Folate is a major component of the one-carbon metabolism (Dhobale and Joshi 2012). It is required for the conversion of methionine into SAM, which provides the methyl groups that will be transferred to histone tail lysine and arginine residues by HMTs during the process of histone methylation (Mentch and Locasale 2016). Both dietary folate deficiency (Garcia et al. 2016; Friso et al. 2016) and reduced SAM levels (Mentch and Locasale 2016) have been associated with reduced histone methylation, leading to changes in gene expression.

Recently, a link between maternal dietary DHA, one of the physiologically relevant omega-3 fatty acids, and histone modifications during pregnancy has been reported (Joss-Moore et al. 2010). Using a rat uteroplacental insufficiency model of IUGR, Joss-Moore et al. observed a decrease in serum DHA levels and in lung peroxisome proliferator-activated receptor gamma (PPARγ) gene expression, Setd8 mRNA levels, and methylation of the lysine residue 20 of histone H4 lysine 20 (H4K20Me) in neonatal pups. These observations imply an importance for DHA maternal dietary status in regulating offspring lung development, since a decrease in Setd8, a histone lysine-methyltransferase, leads to a reduction in histone methylation of the promoter of the gene PPARγ and, consequently, downregulates its expression. Interestingly, maternal DHA supplementation during pregnancy prevented the decrease of all these factors.

Vitamin D has also been shown to be associated with histone modifications through the transcriptional activities mediated by its receptor (Fig. 3). After the active vitamin D binds and activates VDR in target tissues, the VDR/RXR dimer interacts with HATS to induce histone acetylation and, therefore, transcriptional activation (Christakos et al. 2007; Fetahu et al. 2014). Since vitamin D induces histone modifications, it is possible that maternal vitamin D status regulates fetal development through this epigenetic mechanism.

MicroRNAs

MicroRNAs are a family of 21–25-nucleotides small, noncoding, RNAs that suppress gene expression at the posttranscriptional level. In several cases, microRNAs negatively regulate target expression at the translational level by binding to the target gene 3′ untranslated regions (UTRs) through imperfect complementarity at multiple sites, but some microRNAs also repress gene translation by directly cleaving target mRNA or by incorporating into the RNA-induced silencing complex (He and Hannon 2004). Moreover, several microRNAs can interact with other epigenetic mechanisms, being able to target DNMTs and chromatin-modifying genes, whereas their expression can be regulated by DNA methylation and histone modifications (Chango and Pogribny 2015).

MicroRNAs are abundant in the placenta, where they have the potential to regulate trophoblast proliferation and physiology and adaptation to challenges such as hypoxia (Mouillet et al. 2011; Baker et al. 2017). Also, microRNA expression is altered in pregnancy disorders such as preeclampsia, preterm labor, and IUGR (Mouillet et al. 2011). Interestingly, microRNA expression is affected by maternal diet, which highlights the importance of nutraceuticals on preventing pregnancy complications through epigenetics (Chango and Pogribny 2015).

The association between placental microRNA expression and diet is becoming clear with recent studies investigating the relationship between maternal dietary nutrient status and microRNA expression during pregnancy. For example, in low folate conditions, microRNAs involved in the folate-mediated one-carbon metabolism are altered, implying a function of microRNAs in regulating fetal development and pregnancy outcomes (Chango and Pogribny 2015). An association between placental dysfunction and altered placental microRNA expression has also been reported in pregnant teens with low dietary folate status (Baker et al. 2017). In addition, the expression of circulating microRNAs in pregnant women is altered in response to dietary vitamin D (Enquobahrie et al. 2011) and omega-3 fatty acid levels (Ortega et al. 2015). Interestingly, the altered microRNAs in response to maternal diets deficient in folate and vitamin D target genes are involved in tissue remodeling, angiogenesis, and placental development, suggesting an important role of microRNAs in pregnancy health in response to maternal nutrient status (Enquobahrie et al. 2011; Baker et al. 2017).

Conclusion

Maternal dietary status during gestation has a potential impact on pregnancy health through epigenetic mechanisms. The use of nutraceuticals should therefore be used to complement nutrient deficiency to prevent pregnancy disorders. Nevertheless, further research should be made to unravel the effects of specific epigenetic alterations on pregnancy outcomes in response to maternal nutrition.

Dictionary of Terms

- **Gene** – A sequence of nucleotides along a segment of DNA that codes for a protein or RNA molecule.
- **Nutraceutical** – A food or part of a food that not only complement the diet but also prevent and/or treat disorders or diseases.
- **Oxidative stress** – A disturbance in the balance between the production of reactive oxygen species and antioxidant defenses.
- **Preeclampsia** – Pregnancy-induced hypertensive disorder characterized by high blood pressure and increased levels of protein in the urine.
- **Trophoblast** – The peripheral cells of the blastocyst, which attach the zygote to the uterine wall and become the placenta and the membranes that nourish and protect the developing organism.

Key Facts of Nutraceuticals

- Nutraceuticals are known as a food or part of a food that not only complement the diet but also prevent and/or treat disorders or diseases.
- The list of nutraceuticals includes vitamins, probiotics, and antioxidants.
- Nutraceutical compounds can be sold as tablets, capsules, powders, potions, or solutions.
- Nutraceuticals can prevent diseases by altering gene/epigene expression.
- During gestation, nutraceuticals may affect the epigenome of the offspring, modulating gene expression in babies.

Summary Points

- *Nutraceutical* is characterized as “a food, or part of a food, that provides medical or health benefits, including the prevention and/or treatment of a disease.”
- The list of nutraceutical compounds includes probiotics, vitamins, and antioxidants.
- Dietary supplementation with omega-3 polyunsaturated fatty acids, minerals, and vitamins, including folic acid and vitamin D, during gestation has been reported

to reduce the risks of disorders such as preeclampsia and prevent impaired fetal development.

- Dietary omega-3 fatty acids reduce placental inflammation and limit placental oxidative stress, preventing pregnancy complications such as preeclampsia, intra-uterine growth restriction, and gestational diabetes mellitus.
- Vitamin D may prevent pregnancy complications such as gestational diabetes and preeclampsia by stimulating insulin production and decreasing pro-inflammatory cytokines, respectively.
- Folate may prevent pregnancy complications such as recurrent pregnancy loss, preeclampsia, preterm birth, and miscarriage by reducing homocysteine levels, which might reduce oxidative stress.
- Maternal nutrition and the use of nutraceuticals may prevent pregnancy complications through epigenetic mechanisms.
- Micronutrients such as folate and docosahexaenoic acid take part in the one-carbon metabolism and affect fetal development and pregnancy outcomes through epigenetic mechanisms.
- Vitamin D synthesis and biological effects are regulated by DNA methylation and histone modifications.
- Nutrients such as folate, omega-3 fatty acids, and vitamin D may affect pregnancy health through alterations in microRNA expression in pregnant women.

Acknowledgments Mahua Choudhury is supported by Morris L Lichtenstein Jr Medical Research Foundation for diabetes and obesity research and Texas A & M Health Science Center Faculty Development Fund

References

Agrawal S, Fledderjohann J, Vellakkal S, Stuckler D (2015) Adequately diversified dietary intake and iron and folic acid supplementation during pregnancy is associated with reduced occurrence of symptoms suggestive of pre-eclampsia or eclampsia in Indian women. PLoS One 10(3): e0119120. https://doi.org/10.1371/journal.pone.0119120

Al-Dughaishi T, Nikolic D, Zadjali F, Al-Hashmi K, Al-Waili K, Rizzo M et al (2016) Nutraceuticals as lipid-lowering treatment in pregnancy and their effects on the metabolic syndrome. Curr Pharm Biotechnol 17(7):614–623

Amhed R, Dunford J, Mehran R, Robson S, Kunadian V (2014) Pre-eclampsia and future cardiovascular risk among women: a review. J Am Coll Cardiol 63(18):1815–1822

Anderson CM, Ralph JL, Johnson L, Scheett A, Wright ML, Taylor JY et al (2015) First trimester vitamin D status and placental epigenomics in preeclampsia among Northern Plains primiparas. Life Sci 129:10–15

Armelagos GJ (2014) Brain evolution, the determinates of food choice, and the omnivore's dilemma. Crit Rev Food Sci Nutr 54(10):1330–1341

Aubard Y, Darodes N, Cantaloube M (2000) Hyperhomocysteinemia and pregnancy – review of our present understanding and therapeutic implications. Eur J Obstet Gynecol Reprod Biol 93(2):157–165

Baker BC, Mackie FL, Lean SC, Greenwood SL, Heazell AE, Forbes K et al (2017) Placental dysfunction is associated with altered microRNA expression in pregnant women with low folate status. Mol Nutr Food Res. https://doi.org/10.1002/mnfr.201600646

Balogun OO, da Silva LK, Ota E, Takemoto Y, Rumbold A, Takegata M et al (2016) Vitamin supplementation for preventing miscarriage. Cochrane Database Syst Rev 5:CD004073. https://doi.org/10.1002/14651858.CD004073.pub4

Bouillon R, Van Cromphaut S, Carmeliet G (2003) Intestinal calcium absorption: molecular vitamin D mediated mechanisms. J Cell Biochem 88(2):332–339

Brower V (1998) Nutraceuticals: poised for a healthy slice of the healthcare market? Nat Biotechnol 16(8):728–731

Brown SB, Reeves KW, Bertone-Johnson ER (2014) Maternal folate exposure in pregnancy and childhood asthma and allergy: a systematic review. Nutr Rev 72(1):55–64

Calvo MS, Whiting SJ, Barton CN (2005) Vitamin D intake: a global perspective of current status. J Nutr 135(2):310–316

Cetin I, Giovannini N, Alvino G, Agostoni C, Riva E, Giovannini M et al (2002) Intrauterine growth restriction is associated with changes in polyunsaturated fatty acid fetal-maternal relationships. Pediatr Res 52(5):750–755

Chango A, Pogribny IP (2015) Considering maternal dietary modulators for epigenetic regulation and programming of the fetal epigenome. Forum Nutr 7(4):2748–2770

Choi SW, Mason JB (2002) Folate status: effects on pathways of colorectal carcinogenesis. J Nutr 132(Suppl 8):2413S–2418S

Christakos S, Dhawan P, Peng X, Obukhov AG, Nowycky MC, Benn BS et al (2007) New insights into the function and regulation of vitamin D target proteins. Steroid Biochem Mol Biol 103(3–5):405–410

Dang J, Arcot J, Shrestha A (2000) Folate retention in selected processed legumes. Food Chem 68(3):295–298

DeLuca HF, Zierold C (1998) Mechanisms and functions of vitamin D. Nutr Rev 56(2 Pt 2):S4–10; discussion S 54–75

De-Regil LM, Palacios C, Lombardo LK, Pena-Rosas JP (2016) Vitamin D supplementation for women during pregnancy. Cochrane Database Syst Rev 1:CD008873. https://doi.org/10.1002/14651858.CD008873.pub3

DeVilbiss EA, Gardner RM, Newschaffer CJ, Lee BK (2015) Maternal folate status as a risk factor for autism spectrum disorders: a review of existing evidence. Br J Nutr 114(5):663–672

Dhobale M, Joshi S (2012) Altered maternal micronutrients (folic acid, vitamin B(12)) and omega 3 fatty acids through oxidative stress may reduce neurotrophic factors in preterm pregnancy. J Matern Fetal Neonatal Med 25(4):317–323

Enquobahrie DA, Williams MA, Qiu C, Siscovick DS, Sorensen TK (2011) Global maternal early pregnancy peripheral blood mRNA and miRNA expression profiles according to plasma 25-hydroxyvitamin D concentrations. J Matern Fetal Neonatal Med 24(8):1002–1012

Fetahu IS, Hobaus J, Kallay E (2014) Vitamin D and the epigenome. Front Physiol 5:164

Fleet JC, Schoch RD (2010) Molecular mechanisms for regulation of intestinal calcium absorption by vitamin D and other factors. Crit Rev Clin Lab Sci 47(4):181–195

Friso S, Udali S, De Santis D, Choi SW (2016) One-carbon metabolism and epigenetics. Mol Asp Med. https://doi.org/10.1016/j.mam.2016.11.007

Gage TB, O'Connor K (1994) Nutrition and the variation in level and age patterns of mortality. Hum Biol 66(1):77–103

Garcia BA, Luka Z, Loukachevitch LV, Bhanu NV, Wagner C (2016) Folate deficiency affects histone methylation. Med Hypotheses 88:63–67

Gladyshev MI, Sushchik NN, Makhutova ON (2013) Production of EPA and DHA in aquatic ecosystems and their transfer to the land. Prostaglandins Other Lipid Mediat 107:117–126

Goh YI, Koren G (2008) Folic acid in pregnancy and fetal outcomes. J Obstet Gynaecol 28(1):3–13

Goyal R, Zhang L, Blood AB, Baylink DJ, Longo LD, Oshiro B et al (2014) Characterization of an animal model of pregnancy-induced vitamin D deficiency due to metabolic gene dysregulation. Am J Physiol Endocrinol Metab 306(3):E256–E266

Gul K, Singh AK, Jabeen R (2016) Nutraceuticals and functional foods: the foods for the future world. Crit Rev Food Sci Nutr 56(16):2617–2627

Handy DE, Castro R, Loscalzo J (2011) Epigenetic modifications: basic mechanisms and role in cardiovascular disease. Circulation 123(19):2145–2156
Hardy G (2000) Nutraceuticals and functional foods: introduction and meaning. Nutrition 16(7–8):688–689
He L, Hannon GJ (2004) MicroRNAs: small RNAs with a big role in gene regulation. Nat Rev Genet 5(7):522–531
Huppertz B, Weiss G, Moser G (2014) Trophoblast invasion and oxygenation of the placenta: measurements versus presumptions. J Reprod Immunol 101–102:74–79
Jones ML, Mark PJ, Waddell BJ (2014) Maternal dietary omega-3 fatty acids and placental function. Reproduction 147(5):R143–R152
Joss-Moore LA, Wang Y, Baack ML, Yao J, Norris AW, Yu X et al (2010) IUGR decreases PPARgamma and SETD8 expression in neonatal rat lung and these effects are ameliorated by maternal DHA supplementation. Early Hum Dev 86(12):785–791
Kalani A, Kamat PK, Givvimani S, Brown K, Metreveli N, Tyagi SC et al (2014) Nutri-epigenetics ameliorates blood-brain barrier damage and neurodegeneration in hyperhomocysteinemia: role of folic acid. J Mol Neurosci 52(2):202–215
Kalra EK (2003) Nutraceutical – definition and introduction. AAPS PharmSci 5(3):E25
Kar S, Wong M, Rogozinska E, Thangaratinam S (2016) Effects of omega-3 fatty acids in prevention of early preterm delivery: a systematic review and meta-analysis of randomized studies. Eur J Obstet Gynecol Reprod Biol 198:40–46
Kemse NG, Kale AA, Joshi SR (2014) A combined supplementation of omega-3 fatty acids and micronutrients (folic acid, vitamin B12) reduces oxidative stress markers in a rat model of pregnancy induced hypertension. PLoS One 9(11):e111902. https://doi.org/10.1371/journal.pone.0111902
Konings EJ, Roomans HH, Dorant E, Goldbohm RA, Saris WH, van den Brandt PA (2001) Folate intake of the Dutch population according to newly established liquid chromatography data for foods. Am J Clin Nutr 73(4):765–776
Kulkarni A, Dangat K, Kale A, Sable P, Chavan-Gautam P, Joshi S (2011) Effects of altered maternal folic acid, vitamin B12 and docosahexaenoic acid on placental global DNA methylation patterns in Wistar rats. PLoS One 6(3):e17706. https://doi.org/10.1371/journal.pone.0017706
Lewis S, Lucas RM, Halliday J, Ponsonby AL (2010) Vitamin D deficiency and pregnancy: from preconception to birth. Mol Nutr Food Res 54(8):1092–1102
Lillycrop KA, Phillips ES, Jackson AA, Hanson MA, Burdge GC (2005) Dietary protein restriction of pregnant rats induces and folic acid supplementation prevents epigenetic modification of hepatic gene expression in the offspring. J Nutr 135(6):1382–1386
Lucock M (2000) Folic acid: nutritional biochemistry, molecular biology, and role in disease processes. Mol Genet Metab 71(1–2):121–138
Mentch SJ, Locasale JW (2016) One-carbon metabolism and epigenetics: understanding the specificity. Ann N Y Acad Sci 1363:91–98
Mouillet JF, Chu T, Sadovsky Y (2011) Expression patterns of placental microRNAs. Birth Defects Res A Clin Mol Teratol 91(8):737–743
Mozurkewich EL, Klemens C (2012) Omega-3 fatty acids and pregnancy: current implications for practice. Curr Opin Obstet Gynecol 24(2):72–77
Novakovic B, Sibson M, Ng HK, Manuelpillai U, Rakyan V, Down T et al (2009) Placenta-specific methylation of the vitamin D 24-hydroxylase gene: implications for feedback autoregulation of active vitamin D levels at the fetomaternal interface. J Biol Chem 284(22):14838–14848
Omotayo MO, Dickin KL, O'Brien KO, Neufeld LM, De Regil LM, Stoltzfus RJ (2016) Calcium supplementation to prevent preeclampsia: translating guidelines into practice in low-income countries. Adv Nutr 7(2):275–278
Ortega FJ, Cardona-Alvarado MI, Mercader JM, Moreno-Navarrete JM, Moreno M, Sabater M et al (2015) Circulating profiling reveals the effect of a polyunsaturated fatty acid-enriched diet on common microRNAs. J Nutr Biochem 26(10):1095–1101

Perez-Lopez FR (2007) Vitamin D: the secosteroid hormone and human reproduction. Gynecol Endocrinol 23(1):13–24

Peterson CL, Laniel MA (2004) Histones and histone modifications. Curr Biol 14(14):R546–R551

Poniedzialek-Czajkowska E, Mierzynski R, Kimber-Trojnar Z, Leszczynska-Gorzelak B, Oleszczuk J (2014) Polyunsaturated fatty acids in pregnancy and metabolic syndrome: a review. Curr Pharm Biotechnol 15(1):84–99

Saini RK, Nile SH, Keum Y (2016) Folates: chemistry, analysis, occurrence, biofortification and bioavailability. Food Res Int 89:1–13

Scorletti E, Byrne CD (2013) Omega-3 fatty acids, hepatic lipid metabolism, and nonalcoholic fatty liver disease. Annu Rev Nutr 33:231–248

Shahbazian N, Jafari RM, Haghnia S (2016) The evaluation of serum homocysteine, folic acid, and vitamin B12 in patients complicated with preeclampsia. Electron Physician 8(10):3057–3061

Stea TH, Johansson M, Jägerstad M, Frølich W (2006) Retention of folates in cooked, stored and reheated peas, broccoli and potatoes for use in modern large-scale service systems. Food Chem 101:1095–1107

Sutton AL, MacDonald PN (2003) Vitamin D: more than a "bone-a-fide" hormone. Mol Endocrinol 17(5):777–791

Tammen SA, Friso S, Choi SW (2013) Epigenetics: the link between nature and nurture. Mol Asp Med 34(4):753–764

Tokunaga M, Takahashi T, Singh RB, De Meester F, Wilson DW (2013) Nutrition and epigenetics. Med Epigenet 1:70–77

Tsukiyama T, Wu C (1997) Chromatin remodeling and transcription. Curr Opin Genet Dev 7(2):182–191

Wiktorowska-Owczarek A, Berezinska M, Nowak JZ (2015) PUFAs: structures, metabolism and functions. Adv Clin Exp Med 24(6):931–941

Winkels RM, Brouwer IA, Siebelink E, Katan MB, Verhoef P (2007) Bioavailability of food folates is 80% of that of folic acid. Am J Clin Nutr 85(2):465–473

Wolffe AP, Matzke MA (1999) Epigenetics: regulation through repression. Science 286(5439):481–486

Zhang MX, Pan GT, Guo JF, Li BY, Qin LQ, Zhang ZL (2015) Vitamin D deficiency increases the risk of gestational diabetes mellitus: a meta-analysis of observational studies. Forum Nutr 7(10):8366–8375

Zhong Y, Tuuli M, Odibo AO (2010) First-trimester assessment of placenta function and the prediction of preeclampsia and intrauterine growth restriction. Prenat Diagn 30(4):293–308

Polyphenols and Histone Acetylation 103

Anna K. Kiss

Contents

Abstract

In this chapter, recent findings concerning the effect of polyphenols on histone acetylation and acetylation as posttranslational proteins modification are summarized. The relevance of an in vitro study to an in vivo situation is discussed, as orally administered polyphenols have a limited bioavailability. Additionally, the effects of polyphenol metabolites produced by gut microbiota metabolism and type I and II phase enzyme activity are described.

Polyphenols such as curcumin, gallic acid, epicatechin, and some flavonoids in relatively low concentrations are able to reduce several proinflammatory responses by modulating the HAT and HDAC activities, and at the epigenetic level may affect the proinflammatory state of the human body. On the other hand, the relevance of the observed in vitro epigenetic modulation of green tea, epigallogatechin-3-gallate, and isoflavones in cancer cells are difficult to extrapolate to in vivo conditions as the concentrations used are far higher than those detected in humans. Furthermore, information about the effects of polyphenols'

A. K. Kiss (✉)
Department of Pharmacognosy and Molecular Basis of Phytotherapy, Faculty of Pharmacy, Medical University of Warsaw, Warsaw, Poland
e-mail: akiss@wum.edu.pl; farmakognozja@wum.edu.pl

V. B. Patel, V. R. Preedy (eds.), *Handbook of Nutrition, Diet, and Epigenetics*,
https://doi.org/10.1007/978-3-319-55530-0_105

metabolites on epigenetic modulation is scarce and should be taken into consideration in future studies.

Keywords
Polyphenols · Curcumin · Green tea · Flavonoids · Flavan-3-ol · Polyphenols bioavailability · Epigenetics · Chronic diseases · Prevention · Regulation of inflammatory state · Chemoprevention · Histone acetylation

List of Abbreviations

ADR	Adriamycin
AR	Androgen receptor
CVD	Cardiovascular diseases
DHT	Dihydrotestosterone
DS	Dahl salt-sensitive rats
EGCG	Epigallogatechin-3-gallate
ER	Estrogen receptor
GATA	Zinc finger transcription factor
HAT	Histone acetyltransferase
HDAC	Histone deacetylase
HG	High glucose
IL-6	Interleukin 6
MCP-1	Monocyte chemoattractant protein-1
MIP-2	Macrophage inflammatory protein
NF-κB	Nuclear factor kappa-light-chain-enhancer of activated B cells
PE	Phenylephrine
SIRT	Sirtuin
Smad2	Intracellular protein transducing extracellular signals from TGF-β ligands to the nucleus where they activate downstream gene transcription
TGF-β	Transforming growth factor β
TNF-α	Tumor necrosis factor α

Introduction

Polyphenols are a group of heterogenous compounds present in food products (fruits, vegetables, nuts, wine, tea, and coffee) and medicinal plants (Table 1). The high consumption of polyphenols is associated with a reduced risk of chronic diseases connected with an elevated inflammatory state particularly cardiovascular diseases (CVD), type-2 diabetes, and cancer. Some of these beneficial effects may be associated with the epigenetic modulatory effect of polyphenols. Until recently, various polyphenolic compounds have been characterized as influencing HDAC, HAT activities (Table 2), and by doing so modulating metabolic disturbances, inflammatory responses, and the development of cancer. Various flavonoids have been identified as activators of class III HDACs (SIRTs). Turmeric and green tea and their constituents

Table 1 Occurrence of selected polyphenols in plants

Compound	Food products	Medicinal plants
Apigenin[a] (flavonoid)	Celery, parsley	Chamomile flower (*Matricaria chamomilla* L.)
Chrysin[a] (flavonoid)	Honey	Poplar bud (*Populus* sp.)
Curcumin	Turmeric, Javanese turmeric, curry spice	Turmeric rhizome (*Curcuma longa* L.), Javenese turmeric rhizome (*Curcuma xanthorrhiza* D. Dietrich)
Delphinidin[a] (anthocyanidin)	Aubergine skin, blackcurrant, blueberry	Blueberry fruit (*Vaccinium myrtillus* L.) Roselle flower (*Hibiscus sabdariffa* L.)
Ellagic acid[a]	Almond, grape, pomegranate, raspberry, strawberry	Leafflower (*Phyllanthus* sp.) Oak bark (*Quercus* sp.) Raspberry leaf (*Rubus* sp.)
Epicatechin (flavan-3-ol)	Apple, cocoa, grape, lentil plum, red wine	Ginkgo leaf (*Ginkgo biloba* L.) Hawthorn leaf and flowers fruits (*Crataegus* sp) Grape seed (*Vitis vinifera* L.)
Epigallogatechin-3-gallate (flavan-3-ol)	Green tea	Tea leaf (*Cammelia sinensis* L.) Witch-hazel bark (*Hamamelis virginiana* L.)
Fisetin[a] (flavonoid)	Apple, Chinese water chestnut, persimmon	Sumac (*Rhus* sp.)
Gallic acid	Green tea, black tea, mango, pomegranate	Tea leaf (*Cammelia sinensis* L.) Witch-hazel bark and leaf (*Hamamelis virginiana* L.) Peony root bark (*Paeonia suffruticosa* Andrews)
Genistein[a] (isoflavone)	Soyabean	Red clover flower (*Trifolium pratense* L.)
Kaempferol[a] (flavonoid)	Broccoli, kale	Ginkgo leaf (*Ginkgo biloba* L.) Lime flower (*Tilia* sp) Blackthorn flower (*Prunus spinosa* L.)
Luteolin[a] (flavonoid)	Artichoke, asparagus, broccoli, carrot, celery,	Artichoke leaf (*Cynara scolymus* L.) Chamomile flower (*Matricaria chamomilla* L.)
Procyanidin B (flavan-3-ol)	Apple, cocoa, grape, lentil, plum, red wine, strawberry	Hawthorn leaf and flowers fruits (*Crataegus* sp) Grape seed (*Vitis vinifera* L.) Oak bark (*Quercus* sp.)
Quercetin[a] (flavonoid)	Apple, broccoli, buckwheat, kale lettuce, onion	*Sophora japonica* L. Buckwheat herb (*Fagopyrum esculentum* MOENCH.) Ginkgo leaf (*Ginkgo biloba* L.) Lime flower (*Tilia* sp) Elder lower (*Sambucus nigra* L.) Blackthorn flower (*Prunus spinosa* L.)
Resveratrol[a]	Grape, red wine, peanut	Grape seed (*Vitis vinifera* L.) Knotweed rhizome (*Polygonum cuspidatum* Siebold & Zucc. and *P. multiflorum* Thunb.)

[a]Compounds present also in glycosylated forms and plants

Table 2 Inhibition of HAT and HCDA enzyme activities by polyphenols

Compound	HAT	HCDA/SIRT	References
Apigenin (flavonoid)	No data	↓ HDAC activity by ~50% at 40 μM	Tseng et al. 2017
Chrysin (flavonoid)	No data	↓ HDAC-8 activity by ~70% at 40 μM	Pal-Bhadra et al. 2012
Curcumin	p300 IC_{50} = ~25 μM; CBP IC_{50} = ~25 μM; ↔ PCAF;	↑ HDAC at 5 μM	Balasubramanyam et al. 2004a Kiss et al. 2012
Delphinidin (anthocyanidin)	↓ HAT activity by ~70% at 100 μM	↔ HDAC and SIRT	Seong et al. 2011
Ellagic acid	↓ HAT activity at 5 μM	↑ HDAC activity at 5 μM	Kiss et al. 2012
Epicatechin (flavan-3-ol)	↓ HAT activity at 5 μM	↓ HDAC-4 activity at 5 μM ↔ HDAC-1,2,3	Cordero-Herrera et al. 2015
Epigallogatechin-3-gallate (flavan-3-ol)	p300 IC_{50} = ~30 μM; CBP IC_{50} = ~50 μM; PCAF IC_{50} = ~60 μM; TIP60 IC_{50} = ~70 μM;	↔ HDAC and SIRT	Choi et al. 2009a
Fisetin (flavonoid)	↓ HAT activity at 10 μM	↑ HDAC activity at 10 μM	Kim et al. 2012
Gallic acid	↓ HAT activity at 25 μM p300 IC_{50} = ~14 μM; CBP IC_{50} = ~24 μM; PCAF IC_{50} = ~25 μM; TIP60 IC_{50} = ~34 μM;	↑ HDAC-2 activity at 25 μM	Lee et al. 2015 Choi et al. 2009b
Genistein (isoflavone)	↑ HAT activity up to ~12 μM	↓ HDAC activity at 25 μM	Hong et al. 2004 Li et al. 2013
Kaempferol (flavonoid)	No data	↓ HDACs activity up to 60% at 50 μM	Berger et al. 2013
Luteolin (flavonoid)	↓ HAT activity at 10 μM	↑ HDAC activity at 10 μM	Kim et al. 2014
Procyanidin B3 (flavan-3-ol)	↓ HAT activity by ~80% at 100 μM (the most specific for p300)	↔ HDAC	Choi et al. 2011
Quercetin (flavonoid)	↓ HAT activity by ~40% at 100 μM	No data	Ruiz et al. 2007
Resveratrol	No data	↑ SIRT activity at 30 μM	Bagul et al. 2015

have been identified as sources of natural inhibitors of p300/CBP HAT. In this chapter, recent findings concerning the effect of polyphenols (Fig. 1) on histone acetylation and acetylation as posttranslational proteins modification will be summarized. The relevance of an in vitro study to an in vivo situation will be discussed, as orally administered polyphenols have a limited bioavailability. Additionally, due to the chemical changes being a result of gut microbiota metabolism, and type I and II phase enzyme activity, the effects of the metabolites produced will be taken into consideration.

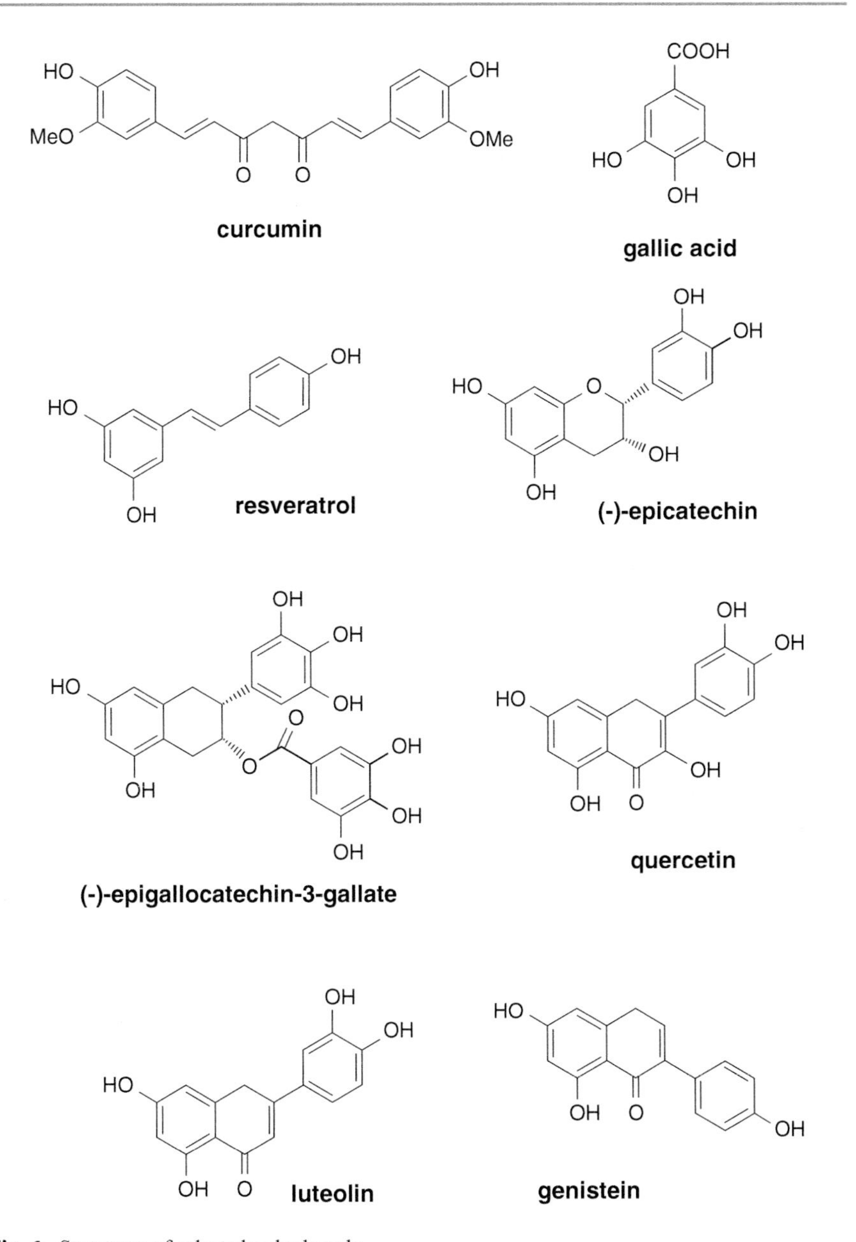

Fig. 1 Structures of selected polyphenols

Curcumin (Cur) (diferuloylmethane; (1E,6E)-1,7-bis (4-hydroxy-3-methoxyphenyl) hepta-1,6-diene-3,5-dione) is a yellow phenolic substance derived from *Curcuma longa* L. rhizomes (turmeric), which is widely used in Indian traditional medicine and in the food industry as a natural food coloring agent and curry spice (Wang and Qiu 2013; Liu et al. 2016).

Curcumin was first described as being able to inhibit the acetylation of histones H3 and H4 by p300/CREB-binding protein (CBP) with an IC_{50} of about 25 μM, whereas the PCAF HAT activity showed no change, nor histone deacetylase and methyltransferase activities, even in the presence of 100 μM of curcumin (Balasubramanyam et al. 2004a). However, the inhibitory activity of Cur was weaker in comparison with that of anacardic acid and garcinol, both isolated from natural products *Anacardium occidentale* L. nuts and *Garcinia indica* (Thouars) Choisy fruits, as potent inhibitors of histone acetyltransferases p300 (IC_{50} ~ 7–8.5 μM) and PCAF (IC_{50} ~ 5 μM) (Balasubramanyam et al. 2003, 2004b). The in vivo significance of HAT inhibition was assessed during myocardial cell hypertrophy, where nuclear acetylation by p300 is a critical event. Cur was tested in two heart failure models: hypertensive heart disease in Dahl salt-sensitive rats (DS) and surgically induced myocardial infarction in rats. In both models, curcumin given orally in a dose of 50 mg/kg/day for 6 weeks prevented the deterioration of systolic function and heart failure–induced increases in both myocardial wall thickness and diameter. Furthermore, curcumin treatment inhibited the hypertension induced acetylation of zinc finger transcription factors GATA4 and formation of p300/GATA4 complex in cardiomyocytes of DS rats (Morimoto et al. 2008). Moreover, the treatment of diabetic Ren-2 rats with curcumin reduced pathological cardiomyocyte hypertrophy (Bugyei-Twum et al. 2014). The authors deduced, from in vitro and in vivo studies, that high levels of glucose enhance the activity of the transcriptional coactivator p300, leading to the activation of transforming growth factor β (TGF-β) via acetylation of Smad2 (intracellular protein transducing extracellular signals from TGF-β ligands to the nucleus where they activate downstream gene transcription), and that Cur by inhibiting p300 activity reduces diabetic cardiomyopathy (Bugyei-Twum et al. 2014). The in vitro study using cultured primary cardiomyocytes from neonatal rats, in which the hypertrophy of cardiomyocytes was induced by phenylephrine, confirmed that Cur inhibited the increase of cardiomyocyte diameter by decreasing the acetylation of GATA4 and GATA4 binding to p300 (Morimoto et al. 2008). In further studies using human primary ventricular cardiomyocytes (HVCM cells), nanocurcumin was shown to prevent hypoxia induced cell hypertrophy by decreasing p300 HAT activity and increasing HDAC and downregulating p-GATA-4 expression (Nehra et al. 2015). What seems to be of importance is that in the cells not induced, the HAT and HDAC activities were not affected by Cur (Nehra et al. 2015). Curcumin encapsulated by carboxymethyl chitosan (CMC) nanoparticle conjugated to a myocyte specific homing peptide at a dose of 5 mg/kg/day applied via the tail vein was able to regress cardiac hypertrophy induced by the ligation of the right renal artery in a rat model (Ray et al. 2016). Similarly, according to the observations of Nehra et al. (2015), some pharmacological effects were correlated with the alteration of cardiomyocytes apoptosis and p53 degradation (Ray et al. 2016). Curcumin treatment also revealed significant inhibition of p53–p300 interaction resulting in reduced acetylation of p53, upregulated during hypertrophy (Ray et al. 2016).

Besides the positive effect on myocardial hypertrophy, in the Dahl salt-sensitive rats used as a model of nephrosclerosis, the administration of 100 mg/kg/day of curcumin suppressed inflammation and fibrosis in the kidneys. Cur decreased acetylation of the interleukin 6 (*IL-6*) gene expression in the kidneys via the suppression of histone acetylation, without affecting other proinflammatory factors such as tumor necrosis factor α (TNF-α) and macrophage inflammatory protein (MIP-2) (Muta et al. 2016). In the adriamycin-induced model of nephropathy, animals after induction received Cur (200 mg/kg/day for 2 weeks), which as above, suppressed the inflammation and fibrosis, which itself was correlated with the decrease of H3 an H4 acetylation in the promoters of monocyte chemoattractant protein-1 (MCP-1) (Liu et al. 2016). The in vivo results were supported by in vitro experiments using monocytes THP-1 cells induced with high glucose (HG). HG significantly induced histone acetylation by induction of p300 mRNA and reduction of HDAC2 mRNA expression, NF-κB activity, and proinflammatory cytokine (IL-6, TNF-α, and MCP-1) release from THP-1 cells. Curcumin at low concentrations of 1.5–6 μM suppressed p65 acetylation and NF-κB transcriptional activity and as a result the release of cytokines. Curcumin treatment also significantly reduced HAT activity, level of p300 and acetylated CBP/p300 gene expression, and induced HDAC2 activity (Yun et al. 2011).

Modulation of p300 and partly HDAC may be a novel strategy to treat heart failure and diabetes induced complications such as fibrosis and inflammation. Curcumin appears to be an interesting molecule for further in vivo and clinical study. However, some limitations should be pointed out. Pharmacological efficacies of Cur may be limited due to its low aqueous solubility and poor bioavailability. When curcumin is orally ingested, the majority is excreted through the feces and only a small proportion is absorbed within the intestine and rapidly converted to water-soluble metabolites (glucuronides and sulfates) (Wang and Qiu 2013; Liu et al. 2016). The activity of those compounds is rarely taken into consideration mainly because of the lack of commercial sources of those compounds. The maximum rat serum concentration of 0.36 ± 0.05 μg/mL (~ 0.13 μM) was reached after i.v. (10 mg/kg) administration of Cur, whereas a maximum plasma concentration of 0.06 ± 0.01 μg/mL (~ 0.022 μM) was achieved for orally (500 mg/kg) administered Cur (Yang et al. 2007). To overcome this limitation some delivery systems including liposomes, nanoparticles and polymeric micelles, phospholipid complexes, and microemulsions are applied. This enables the curcumin half-life to increase by 1.5–5.6 times (Liu et al. 2016) (Table 3).

Resveratrol (Res) (3,5,4′-trihydroxy-*trans*-stilbene) is found in grape products (red wine and grape juice) and peanuts as a major source for human consumption. Its potential role in disease was described, such as chemopreventive activity in various cancers, cardioprotection, diabetes prevention, and a capacity to increase longevity (Weiskirchen and Weiskirchen 2016). Res also appears as an epigenetic modulator.

Resveratrol by affecting metastasis-associated protein 1 (MTA1) level in PCa cell lines (LNCaP and Du145) was shown to cause destabilization of MTA1/HDAC1 interactions in NuRD complexes, whose function is to maintain the locked chromatin conformation, which results in the accumulation of p53 available for acetylation. In parallel, Res switched on acetylation mechanisms by HATs increasing the global protein acetylation, including p53, and favor transcriptional activation of genes

Table 3 Effects of curcumin on in vivo and in vitro models connected with heart failure and diabetes induced complications

Compound	Test model	Effect	Concentration/ dose	References
Curcumin	DS rats with developed hypertension and nephrosclerosis in response to high-salt diet	↓ of high salt diet induced number of acetylated H3K9-positive cells in the kidneys and connected ↓of urinary IL-6 protein, ↓ of fibrosis and glomerular sclerosis	Orally; 100 mg/kg/day; 6 weeks	Muta et al. 201 6
	Mouse podocytes	↓ adriamycin (ADR) induced injury and ↓ ADR induced hyoeracetylation of MCP-1 promoter	0.5–1 μg/mL	Liu et al. 2016
	Rats with adriamycin (ADR) induced nephropathy	↓ ADR induced renal pathology, ↓ histone acetylation in kidney cortex tissue	200 mg/kg/day	
	Cardiomyoblast cells H9c2	↓ of HG induced p300 activityand ↓ Smad2 acetylation	25 μM	Bugyei-Twum et al. 2014
	Transgenic (mRen-2)27 rat with induced diabetes	↓ cardiac hypertrophy	2% in chow; 6 weeks	
	Rat neuropathic pain model	↓ p300/CBP mediated acetylation of histone at BDNF brain-derived neurotrophic factor (BDNF) and cyclooxygenase-2 (COX-2) promoter; ↓ of thermal hyperalgesia and mechanical allodynia	i.p. 20, 40, and 60 mg/kg/day; 2 weeks	Zhu et al. 2014
	Rat cardiomyocytes	↓ hypertrophy induced by PE *via* ↓ of overexpress p-300 and ↓ GATA4 acetylation and DNA binding	5 and 10 μM	Morinoto et al. 2018
	DS rats with induced hypertension	↓ hypertension induced GATA4 acetylation, restore systolic function	50 mg/kg/day; weeks	
Nanocurcumin	Adult human primary ventricular cardiomyocytes (HVCM)	↓ hypoxia induced stress by ↓ p-300 HAT activity	100-1000 ng/mL	Nehra et al. 2015
Curcumin loaded carboxymethyl chitosan nanoparticle (CMCconjugated) with peptid	Left ventricular hypertrophy rat model	↓ cardiac hypertrophy	i.p. 5 mg/kg/day	Ray et al. 2016
	Neonatal myocyte angiotensin II induced hypertrophy	↓ acetylation of p53 and ↓ of cells apoptosis	20 μM	
Curcumin analogue C66	C57BL/6J mice with induced diabetes by streptozotocin	↓ renal histone acetylation by ↓ p300/CBP expression and HAT's activity, ↓ of renal fibrosis and hypertrophy	Orally; 5mg/kg/day, 3 months	Wang et al. 2015

implicated in cell cycle arrest and apoptosis. The author also underlined the increased effect of Res with HADCi suberoylanilide hydroxamic acid (SAHA), which is used clinically (Kai et al. 2010). Furthermore, in the cells unresponsive to androgen (line DU145) Res was found to decrease HDAC1 and HDAC2 levels, but less significantly in comparison with the LNCaP cell (Dhar et al. 2015). The decrease of HDACs levels correlated with the inhibition of deacetylation and inactivation of tumor suppressor PTEN via the inhibition of the MTA1/HDAC complex, and thereby the inactivation of downstream Akt cell survival and migration pathways (Dhar et al. 2015). Res also induced the acetylated histone H3 level in MCF-7 and Hela cells, although at a rather irrelevant in vivo concentration of 50–150 μM (Saenglee et al. 2016). However, such an effect was not observed in colon cancer cells NCM46 after incubation with 50 μM of res, in comparison with a positive control valproate acid (Lea et al. 2010). In most cases, the positive effect on cancer cells was seen at concentrations of 50 and 100 μM. The in vivo relevance of the observed effects is severely limited by low bioavailability in humans, as the level of res in the plasma varies from 0.3 to 2.4 μM, while res metabolites may achieve 14 μM (Howells et al. 2007).

The epigenetic modulation by resveratrol may also affect metabolic and cardiac diseases by sirtuin-1 (SIRT-1) activation which deacetylates NFκB-p65 at lysine 310 and H3 at lysine 9 position (seen in vitro and in vivo) (Bagul et al. 2015). SIRT1 activation leads to the decreased binding of NFkB-p65 to DNA and attenuated cardiac hypertrophy and oxidative stress in an in vivo animal model of diabetes, which received a quite low dose of 10 mg/kg/day of resveratrol orally (Bagul et al. 2015). Res at concentrations of 2.5 and 10 μM was also able to upregulate hypoxia-inducible factor-1 (HIF-1α) via SIRT1 activation and downregulate the c-Myc transcription factor and β-catenin expressions by deacetylation, in HepG2 cells, in a manner that mimics hypoxic preconditioning, and thus may play a beneficial role in response to damage to organs and tissues (Hong et al. 2012). Resveratrol at a dose of 50 mg/kg/day increased the expression of *Hdac4* and decreased the expression of *Hat1* in the hippocampus of a streptozotocin-induced C57BL/6 diabetic mice (Thomas et al. 2014). **Res seems to act appositively on HAT and HDAC/SIRT activity depending on the type of cells (cancerous or not)** (Fig. 2).

However, this may be a matter of the concentrations used. Nonetheless, it is not clear if resveratrol does act as a direct inhibitor/inducer of HAT, HDAC, or SIRT activity or the acetylation/deacetylation is an effect of other mechanisms of action. At least, in an inflammatory state the mechanism of action may be connected with an antioxidant effect (Rahman et al. 2004). In healthy subjects, a single dose of resveratrol (100 mg) combined with polyphenols from muscadine grape extract (75 mg) was shown to suppress oxidative and inflammatory markers of a high-fat, high-carbohydrate meal induced stress response (Ghanim et al. 2011).

Gallic acid (GA) (3,4,5-trihydroxybenzoic acid) a low molecular weight phenol present in grape seed extract (GSE), but also in tea and red wine, was shown similarly as curcumin to significantly reduce HAT activity and to induce HDAC2 (class I) activity in THP-1 cells treated with high glucose (HG), although at a higher concentration of 25 μM. GA at the same concentration also significantly reduced the level of acetylated CBP/p300 and acetylated NF-κBp65, which resulted in a significant reduction of IL-6 and TNF-α mRNA expression (Lee et al. 2015). GA at a concentration of 5 μM also reduced HAT activity in TNF-α stimulated THP-1 cells (Choi et al. 2009b; Kiss et al. 2012) and also inhibited the activity of isolated HAT enzyme but not HDAC activity (Kiss et al. 2012). The HAT activities of p300, CBP, PCAF, and Tip60 were inhibited by GA with an IC_{50} of ~14, ~24, ~25, and ~34 μM, respectively (Choi et al. 2009a). In an in vivo condition, oral pretreatment

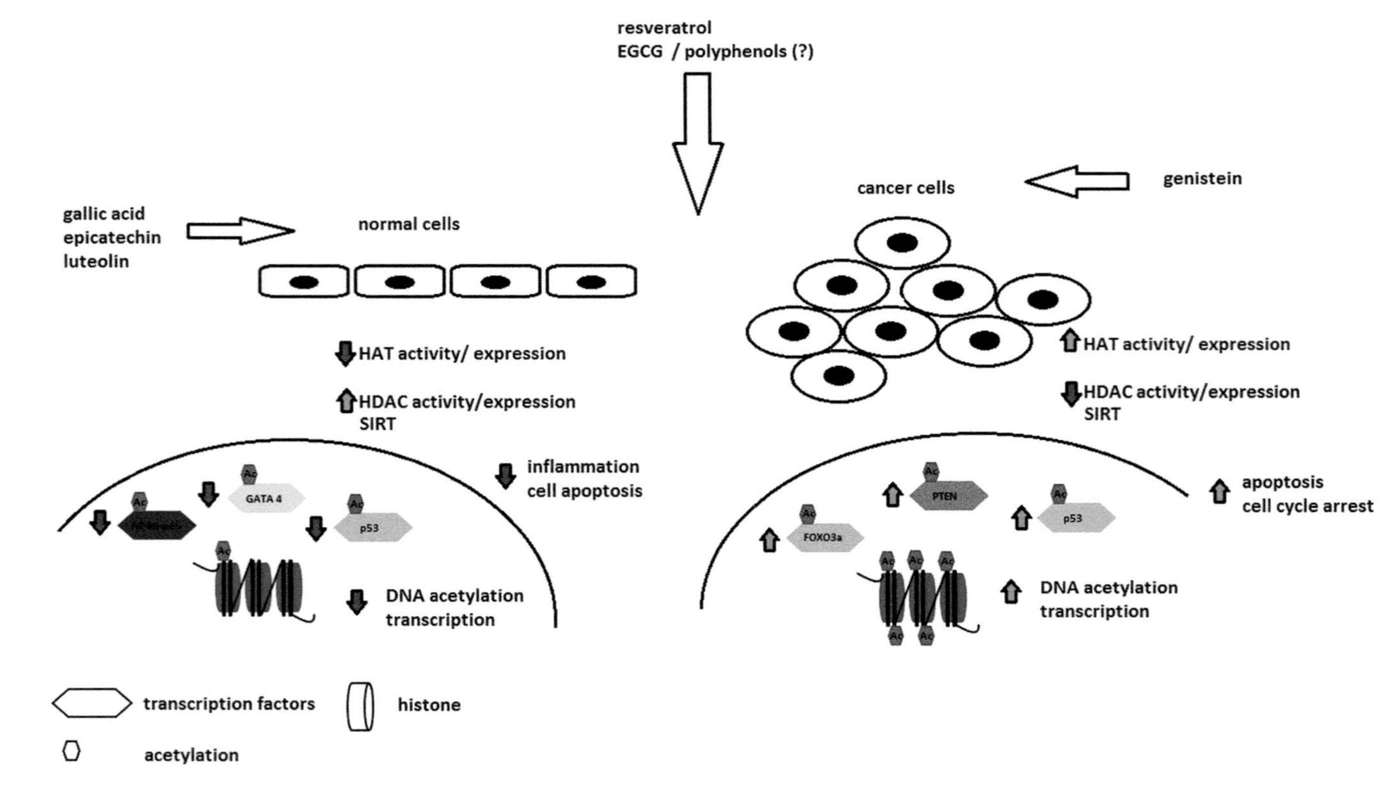

Fig. 2 Proposition of epigenetic modulation by polyphenols depending on the type of cells (normal or cancerous)

of mice with 40 mg of gallic acid/kg/day for 4 weeks prevented the IL-6 from increasing after the i.p. injection of LPS, which was consistent with the observed reduced levels of p65 acetylation in primary peritoneal macrophages following GA treatment (Choi et al. 2009b). In humans, the plasma level of gallic acid and its 4-*O*-methylated metabolite reached 4.7 μM after the consumption of black tea beverages containing 50 mg of gallic acid, and 1.57 μM after red wine containing 4 mg of GA (Lafay and Gil-Izquierdo 2008).

(−)-Epicatechin (EC) – (2R,3R)-2-(3,4-dihydroxyphenyl)-3,4-dihydro-2H-chromene-3,5,7-triol – also present in GSE but also in cocoa, tea, fruits, vegetables, and nuts was able to restore the cell physiology to control level in THP-1 cells treated with high glucose (HG) at a concentration of 5 μM. EC significantly reduced the level of acetylated CBP/p300 and counteracted the increased acetylation of H3K9 and inhibited TNF-α release (Cordero-Herrera et al. 2015). The bioavailability of EC is higher than that of EGCG (see below) and may reach a plasma concentration of 0.13–5.9 μM after the consumption of a cocoa preparation (Manach et al. 2005). **Evaluation of the data suggests that plant extracts and food containing GA and EC may affect the proinflammatory state of the human body.**

Green tea (GT) contains gallic acid and catechins with the dominating epicatechin, epicatechin-3-gallate, epigallocatechin, and epigallogatechin-3-gallate (EGCG). GT and its preparations are used for a wide range of indications but especially in the prevention such as cancer and CVD (Edwards et al. 2015).

Most studies were related with the epigenetic regulation of prostate cancer cells apoptosis. Treatment of hormone-dependent LNCaP cells with GT (2.5–10 μg/mL) and its main compounds EGCG (5–20 μM) resulted in dose- and time-dependent inhibition of protein levels of class I HDACs (HDAC1, 2, 3, and 8), albeit mostly 1–3 (Thakur et al. 2012a). The HDACs inhibition was also observed in hormone-independent PC-3 cancer cells at a higher concentration of GT of 80 μg/mL (Thakur et al. 2012b) and in human breast cells at a concentration 20 μM of EGCG and 20 μg/mL of GT (Deb et al. 2015). GT/EGCG treatment resulted in the stabilization and acetylation of p53 at Lys373 and Lys382 and enhanced its binding on the promoters of p21/wafl and Bax, which was associated with increased accumulation of cells in the G0/G1 phase of the cell cycle and induction of apoptosis of LNCaP cells (Thakur et al. 2012a). Li et al. (2010) demonstrated that EGCG at a concentration of 10 μM induced HAT protein level, and slightly decreased HDAC protein level in ERα-negative MDA-MB-231 breast cancer cells, and was able to induce a pronounced ERα re-expression.

However, the group of Lee et al. (2012) referred to the ability of EGCG and other green tea catechins to inhibit the HAT activity within a concentration range of 25–100 μM, without affecting the HDCA or SIRT activity. It should be emphasized that Thakur et al. (2012a) and Li et al. (2010) investigated the level of the proteins, while Lee et al. (2012) used the HeLa cells extract as a source of enzymes. It seems probable that the decreased level of HDAC protein is due to the increased proteasomal degradation by GT (Thakur et al. 2012b).

The relevance of the observed in vitro epigenetic modulation of GT and EGCG in cancer cells are difficult to extrapolate to in vivo conditions as the

concentrations used are far higher than those detected in humans. The maximal plasma level (C_{max}) of EGCG in humans after green tea preparations consumption vary depending on the dose from 0.052–0.326 μg/mL (~ 0.02–0.15 μM), while the total catechins may reach about 1 μg/mL (Yang et al. 1998; Lee et al. 2002). Many metabolites of catechins were also identified in human plasma such as methyl-(epi) gallocatechin-glucuronide ($C_{max} = 24.6 \pm 13.2$ nM), (epi)gallocatechin-glucuronide ($C_{max} = 40.5 \pm 23.6$ nM), (epi)catechin-glucuronide ($C_{max} = 29.3 \pm 8.7$ nM), methyl-(epi)gallocatechin-sulfate ($C_{max} = 38.0 \pm 10.4$ nM), and methyl-(epi) catechin-sulfate ($C_{max} = 14.2 \pm 10.4$ nM) (Del Rio et al. 2010) and their activities have rarely been tested in in vitro studies (Lambert et al. 2007). Interestingly, in higher micromolar concentrations catechins' metabolites 3′,4′- di- and 3′,4′, 5′-trihydroxyphenyl-γ-valerolactone glucuronides and sulfates were detected in human urine, with strong individual variation (van der Hooft et al. 2012), and their contribution to their potential beneficial effects is unknown. Additionally, current meta-analysis suggests that available evidence from epidemiological studies are not sufficient to conclude that tea consumption could reduce the risk of prostate cancer (Lin et al. 2014).

Flavonoids. **Quercetin** (3, 5, 7, 3′, 4′-pentahydroxyflavone) (Q) is the most common flavonoid in nature and is often linked to sugars such as rutin (quercetin-3-rutinoside) and quercitrin (quercetin-3-rhamnoside). A rich source of this compound are medicinal plants, onions, apple, and buckwheat.

The well-known anti-inflammatory activity of Q may be partly connected with its influence on the epigenetic control on inflammatory states. Q in dose of 40 μM inhibited TNF-induced interferon-γ-inducible protein 10 (IP-10) and macrophage inflammatory protein 2 (MIP-2) expression in primary ileal epithelial cells. The mechanism of action was connected with the inhibition of NF-κB and cofactor recruitment to the IP-10 and MIP-2 gene promoters and consistent with the reduced HAT activity and inhibition of TNF-induced acetylation/phosphorylation of H3 and CBP/p300 binding at those gene promoters. However, the inhibition of HAT activity by Q was quite modest (~ 40% reduction at concentration of 100 μM). In this rare case, the effect was also investigated for Q metabolites produced by gut microbiota (taxifolin, alphitonin, and 3,4-dihydroxy-phenylacetic acid), but their activity was negligible (Ruiz et al. 2007). Quercetin treatment (10 mg/kg p.o) decreased lung inflammation by decreasing levels of metalloproteinases MMP9 and MMP12, which was connected with the increased expression of SIRT1 in elastase/LPS-exposed mice (Ganesan et al. 2010). SIRT1 induction was also observed in HepG2 cells, though treated with a quite high concentration of Q of 50 and 100 μM (Hong et al. 2012). The observed plasma quercetin level in this study was 0.131 ± 0.038 μM (Ganesan et al. 2010), which was consistent with other studies in which the quercetin plasma concentration varied between 0.14 and7.6 μM depending on the sources of compounds (Manach et al. 2005).

Luteolin (5, 7, 3′, 4′-tetrahydroxyflavone) present in significant amounts in medicinal plant and food products derived from plants of the Asteraceae family (artichoke, chicory, chamomile flower tea, etc.), but also in celery, carrot, and broccoli, and **fisetin** (3, 7, 3′, 4′-tetrahydroxyflavone), a much less common flavonoid, found in a bound form (glycosides) in strawberries, apple, and persimmon (Khan et al. 2013).

Both compounds, similarly to other polyphenols described above, at the concentration of 3–10 μM significantly reduced HAT activity and induced HDAC activity in HG-treated THP-1 cells. The activity observed has been correlated with the decreased level of CBP/p300 and NF-κB p65 acetylation, which resulted in a decrease of release/expression of TNF-α and IL-6 (Kim et al. 2012, 2014). The concentration of luteolin in plasma after p.o. application of luteolin (100 mg/kg) or luteolin-7-glucoside (1 g/kg) reached a level of about 3 μg/mL (~ 0.9 μM) (Lin et al. 2015). However, in the case of flavonoids and catechins, which are present in the plasma as glucuronides, it has been shown in vivo and in vitro that during the inflammatory state the deglucuronization takes place and that the local concentration of native compounds at the site of inflammation may reach higher levels (Shimoi et al. 2001; Kawai 2014) (Fig. 3).

Isoflavones. Genistein (5,7,4′-trihydroxyisoflavone) (Gen) and **daidzein** (7,4′-dihydroxyisoflavone) are compounds occurring mainly in soya products, which have been identified as inhibitors of protein tyrosine kinases, which play key roles in cell growth and apoptosis (Akiyama et al. 1987). Gen has also been reported to have estrogenic properties and antineoplastic activity in multiple tumor types (Zava and Duwe 1997).

Incubation of prostate cancer cell lines LNCaP and PC-3 with Gen at a concentration of 50 μM induced mRNA expression of phosphatase and tensin homolog (PTEN), p53, and fork head transcription factor (FOXO3a), which was correlated with a decrease in SIRT1 expression and accumulation of SIRT1 in the cytoplasm from the nucleus, which further resulted in the increased acetylation of H3-K9 at the PTEN, p53, and the FOXO3a promoters (Kikuno et al. 2008). Gen at concentration of 20 μM was also shown to induce histone H3K9 acetylation in DU145, PC3, and LNCaP cells, by inducing the expression of histone acetyltransferase 1 (HAT1) (Phillip et al. 2012). Moreover, in breast cancer cell lines, MCF-7 and MDA-MB 231 genistein, daidzein,, and its metabolite equol induced a decrease in trimethylated marks and an increase in acetylating marks studied, at concentrations of 18.5, 78.5, and 12.8 μM, at six selected genes (histone-lysine N-methyltransferase *EZH2*, tumor suppressor gene *BRCA1*, *ERα*, *ERβ*, nuclear receptor coactivator 3*SRC3,* and *P300*) in a comparable manner to 17β-estradiol (10 nM) and HADC inhibitor suberoylanilide hydroxamic acid (1 μM) (Dagdemir et al. 2013). Gen (25 μM) can also reactivate in ERα-negative MDA-MB-231 and MDA-MB-157 cells the ERα expression via histone modification changes in the ERα promoter (induction of H3 acetylation), and this effect was synergistically enhanced when combined with a histone deacetylase (HDAC) inhibitor, trichostatin A (TSA). Gen treatment also resensitized ERα-dependent cellular responses to the activator 17β-estradiol and antagonist tamoxifen (Li et al. 2013). Importantly, in an in vivo xenograft model of ERα-negative breast cancer, supplementation with Gen (250 mg/kg/day) alone or with tamoxifen resulted in preventing the development of cancer and reduced the growth of mouse breast tumors by induction of percentage of ERα-positive cells, which correlated with the decrease of HDAC1 protein and mRNA expression (Li et al. 2013).

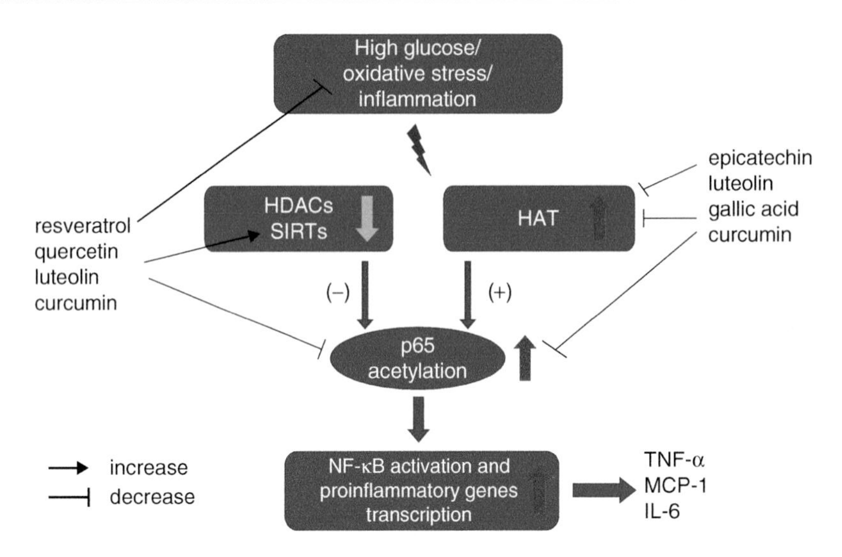

Fig. 3 Proposed epigenetic regulation of inflammatory state by selected polyphenols. Designed based on Yun et al. 2011. IL-6- interleukin 6; MCP-1- monocyte chemoattractant protein-1; NF-κB- nuclear factor kappa-light-chain-enhancer of activated B cells; TNF-α- tumor necrosis factor α

These results support the fact that isoflavones act also as epigenetic modulators in prostate and breast cancer cells. The main question that arises is, are the concentrations used in vitro relevant to the in vivo situation? Mean plasma levels of deconjugated isoflavones after the consumption of soya preparations are: 0.3–4.1 μM for genistein, 0.5–3.1 μM for daidzein,, and 0.39–1.2 μM for equol, in the case of so-called equol producers (Andersen and Markham 2006; van der Velpen et al. 2014a, b). **Taken together, if the total concentration of isoflavone does not exceed 10 μM, though still much higher than the concentration of other flavonoids and catechins, then it is still difficult to predict the long-term effect of the consumption of isoflavones on the epigenetic state of hormone dependent cells.** However, the meta-analysis of clinical studies support the beneficial role of isoflavones in prostate and breast cancer risk reduction (Fritz et al. 2013; van Die et al. 2014).

Key Facts

Key Facts of Polyphenols

- A broad term for structurally diverse, naturally occurring compounds bearing aromatic hydroxy substituted ring.
- Major classes of polyphenols in plants are simple phenols, phenolic acids (e.g., gallic acid), lignans, stilbenes (e.g., resveratrol), flavan-3-ol (e.g., epicatechin), flavonoids (e.g., quercetin), anthocyanins, isoflavonoids (e.g., genistein), gallotannins, and ellagotannins.

- Several modifications of core structures (so-called aglycon) such as glycosylation, *O*-methylation, esterification, and dimerization are responsible for such a huge diversity of this class of compound in the plant kingdom.
- Generally, polyphenols have limited bioavailability in humans. Their structure may be significantly changed by gut microbiota enzymes, and due to this their absorption and activity may be also enhanced (e.g., daidzein is transformed to equol). Once absorbed, polyphenols may undergo I and II phase transformation (methylation, glucuronidation, and sulfation).
- Polyphenols are regarded as strong antioxidants (at least), and there is a general international consensus that a diet rich in fruits and vegetables (also meaning rich in polyphenols) may have a protective effect against cancer, cardiovascular diseases, type-2 diabetes, and inflammatory bowel diseases.

Key Facts of Inflammation

- Inflammation is an important host defense response, and acute inflammation involves neutrophils, basophils, eosinophils, and mononuclear cells, which can be beneficial, as it stimulates the healing process and homeostasis restoration.
- Chronic inflammation lasting months or even years is an important background of many diseases: atherosclerosis, diabetes mellitus, rheumatoid arthritis, cancer, asthma, skin diseases, periodontal inflammation, etc.
- Chronic inflammation is linked to uncontrolled oxidative stress, inflammatory cytokine, chemokine production (e.g., Il-6, TNF-α) due to deregulation of signaling pathways, transcription factors (e.g., NF-κB), and epigenetic control of proinflammatory gene expression.
- The NF-κB nuclear factor plays a central role in the control of inflammation, thus this transcriptional factor serves as a drug/phytonutrients target.
- A diet containing a large amount of refined sugars, saturated fats, red meat, and small amount of fiber, fish, fruits, vegetables, and whole grains may promote inflammation.

Dictionary of Terms

- **Bioavailability** – Ability of substance (here polyphenols) to reach the circulation and target organ after oral administration.
- **Cardiomyocyte/cardiac hypertrophy** – Increase of the side of the heart muscle cells (cardiomyocytes) mostly due to the high tension.
- **Diabetic cardiomyopathy and nephropathy** – Long-term damage of heart and kidney blood vessels due to systemic hyperglycemia in diabetes (diabetes complications).
- **Gut microbiota** – Microorganism (bacteria, yeasts, and fungus) living in human intestine (previously known as gut flora).

- **Heart failure** – Pathophysiological state when heart muscles are not able to sufficiently pump the blood which may cause the depletion of oxygen.
- **Hormone-dependent cancers of breast and prostate** – Type of cancer when uncontrolled cell growth is stimulated by sex hormones estrogens or androgens.
- **Polyphenols metabolites** – Products of enzymatic transformation of genuine polyphenols, which may occur in the intestine and/or liver.

Summary Points

- This chapter focuses on influencing by polyphenolic compounds HDAC and HAT activities, and by doing so modulating/preventing metabolic disturbances, inflammatory responses, and the development of cancer.
- Depending on the type of cells (cancerous or not) some compounds (e.g., resveratrol) may act appositively on HAT and HDAC/SIRT activity.
- Curcumin by modulation of p300 and partly HDAC may be a novel strategy for treating heart failure and diabetes induced complications such as fibrosis and inflammation. Curcumin appears to be an interesting molecule for further in vivo and clinical study.
- Evaluation of the data suggests that plant extracts and food containing a significant amount of gallic acid, epicatechin, and flavonoids may affect at the epigenetic level the proinflammatory state of the human body.
- The relevance of the observed in vitro epigenetic modulation of green tea and epigallogatechin-3-gallate in cancer cells are difficult to extrapolate to in vivo conditions as the concentrations used are far higher than those detected in humans.
- Isoflavones may act as epigenetic modulators in prostate and breast cancer cells. The total concentration of isoflavones in human does not exceed 10 μM, though still much higher than the concentration of green tea catechins. However, it is still difficult to predict the long-term effect of the consumption of isoflavones on the epigenetic state of hormone dependent cancer cells.
- As orally administered polyphenols generally have limited bioavailability and undergo changes due to gut microbiota, type I and II phase enzyme metabolism, the effects of those metabolites on the epigenetic modulation should be taken into consideration as well.

References

Akiyama T, Ishida J, Nakagawa S, Ogawara H, Watanabe S, Itoh N, Shibuya M, Fukami Y (1987) Genistein, a specific inhibitor of tyrosine-specific protein kinases. J Biol Chem 262:5592–5595

Andersen ØM, Markham KR (2006) Flavonoids. Chemistry, biochemistry and applications. Taylor & Francis Group, Boca Raton, p 328

Bagul PK, Deepthi N, Sultana R, Banerjee SK (2015) Resveratrol ameliorates cardiac oxidative stress in diabetes through deacetylation of NFkB-p65 and histone 3. J Nutr Biochem 26:1298–1307. https://doi.org/10.1016/j.jnutbio.2015.06.006

Balasubramanyam K, Swaminathan V, Ranganathan A, Kundu TK (2003) Small molecule modulators of histone acetyltransferase p300. J Biol Chem 278:19134–19140. https://doi.org/10.1074/jbc.M301580200

Balasubramanyam K, Varier RA, Altaf M, Swaminathan V, Siddappa NB, Ranga U, Kundu TK (2004a) Curcumin, a novel p300/CREB-binding protein-specific inhibitor of acetyltransferase, represses the acetylation of histone/nonhistone proteins and histone acetyltransferase-dependent chromatin transcription. J Biol Chem 279:51163–51171. https://doi.org/10.1074/jbc.M409024200

Balasubramanyam K, Altaf M, Varier RA, Swaminathan V, Ravindran A, Sadhale PP, Kundu TK (2004b) Polyisoprenylated benzophenone, garcinol, a natural histone acetyltransferase inhibitor, represses chromatin transcription and alters global gene expression. J Biol Chem 279:33716–33726. https://doi.org/10.1074/jbc.M402839200

Berger A, Venturelli S, Kallnischkies M, Böcker A, Busch C, Weiland T, Noor S, Leischner C, Weiss TS, Lauer UM, Bischoff SC, Bitzer M (2013) Kaempferol, a new nutrition-derived pan-inhibitor of human histone deacetylases. J Nutr Biochem 24:977–985. https://doi.org/10.1016/j.jnutbio.2012.07.001

Bugyei-Twum A, Advani A, Advani SL, Zhang Y, Thai K, Kelly DJ, Connelly KA (2014) High glucose induces Smad activation via the transcriptional coregulator p300 and contributes to cardiac fibrosis and hypertrophy. Cardiovasc Diabetol 13:89. https://doi.org/10.1186/1475-2840-13-89

Choi K-C, Jung MG, Lee Y-H, Yoon JC, Kwon SH, Kang H-B, Kim M-J, Cha J-H, Kim YJ, Jun WJ, Lee JM, Yoon H-G (2009a) Epigallocatechin-3-gallate, a histone acetyltransferase inhibitor, inhibits EBV-induced B lymphocyte transformation via suppression of RelA acetylation. Cancer Res 69(2):583–592

Choi K-C, Lee Y-H, Jung MG, Kwon SH, Kim M-J, Jun WJ, Lee J, Lee JM, Yoon H-G (2009b) Gallic acid suppresses lipopolysaccharide-induced nuclear factor-κB signaling by preventing RelA acetylation in A549 lung cancer cells. Mol Cancer Res 7:2011–2021. https://doi.org/10.1158/1541-7786.MCR-09-0239

Choi K-C, Park S, Lim BJ, Seong AR, Lee YH, Shiota M, Yokomizo A, Naito S, Na Y, Yoon HG (2011) Procyanidin B3, an inhibitor of histone acetyltransferase, enhances the action of antagonist for prostate cancer cells via inhibition of p300-dependent acetylation of androgen receptor. Biochem J 433:235–244. https://doi.org/10.1042/BJ20100980

Cordero-Herrera I, Chen X, Ramos S, Devaraj S (2015) (-)-Epicatechin attenuates high-glucose-induced inflammation by epigenetic modulation in human monocytes. Eur J Nutr. https://doi.org/10.1007/s00394-015-1136-2

Dagdemir A, Durif J, Ngollo M, Bignon YJ, Bernard-Gallon D (2013) Histone lysine trimethylation or acetylation can be modulated by phytoestrogen, estrogen or anti-HDAC in breast cancer cell lines. Epigenomics 5:51–63. https://doi.org/10.2217/epi.12.74

Deb G, Thakur VS, Limaye AM, Gupta S (2015) Epigenetic induction of tissue inhibitor of matrix metalloproteinase-3 by green tea polyphenols in breast cancer cells. Mol Carcinog 54:485–499

Del Rio D, Calani L, Cordero C, Salvatore S, Pellegrini N, Brighenti F (2010) Bioavailability and catabolism of green tea flavan-3-ols in humans. Nutrition 26:1110–1116. https://doi.org/10.1016/j.nut.2009.09.021

Dhar S, Kumar A, Li K, Tzivion G, Levenson AS (2015) Resveratrol regulates PTEN/Akt pathway through inhibition of MTA1/HDAC unit of the NuRD complex in prostate cancer. Biochim Biophys Acta 1853:265–275. https://doi.org/10.1016/j.bbamcr.2014.11.004

Edwards SE, da Costa RI, Williamson EM, Heinrich M (2015) Phytopharmacy an evidence-based guide to herbal medicinal products. Wiley, Chichester, p 186, 191

Fritz H, Seely D, Flower G, Skidmore B, Fernandes R, Vadeboncoeur S, Kennedy D, Cooley K, Wong R, Sagar S, Sabri E, Fergusson D (2013) Soy, red clover, and isoflavones and breast cancer: a systematic review. PLoS One 8:e81968. https://doi.org/10.1371/journal.pone.0081968

Ganesan S, Faris AN, Comstock AT, Chattoraj SS, Chattoraj A, Burgess JR, Curtis JL, Martinez FJ, Zick S, Hershenson MB, Sajjan US (2010) Quercetin prevents progression of disease in elastase/LPS-exposed mice by negatively regulating MMP expression. Respir Res 11:131. https://doi.org/10.1186/1465-9921-11-131

Ghanim H, Sia CL, Korzeniewski K, Lohano T, Afbuaysheh S, Marumganti A, Chaudhuri A, Dandona P (2011) A resveratrol and polyphenol preparation suppresses oxidative and inflammatory stress response to a high-fat, high-carbohydrate meal. J Clin Endocrinol Metab 96:1409–1414. https://doi.org/10.1210/jc.2010-1812

Hong T, Nakagawa T, Pan W, Kim MY, Kraus WL, Ikehara T, Yasui K, Aihara H, Takebe M, Muramatsu M, Ito T (2004) Isoflavones stimulate estrogen receptor-mediated core histone acetylation. Biochem Biophys Res Commun 317:259–264

Hong KS, Park JI, Kim MJ, Kim HB, Lee JW, Dao TT, Oh WK, Kang CD, Kim SH (2012) Involvement of SIRT1 in hypoxic down-regulation of c-Myc and β-catenin and hypoxic preconditioning effect of polyphenols. Toxicol Appl Pharmacol 259:210–218. https://doi.org/10.1016/j.taap.2011.12.025

Howells LM, Moiseeva EP, Neal CP, Foreman BE, Andreadi CK, Sun YY, Hudson EA, Manson MM (2007) Predicting the physiological relevance of *in vitro* cancer preventive activities of phytochemicals. Acta Pharm Sin 28:1274–1304. https://doi.org/10.1111/j.1745-7254.2007.00690.x

Kai L, Samuel SK, Levenson AS (2010) Resveratrol enhances p53 acetylation and apoptosis in prostate cancer by inhibiting MTA1/NuRD complex. Int J Cancer 126:1538–1548. https://doi.org/10.1002/ijc.24928

Kawai Y (2014) β-Glucuronidase activity and mitochondrial dysfunction: the sites where flavonoid glucuronides act as anti-inflammatory agents. J Clin Biochem Nutr 54:145–150. https://doi.org/10.3164/jcbn.14-9

Khan N, Syed DN, Ahmad N, Mukhtar H (2013) Fisetin: a dietary antioxidant for health promotion. Antioxid Redox Signal 19:151–162. https://doi.org/10.1089/ars.2012.4901

Kikuno N, Shiina H, Urakami S, Kawamoto K, Hirata H, Tanaka Y, Majid S, Igawa M, Dahiya R (2008) Genistein mediated histone acetylation and demethylation activates tumor suppressor genes in prostate cancer cells. Int J Cancer 123:552–560. https://doi.org/10.1002/ijc.23590

Kim HJ, Kim SH, Yun JM (2012) Fisetin inhibits hyperglycemia-induced proinflammatory cytokine production by epigenetic mechanisms. Evid Based Complement Alternat Med 2012:639469. https://doi.org/10.1155/2012/639469

Kim HJ, Lee W, Yun JM (2014) Luteolin inhibits hyperglycemia-induced proinflammatory cytokine production and its epigenetic mechanism in human monocytes. Phytother Res 28:1383–1391. https://doi.org/10.1002/ptr.5141

Kiss AK, Granica S, Stolarczyk M, Melzig MF (2012) Epigenetic modulation of mechanisms involved in inflammation: influence of selected polyphenolic substances on histone acetylation state. Food Chem 131:1028–1033. https://doi.org/10.1016/j.foodchem.2011.09.109

Lafay S, Gil-Izquierdo A (2008) Bioavailability of phenolic acids. Phytochem Rev 7:301–311

Lambert JD, Sang S, Yang CS (2007) Biotransformation of green tea polyphenols and the biological activities of those metabolites. Mol Pharm 4:819–825. https://doi.org/10.1021/mp700075m

Lea MA, Ibeh C, Han L, des Bordes C (2010) Inhibition of growth and induction of differentiation markers by polyphenolic molecules and histone deacetylase inhibitors in colon cancer cells. Anticancer Res 30:311–318

Lee MJ, Maliakal P, Chen L, Meng X, Bondoc FY, Prabhu S, Lambert G, Mohr S, Yang CS (2002) Pharmacokinetics of tea catechins after ingestion of green tea and (−)-epigallocatechin-3-gallate by humans: formation of different metabolites and individual variability. Cancer Epidemiol Biomark Prev 11:1025–1032

Lee YH, Kwak J, Choi HK, Choi KC, Kim S, Lee J, Jun W, Park HJ, Yoon HG (2012) EGCG suppresses prostate cancer cell growth modulating acetylation of androgen receptor by anti-histone acetyltransferase activity. Int J Mol Med 30:69–74. https://doi.org/10.3892/ijmm.2012.966

Lee W, Lee SY, Son Y-J, Yun J-M (2015) Gallic acid decreases inflammatory cytokine secretion through histone acetyltransferase/histone deacetylase regulation in high glucose-induced human monocytes. J Med Food 18:793–801. https://doi.org/10.1089/jmf.2014.3342

Li Y, Yuan YY, Meeran SM, Tollefsbol TO (2010) Synergistic epigenetic reactivation of estrogen receptor-α (ERα) by combined green tea polyphenol and histone deacetylase inhibitor in ERα-negative breast cancer cells. Mol Cancer 9:274. https://doi.org/10.1186/1476-4598-9-274

Li Y, Meeran SM, Patel SN, Chen H, Hardy TM, Tollefsbol TO (2013) Epigenetic reactivation of estrogen receptor-α (ERα) by genistein enhances hormonal therapy sensitivity in ERα-negative breast cancer. Mol Cancer 12:9. https://doi.org/10.1186/1476-4598-12-9

Lin Y, Hu Z, Wang X, Mao Q, Qin J, Zheng X, Xie L (2014) Tea consumption and prostate cancer: an updated meta-analysis. World J Surg Oncol 12:38. https://doi.org/10.1186/1477-7819-12-38

Lin LC, Pai YF, Tsai TH (2015) Isolation of luteolin and luteolin-7-*O*-glucoside from *Dendranthema morifolium* Ramat Tzvel and their pharmacokinetics in rats. J Agric Food Chem 63:7700–7706. https://doi.org/10.1021/jf505848z

Liu J, Zhong F, Dai Q, Xu L, Wang W, Chen N (2016) Curcumin prevents adriamycin-induced nephropathy MCP-1 expression through blocking histone acetylation. Int J Clin Exp Med 9:12696–12704

Manach C, Williamson G, Morand C, Scalbert A, Rémésy C (2005) Bioavailability and bioefficacy of polyphenols in humans. I. Review of 97 bioavailability studies. Am J Clin Nutr 81:230S–242S

Morimoto T, Sunagawa Y, Kawamura T, Takaya T, Wada H, Nagasawa A, Komeda M, Fujita M, Shimatsu A, Kita T, Hasegawa K (2008) The dietary compound curcumin inhibits p300 histone acetyltransferase activity and prevents heart failure in rats. J Clin Invest 118:868–878. https://doi.org/10.1172/JCI33160

Muta K, Obata Y, Oka S, Abe S, Minami K, Kitamura M, Endo D, Koji T, Nishino T (2016) Curcumin ameliorates nephrosclerosis via suppression of histone acetylation independent of hypertension. Nephrol Dial Transplant 31:1615–1623. https://doi.org/10.1093/ndt/gfw036

Nehra S, Bhardwaj V, Ganju L, Saraswat D (2015) Nanocurcumin prevents hypoxia induced stress in primary human ventricular cardiomyocytes by maintaining mitochondrial homeostasis. PLoS One 10:e0139121. https://doi.org/10.1371/journal.pone.0139121

Pal-Bhadra M, Ramaiah MJ, Reddy TL, Krishnan A, Pushpavalli SN, Babu KS, Tiwari AK, Rao JM, Yadav JS, Bhadra U (2012) Plant HDAC inhibitor chrysin arrest cell growth and induce p21WAF1 by altering chromatin of STAT response element in A375 cells. BMC Cancer 12:180. https://doi.org/10.1186/1471-2407-12-180

Phillip CJ, Giardina CK, Bilir B, Cutler DJ, Lai YH, Kucuk O, Moreno CS (2012) Genistein cooperates with the histone deacetylase inhibitor vorinostat to induce cell death in prostate cancer cells. BMC Cancer 12:145. https://doi.org/10.1186/1471-2407-12-145

Rahman I, Marwick J, Kirkham P (2004) Redox modulation of chromatin remodeling: impact on histone acetylation and deacetylation, NF-kB and pro-inflammatory gene expression. Biochem Pharmacol 68:1255–1267. https://doi.org/10.1016/j.bcp.2004.05.042

Ray A, Rana S, Banerjee D, Mitra A, Datta R, Naskar S, Sarkar S (2016) Improved bioavailability of targeted curcumin delivery efficiently regressed cardiac hypertrophy by modulating apoptotic load within cardiac microenvironment. Toxicol Appl Pharm 290:54–65. https://doi.org/10.1016/j.taap.2015.11.011

Ruiz PA, Braune A, Hölzlwimmer G, Quintanilla-Fend L, Haller D (2007) Quercetin inhibits TNF-induced NF-kappaB transcription factor recruitment to proinflammatory gene promoters in murine intestinal epithelial cells. J Nutr 137:1208–1215

Saenglee S, Jogloy S, Patanothai A, Leid M, Senawong T (2016) Cytotoxic effects of peanut phenolics possessing histone deacetylase inhibitory activity in breast and cervical cancer cell lines. Pharmacol Rep 68:1102–1110. https://doi.org/10.1016/j.pharep.2016.06.017

Seong A-R, Yoo J-Y, Choi KC, Lee M-H, Lee Y-H, Lee J, Jun W, Kim S, Yoon H-G (2011) Delphinidin, a specific inhibitor of histone acetyltransferase, suppresses inflammatory signaling via prevention of NF-κB acetylation in fibroblast-like synoviocyte MH7A cells. Biochem Biophys Res Commun 410:581–586. https://doi.org/10.1016/j.bbrc.2011.06.029. Cover image

Shimoi K, Saka N, Nozawa R, Sato M, Amano I, Nakayama T, Kinae N (2001) Deglucuronidation of a flavonoid, luteolin monoglucuronide, during inflammation. Drug Metab Dispos 29:1521–1524

Thakur VS, Gupta K, Gupta S (2012a) Green tea polyphenols increase p53 transcriptional activity and acetylation by suppressing class I histone deacetylases. Int J Oncol 41:353–361. https://doi.org/10.3892/ijo.2012.1449

Thakur VS, Gupta K, Gupta S (2012b) Green tea polyphenols causes cell cycle arrest and apoptosis in prostate cancer cells by suppressing class I histone deacetylases. Carcinogenesis 33:377–384. https://doi.org/10.1093/carcin/bgr277

Thomas J, Garg ML, Smith DW (2014) Dietary resveratrol supplementation normalizes gene expression in the hippocampus of streptozotocin-induced diabetic C57Bl/6 mice. J Nutr Biochem 25:313–318. https://doi.org/10.1016/j.jnutbio.2013.11.005

Tseng TH, Chien MH, Lin WL, Wen YC, Chow JM, Chen CK, Kuo TC, Lee WJ (2017) Inhibition of MDA-MB-231 breast cancer cell proliferation and tumor growth by apigenin through induction of G2/M arrest and histone H3 acetylation-mediated p21WAF1/CIP1 expression. Environ Toxicol 32:434–444. https://doi.org/10.1002/tox.22247

van der Hooft JJ, de Vos RC, Mihaleva V, Bino RJ, Ridder L, de Roo N, Jacobs DM, van Duynhoven JP, Vervoort J (2012) Structural elucidation and quantification of phenolic conjugates present in human urine after tea intake. Anal Chem 84:7263–7271. https://doi.org/10.1021/ac3017339

van der Velpen V, Hollman PC, van Nielen M, Schouten EG, Mensink M, Van't Veer P, Geelen A (2014a) Large inter-individual variation in isoflavone plasma concentration limits use of isoflavone intake data for risk assessment. Eur J Clin Nutr 68:1141–1147. https://doi.org/10.1038/ejcn.2014.108

van der Velpen V, Geelen A, Hollman PC, Schouten EG, Van't Veer P, Afman LA (2014b) Isoflavone supplement composition and equol producer status affect gene expression in adipose tissue: a double-blind, randomized, placebo-controlled crossover trial in postmenopausal women. Am J Clin Nutr 100:1269–1277. https://doi.org/10.3945/ajcn.114.088484

van Die MD, Bone KM, Williams SG, Pirotta MV (2014) Soy and soy isoflavones in prostate cancer: a systematic review and meta-analysis of randomized controlled trials. BJU Int 113: E119–E130. https://doi.org/10.1111/bju.12435

Wang K, Qiu F (2013) Curcuminoid metabolism and its contribution to the pharmacological effects. Curr Drug Metab 14:791–806. https://doi.org/10.2174/13892002113149990102

Wang Y, Wang Y, Luo M, Wu H, Kong L, Xin Y, Cui W, Zhao Y, Wang J, Liang G, Miao L, Cai L (2015) Novel curcumin analog C66 prevents diabetic nephropathy via JNK pathway with the involvement of p300/CBP-mediated histone acetylation. Biochim Biophys Acta 1852:34–46. https://doi.org/10.1016/j.bbadis.2014.11.006

Weiskirchen S, Weiskirchen R (2016) Resveratrol: how much wine do you have to drink to stay healthy? Adv Nutr 7:706–718. https://doi.org/10.3945/an.115.011627

Yang CS, Chen L, Lee MJ, Balentine D, Kuo MC, Schantz SP (1998) Blood and urine levels of tea catechins after ingestion of different amounts of green tea by human volunteers. Cancer Epidemiol Biomark Prev 7:351–354

Yang KY, Lin LC, Tseng TY (2007) Oral bioavailability of curcumin in rat and the herbal analysis from Curcuma Longa by LC-MS/MS. J Chromatogr B 853:183–189. https://doi.org/10.1016/j.jchromb.2007.03.010

Yun J-M, Jialal I, Devaraj S (2011) Epigenetic regulation of high glucose-induced proinflammatory cytokine production in monocytes by curcumin. J Nutr Biochem 22:450–458. https://doi.org/10.1016/j.jnutbio.2010.03.014

Zava DT, Duwe G (1997) Estrogenic and antiproliferative properties of genistein and other flavonoids in human breast cancer cells in vitro. Nutr Cancer 27:31–40

Zhu X, Li Q, Chang R, Yang D, Song Z, Guo Q, Huang C (2014) Curcumin alleviates neuropathic pain by inhibiting p300/CBP histone acetyltransferase activity-regulated expression of BDNF and cox-2 in a rat model. PLoS One 9:e91303. https://doi.org/10.1371/journal.pone.0091303

Ginkgo biloba, DNA Damage and DNA Repair: Overview

104

Daniela Oliveira, Bjorn Johansson, and Rui Oliveira

Contents

D. Oliveira
Centre for the Research and Technology of Agro-Environmental and Biological Sciences (CITAB), Department of Biology, University of Minho, Braga, Portugal
e-mail: danielasoliveira@outlook.pt; danielaoliveira65623@gmail.com

B. Johansson
Centre of Molecular and Environmental Biology (CBMA), Department of Biology, University of Minho, Braga, Portugal
e-mail: bjorn_johansson@bio.uminho.pt; bjornjobb@gmail.com

R. Oliveira (✉)
Centre for the Research and Technology of Agro-Environmental and Biological Sciences (CITAB), Department of Biology, University of Minho, Braga, Portugal

Centre of Biological Engineering (CEB), Department of Biology, University of Minho, Braga, Portugal
e-mail: ruipso@bio.uminho.pt; ruipso@gmail.com

V. B. Patel, V. R. Preedy (eds.), *Handbook of Nutrition, Diet, and Epigenetics*,
https://doi.org/10.1007/978-3-319-55530-0_11

Abstract

Despite the ancient use in Chinese popular medicine and, more recently, in western modern medicine in many European countries, the biological effects of extracts of *G. biloba* (GBE) are still not clearly known. In modern medicine GBE has been used for tinnitus, to reverse memory loss, for dementia, and Alzheimer's and Parkinson's diseases in elderly people. Besides reports on improvement of blood circulation in the brain, there are a number of studies pointing to complex cellular effects, involving signal transduction pathways and epigenetic modifications. Evidence are presented from recent reports concerning genotoxic and antigenotoxic properties and the corresponding mechanisms underlying such activities, mostly regarding the prooxidant and antioxidant activities of the extract. However, several examples of direct interaction of the extract and its components with specific proteins are provided, especially for DNA damage repair, contributing for antigenotoxicity. Evidence of epigenetic effects of GBE are also presented from approaches involving transcriptomics, detection of activity of histone deacetylases, and screening of plant extracts with cell-based systems for detection of posttranslational modifications. The modulation of chromatin-remodeling enzymes by GBE and their interaction with proteins involved in DNA damage repair, apoptosis, and signal transduction are discussed in the context of neurodegeneration.

Keywords

Ginkgo biloba · Antioxidant · Antigenotoxicity · Genotoxicity · Flavonoids · DNA repair · DNA damage · Epigenetics · Neuroprotection · Alzheimer's disease · EGb 761

List of Abbreviations

AD	Alzheimer's disease
BER	Base excision repair
DSBs	Double-strand breaks
EGb 761	Standardized leaf extract of *Ginkgo biloba*
GBE	*Ginkgo biloba* extract
HDAC	Histone deacetylase
MAPK	Mitogen-activated protein kinases
NER	Nucleotide excision repair
ROS	Reactive oxygen species
SUMO	Small ubiquitin-like modifier
Topo II	Topoisomerase II

Introduction

The *Ginkgo biloba* L. tree of the family of Ginkgoaceae, commonly known as maidenhair tree, is the last surviving species of the order Ginkgoales, being considered a living fossil because no living close relative species are known. For thousands of years, parts of the plant, mainly seeds and leaves, have been used in traditional Chinese medicine for health benefits. Nowadays, a wide variety of products of *G. biloba* is sold worldwide in nutritional supplements and under forms for oral administration.

The standardized leaf extract of *G. biloba* (EGb 761) is a popular herbal supplement developed by Beaufour-Ipsen Pharma (Paris, France) and Dr. Willmar Schwabe Pharmaceuticals (Karlsruhe, Germany) commonly used in clinical practice and trials for prevention and treatment of neurodegenerative disorders, including Parkinson's and Alzheimer's (AD) diseases (reviewed by Shi et al. 2010). EGb 761 has been considered as an antidementia drug due to evidence of its efficacy in the central nervous system, which has been recognized by the World Health Organization (World Health Organization 1999). EGb 761 is composed of 24% of flavonoid glycosides (e.g., quercetin, isorhamnetin, and kaempferol), 6% of terpene trilactones (ginkgolides A, B, C, J, and M and bilobalide), and less than 5 ppm of ginkgolic acids due to their allergenic properties (Fig. 1; Baron-Ruppert and Luepke 2001). Therefore, the extract is mainly rich in flavonoids which are thought to be the major responsible constituents for the protective effects, especially the strong antioxidant and reactive oxygen species (ROS) scavenging activities (Boghdady 2013; He et al. 2014). The extract has also been associated to activities such as antimutagenic (Križková et al. 2008), antiproliferative at high concentrations (Bahri et al. 2014), neuroprotective (Zhang et al. 2012), antiapoptotic, (Schindowski et al. 2001), cardioprotective against doxorubicin (Liu et al. 2008; Boghdady 2013), antigenotoxic and DNA repair-stimulating activity (Vilar et al. 2009; Marques et al. 2011), and antiaging (Osman et al. 2016). Properties such as the abovementioned antiproliferative, antiapoptotic, and stimulation of DNA repair, among others, suggest that the extract interacts with cellular pathways, possibly by modulating the activity of specific proteins. Recently, several evidence have been reported on the epigenetic effects of *G. biloba* extract (GBE) and the consequences in human health. This review will present recent data and discuss direct and indirect evidence pointing to the effects of GBE on DNA, including damage, protection, damage repair, and epigenetics.

Genotoxicity of *G. biloba* Extract

The ability of GBE to damage DNA has been attributed to the flavonoid fraction, mainly quercetin (Lin et al. 2014; Zhang et al. 2015), kaempferol (Lin et al. 2014; Wu et al. 2015; Zhang et al. 2015), and isorhamnetin (Zhang et al. 2015). Some of the studies were done with pure flavonoids and terpene trilactone constituents (Lin et al. 2014; Wu et al. 2015; Zhang et al. 2015), where the latter did not show genotoxicity as assessed with the comet assay in human HepG2 hepatocytes (Fig. 2; Zhang et al. 2015). The finding of markers of oxidative stress exposure, such as increased ROS, decreased glutathione, and methaemoglobin formation (He et al. 2009), supports a prooxidant activity underlying DNA damage induced by kaempferol, observed in human leukemia HL-60 cells (Wu et al. 2015), and DNA

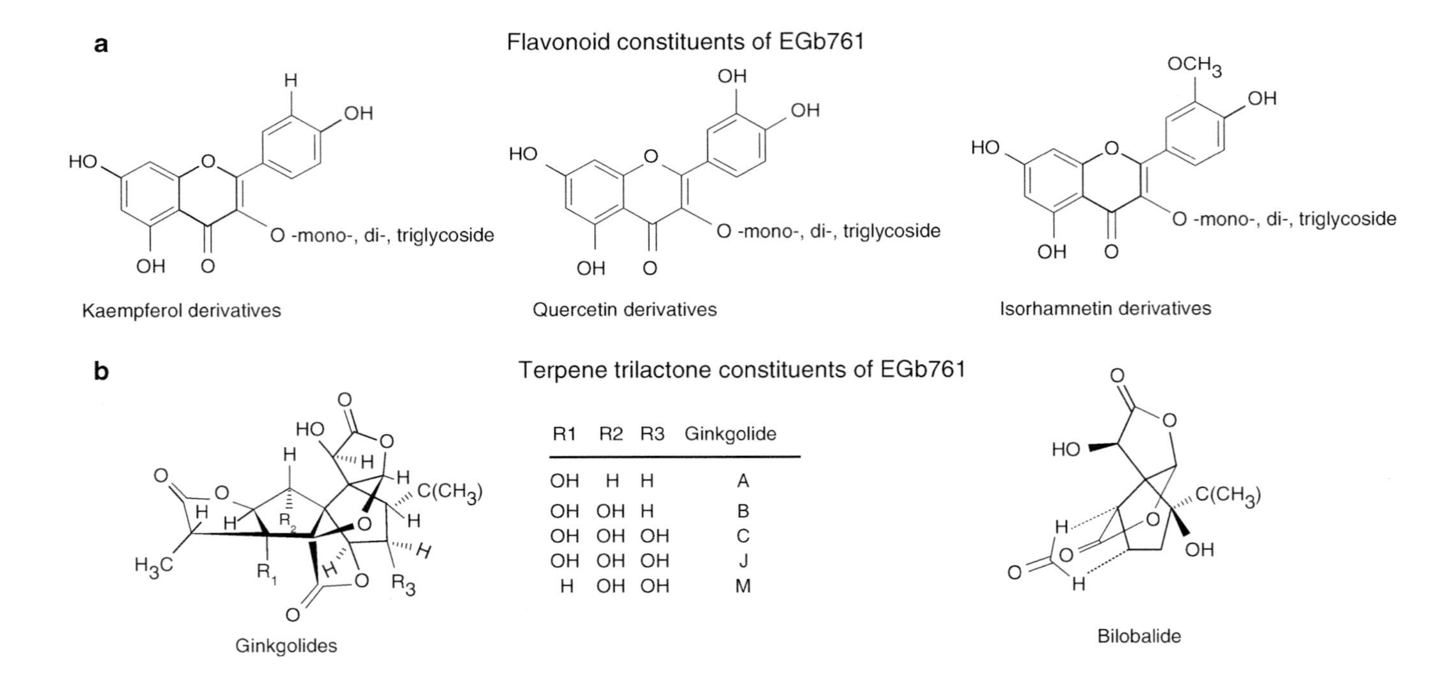

R1	R2	R3	Ginkgolide
OH	H	H	A
OH	OH	H	B
OH	OH	OH	C
OH	OH	OH	J
H	OH	OH	M

Fig. 1 Chemical structure of the main components of the standardized extract of *G. biloba* leaves (EGb 761). Main flavonoid constituents and glycoside derivatives (**a**) and the terpene trilactones ginkgolides and bilobalide (**b**) (Reprinted from Shi et al., copyright (2010), available under a Creative Commons Attribution 3.0 Unported License)

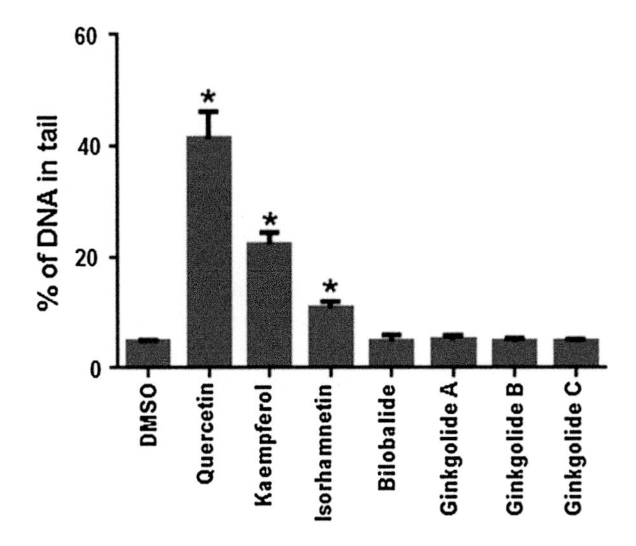

Fig. 2 Genotoxicity of the main constituents of *G. biloba* extract. DNA damage, assessed with the comet assay, of HepG2 cells exposed for 4 h to flavonoids or terpene trilactones (50 μM). Note the higher amount of DNA in the "tails" in the samples treated with the flavonoids, indicating their genotoxicity. DMSO represents the negative control of cells treated only with the solvent dimethyl sulfoxide. Data are the mean ± SD from three independent experiments and * indicates $p < 0.05$ (Adapted from Zhang et al., copyright (2015), available under a Creative Commons Attribution 4.0 International License)

double-strand breaks (DSBs) and increased mutation frequency in a mouse lymphoma cell line by GBE and its constituents quercetin and kaempferol (Lin et al. 2014).

Besides the prooxidant activity, direct influence of the extract and some of the flavonoids in the expression of genes and activity of specific proteins has also been reported. Wu et al. (2015) observed in kaempferol-treated HL-60 human leukemia cells a decrease of expression of proteins involved in DNA damage signaling and repair, proposing that the overall effect was a marked inhibition of DNA damage repair together with a genotoxic activity (Wu et al. 2015). Topoisomerase II (Topo II) inhibition has also been proposed as a DNA-damaging mechanism by GBE, quercetin, kaempferol, and isorhamnetin in human hepatoma cell line HepG2 (Zhang et al. 2015). When Topo II-knockdown cells were used, the DNA damaging effect was dramatically decreased, which correlated with an observed *in vitro* inhibition of Topo II.

Antigenotoxicity of *G. biloba* Extract

Antioxidant properties of the flavonoid fraction determine, at least partly, the antigenotoxic activity of GBE. There are several reports where indirect evidence suggest protection from DNA damage and that this activity is associated to neuronal protection, emphasizing the importance of formulations based on *G. biloba* in the treatment and prophylaxis of neurodegenerative diseases.

Under abnormal physiological conditions that cause oxidative stress, such as intermittent hypoxia typical of the syndrome sleep-disordered breathing, EGb 761 reversed memory deficits and decreased the levels of 8-hydroxy-2′-deoxyguanosine, a common oxidation damage of DNA (Abdel-Wahab and Abd El-Aziz 2012). In addition, in intermittent high glucose, common in diabetes, EGb 761 suppressed elevated ROS, DNA damage, and the levels of 8-hydroxy-2′-deoxyguanosine in DNA in a dose-dependent manner (He et al. 2014). These results are also in line with the decreased DNA single-strand breaks in circulating leukocytes of GBE-treated glaucoma patients, who are affected by systemic oxidative stress (Fang et al. 2015).

Results from a number of studies with exogenous supplementation of H_2O_2 as stressor, an important endogenous ROS, and treatment with GBE suggest antioxidant-mediated antigenotoxic activity of the extract. Among the constituents of GBE, flavonoids are the main compounds implicated in this activity. Accordingly, quercetin was found to have a marked antigenotoxic activity concomitant with increased mRNA expression of human 8-oxoguanine DNA glycosylase, an enzyme that recognizes 8-hydroxy-2′-deoxyguanosine and triggers the repair process (Min and Ebeler 2009). In the animal model adult male Wistar rats, GBE decreased significantly the basal DNA damage in prefrontal cortex cells as well as upon challenge with H_2O_2 (Ribeiro et al. 2016). Although these authors have not observed the protective effect in hippocampal cells, the improved short-term memory in the rats treated with the extract is consistent with a neuroprotective effect. In accordance with the previous studies, in pre- and co-incubation experiments with H_2O_2 and GBE in the simple eukaryotic cell model, *Saccharomyces cerevisiae*, DNA damage was lower than in cells treated only with H_2O_2, which was concomitant with less intracellular oxidation assessed by flow cytometry with an intracellular redox-sensitive fluorochrome, dichlorofluorescein diacetate, and lower cellular loss of viability (Marques et al. 2011).

The antioxidant activity-mediated antigenotoxicity can be observed also against xenobiotics. Roundup®, a herbicide containing the aromatic amino acids synthesis inhibitor glyphosate, induces oxidative stress by increasing the levels of malondialdehyde, a product of lipid peroxidation, and by decreasing the levels of glutathione together with increased genotoxicity markers such as the presence of chromosomal aberrations and micronuclei, as observed in Swiss albino mice (Çavuşoğlu et al. 2011). Interestingly all these markers were reversed in the experimental groups treated with the herbicide and GBE. N-nitrosodiethylamine, a nitrosamine found in tobacco smoke, water, food, cosmetics, and pharmaceuticals, is a hepatocarcinogenic compound that affects DNA, possibly by alkylation (Liu et al. 2010). Nevertheless, there are evidence suggesting that this compound mediates genotoxicity also by an oxidative activity since the administration to male Wistar albino rats increased oxidation markers, namely, higher malondialdehyde and decreased glutathione, superoxide dismutase, glutathione peroxidase, and glutathione reductase levels, concomitant with higher DNA damage as assessed with the comet assay (El Mesallamy et al. 2011). As in the previous case, the experimental group treated with N-nitrosodiethylamine and GBE displayed reversion of all stress markers, further supporting the antioxidant-mediated antigenotoxicity of the extract.

The decay of radioactive isotopes is a source of ionizing radiation that affects biological systems since ionization can yield ROS. While natural radioactivity is negligible in most of human environments, nuclear medicine is based on radioactive isotopes such as ^{99m}Tc as radioactive tracer or ^{131}I used in the treatment of hyperthyroidism. Protein carbonylation, a product of oxidative degradation of proteins, and DNA fragmentation induced by ^{99m}Tc in male Wistar rats were significantly decreased by previous administration of GBE (Alam et al. 2013). Similarly, in a blind clinical trial, micronuclei induction by ^{131}I in Graves' disease patients was significantly decreased in the group treated with EGb 761 when compared with the placebo group (Dardano et al. 2007). These results suggest that the antigenotoxicity of GBE could be used in the control of undesired effects in nuclear medicine.

Antigenotoxicity of GBE has been also observed against genotoxic compounds that specifically react with DNA. Reversion of the mutagenic effect of ofloxacin and acridine orange in the unicellular eukaryotic flagellate *Euglena gracilis* (Križková et al. 2008) and a decrease of micronucleus formation induced by mitomycin C and cyclophosphamide in bone marrow cells from treated mice (Vilar et al. 2009) have been reported. As ofloxacin and mitomycin C trigger the formation of ROS, besides inhibition of topoisomerase and induction of DNA alkylation and cross-links, respectively, the antigenotoxicity of GBE can be attributed to its antioxidant properties. Redox activity of the extract could also explain the activity against cyclophosphamide since this is a prodrug, which bioactivation requires hydroxylation by cytochrome P450. On the other hand, for acridine orange, a DNA intercalating agent that induces frameshift mutations, the proposed antigenotoxic mechanism involves adsorption of the compound to glycosidic moieties of glycosylated flavonoids of the extract (Križková et al. 2008). So, the antigenotoxic activity against acridine orange constitutes an example of a protection mechanism that does not involve the antioxidant properties of GBE.

Effects of *G. biloba* Extract on DNA Repair

Although the literature concerning the effects of GBE on DNA damage repair is not abundant, several clear evidence indicate direct interaction of some constituents with DNA damage repair pathways or with regulatory pathways that modulate DNA repair. Ginkgolide B, one of the exclusive constituents of GBE, was suggested to have chemopreventive properties toward human ovarian cancer (Jiang et al. 2011). The isolated compound was tested in a human ovarian cell line obtained from a BRCA1-mutant carrier, which is defective in BRCA1, a tumor suppressor gene involved in DNA damage repair response (important for DNA DSBs repair by homologous recombination) associated to suppression of breast and ovarian cancers. Analysis of the pattern of protein expression by antibody microarrays upon treatment showed that ginkgolide B increased the expression of DNA damage repair proteins such as the tumor suppressor protein p53, the DNA damage and repair factor xeroderma pigmentosum, complementation group A (XPA), and serine-/arginine-rich protein-specific kinase 1 (SRPK1; the second most induced protein in this study, which is involved in constitutive and alternative splicing and in transcription-related DNA damage response; Boeing et al. 2016). As the authors suggested, ginkgolide B might induce DNA repair response in women carrying BRCA1 mutations, preventing ovarian cancer risk.

Besides the chemopreventive role, GBE has also been referred for its ability to affect DNA repair rate after induction of oxidative damage in budding yeast (Marques et al. 2011). To evaluate the DNA repair along time, yeast spheroplasts were challenged with H_2O_2 to cause damage, followed by incubation with or without GBE to allow DNA repair, with subsequent measurement of the decrease in the tail length at different time points, using the comet assay (Fig. 3). Although the tail length decreased with or without the presence of the extract, the rate was higher in

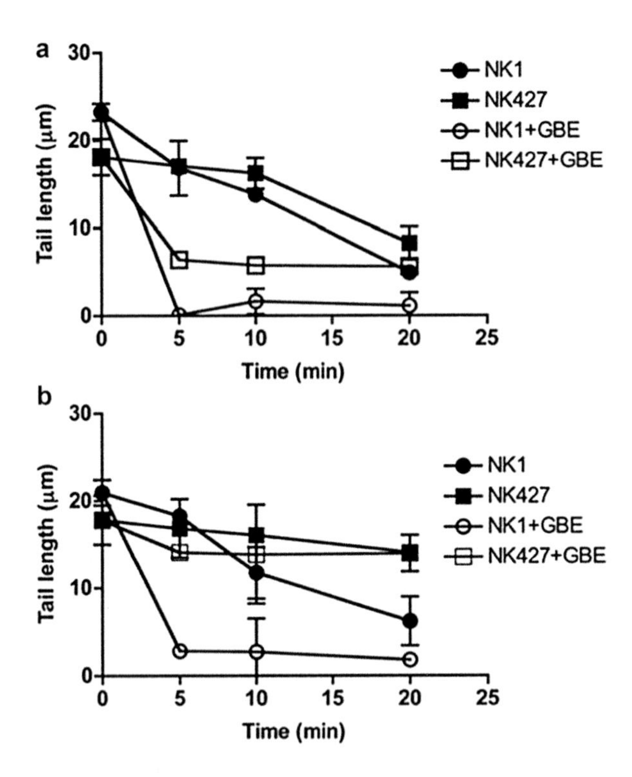

Fig. 3 DNA repair-stimulating activity of *G. biloba* extract and dependency on DNA damage repair pathways. DNA damage repair was measured as comet tail length, with the comet assay, along time of recovery of cells upon an oxidative insult. Yeast spheroplasts were treated with H_2O_2, washed, and incubated to allow DNA repair (a: 23 °C; b: 37 °C), and DNA damage was assessed at different time points. The *G. biloba* extract (GBE) improved the repair of DNA damage in the temperature-sensitive strain NK427, affected in *CDC9*, only under the permissive temperature (23 °C), whereas the wild-type strain NK1 exhibited improved DNA repair under both temperatures. Data are the mean ± SD calculated from three independent experiments (Adapted from Marques et al., copyright (2011), with permission from Elsevier)

treated cells, indicating that GBE may have the ability to stimulate DNA repair (Fig. 3a). Interestingly the temperature-sensitive mutant (NK427), bearing a conditional mutation in *CDC9* (affected in DNA ligase I, which intervenes in base excision repair (BER) and nucleotide excision repair (NER) pathways and is essential to join Okazaki fragments during DNA replication), displayed increased DNA damage recovery upon treatment with the extract under the permissive temperature (Fig. 3a) but failed to be protected only under the restrictive temperature (Fig. 3b), which constitutes one of the few genetic evidence on the DNA repair-stimulating activity of GBE.

Additional evidence on interaction of GBE with DNA damage repair pathways has been also found in the context of neurodegenerative disorders, which highlights this mechanism in the neuroprotective activity of the extract. Human neuroblastoma SH-SY5Y cells and IMR-32 neuroblastoma cells were protected from ROS and nitrogen reactive species induced by β-amyloid peptide, a component of amyloid plaques in AD, upon pretreatment with GBE (Kaur et al. 2015). Remarkably, this antioxidant effect was concomitant, in a similar experiment, with the reversion of the decrease of the levels of APE1 protein, an apurinic/apyrimidinic endonuclease of the BER pathway. So, the decreased capacity to repair oxidative DNA damage in neurodegenerative diseases may be overcome by GBE-mediated induction of expression of proteins involved in BER.

Epigenetic Effects Induced by *G. biloba* Extract

Aberrations in epigenetic mechanisms are considered to contribute to the development of several diseases, including cancer and neurodegenerative disorders (Sadikovic et al. 2008; Kwok 2010). Most epigenetic changes are potentially reversible and, thus, represent a promising target for prevention and treatment of diseases caused by these abnormalities, using epigenetically active compounds such as polyphenols present in plant-based products. Among these, flavonoids have been indicated as exerting an important effect in the modulation of DNA methylation and histone acetylation (Ayissi et al. 2014; Busch et al. 2015). In line with this assumption, in a transcriptomics approach using skeletal muscle cells from male Wistar rats treated with EGb 761, 618 genes were affected with nearly equivalent proportions of upregulated and downregulated genes (Bidon et al. 2009). Functions of the genes with altered expression included transcription-coupled chromatin remodeling (ARID1A), DNA damage repair (MPG, XPC, and RAD23), cell cycle regulation (MAPK14 and CDK7), and the class IV histone deacetylase HDAC11.

From the flavonoids mainly found in GBE, kaempferol is one for which evidence has been reported on inducing epigenetic effects. By *in silico*, *in vivo*, and *in vitro* approaches, Berger et al. (2013) showed that kaempferol has a broad range of histone deacetylase (HDAC) inhibitory spectrum (Fig. 4) and affects viability and proliferation rate of human hepatoma cell lines and colon cancer cells. Another flavonoid found in GBE, quercetin, has been implicated also in HDAC inhibition by potentiation of the senescence of human and rat glioma cell lines by the HDAC-inhibiting

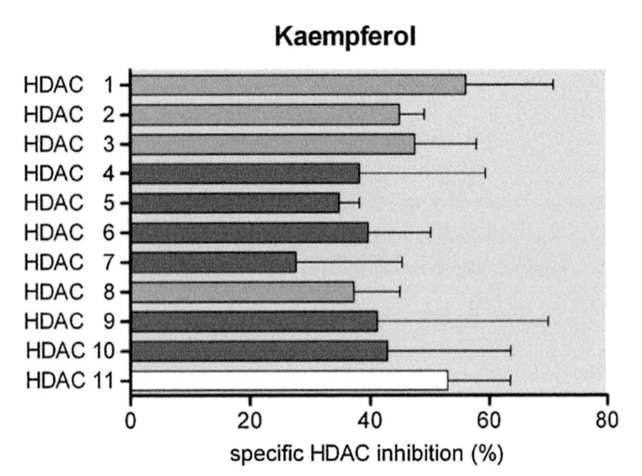

Fig. 4 Profile of histone deacetylase (HDAC) inhibition by kaempferol. Inhibition of HDAC activity was assessed *in vitro* with the commercial HDAC assay kit (Active Motif, Belgium) using recombinant human HDACs and 50 μM kaempferol. Data are the mean ± SD of four replicas from two independent experiments (Adapted from Berger et al., copyright (2013), with permission from Elsevier)

drug sodium butyrate (Vargas et al. 2014). Further evidence was reported by Priyadarsini et al. (2011) on the inhibition of the DNA methylating enzyme DNA (cytosine-5) methyltransferase (DNMT1) and HDAC-1 in male Syrian hamsters.

EGb 761 is commonly used in age-related neurodegenerative diseases, including AD (reviewed by Ude et al. 2013). This condition is characterized by the presence of the toxic peptide β-amyloid, which has been associated to cell death induction mediated by oxidative stress, causing serious damage in many cellular components, such as proteins, lipids, and DNA, and activation of the redox-sensitive nuclear factor kappa B (NF-kB) and MAPK pathways. It has been proposed that β-amyloid-driven overactivation of NF-kB increases inflammatory mediators such as interleukin-1β (IL-1β), tumor necrosis factor-α (TNF-α) and cyclooxygenase-2 (COX-2; Carrero et al. 2012), and subsequently prostaglandins and that these are involved in AD pathogenesis (Fig. 5; Medeiros and LaFerla 2013).

Besides the protective effect against AD driven by free radical scavenging activity of GBE, direct evidence on activation of the class III histone deacetyltransferase SIRT1 has been reported by Longpré et al. (2006). These authors found that preincubation of cells from N2a neuroblastoma cell line with EGb 761 simultaneously increased SIRT1, as detected by immunoassay, and prevented the β-amyloid-mediated activation of the redox-sensitive NF-kB pathway and its upstream activating kinases, such as extracellular signal-regulated protein kinases (ERK1/2) and c-Jun N-terminal kinase (JNK). Since SIRT1 is also able to deacetylate NF-kB, inhibiting this pathway (Yeung et al. 2004), the protective activity against AD could also be explained by this interaction. Neuroprotection by EGb 761 therefore seems to result from a complex interplay of different mechanisms, including free radical scavenging activity, activation of SIRT1 with inhibition of NF-kB, and inhibition of ERK1/2 and JNK (Fig. 6).

The involvement of histone deacetylases and histone acetyltransferases in neurodegeneration is very complex with reports pointing to increased activity of histone deacetylases as beneficial, as reported above for SIRT1 (Longpré et al. 2006), and reports where histone deacetylase (HDAC) inhibitors have been regarded

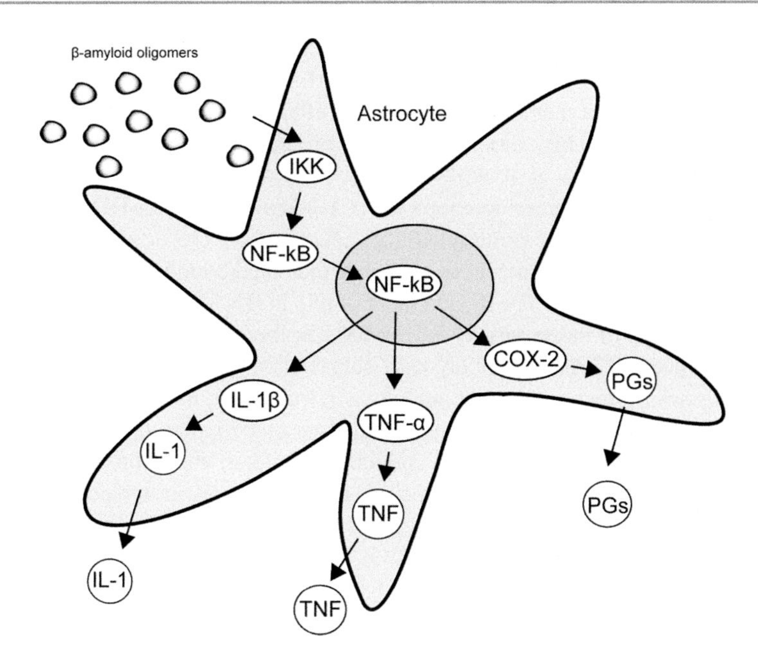

Fig. 5 Putative mechanism of toxicity of β-amyloid in Alzheimer's disease. β-amyloid oligomers activate NF-kB in astrocytes, through IKK kinase, and translocates to the nucleus, leading to the production of the pro-inflammatory mediators interleukin-1β (IL-1β), tumor necrosis factor-α (TNF-α), and prostaglandins (PGs), mediated by the activation of cyclooxygenase-2 (COX-2) (Adapted from Medeiros and LaFerla, copyright (2013), with permission from Elsevier)

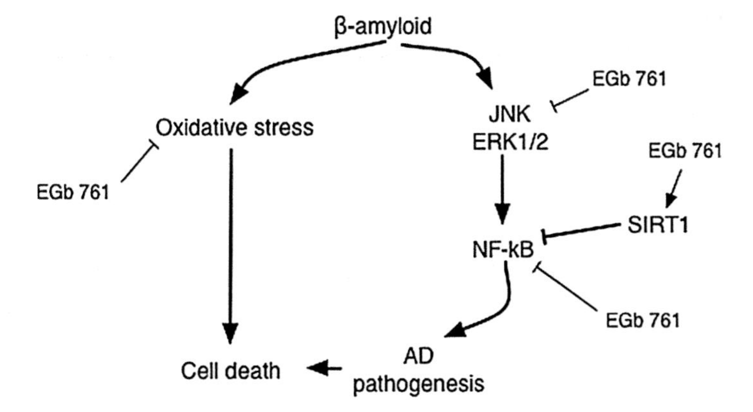

Fig. 6 Mechanisms of action of *G. biloba* extract (EGb 761) against neurotoxicity induced by the β-amyloid peptide. Pathogenicity of β-amyloid involves oxidative stress that may cause cell death and activation of the extracellular signal-regulated kinases (ERK1/2) and c-Jun N-terminal kinase (JNK) pathways that activate the nuclear factor kappa B (NF-kB), leading to Alzheimer's disease (AD) pathogenesis. Neuroprotection by EGb 761 involves the reactive oxygen species-scavenging activity; inhibition of JNK, ERK1/2, and NF-kB; and activation of the histone deacetyltransferase SIRT1. SIRT1 is able to deacetylate NF-kB, contributing to inhibition of this pathway (Adapted from Longpré et al., copyright (2006), with permission from Elsevier)

as a therapeutic strategy with great potential in the treatment of neurodegenerative disorders (reviewed by Kukucka et al. 2013). Accordingly, HDAC inhibitors enhanced learning and memory (Govindarajan et al. 2013) and axonal regeneration (Rivieccio et al. 2009). In addition, nonhistone proteins are also targeted by HDACs, such as p53 (Choudhary et al. 2009), which is involved in the activation of DNA damage repair and in triggering apoptosis. Trichostatin A, a pan-HDAC inhibitor, efficiently protected mouse primary cortical neurons from cell death induced by the DSB-inducing drug camptothecin in a process mediated by increased acetylation of p53 (Brochier et al. 2013). The involvement of p53 in DNA damage repair is known, and the regulation of its activity by SIRT1 has also been demonstrated (Vaziri et al. 2001). However, HDAC inhibition with subsequent p53 acetylation in neuronal tumor cells was found to promote apoptosis (Brochier et al. 2013). The fact that HDAC inhibition, with concomitant acetylation of p53, has opposite effects in proliferating tumor neuronal cells, by inducing apoptosis, and in postmitotic neuronal cells, by inhibiting apoptosis (Brochier et al. 2013), is in agreement with the complex events involved in the epigenetics of AD. Moreover, in the role of inhibiting p53 and NF-kB, SIRT1 has been described as having both antiapoptotic and apoptotic roles, respectively (Yeung et al. 2004). Since p53 acetylation is decreased by SIRT1 and is involved in AD pathogenesis (Vaziri et al. 2001; Tramutola et al. 2016), activation of SIRT1 by EGb 761 (Longpré et al. 2006) could be promoting neuropathology. However, the SIRT1-driven effect of EGb 761 on NF-kB, in agreement with the observed effects of neuroprotection (Longpré et al. 2006), argues in favor of the complexity of the mechanism of action of GBE and for a dual role of p53 in AD pathogenesis.

More indications of the involvement of GBE on epigenetic regulation were reported by Fukuda et al. (2009) who identified GBE in a screen of plant extracts for SUMOylation inhibition in a cell-based system with SUMO (small ubiquitin-like modifier)-tagged GFP. In a similar experiment, it was found that ginkgolic acid also inhibited SUMOylation, suggesting that this compound is responsible for the SUMOylation-inhibiting activity of the extract. Posttranslational modification by conjugation with SUMO is part of the regulation of important cellular processes such as transcription, replication, cell signaling, and DNA repair and was found to affect the function of HDACs and p53 (Johnson 2004). Hence, the inhibition of SUMOylation by specific binding of ginkgolic acid to the SUMO-activating enzyme E1 (Fukuda et al. 2009) is likely to affect p53 transcriptional activity (Stindt et al. 2011) and HDAC-mediated repression of transcription (Yang and Sharrocks 2004). In fact, recently the E1 and E2 enzymes of the SUMO pathway were found to be essential in epigenetic gene silencing (Poleshko et al. 2014) in a genome-wide siRNA knockdown screen with human cells harboring an epigenetically silent GFP reporter. As in the case of SIRT1, effects on SUMOylation can have consequences in neuronal cells as SUMOylation has been proposed to have a regulatory role of proteins such as tau and α-synuclein involved in neurodegenerative diseases (Dorval and Fraser 2006) and to be implicated in normal neuronal function by rescuing amyloid-β-induced neuronal degeneration (Lee et al. 2014).

Conclusion

The complex nature of GBE, the experimental model used in the different studies, and the dual effect, antioxidant and prooxidant, which is reported frequently, contribute to some disparity in the literature. Most of the evidence are related with flavonoids as the main responsible for the antioxidant/prooxidant and antigenotoxic/genotoxic activities. The latter seem to be a consequence of the former; however, there are also evidence pointing to direct interaction with DNA damage repair proteins, p53 and Topo II (Wu et al. 2015; Zhang et al. 2015), for the genotoxic effect, and with DNA damage repair proteins, p53 and with BER and/or NER pathways (Marques et al. 2011; Jiang et al. 2011), for the antigenotoxic effect.

Epigenetic effects are also present in the literature with reports pointing to both improvement and inhibition of histone deacetylases. Regardless of the contradictory reports in the literature, GBE is pointed as influencing chromatin remodeling by altering the acetylation and SUMOylation pattern of proteins. The cellular effects seem to be associated with neuroprotection, since NF-kB overactivation by β-amyloid is reverted by GBE where the role of p53 is central as it may be the target of HDAC and the SUMOylation pathway, which are modulated by GBE. The key role of p53 in linking DNA damage response with cell cycle regulation, DNA damage repair, and apoptosis can be crucial in the neuroprotective effect of GBE and the reported anticancer properties. Therefore, future studies should focus on the effect of GBE in posttranslational pathways, like acetylation and SUMOylation, and the consequences of p53 activity in proliferating tumor neuronal cells and in postmitotic neuronal cells as the effects were found to be opposite (Brochier et al. 2013).

Dictionary of Terms

- **Antigenotoxicity** – The property of Protecting DNA integrity from physical and chemical damage induced by genotoxic agents.
- **Antimutagenicity** – The ability of protecting DNA from spontaneous alterations or the effect of mutagens, reducing the frequency of modifications in the genetic sequence.
- **EGb 761** – Standardized extract of *Ginkgo biloba* leaves, a blend of batches from different harvests in order to achieve a composition of approximately 24% flavonoid glycosides, 6% terpene trilactones, and less than 5 ppm of ginkgolic acids.
- **Flavonoids** – Class of polyphenols commonly found in plants as secondary metabolites, being known for their antioxidant and free radical scavenger activities. These compounds are synthesized via the acetate and shikimate pathways, showing a common basic structure of two aromatic rings linked by a three-carbon chain (C6-C3-C6), and can be classified as flavonols, flavan-3-ols, flavones, isoflavones, flavanones, and anthocyanins.

- **Histone deacetylases** – Proteins responsible for removing acetyl groups from lysines of chromatin proteins and nonhistone proteins, modulating DNA transcription and cellular processes such as differentiation and cell cycle progression.
- **Homologous recombination** – Cellular mechanism to repair double-strand DNA breaks, performed during late-S/G2 cell cycle phases, involving strand invasion and using a sister chromatid as homologous template for DNA repair.
- **NER** – The main DNA repair pathway for removal of bulky DNA lesions responsible for distortions in DNA structure, commonly induced by mutagenic compounds and exposure to ultraviolet light.
- **BER** – Cellular mechanism responsible for repairing lesions in DNA bases, frequently induced by alkylation, oxidation, and deamination, by removing and replacing damaged (e.g., 8-hydroxy-2′-deoxyguanosine) or incorrect bases.

Key Facts

Key Facts of Comet Assay

- Comet assay is a technique that allows the detection and measurement of DNA strand breaks in eukaryotic cells, being commonly used to assess DNA damage and DNA repair.
- Cells are embedded in agarose and lysed with detergent and high salt, which unwinds genomic DNA.
- Upon electrophoresis, unwound DNA will migrate toward the anode, the rate being proportional to the level of strand breaks.
- DNA-labeling with a fluorescent dye will reveal structures similar to comets in which the head corresponds to the nucleoid and the tail to the dragged DNA.
- DNA damage is measured as comet tail length or as percentage of DNA in the tail, among other parameters.

Key Facts of Oxidative Stress

- Reactive oxygen species (ROS) are produced endogenously by normal metabolism in mitochondria and peroxisomes and can also have an exogenous origin such as environmental pollutants and radiation.
- Organisms have evolved antioxidant defenses, including enzymatic and non-enzymatic, which keep redox homeostasis.
- Oxidative stress results from a disturbance of the redox homeostasis by excessive amount of ROS and/or decreased efficiency in antioxidant defenses.
- Free radicals are ROS with one or more unpaired electrons, becoming highly reactive and causing serious damage in structures of high biological value such as DNA.
- Oxidative stress contributes to the development of diseases including cancer, neurodegenerative, and cardiovascular disorders.

Key Facts of SUMOylation

- The conjugation of small ubiquitin-like modifier (SUMO) proteins with the substrate (SUMOylation) represents an important posttranslational regulatory mechanism of proteins.
- The covalent attachment of SUMO to its target protein leads to reversible posttranslational modifications, including the alteration of protein surface, affecting the interaction with other proteins, and the control of subcellular targeting.
- Similarly to ubiquitination, SUMOylation relies on a cascade of enzymatic reactions (SUMO-activating enzyme E1, SUMO-conjugating enzyme E2, and SUMO E3 ligases) to form an isopeptide bond between SUMO and the ε-amino group of a lysine residue of the target protein.
- SUMOylation affects activity, localization, and stability of target proteins and participates in the regulation of cellular events such as DNA repair, transcription, and replication.
- Abnormalities in SUMOylation have been associated to the development of neurodegenerative diseases and cancer.

Key Facts of Tumour Suppressor Protein p53

- p53, the so-called guardian of the genome, is a nuclear transcription factor whose levels are increased in response to situations of cellular stress such as oncogenic activation and DNA damage.
- The activity of p53 is regulated through posttranslational modifications such as phosphorylation, acetylation, and SUMOylation, controlling stability and subcellular localization.
- p53 regulates cell cycle progression, DNA damage repair, and apoptosis, key processes for maintenance of integrity of the genome and in the avoidance of tumorigenesis.
- Deregulation of p53 levels lead to uncontrolled mitosis with accumulation of DNA damage, which can trigger cell death and/or malignancy.
- In Alzheimer's disease, the p53 promoter is activated in response to intracellular accumulation of β-amyloid oligopeptide, suggesting that p53 might promote neurodegeneration.

Summary Points

- This chapter focuses on the activity of *Ginkgo biloba* leaf extract (GBE), which has been associated to many health benefits over the years. The standardized extract of *G. biloba* leaves EGb 761 is commonly used to ameliorate the symptoms of neurodegenerative diseases, including Alzheimer's disease (AD).
- GBE has many biological properties (e.g., antioxidant, antigenotoxic, antimutagenic, cardioprotective, antiproliferative, genotoxic, stimulation of DNA

repair and antiaging), modulating signaling pathways and gene transcription and inducing epigenetic alterations.

- EGb 761 is mainly composed of terpene trilactones and flavonoids, the latter contributing significantly to the concentration-dependent dual response induced by the extract: protective (antioxidant) or damaging (prooxidant) effect.
- Genotoxicity of GBE includes induction of DNA double-strand breaks by a prooxidant mechanism, associated to activation of DNA damage signaling pathways, and inhibition of topoisomerase II.
- EGb 761 has protective effect against oxidative damage, decreasing DNA breaks and the levels of characteristic biomarkers such as 8-hydroxy-2′-deoxyguanosine (DNA damage) and malondialdehyde (lipid peroxidation) and increasing the antioxidant defenses (glutathione and glutathione peroxidase).
- Apart from the antioxidant-mediated antigenotoxicity, GBE protects DNA from the genotoxic effects of intercalating agents (e.g., acridine orange), probably due to direct interaction with the compound which is adsorbed to glycosylated flavonoids of the extract.
- GBE and its constituents can produce effects on DNA damage repair, increasing the expression of DNA repair proteins (ginkgolide B) and increasing the rate of repair (GBE) after oxidative stress probably due to stimulation of nucleotide excision repair and/or base excision repair pathways, along with the reversion of decreased levels of proteins involved in base excision repair under neurodegenerative conditions.
- EGb 761 has neuroprotective properties against AD through free radical scavenging activity and activation of histone deacetyltransferase SIRT1, which inhibits the β-amyloid peptide-driven activation of the redox-sensitive NF-kB pathway.
- GBE inhibits SUMOylation, in which ginkgolic acid seems to be main active component as a result of its direct interaction with SUMO-activating enzyme E1, affecting p53 transcriptional activity and repression of transcription mediated by histone deacetylases (HDACs).
- HDACs and SUMOylation modulation by GBE could target p53 and help in the prevention of neurodegeneration.

References

Abdel-Wahab BA, Abd El-Aziz SM (2012) *Ginkgo biloba* protects against intermittent hypoxia-induced memory deficits and hippocampal DNA damage in rats. Phytomedicine 19:444–450

Alam SS, Hassan NS, Raafat BM (2013) Evaluation of oxidatively generated damage to DNA and proteins in rat liver induced by exposure to technetium radioisotope 99m and protective role of *Angelica archangelica* and *Ginkgo biloba*. WASJ 2:7–17

Ayissi VBO, Ebrahimi A, Schluesenner H (2014) Epigenetic effects of natural polyphenols: a focus on SIRT1-mediated mechanisms. Mol Nutr Food Res 58:22–32

Bahri GG, Lamuki MS, Rezae-Raad MS (2014) Anti-proliferative effects of alcoholic and aqueous extract of *Ginkgo biloba* green leaves on MCF-7 cell line. IJMPR 2:8–11

Baron-Ruppert G, Luepke N-P (2001) Evidence for toxic effects of alkylphenols from *Ginkgo biloba* in the hen's egg test (HET). Phytomedicine 8:133–138

Berger A, Venturelli S, Kallnischkies M et al (2013) Kaempferol, a new nutrition-derived pan-inhibitor of human histone deacetylases. J Nutr Biochem 24:977–985

Bidon C, Lachuer J, Molgo J et al (2009) The extract of *Ginkgo biloba* EGb 761 reactivates a juvenile profile in the skeletal muscle of sarcopenic rats by transcriptional reprogramming. PLoS One 4:e7998

Boeing S, Williamson L, Encheva V et al (2016) Multiomic analysis of the UV-induced DNA damage response. Cell Rep 15:1597–1610

Boghdady NAE (2013) Antioxidant and antiapoptotic effects of proanthocyanidin and *ginkgo biloba* extract against doxorubicin-induced cardiac injury in rats. Cell Biochem Funct 31:344–351

Brochier C, Dennis G, Rivieccio MA et al (2013) Specific acetylation of p53 by HDAC inhibition prevents DNA damage-induced apoptosis in neurons. J Neurosci 33:8621–8632

Busch C, Burkard M, Leischner C et al (2015) Epigenetic activities of flavonoids in the prevention and treatment of cancer. Clin Epigenetics 7:64

Carrero I, Gonzalo MR, Martin B et al (2012) Oligomers of beta-amyloid protein (Aβ1-42) induce the activation of cyclooxygenase-2 in astrocytes via an interaction with interleukin-1beta, tumour necrosis factor-alpha, and a nuclear factor kappa-B mechanism in the rat brain. Exp Neurol 236:215–227

Çavuşoğlu K, Yapar K, Oruç E et al (2011) Protective effect of *Ginkgo biloba* L. leaf extract against glyphosate toxicity in Swiss albino mice. J Med Food 14:1263–1272

Choudhary C, Kumar C, Gnad F et al (2009) Lysine acetylation targets protein complexes and co-regulates major cellular functions. Science 325:834–840

Dardano A, Ballardin M, Ferdeghini M et al (2007) Anticlastogenic effect of *Ginkgo biloba* extract in graves' disease patients receiving radioiodine therapy. J Clin Endocrinol Metab 92:4286–4289

Dorval V, Fraser PE (2006) Small ubiquitin-like modifier (SUMO) modification of natively unfolded proteins tau and α-synuclein. J Biol Chem 281:9919–9924

El Mesallamy HO, Metwally NS, Soliman MS et al (2011) The chemopreventive effect of *Ginkgo biloba* and *Silybum marianum* extracts on hepatocarcinogenesis in rats. Cancer Cell Int 11:38

Fang L, Neutzner A, Turtschi S et al (2015) The effect of *Ginkgo biloba* and Nifedipine on DNA breaks in circulating leukocytes of glaucoma patients. Expert Rev Ophthalmol 10:313–318

Fukuda I, Ito A, Hirai G et al (2009) Ginkgolic acid inhibits protein SUMOylation by blocking formation of the E1-SUMO intermediate. Chem Biol 16:133–140

Govindarajan N, Rao P, Burkhardt S et al (2013) Reducing HDAC6 ameliorates cognitive deficits in a mouse model for Alzheimer's disease. EMBO Mol Med 5:52–63

He J, Lin J, Li J et al (2009) Dual effects of *Ginkgo biloba* leaf extract on human red blood cells. Basic Clin Pharmacol Toxicol 104:138–144

He Y-T, Xing S-S, Gao L et al (2014) *Ginkgo biloba* attenuates oxidative DNA damage of human umbilical vein endothelial cells induced by intermittent high glucose. Pharmazie 69:203–207

Jiang W, Qiu W, Wang Y et al (2011) Ginkgo may prevent genetic-associated ovarian cancer risk: multiple biomarkers and anticancer pathways induced by ginkgolide B in BRCA1-mutant ovarian epithelial cells. Eur J Cancer Prev 20:508–517

Johnson ES (2004) Protein modification by SUMO. Annu Rev Biochem 73:355–382

Kaur N, Dhiman M, Perez-Polo JR et al (2015) Ginkgolide B revamps neuroprotective role of apurinic/apyrimidinic endonuclease 1 and mitochondrial oxidative phosphorylation against Aβ25–35-induced neurotoxicity in human neuroblastoma cells. J Neurosci Res 93:938–947

Križková L, Chovanová Z, Ďuračková Z et al (2008) Antimutagenic in vitro activity of plant polyphenols: Pycnogenol® and *Ginkgo biloba* extract (EGb 761). Phytother Res 22:384–388

Kukucka J, Wyllie T, Read J et al (2013) Human neuronal cells: epigenetic aspects. Biomol Concepts 4:319–333

Kwok JBJ (2010) Role of epigenetics in Alzheimer's and Parkinson's disease. Epigenomics 2:671–682

Lee L, Dale E, Staniszewski A et al (2014) Regulation of synaptic plasticity and cognition by SUMO in normal physiology and Alzheimer's disease. Sci Report 4:7190

Lin H, Guo X, Zhang S et al (2014) Mechanistic evaluation of *Ginkgo biloba* leaf extract-induced genotoxicity in L5178Y cells. Toxicol Sci 139:338–349

Liu T-J, Yeh Y-C, Ting C-T et al (2008) *Ginkgo biloba* extract 761 reduces doxorubicin-induced apoptotic damage in rat hearts and neonatal cardiomyocytes. Cardiovasc Res 80:227–235

Liu W-B, Ao L, Zhou Z-Y et al (2010) CpG island hypermethylation of multiple tumor suppressor genes associated with loss of their protein expression during rat lung carcinogenesis induced by 3-methylcholanthrene and diethylnitrosamine. Biochem Biophys Res Commun 402:507–514

Longpré F, Garneau P, Christen Y et al (2006) Protection by EGb 761 against β-amyloid-induced neurotoxicity: involvement of NF-κB, SIRT1, and MAPKs pathways and inhibition of amyloid fibril formation. Free Radic Biol Med 41:1781–1794

Marques F, Azevedo F, Johansson B et al (2011) Stimulation of DNA repair in *Saccharomyces cerevisiae* by *Ginkgo biloba* leaf extract. Food Chem Toxicol 49:1361–1366

Medeiros R, LaFerla FM (2013) Astrocytes: conductors of the Alzheimer disease neuroinflammatory symphony. Exp Neurol 239:133–138

Min K, Ebeler SE (2009) Quercetin inhibits hydrogen peroxide-induced DNA damage and enhances DNA repair in Caco-2 cells. Food Chem Toxicol 47:2716–2722

Osman NMS, Amer AS, Abdelwahab S (2016) Effects of *Ginkgo biloba* leaf extract on the neurogenesis of the hippocampal dentate gyrus in the elderly mice. Anat Sci Int 91:280–289

Poleshko A, Kossenkov AV, Shalginskikh N et al (2014) Human factors and pathways essential for mediating epigenetic gene silencing. Epigenetics 9:1280–1289

Priyadarsini RV, Vinothini G, Murugan RS et al (2011) The flavonoid quercetin modulates the hallmark capabilities of hamster buccal pouch tumors. Nutr Cancer 63:218–226

Ribeiro ML, Moreira LM, Arçari DP et al (2016) Protective effects of chronic treatment with a standardized extract of *Ginkgo biloba* L. in the prefrontal cortex and dorsal hippocampus of middle-aged rats. Behav Brain Res 313:144–150

Rivieccio MA, Brochier C, Willis DE et al (2009) HDAC6 is a target for protection and regeneration following injury in the nervous system. PNAS 106:19599–19604

Sadikovic B, Al-Romaih K, Squire JA et al (2008) Cause and consequences of genetic and epigenetic alterations in human cancer. Curr Genomics 9:394–408

Schindowski K, Leutner S, Kressmann S et al (2001) Age-related increase of oxidative stress-induced apoptosis in mice prevention by *Ginkgo biloba* extract (EGb761). J Neural Transm 108:969–978

Shi C, Liu J, Wu F et al (2010) *Ginkgo biloba* extract in Alzheimer's disease: from action mechanisms to medical practice. Int J Mol Sci 11:107–123

Stindt MH, Carter SA, Vigneron AM et al (2011) MDM2 promotes SUMO-2/3 modification of p53 to modulate transcriptional activity. Cell Cycle 10:3176–3188

Tramutola A, Pupo G, Di Domenico F et al (2016) Activation of p53 in Down syndrome and in the Ts65Dn mouse brain is associated with a pro-apoptotic phenotype. J Alzheimers Dis 52:359–371

Ude C, Schubert-Zsilavecz M, Wurglics M (2013) *Ginkgo biloba* extracts: a review of the pharmacokinetics of the active ingredients. Clin Pharmacokinet 52:727–749

Vargas JE, Filippi-Chiela EC, Suhre T et al (2014) Inhibition of HDAC increases the senescence induced by natural polyphenols in glioma cells. Biochem Cell Biol 92:297–304

Vaziri H, Dessain SK, Eaton EN et al (2001) hSIR2^{SIRT1} functions as an NAD-dependent p53 deacetylase. Cell 107:149–159

Vilar JB, Leite KR, Chen LC (2009) Antimutagenicity protection of *Ginkgo biloba* extract (Egb 761) against mitomycin C and cyclophosphamide in mouse bone marrow. Genet Mol Res 8:328–333

World Health Organization (1999) WHO monographs on selected medicinal plants, vol 1. World Health Organization, Geneva, pp 154–167

Wu L-Y, Lu H-F, Chou Y-C et al (2015) Kaempferol induces DNA damage and inhibits DNA repair associated protein expressions in human promyelocytic leukemia HL-60 cells. Am J Chin Med 43:365–382

Yang SH, Sharrocks AD (2004) SUMO promotes HDAC-mediated transcriptional repression. Mol Cell 13:611–617

Yeung F, Hoberg JE, Ramsey CS et al (2004) Modulation of NF-κB-dependent transcription and cell survival by the SIRT1 deacetylase. EMBO J 23:2369–2380

Zhang Z, Peng D, Zhu H et al (2012) Experimental evidence of *Ginkgo biloba* extract EGB as a neuroprotective agent in ischemia stroke rats. Brain Res Bull 87:193–198

Zhang Z, Chen S, Mei H et al (2015) *Ginkgo biloba* leaf extract induces DNA damage by inhibiting topoisomerase II activity in human hepatic cells. Sci Report 5:14633

Plant Monoterpenes Camphor, Eucalyptol, Thujone, and DNA Repair

105

Biljana Nikolić, Dragana Mitić-Ćulafić, Branka Vuković-Gačić, and Jelena Knežević-Vukčević

Contents

Abstract

Genotoxic and genoprotective effects of monoterpenes camphor, eucalyptol, and thujone were comparatively studied in bacterial and mammalian cells. In *E. coli* test system, low doses were antimutagenic against UV and 4NQO in the repair-proficient strain, but co-mutagenic in NER-deficient mutant. Additionally, they enhanced UV-induced SOS response and homologous recombination. However, high doses were mutagenic in NER- and MMR-deficient strains. Similarly, low doses decreased genotoxic effect of 4NQO in Vero cell line, while high doses

B. Nikolić (✉) · D. Mitić-Ćulafić · B. Vuković-Gačić · J. Knežević-Vukčević
Microbiology, Center for Genotoxicology and Ecogenotoxicology, Faculty of Biology, University of Belgrade, Belgrade, Serbia
e-mail: biljanan@bio.bg.ac.rs; mdragana@bio.bg.ac.rs; brankavg@bio.bg.ac.rs; jelenakv@bio.bg.ac.rs

V. B. Patel, V. R. Preedy (eds.), *Handbook of Nutrition, Diet, and Epigenetics*, https://doi.org/10.1007/978-3-319-55530-0_106

were genotoxic. Genotoxicity was confirmed in human cell lines: fetal fibroblasts MRC-5 and colon carcinoma HT-29 and HCT116 cells.

Obtained results were consistent with hormesis phenomenon and indicated genotoxin-induced adaptive response provoked by low doses of monoterpenes: small amounts of DNA lesions evoked error-free DNA repair pathways, mainly NER, and provided protection against more potent genotoxic agents, such as UV and 4NQO. Adaptive response in *E. coli* is mediated by enhanced efficiency of NER during SOS induction. On the other hand, adaptive response in mammalian cells may involve transcriptional upregulation of NER genes *DDB2*, *XPC*, *ERCC1*, *XPF*, *XPG*, and *LIG1* previously reported to be induced by UV. In addition, promotion of NER could involve UV-specific histone modifications, such as acetylation of H2A, H2B, H3, and H4, methylation of H3 and H4, and ubiquitination of H2A, H2B, H3, and H4.

Taking into account that numerous genotoxic agents induce DNA lesions repairable by NER, adaptive response provoked by camphor, eucalyptol, and thujone could be important for protection against environmental mutagens and carcinogens.

Keywords

Monoterpenes · Camphor · Eucalyptol · Thujone · *Escherichia coli* model · Mammalian cells · Genotoxicity/antigenotoxicity · DNA repair · Hormesis phenomenon · Genotoxin-induced adaptive response · Nucleotide excision repair · Transcriptional upregulation · Histone modification

List of Abbreviations

4NQO	4-Nitroquinoline 1-oxide
BER	Base excision repair
DSB	Double-strand break
HR	Homologous recombination
MMR	Mismatch repair
NER	Nucleotide excision repair
NHEJ	Nonhomologous end joining
TLS	Translesion synthesis

Introduction

Etiology of different pathological conditions, including cancer, neurodegenerative and cardiovascular diseases, diabetes, and obesity is determined by both genetic and epigenetic factors. As opposed to genetic alterations, epigenetic modifications are potentially reversible, which make them attractive and promising targets in treatment of these pathological conditions (Busch et al. 2015).

Different environmental factors can alter the chromatin conformation, function of microRNA, and methylation status of cytosine in CpG dinucleotide islands in the promoter regions, attenuating the expression of harmful genes, such as oncogenes, or promoting the expression of protective genes, such as tumor suppressor genes. Nutritional pattern is considered to be an extremely important environmental factor affecting epigenetic status. Nutraceuticals proved to be active in epigenetic changes include vitamins and minerals, ω-3 fatty acids and essential and nonessential amino acids, as well as antioxidative phytochemicals, such as quercetin, epigallocatechin gallate, curcumin, and lycopene (Aggarwal et al. 2015).

Terpenes are the largest and highly diverse group of natural substances. They consist of five-carbon isoprene units (C_5H_8) and depending on their number are denoted as monoterpenes (C_{10}), sesquiterpenes (C_{15}), diterpenes (C_{20}), triterpenes (C_{30}), tetraterpenes (C_{40}) , and polyterpenes ($C_{>40}$). They act as signal molecules, participating in communications between two organisms, namely, plants, animals, and microorganisms. Their main function in plants is to provide a chemical defense against environmental stress; they are involved in protection against infections and parasites and in repair mechanism for wounds and injuries. In addition, biological effects of terpenes, like antioxidant, antimicrobial, antiparasitic, spasmolytic, hypoglycemic, anti-allergenic, anti-inflammatory, immune-modulatory, antimutagenic, and anticancer, are well established (Bakkali et al. 2008).

The monoterpenes camphor, eucalyptol, and thujone are widely distributed in essential oils of many medicinal and aromatic plants used worldwide. Camphor is a dominant constituent of camphor tree (*Cinnamomum camphora*), but it is also found in essential oils of *Rosmarinus*, *Artemisia*, *Salvia*, and *Chrysanthemum* species. Eucalyptol and thujone are abundant in *Eucalyptus globulus* and *Thuja occidentalis*, respectively, but they are constituents of *Salvia*, *Artemisia*, and *Juniperus* essential oils and extracts. Camphor and eucalyptol possess strong antimicrobial, analgesic, and antirheumatic properties, while eucalyptol additionally possesses cytotoxic, anti-inflammatory, anti-exudant, gastroprotective, and hepatoprotective properties (Bakkali et al. 2008). Although thujone induces neurotoxicity in experimental animals, the current view based on more accurate safety evaluation downgrades its risk to humans. Thujone-containing essential oils and extracts of *Artemisia*, *Salvia*, and *Thuja* species are used for medical purposes due to their antimicrobial, spasmolytic, anti-inflammatory, and hepatoprotective properties. Furthermore, thujone is a constituent of some food additives used as flavor enhancers (Pelkonen et al. 2013).

The review of the literature data concerning genotoxic/antigenotoxic effect of terpenoid compounds indicates that they could induce genotoxic and mutagenic effects, but numerous studies also indicate their protective capacity against endogenous and environmentally induced genotoxicity (Berić et al. 2008; Bugarin et al. 2014; Di Sotto et al. 2011; Kocaman et al. 2011; Mimica-Dukić et al. 2010; Mitić-Ćulafić et al. 2009; Nikolić et al. 2012; Stajković et al. 2007). Similarly, camphor, eucalyptol, and thujone have been reported to possess both genotoxic and antigenotoxic properties, depending on the cell type, genetic background, experimental setup, and concentrations applied (Mitić-Ćulafić et al. 2009; Nikolić et al. 2011a, b; Vuković-Gačić et al. 2006) and could be considered as phytochemicals with modulating effect on DNA repair.

Regulation of DNA Repair in Response to Genotoxic Stress

DNA is constantly exposed to numerous endogenous and exogenous factors that disturb its structure; DNA lesions that are unrepaired or repaired incorrectly may result in chromosomal changes, gene mutations, cancer promotion and progression, or cell death. DNA damage is associated with numerous diseases, including cancer, cardiovascular and neurodegenerative disorders, immune deficiency, and infertility (Ciccia and Elledge 2010). In order to neutralize detrimental effects of genotoxic factors, and especially to counteract disastrous consequences of genotoxic insult, cells have evolved DNA repair mechanisms that remove or tolerate DNA lesions: reversal of a lesion; base excision repair (BER); nucleotide excision repair (NER); mismatch repair (MMR); double-strand break (DSB) repair, which in eukaryotic cells may proceed by homologous recombination (HR) or nonhomologous end joining (NHEJ); and translesion synthesis (TLS) (Sancar et al. 2004). Multiple and complex DNA repair mechanisms form large networks safeguarding the genomic stability. Taking into account the complexity of DNA repair pathways, as well as the fact that most of them involve nucleases, which by themselves represent a threat to the genome integrity, it is clear that DNA repair network has to be tightly regulated and appropriately activated in the case of genotoxic insults.

Transcriptional Regulation of DNA Repair Genes

DNA repair in eukaryotes is controlled at multiple levels. In addition to regulation by transcription factors, epigenetic changes in promoter regions and posttranslational modifications of histones are very important for DNA repair regulation (Christmann and Kaina 2013). More than 130 different proteins involved in DNA repair have been identified so far in mammals, and among them, 25 have been demonstrated to be inducible by DNA damage (Christmann and Kaina 2013). Their transcriptional upregulation is primarily based on the activities of phosphatidylinositol-3-kinases ATM (*ataxia-telangiectasia mutated*) and ATR (*ATM and Rad3 related*), as well as PARP1 (*poly[ADP-ribose] polymerase 1*). They are the most important sensors of DNA damage that signalize the enzymatic machinery involved in the repair of lesions. While ATM and PARP1 are responsible for the recognition of double-strand breaks (DSB) and single-strand breaks (SSB), respectively, ATR recognizes replication blocking lesions (Ciccia and Elledge 2010; Ko and Ren 2012). ATM and ATR proteins activated at DNA lesions become phosphorylated and subsequently phosphorylate and thus activate other proteins involved in DNA repair, cell cycle control, apoptosis, and autophagy. Proteins activated by ATM and ATR include transcription factors responsible for enhancing DNA repair, such as p53, BRCA1, NF-κB, and AP-1. According to available literature data, p53 and AP-1 together regulate activity of 19 DNA repair genes and thereby appear to have the most important roles in their transcriptional activation. Furthermore, p53 and BRCA1 are also involved in chromatin remodeling, acting at the level of epigenetic regulation of DNA repair (Christmann and Kaina 2013).

Epigenetic Modifications and DNA Repair

Epigenetic changes primarily involve modifications of 5′-carbon of cytosine in CpG islands of DNA in promoter regions, posttranslational modifications of histones, and deposition of certain histone variants in specific regions of DNA. The fact that DNA repair genes are epigenetically silenced in numerous cancer cell lines clearly indicates the influence of epigenetic changes on functionality of DNA repair pathways (Lahtz and Pfeifer 2011). In addition, numerous literature data show that DNA repair is not involved only in protection of DNA from genotoxic insults but also contributes to epigenetic gene regulation.

DNA methylation is associated with transcriptional silencing and several proposed mechanisms, all of them involving DNA repair, and may account for DNA demethylation and consequent transcriptional activation. The anticipated mechanisms appear to use DNA repair-like processes to remove the 5-methylcytosine by BER, or a stretch of nucleotides by NER, rather than the methyl group directly. Members of Gadd45 protein family are considered to be key regulators that recruit BER and NER factors to gene-specific loci and promote the process of active DNA demethylation (Niehrs and Schäfer 2012).

DNA methylation is often preceded by modifications of histones and chromatin structure and this process also involves Gadd45. Together with the histone chaperone nucleophosmin, Gadd45 is involved in chromatin decondensation necessary to enhance the accessibility to DNA demethylation complexes and subsequent transcription activation (Carrier et al. 1999). In addition, Smith et al. (2000) showed that Gadd45 binds to UV-damaged chromatin and facilitates access of NER enzymes to DNA lesion, thus participating in the coupling between chromatin assembly and DNA repair.

Taking into account the link between epigenetic modifications and DNA repair phytochemicals with potential to interfere with DNA repair could be considered as potential modulators of epigenetic status. In previous studies, camphor, eucalyptol, and thujone have been shown to modulate DNA repair and mutagenesis in both prokaryotic and mammalian cells (Nikolić et al. 2011b, 2015). The underlying molecular mechanism(s) and perspectives of monoterpenes to affect epigenetic changes will be discussed.

Mutagenic Versus Antimutagenic Effect of Camphor, Eucalyptol, and Thujone in Bacteria

Mutagenic/antimutagenic potential of monoterpenes was studied in *E. coli* K12 reversion assay performed on isogenic strains differing in DNA repair capacities (Simić et al. 1998). Antimutagenic effect against UV and UV-mimetic chemical agent 4NQO was examined in repair-proficient and nucleotide excision repair (NER)-deficient strain by monitoring *argE3* → Arg^+ reversions. Application of low doses of camphor, eucalyptol, and thujone before or following mutagen treatment induced antimutagenic effect in repair-proficient bacteria (Fig. 1).

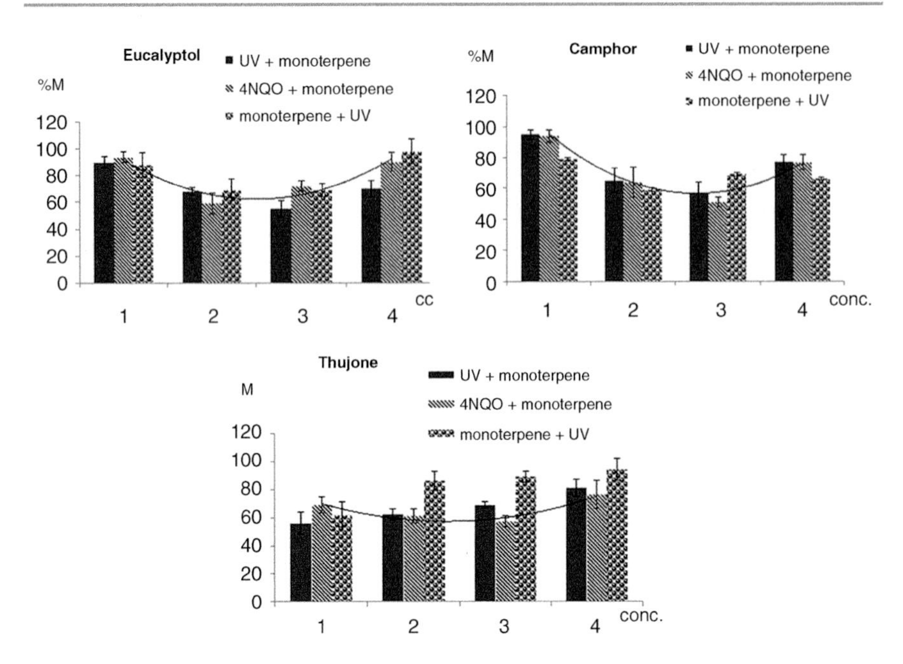

Fig. 1 Antimutagenic effect of monoterpenes in repair-proficient *E. coli* K12 strain. Results are presented as % of mutagenesis (%M) relative to control. Monoterpenes were applied in concentrations up to 100 μM for camphor and up to 200 μM for eucalyptol and thujone

Antimutagenic response was lost with nucleotide excision repair (NER) deficiency, indicating participation of NER pathway in antimutagenic effect.

In spite of being antimutagenic, low doses of monoterpenes increased SOS induction and/or homologous recombination (HR) following UV, indicating their genotoxicity. In repair-proficient and NER-deficient strains treated with camphor and eucalyptol, the level of UV-induced β-galactosidase from SOS controlled *sfiA*::*lacZ* fusion was higher and/or persisted longer, indicating additional DNA damage. Thujone decreased the level of UV-induced β-galactosidase but also significantly reduced general protein synthesis, as evidenced by decreased amounts of alkaline phosphatase constitutively expressed in tester strains (Nikolić et al. 2011b). Moreover, monoterpenes stimulated UV-induced intrachromosomal recombination in *recA*$^{+}$ strain and *recA730* mutant constitutively expressing high level of RecA protein (Fig. 2). In the absence of mutagen treatment, camphor and eucalyptol induced recombination in both strains, while thujone was effective only in *recA730*, which again could be attributed to its inhibitory effects on general protein synthesis.

Several lines of evidence further supported genotoxicity of monoterpenes: (1) decreased antimutagenic effect with increased concentrations (U-shaped concentration-response curves, Fig. 1), (2) co-mutagenic response with UV in NER-deficient cells treated with low doses of monoterpenes (data not shown), (3) mutagenic response to high doses in NER and mismatch repair (MMR)-deficient

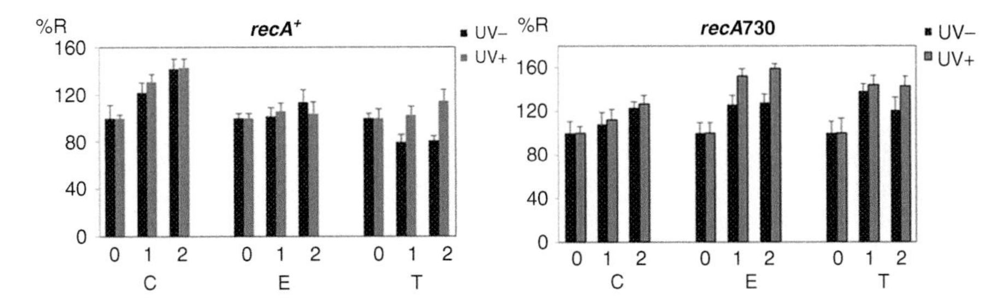

Fig. 2 Recombinogenic and co-recombinogenic effect of monoterpenes in *recA*$^{+}$ and *recA730* *E. coli* K12 cells. Results are presented as % of recombination (%R). C, camphor; E, eucalyptol; T, thujone. Concentrations: 2 and 10 mM for camphor and 2 and 4 mM for eucalyptol and thujone

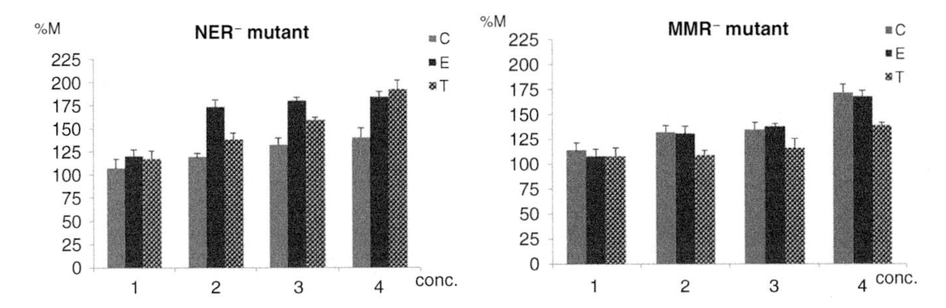

Fig. 3 Mutagenic effect of monoterpenes in NER-deficient (**a**) and MMR-deficient (**b**) strain of *E. coli* K12. Results are presented as % of mutagenesis (%M) relative to control. C, camphor; E, eucalyptol; T, thujone. Monoterpenes were applied in concentrations up to 3 mM for camphor and eucalyptol and 1.2 mM for thujone

strains (Fig. 3), and (4) decreased survival of *uvrA*, *mutS*, and *recB* mutants in the presence of high doses of monoterpenes (Nikolić et al. 2015).

Obtained results clearly demonstrated genotoxic properties of camphor, eucalyptol, and thujone and the involvement of error-free NER, MMR, and HR in the repair of lesions induced by monoterpenes. They also indicated that monoterpenes, although genotoxic, could protect bacteria from UV- and 4NQO-induced mutations if applied at low concentrations.

Genotoxic Versus Antigenotoxic Effect of Camphor, Eucalyptol, and Thujone in Mammalian Cells

Genotoxic/antigenotoxic effects of different doses of monoterpenes were confirmed in eukaryotic models using alkaline comet assay on Vero cells treated with 4NQO. When camphor, eucalyptol, and thujone were applied before (pretreatment) or following 4NQO (posttreatment), significant reduction of DNA single-strand breaks

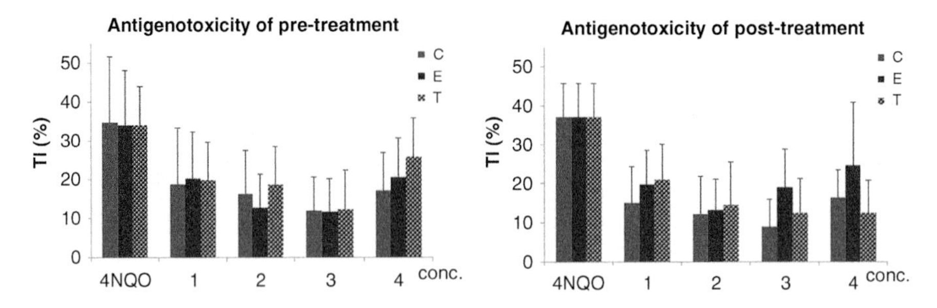

Fig. 4 Antigenotoxic effect of monoterpenes against 4NQO in Vero cell line. Results are presented as % of DNA in comet tail (tail intensity, TI). C, camphor; E, eucalyptol; T, thujone. Monoterpenes were applied in concentrations 0.5–100 μM for camphor and thujone and 0.05–10 μM for eucalyptol

was detected. As in bacteria, attenuation of protective potential with increased doses of monoterpenes was noticed (Fig. 4).

In contrast, when Vero cells were treated with high doses of camphor, eucalyptol, and thujone, genotoxic response was obtained. Comparison with three human cell lines: normal fetal lung fibroblasts (MRC-5) and colorectal cancer cell lines HCT116 and HT-29 demonstrated that the capacity of monoterpenes to induce DNA lesions depended on cell line, the most sensitive being HCT116 (Fig. 5).

Furthermore, evaluation of cytotoxic (Fig. 6) and antiproliferative (Fig. 7) effects of monoterpenes against HCT116 and MRC-5 cells also indicated higher susceptibility of cancer cell line. HCT116 cells are characterized by the lack of hMLH1 protein and consequent MMR deficiency and by mutations in hRAD50 and hMRE11, both being involved in the repair of DSB by HR and NHEJ (Koh et al. 2005).

Similarly as in prokaryotic cells, obtained high vulnerability of cells deficient in MMR and HR pointed at genotoxic potential of camphor, eucalyptol, and thujone, as well as at the involvement of the abovementioned repair mechanisms in the repair of DNA lesions induced by monoterpenes.

Anticipated Mechanisms of Adaptive DNA Repair Response Induced by Camphor, Eucalyptol, and Thujone in Bacteria and Mammalian Cells

Taken together, the results were consistent with hormesis phenomenon, defined as a low-dose beneficial and a high-dose harmful response to a stressor agent. Hormesis is nowadays accepted as a real biological phenomenon, involved in different adaptive responses (Calabrese 2010). Literature data indicate that numerous health beneficial effects of different phytochemicals, including flavonoids, stilbenes, organosulfur compounds, and diferuloyl methanes, are obtained only if they were applied in low concentrations, indicating that hormetic pathways are involved (Surh 2011). In genotoxicology, hormesis is defined as an adaptive response of cells

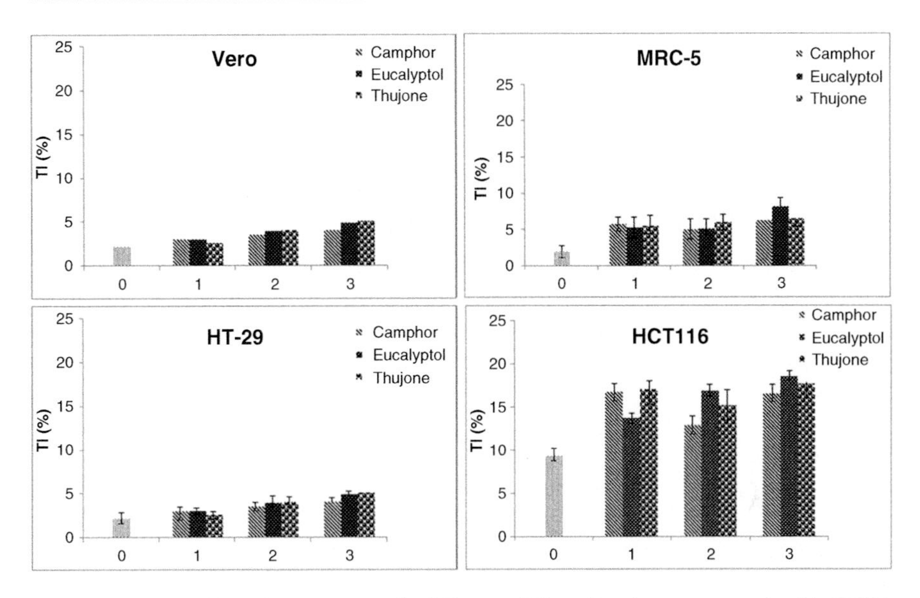

Fig. 5 Genotoxic effect of monoterpenes in different cell lines. Results are presented as % of DNA in comet tail (tail intensity, TI). Concentration range of monoterpenes: Vero, 5–500 mM; MRC-5 and HCT116, camphor and eucalyptol 5–500 mM and thujone 5–250 mM; HT-29, camphor and eucalyptol 50–1000 mM and thujone 5–500 mM

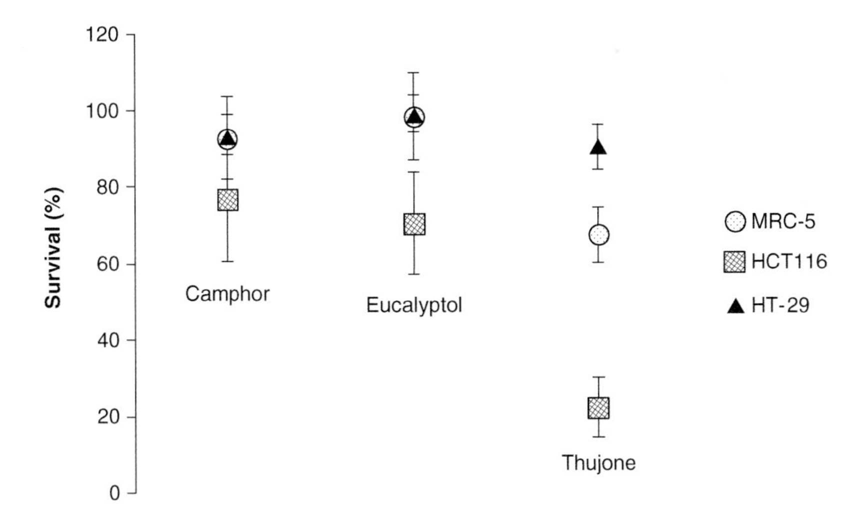

Fig. 6 Cytotoxic effects of monoterpenes in human cell lines. The presented results show survival of cells treated for 24 h with 1 mM of monoterpenes. Cytotoxicity was monitored in MTT assay

exposed to low doses of genotoxin, which leads to their enhanced protection against subsequent higher doses of the same agent. In a broader sense, the genotoxic adaptive response is characterized by protection of pretreated cells against a wider

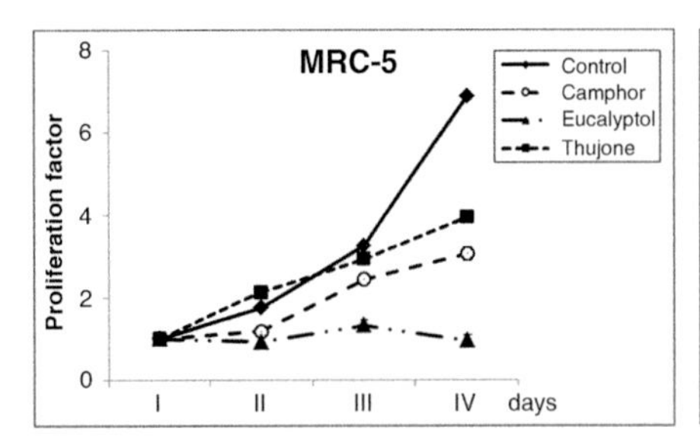

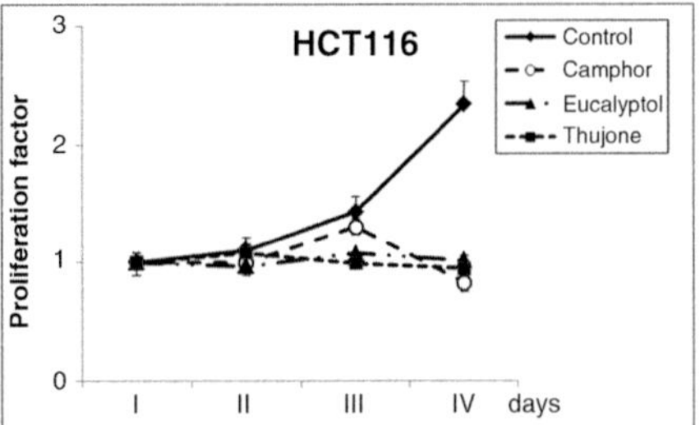

Fig. 7 Antiproliferative effect of monoterpenes in human cell lines. Cells were incubated for 96 h with 2.5 mM of camphor and eucalyptol and 0.05 mM of thujone. Proliferation factor was calculated relative to cell viability at the beginning of incubation

range of genotoxicants (Christmann and Kaina 2013), which is observed in described experiments. It is proposed that, by making small amounts of DNA lesions, low doses of camphor, eucalyptol, and thujone can evoke error-free DNA repair pathways, mainly NER, and incite adaptive response which provides protection against more potent genotoxic agents, such as UV and 4NQO (Nikolić et al. 2011b, 2015).

NER is the major pathway involved in error-free repair of pyrimidine dimers and 4NQO DNA adducts. In *E. coli*, efficiency of DNA repair following UV, 4NQO, and many carcinogens is enhanced due to induction of so-called SOS response. During SOS response, various DNA repair genes become simultaneously upregulated, including *uvrA*, *uvrB*, and *uvrD* participating in NER (Truglio et al. 2006). Since camphor and eucalyptol enhanced SOS induction following UV, increased efficiency of NER could explain their antimutagenic effect. On the other hand, monoterpenes might reduce the rate of DNA replication and delay cell division, giving more time for NER to repair DNA lesions in an error-free manner. This could mainly account for antimutagenic effect of thujone, since it significantly inhibited protein synthesis and severely reduced growth rate of *E. coli* (Nikolić et al. 2011b).

In mammalian cells, camphor, eucalyptol, and thujone also induced genotoxic adaptive response that gave protection against 4NQO. Analysis of underlying mechanisms, using literature data, directs us to transcriptional upregulation of DNA repair genes but also to epigenetic histone modifications. Out of more than hundred DNA repair proteins identified so far, 25 are upregulated following genotoxic stress. Transcription factors AP-1 and p53 regulate nine (*APEX1*, *ERCC1*, *MGMT*, *MSH2*, *NEIL1*, *PCNA*, *TREX1*, *XPF*, and *XPG*) ten DNA repair genes (*DDB2*, *FEN1*, *KARP1*, *MGMT*, *MLH1*, *MSH2*, *PCNA*, *PMS2*, *POLH*, and *XPC*), respectively (Christmann and Kaina 2013). Adaptive responses resulting from upregulation of DNA repair functions have been observed in normal and cancer human cells following UV light (Ye et al. 1999), ionizing radiation (Wolff et al. 1988), and many mutagens and cytostatic drugs (Schlade-Bartusiak et al. 2002).

For UV irradiation, an adaptive response which resulted from amelioration of NER has been demonstrated in human fibroblasts (Ye et al. 1999). Following UV exposure, six members of NER were induced, *DDB2*, *XPC*, *ERCC1*, *XPF*, *XPG*, and

LIG1, leading to enhanced DNA repair and an adaptive response following repeated treatments (Tomicic et al. 2011). Involvement of NER in adaptive response provoked by monoterpenes in *E. coli* is supported by results presented above, but whether camphor, eucalyptol, and thujone upregulate NER genes in Vero and human cells remains to be determined. If so, it could be promising for protection against numerous environmental and synthetic carcinogens inducing DNA damage repairable by NER.

In addition, histone modifications are reported to promote DNA repair by increasing accessibility to lesions and enabling recruitment of factors involved in DNA repair pathways (Cao et al. 2016). Histone modifications, including phosphorylation, methylation, acetylation, and ubiquitination, change the positive charge of the amino acid residues and thereby regulate the tightness of DNA-histone interaction, but also serve as binding sites for the specific proteins (Bannister and Kouzarides 2011). Although DNA damage response has been studied for years, the timing and exact order of particular histone modifications occurring during DNA damage response are still not completely elucidated. Moreover, the exact role of particular histone modification is difficult to define, because one histone modification can influence another (Lee et al. 2010). Nevertheless, histone modifications are considered as promising targets for new chemotherapeutics, and inhibitors of histone deacetylases are already used in cancer treatment (Camphausen and Tofilon 2007).

The important histone modification involved in DNA damage response is phosphorylation of the variant of H2A histone denoted as H2AX. Phosphorylation at serine 139 site (S139) in humans, referred as γH2AX, occurs immediately after exposure to ionizing radiation via ATR or ATM kinase. Phosphorylation of H2AX plays a role in both NHEJ and HR repair pathways, the choice being dependent on additional modifications of other histone proteins (Escargueil et al. 2008). The γH2AX is required to concentrate chromatin remodeling complexes and to promote histone acetylation in order to make the chromatin more accessible for DNA repair proteins. After the repair process is completed, dephosphorylation of H2AX is important for efficient recovery from the DNA damage checkpoint (Douglas et al. 2010).

Other histones, such as H2B at position serine 14 (Fernandez-Capetillo et al. 2004) and H4 at position N-terminal serine 1, are also subjected to phosphorylation in response to DNA damage; the latter is important to promote repair of double-strand breaks by NHEJ and to stabilize nucleosomes during chromatin restoration (Utley et al. 2005). On the other hand, upon DNA damage, H3 is dephosphorylated (Tjeertes et al. 2009).

The second most common histone modification connected with response to DNA damage is methylation, mainly on H3 and H4 histones. It is catalyzed by histone methyltransferases, while histone demethylases are involved in the reverse process. Dynamic profiles of methylation status are determined by methylation of lysines (K) at different positions: H3K4, H3K9, H3K27, H3K36, H3K79, and H4K20. H3K9 methylation is critical for genome stability, its variant H3K9me3 is an essential histone marker of heterochromatin. The level of H3K9me3 nearby DSB is at first upregulated and that is required for γH2AX accumulation and localization at the early phase of DNA damage response (Sasaki et al. 2014). However, since chromatin in the region of lesion should be opened in order to increase the accessibility to repair proteins, appropriate demethylases are recruited and demethylation of H3K9me3 quickly occurs. H3K9me3 is then replaced with H3K9me2 at the damaged DNA, since H3K9me2 is required to retain proteins involved in HR (BARD1 and BRCA1) near the lesion (Wu et al. 2015). Another methylation that is induced in response to DSB and is necessary for recruiting repair factors is H3K36me2, which enhances the activity of NHEJ (Fnu et al. 2011).

Additional histone methylation implicated in facilitated localization of p53-binding protein 1 (53BP1) at the DSB is H4K20 methylation. The binding of 53BP1 at the lesion is indispensable, since it is a key player in the repair of DSB, but is also crucial to recruit other signaling and repair proteins and to promote synapsis of distal DNA ends during NHEJ (Panier and Boulton 2014). On the other hand, H3K79 which is constitutively methylated is important in DNA damage response, by promoting NER (Tatum and Li 2011) and facilitating HR (Conde et al. 2009).

Acetylation of histones neutralizes the positively charged lysine residues and weakens their interaction with DNA; this facilitates chromatin decondensation and thus increases the accessibility to nucleosomal DNA. As mentioned above, γH2AX drives up histone acetyltransferases to the damaged DNA in order to acetylate histones, relax chromatin, and enable access of repair factors. While H4K16 and H3K56 acetylation is important for the repair of DNA strand breaks (Hunt et al. 2013), H3K9 acetylation recruits the NER enzymes to the lesion (Guo et al. 2011). Reverse process of histone deacetylation is also involved in DNA repair and cell cycle progression. This is evidenced by the fact that for 53BP1 binding to the damaged site, both methylation of H4K20 and deacetylation of H4K16 are required (Hunt et al. 2013).

Ubiquitination is a process of addition of small, 76-residue polypeptide ubiquitin to the target proteins. It requires successive actions of three enzymes: activating E1, conjugating E2, and ligating E3, finally resulting in covalently conjugated ubiquitin to the ε-amino group of lysine residue. Ubiquitination regulates many cellular processes by degradation of targeted proteins, but in the case of histones, it functions in the reorganization of chromatin, increasing its accessibility to a number of regulatory factors (Bennett and Harper 2008).

The first ubiquitination event upon ionizing or UV irradiation is RNF8-mediated ubiquitination of H2AX at K63 site, which recruits another E3 ligase, RNF168, to mediate ubiquitination of H2AX at K15 site. The ubK15 is then directly recognized by 53BP1 (Fradet-Turcotte et al. 2013). Ubiquitination events at positions K27 and K119/K120 are also involved in activation of DNA damage response, by recruiting 53BP1, BRCA1, or other repair factors. However, histone ubiquitination is a dynamic process, finely tuned during the cellular response to DNA damage. Among other factors, deubiquitination enzymes can control the rate of H2AX ubiquitination at the damaged site, determining which DNA repair pathway will be selected (Bennett and Harper 2008).

Histone modifications are reported to facilitate DNA repair following UV irradiation by promoting different stages of NER. UV-specific modifications involved in mammalian NER include acetylation of H2A, H2B, H3, and H4, methylation of H3 (S10, T11) and H4 (K20), and ubiquitination of H2A (K118, K119), H2B, H3, and H4. They are implicated in chromatin relaxation prior to lesion detection, recruitment of lesion recognition factors, repair synthesis, and chromatin restoration after NER (for review, see Li 2012). Although we are not aware of any data concerning epigenetic modifications induced by camphor, eucalyptol, and thujone, it was recently shown that *Thymus serpyllum* extract, containing camphor and eucalyptol, interferes with epigenetic events by inhibition of DNA methyltransferase and histone deacetylase in human breast cancer cells (Bozkurt et al. 2012).

Conclusion

Adaptive response provoked by natural compounds, such as camphor, eucalyptol, and thujone, could be extremely important for both cancer prevention and therapy. Upregulation of DNA repair increases overall repair capacity, but the consequences observed in normal and cancer cells are quite different – while increased DNA repair ameliorates protection against genotoxic agents in normal cells, in cancer cells, it could promote acquired drug resistance. Another important point is that adaptive response is initiated in a narrow range of doses: extremely low doses might not be high enough to activate DNA damage response and enhance repair capacity, while doses above upper limits could inhibit transcription counteracting upregulation of DNA repair. These extremely important points need to be carefully examined in further study of camphor, eucalyptol, and thujone as possible modulators of DNA repair in humans.

Dictionary of Terms

- **Antimutagen** – The agent that can reduce the frequency of mutations by inactivating mutagens or modulating DNA repair.
- **DNA repair** – Set of metabolic processes safeguarding cellular genome by recognizing and correcting different lesions on DNA.
- **Genotoxicity** – The ability of physical or chemical agent to induce DNA damage that may lead to genetic alterations (mutations).
- **Histones** – The alkaline proteins that package DNA of eukaryotic cells into chromatin.
- **Hormesis** – A term describing biphasic dose-response of cell or organism to a toxic agent that leads to a low-dose beneficial and a high-dose harmful effect.
- **Monoterpenes** – Linear or cyclic natural substances consisting of two isoprene units that are synthesized mainly by plants to serve as signal molecules; however, they are endowed with numerous biological activities.
- **Mutagenicity** – The capacity of genotoxic agent to cause mutations which may lead to cancer.

- **NER (nucleotide excision repair)** – Error-free DNA repair process that excises a stretch of nucleotides containing lesion from damaged DNA strand and fills the gap using undamaged complementary strand as a template.

Key Facts

- Eukaryotic chromatin is organized into repeating units of nucleosomes that are connected by linker DNA.
- Histone octamers composed of two of each H2A, H2B, H3, and H4 histones are core components of nucleosomes, while histone H1 binds to the linker DNA.
- In order to allow DNA repair, chromatin must be remodeled by epigenetic modifications of histones, such as phosphorylation, methylation, acetylation, and ubiquitination.
- Genotoxin-induced adaptive response refers to a situation when cells pretreated with low doses of genotoxic agent show enhanced protection against subsequent exposure to higher doses of the same agent or, in a broader sense, to a range of genotoxic agents including environmental mutagens and carcinogens.
- Genotoxin-induced adaptive response is based on amelioration of DNA repair processes.
- Amelioration of DNA repair involves enhanced expression of DNA repair genes and histone modifications that lead to improved accessibility of repair proteins to DNA lesions.
- Adaptive response could be provoked by nutraceuticals and other factors.
- While adaptive response is beneficial for normal cells and protects them from genotoxic insults, it can lead to enhanced resistance of cancer cells to cytostatic drugs.

Summary Points

- Effect of monoterpenes camphor, eucalyptol, and thujone on DNA damage and mutagenesis in bacteria and mammalian cells depends on applied concentration and genetic background.
- Low doses of monoterpenes are antimutagenic against UV and 4NQO in repair-proficient *E. coli* strain, while in NER-deficient strain, are co-mutagenic.
- High doses of monoterpenes are mutagenic in NER- and MMR-deficient *E. coli* strains.
- In *E. coli* cells, monoterpenes increase SOS response following UV and enhance recombination.
- The survival of repair-deficient *E. coli* mutants is lower compared to repair-proficient strain in the presence of high doses of monoterpenes.
- Low doses of monoterpenes decrease genotoxic effect of 4NQO in Vero cells.
- High doses of monoterpenes are genotoxic in normal and cancer cell lines.

- Genotoxic, cytotoxic, and antiproliferative effects of monoterpenes are prominently higher in HCT116 cell line characterized by the lack of hMLH1 and mutations in hRAD50 and hMRE11.
- Opposite effect of low and high doses of monoterpenes in both bacteria and mammalian cells is consistent with hormesis phenomenon and indicates genotoxin-induced adaptive response.
- Adaptive response is provoked by small amounts of DNA lesions induced by low doses of monoterpenes which evoke error-free DNA repair pathways, mainly NER, and provide protection against more potent genotoxic agents, such as UV and 4NQO.
- Adaptive response in *E. coli* is mediated by enhanced efficiency of NER during SOS induction.
- Adaptive response in mammalian cells may involve enhanced efficiency of NER by transcriptional upregulation of inducible NER genes and epigenetic modifications of histones.
- Taking into account that numerous genotoxic agents induce DNA lesions repairable by NER, adaptive response provoked by camphor, eucalyptol, and thujone could be important for protection against environmental mutagens and carcinogens.

Acknowledgments This work was supported by the Ministry of Education, Science and Technological Development of Republic of Serbia, Project No. 172058

References

Aggarwal R, Jha M, Shrivastava A et al (2015) Natural compounds: role in reversal of epigenetic changes. Biochem Mosc 80:972–989

Bakkali F, Averbeck S, Averbeck D et al (2008) Biological effects of essential oils – a review. Food Chem Toxicol 46:446–475

Bannister AJ, Kouzarides T (2011) Regulation of chromatin by histone modifications. Cell Res 21:381–395

Bennett EJ, Harper JW (2008) DNA damage: ubiquitin marks the spot. Nat Struct Mol Biol 15:20–22

Berić T, Nikolić B, Stanojević J et al (2008) Protective effect of basil (*Ocimum basilicum* L.) against oxidative DNA damage and mutagenesis. Food Chem Toxicol 46:724–732

Bozkurt E, Atmaca H, Kisim A et al (2012) Effects of *Thymus serpyllum* extract on cell proliferation, apoptosis and epigenetic events in human breast cancer cells. Nutr Cancer 64:1245–1250

Bugarin D, Grbović S, Orčič D et al (2014) Essential oil of *Eucalyptus gunnii* hook. As a novel source of antioxidant, antimutagenic and antibacterial agents. Molecules 19:19007–19020

Busch C, Burkard M, Leischner C et al (2015) Epigenetic activities of flavonoids in the prevention and treatment of cancer. Clin Epigenetics 7:64. https://doi.org/10.1186/s13148-015-0095-z

Calabrese EJ (2010) Hormesis is central to toxicology, pharmacology and risk assessment. Hum Exp Toxicol 29:249e261

Camphausen K, Tofilon PJ (2007) Inhibition of histone deacetylation: a strategy for tumor radiosensitization. J Clin Oncol 25:4051–4056

Cao LL, Shen C, Zhu WG (2016) Histone modifications in DNA damage response. Sci China Life Sci 59:257–270

Carrier F, Georgel PT, Pourquier P et al (1999) Gadd45, a p53-responsive stress protein, modifies DNA accessibility on damaged chromatin. Mol Cell Biol 19:1673–1685

Christmann M, Kaina B (2013) Transcriptional regulation of human DNA repair genes following genotoxic stress: trigger mechanisms, inducible responses and genotoxic adaptation. Nucleic Acids Res 41:8403–8420

Ciccia A, Elledge SJ (2010) The DNA damage response: making it safe to play with knives. Mol Cell 40:179–204

Conde F, Refolio E, Cordon-Preciado V et al (2009) The Dot1 histone methyltransferase and the Rad9 checkpoint adaptor contribute to cohesin-dependent double-strand break repair by sister chromatid recombination in *Saccharomyces cerevisiae*. Genetics 182:437–446

Di Sotto A, Mazzanti G, Carbone F et al (2011) Genotoxicity of lavender oil, linalyl acetate, and linalool on human lymphocytes *in vitro*. Environ Mol Mutagen 52:69–71

Douglas P, Zhong J, Ye R et al (2010) Protein phosphatase 6 interacts with the DNA-dependent protein kinase catalytic subunit and dephosphorylates gamma-H2AX. Mol Cell Biol 30:1368–1381

Escargueil AE, Soares DG, Salvador M et al (2008) What histone code for DNA repair? Mutat Res 658:259–270

Fernandez-Capetillo O, Allis CD, Nussenzweig A (2004) Phosphorylation of histone H2B at DNA double-strand breaks. J Exp Med 199:1671–1677

Fnu S, Williamson EA, De Haro LP et al (2011) Methylation of histone H3 lysine 36 enhances DNA repair by nonhomologous end-joining. P Natal Acad Sci USA 108:540–545

Fradet-Turcotte A, Canny MD, Escribano-Diaz C et al (2013) 53BP1 is a reader of the DNA-damage-induced H2A Lys 15 ubiquitin mark. Nature 499:50–54

Guo R, Chen J, Mitchell DL et al (2011) GCN5 and E2F1 stimulate nucleotide excision repair by promoting H3K9 acetylation at sites of damage. Nucleic Acids Res 39:1390–1397

Hunt CR, Ramnarain D, Horikoshi N et al (2013) Histone modifications and DNA double-strand break repair after exposure to ionizing radiations. Radiat Res 179:383–392

Ko HL, Ren EC (2012) Functional aspects of PARP1 in DNA repair and transcription. Biomol Ther 2:524–548

Kocaman AY, Rencüzoğullari E, Topaktaş M et al (2011) The effects of 4-thujanol on chromosome aberrations, sister chromatid exchanges and micronucleus in human peripheral blood lymphocytes. Cytotechnology 63:493–502

Koh KH, Kang HJ, Li LS et al (2005) Impaired nonhomologous end-joining in mismatch repair-deficient colon carcinomas. Lab Investig 85:1130–1138

Lahtz C, Pfeifer GP (2011) Epigenetic changes of DNA repair genes in cancer. J Mol Cell Biol 3:51–58

Lee JS, Smith E, Shilatifard A (2010) The language of histone crosstalk. Cell 142:682–685

Li S (2012) Implication of posttranslational histone modifications in nucleotide excision repair. J Mol Sci 13:12461–12486

Mimica-Dukić N, Bugarin D, Grbović S et al (2010) Essential oil of *Myrtus communis* L. as a potential antioxidant and antimutagenic agents. Molecules 15:2759–2770

Mitić-Ćulafić D, Žegura B, Nikolić B et al (2009) Protective effect of linalool, myrcene and eucalyptol against *t*-butyl hydroperoxide induced genotoxicity in bacteria and cultured human cells. Food Chem Tox 47:260–266

Niehrs C, Schäfer A (2012) Active DNA demethylation by Gadd45 and DNA repair. Trends Cell Biol 22:220–227

Nikolić B, Jovanović B, Mitić-Ćulafić D et al (2015) Comparative study of genotoxic, anti-genotoxic and cytotoxic activities of monoterpenes camphor, eucalyptol and thujone in bacteria and mammalian cells. Chem Biol Interact 242:263–271

Nikolić B, Mitić-Ćulafić D, Stajković-Srbinović O et al (2012) Effect of metabolic transformation of monoterpenes on antimutagenic potential in bacterial tests. Arch Biol Sci 64:885–894

Nikolić B, Mitić-Ćulafić D, Vuković-Gačić B et al (2011a) The antimutagenic effect of monoterpenes against UV-irradiation-, 4NQO- and *t*-BOOH-induced mutagenesis in *E. coli*. Arch BiolSci 63:117–128

Nikolić B, Mitić-Ćulafić D, Vuković-Gačić B et al (2011b) Modulation of genotoxicity and DNA repair by plant monoterpenes camphor, eucalyptol and thujone in *Escherichia coli* and mammalian cells. Food Chem Toxicol 49:2035–2045

Panier S, Boulton SJ (2014) Double-strand break repair: 53BP1 comes into focus. Nat Rev Mol Cell Bio 15:7–18

Pelkonen O, Abass K, Wiesner J (2013) Thujone and thujone-containing herbal medicinal and botanical products: toxicological assessment. Regul Toxicol Pharmacol 65:100–107

Sancar A, Lindsey-Boltz LA, Unsal-Kacmaz K et al (2004) Molecular mechanisms of mammalian DNA repair and the DNA damage checkpoints. Annu Rev Biochem 73:39–85

Sasaki T, Lynch KL, Mueller CV et al (2014) Heterochromatin controls gammaH2A localization in *Neurospora crassa*. Eukaryot Cell 13:990–1000

Schlade-Bartusiak K, Stembalska-Kozlowska A, Bernady M et al (2002) Analysis of adaptive response to bleomycin and mitomycin C. Mutat Res 513:75–81

Simić D, Vuković-Gačić B, Knežević-Vukčević J (1998) Detection of natural bioantimutagens and their mechanisms of action with bacterial assay-system. Mutat Res 402:51–57

Smith ML, Ford JM, Hollander MC et al (2000) p53-mediated DNA repair responses to UV radiation: studies of mouse cells lacking p53, p21, and/or gadd45 genes. Mol Cell Biol 20:3705–3714

Stajković O, Berić-Bjedov T, Mitić-Ćulafić D et al (2007) Antimutagenic properties of basil (*Ocimum basilicum* L.) in *Salmonella typhimurium* TA100. Food Technol Biotehnol 45:213–217

Surh Y-J (2011) Xenohormesis mechanisms underlying chemopreventive effects of some dietary phytochemicals. Ann N Y Acad Sci 1229:1–6

Tatum D, Li S (2011) Evidence that the histone methyltransferase Dot1 mediates global genomic repair by methylating histone H3 on lysine 79. J Biol Chem 286:17530–17535

Tjeertes JV, Mille KM, Jackson SP (2009) Screen for DNA-damage-responsive histone modifications identifies H3K9Ac and H3K56Ac in human cells. EMBO J 28:1878–1889

Tomicic MT, Reischmann P, Rasenberger B et al (2011) Delayed c-Fos activation in human cells triggers XPF induction and an adaptive response to UVC-induced DNA damage and cytotoxicity. Cell Mol Life Sci 68:1785–1798

Truglio JJ, Croteau DL, VanHouten B et al (2006) Prokaryotic nucleotide excision repair: the UvrABC system. Chem Rev 106:233–252

Utley RT, Lacoste N, Jobin-Robitaille O et al (2005) Regulation of NuA4 histone acetyltransferase activity in transcription and DNA repair by phosphorylation of histone H4. Mol Cell Biol 25:8179–8190

Vuković-Gačić B, Nikčević S, Berić-Bjedov T et al (2006) Antimutagenic effect of essential oil of sage (*Salvia officinalis* L.) and its monoterpenes against UV-induced mutations in *Escherichia coli* and *Saccharomyces cerevisiae*. Food Chem Toxicol 44:1730–1738

Wolff S, Afzal V, Wiencke JK et al (1988) Human lymphocytes exposed to low doses of ionizing radiations become refractory to high doses of radiation as well as to chemical mutagens that induce double-strand breaks in DNA. Int J Radiat Biol Relat Stud Phys Chem Med 53:39–47

Wu W, Nishikawa H, Fukuda T et al (2015) Interaction of BARD1 and HP1 is required for BRCA1 retention at sites of DNA damage. Cancer Res 75:1311–1321

Ye N, Bianchi MS, Bianchi NO et al (1999) Adaptive enhancement and kinetics of nucleotide excision repair in humans. Mutat Res 435:43–61

Modulatory Role of Curcumin in miR-Mediated Regulation in Cancer and Non-cancer Diseases

106

Sayantani Chowdhury, Jyotirmoy Ghosh, and Parames C. Sil

Contents

Abstract

The dietary polyphenol curcumin imparts its pharmacological effects through anticancer, anti-inflammatory, antioxidant, and other mechanisms by inhibiting/modulating aberrant signaling molecules and transcription factors. MicroRNAs (miRs) modulate gene expression by regulating the degradation or translation repression of mRNA. The plethora of evidences over the past few years reflect the disruption of several fundamental regulatory mechanisms, such as carcinogenesis, cell proliferation, differentiation, programed cell death, angiogenesis, migration, invasion, etc., concerning miRs. Curcumin-mediated epigenetic alterations are the regulation of the expression of various pathogenic miRs in liver fibrosis, neurodegenerative diseases, diabetic nephropathy, ocular diseases, etc., on one

S. Chowdhury · P. C. Sil (✉)
Division of Molecular Medicine, Bose Institute, Kolkata, India
e-mail: sayantani@jcbose.ac.in; parames@jcbose.ac.in

J. Ghosh
Department of Chemistry, Banwarilal Bhalotia College, Ushagram Asansol, West Bengal, India
e-mail: bichemjyoti@gmail.com

V. B. Patel, V. R. Preedy (eds.), *Handbook of Nutrition, Diet, and Epigenetics*,
https://doi.org/10.1007/978-3-319-55530-0_64

hand and modulation of several tumor suppressor and oncogenic and epithelial-mesenchymal transition-suppressor microRNAs on the other hand. Based on recent evidences, miRs from miR-21, miR-26, miR-27, miR-28, miR-143, miR-199, miR-200 family, the let-7 family, etc., contribute to anomalies in both cancer and non-cancer diseases through aberrant signaling, tumor formation, and chemoresistance. In context to the significant role of miR homeostasis we summarize, in this book chapter, the findings based on in vitro and in vivo evidences on the regulatory role of curcumin on miR expression involved in cancer and non-cancer diseases.

Keywords
Curcumin · Epigenetics · miR · Cancer · Non-cancer diseases

List of Abbreviations

Akt/mTOR signaling	Protein kinase B (PKB), also known as Akt, is a serine-/threonine-specific protein kinase/mechanistic target of rapamycin signaling
AOF1/2	Amine oxidase domain-containing protein 1/2
ARPE-19 cells	Human retinal pigment epithelial cell line
BAG2	BAG family molecular chaperone regulator 2
BMI1	B lymphoma Mo-MLV insertion region 1 homolog
CAM	Chick chorioallantoic membrane
CDKN1A	Cyclin-dependent kinase inhibitor 1A
COL1A1	Collagen type I alpha 1 chain
COX2	Cyclooxygenase-2
CML xenograft	Chronic myeloid leukemia xenograft
DNMT3b	DNA (cytosine-5-)-methyltransferase 3 beta
Dkk-3/SMAD4	DKK3; dickkopf WNT signaling pathway inhibitor 3/SMAD family member n°4
2D–DIGE	Two-dimensional difference gel electrophoresis
DNMT1	DNA (cytosine-5)-methyltransferase 1
DU145 cell	Human Caucasian prostate cell; derived from metastatic site: brain
EMT-suppressive miRs	Epithelial-mesenchymal transition-suppressive micro-RNAs
EZH2	Enhancer of zeste 2 polycomb repressive complex 2 subunit
Erbb3	V-erb-b2 avian erythroblastic leukemia viral oncogene homolog 3
HCT116 cell	Colon cancer cell line
HCT116-5FUR cell	5-fluorouracil-resistant cellosaurus cell line
HL-60 cell	Human promyelocytic leukemia cell
HAG cells	Human astroglial cells

HSC cells	Hematopoietic stem cells
HNG cells	Human neuronal glial cells
HUVECs	Human umbilical vein endothelial cells
IL-1β	Interleukin-1 beta
K562 cell	Chronic myelogenous leukemia cell
LAMA84 cell	Human leukocytic cell line
Let-7 family miRs	Lethal-7 family microRNAs
mRNA	Messenger RNA
MMP13	Matrix metallopeptidase 13
MAPK signaling	Mitogen-activated protein kinase signaling
MMP-9	Matrix metallopeptidase 9
mmu-miR	Mus musculus microRNA
NF-κB	Nuclear factor kappa-light-chain-enhancer of activated B cells
NKAP	NF-kappa-B-activating protein
Oct4	Octamer-binding transcription factor 4
PTEN	Phosphatase and tensin homolog
PTP1B protein	*Protein*-tyrosine phosphatase 1B protein
PDGFβ	Platelet-derived growth factor beta
PKCs	Protein kinase, catalytic subunit
PDCD4	Programmed Cell Death 4 (Neoplastic Transformation Inhibitor)
PRCs	Polycomb repressive complexes
PI3K	Phosphoinositide 3-kinase
PC cells	Prostate cancer cell
Rko cell	Rectal carcinoma cell line
SCID mice	Severe combined immunodeficient mice
Sox2	SRY (sex determining region Y)-box 2
Sema6a protein	Semaphorin 6A protein
STAT-3	Signal transducer and activator of transcription 3
SUZ12	SUZ12 Polycomb repressive complex 2 subunit
SW480-5FUR cell	5-fluorouracil-resistant cellosaurus cell line
Sp proteins	Surfactant proteins
TIMP-1	Tissue inhibitors of metalloproteinases
TGIF	TG-interacting factor
TGF-β	Transforming growth factor beta
Trps1	Transcriptional repressor GATA binding 1
VEGF	Vascular endothelial growth factor
VEGFR2	Vascular endothelial growth factor receptor 2
VEGFB gene	Vascular endothelial growth factor B gene
WT1 gene	Wilms tumor 1 gene
ZBTB10	Zinc finger and BTB domain-containing 10
ZBTB4	Zinc finger and BTB domain-containing 4
Y79 RB cell	Human Caucasian retinoblastoma cell line

Introduction

Curcumin, a polyphenolic compound, is the predominant curcuminoid present in *Curcuma longa* (turmeric) (Momtazi et al. 2016b). Although the main focus of curcumin research is its anticancer property, but in addition to that curcumin possesses a multitude of its biological function both in vitro and in vivo. Among its numerous medicinal and pharmacological properties, some important biological functions exhibited by curcumin are anti-inflammatory (Momtazi et al. 2016b), lipid-lowering (Mohammadi et al. 2013), immunomodulatory (Seyedzadeh et al. 2014), antioxidant (Chen et al. 2015), neuroprotective, antiarthritic, antidepressant, analgesic, hypoglycemic, anti-ischemic, anti-atherosclerotic, wound healing, and antimicrobial effects (Hasan et al. 2014; Reddy et al. 2005; Sharma et al. 2006; Sidhu et al. 1998). The reason behind this versatile biological application is its interaction with numerous molecular targets, several growth factors, growth receptors, nuclear factors, transcription factors, hormones, and its receptors. Recent research has suggested epigenetic modulators such as microRNAs (miRs) as emerging targets, and it can modulate the pharmacological activities of curcumin in several cancerous and non-cancer diseases (Momtazi et al. 2016a,b).

Extensive researches have revealed growth-inhibiting property of curcumin in different cancer cells by targeting multiple genes and cell-signaling pathways at varied levels, viz., nuclear factors, growth factors, transcription factors, gene expression, and hormone receptors along with several other potent regulators such as miR and various other epigenetic mechanisms (Momtazi et al. 2016a,b). It is well cited in cancer biology that curcumin modulates the expression of gene through the regulation of epigenetic events. Curcumin selectively activates or inactivates genes implicated in cancer mainly through three epigenetic mechanisms. These mechanisms include DNA methylation, histone modifications, and miRs (Sawan et al. 2008). miRs are usually considered as endogenous, single-stranded, noncoding 22-nucleotide RNAs that control and regulate gene expression through posttranscriptional regulation (Chen et al. 2012). miRs regulate the expression of gene at mRNA translation and degradation level. The mature miR on binding perfectly to the 30- untranslated region of mRNA induces targeted cleavage of mRNA, whereas miR on binding imperfectly to mRNA results in translational repression (Momtazi et al. 2016a; Williams and Spencer 2012) (Fig. 1). In this scenario curcumin can mitigate various types of disease through several mechanisms, including modulation of inflammation, apoptosis, oxidative stress, and angiogenesis. It is strongly believed that a complex network of epigenetic events regulate all of these processes wherein miRs play a key role. Although most epigenetic studies (especially miR studies) have been conducted on cancer, few studies regarding noncancerous diseases are also reported in the literature. Recent investigation (Momtazi et al. 2016b) suggests that curcumin has the capacity to modulate the expression of several pathogenic miRs in noncancerous diseases. In this book chapter, we have discussed the new findings regarding to this emerging effect of curcumin.

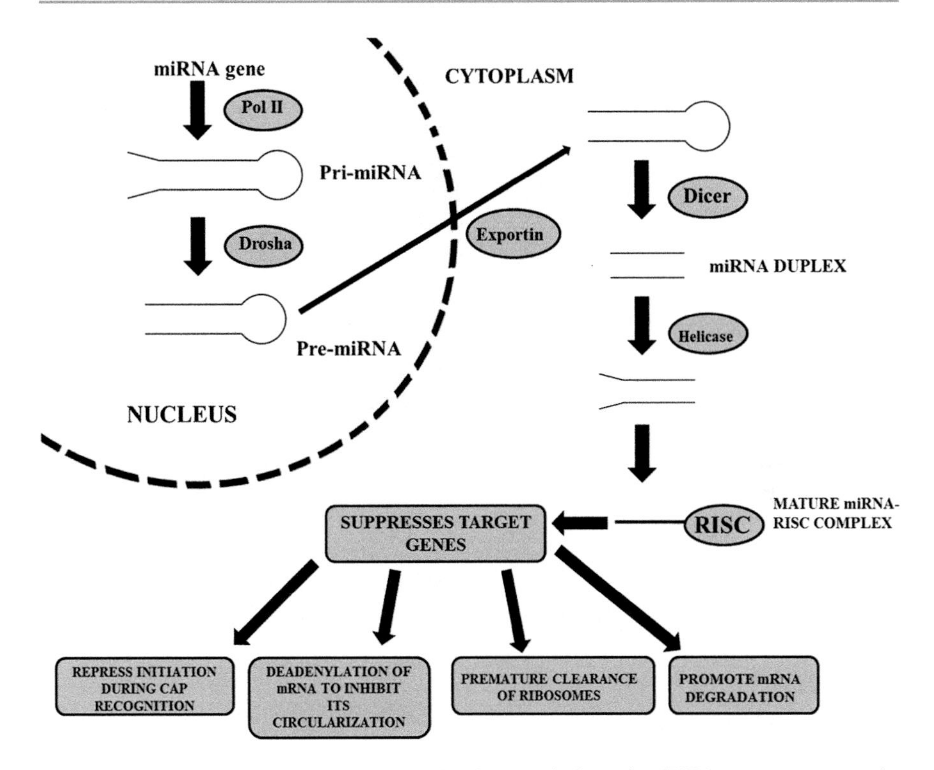

Fig. 1 Biogenesis of miRNA. Pol II-mediated transcription of miRNA gene generates the pri-miRNA which, in turn, gets trimmed to pre-miRNA by Drosha. This pre-miRNA gets translocated from the nucleus to the cytoplasm and is converted to the mature duplex by Dicer. The single-stranded miRNA obtained gets complexed with RISC and suppresses target genes by various mechanisms

Cancer and the Effect of Curcumin on miRs Expression

Over the decades studies have shown the role of curcumin in tumor progression by suppressing/modulating different target molecules like AP-1, STAT-3, NF-κB, several growth factors, enzymes, cytokines, etc., by binding to proteins, viz., PKC, COX-2, thioredoxin reductase, etc., interfering with the Akt/mTOR signaling. Curcumin, an anti-tumorigenic molecule, possesses chemotherapeutic properties like suppressing cell proliferation, inflammation, angiogenesis, and invasion in cancer, inducing apoptosis, and has been included in clinical trials (Khor et al. 2006; Kunnumakkara et al. 2008). Evidence suggests the potential capability of curcumin in regulating different miRs (Yang et al. 2010; Zhang et al. 2010). In this section, we summarize the modulatory role of curcumin on miR expression and the associated mechanisms in cancer (Fig. 2).

Colorectal Cancer Resistance to chemotherapy results in difficulty to obtain a cure for this life-threatening colorectal malignancy. In colon cancer cells, curcumin

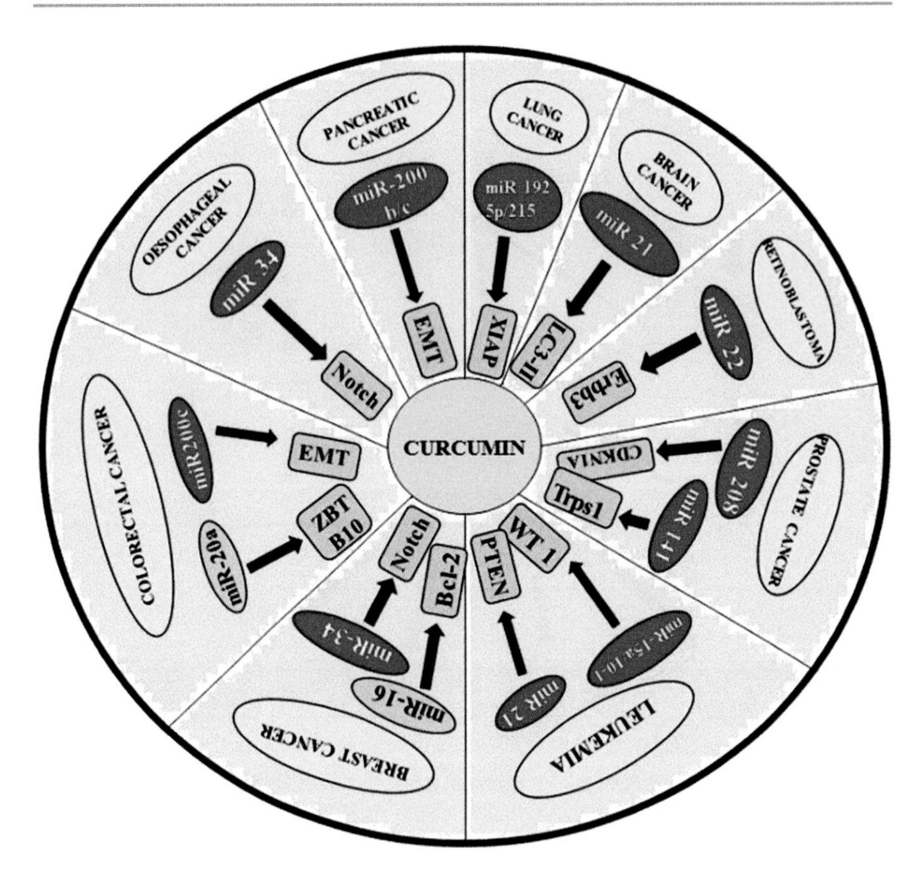

Fig. 2 The role of different miRNAs in targeting specific genes in various types of cancer. A pictorial depiction of protective role of curcumin by regulating different signaling pathways against various types of cancer. "*Red*" *tablets* indicate overexpression, whereas "*green*" indicates under-expression of miRNAs

modulates miRs targeting tumor suppressor genes and transcription factors in colon cancer cells. miR-429, miR-21, miR-17-5p, miR-101, miR-34a/c, miR-20a, miR-200b/c, and miR-141 are involved in curcumin-administered colon cancer treatment. A finding by Mudduluru et al. showed that curcumin inhibited proliferation and triggered apoptosis dose dependently in HCT116 and Rko human colon cancer cells and CAM of chicken embryo by significantly suppressing the expression of miR-21. The group suggested that curcumin-mediated tumor and invasion inhibition, migration, and metastasis (in vivo) might be due to suppression resulted from binding of AP-1 to its promoter inhibition and PDCD4 (a tumor suppressor gene and a target of miR-21) induction (Mudduluru et al. 2011). In colon cancer PRCs, viz., BMI1, EZH2, SUZ12, play a crucial role in silencing of epigenetic events and is the target of EMT-suppressive miRs such as miR-34a, miR-200b, miR-200c, miR-101, miR-141, and miR-429. In HCT116-5FUR cell, following curcumin treatment, the expression of miR-34a was found to increase but not in

SW480-5FUR cells. From this observation it was concluded that miR-34a is upregulated through p53 activation because SW480-5FUR is p53 mutated. The expressions of previously mentioned EMT-suppressive miRs apart from miR-34a were upregulated in both the cells. Similarly, in a mouse xenograft tumor model, miR-200c, an EMT-suppressive miR, was upregulated and resulted in a reduction in the tumor growth following curcumin treatment (Toden et al. 2015). Curcumin repressed the growth of SW480 and Rko colon cancer cells through ROS induction and repression of transcription factors Sp proteins. The proteins are overexpressed in tumor and regulate genes involved in angiogenesis and cell death. Gandhy et al. observed that curcumin suppressed miR-27a and miR-20a and miR-17-5p expressions which in turn inhibited the expression of respective transcriptional Sp repressors, i.e., ZBTB10 and ZBTB4, and effectively interfered with cell survival, proliferation, metabolism, drug transport, and inflammation (Gandhy et al. 2012).

Leukemia WT1 gene encodes a DNA-binding transcription factor responsible for embryonic development and is overexpressed in chronic myelogeous and acute human leukemia. Gao et al. identified miR-mediated molecular mechanism which is associated with the downregulation of WT1 following curcumin treatment. The study reported upregulation of miR-15a/10-1 and downregulation of WT1 expression in curcumin-treated primary AML cells and K562 and HL-60 leukemic cells (Gao et al. 2012). MiR-15a/16-1 overexpression decreased WT1 expression in leukemic cells, but siRNA-WT1-induced downregulation of WT1 failed to increase miR-15a/16-1 expression, thus indicating that curcumin-induced miR-15a/16-1 overexpression is an upstream event to WT1 downregulation. MiR-21 targets several tumor suppressor genes, viz., PTEN, which modulates VEGF-triggered angiogenesis by PI3K/AKT downregulation in solid tumors (Meng et al. 2007). Following investigation on a CML xenograft in SCID mice, Taverna et al. suggested a selective packaging of miR-21 in exosome vesicles that are released from cancer cells containing gene expression regulating miRs. Curcumin imparted its anticancer effects and inhibited growth in both in vivo xenograft CML tumor model and in vitro CML cells by eliminating miR-21 via exosomes. Curcumin treatment decreased miR-21 in CML cells, but the same was increased in the exosomes released by CML cells, K562, and LAMA84. The expression of PTEN, a target of miR-21, on the other hand, was upregulated which in turn modulated AKT phosphorylation. The inverse relation between miR-21 and PTEN supported the possible effect of curcumin on regulation of PTEN expression in CML exosomes via selective packaging of miR-21 (Taverna et al. 2015).

Pancreatic Cancer Due to its high morbidity and mortality rate, early diagnosis of pancreatic cancer (one of the most aggressive malignancies) and development of new preventive and therapeutic agents are highly warranted. Conventional chemotherapeutic agents produce deleterious side effects that augment the mortality and morbidity. In this regard, specific aberrations in the expression of miRs have been identified in pancreatic cancer. The use of miRs (e.g., miR-16, miR-21, miR-22, miR-155, miR-181a, miR-181b, miR-196a, miR-200, and miR-210 family) as potential therapeutic targets and early biomarkers for pancreatic cancer detection has been reported (Liu et al. 2012). Curcumin decreases the expression of miR-21, and as a result tumor suppressor PTEN expression increases in pancreatic cancer cell lines. In the mesenchymal phenotype of gemcitabine-resistant pancreatic cancer cells, curcumin upregulates both miR-200b and miR-200c expressions. It suggests that the reversal of epithelial-mesenchymal transition plays a role after treatment with either of these biomolecules. Curcumin regulates multiple other miRs such as miR-200, miR-34, miR-19, etc. (Ma et al. 2014).

Lung Cancer Lung cancer is one of the major causes of cancer-related deaths worldwide, and its incidence is continually rising. Non-small cell lung cancer accounts for approximately 80% of lung cancers. Zhang et al. showed a novel molecular mechanism in A549 cells in which miR-21 can suppress the anticancer effect of curcumin by upregulation of PTEN gene. So, in this context, suppression of miR-21 may have therapeutic benefits and opens up a new pathway against this malignancy (Zhang et al. 2014). A recent report by Ye et al. suggests that curcumin treatment on non-small cell lung cancer upregulates anti-oncogenic miR-192-5p/215 via activation of p53. This observation further highlights the epigenetic activity of curcumin in cancer treatment. In this regard they also showed a lack of functioning of p53-miR-192-5p/215-XIAP pathway in p53-deficient and missense mutant cancer cells. Among many other possible targets, curcumin can preferentially target multiple cancer-related signaling, and it is very likely that curcumin promotes apoptosis independent of that pathway in p53−/− colon cancer cells (Ye et al. 2015).

Brain Cancer Yeh et al. studied the role of curcumin in migration-prone glioma cells. They observed that this cell line is resistant to curcumin and this resistance is associated with enhanced expression of miR-21 and invasion/antiapoptosis-related proteins. miR-21 was also associated with more advanced clinical pathological stages in the patient tissue specimens. In U251 cells, treatment with curcumin decreased the miR-21 level and antiapoptotic protein expression and increased the expression of proapoptotic proteins and microtubule-associated protein light chain 3-II (LC3-II) in U251 cells (Yeh et al. 2015). Mirgani et al. found that dendrosomal curcumin nanoformulation can suppress pluripotent genes via miR-145 activation in U87MG glioblastoma cells. Dendrosomal curcumin treatment significantly reduced OCT4B1, OCT4A, Nanog and SOX-2 gene expression, and overexpression of miR-145 as the upstream regulator. Combining it can be proposed that dendrosomal curcumin reduces the proliferation of U87MG cells through the downregulation of SOX-2 and OCT4 variants in a miR-145-regulated manner (Tahmasebi Mirgani et al. 2014).

Breast Cancer and Esophageal Cancer Yang et al. observed that both miR-15a and miR-16 expressions were upregulated and that of Bcl-2 was downregulated in curcumin-administered MCF-7 cells. Silencing miR-15a and miR-16 by specific inhibitors restored Bcl-2 expression. The regulation of Bcl-2 expression is controlled by many other regulating factors in addition to the miR-15a and miR-16 expression, but these two miRs can regulate the expression of Bcl-2 in MCF-7 cells. Therefore, they concluded that curcumin can increase the expression of miR-15a and miR-16 and that the miR-15a/16 family may induce apoptosis by downregulating Bcl-2. In this context, in Bcl-2-overexpressing tumors, both miR-15a and miR-16 can probably be regarded as important gene therapy targets (Yang et al. 2010). Another report by Kronski et al. suggested that curcumin affects miR expression in primary tumors inducing miR181b expression, which translates into a pro-inflammatory cytokine CXCL1 downregulation and subsequent loss of metastatic potential (Kronski et al. 2014).

Subramaniam et al. checked the effect of curcumin on esophageal cancer cells and observed that the mechanism is mediated through Notch signaling pathway (Subramaniam et al. 2012). miR-34 modulates Notch pathway and p53 resulting interference with tumor suppression. In addition, a recent report suggests a cross talk between miR and Notch signaling pathways in tumor development and progression (Wang et al. 2010).

Prostate Cancer Recent studies have shown that curcumin effectively blocks prostate carcinoma, one of the present cancers often seen in aged man, via miR-208-triggered activation of CDKN1A. This inhibition of prostate cell proliferation was brought about dose dependently. The study suggested that curcumin regulated the translation of cell cycle suppressor CDKN1A. Moreover, miR-208 was inhibited by curcumin dose dependently in PC cells whereas, in PC cells where miR-208 was overexpressed, failed to decrease cell proliferation following curcumin administration (Guo et al. 2015). In silico analyses highlighted the possible involvement of miR-141, an androgen-mediated miR which interferes with apoptosis in prostate cancer and its modulation by curcumin. miR-141 is overexpressed in cancer via Trps1 which is involved in cellular stress and cellular apoptosis in DU145 cell. Overexpression of miR-183 is observed in prostate cancer, and Dkk-3/SMAD4 has been reported to be the former's potential target. MiR-151 has been identified as a tumor-suppressive miR through 2D-DIGE and modulates DNMT1 (Ueno et al. 2013).

Retinoblastoma Retinoblastoma, a form of intraocular malignant tumor, is involved in the downregulation of miR-22. Recent evidences suggested the possible role of curcumin in modulating miR expressions in tumorigenesis. In human RB cells, miR-22 targets Erbb3 (a molecule which is associated with differentiation, proliferation, cell survival, and mortality). In 2012, a study suggested the anticancer effect of curcumin by evaluating its role on the expression profile of miRs through qRT-PCR in Y79 RB cells (Sreenivasan et al. 2012). Of the upregulated tumor suppressor miRs (miR-22, miR-200c, miR-503, let-7 g*) and downregulated oncomiRs (miR-25*, miR-34c-3p, miR-135b, miR-210, miR-514, miR-95, mir-106*, miR-92a-1*), miR-22 upregulation resulted in decreased expression of its target protein Erbb3, following curcumin treatment. Cells when transfected with miR-22 were found to reduce the migration and inhibit cell proliferation (Momtazi et al. 2016b; Sreenivasan et al. 2012).

Noncancerous Diseases and the Effect of Curcumin on miR Expression

Apart from the involvement of miRs in cancer (Lu et al. 2005), literature reveals the possible role of miRs as biomarkers for diagnosis and prognosis of non-cancer diseases such as liver fibrosis, immune-related diseases, viral diseases, ischemic heart diseases, diabetic nephropathy, polycystic kidney, hepatitis, hematological disorder, psoriasis, etc. (Li and Kowdley 2012; Sayed and Abdellatif 2011). It has been reported that curcumin combats diseases by modulating oxidative stress, inflammation, angiogenesis, and apoptosis. Most of these pathophysiological states are modulated through a complex network of epigenetic events where miRs are known to impart a key role in controlling the expression of gene either by bringing in translational repression or binding to target genes for degradation. Here, in this column, we summarize the evidences suggesting the emerging capacity of curcumin to modulate/regulate several pathogenic miRs expression in non-cancerous diseases (Fig. 3).

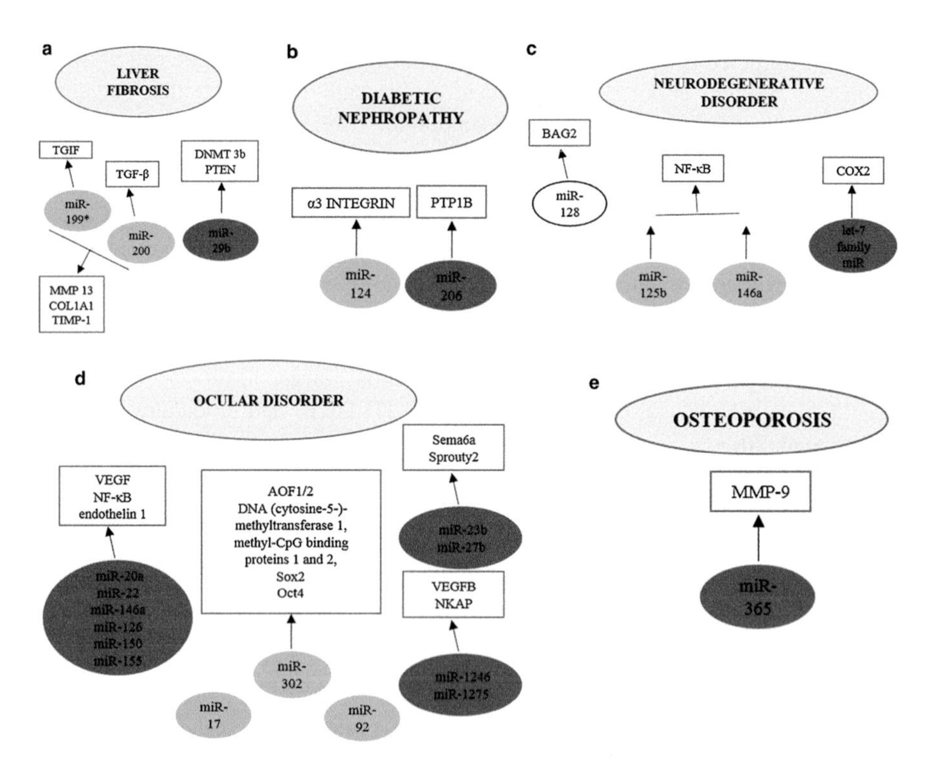

Fig. 3 Ameliorative role of curcumin in modulating miRNAs mediated targeting specific genes in non-cancer diseases: (**a**) liver fibrosis; (**b**) diabetic nephropathy; (**c**) neurodegenerative diseases; (**d**) ocular disorders; (**e**) osteoporosis. A pictorial depiction of curcumin regulating miRNAs induced target specific genes and thereby combating against non-cancer diseases. "*Red*" indicates upregulated miRNAs, whereas "*green*" indicates downregulated miRNAs

Liver Fibrosis Tissue fibrosis is associated with the dysfunction of hepatic system. In this context, literature provides evidences depicting the important changes in the expression of miR, which in concert with various other epigenetic factors influences fibrosis. In 2012, Hassan et al. have shown that curcumin plays an important role in reorganizing miR expression during liver fibrosis (Hassan and Al-Olayan 2012). MiRs, especially miR-199, miR-200*, miR-200a/b, and miR-29 family, are considered as novel biomarkers of liver fibrosis and have the potential to regulate the same (Hassan and Al-Olayan 2012; Murakami et al. 2011). Moreover, TGIF-mediated TGF-β signaling pathways are targeted by miR-199* and miR-200, respectively (Yang and Mahato 2011). Administration of curcumin (5 mg kg^{-1} body wt., i.p., daily for 4 weeks) in CCl_4-induced liver fibrosis in mice significantly downregulated miR-199, miR-199*, miR-200a, and miR-200b expression which were profoundly upregulated in liver fibrotic tissues. Targeting these miRs could significantly downregulate the altered fibrosis-related genes, viz., MMP13, COL1A1, and TIMP-1, as determined by q-PCR (Hassan and Al-Olayan 2012). Zheng et al. showed that in HSC cells curcumin significantly upregulated miR-29b expression resulting in the silencing of DNMT3b which is essential for de novo methylation and hypomethylation of PTEN which is important in the development of liver fibrosis. Thus miR-29b contributes to the suppression of fibrosis (Zheng et al. 2014).

Diabetic Nephropathy Diabetic nephropathy results in glomerular capillary pressure which in turn results in stress on podocytes, the key cells involved in ultrafiltration that surrounds the glomerulus capillaries. Evidence suggests the involvement of miRs in regulating cell adhesion molecules, viz., miR-124, targets integrin α3 which in turn induces damage to podocyte adhesion capacity by downregulating the expression of the same (Moini Zanjani et al. 2014). In this context, a study showed that curcumin could downregulate the altered expression of miR-124 and thus restore the adhesion of podocytes against stress in diabetic rats (Li et al. 2013). Ding et al. showed that curcumin increased miR-206 expression which targets PTP1B protein, a negative regulator of insulin signaling pathway both in vivo and in vitro, thereby exerting protection against fructose-mediated podocyte insulin signaling impairment (Ding et al. 2015).

Neurodegenerative Diseases Endogenous proteins, viz., BAG2, are capable of clearing off tau protein, a target for Alzheimer's disease treatment, from neurons. Evidence showed that curcumin upregulated BAG2 expression, a validated target of miR-128, and hence decreased the expression of phosphorylated tau in primary rat cortical neuronal cell but miR-128 itself showed no significant change in its expression, thus making mechanism of BAG2 upregulation obscure (Patil et al. 2013). Two different studies have shown that curcumin effectively suppressed the expression of miR-125b and/or miR-146a by inhibiting NF-κB in metal sulfate-mediated HAG cells and IL-1β-induced HNG cells (Li et al. 2011). In contrast, Angel-Morales et al. observed inflammation-related miRs, such as miR-126 and miR-146a, had no anti-inflammatory effect on endothelial cells treated with curcumin in respect to inhibition of NF-κB-targeted genes (Angel-Morales et al. 2012). Zaky et al. showed that co-treatment of curcumin with valproic upregulated the suppressed expression of let-7 family miRs (a, b, c, e, and f) during LPS-mediated neuroinflammation, whereas COX2 gene, which is both a pro-inflammatory mediator and a target for let-7 family, was found to be downregulated (Zaky et al. 2014). All these studies cumulatively showed that curcumin imparts neuroprotection by modulating miR-125b, miR-146a, and let-7 family miRs which alter molecules associated with oxidative stress and inflammation.

Ocular Disorders RPE is a primary target in oxidative stress-mediated age-related macular degeneration. Howell et al. investigated the effect of curcumin (known to exert antioxidant effect by modulating miRs in oxidative stress in ocular disorder) in respect to regulating the expression of miRs in H_2O_2-induced ARPE-19 cells (Howell et al. 2013). They showed that curcumin downregulated miR-302 which has been reported to interfere with global DNA demethylation by inhibiting epigenetic regulators, AOF1/2, DNA (cytosine-5-)-methyltransferase 1, and methyl-CpG-binding proteins 1 and 2, and subsequently transcription factors like Sox2, Oct4, etc., are activated. Curcumin-modulated miR-17 and miR-92, both of which when overexpressed following stress, lead to retinoblastoma development (Conkrite et al. 2011). The expressions of miRs (miR-22, miR-20a, miR-146a, miR-150, miR-126, and miR-155) targeting VEGF, NF-κB, endothelin 1, PDGFβ, and other downstream molecules were significantly induced following curcumin administration. Although the exact mechanism through which curcumin modulates the expression of miRs is unclear in this regard, molecular mechanisms, viz., transcriptional processing, posttranscriptional regulation, involvement of epigenetic factors, and genetic abnormalities, may be involved. In response to angiogenic factors, miR-23-27-24 clusters promote choroidal neovascularization and angiogenesis in a murine model by repressing Sema6a and sprouty2 proteins which negatively modulate VEGFR2 and MAPK signaling (Croce 2009). Howell et al. showed that curcumin upregulated the oxidative stress-induced altered expression of miR-23b and miR-27b (Howell et al. 2013). In context to corneal neovascularization-related eye diseases, HUVEC cells are used as in vitro model of pathogenesis for corneal neovascularization since angiogenesis is believed to be the key driving force in several severe disorders. Bai et al. showed that curcumin dramatically regulated the expression profile of miR-1275 and miR-1246 which synergistically played a vital role in the anti-angiogenic property of curcumin by targeting VEGFB and NKAP genes, respectively, the downstream effectors of NF-κB signaling pathway (Bai et al. 2016).

Osteoporosis The wide application of glucocorticoids for inflammatory diseases and immunosuppression following organ transplantation is often associated with the complication, namely, osteoporosis, which results in a decreased bone formation and increased bone fragility and resorption (Yao et al. 2016). Literature suggests the potential antiarthritic and anti-inflammatory effects of curcumin on bone cells. In bone biology, MMP-9, which is highly expressed in osteoclasts, plays a major role in extracellular matrix degradation and hence is regarded as a possible drug target of curcumin (He et al. 2013). Li et al. showed that miR-365, an upstream regulator of

MMP-9, was targeted by curcumin following its supplementation in glucocorticoid-mediated secondary osteoporosis in a murine model and the molecule effectively circumvented bone deterioration at least partially by activating miR-365 via MMP-9 suppression (Li et al. 2015).

Conclusion

Over the past few years, various studies confirmed the pleiotropic effect of the natural bioactive molecule, curcumin, on miRs and their targets in both cancer therapy and other non-cancer diseases. In cancer biology, several of these miRs target tumor suppressor genes or oncogenes that initiate and/or lead to carcinogenesis, whereas curcumin can be used to upregulate tumor-suppressive miRs and downregulate oncogenic miRs to restore drug sensitivity, holds a potential as miR regulator in regard to therapeutic intervention, and can be considered beyond its use just as chemotherapeutic supplements. The regulation of miR by this molecule is disease specific. It can downregulate oncomiR, miR-21, by inhibiting miR-21 gene transcription and increasing exosome exclusion. The progression and metastasis of colon cancer cells can be suppressed by curcumin through miR-20a, miR-27a, and miR-17-7p downregulation and miR-34a, miR-101, miR-141, miR-200b/c, and miR-429 upregulation. Curcumin effectively modulates miR-199a*, miR-7, and miR-22 which in turn regulates pancreatic cancer cells proliferation, apoptosis, and migration. On one hand, overexpressed miR-199a, miR-199a*, and miR-200a/b in liver fibrosis are downregulated by curcumin, while on the other hand miR-146a and miR-125b are modulated by curcumin in Alzheimer's disease. In context to curcumin and neuroinflammation mitigation, let-7 family can be targeted. miR-30 family, especially miR-30e which modulated antioxidant defense, is regulated by curcumin in oxidative stress-induced ocular diseases. miR-124 which regulates cell adhesion molecule and miR-206 which interferes with insulin signaling are regulated by curcumin in diabetic nephropathy together with mmu-miR-374*, mmu-miR-30c-1*, mmu-miR-291b-5p, mmu-miR-497b*, and mmu-miR-296-5p. Varied evidences provide significant elucidation into the interesting changes in miR profile following curcumin treatment. In this regard, extensive and strategic studies are recommended to provide insights on the regulation of curcumin on miR expression in clinical trials. The modulatory role of miRs affecting cancer and several other non-cancer diseases can shed light on curcumin treatment and its potential importance in disease aspect that are linked to miR deregulation.

Dictionary of Terms

- **oncomiR** – microRNA associated with cancer.
- **Epithelial-mesenchymal transition** – Epithelial cells lose cell to cell adherence and polarity and acquire invasive as well as migratory properties to become mesenchymal stem cells.

- **Tumor suppressor** – Antioncogene that regulates cell growth and protects a cell from entering the cancer pathway but when mutated can lead to cancer.
- **Osteoporosis** – A condition accompanied by fragile and brittle bones due to loss of tissue that result due to vitamin D and/or calcium deficiency, hormonal changes, etc.
- **Ocular disorders** – Eye disease and anomalies affecting the visual system, i.e., human eye.

Key Facts of Angiogenesis

- The growth of new capillary blood vessels from existing vasculature is known as angiogenesis which begins in utero and continues through old age, both in health and disease.
- The process plays a pivotal role in reproduction and healing through a precise balance between growth and inhibitory factors.
- Insufficient or excessive growth of blood vessels underlies in several debilitating diseases, viz., cardiovascular disorders, cancer, diabetic ulcers, age-related blindness, stoke, etc.
- Of the two types, the first identified form of angiogenesis was sprouting angiogenesis which is marked by sprouts comprising of endothelial cells which proliferate in response to angiogenic stimuli, viz., VEGF-A.
- Sprouting angiogenesis has the ability to grow blood vessels in regions of tissues devoid of blood vessels previously.
- Intussusceptive angiogenesis comprises of blood vessel formation through splitting mechanism wherein factors of interstitial tissues invade vessels already existing and as a consequence transvascular tissue pillars are formed which grows.
- Angiogenesis is an important step in the transition of benign tumors to malignant form which lead to the implication of angiogenesis inhibitors in treating cancer.
- Judah Folkman, in the year 1971, first proposed the involvement of angiogenesis in tumor growth.

Summary Points

- *Curcuma longa* of Zingiberaceae family has been used as Ayurvedic and traditional Chinese medicine for thousands of years apart from its use as dietary spice and food additive.
- This medicinal plant is cultivated in Indian subcontinent and tropical countries of Southeast Asia.
- The enol form of curcumin (polyphenolic compound extracted from *Curcuma longa*) is stable.
- miRs modulate and/disrupt multiple regulatory mechanisms by interfering with the translation repression or degradation of mRNA.

- MicroRNAs are transcribed by RNA polymerase II as primary transcripts that undergo posttranslational modifications.
- Microprocessor complex that includes cellular RNAse III type endonuclease, Drosha, and DGCR8/Pasha protein crops the primary transcripts to form pre-miRs.
- Ran-GTP-dependent transporter exportin-5 recognizes the 2-nt 3′ overhang of pre-miRs and exports the same into cytoplasm from nucleus.
- Dicer, a RNAIII enzyme along with Argonaute and TAR RNA-binding protein further cleave pre-miR to generate a double-stranded duplex.
- Argonaute then loads the duplex onto miR-associated RNA-induced silencing complex and is delivered to the target mRNA.
- Helicase then unwinds the guiding strand, and the miR is now referred to mature miR.
- Drosha- and DICER-independent pathways as well as shRNA-, tRNA-, and snoRNA-derived pathways are the alternative pathways from miR generation.
- A plethora of evidences have suggested the emerging role of miRs as molecular targets in the therapeutic function of curcumin under several pathophysiological conditions.
- Curcumin-induced epigenetic alterations involve the modulation of several pathogenic miR expressions in non-cancer diseases.
- In context to cancer, curcumin regulates several oncogenic, EMT suppressor, and tumor suppressor miRs.

References

Angel-Morales G, Noratto G, Mertens-Talcott SU (2012) Standardized curcuminoid extract (Curcuma longa l.) decreases gene expression related to inflammation and interacts with associated microRNAs in human umbilical vein endothelial cells (HUVEC). Food Funct 3:1286–1293

Bai Y, Wang W, Sun G, Zhang M, Dong J (2016) Curcumin inhibits angiogenesis by up-regulation of microRNA-1275 and microRNA-1246: a promising therapy for treatment of corneal neovascularization. Cell Prolif:751–762. 49P

Chen B, Li H, Zeng X, Yang P, Liu X, Zhao X, Liang S (2012) Roles of microRNA on cancer cell metabolism. J Transl Med 10:228

Chen FY, Zhou J, Guo N, Ma WG, Huang X, Wang H, Yuan ZY (2015) Curcumin retunes cholesterol transport homeostasis and inflammation response in M1 macrophage to prevent atherosclerosis. Biochem Biophys Res Commun 467:872–878

Conkrite K, Sundby M, Mukai S, Thomson JM, Mu D, Hammond SM, MacPherson D (2011) miR-17~92 cooperates with RB pathway mutations to promote retinoblastoma. Genes Dev 25:1734–1745

Croce CM (2009) Causes and consequences of microRNA dysregulation in cancer. Nat Rev Genet 10:704–714

Ding XQ, Gu TT, Wang W, Song L, Chen TY, Xue QC, Zhou F, Li JM, Kong LD (2015) Curcumin protects against fructose-induced podocyte insulin signaling impairment through upregulation of miR-206. Mol Nutr Food Res 59:2355–2370

Gandhy SU, Kim K, Larsen L, Rosengren RJ, Safe S (2012) Curcumin and synthetic analogs induce reactive oxygen species and decreases specificity protein (Sp) transcription factors by targeting microRNAs. BMC Cancer 12:564

Gao SM, Yang JJ, Chen CQ, Chen JJ, Ye LP, Wang LY, Wu JB, Xing CY, Yu K (2012) Pure curcumin decreases the expression of WT1 by upregulation of miR-15a and miR-16-1 in leukemic cells. J Exp Clin Cancer Res CR 31:27

Guo H, Xu Y, Fu Q (2015) Curcumin inhibits growth of prostate carcinoma via miR-208-mediated CDKN1A activation. Tumour Biol: J Int Soc Oncodev Biol Med 36:8511–8517

Hasan ST, Zingg JM, Kwan P, Noble T, Smith D, Meydani M (2014) Curcumin modulation of high fat diet-induced atherosclerosis and steatohepatosis in LDL receptor deficient mice. Atherosclerosis 232:40–51

Hassan ZK, Al-Olayan EM (2012) Curcumin reorganizes miRNA expression in a mouse model of liver fibrosis. Asian Pac J Cancer Prev 13:5405–5408

He B, Hu M, Li SD, Yang XT, Lu YQ, Liu JX, Chen P, Shen ZQ (2013) Effects of geraniin on osteoclastic bone resorption and matrix metalloproteinase-9 expression. Bioorg Med Chem Lett 23:630–634

Howell JC, Chun E, Farrell AN, Hur EY, Caroti CM, Iuvone PM, Haque R (2013) Global microRNA expression profiling: curcumin (diferuloylmethane) alters oxidative stress-responsive microRNAs in human ARPE-19 cells. Mol Vis 19:544–560

Khor TO, Keum YS, Lin W, Kim JH, Hu R, Shen G, Xu C, Gopalakrishnan A, Reddy B, Zheng X, Conney AH, Kong AN (2006) Combined inhibitory effects of curcumin and phenethyl isothiocyanate on the growth of human PC-3 prostate xenografts in immunodeficient mice. Cancer Res 66:613–621

Kronski E, Fiori ME, Barbieri O, Astigiano S, Mirisola V, Killian PH, Bruno A, Pagani A, Rovera F, Pfeffer U, Sommerhoff CP, Noonan DM, Nerlich AG, Fontana L, Bachmeier BE (2014) miR181b is induced by the chemopreventive polyphenol curcumin and inhibits breast cancer metastasis via down-regulation of the inflammatory cytokines CXCL1 and –2. Mol Oncol 8:581–595

Kunnumakkara AB, Anand P, Aggarwal BB (2008) Curcumin inhibits proliferation, invasion, angiogenesis and metastasis of different cancers through interaction with multiple cell signaling proteins. Cancer Lett 269:199–225

Li D, Lu Z, Jia J, Zheng Z, Lin S (2013) MiR-124 is related to podocytic adhesive capacity damage in STZ-induced uninephrectomized diabetic rats. Kidney Blood Press Res 37:422–431

Li G, Bu J, Zhu Y, Xiao X, Liang Z, Zhang R (2015) Curcumin improves bone microarchitecture in glucocorticoid-induced secondary osteoporosis mice through the activation of microRNA-365 via regulating MMP-9. Int J Clin Exp Pathol 8:15684–15695

Li Y, Kowdley KV (2012) MicroRNAs in common human diseases. Genomics Proteomics Bioinformatics 10:246–253

Li YY, Cui JG, Hill JM, Bhattacharjee S, Zhao Y, Lukiw WJ (2011) Increased expression of miRNA-146a in Alzheimer's disease transgenic mouse models. Neurosci Lett 487:94–98

Liu J, Gao J, Du Y, Li Z, Ren Y, Gu J, Wang X, Gong Y, Wang W, Kong X (2012) Combination of plasma microRNAs with serum CA19-9 for early detection of pancreatic cancer. Int J Cancer 131:683–691

Lu J, Getz G, Miska EA, Alvarez-Saavedra E, Lamb J, Peck D, Sweet-Cordero A, Ebert BL, Mak RH, Ferrando AA, Downing JR, Jacks T, Horvitz HR, Golub TR (2005) MicroRNA expression profiles classify human cancers. Nature 435:834–838

Ma J, Fang B, Zeng F, Pang H, Zhang J, Shi Y, Wu X, Cheng L, Ma C, Xia J, Wang Z (2014) Curcumin inhibits cell growth and invasion through up-regulation of miR-7 in pancreatic cancer cells. Toxicol Lett 231:82–91

Meng F, Henson R, Wehbe-Janek H, Ghoshal K, Jacob ST, Patel T (2007) MicroRNA-21 regulates expression of the PTEN tumor suppressor gene in human hepatocellular cancer. Gastroenterology 133:647–658

Mohammadi A, Sahebkar A, Iranshahi M, Amini M, Khojasteh R, Ghayour-Mobarhan M, Ferns GA (2013) Effects of supplementation with curcuminoids on dyslipidemia in obese patients: a randomized crossover trial. Phytother Res 27:374–379

Moini Zanjani T, Ameli H, Labibi F, Sedaghat K, Sabetkasaei M (2014) The attenuation of pain behavior and serum COX-2 concentration by Curcumin in a rat model of neuropathic pain. Korean J Pain 27:246–252
Momtazi AA, Derosa G, Maffioli P, Banach M, Sahebkar A (2016a) Role of microRNAs in the therapeutic effects of Curcumin in non-cancer diseases. Mol Diagn Ther 20:335–345
Momtazi AA, Shahabipour F, Khatibi S, Johnston TP, Pirro M, Sahebkar A (2016b) Curcumin as a MicroRNA regulator in cancer: a review. Rev Physiol Biochem Pharmacol 171:1–38
Mudduluru G, George-William JN, Muppala S, Asangani IA, Kumarswamy R, Nelson LD, Allgayer H (2011) Curcumin regulates miR-21 expression and inhibits invasion and metastasis in colorectal cancer. Biosci Rep 31:185–197
Murakami Y, Toyoda H, Tanaka M, Kuroda M, Harada Y, Matsuda F, Tajima A, Kosaka N, Ochiya T, Shimotohno K (2011) The progression of liver fibrosis is related with overexpression of the miR-199 and 200 families. PLoS One 6:e16081
Patil SP, Tran N, Geekiyanage H, Liu L, Chan C (2013) Curcumin-induced upregulation of the anti-tau cochaperone BAG2 in primary rat cortical neurons. Neurosci Lett 554:121–125
Reddy RC, Vatsala PG, Keshamouni VG, Padmanaban G, Rangarajan PN (2005) Curcumin for malaria therapy. Biochem Biophys Res Commun 326:472–474
Sawan C, Vaissiere T, Murr R, Herceg Z (2008) Epigenetic drivers and genetic passengers on the road to cancer. Mutat Res 642:1–13
Sayed D, Abdellatif M (2011) MicroRNAs in development and disease. Physiol Rev 91:827–887
Seyedzadeh MH, Safari Z, Zare A, Gholizadeh Navashenaq J, Razavi SA, Kardar GA, Khorramizadeh MR (2014) Study of curcumin immunomodulatory effects on reactive astrocyte cell function. Int Immunopharmacol 22:230–235
Sharma S, Kulkarni SK, Chopra K (2006) Curcumin, the active principle of turmeric (Curcuma longa), ameliorates diabetic nephropathy in rats. Clin Exp Pharmacol Physiol 33:940–945
Sidhu GS, Singh AK, Thaloor D, Banaudha KK, Patnaik GK, Srimal RC, Maheshwari RK (1998) Enhancement of wound healing by curcumin in animals. Wound Repair Regen 6:167–177. Official publication of the Wound Healing Society [and] the European Tissue Repair Society
Sreenivasan S, Thirumalai K, Danda R, Krishnakumar S (2012) Effect of curcumin on miRNA expression in human Y79 retinoblastoma cells. Curr Eye Res 37:421–428
Subramaniam D, Ponnurangam S, Ramamoorthy P, Standing D, Battafarano RJ, Anant S, Sharma P (2012) Curcumin induces cell death in esophageal cancer cells through modulating Notch signaling. PLoS One 7:e30590
Tahmasebi Mirgani M, Isacchi B, Sadeghizadeh M, Marra F, Bilia AR, Mowla SJ, Najafi F, Babaei E (2014) Dendrosomal curcumin nanoformulation downregulates pluripotency genes via miR-145 activation in U87MG glioblastoma cells. Int J Nanomedicine 9:403–417
Taverna S, Giallombardo M, Pucci M, Flugy A, Manno M, Raccosta S, Rolfo C, De Leo G, Alessandro R (2015) Curcumin inhibits in vitro and in vivo chronic myelogenous leukemia cells growth: a possible role for exosomal disposal of miR-21. Oncotarget 6:21918–21933
Toden S, Okugawa Y, Jascur T, Wodarz D, Komarova NL, Buhrmann C, Shakibaei M, Boland CR, Goel A (2015) Curcumin mediates chemosensitization to 5-fluorouracil through miRNA-induced suppression of epithelial-to-mesenchymal transition in chemoresistant colorectal cancer. Carcinogenesis 36:355–367
Ueno K, Hirata H, Shahryari V, Deng G, Tanaka Y, Tabatabai ZL, Hinoda Y, Dahiya R (2013) microRNA-183 is an oncogene targeting Dkk-3 and SMAD4 in prostate cancer. Br J Cancer 108:1659–1667
Wang Z, Li Y, Kong D, Ahmad A, Banerjee S, Sarkar FH (2010) Cross-talk between miRNA and Notch signaling pathways in tumor development and progression. Cancer Lett 292:141–148
Williams RJ, Spencer JP (2012) Flavonoids, cognition, and dementia: actions, mechanisms, and potential therapeutic utility for Alzheimer disease. Free Radic Biol Med 52:35–45
Yang J, Cao Y, Sun J, Zhang Y (2010) Curcumin reduces the expression of Bcl-2 by upregulating miR-15a and miR-16 in MCF-7 cells. Med Oncol (Northwood, London, England) 27:1114–1118

Yang N, Mahato RI (2011) GFAP promoter-driven RNA interference on TGF-beta1 to treat liver fibrosis. Pharm Res 28:752–761

Yao W, Dai W, Jiang L, Lay EY, Zhong Z, Ritchie RO, Li X, Ke H, Lane NE (2016) Sclerostin-antibody treatment of glucocorticoid-induced osteoporosis maintained bone mass and strength. Osteoporos Int 27:283–294. A journal established as result of cooperation between the European Foundation for Osteoporosis and the National Osteoporosis Foundation of the USA

Ye M, Zhang J, Zhang J, Miao Q, Yao L, Zhang J (2015) Curcumin promotes apoptosis by activating the p53-miR-192-5p/215-XIAP pathway in non-small cell lung cancer. Cancer Lett 357:196–205

Yeh WL, Lin HY, Huang CY, Huang BR, Lin C, Lu DY, Wei KC (2015) Migration-prone glioma cells show curcumin resistance associated with enhanced expression of miR-21 and invasion/anti-apoptosis-related proteins. Oncotarget 6:37770–37781

Zaky A, Mahmoud M, Awad D, El Sabaa BM, Kandeel KM, Bassiouny AR (2014) Valproic acid potentiates curcumin-mediated neuroprotection in lipopolysaccharide induced rats. Front Cell Neurosci 8:337

Zhang J, Du Y, Wu C, Ren X, Ti X, Shi J, Zhao F, Yin H (2010) Curcumin promotes apoptosis in human lung adenocarcinoma cells through miR-186* signaling pathway. Oncol Rep 24:1217–1223

Zhang W, Bai W, Zhang W (2014) MiR-21 suppresses the anticancer activities of curcumin by targeting PTEN gene in human non-small cell lung cancer A549 cells. Clin Transl Oncol 16:708–713. Official publication of the Federation of Spanish Oncology Societies and of the National Cancer Institute of Mexico

Zheng J, Wu C, Lin Z, Guo Y, Shi L, Dong P, Lu Z, Gao S, Liao Y, Chen B, Yu F (2014) Curcumin up-regulates phosphatase and tensin homologue deleted on chromosome 10 through microRNA-mediated control of DNA methylation – a novel mechanism suppressing liver fibrosis. FEBS J 281:88–103

Epigenetic Impact of Indoles and Isothiocyanates on Cancer Prevention

107

Pushpinder Kaur and Jaspreet Kaur

Contents

Abstract

Cruciferous vegetables are an excellent source of sulfur-containing compounds called glucosinolates, and their hydrolysis products, such as indoles and isothiocyanates, exert their chemopreventive effect by altering epigenetic mechanisms. These bioactive food components have gained more widespread attention as the potential agent for cancer prevention among other phytochemicals. Many studies have demonstrated the impact of indoles and isothiocyanates in modulating the epigenome in diverse human cancers and in the fetus. This review

P. Kaur (✉)
USC Keck School of Medicine, University of Southern California, Los Angeles, CA, USA
e-mail: Pushpinder.Bains@med.usc.edu; pushpinderkaur2006@gmail.com

J. Kaur
Department of Biotechnology, University Institute of Engineering and Technology, Panjab University, Chandigarh, India
e-mail: Jaspreet_uiet@pu.ac.in

V. B. Patel, V. R. Preedy (eds.), *Handbook of Nutrition, Diet, and Epigenetics*,
https://doi.org/10.1007/978-3-319-55530-0_118

summarizes the evidence from preclinical and clinical studies on the impact of indole and isothiocyanates on epigenetic events and how they relate to chemopreventive effects in the lung, prostate, breast, cervix, and transplacental carcinogenesis. We also discussed the remaining challenges and the research required to gather additional necessary information to identify the promising therapeutical potential of indoles and isothiocyanates in treating and preventing cancer.

Keywords
Indole · Isothiocyanate · Diet · Transplacental · Cancer chemoprevention · Epigenetics

List of Abbreviations

AhR	Aryl hydrocarbon receptor
BaP	Benzo[a]pyrene
BII	2,3-bis[3-indoylmethyl] indole
BR-DIM	BioResponse DIM
BRCA1	Breast cancer 1
CDC2	Cell division cycle 2
CDK2	Cyclin-dependent kinase 2
CIN	Cervical intraepithelial neoplasia
CTAs	Cancer testis antigens
CTR	Cyclic trimer
CYP1A1	Cytochrome P450 1A1
CYP1A2	Cytochrome P450 1A2
DBP	dibenzo[a,l]pyrene
DCIS	Ductal carcinoma in situ
DIM	Diindolylmethane
DNA	Deoxyribonucleic acid
DNMT3B	DNA methyltransferase 3B
DNMTi	DNMT inhibitors
DNMTs	DNA methyltransferases
EMT	Epithelial-to-mesenchymal transition
ERK1/2	Extracellular signal-regulated kinase 1/2
ERVs	Endogenous retroviruses
GD	Gestation day
GSTM1	Glutathione-*S*-transferase 1
H3	Histone H3
H4	Histone H4
HATs	Histone acetylases
HDAC	Histone deacetylase
HDAC1	Histone deacetylase 1
I3C	Indole-3-carbinol
ICZ	Indolocarbazole
LTR	Linear trimer
MI	*Myo*-inositol

miRNA	microRNA
ncRNA	noncoding RNA
NF-kB	Nuclear factor kappa B
NNK	Nitrosamine 4-(methylnitrosamino)-1-(3-pyridyl)-1-butanone
NQO1	NAD(P)H dehydrogenase [quinone] 1
Nrf2	Nuclear factor erythroid 2 (NF-E2)-related factor 2
NSCLC	Non-small cell lung cancer
PBMCs	Peripheral blood mononuclear cells
PEITC-NAC	*N*-acetyl-*S*-(*N*-2-phenethylthiocarbamoyl)-1-cysteine
PTEN	Phosphatase and tensin homolog
RNA	Ribonucleic acid
SAM	*S*-adenosyl methionine
SFN	Sulforaphane
SOX4	(Sex-determining region Y)-box 4
UGT1A1	UDP-glucuronosyl-transferase 1A1
UK	United Kingdom
US	United States

Introduction

The term epigenetics can be explained as changes in gene expression without a change in DNA sequence. Hence, it can be seen as a bridge between genotype and phenotype. In other words, all the cell types in an organism have the same DNA sequence, yet they have high variable phenotypes because of differential expression of genes, regulated by epigenetics. The mechanisms of regulation of gene expression through epigenetics include the process of DNA methylation, posttranscriptional histone modifications, microRNAs (miRNAs)-associated gene silencing, and chromosome inactivation. Mutations in DNA sequences in the genome may be responsible for the plethora of diseases or predisposition to diseases, but epigenetic factors also alter gene expression and are associated with susceptibility to disease. Over the years, researchers have elucidated the association between epigenetic modifications and various human conditions such as aging, cancer, and other diseases like neurological conditions, cardiovascular diseases, immunological disorders, etc. Various other biological processes like genomic imprinting, X-chromosome inactivation, and metastable epialleles have underlying epigenetic mechanisms.

Epigenetic modifications occurring on DNA and its associated histones, chromatin-remodeling enzymes, and miRNAs play a fundamental role during differentiation, fertilization, and development, and the process continues till late adulthood. The epigenetic signatures triggered by environmental stimuli, dietary constituents, and lifestyle factors lead to genomic instability. Diet is one of the important environmental factors that are capable of inducing and/or preventing tumor growth by eliciting epigenetic modifications that affect the transcriptional activity and gene function. Nutrients interfere with epigenetic mechanisms by influencing the supply of methionine for S-adenosyl methionine (SAM) (a product of one-carbon metabolism cycle)

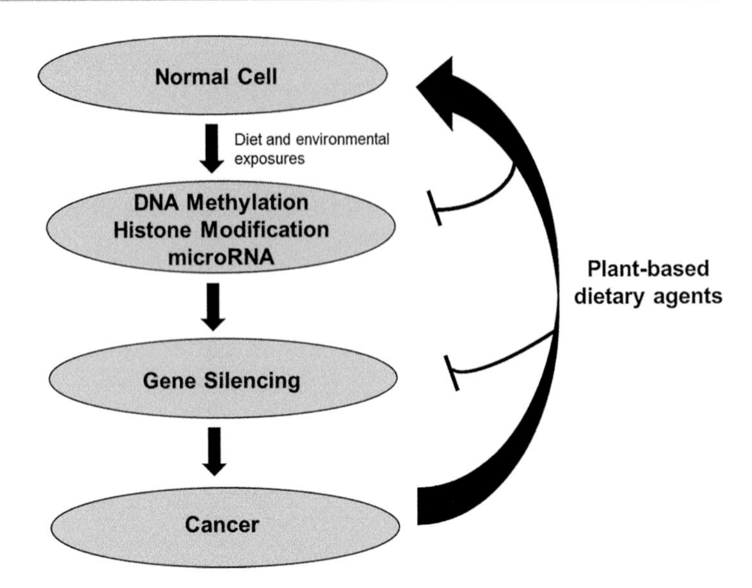

Fig. 1 Schematic view showing the diet and environmental effects on epigenetic modifications, such as hyper/hypomethylation of CpG islands, modification of the histone marks and microRNAs leading to abnormal gene function. Strategy to prevent and overturn the adverse epigenetic alterations through plant-based dietary agents in cancer. Deoxyribonucleic Acid (DNA)

formation and thus modulate the activity of DNA methyltransferases (DNMTs). Dietary components also influence histone modifications, miRNA expression that results in altered epigenetic reprogramming and modulates gene expression (Fig. 1). These epigenetic changes can be inherited and thus influence the health of future generations.

The epigenetic modifications do not alter the sequence of DNA and are reversible and provide promising therapies to restore gene function. Several studies have demonstrated that plant-based dietary agents modify the epigenome by altering the normal epigenetic state and reversing the abnormal gene silencing in diverse human cancers (Fig. 1). Most interestingly, studies have shown that the exposure of dietary agents during pregnancy and lactation could reshape the fetal epigenome providing protection to the offspring from epigenetic-related derangements. Recently, isothiocyanate (sulforaphane (SFN)) and indole glucosinolates (indole-3-carbinol (I3C) and diindolylmethane (DIM)), the major abundant bioactive food components of cruciferous vegetables, have gained substantial attention due to their chemopreventive and chemotherapeutic properties against childhood and adult cancers. In this review, dietary components, with special reference to I3C, DIM, and SFN, will be discussed, which target the epigenome in reducing the risk and provide new insight into clinical research for better disease prevention and therapy. We also discuss the metabolism of indoles and isothiocyanates, their mechanism of action, protective effects on different cancers, and effect of maternal I3C, DIM, and SFN diet on fetal epigenetic mechanisms.

Metabolism and Distribution of Indoles and Isothiocyanates

The cruciferous vegetable family (Brassicaceae) includes broccoli, brussels sprouts, kale, radish, cabbage, watercress, and turnip and contains a number of sulfur-containing compounds called glucosinolates and products of their decomposition, such as indoles and isothiocyanates. SFN is derived from the enzymatic breakdown of glucosinolate glucoraphanin which is abundant in broccoli and broccoli sprouts. SFN is primarily metabolized through the mercapturic acid pathway, which gives rise to several metabolic products. Sulforaphane bioavailability is three times higher in fresh/raw broccoli in comparison to cooked broccoli. However, DIM concentration in cooked cabbage has been reported to be sixfold higher in comparison to uncooked cabbage (Ciska et al. 2009). I3C is formed from the breakdown of glucosinolate glucobrassicin by the enzyme myrosinase and has found to be very unstable compound in both *in vitro* and *in vivo* studies (De Kruif et al. 1991). I3C undergoes self-condensation reactions by the stomach acid to produce a number of metabolites such as DIM, linear trimer (LTR) (2-(indol-3-ylmethyl)-3,3′-diindolylmethane), cyclic trimer (CTR) (5,6,11,12,17,18-hexahydrocyclonona [1,2-b: 4,5-b′:7,8-b″]triindole), 2,3-bis[3-indoylmethyl] indole (BII), indolocarbazole (ICZ), ascorbigen, and ascorbic acid (De Kruif et al. 1991; Preobrazhenskaya et al. 1993; Bradlow et al. 1999). Among these agents, DIM is the most stable and predominant active agent. *In vitro* studies have shown that I3C undergoes self-condensation greater than 60% to produce DIM at a ratio of 2:1 within 24 h of incubation (Bradlow and Zeligs 2010; Reed et al. 2006). I3C is poorly absorbed and has been detected in the bloodstream of humans after its oral ingestion to humans (Reed et al. 2006). In a pharmacokinetic study of single and multiple doses of I3C administration to humans, DIM has been found to be the primary circulating product with the highest concentration at 1000 mg dose (Reed et al. 2006). Anderton et al. (2004) characterized the pharmacokinetic properties and biodistribution of crystalline and BioResponse-DIM (BR-DIM) in a mouse model. The bioavailability of BR-DIM was found to be 50% higher than the crystalline formulation.

DIM was found in gastric contents, stomach tissue, small intestine, and liver in *in vivo* rat models after 1 h of oral exposure of I3C (De Kruif et al. 1991). In another *in vivo* study, 40% of the total radioactivity was detected in the liver extracts as DIM in rainbow trout models fed with radiolabeled I3C (Dashwood et al. 1994). The concentration of DIM was also shown to be pH dependent, with the highest amount at a pH level between 4 and 5. Moreover, DIM has been detected in human urine (Fujioka et al. 2014) and plasma (Reed et al. 2006) samples after ingestion of *Brassica* vegetables. The average intake of glucosinolates in the UK is around 0.2 mg/kg/day, whereas its intake is relatively low in the US (Abdull Razis and Noor 2013). Although research on plant-based dietary agents as a preventive agent is very promising, however, there is limited data about daily intake according to dietary guidelines.

Mechanism and Therapeutic Action of Indoles and Isothiocyanates

SFN has robust antioxidant defense system as it induces phase II detoxifying genes via Nrf2 (nuclear factor erythroid 2 (NF-E2)-related factor 2) signaling (Thimmulappa et al. 2002; Hu et al. 2006) and inhibits procarcinogenic phase I enzymes (Barcelo et al. 1996). The anticancer properties of I3C and DIM and their mechanism(s) of action in cell cultures and animal models have been extensively studied. It is difficult to dissect the response rate of a single agent alone since the protective effects are due to the combinatorial therapy of I3C and DIM. Both compounds have been reported to exert their anticancer efficacy through cellular (apoptosis and cell cycle arrest) mechanism in many cancer cell lines (Li et al. 2005; Rahman et al. 2009; Banerjee et al. 2009) by inhibiting the activity of multiple signaling pathways. Banerjee et al. (2009) have reported that DIM pretreatment to the pancreatic cancer cells, evading the effects of chemotherapeutic drugs such as gemcitabine, cisplatin, and oxaliplatin, inactivates nuclear factor kappa B (NF-kB) signaling and its downstream genes by inducing apoptosis. I3C-treated MCF-7 human breast cancer cells have inhibition of cyclin-dependent kinase 2 activity (CDK2) with associated alterations in cyclin E and subcellular localization of CDK2 complex (Garcia et al. 2005). In HT-29 human colon cancer cells, DIM is reported to inhibit cell proliferation and induces G1 and G2/M phase cell cycle arrest by reduced CDK2 and cell division cycle 2 (CDC2) activity (Choi et al. 2009). Research seems to indicate that I3C and DIM also inhibit angiogenesis via inactivation of Akt and at least in part of extracellular signal-regulated kinase 1/2 (ERK1/2) in human umbilical vein endothelial cells (Kunimasa et al. 2010). Studies have revealed that I3C and DIM supplementation stimulates the cellular detoxifying machinery and facilitates elimination of carcinogens (Li et al. 2003; Loub et al. 1975; McDanell et al. 1987).

An increasing body of evidence has shown that epigenetic deregulation is involved in cancer development. To date, a number of therapeutic drugs and dietary components that target histone deacetylase (HDAC) and/or DNA methylation are very promising in treating and preventing diseases, including cancer. However, there are many unknown dietary agents that play a role in the cross talk between various epigenetic modifications. The best-known therapeutic agents, which have raised hope for cancer therapy and especially for myeloid leukemias, are DNMT inhibitors (DNMTi) decitabine and 5-azacytidine. These agents are potent hypomethylating agents that incorporate into DNA and arrest the release of DNA methyltransferases (DNMTs) during replication and consequently reduce the overall methylation level by reexpressing the genes silenced by promoter methylation. The treatment with DNMTis upregulates the expression of cancer testis antigens (CTAs) and endogenous retroviruses (ERVs) which activates the immune signaling in tumor cells (Siebenkas et al. 2017) as well as induce apoptosis and thus reducing cell cycle activity of cancer stem cells (Siebenkas et al. 2017). At higher concentrations, these agents are cytotoxic via DNA adduct formation and induce chromosomal aberrations. Another major issue with DNMT inhibitors is the return of aberrant DNA methylation pattern near to pretreatment levels (Mossman et al. 2010). In addition to

DNA methylation, histone acetylases (HATs), HDACs, and miRNAs also play a major role in gene regulation and have gained attention as potential therapeutic targets. However, studies have also shown that DNMT and HDAC inhibitors can potentially activate oncogenes due to lack of specificity (Sato et al. 2003). Thus, there is a need for the development of safe and effective dietary epigenetic modulators for cancer prevention. Epigenetic intervention using dietary compounds is now an emerging area of research and due to its "reversible" character, DNA methylation machinery, chromatin modifications, and miRNAs represent a promising target for therapy. Several dietary agents have been explored for their chemopreventive properties, among which indoles and isothiocyanates exhibit significant anticancer activity. The present chapter will provide an insight on the role of I3C, DIM, and SFN in cancer prevention and discuss their effect on the epigenome.

Lung Cancer

Lung cancer is the leading cause of cancer-related death worldwide (http://globocan.iarc.fr). In addition to smoking, diet is a modifiable risk factor in lung carcinogenesis (World Cancer Research Fund/American Institute for Cancer Research). Studies also provide evidence that the consumption of cruciferous vegetables reduces the risk of lung cancer beyond the potential confounding effects of cigarette smoking (Tang et al. 2010; Lam et al. 2010). Arif and coworkers have shown that I3C inhibits the smoke-related lipophilic DNA adduct formation in the lung of the rats exposed to sidestream cigarette smoke (Arif et al. 2000; Tang et al. 2010; Lam et al. 2010). Although evidence indicates that I3C modulate the epigenetic components and is able to reverse altered gene expression in different cancers, only one study has investigated the ability of I3C to target epigenetic mechanism in lung cancer. Morse et al. (1990) examined the mechanism of action of I3C on lung carcinogenesis and DNA methylation induced by the tobacco-specific nitrosamine 4-(methylnitrosamino)-1-(3-pyridyl)-1-butanone (NNK) in the lungs and liver of mice. The authors observed a reduction in DNA methylation level due to decreased bioavailability of NNK and its carcinogenic metabolite 4-(methylnitrosamino)-1-(3-pyridyl)-1-butanol in the lungs of I3C-treated mice. Further detailed research is required to understand the effect of I3C on DNA methylation and other epigenetic mechanisms in lung cancer area. So far, very few studies have looked at the effect of SFN on lung cancer. Jiang et al. (2016) reported that SFN suppresses HDAC activity and increased the acetylation state of histone H3 and H4 in lung cancer cells. The authors also observed that SFN induces apoptosis and cell cycle arrest in *in vitro* and suppresses the lung cancer growth in *in vivo* models, which suggest that it has a chemopreventive potential for lung cancer. SFN is also known to regulate the expression of miR-616-5p, associated with therapy resistance and tumor recurrence, and suppresses the epithelial-to-mesenchymal transition (EMT) process in non-small cell lung cancer (NSCLC) cells (Wang et al. 2017). More interestingly, the combinatorial treatment of *N*-acetyl-*S*-(*N*-2-phenethylthiocarbamoyl)-1-cysteine (PEITC-NAC), I3C/DIM, and *myo*-inositol (MI) were found to be more effective than treatment with a single agent in inhibiting NNK plus benzo[a]pyrene (BaP)-induced lung tumor multiplicity through attenuating Akt, ERK, and NF-kB signaling pathways (Kassie et al. 2010). Although I3C, DIM, and SFN have driven considerable interest toward preventing the progression of lung cancer, it remains to be studied in detail whether these indoles and isothiocyanates exert their anticancer effects through epigenetic mechanisms.

Prostate Cancer

Prostate cancer is the second most common cancer in men worldwide. Androgen signaling and insensitivity play an important role in driving the progression of prostate cancer in men, and this cancer is hypothesized as hormone-dependent cancer. Although androgen ablation is the most common prostate cancer therapy in approximately 80% of the patients, however, prostate cancer often progresses toward the lethal and incurable stage of androgen independence (Schulz et al. 2003). Many investigators are searching for safer and more effective dietary compounds that may manifest protective action against prostate cancer. SFN has been shown to be effective against prostate cancer and to act through cellular and epigenetic mechanisms in *in vivo* and cell culture models. Dietary administration of SFN inhibited HDAC activity in the peripheral blood mononuclear cells (PBMCs) and prostates in xenograft mice that consumed a daily dose of 7.5 μmol for 3 weeks (Myzak et al. 2007). Additionally, SFN and DIM treatment reverses cancer-associated DNA methylation that deregulates gene expression in prostate epithelial cells, and androgen-dependent and androgen-independent prostate cancer cells (Wong et al. 2014). DIM has also shown to inhibit HDAC activity at lower concentration in androgen-sensitive prostate cancer cells (Wong et al. 2014). DIM treatment has also shown to demethylate the nuclear factor erythroid 2 (Nrf2) gene promoter region in association with enhanced mRNA expression of Nrf2 and Nrf2 target genes such as NQO1 and GSTM1 and also reduce the global DNA methylation level in *in vivo* prostate tumors (Wu et al. 2013). Evidence suggests that these bioactive dietary compounds can affect another component of epigenetic machinery such as noncoding RNAs (ncRNAs) that regulates the translation of mRNA into proteins. Treatment of prostate cancer cells with SFN alters the long ncRNA expression of the genes regulating cell cycle, signal transduction, and metabolism (Beaver et al. 2017). These findings could have a significant impact on the understanding of molecular mechanism of action of indoles and isothiocyanates against prostate cancer and will lead to identifying the candidate biomarkers for therapeutic innovation.

Breast Cancer

Epidemiological studies have demonstrated a strong association between cruciferous vegetable intake and breast cancer prevention (Terry et al. 2001; Ambrosone et al. 2004; Steinmetz and Potter 1996). Both I3C and DIM are reported to activate the aryl hydrocarbon receptor (AhR) and suppressed proliferation in ERα-positive MCF-7 cells (Chen et al. 1998; Caruso et al. 2014). Michnovicz et al. have observed that I3C alters endogenous estrogen metabolism in humans that consumed 6–7 mg/kg/day over 7 days (Michnovicz and Bradlow 1991). In combination with dietary lignans, I3C alters estrogen metabolism with increased estrogen C-2 hydroxylation and also reduces breast cancer risk in pre-and postmenopausal women (Laidlaw et al. 2010). It has been reported that I3C supplementation increases the expression of CYP1A1 and CYP1A2 enzymes, involved in estrogen metabolism, and inhibits mammary tumorigenesis in mice (Bradlow et al. 1991). Supplementation of 300 mg BR-DIM for 4–6 weeks significantly upregulates BRCA1 mRNA expression in women with a BRCA1 mutation. I3C, DIM, and SFN have also the potential to influence the epigenetic state and also in preventing and reducing the risk of breast cancer. In *in vitro* studies, I3C treatment was reported to induce demethylation and reactivation of the enzyme UDP-glucuronosyl-transferase 1A1 (UGT1A1) in breast cancer cells (http://cancerres.aacrjournals.org/content/69/9_Supplement/2863). DIM treatment upregulates miR-21 expression in estrogen-dependent MCF-7 cells and also resulted in inhibition of breast cancer cell proliferation (Jin 2011). Also, DIM was shown to downregulate SOX4 expression in MDA-MB-231 and T47D cells by increasing miR-212/132 expression (Hanieh 2015). Li et al. observed that the effect of sulforaphane on DNA methylation in MCF10DCIS, a cell line of poorly differentiated basal-like ductal carcinoma in situ (DCIS), activates miR-140 expression by restoring the normal methylation level and reduced tumor growth in *in vivo* models (Li et al. 2014). These evidence suggest that indoles and isothiocyanates provide an avenue for epigenetic therapy to fight breast cancer.

Cervix Cancer

Cervical cancer is the most common cancer affecting women. Several studies have demonstrated the impact of I3C, DIM, and sulforaphane in *in vivo* and cell culture models of cervix cancer. SFN treatment was found to modulate altered epigenome by restoring the expression and function of epigenetically silenced tumor suppressor genes in cervical cancer cell line. The *in silico* molecular modeling study suggests that SFN interacts with DNMT3B and HDAC1 and mediates the inhibition of epigenetic modulator enzymes (Ali Khan et al. 2015). In cervical cancer cell lines, DIM has been shown to be more potent than I3C in inducing apoptosis (Chen et al. 2001). Qi et al. (2005) found a strong effect of I3C on upregulation of PTEN gene in cervical epithelium of the transgenic mice, suggesting that I3C may exert its anticancer effects by increasing the expression of PTEN gene. Another study by Jin et al. (1999) showed that I3C-supplemented diet inhibits the estrogen-promoted and papillomavirus-initiated cervical-vaginal cancer in K14-HPV16 transgenic mouse models. Additionally, I3C also appears to be chemopreventive for skin cancer in K14-HPV16 mice. A small placebo-controlled trial of 30 patients with cervical intraepithelial neoplasia (CIN) treated with oral I3C supplementation (200 mg/day and 400 mg/day) had complete regression of CIN in comparison to the women who took placebo (Bell et al. 2000). Although these studies suggest that dietary I3C supplementation prevents the progression of cervix cancer, however, the mechanisms behind this I3C chemoprotection is still unclear.

Transplacental Cancer Prevention by Indoles and Isothiocyanates

I3C, DIM, and SFN have been shown to be chemoprotective in *in vitro* and *in vivo* models (Myzak et al. 2006, 2007; Grubbs et al. 1995; Li et al. 2003). However, in some studies, I3C has been shown to act as a modulator of liver carcinogenesis in rats (Stoner et al. 2002) and rainbow trout (Stoner et al. 2002). Very few studies have examined the risk/benefit of these chemoprotective agents in transplacental carcinogenesis models. Yu et al. (2006a) showed that *in utero* exposure to dibenzo[a,l] pyrene (DBP) to mice (15 mg/kg) on gestation day (GD) 17 results in higher incidence of mortality in offsprings due to T-cell acute lymphoblastic lymphoma. All surviving offsprings to 10 months of age exhibited lung tumors and liver tumors. I3C and DIM supplementation increases the activity of certain detoxification enzymes and facilitates elimination of carcinogens (Li et al. 2003). Interestingly, maternal dietary I3C consumption from GD9 till weaning significantly reduced lung cancer mortality in offspring to 10 months of age, suggesting offspring born to mothers consuming chemoprotective agents exhibit significant protection against polyaromatic hydrocarbon-dependent cancer (Yu et al. 2006b). Another study by Oganesian et al. (1997) observed a protective effect against diethylnitrosamine-initiated liver carcinogenesis by long-term dietary administration of I3C in the infant mouse model. Many studies have shown that altered epigenetic marks exist in cancer and research efforts have also been focused on the prevention of cancer by various dietary agents, including indoles and isothiocyanates. However, the research in this field is still curtailed as the questions regarding the mechanism of conferring this protection to the fetus is still unanswered. Although research on SFN as a chemopreventive agent is very promising for many cancers, determining the degree of chemoprotection by sulforaphane for the fetus *in utero* and lactational exposure to carcinogens is still also a challenging task. Other challenges regarding the "ideal dosage," the "combinatorial therapy" of indole and isothiocyanates with other dietary components to improve the fetus development, and whether the transmission of these dietary agents across multiple generations would be able to prevent the successive generation from diseases and cancer still need to be addressed.

Remaining Challenges and Future Prospectives

Nutriepigenomics studies are particularly concerned with how nutrients and diet interact with each other and maintain the epigenome by directly or indirectly affecting DNA methylation, histone modifications, or miRNA expression. Although research has identified numerous plant-based dietary agents, only a relatively few of them have been scrutinized extensively for their long-term prevention effects. Epigenetic dysregulation is recognized as a hallmark of cancer and research efforts have been focused on the prevention and treatment of various diseases by many dietary supplements, including indoles and isothiocyanates. However, the underlying mechanism of conferring this protection is unknown. The field of epigenetic-targeted therapy is the promising area for prevention and treatment of complex diseases, including cancer. We summarized here in this chapter the major epigenetic alterations in various cancers with special emphasis on dietary interventions by indoles and isothiocyanates for cancer chemoprevention. Although some studies have shown that I3C, DIM, and SFN have antitumor and epigenetic properties in the disease risk management, many aspects of the mechanism of action including epigenetics still remain unclear.

Besides this, there are many questions that remain to be answered such as (a) what dosing regimens for sulforaphane will be effective in humans? (b) whether these dietary agents contribute to cytotoxicity at higher/lower concentrations? (c) whether certain pathway or subset of genes modulated by indoles and isothiocyanates? (d) whether these dietary compounds may be able to prevent the successive generation from disease and cancer if given in maternal diet? The challenges are (a) whether the cancer protection comes from isolated dietary supplements or from a combination of phytonutrients, (b) assessing the epigenetic changes induced by phytonutrients and their long-term efficacy and safety of these agents, and (c) identification of other dietary agents that are still untested as to their effectiveness in cancer chemoprevention and transplacental studies.

Dictionary of Terms

- **Apoptosis** – It is a process of programmed cell death for eliminating old and unhealthy cells that may result in uncontrolled cell growth and tumor formation.
- **Cancer chemoprevention** – Cancer chemoprevention is the use of natural substances or other agents to prevent and slow the development or recurrence of cancer.
- **Combinatorial therapy** – The treatment where the patient is treated with more than one therapy.
- **Deoxyribonucleic acid (DNA) methylation** – It is one of the epigenetic mechanisms that modify the DNA through the addition of methyl group which can alter gene expression.
- **Histone modifications** – These are the modifications to the histone proteins by methylation, phosphorylation, acetylation, ubiquitylation, and sumoylation that regulate gene expression.
- **microRNAs** – These are small ribonucleic acid (RNA) molecules that alter or prevent the gene expression by binding to messenger RNA.
- **Transplacental carcinogenesis** – Transplacental carcinogenesis is the initiation of cancer in the fetus by gestational exposure to a carcinogen.

Key Facts of Cancer

- Cancer is the leading cause of death worldwide.
- The lung, prostate, breast, and cervix are the most common malignancy among men and women.
- A single abnormal cell can give rise to cancer.
- Environmental risk factors, lifestyle (including dietary factors) have a significant effect on cancer risk.
- Plant-based dietary agents have chemopreventive properties and can reduce the cancer risk.

Summary Points

- This chapter focuses on bioactive food components derived from cruciferous vegetables that have potential for cancer prevention.
- Cruciferous vegetables contain sulfur-containing compounds called glucosinolates and their hydrolysis products, such as indoles and isothiocyanates, that have been shown to prevent tumorigenesis.
- Many studies have attempted to explore the underlying mechanistic process of the protective role of indoles and isothiocyanates; however, the biological mechanism(s) of action are still under investigation.
- We summarized here the evidence from human subjects, animal, and cell culture models on the impact of indole and isothiocyanates on epigenetic events and how they relate to chemopreventive effects in the lung, prostate, breast, cervix, and transplacental carcinogenesis.
- We also discussed the remaining challenges, questions that are unanswered, and the research required to gather additional necessary information regarding epigenetic changes induced by indoles and isothiocyanates in treating and preventing disease.

References

Abdull Razis AF, Noor NM (2013) Cruciferous vegetables: dietary phytochemicals for cancer prevention. Asian Pac J Cancer Prev 14:1565–1570

Ali Khan M, Sundaram MK, Hamza A et al (2015) Sulforaphane reverses the expression of various tumor suppressor genes by targeting DNMT3B and HDAC1 in human cervical cancer cells. Evid Based Complement Alternat Med 2015:412149

Ambrosone CB, McCann SE, Freudenheim JL et al (2004) Breast cancer risk in premenopausal women is inversely associated with consumption of broccoli, a source of isothiocyanates, but is not modified by GST genotype. J Nutr 134:1134–1138

Anderton MJ, Manson MM, Verschoyle R et al (2004) Physiological modeling of formulated and crystalline 3,3′-diindolylmethane pharmacokinetics following oral administration in mice. Drug Metab Dispos 32:632–638

Arif JM, Gairola CG, Kelloff GJ et al (2000) Inhibition of cigarette smoke-related DNA adducts in rat tissues by indole-3-carbinol. Mutat Res 452:11–18

Banerjee S, Wang Z, Kong D et al (2009) 3,3′-Diindolylmethane enhances chemosensitivity of multiple chemotherapeutic agents in pancreatic cancer. Cancer Res 69:5592–5600

Barcelo S, Gardiner JM, Gescher A et al (1996) CYP2E1-mediated mechanism of anti-genotoxicity of the broccoli constituent sulforaphane. Carcinogenesis 17:277–282

Beaver LM, Kuintzle R, Buchanan A et al (2017) Long noncoding RNAs and sulforaphane: a target for chemoprevention and suppression of prostate cancer. J Nutr Biochem 42:72–83

Bell MC, Crowley-Nowick P, Bradlow HL et al (2000) Placebo-controlled trial of indole-3-carbinol in the treatment of CIN. Gynecol Oncol 78:123–129

Bradlow HL, Zeligs MA (2010) Diindolylmethane (DIM) spontaneously forms from indole-3-carbinol (I3C) during cell culture experiments. In Vivo 24:387–391

Bradlow HL, Michnovicz J, Telang NT et al (1991) Effects of dietary indole-3-carbinol on estradiol metabolism and spontaneous mammary tumors in mice. Carcinogenesis 12:1571–1574

Bradlow HL, Sepkovic DW, Telang NT et al (1999) Multifunctional aspects of the action of indole-3-carbinol as an antitumor agent. Ann N Y Acad Sci 889:204–213

Caruso JA, Campana R, Wei C et al (2014) Indole-3-carbinol and its *N*-alkoxy derivatives preferentially target ERalpha-positive breast cancer cells. Cell Cycle 13:2587–2599

Chen I, McDougal A, Wang F et al (1998) Aryl hydrocarbon receptor-mediated antiestrogenic and antitumorigenic activity of diindolylmethane. Carcinogenesis 19:1631–1639

Chen DZ, Qi M, Auborn KJ et al (2001) Indole-3-carbinol and diindolylmethane induce apoptosis of human cervical cancer cells and in murine HPV16-transgenic preneoplastic cervical epithelium. J Nutr 131:3294–3302

Choi HJ, Lim DY, Park JH (2009) Induction of G1 and G2/M cell cycle arrests by the dietary compound 3,3′-diindolylmethane in HT-29 human colon cancer cells. BMC Gastroenterol 9:39

Ciska E, Verkerk R, Honke J (2009) Effect of boiling on the content of ascorbigen, indole-3-carbinol, indole-3-acetonitrile, and 3,3′-diindolylmethane in fermented cabbage. J Agric Food Chem 57:2334–2338

Dashwood RH, Fong AT, Arbogast DN et al (1994) Anticarcinogenic activity of indole-3-carbinol acid products: ultrasensitive bioassay by trout embryo microinjection. Cancer Res 54:3617–3619

De Kruif CA, Marsman JW, Venekamp JC et al (1991) Structure elucidation of acid reaction products of indole-3-carbinol: detection in vivo and enzyme induction in vitro. Chem Biol Interact 80:303–315

Fujioka N, Ainslie-Waldman CE, Upadhyaya P et al (2014) Urinary 3,3′-diindolylmethane: a biomarker of glucobrassicin exposure and indole-3-carbinol uptake in humans. Cancer Epidemiol Biomarkers Prev 23:282–287

Garcia HH, Brar GA, Nguyen DH et al (2005) Indole-3-carbinol (I3C) inhibits cyclin-dependent kinase-2 function in human breast cancer cells by regulating the size distribution, associated cyclin E forms, and subcellular localization of the CDK2 protein complex. J Biol Chem 280:8756–8764

Grubbs CJ, Steele VE, Casebolt T et al (1995) Chemoprevention of chemically-induced mammary carcinogenesis by indole-3-carbinol. Anticancer Res 15:709–716

Hanieh H (2015) Aryl hydrocarbon receptor-microRNA-212/132 axis in human breast cancer suppresses metastasis by targeting SOX4. Mol Cancer 14:172

Hu R, Xu C, Shen G et al (2006) Gene expression profiles induced by cancer chemopreventive isothiocyanate sulforaphane in the liver of C57BL/6J mice and C57BL/6J/Nrf2 (–/–) mice. Cancer Lett 243:170–192

Jiang LL, Zhou SJ, Zhang XM et al (2016) Sulforaphane suppresses in vitro and in vivo lung tumorigenesis through downregulation of HDAC activity. Biomed Pharmacother 78:74–80

Jin Y (2011) 3,3′-Diindolylmethane inhibits breast cancer cell growth via miR-21-mediated Cdc25A degradation. Mol Cell Biochem 358:345–354

Jin L, Qi M, Chen DZ et al (1999) Indole-3-carbinol prevents cervical cancer in human papilloma virus type 16 (HPV16) transgenic mice. Cancer Res 59:3991–3997

Kassie F, Melkamu T, Endalew A et al (2010) Inhibition of lung carcinogenesis and critical cancer-related signaling pathways by *N*-acetyl-*S*-(*N*-2-phenethylthiocarbamoyl)-L-cysteine, indole-3-carbinol and myo-inositol, alone and in combination. Carcinogenesis 31:1634–1641

Kunimasa K, Kobayashi T, Kaji K et al (2010) Antiangiogenic effects of indole-3-carbinol and 3,3′-diindolylmethane are associated with their differential regulation of ERK1/2 and Akt in tube-forming HUVEC. J Nutr 140:1–6

Laidlaw M, Cockerline CA, Sepkovic DW (2010) Effects of a breast-health herbal formula supplement on estrogen metabolism in pre- and post-menopausal women not taking hormonal contraceptives or supplements: a randomized controlled trial. Breast Cancer (Auckl) 4:85–95

Lam TK, Ruczinski I, Helzlsouer KJ et al (2010) Cruciferous vegetable intake and lung cancer risk: a nested case-control study matched on cigarette smoking. Cancer Epidemiol Biomarkers Prev 19:2534–2540

Li Y, Li X, Sarkar FH (2003) Gene expression profiles of I3C- and DIM-treated PC3 human prostate cancer cells determined by cDNA microarray analysis. J Nutr 133:1011–1019

Li Y, Chinni SR, Sarkar FH (2005) Selective growth regulatory and pro-apoptotic effects of DIM is mediated by AKT and NF-kappaB pathways in prostate cancer cells. Front Biosci 10:236–243

Li Q, Yao Y, Eades G et al (2014) Downregulation of miR-140 promotes cancer stem cell formation in basal-like early stage breast cancer. Oncogene 33:2589–2600

Loub WD, Wattenberg LW, Davis DW (1975) Aryl hydrocarbon hydroxylase induction in rat tissues by naturally occurring indoles of cruciferous plants. J Natl Cancer Inst 54:985–988

McDanell R, McLean AE, Hanley AB et al (1987) Differential induction of mixed-function oxidase (MFO) activity in rat liver and intestine by diets containing processed cabbage: correlation with cabbage levels of glucosinolates and glucosinolate hydrolysis products. Food Chem Toxicol 25:363–368

Michnovicz JJ, Bradlow HL (1991) Altered estrogen metabolism and excretion in humans following consumption of indole-3-carbinol. Nutr Cancer 16:59–66

Morse MA, LaGreca SD, Amin SG et al (1990) Effects of indole-3-carbinol on lung tumorigenesis and DNA methylation induced by 4-(methylnitrosamino)-1-(3-pyridyl)-1-butanone (NNK) and on the metabolism and disposition of NNK in A/J mice. Cancer Res 50:2613–2617

Mossman D, Kim KT, Scott RJ (2010) Demethylation by 5-aza-2′-deoxycytidine in colorectal cancer cells targets genomic DNA whilst promoter CpG island methylation persists. BMC Cancer 10:366

Myzak MC, Dashwood WM, Orner GA et al (2006) Sulforaphane inhibits histone deacetylase in vivo and suppresses tumorigenesis in Apc-minus mice. FASEB J 20:506–508

Myzak MC, Tong P, Dashwood WM et al (2007) Sulforaphane retards the growth of human PC-3 xenografts and inhibits HDAC activity in human subjects. Exp Biol Med (Maywood) 232:227–234

Oganesian A, Hendricks JD, Williams DE (1997) Long term dietary indole-3-carbinol inhibits diethylnitrosamine-initiated hepatocarcinogenesis in the infant mouse model. Cancer Lett 118:87–94

Preobrazhenskaya MN, Bukhman VM, Korolev AM et al (1993) Ascorbigen and other indole-derived compounds from *Brassica* vegetables and their analogs as anticarcinogenic and immunomodulating agents. Pharmacol Ther 60:301–313

Qi M, Anderson AE, Chen DZ et al (2005) Indole-3-carbinol prevents PTEN loss in cervical cancer in vivo. Mol Med 11:59–63

Rahman KM, Banerjee S, Ali S et al (2009) 3,3′-Diindolylmethane enhances taxotere-induced apoptosis in hormone-refractory prostate cancer cells through survivin down-regulation. Cancer Res 69:4468–4475

Reed GA, Arneson DW, Putnam WC et al (2006) Single-dose and multiple-dose administration of indole-3-carbinol to women: pharmacokinetics based on 3,3′-diindolylmethane. Cancer Epidemiol Biomarkers Prev 15:2477–2481

Sato N, Maitra A, Fukushima N et al (2003) Frequent hypomethylation of multiple genes overexpressed in pancreatic ductal adenocarcinoma. Cancer Res 63:4158–4166

Schulz WA, Burchardt M, Cronauer MV (2003) Molecular biology of prostate cancer. Mol Hum Reprod 9:437–448

Siebenkas C, Chiappinelli KB, Guzzetta AA et al (2017) Inhibiting DNA methylation activates cancer testis antigens and expression of the antigen processing and presentation machinery in colon and ovarian cancer cells. PLoS One 12:e0179501

Steinmetz KA, Potter JD (1996) Vegetables, fruit, and cancer prevention: a review. J Am Diet Assoc 96:1027–1039

Stoner G, Casto B, Ralston S et al (2002) Development of a multi-organ rat model for evaluating chemopreventive agents: efficacy of indole-3-carbinol. Carcinogenesis 23:265–272

Tang L, Zirpoli GR, Jayaprakash V et al (2010) Cruciferous vegetable intake is inversely associated with lung cancer risk among smokers: a case-control study. BMC Cancer 10:162

Terry P, Wolk A, Persson I et al (2001) Brassica vegetables and breast cancer risk. JAMA 285:2975–2977

Thimmulappa RK, Mai KH, Srisuma S et al (2002) Identification of Nrf2-regulated genes induced by the chemopreventive agent sulforaphane by oligonucleotide microarray. Cancer Res 62:5196–5203

Wang DX, Zou YJ, Zhuang XB et al (2017) Sulforaphane suppresses EMT and metastasis in human lung cancer through miR-616-5p-mediated GSK3beta/beta-catenin signaling pathways. Acta Pharmacol Sin 38:241–251

Wong CP, Hsu A, Buchanan A et al (2014) Effects of sulforaphane and 3,3′-diindolylmethane on genome-wide promoter methylation in normal prostate epithelial cells and prostate cancer cells. PLoS One 9:e86787

World Cancer Research Fund/American Institute for Cancer Research. Food, nutrition, physical activity, and the prevention of cancer: a global perspective. Available at http://www.aicr.org/assets/docs/pdf/reports/Second_Expert_Report.pdf

Wu TY, Khor TO, Su ZY et al (2013) Epigenetic modifications of Nrf2 by 3,3′-diindolylmethane in vitro in TRAMP C1 cell line and in vivo TRAMP prostate tumors. AAPS J 15:864–874

Yu Z, Loehr CV, Fischer KA et al (2006a) In utero exposure of mice to dibenzo[a,l]pyrene produces lymphoma in the offspring: role of the aryl hydrocarbon receptor. Cancer Res 66:755–762

Yu Z, Mahadevan B, Lohr CV et al (2006b) Indole-3-carbinol in the maternal diet provides chemoprotection for the fetus against transplacental carcinogenesis by the polycyclic aromatic hydrocarbon dibenzo[a,l]pyrene. Carcinogenesis 27:2116–2123

Epigenetic Phenomena of Arsenic and Histone Tail Modifications: Implications for Diet and Nutrition

108

Qiao Yi Chen and Max Costa

Contents

Abstract

Naturally occurring inorganic arsenic has been identified as the causal agent in human skin, lung, bladder, liver, and prostate cancers. Furthermore, arsenic exposure has also been associated with noncarcinogenic health outcomes, including cardiovascular disease, neurologic deficits, neurodevelopmental deficits in childhood, and hypertension. According to the Agency for Toxic Substances and Disease Registry, arsenic is considered number one on the substance priority list. However, the overall risks on human health may exceed the documented levels due to lack of a comprehensive consideration of exposure through diet and anthropogenic factors. Arsenic permeates through water and soil, and related health issues elicit global concerns for the mass public. The exact mechanism of arsenic toxicity is still not fully understood, although convincing evidence and recent advance in epigenetic research such as DNA methylation and histone

Q. Y. Chen · M. Costa (✉)
Department of Environmental Medicine, New York University School of Medicine, Tuxedo, NY, USA
e-mail: qyc203@nyu.edu; Max.Costa@nyumc.org

V. B. Patel, V. R. Preedy (eds.), *Handbook of Nutrition, Diet, and Epigenetics*,
https://doi.org/10.1007/978-3-319-55530-0_17

posttranslational modifications have broadened our scope in understanding the mechanism of arsenic toxicity and carcinogenicity. This chapter will present the most recent literatures on the effect of arsenic on histone tail modifications as well as implications on food and diet.

Keywords
Inorganic arsenic · Histone tail modifications · Methylation · Acetylation · H3.1 · Stem-loop binding protein · Nutrition

List of Abbreviations

AcCOA	Acetyl coenzyme A
As	Arsenic
As (III)	Trivalent As
As (V)	Pentavalent As
As3MT1	Arsenic methyltransferase 1
BL-41	Human lung carcinoma cells
DMA5+	Dimethylarsinic acid
HACAT	Male-derived human keratinocytes
HAT	Histone acetyltransferase
HEK293	Female-derived human embryonic kidney cells
H(X)K(X)	Histone (X) Lysine (X)
iAs	Inorganic arsenic
K	Lysine
MMA5+	Monomethylarsonic acid
PBMC	Peripheral blood mononuclear cells
Ppb	Parts per billion
PTMs	Histone posttranslational modifications
R	Arginine
SAM	Cofactor S-Adenosyl methionine

Introduction

Arsenic (As) is a naturally occurring ubiquitous metalloid found in the Earth's crust, sediments rich in organic matter, as well as volcanic terranes. Corrosion of rocks and minerals coupled with anthropogenic sources of contamination such as application of arsenical pesticides, mining, and burning of fossil fuels exacerbates arsenic contamination in the groundwater. Areas of the world suffering most heavily include Bangladesh and West Bengal. In the 1960s and 1970s, international aids and governmental institutions initiated the digging of tube wells in effort to provide alternatives for people suffering from low-quality surface drinking water and related waterborne diseases. However, due to the lack of pretesting for underground impurities, the hundreds of thousands of tube wells providing arsenic-contaminated drinking water became the sources of one of the most devastating mass poisoning in human history. However, the extent of arsenic contamination is not localized to these two particular regions. In fact, alarming levels found in the French Mediterranean coastal areas, the United States, China, Ghana, and Mexico illustrate the wide-ranging presence of this hazardous carcinogen around the globe. Beginning in the 1990s, arsenic elicited health concerns which began to gain recognition as a global public health issue; today a staggering amount of nearly 200 million people are exposed to unacceptable levels of this class one human carcinogen. Although the World Health Organization's recommended limit is 10 parts per billion (ppb), arsenic concentrations documented in approximately 70 countries can range anywhere from 0.5 to 5000 ppb (Shankar et al. 2014).

Metabolism

The various chemical forms and oxidation states of arsenic dictate its availability and toxicity. Inorganic arsenic is known to be more toxic and prevalent in terrestrial environments. Out of the four types of valence states, inorganic arsenic (iAs) chiefly exists as trivalent As (III) or pentavalent As (V). As the dominant form found in groundwater, the pernicious effect of As (III) is exacerbated by its intrinsically higher toxicity than As (V) as well as its uncharged state, which impedes the effective removal from water. Depending on the regional characteristics and type of human activities, the public can be exposed to arsenic either through ingestion, which is the main route of exposure due to pervasive contamination in drinking water and various foods, and/or inhalation. Once ingested, inorganic arsenic will go through two major steps of metabolism: biotransformation and methylation. Biotransformation refers to the reduction of As (V) to As (III). Methylation, which is also considered a detoxification mechanism, involves a two-step process in facilitating the excretion of methylate inorganic As (III) from the body via the kidneys. Arsenic methyltransferase 1 (As3MT1) catalyzes iAs into monomethylarsonic acid (MMA5+) and dimethylarsinic acid (DMA5+), respectively. The end products are more readily excreted from the body probably due to less protein binding from the reduction in overall charge caused by methylation. However, during the sequential methylation steps, MMA3+ and DMA3+ may form and remain in the tissues. Low levels of these reactive and cytotoxic intermediates can often be detected in people who are chronically exposed to arsenic-contaminated drinking water, and the methylated form of As is inherently more toxic to the body than the non-methylated form with DMA3+ being the most toxic species. It would be of interest to understand the role of AS3MT1 in human carcinogenesis and whether inhibition of this enzyme might be protective against As-induced cancer even though it increases the rate of As excretion.

Health Effects

Depending on the chemical form, oxidation state, amount, and length of exposure, arsenic can elicit a wide range of acute and chronic health concerns. As a class I human carcinogen, arsenic has long been reported to cause cancers of the lung, urinary bladder, kidney, skin, and prostate. Approximately 100 million people are chronically exposed to arsenic in drinking water at levels higher than 50 ppb (Moon et al. 2012). As Chen et al. reported in 2011, arsenic exposure via drinking water at levels between 10 and 300 ppb could induce inimical ramifications including neurological complications, respiratory, hepatic and renal dysfunctions, skin lesions, diabetes, cardiovascular diseases, etc. One of the most distinguishable manifestations of arsenic exposure is skin abnormality. Generally, skin lesions such as melanosis, keratosis, and black foot disease may develop 5–10 years after chronic arsenic exposure. Furthermore the ability of arsenic to readily cross the blood-brain barrier makes the brain especially vulnerable to its toxic effects. Exposed populations may experience difficulties in learning and maintaining concentration. Although the pituitary gland is especially susceptible, arsenic has the ability to deposit in all parts of the brain, thus negatively impacting brain development, learning, and even sustained concentration (Sanchez-Pena et al. 2010).

Epigenetics

The exact mechanism of arsenic toxicity is still not fully understood, although convincing evidence supporting the role of genotoxicity and cytotoxicity has long been studied. Since arsenic lacks the characteristics of a traditional mutagen, recent advances in elucidating the mechanism of arsenic toxicity have shifted toward epigenetic modifications, including DNA methylation and histone posttranslational modifications (PTMs). Epigenetics is broadly defined as the heritable and sometimes reversible change in gene expression and function in the absence of any change in the DNA sequence (Bitto et al. 2014). In order for gene expression to occur, DNA must unravel and become accessible to transcriptional factors. The accessibility of the DNA may be influenced by various mechanisms; histone methylation and acetylation are instrumental in impinging the chromatin structure. Histone methylation is an important tool in distinguishing genes that are or are not transcribed. Lysine (K) and arginine (R) are the only two residues capable of being methylated although lysine methylation is most commonly observed for H3 and H4. Both lysine and arginine are positively charged and retain hydrophobic/basic features. While arginine can only be mono- and di-methylated, lysine can be mono-, di-, and tri-methylated. Methylation of lysines or arginines does not alter the positive charge of the amino acid, while acetylation of lysines neutralizes the positive charge.

Although most literature emphasizes the effect of arsenic on DNA methylation, a growing body of evidence is pointing toward understanding arsenic's ability to induce alterations in histone methylation both in vivo and in vitro. In order for the massive amount of DNA (the length of a car) to fit inside a nucleus which is typically 6 um in diameter, the DNA must be tightly compressed into structures called chromatin. The basic unit of a chromatin is the nucleosome, which consists of 147 base pairs of double-stranded DNA circled 1.65 times around an octamer of alkaline proteins called histones (Luger et al. 1997). Histones are composed of five major groups: H1, H2A, H2B, H3, and H4. Two copies of each of the four canonical histones are required to make up the octamer core, while H1 stabilizes and serves as the linkage between nucleosomes and is involved in the higher-order chromatin structures. Besides the globular core, N-terminal tails that protrude from the nucleosome can have up to 60 different residues, each of which can be readily modified through methylation, acetylation, biotinylation, ubiquitination, phosphorylation, etc. Although globular core modifications have also been identified through mass spectrometry, the N-terminal tail alterations, especially methylation and acetylation, have been more comprehensively studied. Similar to DNA methylation, cofactor S-adenosyl methionine (SAM) acts as the methyl donor. Histone methyltransferases specific for lysine and arginine serve to replace each of the hydrogen on the NH2, NH2+, or NH3 groups with a methyl group. The methylation state (mono-, di-, or tri-) on each residue will prompt different responses for the chromatin structure and the recruitment of transcriptional modifiers (Bannister and Kouzarides 2011). Interestingly, current studies based on the epigenetic effects of arsenic have only reported on the methylation of lysine residues on H3K4, H3K9, H3K27, and H3K36, yet results have been largely inconsistent; see Table 1 (Howe and Gamble 2016).

Table 1 Arsenic's effect on histone methylation in different tissues

Methylation marker	Direction of change	Specific residue	Type of tissue	References
H3K4	Increase	H3K4me2, H3K4me3	A549 cells, steel workers, RWPE1 cells, female Bangladeshi adults	Zhou et al. (2008, 2009) and Chervona et al. (2012)
	Decrease	H3K4me3	Male Bangladeshi adults	Chervona et al. (2012)
H3K9	Increase	H3K9me2, H3K9me3	A549 cells, male and female Bangladeshi adults, Jurkat and CCRF-CEM	Zhou et al. (2008, 2009), Chervona et al. (2012), and Pournara et al. (2016)
	Decrease	H3K9me3	CD4+ cells from Argentinian women	Pournara et al. (2016)
H3K27	Increase	H3K27me3	PMBC of Bangladeshi women, female-derived mouse embryonic fibroblasts	Chervonat et al. (2012) and Kim et al. (2012)
	Decrease	H3K27me3	A549 cells, PBMC of Bangladeshi men	Zhou et al. (2008) and Chervona et al. (2012)
H3K36	Increase	H3K36me2, H3K27me3	PBMCs of Bangladeshi men, A549 cells, lymphocytes of Chinese adults	Howe et al. (2016), Zhou et al. (2008), and Ma et al. (2016)
	Decrease	H3K36me2, H3K27me3 among men	A549 cells, male Bangladeshi adults, HEK293T and HaCaT	Chervona et al. (2012), Ma et al. (2016), and Zhou et al. (2008)

The table summarizes the inconsistencies found in histone methylation after treatment with arsenic. Different tissues induce different posttranslational modifications

H3K4 and H3K36 are associated with transcriptional activation (Bannister and Kouzarides 2011; Wagner and Carpenter 2012). As (III) increased global levels of H3K4me2 in A549 cells and H3K4me3 in RWPE1 cells (Zhou et al. 2008, 2009). A positive correlation between arsenic and H3K4me3 was found in peripheral blood mononuclear cells (PBMC) of Bangladeshi adult women (Chervona et al. 2012). Due to their role in DNA repair, dysregulation in H3K36me2 and H3K36me3 has been implicated in several cancer types (Jha et al. 2014; Kuo et al. 2011; Pfister et al. 2015). Arsenic (III) led to an increase in H3K36me3 in both A549 cells and lymphocytic cells from Chinese adult participants (Zhou et al. 2008; Ma et al. 2016). In contrast to H3K4 and H3K36, H3K9 and H3K27 are typically associated with transcriptional repression, although all four are important for genomic stability (Rivera et al. 2015; Yuan et al. 2011). In A549 cells, As (III) boosted H3K9me2 and H3K9me3 levels. Similarly positive correlations in H3K9me2 and H3K9me3 were detected in the PBMC of adult Bangladeshi participants and CD4+ cells from acute lymphoblastic leukemia patients, respectively (Chervona et al. 2012; Pournara et al. 2016). As for H3K27, two studies involving adult females and in vitro female mouse embryonic fibroblasts showed increase in H3K27me3 after As (III) treatment (Chervona et al. 2012; Kim et al. 2012).

Despite seemingly sufficient evidence supporting the above findings, there are also plenty of studies suggesting the opposite. Specifically, two studies have reported that arsenic differentially influences the level of H3K4me3 based on sex (Chervona et al. 2012; Tyler et al. 2015). Although, as pointed out earlier, female participants showed positive correlation between H3K4me3 and As (III), male participants showed the exact opposite (Chervona et al. 2012). Similarly for H3K9 methylation, despite increase in H3K9me2 and H3K9me3 found in in vivo and in vitro data, a study based on Argentinian women showed an inverse association between H3K9me3 and urinary arsenic (Howe and Gamble 2016; Pournara et al. 2016). Another epidemiology study based in China also observed negative association between urinary As and H3K9me2 in lymphocytic cells (Ma et al. 2016). Interestingly, the level of H3K9me2 seems to change in response to the duration of exposure. The same study by Ma et al. (2016) assessed the response in male-derived human keratinocytes (HaCat) and female-derived human embryonic kidney cells (HEK293) and found that after initial reduction in H3K9me2, both cell lines showed elevation after 8–12 h. This may be an important insight into the discrepancies between in vitro and epidemiological findings. Additionally, the differential effect of arsenic based on sex is also shown in H3K27me3. In contrast to the female participants and female-derived mouse embryonic fibroblasts, H3K27me3 was reduced in A549 cells and male participants in the same study group (Chervona et al. 2012; Zhou et al. 2008). In another study using HepG2 cells, H3K27me3 levels stayed constant after As (III) treatment (Ramirez et al. 2008). Lastly, like the other three posttranslational modifications, H3K36 methylation also demonstrated controversial results. Unlike the positive correlation found in A549 cells and Chinese adult lymphocytes, H3K36me3 levels decreased in a dose-dependent manner in HEK293 and HaCat cells (Ma et al. 2016; Zhou et al. 2008).

Histone acetylation is one of the most extensively studied posttranslational modifications. The negatively charged DNA is tightly bound to the positively charged histone side chain. And upon acetylation, histone acetyltransferase (HAT) transfers an acetyl group from acetyl coenzyme A (AcCOA) to a lysine residue thereby removing the positive charge, loosens bound DNA, and stimulates the accessibility of the promoter region. Unlike histone methylation, less histone acetylation modifications are observed after arsenic treatment. The two most prominent examples are H4K16 acetylation (H4K16ac) and H3K9 acetylation (H3K9ac); see Table 2. Not only is H4K16ac critical for the compact state of higher-order chromatin structure, but this PTM also plays important roles in the interaction between chromatin fibers and nonhistone proteins, DNA damage response, gene expression, and cell cycle control (Chen et al. 2015; Shogren-Knaak et al. 2006). Due to its extensive functions in maintaining genomic stability, dysregulation in H4K16ac is implicated in many cancers (Shogren-Knaak et al. 2006). The effect of arsenic on H4K16ac and H3K9ac levels is analogous to the inconsistencies found in histone methylation. H4K16ac has been shown to be stagnant in female-derived human embryonic kidney and UROTsa cells, elevated in human neonatal keratinocytes, and reduced in another set of studies using UROTsa cells (Herbert et al. 2014; Rahman et al. 2015; Shogren-Knaak et al. 2006). H3K9ac is vital for immune response and recruitment of nucleotide excision repair (NER) factors for DNA repair. Findings on H3K9ac level after arsenic exposure are similarly dispersed as the H4K16ac results. In HepG2, Jurkat, and CCRF-CEM cells, As (III) triggered increase in this PTM (Pournara et al. 2016; Ramirez et al. 2008). On the other hand, H3K9ac was decreased in UROTsa and human embryonic kidney cells, as well as in PMBCs of

Table 2 Arsenic's effect on histone acetylation in different tissues

Acetylation marker	Direction of change	Type of tissue	References
H4K16	No change	Female-derived human embryonic kidney cells	Rahman et al. (2015)
	Increase	Used primary human neonatal keratinocytes	Herbert et al. (2014)
	Decrease	UROTsa cells	Chervona et al. (2012) and Rahman et al. (2015)
H3K9	No change	Lymphocytes from Chinese adults, CD4+ and CD8+ cells from Argentinian women	Ma et al. (2016) and Pournara et al. (2016)
	Increase	HepG2 cells, Jurkat and CCRF-CEM cells	Ramirez et al. (2008) and Pournara et al. (2016)
	Decrease	UROTsa cells from female human ureter, PBMCs of Bangladeshi adults	Rahman et al. (2015) aand Chervona et al. (2012)

The table summarizes the inconsistencies found in histone acetylation after treatment with arsenic. Different tissues induce different posttranslational modifications

adult participants from Bangladesh (Chervona et al. 2012; Rahman et al. 2015). Furthermore, epidemiological studies in both China and Argentina showed no significant correlation between with H3K9ac and As in lymphocytes and immune cells (Howe and Gamble 2016; Ma et al. 2016; Pournara et al. 2016).

The discrepancies found in the canonical histone posttranslational modifications may be attributed by several factors. First, the content and dosage for the studies are varied. Although As (III) can be generated from both As2O3 and NaAsO2, As2O3 may be more toxic. Furthermore, as shown in the H3K9me2 example, PTM levels drastically changed direction after 8–12 h of exposure. In other words, short-term in vitro data may not be an accurate reflection of chronically exposed patients. Second, the degree of arsenic metabolism may differ between cell lines. Differences in methodology such as use of different cell lines, treatment sources, and exposure times all contribute to the lack of consistency among the featured studies. Third, not only are the sources of treatment for in vitro and epidemiological studies different, exposure pathways are also limited to inhalation and drinking water. However, an important aspect of arsenic exposure comes from everyday diet. The method of ingestion may greatly affect the outcome in histone PTMs due to differences in metabolic mechanisms. The lack of consistency in experimental design undermines the precision of the research results and hinders direct comparisons between the studies. On the other hand, the use of different cell lines, participants, dosage, total treatment time, and route of exposure provides us with an array of information which is an important basis for deepening our understanding of arsenic-induced histone posttranslational modifications.

Besides inducing posttranslational histone modifications, arsenic has recently been shown to alter the expression of histone genes themselves. There are three distinct groups of histone genes: replication dependent, replication independent, and those that encode tissue-specific isotypes (Lanzotti et al. 2002; Marzluff 2005). Replication-dependent histone genes encode for canonical histones and are the only genes that form a stem-loop structure at the 3′ end instead of with a poly (A) tail (Dominski and Marzluff 1999). The stem loop consists of 26 highly conserved nucleotides and serves as the binding site for the stem-loop binding protein (SLBP). SLBP engages in pre-mRNA processing and accompanies mature histone mRNA to the cytoplasm to ensure efficient translation (Marzluff 2005; Whitfield et al. 2004). In consideration to its many important roles, the loss of SLPB would nonetheless invoke serious consequences such as abnormal processing of canonical histone mRNAs. Aberrant levels of histone gene expression induced by arsenic were first identified in PBMCs; in this study, 8% of all altered histone genes were canonical histones (Brocato et al. 2014). Brocato et al. specifically looked at the changes in H3.1 gene expression. As predicted, in both human lung carcinoma cells (BL-41) and PMBCs, 1uM of arsenic exposure prompted a double-fold increase in poly-adenylated H3.1 mRNA level (Brocato et al. 2014, 2015). Previous studies have suggested that the loss of stem-loop binding protein may contribute to the poly-adenylation of canonical histones (Brocato et al. 2015; Lanzotti et al. 2002). A sample pathway is presented in Fig. 1. This hypothesis was confirmed when arsenic-induced SLBP reduction due to promoter repression resulted in increased

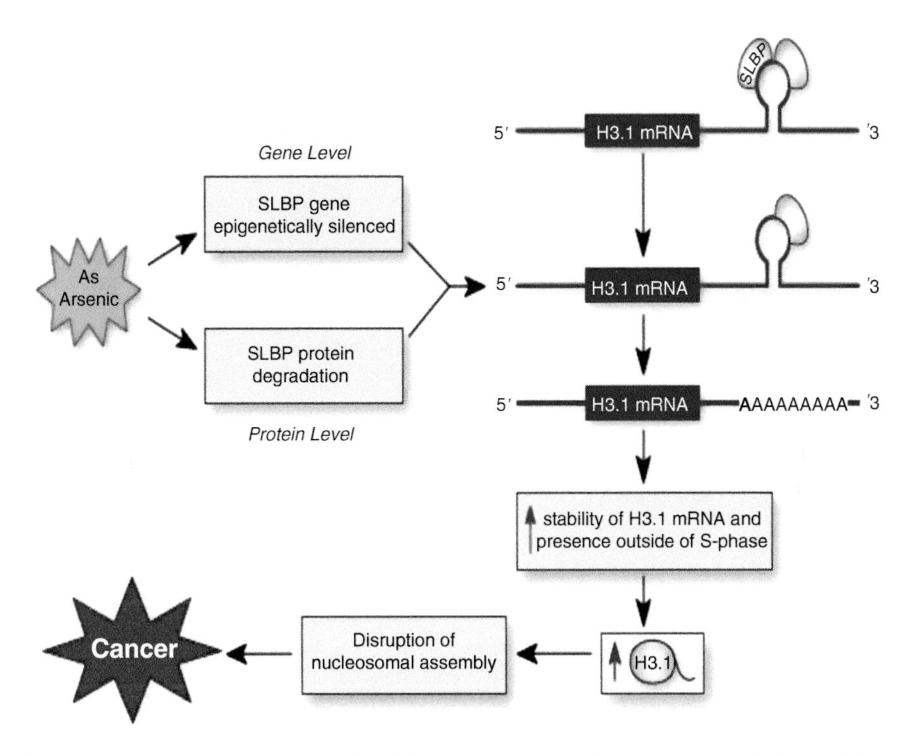

Fig. 1 Arsenic-induced depletion of SLBP and subsequent aberrant polyadenylation of H3.1 mRNA and its effects (Source comes from Brocato et al. 2015. The pathway illustrates the potential mechanism of arsenic-induced SLBP reduction and subsequent polyadenylation of H3.1 mRNA)

expression of polyadenylated H3.1 mRNA (Brocato et al. 2014). Because poly (A) tail provides stability and prolonged half-life, the initial increase in canonical histone genes may be due to increased histone mRNA stability through replacing the stem-loop structure with a poly (A) tail. More importantly, studies are finding that polyadenylation of H3.1 mRNA may be a potential mechanism for arsenic carcinogenesis as H3.1-transfected cells showed significant increase in colony formation/ anchorage-independent growth, hence cell transformation (Brocato et al. 2014). Despite convincing evidence of arsenic-induced SLBP reduction and subsequent increase in H3.1 mRNA polyadenylation, the exact mechanism still requires further examination.

Nutrition

Despite natural occurrences in groundwater, advancement in technology and the increase in the usage of arsenic containing drugs, fertilizers, insecticides, herbicides, etc., fuel the increase in human exposure to this detrimental carcinogen. In addition to drinking water, exposure to vast amount of inorganic arsenic also occurs through

Table 3 Arsenic concentration in contaminated soil, water, and food crops

Country	Soil (mgkg^{-1})	Groundwater (μgL^{-1})	Rice (mgkg^{-1})	Vegetables (mgkg^{-1})	References
Bangladesh	5.64–29.5	290–710	0.02–3.40	0.09–2.03	Das et al. (2002)
India	5.9–9.7	320–640	0.33–0.45	NA	Bhattacharya et al. (2010)
China	129	329	1.09	2.38	Liu et al. (2010)
Nepal	6.1–16.7	ND-1014	0.18	0.33	Dahal et al. (2008)
Taiwan	7.92–12.7	13.8–881	NA	0.01–0.15	Kar et al. (2013)
Limits	20[a]	100[b]	1[c]	1[c]	Bhattacharya et al. (2010)

Source comes from Fayiga et al. (2016). Amount of arsenic found in each four different categories. The limits for each category are set by:
[a]European Union limit for agricultural soil
[b]FAO limit for irrigation water
[c]WHO limit for food crops

food contamination, which is especially prominent in grains, rice, and vegetables. Contaminated groundwater used to irrigate crops and soil will pollute the topsoil, which will eventually lead to the accumulation of significant amounts of arsenic in cultivated vegetables and crops (Fayiga and Saha 2016). One of the most efficient accumulators of arsenic is rice. Rice serves as a staple food for approximately half the world's population, mainly in Southeast Asia, Latin America, and sub-Saharan Africa (Azam et al. 2016). According to an investigation carried out by FAO, the average consumption of rice in China, India, and the United States is approximately 212 g/day, 195 g/day, and 25 g/day, respectively (FAO 2010). As the second most commonly cultivated crop, extensive arsenic contamination may elicit massive public health issues. Table 3 from Fayiga and Saha documents countries that use arsenic-contaminated irrigation water above the FAO limit. Not only does arsenic-contaminated water reduce plant growth and yield, but also accumulation of the carcinogenic compound poses devastating threat to the food consumers, which include both humans and animals. Figure 2 illustrates the increasing emphasis on arsenic and nutrition based on the escalating number of publications throughout the years.

Another anthropogenic source of arsenic contamination comes from arsenic-based drugs such as roxarsone and nitarsone, which release unnecessary sources of inorganic arsenic, MMA, and DMA (Nigra et al. 2016). These drugs have been deliberately and consistently fed to chicken and turkey in the United States for decades in effort to improve weight gain and meat pigmentation and prevent diseases such as histomoniasis and coccidiosis (Abraham et al. 2013; Chapman and Johnson 2002). In fact, 88% of chickens on the market had been treated with roxarsone back in 2010 (Nachman et al. 2012). A study in 2013 reported that based on the average poultry consumption in the United States, the consumers would be receiving a 1.44^10–6 mg/kg daily dose of iAs, which would consequently result in 124 excess bladder and lung cancer cases (Nachman et al. 2013). After

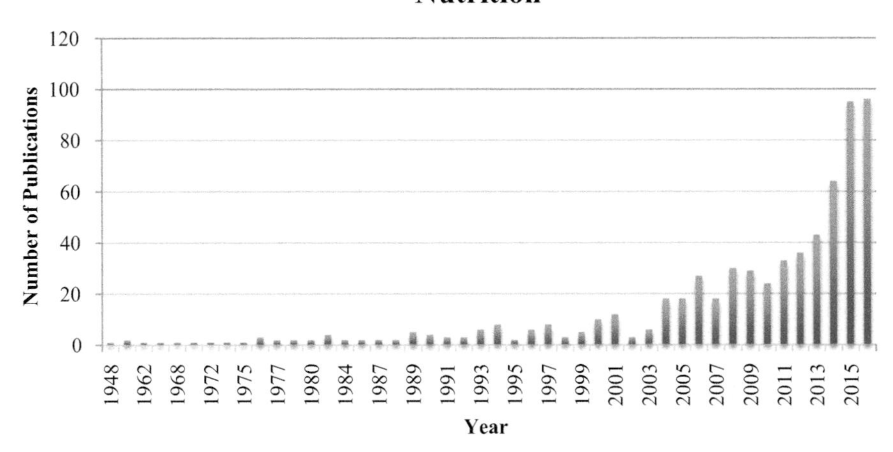

Fig. 2 Increasing number of publications on arsenic and nutrition. The graph is generated based on key word searches on PubMed between 1848 and 2016. The graph illustrates increasing importance placed on arsenic and nutrition over the years

concluding that increase of iAs in poultry was directly induced by feed additives, the FDA subsequently eliminated marketing support for the use of roxarsone and nitarsone in 2013 and 2015, respectively. However, due to decades of chronic arsenic exposure, the extent of the public health issue remain unknown as studies have shown that cancer risks remain elevated years after the exposure was eliminated.

Besides rice and poultry, arsenic tarnishes a wide array of foods such as fish, milk, eggs, fruits, vegetables, etc.; see Fig. 3. Although results are somewhat inconsistent, studies have suggested that selenium, vitamins, tea, Zn, etc., may reduce the damaging health effects of arsenic (Yu et al. 2016). On the other hand, unhealthy diets such as high sugar and high fat have been shown to exacerbate the effect of arsenic exposure. One study exposed pregnant mice to arsenic-contaminated drinking water. After giving birth, a subset of offspring continued to receive the As treatment, while high-sugar and high-fat foods were provided for all mice throughout the experiment (Ditzel et al. 2016). Overall, mice that were exposed to arsenic-contaminated drinking water before and after birth demonstrated more adverse outcomes, as illustrated by insulin resistance, obesity, and high triglycerides in blood compared to the other groups. This study not only illustrated the synergistic effect of arsenic and other food sources, which prompts alarming attention regarding lifestyle factors for those already exposed, but also indicated that neonatal exposure to arsenic may lead to health consequences later on. The prevalence of arsenic in drinking water coupled with prevailing risk through everyday diet further complicates public health outlook for those exposed to arsenic.

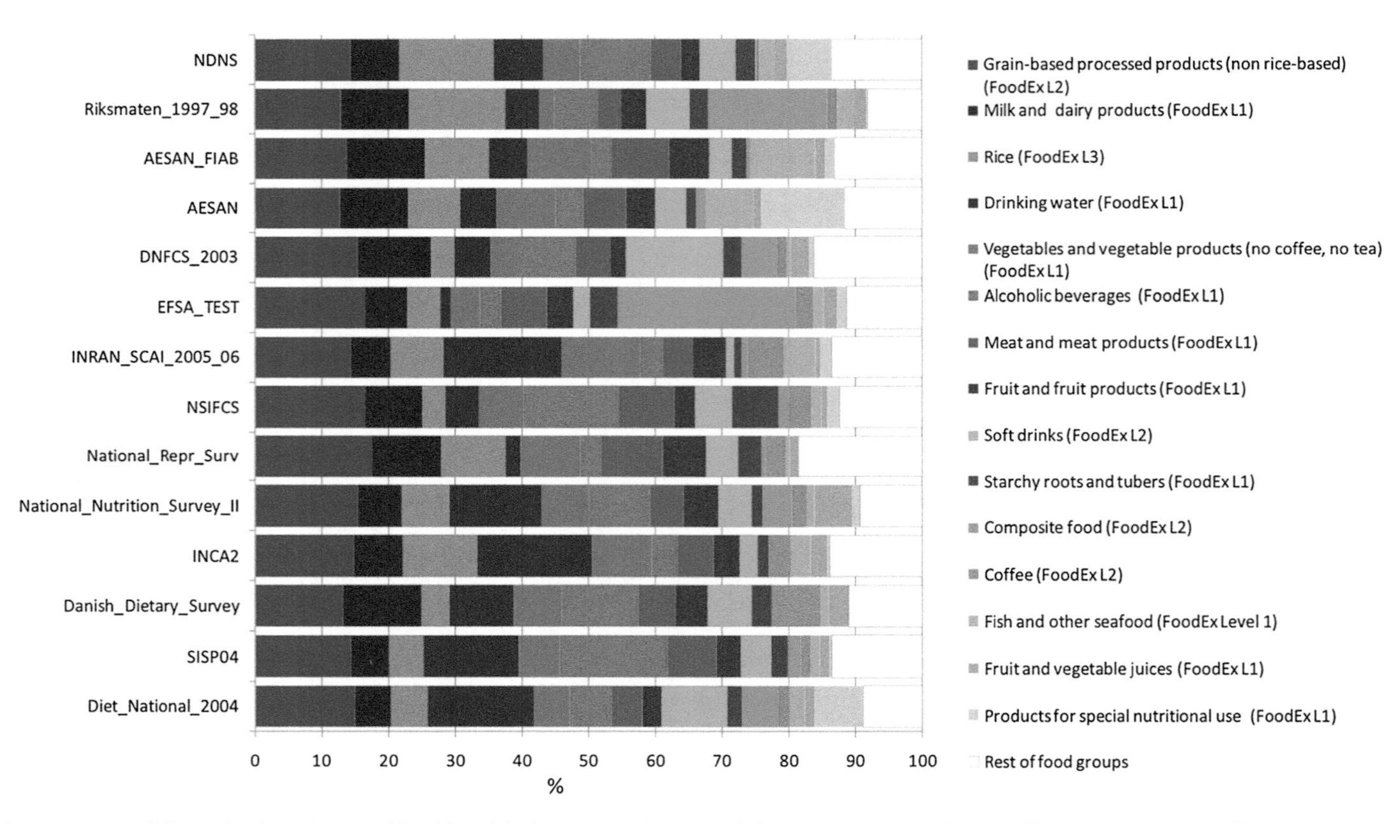

Fig. 3 Percent of dietary iAs in each type of food for adults in Europe. The source is from EFSA (2014). The table illustrates the percent of iAs in various food groups for adult Europeans. The *left* of the graph depicts the study used to collect the data

Conclusion

More than 200 million people are exposed to arsenic either through drinking water, inhalation, or diet. As a class I human carcinogen, exposure to arsenic has been documented to cause neurodevelopmental deficits; cardiovascular disease; human skin, lung, and bladder cancers; etc. Although the exact mechanism of arsenic toxicity and carcinogenicity is still unclear, evidence supporting the role of histone posttranslational modification and histone gene expression reviewed in this chapter provides new insights for our understanding and further investigation. Due to the pervasive presence of arsenic in food and water, policy makers as well as individuals must actively seek preventative ways of reducing chronic exposure. Responsible agencies such as the FDA, lawmakers, and food producers must engage in rigorous monitoring of arsenic contamination and enforce the removal from the market if standards are violated. Physicians and medical professionals should inform the public of potential sources of arsenic and recommend safe substitutions, especially for pregnant women and infants. And most importantly, individuals must take up the responsibility of educating themselves regarding the source in effort to avoid and reduce exposure and further contamination.

Dictionary of Terms

- **Canonical histones** – Most common histones found in the nucleus; H3 is an example of canonical histone.
- **Histone methyltransferase** – An enzyme that carries a methyl group from SAM to a histone lysine or arginine reside.
- **Histomoniasis** – Also called blackhead disease. It is a parasitic disease found in birds.
- **Coccidiosis** – Parasitic infection in the intestinal tract of animals, commonly seen in turkeys.
- **Anthropogenic activity** – Most commonly refers to environmental issues related to human activity such as the use of pesticides, herbicides, etc.
- **Histone N-terminal tail** – N-terminus is the beginning of a protein or peptide and terminated with –NH2. The histone N-terminal tail is the location of posttranslational modifications.

Key Facts of Arsenic Exposure and Application

- Albertus Magnus was the first person to isolate arsenic in 1250.
- Arsenic is a class I human carcinogen, categorized by the International Agency for Cancer Research.
- Inorganic As (III) is the most toxic form for human exposure.
- Set by the US Environmental Protection Agency, the maximum As concentration in drinking water is 10 parts per billion.

- Degree of As toxicity is based on individual metabolic rate/ability.
- Inorganic arsenic can be found in various food sources such as fish, milk, eggs, fruits, and vegetables but mainly in rice and poultry.
- Arsenic can be used in a variety of areas including medicine, alloys, and pesticides.

Summary Points

- Around 200 million people worldwide are exposed to arsenic either through inhalation, drinking water, and/or food.
- Besides DNA methylation which has been extensively studied, arsenic has been shown to effectively alter histone tail modifications such as histone methylation and acetylation.
- Canonical histone gene expression may be directly altered by arsenic exposure.
- Studies have shown that after arsenic exposure, total canonical histone gene expression can go up by twofold. H3.1 is one of these canonical histones.
- Arsenic may degrade stem-loop binding protein as a mechanism of inducing polyadenylated histone H3.1 mRNA.
- Key regulators, lawmakers, food producers, medical professionals, as well as the public must rigorously monitor arsenic contamination in food and enforce the removal from the market if standards are violated.

References

Abraham M, McDougald L, Beckstead R (2013) Blackhead disease: reduced sensitivity of Histomonas meleagridis to nitarsone in vitro and in vivo. Avian Dis 58:60–63

Azam S, Sarkar T, Naz S (2016) Factors affecting the soil arsenic bioavailability, accumulation in rice and risk to human health: a review. Toxicol Mech Methods. https://doi.org/10.1080/15376516.2016.1230165

Bannister AJ, Kouzarides T (2011) Regulation of chromatin by histone modifications. Cell Res 21:381–395

Bhattacharya P, Samal Ac, Majumdar J et al (2010) Arsenic contamination in rice, wheat, pulses, and vegetables: a study in an arsenic affected area of West Bengal, India. Water Air Soil Pollut 213:313

Bitto A, Pizzino G, Irrera N et al (2014) Epigenetic modifications due to heavy metals exposure in children living in polluted areas. Curr Genomics 15:464–468

Brocato J, Fang L, Chervona Y et al (2014) Arsenic induces polyadenylation of canonical histone mRNA by down-regulating stem-loop-binding protein gene expression. J Biol Chem 289:31751–31764

Brocato J, Chen D, Liu J et al (2015) A potential new mechanism of arsenic carcinogenesis: depletion of stem loop binding protein and increase in polyadenylated canonical histone H3.1 mRNA. Biol Trace Elem Res 166:72–81

Chapman HD, Johnson ZB (2002) Use of antibiotics and roxarsone in broiler chickens in the USA: analysis for the years 1995 to 2000. Poult Sci 81:356–364

Chen Y, Graziano J, Parvez F et al (2011) Arsenic exposure from drinking water and mortality from cardiovascular disease in Bangladesh: prospective cohort study. Br Med J 342:2431

Chen Q, Costa M, Sun H (2015) Structure and function of histone acetyltransferase MOF. Aims Biophysics 2:555–569

Chervona Y, Hall MN, Arita A et al (2012) Associations between arsenic exposure and global posttranslational histone modifications among adults in Bangladesh. Cancer Epidemiol Biomark Prev 21:2252–2260

Dahal BM, Fuerhacker M, Mentler A et al (2008) Arsenic contamination of soils and agricultural plants through irrigation water in Nepal. Environ Pollut 155:157–163

Das HK, Chowdhury DA, Rahman S et al (2002) Arsenic contamination of soil and water and related bio-hazards in Bangladesh. Environ Int 30:383–387

Ditzel E, Nguyen T, Parker P et al (2016) Effects of arsenite exposure during fetal development on energy metabolism and susceptibility to diet-induced fatty liver disease in male mice. Environ Health Perspect 124:2

Dominski Z, Marzluff WF (1999) Formation of the 3′ end of histone mRNA. Gene 239:1–14

EFSA (2014) Dietary exposure to inorganic arsenic in the European population. EFSA J 12:3597. https://doi.org/10.2903/j.efsa.2014.3597

FAO (2010) Available from: http://faostat.fao.org/site/339/default.aspx. Last accessed 20 Oct 2016

Fayiga A, Saha U (2016) Arsenic contamination, exposure routes and public health. Evidence-Based Med Pub Health 2:e1392

Herbert KJ, Holloway A, Cook AL et al (2014) Arsenic exposure disrupts epigenetic regulation of SIRT1 in human keratinocytes. Toxicol Appl Pharmacol 281:136–145

Howe C, Gamble M (2016) Influence of arsenic on global levels of histone posttranslational modifications: a review of the literature and challenges in the field. Curr Environ Health Rep 3:225–237

Jha DK, Pfister SX, Humphrey TC et al (2014) SET-ting the stage for DNA repair. Nat Struct Mol Biol 21:655–657

Kar S, Das S, Jean J et al (2013) Arsenic in the water-soil-plant system and the potential health risks in the coastal part of Chianan Plain, Southwestern Taiwan. J Asian Earth Sci 77:195–302

Kim HG, Kim DJ, Li S et al (2012) Polycomb (PcG) proteins, BMI1 and SUZ12, regulate arsenic-induced cell transformation. J Biol Chem 287:31920–31928

Kuo AJ, Cheung P, Chen K et al (2011) NSD2 links dimethylation of histone H3 at lysine 36 to oncogenic programming. Mol Cell 44:609–620

Lanzotti DJ, Kaygun H, Yang X et al (2002) Developmental control of histone mRNA and dSLBP synthesis during Drosophila embryogenesis and the role of dSLBP in histone mRNA 3′ end processing in vivo. Mol Cell Biol 22:2267–2282

Liu CP, Luo CL, Gao Y et al (2010) Arsenic contamination and potential health risk implications at an abandoned tungsten mine, southern China. Envrion Pollut 158:820–826

Luger K, Mader AW, Richmond RK et al (1997) Crystal structure of the nucleosome core particle at 2.8 Å resolution. Nature 389:251–260

Ma L, Li J, Zhan Z et al (2016) Specific histone modification responds to arsenic-induced oxidative stress. Toxicol Appl Pharmacol 302:52–61

Marzluff W (2005) Metazoan replication-dependent histone mRNAs: a distinct set of RNA polymerase II transcripts. Curr Opin Cell Biol 17:274–280

Moon K, Guallar E, Navas-Acien A (2012) Arsenic exposure and cardiovascular disease: an updated systematic review. Curr Atheroscler Rep 14:542–555

Nachman K, Raber G, Francesconi K et al (2012) Arsenic species in poultry feather meal. Sci Total Environ 417:183–188

Nachman K, Baron PA, Raber G et al (2013) Roxarsone, inorganic arsenic, and other arsenic species in chicken: a U.S.-based market basket sample. Environ Health Perspect 121:818–824

Naujokas M, Anderson B, Ahsan H et al (2013) The broad scope of health effects from chronic arsenic exposure: update on worldwide public health problem. Environ Health Perspect 121(3):295–302. https://doi.org/10.1289/ehp.1205875

Nigra A, Nachman K, Love D et al (2016) Consumption and arsenic exposure in the U.S. population. Environ Health Perspect. https://doi.org/10.1289/EHP351

Pfister SX, Markkanen E, Jiang Y et al (2015) Inhibiting WEE1 selectively kills histone H3K36me3-deficient cancers by dNTP starvation. Cancer Cell 28:557–568
Pournara A, Kippler M, Holmlund T et al (2016) Arsenic alters global histone modifications in lymphocytes in vitro and in vivo. Cell Biol Toxicol 32:275
Rahman S, Housein Z, Dabrowska A et al (2015) E2F1-mediated FOS induction in arsenic trioxide-induced cellular transformation: effects of global H3K9 hypoacetylation and promoter-specific hyperacetylation in vitro. Environ Health Perspect 123:484–492
Ramirez T, Brocher J, Stopper H et al (2008) Sodium arsenite modulates histone acetylation, histone deacetylase activity and HMGN protein dynamics in human cells. Chromosoma 117:147–157
Rivera C, Saavedra F, Alvarez F et al (2015) Methylation of histone H3 lysine 9 occurs during translation. Nucleic Acids Res 43:9097–9106
Sanchez-Pena L, Petrosyan P, Morales M et al (2010) Arsenic species, AS3MT amount, and AS3MT gene expression in different brain regions of mouse exposed to arsenite. Environ Res 110:428–434
Shankar S, Shankaer U, Shikha (2014) Arsenic contamination of groundwater: a review of sources, prevalence, health risks, and strategies for mitigation. Sci World J 2014:1–18. https://doi.org/10.1155/2014/304524
Shogren-Knaak M, Ishii H, Sun JM et al (2006) Histone H4-K16 acetylation controls chromatin structure and protein interactions. Science 311:844–847
Tyler CR, Hafez AK, Solomon ER et al (2015) Developmental exposure to 50 parts-per-billion arsenic influences histone modifications and associated epigenetic machinery in a region-and sex-specific manner in the adult mouse brain. Toxicol Appl Pharmacol 288:40–51
Wagner EJ, Carpenter PB (2012) Understanding the language of Lys36 methylation at histone H3. Nat Rev Mol Cell Biol 13:115–126
Whitfield ML, Kaygun H, Erkmann JA et al (2004) SLBP is associated with histone mRNA on polyribosomes as a component of the histone mRNP. Nucleic Acids Res 32:4833–4842
Yu H, Liu S, Li M et al (2016) Influence of diet, vitamin, tea, trace elements and exogenous antioxidants on arsenic metabolism and toxicity. Environ Geochem Health 38:339–351
Yuan W, Xu M, Huang C et al (2011) H3K36 methylation antagonizes PRC2-mediated H3K27 methylation. J Biol Chem 286:7983–7989
Zhou X, Sun H, Ellen TP et al (2008) Arsenite alters global histone H3 methylation. Carcinogenesis 29:1831–1836
Zhou X, Li Q, Arita A et al (2009) Effects of nickel, chromate, and arsenite on histone 3 lysine methylation. Toxicol Appl Pharmacol 236:78–84

Arsenic and microRNA Expression 109

Elena Sturchio, Miriam Zanellato, Priscilla Boccia, Claudia Meconi, and Silvia Gioiosa

Contents

Abstract

Arsenic is a naturally occurring metalloid that poses a major threat to worldwide human health. The most toxic form of arsenic is inorganic arsenic, which has been classified by the International Agency for Research on Cancer as a group 1

E. Sturchio (✉) · M. Zanellato · P. Boccia
Department of Technological Innovation and Safety of Plants, Product and Anthropic Settlements (DIT), Italian Workers' Compensation Authority (INAIL), Rome, Italy
e-mail: e.sturchio@inail.it; m.zanellato@inail.it; p.boccia@inail.it

C. Meconi
Research Organization CRF (Cooperativa Ricerca Finalizzata Sc), Tor Vergata University Science Park, Rome, Italy
e-mail: claudia.meconi@libero.it

S. Gioiosa
Institute of Biomembranes and Bioenergetics, National Research Council, Bari, Italy
e-mail: silvia.gioiosa84@gmail.com

V. B. Patel, V. R. Preedy (eds.), *Handbook of Nutrition, Diet, and Epigenetics*,
https://doi.org/10.1007/978-3-319-55530-0_73

carcinogenic to humans. This classification is based on the increased incidence of primary skin cancer, as well as lung and urinary bladder cancer after exposure to arsenic. Exposure to arsenic typically occurs by oral consumption of contaminated drinking water, soil, and food or by inhalation in an industrial work setting. The main exposure route to inorganic arsenic remains dietary, particularly in infants. This review describes our current understanding of the molecular mechanisms through which arsenic causes harm, although the toxic effects associated with inorganic arsenic exposure are not well understood. Arsenic toxicokinetics varies depending on its form and on several factors such as life-stage, gender, nutritional status, and genetic polymorphisms. MicroRNAs play a key role in many physiological and pathological cellular processes, and they are powerful regulators of gene expression under inorganic arsenic exposure. Several *in vitro* and *in vivo* studies on the effect of inorganic arsenic exposure on the microRNA expression profile showed that micro-RNAs misregulation is involved in a variety of human tumors and in angiogenesis.

Keywords
Arsenic · Inorganic arsenic · Metabolic pathway · Carcinogenesis · Human exposure · Mechanism of action · Toxicity · microRNA expression · Contaminated water · Dietary intake

List of Abbreviations

As	As
AS3MT	Human As methyltransferase
AsIII	Arsenite
AsT cell	As transformed cell
AsV	Arsenate
CCA	Chromated copper arsenate
DMAA	Dimethylarsinic acid
DNMT	DNA methyltransferase
EMT	Epithelial–mesenchymal transition
GSH	Glutathione
HBEC	Human bronchial epithelial cell
HELF	Human embryo lung fibroblast cell
iAs	Inorganic arsenic
MCL	Maximum contaminant level
mESCs	Mouse embryonic stem cells
miRNA	microRNA
MMAA	Monomethylarsonic acid
PHLPP	PH domain leucine-rich repeat protein phosphatase
ROS	Reactive oxygen species
RT-qPCR	Quantitative reverse transcription PCR
SAM	S-adenosyl-methionine
SUMO	Small ubiquitin-like modifier
VEGF	Vascular endothelial growth factor protein

Introduction

Inorganic arsenic (iAs) poses a major threat to worldwide human health. Arsenic (As) is a metalloid, occurring naturally in the Earth's crust generally as one of the several sulfides or as metal arsenates (WHO 2001). As is also a product of anthropogenic activities such as smelting, mining, or waste combustion. Both inorganic and organic species of As occur in the environment. Generally, the iAs forms are more toxic than the organic ones (oAs) for human, animal, and plant health. The chemical structure of As, its electronic status, and its bonding properties give rise to a variety of forms in the solid, aqueous, and gaseous states. Inorganic arsenic is present in different oxidation states: −3, 0, +3, and +5 but the prevalent As forms are arsenite (AsIII) and arsenate (AsV). The most common organic forms are monomethylarsonic acid (MMAA) and dimethylarsinic acid (DMAA).

As dispersion into soil and water through the disintegration of rocks and minerals causes As to enter the food chain (Fig. 1).

Arsenic can be released in the atmosphere by natural processes, such as volcanic activity and dissolution of minerals, or anthropogenic causes such as mining, metal smelting, and pesticide use. It is mainly adsorbed on particles, which are dispersed by winds and deposited on land and water, reaching high concentration mainly in industrial areas (WHO 2001).

The mean concentration of As in soil depends on the physicochemical structure of the soil and it is higher in igneous rock compared to sedimentary rock (Smith et al. 1998). Furthermore, the interaction among soil, plant, and microorganisms may influence the As mobility and bioavailability. Under oxidizing conditions, arsenic is present in the +5 oxidation state and does not coprecipitate nor is adsorbed with the exception of the coprecipitation induced by iron hydroxides (O'Day 2006).

Arsenic water contamination is one of the major health concerns in most countries, in particular in areas, such as Bangladesh, where the major risk of contamination is represented by drinking water from tube wells (Flanagan et al. 2012). Arsenic in water exists in different oxidized or reduced forms depending on the redox potential, pH, temperature, salinity, and on the presence and distribution of microorganisms and biotic component in general (EPA 2001; EPA 2002; Wakao et al. 1988).

The World Health Organization (WHO) guideline value for iAs in drinking water was provisionally reduced in 1993 from 50 to 10 $\mu g\ l^{-1}$ and the same threshold limit has been established for natural mineral waters by the Commission Directive 2003/40/EC (EFSA 2014). The new recommended value was based on the increasing awareness of the toxicity of As, particularly its carcinogenicity, and on the development of sensitive methods to measure it quantitatively. In recent years, there has been a disparity between WHO guideline values and current national standards for drinking water sources. For example, the iAs drinking water limit for Bangladesh is 50 $\mu g\ l^{-1}$, while the American state of New Jersey has enforced an iAs drinking water standard, *Maximum Contaminant Level* (MCL), of 5 $\mu g\ l^{-1}$ instead of the federal MCL of 10 $\mu g\ l^{-1}$. Australia also has a stringent standard for iAs in drinking

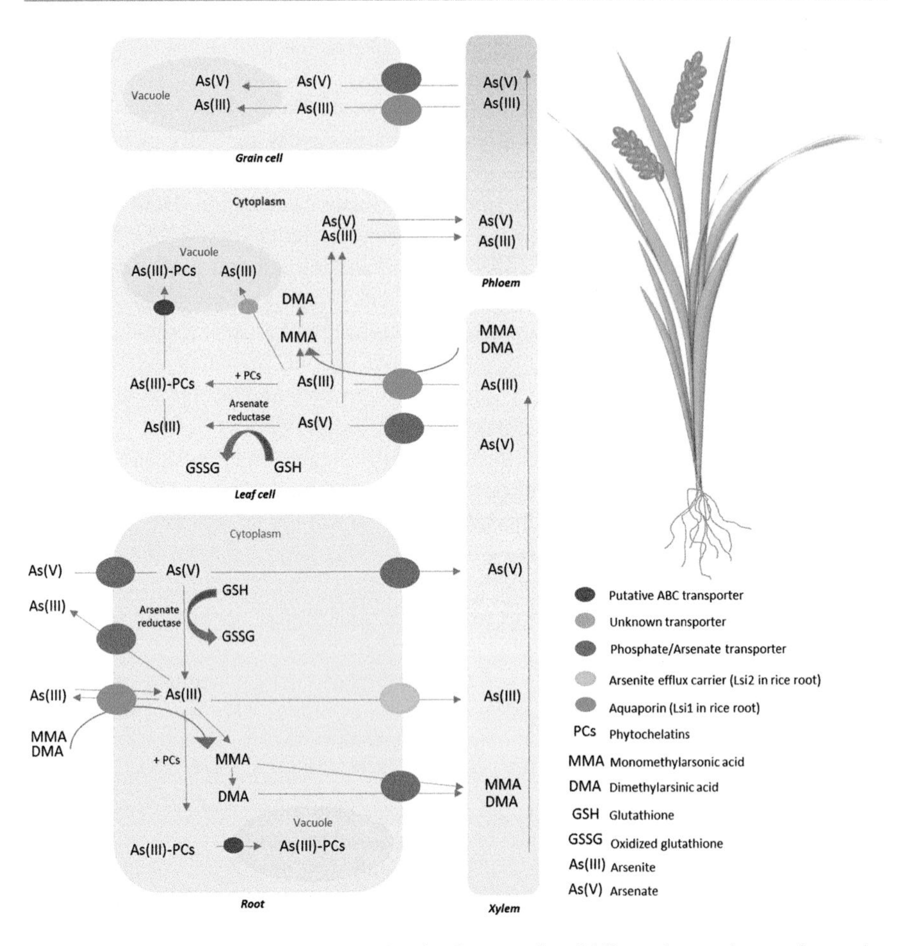

Fig. 1 Arsenic uptake and translocation in *Oryza sativa*. Different inorganic arsenic species uptake from contaminated soil and groundwater in a hyperaccumulating plant and their translocation from roots to grains (Modified by Zhao et al. 2010)

water. In Europe, there is no common recommended maximum level established for As in food, although some member states have their own national guidelines. In the United States, As is used in some veterinary drugs, including those used to treat animals used for commercial food products.

Human Exposure

The International Agency for Research on Cancer (IARC) classified iAs as "carcinogenic to humans" (Group 1) based on sufficient evidence of carcinogenicity in humans (IARC 2012).

Worldwide the most common route of iAs exposure is through the consumption of drinking water contaminated with natural sources of this metalloid.

Human exposure to As typically results from either oral consumption through contaminated drinking water, soil, and food or As inhalation and skin contact in industrial work-settings. Many industrial processes involve the use of arsenic. For instance, chromated copper arsenate (CCA) was intensively used to treat wood as a preservative agent and represents an important source of exposure to arsenic, in particular in the United States. Arsenic contamination may also occur in soil surrounding glass and pharmaceutical industries or metal smelters and mining activities (Ramirez-Andreotta et al. 2013) (Fig. 2). One of the main nonoccupational sources of iAs for human exposure in most European and non-European countries is represented by drinking water and by food, especially cereals, milk, and seafood. For instance, in the Bengal Delta Plain of Bangladesh and West Bengal, India, As in groundwater has emerged as the largest environmental health disaster, putting at least 100 million people at risk of cancer and other As-related diseases (Smith et al. 2000). In Northern Chile, the limited drinking water and irrigation water sources are contaminated by As leading to chronic arsenicosis in resident population. Dietary intake is another common route of exposure to iAs, especially for rice consumers belonging to specific ethnic groups.

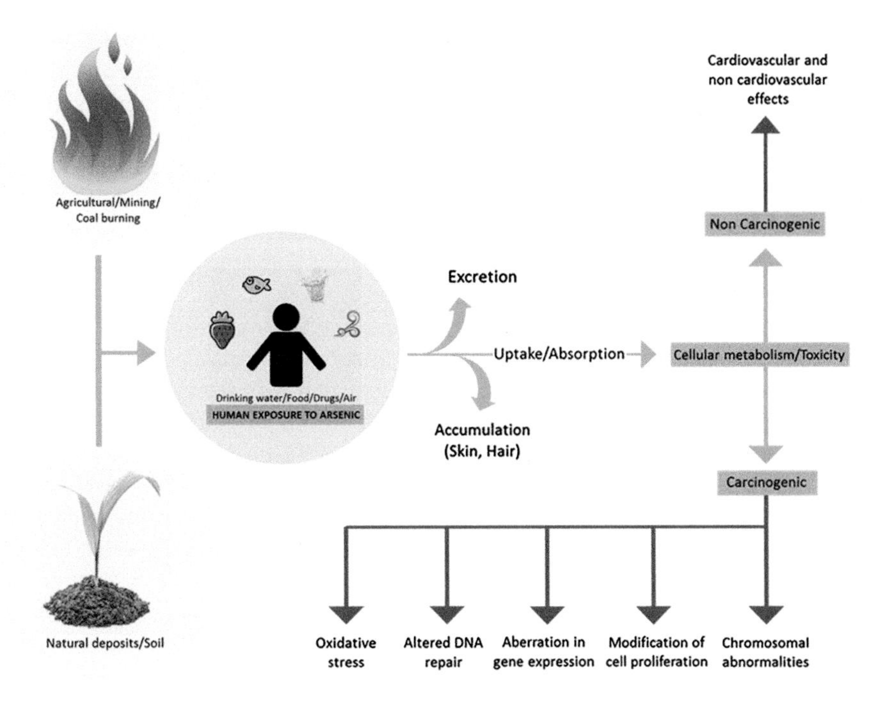

Fig. 2 Source and effects of arsenic human exposure. Human health effects of As are directly correlated with contaminated drinking water consumption and consist of cancer and several nonmalignant diseases

In Italy, high As concentrations were determined on water, marine and river sediments, and on soils only in Lazio and Campania regions. In hydrothermal conditions, As can be selectively mobilized reaching concentrations of the order of mg/l.

In 2009, the EFSA highlighted the toxicological risk associated with young children eating food contaminated with high levels of iAs, exacerbated by the fact that infants consume a large amount of food compared to their body mass. The Joint FAO/WHO Expert Committee on Food Additives (JECFA) established a limit of between 0.3 and 8 μg/kg b.w. per day for an increased risk of skin lesions and cancer of the lung, skin, and bladder (European Food Safety Authority EFSA 2009), and introduced a provisional tolerable weekly intake (PTWI) of 15 μg/kg b.w.

Most of the occurrence data for As in food collected for official food control are reported as total As (tAs) without differentiating between the various As species. The need for speciation data is evident, especially in food associated with emerging risks, such as rice-based infant food. The food subclasses of cereal grains and cereal-based products, followed by food for special dietary uses, bottled water, coffee and beer, rice grains and rice-based products, fish, and vegetables were identified as the main contributors to iAs exposure in the general European population. Meharg et al. (2008) reported that the risk for adverse health effects is higher for exposure occurring during childhood compared to adulthood.

Arsenic toxicity has been well characterized and the cancerogenic effect of human exposure to As has been demonstrated for cancer of skin, lungs, bladder, kidney, and liver (Argos 2015). The molecular mechanisms associated with these effects have been hypotesized to be epigenetic processes. The polymorphism of the genotype may affect the individual response to iAs toxicity and its metabolism within the population (Tantry et al. 2015). Exposure to As has been correlated to epigenetic changes, such as DNA methylation, hypo- and hypermethylation at different loci, and changes in microRNA (miRNA) expression profiling (Reichard and Puga 2010), although *in vivo* studies in human populations exposed to As are necessary for better understanding these correlations (Ren et al. 2011). Arsenic can cause suppression of specific genes involved in the DNA repair pathway and it can interfere with the control of the cell cycle. Furthermore, As is a powerful endocrine disruptor: it can affect gene regulation by interacting with steroid hormones receptors that can result in adverse effects on human development (Bodwell 2006) with different modulations on gene transcription at low or high doses.

The effects of high-dose exposure to As on human health include severe consequences such as neurological disorders; cancer; and liver, renal, fertility, and cardiovascular diseases, including systemic toxicity and death. Chronic exposure to a low dose of As can result in cardiovascular diseases and other disorders such as skin changes and skin cancer, peripheral sensorimotor neuropathy, and diabetes mellitus (Naujokas et al. 2013). Results have been reported that evidence a correlation between exposure to a low level of As in utero and a reduced birth weight, and other authors have linked exposure to As to spontaneous or neonatal death (Marsit 2015). A study by Rebuzzini et al. (2015) regarding the understanding of specific human cardiac diseases caused by in utero exposure to As demonstrated a disruption at each step of developmental cellular process by As trioxide. Such work is the first example of *in vitro* model study performed using the whole 15 days of *in vitro* differentiated mouse embryonic stem cells (mESCs) into cardiomyocytes.

Arsenic Mechanism of Action

The molecular mechanisms underlying the toxic effects associated with iAs exposure are not well understood. Arsenic toxicokinetics varies depending on the As form and on different factors such as life stage, gender, nutritional status, and genetic polymorphisms.

Oral absorption through drinking water and food is the main form of exposure to iAs, with a high rate of absorption in the gastrointestinal tract (typically $>70\%$). Absorption via inhalation is relatively limited and via dermal exposure is even more limited (NRC 1999; WHO 2001).

The absorption of iAs is influenced by the solubility of the arsenic compound, the presence of other food constituents and nutrients in the gastrointestinal tract, and by the food matrix itself [As(III) and As(V) in drinking water are almost completely and rapidly absorbed] (EFSA CONTAM Panel 2009).

The metabolic pathway of iAs is a complex process that gives rise to different organic species.

In humans, the product of the first methylation is monomethylarsonic acid (MMA) and the product of the second and final methylation is dimethylarsinic acid (DMA). DMA and MMA were classified as potentially carcinogenic (IARC Group 2B). Arsenobetaine and other organic As compounds not metabolized in humans are "not classifiable as to their carcinogenicity to humans" (Group 3) (IARC 2012).

After oral ingestion, iAs is absorbed in the intestine, primarily methylated in the liver, and excreted in urine by most species, mainly as DMA and smaller amounts of inorganic and organic AsV and AsIII species.

The metabolic pathway of iAs consists of two sequential reactions: first the inorganic pentavalent As (AsV) is converted to trivalent As (AsIII), which is then methylated to monomethylated and dimethylated arsenicals (MMA, DMA, respectively) (Fig. 3).

Another recently discovered pathway is based on the formations of As-GSH complexes (Hayakawa 2005), which are methylated by AS3MT via transfer of the methyl group from S-adenosyl-methionine (SAM).

Metabolism of inorganic As varies between human populations (Hughes 2006). Such variability is due to genetic polymorphisms in the regulation of the enzyme(s) that metabolize arsenic, which may lead to differences in the toxicity related to As exposure (De Chaudhuri 2008; Schläwicke Engström et al. 2008).

Our understanding of As metabolism and kinetics in infants is very limited, although studies have been conducted on infants exposed to iAs through drinking water in South America and Asia (Vahter 2009). Vahter and coworkers were among the first to raise concerns that children may be more vulnerable to iAs exposure

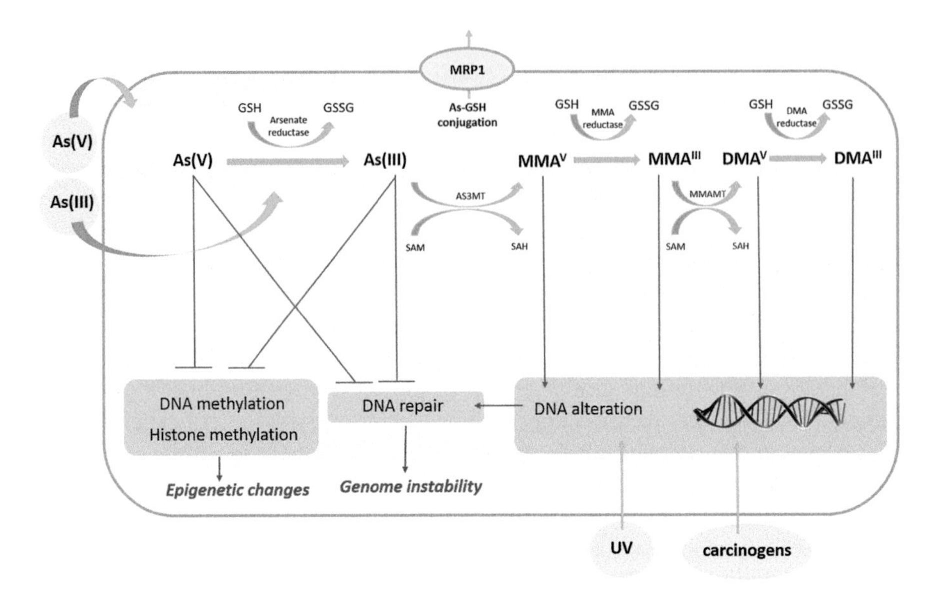

Fig. 3 Metabolic pathway of inorganic arsenic in the liver cell. The main processes, altered after exposure to As, affect the DNA repair and epigenetic changes

compared to adults. In 2009, Fängström et al. demonstrated that the urinary metabolite pattern of children (18 months of age) in Bangladesh is indicative of a decreased As methylation efficiency during weaning. Similar studies that take into account infant exposure to iAs through consumption of rice-based products or drinking water are not available within the EU or any other part of the world. The most important conclusion of the EFSA Scientific Opinion on As in Food (2009) was that the dietary exposure to iAs should be reduced. It is necessary to produce speciation data for different food commodities to obtain dietary exposure assessment and dose-response data for the possible health effects.

According to several toxicological studies, As exposure leads to global changes in gene expression, in particular in epigenetic processes. Epigenetic alterations do not cause a genotoxic effect as they do not involve modification of the DNA sequence, but they can lead to heritable changes in the regulation of gene expression (Feinberg and Tycko 2004). Arsenic induces cancer and changes in gene expression, though maintaining genome stability; changes occur predominantly in CpG islands that are cytosine-rich gene regions (Yoder et al. 1997).

Arsenic exposure has been associated to global genomic DNA hypomethylation that causes illegitimate recombination events and transcriptional deregulation of affected genes. The hypomethylation results from insufficiency of the unique methyl group donor in each conversion step of biomethylation of arsenic, the SAM and from gene expression reduction of the enzyme that catalyzes the As methylation, the DNA methyltransferase (DNMT). Arsenic can affect both people directly exposed and those of future generations in a heritable manner (Ren et al. 2011).

Different epigenetic mechanisms, such as altered DNA methylation, histone modification, and miRNAs expression, have been proposed to play a role in arsenic-induced carcinogenesis. In mammals, DNA methylation is the basic process of gene expression regulation and may cause a reduction of mRNA levels and DNMTs activities.

Arsenic and miRNA Expression Profiles

MiRNAs are endogenous, single-stranded, noncoding RNA molecules ~ 21 nucleotides long which act as negative regulators of gene expression at the post-transcriptional level (Bartel 2009). They exert modulatory roles in crucial cellular processes such as development, differentiation, and apoptosis.

Due to their broad impact on gene regulation, individual miRNAs, or a combination of them, have a potential involvement in a wide range of human diseases such as inflammatory, autoimmune, and metabolic diseases (Sonkoly and Pivarcsi 2009).

It is widely recognized that miRNAs are misregulated in a variety of human tumors, and they act as either oncogenes or tumor suppressors (Yin et al. 2012). In 2013, Xu et al. demonstrated the biphasic effects of arsenite, showing that levels of arsenite in the range of 1.0–2.5 μM induce cell proliferation, while higher levels of arsenite (10–100 μM) induce DNA damage and apoptosis (Xu et al. 2013). In the last years, several experiments based on microarray and RT-qPCR analysis have been conducted both *in vivo* and *in vitro* in order to study miRNA expression changes triggered by iAs exposure.

Despite the significant progress made toward understanding the mechanisms of action of miRNAs, much less is known about the role played by each individual miRNA, and there is a lack of information regarding the effects of chronic As exposure on miRNA expression *in vivo*.

According to Ren et al. (2015), miRNAs expression in rat liver was altered by As exposure in a concentration-dependent manner. Among the identified arsenic-responsive miRNAs, miR-151 and miR-183 were significantly upregulated. miR-151 has been suggested to play a role in breast cancer metastasis and tumor cell proliferation. miR-183 upregulation was observed in lung, breast, prostate, and cervical cancers, and it seems to be associated with tumor cell migration, metastasis, and invasion (Ren et al. 2015). Furthermore, in human urine, miR-200c and miR-205 are inversely associated with As exposure (Michailidi et al. 2015).

These results suggest that the biologic consequences of altered miRNAs induced by As are complex. Further studies are required in order to understand how changes in individual miRNA contribute to arsenic-induced toxicity.

Specific miRNAs Are Implicated in iAs-Mediated Carcinogenesis

Induction of reactive oxygen species (ROS) in cells is closely related with exposure to heavy metals including arsenic, chromium, and cadmium (Wang et al. 2012). In 2012, Ling et al. highlighted that the upregulation of **miR-21** is an important event in the arsenite-induced malignant transformation of human embryo lung fibroblast cells (HELF) mediated by ROS activation of the ERK/NF-κB pathway. It was demonstrated that miR-21 upregulation induces angiogenesis and enhances the invasive potential of transformed cells. Subsequent studies in 2014 by Luo et al. showed that miR-21 acts on PDCD4, an inhibitor of neoplastic transformation, contributing to the EMT (epithelial–mesenchymal transition) induced by arsenite. Wang et al. (2011) treated immortalized p53-knocked down human bronchial epithelial cells ($p53^{low}$HBECs) with low levels of arsenite for 16 weeks ($NaAsO_2$, 2.5 μM). They reported for the first time a causal role for reduction of **miR-200b** expression in human cell malignant transformation and tumor formation resulting from As exposure. Chronic arsenite exposure causes downregulation of miR-200b. This phenomenon can trigger the expression of the EMT-inducing transcription factors ZEB1 and ZEB2, two direct targets of this miRNA. Moreover, the authors highlighted a potential interplay between multiple forms of epigenetic regulations: ZEB1 and ZEB2 are not only targets of miR-200, but they also repress the expression of the miR-200 genes by increasing the methylation of its promoter (Vrba et al. 2010), resulting in a double-negative feedback loop (Bracken et al. 2008; Burk et al. 2008). iAs was also found to trigger a fivefold induction of **miR-190** in a dose-dependent manner in human lung epithelial (BEAS-2B) cells exposed to 20 μM $As3^{+}$ (Beezhold et al. 2011). miR-190 overexpression triggers the downregulation of its target gene PHLPP (PH domain leucine-rich repeat protein phosphatase), which in turns results in an enhanced activation of Akt signaling and in an increased expression of vascular endothelial growth factor protein (VEGF). These events result in enhanced cell proliferation and carcinogenic transformation (Table 1).

Table 1 MicroRNAs implicated in inorganic arsenic-mediated carcinogenesis

Carcinogenesis				
MicroRNA	Up/down-regulated	Cell type	Effects	References
miR-21	Up	Human embryo lung fibroblast cells (HELF)	ROS activation of the ERK/ NF-κB path way	Ling et al. 2012
	Up	Human bronchial epithelial (HBE) cells	Contributes to the epithelial-mesenchymal transition acting on PDCD4	Luo et al. 2014
miR-200b	Down	Immortalized p5 3-knocked down human bronchial epithelial cells	Expression of the EMT-inducing transcription factors ZEB1 and ZEB2	Wang et al. 2011
miR-190	Up	Human lung epithelial (BEAS-2B) cells	Downregulation of its target gene PHLPP, activation of Akt signaling and increased expression of vascular endothelial growth factor protein (VEGF)	Beezhold et al. 2011

Specific miRNAs are up- or downregulated in the arsenite-induced malignant transformation

iAS-Induced miRNA Dysregulation Promotes Angiogenesis

The process of angiogenesis normally occurs in the embryo, in the placenta, during the menstrual cycle, and during wound-healing (Patella and Rainaldi 2012). Under pathological conditions, such as cancer, angiogenic signaling pathways are induced in order to form new blood vessels from existing vasculatures and this induces tumor growth. A number of miRNAs have been reported to be involved in blood vessel development and angiogenesis by directly or indirectly regulating proangiogenic factors or antiangiogenic factors (Urbich et al. 2008). Some studies have indicated that abnormal angiogenesis is fundamental in the pathogenesis of a number of diseases caused by environmental As exposure (Navas-Acien et al. 2005; Straub et al. 2009). In 2012, an *in vivo* experiment by Cui Y. and colleagues demonstrated that **miR-9 and miR-181b** exhibited a massive decrease of expression upon sodium arsenite injection and that this reduction was reflected in increased angiogenesis levels. The experiment was performed by injecting fertilized eggs with 100 nM sodium arsenite at Hamburger–Hamilton (HH) stages 6, 9, and 12, and harvesting them at HH stage 18. miR-9 and miR-181b are implicated in promoting abnormal angiogenesis in iAs-exposed chick embryos by targeting NRP1, a transmembrane receptor implicated in vascular development (Bielenberg and Klagsbrun 2007). Cui et al (Cui et al. 2012) also investigated the role of miR-9 and miR-181b in sodium arsenite-mediated angiogenesis in the human umbilical vein endothelial cell line EA hy926. Their results indicated that overexpression of miR-9 or miR-181b decreased NRP1 expression, cell migration, and tube formation, supporting involvement of miR-9 and miR-181b in arsenite-induced NRP1 expression and angiogenesis *in vitro* in human. In 2011, Carpenter et al. (2011) established an *in vitro* model of arsenic-induced carcinogenesis by transforming immortalized human lung epithelial BEAS-2B cells via chronic exposure to 1 μM sodium As for 26 weeks (Arsenic-transformed BEAS-2B, AsT). They observed that AsT cells produced higher levels of ROS (Carpenter et al. 2011). In 2014, He et al. observed that basal levels of HIF-1α and COX-2 under normoxia were markedly upregulated in AsT cells. A miRNA microarray experiment analysis was performed to compare the miRNA profiles between BEAS-2B and AsT cells, and **miR-199a-5p** resulted to be the most downregulated miRNA. Previously, He et al. (2012) demonstrated that ROS inhibit miR-199a expression through increase of the promoter methylation of miR-199a gene by DNA methyltransferase-1 (He et al. 2012). Taqman RT-qPCR analysis showed MiR-199a-5p expression level was 100-fold lower in AsT cells which, accordingly, brought to overexpression of HIF-1α, a known direct target of mir-199a (Rane et al. 2009). HIF-1α is one of the major proangiogenic factors through inducing transcriptional activation of vascular endothelial growth factor (VEGF) (Semenza 2000). The authors also validated that COX-2, another potent angiogenic activator for tumor angiogenesis (Xue and Shah 2013), is a novel target of miR-199a. Taken together these data indicate that miRNA dysregulation upon As exposure promotes tumor growth and angiogenesis both *in vitro* and *in vivo* (Table 2).

Table 2 Inorganic arsenic-induced microRNA dysregulation promotes angiogenesis

Angiogenesis				
MicroRNA	Up/down-regulated	Cell type	Effects	References
miR-9	Down	*in vivo* on fertilized eggs	Increased angiogenesis levels by targeting NRP1, a transmembrane receptor implicated in vascular development	Cui et al. 2012; Bielenberg et al. 2007; Staton et al. 2007
miR-181b	Down	*in vivo* on fertilized eggs	Increased angiogenesis levels by targeting NRP1, a transmembrane receptor implicated in vascular development	Cui et al. 2012; Bielenberg et al. 2007; Staton et al. 2007
miR-199a-5p	Down	Arsenic-transformed BEAS-2B cells (As T cells)	Up-regulation of HIF-1α and COX-2, two angiogenic activators	He et al. 2014

The downregulation of different miRNAs cause abnormal angiogenesis, a fundamental process in the pathogenesis of a number of diseases caused by environmental As exposure

Arsenic Exposure Impairs DNA Damage Repair and Inhibits Apoptosis Through Specific miRNA Dysregulation

MiR-222, a known oncogene, can increase migration and proliferation of hepatocellular carcinoma (Zhang et al. 2015) and inhibit apoptosis by regulating different targets such as PTEN (Chun-Zhi et al. 2010; Yang et al. 2014), and TIMP3 (Lu et al. 2011). A microarray experiment analysis, also confirmed using RT-PCR and RT-qPCR, showed that miR-222 expression is upregulated the most in arsenic-transformed BEAS-2B cells (As-T cells) (Carpenter et al. 2011), as evidenced by a fourfold higher levels of miR-222 in As-T cells compared to the levels present in B2B cells (Wang et al. 2016). The authors demonstrated that miR-222 directly targets PTEN for inducing the activation of its downstream molecules AKT and ERK in arsenic-transformed cells, inhibiting apoptosis. It is known that upregulation of PI3K/AKT/mTOR pathway is involved in the tumorigenesis of several cancers, particularly through mutations and inactivation of PTEN (Jin et al. 2014; Lavorato-Rocha et al. 2015). In this study, ARID1A was demonstrated to be a new direct target of miR-222. ARID1A, a subunit of the SWItch/sucrose nonfermentable (SWI/SNF) chromatin remodeling complex, has been found to be expressed at low levels in many cancers (Nagymanyoki et al. 2015). An overexpression of miR-222 (twofold), **miR-221** (threefold), and **miR-638** (2.6 fold) was reported by Sturchio et al. (2014) in Jurkat cell line upon 144h iAs exposure. Accordingly, the overexpression of these miRNAS corresponded to the decreasing expression of their putative target gene, the ring finger protein 4 (RNF4). RNF4 is a member of the family of SUMO targeted ubiquitin ligases, and it has a role in active DNA demethylation involving enzymes implicated in DNA repair (Hu et al. 2010). Previously, Marsit and colleagues (Marsit et al. 2006) evidenced the same trend in human immortalized lymphoblast cell line TK-6 and hypothesized that this upregulation could be due to a change in As methylation patterns as a result of reduced levels of S-adenosylmethionine. Moreover, miR-638 is known to be upregulated in oxidative stress conditions, which suggests that miR-638 may play a role in the generalized cellular response to iAs-induced oxidative stress (Simone et al. 2009). Sturchio et al. also showed a time-dependent upregulation of the miR-663 gene expression and at the same time a downregulation of TGFβ1 expression, a validated target of **miR-663** (Tili et al. 2010)

Table 3 Inorganic arsenic-induced microRNA dysregulation impairs DNA damage repair and inhibits apoptosis

Impairs DNA damage repair and inhibits apoptosis				
MicroRNA	Up/down-regulated	Cell type	Effects	References
miR 222	Up	Hepatocellular carcinoma	Inhibit apoptosis by regulating different targets such as FTEN	Chun-Zhi et al. 2010
	Up	Human hepatocellular carcinoma HepG2 cells	Inhibit apoptosis by regulating different targets such as p27	Yang et al. 2014
	Up	Tamoxifen-resistant MCF-7 (OHT(R)) cells and Her2-positive human breast tumors	Inhibit apoptosis by regulating different targets such as TIMFE	Lu et al. 2011
miR-222	Up	Ars enic-transformed BEAS-2B (As-T cells)	Inhibit apoptosis by regulating different targets such as FTEN	Carpenter et al. 2011; Wang et al. 2016
	Up	Jurkat cell line	Targets the Ring Finger Protein 4 (RNF4), that has a role in DNA repair	Sturchio et al. 2014
	Up	Human immortalized lymphoblast cell line TK-6	Targets the Ring Finger Protein 4 (RNF4), that has a role in DNA repair	Marsit et al. 2006
miR-221	Up	Jurkat cell line	Targets the Ring Finger Protein 4 (RNF4), that has a role in DNA repair	Sturchio et al. 2014
miR-638	Up	Jurkat cell line	Targets the Ring Finger Protein 4 (RNF4), that has a role in DNA repair	Sturchio et al. 2014
miR-663	Up	Jurkat cell line	Down-regulation of TGFβl expression	Sturchio et al. 2014

miRNAs dysregulation upon iAs exposure triggers pathways which bring to inhibited cell apoptosis and impaired DNA damage repair, thus prompting tumor growth

and one of the most commonly altered cellular signaling pathways in human cancers. In physiological conditions, TGFβ1 inhibits cell growth, migration, differentiation, apoptosis, and matrix organization (Inman and Allday 2000). Taken together, these results indicate that miRNA dysregulation upon iAs exposure triggers pathways which bring to inhibited cell apoptosis and impaired DNA damage repair, prompting tumor growth (Table 3).

Conclusion

This review evidences that several recommendations regarding As exposure (including knowledge and data gaps) are still to be addressed:

1. Dietary exposure to iAs should be reduced.
2. In order to refine risk assessment of iAs, there is a need to produce speciation data for different food to support dietary exposure assessment and dose-response data for the possible health effects.
3. Future epidemiological studies should incorporate better characterization of exposure to iAs, including food sources.
4. There is a need for more information on critical age periods of As exposure, in particular in early life.
5. There is a need for improved understanding of the human metabolism of organoarsenicals in foods (arsenosugars, arsenolipids, etc.) and the corresponding human health implications.
6. It should be necessary to grow knowledge about coexposure to different factors (i.e., As and uv) on human to develop putative therapeutics for arsenic-induced cancer in occupational and non occupational exposure.
7. Studies regarding miRNA expression profiling highlighted that iAs induced miRNA dysregulation triggers pathways which bring to inhibited cell apoptosis and impaired DNA damage repair, prompting tumor growth.

Future research should focus on the development of new miRNA profile for preventive or therapeutic use.

Key Facts of Arsenic Adverse Effects on Health

- Inorganic arsenic poses a major threat to worldwide human health.
- International Agency for Research on Cancer classified arsenic and inorganic arsenic compounds as "carcinogenic to humans" (Group 1) based on sufficient evidence of carcinogenicity in humans.
- One of the major routes of human exposure is represented by dietary intake in most European and non-European countries.

- The molecular mechanisms underlying the toxic effects of inorganic arsenic exposure are not well understood.
- A chronic arsenic exposure may alter miRNAs expression profiling.

Dictionary of Terms

- **DNA repair pathway** – Set of chemical reactions involved in the processes to correct DNA damage.
- **Endocrine disruptor** – A molecule that mimes a natural hormone, interfering with hormonal equilibrium.
- **Metabolic pathway of inorganic arsenic** – Cellular detoxification process under inorganic arsenic exposure.
- **Reactive oxygen species** – Chemical species containing oxygen that have a crucial role in cell signaling.
- **Apoptosis** – Programmed cellular death.
- **Angiogenesis** – Normal physiological process in growth and development for granulation tissue formation.

Summary Points

- This chapter focuses on inorganic arsenic and its effects on human health.
- Arsenic is a ubiquitous metalloid, widely distributed in the environment.
- Exposure to inorganic arsenic typically results from either oral arsenic consumption through contaminated drinking water, soil, and food or arsenic inhalation in an industrial work setting.
- The data show that the main exposure route to inorganic arsenic remains dietary, particularly for young infants. Therefore, the human exposure to food and drinking water As contamination is a real concern.
- International Agency for Research on Cancer (IARC) classified inorganic arsenic as "carcinogenic to humans" (Group 1) based on sufficient evidence of carcinogenicity in humans.
- Many of the human health effects of inorganic arsenic may be considered: cancers of the skin, lung, bladder, liver, and kidney, neurologic disease, cardiovascular disease, as well as other nonmalignant diseases.
- We examined the effects of inorganic arsenic exposure on miRNAs expression profile through *in vitro* and *in vivo* studies.
- It is widely recognized that miRNAs are misregulated in a variety of human tumors acting as both oncogenes and tumor suppressors.
- Several authors showed that inorganic arsenic-induced microRNA dysregulation promotes angiogenesis and carcinogenesis.
- However, we argue that a list of miRNAs may be considered as potential biomarkers of inorganic arsenic effects.

References

Argos M (2015) Arsenic exposure and epigenetic alterations: recent findings based on the illumina 450K DNA methylation array. Curr Environ Health Rep 2(2):137–144. https://doi.org/10.1007/s40572-015-0052-1

Bartel DP (2009) MicroRNAs: target recognition and regulatory functions. Cell 136:215–233. https://doi.org/10.1016/j.cell.2009.01.002

Beezhold K, Liu J, Kan H, Meighan T, Castranova V, Shi X, Chen F (2011) miR-190-mediated downregulation of PHLPP contributes to arsenic-induced Akt activation and carcinogenesis. Toxicol Sci 123(2):411–420. https://doi.org/10.1093/toxsci/kfr188

Bielenberg DR, Klagsbrun M (2007) Targeting endothelial and tumor cells with semaphorins. Cancer Metastasis Rev 26:421–431. https://doi.org/10.1007/s10555-007-9097-4

Bodwell JE (2006) Arsenic disruption of steroid receptor gene activation: complex dose-response effects are shared by several steroid receptors. Chem Res Toxicol 19:1619–1629. https://doi.org/10.1021/tx060122q

Bracken CP, Gregory PA, Kolesnikoff N, Bert AG, Wang J, Shannon MF, Goodall GJ (2008) A double-negative feedback loop between ZEB1-SIP1 and the microRNA-200 family regulates epithelial-mesenchymal transition. Cancer Res 68:7846–7854. https://doi.org/10.1158/0008-5472.CAN-08-1942

Burk U, Schubert J, Wellner U, Schmalhofer O, Vincan E, Spaderna S, Brabletz T (2008) A reciprocal repression between ZEB1 and members of the miR-200 family promotes EMT and invasion in cancer cells. EMBO Rep 9:582–589. https://doi.org/10.1038/embor.2008.74

Carpenter RL, Jiang Y, Jing Y, He J, Rojanasakul Y, Liu LZ, Jiang BH (2011) Arsenite induces cell transformation by reactive oxygen species, AKT, ERK1/2, and p70S6K1. Biochem Biophys Res Commun 414:533–538. https://doi.org/10.1016/j.bbrc.2011.09.102

Chun-Zhi Z, Lei H, An-Ling Z, Yan-Chao F, Xiao Y, Guang-Xiu W, Zhi-Fan J, Pei-Yu P, Qing-Yu Z, Chun-Sheng K (2010) MicroRNA-221 and microRNA-222 regulate gastric carcinoma cell proliferation and radioresistance by targeting PTEN. BMC Cancer 10:367. https://doi.org/10.1186/1471-2407-10-367

Cui Y, Han Z, Hu Y, Song G, Hao C, Xia H, Ma X (2012) MicroRNA-181b and microRNA-9 mediate arsenic-induced angiogenesis via NRP1. J Cell Physiol 227(2):772–783. https://doi.org/10.1002/jcp.22789

De Chaudhuri S (2008) Genetic variants associated with arsenic susceptibility: study of purine nucleoside phosphorylase, arsenic (+3) methyltransferase, and glutathione s-transferase omega genes. Environ Health Perspect 116:501–505. https://doi.org/10.1289/ehp.10581

European Food Safety Authority (EFSA) (2009) Scientific opinion on arsenic in food; EFSA panel on contaminants in the food chain (CONTAM). EFSA J 7(10):1351

European Food Safety Authority (EFSA) (2014) Dietary exposure to inorganic arsenic in the European population. EFSA J 12(3):3597. 68 pp

Feinberg AP, Tycko B (2004) The history of cancer epigenetics. Nat Rev Cancer 4(2):143–153. https://doi.org/10.1038/nrc1279

Flanagan SV, Johnston RB, Zheng Y (2012) Arsenic in tube well water in Bangladesh: health and economic impacts and implications for arsenic mitigation. B World Health Organ 90:839–846. https://doi.org/10.2471/BLT.11.101253

Hayakawa T (2005) A new metabolic pathway of arsenite: arsenic-glutathione complexes are substrates for human arsenic methyltransferase Cyt19. Arch Toxicol 79:183–191. https://doi.org/10.1007/s00204-004-0620-x

He J, Xu Q, Jing Y, Agani F, Qian X, Carpenter R et al (2012) Reactive oxygen species regulate ERBB2 and ERBB3 expression via miR-199a/125b and DNA methylation. EMBO Rep 13:1116–1122. https://doi.org/10.1038/embor.2012.162

He J, Wang M, Jiang Y, Chen Q, Xu S, Xu Q, Jiang BH, Liu LZ (2014) Chronic arsenic exposure and angiogenesis in human bronchial epithelial cells via the ROS/miR-199a-5p/HIF-1α/COX-2 pathway. Environ Health Perspect 122(3):255–261. https://doi.org/10.1289/ehp.1307545

Hu XV, Rodrigues TM, Tao H, Baker RK, Miraglia L, Orth AP, Lyons GE, Schultz PG, Wu X (2010) Identification of RING finger protein 4 (RNF4) as a modulator of DNA demethylation through a functional genomics screen. Proc Natl Acad Sci U S A 107(34):15087–15092. https://doi.org/10.1073/pnas.1009025107

Hughes MF (2006) Biomarkers of exposure: a case study with inorganic arsenic. Environ Health Perspect 114(11):1790–1796

Inman GJ, Allday MJ (2000) Apoptosis induced by TGF-b1 in Burkitt's lymphoma cells is caspase 8 dependent but is death receptor independent. J Immunol 165:2500–2510

International Agency for Research on Cancer (IARC) (2012) Arsenic, metals, fibres and dusts. IARC monographs on the evaluation of carcinogenic risk to humans, vol. 100 C:11–465. IARC Working Group on the Evaluation of Carcinogenic Risks to Humans

Jin Y, Li Y, Pan L (2014) The target therapy of ovarian clear cell carcinoma. Oncol Targets Ther 7:1647–1652. https://doi.org/10.2147/OTT.S49993

Lavorato-Rocha AM, Anjos LG, Cunha IW, Vassallo J, Soares FA, Rocha RM (2015) Immunohistochemical assessment of PTEN in vulvar cancer: best practices for tissue staining, evaluation, and clinical association. Methods 77–78:20–24. https://doi.org/10.1016/j.ymeth.2014

Ling M, Li Y, Xu Y, Pang Y, Shen L, Jiang R, Zhao Y, Yang X, Zhang J, Zhou J, Wang X, Liu Q (2012) Regulation of miRNA-21 by reactive oxygen species-activated ERK/NF-κB in arsenite-induced cell transformation. Free Radic Biol Med 52(9):1508–1518. https://doi.org/10.1016/j.freeradbiomed.2012.02.020

Lu Y, Roy S, Nuovo G, Ramaswamy B, Miller T, Shapiro C, Jacob ST, Majumder S (2011) Anti-microRNA-222 (anti-miR-222) and −181B suppress growth of tamoxifen-resistant xenografts in mouse by targeting TIMP3 protein and modulating mitogenic signal. J Biol Chem 286:42292–42302. https://doi.org/10.1074/jbc.M111.270926

Luo F, Ji J, Liu Y, Xu Y, Zheng G, Jing J, Wang B, Xu W, Shi L, Lu X, Liu Q (2014) MicroRNA-21, up-regulated by arsenite, directs the epithelial-mesenchymal transition and enhances the invasive potential of transformed human bronchial epithelial cells by targeting PDCD4. Toxicol Lett 232:301–309. https://doi.org/10.1016/j.toxlet.2014.11.001

Marsit CJ (2015) Influence of environmental exposure on human epigenetic regulation. J Exp Biol 218:71–79. https://doi.org/10.1242/jeb.106971

Marsit CJ, Eddy K, Kelsey KT (2006) MicroRNA responses to cellular stress. Cancer Res 66(22):10843–10848. https://doi.org/10.1158/0008-5472.CAN-06-1894

Meharg AA, Sun G, Williams PN, Adamako E, Deacon C, Zhu YG, Feldmann J, Raab A (2008) Inorganic arsenic levels in baby rice are of concern. Environ Pollut 152:746–749. https://doi.org/10.1016/j.envpol.2008.01.043

Michailidi C, Hayashi M, Datta S, Sen T, Zenner K, Oladeru O, Brait M, Izumchenko E, Baras A, VandenBussche C, Argos M, Bivalacqua TJ, Ahsan H, Hahn NM, Netto GJ, Sidransky D, Hoque MO (2015) Involvement of epigenetics and EMT-related miRNA in arsenic-induced neoplastic transformation and their potential clinical use. Cancer Prev Res 8(3):208–221

Nagymanyoki Z, Mutter GL, Hornick JL, Cibas ES (2015) ARID1A is a useful marker of malignancy in peritoneal washings for endometrial carcinoma. Cancer Cytopathol 123:253–257. https://doi.org/10.1002/cncy.21514

Naujokas MF, Anderson B, Ahsan H, Aposhian HV, Graziano JH, Thompson C, Suk WA (2013) The broad scope of health effects from chronic arsenic exposure: update on a worldwide public health problem. Environ Health Perspect 121(3):295–302. https://doi.org/10.1289/ehp.1205875

Navas-Acien A, Sharrett AR, Silbergeld EK, Schwartz BS, Nachman KE, Burke TA, Guallar E (2005) Arsenic exposure and cardiovascular disease: a systematic review of the epidemiologic evidence. Am J Epidemiol 162:1037–1049. https://doi.org/10.1093/aje/kwi330

NRC National Research Council (1999) Arsenic in drinking water. The National Academies Press, Washington, DC

O'Day PA (2006) Chemistry and mineralogy of arsenic. Elements 2:77–83. https://doi.org/10.2113/gselements.2.2.77.

Patella F, Rainaldi G (2012) MicroRNAs mediate metabolic stresses and angiogenesis. Cell Mol Life Sci 69:1049–1065. https://doi.org/10.1007/s00018-011-0775-6

Ramirez-Andreotta MD, Brusseau ML, Artiola JF, Maier RM (2013) A greenhouse and field-based study to determine the accumulation of arsenic in common homegrown vegetables grown in mining-affected soils. Sci Total Environ 443:299–306. https://doi.org/10.1016/j.scitotenv.2012.10.095

Rane S, He M, Sayed D, Vashistha H, Malhotra A, Sadoshima J, Vatner DE, Vatner SF, Abdellatif M (2009) Downregulation of MiR-199a derepresses hypoxia-inducible factor-1α and Sirtuin 1 and recapitulates hypoxia preconditioning in cardiac myocytes. Circ Res 104(7):879–886. https://doi.org/10.1161/CIRCRESAHA.108.193102

Rebuzzini P, Cebral E, Fassina L, Redi CA, Zuccotti M, Garagna S (2015) Arsenic trioxide alters the differentiation of mouse embryonic stem cell into cardiomyocytes. Sci Rep 5:14993. https://doi.org/10.1038/srep14993

Reichard JF, Puga A (2010) Effects of arsenic exposure on DNA methylation and epigenetic gene regulation. Epigenomics 2(1):87–104. https://doi.org/10.2217/epi.09.45

Ren X, McHale CM, Skibola CF, Smith AH, Smith MT, Zhang L (2011) An emerging role for epigenetic dysregulation in arsenic toxicity and carcinogenesis. Environ Health Perspect 119(1):11–19

Ren X, Gailec DP, Gonga Z, Qiud W, Gea Y, Zhangd C, Huangd C, Yand H, Olsona JR, Kavanaghe TJ, Wud H (2015) Arsenic responsive microRNAs in vivo and their potential involvement in arsenic-induced oxidative stress. Toxicol Appl Pharmacol 283(3):198–209. https://doi.org/10.1289/ehp.1002114

Schläwicke Engström K, Nermell B, Concha G, Strömberg U, Vahter M, Broberg K (2008) Arsenic metabolism is influenced by polymorphisms in genes involved in one-carbon metabolism and reduction reactions. Mutat Res 667:4–14. https://doi.org/10.1016/j.mrfmmm.2008.07.003. Epub 2008 Jul 17

Semenza GL (2000) HIF-1: mediator of physiological and pathophysiological responses to hypoxia. J Appl Physiol 88:1474–1480

Simone NL, Soule BP, Ly D, Saleh AD, Savage JE, Degraff W, Cook J, Harris CC, Gius D, Mitchell JB (2009) Ionizing radiation-induced oxidative stress alters miRNA expression. PLoS One 4(7): e6377. https://doi.org/10.1371/journal.pone.0006377

Smith AH, Goycolea M, Haque R, Biggs ML (1998) Marked increase in bladder and lung cancer mortality in a region of northern Chile due to arsenic in drinking water. Am J Epidemiol 147:660–669

Smith AH, Lingas EO, Rahman M (2000) Contamination of drinking-water by arsenic in Bangladesh: a public health emergency. Bull World Health Organ 78(9):1093–1103

Sonkoly E, Pivarcsi A (2009) Advances in microRNAs: implications for immunity and inflammatory diseases. J Cell Mol Med 13(1):24–38. https://doi.org/10.1111/j.1582-4934.2008.00534.x

Staton CA, Kumar I, Reed MWR, Brown NJ (2007) Neuropilins in physiological and pathological angiogenesis. J Pathol 212(3):237–248

Straub AC, Klei LR, Stolz DB, Barchowsky A (2009) Arsenic requires sphingosine-1-phosphate type 1 receptors to induce angiogenic genes and endothelial cell remodeling. Am J Pathol 174:1949–1958. https://doi.org/10.2353/ajpath.2009.081016

Sturchio E, Colombo T, Boccia P, Carucci N, Meconi C, Minoia C, Macino G (2014) Arsenic exposure triggers a shift in microRNA expression. Sci Total Environ 472:672–680. https://doi.org/10.1016/j.scitotenv.2013.11.092. Epub 2013 Dec 7

Tantry BA, Shrivastava D, Taher I, Tantry MN (2015) Arsenic exposure: mechanisms of action and related health effects. J Environ Anal Toxicol 5:327

Tili E, Michaille JJ, Alder H, Volinia S, Delmas D, Latruffe N, Croce CM (2010) Resveratrol modulates the levels of microRNAs targeting genes encoding tumor-suppressors and effectors of TGFβ signaling pathway in SW480 cells. Biochem Pharmacol 80(12):2057–2065. https://doi.org/10.1016/j.bcp.2010.07.003. Epub 2010 Jul 15

Urbich C, Kuehbacher A, Dimmeler S (2008) Role of microRNAs in vascular diseases, inflammation, and angiogenesis. Cardiovasc Res 79:581–588. https://doi.org/10.1093/cvr/cvn156. Epub 2008 Jun 11

U.S. EPA (2001) National primary drinking water regulations: arsenic and clarifications to compliance and new source contaminants monitoring. Final rule. 6976–7066

U.S. EPA (2002) Arsenic treatment technologies for soil, waste and water, EPA-542-R-02-004

Vahter M (2009) Effects of arsenic on maternal and fetal health. Annu Rev Nutr 29:381–399. https://doi.org/10.1146/annurev-nutr-080508-141102

Vrba L, Jensen TJ, Garbe JC, Heimark RL, Cress AE, Dickinson S, Stampfer MR, Futscher BW (2010) Role for DNA methylation in the regulation of miR-200c and miR-141 expression in normal and cancer cells. PLoS One 5:e8697. https://doi.org/10.1371/journal.pone.0008697. Published online 2010 Jan 13

Wakao N, Koyatsu H, Komai Y, Shimokawara H, Sakurai Y, Shiota H (1988) Microbial oxidation of arsenite and occurrence of arsenite-oxidizing bacteria in acid-mine water from a sulfur-pyrite mine. Geomicrobiol J 6:11–24. https://doi.org/10.1080/01490458809377818

Wang Z, Zhao Y, Smith E, Goodall GJ, Drew PA, Brabletz T, Yang C (2011) Reversal and prevention of arsenic-induced human bronchial epithelial cell malignant transformation by microRNA-200b. Toxicol Sci 121(1):110–122. https://doi.org/10.1093/toxsci/kfr029. Epub 2011 Feb 2

Wang X, Mandal AK, Saito H, Pulliam JF, Lee EY, Ke ZJ, Lu J, Ding S, Li L, Shelton BJ, Tucker T, Evers BM, Zhang Z, Shi X (2012) Arsenic and chromium in drinking water promote tumorigenesis in a mouse colitis-associated colorectal cancer model and the potential mechanism is ROS-mediated Wnt/β-catenin signaling pathway. Toxicol Appl Pharmacol 262:11–21. https://doi.org/10.1016/j.taap.2012.04.014. Epub 2012 Apr 19

Wang M, Ge X, Zheng J, Li D, Liu X, Wang L, Jiang C, Shi Z, Qin L, Liu J, Yang H, Liu LZ, He J, Zhen L, Jiang BH (2016) Role and mechanism of miR-222 in arsenic-transformed cells for inducing tumor growth. Oncotarget 7(14):17805–17814. https://doi.org/10.18632/oncotarget.7525. Published online 2016 Feb 20

WHO World Health Organization (2001) Arsenic and arsenic compounds. Environmental Health Criteria 224. World Health Organization, Geneva

Xu Y, Li Y, Li H, Pang H, Zhao Y, Jiang R, Shen L, Zhou J, Wang X, Liu Q (2013) The accumulations of HIF-1α and HIF-2α by JNK and ERK are involved in biphasic effects induced by different levels of arsenite in human bronchial epithelial cells. Toxicol Appl Pharmacol 266(2):187–197. https://doi.org/10.1016/j.taap.2012.11.014. Epub 2012 Nov 27

Xue X, Shah YM (2013) Hypoxia-inducible factor-2α is essential in activating the COX2/mPGES-1/PGE2 signaling axis in colon cancer. Carcinogenesis 34(1):163–169. https://doi.org/10.1093/carcin/bgs313. Epub 2012 Oct 5

Yang YF, Wang F, Xiao JJ, Song Y, Zhao YY, Cao Y, Bei YH, Yang CQ (2014) MiR-222 overexpression promotes proliferation of human hepatocellular carcinoma HepG2 cells by downregulating p27. Int J Clin Exp Med 7:893–902. PMC4057838

Yin Y, Li J, Chen S, Zhou T, Si J (2012) MicroRNAs as diagnostic biomarkers in gastric cancer. Int J Mol Sci 13(10):12544–12555. https://doi.org/10.3390/ijms131012544. Published online 2012 Oct 1

Yoder JA, Walsh CP, Bestor TH (1997) Cytosine methylation and the ecology of intragenomic parasites. Trends Genet 13(8):335–340

Zhang Y, Yao J, Huan L, Lian J, Bao C, Li Y, Ge C, Li J, Yao M, Liang L, He X (2015) GNAI3 inhibits tumor cell migration and invasion and is post-transcriptionally regulated by miR-222 in hepatocellular carcinoma. Cancer Lett 356(2 Pt B):978–984. https://doi.org/10.1016/j.canlet.2014.11.013. Epub 2014 Nov 13

Zhao FJ, McGrath SP, Meharg AA (2010) Arsenic as a food chain contaminant: mechanisms of plant uptake and metabolism and mitigation strategies. Annu Rev Plant Biol 61:535–559. https://doi.org/10.1146/annurev-arplant-042809-112152

Epigenetic Effects of Bisphenol A (BPA): A Literature Review in the Context of Human Dietary Exposure

110

Luísa Camacho and Igor P. Pogribny

Contents

Abstract

Bisphenol A (BPA) is a high production volume industrial chemical used widely in the production of polycarbonate plastics and epoxy resins for manufacturing

---

Disclaimer: The views expressed in this manuscript do not necessarily represent those of the US Food and Drug Administration.

L. Camacho (✉) · I. P. Pogribny
Division of Biochemical Toxicology, National Center for Toxicological Research, Food and Drug Administration, Jefferson, AR, USA
e-mail: Luisa.Camacho@fda.hhs.gov; Igor.Pogribny@fda.hhs.gov

V. B. Patel, V. R. Preedy (eds.), *Handbook of Nutrition, Diet, and Epigenetics*,
https://doi.org/10.1007/978-3-319-55530-0_32

food and drink storage containers, the lining of food cans, dental sealants, medical devices, and thermal paper. Biomonitoring studies show that there is widespread exposure of the human population to BPA, mainly via the diet. The median daily intake of the overall US population is approximately 25 ng BPA/kg body weight (bw)/day, which is estimated to result in serum levels of BPA in the pM range. BPA is an endocrine-disrupting chemical, with known estrogenic activity; more recently, it has been reported to induce epigenetic changes in vitro and in vivo, including effects in DNA methylation, differential histone modifications, and modulation of the levels of noncoding RNAs. This chapter summarizes the studies reporting epigenetic effects associated with human exposure to BPA or induced by the in vitro or in vivo exposure to the chemical. The literature suggesting an association between human exposure to BPA and epigenetic changes is limited. Several studies have assessed the effects of BPA in in vitro cell systems and suggest treatment-related epigenetic effects; the lowest effective BPA dose in vitro was in the nM dose range and the dose-response in these studies tended to be linear. In vivo animal studies suggest epigenetic effects of BPA in a wide range of organs, dose levels, and epigenetic endpoints; however, many studies have limitations, including the use of a single dose level that precludes the characterization of the dose-response of the reported effects. Comprehensive well-designed and well-controlled studies, which include both sexes, multiple and properly spaced BPA doses, multiple time points, and integration of the epigenetic endpoints with other molecular, physiological, and morphological endpoints should provide a better understanding of the potential of BPA to act as an epigenetic modulator.

Keywords
Bisphenol A (BPA) · Endocrine-disrupting chemical · DNA methylation · Histone modification · Noncoding RNA · microRNA · Human · In vitro · Liver · Heart · Brain · Mammary gland · Ovary · Uterus · Placenta · Prostate · Testis · Sperm

List of Abbreviations

BPA	Bisphenol A
bw	Body weight
Casq2	Calsequestrin 2
CLARITY-BPA	Consortium Linking Academic and Regulatory Insights on BPA Toxicity
DNMT	DNA methyltransferase
EFSA	European Food Safety Agency
ER	Estrogen receptor
Esrrg	Estrogen-related receptor-γ
Ezh2	Enhancer of zeste 2 polycomb repressive complex 2 subunit
FDA	Food and Drug Administration
Fkbp5	FK506-binding protein 5
GCK	Glucokinase

GD	Gestational day
GPER	G protein-coupled estrogen receptor
H19	Imprinted maternally expressed transcript
Hdac1	Histone deacetylase 1
HOMA	Homeostatic model assessment
HOTAIR	HOX antisense intergenic RNA
HOX	Homeobox
Hpcal1	Hippocalcin-like 1
IAP	Intracisternal A particle
Igf1	Insulin-like growth factor 1
Igf2r	Insulin-like growth factor 2 receptor
LAMP3	Lysosomal-associated membrane protein 3
LINE-1	Long interspersed nuclear elements-1
lncRNA	Long noncoding RNA
miRNA	microRNA
Nsbp1	Nucleosome-binding protein 1
Pde4d4	Phosphodiesterase type 4D variant 4
Peg3	Paternally expressed gene 3
Pgc-1α	Peroxisome proliferator activated receptor-γ, coactivator 1α
PND	Postnatal day
Scgb2a1	Secretaglobin family 2A member 1
SNORD	Small nucleolar RNAs with C/D motif
Sox2	Sex determining region Y box 2
SREBF	Sterol regulatory element-binding transcription factor
StAR	Steroidogenic acute regulatory protein
Stra8	Stimulated by retinoic acid gene 8
Tpd52	Tumor protein D52
β–Cas	β-casein

Introduction

Bisphenol A (BPA; 2,2-bis(4-hydroxyphenyl)propane; CAS no. 80–05-7) is a high production volume industrial chemical, with an annual production of over 7 billion kilograms (PRNewsWire 2016). BPA is used widely in the production of polycarbonate plastics and epoxy resins for manufacturing food and drink storage containers, the lining of food cans, dental sealants, medical devices, thermal paper, and optical disks (Shelby 2008). Biomonitoring studies suggest widespread exposure of the human population to BPA (Calafat et al. 2008; Teeguarden et al. 2013). The primary route of human exposure to BPA is through ingestion of food containing BPA, due to the migration of the chemical from the packaging material into the food matrix (Shelby 2008). Secondary routes of human exposure to BPA include dermal (Thayer et al. 2016) and inhalation (Loganathan and Kannan 2011). The current estimates of human exposure to BPA indicate that the median daily intake of the overall US population is approximately 25 ng BPA/kg body weight (bw)/day

(LaKind and Naiman 2015). Regulatory agencies across the world, including the US Food and Drug Administration (FDA), European Food Safety Authority (EFSA), and Health Canada, are in agreement that the current exposure levels of the human population to BPA are safe (FDA 2014; EFSA 2015; Health Canada 2012).

The pharmacokinetics of BPA has been studied extensively in multiple species, including rodents (Doerge et al. 2010a, b, 2011a, b, 2012), nonhuman primates (Doerge et al. 2010c; Patterson et al. 2013; Taylor et al. 2011), and humans (Völkel et al. 2002; Teeguarden et al. 2011; Thayer et al. 2015). These studies have shown that, upon oral exposure, BPA is rapidly and efficiently conjugated in the gastrointestinal tract before reaching the systemic circulation. The resulting phase II metabolites, primarily BPA glucuronide, are excreted mainly via urine (primates) or feces (rodents). In rodents, the metabolic capacity of newborns is not fully developed, resulting in a higher level of circulating aglycone BPA in the perinatal period compared to adulthood (Doerge et al. 2010a, 2011a). This developmental metabolic difference is not observed in primates, where newborns are as capable of conjugating and excreting BPA as adults (Doerge et al. 2010c, 2011b). Because of the fast and efficient conjugation and elimination of BPA, the levels of aglycone BPA available to be transferred from the pregnant female to the fetus or via the breast milk to the offspring are attenuated significantly (Doerge et al. 2010a, b, 2011a, b). Animal studies have shown also that, for the same exposure dose of BPA, its internal levels are significantly higher when the chemical is administered via intravenous or subcutaneous routes than via an oral route, because the pre-systemic first-pass metabolism of BPA is bypassed upon parenteral exposure (Doerge et al. 2010a, 2011a, 2012; Fisher et al. 2011). Based on the data from pharmacokinetic studies and computational simulations using physiologically based pharmacokinetic models (Fisher et al. 2011; Yang et al. 2015), the BPA serum concentrations that result from the current levels of human exposure are estimated to be in the pM range (Teeguarden et al. 2013).

Unlike its glucuronide conjugate (Matthews et al. 2001), aglycone BPA was shown to be capable of binding and activating estrogen receptors (ER). The relative binding affinity of aglycone BPA to ERα and ERβ is three to four orders of magnitude lower than that of estradiol (Kuiper et al. 1997, 1998). In agreement, the estrogenic effects of BPA in animals, as evidenced, for example, by increased uterine weight or upregulation of the calbindin D9k (*S100 g*) gene, have been consistently reported in the dose range of hundreds of mg/kg bw/day across animal models and tissues (Chapin et al. 2008). Because of its capacity to interfere with the estrogenic pathway, BPA has been classified as an endocrine-disrupting chemical; however, given its relatively low binding affinity for the ERs, it is unlikely that the internal levels of BPA would be sufficient to induce ER-mediated responses in humans at the current estimated levels of exposure (Teeguarden et al. 2013). In vitro, BPA was shown to bind also to the ER-related receptor-γ (Takayanagi et al. 2006) and to compete with 17β-estradiol for binding to the G protein-coupled ER (GPER) (Thomas and Dong 2006), with a dissociation coefficient of about 6 nM (Okada et al. 2008; Teeguarden et al. 2013). More recently, BPA has been reported to induce epigenetic changes in vitro and in vivo, including effects in DNA methylation, differential histone modifications, and modulation of the levels of noncoding RNAs. These studies will be reviewed here, with a particular focus on important elements of study design, including the BPA dose levels and the window and route of exposure tested. For ease of comparison across studies, the dose level units reported in the cited literature were converted to nM BPA or μg BPA/kg bw/day, as appropriate.

Epigenetic Effects of BPA: A Literature Review

Studies in Exposed Humans

The number of reports on the potential association between human exposure to BPA and epigenetic modifications is limited. Kim et al. (2013) have analyzed the status of DNA methylation in the saliva of Egyptian girls and reported that higher urinary BPA concentrations were generally associated with reduced genome-wide methylation levels of the genomic DNA CpG sites. In a separate study, Faulk et al. (2015) reported that the BPA levels in human fetal liver were associated with increased methylation of CpG islands and decreased DNA methylation at repetitive sequences and CpG shores. Miao et al. (2014) reported that BPA-exposed factory male workers in China had lower levels of methylation in the repetitive element long interspersed nuclear elements-1 (LINE-1) in their sperm compared to workers not exposed to BPA occupationally. The highest tertile of urinary BPA level was also associated with the lowest methylation rate of sperm LINE-1; however, neither of these associations was observed with peripheral blood LINE-1. Additionally, Kundakovic et al. (2015) reported increased DNA methylation of the transcriptionally relevant CpG1A and CpG1B sites in the brain-derived neurotrophic factor (*BDNF*) gene in human cord blood in women with urinary BPA levels >4 μg/L versus those with <1 μg/L; this association was found in the male, but not female, cord blood. By screening the human placenta miRNome, De Felice et al. (2015) reported a positive association between BPA and miR-146a levels in the placenta; however, such association was not confirmed in a separate study by Li et al. (2015). In addition, increased LINE-1 methylation was reported to be associated with increasing levels of BPA in human placenta (Nahar et al. 2015).

Studies in Human Cells In Vitro

Several studies have reported epigenetic effects of BPA in human breast cells in vitro. Weng et al. (2010) reported that treatment of primary human breast epithelial cells with 4 nM BPA for 3 weeks resulted in the silencing of the lysosomal-associated membrane protein 3 (*LAMP3*) gene through CpG island hypermethylation, and Qin et al. (2012) showed that exposure of normal human mammary epithelial cells to 10 nM BPA for 1 week caused hypermethylation of several cancer-related tumor suppressor genes, including *BRCA1*, *CCNA*, *CDKN2A*, *THBS1*, and *TNFRSF*. In addition to alterations in DNA methylation, BPA was reported to affect other components of the cellular epigenome. For example, Doherty et al. (2010) showed that the mRNA levels of the enhancer of zeste 2 polycomb repressive complex 2 subunit (*Ezh2)*, a critical histone methyltransferase that regulates gene expression by methylating histone H3K27, were increased in human breast cancer MCF-7 cells exposed for 2 days to 2,500 and 25,000 nM BPA; the lower BPA levels tested (25 and 250 nM BPA) were not effective. In a separate study, Hussain et al. (2015) showed that treatment of MCF-7 cells with 10–1000 nM BPA for 4 h resulted in a dose-dependent upregulation of the homeobox C6 (*HOXC6)* gene expression, an effect suggested to be mediated by increased histone acetylation and histone H3K4 trimethylation. In a follow-up study, the authors reported a similar overexpression of the homeobox B9 (*HOXB9)* gene after exposure of MCF-7 cells to 100 nM BPA for 6 h (Deb et al. 2016). Also, it has been reported that an 18 h exposure to 10,000 nM BPA altered the expression of multiple miRNAs in MCF-7 cells, including down-regulation of miR-21 (Tilghman et al. 2012). Additionally, Bhan et al. (2014) reported that long noncoding RNA (lncRNA) HOX antisense intergenic RNA (HOTAIR) was induced in MCF-7 cells upon a 4 h exposure to 10-1,000 nM BPA, in a linear dose-response manner.

Avissar-Whiting et al. (2010) reported that exposure of human placental first trimester villous 3A cells and first trimester extravillous HTR-8 cells to 110 nM BPA for 6 days induced the expression of miR-146a; this BPA effect was not observed in third-trimester extravillous TCL-1 cells. Overexpression of miR-146a in 3A placental cells decreased their proliferation rate, but did not affect the survival rate of the cells in response to 110 or 1,100 nM BPA (Avissar-Whiting et al. 2010). In a separate study, Huang et al. (2017) exposed H9 human embryonic stem cell-derived embryoid bodies to 1,000 nM BPA during their differentiation into dopaminergic neurons. The promotor regions of genes involved in the transcriptional regulation of the insulin-like growth factor 1 (*Igf1*) gene were hypermethylated after 7 and 14 days of exposure to the BPA, while the expression of the *Igf1* gene was downregulated upon 3 or 7 days of treatment; the effect of the 14-day BPA exposure on the expression levels of *Igf1* was not reported (Huang et al. 2017). In addition, treatment of human prostaspheres with 10–1,000 nM BPA for 7 days has been reported to alter the expression of 91 genes, including small nucleolar RNAs with C/D motif (SNORDs); this effect was associated with altered recruitment of trimethylated histone H3K9, H3K4, and H3K27 at their gene regulatory sequences (Ho et al. 2015).

Studies in Exposed Experimental Animals

Liver

One of the first reports examining epigenetic effects of BPA in an animal model suggested that dietary exposure of agouti viable yellow female mice to approximately 10,000 μg BPA/kg bw/day from 2 weeks prior to mating through gestation and lactation shifted the coat color distribution toward yellow in the F_1 pups (Dolinoy et al. 2007). The authors proposed that the shift in coat color was due to a BPA-induced decrease in the CpG methylation in an intracisternal A particle (IAP) retrotransposon located upstream of the *agouti* gene. In a follow-up study, this research group confirmed the original finding and showed further that two lower doses of BPA (approximately 0.01 and 10 μg BPA/kg bw/day) induced also color shifts versus control, although these changes were not associated with altered DNA methylation status of CpG sites in the IAP upstream of the *agouti* gene (Anderson et al. 2012). A later report by Rosenfeld et al. (2013) did not support the finding that BPA shifts the coat color of the F_1 pups upon dietary exposure of agouti viable yellow dams to approximately 6.5, 650, or 6,500 μg BPA/kg bw/day from 2 weeks prior to mating through gestation and lactation.

Since the first report of BPA-induced DNA methylation alterations, several studies have investigated the potential of BPA exposure to induce epigenetic effects in multiple tissues of exposed animals. Ma et al. (2013) reported that the oral exposure of Wistar rats to 50 μg BPA/kg bw/day from gestation day (GD) 0 to postnatal day (PND) 21 induced global DNA hypomethylation in the livers of 3- and 21-week-old rats, while the promotor of the glucokinase (*Gck*) gene was hypermethylated and the expression of *Gck* was inhibited. No histological changes were observed in the liver, but hepatic glycogen content was slightly reduced in the older animals, as were the homeostatic model assessment (HOMA)-insulin resistance and insulin sensitivity indices. In a follow-up study, the same research group reported that oral exposure of Sprague-Dawley rats to 40 μg BPA/kg bw/day from GD 20 through PND 21 deregulated glucose and insulin homeostasis in 20-week-old F_2 rats, an effect not observed at 9 weeks of age (Li et al. 2014). These metabolic effects were associated with global DNA hypomethylation and increased DNA methylation of the *Gck* gene promotor in F_1 sperm and F_2 liver; however, the DNA methylation pattern differed across tissues.

Exposure of male ICR (CD-1) mice from PND 0 until sacrifice at 8 weeks or 10 months of age to approximately 0.5 μg BPA/kg bw/day via the drinking water was reported to decrease the extent of DNA methylation of the sterol regulatory element-binding transcription factor 1 (*Srebf1*) and *Srebf2* genes and to increase their expression in the liver (Ke et al. 2016). These effects were accompanied by decreased mRNA and protein levels of hepatic DNA methyltransferase 1 (*Dnmt1*) and *Dnmt3a* in 8-week-old mice; however, at 10 weeks of age, only *Dnmt3b* was found to be downregulated. The accumulation of triglycerides and cholesterol in the livers of the older animals was also reported (Ke et al. 2016). In contrast, van Esterik et al. (2015) reported no changes in global or region-specific DNA methylation in the livers of 23-week-old F_1 C57BL/6JxFVP female offspring of C57BL/6 J female mice exposed through the diet from 2 weeks prior to mating until weaning of their pups to approximately 3, 10, 30, 100, 300, 1,000, or 3,000 μg BPA/kg bw/day.

Heart

Jiang et al. (2015) investigated the effect of BPA exposure on the hearts of male Wistar rats and reported that dietary exposure to 50 μg BPA/kg bw/day from PND 21 to 24 or 48 weeks of age resulted in hypermethylation of the promotor of the mitochondrial peroxisome proliferator activated receptor-γ, coactivator 1α (*Pgc-1α)* gene and downregulation of its expression. These changes were accompanied by an increased expression of *Dnmt1* and *Dnmt3b*. Additionally, decreased mitochondrial function was observed at both time points, while myocardium hypertrophy was reported in the older animals (Jiang et al. 2015). In addition, Patel et al. (2013) reported that lifelong oral exposure of male and female C57BL/6 N mice to 0.5 or 5 μg BPA/kg bw/day from GD 11.5 to PND 120, or to 200 μg BPA/kg bw/day from GD 11.5 to PND 21, increased the global DNA methylation in the hearts of male mice upon exposure to the lowest dose; in female mice, the opposite effect was observed. The expression level of *Dnmt3a* was increased by all BPA doses in male mice; in female mice, the lowest BPA dose increased the *Dnmt3a* gene expression, while 200 μg BPA/kg bw/day decreased it. Additionally, BPA exposure affected the methylation status of different CpG sites in the calsequestrin 2 (*Casq2)* gene promotor, but there was no consistency on the direction of the effect or on the CpG sites affected across BPA level or sex (Patel et al. 2013). Similarly, the reported BPA effects on the heart structure and function of the same animals were not consistent across BPA dose groups or sex (Patel et al. 2013).

Brain

Several studies have assessed the effect of BPA exposure on the brain epigenome of rodent models. In particular, it has been reported that gestational exposure of ICR/Jcl mice to 20 μg BPA/kg bw/day via subcutaneous injection from GD 0 to GD 12.5 or 14.5 affected the methylation status in 48 unique NotI sites of gene promotor-associated CpG islands in the forebrain of the F_1 fetus (Yaoi et al. 2008). The transcriptional levels of vacuolar protein sorting 52 (*Vps52*) and *LOC72325* genes were increased versus control, which correlated with the de novo demethylation at the NotI site in the promotor-associated CpG islands. In a separate study, Kundakovic et al. (2013) reported that oral exposure of pregnant BALB/c mice to 2 or 20 μg BPA/kg bw/day on GDs 0–19 induced changes in the expression of genes encoding ERs (*Esr1* and *Esr2*) and the estrogen-related receptor-γ (*Esrrg)* in the offspring, while the highest dose tested (200 μg BPA/kg bw/day) had minimal effect. Twenty micrograms BPA/kg bw/day, the only dose assessed in the DNA methylation studies, was reported to alter the methylation status of the *Esr1* gene, but the direction of this effect varied across sex and brain subregion (prefrontal cortex and hippocampus) (Kundakovic et al. 2013). Exposure of CD-1 mice on PNDs 28–56 to approximately 250 or 25,000 μg BPA/kg bw/day via drinking water exacerbated the age-related progressive decrease in the hippocampal acetylation of histones H3K9 and H4K8, in particular at the high BPA dose (Jiang et al. 2016). In addition, BPA exposure exacerbated the age-related decreases in the serum levels of free thyroxine and in spatial memory abilities. Although significant correlations were observed between the acetylation levels of histone H3K9 and H4K8 and specific cognitive performances, these were lost when treatment group was taken into account in the correlation analysis (Jiang et al. 2016). Also, although behavior effects were observed in both BPA dose groups, the acetylation effects were mainly restricted to the high BPA dose group. Monje et al. (2007) reported the overexpression of the *Esr1* gene in the hypothalamic preoptical region of 21-day-old female Wistar rats treated by subcutaneous injection with 500 or 20,000 μg BPA/kg bw/day on PNDs 1, 3, 5, and 7; however, this change was not associated with modification of the CpG methylation status in the *Esr1* gene promotor.

Exposure of 10-week-old ICR male mice for 90 days led to an increased acetylation of histone H3K14 in the hippocampus of animals exposed orally to 4,000 and 40,000 μg BPA/kg bw/day, which was associated with an increase in the freezing time 1 and 24 h after fear conditioning; a lower dose (400 μg BPA/kg bw/day) was not effective (Zhang et al. 2014). In addition, Kitraki et al. (2015) reported that oral exposure of Wistar rats to 40 μg BPA/kg bw/day from GD 0 to PND 22 led to hypermethylation of the FK506-binding protein 5 *(Fkbp5)* gene in the hippocampus of 46-day-old male, but not female, rats. The levels of *Fkbp5* mRNA were decreased in males treated with BPA, while the effect of BPA on females was not reported (Kitraki et al. 2015).

Mammary Gland

Several reports have suggested epigenetic effects of BPA in the mammary gland of experimental animals. For instance, intraperitoneal injection of pregnant CD-1 mice with 5,000 μg BPA/kg bw/day on GDs 9–26 increased the protein level of histone methyltransferase EZH2 in the mammary gland of F_1 females at 6 weeks of age (Doherty et al. 2010). Similarly, Altamirano et al. (2016) reported an increased expression of the *Ezh2* and histone deacetylase 1 (*Hdac1*) genes in the mammary gland of lactating Wistar-derived rats that had been exposed orally to 0.6 or 52 μg BPA/kg bw/day from GD 9 through weaning. The authors reported further modifications to the status of DNA methylation and histone modifications in the regulatory regions of the β-casein (*β–Cas)* gene. Additionally, Bhan et al. (2014) reported an upregulation of the lncRNA HOTAIR in the mammary gland of ovariectomized 90-day-old Sprague-Dawley rats exposed acutely to 25 μg BPA/kg bw/day by subcutaneous injection. In a separate study, Dhimolea et al. (2014) reported alterations in the DNA methylation status of the mammary gland of Wistar-Furth rats exposed from GD 9 through birth to 250 μg BPA/kg bw/day via an implanted osmotic pump; however, these effects were not consistent over the three time points assessed (PND 4, 21, and 50) and did not correlate with the gene expression changes observed in the same animals.

Ovary

Neonatal exposure of CD-1 female mice to 20 or 40 μg BPA/kg bw/day by subcutaneous injection from PNDs 7–14 or on PNDs 5, 10, 15, and 20, resulted in reduced methylation of the insulin-like growth factor 2 receptor (*Igf2r*) and paternally expressed gene 3 (*Peg3*) genes and decreased expression of *Dnmt1*, *Dnmt3a*, *Dnmt3b*, and *Dnmt3l* in oocytes (Chao et al. 2012). An abnormal ratio of spindle assembling in meiosis I was also observed (Chao et al. 2012). Zhang et al. (2012a) reported further that oral exposure of CD-1 mice to 80 μg BPA/kg bw/day on GDs 12.5–18.5 induced methylation of CpG sites of the promotor of stimulated by retinoic acid gene 8 (*Stra8*) gene in fetal oocytes collected at GDs 13.5, 15.5, and 17.5. At GD 17.5, the expression levels of *Stra8* gene were decreased by 30% in the BPA-treated oocytes relative to control (Zhang et al. 2012a). In addition, fetal ovaries collected at GD 65 and 90 from ewes that had been exposed from GD 30–90 to 500 μg BPA/kg bw/day subcutaneously were reported to have decreased levels of miRNAs; a single miRNA (miR-203) was found to be downregulated in common at both ages (Veiga-Lopez et al. 2013).

Uterus

In utero exposure of CD-1 pregnant mice on GDs 9–16 to 5,000 μg BPA/kg bw/day via intraperitoneal injection was reported to increase the expression of homeobox A10 (*Hoxa10*) gene and decrease the methylation of its promoter and intron regions in the uterus of BPA-exposed 2-week-old mice; however, this effect did not persist into adulthood (Bromer et al. 2010). In addition, no changes were observed in the uterine expression of *Dnmt1, Dnmt3a,* or *Dnmt3b* (Bromer et al. 2010). These data are consistent with the findings by Camacho et al. (2015), who reported no BPA-induced changes in the expression level of these genes in the uterus of Sprague-Dawley rats orally exposed to 2.5, 8, 25, 80, 260, 840, 2,700, 100,000, or 300,000 μg BPA/kg bw/day from GD 6 to PND 90, despite an increased incidence of uterine cystic endometrial hyperplasia in the highest BPA dose group (Delclos et al. 2014). In contrast, Vigezzi et al. (2016) reported an increased expression of *Dnmt3a* and *Dnmt3b* genes in the uterus of Wistar-derived rats exposed orally to 50 μg BPA/kg bw/day from GD 9 to PND 21; the lowest dose tested (0.5 μg BPA/kg bw/day) was not effective. In a separate study, Hiyama et al. (2011) reported hypomethylation of the intron region of *Hoxa10* gene in the uterus of 8-week-old F_2 generation ICR mice that had been exposed by subcutaneous injection to 100,000 or 200,000 μg BPA/kg bw/day from GD 12–16. On the other hand, although the expression level of *Hoxa10* gene was decreased in the uterus of PND 8 and adult Wistar-derived rats that had been injected subcutaneously with 20,000 μg BPA/kg bw/day on PNDs 1, 3, 5, and 7, the *Hoxa10* gene promotor methylation status was not affected by the BPA exposure (Varayoud et al. 2008). In CD-1 mice exposed to 5,000 μg BPA/kg bw/day from GD 9 to PND 14 via an Alzet osmotic minipump, an altered methylation status of several genes, with preferential hypomethylation of ERα-binding genes, was found at 2 weeks of age and, more markedly, at adulthood (Jorgensen et al. 2016).

Placenta

Exposure of mice to approximately 10 or 10,000 μg BPA/kg bw/day from 2 weeks prior to mating until GD 9.5 or 12.5 was reported to decrease genome-wide DNA methylation in the placenta, but not the embryo. These changes were associated with a loss of genomic imprinting and the differential expression of several genes in the placenta, as well as with abnormal placentation in the highest BPA dose group (Susiarjo et al. 2013).

Prostate

Modification of the DNA methylation status by BPA was reported also in the prostate of adult rats that had been exposed during the neonatal period. In the study by Ho et al. (2006), Sprague-Dawley rats were injected subcutaneously on PNDs 1, 3, and 5 with approximately 10 μg BPA/kg bw/day. The phosphodiesterase type 4D variant 4 (*Pde4d4*) gene, which encodes an enzyme responsible to cyclic AMP breakdown, was found to be hypomethylated in the regulatory CpG island and the expression of the gene was upregulated in the dorsal lobe of the prostate at PND 90 and 200, but not at PND 10. This BPA exposure also led to an increased incidence and severity of prostate intraepithelial hyperplasia in 28-week-old rats that had been implanted with Silastic capsules containing estradiol and testosterone at PND 90 (Ho et al. 2006). In a follow-up study using the same animal model, BPA dose, and dosing paradigm, Tang et al. (2012) reported that the promoter region of the nucleosome-binding protein 1 (*Nsbp1*) gene was hypomethylated and its expression upregulated upon the BPA exposure in the dorsal prostate at PND 10, 90, and 200. In addition, the promoter of the hippocalcin-like 1 (*Hpcal1*) gene was hypermethylated in the dorsal prostate at PND 10, while its expression was decreased at PND 10 and 90 in the BPA-exposed group. Age-specific changes in the expression of *Dnmt3b* and *Dnmt3a* were reported as well (Tang et al. 2012). In contrast, Camacho et al. (2015) did not observe changes in the expression levels of *Dnmt1*, *Dnmt3a*, or *Dnmt3b* in the whole prostate of 90-day-old Sprague-Dawley rats orally exposed to 2.5–300,000 μg BPA/kg bw/day from GD 6 to PND 90; no BPA-induced histopathological changes were observed in the prostates of these animals (Delclos et al. 2014). Cheong et al. (2016) reported that subcutaneous injection of Sprague-Dawley rats with 10 μg BPA/kg bw/day on PNDs 1, 3, and 5 altered the methylation status of the promoter region of 86 genes in the PND 90 dorsal prostate; two of these genes (*Sox2*, sex determining region Y box 2, and *Tpd52*, tumor protein D52) were confirmed to have inverse gene expression changes in the BPA-treated group relative to the vehicle control (Cheong et al. 2016). In addition, Wong et al. (2015) reported that exposure of Sprague-Dawley rats to 50 μg BPA/kg bw/day subcutaneously on PNDs 1, 3, and 5 induced the secretaglobin family 2A member 1 (*Scgb2a1*) gene in the prostate of PND 70 rats; this effect was specific to the anterior lobe of the prostate and was concomitant with increased enrichment of acetylated histone H3K9 and DNA hypomethylation within the CpG island 2 upstream of the transcription start site of the *Scgb2a1* gene (Wong et al. 2015).

Testis and Sperm

Neonatal male Holtzman rats were exposed to approximately 400 μg BPA/kg bw/day via subcutaneous injection on PNDs 1–5 and mated at PND 75 with unexposed females. This treatment led to significant hypomethylation at the imprinting control region of the lncRNA imprinted maternally expressed transcript (*H19*) gene in the spermatozoa of F_0 sires and in the F_1 resorbed embryos in the BPA group versus control viable and resorbed embryos (Doshi et al. 2013); however, the role of the BPA exposure on this effect is confounded by the fact that the control resorbed embryos also had hypomethylated *H19* gene compared to the control viable embryos and no comparisons between the BPA-exposed resorbed and viable embryos were reported. This research group reported also that this exposure paradigm led to hypermethylation of the promoter regions of the *Esr1* and *Esr2* genes and to a slight decrease in their expression levels in the testis of 125-day-old Holtzman rats, accompanied by increased levels of *Dnmt3a* and *Dnmt3b* mRNA and protein (Doshi et al. 2011).

Zhang et al. (2012b) reported that oral exposure of pregnant CD-1 mice on GDs 0.5–12.5 to 40, 80, or 160 μg BPA/kg bw/day led to hypomethylation of the *Peg3* gene in the germ cells of males and females, while the two higher BPA doses led to hypomethylation of *Igf2r* and *H19*. The *Esr1* gene was upregulated in germ cells of male and female mice exposed to 160 μg BPA/kg bw/day versus control. In contrast, treatment of CD-1 mice from PND 3 to PND 21, 35, or 49 with 20 or 40 μg BPA/kg bw/day via subcutaneous injection did not change the DNA methylation status of *H19*, *Igf2*, *Igf2r*, or *Peg3* in germ cells of male mice, despite the reported negative effects of both BPA doses upon spermatogenesis (Zhang et al. 2013). Oral exposure of Long-Evan rats on GDs 12–21 with 25 μg BPA/kg bw/day was reported to increase the testicular levels of 5-hydroxymethyl-2′-deoxycytidine, but not of 5-methyl-2′-deoxycytidine, in 90-day-old rats; no effects were observed upon exposure to 2.5 μg BPA/kg bw/day (Abdel-Maksoud et al. 2015). In addition, oral exposure of pregnant ICR mice to 20 μg BPA/kg bw/day on GDs 1–5 was suggested to downregulate the testicular expression of steroidogenic acute regulatory protein (*StAR*) gene in F_1 pups at PNDs 35 and 50, and to decrease the levels of histone H3 and H3K14 acetylation in the *StAR* promotor; only the older animals were tested for the latter endpoint (Hong et al. 2016).

Discussion, Challenges, and Perspectives

Due to the ubiquitous presence of BPA in the environment and the widespread exposure of the human population to this chemical, mainly via ingestion of food with BPA due to the migration of the chemical from the packaging material into the food matrix, there has been a general interest on the toxicological potential of BPA and its molecular mechanisms of action. Regulatory agencies across the globe are in agreement that the current levels of exposure to BPA are safe to the human population, based on the consistency and robustness of the body of literature that reports BPA can exert biological effects, but only at exposure levels several orders of magnitude higher than those found in the environment and exposed human population. At odds with these assessments are numerous experimental studies that suggest that BPA can exert biological effects at lower dose levels. Among the potential molecular mechanisms proposed to explain these putative BPA low-dose effects are epigenetic mechanisms, including at the DNA methylation, histone modification, and noncoding RNA levels. In this chapter, we attempted to summarize these reports, with particular emphasis on important elements of study design, including the doses of BPA, the window of exposure, and the route of exposure tested. By doing so, we hoped to contribute to a better understanding of the potential of BPA to act as an epigenetic modulator.

One of the fundamental principles of toxicology is the characterization of the dose-response of an effect upon exposure to a compound; however, many of the studies reporting epigenetic effects of BPA used a single dose level, precluding the characterization of the dose-response of the effects being reported. Additionally, the effective dose of BPA varies extensively across studies, in particular when comparing the in vitro and in vivo findings. For example, regardless of the cell model used, the exposure length tested, or the epigenetic endpoint analyzed, the lowest effective BPA dose in vitro was in the range of 1–10 nM, sometimes in the μM dose range. These levels are several orders of magnitude greater than the estimated pM circulating levels of BPA in humans. In contrast, many of the in vivo animal studies, including some using the oral route of exposure, reported epigenetic effects of BPA in the μg/kg bw/day dose range, a dose level that would result in lower internal doses of circulating BPA. In addition, while the dose-response curves reported in the in vitro studies tended to be linear, this was not always the case in the in vivo studies. Also of note, there is a lack of knowledge on the functional consequences of the reported molecular changes, their relation to adverse effects, and their temporality or persistence. Comprehensive well-designed and well-controlled studies, which include both sexes, multiple and properly spaced doses of BPA, multiple time points, and integration of multiple epigenetic endpoints with other molecular, physiological, and morphological endpoints should be informative to address these and other data gaps. An ongoing research program, named *Consortium Linking Academic and Regulatory Insights on BPA Toxicity* (CLARITY-BPA), is expected to contribute to the assessment of the toxicity of BPA upon a lifetime exposure and to a better characterization of the molecular changes induced by this chemical upon sub-chronic and chronic exposures (Schug et al. 2013). In addition, the use of genome-wide approaches to assess the potential of BPA to modulate molecular mechanisms, including at the epigenetic level, may be more informative than targeted ones. The interpretation of this body of data in light of the internal levels of BPA that result from such exposures should provide a better understanding of the potential of BPA to induce epigenetic effects, including in the human population.

Dictionary of Terms

- **Biomonitoring** – Assessment of the exposure to a chemical by measuring it or its metabolites in a biological sample, such as blood, urine, saliva, or tissue.
- **DNA methylation** – Epigenetic mechanism mediated by DNA methyltransferases, by which methyl groups are added to the DNA molecule. These modifications often result in altered gene expression levels.
- **Dose-response** – Relationship between the level of exposure to a compound and the magnitude of its effect.
- **Endocrine-disrupting chemical** – An exogenous chemical that interferes with the endocrine system, leading to adverse effects, such as developmental, reproductive, neurological, metabolic, or immune effects.
- **Histones** – Group of highly alkaline proteins found in eukaryotic cell nuclei that package DNA into structural units called nucleosomes. Histones can be modified posttranslationally by methylation, phosphorylation, acetylation, ubiquitylation, and sumoylation, which may alter the gene expression levels.
- **Noncoding RNA** – Functional RNA molecule that is not translated into a protein. Noncoding RNAs can be short (<30 nucleotides; includes miRNAs, short interfering RNAs, and piwi-interacting RNAs) or long (>200 nucleotides), and can regulate gene expression at the transcriptional or posttranslational level.
- **Pharmacokinetics** – Assessment of how a compound is absorbed, distributed, metabolized, and excreted by an organism.

Key Facts of Bisphenol A

- Bisphenol A (BPA) is a high production volume industrial chemical used widely in the production of polycarbonate plastics and epoxy resins for manufacturing food and drink storage containers, the lining of food cans, dental sealants, medical devices, and thermal paper.
- There is widespread exposure of the human population to BPA; the primary route of exposure is through diet, but there is also exposure via inhalation and skin.
- BPA is an endocrine-disrupting chemical, with known estrogenic activity; the relative binding affinity of BPA to the estrogen receptors is three to four orders of magnitude lower than that of estradiol.
- More recently, BPA has been reported to induce epigenetic changes, including effects in DNA methylation, differential histone modifications, and modulation of the levels of noncoding RNAs.
- Regulatory agencies across the globe are in agreement that the current levels of exposure to BPA are safe to the human population, based on the consistency and robustness of the body of literature that reports BPA can exert biological effects, but only at exposure levels several orders of magnitude higher than those found in the environment and exposed human population.

Summary Points

- Bisphenol A (BPA) is a high production volume industrial chemical used widely in the production of polycarbonate plastics and epoxy resins for manufacturing food and drink storage containers, the lining of food cans, dental sealants, medical devices, and thermal paper.
- Biomonitoring studies show that there is widespread exposure of the human population to BPA, mainly via the diet. The median daily intake of the overall US population is approximately 25 ng BPA/kg body weight (bw)/day, which is estimated to result in serum levels in the pM range.
- BPA is an endocrine-disrupting chemical, with known estrogenic activity; more recently, it has been reported to induce epigenetic changes in vitro and in vivo, including effects in DNA methylation, differential histone modifications, and modulation of the levels of noncoding RNAs.
- This chapter summarizes the studies reporting epigenetic effects associated with human exposure to BPA or induced by the in vitro or in vivo exposure to the chemical.
- The literature suggesting an association between human exposure to BPA and epigenetic changes is limited.
- Several studies have assessed the effects of BPA in in vitro cell systems and suggest treatment-related epigenetic effects; the lowest effective BPA dose in vitro was in the nM dose range and the dose-response in these studies tended to be linear.
- In vivo animal studies suggest epigenetic effects of BPA in a wide range of organs, dose levels, and epigenetic endpoints; however, many studies have limitations, including the use of a single dose level that precludes the characterization of the dose-response of the reported effects.
- Comprehensive well-designed and well-controlled studies, which include both sexes, multiple and properly spaced doses of BPA, multiple time points, and integration of the epigenetic endpoints with other molecular, physiological, and morphological endpoints should provide a better understanding of the potential of BPA to act as an epigenetic modulator.

References

Abdel-Maksoud FM, Leasor KR, Butzen K et al (2015) Prenatal exposures of male rats to the environmental chemicals bisphenol A and di(2-Ethylhexyl) phthalate impact the sexual differentiation process. Endocrinology 156(12):4672–4683

Altamirano GA, Ramos JG, Gomez AL et al (2016) Perinatal exposure to bisphenol A modifies the transcriptional regulation of the beta-casein gene during secretory activation of the rat mammary gland. Mol Cell Endocrinol 439:407–418

Anderson OS, Nahar MS, Faulk C et al (2012) Epigenetic responses following maternal dietary exposure to physiologically relevant levels of bisphenol A. Environ Mol Mutagen 53(5):334–342

Avissar-Whiting M, Veiga KR, Uhl KM et al (2010) Bisphenol A exposure leads to specific microRNA alterations in placental cells. Reprod Toxicol 29(4):401–406

Bhan A, Hussain I, Ansari KI et al (2014) Bisphenol-A and diethylstilbestrol exposure induces the expression of breast cancer associated long noncoding RNA HOTAIR in vitro and in vivo. J Steroid Biochem Mol Biol 141:160–170

Bromer JG, Zhou Y, Taylor MB et al (2010) Bisphenol-A exposure in utero leads to epigenetic alterations in the developmental programming of uterine estrogen response. FASEB J 24(7):2273–2280

Calafat AM, Ye X, Wong LY et al (2008) Exposure of the U.S. population to bisphenol A and 4-tertiary-octylphenol: 2003–2004. Environ Health Perspect 116(1):39–44

Camacho L, Basavarajappa MS, Chang CW et al (2015) Effects of oral exposure to bisphenol A on gene expression and global genomic DNA methylation in the prostate, female mammary gland, and uterus of NCTR Sprague-Dawley rats. Food Chem Toxicol 81:92–103

Chao HH, Zhang XF, Chen B et al (2012) Bisphenol A exposure modifies methylation of imprinted genes in mouse oocytes via the estrogen receptor signaling pathway. Histochem Cell Biol 137(2):249–259

Chapin RE, Adams J, Boekelheide K et al (2008) NTP-CERHR expert panel report on the reproductive and developmental toxicity of bisphenol A. Birth Defects Res B Dev Reprod Toxicol 83(3):157–395

Cheong A, Zhang X, Cheung YY et al (2016) DNA methylome changes by estradiol benzoate and bisphenol A links early-life environmental exposures to prostate cancer risk. Epigenetics 11(9):674–689

De Felice B, Manfellotto F, Palumbo A et al (2015) Genome-wide microRNA expression profiling in placentas from pregnant women exposed to BPA. BMC Med Genet 8:56

Deb P, Bhan A, Hussain I et al (2016) Endocrine disrupting chemical, bisphenol-A, induces breast cancer associated gene HOXB9 expression in vitro and in vivo. Gene 590(2):234–243

Delclos KB, Camacho L, Lewis SM et al (2014) Toxicity evaluation of bisphenol A administered by gavage to Sprague Dawley rats from gestation day 6 through postnatal day 90. Toxicol Sci 139(1):174–197

Dhimolea E, Wadia PR, Murray TJ et al (2014) Prenatal exposure to BPA alters the epigenome of the rat mammary gland and increases the propensity to neoplastic development. PLoS One 9(7): e99800

Doerge DR, Twaddle NC, Vanlandingham M et al (2010a) Pharmacokinetics of bisphenol A in neonatal and adult Sprague-Dawley rats. Toxicol Appl Pharmacol 247(2):158–165

Doerge DR, Vanlandingham M, Twaddle NC et al (2010b) Lactational transfer of bisphenol A in Sprague-Dawley rats. Toxicol Lett 199(3):372–376

Doerge DR, Twaddle NC, Woodling KA et al (2010c) Pharmacokinetics of bisphenol A in neonatal and adult rhesus monkeys. Toxicol Appl Pharmacol 248:1–11

Doerge DR, Twaddle NC, Vanlandingham M et al (2011a) Distribution of bisphenol A into tissues of adult, neonatal, and fetal Sprague-Dawley rats. Toxicol Appl Pharmacol 255(3):261–270

Doerge DR, Twaddle NC, Vanlandingham M et al (2011b) Pharmacokinetics of bisphenol A in neonatal and adult CD-1 mice: inter-species comparisons with Sprague-Dawley rats and rhesus monkeys. Toxicol Lett 207(3):298–305

Doerge DR, Twaddle NC, Vanlandingham M et al (2012) Pharmacokinetics of bisphenol A in serum and adipose tissue following intravenous administration to adult female CD-1 mice. Toxicol Lett 211(2):114–119

Doherty LF, Bromer JG, Zhou Y et al (2010) In utero exposure to diethylstilbestrol (DES) or bisphenol-A (BPA) increases EZH2 expression in the mammary gland: an epigenetic mechanism linking endocrine disruptors to breast cancer. Horm Cancer 1(3):146–155

Dolinoy DC, Huang D, Jirtle RL (2007) Maternal nutrient supplementation counteracts bisphenol A-induced DNA hypomethylation in early development. Proc Natl Acad Sci USA 104(32):13056–13061

Doshi T, Mehta SS, Dighe V et al (2011) Hypermethylation of estrogen receptor promoter region in adult testis of rats exposed neonatally to bisphenol A. Toxicology 289(2–3):74–82

Doshi T, D'Souza C, Vanage G (2013) Aberrant DNA methylation at Igf2-H19 imprinting control region in spermatozoa upon neonatal exposure to bisphenol A and its association with post implantation loss. Mol Biol Rep 40(8):4747–4757

EFSA Panel on Food Contact Materials, Enzymes, Flavourings and Processing Aids (CEF) (2015) Scientific opinion on the risks to public health related to the presence of bisphenol A (BPA) in foodstuffs. EFSA J 13(1):3978

Faulk C, Kim JH, Jones TR et al (2015) Bisphenol A-associated alterations in genome-wide DNA methylation and gene expression patterns reveal sequence-dependent and non-monotonic effects in human fetal liver. Environ Epigenet. 1(1): dvv006

FDA (2014) http://www.fda.gov/downloads/NewsEvents/PublicHealthFocus/UCM424266.pdf. Accessed 15 Dec 2016

Fisher JW, Twaddle NC, Vanlandingham M et al (2011) Pharmacokinetic modeling: prediction and evaluation of route dependent dosimetry of bisphenol A in monkeys with extrapolation to humans. Toxicol Appl Pharmacol 257(1):122–136

Health Canada (2012) http://www.hc-sc.gc.ca/fn-an/securit/packag-emball/bpa/bpa_hra-ers-2012-09-eng.php. Accessed 15 Dec 2016

Hiyama M, Choi EK, Wakitani S et al (2011) Bisphenol-A (BPA) affects reproductive formation across generations in mice. J Vet Med Sci 73(9):1211–1215

Ho SM, Tang WY, Belmonte de Frausto J et al (2006) Developmental exposure to estradiol and bisphenol A increases susceptibility to prostate carcinogenesis and epigenetically regulates phosphodiesterase type 4 variant 4. Cancer Res 66(11):5624–5632

Ho SM, Cheong A, Lam HM et al (2015) Exposure of human prostaspheres to bisphenol A epigenetically regulates SNORD family noncoding RNAs via histone modification. Endocrinology 156(11):3984–3995

Hong J, Chen F, Wang X et al (2016) Exposure of preimplantation embryos to low-dose bisphenol A impairs testes development and suppresses histone acetylation of StAR promoter to reduce production of testosterone in mice. Mol Cell Endocrinol 427:101–111

Huang B, Ning S, Zhang Q et al (2017) Bisphenol A represses dopaminergic neuron differentiation from human embryonic stem cells through downregulating the expression of insulin-like growth factor 1. Mol Neurobiol 54(5):3798–3812

Hussain I, Bhan A, Ansari KI, Deb P, Bobzean SA, Perrotti LI, Mandal SS (2015) Bisphenol-A induces expression of HOXC6, an estrogen-regulated homeobox-containing gene associated with breast cancer. Biochim Biophys Acta 1849(6):697–708

Jiang Y, Xia W, Yang J et al (2015) BPA-induced DNA hypermethylation of the master mitochondrial gene PGC-1alpha contributes to cardiomyopathy in male rats. Toxicology 329:21–31

Jiang W, Cao L, Wang F et al (2016) Accelerated reduction of serum thyroxine and hippocampal histone acetylation links to exacerbation of spatial memory impairment in aged CD-1 mice pubertally exposed to bisphenol-A. Age (Dordr) 38(5–6):405–418

Jorgensen EM, Alderman MH 3rd, Taylor HS (2016) Preferential epigenetic programming of estrogen response after in utero xenoestrogen (bisphenol-A) exposure. FASEB J 30(9):3194–3201

Ke ZH, Pan JX, Jin LY et al (2016) Bisphenol A exposure may induce hepatic lipid accumulation via reprogramming the DNA methylation patterns of genes involved in lipid metabolism. Sci Rep 6:31331

Kim JH, Rozek LS, Soliman AS et al (2013) Bisphenol A-associated epigenomic changes in prepubescent girls: a cross-sectional study in Gharbiah. Egypt Environ Health 12:33

Kitraki E, Nalvarte I, Alavian-Ghavanini A et al (2015) Developmental exposure to bisphenol A alters expression and DNA methylation of Fkbp5, an important regulator of the stress response. Mol Cell Endocrinol 417:191–199

Kuiper GG, Carlsson B, Grandien K et al (1997) Comparison of the ligand binding specificity and transcript tissue distribution of estrogen receptors alpha and beta. Endocrinology 138(3):863–870

Kuiper GG, Lemmen JG, Carlsson B et al (1998) Interaction of estrogenic chemicals and phytoestrogens with estrogen receptor beta. Endocrinology 139(10):4252–4263

Kundakovic M, Gudsnuk K, Franks B et al (2013) Sex-specific epigenetic disruption and behavioral changes following low-dose in utero bisphenol A exposure. Proc Natl Acad Sci USA 110(24):9956–9961

Kundakovic M, Gudsnuk K, Herbstman JB et al (2015) DNA methylation of BDNF as a biomarker of early-life adversity. Proc Natl Acad Sci USA 112(22):6807–6813

LaKind JS, Naiman DQ (2015) Temporal trends in bisphenol A exposure in the United States from 2003–2012 and factors associated with BPA exposure: spot samples and urine dilution complicate data interpretation. Environ Res 142:84–95

Li G, Chang H, Xia W et al (2014) F0 maternal BPA exposure induced glucose intolerance of F2 generation through DNA methylation change in Gck. Toxicol Lett 228(3):192–199

Li Q, Kappil MA, Li A, Dassanayake PS et al (2015) Exploring the associations between microRNA expression profiles and environmental pollutants in human placenta from the National Children's study (NCS). Epigenetics 10(9):793–802

Loganathan SN, Kannan K (2011) Occurrence of bisphenol A in indoor dust from two locations in the eastern United States and implications for human exposures. Arch Environ Contam Toxicol 61(1):68–73

Ma Y, Xia W, Wang DQ et al (2013) Hepatic DNA methylation modifications in early development of rats resulting from perinatal BPA exposure contribute to insulin resistance in adulthood. Diabetologia 56(9):2059–2067

Matthews JB, Twomey K, Zacharewski TR (2001) In vitro and in vivo interactions of bisphenol A and its metabolite, bisphenol A glucuronide, with estrogen receptors alpha and beta. Chem Res Toxicol 14(2):149–157

Miao M, Zhou X, Li Y et al (2014) LINE-1 hypomethylation in spermatozoa is associated with bisphenol A exposure. Andrology 2(1):138–144

Monje L, Varayoud J, Luque EH et al (2007) Neonatal exposure to bisphenol A modifies the abundance of estrogen receptor alpha transcripts with alternative 5′-untranslated regions in the female rat preoptic area. J Endocrinol 194(1):201–212

Nahar MS, Liao C, Kannan K et al (2015) In utero bisphenol A concentration, metabolism, and global DNA methylation across matched placenta, kidney, and liver in the human fetus. Chemosphere 124:54–60

Okada H, Tokunaga T, Liu X et al (2008) Direct evidence revealing structural elements essential for the high binding ability of bisphenol A to human estrogen-related receptor-gamma. Environ Health Perspect 116(1):32–38

Patel BB, Raad M, Sebag IA et al (2013) Lifelong exposure to bisphenol A alters cardiac structure/function, protein expression, and DNA methylation in adult mice. Toxicol Sci 133(1):174–185

Patterson TA, Twaddle NC, Roegge CS et al (2013) Concurrent determination of bisphenol A pharmacokinetics in maternal and fetal rhesus monkeys. Toxicol Appl Pharmacol 267(1):41–48

PRNewWire (2016) http://www.prnewswire.com/news-releases/global-bisphenol-a-market-overview-2016-2022—market-is-projected-to-reach-us225-billion-by-2022-up-from-156-billion-in-2016—research-and-markets-300303934.html. Accessed 15 Dec 2016

Qin XY, Fukuda T, Yang L et al (2012) Effects of bisphenol A exposure on the proliferation and senescence of normal human mammary epithelial cells. Cancer Biol Ther 13(5):296–306

Rosenfeld CS, Sieli PT, Warzak DA et al (2013) Maternal exposure to bisphenol A and genistein has minimal effect on A(vy)/a offspring coat color but favors birth of agouti over nonagouti mice. Proc Natl Acad Sci USA 110(2):537–542

Schug TT, Heindel JJ, Camacho L et al (2013) A new approach to synergize academic and guideline-compliant research: the CLARITY-BPA research program. Reprod Toxicol 40:35–40

Shelby MD (2008) NTP-CERHR monograph on the potential human reproductive and developmental effects of bisphenol A. NTP CERHR MON 2008:1–64

Susiarjo M, Sasson I, Mesaros C et al (2013) Bisphenol A exposure disrupts genomic imprinting in the mouse. PLoS Genet 9(4):e1003401

Takayanagi S, Tokunaga T, Liu X et al (2006) Endocrine disruptor bisphenol A strongly binds to human estrogen-related receptor gamma (ERRgamma) with high constitutive activity. Toxicol Lett 167(2):95–105

Tang WY, Morey LM, Cheung YY et al (2012) Neonatal exposure to estradiol/bisphenol A alters promoter methylation and expression of Nsbp1 and Hpcal1 genes and transcriptional programs of Dnmt3a/b and Mbd2/4 in the rat prostate gland throughout life. Endocrinology 153(1):42–55

Taylor JA, Vom Saal FS, Welshons WV et al (2011) Similarity of bisphenol A pharmacokinetics in rhesus monkeys and mice: relevance for human exposure. Environ Health Perspect 119(4):422–430

Teeguarden JG, Calafat AM, Ye X et al (2011) Twenty-four hour human urine and serum profiles of bisphenol A during high-dietary exposure. Toxicol Sci 123(1):48–57

Teeguarden J, Hanson-Drury S, Fisher JW et al (2013) Are typical human serum BPA concentrations measurable and sufficient to be estrogenic in the general population? Food Chem Toxicol 62:949–963

Thayer KA, Doerge DR, Hunt D et al (2015) Pharmacokinetics of bisphenol A in humans following a single oral administration. Environ Int 83:107–115

Thayer KA, Taylor KW, Garantziotis S et al (2016) Bisphenol A, bisphenol S, and 4-hydroxyphenyl 4-isoprooxyphenylsulfone (BPSIP) in urine and blood of cashiers. Environ Health Perspect 124(4):437–444

Thomas P, Dong J (2006) Binding and activation of the seven-transmembrane estrogen receptor GPR30 by environmental estrogens: a potential novel mechanism of endocrine disruption. J Steroid Biochem Mol Biol 102(1–5):175–179

Tilghman SL, Bratton MR, Segar HC et al (2012) Endocrine disruptor regulation of microRNA expression in breast carcinoma cells. PLoS One 7(3):e32754

van Esterik JC, Vitins AP, Hodemaekers HM et al (2015) Liver DNA methylation analysis in adult female C57BL/6JxFVB mice following perinatal exposure to bisphenol A. Toxicol Lett 232(1):293–300

Varayoud J, Ramos JG, Bosquiazzo VL et al (2008) Developmental exposure to Bisphenol A impairs the uterine response to ovarian steroids in the adult. Endocrinology 149(11):5848–5860

Veiga-Lopez A, Luense LJ, Christenson LK et al (2013) Developmental programming: gestational bisphenol-A treatment alters trajectory of fetal ovarian gene expression. Endocrinology 154(5):1873–1884

Vigezzi L, Ramos JG, Kass L et al (2016) A deregulated expression of estrogen-target genes is associated with an altered response to estradiol in aged rats perinatally exposed to bisphenol A. Mol Cell Endocrinol 426:33–42

Völkel W, Colnot T, Csanády GA et al (2002) Metabolism and kinetics of bisphenol A in humans at low doses following oral administration. Chem Res Toxicol 15(10):1281–1287

Weng YI, Hsu PY, Liyanarachchi S et al (2010) Epigenetic influences of low-dose bisphenol A in primary human breast epithelial cells. Toxicol Appl Pharmacol 248(2):111–121

Wong RL, Wang Q, Trevino LS et al (2015) Identification of secretaglobin Scgb2a1 as a target for developmental reprogramming by BPA in the rat prostate. Epigenetics 10(2):127–134

Yang X, Doerge DR, Teeguarden JG et al (2015) Development of a physiologically based pharmacokinetic model for assessment of human exposure to bisphenol A. Toxicol Appl Pharmacol 289(3):442–456

Yaoi T, Itoh K, Nakamura K et al (2008) Genome-wide analysis of epigenomic alterations in fetal mouse forebrain after exposure to low doses of bisphenol A. Biochem Biophys Res Commun 376(3):563–567

Zhang HQ, Zhang XF, Zhang LJ, Chao HH, Pan B, Feng YM, Li L, Sun XF, Shen W (2012a) Fetal exposure to bisphenol A affects the primordial follicle formation by inhibiting the meiotic progression of oocytes. Mol Biol Rep 39(5):5651–5657

Zhang XF, Zhang LJ, Feng YN et al (2012b) Bisphenol A exposure modifies DNA methylation of imprint genes in mouse fetal germ cells. Mol Biol Rep 39(9):8621–8628
Zhang GL, Zhang XF, Feng YM et al (2013) Exposure to bisphenol A results in a decline in mouse spermatogenesis. Reprod Fertil Dev 25(6):847–859
Zhang Q, Xu X, Li T et al (2014) Exposure to bisphenol-A affects fear memory and histone acetylation of the hippocampus in adult mice. Horm Behav 65(2):106–113

Ochratoxin A and Epigenetics

111

Alessandra Mezzelani

Contents

Abstract
Ochratoxin A is a thermoresistant mycotoxin produced by ubiquitous molds of *Aspergillus* and *Penicillium* genera. It contaminates foodstuffs and feedstuffs worldwide and therefore is of human and animal concern.

A. Mezzelani (✉)
National Research Council, Institute of Biomedical Technologies (CNR-ITB), Segrate (MI), Italy
e-mail: alessandra.mezzelani@itb.cnr.it

V. B. Patel, V. R. Preedy (eds.), *Handbook of Nutrition, Diet, and Epigenetics*,
https://doi.org/10.1007/978-3-319-55530-0_33

Ochratoxin A induces oxidative stress, inflammation, and fibrosis, and is nephrotoxic, hepatotoxic, and neurotoxicin particularly in male subjects. Toxicity is mainly exerted through epigenetic mechanisms.

Nephrotoxicity is probably due to ochratoxin A-induced suppression of the collagen regulator mir-29b that results in an increase of translated collagen, fibrotic alteration, and nephropathy. Alternatively, ochratoxin A induces mir-132 upregulation that occurs in neurologic and psychiatric conditions as well as in oxidative stress. Undeniably, mir-132 acts in the reciprocal regulation of autism-related genes *MeCP2* and *PTEN* decreasing the antioxidant *Nrf2* that leads to the formation of high levels of reactive oxygen species. Reactive oxygen species, in turn, enhance the expression of mir-200c that impairs antioxidative mechanisms and synaptic plasticity through the reduction of HO-1 and *NLGN4X*. As for apoptosis, OTA exposure increases mir-122 that suppresses the anti-apoptotic genes Bcl-w and caspase-3 leading to cell death and hepatic damage.

Interestingly, both *MECP2* and *NLGN4X* are involved in neurodevelopmental disorders, including autism, and are mapped on the X chromosome. As autism is a male predominant disorder, a possible contribution of ochratoxin A in its pathogenesis and in its strong male bias can be suggested.

Very few papers report about ochratoxin A-induced deacetylation:cells exposed to OTA underwent to a dramatic block of histone acetyltransferases leading to mitotic arrest and Nrf2 inhibition that, again, lead to reactive oxygen species formation.

Further studies are needed to obtain a complete picture of ochratoxin A-dependent epigenetic effects and to prevent or to counteract them.

Keywords

Ochratoxin A · Epigenetics · microRNA · Phenylalanine hydroxylase · Nephrotoxic · Neurotoxic · Hepatotoxic · Fibrosis · Oxidative stress · Neurodegenerative diseases · Autism · Nrf2 · MECP2 · NLGN4X · Sex-dependent

List of Abbreviations

AREs	Antioxidant responsive elements
ASD	Autism spectrum disorder (ASD)
BACE1	β-Secretase-1 enzyme
BBB	Blood brain barrier
BDNF	Brain-derived neurotropic factor
BEN	Balkan endemic nephropathy
CASP3	Caspase3
CBP	CREB-binding protein
CNS	Central nervous system
DGCR8	DiGeorge syndrome critical region gene 8
FMRP	Fragile X mental retardation protein
HAT	Histone acetyltransferase
HDAC	Histone deacetylase

HO-1 Heme oxygenase-1
MeCP2 Methyl-CpG-binding protein 2
NLGN4X Neuroligin4x
OTA Ochratoxin A
p300 Adenoviral E1A-associated protein
PAH Phenylalanine hydroxylase
phe Phenylalanine
PKU Phenylketonuria
PTEN Phosphatase and tensin homolog
TGFβ Transforming-growth factor-beta
Nrf2 Nuclear factor erythroid 2-like 2
ROS Reactive oxygen species
tyr tyrosine
ZEB1 Zinc finger E-box binding homeobox 1

Introduction

The mycotoxin Ochratoxin A (OTA) is a small organic molecule composed of anisocumarin nucleus that is amide-inked to a phenylalanine (phe) unit (Fig. 1).

It is a toxic metabolite produced by ubiquitous molds (microfungi) of *Aspergillus* and *Penicillium* genera that contaminate foodstuffs and feedstuffs worldwide (Malir et al. 2016). Indeed, *A. ochraceus* produces OTA within the temperature range 15–37 °C, while *P. verrucosum* can produce significant quantities of toxin at temperatures as low as 4 °C. Since this mycotoxin is thermo stable (Boudra et al. 1995), food decontamination through cooking is practically impossible.

Although cereals are considered the main source of OTA in human diet, it has also been found in legumes, grapes, coffee, nuts, figs, and all derivative products. Due to feedstuffs contamination, animal products can be polluted too and OTA has been found in meat, milk, eggs, and derivative products (Denli and Perez 2010).

OTA toxicity is of human and animal health concern since the exposure arises from food and feed ingestion and, more moderately, by inhalation or dermal exposure.

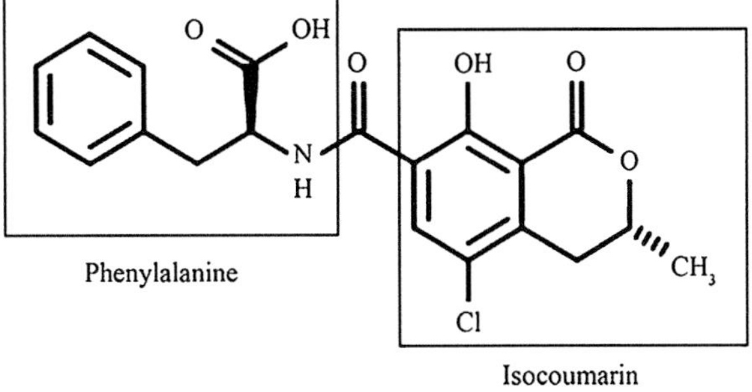

Fig. 1 Chemical structure of OTA. The small organic molecule of OTA consists of an isocumarin nucleus that is amide-inked to a phenylalanine (phe) unit (San Román and Holgado 2015)

After ingestion, the mycotoxin is absorbed through the intestine where it affects the integrity of the walls causing intestinal permeability with molecule trafficking through the intestinal layers, inflammation, gut disorders, and increased bacterial translocation. A study performed on an intestinal cell line, demonstrated that OTA provokes “leaky gut” by removing Claudin-3 and -4 (members of the Claudin family), from the tight junction strands responsible of cell-to-cell adhesion in epithelial or endothelial cell sheets (McLaughlin et al. 2004). Moreover, the composition of gut microbiota from rats subjected to OTA administration revealed that OTA induces changes in microbe fecal profile resembling those associated with aging (Guo et al. 2014) and that induce negative effects to the health of the host (Woodmansey 2007).

Once adsorbed, in humans, OTA has high affinity for *serum* albumin, the most abundant plasma protein. Therefore, about 99.8% of OTA circulates in the blood stream in albumin-bound form that prolongs its persistence in the organism to ~3–5 weeks of half-life (Hagelberg et al. 1989).

OTA Metabolism

In humans xenobiotics are metabolized by the cytochrome P450 and its activity depends on several factors. First of all, age; toddlers, because of their immature cytochrome P450, are much more susceptible to exogenous toxicants than adults (Faustman et al. 2000). Cytochrome P450 also displays a sex dependent activity (Waxman and Holloway 2009), and a sex bias in OTA degradation has been described in humans and rats. Indeed, the liver CYP2C9, CYP2C12, and CYP3A4 enzymes, owing to cytochrome P450 system, that specifically metabolizes OTA (Ringot et al. 2006), are expressed mainly in females while males express more CYP2C11 and CYP3A2 enzymes that, on the contrary, toxify OTA (Mor et al. 2014). These sex differences in OTA metabolism make males more prone to the deleterious effects of OTA than females. As for genetic variability, hundreds of variations in cytochrome P450 such as mutations, single nucleotide polymorphisms, and copy number variants have been described and correlate to the ability of P450 to react to xenobiotics (Preissner et al. 2013). OTA metabolites, such as OTa, (Wu et al. 2011) are excreted in urine, fecal excretion, and through lacteal secretion.

Targets and Toxicity of OTA

In experimental studies OTA has shown a great toxicological variability, and the severity of its toxicity depends on quantity and time exposure as well as on the host species, its genetic variability, age, organs, health conditions and sex. Indeed, as previously reported, toddlers are more sensitive to toxicants and females metabolize OTA better than males.

In many species, including humans, OTA results nephrotoxic, hepatotoxic, neurotoxic, and carcinogenic (Fig. 2). OTA-induced cytotoxicity essentially increases

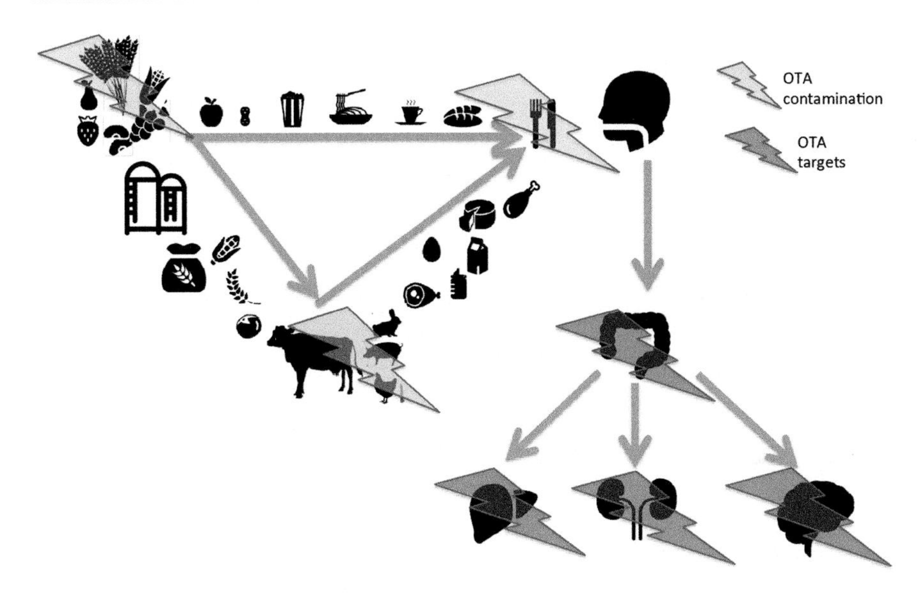

Fig. 2 Mode of OTA food contamination and human exposure. The toxin contaminates foodstuffs and feedstuffs; after ingestion, it passes through the intestinal wall, enters into the bloodstream, and impairs the target organs (icons: http://www.freepik.com/free-icons/)

oxidative stress and lipid peroxidation and interferes with protein biosynthesis. Recent findings suggest that OTA exerts its deleterious effects through epigenetic mechanisms (Schilter et al. 2005).

Nephrotoxicity

In kidney, OTA causes a strong nephrotoxicity with a higher incidence of malignancy of the upper urinary tract (EFSA 2006). The mycotoxin is considered responsible of the chronic tubule-interstitial kidney disease called Balkan Endemic Nephropathy (BEN). In kidney, BEN provokes fibrotic alterations with extracellular matrix accumulation and altered collagen homeostasis (crucial in the onset of renal diseases) leading to chronic pathology such as interstitial nephropathy. OTA has been found in *serum* and urine in patients suffering from BEN as well as OTA-DNA adducts have been detected in tumor samples of patients affected by urinary tract tumors (Castegnaro et al. 2006).

Hepatotoxicity

In liver, OTA induces fibrotic alteration and extracellular matrix accumulation predisposing to hepatic cirrhosis. A recent study has revealed that in HepG2 cell line from liver, OTA alters the mitochondrial transmembrane potential leading to oxidative stress. Reactive oxygen species (ROS), in turn, cause deleterious effect by reacting with cellular macromolecules such as lipids and proteins. ROS induction, together with alteration of the mitochondrial transmembrane potential, results in apoptotic cell death and hepatic damage (Gayathri et al. 2015).

Neurotoxicity

In vivo OTA also passes through the blood brain barrier (BBB)and accumulates in mice brain after a single administration (Sava et al. 2006). In the brain, OTA provokes striatal dopamine acute depletion, a global increase of oxidative stress and consequent cell apoptosis. For this reason, OTA has been proposed as the cause for the pathogenesis of neurodegenerative diseases based on apoptotic processes such as Alzheimer's and Parkinson's disease (Zhang et al. 2009; Hope and Hope 2012). To this purpose, in Neuro-2a cells OTA strongly increases the levels of reactive oxygen species (ROS) and malondialdehyde (resulting from lipid peroxidation) and causes loss of mitochondrial membrane potential in a dose-dependent manner. In these cells, a pretreatment with the antioxidant *N*-acetylcysteine (NAC), before OTA treatment, prevents ROS formation and cellular damage (Bhat et al. 2016). Interestingly, favorable preliminary results have been obtained by NAC administration to a wide range of psychiatric and neurological conditions such as Alzheimer's disease, schizophrenia, autism, and depression, for which oxidative damage and mitochondrial dysfunction have been described (Deepmala et al. 2015).

The role of mycotoxins in the aggravation of autism spectrum disorder (ASD) has also been supposed and, very recently, a link between OTA exposure and ASD has been demonstrated in two observational studies (De Santis et al. 2017a, b).

ASD is a neurodevelopmental disorder that affects more males than females and often presents comorbid conditions (inflammation, oxidative stress, and gastrointestinal disorders) that coincide with those symptoms caused by OTA exposure. Interestingly, even OTA affects more males than females, as already reported in studies about OTA effects on kidney and neural tube development (Pastor et al. 2016; Ueta et al. 2010).

To this purpose, some authors have demonstrated that the OTA-linked phe can act as a substrate for phenylalanine hydroxylase enzyme (PAH). In physiological conditions, PAH converts phe into the amino acid tyrosine (tyr) necessary for the biosynthesis of L-dopa the precursor of the catecholamine neurotransmitters dopamine, noradrenaline, and adrenaline. Dopamine, in turn, interacting with the pro-social neuropeptide oxytocin leads to positive effects on mood and social behavior; interferences in dopaminergic neurotransmission are implicated in some neuropsychiatric conditions including autism and depression (Baskerville and Douglas 2010). Following OTA exposure, the OTA-linked phe competes with phe leading to a significant inhibition of PAH (Creppy et al. 1990; Zanic-Grubisić et al. 2000) and synthesis of OTA-linked tyr that, in turn, interferes with the production of catecholamines and oxytocin. To confirm these findings, simultaneous administration of OTA and its antagonist aspartame (containing high level of phe) reduces the inhibition of PAH and restores the physiological activity of PAH. Interestingly, in humans, phenylketonuria (PKU), the metabolic genetic disorder causing *PAH* knockdown, leads to intellectual disabilities, behavioral problems, and, in some cases, to autism (Baieli et al. 2003) (Fig. 3).

Here the OTA-induced epigenetic effects causing oxidative stress, fibrosis and apoptosis are described. Since several symptoms and epigenetic mechanisms of some neurologic and psychiatric conditions correspond to those induced by OTA, a speculative association between neuropsychiatric disorders and OTA exposure are proposed and well substantiated.

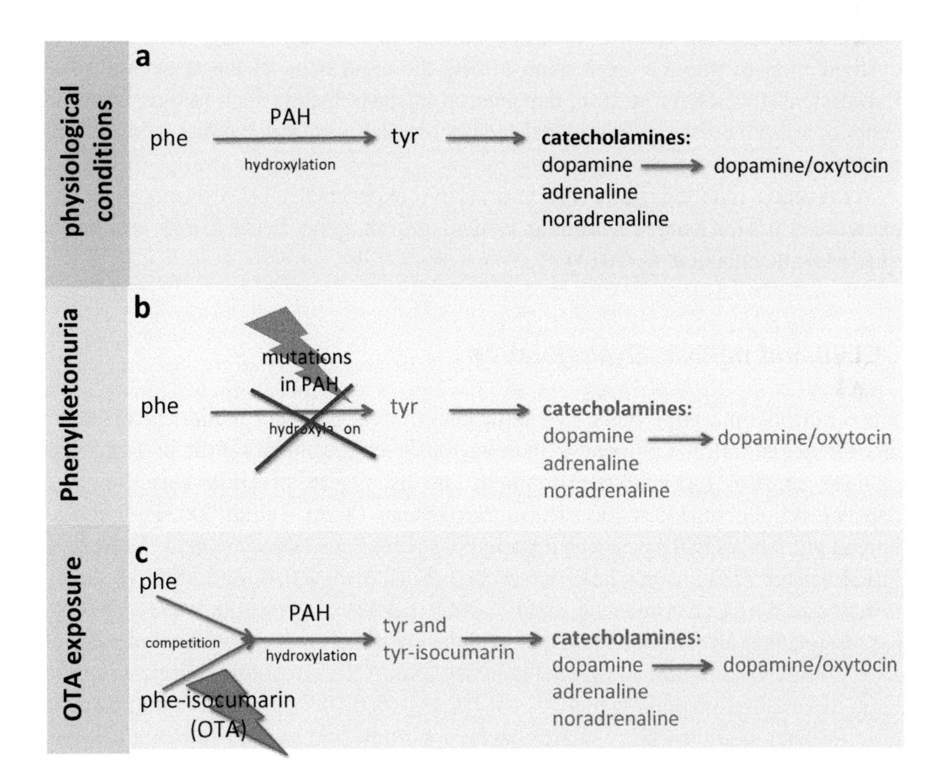

Fig. 3 OTA-induced inhibition of PAH. Comparison between physiologic metabolism of phe and two pathologic conditions: phenylketonuria (PKU) and OTA intoxication. (**a**) Physiological condition: *PAH* hydrolyses phe to tyr necessary for the synthesis of the neurotransmitters catecholamines. (**b**) **PKU**: metabolic disorder in which genetic mutations of *PAH* cause its knockdown, lacking of tyr and, consequently, of catecholamines. This leads to intellectual disabilities, behavioral problems and in some cases autism. (**c**) **OTA exposure**: the phe moiety of OTA competes with phe reducing the PAH activity and the production of catecholamines resembling the PKU pathogen situation

OTA and Epigenetics

OTA toxicity is mainly expressed through epigenetic mechanisms (Schilter et al. 2005); most of the studies refer to the effects of OTA in modulation of microRNAs (abbreviated miRNA) although very little is known about DNA acetylation and methylation.

miRNAs

miRNAs are small, endogenous, single stranded noncoding RNAs that epigenetically orchestrate gene expression through the bond and degradation of target mRNAsat posttranscriptional level (Lu and Clark 2012).

miRNAs are directly involved in physiological and pathological processes in individuals including development and cancer. Indeed, and widely demonstrated, dysregulation of miRNA expression affects the expression of the target mRNAs. Expression of miRNAs, in turn, depends on intrinsic factors such as age, sex, and genetics as well as on environmental factors especially infections, diet, and adsorption of xenobiotics.

As recently reported, in in vitro and in vivo experiments, OTA dysregulates the expression of DiGeorge syndrome critical region gene 8 (*DGCR8*) and some miRNAs Stachurska et al. 2013).

DGCR8 and miRNAs Dysregulation

Since miRNAome is globally dysregulated in many different malignant diseases, interference in miRNA biogenesis pathway has been presumed for the pathogenesis of these cancers. Indeed, knockdown of Drosha, Dgcr8, or Dicer was shown to enhance cellular transformation and tumorigenesis (Kumar et al. 2007) as well as upregulation of DGCR8 has been reported in several tumors such as skin (Sand et al. 2012), gastric (Jafari et al. 2013) colorectal (Kim et al. 2014), and invasive ductal breast carcinoma (Fardmanesh et al. 2016). Recently, dysregulation of *DGCR8* expression has also been associated with psychiatric conditions. An increase of *DGCR8* mRNA, leading to a global increase of miRNAs, was found in postmortem brain tissue of schizophrenic patients as well as genetic deletions and duplications in *DGCR8* lead to DiGeorge/velocardiofacial syndrome and to schizophrenia or other behavioral conditions and cognitive deficits (Beveridge et al. 2010) including autism (Cai et al. 2008).

Interestingly, OTA dysregulates the expression of mir-132, -200c, -200b (Stachurska et al. 2013), -29b (Hennemeier et al. 2014), and -122 (Zhu et al. 2016). In physiological conditions, these miRNAs participate in oxidative stress response, neurophysiology, and regulation of matrix genes and apoptosis and their perturbation is associated with several disorders and diseases (Lyu et al. 2016; Lugli et al. 2008; Jin et al. 2012; Stachurska et al. 2013; Loboda et al. 2016; Hennemeier et al. 2014).

OTA Upregulates Mir-132

Mir-132 is involved in the reciprocal regulation of the autism-related genes methyl-CpG-binding protein 2 (*MeCP2*) and phosphatase and tensin homolog (*PTEN*) and is overexpressed in some neuropsychiatric conditions such as schizophrenia and autism (Mellios and Sur 2012; Lyu et al. 2016). Mir-132 downregulates the expression of *MeCP2* that physiologically binds to methylated CpG islands and, by recruiting histone deacetylase (HDAC) complex (Eden et al. 1998), transcriptionally represses gene expression (Lyu et al. 2016). MECP2 also modulates the expression of a wide range of target genes, including those necessary for neural development, and is involved in miRNA processing by competing, in a dose-dependent manner, with Drosha for binding to DGCR8 (Cheng et al. 2014). Mutations in *MECP2*, that maps on X-chromosome, cause Rett syndrome, severe autism, and X-linked syndromic mental retardation 13. The knockdown of *MeCP2* prevents its interaction with DGCR8 leading to an increase of miRNA synthesis and consequent general inhibition of gene expression (Cheng et al. 2014).

Then again, MECP2 depletion increases the expression of mir-137 that, in turn, targets and suppresses *PTEN*, a ubiquitously expressed tumor suppressor gene whose abnormalities cause tumor progression to more severe stages. Mutations in *PTEN* also cause neurological and psychiatric conditions (Fig. 3b) (Lyu et al. 2016) including macrocephaly/autism syndrome. Decrease of PTEN, in turn, enhances the expression of mir-132 that, downregulating MeCP2, creates a deleterious loop: overexpression of mir-132 downregulates MeCP2; lacking of MeCP2 increases mir-137 that, in turn, downregulates PTEN that lead to mir-132 overexpression (Fig. 3a).

In vitro, miRNA downregulation of MeCP2 also decreases the brain-derived neurotropic factor (*BDNF*) (Manners et al. 2015) a key gene involved in the development and plasticity of synapsis, learning and memory and whose expression is downregulated in Rett disease progression as well as in neuropsychiatric and neurodegenerative diseases such as Parkinson's, Alzheimer's, Huntington's disease, and schizophrenia (Murer et al. 2001; Caccamo et al. 2010). Interestingly, the effect of OTA administration on *BDNF* expression has been recently analyzed in vitro and the results demonstrated that OTA downregulates *BDNF* expression in a dose-dependent manner (Bhat et al. 2016) (Fig. 4).

It is noteworthy a correspondence between the dysregulation of some micro-RNAs and genes (such as mir-132 and BDNF) involved in some genetic neurologic conditions and those dysregulated because of OTA exposure. This leads to speculate that genetic or OTA-induced epigenetic inactivation (through upregulation of mir-132) of MeCP2 triggers the same loop of molecular dysregulation with a deleterious impact on neurodevelopment (Fig. 3c).

Role of mir-132 on Fibrosis

In renal proximal tubular cells from pigs, upregulation of mir-132 also results in a decrease of the nuclear factor erythroid 2-like 2 (*Nrf2*) mRNA (Stachurska et al. 2013). The *Nrf2* pathway activates a robust transcriptional oxidative stress response

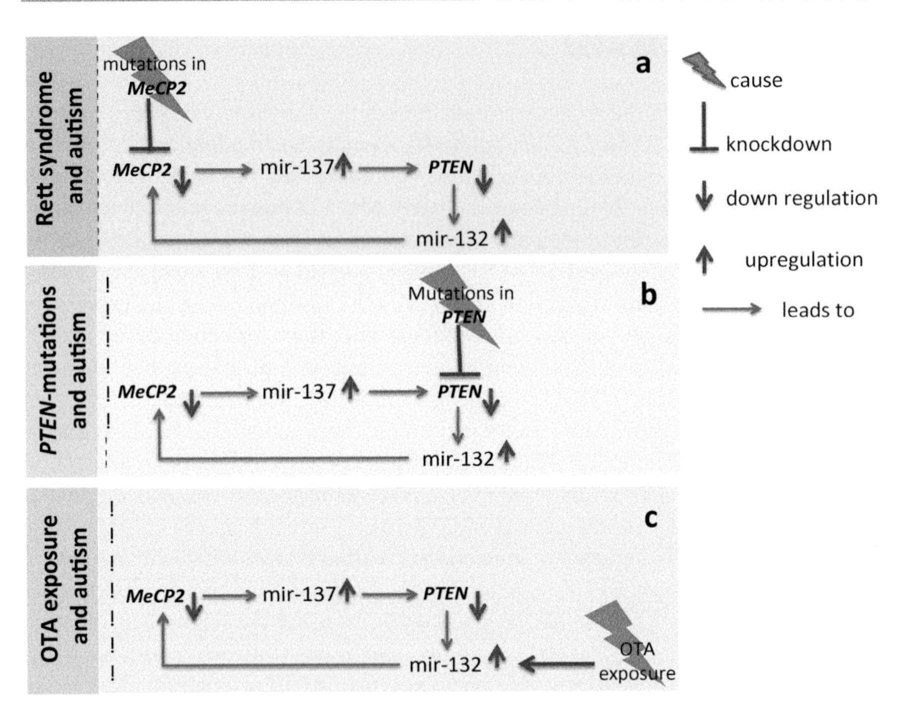

Fig. 4 Causes that alter the reciprocal regulation of autism-related genes. ***MeCP2* and *PTEN***. (**a**) **Rett syndrome and autism**: genetic mutations in *MeCP2* trigger the loop that lead to autism; (**b**) ***PTEN* mutations and autism**: mutations in *PTEN* are the causes of the dysregulation of the loop leading to the neurodevelopmental disorder; (**c**) **OTA exposure and autism**: OTA induces upregulation of mir-132 that triggers the loop resembling cases (**a**) and (**b**) and probably involved in the autism pathogenesis

by binding to the promoter region of many anti-oxidative genes and initiating their transcription. These genes are involved in glutathione synthesis and recycling, in reducing oxygen species and quinones, and in xenobiotic metabolism and transport (Jennings et al. 2013; Limonciel and Jennings 2014). A decrease of Nrf2 results in an attenuation of all these antioxidant functions leading to proteotoxic stress, lipid peroxidation, and oxidative DNA damage. Since no activation of *Nrf2* and *Nrf2*-dependent genes has been found in cells exposed to OTA (Jennings et al. 2012), presumably OTA inhibits *Nrf2* and the cascade of antioxidant genes it activates. Thus, the OTA-dependent upregulation of mir-132 may be the epigenetic key factor that decreases the *Nrf2* protein pool and increases ROS.

Downregulation of the Antifibrotic Mir-29b

In human embryonic kidney cells OTA exposure increases the production of collagen proteins but does not change collagen mRNA expression, thus suggesting that OTA acts by a posttranscriptional mechanism. In these cells, OTA also provokes a shift of intracellular localization of miR-29b that reduces its availability in the cytoplasm. Interestingly, mir-29b belongs to the mir-29 family that, targeting different extracellular matrix genes (including collagen coding genes), modulates fibrotic processes at posttranscriptionally level (Kriegel et al. 2012). Hence, a reduction of mir-29b activity results in a higher amount of translated collagen and fibrotic alterations (Hennemeier et al. 2014).

Moreover, OTA decreases mir-200b that, in physiological conditions, counteracts fibrosis through inhibition of TGFβ, the master cytokine/growth factor in fibrosis. Indeed, in animal models, it has been demonstrated that after ureter obstruction with consequent tubule-interstitial fibrosis, mir-200b injection decreases TGFβ expression and ameliorates fibrosis (Oba et al. 2010).

Downregulation of miR-29a/b also correlates with neurodegenerative diseases such as Huntington's and Alzheimer's (Roshan et al. 2014) with an increase in the level of β-secretase-1 enzyme (BACE1) that, in turn, leads to the formation of amyloid plaques (Xue et al. 2015), the hallmark of several neurodegenerative disease.

Alternatively, in mice, brain-specific knockdown of miR-29 provokes a significant upregulation of its target voltage-dependent anion channel 1 (VDAC1). VDAC1 is a mediator of apoptosis and its dysregulation provokes neuronal cell death and ataxic phenotype. VDAC1 is also upregulated in the brain of transgenic mice models of Alzheimer's as well as in postmortem brain tissue from Alzheimer's patients (Cuadrado-Tejedor et al. 2011).

OTA Increases Mir-122

Mir-122 is highly expressed in the liver and notably acts as a regulator of fatty-acid metabolism. As for pathologic situations, decrease of miR-122 is associated with hepatocellular carcinoma. Experimental results revealed that, in hepatocytes, OTA enhance mir-122 in a time- and dose-dependent manner. Mir-122 upregulation causes inhibition of cell proliferation and induces apoptosis in vivo and in hepatic cell lines. Indeed, mir-122 targets and suppresses Bcl-w (owing to bcl-2 family) and its downstream gene caspase-3 (CASP3). Both the genes are antiapoptotic and their downregulation leads to cell death. Mir-122 also targets Wnt/β-catenin-TCF signaling pathway (Xu et al. 2012) (MacDonald et al. 2009). Wnt family strongly directs cell proliferation and variations in the Wnt pathway has been associated to many diseases including cancer (Clevers 2006).

OTA Increases Mir-200c

In vitro and in vivo OTA increases the expression of mir-200c (Stachurska et al. 2013; Dai et al. 2014) probably through ROS induction. Indeed, a recent study demonstrated that the effect of ROS on miRNA expression results in a dramatic increase of mir-200c that, in turn, provokes cell growth arrest, senescence, and apoptosis through the suppression of the transcriptional repressor zinc finger E-box binding homeobox 1 (*ZEB1)* mRNA (Magenta et al. 2011).

Moreover, the upregulation of mir-200c decreases the expression of the oxidative stress protective HO-1 (Stachurska et al. 2013); then OTA may exert its toxicity by inhibition of the antioxidant response system Nrf2/HO-1 by means of mir-132 and -200c upregulation. This leads to ROS production and, consequently, to an increase of TGFβ expression.

Mir-200c also plays a key role in synaptic plasticity: in synaptoneurosomes (small vesicular structures enriched in dendritic spines and isolated from total forebrain homogenate) mir-200c is ~5 fold more abundant than in the total forebrain homogenate (Lugli et al. 2008). To this purpose, mir-200c is predicted, by bioinformatic tools, and validated, by next generation sequencing experiments, to target, and then down-regulate, neuroligin4X (NLGN4X) (http://mirtarbase.mbc.nctu.edu.tw) (Chou et al. 2015) a protein involved in the formation and plasticity of synapses. Causative point mutations in *NLGN4X* gene have been described in ~1% of patients suffering from ASD demonstrating that functional loss of this protein leads to this pathology; indeed, a mere impairment of a protein at the synapse can provoke a dramatic perturbation in synaptic transmission (Bemben et al. 2015). Interestingly, *NLGN4X* is mapped on X chromosome making males more prone to the effects of its inactivation.

OTA Induces Deacetylation

Recently has been proposed that OTA provokes genomic instability and renal tumorigenesis through interfering with mitotic processes that prevent cell division and lead to cell death or generation of polyploidy and malignant phenotype. To this purpose, Czakai and collaborators (Czakai et al. 2011) demonstrated that immortalized human kidney epithelial cells, treated with OTA at carcinogenic concentration, were undergoing to mitotic arrest due to chromosome aberrant condensation, premature loss of sister chromatid pairing, and altered histone acetylation and phosphorylation. In these cells, OTA dramatically blocked the activity of histone acetyltransferases (HATs) in a dose-dependent manner. HATs are a group of ~30 enzymes that catalyze acetylation of conserved lysine residues on histone proteins leading to chromatin relaxation and a consequent increase of gene expression. HATs are also crucial regulators of cell division and cohesion of sister chromatid during mitosis (Choudhary et al. 2009). Depletion of some HATs, such as Adenoviral E1A-associated protein (p300) and CREB binding protein (CBP), results in mitotic aberration and relative arrest of mitosis. Since HATs seem to be OTA primary cellular targets, it is possible that the mycotoxin causes mitotic catastrophe through depletion of HATS and consequent chromosome hypercondensation.

A plethora of nonhistone proteins involved in antioxidation, cell cycle control, or DNA repair are also susceptible to OTA posttranslational acetylation interfering that leads to their functional loss. Indeed, the transcription coactivator p300/CBP has been demonstrated to directly acetylate *Nrf2* in response to oxidative stress (Choudhary et al. 2009). Thus, OTA-dependent p300/CBP block could reduce Nrf2 acetylation and

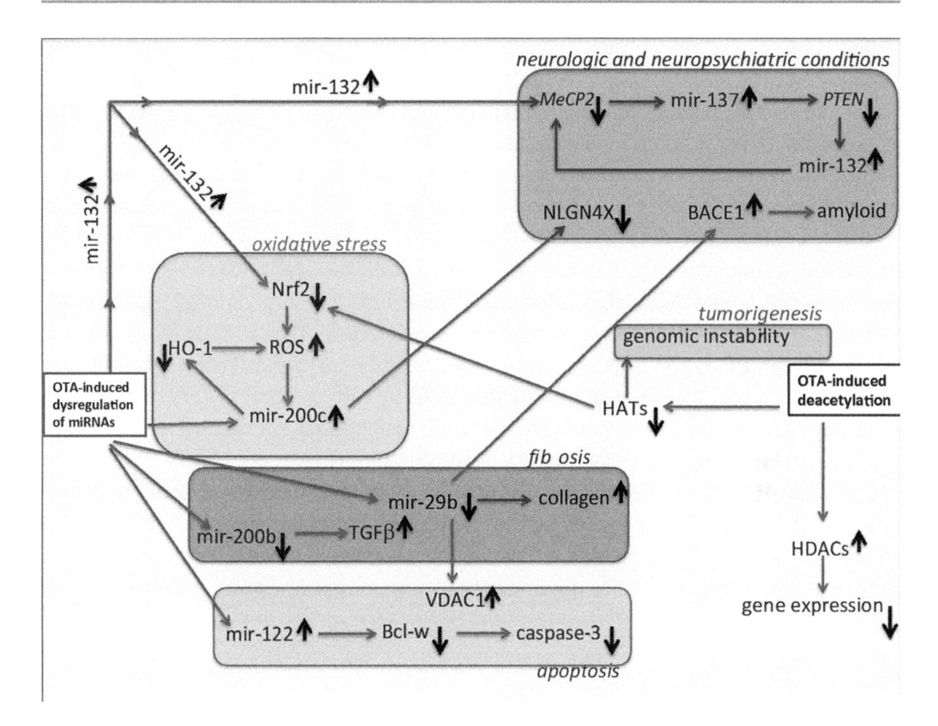

Fig. 5 Summary of the OTA-induced epigenetic effects. OTA-induce epigenetic effects lead to oxidative stress, fibrosis, apoptosis, genomic instability, and neurologic and psychiatric conditions

its consequent activation (Sun et al. 2009; Czakai et al. 2011). In this way, OTA could interfere with the antioxidant response system by two strategies: mir-132 increase and p300/CBP depletion that, in turn, lead to Nrf2 inactivation (Fig. 5). Alternatively, has been proposed that OTA-dependent hypoacetylation also involves p53 and its network proteins as well as proteins involved in DNA repair. The downregulation of these proteins could affect DNA replication explaining the persistence of DNA damages found in cells and animals exposed to OTA (Mally et al. 2005).

OTA, in addition, modulates histone deacetylases (HDACs) that, enhancing chromatin condensation, reduce gene expression; indeed, in kidney of rats exposed to OTA, a significant increase of HADCs has been found (Marin-Kuan et al. 2007). Since HATs and HDACs activities are inversely proportional, OTA-induced HATs reduction indirectly results in HDCAs increase and in consequent global gene expression down-regulation.

Conclusion

In conclusion OTA toxicity acts especially by means of epigenetic mechanisms and in many different ways: blocking the oxidative stress response, dysregulating the expression of genes involved in neurophysiology, and increasing fibrosis and malignant phenotypes. These effects are associated with some pathology such as BEN and hepatic-carcinogenesis and, in emerging way, neurological and psychiatric conditions.

As the mycotoxin contaminates foodstuff and feedstuff worldwide, OTA toxicity is of human and animal concern. So the effects of the mycotoxin on general population should be redefined to establish the risk assessment and the daily tolerable intake.

Then again, OTA detoxification should also be considered for those subjects displaying symptoms of OTA toxicosis. First of all, the antioxidant NAC prevents the OTA-dependent ROS formation and consequent cellular damage (Bhat et al. 2016). Then, in order to counteract the OTA deleterious effect on intestinal microbiota, the administration of probiotics such as *Lactobacillus Plantarum* that is able to adsorb the mycotoxin (Guo et al. 2014) should be tested. As for OTA competition with phe, administration of the OTA antagonist aspartame counteracts OTA inhibition of PAH restoring its physiological activity (Baudrimont et al. 2001; McMaster and Vadani 1999). More recently, the use of layered double hydroxides nano compound has been suggested for removing OTA from contaminated food and feedstuff. Indeed, OTA is intercalated and adsorbed in this compound that is nontoxic, low cost, and easily obtained (San Román and Holgado 2015).

Dictionary of Terms

Autism spectrum disorder	Neurodevelopmental disorder with uncertain pathogenesis, characterized by difficulty in communication and social interaction, limited interests, and repetitive behaviors. About 30% of patients display causative genetic variants while, for the remaining 70%, a gene-environmental interaction has been proposed. Autism is a male-typical disorder with males/females ratios ranging from 2:1 to 16:1.
Apoptosis	Physiological process of programmed cell death to keep cells in tissues that occurs during development and aging. Apoptosis can also act as a defense when diseases or noxious factors impair cells.
Balkan endemic nephropathy	Is a form of fatal familial chronic nephropathy strictly associated to upper urothelial cancer with a high-prevalence rate in rural population along the Danube River. It is characterized by progressive and extensive atrophy and sclerosis of

	kidney, interstitial fibrosis, and glomerular and vascular lesions.
Blood brain barrier	The barrier with high selective permeability that limits the paracellular transport into the central nervous system.
Cytochrome P450	Is a family of ~30 hepatic enzymes responsible for the biotransformation of xenobiotics.
DiGeorge syndrome critical region gene 8 (DGCR8)	Double-stranded-RNA-binding protein part of the miRNA biogenesis machinery components including Drosha-DGCR8 and Dicer-TRBP complexes. Upregulation of DiGeorge syndrome critical region gene 8 increases global microRNA expression provoking an extensive gene silencing whereas its downregulation results in a general induction of gene expression.
Fibrosis	The pathological deposition of extracellular matrixproteins, such as collagen, and connective tissue in response to cellular damage. Fibrosis can destroy the architecture and function of the organs involved.
Gut microbiota	The complex community of microbes residing in the intestine containing up to trillions of microorganisms. It plays a key role in health and disease of the host.
Oxidative stress	A disturbance in the *equilibrium* between oxidants (reactive oxygen species resulted from cellular metabolism or exogenous sources), and antioxidant defenses in favor of oxidants. Oxidative stress produces negative effects on DNA, lipids, proteins, and signal transduction leading to cell and tissue damage.
Single-nucleotide polymorphism	The most common type of genetic variation in a single nucleotide occurring at a specific position in the genome. Most of them have no effect on health while some of them are associated to specific conditions.
Synapsis	A structure across which nerve impulses pass from a neuron to another.

Key Facts About Ochratoxin A Toxicity

- Intestinal permeability allows a plethora of xenobiotics to enter the bloodstream.
- Ochratoxin A provokes leaky gut and also passes through the blood brain barrier.
- The enzymes that metabolize Ochratoxin A are expressed mainly in females making males more prone to the deleterious effects of Ochratoxin A.
- Ochratoxin A causes nephrotoxicity, hepatotoxicity, and neurotoxicity.
- Ochratoxin A exerts its toxicity essentially through epigenetic mechanisms.
- Most of the studies regarding Ochratoxin A-induced epigenetic effects refer to micro-RNA modulation.
- Ochratoxin A epigenetic effects involve oxidative stress response, fibrosis, apoptosis, and tumorigenesis.
- Ochratoxin A modulates genes and micro-RNAs that are dysregulated in neurologic and psychiatric conditions.
- The phenylalanine moiety of Ochratoxin A competes with phe reducing the phenylalanine hydroxylase activity and the production of catecholamines.
- The Ochratoxin A risk assessment and the daily tolerable intake as well as mode of Ochratoxin A detoxification should be investigated.

Summary Points

- Ochratoxin A (OTA) is a thermoresistant mycotoxin produced by ubiquitous molds. Since it contaminates foodstuffs and feedstuffs worldwide and therefore of human and animal concern.
- OTA induces oxidative stress, inflammation, and fibrosis, and is nephrotoxic, hepatotoxic, and neurotoxic especially in males.
- It exerts its toxicity essentially through epigenetic mechanisms although most studies refer only to OTA dysregulation of microRNAs.
- Nephrotoxicity is due to OTA-induced suppression of the collagen regulator mir-29b that results in an increase of fibrosis
- OTA induced mir-132 upregulation that is implicated in neurologic and psychiatric conditions as well as in increasing level of oxidative stress.
- OTA induces overexpression of mir-200c that reduces antioxidative response and alters synaptic plasticity.
- In liver, OTA exposure increases mir-122 that targets and suppresses anti-apoptotic genes leading to cell death and hepatic damage.
- OTA dysregulates some mi-RNAs and target genes involved in neurologic and psychiatric conditions, including autism, suggesting a possible contribution of OTA in the pathogenesis of the disorder.
- OTA induces block of histone acetyltransferases leading to mitotic arrest and inhibition of antioxidation.

References

Baieli S, Pavone L, Meli C et al (2003) Autism and phenylketonuria. J Autism Dev Disord 33(2):201–204

Baskerville TA, Douglas AJ (2010) Dopamine and oxytocin interactions underlying behaviors: potential contributions to behavioral disorders. CNS Neurosci Ther 16(3):e92–123. https://doi.org/10.1111/j.1755-5949.2010.00154.x. Review. PubMed PMID: 20557568

Baudrimont I, Sostaric B, Yenot C, Betbeder AM, Dano-Djedje S, Sanni A, Steyn PS, Creppy EE (2001) Aspartame prevents the karyomegaly induced by ochratoxin A in rat kidney. Arch Toxicol 75(3):176–83

Bemben MA, Nguyen QA, Wang T et al (2015) Autism-associated mutation inhibits protein kinase C-mediated neuroligin-4X enhancement of excitatory synapses. Proc Natl Acad Sci U S A 112 (8):2551–2556. https://doi.org/10.1073/pnas.1500501112

Beveridge NJ, Gardiner E, Carroll AP et al (2010) Schizophrenia is associated with an increase in cortical microRNA biogenesis. Mol Psychiatry 15(12):1176–1189. https://doi.org/10.1038/mp.2009.84

Bhat PV, Md P, Khanum F et al (2016) Cytotoxic effects of ochratoxin A in neuro-2a cells: role of oxidative stress evidenced by N-acetylcysteine. Front Microbiol 7:1142. https://doi.org/10.3389/fmicb.2016.01142

Boudra H, Le Bars P, Le Bars J (1995) Thermostability of ochratoxin A in wheat under two moisture conditions. Appl Environ Microbiol 61:1156–1158

Caccamo A, Maldonado MA, Bokov AF et al (2010) CBP gene transfer increases BDNF levels and ameliorates learning and memory deficits in a mouse model of Alzheimer's disease. Proc Natl Acad Sci U S A 107:22687–22692. https://doi.org/10.1073/pnas.1012851108

Cai G, Edelmann L, Goldsmith JE et al (2008) Multiplex ligation-dependent probe amplification for genetic screening in autism spectrum disorders: efficient identification of known microduplications and identification of a novel microduplication in ASMT. BMC Med Genet 1:50. https://doi.org/10.1186/1755-8794-1-50

Castegnaro M, Canadas D, Vrabcheva T et al (2006) Balkan endemic nephropathy: role of ochratoxins A through biomarkers. Mol Nutr Food Res 50(6):519–529

Cheng TL, Wang Z, Liao Q et al (2014) MeCP2 suppresses nuclear microRNA processing and dendritic growth by regulating the DGCR8/Drosha complex. Dev Cell 28(5):547–560. https://doi.org/10.1016/j.devcel.2014.01.032

Chou C, Chang N, Shrestha S, Hsu S, Lin Y, Lee W et al (2015) miRTarBase 2016: updates to the experimentally validated miRNA-target interactions database. Nucleic Acids Res 44(D1): D239–D247

Choudhary C, Kumar C, Gnad F et al (2009) Lysine acetylation targets protein complexes and co-regulates major cellular functions. Science 325(5942):834–840. https://doi.org/10.1126/science.1175371

Clevers H (2006) Wnt/beta-catenin signaling in development and disease. Cell 127:469–480

Creppy EE, Chakor K, Fisher MJ et al (1990) The myocotoxin ochratoxin A is a substrate for phenylalanine hydroxylase in isolated rat hepatocytes and in vivo. Arch Toxicol 64(4):279–284

Cuadrado-Tejedor M, Vilariño M, Cabodevilla F et al (2011) Enhanced expression of the voltage-dependent anion channel 1 (VDAC1) in Alzheimer's disease transgenic mice: an insight into the pathogenic effects of amyloid-β. J Alzheimers Dis 23(2):195–206. https://doi.org/10.3233/JAD-2010-100966

Czakai K, Müller K, Mosesso P et al (2011) Perturbation of mitosis through inhibition of histone acetyltransferases: the key to ochratoxin a toxicity and carcinogenicity? Toxicol Sci 122 (2):317–329. https://doi.org/10.1093/toxsci/kfr110

Dai Q, Zhao J, Qi X et al (2014) MicroRNA profiling of rats with ochratoxin A nephrotoxicity. BMC Genomics 15:333. https://doi.org/10.1186/1471-2164-15-333

De Santis B, Brera C, Mezzelani A et al (2017a) Role of mycotoxins in the pathobiology of autism: a first evidence. Nutr Neurosci 1–13. https://doi.org/10.1080/1028415X.2017.1357793

De Santis B, Raggi ME, Moretti G et al (2017b) Study on the association among mycotoxins and other variables in children with autism. Toxins 29;9(7). pii: E203. https://doi.org/10.3390/toxins9070203

Deepmala, Slattery J, Kumar N et al (2015) Clinical trials of N-acetylcysteine in psychiatry and neurology: a systematic review. Neurosci Biobehav Rev 55:294–321. https://doi.org/10.1016/j.neubiorev.2015.04.015

Denli M, Perez JF (2010) Ochratoxins in feed, a risk for animal and human health: control strategies. Toxins (Basel) 2(5):1065–1077. https://doi.org/10.3390/toxins2051065

Eden S, Hashimshony T, Keshet I et al (1998) DNA methylation models histone acetylation. Nature 394(6696):842

EFSA, European Food Safety Authority (2006) EFSA Opinion of the scientific panel on contaminants in the food chain on a request from the commission related to ochratoxin A (OTA) in food Quest. N° EFSA-Q-2005-154. EFSA J 365(2006):1–56. https://doi.org/10.2903/j.efsa.2006.365

Fardmanesh H, Shekari M, Movafagh A et al (2016) Upregulation of the double-stranded RNA binding protein DGCR8 in invasive ductal breast carcinoma. Gene 581(2):146–151. https://doi.org/10.1016/j.gene.2016.01.033

Faustman EM, Silbernagel SM, Fenske RA, Burbacher TM, Ponce RA (2000) Mechanisms underlying Children's susceptibility to environmental toxicants. Environ Health Perspect 108 (Suppl 1):13–21

Gayathri L, Dhivya R, Dhanasekaran D et al (2015) Hepatotoxic effect of ochratoxin A and citrinin, alone and in combination, and protective effect of vitamin E: in vitro study in HepG2 cell. Food Chem Toxicol 83:151–163. https://doi.org/10.1016/j.fct.2015.06.009

Guo M, Huang K, Chen S, Qi X, He X, Cheng WH, Luo Y, Xia K, Xu W (2014) Combination of metagenomics and culture-based methods to study the interaction between ochratoxin a and gut microbiota. Toxicol Sci 141(1):314–323. https://doi.org/10.1093/toxsci/kfu128

Hagelberg S, Hult K, Fuchs R (1989) Toxicokinetics of ochratoxin A in several species and its plasma-binding properties. J Appl Toxicol 9(2):91–96

Hennemeier I, Humpf HU, Gekle M et al (2014) Role of microRNA-29b in the ochratoxin A-induced enhanced collagen formation in human kidney cells. Toxicology 3(324):116–122. https://doi.org/10.1016/j.tox.2014.07.012

Hope JH, Hope BE (2012) A review of the diagnosis and treatment of ochratoxin A inhalational exposure associated with human illness and kidney disease including focal segmental glomerulosclerosis. J Environ Public Health 2012:835059. https://doi.org/10.1155/2012/835059

Jafari N, Dogaheh HP, Bohlooli S et al. (2013) Expression levels of microRNA machinery components Drosha, Dicer and DGCR8 in human (AGS, HepG2, and KEYSE-30) cancer cell lines. Int J Clin Exp Med 6(4):269–274

Jennings P, Weiland C, Limonciel A et al (2012) Transcriptomic alterations induced by ochratoxin A in rat and human renal proximal tubular in vitro models and comparison to a ratinvivo model. Arch Toxicol 86:571–589

Jennings P, Limonciel A, Felice L, Leonard MO (2013) An overview of transcriptional regulation in response to toxicological insult. Arch Toxicol 87:49–72

Jin J, Cheng Y, Zhang Y et al (2012) Interrogation of brain miRNA and mRNA expression profiles reveals a molecular regulatory network that is perturbed by mutant huntingtin. J Neurochem 123 (4):477–490. https://doi.org/10.1111/j.1471-4159.2012.07925.x

Kim B, Lee JH, Park JW et al (2014) An essential microRNA maturing microprocessor complex component DGCR8 is up-regulated in colorectal carcinomas. Clin Exp Med 14(3):331–336. https://doi.org/10.1007/s10238-013-0243-8

Kriegel AJ, Liu Y, Fang Y et al (2012) The miR-29 family: genomics, cell biology, and relevance to renal and cardiovascular injury. Physiol Genomics 44:237–244

Kumar MS, Lu J, Mercer KL et al (2007) Impaired microRNA processing enhances cellular transformation and tumorigenesis. Nature Genet 39:673–677

Limonciel A, Jennings P (2014) A review of the evidence that ochratoxin A is an Nrf2 inhibitor: implications for nephrotoxicity and renal carcinogenicity. Toxins (Basel) 6(1):371–379. https://doi.org/10.3390/toxins6010371

Loboda A, Damulewicz M, Pyza E et al (2016) Role of Nrf2/HO-1 system in development, oxidative stress response and diseases: an evolutionarily conserved mechanism. Cell Mol Life Sci 73(17):3221–3247. https://doi.org/10.1007/s00018-016-2223-0

Lu J, Clark AG (2012) Impact of microRNA regulation on variation in human gene expression. Genome Res 22(7):1243–1254

Lugli G, Torvik VI, Larson J, Smalheiser NR (2008) Expression of microRNAs and their precursors in synaptic fractions of adult mouse forebrain. J Neurochem 106(2):650–661. https://doi.org/10.1111/j.1471-4159.2008.05413.x

Lyu JW, Yuan B, Cheng TL et al (2016) Reciprocal regulation of autism-related genes MeCP2 and PTEN via microRNAs. Sci Rep 6:20392. https://doi.org/10.1038/srep20392

MacDonald BT, Tamai K, He X (2009) Wnt/beta-catenin signaling: components, mechanisms, and diseases. Dev Cell 17(1):9–26. https://doi.org/10.1016/j.devcel.2009.06.016

Magenta A, Cencioni C, Fasanaro P et al (2011) MC. miR-200c is upregulated by oxidative stress and induces endothelial cell apoptosis and senescence via ZEB1 inhibition. Cell Death Differ 18 (10):1628–1639. https://doi.org/10.1038/cdd.2011.42

Malir F, Ostry V, Pfohl-Leszkowicz A et al (2016) Ochratoxin A: 50 years of research. Toxins (Basel) 8(7):191. https://doi.org/10.3390/toxins8070191

Mally A, Pepe G, Ravoori S et al (2005) Ochratoxin a causes DNA damage and cytogenetic effects but no DNA adducts in rats. Chem Res Toxicol 18(8):1253–1261. PubMed PMID: 16097798

Manners MT, Tian Y, Zhou Z, Ajit SK (2015) MicroRNAs downregulated in neuropathic pain regulate MeCP2 and BDNF related to pain sensitivity. FEBS Open Bio 5:733–740

Marin-Kuan M, Nestler S, Verguet C et al (2007) MAPK-ERK activation in kidney of male rats chronically fed ochratoxin A at a dose causing a significant incidence of renal carcinoma. Toxicol Appl Pharmacol 224(2):174–181

McLaughlin J, Padfield PJ, Burt JP, O'Neill CA (2004) Ochratoxin A increases permeability through tight junctions by removal of specific claudin isoforms. Am J Physiol Cell Physiol 287(5):C1412–C1417

McMasters DR, Angelo Vedani A (1999) Ochratoxin Binding to Phenylalanyl-tRNA Synthetase: Â Computational Approach to the Mechanism of Ochratoxicosis and Its Antagonism. Journal of Medicinal Chemistry 42(16):3075–3086

Mellios N, Sur M (2012) The emerging role of microRNAs in schizophrenia and autism spectrum disorders. Front Psych 3:39. https://doi.org/10.3389/fpsyt.2012.00039

Mor F, Kilic MA, Ozmen O et al (2014) The effects of orchidectomy on toxicological responses to dietary ochratoxin A in Wistar rats. Exp Toxicol Pathol 66(5-6):267–275. https://doi.org/10.1016/j.etp.2014.04.002

Murer MG, Yan Q, Raisman-Vozari R (2001) Brain-derived neurotrophic factor in the control human brain, and in Alzheimer's disease and Parkinson's disease. Prog Neurobiol 63:71–124. https://doi.org/10.1016/S0301-0082(00)00014-9

Oba S, Kumano S, Suzuki E et al (2010) miR-200b precursor can ameliorate renal tubulointerstitial fibrosis. PLoS One 5(10):e13614. https://doi.org/10.1371/journal.pone.0013614

Pastor L, Vettorazzi A, Campión J, Cordero P, López de Cerain A (2016) Gene expression kinetics of renal transporters induced by ochratoxin A in male and female F344 rats. Food Chem Toxicol 98(Pt B):169–178. https://doi.org/10.1016/j.fct.2016.10.019

Preissner SC, Hoffmann MF, Preissner R, Dunkel M, Gewiess A, Preissner S (2013) Polymorphic cytochrome P450 enzymes (CYPs) and their role in personalized therapy. PLoS One 8(12): e82562. https://doi.org/10.1371/journal.pone.0082562

Ringot D, Chango A, Schneider YJ, Larondelle Y (2006) Toxicokinetics and toxicodynamics of ochratoxin A, an update. Chem Biol Interact 159(1):18–46

Roshan R, Shridhar S, Sarangdhar MA et al (2014) Brain-specific knockdown of miR-29 results in neuronal cell death and ataxia in mice. RNA 20(8):1287–1297. https://doi.org/10.1261/rna.044008.113

San Román MS, Holgado MJ (2015) Intercalation of phenylalanine, isocoumarin and ochratoxin A (OTA) into LDH's. Open Journal of Inorganic Chemistry 5:52–62. https://doi.org/10.4236/ojic.2015.53007

Sand M, Skrygan M, Georgas D, Arenz C, Gambichler T, Sand D, Altmeyer P, Bechara FG (2012) Expression levels of the microRNA maturing microprocessor complex component DGCR8 and the RNA-induced silencing complex (RISC) components argonaute-1, argonaute-2, PACT, TARBP1, and TARBP2 in epithelial skin cancer. Mol Carcinog 51(11):916–922. https://doi.org/10.1002/mc.20861

Sava V, Reunova O, Velasquez A, Harbison R, Sanchez-Ramos J (2006) (2006a). Acute neurotoxic effects of the fungal metabolite ochratoxin-A. Neurotoxicology 27:82–92. https://doi.org/10.1016/j.neuro.2005.07.004

Schilter B, Marin-Kuan M, Delatour T et al (2005) Ochratoxin A: potential epigenetic mechanisms of toxicity and carcinogenicity. Food Addit Contam 22(Suppl 1):88–93

Stachurska A, Ciesla M, Kozakowska M et al (2013) Cross-talk between microRNAs, nuclear factor E2-related factor 2, and heme oxygenase-1 in ochratoxin A-induced toxic effects in renal proximal tubular epithelial cells. Mol Nutr Food Res 57(3):504–515. https://doi.org/10.1002/mnfr.201200456

Sun Z, Chin YE, Zhang DD (2009) Acetylation of Nrf2 by p300/CBP augments promoter-specific DNA binding of Nrf2 during the antioxidant response. Mol Cell Biol 29(10):2658–2672. https://doi.org/10.1128/MCB.01639-08

Ueta E, Kodama M, Sumino Y et al (2010) Gender-dependent differences in the incidence of ochratoxin A-induced neural tube defects in the Pdn/Pdn mouse. CongenitAnom (Kyoto) 50(1):29–39. https://doi.org/10.1111/j.1741-4520.2009.00255.x. PubMed PMID: 20201966

Waxman DJ, Holloway MG (2009) Sex differences in the expression of hepatic drug metabolizing enzymes. Mol Pharmacol 76(2):215–228. https://doi.org/10.1124/mol.109.056705

Woodmansey EJ (2007) Intestinal bacteria and ageing. Appl Microbiol 102(5):1178–1186

Wu Q, Dohnal V, Huang L et al (2011) Metabolic pathways of ochratoxin A. Curr Drug Metab 12(1):1–10

Xu J, Zhu X, Wu L et al (2012) MicroRNA-122 suppresses cell proliferation and induces cell apoptosis in hepatocellular carcinoma by directly targeting Wnt/β-catenin pathway. Liver Int 32(5):752–760. https://doi.org/10.1111/j.1478-3231.2011.02750.x

Xue ZQ, He ZW, Yu JJ et al (2015) Non-neuronal and neuronal BACE1 elevation in association with angiopathic and leptomeningeal β-amyloid deposition in the human brain. BMC Neurol 15:71. https://doi.org/10.1186/s12883-015-0327-z

Zanic-Grubisić T, Zrinski R, Cepelak I, Petrik J, Radić B, Pepeljnjak S (2000) Studies of ochratoxin A-induced inhibition of phenylalanine hydroxylase and its reversal by phenylalanine. Toxicol Appl Pharmacol 167(2):132–139

Zhang X, Boesch-Saadatmandi C, Lou Y, Wolffram S, Huebbe P, Rimbach G (2009) Ochratoxin A induces apoptosis in neuronal cells. Genes Nutr 4(1):41–48. https://doi.org/10.1007/s12263-008-0109-y

Zhu L, Yu T, Qi X, Yang B, Shi L, Luo H, He X, Huang K, Xu W (2016) miR-122 plays an important role in ochratoxin A-induced hepatocyte apoptosis in vitro and in vivo. Toxicol Res 5:160–167

Silver and Histone Modifications 112

Yuko Ibuki

Contents

Abstract

Silver (Ag) is used in a wide range of industries due to its antibacterial properties. The daily intake of Ag from the diet is estimated to be approximately 7 μg per person, and this is predicted to increase due to artificial treatments in food production chains and drinking water. Colloidal Ag and nano-sized Ag (AgNPs) have recently been used in dietary supplements and food packaging materials, respectively, and concerns have been raised over their safety. Ag is

Y. Ibuki (✉)
Graduate Division of Nutritional and Environmental Sciences, University of Shizuoka, Shizuoka, Japan
e-mail: ibuki@u-shizuoka-ken.ac.jp

V. B. Patel, V. R. Preedy (eds.), *Handbook of Nutrition, Diet, and Epigenetics*,
https://doi.org/10.1007/978-3-319-55530-0_74

regarded as a safe metal; however, in vitro experiments recently revealed the genotoxicity of AgNPs. The toxic effects of AgNPs have mainly been attributed to the binding of released Ag ions with functional proteins and the formation of reactive oxygen species. Post-translational modifications in histones have been linked to a number of biological and toxicological processes as well as diseases, and carcinogenic metals are known to alter histone modifications, leading to changes in gene expression. Although few Ag-induced histone modifications have been reported to date, we demonstrated that AgNPs induced the phosphorylation of histone H2AX at serine 139 based on DNA damage and the phosphorylation of histone H3 at serine 10. In this review, these histone modifications are mainly introduced with other related modifications. Histone modifications have been suggested to change the chromatin structure, followed by alterations in the formation and repair of DNA damage. These changes may play a role in the enhanced genotoxicity and carcinogenicity of Ag when it is coexposed with other genotoxic factors.

Keywords

Silver · Nanoparticles · Disinfection · Drinking water · Food packaging · Histone · Phosphorylation · Acetylation · Chromatin · DNA damage · Ultraviolet rays

List of Abbreviations

DSBs	Double-strand breaks
GSH	Glutathione
HDACs	Histone deacetylases
HDACI	HDAC inhibitor
AgNPs	Nano-sized Ag
NOAEL	No observable adverse effect level
γ-H2AX	Phosphorylation of histone H2AX
p-H3S10	Phosphorylation of histone H3 at serine 10
ROS	Reactive oxygen species
Ag	Silver
SB	Sodium butyrate
UV	Ultraviolet
WHO	World Health Organization

Introduction

Silver (Ag) is a naturally occurring element that mainly exists in the form of very insoluble and immobile oxides, sulfides, and some salts. It is found in sea and fresh water as well as in soil and rocks. It is even present in the air as a result of industrial pollution. Ag is used to make a number of products including mirrors, coins, table silver, jewelry, and dental fillings. The antibiotic characteristics of soluble Ag compounds have recently been applied to the disinfection of drinking water and water in swimming pools as well as antibacterial agents (Fewtrell 2014; WHO 2003, 2008). Ag levels in groundwater, surface water, and drinking water are less than 5 μg/L, whereas those in drinking water treated with Ag for disinfection are greater than 50 μg/L. Ag is now being increasingly used in a wide range of applications, such as in food, dietary supplements, and cosmetics. A shift in human and environmental exposure is expected in conjunction with the growing use of Ag.

Ag is regarded as a safe metal. A total lifetime oral intake of 10 g of Ag is considered to be the human no observable adverse effect level (NOAEL) (WHO 2003). Exposure to high levels of Ag for a long period of time results in a condition called argyria, a blue-gray discoloration of the skin and other body tissues (Chang et al. 2006; Drake and Hazelwood 2005). Previous studies reported that argyria was caused by the soluble form of Ag. Lower-level exposure to Ag may also result in its deposition in the skin and other parts of the body; however, this is not known to be so harmful. The World Health Organization (WHO) established first the no health-based guideline value for Ag in drinking water in 1993 (WHO 2003, 2008). Ag has been used in the form of salts, oxides, and halides or as the element, whereas colloidal Ag (small Ag particles in liquid) and nano-sized Ag particles (AgNPs) have recently been used in dietary supplements and food packaging materials for antibacterial activity, respectively, and their safety has been investigated (European Commission 2014; NIH 2004). AgNPs may have different physicochemical properties from micro-sized Ag particles; therefore, concerns have been raised regarding toxic effects in humans (European Commission 2014). The release of Ag ions from AgNPs has been identified as the main cause of toxicity; nevertheless, an increasing number of studies demonstrated that this release alone does not account for the toxic effects observed (Beer et al. 2012).

Post-translational modifications in histones have recently attracted attention because they have been linked to a number of biological and toxicological processes as well as disease states such as cancer (Hake et al. 2004; Kouzarides 2007). Carcinogenic metals, including nickel, arsenic, and chromium, alter histone modifications, which change the programs of gene expression and carcinogenesis (Chervona et al. 2012; Paul and Giri 2015; Wang et al. 2012). Few Ag-induced histone modifications have been reported to date. We recently noted that AgNPs induced some histone modifications, which may be different from the modifications induced by bulky Ag.

The usefulness and toxicological aspects of Ag in food and drinking water are summarized in this review. AgNP-induced histone modifications are subsequently introduced. Furthermore, the relationship between Ag-enhanced DNA damage following ultraviolet (UV) irradiation and histone modifications is discussed.

Ag in Food and Water

A recent estimate of the daily intake of Ag from the diet including drinking water was approximately 7 μg per person (WHO 2003). Higher intake has been reported ranging from 20 to 80 μg per day. Most foods contain traces of Ag in the 10–100 μg/kg range. Refined wheat flour, for example, contains 0.3 μg/g, wheat bran 0.9 μg/g, milk 27–54 μg/L, and mushrooms hundreds of μg/g (European Commission 2000). On the other hand, treatment of food and drinking water for bacteriostasis will increase Ag uptake into humans. In addition, the effects of AgNPs are a concern because of their specific properties. The application of AgNPs to the food production chain was summarized by Bouwmeester et al. (2009). AgNPs may migrate to humans during agricultural production, food procession, and conservation. The Woodrow Wilson International Center for Scholars published a review on the application of and regulatory issues related to NPs incorporated in food packaging materials (Taylor 2008). The direct contact of AgNPs with food, when they are used in packaging materials for preservation, provides a route of AgNPs into food (Bouwmeester et al. 2009; Taylor 2008). The migration limit for Ag from packaging products into food is 0.05 mg/kg (European Commission 2014).

The relative contribution of drinking water is generally very low. Ag levels in drinking water in the USA that had not been treated with Ag for disinfection purposes varied between "nondetectable" and 5 μg/L (WHO 2003). Water treatments with Ag may increase these levels. In terms of water disinfection-related applications, Ag is used in domestic water filters. AgNPs are currently being tested in a number of experimental point-of-use treatment systems, and ionic Ag has been examined for its potential use as a secondary disinfectant (to reduce chlorine levels) in drinking water supplies (Fewtrell 2014). A large number of studies have been conducted on the disinfection efficacies of Ag and AgNP applications against a range of microorganisms and viruses found in food and water.

Some food supplements containing Ag particles make a number of claims such as that they "support the immune system" and "are helpful against severe illnesses" (Wijnhoven et al. 2009). However, these supposed effects have not been examined by national evaluation organizations. Furthermore, the daily intake of supplements may contribute to an excess uptake of Ag.

Benefits of Ag–Antibiotics Against a Broad Spectrum of Microorganisms

Ag including AgNPs have been shown to possess antibacterial properties against a broad spectrum of Gram-negative and Gram-positive bacteria (Franci et al. 2015; Wijnhoven et al. 2009). Previous studies demonstrated that they were also effective against fungi and viruses (Marambio-Jones and Hoek 2010; Wijnhoven et al. 2009).

The mechanisms by which Ag exhibits antibiotic activity have not yet been elucidated in detail (Franci et al. 2015; Marambio-Jones and Hoek 2010); however, reactions between Ag ions and functional proteins and antioxidant defense molecules have been suggested. Ag ions are released at an amount that is equivalent to approximately 0.1% of the total amount of Ag loaded. We also found that AgNPs (1 mg/mL) released approximately 0.5 μg/mL of Ag ions during an incubation for 24 h (Zhao et al. 2013). Ag ions interact with enzymes of the respiratory chain reaction and transport proteins. This has been attributed to the high affinity of Ag ions for the –SH groups present in the cysteine residues of these proteins. Furthermore, the Ag ion-mediated perturbation of the bacterial respiratory chain suggests the generation of reactive oxygen species (ROS) (Park et al. 2009; Xu et al. 2012). The reaction of intracellular antioxidant molecules such as glutathione (GSH) having a –SH group with Ag ions also increases intracellular oxidation (Carlson et al. 2008). ROS may induce cellular membrane and DNA damage. On the other hand, an increasing number of studies on AgNPs found that this release of Ag ions alone does not account for the toxic effects observed (Beer et al. 2012; Eom and Choi 2010; Kawata et al. 2009). AgNPs interact with and directly damage bacterial membranes (Franci et al. 2015; Marambio-Jones and Hoek 2010). In any event, the effective antibacterial activities of Ag and AgNPs have potential in a wide range of applications, with a simultaneous increase in exposure to humans being expected.

Risks of Ag to Humans

Since the risk of Ag to humans is considered to be low, the antibacterial properties of Ag have been utilized for various products, and have been actively investigated for their applications to disinfect drinking water. Acute symptoms of overexposure to Ag nitrate include decreased blood pressure, stomach irritation, and decreased respiration (Drake and Hazelwood 2005). The only known clinical symptom of chronic silver intoxication is argyria, a blue-gray discoloration of the skin and other body tissues (Chang et al. 2006). The NOAEL in the total lifetime oral intake has been established as 10 g of Ag (WHO 2003). In several animal toxicity studies, an increase in various liver enzymes was observed following the administration of AgNPs (European Commission 2014). Intravenous exposure revealed that the immune system is the most sensitive target for AgNPs. Ag salts do not exhibit mutagenic or carcinogenic activity (WHO 2003), whereas DNA damage induced by AgNPs was recently demonstrated using single cell gel electrophoresis and micronucleus assays (AshaRani et al. 2009; European Commission 2014; Kim and Ryu 2013). We also found that AgNPs induced the phosphorylation of histone H2AX (γ-H2AX), a sensitive genotoxic marker in human lung cells (Zhao et al. 2014). Furthermore, epigenetic histone modifications that may mediate gene transcription have been detected (Zhao and Ibuki 2015). I will describe histone modifications reported by us and other researchers in the next section.

Histone Modifications and Ag

Histone Modifications

In eukaryotic cells, genomic DNA winds around histones to form nucleosomes, which are packaged into the condensed structure of chromatin. The fundamental unit of chromatin, the nucleosome, consists of an octamer of four core histones (two each of H2A, H2B, H3, and H4), around which approximately 146 base pairs of DNA are wrapped. Histones contain a large proportion of the positively charged amino acids, lysine and arginine, in their structure. DNA is negatively charged due to the presence of phosphate groups. These opposite charges result in high binding affinity between histones and DNA, and condense the structure of chromatin.

The chromatin structure is needed in order to fit a genome into the small volume of the nucleus in its condensed form; however, it also needs to allow the proteins involved in transcription, replication, and repair to access DNA. Therefore, the condensed structure of chromatin has to be temporarily relaxed (Fischle et al. 2003). Histone modifications are a method by which relaxation is achieved. More than 100 distinct post-translational histone modifications have been identified to date, and include lysine methylation, lysine acetylation, and serine/threonine phosphorylation (Kouzarides 2007; Strahl and Allis 2000). These modifications, which mainly occur on the N-terminal tails that protrude from the nucleosome, may affect the interactions of nucleosomes, and a combination of these modifications changes the chromatin structure, resulting in the regulation of cell functions and responses.

Carcinogenic metals, such as nickel, arsenic, and chromium, alter histone modifications, thereby inducing changes in gene expression, and these epigenetic changes have been suggested to play a role in the malignant transformation of cells (Chervona et al. 2012; Paul and Giri 2015; Wang et al. 2012). These metals alter histone methylation, which is a critical marker for DNA methylation and gene silencing. The arsenic metabolite, monomethylarsonous acid, has been shown to decrease the level of global histone acetylation by activating histone deacetylases (HDACs) (Ge et al. 2013). The dysregulation of histone acetylation may be a key mechanism in monomethylarsonous acid-induced malignant transformation and carcinogenesis. Nickel also induced the loss of the acetylation of histones (Ke et al. 2006). Nickel showed other histone modifications such as ubiquitination and methylation, indicating a relationship with nickel-mediated carcinogenesis. Few studies have described Ag and AgNP-induced histone modifications. We found that AgNPs induced γ-H2AX and H3.

Phosphorylation of Histone H2AX

H2AX is a minor component of histone H2A. γ-H2AX was originally identified as an early event after the direct formation of DNA double-strand breaks (DSBs) by ionizing radiation (Rogakou et al. 1998). However, the generation of γ-H2AX is now considered to occur after the indirect formation of DSBs caused by the collision of replication forks at the sites of DNA damage including oxidative bases, DNA adducts, single-strand breaks, and crosslinking, and the repair of these damages (Bonner et al. 2008) (Fig. 1). Therefore, this modification of histone is a marker of DNA damage and differs from other modifications that are directly related to changes in the chromatin structure and gene expression.

Ag has been shown to be a nonmutagenic metal. On the other hand, the genotoxic effects of AgNPs have recently been reported; however, the effects are weaker than

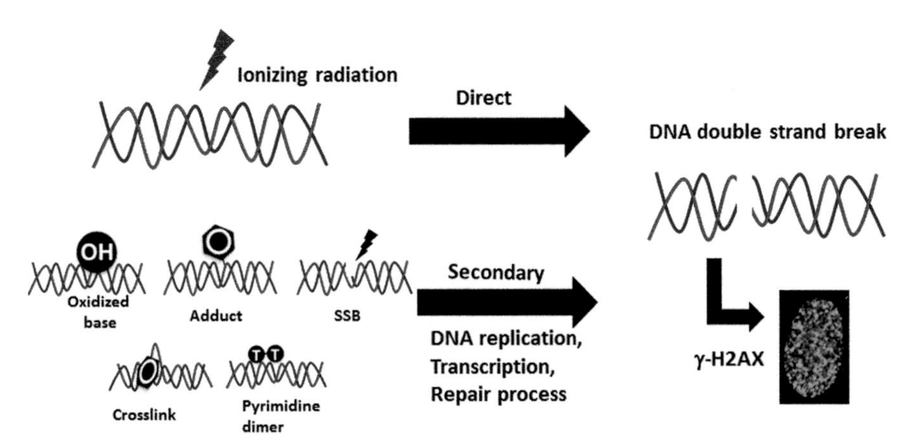

Fig. 1 Direct and indirect formation of DSBs and γ-H2AX. γ-H2AX is generated following formation of DSBs. DSBs are caused secondly by the collision of replication forks at sites of several kinds of DNA damage, and also by the repair of such damage. Therefore, H2AX could be phosphorylated by many mutagens and carcinogens, not only by direct DSBs formation after exposure to ionizing radiation

those of other metal particles such as CuO, ZnO, and NiO, and previous studies demonstrated that genotoxicity did not occur (Karlsson et al. 2014). Two different sizes of AgNPs did not increase γ-H2AX levels in three cell lines (Kruszewski et al. 2013). On the other hand, other studies have shown the genotoxicity of AgNPs, in which AgNP-induced oxidative stress has been reported in detail. Kim and Ryu (2013) summarized AgNP-induced changes related to oxidative stress, genotoxicity, and apoptosis in vitro. AgNP-induced γ-H2AX has been observed in the cells of mammals and fish (Ahamed et al. 2008; Choi et al. 2010; Eom and Choi 2010; Kim et al. 2009). Ag ions have been identified as the cause of the toxicity of AgNPs including the generation of γ-H2AX; however, Ag ions derived from AgNPs do not fully explain their toxicity (Kawata et al. 2009). In T cells, AgNPs clearly phosphorylated H2AX, whereas $AgNO_3$ did not (Eom and Choi 2010). ROS produced by the incorporation of AgNPs were suggested to be due to another cause (Kim et al. 2009). Since γ-H2AX is generated by several types of DNA damage, as described above, ROS-induced oxidative damage in DNA may contribute to AgNP-induced γ-H2AX (Fig. 2). Ag ions released from AgNPs bind with antioxidant functional molecules, leading to an oxidation status, which may form γ-H2AX.

Although γ-H2AX follows DNA damage, chromatin remodeling with histone modifications such as acetylation has been reported with γ-H2AX. Histone acetyltransferases bind to γ-H2AX nucleosomes and acetylate histone H3 (Lee et al. 2010). This acetylation relaxes chromatin and DNA damage may be easily repaired. However, AgNPs did not change the acetylation of H4K5, H4K8, or H3K23, whereas a decrease in the trimethylation of H3K9 was detected (Igaz et al. 2016). We have not yet reached a conclusion regarding Ag-induced DNA damage-mediated histone modifications accompanying γ-H2AX.

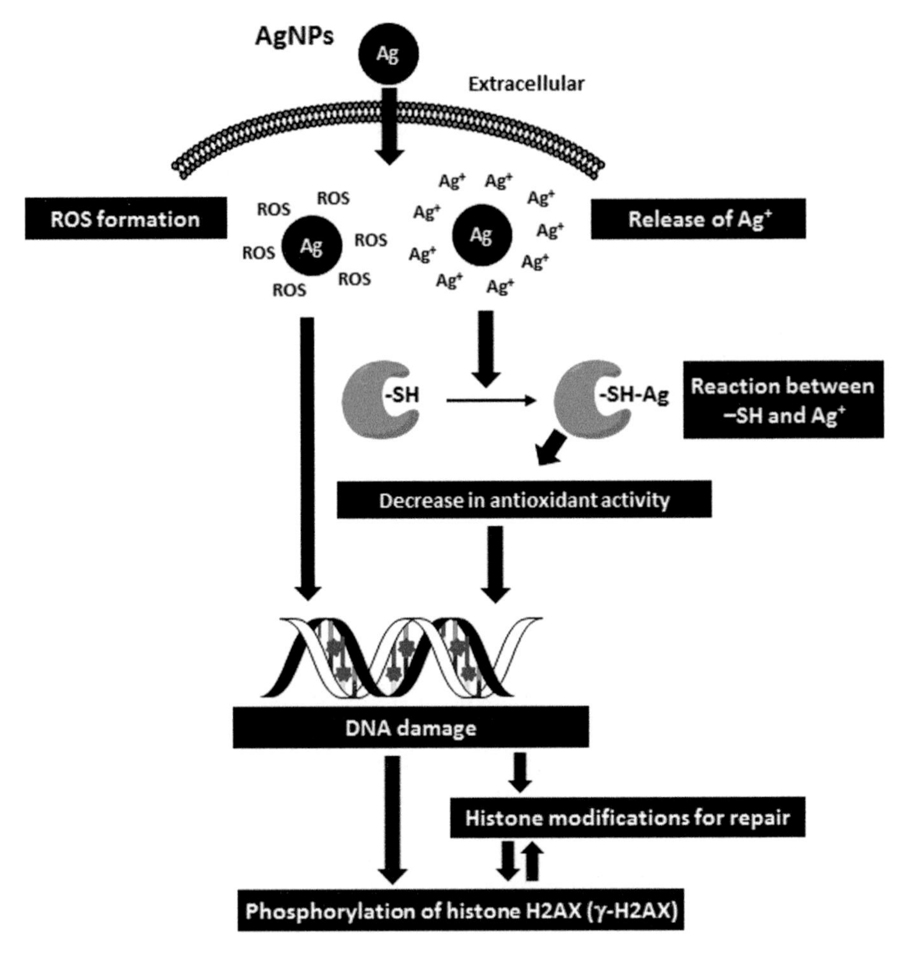

Fig. 2 AgNP-induced γ-H2AX. Incorporated AgNPs produce ROS, followed by the formation of oxidative DNA damage. On the other hand, incorporated AgNPs release Ag ions, which react with antioxidant molecules and increase intracellular oxidation, leading to DNA damage. This DNA damage may phosphorylate histone H2AX in the process of DNA replication or DNA damage repair

Phosphorylation of Histone H3

We previously demonstrated that AgNPs significantly induced the phosphorylation of histone H3 at serine 10 (p-H3S10) (Zhao and Ibuki 2015). p-H3S10 has been associated with mitotic chromosome condensation (Wei et al. 1999); however, AgNP-induced p-H3S10 differed from that involving chromosome condensation. Immunostaining revealed that phosphorylation occurred throughout the nucleus and was induced soon after a treatment with AgNPs. Ke et al. (2008) also reported that increases in the levels of p-H3S10 induced by nickel chloride were not simply due to a larger number of nickel-exposed cells undergoing mitosis. Epidermal growth factor and 12-*O*-tetradecanoylphorbol 13-acetate (Choi et al. 2005; Kim et al. 2008b) as well as carcinogenic environmental factors, such as cigarette sidestream smoke, UVB, arsenite, and nickel (Ibuki et al. 2014; Keum et al. 2013; Li et al. 2003), enhance the similar induction of p-H3S10. They have been associated with the induction of immediate-early genes, including the proto-oncogenes, *c-fos* and *c-jun*. Cell-transforming activity was previously reported to be weaker in histone H3S10A mutant cells than in H3 wild-type cells (Choi et al. 2005). AgNP-induced p-H3S10 was increased in the promoter sites of *c-jun* (Zhao and Ibuki 2015). Furthermore, this phosphorylation was due to Ag ions incorporated into inner cells, and not to the production of ROS.

p-H3S10 has been reported to accelerate other histone modifications. For example, p-H3S10 enables the recruitment of acetyltransferases, which acetylate either H3K9 or H3K14 (Sawicka and Seiser 2012). As described in the previous chapter, it currently remains unclear whether Ag induces histone acetylation, which then changes the chromatin structure. Furthermore, the roles of p-H3S10 other than its cell-transforming activity have not yet been elucidated; however, AgNP-induced p-H3S10 is attracting increasing attention and warrants further study.

Other Histone Modifications

Limited information is available on other Ag-induced histone modifications. In in vitro experiments, an exposure to polyvinyl pyrrolidone-coated AgNPs induced higher levels of global DNA methylation and histone H3 deacetylation (Blanco et al. 2016). In contrast, other studies showed significant decreases in methylation in the lysine residues of some histones (Igaz et al. 2016; Qian et al. 2015). AgNPs induced hemoglobin and the methylation of H3 at lysines 4 and 79 on the β-globin locus was markedly reduced (Qian et al. 2015). The global hypoacetylation of histones was detected after an exposure to other types of NPs, quantum dots, and SiO_2 (Choi et al. 2008; Gong et al. 2010). Acetylation and methylation interact with each other and change the chromatin structure, which regulates the transcriptional activities of some genes. Epigenetic changes may influence the expression of DNA damage-related genes. This may be linked to the prominent formation of γ-H2AX.

Combined Exposure to Ag and Ultraviolet Rays

Enhanced Antibacterial Activity

The combined use of metal NPs and UV has recently been applied to antibiotic treatments. TiO_2 NPs are photoactive and have been used in water and air purification as well as in several antibacterial products due to the formation of ROS. ZnO NPs were found to be more toxic to bacteria when preilluminated with UVA and UVB light than visible light (Kim and An 2012). This was attributed to the photodissolution of ZnO particles in UV light (Han et al. 2010). AgNPs also exhibited enhanced antibiotic activity under longer wavelength UVA light, not shorter wavelength UVB (Zhao et al. 2013). This may be due to an increase in Ag ion release by AgNPs oxidized by UVA irradiation. Furthermore, several studies showed that the use of $AgNO_3$, not AgNPs, plus UVA or visible light enhanced the inactivation of microorganisms such as viruses and bacteria (Butkus et al. 2004; Kim et al. 2008a).

Enhanced Toxicity in Humans

The toxicity of the combined effects of metals and UV in mammals has been reported. Several types of metals such as arsenite, chromium, nickel, cadmium, and lead were found to enhance UV-induced mutagenesis (in vitro) and/or carcinogenesis (in vivo) (Burns et al. 2004; Hartwig 1994; Rossman et al. 2001; Uddin et al. 2007). Rossman et al. (2001) demonstrated that the incidence of tumors was significantly higher (2.4-fold) in Skh1 mice treated with arsenite and UV than in those treated with UV or arsenite alone. We previously reported that UVA enhanced γ-H2AX induced by AgNPs (Zhao et al. 2016). This may be caused by Ag ions, which is consistent with the mechanisms underlying AgNP-induced antibacterial activity. Furthermore, Ag ions enhanced UVB-induced γ-H2AX in human cells (Zhao et al. 2014), whereas UVB did not significantly enhance antibacterial activity (Zhao et al. 2013). The formation of UVB-specific DNA damage was increased by Ag ions.

Histone Modifications Induced by Ag and DNA Damage Repair

Histone modifications change the chromatin structure and may also have an effect on the formation and repair of UV-induced DNA damage (Fig. 3). The condensed

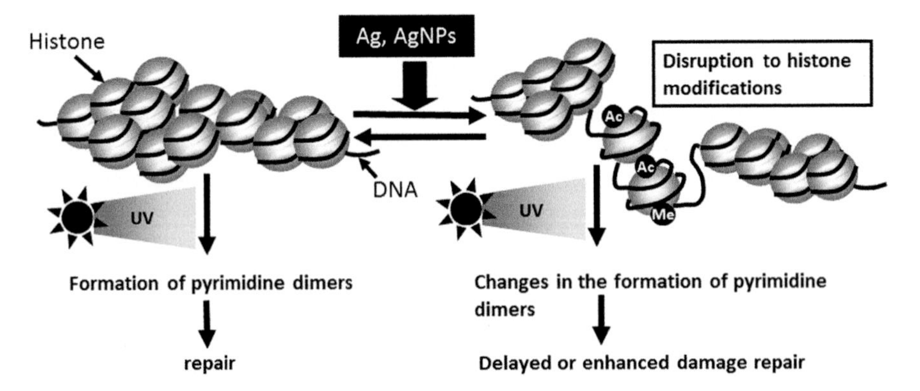

Fig. 3 Histone modifications disrupted by Ag and DNA damage repair. Ag and AgNPs change histone modifications, which may change chromatin conformation. The condensed chromatin environment specifically interferes with the formation of DNA lesions. Furthermore, the repair of DNA damage may be more profoundly affected by the chromatin environment. The repair of DNA damage may be easily predicted by the accessibility of repair molecules to damaged sites in relaxed chromatin

chromatin environment specifically interferes with the formation of the DNA lesions that cause major distortions in the DNA helix, e.g., 6–4 photoproducts may be detected in euchromatin, but not in heterochromatin (Han et al. 2016). We previously showed that Ag ions increased the formation of pyrimidine dimers following UVB exposure and cytotoxicity was increased by concomitant exposure (Zhao et al. 2013). Since Ag ions released from AgNPs change histone modifications, DNA damage after UVB irradiation is predicted to be affected and toxicity may be augmented.

The repair of DNA damage may be more profoundly affected by histone modifications (Fig. 3). The repair of DNA damage may be easily predicted by the accessibility of repair molecules to damaged sites in relaxed chromatin (Palomera-Sanchez and Zurita 2011). Smerdon et al. (1982) indicated that the histone deacetylase inhibitor, sodium butyrate (SB), stimulated the initial rate of nucleotide excision repair in normal and repair-deficient human cells at concentrations at which histones were maximally hyperacetylated. Repair synthesis was approximately 1.8-fold higher in nucleosomes with an average 2.3 acetyl residues/H4 molecules than in those with 1.5 or 1.0 acetyl residues/H4 molecules (Ramanathan and Smerdon 1989). On the other hand, we obtained controversial findings showing that histone acetylation induced by a HDAC inhibitor (HDACI) decreased the accessibility of DNA repair molecules to DNA-damaged sites and enhanced sensitivity to UV (Toyooka and Ibuki 2009). SB and another HDACI, trichostatin A, enhanced cell death following UV irradiation (Kim et al. 2005; Toyooka and Ibuki 2009). The use of HDACI is effective in combination with ionizing radiation due to its suppression of DNA repair activity (Munshi et al. 2005, 2006). In either case, the dysregulation of histone modifications may affect the formation or repair of DNA damage. Therefore, Ag-induced histone modifications may change toxicity when simultaneously exposed with other DNA damage agents.

Dictionary of Terms

- **Nanoparticles** – Particles with a diameter of less than 100 nm are called nanoparticles. They easily enter the inner body and pass through cell membranes in organisms. The high surface to volume ratio of these particles makes them very reactive or catalytic, which has led to toxicity concerns.
- **Ultraviolet rays** – Ultraviolet (UV) rays are divided into UVA (320 ~ 400 nm), UVB (280 ~ 320 nm), and UVC (200 ~ 320 nm). UVC and short-wavelength UVB do not reach the surface of the earth. UVB effectively forms pyrimidine dimers in DNA, which is the main cause of skin cancer.
- **Histone deacetylase inhibitors** – The acetylation status of histones is regulated by histone deacetylases (HDACs) and histone acetyl-transferases (HATs). The acetylation of histone generally promotes a more relaxed chromatin structure, allowing transcriptional activation. HDAC inhibitors (HDACI) generate histone hyperacetylation. HDACI are a relatively new class of anticancer agents that target epigenetic or nonepigenetic regulation.

Key Facts of Silver

- Silver (Ag) is an element found naturally in the environment.
- Although high Ag levels cause argyria, a blue-gray discoloration of the skin and other organs, normal Ag levels are considered to be safe.
- The broad and strong antimicrobial properties of Ag have been used in food and water disinfection treatments.
- Colloidal Ag and nano-sized Ag (AgNPs) have recently been used in dietary supplements and food packaging materials, respectively, and their safeties are being examined.
- Ag mainly causes toxicity through the binding of released Ag ions to functional proteins or the formation of reactive oxygen species.
- Few reports have been published on epigenetic changes, such as Ag-induced histone modifications.

Summary Points

- Epigenetics including histone modifications is a novel concept in toxicology, and Ag-induced epigenetic changes have not yet been examined in detail.
- This chapter focuses on histone modifications induced by Ag or AgNPs.
- AgNP-induced DNA damage phosphorylated histone H2AX at serine 139 (γ-H2AX), and this is a histone modification based on a genetic change.
- γ-H2AX may trigger other histone modifications in order to relax the chromatin structure for DNA damage repair.
- Epigenetic histone modifications such as acetylation and methylation change the chromatin structure, leading to specific gene expression. The phosphorylation of histone H3 at serine 10 (p-H3S10) was clearly detected following an exposure to AgNPs.
- Ag enhanced sensitivity to UV and increased formation of pyrimidine dimers. The altered chromatin structure may change the formation and repair of UV-induced DNA damage.
- The increased use of Ag and AgNPs is predicted for bacteriostatic purposes in food and drinking water. More detailed research concerning epigenetic changes is required in the future.

References

Ahamed M, Karns M, Goodson M, Rowe J, Hussain SM, Schlager JJ, Hong Y (2008) DNA damage response to different surface chemistry of silver nanoparticles in mammalian cells. Toxicol Appl Pharmacol 233(3):404–410. https://doi.org/10.1016/j.taap.2008.09.015

AshaRani PV, Low Kah Mun G, Hande MP, Valiyaveettil S (2009) Cytotoxicity and genotoxicity of silver nanoparticles in human cells. ACS Nano 3(2):279–290. https://doi.org/10.1021/nn800596w

Beer C, Foldbjerg R, Hayashi Y, Sutherland DS, Autrup H (2012) Toxicity of silver nanoparticles – nanoparticle or silver ion? Toxicol Lett 208:286–292. https://doi.org/10.1016/j.toxlet.2011.11.002

Blanco J, Lafuente D, Gómez M, García T, Domingo JL, Sánchez DJ (2016) Polyvinyl pyrrolidone-coated silver nanoparticles in a human lung cancer cells: time- and dose-dependent influence over p53 and caspase-3 protein expression and epigenetic effects. Arch Toxicol. [Epub ahead of print]

Bonner WM, Redon CE, Dickey JS, Nakamura AJ, Sedelnikova OA, Solier S, Pommier Y (2008) GammaH2AX and cancer. Nat Rev Cancer 8(12):957–967. https://doi.org/10.1038/nrc2523

Bouwmeester H, Dekkers S, Noordam MY, Hagens WI, Bulder AS, de Heer C, ten Voorde SE, Wijnhoven SW, Marvin HJ, Sips AJ (2009) Review of health safety aspects of nanotechnologies in food production. Regul Toxicol Pharmacol 53(1):52–62. https://doi.org/10.1016/j.yrtph.2008.10.008

Burns FJ, Uddin AN, Wu F, Nádas A, Rossman TG (2004) Arsenic-induced enhancement of ultraviolet radiation carcinogenesis in mouse skin: a dose-response study. Environ Health Perspect 112(5):599–603

Butkus MA, Labare MP, Starke JA, Moon K, Talbot M (2004) Use of aqueous silver to enhance inactivation of coliphage MS-2 by UV disinfection. Appl Environ Microbiol 70(5):2848–2853

Carlson C, Hussain SM, Schrand AM, Braydich-Stolle LK, Hess KL, Jones RL, Schlager JJ (2008) Unique cellular interaction of silver nanoparticles: size-dependent generation of reactive oxygen species. J Phys Chem B 112(43):13608–13619. https://doi.org/10.1021/jp712087m

Chang AL, Khosravi V, Egbert B (2006) A case of argyria after colloidal silver ingestion. J Cutan Pathol 33:809–811. https://doi.org/10.1111/j.1600-0560.2006.00557.x

Chervona Y, Arita A, Costa M (2012) Carcinogenic metals and the epigenome: understanding the effect of nickel, arsenic, and chromium. Metallomics 4(7):619–627. https://doi.org/10.1039/c2mt20033c

Choi HS, Choi BY, Cho YY, Mizuno H, Kang BS, Bode AM, Dong Z (2005) Phosphorylation of histone H3 at serine 10 is indispensable for neoplastic cell transformation. Cancer Res 65 (13):5818–5827. https://doi.org/10.1158/0008-5472.CAN-05-0197

Choi AO, Brown SE, Szyf M, Maysinger D (2008) Quantum dot-induced epigenetic and genotoxic changes in human breast cancer cells. J Mol Med (Berl) 86(3):291–302

Choi JE, Kim S, Ahn JH, Youn P, Kang JS, Park K, Yi J, Ryu DY (2010) Induction of oxidative stress and apoptosis by silver nanoparticles in the liver of adult zebrafish. Aquat Toxicol 100 (2):151–159. https://doi.org/10.1016/j.aquatox.2009.12.012

Drake PL, Hazelwood KJ (2005) Exposure-related health effects of silver and silver compounds: a review. Ann Occup Hyg 49:575–585. https://doi.org/10.1093/annhyg/mei019

Eom HJ, Choi J (2010) p38 MAPK activation, DNA damage, cell cycle arrest and apoptosis as mechanisms of toxicity of silver nanoparticles in Jurkat T cells. Environ Sci Technol 44 (21):8337–8342. https://doi.org/10.1021/es1020668

European Commission (2000) Opinion on Toxicological dana on colouring agents for medicinal products: E174 silver. http://ec.europa.eu/health/ph_risk/committees/scmp/documents/out30_en.pdf

European Commission (2014) Opinion on Nanosilver: safety, health and environmental effects and role in antimicrobial resistance. http://ec.europa.eu/health/scientific_committees/emerging/docs/scenihr_o_039.pdf

Fewtrell L (2014) Silver: water disinfection and toxicity. http://www.who.int/water_sanitation_health/dwq/chemicals/Silver_water_disinfection_toxicity_2014V2.pdf

Fischle W, Wang Y, Allis CD (2003) Histone and chromatin cross-talk. Curr Opin Cell Biol 15 (2):172–183

Franci G, Falanga A, Galdiero S, Palomba L, Rai M, Morelli G, Galdiero M (2015) Silver Nanoparticles as potential antibacterial agents. Molecules 20(5):8856–8874. https://doi.org/10.3390/molecules20058856

Ge Y, Gong Z, Olson JR, Xu P, Buck MJ, Ren X (2013) Inhibition of monomethylarsonous acid (MMA(III))-induced cell malignant transformation through restoring dysregulated histone acetylation. Toxicology 312:30–35. https://doi.org/10.1016/j.tox.2013.07.011

Gong C, Tao G, Yang L, Liu J, Liu Q, Zhuang Z (2010) SiO(2) nanoparticles induce global genomic hypomethylation in HaCaT cells. Biochem Biophys Res Commun 397(3):397–400. https://doi.org/10.1016/j.bbrc.2010.05.076

Hake SB, Xiao A, Allis CD (2004) Linking the epigenetic 'language' of covalent histone modifications to cancer. Br J Cancer 90:761–769. https://doi.org/10.1038/sj.bjc.6601575

Han J, Qiu W, Gao W (2010) Potential dissolution and photo-dissolution of ZnO thin films. J Hazard Mater 178(1–3):115–122. https://doi.org/10.1016/j.jhazmat.2010.01.050

Han C, Srivastava AK, Cui T, Wang QE, Wani AA (2016) Differential DNA lesion formation and repair in heterochromatin and euchromatin. Carcinogenesis 37(2):129–138. https://doi.org/10.1093/carcin/bgv247

Hartwig A (1994) Role of DNA repair inhibition in lead- and cadmium-induced genotoxicity: a review. Environ Health Perspect 102(Suppl 3):45–50

Ibuki Y, Toyooka T, Zhao X, Yoshida I (2014) Cigarette sidestream smoke induces histone H3 phosphorylation via JNK and PI3K/Akt pathways, leading to the expression of proto-oncogenes. Carcinogenesis 35(6):1228–1237. https://doi.org/10.1093/carcin/bgt492

Igaz N, Kovács D, Rázga Z, Kónya Z, Boros IM, Kiricsi M (2016) Modulating chromatin structure and DNA accessibility by deacetylase inhibition enhances the anti-cancer activity of silver nanoparticles. Colloids Surf B Biointerfaces 146:670–677. https://doi.org/10.1016/j.colsurfb.2016.07.004

Karlsson HL, Gliga AR, Calléja FM, Gonçalves CS, Wallinder IO, Vrieling H, Fadeel B, Hendriks G (2014) Mechanism-based genotoxicity screening of metal oxide nanoparticles using the ToxTracker panel of reporter cell lines. Part Fibre Toxicol 11:41. https://doi.org/10.1186/s12989-014-0041-9

Kawata K, Osawa M, Okabe S (2009) In vitro toxicity of silver nanoparticles at noncytotoxic doses to HepG2 human hepatoma cells. Environ Sci Technol 43(15):6046–6051

Ke Q, Davidson T, Chen H, Kluz T, Costa M (2006) Alterations of histone modifications and transgene silencing by nickel chloride. Carcinogenesis 27(7):1481–1488. https://doi.org/10.1093/carcin/bgl004

Ke Q, Li Q, Ellen TP, Sun H, Costa M (2008) Nickel compounds induce phosphorylation of histone H3 at serine 10 by activating JNK-MAPK pathway. Carcinogenesis 29(6):1276–1281. https://doi.org/10.1093/carcin/bgn084

Keum YS, Kim HG, Bode AM, Surh YJ, Dong Z (2013) UVB-induced COX-2 expression requires histone H3 phosphorylation at Ser10 and Ser28. Oncogene 32(4):444–452. https://doi.org/10.1038/onc.2012.71

Kim SW, An YJ (2012) Effect of ZnO and TiO_2 nanoparticles preilluminated with UVA and UVB light on *Escherichia coli* and Bacillus Subtilis. Appl Microbiol Biotechnol 95(1):243–253. https://doi.org/10.1007/s00253-012-4153-6

Kim S, Ryu DY (2013) Silver nanoparticle-induced oxidative stress, genotoxicity and apoptosis in cultured cells and animal tissues. J Appl Toxicol 33(2):78–89. https://doi.org/10.1002/jat.2792

Kim MS, Baek JH, Chakravarty D, Sidransky D, Carrier F (2005) Sensitization to UV-induced apoptosis by the histone deacetylase inhibitor trichostatin a (TSA). Exp Cell Res 306(1):94–102

Kim JY, Lee C, Cho M, Yoon J (2008a) Enhanced inactivation of E. coli and MS-2 phage by silver ions combined with UV-A and visible light irradiation. Water Res 42(1–2):356–362

Kim HG, Lee KW, Cho YY, Kang NJ, SM O, Bode AM, Dong Z (2008b) Mitogen- and stress-activated kinase 1-mediated histone H3 phosphorylation is crucial for cell transformation. Cancer Res 68(7):2538–2547. https://doi.org/10.1158/0008-5472.CAN-07-6597

Kim S, Choi JE, Choi J, Chung KH, Park K, Yi J, Ryu DY (2009) Oxidative stress-dependent toxicity of silver nanoparticles in human hepatoma cells. Toxicol In Vitro 23(6):1076–1084. https://doi.org/10.1016/j.tiv.2009.06.001

Kouzarides T (2007) Chromatin modifications and their function. Cell 128:693–705. https://doi.org/10.1016/j.cell.2007.02.005

Kruszewski M, Grądzka I, Bartłomiejczyk T, Chwastowska J, Sommer S, Grzelak A, Zuberek M, Lankoff A, Dusinska M, Wojewódzka M (2013) Oxidative DNA damage corresponds to the

long term survival of human cells treated with silver nanoparticles. Toxicol Lett 219 (2):151–159. https://doi.org/10.1016/j.toxlet.2013.03.006
Lee HS, Park JH, Kim SJ, Kwon SJ, Kwon J (2010) A cooperative activation loop among SWI/SNF, gamma-H2AX and H3 acetylation for DNA double-strand break repair. EMBO J 29 (8):1434–1445. https://doi.org/10.1038/emboj.2010.27
Li J, Gorospe M, Barnes J, Liu Y (2003) Tumor promoter arsenite stimulates histone H3 phosphoacetylation of proto-oncogenes c-fos and c-jun chromatin in human diploid fibroblasts. J Biol Chem 278(15):13183–13191. https://doi.org/10.1074/jbc.M300269200
Marambio-Jones C, Hoek EMV (2010) A review of the antibacterial effects of silver nanomaterials and potential implications for human health and the environment. J Nanopart Res 12:1531–1551
Munshi A, Kurland JF, Nishikawa T, Tanaka T, Hobbs ML, Tucker SL, Ismail S, Stevens C, Meyn RE (2005) Histone deacetylase inhibitors radiosensitize human melanoma cells by suppressing DNA repair activity. Clin Cancer Res 11(13):4912–4922
Munshi A, Tanaka T, Hobbs ML, Tucker SL, Richon VM, Meyn RE (2006) Vorinostat, a histone deacetylase inhibitor, enhances the response of human tumor cells to ionizing radiation through prolongation of gamma-H2AX foci. Mol Cancer Ther 5(8):1967–1974
National Institutes of Health (2004) Colloidal Silver. http://nccih.nih.gov/health/silver
Palomera-Sanchez Z, Zurita M (2011) Open, repair and close again: chromatin dynamics and the response to UV-induced DNA damage. DNA Repair (Amst) 10(2):119–125. https://doi.org/10.1016/j.dnarep.2010.10.010
Park HJ, Kim JY, Kim J, Lee JH, Hahn JS, MB G, Yoon J (2009) Silver-ion-mediated reactive oxygen species generation affecting bactericidal activity. Water Res 43(4):1027–1032. https://doi.org/10.1016/j.watres.2008.12.002
Paul S, Giri AK (2015) Epimutagenesis: a prospective mechanism to remediate arsenic-induced toxicity. Environ Int 81:8–17. https://doi.org/10.1016/j.envint.2015.04.002
Qian Y, Zhang J, Hu Q, Xu M, Chen Y, Hu G, Zhao M, Liu S (2015) Silver nanoparticle-induced hemoglobin decrease involves alteration of histone 3 methylation status. Biomaterials 70:12–22. https://doi.org/10.1016/j.biomaterials.2015.08.015
Ramanathan B, Smerdon MJ (1989) Enhanced DNA repair synthesis in hyperacetylated nucleosomes. J Biol Chem 264(19):11026–11034
Rogakou EP, Pilch DR, Orr AH, Ivanova VS, Bonner WM (1998) DNA double-stranded breaks induce histone H2AX phosphorylation on serine 139. J Biol Chem 273(10):5858–5868
Rossman TG, Uddin AN, Burns FJ, Bosland MC (2001) Arsenite is a cocarcinogen with solar ultraviolet radiation for mouse skin: an animal model for arsenic carcinogenesis. Toxicol Appl Pharmacol 176(1):64–71
Sawicka A, Seiser C (2012) Histone H3 phosphorylation – a versatile chromatin modification for different occasions. Biochimie 94(11):2193–2201. https://doi.org/10.1016/j.biochi.2012.04.018
Smerdon MJ, Lan SY, Calza RE, Reeves R (1982) Sodium butyrate stimulates DNA repair in UV-irradiated normal and xeroderma pigmentosum human fibroblasts. J Biol Chem 257 (22):13441–13447
Strahl BD, Allis CD (2000) The language of covalent histone modifications. Nature 403 (6765):41–45. https://doi.org/10.1038/47412
Taylor MR (2008) Assuring the safety of Nanomaterials in food packaging. Woodrow Wilson International Center For Scholars. http://www.nanotechproject.org/publications/archive/nano_food_packaging/
Toyooka T, Ibuki Y (2009) Histone deacetylase inhibitor sodium butyrate enhances the cell killing effect of psoralen plus UVA by attenuating nucleotide excision repair. Cancer Res 69 (8):3492–3500. https://doi.org/10.1158/0008-5472.CAN-08-2546
Uddin AN, Burns FJ, Rossman TG, Chen H, Kluz T, Costa M (2007) Dietary chromium and nickel enhance UV-carcinogenesis in skin of hairless mice. Toxicol Appl Pharmacol 221(3):329–338
Wang B, Li Y, Shao C, Tan Y, Cai L (2012) Cadmium and its epigenetic effects. Curr Med Chem 19 (16):2611–2620

Wei Y, Yu L, Bowen J, Gorovsky MA, Allis CD (1999) Phosphorylation of histone H3 is required for proper chromosome condensation and segregation. Cell 97(1):99–109

Wijnhoven SWP, Peijnenburg WJGM, Herberts CA, Hagens WI, Oomen AG, Heugens EHW et al (2009) Nano-silver – a review of available data and knowledge gaps in human and environmental risk assessment. Nanotoxicology 3(2):109–138

World Health Organization (2003) Silver in Drinking-water. Background document for development of WHO Guidelines for Drinking-water Quality

World Health Organization (2008) Guideline for drinking-water quality, pp 434–5. http://www.who.int/water_sanitation_health/dwq/chemicals/silversum.pdf

Xu H, Qu F, Xu H, Lai W, Andrew Wang Y, Aguilar ZP, Wei H (2012) Role of reactive oxygen species in the antibacterial mechanism of silver nanoparticles on Escherichia coli O157:H7. Biometals 25(1):45–53. https://doi.org/10.1007/s10534-011-9482-x

Zhao X, Ibuki Y (2015) Evaluating the toxicity of silver nanoparticles by detecting phosphorylation of histone H3 in combination with flow cytometry side-scattered light. Environ Sci Technol 49 (8):5003–5012. https://doi.org/10.1021/acs.est.5b00542

Zhao X, Toyooka T, Ibuki Y (2013) Synergistic bactericidal effect by combined exposure to ag nanoparticles and UVA. Sci Total Environ 458-460:54–62. https://doi.org/10.1016/j.scitotenv.2013.03.098

Zhao X, Toyooka T, Ibuki Y (2014) Silver ions enhance UVB-induced phosphorylation of histone H2AX. Environ Mol Mutagen 55(7):556–565. https://doi.org/10.1002/em.21875

Zhao X, Takabayashi F, Ibuki Y (2016) Coexposure to silver nanoparticles and ultraviolet a synergistically enhances the phosphorylation of histone H2AX. J Photochem Photobiol B 162:213–222. https://doi.org/10.1016/j.jphotobiol.2016.06.046

High-Fructose Consumption and the Epigenetics of DNA Methylation

113

Hiroya Yamada, Eiji Munetsuna, and Koji Ohashi

Contents

Abstract

Epidemiological studies have been demonstrated that fructose, which are used for beverages, are associated with the incidence of metabolic disorders. However, the pathological mechanism of fructose effect remains unclear. Recently, there are accumulating evidences that nutrition status may induce epigenetic modification, which lead to cause several diseases such as diabetes. Interestingly, it is becoming

H. Yamada (✉)
Department of Hygiene, Fujita health University School of Medicine, Toyoake, Aichi, Japan
e-mail: hyamada@fujita-hu.ac.jp

E. Munetsuna
Department of Biochemistry, Fujita Health University School of Medicine, Toyoake, Aichi, Japan
e-mail: mntneiji@fujita-hu.ac.jp

K. Ohashi
Department of Clinical Biochemistry, Fujita Health University School of Health Sciences, Toyoake, Aichi, Japan
e-mail: ohashi@fujita-hu.ac.jp

V. B. Patel, V. R. Preedy (eds.), *Handbook of Nutrition, Diet, and Epigenetics*,
https://doi.org/10.1007/978-3-319-55530-0_49

clear that the adverse effect of fructose is mediated by epigenetic modifications. Here, this chapter describes the epigenetic effect of high-fructose consumption.

Keywords
Fructose metabolism · High-fructose corn syrup · Developmental origins of health and Disease (DOHaD) · Pregnancy · Programming · Nonalcoholic fatty liver disease · Metabolic syndrome · DNA methylation · Carnitine palmitoyltrasferase 1A · Peroxisome proliferator-activated receptor-α · Uncoupling proteins · Mitochondrial DNA

List of Abbreviations

ChREBP	Carbohydrate response element-binding protein
CPT1A	Carnitine palmitoyltransferase 1A
DHAP	Dihydroxyacetone phosphate
DOHaD	Developmental Origins of Health and Disease
F1P	Fructose 1-phosphate
GA	Glyceraldehyde
GA3P	Glyceraldehyde 3-phosphate
GK	Glucokinase
GLUT	Glucose transporter
HFCS	High-fructose corn syrup
LDLR	Low-density lipoprotein
MTTP	Microsomal triglyceride transfer protein large subunit
NAFLD	Nonalcoholic fatty liver disease
NASH	Nonalcoholic steatohepatitis
PFK	Phosphofructokinase
PKC	Protein kinase C
PPAR	Peroxisome proliferator-activated receptor
SREBP	Sterol regulatory element-binding protein
TG	Triglyceride

Fructose Consumption

Fructose is a dietary monosaccharide which is naturally present in fruits and vegetables, either as free fructose or as part of the disaccharide sucrose. It is also present in the form of refined sugars, including granulated sugars such as white crystalline table sugar. The development of a quick and convenient method to synthesize fructose from glucose some 50 years ago (Douard and Ferraris 2008) led to an increase in consumption of high-fructose corn syrup (HFCS). Fructose is much sweeter than other natural sugars such as glucose, and large amounts of HFCS are now used in the manufacture of sweetened beverage products (Douard and Ferraris 2008). As shown in Fig. 1, consumption of fructose has increased since the mid-1970s, particularly that of free fructose, likely due to the increased use of HFCS.

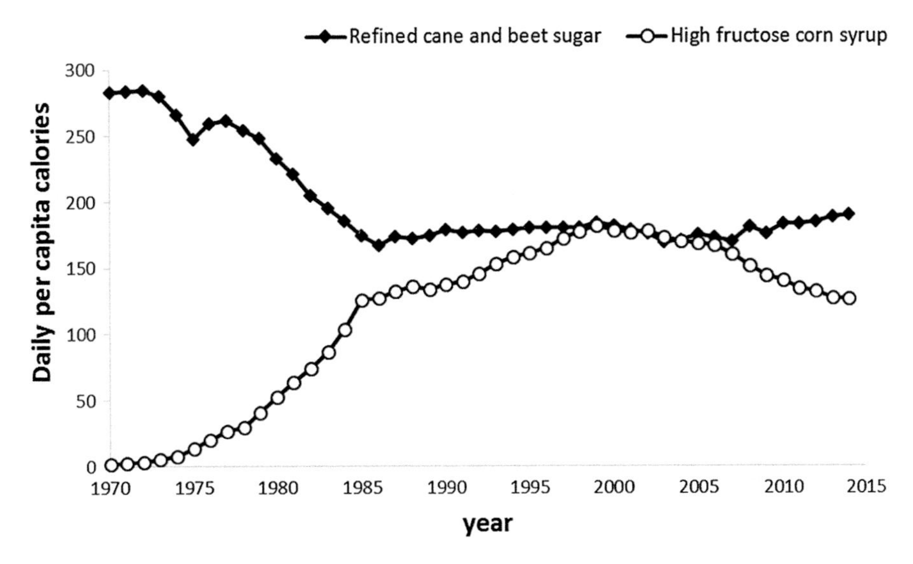

Fig. 1 Daily per capita calories of high-fructose corn syrup in the United States (Data source is the US Department of Agriculture – Economic Research Service)

Sugar intake has considerably increased over the past four decades (Goran et al. 2013; Herman and Samuel 2016). Of note, the rapid increase in the intake of calorically sweetened soft drinks might be a contributing factor to the epidemic of overweight and obesity (Dornas et al. 2015). Sweetened beverages, originally developed about a hundred years ago, may account for a significant portion of total energy intake and added sugars. Ventura et al. reported that fructose accounts for >55% of sugar content in most sweetened beverages (Ventura et al. 2011). Bray et al. noted that increased sugar intake has been observed since the latter years of the eighteenth century (the time of the American Revolution) and that sugar-sweetened beverages have been a main source of this increased sugar consumption in the twentieth century (Bray 2013). The average American consumed 4 lbs (1.81 kg) of sugar per year in the middle of eighteenth century. By 1850, however, this had increased about fivefold, to 20 lbs (9.1 kg) per capita, and has consistently increased since then, reaching 120 lbs (54.4 kg) by 1994 and exceeding 160 lbs(72.6 kg) in the early twenty-first century (Bray 2013).

In other studies, Elliott et al. (2002) reported that the per capita consumption of HFCS drastically increased by about 100 times from 1970 (0.23 kg) to 1997 (28.4 kg) (Elliott et al. 2002). Several other reports have indicated a drastic increase in fructose consumption during these several decades. However, as previously suggested (Goran et al. 2013), there is no information about other foods that contain substantial amounts of fructose, such as breakfast cereals with something like fresh or dried fruits. Given the fact that people regularly consume many foods that contains relatively high amounts of natural fructose, the actual level of fructose consumption in the population might be higher than predicted on the basis of common assumptions regarding HFCS consumption.

Fructose Metabolism

Although the metabolism of glucose through glycolysis uses many of the same enzymes and intermediate structures as in fructolysis, the two monosaccharides are metabolized differentially in human cells (Tappy and Le 2010). The two monosaccharides differ in structure, with glucose containing an aldehyde and fructose containing a ketone group. This difference in structure results in differences in their metabolism and in the biological processes to which they contribute. More specifically, fructose is preferentially metabolized in the liver, while glucose is consumed by cells in a wide variety of organs, including the brain.

Intracellularly, the phosphorylation of glucose by glucokinase is the rate-determining step in glucose metabolism, while fructose phosphorylation in the liver occurs via the enzyme fructokinase. This enzyme catalyzes the conversion of fructose into fructose 1-phosphate (F1P). Because glucokinase has a higher affinity for glucose, phosphorylation of fructose to fructose 6-phosphate is inhibited by glucose. Therefore, almost all fructose is metabolized in the liver via the F1P pathway. F1P is converted by aldolase B into glyceraldehyde and dihydroxyacetone phosphate. Subsequently, dihydroxyacetone kinase-2 phosphorylates glyceraldehyde, which leads to the synthesis of glyceraldehyde 3-phoshate. The products synthesized in fructolysis, glyceraldehyde 3-phosphate and dihydroxyacetone phosphate, can be substrates for glycogen synthesis or lipogenesis (Fig. 2). When fructose consumption is high, large amounts of fructose metabolite are synthesized, which may explain the increases in plasma triglyceride concentrations seen after fructose consumption. Additionally, the metabolism of F1P in the liver occurs independently of phosphofructokinase, a second rate-determining step in glucose metabolism. Fructokinase has no negative feedback system, so ATP is used for this process unless intracellular fructose exits, which results in phosphate depletion. Overall, fructose has greater potential for lipogenesis than glucose. Fructose consumption favors triglyceride synthesis, particularly when glucose is also present (Sloboda et al. 2014). Fructose consumption typically leads to an increase in plasma triglycerides. The liver is the primary site of fructose extraction and metabolism, with extraction approaching 50–70% of fructose delivery. Given that nutritional status may cause epigenetic alterations (Chang et al. 2010), chronic fructose intake may trigger signaling events that further enhance lipogenesis through epigenetic modification.

Epidemiological Studies of Fructose Consumption Metabolism

Several studies have shown the adverse effects of sugar intake on fat accumulation and the metabolic risk involved in fructose consumption (Stanhope and Havel 2009; Teff et al. 2009). For example, Teff et al. examined the effect of consuming fructose-sweetened beverage with meals in obese men and women. They demonstrated that consumption of fructose-sweetened beverages increases postprandial triglycerides in obese subjects, potentially exacerbating the known adverse

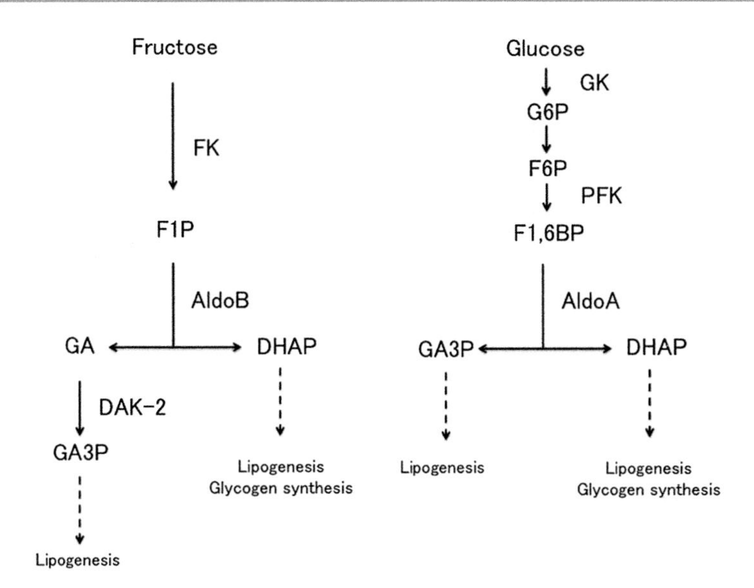

Fig. 2 Metabolism of glucose (*left*) and fructose (*right*). Fructose is metabolized by FK. FK synthesizes F1P from fructose. Fructose is metabolized to GA and DHAP by Aldo B. These metabolites are used for lipogenesis and ultimately for glycogen synthesis. It should be noted that there is no negative feedback system for fructokinase activity. Unless intracellular fructose is present, ATP is used for this process, which results in phosphate depletion. In glucose metabolism, the phosphorylation of glucose by GK is the rate-determining step, which is regulated by glucagon. Also, PFK activity is regulated by ATP and citric acid. Overall, metabolism of fructose is less regulated than that of glucose

metabolic profile associated with obesity, while glucose-sweetened beverages had less effect on postprandial triglycerides (Teff et al. 2009). In a study by Maersk, overweight subjects ($n = 47$) were randomly assigned to groups that were instructed to consume 1 l/day of regular cola, isocaloric semi-skim milk, aspartame-sweetened diet cola, or water (Maersk et al. 2012). They found that daily intake of regular cola for 6 months increased ectopic fat accumulation and lipids compared with milk, diet cola, and water. Another study reported a similar effect. Schwarz and colleagues compared the effects of short-term feeding of a diet high in fructose with that of an isocaloric diet in which complex carbohydrates were substituted for fructose. Only high-fructose consumption was found to stimulate lipogenesis and attenuate the suppression of insulin-mediated hepatic glucose production (Schwarz et al. 2015). It should be noted that moderate amounts of fructose intake attenuate insulin sensitivity in healthy men, whereas no such change has been observed for glucose consumption.

Fructose-induced metabolic syndrome has also been confirmed in epidemiological studies. Accumulating evidence suggests that consumption of soft drinks has led to increased prevalence of metabolic syndrome among adults and children in industrial countries (Nelson et al. 2009; Popkin 2010). Consistent with this notion, many cross-sectional and longitudinal epidemiological studies have reported associations of high intake of sugar-sweetened beverages with risk of obesity, insulin resistance, metabolic syndrome, and diabetes (Schulze et al. 2004; Dhingra et al. 2007). A meta-analysis by Malik et al. (2010) also indicated an effect of consumption of sugar-sweetened beverages on chronic metabolic diseases. They examined 11 prospective cohort studies including 310,819 participants and 15,043 cases of type 2 diabetes. Individuals in the highest quartile of sugar-sweetened beverage intake (most often, 1–2 servings/day) had a 26% greater risk of developing type 2 diabetes than those in the lowest quantile (none or <1 serving/month) (relative risk [RR] 1.26; 95% confidence interval [CI], 1.12–1.41). Among studies evaluating metabolic syndrome, which included 19,431 participants and 5803 cases, the pooled RR was 1.20 (95% CI, 1.02–1.42) (Malik et al. 2010). A cross-sectional study conducted by Ledoux et al. examined the components of a diet associated with child obesity (Ledoux et al. 2011). In that study, which included 342 children (aged 9–10 years) and 323 adolescents (aged 17–18 years), child obesity was positively associated with the consumption of sweetened beverages. According to prospective cohort analyses conducted from 1991 to 1999 among women in the Nurses' Health Study II, Schulze et al. demonstrated associations between the consumption level of sugar-sweetened beverages with the risk of type 2 diabetes and weight gain (Schulze et al. 2004). An association between fructose intake and hypertension has also been reported (Perez-Pozo et al. 2010). In a randomized controlled trial of 74 adult men administered 200 g fructose daily for 2 weeks with or without allopurinol, researchers demonstrated that high doses of fructose raised blood pressure (BP) and caused the features of metabolic syndrome, while lowering uric acid levels prevented an increase in mean arterial BP (Perez-Pozo et al. 2010). Collectively, these findings suggest that increased consumption of fructose in beverages may play a role in the epidemic of obesity and metabolic syndrome.

Although some reports indicate a relationship between fructose consumption and several disorders, the sole effect of fructose consumption cannot be assessed in humans, even in the short term. Given this fact, the effect of chronic fructose consumption is extremely hard to evaluate. A single administration of fructose may have a different effect than chronic fructose intake (Sloboda et al. 2014). As noted by Sloboda et al., there is no clear consensus on the long-term effects of fructose intake in humans; studies are limited in sample size and power analysis and are often retrospective (Sloboda et al. 2014).

Fructose-Induced Metabolic Syndrome and Its General Mechanisms

As mentioned above, the association of increments of fructose consumption with the incidence of several diseases, such as insulin resistance, diabetes, and obesity, has been suggested (Goran et al. 2013; Herman and Samuel 2016; Tappy and Le 2010). Fructose consumption also causes hyperuricemia in humans (Goran et al. 2013; Herman and Samuel 2016; Tappy and Le 2010). Because there is no negative feedback system in the reaction catalyzed by fructokinase, which converts fructose to fructose-1-phosphate, the reaction wastes cellular ATP (discussed above in the "Fructose Metabolism" section), resulting in the activation of enzymes involved in the degradation of AMP and formation of uric acid (Tappy and Le 2010). Uric acid induces adipocyte dysfunction, which leads to metabolic syndrome. Additionally, stimulated expression of transcription factors and enzymes involved in lipogenesis, including sterol regulatory element-binding protein 1c (SREBP1c), has been observed following fructose consumption.

The mechanisms of the adverse effects of fructose consumption have been mainly studied in animals. ChREBP is important for the enhanced expression of liver enzymes related to lipogenesis. When intracellular carbohydrate is low, ChREBPα is phosphorylated (Xu et al. 2013). When the concentration of intracellular carbohydrate is high, phosphorylated ChREBPα is dephosphorylated, resulting in transcriptional activation of key lipogenesis enzymes, such as fatty acid synthase. Like ChREBP, SREBP1c plays an important role in lipogenesis and is considered to be a major player in regulating lipid metabolism. SREBP1c also stimulates transcriptional activation of lipogenesis-related enzymes. In the fasted state, SREBP1c is inactivated. Insulin-stimulated activation of the transcriptional factor is observed after feeding (Xu et al. 2013). After insulin signaling occurs, activated SREBP1c binds to specific DNA sequences called sterol regulatory elements (SRE), which results in the upregulation of fatty acid synthesis. Liver-specific insulin receptor knockout mice did not show the upregulation of hepatic triglyceride levels mediated through high-fructose feeding, suggesting the physiological importance of SREBP1c in high-fructose feeding (Haas et al. 2012). An excellent review article on fructose-induced insulin resistance published by Herman et al. reported that fructose-mediated lipogenesis increases hepatic diacylglycerol content, which activates protein kinase C (PKC) (Herman and Samuel 2016). Activation of PKC subsequently suppresses insulin receptor kinase activation, leading to the deceased activation of downstream kinases, which limits the gluconeogenesis-suppressing effect of insulin. In earlier papers, Nagai et al. reported that fructose-induced impaired insulin sensitivity was improved using peroxisome proliferator-activated receptor-α (PPARα) ligand fibrate, suggesting that PPARα signaling may be altered by fructose consumption (Nagai et al. 2002). Expression of carnitine palmitoyltrasferase 1A (CPT1A) mRNA was also decreased (Nagai et al. 2002). Considering the fact that CPT1A expression is regulated by PPARα, reduced receptor expression may lead to subsequently decreased level of CPT1A. These phenomena may be associated with attenuated β-oxidation activity, which may contribute to lipid accumulation. PPARα also plays an important role in regulating insulin sensitivity (Nagai et al. 2002). Consistent with the results of Nagai's group, most experimental studies in rodents have found that activation of PPARα improves insulin sensitivity (Guerre-Millo et al. 2000; Moller and Berger, 2003; Ye et al. 2001). Taken together, this evidence suggests that the unfavorable effect of fructose consumption may be at least partially mediated by the alteration of PPARα signaling. The mechanism of the fructose effect has been clarified mainly from a biochemical perspective, while there have been no reports concerning the epigenetic effect of fructose.

Epigenetics and Fructose: A New Mechanism of Fructose Effects

It is well known that nutrients can modify epigenetic phenomena and regulate gene expression. Inadequate nutrient intake may cause inappropriate epigenetic changes and lead to metabolic syndrome. For example, Kim et al. demonstrated that high-fat-diet feeding for 20 weeks reduced adiponectin mRNA levels in mice adipocytes (Kim et al. 2015). Transcriptional repression, which is mediated by upregulated DNA methylation in promoter regions of adiponectin, may cause metabolic diseases and obesity expression of the gene. In addition, several nutrients, such as folate and vitamin B12, play important roles in maintaining epigenetic conditions (Ba et al. 2011; Crider et al. 2012).

Based on these findings, we hypothesize that excess fructose intake can cause epigenetic alterations. Previously, we found that melatonin improves fructose-induced metabolic syndrome (Kitagawa et al. 2012). High-fructose feeding for 6 weeks increased serum tumor necrosis factor-α concentration and hepatic lipid peroxide concentration and lowered hepatic glutathione concentration. Daily intraperitoneal administration of melatonin (1 or 10 mg/kg body weight) attenuated these changes at 6 weeks. In an oral glucose tolerance test, high-fructose-fed rats showed higher serum insulin response curves and normal serum glucose response curves compared with rats fed a control diet. Insulin resistance was observed after 4 or 6 weeks of feeding, but daily administration of melatonin ameliorated this insulin resistance and the higher serum insulin response curve seen in the oral glucose tolerance test. These results indicate that melatonin improves metabolic syndrome induced by high-fructose intake in rats. Given the potential role of melatonin in regulating DNA methyltransferase activity (Korkmaz and Reiter 2008), the beneficial effect of melatonin on fructose-induced disease might be mediated by epigenetic alteration. Therefore, the effect of fructose consumption on epigenetic regulation was examined in our previous study in which rats were fed a 20% fructose solution for 14 weeks (Ohashi et al. 2015b). Enhanced serum triglyceride and total cholesterol levels were observed after feeding, indicating the induction of fructose-mediated metabolic syndrome. Upregulation of global DNA methylation levels was also observed, suggesting that fructose feeding resulted in epigenetic regulation (Fig. 3a). These data promoted us to examine the association between fructose-induced metabolic syndrome and DNA methylation levels. We selected four genes involved in lipid metabolisms (CPT1A, PPARα, LDLR, and MTTP) based on previous studies. Of these genes, hepatic mRNA expression of PPARα and CPT1A were decreased (Fig. 3b; data for LDLR and MTTP not shown). It has been reported that the transcription of PPARα is regulated epigenetically (Gluckman et al. 2007; Lillycrop et al. 2005). In our study, administration of fructose solution for 14 weeks resulted in a doubling of the upregulation of DNA methylation levels in promoter regions of PPARα gene (from −20 to −138 bp upstream from transcription sites) (Fig. 3c). Similarly, CPT1A promoter regions were hypermethylated after fructose intake (Fig. 3d). Considering that CPT1A is a target gene for PPARα (Shi et al. 2013), reduced expression of PPARα may lead to subsequently decreased levels of CPT1A. In addition, the promoter region of CPT1A was hypermethylated. In general, DNA

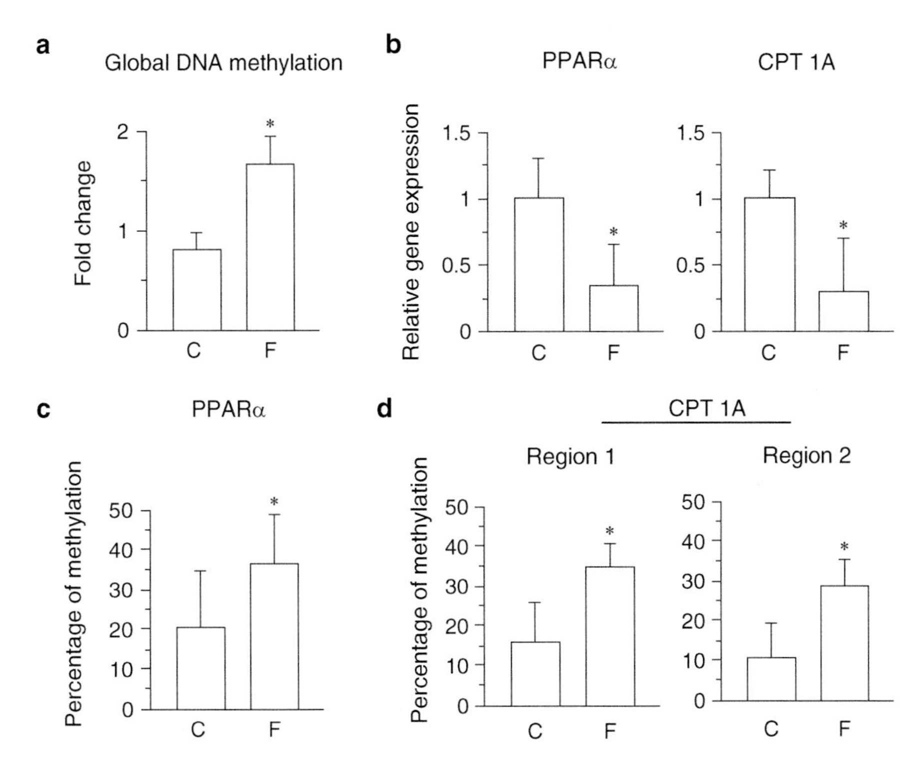

Fig. 3 Epigenetic effect of fructose consumption in rat liver. The effect of fructose consumption on hepatic global DNA methylation. After fructose feeding for 14 weeks, global DNA methylation was measured. A significant increase in DNA methylation was observed after feeding (**a**). The effect of fructose consumption on PPARα and CPT1a mRNA levels. After feeding, PPARα and CPT1a mRNA levels were determined using quantitative real-time PCR. Both mRNAs were decreased following fructose intake (**b**). The effect of fructose consumption on PPARα promoter (20–138 bp upstream of the transcription start site) DNA methylation level. Upregulated DNA methylation was observed (**c**). Similarly, CPT1A promoter (1118–1202 and 1052–1136 bp upstream) methylation levels were stimulated by fructose consumption (**d**). $^{*}P < 0.05$

methylation decreases the accessibility of transcriptional factors, and reduced expression of PPARα and reduced accessibility of the receptor to the promoter may cause transcriptional repression of CPT1A, leading to hepatic lipid accumulation.

We have shown that fructose consumption leads to increased DNA methylation in promoter regions of the PPARα and CPT1A genes, phenomena which may be associated with reduced mRNA levels of these genes, leading to downregulated β-oxidation activity. PPARα is a master regulator of fatty acid oxidation, and its reduction may lead to the development of metabolic syndrome. Accordingly, PPARα agonists have been shown to ameliorate metabolic syndrome in rats (Nagai et al. 2002). Consistent with our findings, a high-fructose diet has been shown to reduce PPARα expression in rat liver. However, while PPARα is known to play a central role in fructose-induced metabolic syndrome, the mechanism for the reduction in PPARα following fructose intake has not been elucidated. Our findings suggest that hypermethylation of the PPARα promoter suppresses its gene expression via a fructose-mediated signal.

Epigenetics and Fructose: The Effect of Fructose on Mitochondrial DNA Methylation

An association between nutritional status and the alteration of mitochondrial function has been demonstrated. Diet-induced insulin-resistant mice showed altered mitochondrial biogenesis, structure, and function in the muscle (Bonnard et al. 2008). Also, Carabelli et al. reported that rat fed a high-fat diet showed significant increases in hepatic mitochondrial DNA (mtDNA) levels, which positively correlated with triglyceride content in the rat liver (Carabelli et al. 2011). Considering these findings, mitochondrial quantity and quality may play important roles in the pathogenesis of metabolic syndrome. Expression of mtDNA may be associated with pathological states in the development of metabolic syndrome, based on the fact that the mitochondrial genome encodes many proteins essential for ATP production. A previous study assessed the effect of fructose consumption on mtDNA content and mitochondrial gene expression, indicating the potential involvement of fructose in adverse effects on mitochondrial function (Yamazaki et al. 2016).

It appears that not only nuclear DNA but also mtDNA is epigenetically regulated. The epigenetic alteration of mtDNA may be involved in the induction of some diseases. Pirola et al. (2013) reported the hypermethylation of mtDNA in nonalcoholic steatohepatitis patients and also found that liver mt-ND6 mRNA expression significantly decreased in these patients, suggesting an association between mtDNA methylation and mitochondrial transcription (Pirola et al. 2013). More recently, the same research group also suggested the possibility that 5-hydroxymethylcystosine, the first oxidative product in the active demethylation of 5-methylcytosine, might be involved in the pathogenesis of nonalcoholic fatty liver disease (NAFLD) through regulation of liver mitochondrial biogenesis (Pirola et al. 2015). These results suggest that epigenetic modification of mtDNA is involved in the pathogenesis of liver disease.

Based on these observations, we analyzed the effect of fructose consumption on mtDNA methylation and its gene expression. As expected, rats fed a 20% fructose solution for 14 weeks showed increases in body weight compared to control rats. Significant increases in serum total cholesterol, triglyceride, and low-density lipoprotein cholesterol/very-low-density lipoprotein cholesterol levels in fructose-fed animals compared to controls were also observed. Consistent with previous studies, we confirmed that fructose intake can induce metabolic syndrome. Previous studies concerning mitochondrial gene expression have investigated the expression levels of very few genes. However, in some cases, the transcription of mitochondrial genes has been found to be independently regulated, despite the fact that mitochondrial transcription uses only three promoters. Thus, a better understanding of the effects of a high-fructose diet on mitochondrial gene expression will require the comprehensive analysis of many mitochondrial genes. Significant upregulation of hepatic mtDNA-

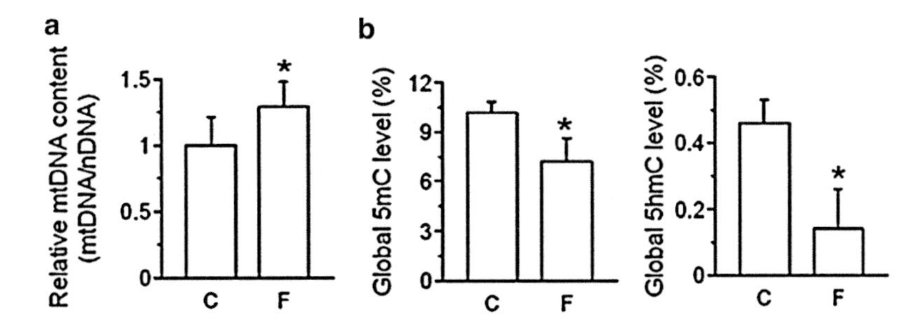

Fig. 4 Relative mtDNA content and mitochondrial epigenetic alteration following fructose consumption. Relative quantification of mtDNA content was determined from the ratio of a mitochondrial gene, mt-ND6, to an intron of the nuclear gene, beta-actin. Amounts of mtDNA in fructose-fed animals are shown relative to those of controls (**a**). Global 5-mC and 5-hmC levels of hepatic mtDNA were significantly decreased in fructose-fed rats (**b**). $^{*}P < 0.05$

encoded genes has been observed in rats fed a high-fructose diet compared to control rats. Specifically, some mitochondrial genes (mt-ND3, mt-ND4, mt-ND5, mt-ND6, mt-CO1, mt-CO2, mt-CO3, mt-ATP6, mt-ATP8, and mt-Cytb) were significantly upregulated in the fructose group relative to the control group, and mt-ND1 and mt-ND2 also showed a trend toward upregulation (data not shown) (Yamazaki et al. 2016). We investigated whether the fructose diet affected mtDNA-encoded gene expression via alteration of mtDNA content and methylation level of mtDNA in the liver. Fructose treatment induced a subtle but significant increase in hepatic mtDNA content compared to that found in control rats (Fig. 4a). These observations suggest the possibility that mitochondrial respiration may be stimulated by excessive fructose consumption. In fact, mitochondria are known to be sensitive to excessive fructose intake (Nagai et al. 2002; Crescenzo et al. 2013), and an association between reduced mtDNA content and metabolic syndrome in leukocytes has been reported (Huang et al. 2011). A reduction of mtDNA content was also observed in various tissues of humans and rodents (e.g., in studies of NAFLD and insulin resistance (Lindinger et al. 2010; Serradas et al. 1995). We observed increased mitochondrial gene expression and mtDNA content in the liver. High-fat-diet-induced increases in mitochondrial biogenesis and function have also been reported (Turner et al. 2007). Thus, mitochondrial dysfunction may be involved in development of metabolic syndrome, resulting in decreased mtDNA content and gene expression.

Interestingly, our study showed that the fructose diet also induced global hypomethylation of mtDNA (Fig. 4b). These results suggest that, while the role of mtDNA methylation is still largely unknown, hypomethylation of mtDNA can upregulate the expression of mtDNA-encoded genes. Given the association between alteration of mtDNA methylation and hepatic disease (Pirola et al. 2013), epigenetic changes in mtDNA might also be associated with pathogenic mechanisms of fructose-induced metabolic disease. In addition to the effect of fructose consumption on 5-methylcytosine, we observed a similar effect on 5-hydroxymethylcytosine, which is the oxidative product of the active demethylation of 5-methylcytosine. Although

5-hydroxymethylcytosine levels were suppressed in fructose-fed rats, we found that the 5-hydroxymethycytosine/5-methylcytosine ratio was not changed, suggesting that the demethylation process is not affected.

Cross Generational Effect of Fructose

Epidemiology formed the basis of "the Barker hypothesis," the concept of "developmental programming," and today's discipline of the Developmental Origins of Health and Disease (DOHaD). Many animal model experimentations have provided proof of the underlying concepts and continue to generate knowledge of underlying mechanisms. It has been suggested that maternal nutrition status can alter the phenotype of offspring (Segovia et al. 2014; Zou et al. 2012). The effects of maternal fructose intake on offspring health remain largely unknown, despite the marked increase in consumption of sweetened beverages that has paralleled the obesity epidemic. In an early study, maternal fructose consumption induced low birth weight, suggesting a potential effect of maternal fructose on fetal development (Jen et al. 1991). Our results were consistent with our previous findings: when dams were fed a 20% fructose solution during the gestational and lactation periods and pups were fed with normal water after lactation, the pups showed decreased insulin sensitivity in adulthood (unpublished data). In another study, we found that maternal fructose intake appears to impact brain function; maternal fructose consumption was found to affect the expression of brain genes involved in steroid metabolism (Ohashi et al. 2015a). These results indicated that maternal fructose consumption may exert systemic effects in offspring. Although several research groups have now indicated a cross generational effect of maternal fructose consumption, the mechanism of this effect remains unknown.

The epigenetic mechanisms (DNA methylation and miRNA) responsible for the adverse effect of maternal fructose consumption during gestation and lactation have been identified. The effect of maternal fructose intake on DNA methylation was assessed via its effect on the hippocampus in offspring: in dams fed a high-fructose diet (50% of energy from fructose), Mortensen et al. observed altered brain uncoupling protein 5 (UCP5) expression in offspring at 15 months (Mortensen et al. 2014). UCP5 is expressed in the inner mitochondrial membrane and transports protons from the intermembrane to the matrix (Ramsden et al. 2012). Consistent with this report, decreased UCP5 expression was observed in hippocampi of offspring (150 days). Based on our data, we conducted a methylation analysis in the putative promoter regions of the UCP5 gene. We observed upregulation of DNA methylation in this region, suggesting that the adverse effects of maternal fructose intake may be mediated by the alteration of DNA methylation levels.

In addition to DNA methylation, accumulating evidence suggests that maternal nutrition status may alter miRNA expression in offspring. For example, Zhang et al. reported that a high-fat diet during pregnancy and lactation alters hepatic expression of IGF-2 and key miRNAs in adult offspring (Zhang et al. 2009). Similarly, we found that miRNA expression changes may play an important role in the cross generational fructose effect. Specifically, we found decreased hepatic expression of IGF-1 in offspring at 160 days. We also found that some miRNAs that might target IGF1 miRNA were also upregulated, suggesting that the adverse effects of maternal fructose intake are mediated by miRNA.

Concluding Remarks

The effect of chronic fructose consumption may differ from that of short-term consumption, based on the fact chronic consumption has a more potent lipogenetic effect. This phenomenon is not fully explained by the genetic effect of fructose consumption.

In addition to its genetic effect, we have also analyzed fructose consumption from the viewpoint of epigenetics. Results have shown that high consumption results in alterations in DNA methylation in both nuclear DNA and mtDNA. We suggest that chronic, but not short-term, fructose consumption may induce DNA methylation, which may affect susceptibility to disease. Fructose-induced epigenetic change appears to be a novel mechanism of disease pathology.

Mini-dictionary of Terms

High-Fructose Corn Syrup (HFCS)	HFCS is a fructose-glucose liquid sweetener that is found in a wide range of processed foods, from ketchup and cereals to crackers and salad dressings.
Insulin Resistance (IR)	IR is a syndrome in which a given concentration of insulin produces a less than expected biological effect.
Nonalcoholic Fatty Liver Disease (NAFLD)	NAFLD is a disease spectrum from fatty liver to nonalcoholic steatohepatitis, fibrosis, and cirrhosis. It is considered to be the most common liver disorder in Western countries.
Mitochondrial DNA (mtDNA)	mtDNA is the double-stranded DNA located in the mitochondria matrix. The mitochondrial DNA encodes genes essential for ATP production (e.g., ATP synthase and NADH dehydrogenase).
Developmental Origins of Health and Disease (DOHaD)	DOHaD is provided new concept into the early origins of health and disease from a lifecycle perspective.

Key Facts of Fructose

- Fructose is a dietary monosaccharide present naturally in fruits and vegetables.
- Fructose is much sweeter than other natural sugar such as glucoses.
- Much amount of HFCS is used for beverage products.
- The consumption of fructose has increased gradually in the industrial countries.
- Chronic fructose consumption causes several diseases such as obesity and diabetes.

Summary Points

- The chapter describes the mechanism of the adverse effect of fructose intake.
- The association between the increment of fructose consumption and the incidence of several diseases such as insulin resistance, diabetes, and obesity has been suggested.
- Recent reports demonstrated the genomic action of fructose involved in some transcriptional factors such as ChREBP, SREBP1c, and PPARα.
- Fructose-mediated attenuated gene expression may be mediated by alterations of DNA methylation status in rat liver.
- Chronic fructose consumption cause twice upregulation of DNA methylation level in promoter regions of PPARα gene and CPT1A.
- Fructose-mediated attenuated gene expression may be mediated by alterations of DNA methylation status in rat liver.
- The Developmental Origins of Health and Disease hypothesis indicates that maternal nutrition in pregnancy has a significant impact on offspring disease risk later in life.
- Maternal fructose intake appears to impact metabolic function and brain function.
- Maternal fructose-induced epigenetic alternation is also observed.
- DNA methylation and small RNA may mediate the effect of the maternal fructose.

References

Ba Y, Yu H, Liu F, Geng X, Zhu C, Zhu Q, Zheng T, Ma S, Wang G, Li Z, Zhang Y (2011) Relationship of folate, vitamin B12 and methylation of insulin-like growth factor-II in maternal and cord blood. Eur J Clin Nutr 65:480–485

Bonnard C, Durand A, Peyrol S, Chanseaume E, Chauvin MA, Morio B, Vidal H, Rieusset J (2008) Mitochondrial dysfunction results from oxidative stress in the skeletal muscle of diet-induced insulin-resistant mice. J Clin Invest 118:789–800

Bray GA (2013) Energy and fructose from beverages sweetened with sugar or high-fructose corn syrup pose a health risk for some people. Adv Nutr 4:220–225

Carabelli J, Burgueno AL, Rosselli MS, Gianotti TF, Lago NR, Pirola CJ, Sookoian S (2011) High fat diet-induced liver steatosis promotes an increase in liver mitochondrial biogenesis in response to hypoxia. J Cell Mol Med 15:1329–1338

Chang X, Yan H, Fei J, Jiang M, Zhu H, Lu D, Gao X (2010) Berberine reduces methylation of the MTTP promoter and alleviates fatty liver induced by a high-fat diet in rats. J Lipid Res 51:2504–2515

Crescenzo R, Bianco F, Falcone I, Coppola P, Liverini G, Iossa S (2013) Increased hepatic de novo lipogenesis and mitochondrial efficiency in a model of obesity induced by diets rich in fructose. Eur J Nutr 52:537–545

Crider KS, Yang TP, Berry RJ, Bailey LB (2012) Folate and DNA methylation: a review of molecular mechanisms and the evidence for folate's role. Adv Nutr 3:21–38

Dhingra R, Sullivan L, Jacques PF, Wang TJ, Fox CS, Meigs JB, D'agostino RB, Gaziano JM, Vasan RS (2007) Soft drink consumption and risk of developing cardiometabolic risk factors and the metabolic syndrome in middle-aged adults in the community. Circulation 116:480–488

Dornas WC, de Lima WG, Pedrosa ML, Silva ME (2015) Health implications of high-fructose intake and current research. Adv Nutr 6:729–737

Douard V, Ferraris RP (2008) Regulation of the fructose transporter GLUT5 in health and disease. Am J Physiol Endocrinol Metab 295:E227–E237

Elliott SS, Keim NL, Stern JS, Teff K, Havel PJ (2002) Fructose, weight gain, and the insulin resistance syndrome. Am J Clin Nutr 76:911–922

Gluckman PD, Lillycrop KA, Vickers MH, Pleasants AB, Phillips ES, Beedle AS, Burdge GC, Hanson MA (2007) Metabolic plasticity during mammalian development is directionally dependent on early nutritional status. Proc Natl Acad Sci USA 104:12796–12800

Goran MI, Dumke K, Bouret SG, Kayser B, Walker RW, Blumberg B (2013) The obesogenic effect of high fructose exposure during early development. Nat Rev Endocrinol 9:494–500

Guerre-Millo M, Gervois P, Raspe E, Madsen L, Poulain P, Derudas B, Herbert JM, Winegar DA, Willson TM, Fruchart JC, Berge RK, Staels B (2000) Peroxisome proliferator-activated receptor alpha activators improve insulin sensitivity and reduce adiposity. J Biol Chem 275:16638–16642

Haas JT, Miao J, Chanda D, Wang Y, Zhao E, Haas ME, Hirschey M, Vaitheesvaran B, Farese RV Jr, Kurland IJ, Graham M, Crooke R, Foufelle F, Biddinger SB (2012) Hepatic insulin signaling is required for obesity-dependent expression of SREBP-1c mRNA but not for feeding-dependent expression. Cell Metab 15:873–884

Herman MA, Samuel VT (2016) The sweet path to metabolic demise: fructose and lipid synthesis. Trends Endocrinol Metab 27:719–730

Huang CH, Su SL, Hsieh MC, Cheng WL, Chang CC, Wu HL, Kuo CL, Lin TT, Liu CS (2011) Depleted leukocyte mitochondrial DNA copy number in metabolic syndrome. J Atheroscler Thromb 18:867–873

Jen KL, Rochon C, Zhong SB, Whitcomb L (1991) Fructose and sucrose feeding during pregnancy and lactation in rats changes maternal and pup fuel metabolism. J Nutr 121:1999–2005

Kim AY, Park YJ, Pan X, Shin KC, Kwak SH, Bassas AF, Sallam RM, Park KS, Alfadda AA, Xu A, Kim JB (2015) Obesity-induced DNA hypermethylation of the adiponectin gene mediates insulin resistance. Nat Commun 6:7585

Kitagawa A, Ohta Y, Ohashi K (2012) Melatonin improves metabolic syndrome induced by high fructose intake in rats. J Pineal Res 52:403–413

Korkmaz A, Reiter RJ (2008) Epigenetic regulation: a new research area for melatonin? J Pineal Res 44:41–44

Ledoux TA, Watson K, Barnett A, Nguyen NT, Baranowski JC, Baranowski T (2011) Components of the diet associated with child adiposity: a cross-sectional study. J Am Coll Nutr 30:536–546

Lillycrop KA, Phillips ES, Jackson AA, Hanson MA, Burdge GC (2005) Dietary protein restriction of pregnant rats induces and folic acid supplementation prevents epigenetic modification of hepatic gene expression in the offspring. J Nutr 135:1382–1386

Lindinger A, Peterli R, Peters T, Kern B, Von Flue M, Calame M, Hoch M, Eberle AN, Lindinger PW (2010) Mitochondrial DNA content in human omental adipose tissue. Obes Surg 20:84–92

Maersk M, Belza A, Stodkilde-Jorgensen H, Ringgaard S, Chabanova E, Thomsen H, Pedersen SB, Astrup A, Richelsen B (2012) Sucrose-sweetened beverages increase fat storage in the liver, muscle, and visceral fat depot: a 6-mo randomized intervention study. Am J Clin Nutr 95:283–289

Malik VS, Popkin BM, Bray GA, Despres JP, Willett WC, Hu FB (2010) Sugar-sweetened beverages and risk of metabolic syndrome and type 2 diabetes: a meta-analysis. Diabetes Care 33:2477–2483

Moller DE, Berger JP (2003) Role of PPARs in the regulation of obesity-related insulin sensitivity and inflammation. Int J Obes Relat Metab Disord 27(Suppl 3):S17–S21

Mortensen OH, Larsen LH, Orstrup LK, Hansen LH, Grunnet N, Quistorff B (2014) Developmental programming by high fructose decreases phosphorylation efficiency in aging offspring brain mitochondria, correlating with enhanced UCP5 expression. J Cereb Blood Flow Metab 34:1205–1211

Nagai Y, Nishio Y, Nakamura T, Maegawa H, Kikkawa R, Kashiwagi A (2002) Amelioration of high fructose-induced metabolic derangements by activation of PPARalpha. Am J Physiol Endocrinol Metab 282:E1180–E1190

Nelson MC, Neumark-Sztainer D, Hannan PJ, Story M (2009) Five-year longitudinal and secular shifts in adolescent beverage intake: findings from project EAT (eating among teens)-II. J Am Diet Assoc 109:308–312

Ohashi K, Ando Y, Munetsuna E, Yamada H, Yamazaki M, Nagura A, Taromaru N, Ishikawa H, Suzuki K, Teradaira R (2015a) Maternal fructose consumption alters messenger RNA expression of hippocampal StAR, PBR, P450(11beta), 11beta-HSD, and 17beta-HSD in rat offspring. Nutr Res 35:259–264

Ohashi K, Munetsuna E, Yamada H, Ando Y, Yamazaki M, Taromaru N, Nagura A, Ishikawa H, Suzuki K, Teradaira R, Hashimoto S (2015b) High fructose consumption induces DNA methylation at PPARalpha and CPT1A promoter regions in the rat liver. Biochem Biophys Res Commun 468:185–189

Perez-Pozo SE, Schold J, Nakagawa T, Sanchez-Lozada LG, Johnson RJ, Lillo JL (2010) Excessive fructose intake induces the features of metabolic syndrome in healthy adult men: role of uric acid in the hypertensive response. Int J Obes 34:454–461

Pirola CJ, Gianotti TF, Burgueno AL, Rey-Funes M, Loidl CF, Mallardi P, Martino JS, Castano GO, Sookoian S (2013) Epigenetic modification of liver mitochondrial DNA is associated with histological severity of nonalcoholic fatty liver disease. Gut 62:1356–1363

Pirola CJ, Scian R, Gianotti TF, Dopazo H, Rohr C, Martino JS, Castano GO, Sookoian S (2015) Epigenetic modifications in the biology of nonalcoholic fatty liver disease: the role of DNA hydroxymethylation and TET proteins. Medicine (Baltimore) 94:e1480

Popkin BM (2010) Patterns of beverage use across the lifecycle. Physiol Behav 100:4–9

Ramsden DB, Ho PW, Ho JW, Liu HF, So DH, Tse HM, Chan KH, Ho SL (2012) Human neuronal uncoupling proteins 4 and 5 (UCP4 and UCP5): structural properties, regulation, and physiological role in protection against oxidative stress and mitochondrial dysfunction. Brain Behav 2:468–478

Schulze MB, Manson JE, Ludwig DS, Colditz GA, Stampfer MJ, Willett WC, Hu FB (2004) Sugar-sweetened beverages, weight gain, and incidence of type 2 diabetes in young and middle-aged women. JAMA 292:927–934

Schwarz JM, Noworolski SM, Wen MJ, Dyachenko A, Prior JL, Weinberg ME, Herraiz LA, Tai VW, Bergeron N, Bersot TP, Rao MN, Schambelan M, Mulligan K (2015) Effect of a high-fructose weight-maintaining diet on lipogenesis and liver fat. J Clin Endocrinol Metab 100:2434–2442

Segovia SA, Vickers MH, Gray C, Reynolds CM (2014) Maternal obesity, inflammation, and developmental programming. Biomed Res Int 2014:418975

Serradas P, Giroix MH, Saulnier C, Gangnerau MN, Borg LA, Welsh M, Portha B, Welsh N (1995) Mitochondrial deoxyribonucleic acid content is specifically decreased in adult, but not fetal, pancreatic islets of the Goto-Kakizaki rat, a genetic model of noninsulin-dependent diabetes. Endocrinology 136:5623–5631

Shi L, Shi L, Zhang H, Hu Z, Wang C, Zhang D, Song G (2013) Oxymatrine ameliorates non-alcoholic fatty liver disease in rats through peroxisome proliferator-activated receptor-alpha activation. Mol Med Rep 8:439–445

Sloboda DM, Li M, Patel R, Clayton ZE, Yap C, Vickers MH (2014) Early life exposure to fructose and offspring phenotype: implications for long term metabolic homeostasis. J Obes 2014:203474

Stanhope KL, Havel PJ (2009) Fructose consumption: considerations for future research on its effects on adipose distribution, lipid metabolism, and insulin sensitivity in humans. J Nutr 139:1236S–1241S

Tappy L, Le KA (2010) Metabolic effects of fructose and the worldwide increase in obesity. Physiol Rev 90:23–46

Teff KL, Grudziak J, Townsend RR, Dunn TN, Grant RW, Adams SH, Keim NL, Cummings BP, Stanhope KL, Havel PJ (2009) Endocrine and metabolic effects of consuming fructose- and glucose-sweetened beverages with meals in obese men and women: influence of insulin resistance on plasma triglyceride responses. J Clin Endocrinol Metab 94:1562–1569

Turner N, Bruce CR, Beale SM, Hoehn KL, So T, Rolph MS, Cooney GJ (2007) Excess lipid availability increases mitochondrial fatty acid oxidative capacity in muscle: evidence against a role for reduced fatty acid oxidation in lipid-induced insulin resistance in rodents. Diabetes 56:2085–2092

Ventura EE, Davis JN, Goran MI (2011) Sugar content of popular sweetened beverages based on objective laboratory analysis: focus on fructose content. Obesity (Silver Spring) 19:868–874

Xu X, So JS, Park JG, Lee AH (2013) Transcriptional control of hepatic lipid metabolism by SREBP and ChREBP. Semin Liver Dis 33:301–311

Yamazaki M, Munetsuna E, Yamada H, Ando Y, Mizuno G, Murase Y, Kondo K, Ishikawa H, Teradaira R, Suzuki K, Ohashi K (2016) Fructose consumption induces hypomethylation of hepatic mitochondrial DNA in rats. Life Sci 149:146–152

Ye JM, Doyle PJ, Iglesias MA, Watson DG, Cooney GJ, Kraegen EW (2001) Peroxisome proliferator-activated receptor (PPAR)-alpha activation lowers muscle lipids and improves insulin sensitivity in high fat-fed rats: comparison with PPAR-gamma activation. Diabetes 50:411–417

Zhang J, Zhang F, Didelot X, Bruce KD, Cagampang FR, Vatish M, Hanson M, Lehnert H, Ceriello A, Byrne CD (2009) Maternal high fat diet during pregnancy and lactation alters hepatic expression of insulin like growth factor-2 and key microRNAs in the adult offspring. BMC Genomics 10:478

Zou M, Arentson EJ, Teegarden D, Koser SL, Onyskow L, Donkin SS (2012) Fructose consumption during pregnancy and lactation induces fatty liver and glucose intolerance in rats. Nutr Res 32:588–598

Part IV

Practical Techniques and Applications

Multilocus Methylation Assays in Epigenetics

114

Thomas Eggermann

Contents

Abstract

Due to the basic significance of DNA methylation patterns and their changes for nearly all physiological processes, the interest in procedures to determine DNA methylation has grown rapidly and expanded. Though different levels of epigenetic regulation are known, it is advantageous to focus on DNA methylation as DNA methylation is very stable, and it is currently assumed that environmental

T. Eggermann (✉)
Institute of Human Genetics, University Hospital, RWTH Technical University Aachen, Aachen, Germany
e-mail: teggermann@ukaachen.de

V. B. Patel, V. R. Preedy (eds.), *Handbook of Nutrition, Diet, and Epigenetics*,
https://doi.org/10.1007/978-3-319-55530-0_50

(and inborn) factors altering epigenetic patterns mainly affect DNA methylation. Furthermore, DNA methylation tests are meanwhile well-established tools in both research and diagnostic laboratories. In the past, determination of the 5-methylcytosin and DNA methylation status was hampered because the methylation profile is not maintained during standard amplification processes. However, with the development of methylation-specific (MS) pretreatment protocols, this problem could be circumvented, and these protocols serve as pretreatment steps before the application of methylation-specific approaches. This chapter focuses on molecular tests determining the DNA methylation from single CpG to genome-wide methylation resolution. As examples for single-locus and multi-loci tests, multiplex ligation probe-dependent amplification (MLPA) and pyrosequencing are described in detail, but this chapter will also introduce the potential of next-generation sequencing (NGS)-based approaches for methylation analysis. In fact NGS-based assays allow the deep and comprehensive analyses of all three mechanisms of epigenetic regulation, i.e., for histone function, analysis of (noncoding) RNA, and characterization of (differentially) methylated DNA:

- The rapid progress in the field of DNA methylation analysis offers unique opportunities to comprehensively analyze the influence of epigenetic regulation on a broad spectrum of biological processes.
- A broad range of methods to determine DNA methylation statuses has been developed, but the decision on the test for DNA methylation will depend on the question to be answered and the sample size which should be investigated.
- In particular for NGS-based approaches, the running costs are currently high per sample, and bioinformatic pipelines for interpretation have to be improved, but these problems will be circumvented in the near future, and standardized software solutions will become available in the near future.

Keywords
5-Hydroxymethyl cytosine · DNA methylation · Enrichment-restriction enzyme · Affinity enrichment · Bisulfite conversion · Array hybridization · Next-generation sequencing · MLPA · Pyrosequencing

List of Abbreviations

5-caC	5-Carboxylcytosine
5-fC	5-Formylcytosine
5-hmC	5-Hydroxymethylcytosine
5-mC	5-Methylcytosine
Bp	Base pair
ChIP	Chromatin immunoprecipitation
CNV	Copy number variation
COBRA	Combined bisulfite restriction analysis
CpG	Cytosine-phosphate-guanosine
DMR	Differentially methylated region
DNMT	DNA methyltransferase

MLPA	Multiplex ligation probe-dependent amplification
MS	Methylation specific
MSRE-PCR	Methylation-sensitive restriction enzyme PCR
ncRNA	Noncoding RNA
NGS	Next-generation sequencing
PCR	Polymerase chain reaction
QAMA	Quantitative analysis of methylated alleles
RE	Restriction endonuclease
RRBS	Reduced representation bisulfite sequencing
SBL	Sequencing by ligation
SBS	Sequencing by synthesis
TET	Ten-eleven translocation methylcytosine dioxygenase
UPD	Uniparental disomy

Introduction

Epigenetics in the sense of a spatial and temporal control of gene activity without alteration of the DNA sequence itself has firstly been suggested in 1990 by Holliday, and meanwhile it is used to describe anything other than DNA sequence that influences the development of an organism. Though this concept has already been suggested more than 25 years ago, it has been in the last decade that the number of studies on the physiological significance of epigenetics has risen exponentially. As a result of these reports, it is now generally accepted that epigenetic regulation influences nearly all cellular and physiological processes and that one major role of this complex mode of gene regulation is the mediation between genome and environment. A specific form of epigenetic regulation is genomic imprinting, resulting in the expression of certain genes in a parent-of-origin-specific manner. Both maternally and paternally inherited alleles can be silenced, and in human more than 100 genes are expected to be imprinted (http://www.geneimprint.com/site/genes-by-species). Disturbed imprinting has meanwhile been reported for several human genes and/or imprinting domains and is associated with congenital disorders, the so-called imprinting disorders (for review: Soellner et al. 2016). Up to now, four different molecular alterations have been identified to result in a disturbed imprinting: uniparental disomies (UPD, i.e., the inheritance of the homologues of a chromosomal pair from only one parent), aberrant methylation itself (epimutations), copy number variations (CNVs, deletions/duplications), and point mutations affecting the imprinted gene itself.

Whereas UPDs, CNVs, and point mutations represent genomic alterations, epimutations are sensu stricto typical epigenetic alterations as the genomic DNA sequence is not altered. The causes of epimutations in patients with imprinting disorders are widely unknown, but there is evidence for a contribution of genomic factors (Mackay et al. 2008; Docherty et al. 2015). However, also environmental factors have been suggested to contribute to altered imprints in these disorders and/or imprinted loci, e.g., assisted reproduction (for review: Uyar and Seli 2014), maternal nutrition during pregnancy (Heijmans et al. 2008; for review: McCullough et al.

2016), and even the paternal and grandparental lifestyles (for review: Stuppia et al. 2015). Environmental factors generally play a significant role in maintenance and modification of the epigenetic status of an individual, and one major factor is nutrition. The consequences of these epigenetic changes affect the development, health status, and even behavior of an individual itself but can also be transferred over the generations (for review: McCarrey et al. 2016).

Due to the basic significance of DNA methylation patterns and their changes for nearly all physiological processes, the interest in procedures to determine DNA methylation has grown rapidly and expanded. In this chapter, a general overview on methylation tests will be given, but due to the rapid technical and bioinformatic progress, only general principles can be presented on the basis of selected examples. For the current state of technical developments and test analyses, the reader is therefore asked to update himself.

DNA Methylation

The main molecular mechanisms which mediate epigenetic signaling include DNA methylation, histone modification, and (small) noncoding RNAs (ncRNAs) (for review: Kelsey and Feil 2013). As assays targeting DNA methylation are meanwhile well-established tools in both research and diagnostic laboratories, this review focuses on this level of epigenetic regulation.

DNA methylation as a covalent modification enables the dynamical reprogramming of cells during ontogenesis. The most prominent epigenetic mark of DNA is 5-methylcytosine (5-mC), but recently further cytosine derivatives have been identified (Fig. 1) (for review: Bochtler et al. 2017). Whereas the methylation of

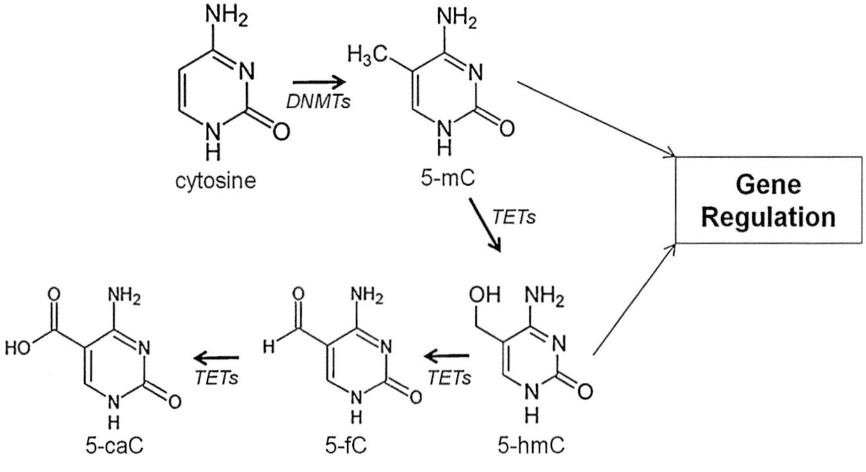

Fig. 1 Methylation of cytosine and its derivatives (*5-mC* 5 methyl cytosine; *5-caC* 5 carboxyl cytosine; *5-fC* 5-formyl cytosine; *5-hmC* 5-hydroxymethyl cytosine; *DNMT* DNA methyltransferase; *TET* ten-eleven translocation methylcytosine dioxygenase)

cytosine to 5-mC is catalyzed by DNA methyltransferases, the modification of 5-mC to 5-hydroxymethylcytosine (5-hmC), 5-formylcytosine (5-fC), and 5-carboxycytosine (5-caC) is mediated by TET proteins (ten-eleven translocation methylcytosine dioxygenase), a group of iron(II)/αKG-dependent dioxygenases. 5-mC and its derivatives constitute only 2–8% of the total cytosines in the human DNA, but all these molecules have crucial roles in physiological mechanisms.

DNA methylation interacts with the condensation status of chromatin, and directly or indirectly regulates gene expression, and thus 5-mC DNA plays an important role in switching genes on and off. The physiological role of 5-hmC is currently unclear, but it has a relatively high abundance in embryonic stem cells and neurons. DNA methylation processes mainly occur in CpG dinucleotide clusters and often in promoter regions. However, intergenic epigenetic marks have also been identified and are of substantial significance for differentially methylated regions (DMRs) in imprinting disorders. Several of these diseases are associated with aberrant methylation of intergenic DMRs which regulate the (hierarchical) expression of several genes. An example is the H19/IGF2:IG-DMR in 11p15.5: its methylation directly silences H19 expression but enhances IGF2 expression by alteration of the chromatin formation (for review: Enklaar et al. 2006).

Pretreatment of Genomic DNA to Determine Its Methylation Status

As already mentioned, histones, RNA, and DNA are targeted in research to enlighten the complex epigenetic mechanisms, and a multitude of assays and strategies have been developed to decipher the molecular basis of epigenetic regulation. However, expression profiling of cells after treatment with DNA methyltransferase or histone deacetylase inhibitors is prone to false results; thus, both in research and diagnostic (screening) studies, it is advantageous to focus on DNA methylation due to two major reasons: (a) DNA methylation is very stable, DNA can easily be obtained, and very small amounts of DNA are required. (b) It is currently assumed that environmental (and inborn) factors altering epigenetic patterns mainly affect DNA methylation; therefore, the determination of the DNA methylation status is an appropriate approach to determine these influences.

However, the determination of the DNA methylation status by standard molecular techniques like PCR or bacterial cloning is hampered because the methylation profile is not maintained during the amplification processes. Hybridization assays are also not suitable to detect methylation groups, as these are located in the major groove of a DNA molecular and do not affect hydrogen bond binding. Thus, the routine analysis of DNA methylation became first possible with the development of methylation-specific (MS) pretreatment protocols which are applied prior to specific MS approaches.

Up to now, three major pretreatment approaches have been established in epigenetic studies (Fig. 2): MS restriction enzyme digestion, affinity enrichment, and bisulfite conversion.

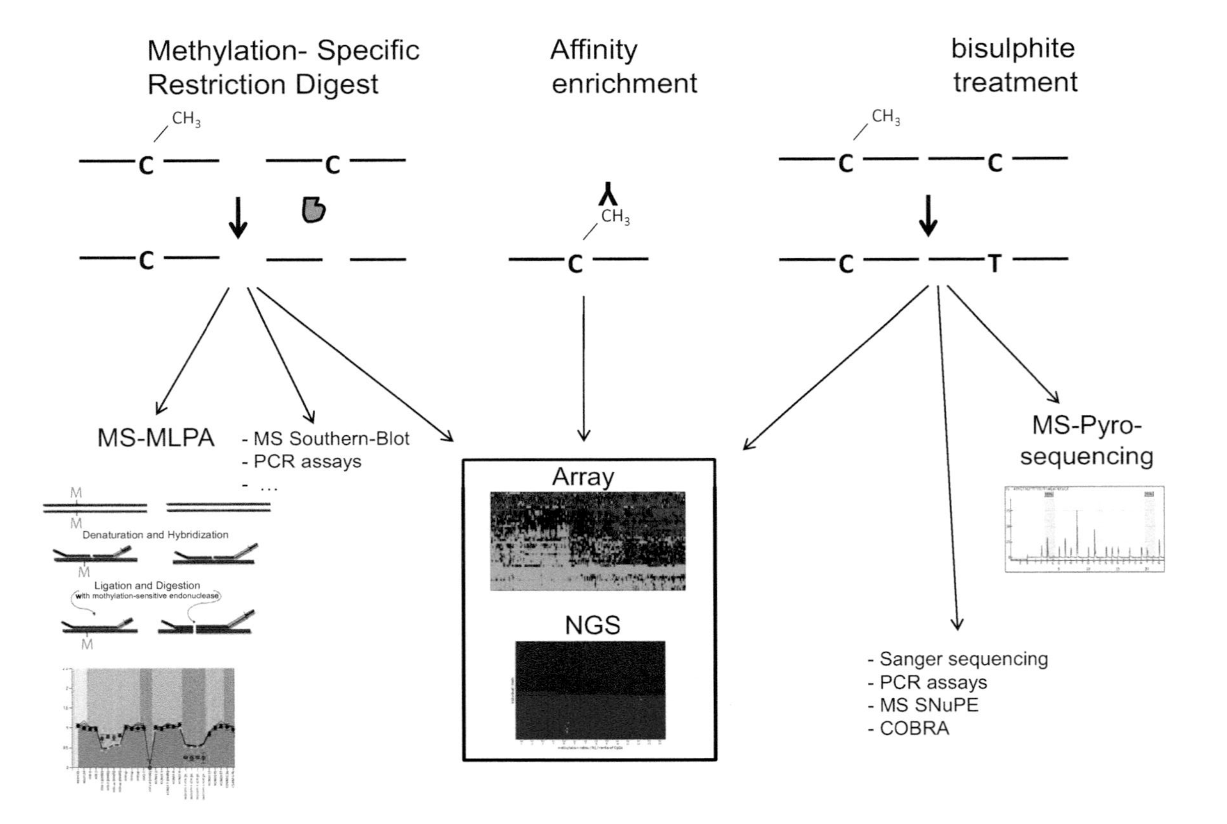

Fig. 2 The three methods for DNA pretreatment as the basis for MS assays and examples for the major tests to determine the methylation of single or multiple loci

MS Restriction Enzymes

MS restriction enzymes (REs) were the first tools which have been applied in locus-specific DNA methylation analyses in the early 1980s. These enzymes are inhibited by 5-mC, so their cutting pattern reflects the methylation of a DNA strand. The most widely used MS REs are *Hpa*II and *Sma*I, and in fact *Hpa*II digestion is the frequently used in MS MLPA (see below), a widely used technique in routine diagnostics of imprinting disorders.

Affinity Enrichment

Affinity enrichment (also named chromatin immunoprecipitation (ChIP)) is based on the use of antibodies specific for 5-mC or methyl-binding proteins, and its power has been proven by genome-wide studies of histone modifications (e.g., Barski et al. 2007). The affinity enrichment is often combined with array hybridization, but a shift to next-generation sequencing (NGS)-based approaches can be observed now. In respect to DNA methylation, affinity enrichment in fact allows a genome-wide assessment, whereas information on methylation on individual CpG level is hampered.

Bisulfite Conversion

The introduction of bisulfite conversion of unmethylated cytosine residues to uracil (Figs. 2 and 3), leaving methylated cytosines unaffected (Frommer et al. 1992), was a revolutionizing step toward a comprehensive analysis of the human methylome, and it is nowadays the major DNA treatment prior to methods enabling the detailed analysis of the methylation status of CpGs. The chemical treatment of genomic DNA by bisulfite by sulfonation, deamination, and subsequent desulfonation causes a conversion of unmethylated cytosine into uracil. Thus, it can be regarded as an induced mutagenesis, translating an epigenetic mark into a genomic alteration and leaving methylated cytosine residues unconverted.

DNA Methylation Studies After DNA Pretreatment

The decision on the technique to analyze DNA methylation depends on the question a study wants to answer. In case only specific CpG sites should be targeted, single-locus tests are the appropriate tools. In fact, several of these tests are feasible to cover more than one locus and can therefore be regarded as multilocus tests. For projects aiming on methylation patterns in the whole epigenome of an individual or cell type, global DNA methylation assays are suitable.

In the last two decades, a multitude of assays have then been developed and implemented in DNA methylation studies (Table 1, Fig. 2) (for review: Laird 2010),

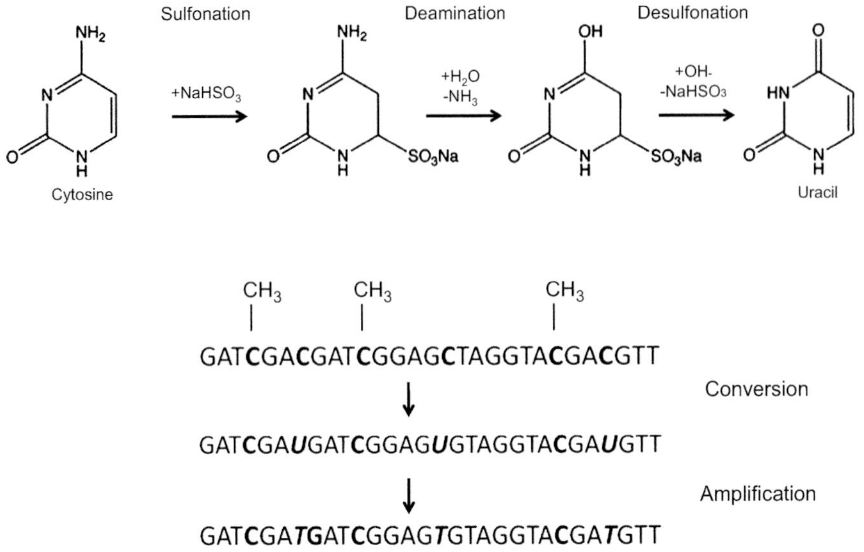

Sulfonation Deamination Desulfonation

Cytosine Uracil

GATCGACGATCGGAGCTAGGTACGACGTT

Conversion

GATCGAUGATCGGAGUGTAGGTACGAUGTT

Amplification

GATCGATGATCGGAGTGTAGGTACGATGTT

Fig. 3 Steps of bisulfite treatment of genomic DNA: unmethylated cytosine residues are modified and result in thymine, whereas methylated cytosines are protected against conversion. In case of amplification (e.g., by PCR or bacterial cloning), uracil is replaced by thymine

and they reflect the evolution of the DNA pretreatment methods. Several of these techniques are strictly linked to the pretreatment method, but others can be flexibly combined.

Due to the complexity of the topic and the huge number of approaches, only some techniques can be exemplified; to get a comprehensive overview, the reader is referred to recent overviews on this topic (e.g., Laird 2010).

MS RE-based DNA methylation assays were the first locus-specific tests established in the early 1980s; they were followed by gel electrophoresis and Southern blotting. After the introduction of PCR, it was also applied in DNA methylation analysis and allowed the differentiation of methylated and unmethylated DNA strands after MS RE digestion (e.g., MSRE-PCR/methylation-sensitive restriction enzyme PCR, Melnikov et al. 2005).

For (epi)genome-wide analysis of DNA methylation after MS RE digestion, both array- and sequencing-based approaches had been developed. In fact, all these MS RE-based approaches are limited by the distribution of their respective digestion sites. However, they are broadly applied in human genetic diagnostic labs as MS MLPA (multiplex ligation-dependent probe amplification) is meanwhile a standard tool for determination of both methylation defects and CNVs.

As already mentioned, affinity enrichment does not provide information on methylation on individual CpG level, but it is rather used for the genome-wide study of histone modification. Another reason for its rare use in routine settings is the extensive experimental and bioinformatic adjustments.

Table 1 Examples of methylation-specific (MS) assays applied in diagnostics and research of imprinted loci

Pretreatment	General comments	Technique[a]	Main application (i.e., single or multilocus testing, genome-wide	Equipment	Discrimination of (epi) mutations	Comments
Restriction digest	Information value depends on the localization of restriction sites; prone to incomplete digestion	Southern blotting	Single-locus testing	Gel electrophoresis	No	Linkage of methylation over large distances; labor-intensive, time-consuming, large amount of DNA, low sensitivity
		PCR		Thermocycler, electrophoresis	No	Fast, cheap, not quantitative
		MLPA	Multilocus testing	Thermocycler, capillary gel electrophoresis	Yes	Commercial assay. Disturbed methylation and CNVs can be detected in one reaction. Not very flexible
		Array	Genome-wide DNA methylation	Array platform	No	Often comparative hybridization of MS and non-MS-digested DNA. Hybridization artifacts might hamper the interpretation
		NGS		NGS platform	(yes)	Requires complex bioinformatics

(*continued*)

Table 1 (continued)

Pretreatment	General comments	Technique[a]	Main application (i.e., single or multilocus testing, genome-wide	Equipment	Discrimination of (epi) mutations	Comments
Affinity enrichment		Array	Genome-wide DNA methylation	Array platform	No	No single CpG sites are targeted; complex bioinformatics
		NGS		NGS platform	(yes)	Requires complex bioinformatics
Bisulfite conversion	Incomplete bisulfite conversion should be considered	Pyrosequencing	Determination of several CpGs at one locus	Pyrosequencing platform	No	Fast, quantitative; 96 well format, screening of large number of samples
		MethyLight; QAMA (real-time PCR-based methylation assay)	Single-locus testing	Real-time thermocycler	No	Fast, quantitative; use of differentially labeled probes
		MS-SNuPE	Single-locus testing	Thermocycler, capillary gel electrophoresis	No	flexible
		Sanger sequencing		Thermocycler, capillary gel electrophoresis	No	Cloning required; time-consuming
		Array	Genome-wide DNA methylation	Array platform	No	Decreased hybridization specificity possible
		NGS		NGS platform	(Yes)	Requires complex bioinformatics

[a]In general, with single exceptions, the general method is named; (yes) in principle, the discrimination of (epi)mutations is possible but requires separate bioinformatic solutions; QAMA real-time PCR-based methylation assay; *MS-SNuPE* MS single-nucleotide primer extension)

Due to its simplicity and reliability, bisulfite conversion (Fig. 3) is meanwhile the major pretreatment method for DNA methylation studies, and numerous assays to determine the grade of methylation have meanwhile been developed. The first characterization of bisulfite-converted DNA comprised PCR amplification, cloning of single PCR products, and Sanger sequencing (Frommer et al. 1992). However, the time-consuming and labor-intensive cloning was rapidly replaced by direct sequencing of amplicons (Paul and Clark 1996). In contrast to MS RE-based strategies, tests based on bisulfite conversion are not hampered by the presence of recognition sites of MS REs. Nevertheless, the bisulfite-treated unmethylated DNA finally comprises only three instead of four different nucleotides. Whereas sequencing-based methods are not influenced by this reduced complexity, it might affect array-based hybridization assays as it might lead to a decreased specificity.

In fact, there are a growing number of methods and strategies aiming on the identification of altered DNA methylation, and the tests summarized in Table 1 are just examples for different optional methods, but not all assays can be covered. In general, these tests can be differentiated in tests aiming on single-loci analyses and assays covering multiple methylated loci, but there is an overlap of applications and capacities. In particular the implementation of bisulfite conversion lead to the development of quantitative and qualitative PCR approaches (e.g., MethyLight, QAMA, COBRA (Eads et al. 2000; Xiong and Laird 1997; Zeschnigk et al. 2008). However, there is a growing demand for multilocus and whole methylome tests; thus, in the following, some of these methods are described in more details as they are currently broadly used or will become important in the (near) future.

MLPA (Multiplex Ligation-Dependent Probe Amplification)

MLPA (multiplex ligation-dependent probe amplification) (Fig. 4a) is a method to simultaneously detect CNVs and altered methylation patterns at up to 46 specific target sequences (Schouten et al. 2002). The respective probes contain these specific sequences but are initially splitted into two oligonucleotides which recognize adjacent target sites. When both oligonucleotides of the same probe are hybridized to their respective targets, they can be ligated to become a complete probe. In the subsequent PCR with primers annealing to the complete probe, only the ligated oligonucleotides, but not the unbound probe oligonucleotides, are amplified. Each complete probe and its resulting PCR product have a unique length, so that its resulting amplicons can be separated and identified by electrophoresis. Thereby, the length and the quantity of the PCR products and their corresponding probes and DNA fragments can be determined. In case of MS MLPA, genomic DNA is digested with an MS RE after hybridization of the MLPA probes. Comparison of peak heights of undigested and digested DNA indicates the methylation level of a particular probe.

A broad portfolio of (MS)MLPA assays has been developed and is distributed by MRC Holland (www.mlpa.com/). (MS) MLPA belongs to the major methods used in

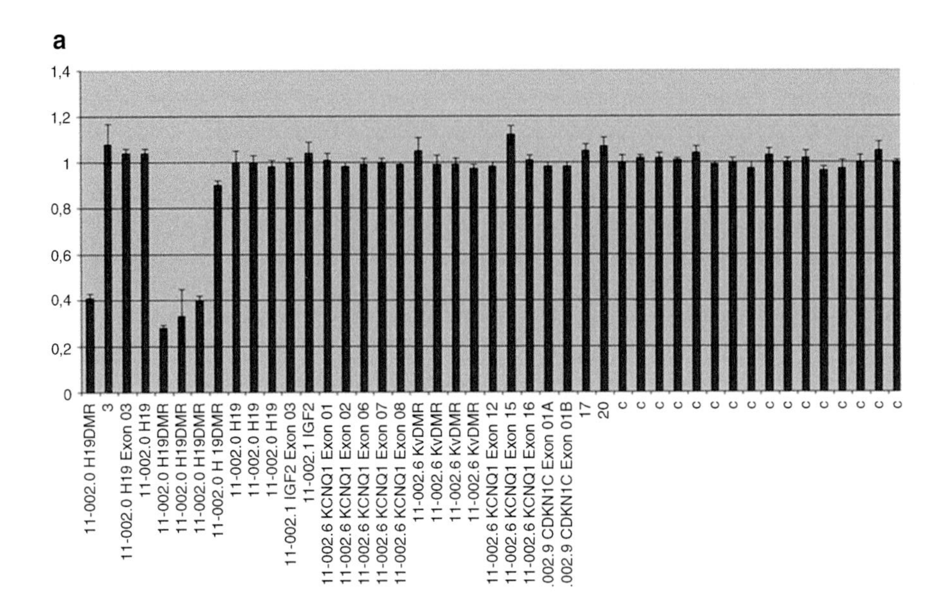

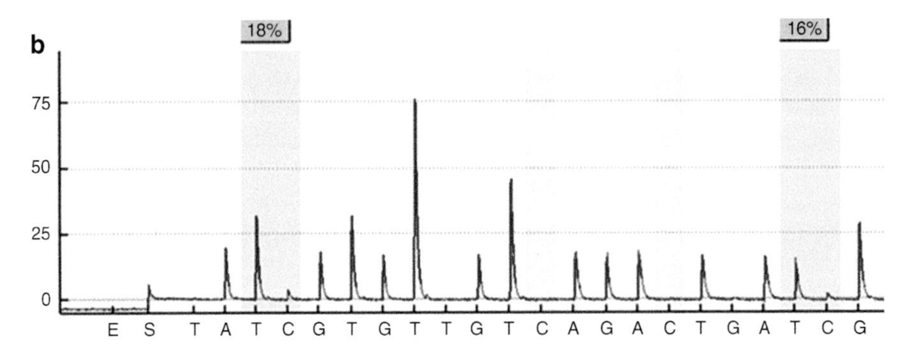

Fig. 4 MS MLPA (**a**) and MS pyrosequencing (**b**) as two major tests for multilocus analysis. (**a**) Bioinformatic analysis of MLPA raw data shows a hypomethylation of cytosines at the H19 differentially methylated region in 11p15.5 (methylation index (y axis) <1), whereas the other probes give a normal methylation index (=1). Each bar represents one probe, targeting one CpG. (**b**) Analysis of a pyrosequenced DNA fragment comprising to CpGs: the comparative analysis of the unconverted and methylated cytosine and the thymine residue resulting from unmethylated cytosine conversion allows the calculation of methylation index (18% in the first and 16% at the second CpG)

diagnostics of imprinting disorders (e.g., Eggermann et al. 2016) because it is robust, easy to handle, and solely requires a standard molecular routine laboratory equipment (e.g., automated Sanger sequencing platform). The data analysis is simple, and several (gratis) software packages are available. In principle, (MS) MLPA can be modified, and custom-designed assays are available, but as the probe composition might be restrictive and in particular MS analysis always requires a restriction site for a MS RE, it is often not applicable for research purposes.

MS Pyrosequencing

In contrast to MS MLPA, MS pyrosequencing (Fig. 4b) does not depend on the presence of a MS RE restriction site and is thus more flexible. Like other MS sequencing-based methods, it detects the bisulfite-induced C to T transition. It allows the analysis of multiple CpG methylation sites in the same DNA strand in the same reaction and provides an accurate quantitation of methylation. After bisulfite treatment, the first step of MS pyrosequencing consists of PCR amplifications using primers not discriminatory for methylation status of the amplicon. The pyrosequencing (Ronaghi et al. 1998) step itself is then targeted on a sequence detection of bisulfite-altered differentially methylated cytosines within an amplicon.

In contrast to single-locus tests, both MS MLPA and MS pyrosequencing can target several CpGs, but in both the number of loci is limited, and in case of MS pyrosequencing, only CpGs in the same small DNA region can be analyzed. In fact, they are sensitive, specific, and relatively cheap, but they are not suitable for the analysis of the whole epigenome. For that purpose, array hybridization and NGS-based assays have been developed.

Array-Based Hybridization

Array-based hybridization assays consisting of a collection of DNA spots attached to a solid surface can be used for further analysis of pretreated DNA from all MS pretreatment procedures. They allow a comprehensive profiling of genome-wide DNA methylation (see Schumacher et al. 2006), but depending on the pretreatment, also custom-designed assays are available targeting specific regions or groups of CpGs, promoters, and genes. However, the spotting of oligos is not that flexible and does not allow an easy modification of the assay. In case of affinity enrichment, using antibodies for 5-mC has been combined with array hybridization, but it does not provide information on individual CpGs. All these reasons lead to a shift to NGS-based approaches. However, due to the relatively low costs of new generations of genome-wide DNA methylation arrays, some formats are broadly used and under permanent improvement. Examples are the genome-wide DNA methylation BeadChips (Bibikova et al. 2011), generations of which have been successfully used in research projects (e.g., Infinium HumanMethylation27, Human-Methylation450, MethylationEPIC; Illumina, Inc., San Diego, CA/USA). The technique is based on a whole-genome amplification of bisulfite-converted DNA, followed by fragmentation and hybridization of the sample to methylation-specific DNA probes that are bound to array surface. Each bead corresponds to a specific CpG and methylation status. An example for the resolution capacity of a genome-wide methylation array is the HumanMethylation450 BeadChip which covers 482,421 CpG sites, 3091 non-CpG sites, and 65 random SNPs each analyzed in parallel at a single-nucleotide resolution per sample.

NGS-Based Assays

In comparison to array-based techniques, NGS-based assays are often more flexible as they do not require a specific oligodesign, are not affected by artificial hybridization, and require less DNA. Thus, NGS allows to understand the methylation status on a large scale and at a single-base resolution. Its high sensitivity, specificity, and scalability have made NGS a powerful tool in research and diagnostic (for review: Goodwin et al. 2016). In addition to the large-scale application, a further advantage of NGS is the possibility of ultradeep sequencing (>1000 reads) allowing a better estimation of (epi) genetic mosaicism (Beygo et al. 2013) and slight differences in methylation in general. Corresponding to the other techniques, a broad spectrum of NGS-based approaches have been developed in the last 10 years, starting with assays targeting a limited number of CpGs on the Roche 454 platform (Taylor et al. 2007).

The majority of wet-lab NGS protocols consist of three major steps: (a) Library preparation aims on breaking genomic DNA randomly into smaller fragments as the prerequisite for an efficient enrichment of the regions of interest (whole genome/exome vs. targeted enrichment). Meanwhile a large number of enrichment assays are available, mainly based on PCR or capture (array or solid) technologies. (b) Template preparation serves to prepare the library fragments and to immobilize them as spatially separated template sites which allow thousands to billions of (c) simultaneous sequencing reactions. At the end of a successful NGS approach, billions of DNA sequences have been generated, and the analysis of these big datasets is a challenge for bioinformatic data analyses. It is comparable to puzzling, and depending on the size of the targeted genomic regions (e.g., selected genes or a whole genome), it might be time-consuming and require large data storage capacities. After alignment of the sequence fragment (primary data analysis), the obtained sequences have to be filtered for platform-specific artifacts and frequent genomic variants (secondary data analysis). Due to the enormous amount of information, the primary and secondary data analyses steps “simply” serve to eliminate artificial and biological irrelevant findings; thus, the tertiary data analysis is the final analysis step which should give the answer to a research or diagnostic question.

A large number of protocols and different NGS platforms are meanwhile (commercially) available, and they differ remarkably in capacity, run time, costs, and output. In particular, the length of fragment and ability to precisely sequence homopolymer stretches are basic aspects for the design of NGS-based projects.

Many NGS applications are based on short-read sequencing approaches (for review: Goodwin et al. 2016), either on sequencing by ligation (SBL) or sequencing by synthesis (SBS). SBL means that a fluorophore-labeled probe hybridizes to a DNA fragment and is then ligated to an adjacent oligonucleotide for imaging. SBS is based on amplification by a polymerase, and the incorporation of a nucleotide is measured, either by the parallel release of a fluorophore signal or a change in ionic concentration. However, for the analysis of the high complexity of whole genomes, including copy number and structural variations, short-read sequencing is often insufficient but requires long-read sequencing which results in reads of several kilobases. For that purpose, several approaches have meanwhile been suggested or are in development (for review: Goodwin et al. 2016).

The application of NGS in epigenetics contributes to an increased knowledge of already known methylated DNA regions, and it leads to the discovery of new epigenetically regulated genomic elements. NGS makes the analysis of the whole genome possible; thus, the methylome can be enlightened on single-based pair level. Furthermore, by comparing methylome with transcriptome data, the cell type-specific significance of changes in DNA methylation can be determined directly.

In case of MS NGS approaches, it has to be considered that the methylation signature as the target of the analysis should not be influenced by any pretreatment and enrichment assay (for review: Goodwin et al. 2016). Thus, PCR-based enrichments are a priori not suitable as PCR does not preserve this signature. However, in respect to PCR, this limitation can be circumvented by bisulfite conversion.

Like array-based methylation analyses, NGS assays have meanwhile been developed for all three pretreatment methods. However, MS RE-based approaches and array-hybridization assays only allow the partial analysis of the methylome, as they are restricted to RE recognition sites, or the probe format does not allow the determination of methylation on single CpG level. Furthermore, biases toward highly methylated regions can be observed.

The use of bisulfite-converted DNA in NGS approaches circumvents these problems, and meanwhile various assays consisting o f a combined use of enrichment and conversion have been established (for review: Soto et al. 2016). In fact, enrichment approaches based on capturing technologies, e.g., reduced representation bisulfite sequencing RRBS (Gu et al. 2010) or SureSelect MethylSeq, are meanwhile widely used, and their advantages comprise cost benefits for many samples and large genomes. However, they harbor the risk to miss DNA methylation marks outside the captured regions, a limitation which can be circumvented by whole-genome bisulfite sequencing. This technique is based on the random fragmentation of DNA, which is then bounded by methylated adapters and then bisulfite treated. The subsequent massive parallel sequencing then provides a single-base resolution-wide DNA methylome profile but is more expensive than the aforementioned capture-based approaches.

Limitations and Challenges of Methylation Tests

The rapid progress in the field of DNA methylation analysis offers unique opportunities to comprehensively analyze the influence of epigenetic regulation on a broad spectrum of biological processes.

The decision on the suitability of a specific DNA methylation strategy has to consider numerous aspects, and some of them are listed in the following, and some of them will be addressed in the following paragraph:

- Should the test be used for diagnostic or research purposes?
- Which tissue/cell type should be investigated?
- Can mosaicism influence the interpretation?
- Which is the region of interest? Is the analysis of one or a small number of CpGs or methylated loci sufficient, or does the study aim on a genome-wide analysis? In the latter, what is the required coverage and resolution?
- Is the test "simply" aiming on methylation differences, or does it discriminate between the basic causes (CNVs, epimutations, UPDs)?
- Quantity and Quality of DNA.
- Which sensitivities and specificities are required?
- How many samples should be tested, is a single sample test needed or a screening test for a large number of samples?
- Which equipment is needed, what are the running costs?
- Is it labor intensive?
- Can the bioinformatic analysis of raw date be managed, is the development of new bioinformatic pipelines required?

In any case, the application of these tests requires to consider the potential results and their interpretation. In case of single/multilocus tests, the results in a defined set of CpGs allow the more or less precise determination of the role and status of these loci in a specific context, but general conclusions on the role of epigenetic alterations in a given context are not possible. Vice versa, global DNA methylation assays allow an overview on (aberrant) methylation of a genome in context with a specific question, but are complex to interpret, often associated with bioinformatic challenges, and might be associated with incidental findings.

As a result, single/multilocus tests are preferably used in diagnostic contexts, e.g., as biomarkers, whereas global (genome-wide) tests should not be applied in diagnostic settings, due to the unforeseeable incidental findings. However, these techniques are indispensable tools in research projects. For diagnostic purposes it has to be noted that technical, biological, and clinical factors can influence the diagnostic yield in IDs in any case and thereby limit the diagnostic detection rates.

Biological Aspects

Factors which might influence the interpretation value of DNA methylation results are tissue and cell types from which the DNA has been isolated. For many questions, it might be difficult to target the tissue which mediates the epigenetic regulation: examples are imprinting disorders where the molecular diagnosis is nearly always based on DNA methylation analysis from peripheral lymphocyte DNA. However, this tissue is relevant for the clinical features, and it might even be not affected by the epimutation. Thus, an epigenetics might be undetected though it is present in another tissue. This so-called mosaicism is a well-known phenomenon in several imprinting disorders and has a significant impact on the diagnostic detection rate (e.g., Alders et al. 2014; Russo et al. 2016).

The sensitivity and suitability to detect even slight differences in DNA methylation are generally major challenges as the variation of the epigenetic pattern between cell types and even cells in the same tissue is remarkable, and they are further influenced by the developmental stage of the organism and environmental factors. Many of the screening tests aim on the characterization of DNA samples from a collection of cells, and the measurements are thus based on average methylation level at selected single genomic locus across many DNA molecules. As a result, aberrant methylation in a specific cell type might escape detection, as it is covered by other cell types in the same sample.

A further aspect is the definition of the region of interest: for the same epigenetically regulated loci, a broad range of molecular techniques with different sensitivities is applied, but they often target different differentially methylated CpG dinucleotides and even different differentially methylated regions. Due to this lack of standardization, it is difficult to compare the molecular results between different studies and diagnostic tests. For that purpose, a consensus on the definition of region of interests is required, and the use of a common nomenclature, e.g., based on approved symbol from the Human Genome Organization (HUGO) Gene Nomenclature Committee (HGNC), is desirable. An example is the recently published consensus on the nomenclature for methylation aberrations in imprinted domains (Monk et al. 2016).

Depending on genomic organization of an epigenetically regulated region, the interpretation of methylation data can be influenced by the presence of copy number variations or SNPs; thus, the design of an assay and the data analysis should consider the consequences of such a genomic alteration. In particular in diagnostic testing of imprinting disorders, the discrimination between the different causes of an apparent aberrant methylation pattern (aberrant methylation itself, CNVs, UPD) should be possible (for review: Soellner et al. 2015).

Technical Aspects

One important aspect for the decision on the way of DNA pretreatment and MS assay is the DNA sample itself. The different methods differ markedly in sample requirements, comprising quality, quantity, and integrity. In particular those assays requiring large DNA fragments need highly integer and pure DNA, whereas others can provide results even on the basis of degraded DNA (Killian et al. 2009).

In addition to the DNA itself, the number of samples and the size of study cohort are other crucial aspects: in fact, whole-genome bisulfite sequencing is the most comprehensive method to provide information both on methylation statuses of single CpGs as on even slight methylation changes. However, currently, it is too cost-intensive for a broad routine application, and bioinformatic challenges are enormous and unsolved yet. Thus, many screening studies targeting large numbers of CpGs are performed by array analysis (i.e., 450 K arrays from Illumina), and this technique has been turned out to be robust and accurate. Its relative low running costs allow the analysis of sample numbers of a size with statistical power. In case of projects aiming on a limited number of CpGs, single or multilocus tests are the preferable techniques in respect to the costs and interpretation and data management.

In case of genome-wide assays, the bioinformatic analyses can be very complex, and in particular in NGS-based approaches targeting large number of CpGs, an enormous set of data ("big data") are generated, and standardized and validated solutions are currently missing. Indeed, massive parallel sequencing in multiple runs generates terabytes of data; thus, powerful tools for data analysis, storing, and managing are required. This is different in array-based approaches for which robust and comparable software solutions have been developed. However, in any case, the use of standard and control samples and the size of the study cohort/number of samples have to be considered.

Conclusion

The rapid development of a broad spectrum of assays aiming on the determination of DNA methylation allows the analysis of selected target sequences, but also genome-wide analysis of CpGs has become possible. In fact, in particular for the NGS-based approaches, the running costs are currently high per sample, and bioinformatic pipelines for interpretation have to be improved. However, comparable to genomic NGS assays, these problems will be circumvented, the prices will decrease, and standardized software solutions will become available.

Despite these technical and logistic aspects, the decision on the test for DNA methylation will depend on the question to be answered, and single or multilocus assays are often sufficient.

For many research projects, it is obvious that methylation analysis alone is not sufficient to enlighten the physiological role of DNA methylation patterns but that gene expression and histone modifications have to be considered. Thus, there are a growing number of studies combining approaches which provide data on interactions between DNA and its methylation, histones, and ncRNAs. Thereby, novel genes and pathomechanisms involving epigenetic regulation as well as biomarkers can be identified (e.g., Ibanez de Caceres et al. 2006; Veeck et al. 2009).

Among other fields, the translational use of DNA methylation assays has already been implemented in routine diagnostics of imprinting disorders and for biomarker determination in tumor surveillance and therapy. Nevertheless, there is a broad field of applications in medicine, i.e., in personalized medicine, including cancer therapy (for review: Soto et al. 2016) or psychopharmacology (Shin et al. 2016).

Dictionary of Terms

- **Affinity enrichment** – By the use of antibodies specific for 5-methylcytosine or methyl-binding proteins, 5-methylcytosine-rich domains can be enriched and characterized and quantified in subsequent assays.

- **Bisulfite conversion** – A DNA pretreatment method to differentiate between methylated 5-methylcytosine and unmethylated cytosine. It is based on a conversion of unmethylated cytosine to uracil, which is then determined in a second assay. Bisulfite conversion can be regarded as a targeted mutagenesis.
- **Genome-wide methylation assays** – Assays aiming on the determination of the DNA methylation of the whole genome and even on every single CpG in the genome. For the latter, NGS-based approaches should be used, as array-hybridization techniques are rather focused on methylation patterns of larger regions.
- **Imprinting disorders** – A group of congenital human disorders caused by molecular alterations affecting imprinted genes, i.e., genes which are epigenetically regulated depending on the parent of origin of the affected allele.
- **Methylome or whole-genome methylation** – The methylation marks of the whole genome.
- **MLPA/multiplex ligation probe depend amplification** – A method which has primarily been developed to determine copy numbers in DNA sample. It is based on the use of specific DNA probes (~60 bp), targeting the region of interest, which in case of hybridization serves as the template for a subsequent PCR. The amount of amplicons is then quantified. A modification of MLPA is methylation-specific MLPA, which also determines the methylation status by the use of methylation-specific restriction endonucleases.
- **Next-generation sequencing** – Term used for modern DNA sequencing platforms and assays which allow a massive parallel sequencing of DNA fragments for relatively low costs (per base pair) and very fast, in comparison to the standard base pair-per-base pair sequencing by the conventional Sanger sequencing.
- **Pyrosequencing** – A sequencing method based on the release of pyrophosphate during incorporation of the sequence-dependent nucleotide. The pyrophosphate is then used for light emission which is then used for sequence determination.
- **Restriction endonucleases** – DNA-digesting enzymes cutting DNA at a specific recognition site, i.e., a specific DNA sequence. This recognition site is unique for each enzyme.
- **Single and multilocus methylation assays** – The different techniques to determine the DNA methylation status can target only single CpGs (single-locus testing) or can analyze multiple CpGs in parallel (multilocus assays). These assays are different from genome-wide methylation assays.

Key Facts

- Due to the basic significance of DNA methylation patterns for biological processes, procedures to determine DNA methylation are rapidly developing.
- Three DNA pretreatment approaches are available for DNA methylation studies: MS restriction enzyme digestion, affinity enrichment, and bisulfite conversion.

- The introduction of bisulfite conversion of unmethylated cytosine residues to uracil was a revolutionizing step toward a comprehensive analysis of the human methylome.
- DNA methylation tests can be differentiated in tests aiming on single-loci analyses and assays covering multiple methylated loci, but there is an overlap of applications and capacities.
- The introduction of NGS in epigenetic research significantly contributes to an increased knowledge of already known methylated DNA regions, and it leads to the discovery of new epigenetically regulated genomic elements.

Summary Points

- This chapter focuses on molecular tests determining the DNA methylation from single CpG to genome-wide methylation resolution.
- Though different levels of epigenetic regulation are known, it is advantageous to focus on DNA methylation as DNA methylation is very stable, and it is currently assumed that environmental (and inborn) factors altering epigenetic patterns mainly affect DNA methylation.
- The determination of the DNA methylation status was hampered in the past because the methylation profile is not maintained during standard amplification processes but could be circumvented by the development of methylation-specific (MS) pretreatment protocols which are applied prior to specific MS approaches.
- The rapid progress in the field of DNA methylation analysis offers unique opportunities to comprehensively analyze the influence of epigenetic regulation on a broad spectrum of biological processes.
- A broad range of methods to determine DNA methylation statuses has been developed, but the decision on the test for DNA methylation will depend on the question to be answered and the sample size which should be investigated.
- In particular for NGS-based approaches, the running costs are currently high per sample, and bioinformatic pipelines for interpretation have to be improved, but these problems will be circumvented in the near future, and standardized software solutions will become available in the near future.

References

Alders M, Maas SM, Kadouch DJ et al (2014) Methylation analysis in tongue tissue of BWS patients identifies the (EPI) genetic cause in 3 patients with normal methylation levels in blood. Eur J Med Genet 57:293–297

Barski A, Cuddapah S, Cui K et al (2007) High-resolution profiling of histone methylations in the human genome. Cell 129:823–837

Beygo J, Ammerpohl O, Gritzan D et al (2013) Deep bisulfite sequencing of aberrantly methylated loci in a patient with multiple methylation defects. PLoS One 9:e76953

Bibikova M, Barnes B, Tsan C et al (2011) High density DNA methylation array with single CpG site resolution. Genomics 98:288–295

Bochtler M, Kolano A, Xu GL (2017) DNA demethylation pathways: additional players and regulators. BioEssays 39:1–13

Docherty LE, Rezwan FI, Poole RL et al (2015) Mutations in NLRP5 are associated with reproductive wastage and multilocus imprinting disorders in humans. Nat Commun 6:8086. https://doi.org/10.1038/ncomms9086

Eads CA, Danenberg KD, Kawakami K et al (2000) MethyLight: a high-throughput assay to measure DNA methylation. Nucleic Acids Res 28:E32

Eggermann K, Bliek J, Brioude F et al (2016) EMQN best practice guidelines for the molecular genetic testing and reporting of chromosome 11p15 imprinting disorders: Silver-Russell and Beckwith-Wiedemann syndrome. Eur J Hum Genet 24:1377–1387

Enklaar T, Zabel BU, Prawitt D (2006) Beckwith-Wiedemann syndrome: multiple molecular mechanisms. Expert Rev Mol Med 8:1–19

Frommer M, McDonald LE, Millar DS et al (1992) A genomic sequencing protocol that yields a positive display of 5-methylcytosine residues in individual DNA strands. Proc Natl Acad Sci U S A 89:1827–1831

Goodwin S, McPherson JD, McCombie WR (2016) Coming of age: ten years of next-generation sequencing technologies. Nat Rev Genet 17:333–351

Gu H, Bock C, Mikkelsen TS, Jäger N et al (2010) Genome-scale DNA methylation mapping of clinical samples at single-nucleotide resolution. Nat Methods 7:133–136

Heijmans BT, Tobi EW, Stein AD et al (2008) Persistent epigenetic differences associated with prenatal exposure to famine in humans. Proc Natl Acad Sci U S A 105:17046–17049

Holliday R (1990) DNA methylation and epigenetic inheritance. Philos Trans R Soc Lond Ser B Biol Sci 326:329–338

Ibanez de Caceres I, Dulaimi E, Hoffman AM et al (2006) Identification of novel target genes by an epigenetic reactivation screen of renal cancer. Cancer Res 66:5021–5028

Kelsey G, Feil R (2013) New insights into establishment and maintenance of DNA methylation imprints in mammals. Philos Trans R Soc Lond Ser B Biol Sci 368:20110336. https://doi.org/10.1098/rstb.2011.0336

Killian JK, Bilke S, Davis S et al (2009) Large-scale profiling of archival lymph nodes reveals pervasive remodeling of the follicular lymphoma methylome. Cancer Res 69:758–764

Laird PW (2010) Principles and challenges of genomewide DNA methylation analysis. Nat Rev Genet 1(1):191–203

Mackay DJ, Callaway JL, Marks SM et al (2008) Hypomethylation of multiple imprinted loci in individuals with transient neonatal diabetes is associated with mutations in ZFP57. Nat Genet 40:949–951

McCarrey JR, Lehle JD, Raju SS (2016) Tertiary epimutations – a novel aspect of epigenetic transgenerational inheritance promoting genome instability. PLoS One 11(12):e0168038

McCullough LE, Miller EE, Mendez MA et al (2016) Maternal B vitamins: effects on offspring weight and DNA methylation at genomically imprinted domains. Clin Epigenetics 8:8. https://doi.org/10.1186/s13148-016-0174-9

Melnikov AA, Gartenhaus RB, Levenson AS et al (2005) MSRE-PCR for analysis of gene-specific DNA methylation. Nucleic Acids Res 33:e93

Monk D, Morales J, den Dunnen JT, et al (2016) Recommendations for a nomenclature system for reporting methylation aberrations in imprinted domains. Epigenetics [Epub ahead of print]

Paul CL, Clark SJ (1996) Cytosine methylation: quantitation by automated genomic sequencing and GENESCAN analysis. BioTechniques 21:126–133

Ronaghi M, Uhlén M, Nyrén PA (1998) Sequencing method based on real-time pyrophosphate. Science 281:363

Russo S, Calzari L, Mussa A et al (2016) A multi-method approach to the molecular diagnosis of overt and borderline 11p15.5 defects underlying Silver-Russell and Beckwith-Wiedemann syndromes. Clin Epigenetics 8:23

Schouten JP, McElgunn CJ, Waaijer R et al (2002) Relative quantification of 40 nucleic acid sequences by multiplex ligation-dependent probe amplification. Nucleic Acids Res e57:30

Schumacher A, Kapranov P, Kaminsky Z et al (2006) Microarray-based DNA methylation profiling: technology and applications. Nucleic Acids Res 34:528–542
Shin C, Han C, Pae CU et al (2016) Precision medicine for psychopharmacology: a general introduction. Expert Rev Neurother 16(7):831–839
Soellner L, Monk D, Rezwan FI et al (2015) Congenital imprinting disorders: application of multilocus and high throughput methods to decipher new pathomechanisms and improve their management. Mol Cell Probes 29:282–290
Soellner L, Begemann M, Mackay DJ et al (2016) Recent advances in imprinting disorders. Clin Genet. https://doi.org/10.1111/cge.12827
Soto J, Rodriguez-Antolin C, Vallespín E et al (2016) The impact of next-generation sequencing on the DNA methylation-based translational cancer research. Transl Res 169:1–18.e1
Stuppia L, Franzago M, Ballerini P et al (2015) Epigenetics and male reproduction: the consequences of paternal lifestyle on fertility, embryo development, and children lifetime health. Clin Epigenetics 7:120
Taylor KH, Kramer RS, Davis JW et al (2007) Ultradeep bisulfite sequencing analysis of DNA methylation patterns in multiple gene promoters by 454 sequencing. Cancer Res 67:8511–8518
Uyar A, Seli E (2014) The impact of assisted reproductive technologies on genomic imprinting and imprinting disorders. Curr Opin Obstet Gynecol 26:210–221
Veeck J, Wild PJ, Fuchs T et al (2009) Prognostic relevance of Wnt-inhibitory factor-1 (WIF1) and Dickkopf-3 (DKK3) promoter methylation in human breast cancer. BMC Cancer 9:217
Xiong Z, Laird PW (1997) COBRA: a sensitive and quantitative DNA methylation assay. Nucleic Acids Res 25:2532–2534
Zeschnigk M, Albrecht B, Buiting K et al (2008) IGF2/H19 hypomethylation in Silver-Russell syndrome and isolated hemihypoplasia. Eur J Hum Genet 16:328–334

Illumina HumanMethylation BeadChip for Genome-Wide DNA Methylation Profiling: Advantages and Limitations

115

Kazuhiko Nakabayashi

Contents

Abstract

HumanMethylation BeadChip is an array platform for highly multiplexed measurement of DNA methylation at individual CpG locus in the human genome based on Illumina's bead technology. It measures the DNA methylation level of individual CpG site by quantitative genotyping of C/T polymorphisms generated in bisulfite-converted and amplified genomic DNA. The current version, HumanMethylationEPIC, measures the DNA methylation level of >850,000 CpG sites, while the previous versions, HumanMethylation450 (HM450) and

K. Nakabayashi (✉)
Division of Developmental Genomics, Department of Maternal-Fetal Biology, National Research Institute for Child Health and Development, Tokyo, Japan
e-mail: nakabaya-k@ncchd.go.jp

V. B. Patel, V. R. Preedy (eds.), *Handbook of Nutrition, Diet, and Epigenetics*,
https://doi.org/10.1007/978-3-319-55530-0_89

HumanMethylation27 (HM27), measured that of >480,000 and >27,000 CpG sites, respectively. HumanMethylation BeadChip requires only 4 days to produce methylome profiles of human samples using 250–500 ng of genomic DNA as a starting material. Because of its time and cost efficiency, high sample output, and overall quantitative accuracy and reproducibility, HM450 has become the most widely used means of large-scale methylation profiling of human samples in recent years. However, it is important to consider potential confounders originating in the technical limitations of HumanMethylation BeadChip such as cross-reactive probes, SNP-affected probes, within-array bias (Infinium I and II bias), and between-array bias (batch effects) especially when subtle methylation differences need to be detected by statistical tests between large numbers of cases and controls. Many integrated analysis packages have been developed by the epigenetics research community as computational solutions for technical and biological confounders associated with HumanMethylation BeadChip data. Considering the substantial increase of the coverage of regulatory regions along with the advantages inherited from HM450, EPIC is expected to maintain its popularity as a platform for epigenome-wide association studies for the foreseeable future.

Keywords
DNA methylation · Bisulfite conversion · Methylome · Infinium assay · Array · Epigenome-wide association study · Genomic imprinting · Data normalization · Batch effects · SNP genotyping

List of Abbreviations

5-mC	5-methylcytosine
CpG	Cytosine-guanine
ddNTP	Dideoxynucleotide triphosphate
DNP	Dinitrophenol
EWAS	Epigenome-wide association study
NGS	Next-generation sequencing
nt	Nucleotide
SNP	Single-nucleotide polymorphism
UTR	Untranslated region

Introduction

DNA methylation plays a critical role in the regulation of gene expression during development and differentiation. Global hypomethylation and regional hypermethylation at specific gene promoters are observed in many cancers (Kelly et al. 2010). Aberrant DNA methylation has also been described in many other diseases such as neurological, autoimmune, and metabolic diseases (Wen et al. 2016; Zhang and Zhang 2015; de Mello et al. 2014). In mammals, DNA methylation primarily occurs at the carbon 5 position of the cytosine of the CpG dinucleotide and produces

5-mC. Whole-genome and reduced representation bisulfite sequencing has up to now been the most common method of genome-wide DNA methylation profiling (Stirzaker et al. 2014). These NGS-based methods work for any species if genomic sequence information of the species is available. Illumina HumanMethylation BeadChip is an alternative method of DNA methylation profiling of human samples. While this array-based method does not interrogate DNA methylation as comprehensively as whole-genome bisulfite sequencing, it has two key advantages over the NGS-based methods: cost-effectiveness and high sample throughput. Such features have made HumanMethylation450 BeadChip and its successor HumanMethylationEPIC the first choice for large-scale epigenome-wide association studies in which hundreds or thousands of human samples are enrolled. This chapter outlines the basic principles and the experimental and analytical procedures of the HumanMethylation BeadChip and covers strengths and limitations of this array-based platform that are widely recognized by the epigenetics research community.

Basic Principles Underlying the Infinium Methylation Assay

Illumina's bead technology (Gunderson et al. 2004) enables highly multiplexed measurement of DNA methylation at individual CpG locus in the human genome. HumanMethylation BeadChip measures DNA methylation levels at the cytosine of CpG sites using quantitative genotyping of C/T polymorphisms in bisulfite-converted genomic DNA followed by whole-genome amplification: unmethylated cytosines are converted to uracils (and amplified as thymines), and methylated cytosines are protected and detected as cytosines (Hayatsu 2008).

HumanMethylation BeadChip uses two assay designs, Infinium I and Infinium II (Fig. 1). Infinium assays consist of the following three major steps: hybridization of the bisulfite-converted and amplified DNA fragments to 50-mer oligonucleotides attached to each bead, single-base extension at the 3′ terminus of the oligonucleotide using fluorescently labeled ddNTPs, and the measurement of fluorescent intensity for each bead. The Infinium I assay uses two bead types per CpG locus: one for measuring the methylated state and the other for the unmethylated state. The 3′ terminus of 50-mer probes is designed to match the interrogated site of either cytosine or thymine. A single-base extension occurs at the base immediately adjacent to the interrogated CpG site. Probes for Infinium I assays are designed on the assumption that methylation is regionally correlated within a 50 base interval. This "comethylation" assumption adopted by Illumina is based on the results of a large-scale bisulfite sequencing study showing that >90% of CpG sites within 50 bases had the same methylation status (Eckhardt et al. 2006) and of another study demonstrating that the methylation status of adjacent CpG sites tends to be correlated (Shoemaker et al. 2010). The Infinium II assay requires only one bead type per CpG locus (Fig. 1). The 3′ terminus of 50-mer probes is cytosine complementary to the guanine of the query CpG site. A single-base extension of labeled G or A base occurs, depending on the complementarity to either the methylated C or unmethylated T. Illumina determined that Infinium II probes can contain up to 3

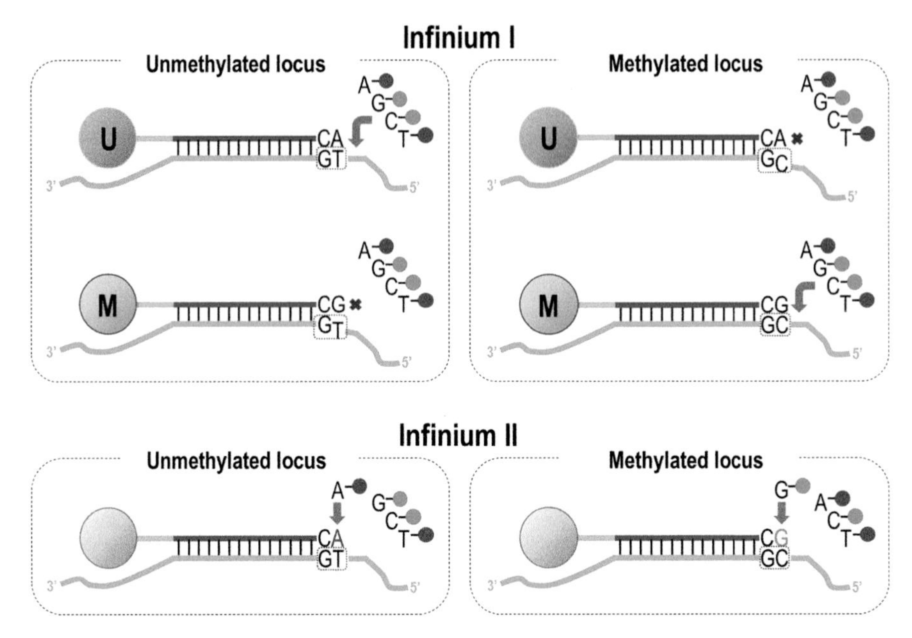

Fig. 1 **Infinium assay designs**. A figure in an Illumina's technical note (http://www.illumina.com/documents/products/technotes/technote_hm450_data_analysis_optimization.pdf) was reproduced with slight modifications. Bisulfite-converted DNA and interrogated CpG sites are indicated by green lines and the boxes outlined in red dots, respectively. DNP and biotin conjugated to ddNTPs are shown as red and green circles. One of the four types of ddNTPs (DNP-labeled ddATP/ddTTP and biotin-labeled ddCTP/ddGTP) is incorporated at the 3′ end of the probe oligonucleotides when the last two bases of the probe oligonucleotides are complementary to the interrogated CpG site in bisulfite-converted DNA. In the Infinium I assay design, single-base extension occurs at the base adjacent to the interrogated CpG site. The same type of dideoxynucleotide is incorporated in both the unmethylated and methylated loci. However, in the Infinium II assay design, the single-base extension occurs at the C/T nucleotides of the interrogated CpG site. Either DNP-labeled ddATP or biotin-labeled ddGTP is incorporated depending on the methylation state of the interrogated CpG site. Incorporated ddNTPs are immunofluorescently detected using Cy5-labeled anti-DNP and Cy3-labeled streptavidin followed by signal amplification. Therefore, while the same color channel (red or green) is used to measure the fluorescent intensities for the unmethylated and methylated status of the interrogated CpG site in the Infinium I assay design, in the Infinium assay II design, two color channels are used, red for the unmethylated status and green for the methylated status of the interrogated CpG site. Dye bias between two color channels can be adjusted using signal intensities of built-in control probes

CpG sites within a 50-mer probe sequence and that the underlying CpG sites may be represented by "degenerate" R-bases (A or G base). This feature makes it possible to assess the methylation status of a query site independently of assumptions on the status of neighboring CpG sites and to increase the number of CpG sites interrogated. While the Infinium II design was applied whenever possible, the Infinium I design was required to measure the DNA methylation levels of CpG sites within CpG-rich regions (i.e., CpG islands).

Manufacture and Decoding of BeadChip Arrays

BeadChip technology is based on 3 micron silica beads that self-assemble in microwells with a uniform spacing of approximately 5.7 microns on planar silica slides (Fan et al. 2005). Each bead in the Infinium assay holds hundreds of thousands of copies of a specific oligonucleotide comprising an address code and a 50-mer probe sequence (Fig. 2). All types of beads are pooled and randomly assembled in the microwells by Van der Waals forces and hydrostatic interactions with the walls of the well on the array slide. The arrays are subsequently subjected to the decoding procedure, which determines the positions of all types of beads in each array (Fan et al. 2005). Sequential hybridizations of fluorescently labeled decoder oligonucleotides (with four distinguishable labels) to the address code of the beads impart a color signature specifying a bead type to each of the beads on the array. The information of the bead locations for each array is provided as a DMAP file. In the case of HumanMethylation BeadChip, a median of 14 beads is randomly distributed on the array. The presence of multiple beads of each type provides quantitative accuracy through multiple measurements with statistical processing. The random distribution

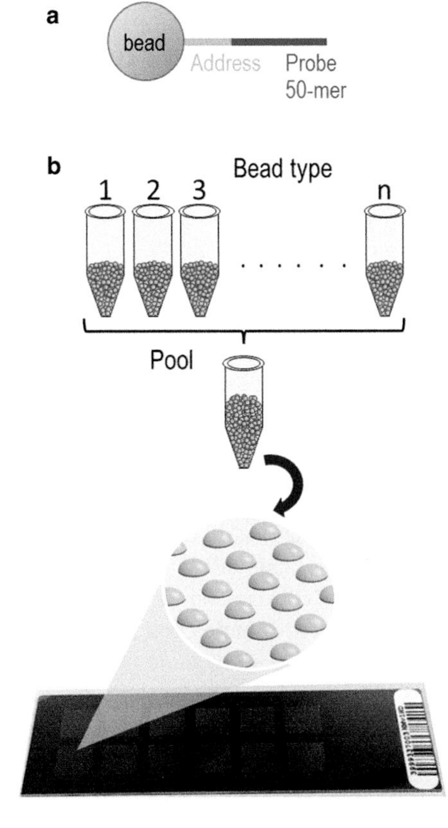

Fig. 2 Manufacture of BeadChip arrays. **(a)** Illumina's BeadChip technology uses 3 micron silica beads attached to hundreds of thousands of copies of a specific oligonucleotide comprising an address code and a 50-mer probe sequence (depicted in orange and blue, respectively). **(b)** In the case of HumanMethylation450 BeadChip, over 620,000 types of beads are generated, pooled, and self-assembled onto the array. On average, 14 copies of each bead type are present on an array. Each array is subjected to the decoding procedure, which determines the positions of each type of beads for each array (Gunderson et al. 2004)

of the beads increases assay robustness by minimizing the chance of any local failure affecting the overall result (Gunderson et al. 2004). In the randomly assembled arrays, the number of each type of bead is a random variable with a Poisson distribution. When the number of a bead is less than three (i.e., 0, 1, or 2), the methylation beta value is not calculated for the corresponding CpG site by the GenomeStudio software (Illumina). In the case of HumanMethylation450 BeadChip, several hundred sites (out of >480,000 CpG sites) per sample inevitably lack the methylation beta value due to the bead number threshold.

Experimental Procedures

Experimental procedures using HumanMethylation BeadChip include sodium bisulfite conversion of genomic DNA (day 1), whole-genome amplification (day 2) followed by DNA fragmentation, hybridization of bisulfite-converted and amplified DNA to BeadChip (day 3), and washing, staining, and scanning (day 4) (Carless 2015). The staining procedure includes a single-base extension using labeled ddNTPs (DNP-labeled ddATP and ddTTP and biotin-labeled ddCTP and ddGTP), fluorescent detection of DNP and biotin with Cy5-labeled anti-DNP and Cy3-labeled streptavidin, and signal amplification. The stained BeadChip is scanned using iScan or HiScan SQ system (Illumina) to measure the intensities of methylated and unmethylated signals for each of the target CpG sites. In this scanning procedure, decoded data (DMAP) files containing bead-type information for each position on the BeadChip are required to obtain raw data files (IDAT files). The GenomeStudio software (methylation module) calculates the methylation level of individual CpG sites from IDAT files as beta value ranging from 0 (completely unmethylated) to 1 (completely methylated).

HumanMethylation27, 450, and EPIC BeadChip Arrays

HumanMethylation27 (HM27) BeadChip, the first version of HumanMethylation BeadChip, was released in 2008 and contained Infinium I probes for 27,578 CpG sites located within the promoter regions of 14,475 consensus coding sequences (CCDS) genes and well-known cancer genes (Bibikova et al. 2009). HumanMethylation450 (HM450) BeadChip was released in 2011. HM450 was designed to include 485,577 assays (482,421 CpG sites, 3091 non-CpG sites, and 65 random SNPs) containing 25,978 (94%) of CpG sites present on HM27. The new content covered a more diverse set of genomic features: three CpG island (CGI)-related categories, CGI, shore (0–2 kb from CGI), and shelf (2–4 kb from CGI), and six gene feature categories, namely, TSS200 (from transcription start site (TSS) to −200 nt upstream of TSS), TSS1500 (−200 to −1500 nt upstream of TSS), 5′UTR, first exon, gene body, and 3′UTR, which were introduced to probe annotation. HM450 provides coverage of a total of 21,231 out of 21,474 UCSC RefGenes (99%) and 26,658 (96%) CGIs (Bibikova et al. 2011). This improved coverage of

HM450 along with the advantages of the HumanMethylation BeadChip platform, time and cost-effectiveness, and high sample throughput resulted in the use of HM450 in many studies of various types. Examples of the HM450 studies include those of genomic imprinting (Court et al. 2014), X inactivation (Cotton et al. 2015), myoblast differentiation (Miyata et al. 2015), aging (Florath et al. 2014), and many types of cancers. Examples of EWAS, in which the DNA methylation profiles of over a thousand subjects were obtained by HM450, are those for rheumatoid arthritis (Liu et al. 2013), metabolic traits (Dick et al. 2014, Petersen et al. 2014), cardiovascular diseases (Rask-Andersen et al. 2016), and diabetes (Florath et al. 2016). Other examples of large-scale projects that used HM450 are the ARIES study, a population-based methylome project that profiled 1000 mother and child pairs (Relton et al. 2015), and the Cancer Genome Atlas Consortium (TCGA, http://cancergenome.nih.gov/), which profiled about 8000 samples from over 200 cancer types.

DNA methylation at distal regulatory elements, particularly enhancers and transcription factor binding sites, has been shown to be dynamically regulated among different tissues and cell types (Ziller et al. 2013; Gu et al. 2016). The latest version, HumanMethylationEPIC BeadChip, was released in 2015. EPIC covers >850,000 CpG sites including >90% of HM450 content and 413,743 additional CpG sites. The newly added CpG sites include 35,000 CpG sites located at potential enhancers identified by the FANTOM5 project (http://fantom.gsc.riken.jp/5/) and the ENCODE project (http://www.encodeproject.org/). Pidsley et al. (2016) reported that EPIC probes cover 58% of FANTOM5 enhancers and 7% distal and 27% proximal ENCODE regulatory elements and that a single EPIC probe does not always represent the methylation level of distal enhancer elements, which tend to show variable methylation levels across a region. However, because of the substantial increase of coverage of regulatory regions along with the advantages it has inherited from HM450, EPIC is expected to maintain its popularity as an EWAS platform. The major features of HM27, HM450, and EPIC BeadChip arrays are summarized in Table 1. The genomic locations of the CpG probes contained in HM27, HM450, and EPIC BeadChip arrays are shown for a 36 kb genomic interval including the *TAL1* gene (Fig. 3) to exemplify the difference in probe distribution and density among the three types of HumanMethylation BeadChip arrays.

Data Analysis

This section describes the chief data processing procedures generally recommended for an HM450/EPIC dataset: probe filtering, within-array normalization, and between-array normalization (batch effect correction). These procedures, if correctly conducted, minimize variance and improve statistical power for detecting small DNA methylation changes. Several bioconductor or R packages that perform all or most of the procedures are available for free (Table 2) including methylumi, minfi, waterRmelon, ChAMP, and RnBeads, which have been reviewed by Morris and Beck (2015). JIllumina is an open-source Java library recently released for the handling and processing of HM450/EPIC raw data (Almeida et al. 2016) (Table 2)

Table 1 Comparison of probe content and the number of use of HM27, HM450 and EPIC

	EPIC (2015)	HM450 (2011)	HM27 (2009)
Target C	867,867	485,512	27,578
Target CpG	864,935	482,421	27,578
Target CpG_Infinium I	143,293	135,476	27,578
Target CpG_Infinium II	721,642	346,945	0
Non-CpG target	2,932	3,091	0
SNP probes	59	65	32
Control probes	638	850	112
Sample number/array	8	12	12
Genomic DNA required	250 ng	500 ng	1000 ng
GEO platform ID	GPL21145	GPL13534 GPL16304	GPL8490
Samples registered in GEO (as of march 2017)	70	56,543 2,268	18,708

GEO: Gene Expression Omnibus (http://www.ncbi.nlm.nih.gov/geo/)
Probe annotation files are available at GEO

and forms the backbone of DiMmer (Almeida et al. 2017), a graphical user interface software that interactively guides EWAS data analysis procedures. Although Illumina's GenomeStudio is useful for analyzing a small number of samples, its functions are limited compared to the software packages listed above.

Step 1: Filtering Problematic Probes

Removal of probes with high detection p-values (probes that have failed to hybridize) is generally recommended. The detection p-value is defined as "1-R/N," where R is the relative rank of the signal intensity relative to the negative controls and N is the number of negative controls (N = 614 in the case of HM450). Illumina recommends the removal of probes with a detection p-value >0.05. In addition, removing probes with missing beta values, probes containing common SNPs, and cross-reactive probes should also be considered. The Infinium methylation assay is based on the quantitative genotyping of C/T polymorphisms generated at CpG sites after bisulfite treatment. When C/T SNP exists at the target CpG sites, the genomic TG allele behaves as unmethylated CpG. Therefore, probes with target CpG sites overlapping a common SNP (minor allele frequency > 0.05) should be treated with caution in studies enrolling multiple human subjects. HumanMethylation BeadChip relies on the hybridization of bisulfite-treated DNA to 50-mer oligos attached to the beads. Bisulfite conversion of the majority of Cs to T generates "three-letter genome sequences," increasing the probability of probe cross-reactivity, i.e., hybridization of a probe to loci other than its primary target. Depending on the criteria used, 8–25% of the HM450 probes were classified as cross-reactive (Price et al. 2013; Chen et al. 2013). Therefore, when differentially methylated probes are extracted by the comparison of HumanMethylation data between groups of interest without prior

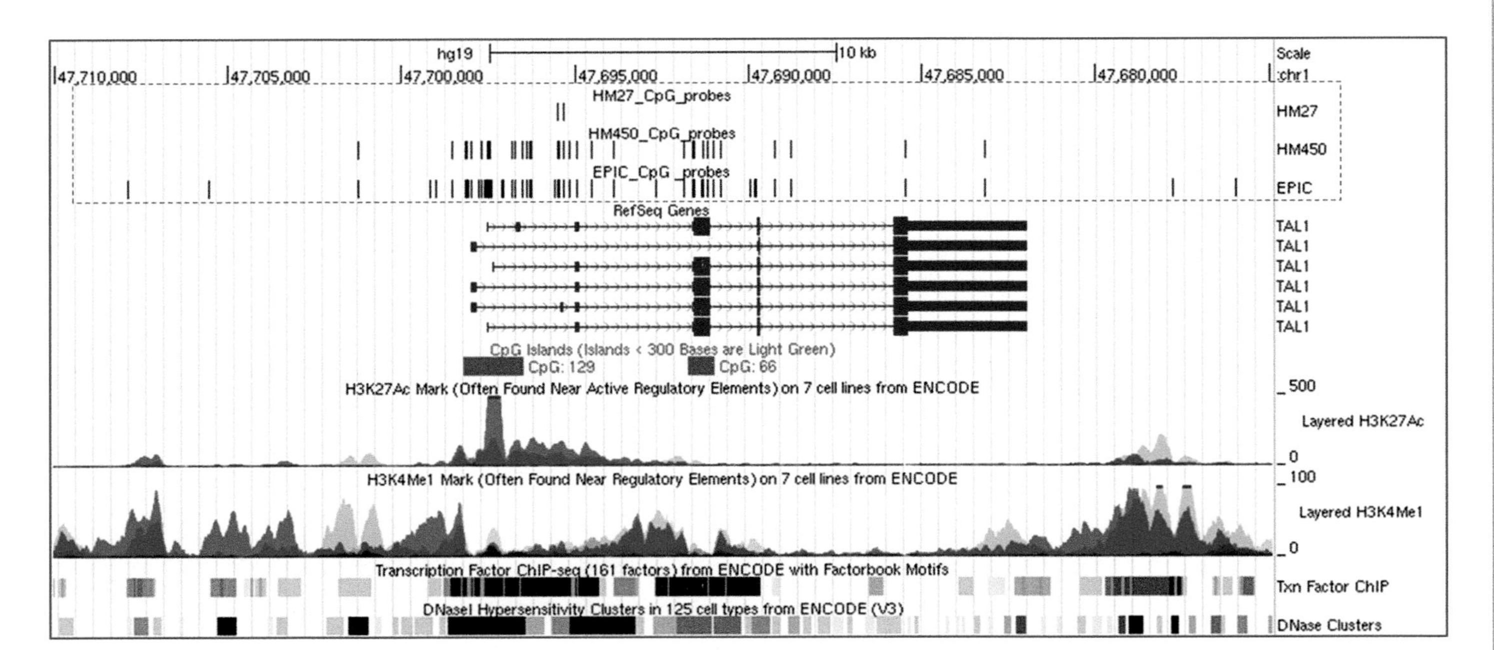

Fig. 3 Distribution of CpG probes included in three types of HumanMethylation BeadChip arrays within a 36 kb genomic interval. The genomic interval of chr1:47,674,001–47,710,000 (hg19) including the *TAL1* gene is shown using UCSC Genome Browser (http://genome.ucsc.edu/) (Kent et al. 2002). The locations of the CpG probes contained in HM27, HM450, and EPIC BeadChip arrays within this interval were shown as vertical black lines by uploading the genomic positions of the probes in the BED (Browser Extensible Data) format using the custom track tool of the Browser. The distribution of the CpG probes in this interval exemplifies the wider coverage of enhancer regions, which are positive for H3K4Me1 signals, by EPIC probes than by HM450 probes

Table 2 Features of data analysis packages for HumanMethylation BeadChip data

	GenomeStudio (Illumina)	Minfi	WaterMelon	ChAMP	RnBeads	JIllumina
Import raw data (IDAT) files	Yes	Yes	Yes	Yes	Yes	Yes
Probe filtering 1. Detection p-value 2. Bead count 3. SNP-containing probes 4. Cross-reactive probes	Yes (1 and 2 only)	Yes	Yes	Yes	Yes	Yes
Within-array normalization: 1. Background correction 2. Type II bias adjustment	Yes (1 only)	Yes	Yes	Yes	Yes	Yes
Between-array normalization: Batch effect correction	No	Yes	No	Yes	Yes	No
Cell composition correction	No	Yes	No	Yes	Yes	Yes
Differentially methylated probes	Yes	Yes	No	Yes	Yes	Yes
Differentially methylated regions	No	Yes	No	Yes	Yes	Yes
Copy number analysis	No	No	No	Yes	No	No

exclusion of SNP-containing and cross-reactive probes, it is important to consider the possibility that such problematic probes may have been falsely identified as differentially methylated.

Step 2: Within-Array Normalization

Removing variations of a technical, rather than a biological, origin is crucial. Within-array normalization is performed for background correction, dye-bias adjustment, and type II bias adjustment (Yousefi et al. 2013; Dedeurwaerder et al. 2014). Type II bias is the foremost issue originating from the two different assay designs, Infinium I and Infinium II, used in HM450 and EPIC. The methylation values derived from these two designs were shown to exhibit different distributions. Infinium II probes showed a lower dynamic range compared with Infinium I probes (Dedeurwaerder et al. 2011; Bibikova et al. 2011). Furthermore, methylation values of the Infinium II probes were shown to be biased and less reproducible in comparison with bisulfite pyrosequencing data (Dedeurwaerder et al. 2011). This issue has led to the development of a variety of algorithms to adjust for type II bias (reviewed in Morris and Beck 2015).

Step 3: Between-Array Normalization

Batch effects represent nonbiological, technical variations observed across different batches of experiments and are common among high-throughput data including HumanMethylation BeadChip data (Leek et al. 2010; Sun et al. 2011). Differences in experiment dates, reagent lots, laboratory conditions, and personnel can cause batch effects. In large-scale studies, in which BeadChip experiments need to be performed with many arrays at different times, batch effects may not always be avoidable. Batch effects are also expected when data from multiple institutes are analyzed together. Several algorithms are available for batch effect correction for HumanMethylation BeadChip data including ComBat, surrogate variable analysis (SVA), RUVm, functional normalization, and BEclear (see the references in Akulenko et al. 2016).

Lie and Siegmund (2016) evaluated the effects of nine processing methods including both within- and between-array normalization methods and found that within-array normalization using Noob (Triche et al. 2013) and BMIQ (Teschendorff et al. 2013) consistently improved signal sensitivity and that RUVm (Maksimovic et al. 2015) outperformed the other batch effect correction algorithms in a combination analysis using Noob and BMIQ. Inclusion of batch effect correction depends on the degree of methylation differences expected between the groups compared; the process is facilitated if variations stemming from a technical origin outnumber those stemming from biological signals, but in the opposite situation, some batch effect correction tools may remove biological signals (Liu and Siegmund 2016).

In addition to the options for probe filtering and within- and between-array normalization, many of the current software packages (Table 2) include options for correcting for cell heterogeneity, another potential confounding factor associated with methylome studies using human population samples. Whole blood is frequently chosen as a source of genomic DNA in EWAS. Because blood is a heterogeneous collection of different cell types, the associations detected in such studies may be confounded by cellular heterogeneity (Jaffe and Irizarry 2014).

After appropriate data preprocessing, differentially methylated probes (DMP) or regions (DMR) between groups of interest can be searched. Probe annotations such as those of Price et al. (2013) and Zhou et al. (2017) are useful for the biological interpretation of identified DMPs/DMRs. The functional annotation provided by Zhou et al. (2017) includes the positions of HM27/HM450/EPIC probes relative to imprinted differentially methylated regions (Court et al. 2014), transcription factor binding sites identified by the ENCODE project, and chromatin states identified by the Roadmap Epigenomics Project (http://www.roadmapepigenomics.org/).

Conclusions

The HumanMethylationEPIC array represents a significant improvement over the HM450 array in terms of genome coverage of regulatory regions, consistent reproducibility, and reliability. Therefore, this platform will serve as a valuable tool in large-scale human methylome studies requiring identification of disease biomarkers and/or etiology and assessment of the impact of environmental exposures. Many integrated analysis packages have been developed as computational solutions for technical confounders originating in the underlying principle and design of HumanMethylation BeadChip as well as biological cofounders inherent in human biology.

Dictionary of Terms

- **Methylome** – The genome-wide profile of DNA methylation.
- **EWAS** – Epigenome-wide association study for uncovering epigenetic variants underlying common diseases.
- **Bisulfite conversion** – Sodium bisulfite conversion of genomic DNA to differentiate and detect unmethylated versus methylated cytosine.
- **Infinium assay** – Illumina's SNP genotyping technology based on on-bead single-base extension and fluorescent detection.
- **Methylation beta value** – Beta = methylated intensity/ (unmethylated intensity + methylated intensity +100).
- **Detection p-value** – Threshold value for filtering out probes that failed to hybridize in HumanMethylation BeadChip experiments.
- **Batch effects** – Nonbiological experimental variation observed across multiple batches of microarray experiments.

Key Facts of HumanMethylation BeadChip

- HumanMethylation BeadChip is an array platform for highly multiplexed measurement of DNA methylation at individual CpG loci in the human genome based on Illumina's bead technology.
- The platform measures the DNA methylation level of individual CpG sites by quantitative genotyping of C/T polymorphisms generated in bisulfite-converted and amplified genomic DNA.
- The current version, HumanMethylation EPIC, measures the DNA methylation level of $>$850,000 CpG sites, while previous versions, HumanMethylation450 and HumanMethylation27, measured that of $>$480,000 and $>$27,000 CpG sites, respectively.
- HumanMethylation BeadChip requires only 4 days to produce methylome profiles of human samples using 250–500 ng of genomic DNA as a starting material.
- Because of its time and cost efficiency, high sample output, and overall quantitative accuracy and reproducibility, HumanMethylation450 was the most widely used means of large-scale methylation profiling of human samples in recent years.
- Considering the substantial increase of the coverage of regulatory regions along with the advantages inherited from HumanMethylation450, HumanMethylation EPIC is expected to maintain its popularity as a platform for epigenome-wide association studies for the foreseeable future.

Summary Points

- Illumina HumanMethylation BeadChip is a popular platform for obtaining DNA methylome data from human samples.
- The current version, HumanMethylationEPIC, measures the DNA methylation levels of over 850,000 CpG sites by highly multiplexed quantitative genotyping of C/T polymorphisms in bisulfite-converted genomic DNA.
- HumanMethylation450 was very widely used in EWAS projects, and its successor, EPIC, is expected to remain the preferred tool for the foreseeable future in studies with large sample numbers due to its cost and time effectiveness and high sample throughput.
- Considering potential confounders originating in the technical limitations of HumanMethylation BeadChip such as cross-reactive probes, SNP-affected probes, within-array bias (Infinium I and II bias), and between-array bias (batch effects) is of particular importance, especially when subtle methylation differences need to be detected by statistical tests between large numbers of cases and controls.
- Many integrated analysis packages have been developed by the epigenetics research community as computational solutions for technical and biological confounders associated with HumanMethylation BeadChip data.

References

Akulenko R, Merl M, Helms V (2016) BEclear: batch effect detection and adjustment in DNA methylation data. PLoS One 11:e0159921

Almeida D, Skov I, Lund J et al (2016) Jllumina - a comprehensive java-based API for statistical Illumina Infinium HumanMethylation450 and MethylationEPIC data processing. J Integr Bioinform 13:294

Almeida D, Skov I, Silva A et al (2017) Efficient detection of differentially methylated regions using DiMmeR. Bioinformatics 33:549–551

Bibikova M, Le J, Barnes B et al (2009) Genome-wide DNA methylation profiling using Infinium® assay. Epigenomics. 1:177–200

Bibikova M, Barnes B, Tsan C et al (2011) High density DNA methylation array with single CpG site resolution. Genomics 98:288–295

Carless MA (2015) Determination of DNA methylation levels using Illumina human methylation BeadChips. In: Chellappan SP (ed) Chromatin protocols, Methods in molecular biology, vol 1288. Springer, New York, pp 143–192

Chen YA, Lemire M, Choufani S et al (2013) Discovery of cross-reactive probes and polymorphic CpGs in the Illumina Infinium HumanMethylation450 microarray. Epigenetics 8:203–209

Cotton AM, Price EM, Jones MJ et al (2015) Landscape of DNA methylation on the X chromosome reflects CpG density, functional chromatin state and X-chromosome inactivation. Hum Mol Genet 24:1528–1539

Court F, Tayama C, Romanelli V et al (2014) Genome-wide parent-of-origin DNA methylation analysis reveals the intricacies of human imprinting and suggests a germline methylation-independent mechanism of establishment. Genome Res 24:554–569

Dedeurwaerder S, Defrance M, Bizet M et al (2014) A comprehensive overview of Infinium HumanMethylation450 data processing. Brief Bioinform 15:929–941

Dedeurwaerder S, Defrance M, Calonne E et al (2011) Evaluation of the Infinium methylation 450K technology. Epigenomics 3:771–784

de Mello VD, Pulkkinen L, Lalli M et al (2014) DNA methylation in obesity and type 2 diabetes. Ann Med 46:103–113

Dick KJ, Nelson CP, Tsaprouni L et al (2014) DNA methylation and body-mass index: a genome-wide analysis. Lancet 383:1990–1998

Eckhardt F, Lewin J, Cortese R et al (2006) DNA methylation profiling of human chromosomes 6, 20 and 22. Nat Genet 38:1378–1385

Fan JB, Hu SX, Craumer WC et al (2005) BeadArray-based solutions for enabling the promise of pharmacogenomics. BioTechniques 39:583–588

Florath I, Butterbach K, Müller H et al (2014) Cross-sectional and longitudinal changes in DNA methylation with age: an epigenome-wide analysis revealing over 60 novel age-associated CpG sites. Hum Mol Genet 23:1186–1201

Florath I, Butterbach K, Heiss J et al (2016) Type 2 diabetes and leucocyte DNA methylation: an epigenome-wide association study in over 1,500 older adults. Diabetologia 59:130–138

Gunderson KL, Kruglyak S, Graige MS et al (2004) Decoding randomly ordered DNA arrays. Genome Res 14:870–877

Gu J, Stevens M, Xing X et al (2016) Mapping of variable DNA methylation across multiple cell types defines a dynamic regulatory landscape of the human genome. G3 (Bethesda) 6:973–986

Hayatsu H (2008) Discovery of bisulfite-mediated cytosine conversion to uracil, the key reaction for DNA methylation analysis – a personal account. Proc Jpn Acad Ser B Phys Biol Sci 84:321–330

Jaffe AE, Irizarry RA (2014) Accounting for cellular heterogeneity is critical in epigenome-wide association studies. Genome Biol 15:R31

Kelly TK, De Carvalho DD, Jones PA (2010) Epigenetic modifications as therapeutic targets. Nat Biotechnol 28:1069–1078

Kent WJ, Sugnet CW, Furey TS et al (2002) The human genome browser at UCSC. Genome Res 12:996–1006

Leek JT, Scharpf RB, Bravo HC et al (2010) Tackling the widespread and critical impact of batch effects in high-throughput data. Nat Rev Genet 11:733–739

Liu J, Siegmund KD (2016) An evaluation of processing methods for HumanMethylation450 BeadChip data. BMC Genomics 17:469

Liu Y, Aryee MJ, Padyukov L et al (2013) Epigenome-wide association data implicate DNA methylation as an intermediary of genetic risk in rheumatoid arthritis. Nat Biotechnol 31:142–147

Miyata K, Miyata T, Nakabayashi K et al (2015) DNA methylation analysis of human myoblasts during in vitro myogenic differentiation: de novo methylation of promoters of muscle-related genes and its involvement in transcriptional down-regulation. Hum Mol Genet 24:410–423

Morris TJ, Beck S (2015) Analysis pipelines and packages for Infinium HumanMethylation450 BeadChip (450k) data. Methods 72:3–8

Maksimovic J, Gagnon-Bartsch JA, Speed TP et al (2015) Removing unwanted variation in a differential methylation analysis of Illumina HumanMethylation450 array data. Nucleic Acids Res 43:e106

Petersen AK, Zeilinger S, Kastenmüller G et al (2014) Epigenetics meets metabolomics: an epigenome-wide association study with blood serum metabolic traits. Hum Mol Genet 23:534–545

Pidsley R, Zotenko E, Peters TJ et al (2016) Critical evaluation of the Illumina MethylationEPIC BeadChip microarray for whole-genome DNA methylation profiling. Genome Biol 17:208

Price ME, Cotton AM, Lam LL et al (2013) Additional annotation enhances potential for biologically-relevant analysis of the Illumina Infinium HumanMethylation450 BeadChip array. Epigenetics Chromatin 6:4

Rask-Andersen M, Martinsson D, Ahsan M, et al (2016) Epigenome-wide association study reveals differential DNA methylation in individuals with a history of myocardial infarction. Hum Mol Genet (Epub ahead of print) PMID: 27634651

Relton CL, Gaunt T, McArdle W, Ho K, Duggirala A, Shihab H, Woodward G, Lyttleton O, Evans DM, Reik W, Paul YL, Ficz G, Ozanne SE, Wipat A, Flanagan K, Lister A, Heijmans BT, Ring SM, Davey SG (2015) Data resource profile: accessible resource for integrated Epigenomic studies (ARIES). Int J Epidemiol 44:1181–1190

Sun Z, Chai HS, Wu Y et al (2011) Batch effect correction for genome-wide methylation data with Illumina Infinium platform. BMC Med Genet 4:84

Shoemaker R, Deng J, Wang W, Zhang K (2010) Allele-specific methylation is prevalent and is contributed by CpG-SNPs in the human genome. Genome Res 20:883–889

Stirzaker C, Taberlay PC, Statham AL, Clark SJ (2014) Mining cancer methylomes: prospects and challenges. Trends Genet 30:75–84

Triche TJ Jr, Weisenberger DJ, Van Den Berg D et al (2013) Low-level processing of Illumina Infinium DNA methylation BeadArrays. Nucleic Acids Res 41:e90

Teschendorff AE, Marabita F, Lechner M et al (2013) A beta-mixture quantile normalization method for correcting probe design bias in Illumina Infinium 450 k DNA methylation data. Bioinformatics 29:189–196

Wen KX, Miliç J, El-Khodor B et al (2016) The role of DNA methylation and histone modifications in neurodegenerative diseases: a systematic review. PLoS One 11:e0167201

Yousefi P, Huen K, Aguilar Schall R et al (2013) Considerations for normalization of DNA methylation data by Illumina 450K BeadChip assay in population studies. Epigenetics 8:1141–1152

Zhang Z, Zhang R (2015) Epigenetics in autoimmune diseases: pathogenesis and prospects for therapy. Autoimmun Rev 14:854–863

Zhou W, Laird PW, Shen H (2017) Comprehensive characterization, annotation and innovative use of Infinium DNA methylation BeadChip probes. Nucleic Acids Res 45:e22

Ziller MJ, Gu H, Müller F, Donaghey J, Tsai LT-Y, Kohlbacher O, De Jager PL, Rosen ED, Bennett DA, Bernstein BE, Gnirke A, Meissner A (2013) Charting a dynamic DNA methylation landscape of the human genome. Nature 500:477–481

Bioinformatics Databases and Tools on Dietary MicroRNA

116

Juan Cui

Contents

Abstract

Empowered by the emerging genomics technology, there has been an increasingly large accumulation of noncoding RNA information, particularly on microRNAs in animals and plants. Research interest in microRNAs originating from food mostly focused on their bioavailability and cross-species transportation and regulation has recently grown due to the potential implication of regulatory role and cofounding impact in human health. This chapter reviews publicly available repositories and bioinformatics tools developed for dietary microRNA research, including the first dietary microRNA database (DMD) that archives microRNA sequence and annotation in various dietary sources, a new small RNA sequencing analytical pipeline focusing on the detection of both endogenous and exogenous microRNAs, and other general computational resources and tools for microRNA target prediction and functional analysis.

J. Cui (✉)
Department of Computer Science and Engineering, University of Nebraska-Lincoln, Lincoln, NE, USA
e-mail: jcui@unl.edu

V. B. Patel, V. R. Preedy (eds.), *Handbook of Nutrition, Diet, and Epigenetics*,
https://doi.org/10.1007/978-3-319-55530-0_90

Keywords
Dietary microRNA · Exogenous microRNA · Bioinformatics · microRNA target prediction · Small RNA sequencing · Dietary microRNA database (DMD) · microRNA function · microRNA transportation · Exosome · Gene regulation network · Sequence motif

List of Abbreviations

DMD	Dietary microRNA database
miRs, miRNAs	microRNAs
pre-miRNA	Precursor microRNA
NGS	Next-generation sequencing
KEGG	Kyoto Encyclopedia of Genes and Genomes
PPI	Protein-protein interaction
RPKM	Reads Per Kilobase of transcript per Million mapped reads
MFE	Minimum free energy

Introduction

Noncoding RNAs, particularly microRNAs (miRNAs), have been emerging as important global gene regulators in most physiological and pathological conditions in human (Pogue et al. 2010; Bueno and Malumbres 2011; Xiong et al. 2011; Khalid et al. 2014). Given the fact that one single miRNA can regulate up to 1,000 genes while miRNAs with different sequences can co-repress the same target gene, it is conceived that small irregularities in miRNA expression or sequence may lead to important cellular changes (Bartel 2004; Krek et al. 2005; Cannell et al. 2008; Seitz 2009). In addition to the well-characterized endogenous miRNAs, e.g., over 2,000 human miRNAs that regulate over 70% of human genes (Friedman et al. 2009), recent research evidence shows that exogenous miRNA from food are bioactive, and both human and animal can absorb miRNAs from a diet of plant or animal source, such as rice, honeysuckle, milk, and egg (Wang et al. 2012; Zhang et al. 2012; Baier et al. 2014; Zhou et al. 2015), and the biogenesis and function of such exogenous miRNAs are evidently health related (Arnold et al. 2012; Izumi et al. 2012; Liu et al. 2012). Furthermore, it has been documented that human can absorb milk exosomes and deliver the miRNA cargos to various peripheral tissues (Kusuma et al. 2016), possibly regulating human genes. There is a skeptical viewpoint that the abundance of exogenous miRNA may be too low to be biologically meaningful, which however does not hinder the enthusiasm and progress on dietary miRNA research. Instead, it promotes studies particularly solving puzzles about dietary bioavailability and regulatory roles of the dietary cofounder in host species.

Recent advances in high-throughput genomic technology on small RNAs has greatly boosted dietary miRNA discovery and led to the profiling of miRNA expression in various common dietary species. For example, 450 dietary miRNAs have been discovered in human milk (Alsaweed et al. 2016) and 213 miRNAs in bovine slim milk (Chen et al. 2010). With a wealth of expression data and genomic information now available, many large-scale analyses have been performed for different discovery and hypothesis-driven studies. Among them, one of particular interest is to analyze dietary miRNAs as bioactive compound in the circulation of humans and animals, with the ultimate goal to gain new insights in innovative strategy for human disease prevention and treatment.

While topics on cross-species bioavailability, transportation, and regulation of dietary miRNA have offered novel directions in the emerging nutrigenomics field, there are exceptional challenges. For example, exogenous dietary animal miRNAs may have identical or similar sequence with their human homologs, making it extremely difficult to distinguish the endogenous and exogenous sources; they may or may not bind to the same targets of identical human miRNAs. On the contrary, plant diet has different miRNAs sequences from human, and the different miRNA-gene interaction mechanisms often lead to different gene regulation. Given the increased appreciation and challenges in this field, there is a pressing bioinformatics need to develop centralized databases and data analytics tools that can facilitate each specific research on dietary miRNA, which will be reviewed in the next section.

The Dietary miRNA Database

To centralize the annotation of dietary miRNA sequences, bioavailability, and related experimental data available in the exiting literature and public databases, Cui's group developed the first system of its kind, named Dietary MiRNA Database (DMD), for archiving and analyzing miRNAs discovered in dietary resources (Chiang et al. 2015). The system currently covers 5,865 miRNAs that have been reported in 16 dietary species from either plant or animal sources such as apple, grape, cow milk, and cow fat. Information about miRNA sequence, structure, experimental and predicted target, involved functional pathways, and gene regulation networks was integrated as cross references from public miRNA databases such as miRBase (Griffiths-Jones et al. 2006), TargetScan (Lewis et al. 2005), and MirTarBase (Hsu et al. 2014). The database can be accessed at http://sbbi.unl.edu/dmd/.

The DMD supports sequence search and comparison, feature generation, and pathway and gene interaction network analysis. For each mature miRNA entry, the annotation covers information of sequences, genome locations, hairpin structures of parental pre-miRNAs, cross-species sequence comparison, disease relevance, and the experimentally validated gene targets. Users can search for specific entry by using mature or pre-miRNA name or view all miRNAs by choosing the species of interest (Fig. 1).

Each entry in the database (indexed as DM0000*) represents a mature sequence, with the information on the genomic location and hairpin sequence of the parental pre-miRNA (indexed as DP0000*). The corresponding miRBase index is given if

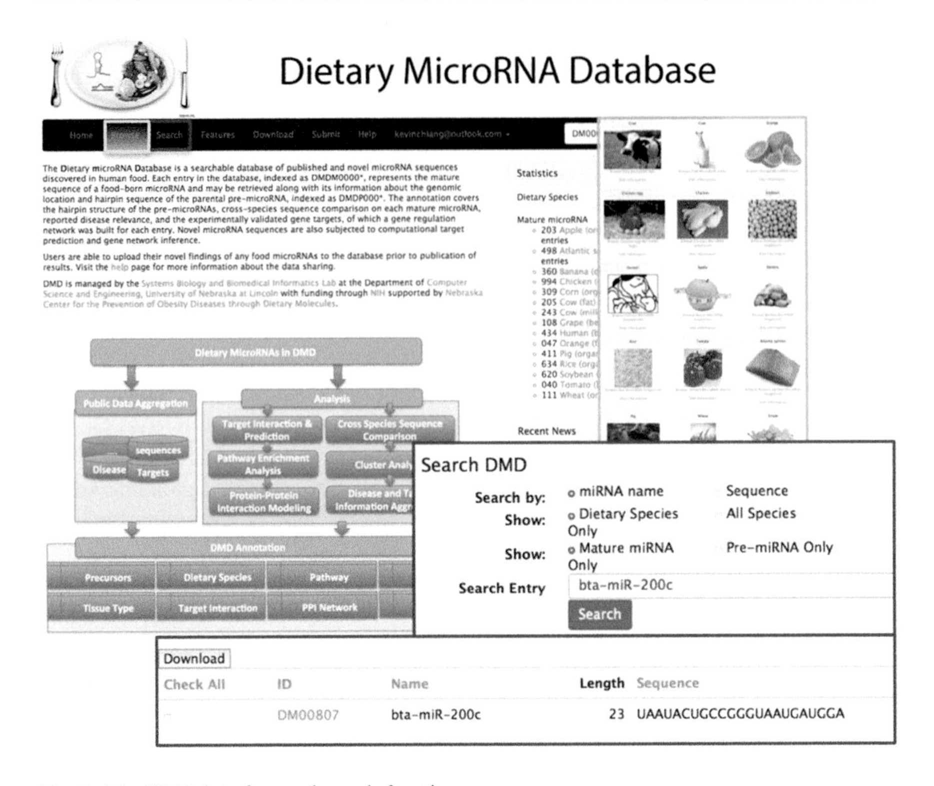

Fig. 1 The DMD interface and search function

entries from both databases are the same. Homologous miRNA loci in different species are assigned the same number, e.g., hsa-mir-100 and dme-mir-100. Paralogous miRNAs have names assigned with lettered and numbered suffixes, depending on whether the derived mature miRNA is identical in sequence or contains sequence differences, e.g., hsa-mir-29a and has-mir-29b. The derived mature miRNAs were named in the following fashion, hsa-miR-100-5p and hsa-miR-100-3p, for sequences derived from the 5′ and 3′ arms of the hsa-miR-100 hairpin precursor. On each annotation page, it provides targets from both experimental studies and computational prediction mainly by MirTarget (Wong and Wang 2015) and psRNAtarget (Bulow et al. 2012). Note that the plant and animal cases were separated when assessing a possible target in human because of the different mechanisms. Furthermore, disease information was extracted from the Human microRNA Disease Database (Li et al. 2014), PhenomiR (Ruepp et al. 2012) and PubMed literature search.

One unique characteristic of DMD is the integration of a feature generator that calculates 411 descriptive attributes (features) that are potentially related to molecular properties key for target interaction of a chosen miRNA. In addition to the mature sequence, the features also include information on the corresponding pre-miRNA sequence, such as palindrome, length, composition of monomers and dimers, and secondary structure, and minimum free energy (MFE) analyzed by RNAfold (Denman 1993), RNAshape (Janssen and Giegerich 2015), and STOAT (Knudsen and Caetano-Anolles 2008). The *feature* page allows users to calculate these features in two categories, namely, sequence based and secondary structure based.

Furthermore, functional analysis has been integrated into the system, enabling users to generate functional discoveries and insights through viewing pathways and building protein-protein interaction networks associated with each miRNA. Users can also run a pathway enrichment analysis on the selected targets, either experimentally validated or predicted. A total of 1,955 pathways from KEGG (Aoki and Kanehisa 2005) are included in the current system. A modified p-value was calculated for each enriched pathway based on Fisher's exact test on queried targets against the whole genome. In addition, protein-protein interaction (PPI) data (Stark et al. 2006) were employed to visualize the miRNA-mediated gene regulation network. First, selecting from the list of targets will display a new list and opting to show the interaction will open a new window, which visualizes the interaction between any two targets with up to three shared intermediates (Fig. 2). With all the above functionalities, DMD would be particularly useful for research groups studying miRNA regulation from a nutrition point of view.

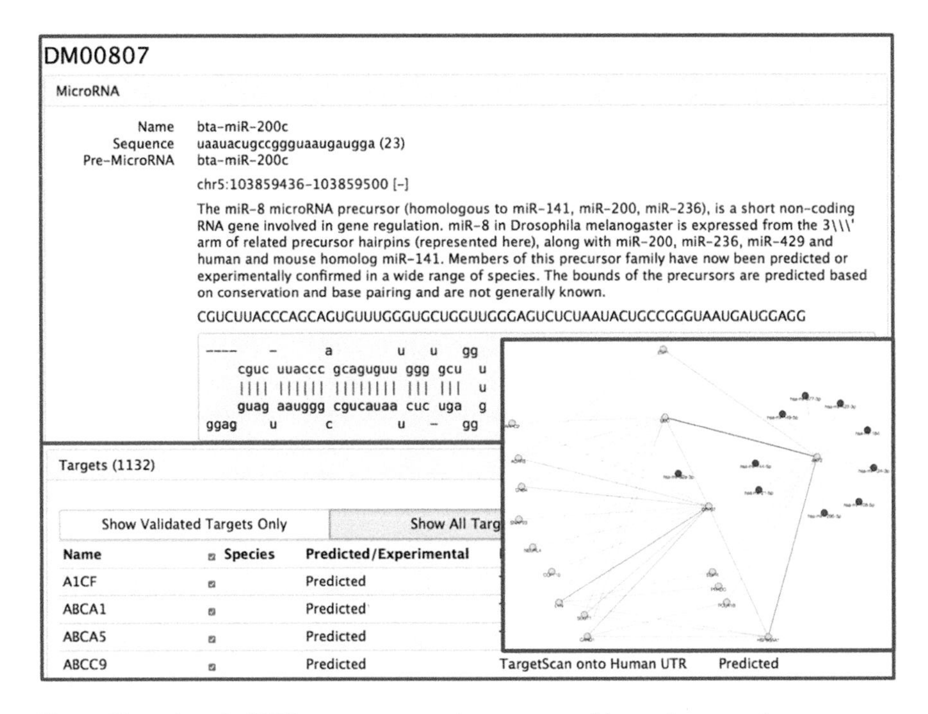

Fig. 2 Illustration of miRNA sequence annotation, targets, and interaction network

The Exogenous miRNA Detection Based on Sequencing Data Analysis

Another fact that warrants the efforts in bioinformatics tool development is that next-generation sequencing (NGS) became the most efficient techniques in the emerging exogenous miRNA discovery, e.g., a large abundance of nonhuman miRNAs have been detected in human circulation using noncoding small RNA sequencing (Lukasik and Zielenkiewicz 2014; Kitchen 2015); however, no existing NGS pipeline exists for exogenous miRNA detection. Tools for small RNA sequencing data analysis such as miRDeep2 (Friedlander et al. 2012), CAP-miRSeq (Sun et al. 2014), DSAP (Huang et al. 2010), DARIO (Fasold et al. 2011), omiRas (Muller et al. 2013), sRNAbench (Barturen et al. 2014), ShortStack (Axtell 2013), miRDeep-P (Yang and Li 2011), and miR-PREFeR (Lei and Sun 2014) all focus on miRNA expression profiling and novel host miRNA detection, whereas the challenges of exogenous sequence detection lie in the fast multigenome mapping, annotation, and the cross-species comparison for detection of subtle sequence differences between endogenous and exogenous miRNAs. Very recently, a new small RNA-seq data analysis pipeline for both endogenous and exogenous miRNA detection, miRDis (**mi**cro**R**NA **Dis**covery), has been developed (Zhang et al. 2017), which allows the detection in the host species of dietary miRNAs that originate from major dietary species. The server can be accessed at http://sbbi-panda.unl.edu/miRDis/.

The miRDis system requires input files containing small RNA-seq read data in *.zip or *.gz format, along with the user-given adapter information. Currently, five host species (human, chimpanzee, dog, rat, and mouse) and eight common dietary species including cow, pig, chicken, tomato, maize, soybean, rice, and grape are available. Every submitted job holds a unique id, e.g., 20160817135554r; once it is finished, the result can be accessed in miRDis within 2 weeks upon email notification.

The schematic flowchart in Fig. 3 showcases the implementation workflow and the functionality of this pipeline, including four main components: read processing, read mapping, annotation, and differential expression analysis. First, the sequencing data quality control (QC) will be performed by FASTQC (Bioinformatics 2011) with a report generated by the system. Cutadapt (Martin 2011) can be used to eliminate the low quality reads and adaptors with the specified sequence. The genome annotation on 13 types of animals and plants are downloaded from the Ensemble database (Kersey et al. 2014). All unique reads are mapped to the genomes of the host or dietary species by bowtie (Langmead et al. 2009) and the mapped regions are identified using BEDtools (Quinlan 2014). Reads that have more than one mapped region are assigned to the loci that have more unique reads or more stable secondary structure inferred by RNAfold (Lorenz et al. 2011). The BED file covers information about all mapped reads and the depth in every single position. After mapping, the consensus sequence from each mapped region is extracted through the following analysis: all peaks (single positions with the highest depth (no less than 5) within a consecutive mapped region) are examined. The expanded region where the depth in the peak is more than 3-fold higher than those at both ends becomes an expressed

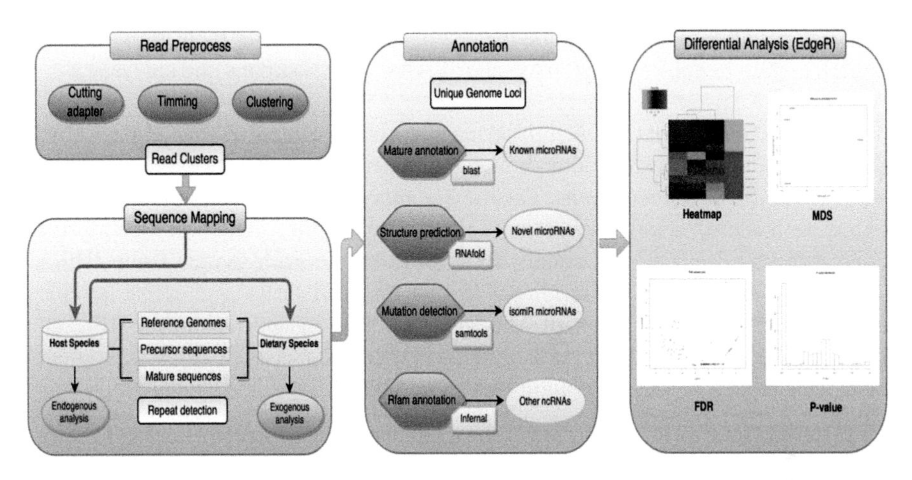

Fig. 3 The pipeline workflow for endogenous and exogenous miRNA discovery based on small RNA sequencing analysis

region. The consensus sequence retrieved from such expressed region will be subject to further annotation as a mature miRNA (either endogenous or exogenous), novel miRNA, or other noncoding RNA, according to the following protocol:

- *Known miRNAs are* expressed regions that overlap with annotated miRNAs in the genome. A new term mature information (I_{mature}) was defined to evaluate the confidence:
  $I_{\text{mature}} = \frac{\text{Identity}^* 100}{\text{length (sequence of the expressed region)}}$, where the identity is calculated between the annotated mature miRNA sequence and the expressed sequence. An expressed region with $I = 100$ indicates a perfectly matched mature miRNA while a lower I (<100 and $> = 85$) may indicate isomiRs that have one sequence variation or one extended base compared to the reported sequence. In addition, the miR-precursor type refers to the expressed sequences that are identical to part of a known precursor but not reported as mature sequences in the databases.
- An exogenous miRNA sequence can be determined based on the following scenarios: (1) the expressed region along with their flanking sequence shows a better match with a known miRNA sequence in a dietary species compared to the host, (e.g., the extended sequence on either side matched to dietary genome but not human although the expressed region may be identical in both genomes, or the similarity is higher in dietary species vs. human), and (2) the expressed region is corresponding to a known miRNA of the dietary species but unrelated to any host sequence (e.g., a cow-milk-specific miRNA that humans do not have).
- *Novel miRNAs and noncoding RNAs* represent the expressed regions that are either annotated as predicted miRNAs in Ensemble (Guttman et al. 2009) or homologs to known miRNA structure in Rfam evaluated by infernal based on CM model (e-value $<$ 1E-5) (Nawrocki and Eddy 2013; Nawrocki et al. 2015) and

$I_{mature} > 40$. Similarly, this pipeline also differentiates other types of noncoding RNAs such as snoRNAs, rRNAs, and tRNAs, based on the same analysis.

The miRDis outputs all identified candidates in categories including known miRNAs, novel miRNAs, other noncoding miRNAs, and exogenous miRNAs, along with detailed annotations. On the summary page, Pie charts are generated based on the read count distribution across all categories. MiRNA expression is quantified based on its read counts normalized within each sample using RPKM (Reads Per Kilobase of transcript per Million mapped reads) and across samples using TMM (Trimmed Mean of M values) (Robinson and Oshlack 2010). The differential analysis is performed using EdgeR (Robinson et al. 2010) which outputs visualized results on the expression Heatmap, MDS (Multidimensional scaling), p-value, and FDR distribution.

Other Bioinformatics Recourse on miRNA

In addition to the aforementioned resources that are specifically designed for dietary miRNA, many tools for studying miRNA in humans, animals, and plants can be used on dietary miRNA research. The examples are as follows.

Target Prediction: Numerous bioinformatics tools have been focused on miRNA target prediction, such as TargetScan (Lewis et al. 2005), miRanda (Miranda et al. 2006), mirSVR (Betel et al. 2010), RFMirTarget (Mendoza et al. 2013), MirTarget (Wong and Wang 2015), miRDB (Wong and Wang 2015), in order to indirectly infer the functional processes that a miRNA can participate in. Some early tools are focused on search for complementary sequences in the seed region, e.g., MirTarget begins with scanning all 3′-UTR sequences for seed paring and therefore suffers from the high false prediction due to the challenges posed by the complex miRNA-mRNA interactions. On the other hand, recent experimental studies have uncovered a large set of genome-wide miRNA-mRNA interactions using novel sequencing techniques. For example, 18,514 miRNA-mRNA interactions were detected through crosslinking, ligation, and sequencing of hybrids (CLASH) analysis (Helwak et al. 2013). This study also unveiled that ~60% of the binding sites are within coding region of mRNA, as opposed to the 3’UTR-centric search by the existing algorithms. Similar observations were made using covalent ligation of endogenous Argonaute-bound RNAs (CLEAR)-CLIP in human hepatoma cells (Moore et al. 2015). By integrating the sequencing-detected interactions into the models, updated version of the computational tools and databases such as TargetScan and MirTarBase can now provide much more reliable information of miRNA targets.

Sequence Motif: Another important topic related to dietary miRNA is to identify molecular features that contribute to sorting, packaging, and transport of miRNAs from cells of the dietary origins to the host circulation. One recent study has reported a motif (GGAG) that is enriched in exosomal miRNAs secreted from T-cells, through which the loading protein, heterogeneous nuclear ribonucleoprotein A2B1 (hnRNPA2B1), can specifically binds to some miRNAs and load them to exosomes (Villarroya-Beltri et al. 2013). To tackle this problem using a systematic approach, a bioinformatics analysis was conducted to generate 1,620 features based on the sequence and structure of miRNA and applied to a support-vector-machine-based classification analysis on circulating miRNA versus other human miRNAs. Eight groups features were identified to be associated with human circulating miRNAs (Shu et al. 2015). Further motif finding analysis based on known exosomal miRNAs from various cells and liquid in human shows significant sequence motifs that might contribute to miRNA-loading to exosome (Cha et al. 2015; Santangelo et al. 2016). Motif finding tools, including MEME (Bailey et al. 2009) and Improbizer (Ao et al. 2004), have been utilized in these analysis. Current efforts are focused on the assessment of motifs within specific groups that have the same cell type or dietary species. Along this line of research, there are databases specific for exosomal cargo including miRNA from various sources. Three examples are listed in Table 1.

Last but not least, tools such as DIANA miRPath (Vlachos et al. 2015) utilize the comprehensive biological pathways annotated in KEGG (Kanehisa and Goto 2000) to provide better understanding of which biological pathways are regulated by miRNAs.

Table 1 Statistics of exosomal miRNAs reported in the public databases

Vesiclepedia (Kalra et al. 2012)		ExoCarta (Mathivanan and Simpson 2009)		EVpedia (Kim et al. 2013)	
Source	Count	Source	Count	Source	Count
Neuronal cells (mmu)	193	Plasma	366	Lung cancer cell (NCI_H69)	46
Urine	176	Colorectal cancer cells	271	Colorectal cancer cell (SW480)	39
T cells	99	Colon carcinoma cells	238	Colorectal cancer cell (SW620)	39
Serum	64	Endothelial cells	228	Seminal plasma	35
Breast cancer cells	47	T cells	78	Caput epithelial cell	30
B cells	38	Mast cells (mmu)	69	Cauda epithelial cell	28
Lung cancer cells	31	Serum	54	Lung cancer cell (DMS563)	27
Mast cells (mmu)	27	B cells	29	Caput luminal fluid	26
Saliva	14	Melanoma cells	24	Cauda luminal fluid	25
Dendritic cells	13	Mesenchymal stem cells	14	Metastatic gastric cancer cell (AZP7a)	24
Plasma	4	Dendritic cells	11	Metastatic gastric cancer cell (AZ521)	13
Epithelial cells	3	Ascites	2	Lung cancer cell (SBC-3)	9
Total	709		1384		341

mmu mus musculus; the rest are from human

Dictionary of Terms

- Next-generation sequencing represents modern sequencing technologies including Illumina, Roche 454, and SOLiD that detect the sequence of DNA and RNA in a more efficient fashion (high-throughput and cheaper) compared to the previously used Sanger sequencing.
- Exosome are small cell-derived vesicles (30–150 nm) present in many (or perhaps all) eukaryotic fluids, including blood, urine, and cultured medium of cell cultures, containing sophisticated RNA, protein, and lipids cargos.
- Kyoto Encyclopedia of Genes and Genomes (KEGG) is an online bioinformatics database that provides molecular-level information related to a large collection of genomes, pathways, disease, and drugs.
- A sequence motif is a short nucleotide sequence pattern in RNA that has a biological significance, which often indicates a sequence-specific biding site and/or a biological function.
- Minimum free energy (MFE) structure refers to, theoretically, the most likely secondary structure of an RNA.

Key Facts

- Animal can acquire exogenous miRNA from diet; however, the mechanism of cross-species transport miRNA has yet to be fully explored.
- Exsomes are important vehicles to transport cellular cargos including miRNA, mRNA, proteins, and lipids to the recipient cells through blood circulation.
- It is now possible to identify exogenous miRNA sequences in host species using computational analysis from small noncoding RNA sequencing data.
- Many computational systems have been developed for the comprehensive discovery of miRNA of all kinds based on sequencing data analysis.
- miRNA-mediated gene regulation is a very complex, semi-stochastic process, which integrates both cooperative and competitive binding mechanisms that are largely ambiguous.

Summary Points

- With increased research efforts in miRNA biology, there is a pressing need for efficient and friendly tools for making promising discoveries in miRNA cross-species transport.
- Detection of exogenous dietary miRNA through sequencing requires efficient systems that have high performance in miRNA detection, expression quantification, and differentiation of the isomiRs and nonhost miRNAs.
- miRDis represents the first pipeline for small noncoding RNA sequencing data analysis that enables the automated detection in host samples of the presence of both endogenous miRNA and exogenous miRNA from dietary species.

- System like miRDis overcomes the scaling challenges for processing large set of miRNA-seq data through parallel computation.
- Continued bioinformatics efforts on miRNA are focused on the following topics: (1) development of efficient methods for miRNA target prediction that can improve the high false positive rate with the existing methods; (2) implementation of new techniques to evaluate the functional association among miRNAs; and (3) study of the miRNA-mediated gene regulation in a dynamic manner through network analysis.

References

Alsaweed M, Lai CT, Hartmann PE, Geddes DT, Kakulas F (2016) Human milk miRNAs primarily originate from the mammary gland resulting in unique miRNA profiles of fractionated milk. Sci Rep 6:20680

Ao W, Gaudet J, Kent WJ, Muttumu S, Mango SE (2004) Environmentally induced foregut remodeling by PHA-4/FoxA and DAF-12/NHR. Science 305(5691):1743–1746

Aoki KF, Kanehisa M (2005) Using the KEGG database resource. Curr Protoc Bioinformatics. Chapter 1: Unit 1.12. https://doi.org/10.1002/0471250953.bi0112s11.

Arnold CN, Pirie E, Dosenovic P, McInerney GM, Xia Y, Wang N, Li X, Siggs OM, Karlsson Hedestam GB, Beutler B (2012) A forward genetic screen reveals roles for Nfkbid, Zeb1, and Ruvbl2 in humoral immunity. Proc Natl Acad Sci U S A 109(31):12286–12293

Axtell MJ (2013) ShortStack: comprehensive annotation and quantification of small RNA genes. RNA 19(6):740–751

Baier SR, Nguyen C, Xie F, Wood JR, Zempleni J (2014) MicroRNAs are absorbed in biologically meaningful amounts from nutritionally relevant doses of cow milk and affect gene expression in peripheral blood mononuclear cells, HEK-293 kidney cell cultures, and mouse livers. J Nutr 144 (10):1495–1500

Bailey TL, Boden M, Buske FA, Frith M, Grant CE, Clementi L, Ren J, Li WW, Noble WS (2009) MEME SUITE: tools for motif discovery and searching. Nucleic Acids Res 37(Web Server issue):W202–W208

Bartel DP (2004) MicroRNAs: genomics, biogenesis, mechanism, and function. Cell 116 (2):281–297

Barturen G, Rueda A, Hamberg M, Alganza A, Lebron R, Kotsyfakis M, Shi B-J, Koppers-Lalic D, Hackenberg M (2014) sRNAbench: profiling of small RNAs and its sequence variants in single or multi-species high-throughput experiments. Methods Next Gener Seq 1:21–31

Betel D, Koppal A, Agius P, Sander C, Leslie C (2010) Comprehensive modeling of microRNA targets predicts functional non-conserved and non-canonical sites. Genome Biol 11:R90

Bioinformatics B (2011) FastQC a quality control tool for high throughput sequence data. Babraham Institute, Cambridge, UK

Bueno MJ, Malumbres M (2011) MicroRNAs and the cell cycle. Biochim Biophys Acta 1812 (5):592–601

Bulow L, Bolivar JC, Ruhe J, Brill Y, Hehl R (2012) 'MicroRNA targets', a new AthaMap web-tool for genome-wide identification of miRNA targets in *Arabidopsis thaliana*. BioData Min 5:7

Cannell IG, Kong YW, Bushell M (2008) How do microRNAs regulate gene expression? Biochem Soc Trans 36(Pt 6):1224–1231

Cha DJ, Franklin JL, Dou Y, Liu Q, Higginbotham JN, Demory Beckler M, Weaver AM, Vickers K, Prasad N, Levy S, Zhang B, Coffey RJ, Patton JG (2015) KRAS-dependent sorting of miRNA to exosomes. elife 4:e07197

Chen X, Gao C, Li H, Huang L, Sun Q, Dong Y, Tian C, Gao S, Dong H, Guan D, Hu X, Zhao S, Li L, Zhu L, Yan Q, Zhang J, Zen K, Zhang CY (2010) Identification and characterization of microRNAs in raw milk during different periods of lactation, commercial fluid, and powdered milk products. Cell Res 20(10):1128–1137

Chiang K, Shu J, Zempleni J, Cui J (2015) Dietary MicroRNA database (DMD): an archive database and analytic tool for food-borne microRNAs. PLoS One 10(6):e0128089

Denman RB (1993) Using RNAFOLD to predict the activity of small catalytic RNAs. BioTechniques 15(6):1090–1095

Fasold M, Langenberger D, Binder H, Stadler PF, Hoffmann S (2011) DARIO: a ncRNA detection and analysis tool for next-generation sequencing experiments. Nucleic Acids Res 39(Web Server issue):W112–W117

Friedlander MR, Mackowiak SD, Li N, Chen W, Rajewsky N (2012) miRDeep2 accurately identifies known and hundreds of novel microRNA genes in seven animal clades. Nucleic Acids Res 40(1):37–52

Friedman RC, Farh KK, Burge CB, Bartel DP (2009) Most mammalian mRNAs are conserved targets of microRNAs. Genome Res 19(1):92–105

Griffiths-Jones S, Grocock RJ, van Dongen S, Bateman A, Enright AJ (2006) miRBase: microRNA sequences, targets and gene nomenclature. Nucleic Acids Res 34(Database issue):D140–D144

Guttman M, Amit I, Garber M, French C, Lin MF, Feldser D, Huarte M, Zuk O, Carey BW, Cassady JP, Cabili MN, Jaenisch R, Mikkelsen TS, Jacks T, Hacohen N, Bernstein BE, Kellis M, Regev A, Rinn JL, Lander ES (2009) Chromatin signature reveals over a thousand highly conserved large non-coding RNAs in mammals. Nature 458(7235):223–227

Helwak A, Kudla g, Dudnakova t, Tollervey D (2013) Mapping the human miRNA interactome by CLASH reveals frequent noncanonical binding. Cell 153(3):654–665. https://doi.org/10.1016/j.cell.2013.03.043. PMCID: PMC3650559

Hsu SD, Tseng YT, Shrestha S, Lin YL, Khaleel A, Chou CH, Chu CF, Huang HY, Lin CM, Ho SY, Jian TY, Lin FM, Chang TH, Weng SL, Liao KW, Liao IE, Liu CC, Huang HD (2014) miRTarBase update 2014: an information resource for experimentally validated miRNA-target interactions. Nucleic Acids Res 42(Database issue):D78–D85

Huang PJ, Liu YC, Lee CC, Lin WC, Gan RR, Lyu PC, Tang P (2010) DSAP: deep-sequencing small RNA analysis pipeline. Nucleic Acids Res 38(Web Server issue):W385–W391

Izumi H, Kosaka N, Shimizu T, Sekine K, Ochiya T, Takase M (2012) Bovine milk contains microRNA and messenger RNA that are stable under degradative conditions. J Dairy Sci 95 (9):4831–4841

Janssen S, Giegerich R (2015) The RNA shapes studio. Bioinformatics 31(3):423–425

Kalra H, Simpson RJ, Ji H, Aikawa E, Altevogt P, Askenase P, Bond VC, Borras FE, Breakefield X, Budnik V, Buzas E, Camussi G, Clayton A, Cocucci E, Falcon-Perez JM, Gabrielsson S, Gho YS, Gupta D, Harsha HC, Hendrix A, Hill AF, Inal JM, Jenster G, Kramer-Albers EM, Lim SK, Llorente A, Lotvall J, Marcilla A, Mincheva-Nilsson L, Nazarenko I, Nieuwland R, Nolte-'t Hoen EN, Pandey A, Patel T, Piper MG, Pluchino S, Prasad TS, Rajendran L, Raposo G, Record M, Reid GE, Sanchez-Madrid F, Schiffelers RM, Siljander P, Stensballe A, Stoorvogel W, Taylor D, Thery C, Valadi H, van Balkom BW, Vazquez J, Vidal M, Wauben MH, Yanez-Mo M, Zoeller M, Mathivanan S (2012) Vesiclepedia: a compendium for extracellular vesicles with continuous community annotation. PLoS Biol 10(12):e1001450

Kanehisa M, Goto S (2000) KEGG: kyoto encyclopedia of genes and genomes. Nucleic Acids Res 28(1):27–30

Kersey PJ, Allen JE, Christensen M, Davis P, Falin LJ, Grabmueller C, Hughes DS, Humphrey J, Kerhornou A, Khobova J, Langridge N, McDowall MD, Maheswari U, Maslen G, Nuhn M, Ong CK, Paulini M, Pedro H, Toneva I, Tuli MA, Walts B, Williams G, Wilson D, Youens-Clark K, Monaco MK, Stein J, Wei X, Ware D, Bolser DM, Howe KL, Kulesha E, Lawson D, Staines DM (2014) Ensembl genomes 2013: scaling up access to genome-wide data. Nucleic Acids Res 42(Database issue):D546–D552

Khalid U, Bowen T, Fraser DJ, Jenkins RH (2014) Acute kidney injury: a paradigm for miRNA regulation of the cell cycle. Biochem Soc Trans 42(4):1219–1223

Kim DK, Kang B, Kim OY, Choi DS, Lee J, Kim SR, Go G, Yoon YJ, Kim JH, Jang SC, Park KS, Choi EJ, Kim KP, Desiderio DM, Kim YK, Lotvall J, Hwang D, Gho YS (2013) EVpedia: an integrated database of high-throughput data for systemic analyses of extracellular vesicles. J Extracell Vesicles, Mar 19;2. https://doi.org/10.3402/jev.v2i0.20384. eCollection 2013.

Kitchen R (2015) A comprehensive method for the analysis of extracellular small RNA-seq data, including characterisation based on cellular expression profiles and exogenous sequence detection [conference abstract]. In: 2015 meeting of the international society of extracellular vesicles, Bethesda

Knudsen V, Caetano-Anolles G (2008) NOBAI: a web server for character coding of geometrical and statistical features in RNA structure. Nucleic Acids Res 36(Web Server issue):W85–W90

Krek A, Grun D, Poy MN, Wolf R, Rosenberg L, Epstein EJ, MacMenamin P, da Piedade I, Gunsalus KC, Stoffel M, Rajewsky N (2005) Combinatorial microRNA target predictions. Nat Genet 37(5):495–500

Kusuma RJ, Manca S, Friemel T, Sukreet S, Nguyen C, Zempleni J (2016) Human vascular endothelial cells transport foreign exosomes from cow's milk by endocytosis. Am J Physiol Cell Physiol 310(10):C800–C807

Langmead B, Trapnell C, Pop M, Salzberg SL (2009) Ultrafast and memory-efficient alignment of short DNA sequences to the human genome. Genome Biol 10(3):R25

Lei J, Sun Y (2014) miR-PREFeR: an accurate, fast and easy-to-use plant miRNA prediction tool using small RNA-Seq data. Bioinformatics 30(19):2837–2839

Lewis BP, Burge CB, Bartel DP (2005) Conserved seed pairing, often flanked by adenosines, indicates that thousands of human genes are microRNA targets. Cell 120(1):15–20

Li Y, Qiu C, Tu J, Geng B, Yang J, Jiang T, Cui Q (2014) HMDD v2.0: a database for experimentally supported human microRNA and disease associations. Nucleic Acids Res 42 (Database issue):D1070–D1074

Liu R, Ma X, Xu L, Wang D, Jiang X, Zhu W, Cui B, Ning G, Lin D, Wang S (2012) Differential microRNA expression in peripheral blood mononuclear cells from Graves' disease patients. J Clin Endocrinol Metab 97(6):E968–E972

Lorenz R, Bernhart SH, Honer Zu Siederdissen C, Tafer H, Flamm C, Stadler PF, Hofacker IL (2011) ViennaRNA Package 2.0. Algorithms Mol Biol 6:26

Lukasik A, Zielenkiewicz P (2014) In silico identification of plant miRNAs in mammalian breast milk exosomes – a small step forward? PLoS One 9(6):e99963

Martin M (2011) Cutadapt removes adapter sequences from high-throughput sequencing reads. EMBnet J 17(1):10–12

Mathivanan S, Simpson RJ (2009) ExoCarta: a compendium of exosomal proteins and RNA. Proteomics 9(21):4997–5000

Mendoza MR, da Fonseca GC, Loss-Morais G, Alves R, Margis R, Bazzan AL (2013) RFMirTarget: predicting human microRNA target genes with a random forest classifier. PLoS One 8:e70153

Miranda KC, Huynh T, Tay Y, Ang YS, Tam WL, Thomson AM, Lim B, Rigoutsos I (2006) A pattern-based method for the identification of MicroRNA binding sites and their corresponding heteroduplexes. Cell 126(6):1203–1217

Moore MJ, Scheel TK, Luna JM, Park CY, Fak JJ, Nishiuchi E, Rice CM, Darnell RB (2015) miRNA-target chimeras reveal miRNA 3′-end pairing as a major determinant of Argonaute target specificity. Nat Commun 6:8864

Muller S, Rycak L, Winter P, Kahl G, Koch I, Rotter B (2013) omiRas: a web server for differential expression analysis of miRNAs derived from small RNA-Seq data. Bioinformatics 29 (20):2651–2652

Nawrocki EP, Burge SW, Bateman A, Daub J, Eberhardt RY, Eddy SR, Floden EW, Gardner PP, Jones TA, Tate J, Finn RD (2015) Rfam 12.0: updates to the RNA families database. Nucleic Acids Res 43(Database issue):D130–D137

Nawrocki EP, Eddy SR (2013) Computational identification of functional RNA homologs in metagenomic data. RNA Biol 10(7):1170–1179

Pogue AI, Cui JG, Li YY, Zhao Y, Culicchia F, Lukiw WJ (2010) Micro RNA-125b (miRNA-125b) function in astrogliosis and glial cell proliferation. Neurosci Lett 476(1):18–22
Quinlan AR (2014) BEDTools: the Swiss-Army tool for genome feature analysis. Curr Protoc Bioinformatics 47:11-12–11-34
Robinson MD, McCarthy DJ, Smyth GK (2010) edgeR: a Bioconductor package for differential expression analysis of digital gene expression data. Bioinformatics 26(1):139–140
Robinson MD, Oshlack A (2010) A scaling normalization method for differential expression analysis of RNA-seq data. Genome Biol 11(3):R25
Ruepp A, Kowarsch A, Theis F (2012) PhenomiR: microRNAs in human diseases and biological processes. Methods Mol Biol 822:249–260
Santangelo L, Giurato G, Cicchini C, Montaldo C, Mancone C, Tarallo R, Battistelli C, Alonzi T, Weisz A, Tripodi M (2016) The RNA-binding protein SYNCRIP is a component of the hepatocyte exosomal machinery controlling microRNA sorting. Cell Rep 17(3):799–808
Seitz H (2009) Redefining microRNA targets. Curr Biol 19(10):870–873
Shu J, Chiang K, Zempleni J, Cui J (2015) Computational characterization of exogenous microRNAs that can be transferred into human circulation. PLoS One 10(11):e0140587
Stark C, Breitkreutz BJ, Reguly T, Boucher L, Breitkreutz A, Tyers M (2006) BioGRID: a general repository for interaction datasets. Nucleic Acids Res 34(Database issue):D535–D539
Sun Z, Evans J, Bhagwate A, Middha S, Bockol M, Yan H, Kocher JP (2014) CAP-miRSeq: a comprehensive analysis pipeline for microRNA sequencing data. BMC Genomics 15:423
Villarroya-Beltri C, Gutierrez-Vazquez C, Sanchez-Cabo F, Perez-Hernandez D, Vazquez J, Martin-Cofreces N, Martinez-Herrera DJ, Pascual-Montano A, Mittelbrunn M, Sanchez-Madrid F (2013) Sumoylated hnRNPA2B1 controls the sorting of miRNAs into exosomes through binding to specific motifs. Nat Commun 4:2980
Vlachos IS, Zagganas K, Paraskevopoulou MD, Georgakilas G, Karagkouni D, Vergoulis T, Dalamagas T, Hatzigeorgiou AG (2015) DIANA-miRPath v3.0: deciphering microRNA function with experimental support. Nucleic Acids Res 43(W1):W460–W466
Wang K, Li H, Yuan Y, Etheridge A, Zhou Y, Huang D, Wilmes P, Galas D (2012) The complex exogenous RNA spectra in human plasma: an interface with human gut biota? PLoS One 7(12): e51009
Wong N, Wang X (2015) miRDB: an online resource for microRNA target prediction and functional annotations. Nucleic Acids Res 43(Database issue):D146–D152
Xiong X, Ren HZ, Li MH, Mei JH, Wen JF, Zheng CL (2011) Down-regulated miRNA-214 induces a cell cycle G1 arrest in gastric cancer cells by up-regulating the PTEN protein. Pathol Oncol Res 17(4):931–937
Yang X, Li L (2011) miRDeep-P: a computational tool for analyzing the microRNA transcriptome in plants. Bioinformatics 27(18):2614–2615
Zhang L, Hou D, Chen X, Li D, Zhu L, Zhang Y, Li J, Bian Z, Liang X, Cai X, Yin Y, Wang C, Zhang T, Zhu D, Zhang D, Xu J, Chen Q, Ba Y, Liu J, Wang Q, Chen J, Wang J, Wang M, Zhang Q, Zhang J, Zen K, Zhang CY (2012) Exogenous plant MIR168a specifically targets mammalian LDLRAP1: evidence of cross-kingdom regulation by microRNA. Cell Res 22(1):107–126
Zhang H, Vieira Resende ESB, Cui J (2017) miRDis: a Web tool for endogenous and exogenous microRNA discovery based on deep-sequencing data analysis. Brief Bioinform, Jan 10. pii: bbw140. https://doi.org/10.1093/bib/bbw140.
Zhou Z, Li X, Liu J, Dong L, Chen Q, Liu J, Kong H, Zhang Q, Qi X, Hou D, Zhang L, Zhang G, Liu Y, Zhang Y, Li J, Wang J, Chen X, Wang H, Zhang J, Chen H, Zen K, Zhang CY (2015) Honeysuckle-encoded atypical microRNA2911 directly targets influenza a viruses. Cell Res 25 (1):39–49

MicroRNAs and Reference Gene Methodology

117

Petra Matoušková

Contents

Abstract

microRNAs (miRNAs), short noncoding RNAs, are posttranscriptional negative regulators, with extracellular circulating miRNAs considered promising biomarkers of various diseases. The relative quantification of miRNA transcripts requires an endogenous reference gene for data normalization, which is the critical step in this process. In the present chapter, the means of normalization and methods for the selection of suitable reference gene(s) are discussed, with a key finding being how common reference genes such as RNU6 and miR-16 can cause bias in data interpretation if used without proper validation.

P. Matoušková (✉)
Department of Biochemical Sciences, Charles University, Faculty of Pharmacy, Hradec Králové, Czech Republic
e-mail: matousp7@faf.cuni.cz

V. B. Patel, V. R. Preedy (eds.), *Handbook of Nutrition, Diet, and Epigenetics*,
https://doi.org/10.1007/978-3-319-55530-0_34

Keywords
microRNA · qPCR · Reference gene · Normalization · RNU6 · miR-16 · Circulating miRNA · geNorm · BestKeeper · NormFinder · miRBase · Absolute quantification · Relative quantification · MIQE guidelines

List of Abbreviations

18S	Ribosomal RNA 18S
5S	Ribosomal RNA 5S
ACT	Actin
Cq	Quantitation cycle
FFPE	Formalin-fixed paraffin-embedded
GAPDH	Glyceraldehyde 3-phosphate dehydrogenase
LNA	Locked nucleic acid
MIQE	Minimum information for publication of quantitative real-time PCR experiments
miRNA	microRNA
qPCR	Quantitative real-time polymerase chain reaction

Introduction

microRNAs (miRNAs) are small (19–23 nucleotides) evolutionary conserved non-coding RNAs which are generally considered as negative regulators of gene expression. miRNAs regulate not only the initially discovered timing of developmental events in *Caenorhabditis elegans* (Lee and Ambros 2001), as well as diverse developmental processes in plants (Dugas and Bartel 2004), but much more. miRNAs have been detected in all kingdoms; miRNA-directed gene regulation processes include control of the cell cycle, cell differentiation, metabolism control, and related functions. Furthermore, the deregulation of miRNA expression alters normal cell functions and participates in the development of human diseases.

miRNAs are transcribed from various regions of the genome (intronic, intergenic, or exonic) initially as primary transcripts (pri-miRNA) that can be monocistronic or polycistronic. These long transcripts are cleaved by the RNase III enzyme Drosha while still in the nucleus into 70-nucleotide-long hairpin loop precursor miRNAs (pre-miRNA) which are actively transported by Ran-GTPase and Exportin-5 to the cytoplasm. Pre-miRNAs are cleaved by the endonuclease Dicer into double-stranded miRNAs; one strand (Fig. 1, guide) mature miRNA is subsequently associated with Argonaute-2 proteins and loaded into the miRNA-induced silencing complex (miRISC), which mediates the interaction between the mature miRNA and target mRNA molecule, in turn causing either translation repression or direct mRNA degradation. The other strand (passenger, *) is usually degraded (Bartel 2004), while the strand considered as the guide is cell type dependent (Ludwig et al. 2016). Therefore, instead of using * for passenger miRNA (miR-X*), novel nomenclature has been adopted using miR-X-3p and miR-X-5p, depending on the arm

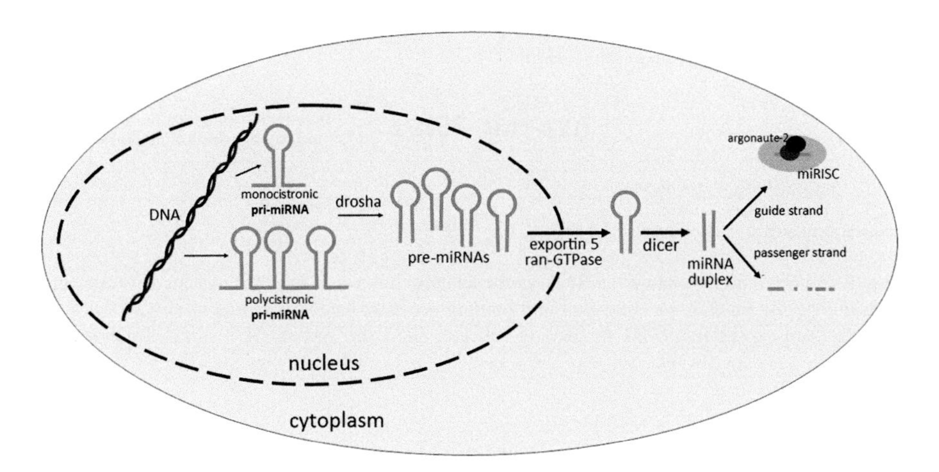

Fig. 1 Annotation of mature miRNAs. miRNA biogenesis starts in the nucleus; long primary miRNA (pri-miRNA) is transcribed and then endonuclease drosha cleaves 70 nt precursor miRNA (pre-miRNA), which is actively transported by Exportin-5 and Ran-GTPase to the cytoplasm. Subsequently, endonuclease dicer excise miRNA duplex, of which one strand (guide; mature miRNA) is incorporated into miRNA-induced silencing complex (miRISC) and the other strand (passenger) is degraded

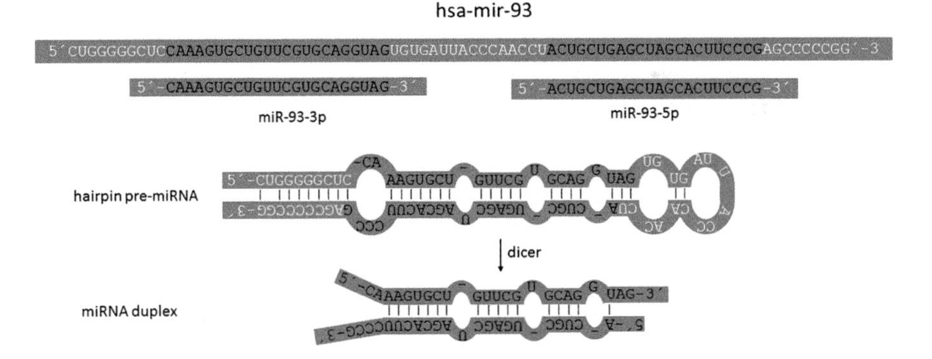

Fig. 2 miRNA biogenesis. Novel annotation of mature miRNAs include $-5'$ and $-3'$ suffixes based on the closer arm of the pre-miR hairpin. Such nomenclature replaces obsolete * sign, which was used for passenger miRNA that was usually degraded

(3′- or 5′-) of the hairpin structure (Fig. 2, (Griffiths-Jones et al. 2006), Fig. 3). The reannotation of miRNAs, necessary for the proper identification of miRNA of interest, however, has caused confusion in literature. Therefore, to simplify the search for miRNAs and to ensure proper annotation, Van Peer et al. have developed the miRBase Tracker (www.mirbasetracker.org) to facilitate the search for correct miRNA names and to help researches locate their miRNAs of interest in "older" literature (Van Peer et al. 2014).

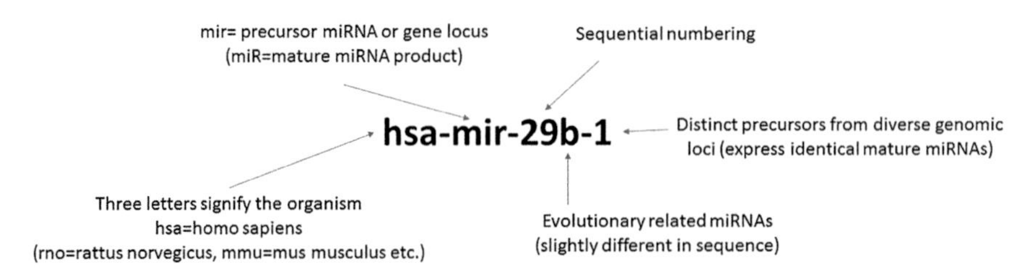

Fig. 3 miRNA nomenclature. miRNA nomenclature follows a uniform system. Criteria and conventions for miRNA identification and naming are described in Ambros et al. (2003). let-7 and lin-4 are exceptions to the numbering scheme, and these names are retained for historical reasons

To date, 35,828 mature miRNAs from 223 species have been registered in the miRBase (release 21, June 2014) (http://www.mirbase.org) (Griffiths-Jones et al. 2006). The field of miRNA research is rapidly growing, as is the number of terms relating to different groups of miRNAs, e.g., oncomiRs for tumors associated with differently expressed miRNAs (Esquela-Kerscher and Slack 2006), apoptomiRs for miRNAs associated with apoptotic process (Vecchione and Croce 2010), myomiRs for muscle-specific miRNAs (McCarthy 2008), redoximiRs for redox-sensitive miRNAs (Cheng et al. 2013), fibromiRs for miRNAs that regulate fibrotic processes (Fierro-Fernandez et al. 2016), purported xenomiRs for plant or animal miRNAs ingested and transferred across the gut (Witwer 2012; Witwer et al. 2013), etc.

Significant amounts of miRNAs have been found in extracellular human body fluids, not only in serum/plasma and urine but also saliva, milk, tears, etc. miRNAs are surprisingly stable despite the ubiquitous presence of extracellular RNases. Extracellular miRNAs – **circulating miRNAs** – are enclosed in vesicles (e.g., in exosomes) or associated with proteins (Hrustincova et al. 2015; Mitchell et al. 2008). A global analysis of the miRNA distribution in 12 human body fluids showed distinct compositions, in which some of the miRNAs were common in multiple fluid types, while some were enriched in specific fluids (Weber et al. 2010). The functions of extracellular miRNAs are not yet fully understood; some of them may represent by-products of dead/dying cells (Turchinovich et al. 2011), but it is likely that at least some miRNAs mediate intercellular communication. Extracellular miRNAs do not simply reflect the repertoire of the cells of origin, although some miRNAs are exported or retained within the cell, a process which is highly selective and tightly regulated (Guduric-Fuchs et al. 2012). Furthermore, the level and composition of circulating miRNAs correlate well with some disease conditions, e.g., various cancers (reviewed by Chakraborty and Das 2016), Alzheimer's disease (Cheng et al. 2015), hepatitis infection (Tan et al. 2015), and amyotrophic lateral sclerosis (Cloutier et al. 2015); therefore the analysis of deregulated specific miRNAs as disease biomarkers is intriguing. At present, several clinical trials are investigating circulating miRNAs as biomarkers (reviewed by D'angelo et al. 2016).

Recently, miRNAs have also been identified as important factors in the response to environmental stress, as they may help to restore homeostasis and are responsive to various chemicals. Accordingly, certain **dietary compounds** – micro- and macronutrients, phytochemicals, etc. – can lead to changes in miRNAs expression and affect their functions (Garcia-Segura et al. 2013). A recent review by Rome sums up the information gathered so far about the possible use of miRNAs as biomarkers of differences regarding various environmental factors such as diet and lifestyle. So far, the most relevant body fluids for miRNA quantification in response to nutrition seem to be plasma and serum (Rome 2015). Tarallo et al., however, have compared the expression of seven human miRNAs in plasma and stool from healthy individuals with various dietary habits of vegans, vegetarians, and omnivores and found that different diet regime may induce differences in miRNA expression; both the investigated specimens reflected such differences. Specifically, miR-92a was reflected well in terms of diet, with vegans and vegetarians having higher levels in plasma and stool samples (Tarallo et al. 2014).

microRNA detection and quantification methods have evolved significantly since the discovery of miRNA, with a wide spectrum of innovative methods reported. A handbook of miRNA expression detection methods was published in 2010 including practical protocols, applications, and the limitations of each method (Wang 2010). Due to the extremely small size of miRNAs, most conventional methods of detection require specific modifications; many of the methods presented difficulties regarding the close similarities among miRNA family members. On the other hand, miRNAs have higher stability compared to mRNAs, even in degraded samples, e.g., even from formalin-fixed paraffin-embedded (FFPE) tissues, it is possible to obtain satisfactory expression data. The detection of miRNA can be performed by several different approaches, e.g., high-throughput hybridization or sequencing-based techniques, medium- to low-throughput amplification-based techniques, nanoparticle-based methods, and various modifications of electrical and optical methods (Wang 2010). In recent years, as sequencing methodologies advance, small RNA-seq (microRNA-seq) has become a key method for global microRNA expression analysis. Witwer and Halushka have thoroughly reviewed microRNA quantification reproducibility and microRNA-seq hindrances in general, including normalization strategies (Witwer and Halushka 2016).

Real-time PCR amplification is the gold standard for low- to mid-throughput mRNA quantification. Nevertheless, some modifications are required for miRNA quantification due to their short length. Several methods have been reported, for example, the use of polymerase-A extension for miRNA prolongation for subsequent use of the modified oligo (dT) primer (Raymond et al. 2005). Chen and colleagues have developed the method of stem-loop RT-PCR for the detection of mature miRNAs using a special stem-loop primer for reverse transcription followed by qPCR with specific Taqman probes (Chen et al. 2005). The insertion of special nucleotides (e.g., LNA-locked-nucleic acids) for primer modification is also possible (Takada and Mano 2007). Bustin et al. formulated guidelines regarding the minimum information necessary for the publication of quantitative real-time PCR experiments (**MIQE**), initially used for the quantification of mRNAs (Bustin et al. 2009). These guidelines ensure a more efficient and accurate experimental practice and interpretation of qPCR results. Furthermore, enable greater integrity and consistency between laboratories in miRNA quantification, as data analysis and normalization are few of the key components of a reliable qPCR assay. Although the **absolute quantification** of miRNAs based on serial dilutions of artificial standards is also possible, this method is used only for detecting a small number of miRNAs, since for each miRNA, a respective standard is needed. A novel approach for the absolute quantification of a DNA copy number is enabled by the droplet digital PCR (dPCR) system through dilution down to single molecules (Vogelstein and Kinzler 1999). dPCR is used mainly for the sensitive detection of rare alleles or mutant DNA (Hindson et al. 2011). Generally, **relative quantification** is the prevailing method; however a normalization strategy needs to be considered before experimentation commences. Normalization using reference genes is the generally accepted method, but such reference genes should be selected initially from a candidate gene set, and stability across each particular experimental setup must be verified.

Although there is an increasing awareness of the importance of the systematic validation of suitable reference genes, common "housekeeping" genes (e.g., GAPDH, ACT, 18S, etc.) are often used for mRNA quantification without any stability validation (Dheda et al. 2005). Similar misleading approaches have also been used for miRNA quantification. Thorough reviews describing a data normalization strategy for miRNA quantification from various sources have recently been published (Occhipinti et al. 2016; Schwarzenbach et al. 2015). One of the first successful attempts at a proper selection of suitable miRNA reference genes was reported by Peltier and Latham, who proposed "horizontal" and "vertical" sample scans (Peltier and Latham 2008). Two miRNAs were suggested (miR-191, miR-103) as suitable reference genes for comparing normal and cancerous solid tissues. However, again as reported by Mestdagh et al., these miRNAs can be used only in verified sample sets (Mestdagh et al. 2009).

Selection of Reference Gene

Suitable reference gene(s) should be expressed in all samples, should not vary across the sample set, and should not be affected by tested experimental conditions or compounds. Ideally, the expression level should be close to the genes of interest. Several programs are available to select reliable reference genes based on candidate genes, four of which – geNorm, NormFinder, Delta Ct, and BestKeeper – are included in **RefFinder**, a user-friendly web-based tool that compares and ranks tested genes based on rankings from each program (Xie et al. 2012). Although the implemented methods were developed for the normalization of mRNA data, they can be used for miRNA reference genes analysis as well. The drawback of this webtool is its use of raw Cq values as an input, which does not enable corrections for the PCR efficiency of each separate assay (De Spiegelaere et al. 2015); thus a normalization strategy is restricted only to fully detected miRNAs across the sample set. **geNorm** calculates the average expression stability value M, the pairwise variation of one gene compared to all the other tested genes (Vandesompele et al. 2002).

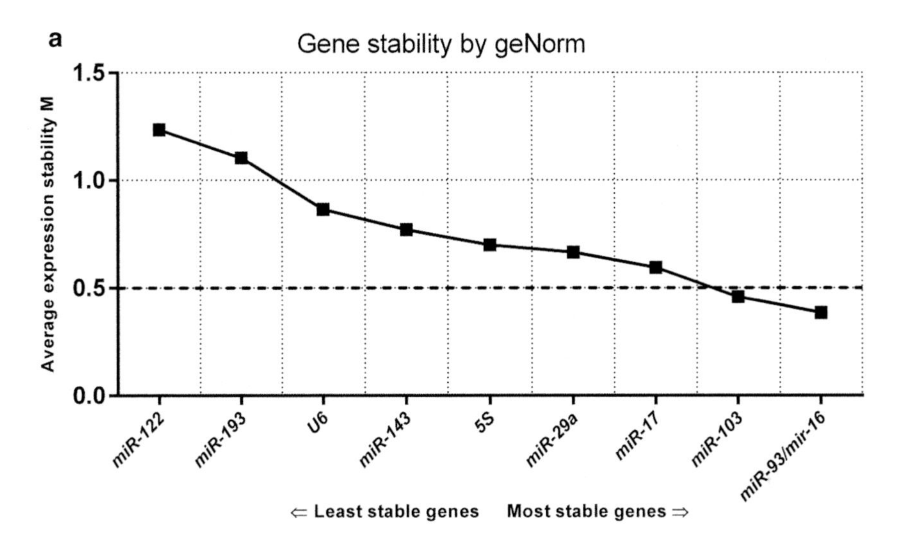

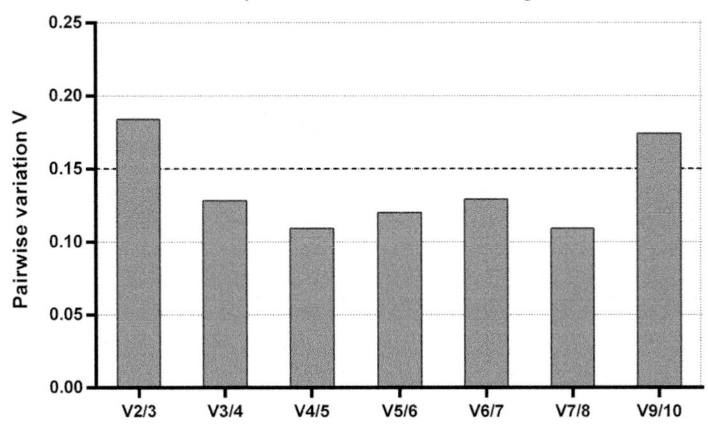

Fig. 4 Example of GeNorm output. GeNorm output consists of gene stability graph, with highlighted 0.5 M value as an ideal cutoff (**a**), and graph of pairwise variation for determination of the optimal number of reference genes for normalization with a cutoff line of 0.15 (**b**). An example of ten tested genes is given with miR-93 and miR-16 identified as the most stable genes followed by miR-103. V3/V4 below the threshold limit (0.15) suggests that three reference genes combined would be sufficient as reference genes for this experiment

Typically, M values lower than 0.5 are observed for homogenous sample panels (Hellemans et al. 2007), but an M value below the threshold of 1.5 is suitable for the identification of reference genes. Furthermore, geNorm enables an estimation of the optimal number of reference genes that should be used for a particular experimental setup through a calculation of pair-wise variation values (Fig. 4). This feature is not available in RefFinder and has to be calculated by geNorm, either implemented in the licensed qbase + software (http://www.qbaseplus.com/) or by an open-source Excel sheet plugin in which the raw Cq values need to be converted to relative quantities, the required data input format. geNorm is not suitable for analyzing a large number of candidate genes, since the possibility exists that testing genes are mutually correlated (e.g., sharing the same pathway and undergoing the same regulation). Such genes would be erroneously evaluated by geNorm as the most stable, since their pairwise variation would be minimal, a result which can cause bias in reference gene selection. **NormFinder**, the model-based approach, calculates overall expression level variation, a parameter which can be easily obtained by RefFinder. In addition, NormFinder itself can differentiate among intragroup variations and therefore can be used when reference genes are to be assessed for different sample groups (Andersen et al. 2004). To obtain such a deeper analysis, the NormFinder functionality can be added directly to Excel, and a version for R is also available. **BestKeeper** calculates the coefficient of variance and standard deviation (SD) of the raw Cq values and establishes a BestKeeper Index (BKI) based on the Pearson correlation of each individual gene with the geometric mean of all candidate reference genes (Pfaffl et al. 2004). The interpretation of the calculated data is not as straightforward as in the two programs above; thus good judgment needs to be employed; the correlation factor (BKI) (the closer to 1, the better) and the SD (the larger, the worse) should both be taken into account. RefFinder ranks the candidate genes based solely on SD, a measure which can differ substantially from the ranking based on BKI, as demonstrated by De Spiegelaere et al. (2015). **The comparative delta-Ct** (ΔCt) is a simple method that ranks genes based on the spread of Cq values, considering the gene with the least variation to be the best reference gene within a given sample set (Silver et al. 2006). Even though differences among the programs bring a different outcome with each one, the most stable and the least stable genes are usually comparable using any of the programs for each sample set. Therefore, a choice of reference gene based only on RefFinder that takes into consideration the output of all four programs can be used, albeit with a degree of caution (De Spiegelaere et al. 2015).

Such a knowledge-driven approach for the normalization of low-throughput qPCRs based on pilot study identifying suitable endogenous reference gene(s) from a set of candidate genes is recommended. On the other hand, several data-driven approaches for the normalization of **larger data sets** have been proposed. A mean-centering method developed by Mestdagh et al. uses the mean expression value of all expressed miRNAs in a given sample as a normalization factor (Mestdagh et al. 2009). Such a method can be used when large number of miRNAs are profiled, e.g., for Megaplex RT-qPCR technology, for microarrays, and recently also for transcriptome sequencing. The mean expression value normalization reduces technical variation and identifies true biological changes in patient samples. Improvement of this method was reported by Wylie et al. as being mean-centering restricted, i.e., including only those miRNAs that are expressed in all samples with no value missing for the "mean" calculation. Such improvement is useful especially for biofluid data, during which a substantial fraction of miRNA data values may be missing (Wylie et al. 2011). Both of these data-driven methods can be used when moving from large-scale to small-scale experiments, where such approaches can be used in pilot experiments for the selection of miRNAs that resemble the mean expression value, followed by a targeted experiment analyzing a limited number of genes (knowledge-driven selection). Other normalization methods specifically developed for microarray data (e.g., scale, quantile, etc.) that are usually included in microarray expression analysis software tend to perform worse than miRNA-specific normalization. These procedures should be used with caution, especially with biofluids, where numerous expression values are missing, which in turn can cause bias in the interpretation.

Other means of normalization without the use of reference genes have also been proposed, e.g., the use of a **fixed starting amount of plasma** for RNA extraction and subsequently using a fixed amount of RNA eluate in the RT reaction (Mitchell et al. 2008) or normalizing by matching the amount of **input RNA** in the RT reaction. This is not an appropriate approach, however, since the RNA content of plasma can vary considerably, in particular with various disease states. Furthermore, synthetic **RNA spike-in** controls are sometimes used in qPCR experiments. Various synthetic nonhuman miRNAs have been used, for example, cel-miR-39 from *C. elegans* or ath-miR-159a *from Arabidopsis thaliana*. The concentration of spike-in controls should be optimized to cover the expected range of intensities of biological samples (Sarkar et al. 2009). Mitchell et al. based normalization upon spiking three synthetic *C. elegans* RNA oligonucleotides into the plasma sample after the addition of a denaturing solution (Mitchell et al. 2008). Such an approach is useful for assay validation and quality control and to correct for extraction efficiency or reverse transcription efficiency, but it does not truly reflect the biological variation.

Examples of Commonly Used Reference Genes for miRNA Quantification

As mentioned above, there is no universally accepted reference gene/normalizer for miRNA data normalization (Schwarzenbach et al. 2015). The choice of an endogenous reference gene is a critical step in order to truly assess biological variations in analyzed samples. Numerous reports on the selection of reference genes have served as pilot experiments, and as can be seen in Table 1, the outcome for each particular experiment can be very different, although the same miRNAs have been analyzed. It should be noted that small nuclear RNA RNU6 and miR-16 have frequently been used in studies of miRNA profiling, often without any validation of their expression stability. The selection of normalizers should be undertaken carefully prior to data interpretation, with a combination of several normalizers possibly being more effective than the use of a single universal normalizer (Shen et al. 2015).

Table 1 Selection of studies focused on reference miRNA identification for the data normalization. Studies that focus on selection of stable reference genes have different number of tested candidate genes (when also tested RNU6 or ribosomal RNA 5S "+U6" or "+5S" is included, respectively). Outcome of such selection can significantly differ, even when the same sources (serum, plasma) are used, because the tested genes can be affected by different disease conditions

No. of miRNAs tested	Identified the most stable miRNAs	Algorithms/ Programs used for evaluation	Source	Disease	References
5	miR-193a, miR-16	geNorm, NormFinder	Serum	Bladder cancer	(Wang et al. 2015b)
6+U6	miR-16, miR-93	RefFinder	Serum	Gastric cancer	(Song et al. 2012)
9+5S+U6	miR-221, miR-191, let-7a, miR-181a, miR-26a	geNorm, NormFinder	Serum exosomes	Hepatitis B or hepatocellular carcinoma	(Li et al. 2015b)
8+U6+5S	miR-221, let-7a, miR-26a	RefFinder	Serum exosomes (pre and post resection)	Liver carcinoma	(Li et al. 2015a)
6	miR-101, miR-93	geNorm, BestKeeper, NormFinder, Delta Ct (RankAggreg)	Plasma	Major depressive disorder	(Liu et al. 2014)
12+U6	miR-93	geNorm, NormFinder	Plasma	Tuberculosis patients	(Barry et al. 2015)
11+U6	miR-25, miR-93	NormFinder	6 human cancer cell lines		(Das et al. 2016)
8+U6+5S	miR-103, miR-106b, miR-26b	geNorm, NormFinder	Differentiating neuronal cell lines		(Lim et al. 2011)
9+U6+5S	miR-191, miR-93	geNorm, NormFinder	Tissue	Lung cancer (+FFPE)	(Peltier and Latham 2008)
15+U6	mir-152, mir-23b	geNormPlus (qbase), NormFinder	Tissue	Liver-healthy male/female	(Lamba et al. 2014)
3+U6	miR-130b, U6	geNorm, NormFinder	Tissue	Prostate carcinoma/ adjacent normal tissue	(Schaefer et al. 2010)
11	miR-16, miR-29a	geNorm, NormFinder	Tissue	FFPE, fresh biopsies	(Rinnerthaler et al. 2016)

RNU6

Small noncoding RNA, e.g., RNU6, is the most frequently used normalizer for miRNA relative quantification. Nevertheless, several reasons can be put forth against the use of RNU6 as reference gene. Technically, small noncoding RNA is not miRNA, and therefore it does not reflect the true biochemical character of miRNA molecules; thus the efficiency of extraction or reverse transcription might differ from that of miRNAs (Schwarzenbach et al. 2015). Longer RNAs are more easily degraded by RNase, and RNU6 is significantly longer than miRNA. Xiang et al. have reported the lower stability of RNU6 compared to miRNAs after repeated freeze-thaw cycles (Xiang et al. 2014). Especially for the normalization of circulating miRNAs, RNU6 proved to be unsuitable. Wang et al. have compared differences between serum and plasma in the miRNA spectrum, revealing that RNU6 were not detectable or had very high concentration variations in both components. In fact, RNU6 is present in plasma or serum only as a result of cell lysis or coagulation, both of which are highly influenced by pre-analytical factors; thus using RNU6 for the normalization of serum and plasma miRNAs is therefore highly inappropriate (Witwer and Halushka 2016). Furthermore, RNU6 has repeatedly been reported as unsuitable for normalization, e.g., in liver tissue (Lamba et al. 2014) and serum (Benz et al. 2013). Apparently, RNU6 is also not suitable as a reference gene for comparison of carcinoma and adjacent normal tissue. Lou et al. compared RNU6 distribution among several carcinoma and corresponding normal tissues and found a high variation in comparison to miR-16. Furthermore, miRNA in situ hybridization used for the detection of RNU6 expression in breast carcinoma sections revealed higher expression levels of RNU6 in epithelial cells than in mesenchymal cells (Lou et al. 2015). RNU6 is possibly suitable for use in the normalization of tissue microRNA but only after thorough validation of its stability across all samples (Schaefer et al. 2010; Liang et al. 2007).

5S Ribosomal RNA

Although the reasons against the use of 5S ribosomal RNA are similar (e.g., it is not miRNA and possesses a longer sequence), some publications revealed its suitability as an endogenous reference gene. The selection of reference genes for qPCR in the tea plant revealed 5S as the most stable gene for the comparison of data from different organs (Song et al. 2016). Reference gene selection in a rat model of acute hepatotoxicity identified 5S (along with miR-16) as the most stable (Lardizabal et al. 2012). In contrast, 5S was among the least stable in serum exosomes from patients with liver carcinoma (Li et al. 2015a) as well as from FFPE lung carcinoma samples (Peltier and Latham 2008). 5S was also proved as unstable for neuronal differentiation studies (Lim et al. 2011). In some cases, 5S was even used as a normalizer without any validation (e.g., Wang et al. 2015a).

miR-16

miR-16 is a frequently used endogenous miRNA for normalization. As revealed in several publications, the miR-16 was indeed among the most stable identified miRNAs, e.g., levels of miR-16 in serum taken from patients with bladder cancer and gastric cancer were stable (Wang et al. 2015b; Song et al. 2012). Similarly, miRNA analysis in FFPE samples taken from breast cancer samples and surrounding tissues as well as a comparison of primary tumors with metastases revealed miR-16 (along with miR-29a) as being the most stably expressed gene (Rinnerthaler et al. 2016). On the contrary, Li et al. identified miR-16 as highly unstable (together with RNU6) in serum exosomes in patients with hepatocellular carcinoma (Li et al. 2015b). Furthermore, miR-16 has been proposed as a differential plasma biomarker in the diagnosis and prognosis of hyperacute cerebral infarction (Tian et al. 2016) and Alzheimer's disease (Cheng et al. 2015).

miR-93

An interesting miRNA candidate seems to be miR-93, as it was identified as ranking among the most stable miRNAs in plasma from patients with major depressive disorder in comparison to healthy controls (Liu et al. 2014) and patients with tuberculosis (Barry et al. 2015). In addition, miRNA analysis of various human cancer cell lines has revealed miR-93 (along with miR-25) as the most stable gene (Das et al. 2016). Peltier has also suggested miR-93 (along with miR-191) as a suitable normalizer for solid cancer tissues (Peltier and Latham 2008). However, miR-93 as a normalizer should also be used with caution, as it has repeatedly been reported as deregulated miRNA in various conditions (e.g., Ansari et al. 2016).

Conclusion

In conclusion, as of this writing, no microRNA has been proved to be the ideal reference gene. However, qPCR experiments especially for the selection of biomarkers of a disease or dietary intervention need to reflect subtler changes in homeostasis. A thorough analysis of the method for referencing is highly desired to avoid any possible bias caused by the misinterpretation of obtained results. Therefore, several microRNAs should be tested across the whole experimental setup to verify their stability and usefulness for particular experiments. As repeatedly recommended, the workflow in proper qPCR analysis should start with a selection of candidate reference genes based on a literature search, analysis, and validation of the most stable genes within a given sample set. To minimize processing bias and technical variation within the sample preparation, the use of spike-in controls is also recommended.

Dictionary of Terms

- **Quantitative real-time polymerase chain reaction (qPCR)** – A method for the quantitative detection of a gene of interest in a tested sample. qPCR is based on the monitoring of the amplification in each cycle of the process and enables the quantitative assessment of the amount of a target gene.
- **Absolute quantification** – Absolute quantity, e.g., the copy number of a gene of interest, can be obtained by a comparison to the synthetic standard of known concentration through the use of a standard curve.
- **Relative quantification** – The amount of a target gene can be relatively compared without knowing the absolute level of expression. Changes of gene expression in a given sample or under certain conditions can be relatively compared to a reference sample, e.g., an untreated control.
- **Normalization** – Normalization is the process of reducing variation between samples of non-biological origin, the purpose being to reduce any differences that may arise in the handling or processing of the samples.
- **Reference gene (RG)** – RG is used for normalization. RG must be stably expressed in all samples tested and should not vary under any experimental conditions.
- **Spike-in control** – To reduce variations in an extraction method, a synthetic spike-in control can be introduced into the sample before isolation. RNA spike-in controls added before RNA isolation can be used for extraction assay validation or reverse transcription efficiency correction.
- **miRBase** – All annotated microRNAs are stored in a miRBase database under a specific accession number.

Key Facts of microRNAs

- microRNAs are short noncoding RNAs first described in 1993.
- microRNAs are negative posttranscriptional regulators which generally cause either mRNA degradation or repression of translation.
- Single microRNA can target several mRNAs, and a single mRNA can be targeted by several microRNAs.
- Circulating microRNAs are surprisingly stable.
- All annotated microRNAs are stored in the database miRBase.

Summary Points

- This chapter provides basic information on microRNAs, their biogenesis, and nomenclature with special focus on circulating microRNAs.
- Methods of microRNA detection and quantification are briefly summarized.
- Real-time PCR amplification enables accurate microRNA quantification; MIQE guidelines outline the minimum information required for publishing qPCR data.

- Relative quantification requires the selection of an endogenous control/reference gene for proper normalization. Means of normalization and available programs are discussed.
- Reference gene(s) should be expressed in all samples, should not vary across the sample set, and should not be affected by tested experimental conditions or compounds. Ideally the expression level should be close to the genes of interest.
- Commonly used reference genes miR-16 and RNU6 should not be used without prior verification of their stability across the experimental setup.

References

Ambros V, Bartel B, Bartel DP, Burge CB, Carrington JC, Chen XM, Dreyfuss G, Eddy SR, Griffiths-Jones S, Marshall M, Matzke M, Ruvkun G, Tuschl T (2003) A uniform system for microRNA annotation. RNA 9:277–279. https://doi.org/10.1261/rna.2183803

Andersen CL, Jensen JL, Ørntoft TF (2004) Normalization of real-time quantitative reverse transcription-PCR data: a model-based variance estimation approach to identify genes suited for normalization, applied to bladder and colon cancer data sets. Cancer Res 64:5245–5250. https://doi.org/10.1158/0008-5472.CAN-04-0496

Ansari MH, Irani S, Edalat H, Amin R, Roushandeh AM (2016) Deregulation of miR-93 and miR-143 in human esophageal cancer. Tumor Biol 37:3097–3103. https://doi.org/10.1007/s13277-015-3987-9

Barry SE, Chan B, Ellis M, Yang YR, Plit ML, Guan GY, Wang XL, Britton WJ, Saunders BM (2015) Identification of miR-93 as a suitable miR for normalizing miRNA in plasma of tuberculosis patients. J Cell Mol Med 19:1606–1613. https://doi.org/10.1111/jcmm.12535

Bartel DP (2004) MicroRNAs: genomics, biogenesis, mechanism, and function. Cell 116:281–297. https://doi.org/10.1016/s0092-8674(04)00045-5

Benz F, Roderburg C, Cardenas DV, Vucur M, Gautheron J, Koch A, Zimmermann H, Janssen J, Nieuwenhuijsen L, Luedde M, Frey N, Tacke F, Trautwein C, Luedde T (2013) U6 is unsuitable for normalization of serum miRNA levels in patients with sepsis or liver fibrosis. Exp Mol Med 45. https://doi.org/10.1038/emm.2013.81

Bustin SA, Benes V, Garson JA, Hellemans J, Huggett J, Kubista M, Mueller R, Nolan T, Pfaffl MW, Shipley GL, Vandesompele J, Wittwer CT (2009) The MIQE guidelines: minimum information for publication of quantitative real-time PCR experiments. Clin Chem 55:611–622. https://doi.org/10.1373/clinchem.2008.112797

Chakraborty C, Das S (2016) Profiling cell-free and circulating miRNA: a clinical diagnostic tool for different cancers. Tumour Biol 37:5705–5714. https://doi.org/10.1007/s13277-016-4907-3

Chen CF, Ridzon DA, Broomer AJ, Zhou ZH, Lee DH, Nguyen JT, Barbisin M, Xu NL, Mahuvakar VR, Andersen MR, Lao KQ, Livak KJ, Guegler KJ (2005) Real-time quantification of microRNAs by stem-loop RT-PCR. Nucleic Acids Res 33. https://doi.org/10.1093/nar/gni178

Cheng XH, Ku CH, Siow RCM (2013) Regulation of the Nrf2 antioxidant pathway by microRNAs: new players in micromanaging redox homeostasis. Free Radic Biol Med 64:4–11. https://doi.org/10.1016/j.freeradbiomed.2013.07.025

Cheng L, Doecke JD, Sharples RA, Villemagne VL, Fowler CJ, Rembach A, Martins RN, Rowe CC, Macaulay SL, Masters CL, Hill AF, Australian Imaging B, Lifestyle Research G (2015) Prognostic serum miRNA biomarkers associated with Alzheimer's disease shows concordance with neuropsychological and neuroimaging assessment. Mol Psychiatry 20:1188–1196. https://doi.org/10.1038/mp.2014.127

Cloutier F, Marrero A, O'connell C, Morin P (2015) MicroRNAs as potential circulating biomarkers for amyotrophic lateral sclerosis. J Mol Neurosci 56:102–112. https://doi.org/10.1007/s12031-014-0471-8

D'angelo B, Benedetti E, Cimini A, Giordano A (2016) MicroRNAs: a puzzling tool in cancer diagnostics and therapy. Anticancer Res 36:5571–5575. https://doi.org/10.21873/anticanres.11142

Das MK, Andreassen R, Haugen TB, Furu K (2016) Identification of endogenous controls for use in miRNA quantification in human cancer cell lines. Cancer Genomics Proteomics 13:63–68

De Spiegelaere W, Dern-Wieloch J, Weigel R, Schumacher V, Schorle H, Nettersheim D, Bergmann M, Brehm R, Kliesch S, Vandekerckhove L, Fink C (2015) Reference gene validation for RT-qPCR, a note on different available software packages. PLoS One 10. https://doi.org/10.1371/journal.pone.0122515

Dheda K, Huggett JF, Chang JS, Kim LU, Bustin SA, Johnson MA, Rook GAW, Zumla A (2005) The implications of using an inappropriate reference gene for real-time reverse transcription PCR data normalization. Anal Biochem 344:141–143. https://doi.org/10.1016/j.ab.2005.05.022

Dugas DV, Bartel B (2004) MicroRNA regulation of gene expression in plants. Curr Opin Plant Biol 7:512–520. https://doi.org/10.1016/j.pbi.2004.07.011

Esquela-Kerscher A, Slack FJ (2006) Oncomirs – microRNAs with a role in cancer. Nat Rev Cancer 6:259–269. https://doi.org/10.1038/nrc1840

Fierro-Fernandez M, Miguel V, Lamas S (2016) Role of redoximiRs in fibrogenesis. Redox Biol 7:58–67. https://doi.org/10.1016/j.redox.2015.11.006

Garcia-Segura L, Perez-Andrade M, Miranda-Rios J (2013) The emerging role of microRNAs in the regulation of gene expression by nutrients. J Nutrigenet Nutrigenomics 6:16–31. https://doi.org/10.1159/000345826

Griffiths-Jones S, Grocock RJ, Van Dongen S, Bateman A, Enright AJ (2006) miRBase: microRNA sequences, targets and gene nomenclature. Nucleic Acids Res 34:D140–D144. https://doi.org/10.1093/nar/gkj112

Guduric-Fuchs J, O'connor A, Camp B, O'neill CL, Medina RJ, Simpson DA (2012) Selective extracellular vesicle-mediated export of an overlapping set of microRNAs from multiple cell types. BMC Genomics 13. https://doi.org/10.1186/1471-2164-13-357

Hellemans J, Mortier G, De Paepe A, Speleman F, Vandesompele J (2007) qBase relative quantification framework and software for management and automated analysis of real-time quantitative PCR data. Genome Biol 8:1–14

Hindson BJ, Ness KD, Masquelier DA, Belgrader P, Heredia NJ, Makarewicz AJ, Bright IJ, Lucero MY, Hiddessen AL, Legler TC, Kitano TK, Hodel MR, Petersen JF, Wyatt PW, Steenblock ER, Shah PH, Bousse LJ, Troup CB, Mellen JC, Wittmann DK, Erndt NG, Cauley TH, Koehler RT, So AP, Dube S, Rose KA, Montesclaros L, Wang SL, Stumbo DP, Hodges SP, Romine S, Milanovich FP, White HE, Regan JF, Karlin-Neumann GA, Hindson CM, Saxonov S, Colston BW (2011) High-throughput droplet digital PCR system for absolute quantitation of DNA copy number. Anal Chem 83:8604–8610. https://doi.org/10.1021/ac202028g

Hrustincova A, Votavova H, Merkerova MD (2015) Circulating microRNAs: methodological aspects in detection of these biomarkers. Folia Biol 61:203–218

Lamba V, Ghodke-Puranik Y, Guan W, Lamba JK (2014) Identification of suitable reference genes for hepatic microRNA quantitation. BMC Res Notes 7:129. https://doi.org/10.1186/1756-0500-7-129

Lardizabal MN, Nocito AL, Daniele SM, Ornella LA, Palatnik JF, Veggi LM (2012) Reference genes for real-time PCR quantification of microRNAs and messenger RNAs in rat models of hepatotoxicity. PLoS One 7. https://doi.org/10.1371/journal.pone.0036323

Lee RC, Ambros V (2001) An extensive class of small RNAs in *Caenorhabditis elegans*. Science 294:862–864. https://doi.org/10.1126/science.1065329

Li Y, Xiang GM, Liu LL, Liu C, Liu F, Jiang DN, Pu XY (2015a) Assessment of endogenous reference gene suitability for serum exosomal microRNA expression analysis in liver carcinoma resection studies. Mol Med Rep 12:4683–4691. https://doi.org/10.3892/mmr.2015.3919

Li Y, Zhang L, Liu F, Xiang G, Jiang D, Pu X (2015b) Identification of endogenous controls for analyzing serum exosomal miRNA in patients with hepatitis B or hepatocellular carcinoma. Dis Markers 2015:893594. https://doi.org/10.1155/2015/893594

Liang Y, Ridzon D, Wong L, Chen C (2007) Characterization of microRNA expression profiles in normal human tissues. BMC Genomics 8:166. https://doi.org/10.1186/1471-2164-8-166

Lim QE, Zhou L, Ho YK, Wan G, Too HP (2011) snoU6 and 5S RNAs are not reliable miRNA reference genes in neuronal differentiation. Neuroscience 199:32–43. https://doi.org/10.1016/j.neuroscience.2011.10.024

Liu XL, Zhang L, Cheng K, Wang X, Ren GP, Xie P (2014) Identification of suitable plasma-based reference genes for miRNAome analysis of major depressive disorder. J Affect Disord 163:133–139. https://doi.org/10.1016/j.jad.2013.12.035

Lou G, Ma N, Xu Y, Jiang L, Yang J, Wang C, Jiao Y, Gao X (2015) Differential distribution of U6 (RNU6-1) expression in human carcinoma tissues demonstrates the requirement for caution in the internal control gene selection for microRNA quantification. Int J Mol Med 36:1400–1408. https://doi.org/10.3892/ijmm.2015.2338

Ludwig N, Leidinger P, Becker K, Backes C, Fehlmann T, Pallasch C, Rheinheimer S, Meder B, Stahler C, Meese E, Keller A (2016) Distribution of miRNA expression across human tissues. Nucleic Acids Res 44:3865–3877. https://doi.org/10.1093/nar/gkw116

Mccarthy JJ (2008) MicroRNA-206: the skeletal muscle-specific myomiR. Biochim Biophys Acta 1779:682–691. https://doi.org/10.1016/j.bbagrm.2008.03.001

Mestdagh P, Van Vlierberghe P, De Weer A, Muth D, Westermann F, Speleman F, Vandesompele J (2009) A novel and universal method for microRNA RT-qPCR data normalization. Genome Biol 10. https://doi.org/10.1186/gb-2009-10-6-r64

Mitchell PS, Parkin RK, Kroh EM, Fritz BR, Wyman SK, Pogosova-Agadjanyan EL, Peterson A, Noteboom J, O'briant KC, Allen A, Lin DW, Urban N, Drescher CW, Knudsen BS, Stirewalt DL, Gentleman R, Vessella RL, Nelson PS, Martin DB, Tewari M (2008) Circulating microRNAs as stable blood-based markers for cancer detection. Proc Natl Acad Sci USA 105:10513–10518. https://doi.org/10.1073/pnas.0804549105

Occhipinti G, Giulietti M, Principato G, Piva F (2016) The choice of endogenous controls in exosomal microRNA assessments from biofluids. Tumor Biol 37:11657–11665. https://doi.org/10.1007/s13277-016-5164-1

Peltier HJ, Latham GJ (2008) Normalization of microRNA expression levels in quantitative RT-PCR assays: identification of suitable reference RNA targets in normal and cancerous human solid tissues. RNA 14:844–852. https://doi.org/10.1261/rna.939908

Pfaffl MW, Tichopad A, Prgomet C, Neuvians TP (2004) Determination of stable housekeeping genes, differentially regulated target genes and sample integrity: BestKeeper – excel-based tool using pair-wise correlations. Biotechnol Lett 26:509–515. https://doi.org/10.1023/b:bile.0000019559.84305.47

Raymond CK, Roberts BS, Garrett-Engele P, Lim LP, Johnson JM (2005) Simple, quantitative primer-extension PCR assay for direct monitoring of microRNAs and short-interfering RNAs. RNA 11:1737–1744. https://doi.org/10.1261/rna.2148705

Rinnerthaler G, Hackl H, Gampenrieder SP, Hamacher F, Hufnagl C, Hauser-Kronberger C, Zehentmayr F, Fastner G, Sedlmayer F, Mlineritsch B, Greil R (2016) miR-16-5p is a stably-expressed housekeeping microRNA in breast cancer tissues from primary tumors and from metastatic sites. Int J Mol Sci 17:156. https://doi.org/10.3390/ijms17020156

Rome S (2015) Use of miRNAs in biofluids as biomarkers in dietary and lifestyle intervention studies. Genes Nutr 10:483. https://doi.org/10.1007/s12263-015-0483-1

Sarkar D, Parkin R, Wyman S, Bendoraite A, Sather C, Delrow J, Godwin AK, Drescher C, Huber W, Gentleman R, Tewari M (2009) Quality assessment and data analysis for microRNA expression arrays. Nucleic Acids Res 37:e17. https://doi.org/10.1093/nar/gkn932

Schaefer A, Jung M, Miller K, Lein M, Kristiansen G, Erbersdobler A, Jung K (2010) Suitable reference genes for relative quantification of miRNA expression in prostate cancer. Exp Mol Med 42:749–758. https://doi.org/10.3858/emm.2010.42.11.07

Schwarzenbach H, Da Silva AM, Calin G, Pantel K (2015) Data normalization strategies for microRNA quantification. Clin Chem 61:1333–1342. https://doi.org/10.1373/clinchem.2015.239459

Shen Y, Tian F, Chen Z, Li R, Ge Q, Lu Z (2015) Amplification-based method for microRNA detection. Biosens Bioelectron 71:322–331. https://doi.org/10.1016/j.bios.2015.04.057

Silver N, Best S, Jiang J, Thein SL (2006) Selection of housekeeping genes for gene expression studies in human reticulocytes using real-time PCR. BMC Mol Biol 7. https://doi.org/10.1186/1471-2199-7-33

Song J, Bai Z, Han W, Zhang J, Meng H, Bi J, Ma X, Han S, Zhang Z (2012) Identification of suitable reference genes for qPCR analysis of serum microRNA in gastric cancer patients. Dig Dis Sci 57:897–904. https://doi.org/10.1007/s10620-011-1981-7

Song H, Zhang X, Shi C, Wang S, Wu A, Wei C (2016) Selection and verification of candidate reference genes for mature microRNA expression by quantitative RT-PCR in the tea plant (Camellia sinensis). Genes (Basel) 7:25. https://doi.org/10.3390/genes7060025

Takada S, Mano H (2007) Profiling of microRNA expression by mRAP. Nat Protoc 2:3136–3145. https://doi.org/10.1038/nprot.2007.457

Tan YW, Ge GH, Pan TL, Wen DF, Gan JH (2015) Serum MiRNA panel as potential biomarkers for chronic hepatitis B with persistently normal alanine aminotransferase. Clin Chim Acta 451:232–239. https://doi.org/10.1016/j.cca.2015.10.002

Tarallo S, Pardini B, Mancuso G, Rosa F, Di Gaetano C, Rosina F, Vineis P, Naccarati A (2014) MicroRNA expression in relation to different dietary habits: a comparison in stool and plasma samples. Mutagenesis 29:385–391. https://doi.org/10.1093/mutage/geu028

Tian CO, Li ZF, Yang ZG, Huang QH, Liu JM, Hong B (2016) Plasma microRNA-16 is a biomarker for diagnosis, stratification, and prognosis of hyperacute cerebral infarction. PLoS One 11. https://doi.org/10.1371/journal.pone.0166688

Turchinovich A, Weiz L, Langheinz A, Burwinkel B (2011) Characterization of extracellular circulating microRNA. Nucleic Acids Res 39:7223–7233. https://doi.org/10.1093/nar/gkr254

Van Peer G, Lefever S, Anckaert J, Beckers A, Rihani A, Van Goethem A, Volders PJ, Zeka F, Ongenaert M, Mestdagh P, Vandesompele J (2014) miRBase tracker: keeping track of microRNA annotation changes. Database (Oxford) 2014. https://doi.org/10.1093/database/bau080

Vandesompele J, De Preter K, Pattyn F, Poppe B, Van Roy N, De Paepe A, Speleman F (2002) Accurate normalization of real-time quantitative RT-PCR data by geometric averaging of multiple internal control genes. Genome Biol 3. https://doi.org/10.1186/gb-2002-3-7-research0034

Vecchione A, Croce CM (2010) Apoptomirs: small molecules have gained the license to kill. Endocr Relat Cancer 17:F37–F50. https://doi.org/10.1677/erc-09-0163

Vogelstein B, Kinzler KW (1999) Digital PCR. Proc Natl Acad Sci USA 96:9236–9241. https://doi.org/10.1073/pnas.96.16.9236

Wang ZYB (2010) MicroRNAs expression detection methods. Springer, Berlin

Wang JY, Mao RC, Zhang YM, Zhang YJ, Liu HY, Qin YL, Lu MJ, Zhang JM (2015a) Serum microRNA-124 is a novel biomarker for liver necroinflammation in patients with chronic hepatitis B virus infection. J Viral Hepat 22:128–136. https://doi.org/10.1111/jvh.12284

Wang LS, Liu YM, Du LT, Li J, Jiang XM, Zheng GX, Qu AL, Wang HY, Wang LL, Zhang X, Liu H, Pan HW, Yang YM, Wang CX (2015b) Identification and validation of reference genes for the detection of serum microRNAs by reverse transcription-quantitative polymerase chain reaction in patients with bladder cancer. Mol Med Rep 12:615–622. https://doi.org/10.3892/mmr.2015.3428

Weber JA, Baxter DH, Zhang S, Huang DY, Huang KH, Lee MJ, Galas DJ, Wang K (2010) The microRNA spectrum in 12 body fluids. Clin Chem 56:1733–1741. https://doi.org/10.1373/clinchem.2010.147405

Witwer KW (2012) XenomiRs and miRNA homeostasis in health and disease evidence that diet and dietary miRNAs directly and indirectly influence circulating miRNA profiles. RNA Biol 9:1147–1154. https://doi.org/10.4161/rna.21619

Witwer KW, Halushka MK (2016) Toward the promise of microRNAs – enhancing reproducibility and rigor in microRNA research. RNA Biol 13:1103–1116. https://doi.org/10.1080/15476286.2016.1236172

Witwer KW, Mcalexander MA, Queen SE, Adams RJ (2013) Real-time quantitative PCR and droplet digital PCR for plant miRNAs in mammalian blood provide little evidence for general uptake of dietary miRNAs limited evidence for general uptake of dietary plant xenomiRs. RNA Biol 10:1080–1086

Wylie D, Shelton J, Choudhary A, Adai AT (2011) A novel mean-centering method for normalizing microRNA expression from high-throughput RT-qPCR data. BMC Res Notes 4:555. https://doi.org/10.1186/1756-0500-4-555

Xiang MQ, Zeng Y, Yang RR, Xu HF, Chen Z, Zhong J, Xie HL, Xu YH, Zeng X (2014) U6 is not a suitable endogenous control for the quantification of circulating microRNAs. Biochem Biophys Res Commun 454:210–214. https://doi.org/10.1016/j.bbrc.2014.10.064

Xie F, Xiao P, Chen D, Xu L, Zhang B (2012) miRDeepFinder: a miRNA analysis tool for deep sequencing of plant small RNAs. Plant Mol Biol. https://doi.org/10.1007/s11103-012-9885-2

Mass Spectrometry and Epigenetics 118

Luciano Nicosia, Roberta Noberini, Monica Soldi,
Alessandro Cuomo, Daniele Musiani, Valeria Spadotto, and
Tiziana Bonaldi

Contents

Abstract

Chromatin is a nucleoprotein complex composed of DNA and histone proteins. The concerted activity of chromatin-associated proteins, histone post-translational modifications, and DNA methylation induces epigenetic variations that regulate most of the physiological processes of eukaryotic cells, ranging from gene expression to DNA replication and repair. Epigenetics has also been shown to be tightly linked to cell metabolism. For instance, histone modifications are highly sensitive to the changes in the microenvironment and the local concentration of specific metabolites. Mass-spectrometry (MS)-based proteomics

L. Nicosia · M. Soldi · A. Cuomo · D. Musiani · V. Spadotto · T. Bonaldi (✉)
Department of Experimental Oncology, European Institute of Oncology, Milan, Italy
e-mail: luciano.nicosia@ieo.it; monica.soldi@ieo.it; alessandro.cuomo@ieo.it; daniele.musiani@ieo.it; valeria.spadotto@ieo.it; tiziana.bonaldi@ieo.it

R. Noberini
Center for Genomic Science of IIT@SEMM, Istituto Italiano di Tecnologia, Milan, Italy
e-mail: roberta.noberini@iit.it

V. B. Patel, V. R. Preedy (eds.), *Handbook of Nutrition, Diet, and Epigenetics*,
https://doi.org/10.1007/978-3-319-55530-0_115

significantly contributed to the recent advances in the epigenetic field, by allowing the comprehensive analysis of histone post-translational modifications as well as the systematic identification of chromatin constituents.

In this chapter, we will provide a general overview of various MS-based experimental strategies developed to boost the epigenetic field, with references to the studies whereby chromatin biology was assessed in relation to cell metabolism.

Keywords
Mass spectrometry · Histone post-translational modifications · Chromatin · Metabolism · Epigenetics · Stable isotope labeling with amino acids in cell culture · Chromatin-associated proteins · Histone modifying enzymes · Proteomics · Global post-translational modification profiling

List of Abbreviations

AF-10	Antisecretor factor 10
ChIP	Chromatin immunoprecipitation
CID	Collision-induced dissociation
CRISPR	Clustered regularly interspaced short palindromic repeats
DDA	Data-dependent acquisition
ECD	Electron capture dissociation
ESC	Embryonic stem cell
ETD	Electron transfer dissociation
HAT	Histone acetyltransferase
HMCV	Cytomegalovirus
hmSILAC	Heavy-methyl SILAC
KAT	Lysine acetyl-transferase
LC	Liquid chromatography
LF	Label-free
MRM	Multiple reaction monitoring
MS	Mass-spectrometry
NPC	Neural progenitor cell
NSC	Neural stem cell
PTM	Post-translational modification
RA	Relative abundance
RP	Reversed-phase
SAM	S-adenosyl-methionine
SILAC	Stable isotope labeling with amino acids in cell culture
SRM	Selected reaction monitoring
TAL	Transcription activator-like
TF	Transcription factor
WCX-HILIC	Weak-cation exchange hydrophilic interaction liquid chromatography
XIC	MS-extracted ion chromatogram

Introduction

Epigenetic mechanisms refer to heritable changes in gene expression that do not involve a change in the nucleotide sequence (Wu and Morris 2001). The main epigenetic mechanisms involve DNA methylation and post-translational modifications on histones, as well as the activity of noncoding RNAs and chromatin remodeling complexes (Tammen et al. 2013).

Chromatin is a composite nucleoprotein complex that mediates the packaging of DNA into the nucleus and through which the eukaryotic cell exerts the regulation of different DNA-mediated processes, such as transcription, replication, and DNA repair. Nucleosomes are the basic units of chromatin and are constituted by repeated units of 147 bp of DNA, wrapped around the histone octamer, which comprises one histone H3-H4 tetramer and two histone H2A-H2B dimers (Kornberg 1974). Histones are subjected to different post-translational modifications (PTMs), mainly occurring at their N-terminal tails (Kouzarides 2007), which vary in number, type, and abundance. The combination of such modifications is hypothesized to create the so-called "histone code" (Strahl and Allis 2000), whereby different PTM combinations generate a code that determines the functional state of the underlying DNA. The regulatory function of histone PTMs is achieved either by controlling chromatin accessibility to different DNA binding proteins that, in turn, mediate downstream processes (Akhtar and Becker 2000), or through the recruitment of "reader" proteins that recognize and bind these marks, initiating the execution of different DNA-based processes (Jenuwein and Allis 2001).

It has recently been shown that histone PTMs are also closely linked with the metabolic state of the cell. Cell metabolism affects the activity of the enzymes responsible for the placement and removal of the histone marks, known as "writers" and "erasers," respectively, and determines the abundance of several metabolites that function as either co-factors or donors in the enzymatic reaction of modification (Fan et al. 2015). For instance, acetyl-coA, the donor of acetyl groups in the histone acetylation reaction catalyzed by histone acetyltransferase (HAT) enzymes, is also a crucial molecule in the metabolism of the eukaryotes, at the cross road of both catabolic and anabolic processes. Its level is dynamically regulated by environmental and genetic factors and impacts on histone acetylation and, consequently, gene expression (Lee et al. 2014; Morrish et al. 2010). Histone methylation is also downstream of various metabolic pathways impinging on S-adenosyl-methionine (SAM), which is used by methyltransferase enzymes as a methyl-donor in the reaction of histone methylation. SAM is the product of methionine degradation, an essential amino acid taken from the diet. Therefore, dietary and metabolic factors strongly influence the levels of SAM and, thus, of histone methylation (Fan et al. 2015; Shyh-Chang et al. 2013).

Mass-spectrometry (MS)-based proteomics has emerged as a very powerful approach to analyze histone PTMs. The main advantages of MS-based techniques include the possibility to study in a comprehensive manner known and novel histone PTMs and their combinations, with a quantitative assessment of their overall relative abundance. Moreover, MS-methods allow comparing histone PTM abundances among different cellular conditions (Sidoli et al. 2012; Zee et al. 2011).

Beyond histone modifications, quantitative MS allows analyzing chromatin-associated proteins and the proteomic composition of specific chromatin regions (Shiio et al. 2003; Soldi and Bonaldi 2013). Among the different strategies employed to study chromatin composition and architecture, the combination of chromatin immunoprecipitation (ChIP) and mass spectrometry has emerged as the most efficient technique, capable of analyzing both histone modifications and binding proteins in parallel (Soldi et al. 2014).

In this chapter, we will provide an overview of how MS-based proteomics can contribute to epigenetic research, illustrating in particular its applications to the analysis of PTMs on histones and beyond them and to the identification of chromatin-associated proteins.

Mass Spectrometry-Based Proteomics for the Analysis of Histone Modifications

Identification of Histone PTMs

Over the past two decades, MS has proven to be the most suitable technique to identify protein PTMs, thanks to the high mass accuracy and resolution now routinely reachable with modern bench-top MS instruments. In particular, peptide-centric (bottom-up) approaches are widely used for the characterization of histone proteins, following a workflow that is similar to standard shotgun proteomic pipelines. In a bottom-up workflow, proteins are first digested with trypsin and the resulting proteolytic peptides are analyzed by reversed-phase (RP) liquid chromatography (LC)/MSMS in a data-dependent acquisition (DDA) mode, where the ion precursors with the highest intensities in the full MS scan are selected for subsequent fragmentation. However, histones are very basic proteins and trypsin digestion produces some very short peptides that bear functionally relevant epigenetic marks (e.g., the peptide 3–8 of histone H3 bearing K4 methylations), which are barely detectable in standard RP nano-chromatography. In order to overcome this limitation, the Arg-C protease, which digests at the C-terminus of arginine residues, is normally used for bottom-up studies focused on histones. Alternatively, when Arg-C digestion cannot be performed (e.g., for in-gel digestions), lysine residues on histones can be modified with chemical agents, such as propionic or deuterated acetic anhydride, that impair trypsin digestion on the derivatized sites, thus generating "Arg-C like" peptide products displaying an optimal size for subsequent RP-LC/MSMS analysis. In addition, the chemical alkylation of lysine residues increases the hydrophobicity of short peptides and augments their retention time in RP chromatography, which can be exploited to better discriminate between isobaric histone modified peptides. More recently, alternative chemicals for histone lysine derivatization were shown to further improve the detection of very short and hydrophilic peptides (Maile et al. 2015).

Despite its popularity, the bottom-up MS strategy is not ideal to study the physical co-existence of long-distance modifications on histones, which can instead be assessed through top-down and middle-down MS methods, whereby intact or large protein fragments, respectively, are ionized and analyzed. The ability to efficiently fragment intact proteins with high molecular weight is critical in top-down MS. Because collision-induced dissociation (CID), typically used in bottom-up approaches, is inefficient with intact proteins and results in poor fragmentation spectra, especially for multicharged proteins as histones, top-down experiments usually use electron capture dissociation (ECD) and electron transfer dissociation (ETD) fragmentation techniques, which lead to improved MSMS spectra and more extensive sequence coverage of histone proteoforms (Syka et al. 2004; Zubarev et al. 2000). While potentially powerful, top-down approaches currently suffer from several limitations. The separation of isobaric histone proteoforms is extremely challenging and, as a consequence, several isoforms are co-isolated and co-fragmented, generating very complex MSMS spectra, difficult to interpret.

Middle-down MS has gained interest for the combinatorial analysis of histones as a "compromise" between bottom-up and top-down approaches. Middle-down allows studying long peptides (>5 kDa) generated by the digestion with proteases that cleave at less frequent sites compared with trypsin, such as GluC (cleaving at the C-terminus of glutamic acid) or AspN (digesting the N-terminus of aspartic acid). In mammalian histone H3, for instance, a GluC digestion generates a peptide spanning amino acids 1–50, allowing the analysis of the co-occurrence and stoichiometry of at least 23 individual PTMs located within this region. Middle-down approaches combined with weak-cation exchange hydrophilic interaction liquid chromatography (WCX-HILIC) analysis have proved to be highly effective in separating large N-terminal histone tails (Sidoli et al. 2014; Young et al. 2009). Similarly to top-down, ECD and ETD are used for polypeptide fragmentation, as they increase the sequence coverage of long basic peptides (Thomas et al. 2006). Middle-down MS-analysis has also proved valuable in distinguishing histone variants, which is a challenging task in bottom-up MS experiment due to their high sequence similarity (Pesavento et al. 2006; Fig. 1).

MS-Based Strategies to Quantify Histone PTMs

The abundance of histone PTMs can be measured by means of different quantitative MS strategies, including label-free (LF) approaches, stable isotope labeling with amino acids in cell culture (SILAC) and SILAC-derived methods, chemical labeling, and targeted-based methods.

The classical label-free approach for the analysis of histone modifications consists in the direct comparison of unlabeled samples through ion intensity-based quantitation and involves the calculation of relative abundance (RA) value, which corresponds to the ratio between the MS-extracted ion chromatogram (XIC) of a certain modified histone peptide and the sum of the XICs of all the observed modified and unmodified isoforms of the same peptide (Fig. 2; Jung et al. 2010). Isobaric isoforms that cannot be separated by liquid chromatography can be quantified based on the

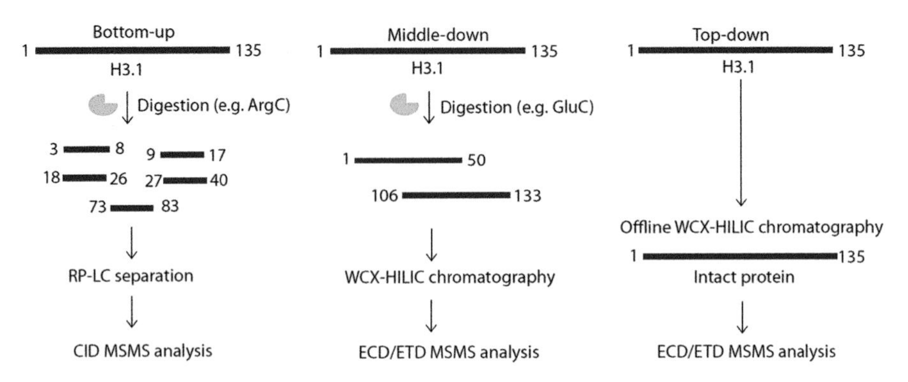

Fig. 1 **MS-based approaches to identify histone PTMs**. The three main strategies for histone PTM identification in MS. "Bottom-up" approach is based on the use of proteases, which produce peptides 6–15 aa long, prior to reversed phase liquid chromatography (RP-LC) separation and MS analysis. "Middle-down" involves the use of proteases, such as Glu-C and Asp-N, which generate longer peptides (around 50 aa), while "Top-Down" consists in the MS- analysis of intact proteins. "Middle-down" and "Top-down" frequently require the use of WCX-HILIC liquid chromatography for peptide/protein separation and ETD for ion fragmentation

relative ratios of fragment ions at the MS2 level (Pesavento et al. 2006). LF- based methods rely on the concept that the XIC of a given modified peptide linearly correlates with its abundance. Although modifications can alter the peptide ionization and consequently their XIC, this assumption is acceptable when the amount of the same modified peptide is compared among distinct samples. Label-free approaches have been used to profile histone modifications in a panel of cancer cells (Leroy et al. 2013), as well as to determine changes in the rate of histone acetylation in the absence of DNA glycosylase (Henry et al. 2016). A great advantage of label-free strategies consists in the possibility to analyze an elevated number of biological conditions without the need of time-consuming and costly labeling, but with the limitation that very high technical and experimental reproducibility is required.

Isotopic-labeling methods have the advantage to minimize the contribution of experimental variations during sample preparation and obtain a more accurate quantitation, even if they are more laborious and, hence, they typically allow lower throughput. The most popular labeling strategy for MS analysis of histones is SILAC, which is based on the metabolic labeling of replicating cells with essential amino acids, such as lysine and arginine, marked with either light or heavy isotopes of various elements, to label up to three different cell populations. When heavy and light labeled protein extracts are mixed and MS-analyzed, the mass difference due to the presence of distinct isotope-encoded tags allows the proteins to remain distinguishable within the same MS spectrum and therefore to be relatively quantified (Ong and Mann 2006). The high accuracy of SILAC-based relative quantitation relies on the fact that protein samples are combined at early stages of the MS-proteomic workflow. SILAC has been employed in various histone-focused studies, for instance, it has been recently used to evaluate the influence of the glycolytic flux on the levels of histone acetylation (Cluntun et al. 2015)

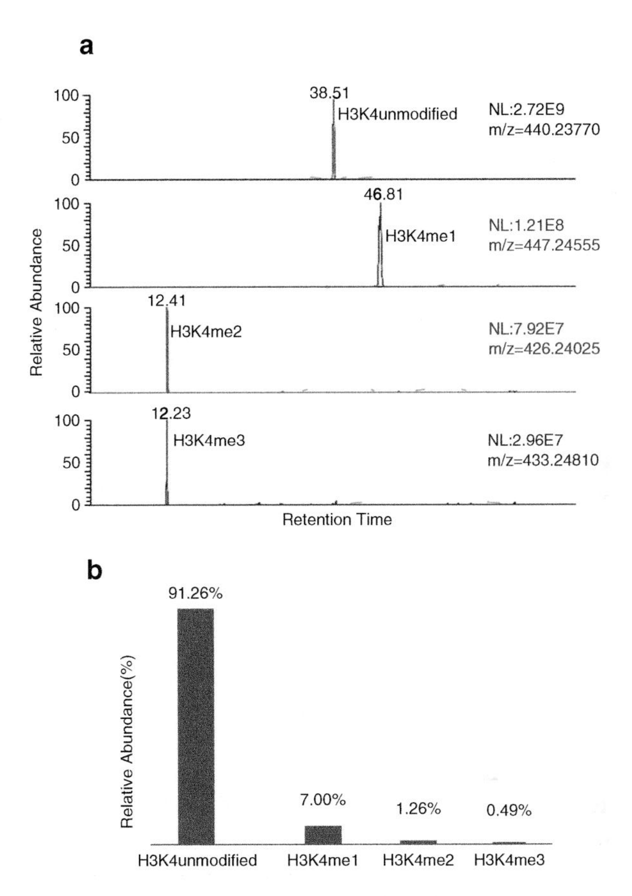

Fig. 2 Label-free approach to quantify histone PTMs. (**a**) Chromatographic peaks associated to the unmodified and modified forms of the H3K4 (mono-, di-, and tri- methylation), extracted from Thermo Xcalibur Qual Browser. The histone peptides showed in this figure have been obtained by means of a "hybrid" chemical labeling method based on an initial propionylation of free and mono-methylated lysines of histones, followed by trypsin digestion and labeling of peptide N-termini with phenyl isocyanate (PIC). The figure shows the retention time, the intensity values (NL), and the mass to charge ratio (m/z) of each histone peptide. m/z is obtained by considering $z = 2$. (**b**) Calculation of the percentage relative abundance (%RA) value of the H3K4 modifications. %RA is given by the ratio between the MS-extracted ion chromatogram (XIC) of each modified peptide over the sum of the XICs of all isoforms of that peptide

and to study modification changes during the differentiation of neural stem cells (NSCs) (An et al. 2016).

Variations of standard SILAC can also be used to quantify histone modifications. For instance, spike-in SILAC strategies can be employed to profile histone PTMs across multiple samples. The rationale behind these approaches is the use of an internal standard, which can derive either from a single cell line or a group of representative cell lines that generate a histone-focused version of the "super-SILAC" mix (Fig. 3). A super-SILAC strategy has been recently used to profile

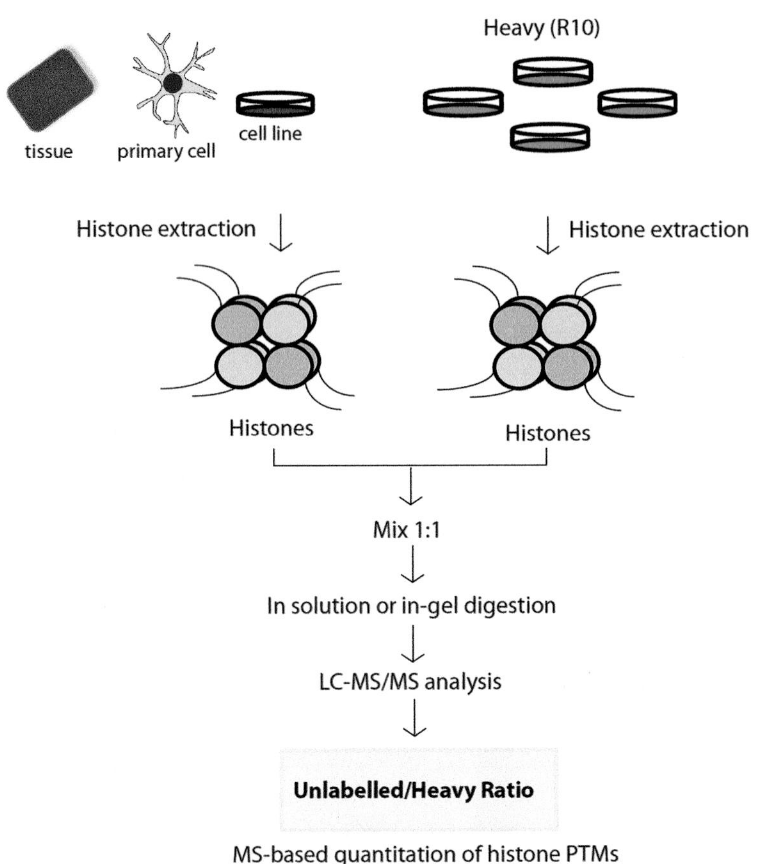

Fig. 3 Analytical workflow of the Super-SILAC strategy for histone PTM quantitation. Histones from unlabeled samples and from cells labeled with stable isotope "heavy-arginine" (R10) are extracted and mixed to a 1:1 ratio prior to protein digestion and MS analysis. Each histone modified peptide is from the unlabeled sample and is then compared to the level of its internal "heavy" counterpart, by calculating the Unlabeled/Heavy SILAC Ratio. Multiple unlabeled samples can be compared by normalizing their abundance to the spike-in SILAC reference

global histone PTMs in human cancer cell lines (Jaffe et al. 2013) and to detect epigenetic differences among breast cancer subtypes (Noberini et al. 2016). This method overcomes two major limitations of standard SILAC: the inability to measure more than three biological conditions and to use samples that cannot be metabolically labeled, such as clinical samples.

Heavy-methyl SILAC (hmSILAC) is another variant of SILAC used for the high confidence identification and quantitation of *in vivo* protein methylations. Cells are grown in a medium supplemented with "heavy" [13CD3]-methionine, which is converted into S-adenosyl methionine (SAM), the sole biological donor of methyl groups during the enzymatic methylation of proteins, leading to specific heavy labeling of all methylations within proteins (Ong et al. 2004). This approach was applied to investigate the changes in the histone methylation flux upon cytomegalovirus (HMCV) infection (O'Connor et al. 2014) and to study the turnover of histone methylation in quiescent cells that were induced to re-enter the cell cycle by means of nutrient feeding. In this latter study, the quiescent cells were also labeled with heavy-glucose to track histone acetylation dynamics (Mews et al. 2014).

Another useful labeling strategy for histone PTM quantitation is based on chemical derivatization of lysines by using different isotope-labeled compounds, such as propionic anhydride, that chemically react with these residues. Similarly to metabolic labeling, after labeling with differentially isotope-encoded reactive compounds, the histone samples are mixed in equal amounts and analyzed by MS. This labeling method was employed to profile histone modifications in a prediabetic mouse model (Nie et al. 2017). Alkylation of lysines through deuterated acetic anhydride has also been used to quantitatively evaluate the acetylation level of histone H4 in *Drosophila melanogaster* (Bonaldi et al. 2004; Fraga et al. 2005).

Last, histone modifications can be quantified through targeted approaches, such as selected and multiple reaction monitoring (SRM and MRM, respectively), where the peptides of interest are selected and single or multiple fragment ions derived from their fragmentation are measured. This analysis is usually carried out using synthetic, isotopically labeled peptides as spike-in internal standards, in order to achieve both relative and absolute quantification. Relative quantitation is obtained by comparing the intensity of each modified peptide with that of the standard, spiked-in at constant concentration across multiple samples. An absolute quantitation, instead, can be achieved by building a calibration curve of the ion intensity of the modified peptide versus the standard, injected at distinct concentrations. Targeted approaches have been successfully employed to compare levels of H2B ubiquitination and H3K79 di-methylation in wild-type U937 human leukemia cell line overexpressing wild-type or mutant antisecretor factor 10 (AF-10) (Darwanto et al. 2010) and to profile 42 differentially modified histone peptides in a panel of cancer cell lines (Gao et al. 2014).

Identification of Chromatin-Associated Proteins by Mass Spectrometry

Transcriptional regulation is mediated by the activity of proteins that alter the chromatin structure by recognizing and binding to *cis*-regulatory elements or by recruiting proteins of the transcriptional machinery. The in-depth characterization of these chromatin-associated proteins is crucial to better understand how genes are regulated in physiological conditions and how such regulatory mechanisms are altered during the transition to a diseased state.

A panel of approaches based on MS-proteomics is currently available for the investigation of the plasticity of chromatin composition. The first attempt to purify chromatin was based on its insolubility in nonionic detergents and led to the identification of chromatin-associated proteins in cells expressing, or not, Myc (Shiio et al. 2003). More recently, partial MNase digestion has been employed to discriminate between proteins associated with eu- and hetero-chromatin, exploiting the differential accessibility of this enzyme to chromatin displaying different condensation levels (Torrente et al. 2011). Similarly, MNase digestion was used to identify proteins differentially expressed between embryonic stem cells (ESCs) and neural progenitor cells (NPCs) (Alajem et al. 2015). An approach based on chromatin cross-linking by formaldehyde, in combination with denaturing washing conditions, was instead employed to capture proteins tightly bound to chromatin (Kustatscher et al. 2014).

Histone readers, which recognize and directly bind to specific histone marks, can be systematically identified by pull-down experiments where nucleosolic extracts are incubated with peptides mimicking modified histone-tails (Eberl et al. 2013; Migliori et al. 2012; Vermeulen et al. 2007). Proteins specifically recognizing the PTMs are thus enriched and analyzed by MS, either through SILAC, which significantly increases the sensitivity and specificity of these types of screening (Migliori et al. 2012; Vermeulen et al. 2007), or label-free quantification (Eberl et al. 2013). In addition, photo-cross-linking-based peptide probes were proven to be useful to improve the detection of weak but specific interactions that may be missed by standard pull-down approaches (Li et al. 2012). Mono- or oligo-nucleosomes bearing differentially modified histone tails (Nikolov et al. 2011), possibly in combination with methylated DNA (Bartke et al. 2010), can also be used as baits to assess protein interactions that require a chromatin context for binding to histone marks.

In addition, DNA sequences with specific binding motifs or single-nucleotide polymorphisms, as well as DNA modifications and mutations, can be used to identify novel DNA binding proteins, such as transcription factors (TF), or to study the impact of various DNA alterations on the recruitment of TFs and co-regulator proteins (Curina et al. 2017; Mittler et al. 2009; Spruijt et al. 2013).

To investigate chromatin-associated complexes within specific genomic regions, various approaches based on the combination of standard chromatin immuno-precipitation (ChIP) with MS-protein profiling have been developed. These methods include strategies to identify interaction partners of different histone variants (Lambert et al. 2009; Sansoni et al. 2014), as well as approaches that employ antibodies against histone PTMs (Ji et al. 2015; Soldi and Bonaldi 2013) or endogenous proteins as baits (Mohammed et al. 2016) and tagged proteins (Wang et al. 2013). Recently, an approach that couples ChIP with DNA biotinylation using TdT and biotin-ddUTP has been described for the identification of chromatin-bound proteins (Rafiee et al. 2016). In the context of metabolism, a TAP-based immuno-purification followed by an *in vitro* lysine acetyl-transferase (KAT) reaction in the presence of isotopically labeled acetyl-CoA allowed identifying the link between individual KATs and their substrates within chromatin-associated complexes (Mitchell et al. 2013).

As an added value, ChIP followed by MS analysis can also be employed to identify co-enriched histone PTMs, thus providing information on the combinatorial marks tagging specific chromatin regions (Soldi and Bonaldi 2013).

Labeling of newly synthetized DNA has been employed to identify proteins associated with nascent DNA at replication forks (Sirbu et al. 2012). Similarly, replicating DNA was tagged using biotin-dUTPs, to study *in vivo* chromatin dynamics during DNA replication (Alabert et al. 2014).

A yet unachieved goal is the purification and characterization of a single chromatin locus. The proteomic characterization of telomeres using a DNA probe

Table 1 Mass spectrometry-based approaches for the identification of chromatin-associated proteins

Aim	Experimental strategy	References
Study of chromatin composition	MNase digestion	Torrente et al. 2011; Alajem et al. 2015
	Formaldehyde cross-linking combined with denaturing washing conditions	Kustatscher et al. 2014
Identification of histone readers	Pull-down; Bait: peptides simulating modified histone tails	Vermeulen et al. 2007; Eberl et al. 2013; Migliori et al. 2012
	Pull-down; Bait: photo-cross-linking based peptide probes	Li et al. 2012
	Pull-down; Bait: Recombinant mono- or oligo- nucleosomes	Nikolov et al. 2011; Bartke et al. 2010
Analysis of chromatin-associated complexes	ChIP-MS; Bait: histone PTMs	Soldi and Bonaldi 2013; Ji et al. 2015
	ChIP-MS; Bait: endogenous proteins	Mohammed et al. 2016
	ChIP-MS; Bait: tagged proteins	Wang et al. 2013
Characterization of proteomic components in specific loci	Bait: DNA probe complementary to telomeric repeats	Dejardin and Kingston 2009
	Bait: tagged protein bound to specific DNA binding sites engineered in specific loci of interest	Byrum et al. 2012
	Bait: proteins such as TAL or dCas9 containing a DNA-binding sequence combined with a guide RNA	Byrum et al. 2013; Fujita et al. 2013

complementary to telomere repeats was the first example of a region-specific analysis (Dejardin and Kingston 2009) and paved the way to other strategies whereby distinct genomic loci are targeted either by an artificial DNA-binding site and the corresponding tagged binding protein (Byrum et al. 2012), or via a protein displaying DNA sequence specificity, such as TAL or dCas9 coupled to a guide RNA (Byrum et al. 2013; Fujita et al. 2013; Table 1).

The development of the CRISPR (clustered regularly interspaced short palindromic repeats) system as a method to easily create engineered DNA-binding molecules made this approach an ideal tool for targeting specific genomic loci, such as the GAL1 locus or the IRF1 promoter, becoming the elective method for single-locus chromatin proteomics.

Global PTM Analysis Beyond Histones

Protein methyltransferases/demethylases and acetyltransferases/deacetylases were initially assumed to target only histones, thus contributing to the epigenetic regulation of gene expression. However, it has now become widely recognized that these enzymes can also modify a large number of nonhistone proteins, playing crucial roles in the regulation of almost every cellular process. Recent advances in high

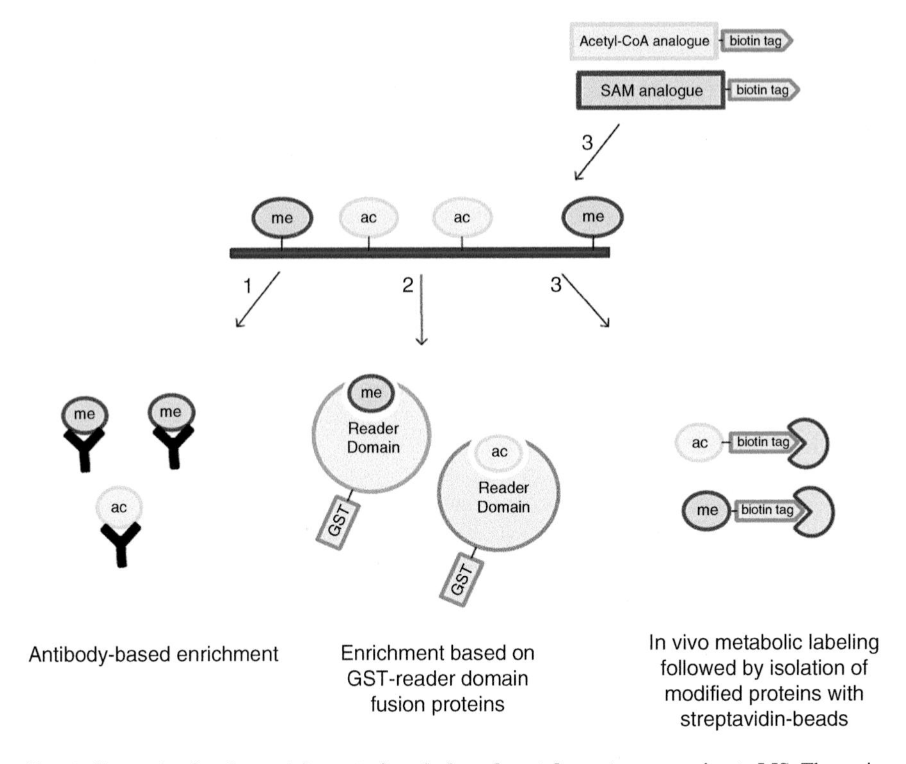

Fig. 4 Strategies for the enrichment of methyl- and acetyl- proteomes, prior to MS. The main strategies developed for the global enrichment of methyl- and acetyl-proteomes prior to MS detection are illustrated. The first method (*1*) is based on the use of antibodies against methylated or acetylated peptides; alternatively (*2*), modification-specific Reader Domains, usually fused with specific tags such as Glutathione S-transferase (GST), can be used to affinity-capture modified peptides. The third strategy (*3*) relies on the use of tagged chemical reporters, such as analogues of acetyl-CoA and SAM, fused with biotin to *in vivo* label acetylated and methylated residues. Enzymatically modified proteins are, then, captured with streptavidin-conjugated beads

resolution MS, in combination with improved protocols for the biochemical enrichment of methylated and acetylated peptides, have greatly contributed to expand our knowledge of the global methyl- and acetyl-proteomes.

Because PTMs on nonhistone proteins are substoichiometric, strategies to affinity-enrich them, both at the peptide- and protein-level, prior to MS are essential to increase their identification rate in complex samples. Typically, lysine-acetylation and lysine- and arginine-methylations are enriched through commercially available antibodies specifically raised against these modified residues (Guo et al. 2014); alternatively, they can also be enriched by using PTM-specific reader domains, such as the chromo- and tudor domains (Kim et al. 2006), PHD domains (Lange et al. 2008), and the MBT domain (Moore et al. 2013). In addition, bio-orthogonal chemical reporters, such as analogues of acetyl-coA and SAM, can be used to *in vivo* label acetylated and methylated residues, respectively, by click-chemistry, tagging them for subsequent isolation with streptavidin-conjugated beads (Fig. 4; Islam et al. 2012; Yang et al. 2010).

In combination with these modification-enrichment strategies, fractionation can also be used to reduce sample complexity. For instance, high pH-based, off-line reversed-phase chromatography of arginine-methylated peptides prior to affinity-enrichment with pan-methyl-antibodies has been recently shown to significantly increase the identification of arginine methylations in human cell lines (Larsen et al. 2016). Isoelectric focusing and ion exchange chromatography are also boosting the identification of lysine-acetylated peptides (Choudhary et al. 2009).

Acetylations and methylations can be quantified in different functional states using different quantitative proteomic approaches. SILAC-based strategies have been employed for global profiling of acetylation and methylation dynamics during the DNA damage response (Beli et al. 2012), or following knockdown of different PRMTs (Larsen et al. 2016). Chemical labeling (TMT) and label-free strategies have been applied for quantitative profiling of global protein-acetylation. These studies highlighted for the first time the crucial role of nonhistone protein modifiers in the regulation of cell metabolism: for instance, the lysine deacetylase SIRT3 was shown to finely tune the acetylation of several mitochondrial proteins involved in various metabolic pathways, thus orchestrating cellular responses to nutrient variation and energy homeostasis (Hebert et al. 2013; Rardin et al. 2013).

Conclusions

Mass-spectrometry-based proteomics is a powerful analytical method to investigate histone modifications and their related enzymes. Here we have reviewed the recent MS-based approaches designed to study chromatin, ranging from modification profiling of histones and non-histone proteins to chromatin-associated protein screening.

Epigenetic processes are tightly intertwined with the metabolic state of the cell. For instance, the activity of histone-modifying enzymes is regulated by changes in the local concentration of key metabolites, so that variations in their concentrations affect histone modifications and chromatin composition and accessibility, leading to profound changes in gene expression patterns and, in extreme circumstances, to pathological disorders. In turn, several metabolic pathways are regulated and controlled by PTMs added or removed by histone modifying enzymes. Understanding the interconnections between metabolism and epigenetics is not only an issue of cell biology, but may have relevant clinical implications, since epigenetic determinants can function as both biomarkers and therapeutic targets for the treatment of metabolic diseases.

MS-proteomics can contribute to the investigation of the impact of the metabolic landscape on the epigenome, thanks to the many analytical options available to profile histone and nonhistone modifications, dynamic chromatin composition, and histone modifier activities.

Dictionary of Terms

- **Bottom-up** – Peptide-centric MS strategy where proteins are digested with proteases, such as trypsin, generating peptides of about 10–12 amino acids, easily detectable by mass spectrometry.
- **Middle-down** – Proteomic approach in which proteins are digested with proteases such as Glu-C or Asp-N, producing long polypeptides around 50–60 amino acids, prior to MS analysis.
- **Top-down** – MS analysis of intact proteins.
- **Isobaric peptides** – Different peptides having the same nominal mass.
- **Label-free proteomics** – Quantitative proteomic approaches based on the quantitation of the relative amount of proteins in biological samples, without the use of labeling methods.
- **Relative Abundance** – In histone PTM quantitation, it corresponds to the ratio of the MS-extracted ion chromatogram (XIC) of each modified histone peptide over the sum of the XICs of all detected isoforms for the same peptide.
- **SILAC** – Quantitative proteomic strategy based on the metabolic incorporation of essential amino acids marked with differentially labeled isotopes that are distinguishable in MS by a specific and univocal deltamass.
- **MNase** – Endo-exonuclease that – in the presence of divalent cations – generates double-strand breaks in nucleosome free-regions.
- **Methyl–/Acetyl- proteome** – The entire set of methylated and acetylated proteins in a cell.

Key Facts of Mass-Spectrometry Based Proteomics

- Mass-spectrometry (MS)-based proteomics is an analytical method employed for identification, quantification, and in-depth characterization of proteins.
- MS allows accurate determination of the molecular mass of proteins/peptides.
- In a classical bottom-up MS-proteomic workflow, proteins are digested with proteases that cleave specific sites and generate a complex mixture of peptides.
- Peptides are typically separated by high performance liquid chromatography (HPLC) prior to MS analysis.
- A mass-spectrometer is composed by three parts: an ion source, a mass analyzer, and a detector.
- The ion source converts the digested peptides into gas-phase ions, the mass analyzer separates the ions according to their mass-to-charge ratio (m/z), and the detector records the number of ions at each m/z value.
- Dedicated algorithms are used to analyze the output generated by the mass spectrometer, in order to identify peptides and proteins from the m/z data list.

Summary Points

- Epigenetic mechanisms are strongly influenced by the metabolic state of the cell.
- Mass-spectrometry (MS)-based proteomics is a powerful method to study histone post-translational modifications (PTMs), thanks to its capability to quantify them and assess their combinations.
- Bottom-up and middle-down are the main MS-based strategies adopted for the identification of histone modifications.
- Label-free based methods for the quantitation of histone PTMs compare unlabeled samples and are mainly based on the calculation of the relative abundance.
- Metabolic and chemical labeling approaches allow quantifying with high accuracy histone PTMs by allowing the combination of different samples at early stages of the proteomic workflow.
- Spike-in methods, such as the Super-SILAC strategy, are particularly useful for the systematic profiling of histone PTMs because they allow higher multiplexing and the analysis of samples that cannot be metabolically labeled, such as clinical specimens.
- Pull-down assays and chromatin immuno-precipitation (ChIP), followed by with quantitative MS analysis allow the identification of histone readers and chromatin-associated complexes, respectively.
- Proteomic components of a certain chromatin locus can be enriched by means of artificial DNA-binding sites or engineered DNA-binding molecules.
- Histone modifying-enzymes target a wide number of nonhistone proteins.
- High-resolution MS, in combination with biochemical enrichment strategies, allowed extending the annotation of global methyl- and acetyl-proteomes.

References

Akhtar A, Becker PB (2000) Activation of transcription through histone H4 acetylation by MOF, an acetyltransferase essential for dosage compensation in drosophila. Mol Cell 5:367–375

Alabert C et al (2014) Nascent chromatin capture proteomics determines chromatin dynamics during DNA replication and identifies unknown fork components. Nat Cell Biol 16:281–293

Alajem A et al (2015) Differential association of chromatin proteins identifies BAF60a/SMARCD1 as a regulator of embryonic stem cell differentiation. Cell Rep 10:2019–2031

An M et al (2016) The alteration of H4-K16ac and H3-K27met influences the differentiation of neural stem cells. Anal Biochem 509:92–99

Bartke T et al (2010) Nucleosome-interacting proteins regulated by DNA and histone methylation. Cell 143:470–484

Beli P et al (2012) Proteomic investigations reveal a role for RNA processing factor THRAP3 in the DNA damage response. Mol Cell 46:212–225

Bonaldi T, Imhof A, Regula JT (2004) A combination of different mass spectroscopic techniques for the analysis of dynamic changes of histone modifications. Proteomics 4:1382–1396

Byrum SD et al (2012) ChAP-MS: a method for identification of proteins and histone posttranslational modifications at a single genomic locus. Cell Rep 2:198–205

Byrum SD, Taverna SD, Tackett AJ (2013) Purification of a specific native genomic locus for proteomic analysis. Nucleic Acids Res 41:e195

Choudhary C et al (2009) Lysine acetylation targets protein complexes and co-regulates major cellular functions. Science 325:834–840
Cluntun AA et al (2015) The rate of glycolysis quantitatively mediates specific histone acetylation sites. Cancer Metab 3:10
Curina A et al (2017) High constitutive activity of a broad panel of housekeeping and tissue-specific cis-regulatory elements depends on a subset of ETS proteins. Genes Dev 31:399–412
Darwanto A et al (2010) A modified "cross-talk" between histone H2B Lys-120 ubiquitination and H3 Lys-79 methylation. J Biol Chem 285:21868–21876
Dejardin J, Kingston RE (2009) Purification of proteins associated with specific genomic Loci. Cell 136:175–186
Eberl HC et al (2013) A map of general and specialized chromatin readers in mouse tissues generated by label-free interaction proteomics. Mol Cell 49:368–378
Fan J et al (2015) Metabolic regulation of histone post-translational modifications. ACS Chem Biol 10:95–108
Fraga MF et al (2005) Loss of acetylation at Lys16 and trimethylation at Lys20 of histone H4 is a common hallmark of human cancer. Nat Genet 37:391–400
Fujita T et al (2013) Identification of telomere-associated molecules by engineered DNA-binding molecule-mediated chromatin immunoprecipitation (enChIP). Sci Rep 3:3171
Gao J et al (2014) Absolute quantification of histone PTM marks by MRM-based LC-MS/MS. Anal Chem 86:9679–9686
Guo A et al (2014) Immunoaffinity enrichment and mass spectrometry analysis of protein methylation. Mol Cell Proteomics 13:372–387
Hebert AS et al (2013) Calorie restriction and SIRT3 trigger global reprogramming of the mitochondrial protein acetylome. Mol Cell 49:186–199
Henry RA et al (2016) Interaction with the DNA repair protein thymine DNA Glycosylase regulates histone acetylation by p300. Biochemistry 55:6766–6775
Islam K et al (2012) Bioorthogonal profiling of protein methylation using azido derivative of S-adenosyl-L-methionine. J Am Chem Soc 134:5909–5915
Jaffe JD et al (2013) Global chromatin profiling reveals NSD2 mutations in pediatric acute lymphoblastic leukemia. Nat Genet 45:1386–1391
Jenuwein T, Allis CD (2001) Translating the histone code. Science 293:1074–1080
Ji X et al (2015) Chromatin proteomic profiling reveals novel proteins associated with histone-marked genomic regions. Proc Natl Acad Sci U S A 112:3841–3846
Jung HR et al (2010) Quantitative mass spectrometry of histones H3.2 and H3.3 in Suz12-deficient mouse embryonic stem cells reveals distinct, dynamic post-translational modifications at Lys-27 and Lys-36. Mol Cell Proteomics 9:838–850
Kim J et al (2006) Tudor, MBT and chromo domains gauge the degree of lysine methylation. EMBO Rep 7:397–403
Kornberg RD (1974) Chromatin structure: a repeating unit of histones and DNA. Science 184:868–871
Kouzarides T (2007) Chromatin modifications and their function. Cell 128:693–705
Kustatscher G et al (2014) Chromatin enrichment for proteomics. Nat Protoc 9:2090–2099
Lambert JP et al (2009) A novel proteomics approach for the discovery of chromatin-associated protein networks. Mol Cell Proteomics 8:870–882
Lange M et al (2008) Regulation of muscle development by DPF3, a novel histone acetylation and methylation reader of the BAF chromatin remodeling complex. Genes Dev 22:2370–2384
Larsen SC et al (2016) Proteome-wide analysis of arginine monomethylation reveals widespread occurrence in human cells. Sci Signal 9:rs9
Lee JV et al (2014) Akt-dependent metabolic reprogramming regulates tumor cell histone acetylation. Cell Metab 20:306–319
Leroy G et al (2013) A quantitative atlas of histone modification signatures from human cancer cells. Epigenetics Chromatin 6:20

Li X et al (2012) Quantitative chemical proteomics approach to identify post-translational modification-mediated protein-protein interactions. J Am Chem Soc 134:1982–1985

Maile TM et al (2015) Mass spectrometric quantification of histone post-translational modifications by a hybrid chemical labeling method. Mol Cell Proteomics 14:1148–1158

Mews P et al (2014) Histone methylation has dynamics distinct from those of histone acetylation in cell cycle reentry from quiescence. Mol Cell Biol 34:3968–3980

Migliori V et al (2012) Symmetric dimethylation of H3R2 is a newly identified histone mark that supports euchromatin maintenance. Nat Struct Mol Biol 19:136–144

Mitchell L et al (2013) mChIP-KAT-MS, a method to map protein interactions and acetylation sites for lysine acetyltransferases. Proc Natl Acad Sci U S A 110:E1641–E1650

Mittler G, Butter F, Mann M (2009) A SILAC-based DNA protein interaction screen that identifies candidate binding proteins to functional DNA elements. Genome Res 19:284–293

Mohammed H et al (2016) Rapid immunoprecipitation mass spectrometry of endogenous proteins (RIME) for analysis of chromatin complexes. Nat Protoc 11:316–326

Moore KE et al (2013) A general molecular affinity strategy for global detection and proteomic analysis of lysine methylation. Mol Cell 50:444–456

Morrish F et al (2010) Myc-dependent mitochondrial generation of acetyl-CoA contributes to fatty acid biosynthesis and histone acetylation during cell cycle entry. J Biol Chem 285:36267–36274

Nie L et al (2017) The landscape of histone modifications in a high-fat diet-induced obese (DIO) mouse model. Mol Cell Proteomics 16:1324–1334

Nikolov M et al (2011) Chromatin affinity purification and quantitative mass spectrometry defining the interactome of histone modification patterns. Mol Cell Proteomics M110(005371):10

Noberini R et al (2016) Pathology tissue-quantitative mass spectrometry analysis to profile histone post-translational modification patterns in patient samples. Mol Cell Proteomics 15:866–877

O'Connor CM et al (2014) Quantitative proteomic discovery of dynamic epigenome changes that control human cytomegalovirus (HCMV) infection. Mol Cell Proteomics 13:2399–2410

Ong SE, Mann M (2006) A practical recipe for stable isotope labeling by amino acids in cell culture (SILAC). Nat Protoc 1:2650–2660

Ong SE, Mittler G, Mann M (2004) Identifying and quantifying in vivo methylation sites by heavy methyl SILAC. Nat Methods 1:119–126

Pesavento JJ, Mizzen CA, Kelleher NL (2006) Quantitative analysis of modified proteins and their positional isomers by tandem mass spectrometry: human histone H4. Anal Chem 78:4271–4280

Rafiee MR et al (2016) Expanding the circuitry of Pluripotency by selective isolation of chromatin-associated proteins. Mol Cell 64:624–635

Rardin MJ et al (2013) Label-free quantitative proteomics of the lysine acetylome in mitochondria identifies substrates of SIRT3 in metabolic pathways. Proc Natl Acad Sci U S A 110:6601–6606

Sansoni V et al (2014) The histone variant H2A.B.bd is enriched at sites of DNA synthesis. Nucleic Acids Res 42:6405–6420

Shiio Y et al (2003) Quantitative proteomic analysis of chromatin-associated factors. J Am Soc Mass Spectrom 14:696–703

Shyh-Chang N et al (2013) Influence of threonine metabolism on S-adenosylmethionine and histone methylation. Science 339:222–226

Sidoli S, Cheng L, Jensen ON (2012) Proteomics in chromatin biology and epigenetics: elucidation of post-translational modifications of histone proteins by mass spectrometry. J Proteome 75:3419–3433

Sidoli S et al (2014) Middle-down hybrid chromatography/tandem mass spectrometry workflow for characterization of combinatorial post-translational modifications in histones. Proteomics 14:2200–2211

Sirbu BM, Couch FB, Cortez D (2012) Monitoring the spatiotemporal dynamics of proteins at replication forks and in assembled chromatin using isolation of proteins on nascent DNA. Nat Protoc 7:594–605

Soldi M, Bonaldi T (2013) The proteomic investigation of chromatin functional domains reveals novel synergisms among distinct heterochromatin components. Mol Cell Proteomics 12:764–780

Soldi M, Bremang M, Bonaldi T (2014) Biochemical systems approaches for the analysis of histone modification readout. Biochim Biophys Acta 1839:657–668

Spruijt CG et al (2013) Dynamic readers for 5-(hydroxy)methylcytosine and its oxidized derivatives. Cell 152:1146–1159

Strahl BD, Allis CD (2000) The language of covalent histone modifications. Nature 403:41–45

Syka JE et al (2004) Peptide and protein sequence analysis by electron transfer dissociation mass spectrometry. Proc Natl Acad Sci U S A 101:9528–9533

Tammen SA, Friso S, Choi SW (2013) Epigenetics: the link between nature and nurture. Mol Asp Med 34:753–764

Thomas CE, Kelleher NL, Mizzen CA (2006) Mass spectrometric characterization of human histone H3: a bird's eye view. J Proteome Res 5:240–247

Torrente MP et al (2011) Proteomic interrogation of human chromatin. PLoS One 6:e24747

Vermeulen M et al (2007) Selective anchoring of TFIID to nucleosomes by trimethylation of histone H3 lysine 4. Cell 131:58–69

Wang CI et al (2013) Chromatin proteins captured by ChIP-mass spectrometry are linked to dosage compensation in drosophila. Nat Struct Mol Biol 20:202–209

Wu C, Morris JR (2001) Genes, genetics, and epigenetics: a correspondence. Science 293:1103–1105

Yang YY, Ascano JM, Hang HC (2010) Bioorthogonal chemical reporters for monitoring protein acetylation. J Am Chem Soc 132:3640–3641

Young NL et al (2009) High throughput characterization of combinatorial histone codes. Mol Cell Proteomics 8:2266–2284

Zee BM, Young NL, Garcia BA (2011) Quantitative proteomic approaches to studying histone modifications. Curr Chem Genomics 5:106–114

Zubarev RA et al (2000) Electron capture dissociation for structural characterization of multiply charged protein cations. Anal Chem 72:563–573

Forward and Reverse Epigenomics in Embryonic Stem Cells

119

Ilana Livyatan and Eran Meshorer

Contents

Abstract

The self-renewing and pluripotent properties of ESCs make them a precious tool for the advancement of general biological research, discerning the process of differentiation and embryonic development, disease modeling, drug discovery, drug testing, and, ultimately, cell- and tissue-based regenerative medicine. To further these goals, it is imperative that a deep and comprehensive understanding of all aspects of ESC biology are attained, particularly the transcriptional program and its regulation.

Chromatin immunoprecipitation (ChIP) followed by next-generation sequencing (NGS) (ChIP-seq) pinpoints the binding locations of factors involved in

I. Livyatan
Department of Genetics, The Hebrew University of Jerusalem, Jerusalem, Israel
e-mail: ilana.forst-livya@mail.huji.ac.il

E. Meshorer (✉)
Department of Genetics and the Edmond and Lily Safra Center for Brain Sciences (ELSC), The Hebrew University of Jerusalem, Jerusalem, Israel
e-mail: meshorer@huji.ac.il

V. B. Patel, V. R. Preedy (eds.), *Handbook of Nutrition, Diet, and Epigenetics*,
https://doi.org/10.1007/978-3-319-55530-0_51

epigenomic regulation of transcription such as transcription factors, modifications on histone proteins, chromatin modifiers and remodelers, and structural and insulator proteins. Each binding map by itself leads to insights into the mechanism of regulation of a specific factor and its downstream target genes upon which it exerts its regulatory effect leading to the discovery of the epigenomic "hallmarks" of ESCs which govern these cells' state and fate.

On the other hand, an integration of a combination of binding maps enables researchers to gain an additional and complementary perspective on epigenomic regulation from the genomic point of view. By combining over 450 ChIP-seq datasets from large consortiums and singleton experimental efforts in our BindDB platform, we discovered a remarkably extensive epigenomic profile at active genes in ESCs. Based on this platform, we generated a robust in silico simulation of a reverse-ChIP protocol via implementation of an easy querying and analysis webtool (http://bind-db.huji.ac.il/) to enable researchers to learn about which epigenomic features bind their genes or genomic regions of interest in ESCs. By querying several gene groups as case studies, we noted the participation of histone modifications, chromatin modifiers, chromatin remodelers, transcription factors, and structural proteins in the regulation of the same pieces of DNA, indicating how crucial and precise the epigenomic mechanism must be in ESCs. This utilization of both a forward and reverse approach to epigenomic research will greatly advance the acquisition of a more complete picture of the mechanisms of transcriptional regulation in ESCs and improve the ability to harness it toward advancing the research and medical potential of these unique cells.

Keywords
Embryonic stem cells · ESCs · Pluripotent · Epigenome · Epigenomic profile · Binding maps · Webtool · Database · Chromatin · Epigenomic signature · Transcription factors

Abbreviations

BED	Browser extensible data format
ChIP	Chromatin immunoprecipitation
ChIP-chip	Chromatin immunoprecipitation followed by hybridization to microarray
ChIP-seq	Chromatin immunoprecipitation followed by sequencing
DNA	Deoxyribonucleic acid
DNMT	DNA Methyltransferase
ENCODE	Encyclopedia of DNA elements (consortium)
ESC	Embryonic stem cell
HAT	Histone acetyl transferase
HDAC	Histone deacetylase
HDM	Histone demethylase
Hi-C	High-throughput chromosome conformation capture
HMT	Histone methyltransferase
MBD	Methyl-binding domain

RNA	Ribonucleic acid
TAD	Topologically associating domain
TF	Transcription factor

Introduction

Embryonic stem cells (ESCs) are derived from the inner cell mass of a blastocyst or early-stage embryo in mouse (Evans and Kaufman 1981; Martin 1981), human (Thomson et al. 1998), and rat (Kawamata and Ochiya 2010). These cells possess the ability to self-renew in cell culture environments, meaning they divide, grow, and spread indefinitely in a culture dish with the appropriate conditions making them suitable for long-term growth. In addition, ESCs are pluripotent possessing the potential and ability to differentiate to cell types of all three germ layers in the embryo and can give rise to any cell type comprising the developing embryo in early and late stages and also the adult organism. These unique cell traits can be harnessed for different types of biological and medical applications (Keller 2005; Mayhall et al. 2004). These include:

(i) Study of early embryonic development: differentiation of ESCs along different differentiation pathways can unveil the mechanisms and their roles in embryonic development and how they may be affected by different perturbations.
(ii) Disease modeling: derivation of ESCs from diseased embryos or genetic and/or other modifications of ESCs can be used as a model of disease, especially for those which affect early embryonic development, but not limited to them.
(iii) Drug screens and testing: the karyotypic and genetic stability of ESCs in culture and the ability to acquire different cell types from these cells by the induction of various differentiation pathways provide an excellent and reproducible platform for testing drug safety and efficacy.
(iv) Cell- and tissue-based therapies: ESCs' expansion in culture and subsequent directed differentiation can potentially provide a source of replacement cells and tissues to treat diseases and injuries, which involve cell degeneration or damage. In order to achieve these four majorly impactful applications of ESCs, the key is to understand, direct, and ultimately be able to control the process of differentiation.

Transcription

While several biological processes affect ESC differentiation, one of the most drastic and impactful changes is at the level of DNA transcription. Almost all cells in a mammalian organism contain the same DNA sequence packaged within their nuclei, but these same DNA copies are expressed or transcribed in very different manners in different cell types. The process of transcription involves the copying of a portion of DNA sequence into an RNA molecule by the macromolecular complex, RNA polymerase. The subsequent RNA molecules or transcripts can then be translated into protein by the ribosomal complexes, which then carry out cellular functionality. Therefore, the functionality of the cell is largely dictated by the portions of expressed DNA, termed the “transcriptional program” of the cell. During differentiation and development, the transcriptional program of a cell changes drastically with portions of DNA (i.e., genes) becoming active while others are deactivated. Transcriptional levels, or number of RNA copies per gene, are also modulated and fine-tuned by transcriptional (RNA polymerase), co-transcriptional (RNA splicing), and posttranscriptional (RNA splicing, RNA modifications, and RNA silencing) factors. The transcriptional activity of a cell is tightly regulated by different factors including proteins, termed “transcription factors” (TFs), which bind DNA to recruit RNA polymerase to sites of transcription. Different cell types having a different composition of such transcription factors and those that are found to be most impactful are considered “hallmarks” of cell state and function. Therefore, transcription factors and other signaling molecules which affect transcription are the key ingredients in any directed ESC differentiation protocol into new cell types, so the identification of factors and their mechanisms of action is key to harnessing the power of ESCs and enabling the aforementioned applications. The early analysis of ESCs’ transcription revealed several such transcriptional “hallmarks” (Heng and Ng 2010): OCT4 (Mountford et al. 1998), SOX2 (Avilion et al. 2003), and NANOG (Chambers et al. 2003), are the core TFs which govern the transcriptional network in ESCs, which confer and maintain self-renewing and pluripotent capabilities (Boyer et al. 2005; Chambers and Tomlinson 2009; Chen and Daley 2008; Kim et al. 2008; Loh et al. 2006; Whyte et al. 2013; Yeo and Ng 2013; Young 2011).

Epigenomic Mechanisms in ESCs: Forward

In addition to the binding of transcription factors to upstream promoter elements of genes, transcription is tightly regulated by chromatin structure itself, or the “epigenome,” which lies beyond the heritable, genomic DNA sequence. The epigenome constitutes an orchestra of chemical modifications within the DNA sequence, DNA packaging proteins and their chemical modifications, and DNA-binding proteins, which recruit or interfere with DNA expressing or maintaining apparatus. Epigenomic mechanisms regulate the expression of genes and comprise the “second dimension” of the transcriptional program of cells.

The next-generation sequencing (NGS)revolution has enabled researchers to determine the location of each of these epigenomic features along the DNA molecule with respect to genes, subgenic territories, and intergenic regions, shedding light on their mode of regulation of transcription. First, cross-linking agents are added to cells in order to capture and stabilize the epigenomic state in cells. Next, a chromatin protein of interest is precipitated out of a whole cell extract using a specific antibody in a process called chromatin immunoprecipitation (ChIP). The millions of DNA pieces bound by this chromatin protein can then be extracted and sequenced on a high-throughput platform (ChIP-seq). In silico alignment of resulting sequences to reference DNA reveals the genomic locations bound by the chromatin protein and creates a binding map for this protein in the examined cell type condition. The ChIP-seq methodology has made it possible for researchers to generate binding maps for a plethora of transcription factors and chromatin-related elements in many cell types (ENCODE 2012; Furey 2012; Roadmap Epigenomics et al. 2015) and especially in pluripotent stem cells (Bernstein et al. 2010; Chen et al. 2008; Kraushaar and Zhao 2013; Morey et al. 2015; Paranjpe and Veenstra 2015) paving the way to understand the multilayered nature of transcriptional regulation.

Being cell-type specific, the epigenome of ESCs is under particular scrutiny (Burton and Torres-Padilla 2014; Morey et al. 2015; Tee and Reinberg 2014). Major fluctuations in genome-wide DNA methylation occur during embryonic development (Brandeis et al. 1993; Smith and Meissner 2013) including inactivation of an entire X chromosome in females. Packaging of DNA around nucleosomes comprised of octamer histone proteins facilitates the unique diffusive and open structure of ESC chromatin (Gaspar-Maia et al. 2011; Melcer and Meshorer 2010). The four types of core histone proteins in the nucleosome H2A, H2B, H3, and H4 in ESCs carry a significantly high percentage of activating histone modifications such as lysine (K) acetylation (Ac) and single (mono-) methylation (me1), double (bi-) methylation (me2), or triple (tri-) methylation (me3) (Bernstein et al. 2005, 2010; Mattout et al. 2011). At the gene level, a unique "bivalent" character of histone modifications in the region of gene promoters was discovered in ESCs: both tri-methylation of lysine 4 (active mark) and tri-methylation of lysine 27 (repressive mark) in histone H3 were found as promoters of developmental genes, imparting a "poised" configuration to these genes which can be activated or deactivated as development progresses and lineages determined (Azuara et al. 2006; Bernstein et al. 2006). These chemical modifications are conferred and regulated by enzymes such as histone acetyl transferases (HATs) and deacetylases (HDACs) and histone methyltransferases (HMTs) and demethylases (HDMs) such that the pattern of their, and other chromatin modifier, binding to the DNA is also an important feature of ESC chromatin. Binding maps of the core (Boyer et al. 2005; Chambers and Tomlinson 2009; Loh et al. 2006) and supportive (Chen et al. 2008; Kim et al. 2008; Young 2011) TFs have established a key role in regulating the transcriptional program of ESCs. Chromatin remodeler protein complexes (Chen and Dent 2014) and other structural (e.g., Cohesin and Mediator complex (Kagey et al. 2010)) and insulator (e.g., CTCF (Barski et al. 2007)) proteins also play roles in gene regulation in ESCs by conferring higher-order/three-dimensional structures to the chromatin such as enhancer-promoter loops (Whyte et al. 2013).

Epigenomic Mechanisms in ESCs: Reverse

Until recently, most epigenomic studies in ESCs were "protein oriented" and were asking questions regarding the factor binding locations and how they can lead to conclusions about factor function. Several studies had begun to look at the crosstalk among a handful of factors such as activating histone modifications alongside the core ESC transcription factors (Boyer et al. 2005; Chen et al. 2008; Mikkelsen et al. 2007; Tsankov et al. 2015; Whyte et al. 2013).

Yet a question that could not be sufficiently answered in ESCs was the opposite one: given a genomic location, which factors bind/exist there? This question reverses the approach to chromatin analysis from a protein point of view to a genomic point of view. Theoretically, a reversed biochemical assay, whereby a region of DNA is captured and then the proteins and chemical modifications which bind it are identified, would enable this kind of shift of focus from protein to DNA, but methods of DNA capture and protein identification are not fully established or very inefficient (Dejardin and Kingston 2009). The sequencing revolution pertains to nucleic acid sequences and not polypeptides. Yet, due to the sufficient number of accumulated forward ChIP-seq profiles in ESCs (Adams et al. 2012; Bernstein et al. 2010; ENCODE 2012), we could simulate such a reverse assay in silico by assembling over 450 ChIP-seq datasets in ESCs into the BindDB database (http://bind-db.huji.ac.il/) and implementing an easy querying and analysis platform to enable all end users, computational and noncomputational alike, to query their gene(s) or genomic regions of interest against it (Livyatan et al. 2015).

BindDB

BindDB is a database of all known binding maps resulting from 100 s of published ChIP-seq or ChIP-chip experiments in over 40 mouse and human ESC types (Fig. 1). It is a user-friendly, interactive, publicly accessible webtool, which enables users to query one (Fig. 2) or more (Fig. 3) genes (defined by symbol or accession) or genomic regions (BED format). The output types vary from raw data requiring additional computational analysis for bioinformaticians to enrichment and statistical analyses of larger query groups. These results are presented visually in the form of interactive tables, bar plots, heatmaps, and line graphs to enable easy interpretation of the results and insights into real biological conclusions.

The larger consortiums (Adams et al. 2012; Bernstein et al. 2010; ENCODE 2012), after publishing a giant data mining analysis, generally left researchers with semi-processed data (e.g., factorbook (Wang et al. 2012)) and not enough tools with which to integrate it. Several tools had come online in recent years with varying data sources and degrees of correlative and integrative analysis capabilities (Table 1). Many took the conventional "factor" point of view, looking for correlations in binding profiles among different factors (e.g., CR Cistrome (Wang et al. 2014), WashU Epigenome Browser (Zhou et al. 2011)). The newer ones, which also offered a "gene"-based view to find epigenomic factors regulating given gene sets, were limited to current gene annotations and the assumption that epigenomic regulation occurs at promoters only (e.g., ESCAPE (Xu et al. 2013), CODEX (Sanchez-Castillo et al. 2015)) or had scalability limitations (WashU Epigenome Browser).

BindDB enabled us to recapitulate previous analysis we performed for histone genes (Gokhman et al. 2013), detecting enrichment of factors in the promoter regions of the histone clusters, including E2F1, E2F4 (shedding light on the

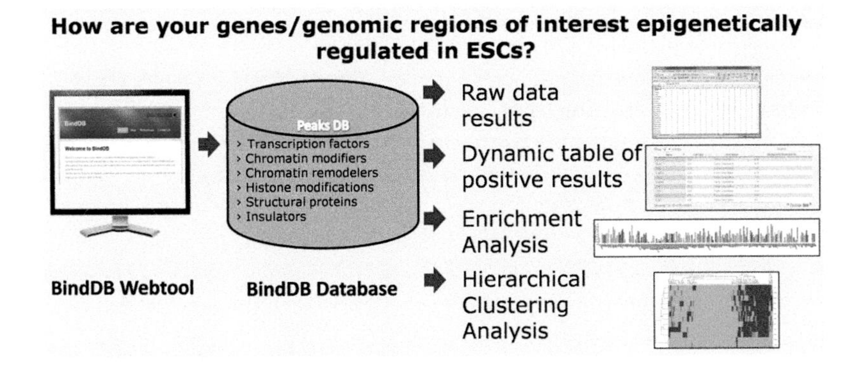

Fig. 1 The BindDB scheme. The BindDB database includes over 450 binding maps generated from ChIP-seq experiments in human and mouse ESCs on various types of transcription- and chromatin-regulating proteins. The BindDB webtool can be accessed via web browsers to query genes or genomic regions of interest and generate several types of views and analyses of query results including enrichment analysis and hierarchical clustering of resulting epigenomic profiles

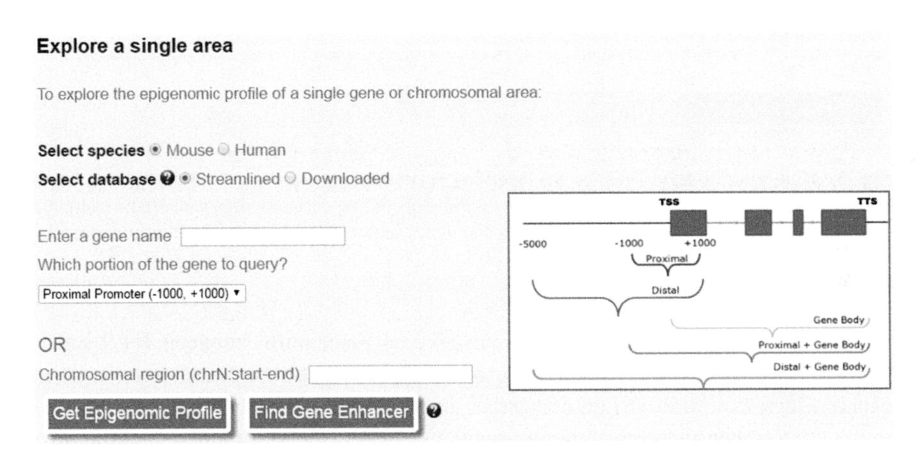

Fig. 2 Single gene/region query. User may choose to query mouse or human data. An Entrez gene symbol, Refseq, or UCSC accession number can be provided and precise area to be queried selected (proximal promoter only, proximal promoter and gene body, etc.). A chromosomal region may be designated for intergenic regions or if a different location (not offered in the dropdown menu) around the gene is desired

regulation of their cell cycle-related expression profile and functionality during DNA replication), SMAD1, SMAD2, P300, MED1 and MED12, and identified several novel factors, including GCN5, AFF4, and CAPH2 (Aaronson et al. 2016). This type of analysis also revealed the complexity of histone gene regulation involving cell-type-specific regulators and differential repressors of specific histone subtypes. The hierarchical clustering analysis was statistically strong enough to confirm that ZFX was indeed preferentially enriched at H1 linker histone genes over core histone genes while adding two more factors with similar behavior: REST and KAT5

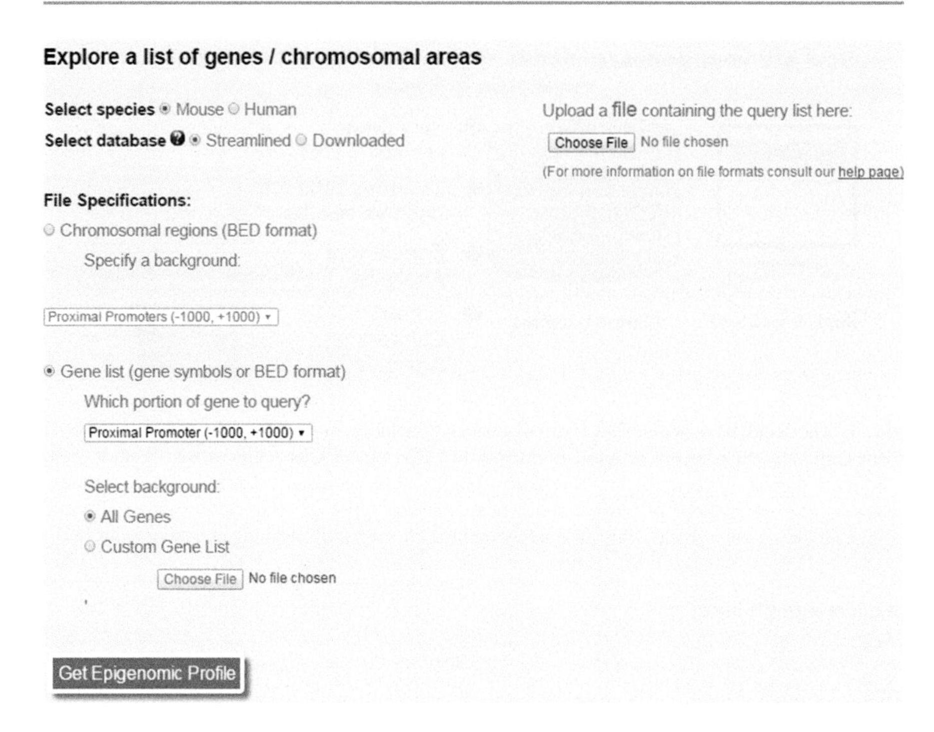

Fig. 3 Multiregion query. Enables the querying of multiple genes or genomic regions, which allows for enrichment analysis and statistical assessments of that enrichment in order to pinpoint the factors that were most likely to be regulating the genes/regions' expression. Backgrounds may be selected for enrichment and statistical analysis

(Aaronson et al. 2016). The recently discovered association between H1.0 linker histone silencing and the maintenance of self-renewing tumor-maintaining cells in cancer (Torres et al. 2016) demonstrates the importance of identifying the players involved in linker histone expression regulation. ZFX, REST, and KAT5 could now be candidates of therapeutic efforts to restore higher levels of H1.0 in cancer cells to enhance differentiation and reduce tumorigenicity. Hierarchical clustering of epigenomic signatures of histone genes first detected a clear difference between canonical replication-dependent and variant replication-independent histone genes. A second iteration of the BindDB on only the variant histones showed another clustered breakdown of these genes, dividing them into those with an active epigenomic signature and those with a very sparse signature involving repressive factors. The latter group comprised of histone genes with a very cell-specific expression pattern: *H1fnt* in testis, *H1foo* in oocyte, and *H2afy3*, *H2afb1*, and *H2bfm* with germ cell expression patterns. Interestingly, evidence of methyl-binding domain protein (MBD) binding is found in the vicinity of these histones implicating DNA methylation as the silencing mechanism in germ-cell types.

The BindDB epigenomic analysis of histones demonstrates how to harness the power of in silico mining of "big" epigenomic data before designing further wetlab

Table 1 Comparison of characteristics of current epigenomic tools in ESCs

Tool	Publication	Cell types	Species	Factors	Data curation	Download	Visualization	Analysis	Upload
ChIPBase v2.0	NAR 2017	Diverse, hESCs, mESCs, tumor samples from TGCA project	Human, Mouse, Dog, Chicken, Drosophila, C.elegans, Rat, Xenopus tropicalis, Zebrafish, Yeast and Arabidopsis thaliana	Transcription factors, transcription cofactors, chromatin-remodeling factors	Downloaded peak datasets from GEO, ENCODE, modENCODE and supplementary info of articles. No standard pipeline. Translated via liftOver. DB of TFBS's from chip-seqs and Jaspar, Transfac, Cistrome, and Uniprobe. LincRNAs were taken from publications and Ensembl, Refseq and UCSC and lncRNAdb. miRNAs were taken from mirBase	Download interactions between TFs, CTFs, CRFs and lincRNAs, miRNAs, protein coding genes	Tables of factors that bind (no scoring or statistical tests). Interactions are limited to ncRNAs	Browsing, narrowing and widening regulatory window. Regulator module for finding TFBS in gene sets of interest. Motif discovery. Functional prediction of DNA-binding proteins. Co-expression analysis of TFs and genes	Upload user tracks and Gene Sets for simultanious viewing

(*continued*)

Table 1 (continued)

Tool	Publication	Cell types	Species	Factors	Data curation	Download	Visualization	Analysis	Upload
Cistrome Map	Genome Biology 2011	Diverse, hESCs, mESCs		571 in stem cells	Reads	Easy download of reads	Table of experiments by cell type	None (DB only)	None
CR Cistrome	NAR 2014	hESCs, mESCs	Human, Mouse	36 CRs, 13 HMs	Reads' data from Cistrome Map, aligned with Bowtie (2 mismatches) and peaks with MACs (Qvalue = 0.01)	Easy download of data as bed or wiggle	Average profiles, motifs, venn diagrams, enrichment bar plots over genic regions	Factor oriented – motif. Gene, TSS and TTS average profile. Enrichment in genic categories. Comparison between CR and HM datasets: Venn of overlapping peaks, scatter plots, average profiles of one peak set over the other	None

ESCAPE	Database 2013	hESCs, mESCs	Human, Mouse	61 TFs, 18 HMs	Not described	MySQL table download of database (not data)	Networks, enrichment bar plots	Among other things, the ChIP- seqs are used to draw connections between factors and target genes. The entire analysis is gene based (assuming TF at promoter of gene makes gene a target). No option for other genomic locations	None
CODEX	NAR 2014	93 cell types	Human, Mouse	159 TF datasets in mouse ESCs	ChIP-seq, RNA-seq, DNASE-seq, Data mining via web crawling for new data from	All raw and processed data	As UCSC tracks, simultaneously or individually	GeneQuest to find which factors are associated to genes or which genes	None

(*continued*)

Table 1 (continued)

Tool	Publication	Cell types	Species	Factors	Data curation	Download	Visualization	Analysis	Upload
					GEO, Array Express. Extract Fastq from SRA, fastQC, align with Bowtie2 to mm10/hg19, Samtools to convert to BED and bigwig, MACs2.0 for peak calling with a series of pvalues which are manually inspected, BED files with peak summit inside a 400 bp window			are targets of a particular factor. Correlation tool (Dice correlation of a binary matrix over "genomic elements"), Gene Set Control Analysis (GSCA) like GO for sets of target genes. Motif discovery	
WashU Epigenome Browser	Nature Methods 2011	Diverse, hESCs, mESCs	Human, Mouse, Rat, Rhesus maqaque Zebrafish, Fruitfly, Guinea Pig, Arabidopsis,	In mouse ESCs: 8 HMs and 10 TFs, In human: Roadmap and ENCODE	Uses read data to generate wiggle plots	None	UCSC browser like, zoom in/ out interactive, addition of tracks from public and custom, gene tracks, focus on	Correlation between tracks. Statistical test to identify regions with differences	None

			Maize, Soybean, Common bean				set of genes or genomic regions, metadata from experiments. Gene Set view for up to 200 genes	between 2+ tracks	
BindDB	Cell Stem Cell 2015	hESCs, mESCs	Human, Mouse	232 in mESCs, 223 in hESCs. Includes Roadmap and ENCODE. HMs, TFs, coactivators, chromatin modifiers, remodelers, structural proteins	ChIP-seq, ChIP-chip, RNA- seq. Data mining of data from GEO, ENCODE, Roadmap. "As is" or Extract Fastq from SRA, align with Bowtie to mm9/ hg19, Samtools to convert to BED and bigwig, MACs2.0 for peak calling	None	Excel tables of raw results, enrichment barplots, interactive, searchable table of positive results, heatmap if clustering, line plots of enhancer scores, integration with UCSC browser and wiggle plots	Enrichment of binding, hierarchical clustering of epigenomic signatures. Correlation with tx levels. Enhancer scoring. Gene perturbation analysis	None

experiments. Implication of MBD proteins in germ cell-specific histone variant silencing can now be checked in MBD or DNA methyltransferase (DNMT) knockout ESC lines and ESC-derived primordial germ cells to test their involvement.

BindDB also enabled the identification of a novel regulator of ribosomal genes in ESCs, NR5A2 (Aaronson et al. 2016). This conclusion was enabled after systematic gene perturbation datasets were integrated with the epigenomic results and probably underestimated for a few reasons: (1) only three out of five potential binding factors had associated perturbation sets that could be tested; (2) BindDB can only detect direct binding, whereas perturbations may be the result of indirect effect via intermediate regulatory genes.

This case study brings out two important points to consider for future analyses: firstly, that direct factor binding, while being extensive and complex and involving crosstalk among chromatin readers, writers, remodelers, TFs, and histone modification, is only the first tier of a transcriptional regulatory network. For example, in ESCs, the epigenomic regulation of the *Oct4* gene has a potential exponential downstream effect on genes regulated by the OCT4 transcription factor protein. Therefore, the incorporation of data regarding regulatory circuits of transcription can greatly enhance our understanding of the more complete picture of the outcomes of epigenomic regulation. Such regulatory networks can even be constructed from the big data in BindDB as shown for MBD proteins (Aaronson et al. 2016). Secondly, in the era of "big data," integration among different types of datasets is key in deciphering the true biological significance of the data. Further integration with datasets representing higher-order chromatin structure (Hi-C data, TAD domains), DNA methylation data, and additional gene perturbations will definitely lead to more insights and conclusions and obviously many more questions.

One common criticism when studying ESC transcription is that its detected "global" (Efroni et al. 2008) or "hypertranscriptive" nature may be the byproduct of certain biological or experimental technicalities, and, hence, the discovery of novel transcription needs to be carefully considered. First, the "open" (Gaspar-Maia et al. 2011), accessible, and hyperdynamic (Melcer and Meshorer 2010) nature of chromatin in ESCs may lead to spurious opportunistic transcription without an intentional function. This could also result from hyperactivity of the transcriptional and even posttranscriptional apparatus such as RNA splicing (Pritsker et al. 2005), but the former is also hypothesized to be a byproduct of a principal method of growing ESCs in fetal bovine/calf serum. Recently revised methods of ESC culturing replace serum with the addition of two inhibitors (2i) or three inhibitors (3i) of signaling pathways which promote differentiation, resulting in the maintenance of the undifferentiated embryonic state (Ying et al. 2008). Epigenomic and transcriptomic data is starting to appear for ESCs grown in "2i" medium and, once amassed, can be added to the BindDB and other databases and tested alongside data from serum conditions to further iron out this point.

Concluding Remarks

The analysis of "big" data gives us an opportunity to look at the bigger picture of transcription and its regulation, but it is not free of confounding variables and requires the integration of different types of big datasets to corroborate results and strengthen biological conclusions. Future acquisition of transcriptional and epigenomic data in single cells, which is now beginning to emerge, will increase the association between these two data types and fine-tune the deciphering of the effects of chromatin regulation on DNA expression and vice versa. The BindDB database and querying platform demonstrate how an in silico approach can complement a biochemical one to show both sides of epigenomic analysis, the factor perspective and the genomic perspective, to generate a more complete understanding of the elements governing transcriptional regulation and their modes of action in ESCs.

Dictionary of Terms

- **Self-renewal** – the ability of cells to multiply and maintain cell state indefinitely in culture conditions.
- **Pluripotent** – the ability of a cell to differentiate into cell types of all three embryonic germ layers: endoderm, mesoderm, and ectoderm.
- **DNA methylation** – a chemical modification involving the addition of a methyl group to cytosine residues in the DNA molecule.
- **Histone modification** – chemical modifications on histone proteins including acetylation and methylation which influence gene expression and/or chromatin structure.
- **Insulator protein** – a DNA-binding protein which buffers between two types of chromatin from (linear) adjacent DNA sections.
- **Chromatin modifier** – enzyme or enzymatic complex which chemically modifies histone proteins via addition or cleavage of chemical groups.
- **Chromatin remodeler** – enzyme or enzymatic complex which changes local structure of chromatin.
- **Binding map** – a list of binding coordinates along the linear representation of DNA of a chromatin-related protein or DNA modification.
- **Cell/tissue therapy** – the transplantation of cells or tissue in order to regenerate damaged tissue.
- **Promoter element** – the adjacent upstream DNA sequence to a gene, where much of the regulation of the transcriptional machinery is applied and epigenomic factors bind.
- **Genic/subgenic/intergenic** – the regions of DNA demarcated as genes (regions which are expressed by the transcriptional machinery in some cell type) are genic regions. Genic regions can be subdivided into subgenic regions according to gene structure. DNA sequences outside of genic regions are termed intergenic.

- **High-throughput sequencing** – also known as next-generation sequencing (NGS) is the physical technology by which millions of short DNA sequences can be determined simultaneously enabling the sequencing of whole genomes or large libraries of genomic sequences with high accuracy in a short amount of time.
- **Hierarchical clustering** – during data analysis, the measurement of distance between the data components is used to create a treelike structure which depicts the similarity/difference between components and component subgroups facilitating their categorization.
- **Histones and variant histones** – histone proteins are generally expressed from histone genes in S-phase of the cell cycle alongside the process of DNA replication and required for the packaging of the new DNA molecule. The high demand for many histone proteins in a short period of time confers a different pathway of histone gene transcription including the clustered assembly of histone genes and alternative stabilization of histone mRNA transcripts to 3′ polyadenylation. Histone variants are proteins generated outside of S-phase, generated from classic, polyadenylated mRNAs, which participate in chromatin maintenance and regulation.
- **Ribosomal genes** – genes which encode ribosomal proteins. Together with ribosomal RNAs (rRNAs), ribosomal proteins comprise the subunits of ribosomal complexes responsible for translating mRNAs into proteins in the cytoplasmic region of the cell. Ribosomal genes reside in both the cellular DNA molecule and the DNA of the cellular mitochondria.
- **Perturbation datasets** – each dataset comprises a list of genes that show significant changes in expression levels after a perturbation is applied to a factor involved in transcriptional regulation. Perturbations include knockdown (lowered expression), overexpression (higher expression), and knockout (canceled expression) of the transcriptional regulator.

Key Facts

Key Facts About Embryonic Stem Cells

- ESCs are derived from the preimplantation embryo, a cluster of 100–200 cells also termed the blastocyst.
- In humans, ESCs are harvested from donated embryos resulting from in vitro fertilization (IVF) at the blastocyst stage.
- In mice (and rats), ESCs are harvested directly from pregnant females around 3.5 days postconception.
- ESCs can be expanded indefinitely in culture, unlike cells derived from primary tissue of fetal or adult organisms.
- They remain karyotypically stable during culturing and expansion unlike cancer cells, which can also self-renew, but acquire genetic abnormalities.

- They can be driven to differentiate into cell types from all three main germ layers of the embryo: ectoderm, endoderm, and mesoderm.
- Theoretically, ESCs can differentiate into any cell type in the adult organism, and there are differentiation protocols to generate many cell types from ESCs in culture conditions.
- From an epigenetic perspective, ESCs have an open and accessible chromatin conformation when compared to differentiated cell types. This allows their plasticity.

Summary Points

- ESCs are a unique cell type with great research and medical potential.
- The transcriptional program of ESCs is governed by the regulation of three core transcription factors: OCT4, SOX2, and NANOG.
- The ChIP-seq methodology generates binding maps for a plethora of transcription factors and chromatin-related elements in many cell types.
- The application of ChIP-seq to ESCs has uncovered an intricate and complex epigenomic regulatory network involving additional transcription factors and epigenomic features.
- The forward application of ChIP-seq can reveal the binding locations, mechanism of action, and role of a given epigenomic factor.
- The reverse application of ChIP-seq, to determine the epigenomic signature of a given genomic region, is not readily available as a biochemical assay but can be simulated in silico in ESCs.
- Several computational tools address both forward and reverse approaches.
- BindDB comprises a collection of over 450 ChIP-seq datasets and a user-friendly webtool application (http://bind-db.huji.ac.il/), which facilitates the reverse application of ChIP-seq and enables users to query the epigenomic profiles in ESCs of genes or any genomic regions of interest.

References

Aaronson Y, Livyatan I, Gokhman D, Meshorer E (2016) Systematic identification of gene family regulators in mouse and human embryonic stem cells. Nucleic Acids Res 44:4080–4089

Adams D, Altucci L, Antonarakis SE, Ballesteros J, Beck S, Bird A, Bock C, Boehm B, Campo E, Caricasole A et al (2012) BLUEPRINT to decode the epigenetic signature written in blood. Nat Biotechnol 30:224–226

Avilion AA, Nicolis SK, Pevny LH, Perez L, Vivian N, Lovell-Badge R (2003) Multipotent cell lineages in early mouse development depend on SOX2 function. Genes Dev 17:126–140

Azuara V, Perry P, Sauer S, Spivakov M, Jorgensen HF, John RM, Gouti M, Casanova M, Warnes G, Merkenschlager M et al (2006) Chromatin signatures of pluripotent cell lines. Nat Cell Biol 8:532–538

Barski A, Cuddapah S, Cui K, Roh TY, Schones DE, Wang Z, Wei G, Chepelev I, Zhao K (2007) High-resolution profiling of histone methylations in the human genome. Cell 129:823–837

Bernstein BE, Kamal M, Lindblad-Toh K, Bekiranov S, Bailey DK, Huebert DJ, McMahon S, Karlsson EK, Kulbokas EJ 3rd, Gingeras TR et al (2005) Genomic maps and comparative analysis of histone modifications in human and mouse. Cell 120:169–181

Bernstein BE, Mikkelsen TS, Xie X, Kamal M, Huebert DJ, Cuff J, Fry B, Meissner A, Wernig M, Plath K et al (2006) A bivalent chromatin structure marks key developmental genes in embryonic stem cells. Cell 125:315–326

Bernstein BE, Stamatoyannopoulos JA, Costello JF, Ren B, Milosavljevic A, Meissner A, Kellis M, Marra MA, Beaudet AL, Ecker JR et al (2010) The NIH Roadmap Epigenomics mapping consortium. Nat Biotechnol 28:1045–1048

Boyer LA, Lee TI, Cole MF, Johnstone SE, Levine SS, Zucker JP, Guenther MG, Kumar RM, Murray HL, Jenner RG et al (2005) Core transcriptional regulatory circuitry in human embryonic stem cells. Cell 122:947–956

Brandeis M, Ariel M, Cedar H (1993) Dynamics of DNA methylation during development. BioEssays: News Rev Mol Cell Dev Biol 15:709–713

Burton A, Torres-Padilla ME (2014) Chromatin dynamics in the regulation of cell fate allocation during early embryogenesis. Nat Rev Mol Cell Biol 15:723–734

Chambers I, Colby D, Robertson M, Nichols J, Lee S, Tweedie S, Smith A (2003) Functional expression cloning of Nanog, a pluripotency sustaining factor in embryonic stem cells. Cell 113:643–655

Chambers I, Tomlinson SR (2009) The transcriptional foundation of pluripotency. Development 136:2311–2322

Chen L, Daley GQ (2008) Molecular basis of pluripotency. Hum Mol Genet 17:R23–R27

Chen T, Dent SY (2014) Chromatin modifiers and remodellers: regulators of cellular differentiation. Nat Rev Genet 15:93–106

Chen X, Xu H, Yuan P, Fang F, Huss M, Vega VB, Wong E, Orlov YL, Zhang W, Jiang J et al (2008) Integration of external signaling pathways with the core transcriptional network in embryonic stem cells. Cell 133:1106–1117

Dejardin J, Kingston RE (2009) Purification of proteins associated with specific genomic loci. Cell 136:175–186

Efroni S, Duttagupta R, Cheng J, Dehghani H, Hoeppner DJ, Dash C, Bazett-Jones DP, Le Grice S, McKay RD, Buetow KH et al (2008) Global transcription in pluripotent embryonic stem cells. Cell Stem Cell 2:437–447

ENCODE (2012) An integrated encyclopedia of DNA elements in the human genome. Nature 489:57–74

Evans MJ, Kaufman MH (1981) Establishment in culture of pluripotential cells from mouse embryos. Nature 292:154–156

Furey TS (2012) ChIP-seq and beyond: new and improved methodologies to detect and characterize protein-DNA interactions. Nat Rev Genet 13:840–852

Gaspar-Maia A, Alajem A, Meshorer E, Ramalho-Santos M (2011) Open chromatin in pluripotency and reprogramming. Nat Rev Mol Cell Biol 12:36–47

Gokhman D, Livyatan I, Sailaja BS, Melcer S, Meshorer E (2013) Multilayered chromatin analysis reveals E2f, Smad and Zfx as transcriptional regulators of histones. Nat Struct Mol Biol 20:119–126

Heng JC, Ng HH (2010) Transcriptional regulation in embryonic stem cells. Adv Exp Med Biol 695:76–91

Kagey MH, Newman JJ, Bilodeau S, Zhan Y, Orlando DA, van Berkum NL, Ebmeier CC, Goossens J, Rahl PB, Levine SS et al (2010) Mediator and cohesin connect gene expression and chromatin architecture. Nature 467:430–435

Kawamata M, Ochiya T (2010) Establishment of embryonic stem cells from rat blastocysts. Methods Mol Biol 597:169–177

Keller G (2005) Embryonic stem cell differentiation: emergence of a new era in biology and medicine. Genes Dev 19:1129–1155

Kim J, Chu J, Shen X, Wang J, Orkin SH (2008) An extended transcriptional network for pluripotency of embryonic stem cells. Cell 132:1049–1061

Kraushaar DC, Zhao K (2013) The epigenomics of embryonic stem cell differentiation. Int J Biol Sci 9:1134–1144

Livyatan I, Aaronson Y, Gokhman D, Ashkenazi R, Meshorer E (2015) BindDB: an integrated database and Webtool platform for "reverse-ChIP" epigenomic analysis. Cell Stem Cell 17:647–648

Loh YH, Wu Q, Chew JL, Vega VB, Zhang W, Chen X, Bourque G, George J, Leong B, Liu J et al (2006) The Oct4 and Nanog transcription network regulates pluripotency in mouse embryonic stem cells. Nat Genet 38:431–440

Martin GR (1981) Isolation of a pluripotent cell line from early mouse embryos cultured in medium conditioned by teratocarcinoma stem cells. Proc Natl Acad Sci USA 78:7634–7638

Mattout A, Biran A, Meshorer E (2011) Global epigenetic changes during somatic cell reprogramming to iPS cells. J Mol Cell Biol 3:341–350

Mayhall EA, Paffett-Lugassy N, Zon LI (2004) The clinical potential of stem cells. Curr Opin Cell Biol 16:713–720

Melcer S, Meshorer E (2010) Chromatin plasticity in pluripotent cells. Essays Biochem 48:245–262

Mikkelsen TS, Ku M, Jaffe DB, Issac B, Lieberman E, Giannoukos G, Alvarez P, Brockman W, Kim TK, Koche RP et al (2007) Genome-wide maps of chromatin state in pluripotent and lineage-committed cells. Nature 448:553–560

Morey L, Santanach A, Di Croce L (2015) Pluripotency and epigenetic factors in mouse embryonic stem cell fate regulation. Mol Cell Biol 35:2716–2728

Mountford P, Nichols J, Zevnik B, O'Brien C, Smith A (1998) Maintenance of pluripotential embryonic stem cells by stem cell selection. Reprod Fert Develop 10:527–533

Paranjpe SS, Veenstra GJ (2015) Establishing pluripotency in early development. Biochim Biophys Acta 1849:626–636

Pritsker M, Doniger TT, Kramer LC, Westcot SE, Lemischka IR (2005) Diversification of stem cell molecular repertoire by alternative splicing. Proc Natl Acad Sci USA 102:14290–14295

Roadmap Epigenomics C, Kundaje A, Meuleman W, Ernst J, Bilenky M, Yen A, Heravi-Moussavi A, Kheradpour P, Zhang Z, Wang J et al (2015) Integrative analysis of 111 reference human epigenomes. Nature 518:317–330

Sanchez-Castillo M, Ruau D, Wilkinson AC, Ng FS, Hannah R, Diamanti E, Lombard P, Wilson NK, Gottgens B (2015) CODEX: a next-generation sequencing experiment database for the haematopoietic and embryonic stem cell communities. Nucleic Acids Res 43:D1117–D1123

Smith ZD, Meissner A (2013) DNA methylation: roles in mammalian development. Nat Rev Genet 14:204–220

Tee WW, Reinberg D (2014) Chromatin features and the epigenetic regulation of pluripotency states in ESCs. Development 141:2376–2390

Thomson JA, Itskovitz-Eldor J, Shapiro SS, Waknitz MA, Swiergiel JJ, Marshall VS, Jones JM (1998) Embryonic stem cell lines derived from human blastocysts. Science 282:1145–1147

Torres CM, Biran A, Burney MJ, Patel H, Henser-Brownhill T, Cohen AS, Li Y, Ben-Hamo R, Nye E, Spencer-Dene B et al (2016) The linker histone H1.0 generates epigenetic and functional intratumor heterogeneity. Science 353:1514

Tsankov AM, Gu H, Akopian V, Ziller MJ, Donaghey J, Amit I, Gnirke A, Meissner A (2015) Transcription factor binding dynamics during human ES cell differentiation. Nature 518:344–349

Wang J, Zhuang J, Iyer S, Lin X, Whitfield TW, Greven MC, Pierce BG, Dong X, Kundaje A, Cheng Y et al (2012) Sequence features and chromatin structure around the genomic regions bound by 119 human transcription factors. Genome Res 22:1798–1812

Wang Q, Huang J, Sun H, Liu J, Wang J, Wang Q, Qin Q, Mei S, Zhao C, Yang X et al (2014) CR Cistrome: a ChIP-Seq database for chromatin regulators and histone modification linkages in human and mouse. Nucleic Acids Res 42:D450–D458

Whyte WA, Orlando DA, Hnisz D, Abraham BJ, Lin CY, Kagey MH, Rahl PB, Lee TI, Young RA (2013) Master transcription factors and mediator establish super-enhancers at key cell identity genes. Cell 153:307–319

Xu H, Baroukh C, Dannenfelser R, Chen EY, Tan CM, Kou Y, Kim YE, Lemischka IR, Ma'ayan A (2013) ESCAPE: database for integrating high-content published data collected from human and mouse embryonic stem cells. Database: J Biol Databases Curation 2013:bat045

Yeo JC, Ng HH (2013) The transcriptional regulation of pluripotency. Cell Res 23:20–32

Ying QL, Wray J, Nichols J, Batlle-Morera L, Doble B, Woodgett J, Cohen P, Smith A (2008) The ground state of embryonic stem cell self-renewal. Nature 453:519–523

Young RA (2011) Control of the embryonic stem cell state. Cell 144:940–954

Zhou X, Maricque B, Xie M, Li D, Sundaram V, Martin EA, Koebbe BC, Nielsen C, Hirst M, Farnham P et al (2011) The human epigenome browser at Washington University. Nat Methods 8:989–990

MiRImpact as a Methodological Tool for the Analysis of MicroRNA at the Level of Molecular Pathways

120

Anton A. Buzdin and Nikolay M. Borisov

Contents

Abstract

Intracellular molecular pathways (IMPs) involve multiple gene products implicated in certain biological functions. The best-known IMPs are metabolic pathways, signaling pathways, DNA repair pathways, and cytoskeleton reorganization pathways. The pathway-level of analysis in molecular biology

A. A. Buzdin (✉)
Centre for Convergence of Nano-, Bio-, Information and Cognitive Sciences and Technologies, National Research Centre "Kurchatov Institute", Moscow, Russia

OmicsWay Corporation, Walnut, CA, USA

Group for Genomic Regulation of Cell Signaling Systems, Shemyakin-Ovchinnikov Institute of Bioorganic Chemistry, Moscow, Russia
e-mail: Buzdin@ponkc.com; bu3din@mail.ru

N. M. Borisov
Centre for Convergence of Nano-, Bio-, Information and Cognitive Sciences and Technologies, National Research Centre "Kurchatov Institute", Moscow, Russia

Department of Personalized Medicine, First Oncology Research and Advisory Center, Moscow, Russia
e-mail: nicolasborissoff@gmail.com

V. B. Patel, V. R. Preedy (eds.), *Handbook of Nutrition, Diet, and Epigenetics*,
https://doi.org/10.1007/978-3-319-55530-0_91

provides a number of advantages compared to the analysis of single genes. First of all, IMPs are more stable biomarkers. This can be explained by the fact that most frequently several or even many individual gene products are involved in a single elementary biological process. For example, the members of RAF family or regulatory protein kinases can be all involved in the same biological process of signal transduction, by acting in an interchangeable way as the MAP kinase kinase kinases downstream to the RAS proteins. The RAS family, in turn, consists of many proteins that may exert basically the same functions, and so on. A variation in the expression of a single family member is hard to interpret, whereas the pathway level of analysis enables obtaining an integral figure for all the nodes and family members. Secondly, the pathway level of data analysis makes it possible to significantly reduce the experimental error of measuring gene expression. This allows to reduce or even eliminate the batch effects and to compare the data obtained using different experimental platforms. Several analytic approaches have been published to digest the mRNA or proteomic data at the level of IMPs, but an approach crosslinking the changes in microRNA (miR) profiles with the activation of molecular pathways was missing. Recently, we proposed a bioinformatic method termed MiRImpact, which enables to link the high-throughput miR expression data with the estimated outcome on the regulation of molecular pathways. MiRImpact was used to establish interactomic signatures for hundreds of molecular pathways, specific to stem cell differentiation, cancer progression, and cytomegalovirus infection. Of note, the impact of miRs appeared orthogonal to pathway regulation at the mRNA level, which stresses the importance of combining all available levels of gene regulation for building more objective models of intracellular molecular processes.

Keywords
Systems biology · Bioinformatics · Intracellular molecular pathways · Gene expression · Transcriptomics · Proteomics · Epigenetics · MicroRNA · miR · Cancer · Biomarkers · Stem cell differentiation

List of Abbreviations

AI	Cells after infection
BC	Bladder cancer
HS	Cells highly sensitive to HCMV infection
IMP	Intracellular molecular pathway
LS	Cells low sensitive to HCMV infection
miPAS	Pathway activation strength, calculated using microRNA expression data
NGS	Next-generation sequencing
PAS	Pathway activation strength, calculated using mRNA or protein expression data
riboPAS	Pathway activation strength, calculated using ribosome profiling gene expression data
WI	Cells without infection

Introduction

Intracellular molecular pathways (IMPs) are involved in all major events in the living organisms. The major groups of IMPs are metabolic, cell signaling, cytoskeleton reorganization, and DNA repair pathways (Blagosklonny 2011, 2013; Demidenko and Blagosklonny 2011). The metabolic pathway implements coordination of biochemical reactions to combine them in a continuously connected network having any biologically significant output. The signaling pathways, in turn, consolidate gene products involved in signal transduction. The transduced signals are very diverse. They may be initiated by the receptor binding on the cell surface or inside the cell, by the concentration changes of certain proteins or low-molecular mass metabolites and ions, by the physical impact such as high or low temperature and ionizing radiation, and by the altered stability of protein complexes (Hanahan and Weinberg 2000, 2011; Sonnenschein and Soto 2013). The signal is initially sensed and transduced to downstream pathway nodes until it reaches one or several effector node/nodes that execute any actionable item (Fig. 1). The final outputs can be also extremely diverse, even for a single signaling pathway. The types of such outputs include affected regulation of gene expression, altered permeability of cellular or mitochondrial membranes to ions and metabolites, biochemical modifications of proteins, such as phosphorylation, assembly or disruption of molecular complexes, activation or downregulation of other molecular pathways. The intermediate steps in molecular signaling are equally variegated and may deal with the protein biochemical modification (primarily phosphorylation), molecular localization (e.g., transfer of a protein molecule from cell membrane to cytoplasm or nucleus), assembly or disaggregation of molecular complexes, release of ions and low-molecular mass metabolites (Marshall 1995; Kholodenko et al. 1999; Kiyatkin et al. 2006; Borisov et al. 2009).

The cytoskeleton reorganization and DNA repair pathways are similarly organized and partly overlap with the signaling pathways. The major distinction is their functional role, which is, respectively, a remodeling of cytoskeleton or repair of damaged DNA. Interestingly, most of the signaling pathways as one of their outputs affect at least one cytoskeleton remodeling pathway, which suggests their tight interconnection in a living cell (Disanza et al. 2009; Filteau et al. 2015). Similarly, DNA repair functions in concert with the regulation at the level of signaling and cytoskeleton remodeling pathways. For example, until DNA lesions are repaired (DNA repair pathways involved), the cell cannot proceed to the cell cycle (signaling and cytoskeleton pathways involved) (Branzei and Foiani 2008; Vermeulen et al. 2003; Malumbres and Barbacid 2009).

The pathways may include tens or hundreds of nodes and aggregate up to several hundreds of different gene products (Elkon et al. 2008; Nakaya et al. 2013; Croft et al. 2014; Nikitin et al. 2003). Importantly, each node in a pathway is typically built not by just a single gene product but rather by their groups. Those can be formed by the homologous families of similarly functionally charged proteins, or by the various protein subunits which may be all needed to execute a function required for the pathway activity (UniProt Consortium 2011; Mathivanan et al. 2006).

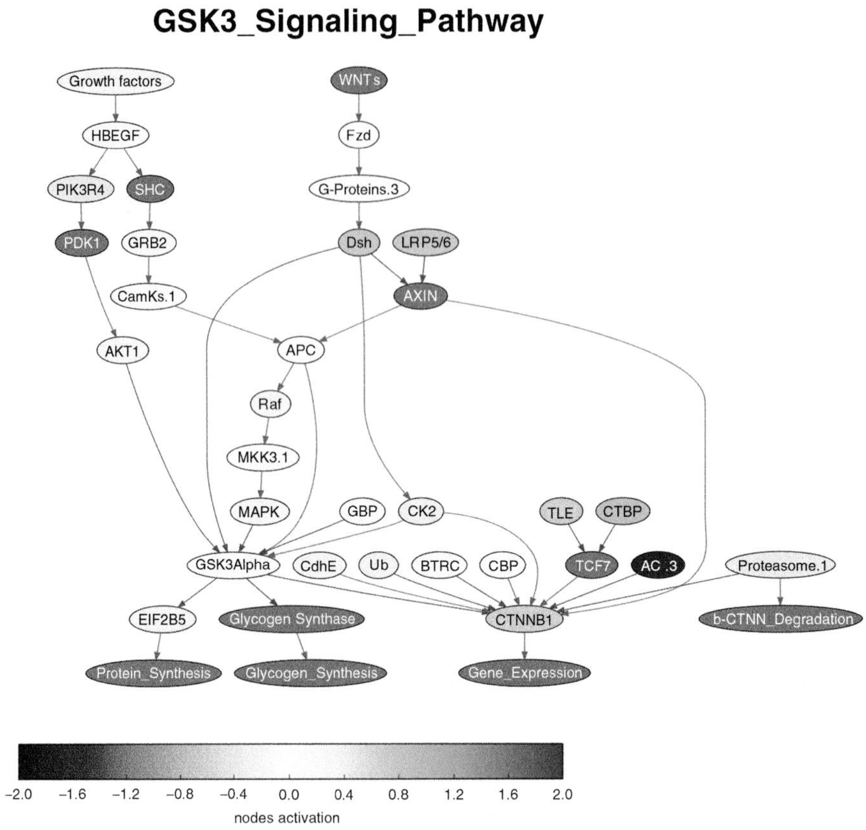

Fig. 1 Schematic representation of a Glycogen synthase kinase 3 (*GSK3*) molecular pathway in one sample of human hepatocellular carcinoma. The pathway includes different input signals such as the activation with growth factors or Wnt proteins, has many intermediate signal transducer nodes and as the output regulates the processes such as the synthesis of protein and glycogen, and induces gene expression shift in a beta-catenin dependent way. The *color bar* represents scale of relative up- or downregulation in the pathway nodes compared to the normal samples

For several decades, the molecular pathways are still on the forefront of biomedical sciences (Marshall 1995; Hanahan and Weinberg 2000, 2011; Borisov et al. 2009).

Hundreds of thousands of molecular interactions and thousands of molecular pathways have been discovered by the molecular biologists and catalogued in different databases. Among those, one can mention the Universal protein resource (UniProt, www.uniprot.org) (UniProt Consortium 2011) QIAGEN SABiosciences (http://www.sabiosciences.com/), Human protein reference database (HPRD) (Mathivanan et al. 2006), Ariadne Pathway Studio (Nikitin et al. 2003), Reactome (Croft et al. 2014), Kyoto Encyclopedia of Genes and Genomes (KEGG) (Nakaya et al. 2013), Signaling Pathways Integrated Knowledge Engine (SPIKE) (Elkon et al. 2008), WikiPathways (Bauer-Mehren et al. 2009), and HumanCyc Encyclopedia of Human Genes and Metabolism (www.humancyc.org) databases. In these resources, one can find the information on the pathway architecture and frequently on the associated functional features.

On the other side, the gene products can be labeled and sorted according to their functional role in the cell and with reference to the molecular or supramolecular processes they are involved. This way of data aggregation does not require knowledge of the particular chains of molecular interactions, mostly protein-protein interactions, as for the previously mentioned group of the databases. For example, the Gene ontology (GO) database provides functional and structural labels to the gene products or their groups (www.geneontology.org). In order to simplify this type of analysis, a specific software has been created (e.g., the annotation system Database for Annotation, Visualization and Integrated Discovery (DAVID; http://david.ncifcrf.gov). By uploading a specific set of gene products (which can be tagged by either official gene or protein names, or by the many other ways), the user can find it out whether this list is statistically significantly enriched in certain types of functions or certain structural groups. For example, this makes it possible to make a quick overview of the differentially expressed gene sets, most frequently mutated groups of genes, et cetera.

However, besides qualitative identification of the molecular pathways, there is still a challenge to quantitatively measure the extent of their activation. In this chapter, we will review several methods thereof and will make a focus on the task of measuring pathway activation using the data on microRNA (miR) expression.

Quantitative Analysis of the Molecular Pathways

In most of the applications, the information about activation of IMPs can be obtained from the high-throughput screenings of gene expression data at the transcriptomic or proteomic levels (Kulesh et al. 1987, Olsvik et al. 1993). Although the proteomic level may seem somewhat closer to the biological function of an IMP, the transcriptomic level of studies for now remains far more cost-effective and feasible in terms of simplicity and reproducibility in obtaining the data. The transcriptomic methods like next-generation sequencing (NGS) or microarray analysis of RNA may be performed for the minute amounts of biological samples and can routinely determine expression levels for all or at least most of all known genes (Vivar et al. 2013).

However, until recently it remained challenging to efficiently do the high-throughput quantification of pathway activation for the individual biological samples. Several biomathematical approaches were published to measure pathway activation based on large-scale gene expression data. For example, Khatri et al. (2012) classified those methods into three major groups: Over-Representation Analysis (ORA), Functional Class Scoring (FCS), and Pathway Topology (PT)-based approaches. ORA-based methods calculate if the pathway is significantly enriched with differentially expressed genes (Khatri et al. 2012; Zeeberg et al. 2003). These methods have many limitations, because they ignore all nondifferentially expressed genes and do not take into account many gene-specific characteristics influencing the nature of interactions between the pathway members. FCS-based approaches partially tackle these limitations by calculating fold change-based scores for each gene and then combining them into a single pathway enrichment score (Tian et al. 2005). Finally, the PT-based analysis also takes into account topological characteristics of each given pathway, assigning additional weights to the enclosed genes (Mitrea et al. 2013). Recently, to account for gene expression variability within a pathway, another set of differential variability methods has been developed (Afsari et al. 2014). Differential variability analysis determines a group of genes with a significant change in variance of gene expression between case and control groups (Ho et al. 2008). This approach was further extended and applied on the pathway level (Zhang et al. 2009). In 2014, we published a new method for pathway analysis, termed OncoFinder (Buzdin et al. 2014a). Based on kinetic models that use the "low-level" approach of mass action law, the OncoFinder performs quantitative and qualitative enrichment analysis of the signaling pathways. For each sample under investigation, it does a case-control pairwise comparison and calculates the Pathway Activation Strength (PAS), a value which serves as a qualitative measure of pathway activation. Unlike most other methods, this approach identifies not simply if the molecular pathway is somehow affected, but also determines if the pathway is significantly up- or downregulated compared to the reference, and returns the extent of this deregulation. Negative and positive PAS values indicate on an inhibited or activated state of each respective molecular pathway (Buzdin et al. 2014a). OncoFinder was also shown to provide the output data with strongly reduced levels of noise introduced by the experimental systems profiling the gene expression at the RNA level, either microarray- or NGS-based (Buzdin et al. 2014b). This method was effective in finding new biomarkers for stimulation by the nutrients, for various chronicle human diseases, e.g., (Lezhnina et al. 2014) (Aliper et al. 2014; Makarev et al. 2014, 2016) in modeling melanoma development (Shepelin et al. 2016), drug sensitivity in cancer (Lebedev et al. 2015; Zhu et al. 2015) and acquiring drug resistance in leukemia (Spirin et al. 2014), in studying immunity and apoptosis (Ram et al. 2016), and in many other applications. A variant of the OncoFinder PAS scoring method termed in silico Pathway Activation Network Decomposition Analysis (iPANDA) has been recently elaborated to analyze the age-related diseases and functional changes (Ozerov et al. 2016). Interestingly, application of the OncoFinder PAS approach enabled us to identify novel potential geroprotector molecules and their compositions (Aliper et al. 2016). One of these compositions including four different substances was further used by the pharmaceutical company LifeExtension to generate a series of supplements termed GEROPROTECT with the first of them termed Ageless Cell already on the market (http://www.lifeextension.com/Vitamins-Supplements/item02119/Ageless-Cell).

The OncoFinder Algorithm

The OncoFinder algorithm operates with calculation of the Pathway Activation Strength (PAS), a value which serves as a quantitative measure of a molecular pathway activation. The formula for PAS calculation requires gene expression data and the information on protein-protein interactions in a pathway in order to be able to attach functional label to each gene product, i.e., its activator or repressor role in a pathway (Buzdin et al. 2014a). Another important aspect is the identification of control sample/group of control samples, which will be used to normalize the gene expression at the first step of the PAS calculation. For microarray gene expression data, previous data normalization may be required, such as the quantile normalization (Bolstad et al. 2003). For the NGS data, another group of prenormalization techniques is recommended, such as the DESeq or DESeq2 methods (Love et al. 2014). The positive value of PAS indicates increased activity of a molecular pathway compared to the normal samples, and the negative value means its downregulation, whereas zero PAS scores denote unaffected pathways which make no difference between the case and the normal samples. The following formula may be used to evaluate pathway activation, where summation is applied for all the gene products that take part in a pathway:

$$PAS_p = \sum_n ARR_{np} \cdot BTIF_n \cdot \lg(CNR_n).$$

Here the *case-to-normal ratio* (*CNRn*) is the ratio of expression levels for a gene n in the sample under investigation to the same average value for the control group of samples. For each gene i, the case to normal ratio (CNR_i) is calculated for the concentrations of mRNA or protein, depending on the input data (most frequently used for mRNA expression data):

$$CNR_i = \frac{Case_{mRNA}Signal_i}{Norm_{mRNA}Signal_i}.$$

To increase stability of the differential expression data, for each CNR value, a Boolean flag of beyond tolerance interval flag (BTIF) may be applied, which equals to 1 when the CNR value has passed, and to 0 when CNR value did not pass the following two criteria of significance for the differential expression. First, the expression level for the sample must lay outside of the tolerance interval for the norms, with $p < 0.05$, and second, the gene expression level in the sample must differ from the mean expression level in the controls by at least 1.5-fold.

Next, the OncoFinder algorithm comprises application of another criterion: discrete value of *activator/repressor role* (*ARR*) that reflects the functional role of a gene product n in a pathway (Buzdin et al. 2014b). For each gene product that participates in a pathway p, its activator-repressor role ($\mathrm{ARR}_{\mathrm{i,p}} ARR_{i,p}$) is defined, which depends on the functional role of this gene product in a pathway:

$$ARR_{i,p} = \begin{cases} -1, & repressor \\ -0.5, & repressor > activator \\ 0, & neither \\ 0.5, & activator > repressor \\ 1, & activator \end{cases}.$$

The structures of the molecular pathways and knowledge of the functional roles of the individual nodes may be obtained by manually or automatically curating the specialized databases. In the previous OncoFinder applications, the abovementioned resources QIAGEN SABiosciences, WikiPathways (Bauer-Mehren et al. 2009), Ariadne Pathway Studio (Nikitin et al. 2003), Reactome (Croft et al. 2014), KEGG (Nakaya et al. 2013), and HumanCyc (www.humancyc.org) have been used to create the pathway structure data matrixes.

This OncoFinder algorithm has been validated using both experimental data and theoretical approaches (Zhavoronkov et al. 2014; Buzdin et al. 2014a). The overall graph for the protein interaction events may include both sequential (pathway-like) and parallel (network-like) edges (Conzelmann et al. 2006; Borisov et al. 2008). The role of each gene product in the signal transduction may depend on whether it works in a sequential or a parallel way. Alternatively, as the raw approximation of this situation, one may propose a simplified method that utilizes only the overall roles of each gene product in the molecular pathway, as in the case of the OncoFinder method. Here, each simplified pathway graph includes only two types of branches of protein interaction chain: one for sequential events that promote the pathway activation and another – for repressor sequential events. Under these conditions, it can be presumed that all activator/repressor members have equal importance for the pathway activation, and there are no additional weight coefficients depending on the pathway architecture. However, this assumption of sequential protein-protein interaction in pathways with equal importance may seem artificial.

At least two ways for determination of relative importance of gene products may be suggested. First, a concept of sensitivity of the ordinary differential equations (ODE) system on the free parameters can be applied (Kholodenko et al. 2002), which is generally applied to kinetic constants such as the dissociation constant, the Michaelis-Menten constants, etc., but also may be used (exactly as here) for the total concentrations of certain proteins in the kinetic model of a pathway.

The second way to calculate the importance function for the gene product in a pathway is related to the stiffness/sloppiness analysis (Daniels et al. 2008) for the effector activation upon total protein concentrations. Following this approach, we calculated alternative line of PAS scores.

Taking into account the above considerations, we come to the following adjusted formula for PAS calculation: $PAS_p^{(1,2)} = \sum_n ARR_{np} \cdot BTIF_n \cdot w_n^{(1,2)} \cdot \log(CNR_n)$.

To check the introduction of the weight (importance) coefficients, either sensitivity-based, $w^{(1)}$, or stiffness-based, $w^{(2)}$, we performed the algorithm validation on the example of the EGFR pathway (Kuzmina and Nikolay 2011). For these two types of weighting factors and without them, we performed a computational analysis using microarray gene expression data for the patients with high-grade glioblastoma (Griesinger et al. 2013). Our results suggest that the cloud of values for the ratio of $\frac{PAS_{\mathrm{EGFR}}^{(1)}}{PAS_{\mathrm{EGFR}}}$ lies within the interval of (0.7 ± 0.3), whereas the ratios of $\frac{PAS_{\mathrm{EGFR}}^{(2)}}{PAS_{\mathrm{EGFR}}}$ is slightly higher and may be assessed as (1.0 ± 0.6); here PAS_{EGFR} is the PAS

value for the EGFR pathway with all importance factors equal to 1, which is the case for the basic version of OncoFinder.

Of course, the proportion of the molecular pathways that have been quantitatively characterized using the kinetic models is small. Thus, for most of the pathways the evaluation of weighting factors $w^{(1,2)}$ was impossible. However, we performed modeling tests for the stochastic robustness analysis of the above PAS calculation formula (Buzdin et al. 2014a). In this model, we introduced the additional random perturbation factors w_n, which were used as multiplication coefficients for each logarithmic gene expression member. In our computational study, we assumed the distributions for w_n as logarithmically normal; they were calculated as follows: $w_n = 2^{x_n}$, where x_n are normally distributed random numbers with the expected value of $M = 0$ and standard deviation $\sigma = 0.5$. The random perturbation factors w_n were applied to one of the glioblastoma samples (Griesinger et al. 2013). Importantly, although the perturbation simulation was made independently 98 times with independent weighting factors w_n for each gene, the values of standard deviation for the set of alternate PAS (APAS) were not big enough to bias the linear correlation between the average perturbed and unperturbed PAS for each signaling pathway (Griesinger et al. 2013) (Fig. 2).

Taken together, these findings suggest that the basic OncoFinder formula for PAS calculation that does not involve weighting coefficients w can be used efficiently to profile the pathway activation in the experimental samples. Furthermore, the same rationale was employed for the pathway activity assessment based on the microRNA (miR) expression data. The related technique, termed MiRImpact, enables to link the total miR profiles with the estimated outcomes on the regulation of molecular pathways (Artcibasova et al. 2016).

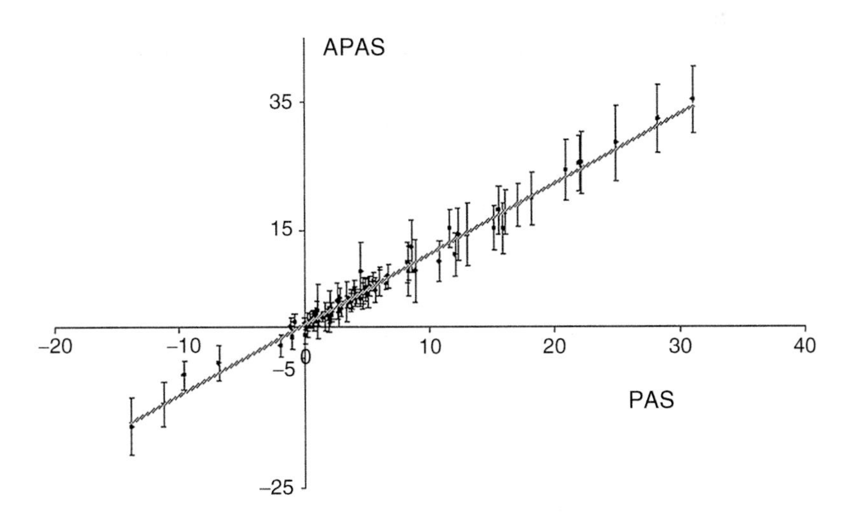

Fig. 2 Alternated values of pathway activation strength (*APAS*) that were calculated for each of 80 signaling pathways in the OncoFinder database in 98 random trials, having random log-normally distributed weighting factors wn, versus nonperturbed PAS. For the perturbed values (*APAS*), both average values (points at the plot) and standard deviation span are shown

MiRImpact: An Algorithm Linking miR Expression Profiles and Pathway Activation

This method was developed to measure the effects of changed miR concentration profiles on the activity of molecular pathways (Artcibasova et al. 2016). MiRImpact was built on the basis of the OncoFinder algorithm (Buzdin et al. 2014a). For each miR, a case to normal ratio is calculated for the respective miR concentrations ($miCNR_j$):

$$miCNR_j = \frac{Case\_microRNA\_Signal_j}{Norm\_microRNA\_Signal_j}.$$

Similarly to the OncoFinder, the miR beyond tolerance interval flag ($miBTIFmiBTIF_j$) determines if the difference between the case and the normal samples is significant:

$$miBTIF_j = \begin{cases} 0, miCNR_j\, belongs\, to\, microRNA\_tolerance\_interval \\ 1, miCNR_j\, doesn't\, belong\, to\, microRNA\_tolerance\_interval \end{cases}$$

The unique coefficient termed miR involvement index (mII) determines if a given expression product of a gene i is a molecular target of a miR j:

$$miII_{j,i} = \begin{cases} 0, & target \\ 1, & not\, target \end{cases}$$

Then the value of miR-based pathway activation strength for a pathway p ($miPAS_p$) is calculated according to the formula:

$$miPAS_p = \sum\nolimits_i (-ARR_{i,p}) \cdot \sum\nolimits_j mII_{j,i} \cdot miBTIF_i \cdot \lg(miCNR_i),$$

The activator-repressor role value ($ARR_{i,p}$) indicates if a gene project has a stimulatory or repressor role in a pathway:

$$ARR_{i,p} = \begin{cases} -1, & repressor \\ -0.5, & repressor > activator \\ 0, & neither \\ 0.5, & activator > repressor \\ 1, & activator \end{cases}$$

MicroRNAs are typically the negative regulators of the expression of their target genes, and the higher is the expression of a particular miR the lower should be the expression of its molecular target gene products. To address this feature, in the miPAS calculation formula, the ARR is multiplied by a factor of (-1). Like in the case of the OncoFider algorithm, a positive value of $miPAS_p$ indicates activation, whereas a negative one means downregulation of a pathway pp, calculated based on the miR expression data.

In the initial application of MiRImpact, we took previously published OncoFinder signaling pathway database featuring 2725 unique genes and 271 signaling pathways (Lezhnina et al. 2014; Spirin et al. 2014). These data were needed to identify genes involved in each pathway and their functional roles expressed by the *ARR* values. In addition, to find out *mII* indexes, another database is required covering the target gene product specificities of miRs.

In this regard, we used two most complete alternative knowledge bases on experimentally confirmed miRs and their molecular targets: miRTarBase (Hsu et al. 2014) and Diana TarBase (Vergoulis et al. 2012). At the time of the method publication, both databases included information on more than 50,000 molecular interactions of miRs with target mRNAs, in the case of miRTarBase – for 18 species, in the case of Diana-TarBase – for 24 species, including human. The most commonly used experimental approaches for validating molecular targets of miRs are luciferase reporter assay, Western blots, and next-generation sequencing approaches. This information is manually curated by the database developers based on published literature reports (Hsu et al. 2014; Vergoulis et al. 2012). The target specificities of miRs catalogued there covered, respectively, 72% and 18% of the genes included in that variant of the OncoFinder database of molecular pathways (Table 1).

We used both databases to calculate the miPAS scores for the human bladder cancer patients in comparison with the normal bladder specimens. We observed a weak, but still statistically significant correlation between the miPAS data calculated for both databases (Fig. 3). The results obtained suggest that the method MiRImpact may be compatible with different miR databases (Artcibasova et al. 2016). We compared the obtained results with the literature data on the impact of particular miRs on the respective signaling pathways. For the data calculated using the miRTarBase, we observed a greater congruence between the experimental and the literature data (in 47% of the cases), whereas for Diana-TarBase, the data were compatible in only 23% of the cases. Based on this finding, we suggested that the miRTarBase was a database of choice for the estimation of molecular pathways regulation by miRs in humans (Artcibasova et al. 2016).

For specific applications, the enclosed information on the molecular pathways may be also updated in a user-definitive way. We hypothesize that in the future, knowledge of the qualitative aspects of molecular interactions between the miRs and their targets, and between the molecules participating in the molecular pathways, may be used to tune the databases and algorithms in order to assign specific weighting coefficients to each miR and/or each gene product.

Table 1 Comparative characteristics of validated miR target databases MiRTarBase and Diana TarBase, at the state of September 2016

Database	MiRTarBase	Diana TarBase
Number of miR records, all species	12,103	3006
Number of miRs targeting gene products from OncoFinder signaling pathway database	596	183
Number of target genes in OncoFinder signaling pathway database	1968	497

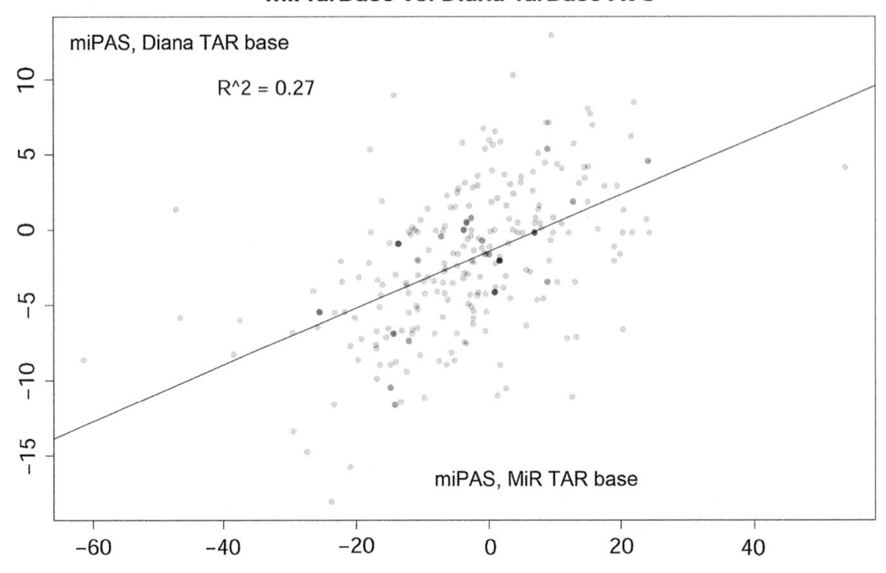

Fig. 3 Bladder cancer data. Comparison of microRNA Pathway Activation Strength (*miPAS*) values calculated using miRTarBase and Diana TarBase databases of miR targets, for an averaged miR expression. The resulting virtual sample is the result of averaging of miR expression measured by deep sequencing for eight bladder cancer samples

Finally, we propose that for the future applications, other types of noncoding RNAs than miRs can be also analyzed using the MiRImpact method, if/when their regulatory roles and their target/effector gene products become known.

Applications of MiRImpact Algorithm

Identification of new bladder cancer biomarkers. Bladder cancer (BC) is the second most frequent urological cancer worldwide, with about 356,000 new BC cases reported annually, and there is a need to develop novel diagnostic tools to efficiently detect BC at the early stages (Lezhnina et al. 2014). Discovery of pathway-based biomarkers may provide direction for moving towards a more efficient, patient-oriented anticancer therapy.

Using a combination of microarray and NGS methods, we profiled mRNA and miR expression in 17 experimental cancer and seven normal bladder tissue samples (Lezhnina et al. 2014; Artcibasova et al. 2016). We screened 271 signaling pathways and found among them 44 pathways that may serve as excellent new biomarkers of BC at the mRNA level, supported by the high AUC values >0.75 (Lezhnina et al. 2014). Interestingly, overall pathway activation profiles obtained using the OncoFinder for mRNA expression data, and using the MiRImpact for miR expression, differed dramatically (Artcibasova et al. 2016). Previously, we identified 44 molecular signaling pathways which may serve as potent biomarkers of BC. For 21 of them, we found literature data connecting miR expression and pathway activation abnormalities in cancer (Artcibasova et al. 2016). Basing on our own experimental assay, for miRTarBase we observed congruence with finding of pathway up/downregulated state in 10/21 molecular pathways (Artcibasova et al. 2016). The remaining pathways that did not coincide with the MiRImpact results were either apparently inconclusively (bidirectionally) regulated in BC or were unchanged according to miPAS data (Artcibasova et al. 2016).

We next compared pathway activation signatures for the 44 above characteristic BC-associated pathways at the mRNA and miR levels. In the case of miRTarBase version, 20 pathways had contrary trends, and only 10 had common trends at the miPAS and PAS levels (Artcibasova et al. 2016). This suggests that the regulation of many characteristic BC-linked pathways differs dramatically at the mRNA and miR levels. Similar figure was seen when looking at the miPAS values for all available pathways. Comparison of mRNA and miR pathway activation also showed quite distinct peculiarities in terms of variation between the individual samples. We observed relatively uniform regulation of pathways at the mRNA level, with relatively small number of pathways showing significant variations between the individual samples (Artcibasova et al. 2016). In contrast, at the level of miR regulation, the apparently observed differences between the samples were significantly higher (Artcibasova et al. 2016). The majority of the pathways were also strongly differential between the normal and cancer samples at the miR level. These peculiarities of miPAS scores suggest that they may be more sensitive compared to the PAS values to discriminate between the individual cancer samples. This may be highly beneficial for finding new diagnostic markers, e.g., linked with the individual sensitivity of a patient to certain type of treatment.

Finally, we demonstrated that at least for the human BC tissues, the intracellular pathway regulation at the miR level differs greatly from that at the mRNA level, thus showing orthogonal dependencies for the extents of pathway activation. This trend could be observed for all individual samples, and also for the averaged samples shown on Fig. 4.

Identification of cytomegalovirus infection-linked biomarkers. We examined two untransformed normal human fibroblast cell lines, HELF-977 and HAF-1608, with dramatically different proliferative activities (Buzdin et al. 2016). The cells were infected with the human cytomegalovirus (HCMV). In parallel, a fraction of each cell culture was mock infected with a buffer solution. The data on time-dependence of viral gene expression allowed to characterize HELF-977 cells as highly sensitive (HS) and HAF-1608 – as low sensitive (LS) to the HCMV infection.

Three hours after infection, the cells were collected and aliquots were taken for RNA isolation and expression profiling. For each cell line, we examined three samples: (i) the preanalysis norm representing cells before infection, (ii) infected cells 3 h after infection (3H AI), and (iii) mock-infected cells 3 h after mock infection (3H WI).

We used NGS to establish miR expression profiles and microarray hybridization to interrogate transcription of protein coding genes. The expression data were next processed to establish the activation features of intracellular molecular pathways, and a total of 271 intracellular signaling pathways were screened. For all calculations, AI and WI expression profiles were normalized on the data obtained for the

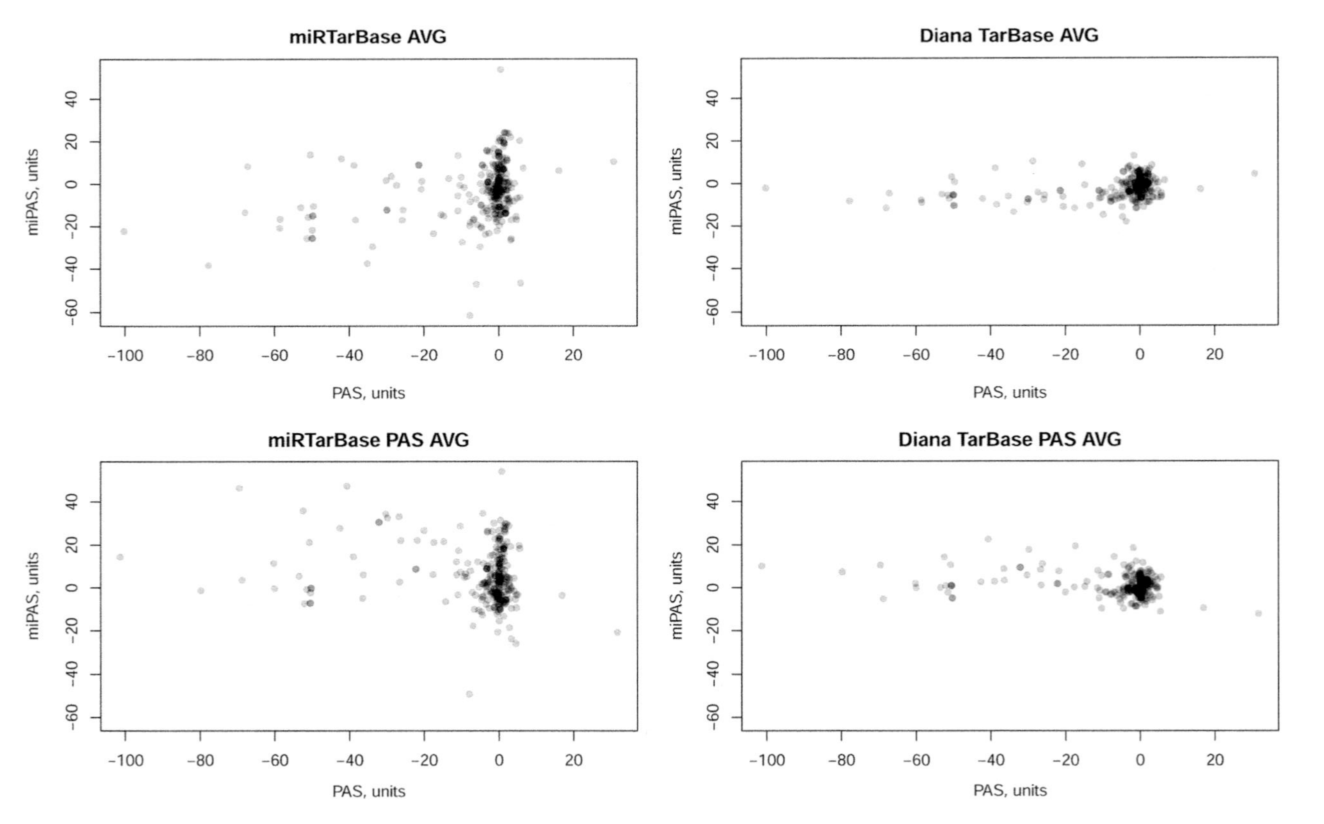

Fig. 4 Cytomegalovirus data. Comparison of microRNA Pathway Activation Strength (*miPAS*) values calculated using miRTarBase and Diana TarBase databases of miR targets, for an averaged miR expression. The resulting virtual sample is the result of averaging of miR expression measured by deep sequencing for eight bladder cancer samples

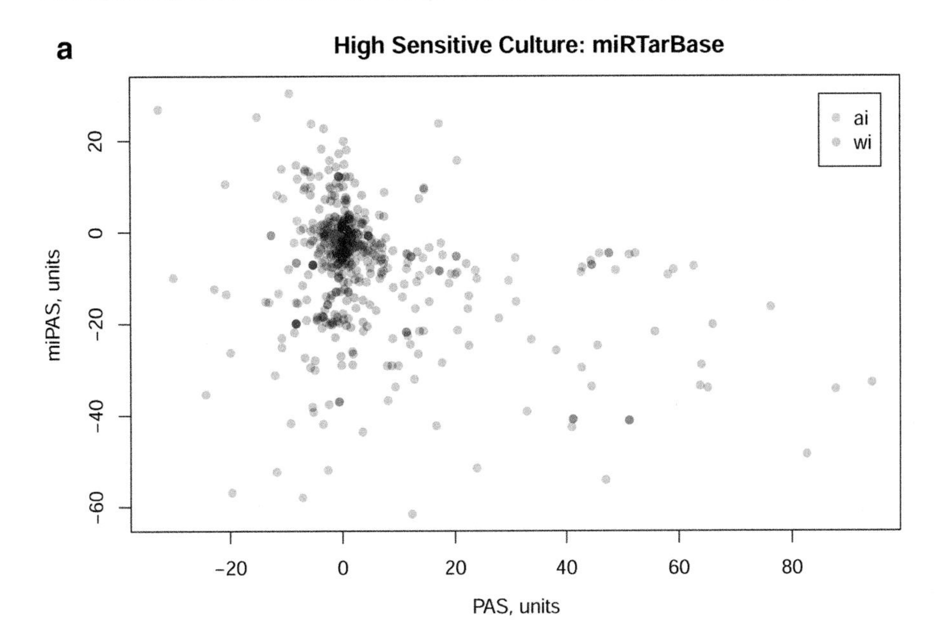

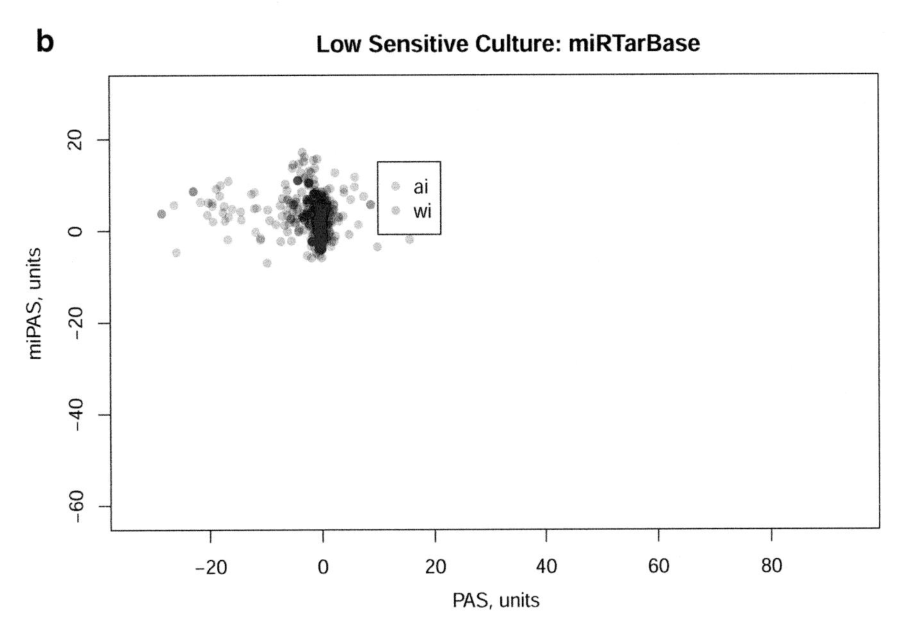

Fig. 5 Dependencies between PAS and miPAS in HCMV-infected fibroblasts. MiPAS data were calculated using miRTarBase database. Each *dot* represents one molecular pathway, for which PAS and miPAS values were calculated. *Blue dots* are given for the samples without HCMV infection, *red dots* – for the samples after infection. All PAS and miPAS values were calculated against preinfection controls. (**a**) comparison for the HS cells; (**b**) comparison for the LS cells

initial preinfection cells. We established large-scale profiles of signaling pathway activation associated with the HCMV infection. These profiles were markedly different for the fibroblasts with high and low sensitivity to infection at both levels of RNA regulation (Fig. 5). Similarly to the previously published data for bladder cancer (Artcibasova et al. 2016), we observed that the intracellular pathway regulation at the miR level differs greatly from that at the mRNA level, thus showing orthogonal dependencies for the extent of pathway activation. In addition to multiple specific features of pathway activation at the mRNA and miR levels, we also detected several trends characterizing large-scale behaviors of the infected versus noninfected HS and LS cells at the molecular level (Buzdin et al. 2016).

First, for the time-control (WI) HS cells, overall sets of RNA regulation features showed significantly greater levels of variation compared to the LS cells: at the mRNA level, at the miR level, and at the signaling pathway regulation levels. This suggests that the HS cells change their molecular landscape more rapidly than the LS cells even in the noninfected conditions. The same was also true for the infected HS cells, but only for the regulation at the mRNA and PAS levels (Buzdin et al. 2016).

Second, for the infected HS cells, we observed a dramatically lower impact of miR regulation of the molecular pathways, as reflected by close-to-zero clouds of CNR and miPAS values compared to the mock-infected controls (Buzdin et al. 2016). It means that the infected HS cells demonstrate little difference from the preinfection controls in their miR expression profile. This was evidenced by the respective distribution curves with approximately fivefold lower dispersion for the AI cells compared to WI cells for the miPAS values. Notably, no significant bias in distribution trends could be seen for the LS cells, and, at the levels of mRNA and PAS – in both cell types. Such an effect shows conservation of miR expression landscape in the highly sensitive fibroblasts at the moment of HCMV invasion and further, at least up to 3 h after infection. We propose that this may be due to a previously unknown property of HCMV proteins to suppress intracellular defensive mechanisms related to miR expression. This finding evidences that at the early stage of HCMV infection, the miR regulation in the host cells is significantly down-regulated. Interestingly, this suppression was not seen for the LS cells, which may suggest its importance for the progression of the HCMV infection.

Dictionary of Terms

- **Metabolic pathway** – Intracellular molecular pathway that coordinates biochemical reactions in a continuously connected network having a biologically significant output.
- **Signaling pathway** – Intracellular molecular pathway that consolidates gene products involved in a sequential signal transduction chain.
- **Cytoskeleton reorganization pathway** – Intracellular molecular pathway that aggregates gene products involved in a sequential events in cytoskeleton remodeling.

- **DNA repair pathway** – Intracellular molecular pathway that consolidates gene products involved in a sequential processes of DNA repair.
- **Pathway activation strength** – Quantitative characteristics of a molecular pathway activation, based on the protein-coding gene expression (PAS) or microRNA expression (miPAS) data.

Key Facts

- All the aspects of the living cell functioning are controlled by the molecular pathways.
- Most important groups are signaling, metabolic, DNA repair, and cytoskeleton remodeling pathways.
- The activity of molecular pathways can be quantitatively profiled using high throughput gene expression data.
- Pathway activation strength is a novel type of biomarkers superior to the individual gene expression.
- MiR expression influences translation of the protein-coding gene products and therefore can alter the mRNA-based pathway homeostasis.
- MiRImpact is the first method to quantitate the effects of the total miR profiles on the activation of molecular pathways.

Summary Points

- Most of the molecular processes in the cell are organized in the pathway-based manner.
- Several thousands of different molecular pathways have been discovered in previous studies.
- The information about the pathway architecture and internal/external functional relationship is stored in the specific databases.
- To quantitatively characterize activation of a molecular pathway, a specific value termed *pathway activation strength (PAS)* can be introduced.
- PAS is a novel type of biomarkers.
- PAS is a stronger biomarker compared to the expression of individual gene products.
- Pathway activation strength can be calculated based on either the protein-coding gene expression (PAS) or microRNA expression (miPAS) data.
- Public databases can be used to extract information on the molecular targets of microRNAs.
- The miPAS values are more variable and thus polymorphic biomarkers, whereas the PAS profiles show a more stable distribution.
- The applications of PAS and miPAS profiles include fundamental science, biomedicine, and clinical applications.

- Harmonization of genomic, transcriptomic, epigenetic, and proteomic gene expression data at the PAS levels is the perspective area of future developments in biomedicine.

References

Afsari B, Geman D, Fertig EJ (2014) Learning dysregulated pathways in cancers from differential variability analysis. Cancer Informat 13:61–67

Aliper A, Belikov AV et al (2016) In search for geroprotectors: in silico screening and in vitro validation of signalome-level mimetics of young healthy state. Aging 8:2127–2152

Aliper AM, Frieden-Korovkina VP et al (2014) Interactome analysis of myeloid-derived suppressor cells in murine models of colon and breast cancer. Oncotarget 5:11345–11353

Artcibasova AV, Korzinkin MB et al (2016) MiRImpact, a new bioinformatic method using complete microRNA expression profiles to assess their overall influence on the activity of intracellular molecular pathways. Cell Cycle 5:689–698

Bauer-Mehren A, Furlong LI, Sanz F (2009) Pathway databases and tools for their exploitation: benefits, current limitations and challenges. Mol Syst Biol 5:290

Blagosklonny MV (2011) The power of chemotherapeutic engineering: arresting cell cycle and suppressing senescence to protect from mitotic inhibitors. Cell Cycle 10:2295–2298

Blagosklonny MV (2013) MTOR-driven quasi-programmed aging as a disposable soma theory: blind watchmaker vs. intelligent designer. Cell Cycle 12:1842–1847

Bolstad BM, Irizarry RA et al (2003) A comparison of normalization methods for high density oligonucleotide array data based on variance and bias. Bioinformatics 19:185–193

Borisov N, Aksamitiene E et al (2009) Systems-level interactions between insulin-EGF networks amplify mitogenic signaling. Mol Syst Biol 5:256

Borisov NM, Chistopolsky AS et al (2008) Domain-oriented reduction of rule-based network models. IET Syst Biol 2:342–351

Branzei D, Foiani M (2008) Regulation of DNA repair throughout the cell cycle. Nat Rev Mol Cell Biol 9:297–308

Buzdin AA, Zhavoronkov AA et al (2014a) Oncofinder, a new method for the analysis of intracellular signaling pathway activation using transcriptomic data. Front Genet 5:55

Buzdin AA, Zhavoronkov AA et al (2014b) The Oncofinder algorithm for minimizing the errors introduced by the high-throughput methods of transcriptome analysis. Front Mol Biosci 1:8

Buzdin AA, Artcibasova AV et al (2016) Early stage of cytomegalovirus infection suppresses host microRNA expression regulation in human fibroblasts. Cell Cycle 15:3378–3389

Conzelmann H, Saez-Rodriguez J et al (2006) A domain-oriented approach to the reduction of combinatorial complexity in signal transduction networks. BMC Bioinformatics 7:34

Croft D, Mundo AF et al (2014) The Reactome pathway knowledgebase. Nucleic Acids Res 42: D472–D477

Daniels BC, Chen YL et al (2008) Sloppiness, robustness, and evolvability in systems biology. Curr Opin Biotechnol 19:389–395

Demidenko ZN, Blagosklonny MV (2011) The purpose of the HIF-1/PHD feedback loop: to limit mTOR-induced HIF-1α. Cell Cycle 10:1557–1562

Disanza A, Frittoli E et al (2009) Endocytosis and spatial restriction of cell signaling. Mol Oncol 3:280–296

Elkon R, Vesterman R et al (2008) SPIKE- a database, visualization and analysis tool of cellular signaling pathways. BMC Bioinformatics 9:110

Filteau M, Diss G, Torres-Quiroz F, Dube AK, Schraffl A, Bachmann VA, Gagnon-Arsenault I et al (2015) Systematic identification of signal integration by protein kinase A. Proc Natl Acad Sci 112(14):4501–4506. https://doi.org/10.1073/pnas.1409938112

Griesinger AM, Birks DK et al (2013) Characterization of distinct immunophenotypes across pediatric brain tumor types. J Immunol 191(9):4880–4888

Hanahan D, Weinberg RA (2000) The hallmarks of cancer. Cell 100:57–70
Hanahan D, Weinberg RA (2011) Hallmarks of cancer: the next generation. Cell 144:646–674
Ho JWK, Stefani M et al (2008) Differential variability analysis of gene expression and its application to human diseases. Bioinformatics 24:i390–i398
Hsu SD, Tseng YT et al (2014) miRTarBase update 2014: an information resource for experimentally validated miRNA-TARGET INTERACTIONS. Nucleic Acids Res 42:D78–D85
Khatri P, Sirota M, Butte AJ (2012) Ten years of pathway analysis: current approaches and outstanding challenges. PLoS Comput Biol 8:e1002375
Kholodenko BN, Demin OV et al (1999) Quantification of short term signaling by the epidermal growth factor receptor. J Biol Chem 274:30169–30181
Kholodenko BN, Kiyatkin A et al (2002) Untangling the wires: a strategy to trace functional interactions in signaling and gene networks. PNAS 99:12841–12846
Kiyatkin A, Aksamitiene A et al (2006) Scaffolding protein Grb2-associated binder 1 sustains epidermal growth factor-induced mitogenic and survival signaling by multiple positive feedback loops. J Biol Chem 281:19925–19938
Kulesh DA, Clive DR et al (1987) Identification of interferon-modulated proliferation-related cDNA sequences. PNAS 84:8453–8457
Kuzmina NB, Nikolay MM (2011) Handling complex rule-based models of mitogenic cell signaling (on the example of ERK activation upon EGF stimulation). Int Proc Chem Biol Environ Eng 5:76–82
Lebedev TD, Spirin PV et al (2015) Receptor tyrosine kinase KIT may regulate expression of genes involved in spontaneous regression of neuroblastoma. Mol Biol 49:1052–1055
Lezhnina K, Kovalchuk O et al (2014) Novel robust biomarkers for human bladder cancer based on activation of intracellular signaling pathways. Oncotarget 5:9022–9032
Love ML, Huber W, Anders A (2014) Moderated estimation of fold change and dispersion for RNA-seq data with DESeq2. Genome Biol 15:550
Makarev E, Cantor C et al (2014) Pathway activation profiling reveals new insights into age-related macular degeneration and provides avenues for therapeutic interventions. Aging 6:1064–1075
Makarev E, Izumchenko E et al (2016) Common pathway signature in lung and liver fibrosis. Cell Cycle 15:1667–1673
Malumbres M, Barbacid M (2009) Cell cycle, CDKs and cancer: a changing paradigm. Nat Rev Cancer 9:153–166
Marshall CJ (1995) Specificity of receptor tyrosine kinase signaling: transient versus sustained extracellular signal-regulated kinase activation. Cell 80:179–185
Mathivanan S, Periaswamy B et al (2006) An evaluation of human protein-protein interaction data in the public domain. BMC Bioinformatics 7(Suppl 5):S19
Mitrea C, Taghavi Z et al (2013) Methods and approaches in the topology-based analysis of biological pathways. Front Physiol 4:278
Nakaya A, Katayama T et al (2013) KEGG OC: a large-scale automatic construction of taxonomy-based ortholog clusters. Nucleic Acids Res 41:D353–DS57
Nikitin A, Egorov S et al (2003) Pathway Studio – the analysis and navigation of molecular networks. Bioinformatics 19:2155–2157
Olsvik O, Wahlberg J et al (1993) Use of automated sequencing of polymerase chain reaction-generated amplicons to identify three types of cholera toxin subunit B in *Vibrio cholerae* O1 strains. J Clin Microbiol 31:22–25
Ozerov IV, Lezhnina LV et al (2016) In silico pathway activation network decomposition analysis (iPANDA) as a method for biomarker development. Nat Commun 7:13427
Ram DR, Ilyukha V et al (2016) Balance between short and long isoforms of cFLIP regulates FAS-mediated apoptosis in vivo. PNAS 113:1606–1611
Shepelin D, Korzinkin M et al (2016) Molecular pathway activation features linked with transition from normal skin to primary and metastatic melanomas in human. Oncotarget 7:656–670
Sonnenschein C, Soto AM (2013) The aging of the 2000 and 2011 *Hallmarks of Cancer* reviews: a critique. J Biosci 38(3):651–663
Spirin PV, Lebedev TD et al (2014) Silencing AML1-ETO gene expression leads to simultaneous activation of both pro-apoptotic and proliferation signaling. Leukemia 28:2222–2228

Tian L, Greenberg SA et al (2005) Discovering statistically significant pathways in expression profiling studies. PNAS 102:13544–13549
UniProt Consortium (2011) Ongoing and future developments at the universal protein resource. Nucleic Acids Res 39:D214–DD19
Vergoulis T, Vlachos IS et al (2012) TarBase 6.0: capturing the exponential growth of miRNA targets with experimental support. Nucleic Acids Res 40:D222–D229
Vermeulen K, van Bockstaele DR, Berneman ZN (2003) The cell cycle: a review of regulation, deregulation and therapeutic targets in cancer. Cell Prolif 36:131–149
Vivar JC, Pemu P et al (2013) Redundancy control in pathway databases (ReCiPa): an application for improving gene-set enrichment analysis in omics studies and 'big data' biology. Omics J Integr Biol 17:414–422
Zeeberg BR, Feng W et al (2003) GoMiner: a resource for biological interpretation of genomic and proteomic data. Genome Biol 4:R28
Zhang J, Li J, Deng HW (2009) Identifying gene interaction enrichment for gene expression data. PLoS One 4:e8064
Zhavoronkov A, Buzdin AA et al (2014) Signaling pathway cloud regulation for in silico screening and ranking of the potential geroprotective drugs. Front Genet 5:49
Zhu Q, Izumchenko E et al (2015) Pathway activation strength is a novel independent prognostic biomarker for cetuximab sensitivity in colorectal cancer patients. Hum Genome Var 2:15009

Resources in Diet, Nutrition, and Epigenetics

121

Rajkumar Rajendram, Vinood B. Patel, and Victor R. Preedy

Contents

Abstract

Epigenetics studies inheritable changes in gene expression and function that occur in the absence of modification in the DNA sequence. In other words, this represents stable phenotypic changes which occur without genomic changes. Epigenetics is still developing as a science. It was only in 2008 that an operational definition of an epigenetics was described in a consensus document by experts at a meeting on chromatin-based epigenetics. Since then there has been an explosion in the knowledge and understanding of epigenetics. It is now difficult even for

R. Rajendram (✉)
Department of Medicine, King Abdulaziz Medical City, Ministry of National Guard Health Affairs, Riyadh, Saudi Arabia

Diabetes and Nutritional Sciences Research Division, Faculty of Life Sciences and Medicine, King's College London, London, UK
e-mail: rajkumarrajendram@doctors.org.uk

V. B. Patel
School of Life Sciences, University of Westminster, London, UK

V. R. Preedy
Diabetes and Nutritional Sciences Research Division, Faculty of Life Sciences and Medicine, King's College London, London, UK

V. B. Patel, V. R. Preedy (eds.), *Handbook of Nutrition, Diet, and Epigenetics*,
https://doi.org/10.1007/978-3-319-55530-0_125

experienced scientists to remain up-to-date. For those new to the field, it is difficult to know which of the myriad of available sources are reliable. To further aid colleagues who are interested in understanding more about epigenetics, we have therefore produced tables containing reliable, up-to-date resources on epigenetics in this chapter. The experts who assisted with the compilation of these tables of resources are acknowledged below.

Keywords
Epigenetics · Evidence · Resources · Books · Journals · Regulatory bodies · Professional societies

Introduction

Epigenetics studies inheritable changes in gene expression and function that occur in the absence of modification in the DNA sequence. In other words, this represents stable phenotypic changes which occur without genomic changes. Epigenetics is still

Table 1 Professional societies and organizations

American Association for the Advancement of Science	www.aaas.org
American Society for Nutrition (ASN)	nutrition.org
Cancer Epigenetic Society	ces.b2sg.org
Clinical Epigenetics Society	www.clinical-epigenetics-society.org/about-us
Epigenetics Interest Group	www.oir.nih.gov/sigs/epigenetics-interest-group
Epigenetics Society	www.epigeneticssocietyint.com
Federation of European Nutrition Societies	www.fensnutrition.eu
International Federation of Cell Biology	www.ifcbiol.com
International Society for Developmental Origins of Health and Disease (DOHaD)	www.dohadsoc.org
International Union of Nutritional Sciences	www.iuns.org
Keystone Symposia	www.keystonesymposia.org
National Institute of Health	www.nih.gov
NOVA	www.pbs.org/wgbh/nova/body/epigenetics.html
The Nutrition Society	www.nutritionsociety.org
The Royal Society of Medicine	www.rsm.ac.uk
US Society for Developmental Origins of Health and Disease (US DOHaD)	www.usdohad.org

This table lists some professional societies and organizations involved with epigenetics. The list is not meant to be exhaustive but some sites provide links to other societies. For example, the Federation of European Nutrition Societies has links to the Finnish Society for Nutrition Research, Austrian Nutrition Society, Belgium Nutrition Society, Dutch Academy of Nutritional Sciences, French Society of Nutrition, and over 20 other societies. Occasionally the location of the websites or web address changes. See also Table 4

Table 2 Journals relevant to epigenetics

PLOS One
Oncotarget
Scientific Reports
Epigenetics
Proceedings of the National Academy of Sciences of the United States of America
Methods in Molecular Biology
Epigenomics
Nucleic Acids Research
Clinical Epigenetics
Nature Communications
International Journal of Molecular Sciences
BMC Genomics
Advances in Experimental Medicine and Biology
Journal of Biological Chemistry
PLOS Genetics
Genome Biology
Oncogene
Cell Reports
Frontiers in Genetics
Nature
Biochemical and Biophysical Research Communications
Cell
Cancer Research
Epigenetics & Chromatin
Cancer Letters

Journals publishing original research and review articles related to epigenetics. Included in this list are the top 25 journals which have published the most number of articles on epigenetics over the past 5 years using epigenetics as the search word in Scopus. Of the 25, *PLOS One* has published the most and the least (ranked 25th) is *Cancer Letters*. Of course other terms and other databases will produce different listings. Contributing authors also identified *Nature Medicine*, *New England Journal of Medicine*, and *PLOS One* which were also captured by publication analysis

developing as a science. It was only within the last decade that an operational definition of an epigenetics was described in a consensus document. In 2008, experts at a meeting on chromatin-based epigenetics hosted by the Banbury Conference Center and Cold Spring Harbor Laboratory defined an epigenetic trait as a "stably heritable phenotype resulting from changes in a chromosome without alteration in the DNA sequence" (Berger et al. 2009).

These changes can persist as the cell divides throughout its life cycle. These changes may persist through multiple generations, maintained through the progeny of the index cell in which the initial change occurred (Berger et al. 2009; Reik 2007). The DNA sequence of the organism remains unchanged (Berger et al. 2009; Reik 2007; Bird 2007). It is exposure to environmental nongenetic factors that influences

Table 3 Relevant books

Bioactive Compounds and Cancer. Milner JA, Romagnolo DF (Editors). Humana Press, Springer, 2010, USA.
Environmental epigenetics Su, LJ, Chiang, T (Editors) Springer *2015 USA*
Epigenetic Technological Applications. Zheng Y (Editor) Academic Press 2015 USA
Epigenetic Toxicology. McCullough S, Dolinoy D. Academic Press. 2018 USA
Epigenetics. Heil R, Seitz SB, König H, Robienski J (Editors). Springer 2017, USA
Epigenetics. Allis CD, Jenuwein T, Reinberg D (Editors); Associate Editor Monika Lachlan, Max-Planck Institute of Immunobiology and Epigenetics Cold Spring Harbor Laboratory Press, 2015. USA
Handbook of Epigenetics, Second Edition, Tollefsbol T. Elsevier, 2017, AL, USA
Medical Epigenetics. Tollefsbol T. Elsevier, 2016, AL, USA
Metabolic control. Herzig S, Springer, 2015, Switzerland
Neuropsychiatric Disorders and Epigenetics. Yasui D, Peedicayi J, Grayson D, (Editor) Academic Press, 2018, USA
Nutrition and Epigenetics. Ho E and Domann F (Editors). CRC Press, 2015, USA.
Personalized Epigenetics, First edition, Tollefsbol T (Editor) Elsevier, 2015, AL, USA
Transgenerational Epigenetics. Tollefsbol T (Editor) Academic Press, 2014, USA

This table lists recommended books on epigenetics

gene expression (Berger et al. 2009; Reik 2007; Bird 2007). Environmental factors which have been found to cause epigenetic phenomena include chemicals, pharmaceuticals, development (in utero and during childhood), aging, and nutrition (Berger et al. 2009; Tabish et al. 2012; Burdge et al. 2011). The latter is of particular relevance to these multi-chaptered works. The mechanisms of these phenotypic changes include methylation of DNA, alteration of histones, and many other modulating influences identified in the *Handbook of Nutrition, Diet, and Epigenetics* (Berger et al. 2009; Tabish et al. 2012; Burdge et al. 2011).

Still in its infancy as a science, epigenetics has captured the imagination of the general public. There has been significant interest and indeed sensationalism of epigenetics in the media and popular culture. Since 2008 when Berger et al. (Berger et al. 2009) reported the operational definition of epigenetics, there has been an explosion in the knowledge and understanding of epigenetics. It is now difficult even for experienced scientists to remain up-to-date. For those new to the field, it is difficult to know which of the myriad of available sources are reliable. To assist colleagues who are interested in understanding more about epigenetics, we have therefore produced tables containing reliable, up-to-date resources on epigenetics in this chapter.

Tables 1, 2, 3, 4 list the most up-to-date information on professional societies and organizations (Table 1), journals on epigenetics (Table 2), books (Table 3), and online resources (Table 4) that are relevant to an evidence-based approach to epigenetics.

Table 4 Relevant online resources or information on emerging techniques and scientific fields

Abcam	www.abcam.com/epigenetics/ultimate-epigenetics-event-calendar
	www.abcam.com/research/epigenetics
Bio-Rad	www.bio-rad.com
BioTechniques	www.biotechniques.com
Blueprint epigenome	www.blueprint-epigenome.eu
DeepBlue	http://deepblue.mpi-inf.mpg.de
Diagenode	www.diagenode.com
DiseaseMeth	http://bio-bigdata.hrbmu.edu.cn/diseasemeth/
EpiGenie	www.epigenie.com/epigenetic-tools-and-databases/
Epigenetics Boot Camp/ Columbia University	www.mailman.columbia.edu/bootcamp
Epigenetics Glossary	www.mailman.columbia.edu/research/laboratory-precision-environmental-biosciences/epigenetics-glossary
European Bioinformatics Institute (EMBL-EBI)	www.ebi.ac.uk
HIstome: The Histone Infobase	www.actrec.gov.in/histome/index.php
Illumina	www.illumina.com/products/by-type/microarray-kits/infinium-methylation-epic.html
Illumina	www.illumina.com/products/by-type/sequencing-kits/library-prep-kits/truseq-methyl-capture-epic.html
miRNEST	www.omictools.com/mirnest-tool
Nature	www.nature.com/subjects/epigenetics-analysis
NIH Roadmap Epigenomics Mapping Consortium	www.roadmapepigenomics.org
Noncode	www.noncode.org
Nuclear receptor Signalling Atlas	www.nursa.org/nursa/index.jsf
QIAGEN	www.qiagen.com/us/resources/molecular-biology-methods/epigenetics
Science	www.sciencemag.org/site/feature/plus/sfg/resources/res_epigenetics.xhtml
ThermoFisher	www.thermofisher.com/us/en/home/brands/thermo-scientific/molecular-biology/thermo-scientific-specialized-molecular-biology-applications/epigenetics-thermo-scientific.html
Zymo Research	www.zymoresearch.com

This table lists some Internet resources on epigenetics. Some pages are direct links. In some cases (e.g., see Bio-Rad), material on epigenetics can be found using the search option. Some websites such as the one relating to EpiGenie will have links to the websites. Occasionally the location of the websites or web address changes

Summary Points

- The significance of epigenetics is currently under recognized in modern medicine.
- This chapter lists the most up-to-date resources on the regulatory bodies, professional bodies, journals, books, and websites that are relevant to an evidence-based approach to epigenetics.
- Epigenetics is still in its infancy as a science.

Acknowledgments This is of vital importance for this chapter. When compiling the resources it was agreed that all contributing authors would be acknowledged. Acknowledging these authors also increases the credibility of this work.

References

Berger SL, Kouzarides T, Shiekhattar R, Shilatifard A (2009) An operational definition of epigenetics. Genes Dev 23:781–783

Bird A (2007) Perceptions of epigenetics. Nature 447:396–398

Burdge GC, Hoile SP, Uller T, Thomas NA, Gluckman PD, Hanson MA, Lillycrop KA (2011) Progressive, transgenerational changes in offspring phenotype and epigenotype following nutritional transition. PLoS One 6:e28282

Reik W (2007) Stability and flexibility of epigenetic gene regulation in mammalian development. Nature 447(7143):425–432

Tabish AM, Poels K, Hoet P, Godderis L (2012) Epigenetic factors in cancer risk: effect of chemical carcinogens on global DNA methylation pattern in human TK6 cells. PLoS One 7:e34674

Index

V. B. Patel, V. R. Preedy (eds.), *Handbook of Nutrition, Diet, and Epigenetics*, https://doi.org/10.1007/978-3-319-55530-0

C

I

O

Q

R